Cancer Sensitizing Agents for Chemotherapy

GLIOBLASTOMA RESISTANCE TO CHEMOTHERAPY: MOLECULAR MECHANISMS AND INNOVATIVE REVERSAL STRATEGIES

VOLUME 15

Cancer Sensitizing Agents for Chemotherapy Series

Series Editor: Benjamin Bonavida, PhD

Volume 1: Nitric Oxide (Donor/Induced) in Chemosensitization
Edited by Benjamin Bonavida, PhD

Volume 2: Role of Nutraceuticals in Cancer Chemosensitization
Edited by Alok Chandra Bharti and Bharat Bhushan Aggarwal

Volume 3: Targeting Cell Survival Pathways to Enhance Response to Chemotherapy
Edited by Daniel E. Johnson, PhD

Volume 4: Protein Kinase Inhibitors as Sensitizing Agents for Chemotherapy
Edited by Zhe-Sheng Chen and Dong-Hua Yang

Volume 5: Breaking Tolerance to Pancreatic Cancer Unresponsiveness to Chemotherapy
Edited by Ganji Purnachandra Nagaraju, PhD, DSc

Volume 6: Improving the Therapeutic Ratio in Head and Neck Cancer
Edited by Randall J. Kimple, MD, PhD

Volume 7: Drug Efflux Pumps in Cancer Resistance Pathways: From Molecular Recognition and Characterization to Possible Inhibition Strategies in Chemotherapy
Edited by Alejandro Sosnik, PhD and Reina Bendayan, PharmD

Volume 8: Drug Resistance in Colorectal Cancer: Molecular Mechanisms and Therapeutic Strategies
Edited by Chi Hin Cho and Tao Hu, PhD

Volume 9: Novel Therapies in Head and Neck Cancer: Beyond the Horizon
Edited by Maie A. St. John, MD, PhD, FACS and No-Hee Park, DMD, PhD

Volume 10: pH-Interfering Agents as Chemosensitizers in Cancer Therapy
Edited by Claudiu T. Supuran, PhD and Simone Carradori, PhD

Volume 11: Overcoming Ovarian Cancer Chemoresistance
Edited by Goli Samimi, PhD, MPH and Christina Annunziata, MD, PhD

Volume 12: Biological Mechanisms and the Advancing Approaches to Overcoming Cancer Drug Resistance
Edited by Andrew Freywald, PhD and Franco J. Vizeacoumar, PhD

Volume 13: Therapeutic Strategies to Overcome ALK Resistance in Cancer
Edited by Luc Friboulet, PhD

Volume 14: New Targeting in the Reversal of Resistant Glioblastomas
Edited by Ali Syed Arbab, MD, PhD

Volume 15: Glioblastoma Resistance to Chemotherapy: Molecular Mechanisms and Innovative Reversal Strategies
Edited by Ramasamy Paulmurugan, PhD and Tarik F. Massoud, MD, PhD

Volume 16: Epigenetic Regulation in Overcoming Chemoresistance
Edited by Chunfu Wu, PhD and Lihui Wang, PhD

Volume 17: Overcoming Drug Resistance in Gynecologic Cancers
Edited by Riyaz Basha, PhD, FAACC, FABAP and Sarfraz Ahmad, PhD, FAACC, FABAP

Upcoming Volumes:

Terpenoids as Chemosensitizers Against Drug-resistant Tumors
Edited by Thomas Efferth

Chemotherapy Sensitizing Agents for Prostate Cancer
Edited by K. Ketola

Chemosensitization by TRK Inhibitors
Edited by Dirk Hempel

Chemosensitization by BTK Inhibitors
Edited by Varinder Kaur

Chemosensitization by Anti-angiogenic Agents
Edited by Lucia Morbidelli

Chemosensitization by IDH1 Inhibitors
Edited by Robert Musiol

Chemosensitization by EGFR Inhibitors
Edited by M. Pancione

Chemosensitization by IDH2 Inhibitors
Edited by Sanjee Singh

Chemosensitization by Bcl2 Inhibitors
Edited by Yu Yu

Chemosensitization by BTK Inhibitors (II)
Edited by Jean Koff

Overcoming Resistance to EGFR Inhibitors in EGFR Mutant NSCLC
Edited by Anthony Faber

Cancer Sensitizing Agents for Chemotherapy

GLIOBLASTOMA RESISTANCE TO CHEMOTHERAPY: MOLECULAR MECHANISMS AND INNOVATIVE REVERSAL STRATEGIES

VOLUME 15

Edited by

RAMASAMY PAULMURUGAN, PHD

Cellular Pathway Imaging Laboratory (CPIL), Molecular Imaging Program at Stanford (MIPS), Department of Radiology, Stanford University School of Medicine, Palo Alto, CA, United States

TARIK F. MASSOUD, MD, PHD

Laboratory of Experimental and Molecular Neuroimaging (LEMNI), Molecular Imaging Program at Stanford (MIPS), Department of Radiology, Stanford University School of Medicine, Stanford, CA, United States

ELSEVIER

ACADEMIC PRESS

An imprint of Elsevier

Academic Press is an imprint of Elsevier
125 London Wall, London EC2Y 5AS, United Kingdom
525 B Street, Suite 1650, San Diego, CA 92101, United States
50 Hampshire Street, 5th Floor, Cambridge, MA 02139, United States
The Boulevard, Langford Lane, Kidlington, Oxford OX5 1GB, United Kingdom

Notices
Knowledge and best practice in this field are constantly changing. As new research and experience broaden our understanding, changes in research methods, professional practices, or medical treatment may become necessary.

Practitioners and researchers must always rely on their own experience and knowledge in evaluating and using any information, methods, compounds, or experiments described herein. In using such information or methods they should be mindful of their own safety and the safety of others, including parties for whom they have a professional responsibility.

To the fullest extent of the law, neither the Publisher nor the authors, contributors, or editors, assume any liability for any injury and/or damage to persons or property as a matter of products liability, negligence or otherwise, or from any use or operation of any methods, products, instructions, or ideas contained in the material herein.

Library of Congress Cataloging-in-Publication Data
A catalog record for this book is available from the Library of Congress

British Library Cataloguing-in-Publication Data
A catalogue record for this book is available from the British Library

ISSN: 2468-3183

ISBN: 978-0-12-821567-8

For information on all Academic Press publications
visit our website at https://www.elsevier.com/books-and-journals

Publisher: Stacy Masucci
Senior Acquisitions Editor: Rafael Teixeira
Editorial Project Manager: Samantha Allard
Production Project Manager: Punithavathy Govindaradjane
Cover Designer: Greg Harris

Typeset by SPi Global, India

Cover Image Legend

Temozolomide (TMZ) chemotherapy induces the methylation of DNA guanine residues at the O6 position (O6-methylguanine). This treatment of glioblastomas can effectively target differentiated glioma cells that are highly proliferative, where defective mismatch repair can lead to DNA strand breaks and cell death. However, slow-dividing or quiescent glioma stem cells remain unaffected because the expression of methylguanine methyltransferase (MGMT) neutralizes the O6-methylguanine adduct and restores DNA integrity. This leads to cell survival, tumor relapse, and chemoresistance (see Fares et al., Chapter 23).

Aims and Scope for Series "Cancer Sensitizing Agents for Chemotherapy"

Current cancer management strategies fail to adequately treat malignancies with chemotherapy, with multivariable dose-restrictive factors such as systemic toxicity and multidrug resistance; hence, limiting therapeutic benefits, quality of life, and complete long-term remission rates. The resistance of cancer cells to anticancer drugs is one of the major reasons for the failure of traditional cancer treatments. Cellular components and dysregulation of signaling pathways contribute to drug resistance. If modulated, such perturbations may restore the drug response and its efficacy. The recent understanding of the molecular mechanisms and targets that are implicated for cancer chemoresistance has paved the way to develop a large battery of small molecules (sensitizing agents) that can target resistant factors and reduce the threshold of resistance, thus allowing their combination with chemotherapeutic drugs to be effective and to reverse the chemoresistance. A large variety of chemotherapy-sensitizing agents has been developed and several have been shown to be effective in experimental models and in cancer patients.

The main objective of the proposed series "Cancer Sensitizing Agents for Chemotherapy" is to publish individualized and focused volumes whereby each volume is edited by an invited expert Editor(s). Each volume will dwell on specific chemosensitizing agents with similar targeting activities. The combination treatment of the sensitizing agent with chemotherapy may result in a synergistic/additive activity and the reversal of tumor cells resistance to drugs.

The Editor(s) will compile nonoverlapping review chapters on reported findings in various cancers, both experimentally and clinically, with particular emphases on underlying biochemical, genetic, and molecular mechanisms of the sensitizing agent and the combination treatment.

The scope of the series is to provide scientists and clinicians with updated basic and clinical information that will be valuable in their quest to investigate, develop, and apply novel combination therapies to reverse drug resistance, thereby prolonging survival and even resulting in a cure in cancer patients.

Dr. Benjamin Bonavida, PhD (Series Editor)

About the Series Editor

Dr. Benjamin Bonavida, PhD, (Series Editor), is currently Distinguished Research Professor at the University of California, Los Angeles (UCLA). He is affiliated with the Department of Microbiology, Immunology and Molecular Genetics, UCLA David Geffen School of Medicine. His research career, thus far, has focused on investigations in the fields of basic immunochemistry and cancer immunobiology. His research investigations have ranged from the biochemical, molecular, and genetic mechanisms of cell-mediated killing and tumor cell resistance to chemo-immuno cytotoxic drugs. The reversal of tumor cell resistance was investigated by the use of various selected sensitizing agents based on molecular mechanisms of resistance. In these investigations, there was the newly characterized dysregulated NF-κB/Snail/YY1/RKIP/PTEN loop in many cancers that was reported to regulate cell survival, proliferation, invasion, metastasis, and resistance. Emphasis was focused on the roles of the tumor suppressor Raf Kinase Inhibitor Protein (RKIP), the tumor promoter Yin Yang 1 (YY1), and the role of nitric oxide as a chemo-immuno-sensitizing factor. Many of the earlier mentioned studies are centered on the clinical challenging features of cancer patients' failure to respond to both conventional and targeted therapies.

The Editor has been active in the organization of regular sequential international miniconferences that are highly focused on the roles of YY1, RKIP, and nitric oxide in cancer and their potential therapeutic applications. Several books edited or coedited by the Editor have been published. In addition, the Editor has been the Series Editor of books (over 23) published by Springer on *Resistance to Anti-Cancer Targeted Therapeutics*. In addition, the Editor is presently the Series Editor of three series published by Elsevier/Academic Press on *Cancer Sensitizing Agents for Chemotherapy*, *Sensitizing Agents for Cancer Resistant to Cell Mediated Immunotherapy*, and *Breaking Tolerance to Anti-Cancer Immunotherapy*. Lastly, the Editor is the Editor-in-Chief of the Journal Critical Reviews in Oncogenesis. The Editor has published over 500 research publications and reviews in various scientific journals of high impact.

Acknowledgments: The Series Editor acknowledges the Department of Microbiology, Immunology and Molecular Genetics and the UCLA David Geffen School of Medicine for their continuous support. The Series Editor also acknowledges the assistance of Mr. Rafael Teixeira, Acquisitions Editor for Elsevier/Academic Press, and the excellent assistance and support of Ms. Samantha Allard, Editorial Project Manager for Elsevier/Academic Press, for their continuous cooperation throughout the development of this book.

Aims and Scope of the Volume

Glioblastoma (GBM) is the most malignant primary brain tumor in humans, with extremely poor survival after diagnosis. The dismal prognosis of GBM is multifactorial but attributable primarily to its resistance to current therapeutic approaches. This contributes to virtually inevitable tumor recurrence in GBM patients following treatment. Thus, it is imperative that molecular and cellular mechanisms at the root of chemotherapy-resistant GBM cells, as well as therapeutic strategies available to counter this resistance are understood. Indeed, a novel tactic in combating this deadly disease is to target its mechanisms of chemoresistance. This unique book consists of 29 chapters providing a comprehensive resource on the latest developments in the field of GBM resistance to chemotherapy, including the implicated biological mechanisms that can inform the creation of innovative therapeutic strategies to reverse it. This information will be valuable to basic and clinical neuroscientists with scientific interest or practical clinical involvement in the management of GBM patients, including neuroscientists, neuro-oncologists, neurosurgeons, radiation therapists, neuropathologists, neurointensivists, and neuroradiologists. We aim to provide a comprehensive core knowledge related to the entire discipline of GBM chemoresistance, from its many etiological mechanisms to specific strategies to reverse this resistance. This book brings together current perspectives from an international team of experts on the basic science, preclinical research, and clinical management of GBM chemoresistance.

Ramasamy Paulmurugan, PhD
Tarik F. Massoud, MD, PhD

About the Volume Editors

Ramasamy Paulmurugan is an Associate Professor in the Department of Radiology, Stanford University School of Medicine, California, USA. He received his PhD in 1997 from the National Environmental Engineering Research Institute (NEERI), University of Madras, Chennai, India. After serving as a scientist for 4 years in the Rajiv Gandhi Center for Biotechnology, Trivandrum, India, he joined the University of California at Los Angeles (UCLA) as a visiting scientist in 2001 to work on developing in vivo molecular imaging assays to study protein-protein interactions and protein folding. In 2003, he joined Stanford University as a senior research scientist to work in the Molecular Imaging Program at Stanford University (MIPS). Since 2009, he has been an academic member of faculty in MIPS and the Department of Radiology at Stanford. Currently, his preclinical research laboratory develops in vivo molecular imaging assays to noninvasively monitor different cancer epigenetic processes, as well as designs and refines novel molecularly targeted therapies using microRNAs and gene therapies against various cancers, such as breast, hepatocellular carcinoma, and glioblastoma. His laboratory developed a clinically translatable ultrasound-mediated microbubble targeted gene delivery system as a novel approach to targeted cancer therapy. He has published several manuscripts in high impact journals such as *PNAS*, *Cancer Research*, *ACS Nano*, *Nature Medicine*, *Advanced Materials*, *Advanced Therapeutics*, *Molecular Therapy*, *Radiology*, *Journal of Controlled Release*, and *Science Translational Medicine*.

Tarik F. Massoud, MD, PhD is a physician-scientist and Professor of Neuroradiology at Stanford University School of Medicine, where he also directs the Laboratory of Experimental and Molecular Neuroimaging (LEMNI) in the Molecular Imaging Program at Stanford, and is a member of the Stanford Cancer Institute. He qualified from the Medical School of the Royal College of Surgeons in Ireland, and trained in Radiology and Neuroradiology in Oxford, UCLA, and the University of Michigan. He holds an additional research doctorate in experimental neuroimaging, as well as a University of Cambridge PhD in molecular imaging of cancer. From 2000 to 2013 he was a neuroradiology faculty member at the University of Cambridge School of Clinical Medicine, UK. His general academic interests include molecular and translational imaging of the brain especially in neuro-oncology, experimental aspects of neuroimaging, and clinical neuroimaging of glioblastoma (GBM). In his LEMNI laboratory he develops novel theranostic applications for GBM to both interrogate fundamental cellular and molecular events, and for use in new anticancer therapeutic strategies. Generally, this includes preclinical in vivo imaging of gene expression using reporter assays, protein-protein interactions, and signal transduction, as well as cellular and nano-imaging. His achievements include development of a novel split reporter gene biosensor to image p53 protein folding in cancer to help discover new antimisfolding drugs, packaging of therapeutic microRNAs in novel nanoparticles delivered to GBM, and development of approaches for combined microRNA and drugs to treat and image GBMs via the intranasal route. His clinical neuroimaging practice affords him particular insights into the challenges of clinical translation of laboratory innovations within neuro-oncology.

Preface

Glioblastoma (GBM) is one of the most lethal human cancers—a scourge of mankind. It has dismal survival despite conventional multimodal therapies that include surgery, radiation, and chemotherapy. Little progress has been made in GBM treatment, with the last key advances being the standardized implementation of temozolomide (TMZ), along with radiotherapy, and the use of tumor-treating fields for new GBMs. New therapeutic approaches are clearly warranted.

GBM is an extremely complex disease that resists therapeutic agents through highly sophisticated mechanisms that help it to recognize and adapt to constant threats. While the GBMs of many patients initially respond to chemotherapies, no current drug regimen can overcome inevitable tumor recurrence, the hallmark of which is chemoresistance. There is currently no standard of care for these recurrent GBMs. Virtually all patients in developed countries die owing to recurrent disease, and not because of primary GBM.

Temozolomide (TMZ) has been used as standard chemotherapy for GBM since 2005. However, despite increasing patient survival, treatment efficacy is limited, as GBMs frequently show innate or acquired resistance to TMZ. Understanding the mechanisms of this chemoresistance and identifying novel strategies to overcome it remain an important goal in GBM patient management. Our understanding of these mechanisms in GBM is rapidly expanding and could provide an important basis for future individualized treatments according to different resistance mechanisms.

The phenomenon of GBM chemoresistance is related to its biological complexity and to different cellular and molecular mechanisms activated during tumorigenesis and progression. Thus, GBM cells (including their stem cells) may alter drug targets through mutation or reduced expression, upregulate the expression of drug pumps, increase the activity or expression of drug detoxifying mechanisms, reduce their susceptibility to senescence and apoptosis, alter their level of proliferation, and increase their ability to repair DNA damage. Hence, the main protagonists in chemoresistance include highly therapy-resistant populations of glioma stem cells, aided and abetted by the enormous intratumoral and intertumoral heterogeneity that drastically reduces the efficacy of targeted agents, and the dysregulated cellular metabolism that has evolved to take advantage of the nutrient-rich surroundings. In addition, the poor permeability of anticancer drugs through the blood-brain barrier and their inability to reach GBMs at therapeutic concentrations decreases drug effectiveness and contributes to treatment failure. Finally, the immunosuppressive feature of the tumor microenvironment allows cancer cells to evade immune surveillance.

A novel tactic in combating this deadly disease is to target its mechanisms of chemoresistance. Many such mechanisms may be concurrent, but there is considerable heterogeneity within and between tumors to enable escape mechanisms from targeted agents. Hence the need for individualized approaches to GBM treatment. Our knowledge of the plethora of cellular pathways and mechanisms involved continues to provide valuable biological and therapeutic insights into the resistance strategies of GBMs. Given these obstacles, it has become clear from mechanistic and preclinical studies that the key to GBM treatment will be multimodal therapies that synergistically attack the cancer on several fronts.

Until now, no single book has been available as a comprehensive source of information and knowledge on this complex topic. This book consists of 29 chapters providing a comprehensive resource of the latest developments in the field of GBM resistance to chemotherapy, including the molecular and cellular mechanisms implicated in this chemoresistance, and which can inform the creation of innovative therapeutic strategies to reverse it.

The book is organized into six sections that combine to provide a broad core knowledge related to the entire discipline of GBM chemoresistance, defined as the inability of current treatments to destroy or arrest the growth of an entire GBM mass through either innate or acquired mechanisms. The authors discuss the many etiological molecular and cellular mechanisms that drive the formation of this innate and acquired resistance, as well as specific therapeutic strategies to reverse it. Innate therapy resistance is characterized by the presence of GBM cells that possess certain genetic and/or epigenetic mechanisms to overcome the actions of chemoradiation, leading to the regrowth of residual tumor cells following treatment. These accumulated GBM cells are selected in the course of standard treatments. Acquired resistance implicates phenotypic alterations of naive GBM cells to become more aggressive tumors after chemoradiation. This resistance emerges over time, possibly owing to stimulated genetic instability, accumulated gene mutations, and/or epigenetic alterations of GBM cells following treatment. Unfortunately, this diminishes the efficacy of subsequent rounds of chemotherapy for recurrent tumors after the initial short-term disease control for primary tumors.

Overall, we hope that the fine collection of up-to-date reviews contained within this book will represent a useful foundation of knowledge for scientists and clinicians dealing with the unique difficulties of GBM and its treatment. The next big challenge in the management of GBM patients will be to scale up these concepts, tools, efforts, and directions. For instance, this will require new strategies in designing future therapies that better sensitize GBMs to TMZ and that target resistant cells. In the midst of tumor heterogeneity, it is essential to identify and characterize which types of GBM cells can evade therapies and become recurrence-initiating cancer cells so that they can be targeted appropriately. Greater knowledge of the genetic evolution of GBM cell populations after therapy, and the ability of GBM cells to adopt stemlike characteristics will also provide us with invaluable new insights to tackle this recalcitrant cancer. Moreover, in this era of precision medicine, many more molecular techniques will need to be developed to allow targeting of each patient's unique tumor, in the hope that this personalized approach will generate useful information to bypass chemoresistance. Each resistance mechanism presents formidable challenges to achieving effective GBM treatment, but as we gain more insights in surmounting these difficulties, future improved drugs may have more promise for GBM patients.

Finally, we are immensely grateful to each one of the international team of expert contributors to this book who have provided readers with the most up-to-date knowledge on the basic science, preclinical research, and clinical management of GBM and its chemoresistance. We ultimately hope that this book will serve as an important and comprehensive source of information for basic and clinical scientists involved in the management of GBM patients, including neuroscientists, neuro-oncologists, neurosurgeons, radiation therapists, neuropathologists, neurointensivists, and neuroradiologists, as well as clinical and research trainees in these disciplines.

We wish to express our gratitude to Dr. Benjamin Bonavida, series editor, and Elsevier for their excellent support throughout the process leading to the completion of this book.

Ramasamy Paulmurugan, PhD
Cellular Pathway Imaging Laboratory (CPIL), Molecular Imaging Program at Stanford (MIPS), Department of Radiology, Stanford University School of Medicine, Palo Alto, CA, United States
Tarik F. Massoud, MD, PhD
Laboratory of Experimental and Molecular Neuroimaging (LEMNI), Molecular Imaging Program at Stanford (MIPS), Department of Radiology, Stanford University School of Medicine, Stanford, CA, United States

Contents

Contributors xxv

1. Overview of glioblastoma biological hallmarks and molecular pathology

Kirsten Ludwig, Sree Deepthi Muthukrishnan, Alvaro G. Alvarado, and Harley I. Kornblum

Introduction 2
Biological hallmarks 3
Molecular pathology 7
Molecular markers 8
Epigenetic markers 10
Conclusions and future perspectives 11
References 11

2. Past and present drug treatments for glioblastoma

Yasmeen Rauf and Manmeet S. Ahluwalia

Introduction 18
Brief history of glioblastoma 18
Chemotherapy 23
Tyrosine kinase pathway 24
Immunotherapy 26
Conclusion 26
References 27

3. Investigational new drugs against glioblastoma

David Olayinka Kamson, Harmon Singh Khela, and John Laterra

Introduction 32
Agents that target DNA integrity 36
Agents that target epigenetic determinants of chromosome structure and gene transcription 40
Agents that directly alter cell cycle or cell death regulation 42
Agents that directly alter protein homeostasis 49
Targeting upstream molecular pathways 51
Synthetic lethality and drug delivery 57
Conclusions 57
Acknowledgments 58
References 58

4. Molecular mechanisms in temozolomide-resistant glioblastoma

Sang Y. Lee and Gaspar J. Kitange

Introduction 81
Molecular events in GBM cells after TMZ treatment 83
Molecular events in TMZ-resistant GBM cells after TMZ treatment 84
Therapeutic approaches 96
Methods for identification of novel targets for sensitizing TMZ 114
Conclusions and future directions 120
Acknowledgment 120
References 121

5. Glioma stem cells and associated molecular mechanisms in Glioblastoma Chemoresistance

Georgios Solomou and Colin Watts

Introduction 136
Current therapies and the role of GSCs 137
Glioma stem-like cell biology aiding survival 138
Specific molecular pathways of GSC resistance to chemotherapy 142
Conclusion 144
References 144

6. Signaling in the tumor microenvironment of therapy-resistant glioblastoma

Marina Nikolopoulos and Sunit Das

Introduction 154
TME 155
Signals from the ECM 156
Tumor/stroma interactions 158

Phospholipase C (PLC/PKC) 164
Effects of therapy on the TME 170
Targeting the TME 171
Conclusion 172
References 172

7. Intratumoral heterogeneity associated with glioblastoma drug response and resistance

Sabra K. Salim, Neil Savage, William T. Maich, Chitra Venugopal, and Sheila K. Singh

Introduction 186
Genetic and transcriptomic ITH 187
Cellular and functional ITH 188
Molecular mechanisms of treatment resistance 188
Cell-intrinsic and -extrinsic mechanisms of resistance 192
Conclusions 194
References 195

8. Role of apoptosis, autophagy, and the unfolded protein response in glioblastoma chemoresistance

Sanaz Dastghaib, Sima Hajiahmadi, Amir Seyfoori, Meitham Amereh, Mozhdeh Zamani, Zahra Shahsavari, Shahla Shojaei, Mohsen Akbari, Pooneh Mokarram, and Saeid Ghavami

Introduction 204
Apoptosis 206
P53 208
Apoptosis and chemoresistance 208
Chemoresistance in glioblastoma 209
Autophagy pathway 210
The dual roles of autophagy in glioblastoma promotion and suppression 213
Endoplasmic reticulum stress, and unfolded protein response (UPR) 219
Cross talk between UPR, autophagy, and apoptosis 221
Role of UPR in glioblastoma chemoresistance 221
Role of UPR branches in GBM-induced chemoresistant to TMZ 223
UPR affects chemotherapy-induced cell death in GBM 224
How simvastatin/TMZ affect GBM chemoresistance 225
Tissue-engineered tumor models 226
Bioengineered glioblastoma and neuroblastoma in vitro models 227
Brain tumor tissue-engineered models for chemotherapeutic evaluations 229
Conclusion 232
References 232

9. Role of multidrug resistance in glioblastoma chemoresistance: Focus on ABC transporters

Mohammed Ahmed, Maite Verreault, Xavier Declèves, and Ahmed Idbaih

Introduction 244
Structure and functions of ABC proteins 246
ABC transporters in glioblastoma 248
Clinical value 254
Conclusion 257
Acknowledgments 257
Authors' contributions 257
References 257

10. Role of mitochondrial reactive oxygen species in glioblastoma drug resistance and strategies for therapeutic intervention

Alessia Lo Dico, Daniela Salvatore, Cecilia Diceglie, Cristina Martelli, and Luisa Ottobrini

Introduction 265
Reactive oxygen species in GBM 271
Conclusions 285
Acknowledgments 286
References 286

11. Glioblastoma cell-induced immunosuppression causing chemoresistance

Peter P. Sordillo and Laura A. Sordillo

Introduction 294
The glioblastoma secretome 295
Cytokine-induced immunosuppression in glioblastoma 295
Indoleamine 2,3-dioxygenase and the kynurenine pathway 302
Programmed death-ligand 1 (PD-L1) (cluster of differentiation, CD 274) 306
References 307

12. Role of cell quiescence in glioblastoma cytotoxic resistance and strategies for therapeutic intervention

Ryan J. Atkins, Stanley S. Stylli, Niall M. Corcoran, Christopher M. Hovens, and Theo Mantamadiotis

Introduction 320
Cell quiescence 323
Quiescence and treatment resistance 324
Molecular drivers of quiescence 324
Targeting quiescent GBM tumor cells 325
Methods to identify and isolate quiescent cells 326
Vital dye retention, proliferative potential, and cytotoxic sensitivity in glioblastoma cells 327
Conclusions 331
Acknowledgments 331
References 331

13. The molecular and cellular effects of radiotherapy-induced microenvironment changes on potential chemoresistance in glioblastoma

Kshama Gupta

Introduction 337
Extracellular microenvironment of the brain in glioblastoma 338
Radiation-induced changes in the brain and GBM microenvironment 345
Therapeutic implications and future directions 351
Conclusions 352
References 352

14. Molecular and cellular mechanisms in recurrent glioblastoma chemoresistance

Sabrina Battista, Laura Cerchia, and Monica Fedele

Introduction 367
Recurrent glioblastoma 368
Glioblastoma stem cells 369
Other intrinsic and extrinsic mechanisms involved in the recurrence of GBM 381
Conclusion 390
Acknowledgment 391
References 391

15. Role of long noncoding RNAs in temozolomide-resistant glioblastoma

M. Janaki Ramaiah, Shaik Mohammad Naushad, Rajasekhar Reddy Manyam, and Vijay Kumar Kutala

Introduction 403
GBM cancer 403
Long noncoding RNA 405
Cancer susceptibility candidate 2 413
Key cellular components in therapy resistance 415
Long noncoding RNA (lncRNA)-mediated gene silencing through polycomb mechanisms 415
Noncoding RNA epigenetics 416
lncRNA classification 418
Functions of LncRNA in the nucleus 419
Functions of LncRNAs in regulating gene expression in the cytoplasm 419
Mechanisms of actions of lncRNA 419
LncRNA classification based on region implicated by the lncRNA 420
Conclusions 420
Acknowledgments 421
References 421

16. Mechanisms of glioblastoma resistance to antiangiogenic agents and reversal approaches

Kartik Angara and Ali S. Arbab

Introduction 430
Tumor neovascularization through angiogenesis, vasculogenesis, and vascular mimicry 431
Antiangiogenic therapies and development of resistance 434
Mechanisms of AAT resistance and possible intervention 436
References 444

17. Treatment and resistance of glioblastoma to CAR T-cell immunotherapies

Stephen J. Bagley

Introduction to CAR T-cell therapy 454
CAR T-cell therapy in GBM: Recent clinical advances 456

Challenges and mechanisms of resistance to CAR T-cell therapy in GBM 458
Future directions to overcome resistance 460
Conclusions 464
References 464

18. On-chip analysis of glioblastoma cell chemoresistance

Chi Yan Wong, Huilin Shao, and Hakho Lee

Introduction 474
Exosomal RNA analysis for monitoring chemoresistance in GBM 475
Outlook 486
Acknowledgments 487
References 488

19. Three-dimensional in vitro models to study potentiation of glioblastoma resistance by hypoxia

Ahmed Musah-Eroje

Introduction 492
Hypoxia as a tumor microenvironment component in glioblastoma 493
In vitro 3D model for glioblastoma 494
Conclusion 494
References 495

20. Chemoresistance mechanisms in mouse models of glioblastoma

Takamasa Kinoshita, Masafumi Miyai, Toru Iwama, Akira Hara, and Hiroyuki Tomita

Introduction 498
Factors involved in GBM chemoresistance 498
Mouse models 500
Conclusion 503
Acknowledgment 503
References 503

21. Therapeutic approaches to overcome temozolomide resistance in glioblastoma

Marina Gergues, Vibha Harindra Savanur, David Prado, Adam Eljarrah, James K. Liu, Mohammed Jaloudi, and Pranela Rameshwar

Introduction 508
Targeting DNA repair mechanisms 509
Targeting receptor tyrosine kinases and cellular pathways 513
Antiangiogenic therapies 526
Targeting cancer stem cells 529
Tumor treating fields 531
Alternative delivery methods 533
Conclusion 535
References 535

22. Understanding signal transduction pathways to overcome targeted therapy resistance in glioblastoma

Arata Tomiyama

Introduction 550
Current status of chemotherapy in GBM treatment 550
Molecular mechanisms of TMZ resistance in GBM cells 551
Chemoresistance of GBM induced by plasma membrane-localizing receptor-mediated survival signaling 557
Chemoresistance of GBM induced by Hippo/Yes associated-protein 1 (YAP)- mediated survival signaling 567
Chemoresistance of GBMs induced by p16INK4a/p14ARF-mediated survival signaling 567
Chemoresistance induced by cancer stem-like cells, TME, heterogeneity, and plasticity in GBMs 569
miRNA-mediated chemoresistance 574
Conclusions and perspectives 574
References 575

23. Targeting the molecular mechanisms of glioma stem cell resistance to chemotherapy

Jawad Fares, Deepak Kanojia, Alex Cordero, Ilya Ulasov, and Maciej S. Lesniak

Introduction 590
The adaptive microenvironment of glioma stem cells 592
Mechanisms of glioma stem cell resistance to chemotherapy 595
Therapeutic strategies to overcome glioma stem cell resistance to chemotherapy 598
Conclusion 620
References 621

24. Immunotherapy for glioblastoma as a means to overcome resistance to standard therapy

Kristen A. Batich, John H. Sampson, Mustafa Khasraw, Henry S. Friedman, and David M. Ashley

Introduction: Evolution of therapy for glioblastoma 637
Immune surveillance and immunotherapy strategies in translation 641
Lymphatic drainage in the CNS to secondary lymphoid organs 642
Overcoming limited migration of DC therapies 645
Immune checkpoint inhibitors in glioma 647
Mechanisms of chemotherapy resistance and implications for immune activation 648
Conclusion 650
References 653

25. Role of microRNA therapy in presensitizing glioblastoma cells to temozolomide treatment

Ramasamy Paulmurugan and Tarik F. Massoud

Introduction 668
Role of microRNAs in cells 669
Regulatory role of miRNAs in cancer 670
Regulatory role of microRNAs in GBM biology 672
MiRNAs role in therapeutic sensitization of GBM to chemotherapy 677
The role of miRNAs in therapeutic sensitization of GBM to TMZ 677
MiRNAs targeting GSCs to achieve GBM therapy 682
MicroRNAs targeting multiple cellular mechanisms as potential future therapies for GBM 683
Conclusion and future perspectives 684
References 685

26. Drug penetration through the blood–brain barrier after radiotherapy: New approaches to bypass glioblastoma chemoresistance

Iacopo Sardi, Milena Guidi, Laura Giunti, Anna Lisa Iorio, Giacomo Casati, Maria Luigia Censullo, and Barbara Rombi

Introduction 690
The molecular mechanisms and genetics of chemoresistance in glioblastoma 691
Alterations in the BBB permeability to increase the action of doxorubicin 692
Consolidated and new treatments for glioblastoma 695
Ionizing radiation interferes with BBB permeability by facilitating the passage of drugs 697
Conclusion 699
Acknowledgment 700
References 700

27. Nanoparticles beyond the blood-brain barrier for glioblastoma

Lynn Jena, Nicholas J. Dunne, and Helen O. McCarthy

Introduction 709
Barriers to delivery 710
Emerging delivery strategies to mitigate the BBB 718
Conclusion 734
References 736

28. Small molecules targeting misfolded mutants of p53 as a rescue strategy to improve glioblastoma chemotherapy

Arutselvan Natarajan, Husam A. Babikir, Ramasamy Paulmurugan, and Tarik F. Massoud

Introduction 750
History of the tumor suppressor p53 750
p53 regulation 751
p53 structure 751
The p53 pathway 751
p53 mutants 753
Distribution and frequency of p53 mutations and their clinical relevance 754
p53 with single allele, both allele and null mutants 754
Therapeutic strategies targeting p53 756
Therapeutic approaches to restore wild-type p53 function of mutant p53 protein via restoration of structural stability 757
Considerations and principles of drugs that rescue mutant p53 structure 757
Assay systems to identify new drugs stabilizing p53 protein structure 758
Drugs that rescue the deleterious effect of p53 structural mutations 759
Limitations of compounds targeting structural mutations of p53 764
Synergizing the effect of drugs targeting mutant p53 765
Conclusion 766
References 766

29. p53 supplementation as a targeted cancer gene therapy for glioblastoma

Uday Kumar Sukumar, Tarik F. Massoud, and Ramasamy Paulmurugan

Role of p53 in cell cycle regulation 774
TP53 mutation and its relevance to glioblastoma development 774
Correlation of TP53 status to grades of gliomas 776
TP53-based therapeutic strategies for glioblastoma 777
p53-targeted cancer gene therapy 779
Clinical status of p53 cancer gene therapy 782
Future perspectives and therapeutic potentials 783
References 784

Index 787

Contributors

Manmeet S. Ahluwalia Burkhardt Brain Tumor and Neuro-Oncology Center, Cleveland Clinic, Cleveland, OH, United States

Mohammed Ahmed Sorbonne Université, Institut du Cerveau - Paris Brain Institute - ICM, Inserm, CNRS, APHP, Hôpital de la Pitié Salpêtrière, Paris; Faculté de Médecine Paris-Sud - Université Paris-Saclay, Saint-Aubin, France

Mohsen Akbari Laboratory for Innovations in Microengineering (LiME), Department of Mechanical Engineering; Centre for Biomedical Research (CBR); Center for Advanced Materials and Related Technologies, University of Victoria, Victoria, BC, Canada

Alvaro G. Alvarado The Intellectual and Developmental Disabilities Research Center, The Semel Institute for Neuroscience and Human Behavior, David Geffen School of Medicine at UCLA, Los Angeles, CA, United States

Meitham Amereh Laboratory for Innovations in Microengineering (LiME), Department of Mechanical Engineering, University of Victoria, Victoria, BC, Canada

Kartik Angara Caris Life Sciences, Phoenix, AZ, United States

Ali S. Arbab Georgia Cancer Center, Augusta University, Augusta, GA, United States

David M. Ashley Preston Robert Tisch Brain Tumor Center at Duke; Department of Neurosurgery; Duke Cancer Institute, Duke University Medical Center, Durham, NC, United States

Ryan J. Atkins Department of Surgery (RMH), The University of Melbourne, Parkville, VIC, Australia

Husam A. Babikir Laboratory of Experimental and Molecular Neuroimaging (LEMNI), Molecular Imaging Program at Stanford (MIPS), Department of Radiology, Stanford University School of Medicine, Stanford, CA, United States

Stephen J. Bagley Division of Hematology/Oncology, Department of Medicine, Perelman School of Medicine at the University of Pennsylvania, Philadelphia, PA, United States

Kristen A. Batich Preston Robert Tisch Brain Tumor Center at Duke; Department of Neurosurgery, Duke University Medical Center, Durham, NC, United States

Sabrina Battista Institute for Experimental Endocrinology and Oncology (IEOS), National Research Council (CNR), Naples, Italy

Giacomo Casati Neuro-Oncology Unit, Department of Pediatric Oncology, Meyer Children's Hospital, Florence, Italy

Maria Luigia Censullo Neuro-Oncology Unit, Department of Pediatric Oncology, Meyer Children's Hospital, Florence, Italy

Laura Cerchia Institute for Experimental Endocrinology and Oncology (IEOS), National Research Council (CNR), Naples, Italy

Niall M. Corcoran Department of Surgery (RMH), The University of Melbourne, Parkville, VIC, Australia

Alex Cordero Department of Neurological Surgery, Feinberg School of Medicine, Northwestern University, Chicago, IL, United States

Sunit Das Institute for Medical Science; Division of Neurosurgery, St. Michael's Hospital, University of Toronto, Toronto, ON, Canada

Sanaz Dastghaib Endocrine and Metabolism Research Center, Shiraz University of Medical Sciences, Shiraz, Iran

Xavier Declèves Inserm, UMR-S 1144, Université de Paris, Therapeutic Optimization in Neuropsychopharmacology; Biologie du Médicament et Toxicologie, Hôpital Cochin,

Assistance Publique-Hôpitaux de Paris, Paris, France

Cecilia Diceglie Department of Pathophysiology and Transplantation, University of Milan, Segrate, MI, Italy

Nicholas J. Dunne School of Mechanical and Manufacturing Engineering, Dublin City University; Centre for Medical Engineering Research, School of Mechanical and Manufacturing Engineering, Dublin City University; Department of Mechanical and Manufacturing Engineering, School of Engineering, Trinity College Dublin; Advanced Manufacturing Research Centre (I-Form), School of Mechanical and Manufacturing Engineering, Dublin City University; Advanced Materials and Bioengineering Research Centre (AMBER), Royal College of Surgeons in Ireland and Trinity College Dublin; Advanced Processing Technology Research Centre, Dublin City University; Trinity Centre for Biomedical Engineering, Trinity Biomedical Sciences Institute, Trinity College Dublin, Dublin, Ireland

Adam Eljarrah Department of Medicine, Rutgers New Jersey Medical School, Newark, NJ, United States

Jawad Fares Department of Neurological Surgery, Feinberg School of Medicine, Northwestern University, Chicago, IL, United States

Monica Fedele Institute for Experimental Endocrinology and Oncology (IEOS), National Research Council (CNR), Naples, Italy

Henry S. Friedman Preston Robert Tisch Brain Tumor Center at Duke; Department of Neurosurgery; Duke Cancer Institute, Duke University Medical Center, Durham, NC, United States

Marina Gergues Department of Medicine; School of Graduate Studies, Rutgers New Jersey Medical School, Newark, NJ, United States

Saeid Ghavami Autophagy Research Center, Shiraz University of Medical Sciences, Shiraz, Iran; Department of Human Anatomy and Cell Science, Max Rady College of Medicine, Rady Faculty of Health Sciences; Research Institute in Oncology and Hematology, CancerCare Manitoba, University of Manitoba, Winnipeg, MB, Canada; Faculty of Medicine, Katowice School of Technology, Katowice, Poland; Children's Hospital Research Institute of Manitoba, Winnipeg, MB, Canada

Laura Giunti Medical Genetics Unit, Meyer Children's Hospital, Florence, Italy

Milena Guidi Neuro-Oncology Unit, Department of Pediatric Oncology, Meyer Children's Hospital, Florence, Italy

Kshama Gupta Departments of Cancer Biology and Radiation Oncology, Mayo Clinic, Jacksonville, FL, United States

Sima Hajiahmadi Department of Clinical Biochemistry, Faculty of Medicine, Shahid Beheshti University of Medical Sciences, Tehran, Iran

Akira Hara Department of Tumor Pathology, Gifu University Graduate School of Medicine, Gifu, Japan

Christopher M. Hovens Department of Surgery (RMH), The University of Melbourne, Parkville, VIC, Australia

Ahmed Idbaih Sorbonne Université, Inserm, CNRS, UMR S 1127, Institut du Cerveau, ICM, AP-HP, Hôpitaux Universitaires La Pitié Salpêtrière - Charles Foix, Service de Neurologie 2-Mazarin, Paris, France

Anna Lisa Iorio Neuro-Oncology Unit, Department of Pediatric Oncology, Meyer Children's Hospital, Florence, Italy

Toru Iwama Department of Neurosurgery, Gifu University Graduate School of Medicine, Gifu, Japan

Mohammed Jaloudi Department of Medicine, Rutgers New Jersey Medical School; Rutgers Cancer Institute of New Jersey, University Hospital, Newark, NJ, United States

M. Janaki Ramaiah Functional Genomics and Disease Biology Laboratory, School of Chemical and Biotechnology, SASTRA Deemed University, Thanjavur, Tamil Nadu, India

Lynn Jena School of Pharmacy, Queen's University Belfast, Belfast, United Kingdom

David Olayinka Kamson Neuro-Oncology Branch, National Cancer Institute, National Institutes of Health; Department of Neurology, Johns Hopkins Hospital, Baltimore, MD, United States

Deepak Kanojia Department of Neurological Surgery, Feinberg School of Medicine, Northwestern University, Chicago, IL, United States

Mustafa Khasraw Preston Robert Tisch Brain Tumor Center at Duke; Department of Neurosurgery; Duke Cancer Institute, Duke University Medical Center, Durham, NC, United States

Harmon Singh Khela Department of Neurology, Johns Hopkins Hospital, Baltimore, MD, United States

Takamasa Kinoshita Department of Neurosurgery; Department of Tumor Pathology, Gifu University Graduate School of Medicine, Gifu, Japan

Gaspar J. Kitange Department of Radiation Oncology, Mayo Clinic, Rochester, MN, United States

Harley I. Kornblum The Intellectual and Developmental Disabilities Research Center, The Semel Institute for Neuroscience and Human Behavior, David Geffen School of Medicine at UCLA, Los Angeles, CA, United States

Vijay Kumar Kutala Department of Clinical Pharmacology and Therapeutics, Nizam's Institute of Medical Sciences, Hyderabad, Telangana, India

John Laterra Department of Neurology, Johns Hopkins Hospital, Baltimore, MD, United States

Hakho Lee Center for Systems Biology; Department of Radiology, Massachusetts General Hospital/Harvard Medical School, Boston, MA, United States

Sang Y. Lee Department of Neurology, The Johns Hopkins University School of Medicine, Baltimore, MD, United States

Maciej S. Lesniak Department of Neurological Surgery, Feinberg School of Medicine, Northwestern University, Chicago, IL, United States

James K. Liu Department of Neurosurgery, Rutgers New Jersey Medical School, Newark, NJ, United States

Alessia Lo Dico Department of Pathophysiology and Transplantation, University of Milan, Segrate, MI, Italy

Kirsten Ludwig The Intellectual and Developmental Disabilities Research Center, The Semel Institute for Neuroscience and Human Behavior, David Geffen School of Medicine at UCLA, Los Angeles, CA, United States

William T. Maich Department of Biochemistry, McMaster University, Hamilton, ON, Canada

Theo Mantamadiotis Department of Surgery (RMH); Department of Microbiology & Immunology, The University of Melbourne, Parkville, VIC, Australia

Cristina Martelli Department of Pathophysiology and Transplantation, University of Milan, Segrate, MI, Italy

Tarik F. Massoud Laboratory of Experimental and Molecular Neuroimaging (LEMNI), Molecular Imaging Program at Stanford (MIPS), Department of Radiology, Stanford University School of Medicine, Stanford, CA, United States

Helen O. McCarthy School of Pharmacy, Queen's University Belfast, Belfast, United Kingdom; School of Chemical Sciences, Dublin City University, Dublin, Ireland

Masafumi Miyai Department of Neurosurgery, Gifu University Graduate School of Medicine, Gifu, Japan

Shaik Mohammad Naushad Sandor Speciality Diagnostics Private Limited, Hyderabad, Telangana, India

Pooneh Mokarram Autophagy Research Center; Department of Biochemistry, Shiraz University of Medical sciences, Shiraz, Iran

Ahmed Musah-Eroje School of Life Sciences, University of Bedfordshire, Luton; School of Science and Technology, Nottingham Trent University, Nottingham, United Kingdom; Department of Anatomy, Niger Delta University, Amassama, Bayelsa State, Nigeria

Sree Deepthi Muthukrishnan The Intellectual and Developmental Disabilities Research Center, The Semel Institute for Neuroscience and Human Behavior, David Geffen School of Medicine at UCLA, Los Angeles, CA, United States

Arutselvan Natarajan Cellular Pathway Imaging Laboratory (CPIL), Molecular Imaging Program at Stanford (MIPS), Department of Radiology, Stanford University School of Medicine, Palo Alto, CA, United States

Marina Nikolopoulos Institute for Medical Science, University of Toronto, Toronto, ON, Canada

Luisa Ottobrini Department of Pathophysiology and Transplantation, University of Milan; Department of Molecular Bioimaging and Physiology (IBFM), CNR, Segrate, MI, Italy

Ramasamy Paulmurugan Cellular Pathway Imaging Laboratory (CPIL), Molecular Imaging Program at Stanford (MIPS), Department of Radiology, Stanford University School of Medicine, Palo Alto, CA, United States

David Prado Department of Medicine, Rutgers New Jersey Medical School, Newark, NJ, United States

Pranela Rameshwar Department of Medicine, Rutgers New Jersey Medical School, Newark, NJ, United States

Yasmeen Rauf Burkhardt Brain Tumor and Neuro-Oncology Center, Cleveland Clinic, Cleveland, OH, United States

Rajasekhar Reddy Manyam Department of Computer Science and Engineering, Koneru Lakshmaiah Education Foundation, Vaddeswaram, Andhra Pradesh, India

Barbara Rombi Trento Proton Therapy Center, Azienda Provinciale per i Servizi Sanitari (APSS), Trento, Italy

Sabra K. Salim Department of Biochemistry, McMaster University, Hamilton, ON, Canada

Daniela Salvatore Department of Pathophysiology and Transplantation, University of Milan, Segrate, MI, Italy

John H. Sampson Preston Robert Tisch Brain Tumor Center at Duke; Department of Neurosurgery; Duke Cancer Institute, Duke University Medical Center, Durham, NC, United States

Iacopo Sardi Neuro-Oncology Unit, Department of Pediatric Oncology, Meyer Children's Hospital, Florence, Italy

Neil Savage Department of Biochemistry, McMaster University, Hamilton, ON, Canada

Vibha Harindra Savanur Department of Medicine; School of Graduate Studies, Rutgers New Jersey Medical School, Newark, NJ, United States

Amir Seyfoori Laboratory for Innovations in Microengineering (LiME), Department of Mechanical Engineering, University of Victoria, Victoria, BC, Canada; Biomaterials and Tissue Engineering Department, Breast Cancer Research Center, Motamed Cancer Institute, ACECR, Tehran, Iran

Zahra Shahsavari Department of Clinical Biochemistry, Faculty of Medicine, Shahid Beheshti University of Medical Sciences, Tehran, Iran

Huilin Shao Institute for Health Innovation and Technology; Department of Biomedical Engineering, Faculty of Engineering; Department of Surgery, Yong Loo Lin School of Medicine, National University of Singapore; Institute of Molecular and Cell Biology, Agency for Science, Technology and Research, Singapore; National Neuroscience Institute, Singapore

Shahla Shojaei Department of Human Anatomy and Cell Science, Max Rady College of Medicine, Rady Faculty of Health Sciences, University of Manitoba, Winnipeg, MB, Canada

Sheila K. Singh Department of Biochemistry; Department of Surgery, McMaster University, Hamilton, ON, Canada

Georgios Solomou Institute of Cancer and Genome Sciences, University of Birmingham, Birmingham; School of Medicine, Keele University, Stoke on Trent, United Kingdom

Laura A. Sordillo Departments of Physics and Electrical Engineering, The City College of New York, New York, NY, United States

Peter P. Sordillo Department of Hematology-Oncology, Lenox Hill Hospital, New York, NY, United States

Stanley S. Stylli Department of Surgery (RMH), The University of Melbourne; Department of Neurosurgery, The Royal Melbourne Hospital, Parkville, VIC, Australia

Uday Kumar Sukumar Cellular Pathway Imaging Laboratory (CPIL), Molecular Imaging Program at Stanford (MIPS), Department of Radiology, Stanford University School of Medicine, Palo Alto, CA, United States

Hiroyuki Tomita Department of Tumor Pathology, Gifu University Graduate School of Medicine, Gifu, Japan

Arata Tomiyama Department of Neurosurgery, National Defense Medical College, Saitama; Department of Brain Disease Translational Research, Juntendo University Faculty of Medicine, Tokyo, Japan

Ilya Ulasov Group of Experimental Biotherapy and Diagnostics, Institute for Regenerative Medicine, Sechenov First Moscow State Medical University, Moscow, Russia

Chitra Venugopal Department of Surgery, McMaster University, Hamilton, ON, Canada

Maite Verreault Sorbonne Université, Institut du Cerveau - Paris Brain Institute - ICM, Inserm, CNRS, APHP, Hôpital de la Pitié Salpêtrière, Paris, France

Colin Watts Institute of Cancer and Genome Sciences, University of Birmingham; Department of Neurosurgery, University Hospitals Birmingham, Birmingham, United Kingdom

Chi Yan Wong Institute for Health Innovation and Technology; Department of Medicine, Yong Loo Lin School of Medicine, National University of Singapore, Singapore

Mozhdeh Zamani Autophagy Research Center, Shiraz University of Medical Sciences, Shiraz, Iran

CHAPTER

1

Overview of glioblastoma biological hallmarks and molecular pathology

Kirsten Ludwig, Sree Deepthi Muthukrishnan, Alvaro G. Alvarado, and Harley I. Kornblum

The Intellectual and Developmental Disabilities Research Center, The Semel Institute for Neuroscience and Human Behavior, David Geffen School of Medicine at UCLA, Los Angeles, CA, United States

Abstract

Gliomas are the most malignant and aggressive form of brain tumors, and account for the majority of brain cancer-related deaths. Malignant gliomas, including glioblastoma are treated with radiation and temozolomide, with only a minor benefit in survival time. A number of advances have been made in understanding glioma biology, including the discovery of molecular heterogeneity between tumors from different patients as well as within tumors from the same patient. This chapter summarizes some of the biological hallmarks including pseudopalisading necrosis, microvascular hyperplasia and angiogenesis, local invasion, altered metabolism, the immunosuppressive microenvironment and heterogeneity, and cancer stem-like cells. We also discuss the molecular hallmarks of glioblastoma, which include 1p/19q codeletion, IDH, EGFR, p53, PTEN/Akt pathway, Rb, Ras/MAPK pathway, extrachromosomal DNA, MGMT, TERT, and ATRX. This chapter is meant to serve as a primer to glioblastoma and provides a brief overview of a very complex disease.

Abbreviations

ADAMS a disintegrin and metalloproteinases
α-KG alpha-ketoglutarate
ALT alternative lengthening of telomeres
ANG-2 angiopoientin-2
ATRX alpha-thalassemia/mental retardation syndrome X-linked
D-2HG D-2-hydroxy-glutarate
ecDNA extrachromosal DNA
ECM extracellular matrix
EGFR epidermal growth factor
EMT epithelial to mesenchymal transition
EPC endothelial progenitor cells
GBM glioblastoma

Glioblastoma Resistance to Chemotherapy: Molecular Mechanisms and Innovative Reversal Strategies
https://doi.org/10.1016/B978-0-12-821567-8.00020-8

GSC	glioma stem cells
HA	hyaluronic acid
HIF1α	hypoxia-inducible factor-1 alpha
IDH	isocitrate dehydrogenase
IL-8	interleukin-8
LOH	loss of heterozygosity
MDSC	myeloid-derived suppressor cells
MGMT	O6-methylguanin-DNA methyltransferase
MMP	matrix metalloproteinases
NF1	neurofibromin 1
PDGFRα	platelet derived growth factor receptor alpha
PD-L1	programmed death ligand-1
PI3K	phosphoinositide 3-kinase
PTEN	phosphate and tensin homolog
Rb	retinoblastoma
RTK	receptor tyrosine kinase
TAM	tumor-associated macrophages and microglia
TCGA	the cancer genome atlas
TERT	telomerase reverse transcription
TMB	tumor mutational burden
TMZ	temozolomide
TNM	tumor, node, metasis
T-regs	T-regulatory cells
VEGF	vascular endothelial growth factor
WHO	World Health Organization

Conflict of interest

No potential conflicts of interest were disclosed.

Introduction

The TNM (tumor, node, metastasis) system is used to stage most types of solid tumors for prognosis and treatment. Brain tumors, however, rarely metastasize, so malignancy is determined by a grading system developed by the World Health Organization (WHO), which is based on histopathological features, including extent of necrosis, microvascular proliferation, and tumor cell invasion. Grade I and II tumors are considered nonmalignant, and Grade III and IV tumors are malignant, Grade IV tumors are also termed as glioblastoma (GBM) [1]. Further subdivisions are based on other features of the tumor cells, including the predominance of oligodendrocytic or astrocytic characteristics, and the location of the tumor. Glioblastomas are the most common form of adult brain tumors and carry a dismal prognosis. Current standard of care treatments, which include radiation and chemotherapy with temozolomide, provide a survival benefit that can be measured in weeks rather than years [2]. Further understanding of basic GBM biology and translation of this understanding into effective treatments are critically needed.

Many recent studies have highlighted key molecular differences among tumors with seemingly similar pathological features. In 2008, The Cancer Genome Atlas (TCGA) consortium - published a now widely used classification of GBM. This study identified four different subclasses of GBM based on molecular markers: classical, mesenchymal, neural, and proneural, with more recent studies eliminating the neural subgroup [3]. These classifications, which were based on gene expression and reflected underlying oncogenic mutations, opened up a new wave of research aimed at defining the cellular and molecular basis of GBM. With ongoing research, new aspects of the disease are constantly being discovered, and we are learning more about the importance of these biological and molecular hallmarks. This chapter discusses the most salient histopathological and molecular features of GBM and provides a brief overview of their contribution to GBM classification, resistance, and recurrence.

Biological hallmarks

Pseudopalisading necrosis

The most prominent histopathological feature used to define and distinguish GBM from lower grade gliomas is the presence of "pseudopalisading necrosis" in the tumor mass. "Coagulative necrosis" is most commonly seen and refers to the degradation of protein in response to hypoxia or ischemic insults resulting in *en masse* cell death. "Pseudopalisades" refers to the hypercellular zone of abnormally stacked nuclei of tumor cells that surround the necrotic core within the tumor [4]. The mechanistic relationship between necrosis and formation of pseudopalisades is not fully understood.

A study by Brat *et al.* in 2004 demonstrated that pseudopalisading cells overexpress hypoxia-inducible factor 1 alpha (HIF1α) and extracellular matrix proteases and show enhanced gelatinase activity, factors that promote invasion and migration [5]. This study also noted degenerating and thrombosed blood vessels within the center of many pseudopalisades, and concluded that the hypoxic, vaso-occlusive event results in a wave of peripherally moving tumor cells that appear microscopically as palisades [5]. The pseudopalisades also secrete soluble proangiogenic factors such as vascular endothelial growth factor (VEGF) and interleukin-8 (IL-8), which induces an angiogenic response and formation of abnormal blood vessels that aid in growth and expansion of the tumor mass [4].

Microvascular hyperplasia and angiogenesis

Along with necrosis, extensive microvascular proliferation or hyperplasia is a major pathological hallmark and diagnostic feature of primary and secondary GBM [6]. In fact, GBM and high-grade gliomas are among the most highly vascularized of all solid tumors. Microvascular density, a measure of microvascular proliferation, is an independent prognostic factor and correlates directly with tumor aggressiveness and recurrence, and inversely with patient survival [7]. Microvascular hyperplasia is defined by the rapid proliferation of endothelial cells from preexisting parent vessels that together with stromal pericytes and basal lamina form microaggregates of "glomeruloid bodies or tufts" of sprouting vessels [7]. This type of sprouting angiogenesis is most common in GBM and is primarily mediated by VEGF signaling, a factor abundantly secreted by tumor cells [8]. The microvasculature provides nutrients, oxygen, and blood supply to tumor cells promoting their growth and infiltration into the surrounding parenchyma. Angiogenesis is therefore considered a rate-limiting step in tumor growth, a factor that led to the development of several antiangiogenic drugs including bevacizumab (Avastin), a recombinant monoclonal antibody targeting VEGF for treating GBM patients [9]. However, bevacizumab showed limited efficacy and survival benefit in clinical studies, with tumors inevitably developing resistance and evolving alternative mechanisms of neovascularization to evade therapy [10, 11]. These neovascularization mechanisms, which also include sprouting angiogenesis, are: (1) vessel cooption, (2) vasculogenesis, (3) vascular mimicry, and (4) endothelial transdifferentiation [12].

The first alternative neovascularization mechanism that was identified and occurs independent of sprouting angiogenesis is "vessel cooption," a process where tumor cells organize themselves in cuffs around normal microvessels [13]. The coopted vessels express angiogenic factors such as angiopoietin-2 (ANG-2) that promotes vessel stabilization in the presence of VEGF [14]. Another mechanism is "vasculogenesis" and involves differentiation of circulating bone marrow-derived cells known as endothelial progenitor cells (EPCs) into mature endothelial cells that incorporate into vasculature. Support for vasculogenesis as a form of GBM neovascularization came from a study which showed that tumor growth was affected by impaired recruitment of EPC [15, 16]. The mechanisms involved in recruitment and differentiation of EPCs and their contribution to vasculogenesis are not fully understood, but it is thought to involve VEGF and ANG-2 signaling. "Vasculogenic or vascular mimicry," another GBM neovascularization mechanism, is defined as the ability of tumor cells to incorporate and form functional vessel-like networks [17]. It is observed immunohistochemically as tumors that contain CD34− PAS+ vessels and are typically found more in aggressive tumors [18, 19].

The most recent neovascularization mechanism identified in GBM is the "transdifferentiation of glioma cells into endothelial cells." A group of studies have demonstrated that in human GBM specimens, a fraction of CD31+ endothelial-like cells carried the same chromosomal aberrations and mutations as tumor cells, and coexpressed both endothelial and glial markers [20–22]. Further studies have demonstrated that these glioma-derived endothelial cells also incorporate into host vasculature, and produce tube-like network formation on Matrigel, a phenotypic characteristic of normal endothelial cells [20–24]. More recently, glioma stem-like cells were also shown to transdifferentiate into pericytes, a major stromal component of tumor microvasculature. These glioma-derived pericytes are involved in the maintenance of the blood-tumor barrier, and selective depletion of these cells results in the disruption of tumor vessels and inhibits tumor growth [25]. Although a substantial number of studies have since validated and identified molecular pathways promoting the process of transdifferentiation, its relative contribution to glioma neovascularization and recurrence, and its clinical relevance, are yet to be fully defined.

Local invasion

Another key hallmark of GBM that makes them universally fatal and hampers surgery and treatment is the local diffuse invasion of tumor cells into adjacent normal brain parenchyma. As a consequence, most GBM recurrence occurs within 1–2 cm from the original tumor site, and very rarely metastasize outside the brain. In some instances, tumors referred to as "butterfly glioma" develop at distant sites by migrating along blood vessels in the perivascular space, white matter tracts, and leptomeningeal space [26].

Glioma cell invasion is governed by a complex interplay of biophysical and biochemical signaling pathways, and cross talk with the microenvironment that regulate cell shape and motility, cytoskeletal and extracellular matrix (ECM) remodeling, and switching from a proliferative to migratory phenotype [27, 28]. The transition to the migratory phenotype in cancer is generally called epithelial to mesenchymal transition (EMT), although glioma cells are not truly epithelial in nature [29]. In glioma, this transition is often referred to as proneural to mesenchymal, based on the notion that glioma cells with proneural gene signatures have less migratory potential than those with mesenchymal gene signatures [30]. The first step in acquiring an invasive and migratory phenotype requires detachment from the primary tumor, a process that involves cell surface adhesion molecules such as neural cell adhesion molecules, cadherins, and integrins, which modify the ECM and regulate the expression of matrix metalloproteinases to promote invasion [28]. The composition of ECM also plays a critical role in mediating tumor cell invasion, and the ECM associated with tumor cells is markedly different from the normal brain parenchyma. For example, GBM tumors are rich in the ECM component hyaluronic acid (HA), which they secrete up to 20% more than normal brain [31]. CD44, the receptor for HA is upregulated in tumor cells, and correlates with poor prognosis [32]. Glioma tumors also upregulate chondroitin sulfate proteoglycans such as brevican and versican, and several matrix-degrading enzymes such as matrix metalloproteinases MMP-2, MMP-9, cysteine proteases, and A disintegrin and metalloproteinases (ADAMS) that stimulate migration [27].

In addition to these biochemical processes, biophysical properties of ECM such as cell density, stiffness, and geometry also regulate glioma migration by promoting focal adhesion and formation of elastic stress fibers [33]. Cytoskeletal remodeling is essential for the formation of invadopodia and lamellipodia, and for passage through extracellular spaces that are smaller than the nuclei of tumor cells. Rho-GTPases play a key role in regulating these cytoskeletal rearrangements and enable amoeboid and mesenchymal pattern of migration [27, 28].

Furthermore, numerous growth factors and chemokines secreted by blood vessels, tumor-associated macrophages and microglia, collectively referred to as TAMs, and stromal components of the microenvironment act as chemotactic cues for glioma invasion [34, 35]. The most comprehensively examined signaling cues that mediate glioma invasion and dissemination involve WNT/β-catenin, PI3K/AKT, Hedgehog, TGFβ, and tyrosine kinase receptors [27, 28, 35]. Aberrant and constitutive activation of these pathways is common in GBM as they are frequently mutated, including EGFRVIII and PTEN mutations. The complexity of the tumor-ECM interactions, cross talk with the microenvironment, and the mechanistic contribution of each signaling pathway to glioma invasion are poorly understood.

Altered metabolism

Like most solid tumors, GBMs also undergo changes in their metabolism to meet the demand for large amounts of energy required for the synthesis of substrates for membranes, proteins, and nucleic acids to support rapid proliferation. Glucose serves as the primary energy substrate for GBM and is required in high amounts for glycolysis. Early studies underscored the importance of glucose and glycolytic pathway to GBM metabolism by genetic knockdown of key glycolytic genes that impaired tumor growth and survival [36]. In normal brain cells, the majority of carbon in acetyl-CoA is generally derived from glucose, whereas in GBM cells, less than 50% of acetyl-CoA comes from glucose. The greater metabolic need of the tumor cells causes a shift in carbon sources away from glucose and more toward using acetate, along with fatty acids and ketones [37].

Glutamine is another important substrate for GBM metabolism, and the total concentration of glutamine is much higher in GBM than in normal surrounding tissue [38]. Rather than taking glutamine from the circulating pool, GBM cells synthesize their own glutamine, and inhibition of this pathway has been shown to decrease growth and tumor volume [39]. More recent studies have highlighted the role for metabolic plasticity in GBM. For example, in times of metabolic stress, GBM cells are capable of oxidizing fatty acids for energy requirements, a feature contradictory to the general dogma of GBM biology [40, 41]. Finally, highly proliferative cells require a large amount of nucleotides and data suggest that GBM relies primarily on the de novo synthesis pathway, as opposed to the salvage pathway that most differentiated cells use [42].

These metabolic shifts in tumor cells are mediated by mutations in oncogenes and tumor suppressors that normally regulate cellular metabolism and nutrient uptake. *PTEN* and *p53*, two of the most highly mutated genes in primary and secondary GBM, are known to control key steps in glycolysis and glucose metabolism such as regulating the expression of glucose transporters and enhancing glucose uptake by hexokinase [42]. Moreover, the standard of care treatment for GBM radiation and chemotherapy, which induce necrosis and tissue inflammation, elevate glucose and glutamate levels promoting tumor survival and growth [43]. A better understanding of the interactions of oncogenes and tumor suppressors with metabolic pathways, as well as treatment-induced metabolic changes, will be important for developing therapies targeting tumor metabolism.

Immunosuppressive microenvironment

Solid tumors are often classified into immunogenic "hot" tumors and nonimmunogenic "cold" tumors based on their ability to evoke an immune response [44]. GBM is considered one of the most "cold tumors" due to their immunosuppressive microenvironment, which is hostile to infiltrating immune effector cells. This is because they exhibit a relatively low rate of tumor mutational burden (TMB), which results in few tumor-specific neoantigens and poor immunogenicity [45]. In fact, hypermutation rarely occurs in newly diagnosed GBM, but increases by 15%–20% in recurrent tumors [46]. GBM tumors also secrete a number of cytokines and factors that block T-cell migration and activation such as TGF-β, VEGF, IL-10, prostaglandin E, nitric oxide, and indoleamine-2,3-dioxygenase [47]. They also express immunosuppressive cell-surface molecules such as CD95 which induce T-cell apoptosis. More recent studies have shown that only 1%–3% of tumor cells express programmed death ligand-1 (PD-L1/CD271), which inhibits cytotoxic T-cell activity, and therefore tend to have a low response to immune checkpoint inhibitors and immunotherapy [48, 49].

Tumors are predominantly infiltrated by myeloid cells such as resident microglia, systemic macrophages, and myeloid-derived suppressor cells (MDSCs), which comprise almost 30%–50% of the tumor mass [50, 51]. Collectively, these myeloid cells function to block the infiltration and penetration of effector immune and T-cells into the tumor. Tumor-associated macrophages (TAMs) secrete factors like TGF-β and IL-10 which inhibit effector immune cells and promote tumor growth [52]. Another key contributor to the immunosuppressive phenotype of GBM are the FoxP3+ T-regulatory cells (T-regs), which constitute up to 60% of the infiltrating lymphocytes. Tumor cell-derived soluble factors such as CCL22 mediate T-reg proliferation and survival in the microenvironment [53, 54]. In addition, the physical and metabolic landscape of the tumor consisting of hypoxia, acidosis, and nutrient deprivation can diminish T-cell activity and function contributing to the nonimmunogenic "cold" phenotype of GBM [55, 56].

Heterogeneity and cancer stem-like cells

Another hallmark and major impediment to designing targeted therapies against GBM is the high degree of cellular and molecular heterogeneity and diversity exhibited by tumors [57, 58]. First, as described above, GBM exhibits *inter*tumoral heterogeneity with the above-described differentiation into three major types, proneural, mesenchymal, and classical, with more recent studies separating out another group; those tumors that have progressed from low-grade glioma and those that are mutated with the enzyme isocitrate dehydrogenase (IDH) [59, 60]. Importantly, GBM also exhibits significant *intra*tumoral heterogeneity [57]. Within a tumor, there are cells with different degrees of self-renewal and tumor-forming capacity. Cells with a greater capacity to self-renew and initiate new tumors in xenografts are often referred to as glioma stem cells (GSCs) [61, 62]. These cells are thought to be major drivers of progression and recurrence and are highly resistant to chemoradiation therapy. Within a tumor, GSCs themselves are also heterogeneous. GSCs at the highly vascular edge of a tumor are often more proneural-like, whereas GSCs at the relatively hypovascular, hypoxic center, or core of a tumor are often more mesenchymal-like, each with its own molecular characteristics [63].

In addition to these key biological and histopathological hallmarks, recent identification of several molecular and epigenetic markers have led to a better characterization and classification of GBM. We have highlighted some key molecular markers and their relevance to GBM pathology in the following section.

Molecular pathology

In 2016, the WHO reclassified tumor types of the central nervous system to better inform patient outcomes [1]. Currently, tumors are classified based on both histopathology and molecular markers. The most common gliomas are classified as either oligodendrogliomas or astrocytomas. Of note, astrocytic tumors are grouped with oligodendroglial tumors in a diffusely infiltrating glioma group. Glioblastoma, grade IV in this group, is divided into IDH-wild-type (comprising the majority of cases at around 90%), and IDHmutant (representing about 10% of cases) [64]. The former, mostly clinically defined as primary or de novo cases, present telomerase reverse transcription (TERT) mutations in over 70% of the cases as well as mutations in epidermal growth factor (EGFR; 35%) and TP53 (27%) [59]. The latter, which correspond to secondary glioblastoma that arise from a previous lower-grade infiltrative glioma, have alpha-thalassemia/mental retardation syndrome X-linked (ATRX) mutations in at least 70% of the cases together with mutations in TP53 (81%) of the cases and in TERT to a lower extent (25%) [65]. Below, we describe some of these genetic, pathway, and epigenetic markers that have been extensively characterized and studied in GBM.

Molecular markers

1p/19q codeletion

In 1998, Cairncross et al. described a genetic alteration that predicted improved response to chemotherapy and prolonged survival in oligodendrogliomas: the codeletion of the entire short arm of chromosome 1 and the entire long arm of chromosome 19 (1p/19q) [66]. This alteration is an early genetic event, and is closely associated with tumors of the oligodendroglial lineage, being found in 80% of oligodendrogliomas [67]. The codeletion is almost never found in any other nonglial malignancy. It is typically associated with mutations in IDH 1 or 2 (IDH1/2), although current evidence suggests that mutations in IDH are the first to occur in this tumor [68, 69].

Isocitrate dehydrogenase

IDH mutations were identified through an exome-wide deep sequencing analysis performed on GBM samples [70]. IDH mutations often occur in the context of either p53 mutations or 1p/19q codeletion, rarely both. Mutations in ATRX can also be found in IDH mutant tumors that are not 1/p/19q codeleted. Mutations and amplifications of EGFR and loss of chromosome 10 are rarely found in IDH mutant tumors [69, 71]. The most common mutation of either IDH 1 or 2, which are highly homologous, is a single residue alteration that substitutes a histidine for an arginine [70]. Normally, this protein catalyzes the oxidative decarboxylation of isocitrate to alpha-ketoglutarate (α-KG), a key molecule in the citric acid cycle. When IDH is mutated, isocitrate is first converted to α-KG, which is then converted to D-2-hydroxy-glutarate (D-2HG) [72]. Many studies suggest that IDH mutation is an early driver event in gliomas; however, how this occurs is still not fully understood. The prevailing theories are based on increased oxidative stress or whole genome promoter methylation. While the identification of IDH mutations has opened new avenues in the study of glioma, and has prognostic significance, it does not appear to have any strong predictive role for treatment response in GBM [73, 74]. However, studies with novel inhibitors of the mutant IDH protein are being carried out, showing promise for the treatment of lower grade gliomas.

EGFR

Epidermal growth factor receptor (EGFR) is a major activator of the MAPK and PI3K pathways that promote proliferation, survival, migration, and tumorigenesis. In GBM, there can be tens of additional copies of EGFR [75]. Additionally, about 20% of GBM patients have a constitutively active mutation due to the deletion of exons 2–7 (EGFRvIII) [76]. This rearrangement is virtually only found in EGFR-amplified tumors, although not all EGFR-amplified tumors exhibit the rearrangement. EGFRvIII is expressed by small extrachromosomal pieces of DNA, sometimes termed "double minutes," which are under dynamic regulation via unknown mechanisms [77]. While it is commonly thought that amplification or mutation of EGFR is an indicator of poor survival, several studies have failed to validate this conclusion [73, 74]. Many believe that this molecular marker could serve as a predictor for response to receptor tyrosine kinase (RTK) inhibitors. While EGFR-amplified tumors initially respond to RTK inhibition, data suggests that they often become resistant to this form of treatment [2].

p53 pathway

p53 is one of the most well-known tumor suppressor proteins being implicated in almost every cancer including GBM [78, 79]. Deletion of the p53 gene, *TP53*, can occur, but in gliomas, this is often modulated by a number of factors, including upstream regulators MDM2, MDM4, and $p14^{ARF}$ as well as downstream effectors such as ATM and ATR. Alterations in the P53 pathway are thought to promote progression of low-grade gliomas to high grade [74, 80].

PTEN and PI3K-AKT pathway

Phosphate and tensin homolog (PTEN) is a negative regulator of the 3-kinase (PI3K)/AKT pathway that is essential for a number of cellular processes involved in tumorigenesis such as survival, proliferation, and migration. The PI3K pathway is normally activated by receptors tyrosine kinases including EGFR [81]. Almost all GBMs show increased activity somewhere in this pathway, although less than 15% of GBM have activating mutations in PI3K itself. PTEN is located on chromosome 10, and while only 40% of GBM have mutations in this protein, around 70% show a loss of heterozygosity (LOH) at the *PTEN* locus, making it the second most common tumor suppressor mutation in GBM, just behind p53 [80]. The value of PTEN loss as a prognostic marker has not been validated and is still somewhat controversial [82].

Rb pathway

Like the p53 and PI3K pathways, the retinoblastoma pathway is also highly deregulated in brain tumors. Rb is a negative regulator of the cell cycle through its binding to the E2F1 transcription factor and was discovered (and named) because of its loss in retinoblastoma [80]. While only about 20% of GBMs are mutated at the *Rb* locus, inactivating mutations of the upstream regulator p16INK4a, or activating mutations in the downstream factors CDK4 or cyclin D resulting in dysregulated control of the E2F1 are very common [83]. In addition, promoter methylation of the *Rb* gene is 43% more prevalent in secondary GBM as compared to primary tumors. This is not, however, found in low-grade or anaplastic astrocytomas, suggesting that it may be a late event in astrocytoma progression [84].

Ras/Raf/MAPK pathway

A major event that occurs in a subset of largely proneural GBM is the amplification of platelet-derived growth factor receptor alpha (PDGFRα) [85]. This receptor primarily activates the Ras/Raf/MAPK pathway, wherein the GTPase Ras activates the Raf-MEK-ERK pathway, which regulates the activity of transcription factors that function in proliferation, survival, differentiation, and apoptosis [86]. Furthermore, the pathway is also activated by EGFR signaling. The pathway can also be directly or indirectly activated through mutations of downstream components. Ras itself is a common focus of activation in many solid tumors, but such mutations are rarely seen in GBM [87, 88]. However, the upstream Ras antagonist neurofibromin 1 (NF1) is either deleted or mutated in about 20% of primary GBMs. Additionally, alterations in the NF1 gene are closely associated with the mesenchymal subtype [86]. BRAF, a downstream target of Ras, is typically activated in pilocytic (Grade I) astrocytoma, through either an activating mutation (V600E) or a fusion oncoprotein with KIAA1549 (KIAA1549:BRAF). Mutations in this pathway have not been associated with reliable effects on survival, although like EGFR alterations, it is thought that they may provide valuable insight regarding response to RTK inhibitors [88, 89].

Extrachromosomal DNA

As mentioned above with respect to EGFR mutations, GBM, as well as many other cancers, can express oncogenes on small pieces of extrachromosomal DNA (ecDNA) that is generated from mutations in chromosomal DNA via unknown mechanisms [77, 90]. Mathematical modeling indicates that ecDNA-expressed oncogenes are likely the most efficient way to achieve rapid amplification of oncogenes in any particular cell type and thus are highly selected for [90]. Remarkably, there is evidence that ecDNA bearing different oncogenic mutations can be differentially regulated [91]. For example, treatment of cells bearing the EGFRvIII mutation with the EGFR inhibitor erlotinib results in a loss of ecDNA-expressed EGFRvIII but does not diminish ecDNA-expressing MDM4 [77]. It is important to note that genetically engineered murine models may not have oncogene-bearing ecDNA making it important to have availability of reliable human models such as the gliomasphere library that we have created as well as several others available for study [92]. While still in its early days of discovery, the role of ecDNA as a major oncogenic driver in GBM is clearly established and novel therapeutic strategies will need to incorporate this concept.

Epigenetic markers

MGMT methylation

Chemotherapy with temozolomide (TMZ), an alkylating agent, is part of the standard care of treatment for GBM along with radiation. TMZ adds an alkyl group to thymine and guanine creating DNA adducts, which trigger damage response and initiate apoptosis [2]. O6-methylguanine-DNA methyltransferase (MGMT) is a DNA repair enzyme that removes TMZ-induced DNA lesions by transferring the alkyl group from guanine to a cysteine residue. Thus, tumor cells that express this protein are resistant to TMZ, whereas loss of MGMT makes tumors susceptible to TMZ-induced apoptosis [93]. Expression of MGMT is tightly regulated by methylation of its promoter, which has 98 CpG islands in the first exon and many parts of the promoter [94]. Increased methylation leads to decreased expression of this protein, and ultimately increased response to TMZ treatment [94]. Interestingly, this phenotype has been closely associated with IDH mutations, which also show increased methylation throughout the entire genome.

Telomerase reverse transcriptase

Telomerase reverse transcriptase (TERT) is a subunit of the enzyme complex telomerase, which adds nucleotides to the ends of telomeres as they become shortened during cell division [95]. While the entire telomerase subunit is known to be activated in cancer cells, the TERT protein itself is often upregulated due to an activating mutation in the promoter region at positions 228 and 250 [96]. TERT mutations are found primarily in oligodendrogliomas and primary GBM and are rarely found in astrocytomas or secondary GBM [97]. Interestingly, the prognostic factor of TERT is dependent on the mutational background of the disease, as this mutation has a negative prognostic value in IDH wild-type tumors, but a positive prognostic value in IDH mutant tumors [98].

ATRX

As tumors cells are rapidly and continuously dividing, their telomeres consistently become shorter. To avoid senescence as a result of shortened telomeres, tumor cells have to devise a way to lengthen their telomeres either by activating telomerase (TERT mutations), or through the alternative lengthening of telomeres (ALT) phenotype [99]. The alpha-thalassemia/mental retardation syndrome X-linked (ATRX) protein is a DNA helicase that functions in DNA remodeling, and its inactivation is strongly correlated with the ALT phenotype, although the exact mechanism is still unknown. Loss of ATRX is independent of 1p/19q deletions and is consistently found in IDH and p53 mutants of the astrocytic lineage [100]. A recent study found that loss of ATRX resulted in more aggressive tumors due to rapid accumulation of oncogenic mutations [101].

Conclusions and future perspectives

While this chapter provides a brief overview of some of the features of GBM, a great deal of work is still needed to understand GBM biology. However, if one then rings through the study of GBM, it is that GBM is not a single disease even within one patient. This remarkable heterogeneity is the major obstacle for treatment. New therapies will not only have to tackle intra- and intertumoral heterogeneity but also need to take into account the complex tumor microenvironment as well as the inevitable treatment-induced plasticity that occurs.

References

[1] Louis DN, et al. The 2016 World Health Organization classification of tumors of the central nervous system: a summary. Acta Neuropathol 2016;131(6):803–20.
[2] Patel M, et al. Molecular targeted therapy in recurrent glioblastoma: current challenges and future directions. Expert Opin Investig Drugs 2012;21(9):1247–66.
[3] Verhaak RG, et al. Integrated genomic analysis identifies clinically relevant subtypes of glioblastoma characterized by abnormalities in PDGFRA, IDH1, EGFR, and NF1. Cancer Cell 2010;17(1):98–110.
[4] Rong Y, et al. 'Pseudopalisading' necrosis in glioblastoma: a familiar morphologic feature that links vascular pathology, hypoxia, and angiogenesis. J Neuropathol Exp Neurol 2006;65(6):529–39.

[5] Brat DJ, et al. Pseudopalisades in glioblastoma are hypoxic, express extracellular matrix proteases, and are formed by an actively migrating cell population. Cancer Res 2004;64(3):920–7.
[6] Das S, Marsden PA. Angiogenesis in glioblastoma. N Engl J Med 2013;369(16):1561–3.
[7] Brat DJ, Van Meir EG. Glomeruloid microvascular proliferation orchestrated by VPF/VEGF: a new world of angiogenesis research. Am J Pathol 2001;158(3):789–96.
[8] Xu L, Fukumura D, Jain RK. Acidic extracellular pH induces vascular endothelial growth factor (VEGF) in human glioblastoma cells via ERK1/2 MAPK signaling pathway: mechanism of low pH-induced VEGF. J Biol Chem 2002;277(13):11368–74.
[9] Presta LG, et al. Humanization of an anti-vascular endothelial growth factor monoclonal antibody for the therapy of solid tumors and other disorders. Cancer Res 1997;57(20):4593–9.
[10] Rubenstein JL, et al. Anti-VEGF antibody treatment of glioblastoma prolongs survival but results in increased vascular cooption. Neoplasia 2000;2(4):306–14.
[11] DeLay M, et al. Microarray analysis verifies two distinct phenotypes of glioblastomas resistant to antiangiogenic therapy. Clin Cancer Res 2012;18(10):2930–42.
[12] Hardee ME, Zagzag D. Mechanisms of glioma-associated neovascularization. Am J Pathol 2012;181(4): 1126–41.
[13] Holash J, et al. Vessel cooption, regression, and growth in tumors mediated by angiopoietins and VEGF. Science 1999;284(5422):1994–8.
[14] Zagzag D, et al. Vascular apoptosis and involution in gliomas precede neovascularization: a novel concept for glioma growth and angiogenesis. Lab Investig 2000;80(6):837–49.
[15] Lyden D, et al. Impaired recruitment of bone-marrow-derived endothelial and hematopoietic precursor cells blocks tumor angiogenesis and growth. Nat Med 2001;7(11):1194–201.
[16] Machein MR, et al. Minor contribution of bone marrow-derived endothelial progenitors to the vascularization of murine gliomas. Brain Pathol 2003;13(4):582–97.
[17] El Hallani S, et al. A new alternative mechanism in glioblastoma vascularization: tubular vasculogenic mimicry. Brain 2010;133(Pt 4):973–82.
[18] Yue WY, Chen ZP. Does vasculogenic mimicry exist in astrocytoma? J Histochem Cytochem 2005;53(8): 997–1002.
[19] Liu XM, et al. Clinical significance of vasculogenic mimicry in human gliomas. J Neuro-Oncol 2011;105(2): 173–9.
[20] Soda Y, et al. Transdifferentiation of glioblastoma cells into vascular endothelial cells. Proc Natl Acad Sci U S A 2011;108(11):4274–80.
[21] Ricci-Vitiani L, et al. Tumour vascularization via endothelial differentiation of glioblastoma stem-like cells. Nature 2010;468(7325):824–8.
[22] Wang R, et al. Glioblastoma stem-like cells give rise to tumour endothelium. Nature 2010;468(7325):829–33.
[23] Zhao Y, et al. Endothelial cell transdifferentiation of human glioma stem progenitor cells in vitro. Brain Res Bull 2010;82(5–6):308–12.
[24] Dong J, et al. Glioma stem/progenitor cells contribute to neovascularization via transdifferentiation. Stem Cell Rev Rep 2011;7(1):141–52.
[25] Cheng L, et al. Glioblastoma stem cells generate vascular pericytes to support vessel function and tumor growth. Cell 2013;153(1):139–52.
[26] Giese A, Westphal M. Glioma invasion in the central nervous system. Neurosurgery 1996;39(2):235–50 [discussion 250-2].
[27] Velásquez C, et al. Molecular and clinical insights into the invasive capacity of glioblastoma cells. J Oncol 2019;2019:1740763.
[28] Mair DB, Ames HM, Li R. Mechanisms of invasion and motility of high-grade gliomas in the brain. Mol Biol Cell 2018;29(21):2509–15.
[29] Iwadate Y. Epithelial-mesenchymal transition in glioblastoma progression. Oncol Lett 2016;11(3):1615–20.
[30] Fedele M, et al. Proneural-mesenchymal Transition: phenotypic plasticity to acquire multitherapy resistance in glioblastoma. Int J Mol Sci 2019;20(11).
[31] Delpech B, et al. Hyaluronan and hyaluronectin in the extracellular matrix of human brain tumour stroma. Eur J Cancer 1993;29A(7):1012–7.
[32] Misra S, et al. Hyaluronan-CD44 interactions as potential targets for cancer therapy. FEBS J 2011;278(9):1429–43.

[33] Ulrich TA, de Juan Pardo EM, Kumar S. The mechanical rigidity of the extracellular matrix regulates the structure, motility, and proliferation of glioma cells. Cancer Res 2009;69(10):4167–74.
[34] Ye XZ, et al. Tumor-associated microglia/macrophages enhance the invasion of glioma stem-like cells via TGF-β1 signaling pathway. J Immunol 2012;189(1):444–53.
[35] Hambardzumyan D, Gutmann DH, Kettenmann H. The role of microglia and macrophages in glioma maintenance and progression. Nat Neurosci 2016;19(1):20–7.
[36] Libby CJ, et al. The pro-tumorigenic effects of metabolic alterations in glioblastoma including brain tumor initiating cells. Biochim Biophys Acta Rev Cancer 2018;1869(2):175–88.
[37] Masui K, et al. Metabolic reprogramming in the pathogenesis of glioma: update. Neuropathology 2019;39(1): 3–13.
[38] Marin-Valencia I, et al. Analysis of tumor metabolism reveals mitochondrial glucose oxidation in genetically diverse human glioblastomas in the mouse brain in vivo. Cell Metab 2012;15(6):827–37.
[39] Bagga P, et al. Characterization of cerebral glutamine uptake from blood in the mouse brain: implications for metabolic modeling of 13C NMR data. J Cereb Blood Flow Metab 2014;34(10):1666–72.
[40] Duman C, et al. Acyl-CoA-binding protein drives glioblastoma tumorigenesis by sustaining fatty acid oxidation. Cell Metab 2019;30(2):274–289.e5.
[41] Sperry J, et al. Glioblastoma utilizes fatty acids and ketone bodies for growth allowing progression during ketogenic diet therapy. iScience 2020;23(9):101453.
[42] Zhou W, Wahl DR. Metabolic abnormalities in glioblastoma and metabolic strategies to overcome treatment resistance. Cancers (Basel) 2019;11(9).
[43] Seyfried TN, et al. Metabolic therapy: a new paradigm for managing malignant brain cancer. Cancer Lett 2015;356(2 Pt A):289–300.
[44] Sampson JH, et al. Brain immunology and immunotherapy in brain tumours. Nat Rev Cancer 2020;20(1):12–25.
[45] Hodges TR, et al. Mutational burden, immune checkpoint expression, and mismatch repair in glioma: implications for immune checkpoint immunotherapy. Neuro-Oncology 2017;19(8):1047–57.
[46] Wang J, et al. Clonal evolution of glioblastoma under therapy. Nat Genet 2016;48(7):768–76.
[47] Nduom EK, Weller M, Heimberger AB. Immunosuppressive mechanisms in glioblastoma. Neuro-Oncology 2015;17(Suppl 7):vii9–vii14.
[48] Roth P, et al. Malignant glioma cells counteract antitumor immune responses through expression of lectin-like transcript-1. Cancer Res 2007;67(8):3540–4.
[49] Nduom EK, et al. PD-L1 expression and prognostic impact in glioblastoma. Neuro-Oncology 2016;18(2): 195–205.
[50] Rossi ML, et al. Immunohistological study of mononuclear cell infiltrate in malignant gliomas. Acta Neuropathol 1987;74(3):269–77.
[51] Chang AL, et al. CCL2 produced by the glioma microenvironment is essential for the recruitment of regulatory T cells and myeloid-derived suppressor cells. Cancer Res 2016;76(19):5671–82.
[52] Hussain SF, et al. The role of human glioma-infiltrating microglia/macrophages in mediating antitumor immune responses. Neuro-Oncology 2006;8(3):261–79.
[53] Crane CA, et al. Soluble factors secreted by glioblastoma cell lines facilitate recruitment, survival, and expansion of regulatory T cells: implications for immunotherapy. Neuro-Oncology 2012;14(5):584–95.
[54] Heimberger AB, et al. Incidence and prognostic impact of FoxP3+ regulatory T cells in human gliomas. Clin Cancer Res 2008;14(16):5166–72.
[55] Barsoum IB, et al. Mechanisms of hypoxia-mediated immune escape in cancer. Cancer Res 2014;74(24):7185–90.
[56] Pilon-Thomas S, et al. Neutralization of tumor acidity improves antitumor responses to immunotherapy. Cancer Res 2016;76(6):1381–90.
[57] Qazi MA, et al. Intratumoral heterogeneity: pathways to treatment resistance and relapse in human glioblastoma. Ann Oncol 2017;28(7):1448–56.
[58] Schäfer N, et al. Longitudinal heterogeneity in glioblastoma: moving targets in recurrent versus primary tumors. J Transl Med 2019;17(1):96.
[59] Karsy M, et al. New molecular considerations for glioma: IDH, ATRX, BRAF, TERT, H3 K27M. Curr Neurol Neurosci Rep 2017;17(2):19.
[60] Ohgaki H, Kleihues P. The definition of primary and secondary glioblastoma. Clin Cancer Res 2013;19(4): 764–72.

[61] Lathia JD, et al. Cancer stem cells in glioblastoma. Genes Dev 2015;29(12):1203–17.
[62] Prager BC, et al. Glioblastoma stem cells: driving resilience through chaos. Trends Cancer 2020;6(3):223–35.
[63] Karsy M, et al. The impact of hypoxia and mesenchymal transition on glioblastoma pathogenesis and cancer stem cells regulation. World Neurosurg 2016;88:222–36.
[64] Gupta A, Dwivedi T. A simplified overview of World Health Organization classification update of central nervous system tumors 2016. J Neurosci Rural Pract 2017;8(4):629–41.
[65] Nonoguchi N, et al. TERT promoter mutations in primary and secondary glioblastomas. Acta Neuropathol 2013;126(6):931–7.
[66] Cairncross JG, et al. Specific genetic predictors of chemotherapeutic response and survival in patients with anaplastic oligodendrogliomas. J Natl Cancer Inst 1998;90(19):1473–9.
[67] Riemenschneider MJ, et al. Molecular diagnostics of gliomas: state of the art. Acta Neuropathol 2010;120(5):567–84.
[68] Horbinski C, et al. Interplay among BRAF, p16, p53, and MIB1 in pediatric low-grade gliomas. Neuro-Oncology 2012;14(6):777–89.
[69] Ichimura K, et al. IDH1 mutations are present in the majority of common adult gliomas but rare in primary glioblastomas. Neuro-Oncology 2009;11(4):341–7.
[70] Parsons DW, et al. An integrated genomic analysis of human glioblastoma multiforme. Science 2008;321(5897):1807–12.
[71] Yan H, et al. IDH1 and IDH2 mutations in gliomas. N Engl J Med 2009;360(8):765–73.
[72] Balss J, et al. Analysis of the IDH1 codon 132 mutation in brain tumors. Acta Neuropathol 2008;116(6):597–602.
[73] Gupta K, Salunke P. Molecular markers of glioma: an update on recent progress and perspectives. J Cancer Res Clin Oncol 2012;138(12):1971–81.
[74] Karsy M, et al. A practical review of prognostic correlations of molecular biomarkers in glioblastoma. Neurosurg Focus 2015;38(3), E4.
[75] Yip S, Iafrate AJ, Louis DN. Molecular diagnostic testing in malignant gliomas: a practical update on predictive markers. J Neuropathol Exp Neurol 2008;67(1):1–15.
[76] Sugawa N, et al. Identical splicing of aberrant epidermal growth factor receptor transcripts from amplified rearranged genes in human glioblastomas. Proc Natl Acad Sci U S A 1990;87(21):8602–6.
[77] Nathanson DA, et al. Targeted therapy resistance mediated by dynamic regulation of extrachromosomal mutant EGFR DNA. Science 2014;343(6166):72–6.
[78] Frankel RH, et al. p53 mutations in human malignant gliomas: comparison of loss of heterozygosity with mutation frequency. Cancer Res 1992;52(6):1427–33.
[79] Zhang Y, et al. The p53 pathway in glioblastoma. Cancers (Basel) 2018;10(9).
[80] Mao H, et al. Deregulated signaling pathways in glioblastoma multiforme: molecular mechanisms and therapeutic targets. Cancer Investig 2012;30(1):48–56.
[81] Zoncu R, Efeyan A, Sabatini DM. mTOR: from growth signal integration to cancer, diabetes and ageing. Nat Rev Mol Cell Biol 2011;12(1):21–35.
[82] Nikiforova MN, Hamilton RL. Molecular diagnostics of gliomas. Arch Pathol Lab Med 2011;135(5):558–68.
[83] McNamara MG, Sahebjam S, Mason WP. Emerging biomarkers in glioblastoma. Cancers (Basel) 2013;5(3):1103–19.
[84] Lam PY, et al. Expression of p19INK4d, CDK4, CDK6 in glioblastoma multiforme. Br J Neurosurg 2000;14(1):28–32.
[85] Cancer Genome Atlas Research Network. Comprehensive genomic characterization defines human glioblastoma genes and core pathways. Nature 2008;455(7216):1061–8.
[86] Aldape K, et al. Glioblastoma: pathology, molecular mechanisms and markers. Acta Neuropathol 2015;129(6):829–48.
[87] Knobbe CB, Reifenberger J, Reifenberger G. Mutation analysis of the Ras pathway genes NRAS, HRAS, KRAS and BRAF in glioblastomas. Acta Neuropathol 2004;108(6):467–70.
[88] Schindler G, et al. Analysis of BRAF V600E mutation in 1,320 nervous system tumors reveals high mutation frequencies in pleomorphic xanthoastrocytoma, ganglioglioma and extra-cerebellar pilocytic astrocytoma. Acta Neuropathol 2011;121(3):397–405.
[89] Jones DT, et al. Oncogenic RAF1 rearrangement and a novel BRAF mutation as alternatives to KIAA1549:BRAF fusion in activating the MAPK pathway in pilocytic astrocytoma. Oncogene 2009;28(20):2119–23.

[90] Turner KM, et al. Extrachromosomal oncogene amplification drives tumour evolution and genetic heterogeneity. Nature 2017;543(7643):122–5.
[91] Verhaak RGW, Bafna V, Mischel PS. Extrachromosomal oncogene amplification in tumour pathogenesis and evolution. Nat Rev Cancer 2019;19(5):283–8.
[92] Laks DR, et al. Large-scale assessment of the gliomasphere model system. Neuro-Oncology 2016;18(10): 1367–78.
[93] Hegi ME, et al. MGMT gene silencing and benefit from temozolomide in glioblastoma. N Engl J Med 2005;352 (10):997–1003.
[94] Nakagawachi T, et al. Silencing effect of CpG island hypermethylation and histone modifications on O6-methylguanine-DNA methyltransferase (MGMT) gene expression in human cancer. Oncogene 2003;22 (55):8835–44.
[95] Arndt GM, MacKenzie KL. New prospects for targeting telomerase beyond the telomere. Nat Rev Cancer 2016;16(8):508–24.
[96] Killela PJ, et al. TERT promoter mutations occur frequently in gliomas and a subset of tumors derived from cells with low rates of self-renewal. Proc Natl Acad Sci U S A 2013;110(15):6021–6.
[97] Koelsche C, et al. Distribution of TERT promoter mutations in pediatric and adult tumors of the nervous system. Acta Neuropathol 2013;126(6):907–15.
[98] Labussiere M, et al. TERT promoter mutations in gliomas, genetic associations and clinico-pathological correlations. Br J Cancer 2014;111(10):2024–32.
[99] Bush NA, Butowski N. The effect of molecular diagnostics on the treatment of glioma. Curr Oncol Rep 2017;19 (4):26.
[100] Eckel-Passow JE, et al. Glioma groups based on 1p/19q, IDH, and TERT promoter mutations in Tumors. N Engl J Med 2015;372(26):2499–508.
[101] Koschmann C, et al. ATRX loss promotes tumor growth and impairs nonhomologous end joining DNA repair in glioma. Sci Transl Med 2016;8(328), 328ra28.

CHAPTER

2

Past and present drug treatments for glioblastoma

Yasmeen Rauf and Manmeet S. Ahluwalia

Burkhardt Brain Tumor and Neuro-Oncology Center, Cleveland Clinic, Cleveland, OH, United States

Abstract

In the pre-Stupp era, BCNU and radiotherapy were standard of care in the United States. Standard treatment for GBM now includes maximal safe resection, followed by concurrent radiation and chemotherapy with TMZ, followed by TMZ alone. We review past and present drug treatments for GBM including TMZ, nitrosourea-based therapy, nonalkylating chemotherapy including bevacizumab, drugs targeting the tyrosine kinase pathway, and immunotherapies. There are several clinical trials currently ongoing, which will hopefully change the course of this dreaded disease in the coming years.

Abbreviations

ABT 414	an anti-EGFR monoclonal antibody drug conjugate
ACNU	nimustine
ASRO	The American Society of Radiation Oncology
BCNU	carmustine
BELOB study	a randomized phase II study on bevacizumab versus bevacizumab plus lomustine versus lomustine single agent in recurrent glioblastoma
CCNU	lomustine
CNS	central nervous system
CTLA-4	cytotoxic T-lymphocyte-associated protein 4
EGFR	epidermal growth factor receptor
EORTC	the European Organization for Research and Treatment of Cancer
FDA	the United States Food and Drug Administration
GBM	glioblastoma
IDH	isocitrate dehydrogenase
INTELLANCE-1 Trial	a randomized, placebo-controlled Phase III study designed to evaluate the efficacy and safety of depatuxizumab mafodotin versus placebo when administered with concurrent radiation and temozolomide and with adjuvant temozolomide in subjects with newly diagnosed EGFR-amplified GBM

Glioblastoma Resistance to Chemotherapy: Molecular Mechanisms and Innovative Reversal Strategies
https://doi.org/10.1016/B978-0-12-821567-8.00007-5

KPS	Karnofsky Performance Scale
MET	gene encoding c-Met, also called tyrosine-protein kinase Met or hepatocyte growth factor receptor
MGMT	methyl guanine methyl transferase
PD1	programmed cell death protein 1
PI3K/mTOR	phosphatidylinositol 3-kinase/mammalian target of rapamycin
TMZ	temozolomide
TTF	tumor treating fields
VEGF	vascular endothelial growth factors

Conflict of interest

No potential conflicts of interest were disclosed.

Introduction

Glioblastoma (GBM) is the most common malignant primary brain tumor [1, 2]. The average life expectancy of a patient with glioblastoma is 15–18 months. The standard of care for a patient with GBM includes maximum surgical resection followed by concurrent radiation and chemotherapy with temozolomide (TMZ) followed by TMZ alone with the option of using tumor treating fields (TTF). Young age, good performance status, MGMT hypermethylation, and IDH 1/2 mutation are favorable prognostic factors for the patients [3–7]. Of note, 90% of patients lack the IDH mutation and are IDH wild type and 55%–60% of patients do not have hypermethylation of MGMT.

Brief history of glioblastoma

This primary CNS tumor was first described by Berns in British scientific reports in 1800 [8]. He described the gross anatomy of the tumor, which he observed on autopsy. Abberty in 1804 described primary CNS tumor as a tumor with no borders and very different from healthy brain tissue. These were called medullary sarcomas by the British at that time. Later the French called it encephaloide and the Germans called it fungus medullare. Over the next 50 years, light microscopy was introduced and histopathology took birth. Rudolf Virchow, a German pathologist, was the first scientist to give a detailed histopathological description of primary brain tumors in 1865. He described them as malignant cells that invaded the neighboring cells and he was first to use the term "glioma." He also grouped brain tumors into "low-grade" and "high-grade" tumors. Almost one hundred years later, the American pathologist Percival Bailey and pioneer neurosurgeon Harvey Cushing gave the modern classification of gliomas. They coined the term spongioblastoma multiforme for what we now know as GBM. They also described astrocytomas. Modern WHO 2016 classification of brain tumors is based on the classification used by Bailey and Cushing in 1926, modified with the use of molecular markers. Controversial German neuropathologist Hans-Joachim Scherer coined the term primary glioblastoma for tumors that formed de novo and secondary GBM for those that arose from low-grade tumors.

Pre-Stupp era

Prior to 2005, the standard of care for GBM was maximal, safe surgical resection followed by radiation therapy in Europe and Canada [9]. In United States, BCNU and radiotherapy were the standard of care. Gross total resection was defined as the removal of all the enhancing parts of the tumor, and this correlated with maximal survival (Fig. 1). It helped to treat the local mass effect, and provided tissue for tissue diagnosis, including histopathological diagnosis and molecular diagnosis. Surgical biopsy was also of value in recurrent disease to distinguish true progression from pseudoprogression which was often seen in patients who underwent radiation treatment of GBM.

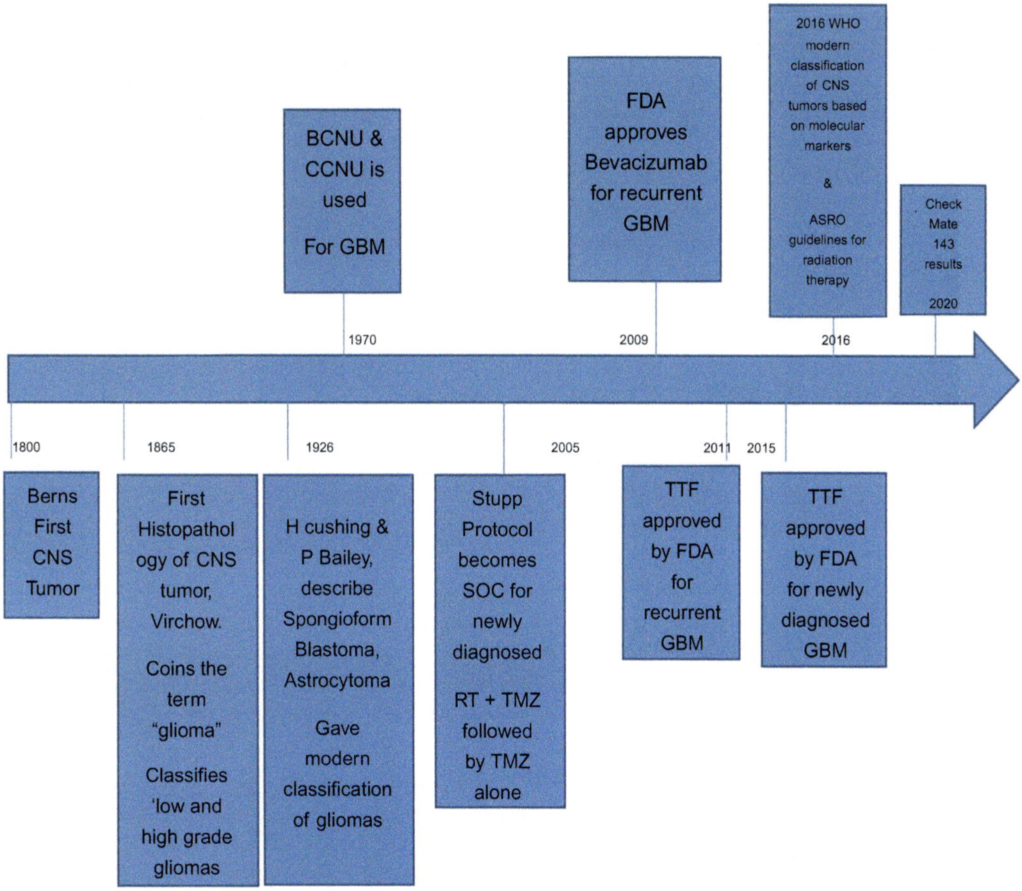

FIG. 1 Historical landmarks in GBM treatment.

Standard of care for glioblastoma (2005 to now)

Standard treatment for GBM now includes maximal safe resection, followed by concurrent radiation and chemotherapy with TMZ, followed by TMZ alone [10–12]. Stupp et al. [9] randomized 573 patients with newly diagnosed GBM, diagnosed by histopathology, to receive the standard of care at that time (focal radiotherapy) for 6 weeks or to receive concomitant radiation and TMZ followed by chemotherapy alone [9, 13, 14]. After 6 weeks of surgery, all patients received focal radiation. The dose used was 2 Gy per fraction over 6 weeks for a total dose of 60 Gy. A dose of 2 Gy per fraction was given once daily Monday to Friday for 6 weeks. The patients in the radiotherapy arm received radiotherapy alone. The patients in the concomitant radiation and chemotherapy arm received TMZ 75 mg/m^2 every single day for 6 weeks including Saturday and Sunday when the patients were not getting radiation therapy. TMZ was started on the first day of radiation and was continued till the last day of radiation. During concomitant chemoradiation, patients received antiemetic prophylaxis for nausea and also prophylaxis for *Pneumocystis carinii* pneumonia (PCP). All patients underwent a follow-up brain MRI 4 weeks after completing radiation. The patients randomized to the concomitant radiation and chemotherapy arm received six cycles of adjuvant TMZ. The dose used in the first cycle was 150 mg/m^2 followed by 200 mg/m^2 in the second to sixth cycle. The dose was given in the first five consecutive days of the 28-day cycle. The study showed that the median survival in the chemoradiation arm was 14.6 months versus 12.1 months in the radiation alone arm. The progression-free survival was 6.9 months in the chemoradiation arm compared to the 5.5 months in the radiation alone arm. The 2-year survival was 26.5% in the chemoradiation arm and 10.4% in the radiation alone arm. In April 2005, based on the results of this study, FDA approved the use of concomitant and adjuvant TMZ for newly diagnosed GBM.

The American Society of Radiation Oncology (ASRO) now recommends that GBMs be treated with 60 Gy of radiation delivered over 30 fractions, 5 days a week for 6 weeks. Radiation therapy improves survival when compared with surgery alone, chemotherapy alone, or surgery with chemotherapy [15]. Society recommends radiation with concurrent chemotherapy with TMZ after surgery for maximal survival benefit in patients with newly diagnosed GBM. These recommendations are for patients with a good performance status (KPS > 60) (Fig. 2). Elderly patients over age 65 years can be treated with 3 weeks of hypofractionated radiation with 40 Gy delivered over 15 fractions, 5 days a week for 3 weeks (Fig. 3). Hypofractionated radiation therapy has been associated with improved survival and lower corticosteroid use in the elderly population [16]. Perry et al. randomized 562 patients with newly diagnosed GBM with a median age of 73 to either receive hypofractionated radiotherapy alone or hypofractionated radiotherapy and TMZ [17]. The median overall survival was 9.3 months and progression-free survival was 5.3 months for those who received combination treatment versus overall survival of 7.6 months and progression-free survival of 3.9 months for the radiotherapy alone arm.

In 2015, the FDA approved the use of tumor treating fields (TTF) for newly diagnosed GBM. TTF delivers low intensity, alternating electric fields to the scalp [18, 19]. The frequency of these electric fields is 200 kHz and is delivered using a transducer. These electric fields are antimitotic and work by interrupting cell division. It is a noninvasive device, where electrodes are placed on a shaved scalp. The electrodes are connected with wires to a battery which is

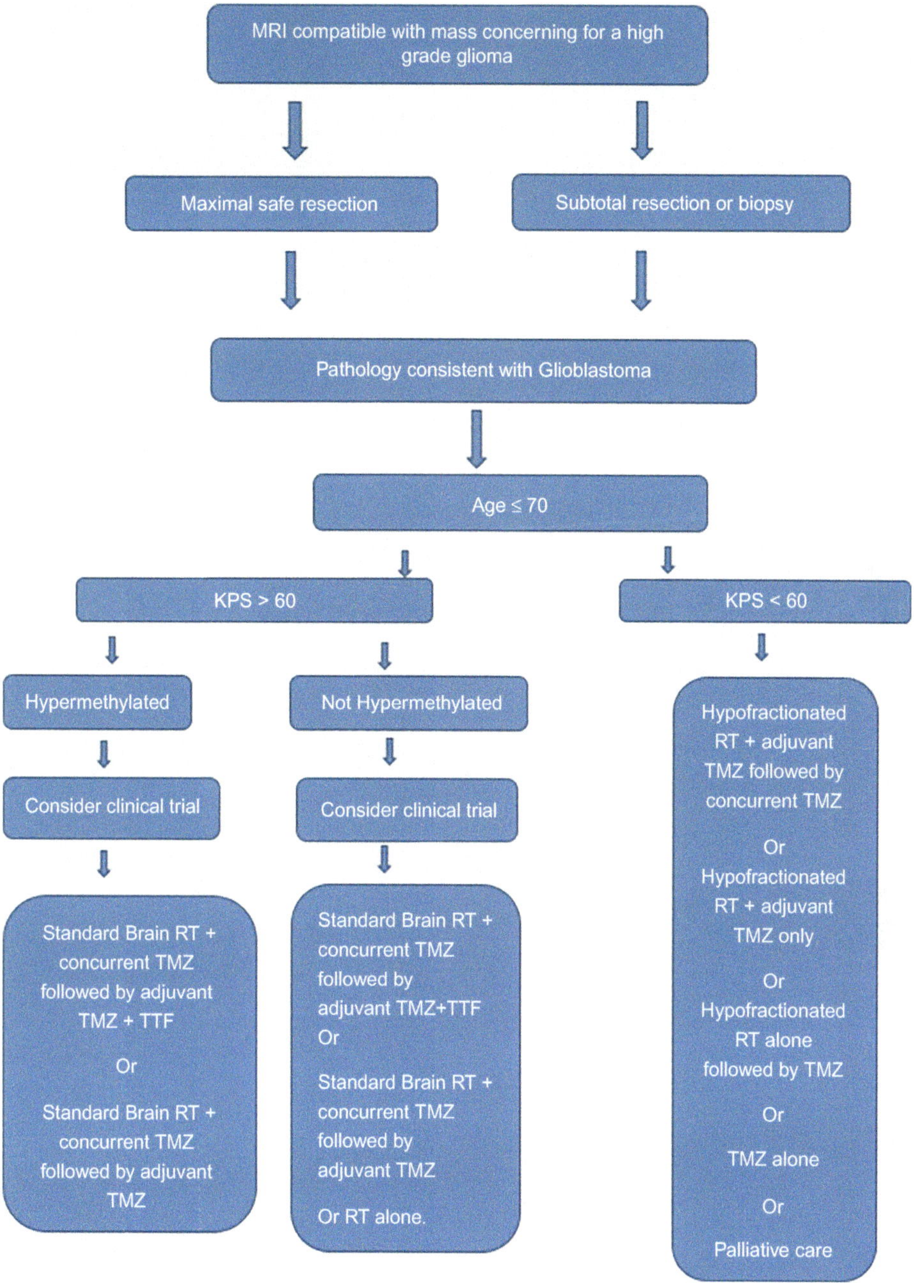

FIG. 2 Standard of care for a patient with GBM aged ≤70 years in the USA based on the NCCN guidelines.

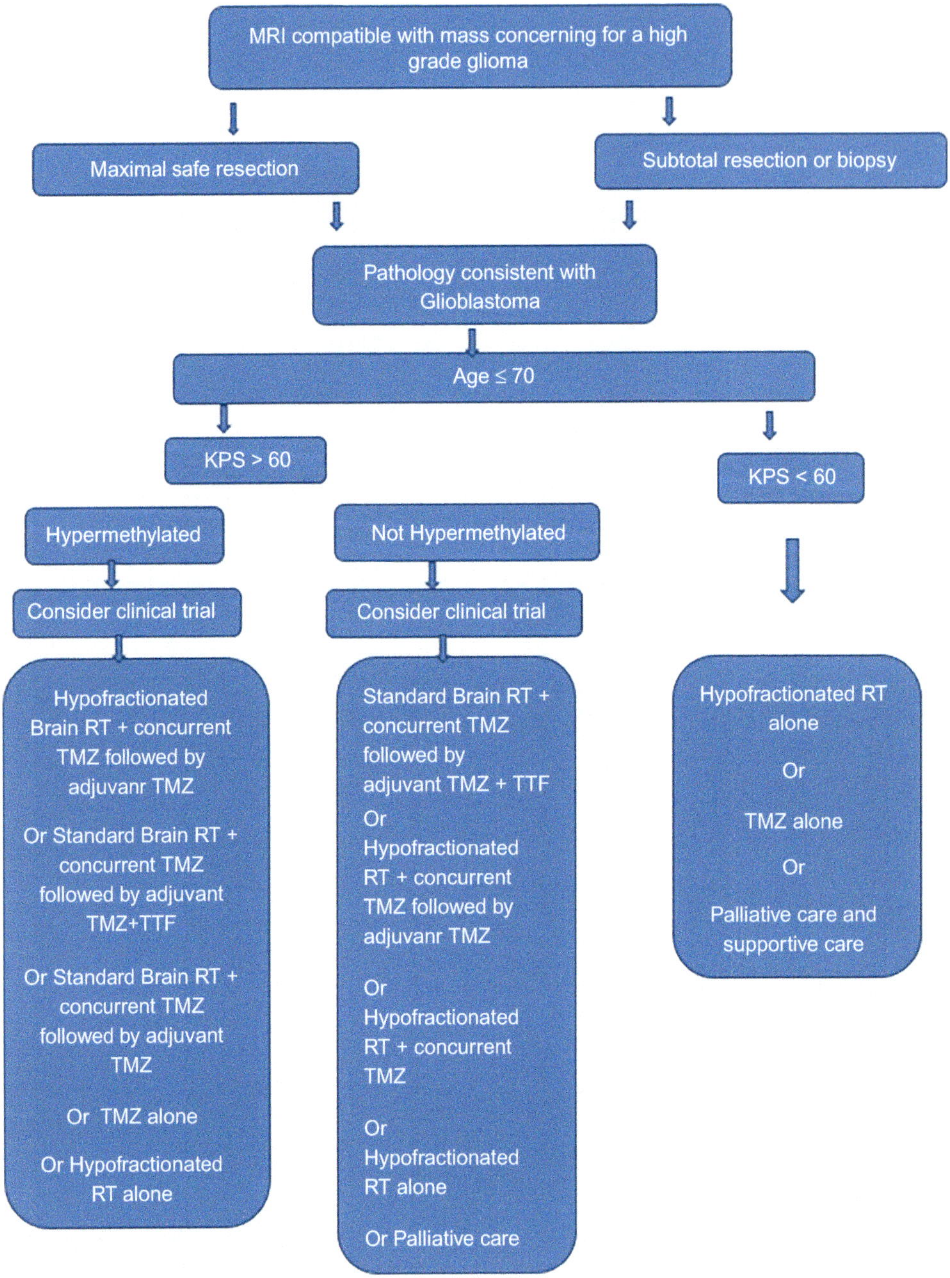

FIG. 3 Standard of care for a patient with GBM aged more than 70 years in the USA based on the NCCN guidelines.

carried by the patient at the time of use. Stupp et al. in 2009 designed a multicenter, phase 3 clinical trial where they used TTF in addition to TMZ in patients with newly diagnosed GBM after they had completed concurrent chemoradiation [18]. Patients were randomized to receive either TTF with adjuvant TMZ or adjuvant TMZ alone. The median progression-free survival increased from 4 to 7.2 months and the median overall survival increased from 16.6 to 19.6 months in patients randomized to receive both TTF and adjuvant TMZ. TTF is used for 18h a day and is associated with benefits.

Chemotherapy

1. Temozolomide

Temozolomide (TMZ) is an oral alkylating agent [9]. Its small size and lipophilic nature help it permeate the blood-brain barrier (BBB) [20, 21]. TMZ adds a methyl group to purine and pyrimidine in DNA. This arrests the cell cycle at G2/M leading to apoptosis or cell death. TMZ synergizes with radiation to increase cell death.

Methyl guanine methyl transferase (MGMT) is an enzyme that counteracts the apoptotic effects of alkylating reagents by transferring a methyl group from guanine to repair DNA damage and prevent cell death [22, 23]. In all, 55% of patients with GBM who are not MGMT hypermethylated have relative resistance to TMZ and derive marginal benefit. MGMT counteracts the apoptotic effects of alkylating reagents [24]. Even patients who have had an initial response to TMZ eventually develop resistance.

As mentioned earlier, Roger Stupp led the randomized clinical trial EORTC showing that TMZ with radiation therapy was superior to RT alone and this established TMZ as a standard of care for GBM. There was a 2.5-month improvement in median survival (Hazard ratio 0.64), increase in 2-year survival of 16%. Patients with hypermethylated MGMT showed greater relative and absolute benefit with concomitant radiation and chemotherapy with TMZ [9]. Patients with MGMT promoter hypermethylation had an 8-month improvement in median survival (HR 0.51), an absolute increase in 2-year survival of 25% and 5-year survival of 9%. Patients with no hypermethylation of the MGMT promoter gene had an improvement in median survival of 1 month (HR 0.69), the absolute increase in 2-year survival of 13%, and 5-year survival of 8%.

The brain tumor concentration of TMZ is only 20% of plasma concentration [25]. Other methods of delivery including intra-arterial delivery of TMZ have shown increased side effects with no improvement in survival.

In terms of dosage, concomitant TMZ at a dose of 75 mg/m^2 is given with RT for 6 weeks. This is followed by a 4-week break. TMZ is increased to 150 mg/m^2 for the first cycle for Day 1–5. This is followed by five cycles of 200 mg/m^2 for Day 1–5 [9]. Regarding side effects of TMZ [25], the most common one is lymphopenia. When given concomitant with radiation, it causes depletion of CD4 T cells and B cells. Lymphopenia is dose dependent. Most patients who develop lymphopenia continue to have low cell counts till about 2 months after stopping the drug. This makes patients susceptible to opportunistic infections. It is standard of care to start PCP prophylaxis during 6 weeks of concomitant radiation and TMZ. Pancytopenia and thrombocytopenia are often conditions that lead to stopping the drug. Other side effects include constipation and nausea. Less common side effects are skin rash, diarrhea, dysphagia, loss of appetite, abdominal pain, headache, alopecia, confusion, lethargy, neuropathy, and gait disturbance. Severe side effects, such as leukemias and other myelodysplastic diseases are seen in patients who receive TMZ for several years.

2. Nitrosourea-based therapy

   Carmustine (BCNU), lomustine (CCNU), and nimustine (ACNU) are DNA alkylating agents that permeate the BBB and therefore have been extensively used in GBM [26–29]. BCNU was one of the first drugs to be used for the treatment of GBM [28]. BNCU is an intravenous drug and is used either at a dose of 80 mg/mq on day 1–3 every 6 weeks or a combined dose of 240 mg/m^2 once every 6 weeks. Unlike BCNU, CCNU is an orally administered drug given at a dose of 90–110 mg/m^2 every 6 weeks. Owing to a similar toxicity profile, but the ease of administration, lomustine became more popular when compared to carmustine for use in recurrent GBM.

   Several clinical trials have compared nitrosoureas as monotherapy and combined with and without radiation therapy [30]. Clinical trials have also used nitrosoureas with and without bevacizumab. However, the ease of use of TMZ and safer toxicity profile with survival benefit has led to the use of nitrosoureas, such as lomustine, for patients who have failed TMZ. Procarbazine, CCNU, and Vincristine, have been tried as a combination therapy before TMZ became standard of care and are now used in some centers for recurrent GBM.

3. BCNU wafer

   A biodegradable polymer of BCNU wafers was implanted in the surgical cavity and this was followed by standard radiation [29, 31]. Randomized controlled trials have shown increased CNS toxicities from local chemotherapy. Therefore, these wafers did not change the care of patients with GBM [32].

4. Nonalkylating chemotherapy

   Carboplatin is a platinum-based drug that crosses the BBB and was used in GBM. Clinical trials with platinum-based chemotherapies alone or in combination with radiation or other chemotherapies have not shown any additional benefit. These trials showed increased toxicities from adding a platinum compound [33].

   Irinotecan was used with bevacizumab in two clinical trials that were both negative for recurrent GBM. One trial (with the addition of bevacizumab) is described later in reference to the tyrosine kinase pathway. Irinotecan is no longer used in GBM treatment owing to the availability of better and safer options.

Tyrosine kinase pathway

EGFR pathway

Epidermal growth factor receptor is an oncogene in IDH wild-type GBM [34]. EGFR tyrosine kinase inhibitors, such as erlotinib and afatinib have not shown to improve survival in patients with GBM so far.

ABT 414 is an anti-EGFR monoclonal antibody-drug conjugate. It was studied in both recurrent and newly diagnosed GBM populations [35, 36]. In a phase I study, patients with recurrent GBM were randomized to get either the ABT 414 drug IV with TMZ or the ABT 414 drug alone. The study showed ABT414 had antitumor activity, more so in patients with amplified EGFR. However, patients had ocular side effects [37]. Phase II study (Intellance 2) showed improved overall survival when ABT 414 was given in combination with TMZ to patients with recurrent GBM [38]. In a phase III study (INTELLANCE- 1), ABT 414 was studied in newly diagnosed GBM patients with an EGFR amplification. The interim analysis based on 639 patients did not show any improvement in overall survival.

VEGF

Vascular endothelial growth factors are critical for vascular proliferation and angiogenesis. The binding of VEGF to VEGF receptors leads to activation of the tyrosine Kinase pathway and results in angiogenesis. This angiogenesis has been implicated in tumor growth and proliferation.

Bevacizumab is a humanized monoclonal antibody that inhibits VEGF. It has been shown to decrease glucocorticoid use and increase progression-free survival in patients with GBM.

Gilbert et al. randomized 637 patients with newly diagnosed GBM, who were receiving standard of care radiation with concomitant TMZ to either receive bevacizumab or placebo starting from week 4 of radiation therapy up to cycle 12 of maintenance chemotherapy. The addition of bevacizumab did not improve overall survival. Progression-free survival was 10.7 months in the bevacizumab group versus 7.3 months in the placebo group. However, the intervention group also had a worse quality of life and increased toxicities including sharper cognitive decline [39].

Chinot et al. randomized 921 patients with newly diagnosed GBM to either receive standard of care or standard of care plus bevacizumab starting treatment with radiation therapy and continuing it till cycle 6 of chemotherapy. The trial did not show any improved overall survival. The progression-free survival was improved in the bevacizumab arm, but these patients also suffered worse toxicities [40].

In the BRAIN trial, patients with a first or second recurrence of GBM, who were previously treated with TMZ were randomized to receive either bevacizumab or bevacizumab plus CPT-11 (irinotecan) [41]. Patients showed tumor shrinkage with a durable response in both arms. Six months PFS was 42.6% in the bevacizumab arm and 50.3% in the Bev plus CPT-11 arm. The trial led to accelerated approval of bevacizumab by the FDA for recurrent GBM in 2009 [42].

In the BELOB study, 153 patients were randomized to receive either lomustine or bevacizumab, or both lomustine and bevacizumab. The OS at 9 and 12 months was higher in the patients who received both lomustine and bevacizumab versus those who received either of these drugs alone [43].

EORTC Phase 3 trial randomized 457 patients to receive lomustine alone or lomustine and bevacizumab after they progressed on standard of care for GBM [43, 44]. The arm with bevacizumab plus lomustine had an improved progression-free survival but did not improve overall survival. The combination of lomustine and bevacizumab was associated with more toxicities.

Small molecule tyrosine kinase inhibitors, including sunitinib which target multiple receptors in the pathway, have also not shown favorable results.

Pl3K/mTOR pathway

Pl3K pathway has been a target for many clinical trials in GBM [45]. The safety and tolerability profile of these drugs have also been questioned.

Met

Met is the receptor for hepatocyte growth factor also known as the scatter factor. It is thought to be responsible for the invasiveness and migration of GBM cells [46]. Increased levels of these growth factors and c-met expression in the CSF are associated with poor survival. In a study of rilotumumab, a human monoclonal antibody that targets HGF/GF, did not show antitumor activity in patients with recurrent glioblastoma [46].

Immunotherapy

Immunotherapy includes using immunomodulatory drugs to activate the immune system to fight cancer. Immune checkpoint inhibition interferes with T cell signaling via programmed death 1 (PD-1), programmed death-1 ligand, and cytotoxic T lymphocyte-associated agent-4 [47]. Side effect profiles of immunotherapies include colitis, cholecystitis, diabetic ketoacidosis, cerebritis, and altered mental status.

Nivolumab, pembrolizumab, and cemiplimab are anti PD1 antibodies. PD-1 inhibition blocks PD-1's engagement with its inhibitory ligands. This leads to cytotoxic T cell-mediated tumor cell death. There are several clinical trials focused on anti PD1 inhibitors in newly diagnosed and recurrent GBM. Nivolumab is a fully human immunoglobulin G4 monoclonal antibody that targets programmed death-1 (PD-1) checkpoint receptor [48].

In the CheckMate 143 phase III trial, 369 patients with recurrent GBM received either nivolumab or bevacizumab. The median OS was similar in both groups, 9.8 months in the nivolumab arm and 10 months in the bevacizumab arm. The study did not meet its primary end point of overall survival. A randomized controlled trial is underway where patients with recurrent GBM are randomized to receive either nivolumab and low-dose bevacizumab or nivolumab and high-dose bevacizumab.

Atezolizumab, avelumab, and durvalumab are anti-PD L1 antibodies approved for solid tumors. Ipilimumab is a CTLA-4 antibody tried in combination with anti-PD-1 inhibitors in clinical trials.

Conclusion

GBM remains the most aggressive malignancy of the central nervous system. With best efforts, the life expectancy of a patient diagnosed with GBM is 14–18 months. The standard of care is maximal surgical resection followed by radiation and chemotherapy with TMZ followed by TMZ alone for six cycle and tumor treating fields. There are several clinical trials currently ongoing, which will hopefully change the course of this dreaded disease in the coming years.

References

[1] Wirsching HG, Galanis E, Weller M. Glioblastoma. Handb Clin Neurol 2016;134:381–97. https://doi.org/10.1016/B978-0-12-802997-8.00023-2.

[2] Malmström A, et al. Temozolomide versus standard 6-week radiotherapy versus hypofractionated radiotherapy in patients older than 60 years with glioblastoma: the Nordic randomised, phase 3 trial. Lancet Oncol 2012;13:916–26. https://doi.org/10.1016/s1470-2045(12)70265-6.

[3] Wick W, et al. Prognostic or predictive value of MGMT promoter methylation in gliomas depends on IDH1 mutation. Neurology 2013;81:1515–22. https://doi.org/10.1212/WNL.0b013e3182a95680.

[4] Yan H, et al. IDH1 and IDH2 mutations in gliomas. N Engl J Med 2009;360:765–73. https://doi.org/10.1056/NEJMoa0808710.

[5] Nobusawa S, Watanabe T, Kleihues P, Ohgaki H. IDH1 mutations as molecular signature and predictive factor of secondary glioblastomas. Clin Cancer Res 2009;15:6002–7. https://doi.org/10.1158/1078-0432.CCR-09-0715.

[6] Lai A, et al. Evidence for sequenced molecular evolution of IDH1 mutant glioblastoma from a distinct cell of origin. J Clin Oncol 2011;29:4482–90. https://doi.org/10.1200/JCO.2010.33.8715.

[7] Cancer Genome Atlas Research, N. Comprehensive genomic characterization defines human glioblastoma genes and core pathways. Nature 2008;**455**:1061–8. https://doi.org/10.1038/nature07385.

[8] Stoyanov GS, Dzhenkov DL. On the concepts and history of glioblastoma multiforme—morphology, genetics and epigenetics. Folia Med (Plovdiv) 2018;60:48–66. https://doi.org/10.1515/folmed-2017-0069.

[9] Stupp R, et al. Radiotherapy plus concomitant and adjuvant temozolomide for glioblastoma. N Engl J Med 2005;352:987–96. https://doi.org/10.1056/NEJMoa043330.

[10] Sanai N, Polley MY, McDermott MW, Parsa AT, Berger MS. An extent of resection threshold for newly diagnosed glioblastomas. J Neurosurg 2011;115:3–8. https://doi.org/10.3171/2011.2.JNS10998.

[11] Oszvald A, et al. Glioblastoma therapy in the elderly and the importance of the extent of resection regardless of age. J Neurosurg 2012;116:357–64. https://doi.org/10.3171/2011.8.JNS102114.

[12] Bloch O, et al. Impact of extent of resection for recurrent glioblastoma on overall survival: clinical article. J Neurosurg 2012;117:1032–8. https://doi.org/10.3171/2012.9.JNS12504.

[13] Stupp R, et al. Effects of radiotherapy with concomitant and adjuvant temozolomide versus radiotherapy alone on survival in glioblastoma in a randomised phase III study: 5-year analysis of the EORTC-NCIC trial. Lancet Oncol 2009;10:459–66. https://doi.org/10.1016/s1470-2045(09)70025-7.

[14] Stupp R, et al. Promising survival for patients with newly diagnosed glioblastoma multiforme treated with concomitant radiation plus temozolomide followed by adjuvant temozolomide. J Clin Oncol 2002;20:1375–82. https://doi.org/10.1200/JCO.2002.20.5.1375.

[15] Cabrera AR, et al. Radiation therapy for glioblastoma: executive summary of an American Society for Radiation Oncology evidence-based clinical practice guideline. Pract Radiat Oncol 2016;6:217–25. https://doi.org/10.1016/j.prro.2016.03.007.

[16] Roa W, et al. Abbreviated course of radiation therapy in older patients with glioblastoma multiforme: a prospective randomized clinical trial. J Clin Oncol 2004;22:1583–8. https://doi.org/10.1200/JCO.2004.06.082.

[17] Perry JR, et al. Short-course radiation plus Temozolomide in elderly patients with glioblastoma. N Engl J Med 2017;376:1027–37. https://doi.org/10.1056/NEJMoa1611977.

[18] Stupp R, et al. NovoTTF-100A versus physician's choice chemotherapy in recurrent glioblastoma: a randomised phase III trial of a novel treatment modality. Eur J Cancer 2012;48:2192–202. https://doi.org/10.1016/j.ejca.2012.04.011.

[19] Stupp R, et al. Maintenance therapy with tumor-treating fields plus Temozolomide vs Temozolomide alone for glioblastoma: a randomized clinical trial. JAMA 2015;314:2535–43. https://doi.org/10.1001/jama.2015.16669.

[20] Karachi A, Dastmalchi F, Mitchell DA, Rahman M. Temozolomide for immunomodulation in the treatment of glioblastoma. Neuro Oncol 2018;20:1566–72. https://doi.org/10.1093/neuonc/noy072.

[21] Gilbert MR, et al. Dose-dense temozolomide for newly diagnosed glioblastoma: a randomized phase III clinical trial. J Clin Oncol 2013;31:4085–91. https://doi.org/10.1200/JCO.2013.49.6968.

[22] Lee SY. Temozolomide resistance in glioblastoma multiforme. Genes Dis 2016;3:198–210. https://doi.org/10.1016/j.gendis.2016.04.007.

[23] Weller M, Wick W, Hegi ME, Stupp R, Tabatabai G. Should biomarkers be used to design personalized medicine for the treatment of glioblastoma? Future Oncol 2010;6:1407–14. https://doi.org/10.2217/fon.10.113.

[24] Esteller M, et al. Inactivation of the DNA-repair gene MGMT and the clinical response of gliomas to alkylating agents. N Engl J Med 2000;343:1350–4. https://doi.org/10.1056/NEJM200011093431901.

[25] Schreck KC, Grossman SA. Role of Temozolomide in the Treatment of Cancers Involving the Central Nervous System. Oncology (Williston Park) 2018;**32**:555–60. 569.

[26] Parney IF, Chang SM. Current chemotherapy for glioblastoma. Cancer J 2003;9:149–56. https://doi.org/10.1097/00130404-200305000-00003.

[27] Aoki T, Hashimoto N, Matsutani M. Management of glioblastoma. Expert Opin Pharmacother 2007;8:3133–46. https://doi.org/10.1517/14656566.8.18.3133.

[28] Brandes AA, Bartolotti M, Tosoni A, Franceschi E. Nitrosoureas in the management of malignant gliomas. Curr Neurol Neurosci Rep 2016;16:13. https://doi.org/10.1007/s11910-015-0611-8.

[29] Stewart LA. Chemotherapy in adult high-grade glioma: a systematic review and meta-analysis of individual patient data from 12 randomised trials. Lancet 2002;359:1011–8. https://doi.org/10.1016/s0140-6736(02)08091-1.

[30] Walker MD, et al. Randomized comparisons of radiotherapy and nitrosoureas for the treatment of malignant glioma after surgery. N Engl J Med 1980;303:1323–9. https://doi.org/10.1056/NEJM198012043032303.

[31] Ashby LS, Smith KA, Stea B. Gliadel wafer implantation combined with standard radiotherapy and concurrent followed by adjuvant temozolomide for treatment of newly diagnosed high-grade glioma: a systematic literature review. World J Surg Oncol 2016;14:225. https://doi.org/10.1186/s12957-016-0975-5.

[32] Bock HC, et al. First-line treatment of malignant glioma with carmustine implants followed by concomitant radiochemotherapy: a multicenter experience. Neurosurg Rev 2010;33:441–9. https://doi.org/10.1007/s10143-010-0280-7.

[33] Roberts NB, et al. Repurposing platinum-based chemotherapies for multi-modal treatment of glioblastoma. Oncoimmunology 2016;**5**. https://doi.org/10.1080/2162402X.2016.1208876, e1208876.

[34] Gan HK, et al. Safety, pharmacokinetics, and antitumor response of depatuxizumab mafodotin as monotherapy or in combination with temozolomide in patients with glioblastoma. Neuro Oncol 2018;20:838–47. https://doi.org/10.1093/neuonc/nox202.

[35] Hui K. Gan, et al. Phase I study of ABT-414 mono- or combination therapy with temozolomide (TMZ) in recurrent glioblastoma (GBM). https://doi.org/10.1200/jco.2015.33.15_suppl.2016 J Clin Oncol 33, no. 15_suppl (May 20, 2015) 2016–2016.

[36] Reardon DA, et al. Efficacy and safety results of ABT-414 in combination with radiation and temozolomide in newly diagnosed glioblastoma. Neuro Oncol 2017;19:965–75. https://doi.org/10.1093/neuonc/now257.

[37] Parrozzani R, et al. Corneal side effects induced by EGFR-inhibitor antibody-drug conjugate ABT-414 in patients with recurrent glioblastoma: a prospective clinical and confocal microscopy study. Ther Adv Med Oncol 2020;12. https://doi.org/10.1177/1758835920907543. 1758835920907543.

[38] van den Bent M, et al. Efficacy of depatuxizumab mafodotin (ABT-414) monotherapy in patients with EGFR-amplified, recurrent glioblastoma: results from a multi-center, international study. Cancer Chemother Pharmacol 2017;80:1209–17. https://doi.org/10.1007/s00280-017-3451-1.

[39] Gilbert MR, et al. A randomized trial of bevacizumab for newly diagnosed glioblastoma. N Engl J Med 2014;370:699–708. https://doi.org/10.1056/NEJMoa1308573.

[40] Chinot OL, et al. Bevacizumab plus radiotherapy-temozolomide for newly diagnosed glioblastoma. N Engl J Med 2014;370:709–22. https://doi.org/10.1056/NEJMoa1308345.

[41] Friedman HS, et al. Bevacizumab alone and in combination with irinotecan in recurrent glioblastoma. J Clin Oncol 2009;27:4733–40. https://doi.org/10.1200/JCO.2008.19.8721.

[42] Cohen MH, Shen YL, Keegan P, Pazdur R. FDA drug approval summary: bevacizumab (Avastin) as treatment of recurrent glioblastoma multiforme. Oncologist 2009;14:1131–8. https://doi.org/10.1634/theoncologist.2009-0121.

[43] Taal W, et al. Single-agent bevacizumab or lomustine versus a combination of bevacizumab plus lomustine in patients with recurrent glioblastoma (BELOB trial): a randomised controlled phase 2 trial. Lancet Oncol 2014;15:943–53. https://doi.org/10.1016/s1470-2045(14)70314-6.

[44] Wick W, et al. Lomustine and bevacizumab in progressive glioblastoma. N Engl J Med 2017; 377:1954–63. https://doi.org/10.1056/NEJMoa1707358.

[45] Li X, et al. PI3K/Akt/mTOR signaling pathway and targeted therapy for glioblastoma. Oncotarget 2016;7:33440–50. https://doi.org/10.18632/oncotarget.7961.

[46] Wen PY, et al. A phase II study evaluating the efficacy and safety of AMG 102 (rilotumumab) in patients with recurrent glioblastoma. Neuro Oncol 2011;13:437–46. https://doi.org/10.1093/neuonc/noq198.

[47] Neagu MR, Reardon DA. An update on the role of immunotherapy and vaccine strategies for primary brain tumors. Curr Treat Options Oncol 2015;16:54. https://doi.org/10.1007/s11864-015-0371-3.

[48] Reardon DA, et al. Effect of nivolumab vs bevacizumab in patients with recurrent glioblastoma: the CheckMate 143 phase 3 randomized clinical trial. JAMA Oncol 2020;6:1003–10. https://doi.org/10.1001/jamaoncol.2020.1024.

CHAPTER

3

Investigational new drugs against glioblastoma

David Olayinka Kamson[a,b], *Harmon Singh Khela*[b], *and John Laterra*[b]

[a]Neuro-Oncology Branch, National Cancer Institute, National Institutes of Health, Bethesda, MD, United States [b]Department of Neurology, Johns Hopkins Hospital, Baltimore, MD, United States

Abstract

This chapter provides a timely overview of new drugs under development for glioblastoma therapy. First, we describe the challenges to drug development that are specific to glioblastoma and inherent issues that explain the high failure rate of late phase clinical trials against these tumors. We identify measures that have the potential to overcome current challenges especially if applied early during drug development, preclinical testing, and clinical translation. We focus on agents that target tumor cell-autonomous mechanisms in primary glioblastoma that represents 90%–95% of glioblastoma diagnoses. Druggable targets in glioblastoma cells are described in a topology-based fashion including agents and strategies directed at (i) DNA, DNA repair, and DNA integrity including novel approaches to DNA alkylation, PARP, DNA protein kinase, and telomerase inhibition; (ii) epigenetic determinants of chromatin structure and gene transcription, including histone deacetylase and methylase inhibitors; (iii) modulators of cell cycle and apoptosis, including CDK4/6 inhibitors, drugs that interfere with the mitotic spindle, MDM2, Wee-1 and protein phosphatase 2A inhibitors and caspase activators; (iv) protein homeostasis, including proteasome and neddylation inhibitors; and (v) upstream cellular signaling pathways, including EGFR, PI3K/AKT/mTOR and MEK inhibitors. We also briefly discuss drug delivery and synthetic lethality approaches that may open new avenues for otherwise potentially effective drugs.

Abbreviations

BBB blood-brain barrier
BER base excision repair
CSF cerebrospinal fluid
EB1 end-binding protein 1
IC_{50} half-maximal inhibitory concentration
IDH isocitrate dehydrogenase

Glioblastoma Resistance to Chemotherapy: Molecular Mechanisms and Innovative Reversal Strategies
https://doi.org/10.1016/B978-0-12-821567-8.00002-6

GBM	glioblastoma
HDAC	histone deacetylase
HR	homologous repair
MGMT	O^6 methyl guanine methyltransferase
MTD	maximum tolerated dose
MTIC	5-(3-methyltriazen-1-yl)imidazole-4-carboxamide
NEDD8	neural precursor cell-expressed developmentally downregulated 8
NHEJ	nonhomologous end joining
PC3	procaspase-3
PI3K	phosphatidyl inositol 3 kinase
PIP_2	phosphatidylinositol-4,5-bisphophate
PIP_3	phosphatidylinositol-3,4,5-trisphophate
PK	pharmacokinetics
PP2A	protein phosphatase 2A
TMZ	temozolomide

Conflict of interest

No potential conflicts of interest were disclosed.

Introduction

This chapter provides a timely overview of new drugs under development for glioblastoma therapy. First, we describe the challenges to drug development, which are specific to glioblastoma and inherent issues that explain the high failure rate of late phase clinical trials against these tumors. We identify measures that have the potential to overcome current challenges especially if applied early during drug development, preclinical testing, and clinical translation. Druggable targets in glioblastoma cells are described in a topology-based fashion including agents and strategies directed at (i) DNA, DNA repair, and DNA integrity; (ii) epigenetic determinants of chromatin structure and gene transcription; (iii) modulators of cell cycle and apoptosis; (iv) protein homeostasis; and (v) upstream cellular signaling pathways. We focus on agents that target tumor cell-autonomous mechanisms in primary glioblastoma that represents 90%–95% of glioblastoma diagnoses. Hence, strategies currently most relevant to lower-grade astrocytoma and secondary glioblastoma (e.g., inhibitors of mutant IDH1/2) are not included. Immunotherapeutics and other strategies that primarily target the tumor microenvironment are covered elsewhere and in considerable detail in this volume. We also briefly discuss drug delivery and synthetic lethality approaches that may open new avenues for otherwise potentially effective drugs.

Special challenges

Glioblastomas (GBM) are diffusely infiltrative gliomas that represent the most common and most aggressive of primary brain tumors [1, 2]. As the recently abandoned adjective "multiforme" implied, GBMs exhibit incredible genetic, epigenetic, and morphologic heterogeneity. This molecular heterogeneity evolves overtime as the tumor adapts in response to treatment. GBMs typically present as a solid, partially necrotic tumor core where there is localized blood-brain barrier (BBB) breakdown. Tumor cells have already invaded diffusely into the surrounded brain at diagnosis. Numerous studies show evidence of GBM cells in the normal-appearing brain with intact BBB even contralateral to the tumor core [3–5]. As a consequence of this extreme infiltration, maximum safe resection is not curative but serves as a diagnostic and cytoreductive measure that lessens neurologic symptoms owing to the core's mass effect and optimizes the patient's ability to tolerate postsurgical therapies [6, 7]. Radiotherapy extends survival but the benefit of radiation therapy alone is measured in months, even if the whole brain receives a high radiation dose to target all infiltrative tumor cells [7, 8]. Such an approach has long been abandoned owing to its neurotoxicity and lack of improved efficacy. Thus, there is more to explain radiotherapy's modest efficacy than inadequate treatment coverage. The promise of systemic drug therapy is best underlined by the Stupp trial proving the utility of the alkylating chemotherapeutic temozolomide (TMZ) against GBM [8]. Subsequent molecular analyses show that the GBM subsets that benefit significantly from TMZ are those with MGMT gene silencing through promoter methylation and those carrying an IDH mutation [9–14]. The former represents ~45%–55% of GBM (including those with intermediate methylation status) and the latter recognized by 2016 WHO classification as secondary GBM (5%–10% of all GBM), developmentally and genetically distinct from primary GBM [2]. There are essentially no chemotherapeutic agents proven to be effective for MGMT unmethylated primary GBM or for GBM that recurs after treatment with radiation and TMZ [15–17]. Thus, novel drugs and drug combinations are desperately needed, and the partial success of TMZ has yet to be surpassed, or at least replicated with other agents.

Despite the enormous demand and effort to develop novel drugs against GBM, the failure rate of phase III trials preceded by positive phase II results between 1991 and 2016 is >91%. Concurrently, the costs of these trials have grown substantially, reaching up to $125K per patient by 2013 [18, 19]. There is clearly a great need for a paradigm shift for GBM trial design. Before covering investigational drugs, it is, therefore, crucial to inspect the intricacies that explain some of this high failure rate of trials.

Concepts to explain failure

Multiple factors have and continue to contribute to the failed translation of promising preclinical to clinical efficacy in GBM and other primary brain malignancies. Yet, as described by Grossman et al. clinical success rates for agents with proven direct on-target effects can be improved by following a few simple principles [20]. First, the drug has to have a concentration (i.e., IC_{50}) that is effective against its target at a cellular level. Second, the effective concentration needs to be achieved in the drug's volume of distribution, specifically tumor tissue. Third, the drug's volume of distribution at the effective concentration has to sufficiently overlap with the entire tumor volume that for GBM represents the tumor core and the diffusely infiltrating tumor cells residing behind an intact BBB (i.e., entire brain). The drug-to-target volume overlap has to cover a clinically meaningful tumor volume and remain within the above-defined parameters for durations compatible with target biochemistry to have a survival impact. These principles are similar to those employed to achieve clinically efficacious bacteriostatic and bactericidal antibiotic concentrations.

In reality, the triad of drug distribution, specific target, and clinically relevant volumes may be highly variable between subjects, who may have great differences in tumor tissue perfusion and BBB integrity. Given the invasiveness of GBM and the lack of cure by the resection of the contrast-enhancing tumor volume, it is reasonable to assume that the clinically relevant tumor volume is the entire brain. Thus, to minimize the risk of drug-to-tumor mismatch, ideal drugs should reach therapeutic concentrations in normal-appearing brains where the BBB is intact. Utilizing advanced methods to evaluate drug pharmacokinetics and bioavailability is important, especially since CSF/plasma drug concentration ratios are poor surrogates of brain accumulation owing in part to the crucial role of efflux pumps and major differences in tissue protein binding [21], Clinical trials are increasing including surgical intervention arms to quantify drug concentration in the resected tissue. Drug concentrations from resected samples that are predominantly from BBB-defective contrast-enhancing tumor regions cannot be extrapolated to drug concentrations within nonenhancing brain that is often the site of treatment failure and tumor recurrence [22]. Thus, developing and integrating methods such as invasive microdialysis and noninvasive imaging to track drug bioavailability spatially and temporally will have great utility [22, 23].

Target-to-tumor mismatch owing to GBM's molecular heterogeneity is an additional critical obstacle to the successful development of new targeted therapies, even for those that otherwise distribute ideally throughout the entire tumor volume. The more targeted the agent (e.g., a drug against mutant EGFR) the higher the risk that its specific target is expressed only in a fraction of the tumor volume. Additional complexities arise from dynamically changing target expression and resistance mechanisms that allow tumor cells to bypass target inhibition. For example, if only a subpopulation of cells expresses mutant EGFR within a volume (i.e., the target penetrance is partial), even excellent spatial overlap of drug and target will likely lead to treatment failure. Emerging resistance can even influence the target volume of drugs that affect ubiquitous cellular processes (e.g., mutagens or agents against the mitotic machinery). Resistance may emerge transiently through cell cycle arrest or upregulation of pathways that bypass the drug target for the duration of drug exposure, or permanently through accelerated selection and expansion of resistant mutants. These issues are as elusive as critical. One solution, albeit impractical, is to continually monitor drug and target biodistributions noninvasively throughout therapy. A more practical approach is to monitor a surrogate signal of biological response (e.g., tumor cell metabolism) to target inhibition. An example is [^{18}F]fluorodeoxy glucose (FDG) PET imaging, where tracking radioactive glucose allows visualization of the Wartburg effect, and thus monitors treatment responses more promptly than that offered by MRI-based changes in tumor volume. Unfortunately, the high baseline glucose utilization in the brain results in low FDG-PET signal-to-noise ratios. Amino acid PET tracers have emerged recently as superior alternatives to FDG for brain tumor imaging. However, clinical trials integrating these tracers and drugs that directly target amino acid metabolism are far and few [24–26].

Unfortunately, no cancer chemotherapeutic exists to date for which all the above information is available. Although the development of such methods would compose significant upfront expense, reducing the number of futile late phase neuro-oncology trials could potentially render the method cost-efficient. Furthermore, characterization of the cause of failure could allow modification of drugs or their delivery accordingly. In the near future, additional advances in BBB opening achieved either selectively (e.g., transcranial ultrasound) or globally (e.g., regadenoson injection) may allow entry of drugs that previously failed against GBM owing to poor drug distribution [27–29]. Nonetheless, methods to achieve consistent response to drugs would revolutionize neuro-oncology. We have yet to formulate strategies for *in vivo* detection and for the eradication of quiescent GBM stem cells that are responsible for the inevitable recurrence of these tumors [30, 31].

Temozolomide, the prototype of antiglioma drugs

Except for modest breakthroughs in the treatment of rare, molecularly distinct subgroups of GBM (e.g., BRAF V600E; NTRK1 gene fusion), no major breakthrough has occurred in the chemotherapy space since the 2005 study confirming improved survival with TMZ added to standard radiotherapy [8, 32, 33]. Subsequent advances in our understanding of molecular biomarkers revealed that it is primarily the 45%–55% of GBM patients with an at least partially methylated MGMT promoter [9–13, 34], or the 5%–10% who have an isocitrate dehydrogenase (IDH) mutation who derive a survival benefit from the use of TMZ [2, 14]. While concurrent IDH mutation and MGMT*p* methylation can nearly triple the otherwise more or less 1-year-median survival of those with neither of these alterations, the benefits offered by TMZ remain short lived and all GBMs recur eventually [10, 13, 35]. Regardless of its limitations, TMZ remains the benchmark other drugs are measured against. Therefore, understanding its action is important to recognize how novel drugs could exceed or improve their activity.

TMZ is a small-molecular-weight (194 Da) lipophilic second-generation imidazotetrazine prodrug that is thought to be uniquely positioned to treat brain tumors owing to its pH-dependent activity, high BBB penetrance, and relatively low systemic toxicity [36–38]. TMZ is stable under the acidic pH of the stomach and has nearly 100% bioavailability. It accumulates well in brain tumors at 1.3-fold higher concentrations compared to the plasma, and even normal brain TMZ concentrations are similar to that of plasma levels (brain-to-plasma ratio 0.9) [38]. Under the alkaline systemic milieu ($pH > 7$), it undergoes nonenzymatic conversion to its active metabolite 5-(3-methyltriazen-1-yl)imidazole-4-carboxamide (MTIC). MTIC has a half-life of 2 min as it is shortly broken down under acidic conditions such as in the tumor microenvironment [39]. The end product is a highly reactive methyl-diazonium cation that preferentially methylates DNA. Cytotoxicity is mainly mediated through the methylation of O^6 on guanine and to a lesser extent N^3 and N^7 on adenine. Methylated O^6 guanine either undergoes stoichiometrically matched demethylation by MGMT or activates the DNA mismatch repair (MMR) system by mispairing guanine with thymine during DNA replication. As MMR only recognizes and excises the mismatch on the daughter strand, futile cycles of reinsertion and excision ensue, which lead to replication fork collapse and subsequent G_2/M cell cycle arrest followed by apoptosis [40–42]. Methylation of N^3 adenine is also lethal if not repaired rapidly by the base excision repair (BER) system [43].

The above-described cascade of events clearly indicates the fragility of the genetic and epigenetic circumstances that enable TMZ to be effective. First and foremost, epigenetic silencing of the MGMT gene promoter translates to 12 months of added survival by TMZ that plateaus for those above a certain level of MGMTp methylation. In contrast, the clinical benefit from TMZ is highly questionable for those below a level of MGMT methylation [13, 34]. Interestingly *in vitro*, DNA hypermethylation such as seen in IDH mutants is associated with even lower MGMT expression, whereas MGMT gene translocations and fusions with new promoters were recently described as an alternative avenue for GBM cells to increase MGMT expression and escape TMZ toxicity [44]. Even if the function of MGMT is persistently lost, defects or disruptions in any of the DNA MMR genes MLH1, MSH2, MSH6, and PMS2 propagate hypermutation and TMZ resistance and is characteristic of post-TMZ GBM recurrence [45, 46]. Loss or inactivating mutations of downstream cell cycle regulators such as RB1 and TP53 are also associated with the recurrence of TMZ resistant GBM owing to the capacity to cycle through DNA damage signals that normally halt cell proliferation. While cytotoxicity to replicating cells is thought to be abrogated mainly by MGMT, the base excision repair (BER) system including poly-ADP-ribose-polymerase 1 (PARP-1) is responsible for the excision and repair of methylated N^3 adenine, which seems to play an important role in protecting nonreplicating cells from TMZ-induced damage [47, 48].

Overall, TMZ pharmacokinetic data matches what is seen both clinically and *in vitro*. TMZ accumulates in humans at concentrations that surpass the IC_{50} (<50 μM) associated with TMZ sensitive GBM cell lines, whereas it falls far below that of the IC_{50} range (>300 μM) associated with TMZ resistance. Therefore, it is clear that approaches that target the above-listed explanations for drug failures are of great interest in neuro-oncology and will be covered in detail in this chapter.

Agents that target DNA integrity

Agents that induce direct DNA damage

Although our understanding of TMZ resistance has greatly expanded over the past decades, attempts to overcome resistance using the MGMT inhibitor **O6-benzylguanine**, or to deplete MGMT using a dose-dense TMZ schedule have consistently failed [49, 50]. The need for an MGMT-resistant alkylating agent has led to a resurrected interest in **dianhydrogalactitol** (VAL-093). Dianhydrogalactitol is a small molecule hexitol epoxide (MW = 146 Da) that first emerged as a potential anti-glioma agent in the 1970s after it was shown to accumulate exceedingly well in the brain [51–53]. In 1973, in a small trial including 42 patients, it was shown to nearly double patient survival when combined with radiotherapy compared to those treated with radiation alone in a cohort of predominantly grade 3 supratentorial high-grade gliomas [54]. Prior treatment histories were not published and as the study took place prior to the era of tumor molecular profiling, it is not known whether the results were skewed by the overrepresentation of a molecularly favorable subgroup. Another randomized controlled phase II study from 1990 showed a 6.5-week-longer survival when added to radiotherapy vs radiotherapy alone in a cohort of high-grade gliomas [55]. Although this survival difference did not reach statistical significance, the study was likely underpowered as it included only 30 patients per group. Preliminary results from more recent trials evaluating the efficacy of VAL-083 combined with radiotherapy in newly diagnosed GBM and as monotherapy for recurrent GBM are also encouraging [56–58]. VAL-083 is thought to target DNA at the N^7 guanine position causing monoalkylation as well as interstrand cross-linking that leads to double-strand breaks as corroborated by γH2AX focus precipitation [59, 60]. Experimental evidence shows persistent antitumor activity even in the setting of an impaired MMR system. A possible explanation is its capacity to activate the double-strand break sensing system through ATM leading to p53 and $p21^{cip1}$ activation and inhibition of G2 to M cell cycle transition. Additional studies suggest that VAL-083 also has antimigration, antiinvasion, and antiangiogenic actions mediated by interfering with VEGF/VEGFR2 and FGF2/FGFR1 signaling pathways [61].

The promising results from early-phase clinical testing need to be viewed in the context of a potential mismatch between effective drug concentrations in vitro and in vivo pharmacokinetics. A number of *in vitro* analyses found that effective anti-GBM VAL-083 concentrations were in the 5–20 μM range [62–65]. Pharmacokinetic studies have found brain tumor/plasma, normal white matter/plasma, and CSF/plasma ratios to be 1.3, 0.3, and 0.2, respectively, with typical plasma peak concentrations of 3.9–6.2 μM. These results suggest that brain drug accumulation may fall short of drug levels required for GBM cytotoxicity [53, 58]. VAL-083 may have synergistic activity with topoisomerase inhibitors, such as etoposide or irinotecan potentially enabling lower drug concentrations to elicit desired tumoristatic or tumoricidal effects. Yet, most topoisomerase inhibitors have relatively high-molecular weight (>500 Da) and poor CNS accumulation as well as poor efficacy against GBM even as a liposomal formulation [66–71].

Nucleotide antimetabolites represent a large group of drugs that inhibit DNA synthesis, DNA repair, or cause direct DNA damage. These agents fell out of favor after initial investigations owing to their lack of apparent clinical efficacy against GBM. Many (e.g., azacytidine, 5FU) have insufficient BBB penetration to reach effective concentrations predicted by *in vitro* and animal model data [72–75]. More recent invasive pharmacokinetic studies showed that **gemcitabine** (MW = 263 Da) does cross the BBB with the potential to act as a radiosensitizer against high-grade gliomas [76, 77]. Yet, the tally includes numerous negative GBM trials vs some potential survival benefits in anaplastic gliomas, including tumors that would now be considered molecular GBM [78–80].

Alternative antinucleotide delivery approaches are currently being studied. Perhaps the most promising of these evaluated **Toca 511** (vocimagene amiretrorepvec) in combination with Toca FC (extended-release 5-fluorocytosine). Toca 511 is a modified tumor cell-selective replication-competent retroviral vector containing a transgene for cytosine deaminase, which converts prodrug 5-fluorocytosine to its active form, 5-fluorouracyl [81, 82]. Animal data showed excellent 5-FU conversion and brain accumulation in transfected mice, and early phase clinical data showed promising responses [83, 84]. Unfortunately, a phase III trial comparing TOCA 511 + Toca FC with standard-of-care in patients with recurrent high-grade glioma did not meet its efficacy target. Post hoc analyses suggested favorable overall survival trends in certain tumor subgroups but not in IDH-wild-type primary GBM [85].

Agents that impair DNA repair

Drugs that target DNA repair are of great interest since they have the potential to enhance the DNA damaging effects of current standard-of-care radiotherapy and TMZ. Inhibitors of the poly ribose-ADP polymerase (PARP) are heavily studied within this group of agents. PARP is part of the BER system and binds single-stranded DNA breaks to which it attaches ADP-ribose polymers and activates a number of DNA damage response proteins. Although the PARP gene family consists of 17 enzymes, PARP-1 is responsible for approximately 90% of cellular PARP activity [86]. PARP-1 is a desirable target as it seems to be consistently expressed in GBM cells, but not in normal brain tissue [87]. PARP can be inhibited by blocking its catalytic activity. However, *in vitro* data suggest that this approach is limited and bypassed by other DNA repair mechanisms [88]. PARP functions by binding to DNA until it undergoes auto-ADP-ribosylation. Blocking auto-ADP-ribosylation "traps" PARP on DNA, impedes the activity of other DNA damage repair enzymes, and requires salvage by the homologous recombination repair (HR) system to avoid replication fork collapse and double-stranded DNA break formation [89]. Monotherapy with "trapping" PARP inhibitors can lead to cell death even at low concentrations in tumors with HR repair defects, such as those with germline BRCA1/2 mutations. Additionally, IDH mutant tumors exhibit a level of "BRCAness" resulting from HR enzyme inhibition by the oncometabolite 2-hydroxyglutarate, and thus seem to be exceptionally sensitive to PARP inhibition [90]. As mentioned before, inhibition of the base excision repair system by PARP inhibitors can synergize with TMZ by enhancing the MGMT-insensitive toxicity of adenine N^3/N^7 methylation [91]. The synergy of PARP inhibitors with radiotherapy is also thought to result from hindering repair of radiation-induced DNA damage [92].

Numerous PARP inhibitors are currently under investigation. Those of the greatest interesting include olaparib, veliparib, niraparib, and pamiparib. These agents differ in molecular weight, polarity, BBB penetrance, PARP inhibition, and trapping potency. **Olaparib** (AZD2281; MW = 435 Da) is the best-studied agent in neuro-oncology to date. Its IC_{50} for tumor cell viability in pediatric high-grade glioma cell lines falls between 1.4 and 7.4 μM [92]. The bulk of preclinical data shows that tissue concentrations within brain xenografts and normal brains are below the 0.2 μM range with very poor accumulation in normal-appearing brain and spine. Brain-plasma ratios are 1%–2% within 2 h of infusion with rapid clearance by 5 h post infusion [93, 94]. Human PK data from the phase I OPARATIC trial studying olaparib combined with TMZ found average olaparib concentrations in contrast-enhancing tumor cores and in nonenhancing tumor margins to be ~0.5 μM. While these concentrations are lower than the expected IC_{50} for tumor cell viability, *in vitro* follow-up experiments showed that olaparib has a significant radiosensitizing effect at both 0.1 and 0.5 μM. Clinical trials testing olaparib with radiotherapy with or without TMZ are currently ongoing [95].

Veliparib (ABT-888; MW = 244 Da) is currently the second most studied PARP inhibitor in glioma trials. Its IC_{50} for killing MGMT*p* unmethylated GBM cell lines ranges from 18 to 80 μM as a monotherapy, and 5–10 μM enabled significant sensitization to TMZ (150 and 300 μM) and to radiotherapy [96]. Animal data shows that veliparib has moderate BBB penetrability with a brain-to-plasma ratio of 0.3 [97, 98]. Unfortunately, veliparib in combination with TMZ failed to prolong survival in adults with recurrent GBM [99]. A similar lack of efficacy led to the premature closure of a phase I/II study of veliparib combined with radiotherapy and TMZ in children with newly diagnosed diffuse midline gliomas [100]. Of note, lack of efficacy in this pediatric study is attributable at least in part to PK data showing mean C_{max} in the highest dosed group to be 4.4 μM in the plasma, which is lower than established effective concentrations, even without taking into account limited BBB penetration.

Xiong et al. compared the brain penetration of four PARP inhibitors in mice and found niraparib and pamiparib to have favorable brain-to-plasma ratios compared to talazoparib and olaparib [93]. Although the data is scarce for **niraparib** (MK4827; MW = 320 Da), it was found to have an IC_{50} above 2 μM for glioma cell line growth inhibition but a low nanomolar range IC_{50} for PARP inhibition against cell lines with impaired HR repair (e.g., BRCA mutants) [101, 102]. Niraparib's peak plasma drug concentrations did not surpass the 4 μM range. This, along with its 15% BBB penetration, predicts that it is unlikely to be effective against cells with intact HR [102]. Preliminary data on **pamiparib** (BGB-290;

MW = 298 Da) seem more promising. It readily traverses the BBB in mice with a brain-to-plasma ratio of 0.19, which is nearly 10-fold that of olaparib [93]. While the *in vitro* IC/EC_{50} of pamiparib for PARP inhibition was in the nanomolar range, more recent data suggests that an IDH mutation may be required to sensitize tumors to pamiparib (IC_{50} 0.67 vs 141 μM in IDH mutant vs IDH wild-type tumors, respectively) [103]. Recently established phase II doses of pamiparib seem to fall in the human-equivalent dose range used in the above mouse study and human pharmacokinetic data are currently being accumulated [104, 105].

DNA protein kinase inhibitors

DNA protein kinase (DNA-PK) has recently emerged as a novel therapeutic target in oncology [106]. DNA-PK is a serine/threonine kinase that belongs in the phosphatidylinositol 3-kinase-related kinase group that contains prominent regulatory enzymes such as ATM, ATR, or mTOR. DNA-PK seems to play a role in determining the type of DNA repair initiated (HR vs nonhomologous end joining [NHEJ]) along with direct regulation of these processes [107]. DNA-PK has also been implicated in the regulation of chromosomal stability, telomere maintenance, cell cycle progression, and maintenance or induction of pluripotency [108–112]. DNA-PK silencing was found to inhibit Chk2-mediated phosphorylation of BRCA1 that ensures appropriate kinetochore-microtubule attachment during mitosis [108]. The c-MYC oncogene and pluripotency factor is also regulated by DNA-PK-dependent phosphorylation [113, 114]. Based on these multiple activities, it is not surprising that DNA-PK expression is upregulated and associated with radioresistance, and poor prognosis, in many forms of cancer including gliomas [109–112, 115].

CC-115 (MW = 336 Da) is a novel dual-action triazole compound with high *in vitro* potency for inhibiting mTOR1/2 and DNA-PK [116]. A human phase I trial revealed that CC-115 accumulates in GBM with an average GBM-to-plasma ratio of 0.7 and average GBM concentration of 116 nM (min-max 12–287 nM) [117]. These concentrations are well above the *in vitro* IC_{50} of 13 nM for DNA-PK and 23 nM for mTOR kinase found against prostate cancer cell lines, but far below the IC_{50} of 510 nM for cell death in CLL samples [116, 118]. IC_{50} data for CC-115 against GBM cells is not currently available. Additional data is currently being collected under the INSIGhT trial [119]. Another DNA-PK inhibitor **VX-984** showed dose-dependent radiosensitizing activity against GBM cell lines at 100–250 nM concentrations, and **nedisertib** is currently in phase I trials [120, 121].

Telomerase inhibition

Altered mechanisms of telomerase maintenance either via alternative lengthening of telomeres in the setting of an ATRX mutation, or mutations in the gene encoding the catalytic component of telomerase (TERT) are oncogenic drivers in the formation of diffuse gliomas [2]. Therefore, targeting telomerase has been a topic of interest in the treatment of GBM. **Imetelstat** (GRN163; MW = 4610 Da) has dose-dependent telomerase inhibitory activity (IC_{50} 0.45–10 μM) and in a pediatric glioma, study it reached peak plasma levels of ~22 μM range with evidence of ~95% telomerase inhibition in tumor tissue [122–124]. Unfortunately, the clinical trial was terminated early owing to two deaths secondary to thrombocytopenia-associated hemorrhages.

Agents that target epigenetic determinants of chromosome structure and gene transcription

Modulators of chromatin organization

Chromatin is organized into nucleosomes consisting of 147 bp DNA segments each wrapped around histone octamers (two H2A-H2B dimers and a H3-H4 tetramer) [125, 126]. The N-terminal "tails" of these core histones undergo a number of covalent modifications, such as acetylation, methylation, and phosphorylation that regulate chromatin structure, protein-DNA interactions, and gene expression. Histone acetyltransferases (HATs) catalyze the transfer of acetyl groups to N-terminal lysine sites neutralizing the positive charge of histones and thus allow unwinding of the negatively charged DNA from the nucleosome and facilitate transcription from the affected DNA region. In contrast, the removal of these acetyl adducts by histone deacetylase enzymes (HDACs) restores the positive histone charge, DNA winding, and transcription silencing. The role of these nonmutational epigenetic changes is being increasingly recognized as important determinants of oncogenesis, tumor heterogeneity, and therapeutic efficacy [127].

Histone deacetylase inhibitors

HDACs are the subject of intense research owing to their increased expression in malignant glioma [128–130]. HDAC1 and HDAC2 have increased expression in GBMs, and are responsible for regulating cell proliferation, as well as additional roles in DNA repair [131, 132].

Currently studied HDAC inhibitors can be classified in broad chemical groups, including benzamides (entinostat), aliphatic acids (e.g., valproic acid), hydroxamic acid analogs (vorinostat, belinostat, panobinostat, etc.), and cyclic tetrapeptides (e.g., romidepsin). Of these, **entinostat** is not covered owing to its poor BBB penetration [133]. The antiepileptic **valproate** (MW = 144 Da) is a class I inhibitor (including HDAC 1 and 2) and is very well-studied [134]. Valproate's IC_{50} for HDAC1/2 inhibition (700–1000 μM) is above the upper limit plasma concentration recommended for antiepileptic therapy (350–700 μM [50–100 mg/L]) which may explain why valproate failed to improve survival when combined with other chemotherapeutic agents in a number of systemic cancer trials [135–137]. Additionally, valproate has a low brain-to-plasma ratio (~11% of total plasma concentration). Therefore, brain levels fall below the above-established IC_{50} for HDAC inhibition even when dosed ~1.5-fold of the recommended antiepileptic dose [138]. Owing potentially to the ability of radiation to increase BBB permeability, a phase I trial testing valproate as a potential radiosensitizer in patients with newly diagnosed GBM yielded promising results that await confirmation [139, 140]. A novel valproic acid prodrug AN446 with *in vitro* anti-GBM cell activity nearly 200-fold higher than valproic acid holds the potential for future development [141].

Vorinostat (SAHA; MW = 264 Da), an HDAC1,2,3,6,10 inhibitor has also been extensively studied. In humans. Its plasma C_{max} is ~1.1 μM, and 1%–4% of the plasma concentration passes the mouse BBB [142–144]. Vorinostat's *in vitro* IC_{50} for inhibiting HDAC1, HDAC2, and in a GBM cell line (G55T2) proliferation were found to be 0.06, 0.25, and 0.8–0.9 μM, respectively [141, 142, 145]. Clinical testing found that vorinostat increased the acetylation of H2A, H2B, and H3 in post-treatment samples but failed as a monotherapy or combination therapy to significantly prolong survival of patients with either newly diagnosed or recurrent GBM [146–148].

Belinostat (PXD101; MW = 318 Da) is a novel pan-HDAC inhibitor that is structurally similar to vorinostat with improved BBB permeability. *In vitro* studies show its IC_{50} for HDAC inhibition and cytotoxicity against GBM cell lines to be 0.3 and 1.3 μM, respectively, with *in vivo* data shows plasma C_{max} 69–100 μM to be tolerable. Animal models using radiolabeled [^{11}C]methyl-belinostat showed a brain-to-plasma ratio of 6%. Therefore, belinostat seems to be a promising agent to treat GBM and human trials are currently underway. The only preliminary clinical efficacy data available to date is a study of newly diagnosed GBM using belinostat 500–750 mg/m^2 i.v. given on days 1–5 of 21-day cycles for a total of three cycles starting 1 week prior to standard radiotherapy with concurrent TMZ [149, 150]. Although OS data is not published yet, there was a nonsignificant trend toward prolonged progression-free survival. Patients treated with belinostat also had improved depression screening scores, which is a known alternate effect of the drug and may indicate successful BBB penetration.

Panobinostat (LBH589; MW = 349 Da), another hydramic acid analog has also been extensively studied and exhibited very high *in vitro* potency against diffuse intrinsic midline glioma (IC_{50} 0.002–0.1 μM), but weaker activity against some GBM cell lines (~1–1.4 μM) and very low toxicity against normal astrocytes (IC50 5.3 μM) [151–154]. In a mouse model, intraperitoneal panobinostat administration (human equivalent dose ~1 mg/kg) was demonstrated to pass the BBB with a brain concentration of 110 nM vs 319 nM in plasma 2 h after injection. In this experiment, plasma drug levels dropped much more rapidly compared to the brain, yielding brain-to-plasma ratios of 0.04, 0.59, and 3.66 over 5 min, 2 h, and 24 h post injection, respectively. Significant reduction in tissue histone acetylation further corroborated the findings in this experiment [155]. A human trial involving a combination of panobinostat and bevacizumab failed to exhibit efficacy against recurrent GBM and was therefore halted early [156, 157]. A phase I study found that combining panobinostat with fractionated stereotactic re-irradiation in recurrent high-grade glioma was well-tolerable. To date, the safety and potential efficacy of this approach have not been evaluated by a phase II trial [158]. Lastly, animal models of convection-enhanced panobinostat delivery had been shown to achieve uniform drug distribution throughout the brains of animals without evidence of toxicity even at concentrations as high as 30 μM [153, 159].

Romidepsin (FK228; MW = 541 Da) is a cyclic tetrapeptide selective class I HDAC inhibitor with high in vitro potency capable of inducing growth arrest at ~2 nM in glioma cell lines [160]. Unfortunately, romidepsin monotherapy failed to show efficacy in patients with recurrent GBM [161]. The investigators of this trial speculated that the plasma concentrations of 0.8–1.4 μM may have yielded therapeutic intratumoral levels based on a nonhuman primate PK study that found 2% of the plasma romidepsin level in the CSF [161–163]. The likelihood of tumor target inhibition remains uncertain in the absence of actual tissue romidepsin levels. Recent in vitro studies in GBM cell lines show synergy between TMZ and romidepsin at concentrations in the 1–10 nM range, PI3K/AKT/mTOR pathway inhibition, and the ability of romidepsin to inhibit AKT activation and deacetylation of histone loci H2AZK4K/K7 and H3K27 by the oncogenic EGFRvIII receptor tyrosine kinase [164]. Combining romidepsin with TMZ for specific GBM molecular subtypes deserves further investigation.

Histone methylase inhibitors

The contextual complexities of histone modification are well reflected in the increasingly recognized importance of histone methylation in carcinogenesis [165]. Unlike histone acetylation that promotes the formation of euchromatin by releasing the electrostatic histone-DNA bond, histone methylation alters the interaction between histone and chromatin-binding proteins. Methylation can either enhance or repress transcription depending on the foci and the number of methyl groups attached [166, 167]. The most extensively studied and likely most important locus in gliomas is at the lysine 27 position on histone-3 (H3K27), the methylation of which is considered a dominant mark of transcriptional repression [2, 165, 168]. Methylation of this locus is catalyzed by EZH2 (enhancer of zeste homologue 2), a subunit of the polycomb repressive complex 2 (PRC2). EZH2 is commonly overexpressed in glioma and has been associated with higher grades and shorter survival [169]. In vitro studies using GBM cell lines also implicate EZH2 activity in the development of TMZ resistance [170]. Canonical H3K27M mutation is considered a driver for the development of diffuse midline gliomas. PRC2-binding affinity to H3K27M is increased 22-fold resulting in PRC2 sequestration, aberrant histone methylation patterns, and altered expression of genes that regulate cell growth and fate determination [168, 171, 172]. Experimental evidence suggests that gliomagenesis in this population remains EZH2 dependent and that inhibiting EZH2 may abolish the growth of these tumors in a $p16^{ink4a}$-dependent fashion [173]. Thus, inhibitors of the EZH2 molecule are of great research interest.

Tazemetostat (EPZ-6438; MW = 573 Da) is an EZH2 inhibitor currently under investigation for the treatment of hematologic and solid malignancies. Tazemetostat has high *in vitro* potency to inhibit EZH2 in an enzyme assay ($IC_{50} < 5$ nM) [174]. Although it seems to require prolonged exposure to elicit its effect, tazemetostat was able to achieve nearly complete depletion of H3K27me2/3 in neural stem cells at a 100 nM concentration after 3 days of incubation [173]. In its first-in-human trial for relapsed refractory non-Hodgkin B-cell lymphoma, tazemetostat was found to be well tolerated at a dose of 800-mg twice a day administered in 28-day cycles [175]. The initial plasma C_{max} at this dose was ~2 μM but levels subsequently decreased by 42% due to autoinduction by CYP3A4. Predose plasma concentrations after the first month of treatment were ~200 nM. Brain accumulation data from mice show a brain-to-plasma ratio of 0.3, 1 h after i.v. injection [176]. Extrapolating this information suggests that tazemetostat may effectively inhibit EZH2 in brain tumors. However, important parameters such as AUC and long-term brain accumulation required to predict potential efficacy are presently unavailable.

Agents that directly alter cell cycle or cell death regulation

In the absence of devastating tumor cell damage and tumor cell necrosis, the efficacy of anticancer drugs relies to a great degree on the function of intracellular mechanisms that disable cell cycle progression and/or initiate orchestrated cell death pathways such as apoptosis and senescence. It is not surprising that the ability to bypass cell cycle stop signals and escape apoptosis is a hallmark of cancer and a major mechanism of chemotherapy resistance [44, 127, 177]. Thus, restoring these regulatory activities or further impairing them in a way to induce devastating damage in cancer cells are desirable pharmacologic goals. Analysis of The Cancer

Genome Atlas (TCGA) suggests that the majority of genetic alterations in GBM cluster around pathways that directly or indirectly inactivate Rb leading to unopposed cell cycle progression (79%), or around the loss of p53 function leading to impaired apoptosis or senescence (86%) [129]. This section focuses on agents that affect these pathways.

Targeting cell cycle regulation

The expansion and survival of cancer cells require successful mitotic activity involving the replication of biochemical machinery (G0/G1), DNA synthesis (S) quality control and DNA error repair (G2), and ultimately the commitment to proceed with mitosis from prophase to cytokinesis. The process of exiting quiescence (G0/G1) and initiating DNA synthesis (S) relies on the activation of the cyclin D-dependent kinases CDK4 and CDK6 [178]. The broad steps of this process are summarized in Fig. 1

TCGA data reveals that ~60% of GBMs have inactivating mutations or deletions in the tumor suppressor genes CDKN2A that codes for $p16^{ink4a}$ and $p14^{ARF}$ or in CDKN2B that codes for $p15^{ink4b}$. As an example of the effect of these mutational backgrounds on personalized therapeutics, the antiproliferative effect of the EZH2 inhibitor tazemetostat seems to rely on the function of $p16^{ink4a}$, and thus loss of $p16^{ink4a}$ function will render tazemetostat ineffective [173]. Activating mutations and amplification of CDK4/6, seen in 15% of GBM, overwhelm antimitogenic signaling and therefore when present, serve as desirable therapeutic targets in GBM. Lastly, about 8% of GBMs either carry an inactivating Rb mutation or deletion, freeing up E2F from its regulatory circuitry, rendering drugs targeting this system ineffective [179–181].

CDK4/6 inhibitors

The most prominent group of drugs that directly target cell cycle regulation are CDK4/6 inhibitors that interrupt G1-S phase transition by preventing downstream Rb phosphorylation. The best studied CDK4/6 inhibitors are ribociclib, palbociclib, and abemaciclib. **Ribociclib** (LEE011; MW = 435 Da), in combination with aromatase inhibitor letrozole is FDA approved for the treatment of ER-positive, Her2-negative breast cancer. Ribociclib has high potency (IC_{50} 10–40 nM) against CDK4/6 enzymatic activity *in vitro* [182–184]. Ribociclib is also able to penetrate the intact BBB in mice with an unbound brain-to-plasma fraction of 0.1 leading to ribociclib tissue levels of ~10 nM for 10 h at the therapeutic dose range [182, 183]. Two early phase human studies of ribociclib in recurrent GBM, one using 900 mg daily for 5 days and the other using 600 mg daily for 14 days prior to tissue sampling produced similar pharmacokinetic findings, with respective unbound drug concentrations of 374 and 632 nM in CSF, 560 and 484 nM in nonenhancing tumor, and 2152 and 1525 nM in contrast-enhancing tumor [184, 185]. Pharmacodynamic analyses of pretreated and on-treatment tumor tissue showed a reduction in phosphorylated Rb in the majority of patients. Tumors that recurred after ribociclib monotherapy showed PI3K/mTOR upregulation suggesting a possible resistance mechanism [185]. Overall survival data from these studies are pending.

Palbociclib (PD-332991; MW = 448 Da) is another potent inhibitor of CDK4/6 (IC_{50} 4–19 nM in enzyme assay) with high *in vitro* efficacy against GBM cell lines (IC_{50}

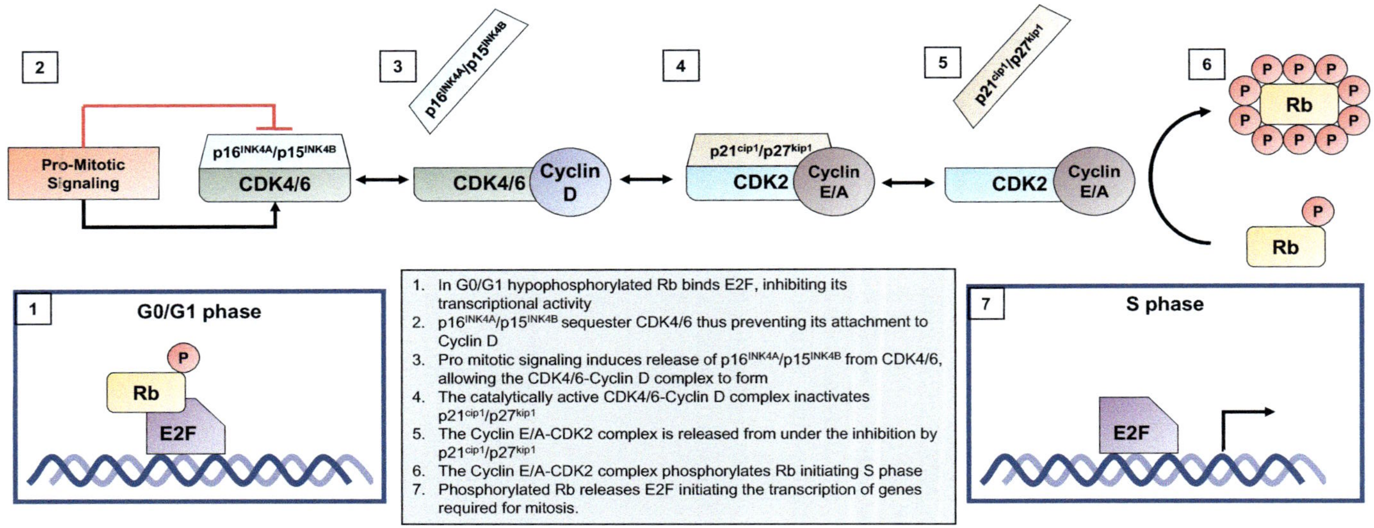

FIG. 1 A simplified overview of cell cycle regulation. *Recreated from Gil J, Peters G. Regulation of the INK4b–ARF–INK4a tumour suppressor locus: all for one or one for all. Nat Rev Mol Cell Biol 2006;7:667–77. https://doi.org/10.1038/nrm1987.*

112–252 nM) and brain-to-plasma ratios of 6.5%–10% [182, 186]. Palbociclib was shown to inhibit GBM xenografts when implanted subcutaneously but this effect did not translate well to mice bearing orthotopic GBM xenografts, which suggest limitations in the CNS bioavailability of the murine models [180]. A phase II study evaluating palbociclib monotherapy for recurrent GBM was halted early due to lack of efficacy [187]. However, PK data from five patients in this study showed an average palbociclib concentration of 1.6 μM in contrast-enhancing tumor tissue, 16-fold higher than plasma levels and well above IC_{50} for in vitro efficacy. Palbociclib is currently included in a Bayesian phase I/II study combining the drug with radiotherapy in a molecularly selected cohort of newly diagnosed MGMT*p* unmethylated GBM [188].

Abemaciclib (LY2835219; MW = 507 Da) is the third CDK4/6 inhibitor approved to treat ER-positive HER2-negative breast cancer. It has a similar IC_{50} against CD4/6 to palbociclib in cell-free assay (2.6–23 nM), but seems to have somewhat lower potency against GBM cell lines (120–540 nM) [179, 186]. Rodent models show a brain-to-plasma ratio of 120%–130% though with an unbound brain drug concentration of only 11%–17% that of plasma [189]. Abemaciclib prolonged survival of mice bearing orthotopic GBM similarly to TMZ and the combination of the two agents showed an additive effect on survival [189]. Early phase human studies with abemaciclib are currently ongoing.

Drugs that interfere with the mitotic spindle

There is a long-standing interest in agents that disrupt microtubule functions owing to their potential to interfere with the mitotic spindle and cell division. Vincristine (MW 825 Da) is commonly used in pediatric protocols for treating CNS malignancies. While it is a component of standard-of-care adjuvant chemotherapy for oligodendroglioma, it is poorly tolerated in adults and its poor brain penetration raises questions about its value in this context [190–195]. Perhaps the most unexpected newly repurposed drug of the group is **mebendazole** (MW = 295 Da), which is a benzimidazole antihelmintic agent. Its activity against GBM was discovered serendipitously after a colony of mice repeatedly failed brain tumor engraftment while receiving treatment for pinworm infection [196]. Mebendazole reduces tubulin polymerization in a dose-dependent fashion in GBM cell lines and significantly inhibits the growth of GBM cell lines *in vitro* (IC_{50} 110–310 nM) and intracranial xenograft growth [194, 196]. Preclinical data shows an additive effect of mebendazole with TMZ or radiotherapy and significant prolongation of survival when used to treat GBM xenograft-bearing mice [196–198]. A pharmacokinetic study comparing mebendazole polymorphs A, B, and C showed an 82% brain-to-plasma ratio in normal mice allowing the accumulation of mebendazole nearly 30-times higher than the IC_{50} in GBM cell lines [199]. A recent phase I study shows that mebendazole (1600 mg three times a day, highest dose tested) is well tolerable when combined with radiotherapy plus concurrent low-dose TMZ (75 mg/m^2) or with adjuvant high-dose TMZ (200 mg/m^2) [200]. While the 4800 mg daily dose can putatively lead to plasma levels close to an order of magnitude higher than the IC_{50} described above [201], confirmatory pharmacokinetic data was not collected. Results from ongoing human mebendazole clinical trials against GBM are to be reported.

BAL101553 (lisavanbulin; MW 516 Da) is a prodrug of BAL27862 (avanbulin, MW 387 Da) that reversibly inhibits microtubule organization by binding to the colchicine site [202]. BAL27862 is potently cytotoxic against GBM neurospheres that are enriched for tumor-propagating stem-like cells (IC_{50} ~ 21 nM), induces their differentiation, and inhibits the migration of these cells at concentrations as low as 6nM [203]. BAL101553 prolongs the survival of mice bearing GBM xenografts. Preclinical antitumor responses seemed to correlate with the expression of end-binding 1 protein (EB1), which is a key protein in microtubule regulation associated with cell stemness, migration, and poor patient survival [204]. Mouse data suggest excellent brain penetration with a brain-to-plasma ratio of 93% [205]. Phase I studies showed plasma drug concentrations at the MTD predict therapeutic brain levels. Phase I studies for GBM are currently ongoing [206–209].

Another target for disrupting the mitotic spindle is Aurora A, a centrosome-associated serine/threonine kinase involved in centrosome duplication, mitotic spindle assembly, and sister chromatid segregation [210, 211]. Aurora A is overexpressed in many malignancies including GBM. Paradoxically, higher Aurora A expression associates with longer patient survival but **alisertib** (MLN8237; MW = 519 Da), an Aurora A kinase inhibitor, is potently cytotoxic to GBM cell lines (IC_{50} 60–225 nM) and potentiates their response to radiation [212]. Pharmacokinetic data from mice show brain-to-plasma ratios of 6% at 1 h, and 4% 24 h following drug administration [213]. Unfortunately, the mouse BBB penetration data predicts that the doses recommended for phase II clinical testing will yield brain drug levels below the IC_{50} of the less sensitive GBM cell lines, potentially compromising the efficacy of alisertib monotherapy [214, 215]. The BBB penetration of other Aurora kinase inhibitors is currently under investigation with **barasertib** (AZD1152; MW = 588 Da) promising more favorable bioavailability [216, 217].

Targeting the apoptotic circuitry

The overwhelming majority (86%) of GBMs have genetic alterations in pathways that converge onto p53 and influence mechanisms of senescence and apoptosis [129]. The activity of p53 is tightly regulated by a number of feedback loops that also overlap with the CDK/Rb1 pathway discussed above [178, 218]. The activity of this $p14^{ARF}$-p53-MDM2 network is summarized in Fig. 2. Direct mutation or deletion of p53 is part of the cascade of events leading to gliomagenesis in IDH mutant tumors, but much less common in IDH-wild-type primary GBMs [2, 15, 129]. Amplification of MDM2 or MDM4 are each detected in 7%–8% of GBM and are mutually exclusive with p53 mutations. Most importantly CDKN2A is deleted in 58% of GBMs, leading to the loss of $p14^{ARF}$ and overactivation of MDM2.

MDM2 has recently emerged as a desirable target. **AMG-232** (MW = 569 Da) is an orally bioavailable novel MDM2 inhibitor with promising *in vitro* data against GBM [219]. AMG-232 exhibits potent growth-inhibitory actions against TP53 wild-type GBM cell lines ($IC_{50} \sim 250$ nM) compared to low potency against TP53 mutant lines (>18 000 nM) [220] and is especially potent against cell lines with MDMD2 gene amplification (IC_{50} 5.3 nM). Sensitivity to AMG232 correlated with elevated $p21^{cip1}$ levels, whereas experimental inactivation of p53 led to at least a 12-fold increase in IC_{50}. Furthermore, AMG-232 concentrations as low as 1 nM were sufficient to abrogate the three-dimensional tumor growth of TP53 wild-type GBM stem cells [220]. Unfortunately, to date brain, pharmacokinetic data has not been published. Phase I clinical trial data from mixed cohorts of hematologic and solid malignancies showed a mean plasma C_{max} at 1.7–2.5 μM following a 240 mg per day dosing regimen [221, 222]. This suggests potentially therapeutic levels against MDM2 amplified GBM even

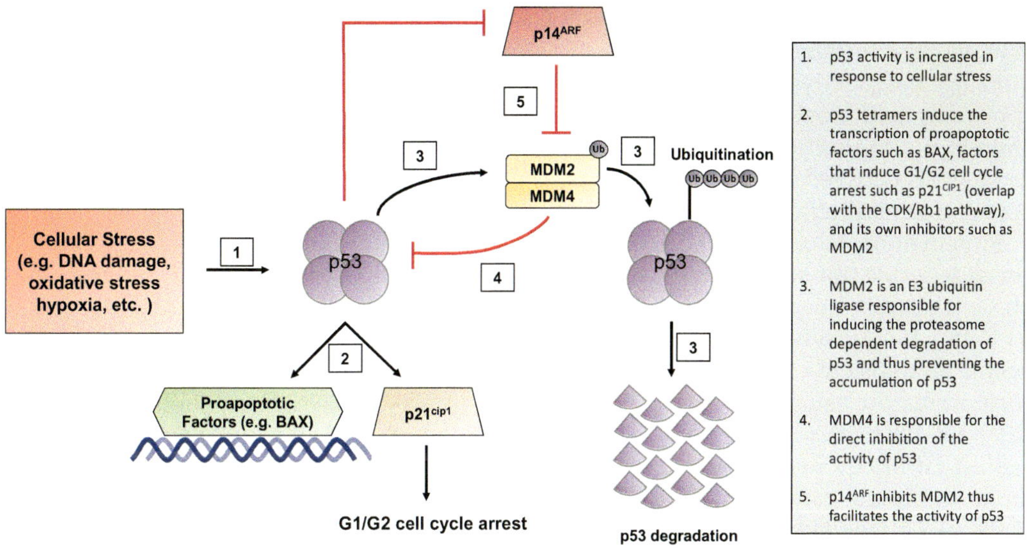

FIG. 2 Schematic summary of the $p14^{ARF}$-p53-MDM2 pathway. *Recreated from Gil J, Peters G. Regulation of the INK4b–ARF–INK4a tumour suppressor locus: all for one or one for all. Nat Rev Mol Cell Biol 2006;7:667–77. https://doi.org/10.1038/nrm1987.*

if the brain-to-plasma ratio is <1%. Ongoing phase 0/I testing dedicated to gliomas should clarify outstanding questions regarding bioavailability in the near future.

Another MDM2 inhibitor, **idasanutlin** (RG7388; MW = 617 Da) seems to have a reasonable potency against TP53-wild-type GBM cell lines (IC_{50} of 46–220 nM) and was shown to cross the mouse BBB (brain-to-plasma ratio 35%) [223, 224]. Idasanutlin was found to cooperate with radiotherapy against GBM cell lines [225]. Drug resistance emerged through the activation of the ERK1/2-IGF-binding protein1 (IGFBP1) signaling cascade and was successfully overcome through inhibition of ERK by trametinib. Idasanutlin is currently being tested in human GBM trials including the N^2M^2 trial [188].

Impairment of the G1 checkpoints (p53 and Rb pathways) increases reliance on G2 checkpoints, such as one mediated by the Wee-1 tyrosine kinase [226]. Wee-1 serves as a gatekeeper that inactivates the G2-M phase transition master switch $CDK1^{CDC2}$ to avert mitotic catastrophe by ensuring adequate DNA replication and repair [227–229]. Consequently, Wee-1 has emerged recently as a potential target. **Adavosertib** (MK1775 or AZD1775; MW = 501 Da) is a novel Wee-1 inhibitor with *in vitro* cytotoxicity against GBM cell lines (IC_{50} 68–695 nM) [230]. *In vitro* and *in vivo* studies show that adavosertib abrogates radiation-induced G2 checkpoint arrest and increases the proportion of mitotic cells leading to increased radiosensitivity. Adavosertib showed efficacy in a heterotopic (flank) but not in an orthotopic (intracranial) mouse model, which was attributed to its poor, ~5% accumulation in the normal brain compared to plasma [231]. The pharmacokinetics of adavosertib has been uniquely well studied in humans as well. In a phase 0 study, a single 400 mg oral dose of the adavosertib led to an unbound drug concentration of ~170 nM in the contrast-enhancing part of recurrent GBM with a median tumor-to-plasma ratio of 3.2 [232, 233]. Interestingly, pharmacodynamic analyses revealed the highest elevations of apoptotic (cleaved caspase-3) and DNA damage (γH2AX) markers at an intermediate, 200 mg dose compared to 100 and 400 mg. Most recently, a study utilizing microdialysis measured unbound adavosertib concentrations in the normal-appearing and noncontrast-enhancing T2/FLAIR hyperintense tumor volume showed brain-to-plasma ratios (AUC_{ECF}/AUC_p) ranging between 3% and 17%, and absolute brain drug levels between 25 and 132 nM [234]. These data again cast doubts about adavosertib's potential clinical efficacy against GBM.

Protein phosphatase 2A (PP2A) is another potential target of cell cycle regulation. PP2A is a versatile serine/threonine phosphatase that prevents the activation of $CDK1^{CDC2}$ and G2/M phase transition in addition to numerous other functions [235, 236]. Similar to adavosertib, inhibiting PP2A leads to premature cell cycle progression causing mitotic catastrophe and cell death. As transient cell cycle arrest is a major cause of radiotherapy resistance, agents that elicit premature cell cycle progression can lead to radiosensitization and mitotic catastrophe [237]. **LB-100** (MW = 268 Da) is a first-in-class inhibitor of PP2A. It has low cytotoxic potency against GBM cell lines (IC_{50} 5–11 μM) and rat models demonstrate poor BBB penetration (brain-to-plasma ratio 1.2%–1.5%). LB-100 is unlikely to reach therapeutic concentrations at the recommended human phase II dose [238–241]. However, in animal models, LB-100 was found to have a synergistic antitumor effect when combined with PD-1 blockade, by enhancing the pool of tumor-infiltrating effector T cells, and reducing the immune-inhibiting regulatory T cells [239, 240]. This combination translated into prolonged survival of study animals, which was not observed with either LB-100 or anti-PD-1 monotherapy [242]. These immune responses suggest interesting and potentially clinically translatable mechanisms of actions distinct from the inhibition of tumor cell PP2A.

The serine/threonine kinases ataxia telangiectasia mutated (ATM) and Rad-3 related protein (ATR/ATM) induce the G2/M checkpoint, cell cycle arrest, and DNA repair by activating CHK1 and are thus involved in the wide range of cell responses to DNA damage and replication stress [243, 244]. Targeting ATR/ATM has been proposed as a radiation sensitizing strategy against GBM, and new drugs are currently being evaluated in these roles. At present, drugs such as **ceralasertib** (AZD6738, MW = 413 Da), **BAY1895344** (MW = 412 Da), **M6620** (VX-970; MW = 464 Da) are actively studied in cancers including brain tumors; however, the data is premature to be reported in detail at this time [245–247].

Direct proapoptotic agents

In cancer, uncontrolled proliferation is complemented by reduced apoptotic cell death through either reduced or lost activity of proapoptotic signaling and machinery or amplification of antiapoptotic molecules [127]. Members of the BCL-2 family (e.g., BCL-XL, BCL-W, MCL-1 A1) are regulatory proteins that bind and inactivate proapoptotic proteins such as Bax and Bak. The latter is a mitochondrial outer membrane protein that when activated, initiates the apoptotic process mediated by the release of mitochondrial cytochrome C. This in turn leads to the activation of a cascade of proteases that culminate in the activation of effector caspases (e.g., caspase 3) that execute apoptotic cell death [248]. BCL-2 is detectable by immunohistochemistry in ~30% of GBM and has been proposed as a therapeutic target of novel drugs [249]. The most prominent of these at the present are **navitoclax** (ABT-263; MW = 975 Da) and **venetoclax** (ABT-199; MW = 868 Da). GBM specific data on these drugs are scarce. However, the size of these molecules and prepublication animal data showing poor CNS bioavailability suggest that these agents may not be applicable to brain tumor therapy [250].

Procaspase 3 (PC3), the zymogen form of caspase 3 was shown to be upregulated in cancer, including 60% of GBMs and high PC3 levels were associated with poor survival in these tumors [251–254]. Consequently, PC3 is a desirable target against GBM. **Procaspase activating compound 1** (**PAC-1** or VO-100; MW = 393 Da) is a chelating agent that sequesters labile Zn^{2+} ions leading to activation of procaspase 3 triggering the apoptotic process [255, 256]. While PAC-1 has a relatively low cytotoxic potency against *in vitro* GBM cell lines (IC_{50} 5.5–27.2 μM after 72H incubation), *in silico* predictions of ~85% brain-to-plasma distribution were surpassed by mouse experiments detecting ~120 μM in the brain vs 70 μM in plasma 5-min post injection of PAC-1 [251, 254, 256, 257]. PAC-1 prolonged survival in rat GBM models when used as monotherapy and exhibited synergistic effects when combined with TMZ [254]. The addition of PAC-1 to TMZ and radiotherapy to treat three companion canine patients with naturally occurring gliomas resulted in significant reductions in tumor burden, including one complete response. Human phase I studies of PAC-1 are currently ongoing, and the maximum tolerated dose (MTD) was established to be 750 mg daily as a monotherapy. Optimal dosing when combined with TMZ is currently being investigated [258].

Agents that directly alter protein homeostasis

Dysfunction of specific protein degradation mechanisms contributes to oncogenesis and therapeutic resistance of malignant cells. The ubiquitin-proteasome system of protein degradation is of considerable relevance in this context. The more recently described NEDD8 protein conjugation system also plays a role [259]. This section covers therapeutic agents that target these systems.

Inhibition of the ubiquitin-proteasome pathway

The ubiquitin-proteasome system is responsible for the degradation of a vast majority of intracellular proteins. This pathway regulates the levels of functional proteins in addition to clearing cells of misfolded or otherwise damaged proteins. Mechanisms influenced by proteasome-dependent protein degradation include oncogenic programs and the emergence of resistance to antineoplastic therapy. Proteins that are critically important to cell growth control and apoptosis (e.g., $p27^{kip1}$, p53, Bax) are regulated by targeted ubiquitination and proteasome degradation [260, 261]. Proteosomes are present both in the cytoplasm and nucleus of eukaryotic cells and thereby contribute to the regulation of proteins with differential subcellular locations and actions. In the ubiquitin-proteasome pathway, proteins are first marked in an ATP-dependent multistep catalytic process resulting in the attachment of ubiquitin moieties. Polyubiquitinated proteins are transferred to the 26S proteasome that consists of one or two 19S ubiquitin cleaving subunits that recycle ubiquitin, and a 20S polycatalytic core with caspase-, trypsin-, and chymotrypsin-like activities [262]. Most proteasome inhibitors target the chymotrypsin-like sites. **Bortezomib** (MW = 384 Da) a boronic acid peptide small molecule is the prototype of the inhibitors of the proteosome chymotrypsin-like activity. It was first-in-class when entered into clinical trials and was approved against multiple myeloma and mantle cell lymphoma [263–265]. Bortezomib has poor brain penetrance in animal models (brain-to-plasma ratio 4%–7%), explaining at least in part its lack of efficacy in human malignant glioma trials [147, 266–268].

Protease inhibitors **carfilzomib** (PR-171; MW = 720 Da) and **ixazomib** (MLN9708; MW = 361 Da) have poor BBB penetration [261, 269, 270]. **Marizomib** (NPI-0052 or salinosporamide A; MW = 314 Da) appears to have more favorable pharmacokinetic and pharmacodynamic profiles for neuro-oncological applications. Marizomib is a beta-lactone class small molecule that has the advantage of inhibiting all proteasome catalytic subunits in an irreversible fashion with high *in vitro* potency against GBM (cytotoxic IC_{50} 9–51 nM; DNA fragmentation at 48H IC_{50} 250 nM) [271–273]. In Sprague-Dawley rats, marizomib was found to have a >31% brain-to-blood concentration ratio, and significant proteasome inhibition was also confirmed in nonhuman primates with both once or twice weekly dosing schedules [272]. Human pharmacokinetic data from a phase I trial of advanced malignancies showed highly variable but low plasma maximum concentrations of intravenously administered marizomib (plasma C_{max} ~10–184 nM if given weekly, 8–90 nM in twice weekly dosing). How this may translate to effective drug levels within the brain is uncertain [274]. Phase I/II trials of marizomib are currently ongoing in combination with standard chemoradiation in newly diagnosed GBM, or in combination with bevacizumab in recurrent GBM, with promising preliminary results [275, 276]. High-throughput drug screening suggests that proteasome inhibitors may synergize with HDAC inhibitors, a potential combinatorial approach for the future [261, 277].

Inhibition of the neddylation pathway

Neddylation is a recently described protein conjugation pathway that shares many similarities with the ubiquitin-proteasome system. It is executed by a set of specific enzymes that do not cross conjugate with ubiquitin or other ubiquitin-like proteins. Neddylation involves ATP-dependent sequential transfer of ubiquitin-like NEDD8 (neural precursor cell-expressed developmentally downregulated 8) moieties to protein substrates [259]. These include numerous regulatory molecules that play a significant role in cancer, such as EGFR, p53, and MDM2 [259, 278]. NEDD8 expression is upregulated in high-grade glioma and associated with poor survival, hence the neddylation pathway poses a promising novel target against GBM [279].

Pevonedistat (MLN4924, MW = 444 Da) is a first-in-class inhibitor of the NEDD8-activating enzyme (NAE), an essential component of the neddylation pathway [280]. Pevonedistat has potent in vitro activity against a GBM cell line (IC_{50} for cytotoxicity: 10–430 nM; IC_{50} for antimigratory: <100 nM), and significantly reduced tumor growth rates with prolonged survival in an orthotopic GBM mouse model [279–281]. Treatment with pevonedistat upregulated $p21^{cip1}$ and $p27^{kip1}$ in conjunction with cell cycle arrest. Preliminary transcriptomic analyses suggest that cell lines enriched for ERK and AKT signaling are more sensitive to pevonedistat, and that this sensitivity may be dependent on wild-type PTEN and induced in PTEN mutants by topoisomerase 2 inhibitors [281, 282]. Pevonedistat was found to be well tolerated in a phase I trial against acute myeloid leukemia, with plasma C_{max} between 1.5 and 2.7 μM depending on the dosing schedule used [283]. This suggests that a brain-to-plasma ratio as low as 5%–10% could potentially yield brain concentrations effective against GBM. However, pevonedistat's BBB permeability has not yet been established. Human GBM trials are yet to be launched.

Targeting upstream molecular pathways

Cell proliferation and survival mechanisms are regulated by a network of cell surface receptors and cytoplasmic kinases and downstream nuclear transcription factors that are hyperactivated in GBM. About 90% of GBMs harbor genetic alterations (e.g., point mutation, deletion, amplification) in the upstream components of these signaling networks. EGFR is hyperactivated in ~50% of GBM and other receptor tyrosine kinases (RTKs) such as PDGFR and c-Met are altered in GBM at somewhat lower frequencies [2, 15, 129]. Hyperactive RTKs in turn hyperactivate cytoplasmic signaling kinase pathways the most critical of which are the PI3K/PTEN/AKT/mTOR and RAS/BRAF/ERK pathways. PI3K/PTEN/AKT signaling can be aberrantly hyperactivated independent of RTK signaling by mutually exclusive activating mutations of phosphatidyl inositol 3 kinase (PI3K) in ~25% of GBM or inactivating mutations in the PI3K inhibitor PTEN in 41% of the cases. Mutations in RAS, BRAF, and ERK are less common in GBM. Loss of function mutations in NF1 (direct inhibitor of RAS) are seen in ~10% of GBM and activating RAS or BRAF mutations are even less common (1%–2% of GBM).

This section presents these three systems separately, however, it is important to appreciate how deeply they are interwoven as RTKs can activate both the PI3K/AKT/mTOR and RAS/BRAF/ERK pathways that also cross talk with each other (Fig. 3). As described in graph theory, these systems represent a highly efficient oncogenic network that may easily recover if a single node is removed [284–286]. Therefore, strategic utilization of pharmacologic targeting is paramount to halt or reverse the oncogenic program. Combination drug therapy is likely required to prevent the emergence of resistance to single agents.

EGFR inhibition

The high frequency of activating EGFR mutations, deletion and amplification in GBM detected by large-scale genomic analyses identified EGFR as a highly desirable therapeutic target [129, 287]. While EGFR inhibition led to major breakthroughs in the treatment of colon and lung cancer, similar successes are not yet realized against GBM [288]. The cause of these failures is likely multifactorial.

Poor BBB penetrance has been an obstacle to efficacy for most EGFR inhibitors, not surprisingly for neutralizing monoclonal antibodies such as cetuximab and nimotuzumab [289–291], but also for many small molecule inhibitors such as erlotinib [292–297], gefitinib [298–301], or lapatinib [302–304]. A common denominator for most of the early studies using these agents was the lack of correlation between levels of tumor EGFR expression and treatment response, potentially explained by drug-to-target distribution mismatch. EGFR alterations are also extremely heterogeneous in GBM with at least four major molecular forms: (i) wild-type EGFR gene amplification, (ii) extracellular domain deletion (i.e., EGFRvIII), (iii) kinase domain duplication (i.e., EGFR-KDD), and (iv) gene fusion with septin-14 (i.e., EGFR-SEPT14) [129, 305,

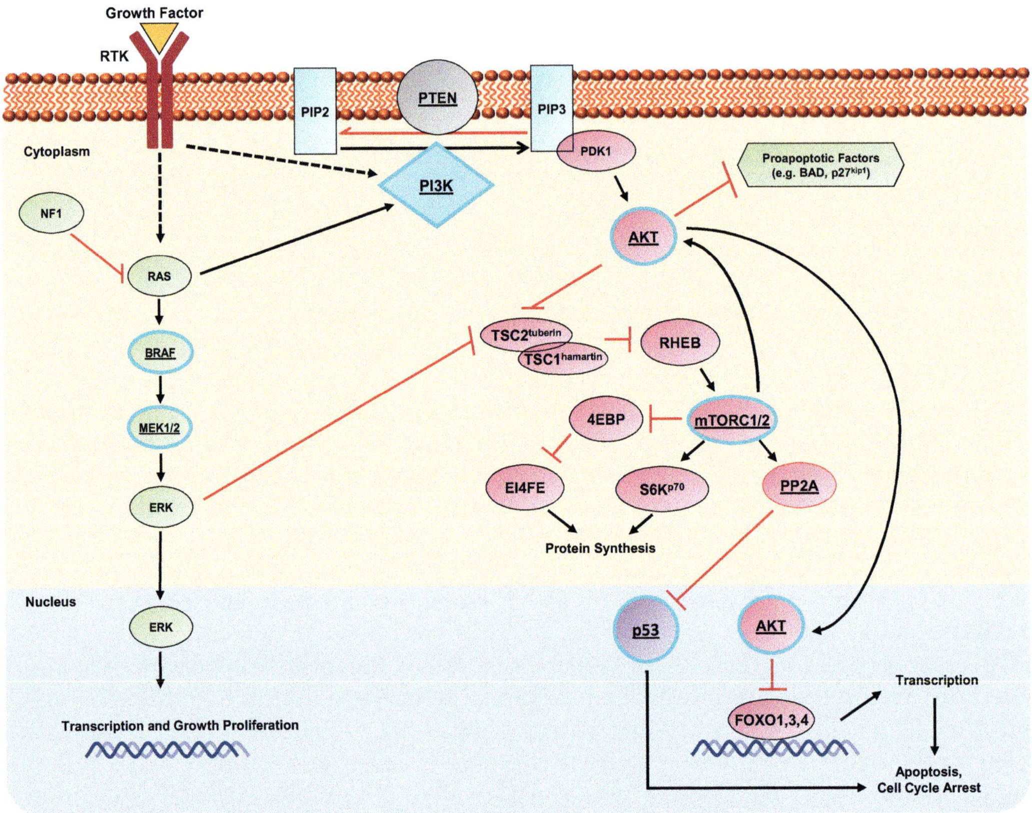

FIG. 3 Schematic overview of downstream signaling pathways most commonly implicated in glioblastoma. *Recreated from Cully M, You H, Levine AJ, Mak TW. Beyond PTEN mutations: the PI3K pathway as an integrator of multiple inputs during tumorigenesis. Nat Rev Cancer 2006;6:184–92. https://doi.org/10.1038/nrc1819.*

306]. These distinct forms have different drug sensitivity profiles and resistance mechanisms. For example, erlotinib is less effective against EGFRvIII and posttranslational modifications that enhance receptor dimerization abrogate the effects of neutralizing antibodies such as cetuximab [307, 308]. Another example is the dynamic modulation of extrachromosomal DNA coding EGFRvIII allowing escape from erlotinib therapy by transiently reducing EGFRvIII expression [309]. Therefore, even if the issue of drug distribution mismatch is solved via the use of BBB penetrating drugs, drug-to-target mismatch may remain a highly relevant concern.

Osimertinib (AZD9291, MW = 500 Da) is a third-generation irreversible EGFR kinase inhibitor with promising activity in the CNS evidenced by efficacy against nonsmall cell lung cancer brain metastases [310–312]. Osimertinib inhibits EGFRvIII autophosphorylation at Y1068 *in vitro* with high potency in an EGFRvIII-positive GBM stem cell line (IC_{50} 50 nM), and inhibited GBM cell proliferation at a higher concentration (IC_{50} ~500 nM) [305]. Another report found that osimertinib is 10–20 times more potent than first-generation EGFR inhibitors such as erlotinib, but still requires relatively high concentrations for a cytotoxic effect (IC_{50} ≤3M) [313]. Both of these studies showed prolonged survival of mice with orthotopically xenografted GBMs. These results are in line with the consistently demonstrated BBB penetration of osimertinib. Animal studies found a brain drug concentration of 1–12 times that of plasma, although the free brain-to-plasma drug concentration may be as low as 30% [304, 305, 314]. PET evaluations using radiolabeled [^{11}C]osimertinib corroborated the excellent brain distribution of this drug including nonhuman primates (brain-to-plasma AUC ratio 2.6) and healthy humans (AUC ratio 3.8) [314, 315]. Spatial distribution of the drug showed twofold higher accumulation in gray matter compared to white matter. At present, there is a case study showing a remarkable radiographic response to osimertinib in a multicentric GBM [316]. The efficacy of osimertinib in humans with glioblastoma is currently undergoing rigorous investigation.

Inhibitors of the PI3K/AKT/mTOR pathway

The PI3K/AKT/mTOR system (Fig. 3) serves as a central integrator of several growth regulating signaling networks. This pathway is responsible for promoting cell survival, proliferation, cellular enlargement, metabolite-dependent integration of insulin signaling among many other functions [284]. PI3K is a major signaling node that is hyperactivated in most GBMs owing to activating PI3K mutations, loss of function of its direct inhibitor PTEN, and aberrant signaling from RTKs. Activated PI3K consists of a regulatory p85 (PIK3R1), and a catalytic p110 (PIK3CA) subunit, which phosphorylates phosphatidylinositol-4,5-bisphophate (PIP_2), to generate phosphatidylinositol-3,4,5-trisphophate (PIP_3). Unless it is broken down to PIP_2 by PTEN, the presence of PIP_3 leads to the activation of AKT (protein kinase B) via phosphorylation. pAKT inactivates $p27^{kip1}$ directly and indirectly through FOXO3 leading to cyclin D transcription and cell cycle progression. AKT also activates mTOR complexes mTORC1 and mTORC2. mTORC1 activates ribosomal protein kinase $S6K^{p70}$ and eukaryotic initiation factor 4E-binding protein (4EBP) that together increase the synthesis of protein, lipid, and nucleotides and decrease autophagy. mTORC2 can activate AKT and thereby amplify cell proliferation and survival [317].

The redundancy of the system leads to numerous possible signaling escape routes. Understanding these is absolutely crucial for the prediction of potential treatment efficacy and for the choice of possible synergistic drug combinations. For example, pharmacologic inhibition of PI3K or AKT may be circumvented via the RAS pathway that can activate mTORC1/2 via ERK in the setting of PI3K inhibition. mTORC2 can then feedback to activate AKT and thereby override the downstream signaling response to PI3K inhibition. Resistance to specific mTORC1 inhibition can similarly occur via AKT activation by mTORC2. A pan-mTORC inhibitor still allows AKT to induce cell cycle progression. Yet the complexity of this system also provides unique opportunities. For example, the anabolic activity promoted by mTOR can be visualized via molecular imaging and provide a promising avenue to noninvasively monitor tumor metabolism as a biomarker of PI3K pathway inhibition using positron emission tomography [318–320]. The following section discusses novel pan-PI3K, AKT, and dual mTOR1/2, and combined mTOR-PI3K inhibitors in detail.

Pan-PI3K inhibitors

Pan-PI3K inhibitors affect all four PI3K class I isoforms (PI3K α, β, γ, and δ) [317, 321]. This drug class has had limited success against most malignancies examined. **Copanlisib** (BAY-80-6946; MW = 481 Da) is the only pan-PI3K inhibitor that received FDA approval for the treatment of follicular lymphoma. Many others have been studied in the context of GBM. **Pictilisib** (GDC-0941; MW = 514 Da) displayed promising *in vitro* efficacy against GBM (IC_{50} for cytostasis 950 nM) [322]. However, animal imaging data showed low brain accumulation compared to the plasma in rats (9%) and in mice (1%–3% for total and 0.4%–0.8% for unbound drug) [323, 324]. Pictilisib is not likely to be clinically effective against GBM in humans since its peak plasma concentration at the maximum tolerated dose is ~2 μM [325]. **Sonolisib** (PX-866; MW = 526 Da), an oral pan-PI3K inhibitor, was studied as a monotherapy against recurrent GBM. Despite a reported durable stable response rate of ~21%, no additional trials for GBM have been published since 2015 [326].

Buparlisib (BKM120; MW = 410 Da), a selective pan-class I PI3K inhibitor that binds to the ATP-binding cleft of PI3K, is perhaps the most rigorously studied in GBM and the most representative of the challenges faced by these agents [321, 327]. Biochemical enzymatic profiling shows high potency against PI3K (IC50 <0.4 nM) with inhibitory effect against mTOR as well (IC50 4.6 nM) [327]. Buparlisib is similarly active against PTEN wild-type and PTEN mutant cell lines (cytostatic IC_{50} ~ 1.7 μM for both). *In vivo* pharmacokinetic analyses of buparlisib in mice showed exceptional BBB penetration with a brain-to-plasma ratio >150% 1 h following both intravenous and oral drug administration. Brain drug concentrations exceeded 1 μM even at the lowest ingested dose [328]. These findings consistently translated to dose-dependent tumor growth reduction and prolonged survival in orthotopic GBM xenograft mouse models [327, 329, 330]. Results from human phase I studies established maximum tolerated doses at 60–80 mg daily and the drug had a prolonged plasma half-life around 40 h with C_{max} measured consistently above 2 μM once a steady state was achieved [331]. The most prominent dose-limiting toxicities were transaminase elevation, hyperglycemia, fatigue, and mood changes, with suicidal ideation in cases. BBB penetration was confirmed in humans, but lower compared to the animal models, with an average brain tumor-to-plasma ratio of 1.1 in the contrast enhancing and 0.7 in the nonenhancing tumor volume [331]. Unfortunately, to date buparlisib studies have been largely unsuccessful in GBM. First, in the newly diagnosed setting, it was found to be intolerable when added to standard chemoradiation, and tolerable, but ineffective when combined with adjuvant TMZ [332]. In the recurrent setting, it had minimal efficacy as a monotherapy, or when combined with lomustine or carboplatin [331, 333]. Pharmacodynamic evaluation via tissue sampling post buparlisib monotherapy showed 42% reduction of pAKT, but a negligible effect on the phosphorylation of $S6K^{p70}$ or on cell proliferation. The latter suggests insufficient target inhibition or a bypass mechanism of drug resistance.

ZSTK474 (MW = 417 Da), a newer pan-class I PI3K inhibitor that crosses the BBB with a brain-to-plasma ratio above 1 [334]. ZSTK474 was shown to be cytotoxic (IC_{50} 630 nM) against a PTEN-null GBM cell line *in vitro*, well tolerated in mice and active against a GBM xenograft model [335]. The maximum tolerated dose in humans was established at 150 mg daily 3 weeks on and 1 week off cycles [336]. Although detailed human pharmacokinetic data was not published yet, the doses used in humans are comparable to those in the mouse studies. To date, all ZSTK474 trials are closed with final results pending publication.

AKT inhibitors

While oncogenic AKT mutations are rare in GBM, AKT remains a desirable pharmacologic target due to its role in promoting cell growth and resistance to cytotoxicity. AKT is a serine-threonine kinase with three isoforms all of which can be inhibited by competitive blockage of its ATP-binding site. Competitive AKT inhibitors have relatively low target specificity and thus high off-target effects [337–339]. In contrast, allosteric inhibitors function by binding to the AKT pleckstrin-homology domain causing a conformational change that prevents AKT translocation to the cell membrane, a requirement for AKT activation. Allosteric inhibitors promise more target specificity and are felt to minimize the emergence of drug-resistant mutations.

MK2206 (MW = 480 Da) is an oral allosteric pan-AKT inhibitor extensively studied in the setting of extra-CNS solid tumors. The limited published *in vitro* data suggest that MK2206 has low potency against GBM cell lines, requiring concentrations >0.5–1 μM to elicit any significant cytotoxic, cytostatic, or even antimigratory effect [340, 341]. Considering that peak plasma concentrations at the maximum tolerated doses are below 0.5 μM, unusually high brain accumulation will be required to reach meaningful concentrations against glioblastoma [339, 342]. Other allosteric AKT inhibitors with sufficient BBB penetration await development.

M2698 (MSC2363318A; MW = 450 Da) is a novel orally bioavailable competitive dual AKT1/3 and $S6K^{P70}$ inhibitor shown to cross the BBB. M2689 levels higher than its cytostatic concentration (IC_{50} 3 μM) were detected in tumor-bearing rat brain and normal mouse brain and drug levels in brain were >fivefold higher than in plasma [343]. M2689 inhibits orthotopic GBM growth and prolongs the survival of tumor-bearing mice. Pharmacodynamic analysis of gene expression changes showed increased pAKT but decreased $pS6K^{P70}$ levels consistent with successful inhibition of these molecules. M2698 is currently studied in humans with advanced cancers with PI3K/AKT/mTOR pathway abnormalities and a preliminary report of the findings showed good tolerability at doses comparable to the animal studies [344].

mTORC1/2 inhibitors

mTOR has been considered a desirable target for anticancer pharmacotherapy as it represents a convergence node for the proliferative and antiapoptotic RAS/ERK and PI3K/AKT pathways. mTORC1 inhibitors such as everolimus, and temsirolimus have been extensively studied with minimal to no survival benefit in newly diagnosed or recurrent GBM [345–350]. This lack of efficacy is predicted since specific mTORC1 inhibition allows for continued AKT and ERK activation by mTORC2 and other mechanisms. Therefore, there is great interest in dual mTORC1/2 inhibitors currently under development.

An ideal example of the pan-mTOR inhibitor is **Vistusertib** (AZD2014; MW = 463 Da). Vistusertib is cytotoxic to everolimus resistant diffuse intrinsic pontine glioma cell lines *in vitro* (IC_{50} <0.6 μM in 3/3 and IC_{90} <2 μM in 2/3) [351]. In an orthotopic GBM xenograft mouse model, vistusertib elicited prolonged survival when combined with radiation, but not as a monotherapy [352]. Phase I trials found average plasma peak and trough concentrations of this drug in the ranges of 3.5–6 μM depending on the dosing schedule [353]. In its only human study in GBM to date, vistusertib showed good tolerability (100-mg twice daily 2 days on 5 days off schedule), when combined with standard TMZ 150 mg/m^2 5 days on 23 days off [354]. Unfortunately, the BBB penetrability of this drug has not been studied in animals or humans. Further studies are pending to establish brain pharmacokinetics and efficacy of this drug.

Multitarget inhibitors if the PI3K pathway

Effective PI3K pathway inhibition requires blockade of multiple nodes to completely abrogate its oncogenic proliferative and cytoprotective effects. The application of promiscuous competitive inhibitors of the ATP-binding sites of multiple kinases comprising this pathway is a rational approach to such multifaceted targeting. The dual pan-PI3K and mTORC1/2 inhibitor **dactolisib** (BEZ325; MW = 470 Da) is an excellent example of such drugs. In chemical assays, dactolisib inhibits its target kinases at low nanomolar concentrations [355]. Dactolisib also exhibits significant tumoristatic activity against GBM cell lines (IC_{50} ≤100 nM) [355–357]. Additional *in vitro* analyses show that dactolisib at concentrations as low as 10 nM synergizes with both radiation and TMZ by facilitating DNA injury inhibiting double-stranded DNA repair and induces autophagy in GBM cells. Dactolisib monotherapy inhibits tumor growth and modestly prolongs the survival of mice bearing GBM xenografts with greater survival prolongation when combined with TMZ or radiotherapy. However, dose-limiting toxicities adversely affecting animal survival are described [355, 357–360]. The limited information on BBB penetration shows a brain-to-plasma C_{max} ratio of 75% after intraperitoneal administration in mice, with additional drug toxicity concerns [245]. In a human phase I trial in solid tumors, dactolisib monotherapy was well tolerated at 300-mg twice daily, which translated to an average steady-state C_{max} of 1.4 µM [361]. Decreased tumor glucose uptake post treatment, a characteristic response to PI3K pathway inhibition, was noted in 74% of tumors. Other phase I/II trials in non-CNS cancers have found dactolisib to be poorly tolerated as monotherapy or in combination with everolimus [362–365]. It has yet to be determined if dactolisib has any efficacy against clinical GBM or if it is tolerated by patients when combined with current standards of care.

Inhibition of the RAS/MEK/ERK pathway

Compared to the other systems covered, the RAS/MEK/ERK (Fig. 3) pathway seems to play only a minor role in adult GBM. In contrast, there is a growing body of evidence regarding the clinical efficacy of inhibiting BRAF (e.g., dabrafenib) and MEK (e.g., trametinib) in pediatric lower-grade gliomas [366–368]. Data from children as well as adults show that mutations in the BRAF or NF1 genes are major determinants of response to BRAF/MEK inhibition [368, 369]. While both of these genetic alterations cause an overdrive of the RAS/MEK/ERK system, a BRAF mutation is only detected in 1% of adult GBM, whereas loss of NF1 function is present in 10% [2, 32, 129]. Experimental evidence from the use of numerous MEK inhibitors, including **selumetinib** (AZD6244, MW = 458 Da) and **PD0325901** (MW = 482 Da) shows that glioma cell lines with competent RAS regulation are inherently resistant to this drug class, whereas those harboring mutant NF1 are sensitive. However, even in NF1 mutant GBM cell lines, the response is usually a transient growth arrest (cytostatic IC_{50} 30–37 and

230–550 nM for PD0325901 and selumetinib, respectively), which is overcome by emerging resistance after prolonged MEK inhibitor exposure [370, 371]. Therefore, it is unlikely for RAF/MEK inhibitors to succeed in monotherapy against GBM. However, they might play an important role as inhibitors of mTOR activation via ERK and thus a brain penetrable, MEK inhibitor such as PD032591 (brain-to-plasma ratio ~60%) may be of future clinical interest [371].

Synthetic lethality and drug delivery

This chapter was written with the feasibility of clinical drug efficacy in mind. While more than 150 investigational drugs were screened, even after excluding those with a postulated primary effect on the tumor microenvironment (e.g., antiangiogenics and immunotherapies) the majority either lacked or had clearly incompatible *in vitro* efficacy, brain accumulation, or human tolerability data. Of the drugs presented in this chapter, a clear pattern is evident: most anti-GBM agents have to be pushed to the highest tolerable doses to approximate biologically significant drug levels in the nonenhancing brain. The BBB is not only a physical but also a pharmacological barrier that effectively reduces the clinical potency of drugs while also narrowing their therapeutic index. Other than the chemical modification of drugs to improve BBB permeability, three additional approaches can be considered to overcome the BBB: (i) direct drug delivery into the brain; (ii) selective or global opening of the BBB to improve the brain entry of systemically administered drugs; and (iii) increasing drug potency by harnessing synthetic lethality.

Convection-enhanced delivery utilizes a catheter placed directly into the brain parenchyma that perfuses the applied drug dissolved in artificial CSF [159]. Other currently tested approaches to open the BBB involve the use of focused ultrasound or vasoactive peptides to transiently increase vascular permeability [27, 29]. Some of these are covered in other sections. The third option, utilizing synthetic lethality is an old concept reinvigorated by new technology. In cancer, it primarily refers to lethality caused by the concurrent loss of function of a pair of genes that individually have minimal effect on cell viability [372]. One member of this pair is usually a cancer-specific mutation such as BRCA (or IDH), whereas the second is a therapeutic intervention that interferes with the function of a second gene (e.g., targeted PARP inhibition) [373]. Ideally, the end result is greatly enhanced drug potency, cancer cell chemosensitivity, and an improved therapeutic index, which culminate in realistically achievable brain drug concentration requirements. This concept has been complemented recently by high-throughput drug combination screening systems, which allows the empirical discovery of synergistic drug pairs rather than relying on individual hypothesis testing or serendipity [277].

Conclusions

Although there is a myriad of available new investigational anticancer drugs, only a fraction of these agents meets the minimum requirements necessary to be clinically efficacious against GBM. Information regarding in vitro activity against GBM, BBB penetrability should be consistently considered when designing clinical trials. Previously failed or promising novel agents with unfavorable brain pharmacokinetics could be reevaluated in combinations in the future, but only if both drugs can feasibly achieve a brain concentration necessary for the combination to be effective. The establishment of methods to safely and consistently open the BBB could eliminate the unique pharmacologic obstacles associated with GBM and is a bypass desperately needed.

Acknowledgments

We would like to sincerely thank Mary Brown, Jan Wemmer NP, Lindsay Blair NP, and Olga Yankulina PharmD of our clinical team at Johns Hopkins for their continuous help caring for our patients. We would also like to thank Brittni Griffin and Joy Fisher, MBA at Johns Hopkins and Lilian Yang, MBA at the Neuro-Oncology Branch for their invaluable administrative help and support.

References

[1] Ostrom QT, Gittleman H, Truitt G, Boscia A, Kruchko C, Barnholtz-Sloan JS. CBTRUS Statistical Report: primary brain and other central nervous system tumors diagnosed in the United States in 2011–2015. Neuro Oncol 2018;20:iv1–86. https://doi.org/10.1093/neuonc/noy131.

[2] World Health Organisation. WHO classification of tumours of the central nervous system. Revised 4th ed; 2016. Lyon: International Agency for Research on Cancer.

[3] Pallud J, Varlet P, Devaux B, Geha S, Badoual M, Deroulers C, et al. Diffuse low-grade oligodendrogliomas extend beyond MRI-defined abnormalities. Neurology 2010;74:1724–31. https://doi.org/10.1212/WNL.0b013e3181e04264.

[4] Capper D. Addressing diffuse glioma as a systemic brain disease with single-cell analysis. Arch Neurol 2012;69:523. https://doi.org/10.1001/archneurol.2011.2910.

[5] Kamson DO, Juhász C, Buth A, Kupsky WJ, Barger GR, Chakraborty PK, et al. Tryptophan PET in pretreatment delineation of newly-diagnosed gliomas: MRI and histopathologic correlates. J Neurooncol 2013;112:121–32. https://doi.org/10.1007/s11060-013-1043-4.

[6] Dandy WE. Physiological studies following extirpation of the right cerebral hemisphere in man. Bull Johns Hopkins Hosp 1933;53:31–35.

[7] Walker MD, Alexander E, Hunt WE, MacCarty CS, Mahaley MS, Mealey J, et al. Evaluation of BCNU and/or radiotherapy in the treatment of anaplastic gliomas: a cooperative clinical trial. J Neurosurg 1978;49:333–43. https://doi.org/10.3171/jns.1978.49.3.0333.

[8] Stupp R, Mason WP, Bent MJ, Weller M, Fisher B, Taphoorn MJ. Radiotherapy plus concomitant and adjuvant temozolomide for glioblastoma. N Engl J Med 2005;352:987–96.

[9] Hegi ME, Diserens A-C, Gorlia T, Hamou M-F, de Tribolet N, Weller M, et al. MGMT gene silencing and benefit from temozolomide in glioblastoma. N Engl J Med 2005;352:997–1003. https://doi.org/10.1056/NEJMoa043331.

[10] Stupp R, Hegi ME, Mason WP, van den Bent MJ, Taphoorn MJ, Janzer RC, et al. Effects of radiotherapy with concomitant and adjuvant temozolomide versus radiotherapy alone on survival in glioblastoma in a randomised phase III study: 5-year analysis of the EORTC-NCIC trial. Lancet Oncol 2009;10:459–66. https://doi.org/10.1016/S1470-2045(09)70025-7.

[11] Hegi ME, Stupp R. Withholding temozolomide in glioblastoma patients with unmethylated MGMT promoter—still a dilemma?: Table 1. Neuro Oncol 2015;17:1425–7. https://doi.org/10.1093/neuonc/nov198.

[12] Xia D, Reardon DA, Bruce JL, Lindeman NI. The clinical implications of inconsistently methylated results from glioblastoma MGMT testing by replicate methylation-specific PCR. J Mol Diagn 2016;18:864–71. https://doi.org/10.1016/j.jmoldx.2016.06.009.

[13] Hegi ME, Genbrugge E, Gorlia T, Stupp R, Gilbert MR, Chinot OL, et al. MGMT promoter methylation cutoff with safety margin for selecting glioblastoma patients into trials omitting temozolomide: a pooled analysis of four clinical trials. Clin Cancer Res 2019;25:1809–16. https://doi.org/10.1158/1078-0432.CCR-18-3181.

[14] Van Den Bent MJ, Erridge S, Vogelbaum MA, Nowak AK, Sanson M, Brandes AA, et al. Second interim and first molecular analysis of the EORTC randomized phase III intergroup CATNON trial on concurrent and adjuvant temozolomide in anaplastic glioma without 1p/19q codeletion. JCO 2019;37:2000. https://doi.org/10.1200/JCO.2019.37.15_suppl.2000.

[15] Brat DJ, Aldape K, Colman H, Holland EC, Louis DN, Jenkins RB, et al. cIMPACT-NOW update 3: recommended diagnostic criteria for "Diffuse astrocytic glioma, IDH-wildtype, with molecular features of glioblastoma, WHO grade IV". Acta Neuropathol 2018;136:805–10. https://doi.org/10.1007/s00401-018-1913-0.

[16] Brat DJ, Aldape K, Colman H, Figrarella-Branger D, Fuller GN, Giannini C, et al. cIMPACT-NOW update 5: recommended grading criteria and terminologies for IDH-mutant astrocytomas. Acta Neuropathol 2020;139:603–8. https://doi.org/10.1007/s00401-020-02127-9.

[17] Louis DN, Wesseling P, Aldape K, Brat DJ, Capper D, Cree IA, et al. cIMPACT-NOW update 6: new entity and diagnostic principle recommendations of the cIMPACT-Utrecht meeting on future CNS tumor classification and grading. Brain Pathol 2020;30:844–56. https://doi.org/10.1111/bpa.12832.

[18] Steensma DP, Kantarjian HM. Impact of cancer research bureaucracy on innovation, costs, and patient care. JCO 2014;32:376–8. https://doi.org/10.1200/JCO.2013.54.2548.

[19] Mandel JJ, Yust-Katz S, Patel AJ, Cachia D, Liu D, Park M, et al. Inability of positive phase II clinical trials of investigational treatments to subsequently predict positive phase III clinical trials in glioblastoma. Neuro Oncol 2018;20:113–22. https://doi.org/10.1093/neuonc/nox144.

[20] Grossman SA, Romo CG, Rudek MA, Supko J, Fisher J, Nabors LB, et al. Baseline requirements for novel agents being considered for phase II/III brain cancer efficacy trials: conclusions from the Adult Brain Tumor Consortium's first workshop on CNS drug delivery. Neuro Oncol 2020;22:1422–4. noaa142. https://doi.org/10.1093/neuonc/noaa142.

[21] Kodaira H, Kusuhara H, Fujita T, Ushiki J, Fuse E, Sugiyama Y. Quantitative evaluation of the impact of active efflux by P-glycoprotein and breast cancer resistance protein at the blood-brain barrier on the predictability of the unbound concentrations of drugs in the brain using cerebrospinal fluid concentration as a surrogate. J Pharmacol Exp Ther 2011;339:935–44. https://doi.org/10.1124/jpet.111.180398.

[22] Blakeley JO, Olson J, Grossman SA, He X, Weingart J, Supko JG, et al. Effect of blood brain barrier permeability in recurrent high grade gliomas on the intratumoral pharmacokinetics of methotrexate: a microdialysis study. J Neurooncol 2009;91:51–8. https://doi.org/10.1007/s11060-008-9678-2.

[23] Li Y, Chen H, Xu J, Yadav NN, Chan KWY, Luo L, et al. CEST theranostics: label-free MR imaging of anticancer drugs. Oncotarget 2016;7:6369–78. https://doi.org/10.18632/oncotarget.7141.

[24] Juhász C, Dwivedi S, Kamson DO, Michelhaugh SK, Mittal S. Comparison of amino acid positron emission tomographic radiotracers for molecular imaging of primary and metastatic brain tumors. Mol Imaging 2014;13. 7290.2014.00015. https://doi.org/10.2310/7290.2014.00015.

[25] Law I, Albert NL, Arbizu J, Boellaard R, Drzezga A, Galldiks N, et al. Joint EANM/EANO/RANO practice guidelines/SNMMI procedure standards for imaging of gliomas using PET with radiolabelled amino acids and [18F]FDG: version 1.0. Eur J Nucl Med Mol Imaging 2019;46:540–57. https://doi.org/10.1007/s00259-018-4207-9.

[26] Lukas RV, Juhász C, Wainwright DA, James CD, Kennedy E, Stupp R, et al. Imaging tryptophan uptake with positron emission tomography in glioblastoma patients treated with indoximod. J Neurooncol 2019;141:111–20. https://doi.org/10.1007/s11060-018-03013-x.

[27] Jackson S, Anders NM, Mangraviti A, Wanjiku TM, Sankey EW, Liu A, et al. The effect of regadenoson-induced transient disruption of the blood–brain barrier on temozolomide delivery to normal rat brain. J Neurooncol 2016;126:433–9. https://doi.org/10.1007/s11060-015-1998-4.

[28] Carman AJ, Mills JH, Krenz A, Kim D-G, Bynoe MS. Adenosine receptor signaling modulates permeability of the blood-brain barrier. J Neurosci 2011;31:13272–80. https://doi.org/10.1523/JNEUROSCI.3337-11.2011.

[29] Curley CT, Mead BP, Negron K, Kim N, Garrison WJ, Miller GW, et al. Augmentation of brain tumor interstitial flow via focused ultrasound promotes brain-penetrating nanoparticle dispersion and transfection. Sci Adv 2020;6, eaay1344. https://doi.org/10.1126/sciadv.aay1344.

[30] Rycaj K, Tang DG. Cancer stem cells and radioresistance. Int J Radiat Biol 2014;90:615–21. https://doi.org/10.3109/09553002.2014.892227.

[31] Lathia JD, Mack SC, Mulkearns-Hubert EE, Valentim CLL, Rich JN. Cancer stem cells in glioblastoma. Genes Dev 2015;29:1203–17. https://doi.org/10.1101/gad.261982.115.

[32] Schreck KC, Guajardo A, Lin DDM, Eberhart CG, Grossman SA. Concurrent BRAF/MEK inhibitors in BRAF V600-mutant high-grade primary brain tumors. J Natl Compr Canc Netw 2018;16:343–7. https://doi.org/10.6004/jnccn.2017.7052.
[33] Ziegler DS, Wong M, Mayoh C, Kumar A, Tsoli M, Mould E, et al. Brief Report: potent clinical and radiological response to larotrectinib in TRK fusion-driven high-grade glioma. Br J Cancer 2018;119:693.
[34] Kamson DO, Grossman SA. The role of temozolomide in patients with newly diagnosed wild-type IDH, unmethylated MGMTp glioblastoma during the COVID-19 pandemic. JAMA Oncol 2020. https://doi.org/10.1001/jamaoncol.2020.6732.
[35] Yang P, Zhang W, Wang Y. IDH mutation and MGMT promoter methylation in glioblastoma: results of a prospective registry. Oncotarget 2015;6:40896–906.
[36] Newlands E, Blackledge G, Slack J, Rustin G, Smith D, Stuart N, et al. Phase I trial of temozolomide (CCRG 81045: M&B 39831: NSC 362856). Br J Cancer 1992;65:287–91. https://doi.org/10.1038/bjc.1992.57.
[37] Saleem A, Brown GD, Brady F, Aboagye EO, Osman S, Luthra SK, et al. Metabolic activation of temozolomide measured in vivo using positron emission tomography. Cancer Res 2003;63:2409–15.
[38] Rosso L, Brock CS, Gallo JM, Saleem A, Price PM, Turkheimer FE, et al. A new model for prediction of drug distribution in tumor and normal tissues: pharmacokinetics of temozolomide in glioma patients. Cancer Res 2009;69:120–7. https://doi.org/10.1158/0008-5472.CAN-08-2356.
[39] Rao JU, Coman D, Walsh JJ, Ali MM, Huang Y, Hyder F. Temozolomide arrests glioma growth and normalizes intratumoral extracellular pH. Sci Rep 2017;7:7865. https://doi.org/10.1038/s41598-017-07609-7.
[40] Mojas N, Lopes M, Jiricny J. Mismatch repair-dependent processing of methylation damage gives rise to persistent single-stranded gaps in newly replicated DNA. Genes Dev 2007;21:3342–55. https://doi.org/10.1101/gad.455407.
[41] Cejka P. Methylation-induced G2/M arrest requires a full complement of the mismatch repair protein hMLH1. EMBO J 2003;22:2245–54. https://doi.org/10.1093/emboj/cdg216.
[42] Roos W, Baumgartner M, Kaina B. Apoptosis triggered by DNA damage O6-methylguanine in human lymphocytes requires DNA replication and is mediated by p53 and Fas/CD95/Apo-1. Oncogene 2004;23:359–67. https://doi.org/10.1038/sj.onc.1207080.
[43] Horton J, Wilson S. Hypersensitivity phenotypes associated with genetic and synthetic inhibitor-induced base excision repair deficiency. DNA Repair 2007;6:530–43. https://doi.org/10.1016/j.dnarep.2006.10.016.
[44] Oldrini B, Vaquero-Siguero N, Mu Q. MGMT genomic rearrangements contribute to chemotherapy resistance in gliomas. Nat Commun 2020;11:3883.
[45] Wang J. Clonal evolution of glioblastoma under therapy. Nat Genet n.d.;48:768–776.
[46] Felsberg J. Promoter methylation and expression of MGMT and the DNA mismatch repair genes MLH1, MSH2, MSH6 and PMS2 in paired primary and recurrent glioblastomas. Int J Cancer 2011;129:659–70.
[47] Kaina B, Christmann M, Naumann S, Roos WP. MGMT: key node in the battle against genotoxicity, carcinogenicity and apoptosis induced by alkylating agents. DNA Repair 2007;6:1079–99.
[48] Kisby GE, Lesselroth H, Olivas A, Samson L, Gold B, Tanaka K, et al. Role of nucleotide- and base-excision repair in genotoxin-induced neuronal cell death. DNA Repair 2004;3:617–27. https://doi.org/10.1016/j.dnarep.2004.02.005.
[49] Quinn JA, Jiang SX, Reardon DA, Desjardins A, Vredenburgh JJ, Rich JN, et al. Phase II trial of temozolomide plus o6-benzylguanine in adults with recurrent, temozolomide-resistant malignant glioma. J Clin Oncol 2009;27:1262–7. https://doi.org/10.1200/JCO.2008.18.8417.
[50] Gilbert MR, Wang M, Aldape KD, Stupp R, Hegi ME, Jaeckle KA. Dose-dense temozolomide for newly diagnosed glioblastoma: a randomized phase III clinical trial. J Clin Oncol 2013;31:4085–91.
[51] Institóris L, Dzurillay É, Pethes G. Comparative studies on the in vivo distribution pattern of dibromodulcitol and diepoxydulcitol. Z Krebsforsch Klin Onkol 1973;79:49–59.
[52] Levin VA, Dove MAF, Maroten CE. Dianhydrogalactitol (NSC-132313): pharmacokinetics in normal and tumor-bearing rat brain and antitumor activity against three intracerebral rodent tumors 2. J Natl Cancer Inst 1976;56:535–9. https://doi.org/10.1093/jnci/56.3.535.
[53] Eckhardt S, Csetenyi J, Horvath I, Kerpel-Fronius S, Szamel I, Institoris L, et al. Uptake of labeled dianhydrogalactitol into human gliomas and nervous. Cancer Treat Rep 1977;61:841–7.
[54] Eagan RT. Dianhydrogalactitol and radiation therapy: treatment of supratentorial glioma. JAMA 1979;241:2046. https://doi.org/10.1001/jama.1979.03290450044023.
[55] Áfra D, Kerpel-Fronius S, Szinai I, Kocsis B, Eckhardt S. Combined treatment of anaplastic astrocytoma (grade 3–4) with diacetyl-dianhydro-galactitol (DADAG). Journal of Neuro Oncol 1990;8:85–91.

[56] O'Brien B, de Groot J, Kamiya-Matsuoka C, Weathers S-P, Bacha J, Brown D, et al. ACTR-27. Phase 2 study of dianhydrogalactitol (VAL-083) in patients with MGMT-unmethylated, bevacizumab-naïve recurrent glioblastoma. Neuro Oncol 2018;20:vi17. https://doi.org/10.1093/neuonc/noy148.060.

[57] Chen Z, Guo C, Yang Q, Lei J-W, Wu S, Johnson G, et al. Abstract CT273: phase 2 trial of dianhydrogalactitol (VAL-083) in patients with newly diagnosed MGMT-unmethylated glioblastoma. In: Tumor biology. American Association for Cancer Research; 2020. p. CT273. https://doi.org/10.1158/1538-7445.AM2020-CT273.

[58] Guo C, Yang Q, Li J, Wu S, Deng M, Du X, et al. Phase 2 clinical trial of VAL-083 as first-line treatment in newly-diagnosed MGMT-unmethylated glioblastoma multiforme (GBM): Halfway report. Glioma 2019;2:167. https://doi.org/10.4103/glioma.glioma_25_19.

[59] Institóris E, Szikla K, Ötvös L, Gál F. Absence of cross-resistance between two alkylating agents: BCNU vs bifunctional galactitol. Cancer Chemother Pharmacol 1989;24:311–3.

[60] Zhai B, Steinø A, Bacha J, Brown D, Daugaard M. Dianhydrogalactitol induces replication-dependent DNA damage in tumor cells preferentially resolved by homologous recombination. Cell Death Dis 2018;9:1016. https://doi.org/10.1038/s41419-018-1069-9.

[61] Jiang X, Huang Y, Wang X, Liang Q, Li Y, Li F, et al. Dianhydrogalactitol, a potential multitarget agent, inhibits glioblastoma migration, invasion, and angiogenesis. Biomed Pharmacother 2017;91:1065–74. https://doi.org/10.1016/j.biopha.2017.05.025.

[62] Berens ME, Giblin JR, Dougherty DV, Hoifodt HK, Tveit K, Rosenblum ML. Comparison of *in vitro* cloning assays for drug sensitivity testing of human brain tumours. BrJ Neurosurg 1988;2:227–34. https://doi.org/10.3109/02688698808992673.

[63] Fotovati A, Hu K, Wakimoto H, Triscott J, Bacha J, Brown D, et al. ET-37. VAL-083, a novel N7 alkylating agent, surpasses temozolomide activity and inhibits cancer stem cells, providing a new potential treatment option for glioblastoma multiforme. Neuro Oncol 2012;14(suppl 6):vi32.

[64] Zhai B, Li Y, Kotapalli SS, Bacha J, Brown D, Steinø A, et al. Dianhydrogalactitol synergizes with topoisomerase poisons to overcome DNA repair activity in tumor cells. Cell Death Dis 2020;11:577. https://doi.org/10.1038/s41419-020-02780-8.

[65] Peng C, Qi X, Miao L, Ren J. 1,2:5,6-dianhydrogalactitol inhibits human glioma cell growth in vivo and in vitro by arresting the cell cycle at G2/M phase. Acta Pharmacol Sin 2017;38:561–70. https://doi.org/10.1038/aps.2016.154.

[66] Hevener KE, Verstak TA, Lutat KE, Riggsbee DL, Mooney JW. Recent developments in topoisomerase-targeted cancer chemotherapy. Acta Pharm Sin B 2018;8:844–61. https://doi.org/10.1016/j.apsb.2018.07.008.

[67] Goldwirt L, Beccaria K, Carpentier A, Farinotti R, Fernandez C. Irinotecan and temozolomide brain distribution: a focus on ABCB1. Cancer Chemother Pharmacol 2014;74:185–93. https://doi.org/10.1007/s00280-014-2490-0.

[68] Prados MD, Lamborn K, Yung WKA, Jaeckle K, Robins HI, Mehta M, et al. A phase 2 trial of irinotecan (CPT-11) in patients with recurrent malignant glioma: a North American Brain Tumor Consortium study1. Neuro Oncol 2006;8:189–93. https://doi.org/10.1215/15228517-2005-010.

[69] Quinn JA, Jiang SX, Reardon DA, Desjardins A, Vredenburgh JJ, Friedman AH, et al. Phase II trial of temozolomide (TMZ) plus irinotecan (CPT-11) in adults with newly diagnosed glioblastoma multiforme before radiotherapy. J Neurooncol 2009;95:393–400. https://doi.org/10.1007/s11060-009-9937-x.

[70] Clarke JL, Molinaro AM, Cabrera JR, DeSilva AA, Rabbitt JE, Prey J, et al. A phase 1 trial of intravenous liposomal irinotecan in patients with recurrent high-grade glioma. Cancer Chemother Pharmacol 2017;79:603–10. https://doi.org/10.1007/s00280-017-3247-3.

[71] Ananda S, Nowak AK, Cher L, Dowling A, Brown C, Simes J, et al. Phase 2 trial of temozolomide and pegylated liposomal doxorubicin in the treatment of patients with glioblastoma multiforme following concurrent radiotherapy and chemotherapy. J Clin Neurosci 2011;18:1444–8. https://doi.org/10.1016/j.jocn.2011.02.026.

[72] Yamashita AS, da Costa RM, Borodovsky A, Festuccia WT, Chan T, Riggins GJ. Demethylation and epigenetic modification with 5-azacytidine reduces IDH1 mutant glioma growth in combination with temozolomide. Neuro Oncol 2019;21:189–200. https://doi.org/10.1093/neuonc/noy146.

[73] Jin Y, Li J, Rong L-F, Lü X-W, Huang Y, Xu S-Y. Pharmacokinetics and tissue distribution of 5-fluorouracil encapsulated by galactosylceramide liposomes in mice. Acta Pharmacol Sin 2005;26:250–6. https://doi.org/10.1111/j.1745-7254.2005.00530.x.

[74] Morikawa A, Peereboom DM, Thorsheim HR, Samala R, Balyan R, Murphy CG, et al. Capecitabine and lapatinib uptake in surgically resected brain metastases from metastatic breast cancer patients: a prospective study. Neuro Oncol 2015;17:289–95. https://doi.org/10.1093/neuonc/nou141.

[75] Miller CR, Williams CR, Buchsbaum DJ, Gillespie GY. Intratumoral 5-fluorouracil produced by cytosine deaminase/5-fluorocytosine gene therapy is effective for experimental human glioblastomas. Cancer Res 2002;62:773–80.

[76] Sigmond J, Honeywell RJ, Postma TJ, Dirven CMF, de Lange SM, van der Born K, et al. Gemcitabine uptake in glioblastoma multiforme: potential as a radiosensitizer. Ann Oncol 2009;20:182–7. https://doi.org/10.1093/annonc/mdn543.
[77] Green AL, Flannery P, Hankinson TC, O'Neill B, Amani V, DeSisto J, et al. Preclinical and clinical investigation of intratumoral chemotherapy pharmacokinetics in DIPG using gemcitabine. Neuro Oncol Adv 2020;2:vdaa021. https://doi.org/10.1093/noajnl/vdaa021.
[78] Wick W, Hermisson M, Kortmann RD, Küker WM, Duffner F, Dichgans J, et al. Neoadjuvant gemcitabine/treosulfan chemotherapy for newly diagnosed glioblastoma. J Neuro Oncol 2002;59:151–5.
[79] Metro G, Fabi A, Mirri MA, Vidiri A, Pace A, Carosi M, et al. Phase II study of fixed dose rate gemcitabine as radiosensitizer for newly diagnosed glioblastoma multiforme. Cancer Chemother Pharmacol 2010;65:391.
[80] Kim MM, Camelo-Piragua S, Schipper M, Tao Y, Normolle D, Junck L, et al. Gemcitabine plus radiation therapy for high-grade glioma: long-term results of a phase 1 dose-escalation study. Int J Radiat Oncol Biol Phys 2016;94:305–11. https://doi.org/10.1016/j.ijrobp.2015.10.032.
[81] Philbrick BD, Adamson DC. Early clinical trials of Toca 511 and Toca FC show a promising novel treatment for recurrent malignant glioma. Expert Opin Investig Drugs 2019;28:207–16. https://doi.org/10.1080/13543784.2019.1572112.
[82] Dalba C, Klatzmann D, Logg CR, Kasahara N. Beyond oncolytic virotherapy: replication-competent retrovirus vectors for selective and stable transduction of tumors. Curr Gene Ther 2005;5:655–67. https://doi.org/10.2174/156652305774964659.
[83] Mitchell LA, Lopez Espinoza F, Mendoza D, Kato Y, Inagaki A, Hiraoka K, et al. Toca 511 gene transfer and treatment with the prodrug, 5-fluorocytosine, promotes durable antitumor immunity in a mouse glioma model. Neuro Oncol 2017;19:930–9. https://doi.org/10.1093/neuonc/nox037.
[84] Cloughesy TF, Landolfi J, Vogelbaum MA, Ostertag D, Elder JB, Bloomfield S, et al. Durable complete responses in some recurrent high-grade glioma patients treated with Toca 511 + Toca FC. Neuro Oncol 2018;20:1383–92. https://doi.org/10.1093/neuonc/noy075.
[85] Cloughesy T, Petrecca K, Walbert T, Butowski N, Salacz M, Perry J, et al. LTBK-08. TOCA 511 & TOCA FC versus standard of care in patients with recurrent high grade glioma. Neuro Oncol 2019;21:vi284.
[86] Beckta JM, Bindra RS, Chalmers AJ. Targeting DNA repair in gliomas. Curr Opin Neurol 2019;32:878–85. https://doi.org/10.1097/WCO.0000000000000760.
[87] Galia A, Calogero AE, Condorelli R, Fraggetta F, La Corte A, Ridolfo F, et al. PARP-1 protein expression in glioblastoma multiforme. Eur J Histochem 2012;56, e9. https://doi.org/10.4081/ejh.2012.e9.
[88] Satoh MS, Lindahl T. Role of poly(ADP-ribose) formation in DNA repair. Nature 1992;356:356–8. https://doi.org/10.1038/356356a0.
[89] Pommier Y, O'Connor MJ, De Bono J. Laying a trap to kill cancer cells: PARP inhibitors and their mechanisms of action. Sci Transl Med 2016;8:362ps17.
[90] Sulkowski PL, Corso CD, Robinson ND, Scanlon SE, Purshouse KR, Bai H, et al. 2-Hydroxyglutarate produced by neomorphic IDH mutations suppresses homologous recombination and induces PARP inhibitor sensitivity. Sci Transl Med 2017;9:eaal2463.
[91] Higuchi F, Nagashima H, Ning J, Koerner MVA, Wakimoto H, Cahill DP. Restoration of temozolomide sensitivity by parp inhibitors in mismatch repair deficient glioblastoma is independent of base excision repair. Clin Cancer Res 2020;26:1690–9. https://doi.org/10.1158/1078-0432.CCR-19-2000.
[92] van Vuurden DG, Hulleman E, Meijer OLM, Wedekind LE, Kool M, Witt H, et al. PARP inhibition sensitizes childhood high grade glioma, medulloblastoma and ependymoma to radiation. Oncotarget 2011;2:984–96. https://doi.org/10.18632/oncotarget.362.
[93] Xiong Y, Guo Y, Liu Y, Wang H, Gong W, Liu Y, et al. Pamiparib is a potent and selective PARP inhibitor with unique potential for the treatment of brain tumor. Neoplasia 2020;22:431–40. https://doi.org/10.1016/j.neo.2020.06.009.
[94] Hanna C, Kurian KM, Williams K, Watts C, Jackson A, Carruthers R, et al. Pharmacokinetics, safety, and tolerability of olaparib and temozolomide for recurrent glioblastoma: results of the phase I OPARATIC trial. Neuro Oncol 2020;noaa104. https://doi.org/10.1093/neuonc/noaa104.
[95] Chalmers AJ, Short S, Watts C, Herbert C, Morris A, Stobo J, et al. Phase I clinical trials evaluating olaparib in combination with radiotherapy (RT) and/or temozolomide (TMZ) in glioblastoma patients: results of

OPARATIC and PARADIGM phase I and early results of PARADIGM-2. JCO 2018;36:2018. https://doi.org/10.1200/JCO.2018.36.15_suppl.2018.
[96] Jue TR, Nozue K, Lester AJ, Joshi S, Schroder LBW, Whittaker SP, et al. Veliparib in combination with radiotherapy for the treatment of MGMT unmethylated glioblastoma. J Transl Med 2017;15:61. https://doi.org/10.1186/s12967-017-1164-1.
[97] Penning TD, Zhu G-D, Gandhi VB, Gong J, Liu X, Shi Y, et al. Discovery of the poly(ADP-ribose) polymerase (PARP) inhibitor 2-[(*R*)-2-methylpyrrolidin-2-yl]-1 *H* -benzimidazole-4-carboxamide (ABT-888) for the treatment of cancer. J Med Chem 2009;52:514–23. https://doi.org/10.1021/jm801171j.
[98] Li X, Delzer J, Voorman R, de Morais SM, Lao Y. Disposition and drug-drug interaction potential of veliparib (ABT-888), a novel and potent inhibitor of poly(ADP-ribose) polymerase. Drug Metab Dispos 2011;39:1161–9. https://doi.org/10.1124/dmd.110.037820.
[99] Robins HI, Zhang P, Gilbert MR, Chakravarti A, de Groot JF, Grimm SA, et al. A randomized phase I/II study of ABT-888 in combination with temozolomide in recurrent temozolomide resistant glioblastoma: an NRG oncology RTOG group study. J Neurooncol 2016;126:309–16. https://doi.org/10.1007/s11060-015-1966-z.
[100] Baxter PA, Su JM, Onar-Thomas A, Billups CA, Li X-N, Poussaint TY, et al. A phase I/II study of veliparib (ABT-888) with radiation and temozolomide in newly diagnosed diffuse pontine glioma: a Pediatric Brain Tumor Consortium study. Neuro Oncol 2020;22:875–85. https://doi.org/10.1093/neuonc/noaa016.
[101] Shen S, Miao Z, Yang C. Efficacy of a novel potent and high water-soluble selective PARP inhibitor CVL218 in glioma. JCO 2019;37:e13539. https://doi.org/10.1200/JCO.2019.37.15_suppl.e13539.
[102] Longoria TC, Tewari KS. Pharmacokinetic drug evaluation of niraparib for the treatment of ovarian cancer. Expert Opin Drug Metab Toxicol 2018;14:543–50. https://doi.org/10.1080/17425255.2018.1461838.
[103] Hong C, Sule A, Beckta J, Sundaram R, Bindra R. Abstract 1389: exploiting inherent DNA damage repair defects in IDH1/2 mutated gliomas with the CNS penetrant PARP inhibitor, pamiparib. In: Molecular and cellular biology/genetics. American Association for Cancer Research; 2020. p. 1389. https://doi.org/10.1158/1538-7445.AM2020-1389.
[104] Friedlander M, Meniawy T, Markman B, Mileshkin L, Harnett P, Millward M, et al. Pamiparib in combination with tislelizumab in patients with advanced solid tumours: results from the dose-escalation stage of a multicentre, open-label, phase 1a/b trial. Lancet Oncol 2019;20:1306–15.
[105] Piotrowski A, Puduvalli V, Wen P, Campian J, Colman H, Pearlman M, et al. ACTR-39. Pamiparib in combination with radiation therapy (RT) and/or temozolomide (TMZ) in patients with newly diagnosed or recurrent/refractory (R/R) glioblastoma (GBM); phase 1B/2 study update. Neuro Oncol 2019;21:vi21–2. https://doi.org/10.1093/neuonc/noz175.081.
[106] Mohiuddin IS, Kang MH. DNA-PK as an emerging therapeutic target in cancer. Front Oncol 2019;9:635. https://doi.org/10.3389/fonc.2019.00635.
[107] Davis AJ, Chen BPC, Chen DJ. DNA-PK: a dynamic enzyme in a versatile DSB repair pathway. DNA Repair 2014;17:21–9. https://doi.org/10.1016/j.dnarep.2014.02.020.
[108] Shang Z, Yu L, Lin Y-F, Matsunaga S, Shen C-Y, Chen BPC. DNA-PKcs activates the Chk2–Brca1 pathway during mitosis to ensure chromosomal stability. Oncogenesis 2014;3:e85. https://doi.org/10.1038/oncsis.2013.49.
[109] Shintani S, Mihara M, Li C, Nakahara Y, Hino S, Nakashiro K, et al. Up-regulation of DNA-dependent protein kinase correlates with radiation resistance in oral squamous cell carcinoma. Cancer Sci 2003;94:894–900. https://doi.org/10.1111/j.1349-7006.2003.tb01372.x.
[110] Lee S, Cho K-J, Park J, Kim SY, Nam SY, Lee B-J, et al. Expressions of Ku70 and DNA-PKcs as prognostic indicators of local control in nasopharyngeal carcinoma. Int J Radiat Oncol Biol Phys 2005;62:1451–7. https://doi.org/10.1016/j.ijrobp.2004.12.049.
[111] Ihara M, Ashizawa K, Shichijo K, Kudo T. Expression of the DNA-dependent protein kinase catalytic subunit is associated with the radiosensitivity of human thyroid cancer cell lines. J Radiat Res 2019;60:171–7. https://doi.org/10.1093/jrr/rry097.
[112] Evert M, Frau M, Tomasi ML, Latte G, Simile MM, Seddaiu MA, et al. Deregulation of DNA-dependent protein kinase catalytic subunit contributes to human hepatocarcinogenesis development and has a putative prognostic value. Br J Cancer 2013;109:2654–64. https://doi.org/10.1038/bjc.2013.606.
[113] Gearhart J, Pashos EE, Prasad MK. Pluripotency redux—advances in stem-cell research. N Engl J Med 2007;357:1469–72. https://doi.org/10.1056/NEJMp078126.

[114] Iijima S, Teraoka H, Date T, Tsukada K. DNA-activated protein kinase in Raji Burkitt's lymphoma cells: phosphorylation of c-Myc oncoprotein. Eur J Biochem 1992;206:595–603. https://doi.org/10.1111/j.1432-1033.1992.tb16964.x.
[115] Kase M, Vardja M, Lipping A, Asser T, Jaal J. Impact of PARP-1 and DNA-PK expression on survival in patients with glioblastoma multiforme. Radiother Oncol 2011;101:127–31. https://doi.org/10.1016/j.radonc.2011.06.024.
[116] Mortensen DS, Perrin-Ninkovic SM, Shevlin G, Elsner J, Zhao J, Whitefield B, et al. Optimization of a series of triazole containing mammalian target of rapamycin (mTOR) kinase inhibitors and the discovery of CC-115. J Med Chem 2015;58:5599–608. https://doi.org/10.1021/acs.jmedchem.5b00627.
[117] Munster P, Mita M, Mahipal A, Nemunaitis J, Massard C, Mikkelsen T, et al. First-in-human phase I study of a dual mTOR kinase and DNA-PK inhibitor (CC-115) in advanced malignancy. CMAR 2019;11:10463–76. https://doi.org/10.2147/CMAR.S208720.
[118] Thijssen R, ter Burg J, Garrick B, van Bochove GGW, Brown JR, Fernandes SM, et al. Dual TORK/DNA-PK inhibition blocks critical signaling pathways in chronic lymphocytic leukemia. Blood 2016;128:574–83. https://doi.org/10.1182/blood-2016-02-700328.
[119] Alexander BM, Trippa L, Gaffey SC, Arrillaga I, Lee EQ, Tanguturi SK, et al. Individualized screening trial of innovative glioblastoma therapy (INSIGhT). JCO 2017;35:TPS2079. https://doi.org/10.1200/JCO.2017.35.15_suppl.TPS2079.
[120] Timme CR, Rath BH, O'Neill JW, Camphausen K, Tofilon PJ. The DNA-PK inhibitor VX-984 enhances the radiosensitivity of glioblastoma cells grown *in vitro* and as orthotopic xenografts. Mol Cancer Ther 2018;17:1207–16. https://doi.org/10.1158/1535-7163.MCT-17-1267.
[121] Pilié PG, Tang C, Mills GB, Yap TA. State-of-the-art strategies for targeting the DNA damage response in cancer. Nat Rev Clin Oncol 2019;16:81–104. https://doi.org/10.1038/s41571-018-0114-z.
[122] Marian CO, Cho SK, Mcellin BM, Maher EA, Hatanpaa KJ, Madden CJ, et al. The telomerase antagonist, imetelstat, efficiently targets glioblastoma tumor-initiating cells leading to decreased proliferation and tumor growth. Clin Cancer Res 2010;16:154–63. https://doi.org/10.1158/1078-0432.CCR-09-2850.
[123] Hashizume R, Ozawa T, Gryaznov SM, Bollen AW, Lamborn KR, Frey WH, et al. New therapeutic approach for brain tumors: intranasal delivery of telomerase inhibitor GRN163. Neuro Oncol 2008;10:112–20. https://doi.org/10.1215/15228517-2007-052.
[124] Salloum R, Hummel TR, Kumar SS, Dorris K, Li S, Lin T, et al. A molecular biology and phase II study of imetelstat (GRN163L) in children with recurrent or refractory central nervous system malignancies: a pediatric brain tumor consortium study. J Neurooncol 2016;129:443–51. https://doi.org/10.1007/s11060-016-2189-7.
[125] Khan O, La Thangue NB. HDAC inhibitors in cancer biology: emerging mechanisms and clinical applications. Immunol Cell Biol 2012;90:85–94. https://doi.org/10.1038/icb.2011.100.
[126] Kouzarides T. Chromatin modifications and their function. Cell 2007;128:693–705. https://doi.org/10.1016/j.cell.2007.02.005.
[127] Hanahan D, Weinberg RA. Hallmarks of cancer: the next generation. Cell 2011;144:646–74. https://doi.org/10.1016/j.cell.2011.02.013.
[128] Lucio-Eterovic AK, Cortez MA, Valera ET, Motta FJ, Queiroz RG, Machado HR, et al. Differential expression of 12 histone deacetylase (HDAC) genes in astrocytomas and normal brain tissue: class II and IV are hypoexpressed in glioblastomas. BMC Cancer 2008;8:243. https://doi.org/10.1186/1471-2407-8-243.
[129] Brennan CW, Verhaak RGW, McKenna A, Campos B, Noushmehr H, Salama SR, et al. The somatic genomic landscape of glioblastoma. Cell 2013;155:462–77. https://doi.org/10.1016/j.cell.2013.09.034.
[130] Was H, Krol SK, Rotili D, Mai A, Wojtas B, Kaminska B, et al. Histone deacetylase inhibitors exert anti-tumor effects on human adherent and stem-like glioma cells. Clin Epigenet 2019;11:11. https://doi.org/10.1186/s13148-018-0598-5.
[131] Marchion DC, Bicaku E, Turner JG, Schmitt ML, Morelli DR, Munster PN. HDAC2 regulates chromatin plasticity and enhances DNA vulnerability. Mol Cancer Ther 2009;8:794–801. https://doi.org/10.1158/1535-7163.MCT-08-0985.
[132] Miller KM, Tjeertes JV, Coates J, Legube G, Polo SE, Britton S, et al. Human HDAC1 and HDAC2 function in the DNA-damage response to promote DNA nonhomologous end-joining. Nat Struct Mol Biol 2010;17:1144–51. https://doi.org/10.1038/nsmb.1899.
[133] Hooker JM, Kim SW, Alexoff D, Xu Y, Shea C, Reid A, et al. Histone deacetylase inhibitor MS-275 exhibits poor brain penetration: pharmacokinetic studies of [^{11}C]MS-275 using positron emission tomography. ACS Chem Neurosci 2010;1:65–73. https://doi.org/10.1021/cn9000268.

[134] Gurvich N, Tsygankova OM, Meinkoth JL, Klein PS. Histone deacetylase is a target of valproic acid-mediated cellular differentiation. Cancer Res 2004;64:1079–86. https://doi.org/10.1158/0008-5472.CAN-03-0799.

[135] Kuendgen A, Schmid M, Schlenk R, Knipp S, Hildebrandt B, Steidl C, et al. The histone deacetylase (HDAC) inhibitor valproic acid as monotherapy or in combination with all-trans retinoic acid in patients with acute myeloid leukemia. Cancer 2006;106:112–9. https://doi.org/10.1002/cncr.21552.

[136] Rocca A, Minucci S, Tosti G, Croci D, Contegno F, Ballarini M, et al. A phase I–II study of the histone deacetylase inhibitor valproic acid plus chemoimmunotherapy in patients with advanced melanoma. Br J Cancer 2009;100:28–36. https://doi.org/10.1038/sj.bjc.6604817.

[137] Chadwick DW. Concentration-effect relationships of valproic acid. Clin Pharmacokinet 1985;10:155–63. https://doi.org/10.2165/00003088-198510020-00003.

[138] Shen DD, Ojemann GA, Rapport RL, Dills RL, Friel PN, Levy RH. Low and variable presence of valproic acid in human brain. Neurology 1992;42:582. https://doi.org/10.1212/WNL.42.3.582.

[139] Krauze AV, Myrehaug SD, Chang MG, Holdford DJ, Smith S, Shih J, et al. A phase 2 study of concurrent radiation therapy, temozolomide, and the histone deacetylase inhibitor valproic acid for patients with glioblastoma. Int J Radiat Oncol Biol Phys 2015;92:986–92. https://doi.org/10.1016/j.ijrobp.2015.04.038.

[140] Krauze AV, Mackey M, Rowe L, Chang MG, Holdford DJ, Cooley T, et al. Late toxicity in long-term survivors from a phase 2 study of concurrent radiation therapy, temozolomide and valproic acid for newly diagnosed glioblastoma. Neurooncol Pract 2018;5:246–50. https://doi.org/10.1093/nop/npy009.

[141] Tarasenko N, Chekroun-Setti H, Nudelman A, Rephaeli A. Comparison of the anticancer properties of a novel valproic acid prodrug to leading histone deacetylase inhibitors. J Cell Biochem 2018;119:3417–28. https://doi.org/10.1002/jcb.26512.

[142] Hanson JE, La H, Plise E, Chen Y-H, Ding X, Hanania T, et al. SAHA enhances synaptic function and plasticity in vitro but has limited brain availability in vivo and does not impact cognition. PLoS ONE 2013;8, e69964. https://doi.org/10.1371/journal.pone.0069964.

[143] Munkacsi AB, Hammond N, Schneider RT, Senanayake DS, Higaki K, Lagutin K, et al. Normalization of hepatic homeostasis in the *Npc1* ^{nmf164} mouse model of Niemann-Pick Type C disease treated with the histone deacetylase inhibitor vorinostat. J Biol Chem 2017;292:4395–410. https://doi.org/10.1074/jbc.M116.770578.

[144] Rubin EH, Agrawal NGB, Friedman EJ, Scott P, Mazina KE, Sun L, et al. A study to determine the effects of food and multiple dosing on the pharmacokinetics of vorinostat given orally to patients with advanced cancer. Clin Cancer Res 2006;12:7039–45. https://doi.org/10.1158/1078-0432.CCR-06-1802.

[145] Reßing N, Marquardt V, Gertzen CGW, Schöler A, Schramm A, Kurz T, et al. Design, synthesis and biological evaluation of β-peptoid-capped HDAC inhibitors with anti-neuroblastoma and anti-glioblastoma activity. Med Chem Commun 2019;10:1109–15. https://doi.org/10.1039/C8MD00454D.

[146] Galanis E, Jaeckle KA, Maurer MJ, Reid JM, Ames MM, Hardwick JS, et al. Phase II trial of vorinostat in recurrent glioblastoma multiforme: a north central cancer treatment group study. J Clin Oncol 2009;27:2052–8. https://doi.org/10.1200/JCO.2008.19.0694.

[147] Friday BB, Anderson SK, Buckner J, Yu C, Giannini C, Geoffroy F, et al. Phase II trial of vorinostat in combination with bortezomib in recurrent glioblastoma: a north central cancer treatment group study. Neuro Oncol 2012;14:215–21. https://doi.org/10.1093/neuonc/nor198.

[148] Galanis E, Anderson SK, Miller CR, Sarkaria JN, Jaeckle K, Buckner JC, et al. Phase I/II trial of vorinostat combined with temozolomide and radiation therapy for newly diagnosed glioblastoma: results of Alliance N0874/ABTC 02. Neuro Oncol 2018;20:546–56. https://doi.org/10.1093/neuonc/nox161.

[149] Gurbani SS, Yoon Y, Weinberg BD, Salgado E, Press RH, Cordova JS, et al. Assessing treatment response of glioblastoma to an HDAC inhibitor using whole-brain spectroscopic MRI. Tomography 2019;5:53–60. https://doi.org/10.18383/j.tom.2018.00031.

[150] MRSI to predict response to RT/TMZ ± belinostat in GBM—full text view - ClinicalTrials.gov n.d. https://clinicaltrials.gov/ct2/show/NCT02137759 (accessed September 18, 2020).

[151] Choi SA, Lee C, Kwak PA, Park C-K, Wang K-C, Phi JH, et al. Histone deacetylase inhibitor panobinostat potentiates the anti-cancer effects of mesenchymal stem cell-based sTRAIL gene therapy against malignant glioma. Cancer Lett 2019;442:161–9. https://doi.org/10.1016/j.canlet.2018.10.012.

[152] Boulet MHC, Marsh LK, Howarth A, Woolman A, Farrer NJ. Oxaliplatin and [Pt(*R*, *R* -DACH)(panobinostat $_{-2H}$)] show nanomolar cytotoxicity towards diffuse intrinsic pontine glioma (DIPG). Dalton Trans 2020;49:5703–10. https://doi.org/10.1039/C9DT04862F.

[153] Kommidi H, Tosi U, Maachani UB, Guo H, Marnell CS, Law B, et al. ^{18}F-radiolabeled panobinostat allows for positron emission tomography guided delivery of a histone deacetylase inhibitor. ACS Med Chem Lett 2018;9:114–9. https://doi.org/10.1021/acsmedchemlett.7b00471.
[154] Meel MH, de Gooijer MC, Metselaar DS, Sewing ACP, Zwaan K, Waranecki P, et al. Combined therapy of AXL and HDAC inhibition reverses mesenchymal transition in diffuse intrinsic pontine glioma. Clin Cancer Res 2020;26:3319–32. https://doi.org/10.1158/1078-0432.CCR-19-3538.
[155] Chopra V, Quinti L, Khanna P, Paganetti P, Kuhn R, Young AB, et al. LBH589, a hydroxamic acid-derived HDAC inhibitor, is neuroprotective in mouse models of Huntington's disease. JHD 2016;5:347–55. https://doi.org/10.3233/JHD-160226.
[156] Drappatz J, Lee EQ, Hammond S, Grimm SA, Norden AD, Beroukhim R, et al. Phase I study of panobinostat in combination with bevacizumab for recurrent high-grade glioma. J Neurooncol 2012;107:133–8. https://doi.org/10.1007/s11060-011-0717-z.
[157] Lee EQ, Reardon DA, Schiff D, Drappatz J, Muzikansky A, Grimm SA, et al. Phase II study of panobinostat in combination with bevacizumab for recurrent glioblastoma and anaplastic glioma. Neuro Oncol 2015;17:862–7. https://doi.org/10.1093/neuonc/nou350.
[158] Shi W, Palmer JD, Werner-Wasik M, Andrews DW, Evans JJ, Glass J, et al. Phase I trial of panobinostat and fractionated stereotactic re-irradiation therapy for recurrent high grade gliomas. J Neurooncol 2016;127:535–9. https://doi.org/10.1007/s11060-016-2059-3.
[159] Singleton WGB, Bienemann AS, Woolley M, Johnson D, Lewis O, Wyatt MJ, et al. The distribution, clearance, and brainstem toxicity of panobinostat administered by convection-enhanced delivery. J Neurosurg Pediatr 2018;22:288–96. https://doi.org/10.3171/2018.2.PEDS17663.
[160] Sawa H, Murakami H, Kumagai M, Nakasato M, Yamauchi S, Matsuyama N, et al. Histone deacetylase inhibitor, FK228, induces apoptosis and suppresses cell proliferation of human glioblastoma cells in vitro and in vivo. Acta Neuropathologica 2004;107:523–31. https://doi.org/10.1007/s00401-004-0841-3.
[161] Iwamoto FM, Lamborn KR, Kuhn JG, Wen PY, Alfred Yung WK, Gilbert MR, et al. A phase I/II trial of the histone deacetylase inhibitor romidepsin for adults with recurrent malignant glioma: North American Brain Tumor Consortium Study 03-03. Neuro Oncol 2011;13:509–16. https://doi.org/10.1093/neuonc/nor017.
[162] Byrd JC. A phase 1 and pharmacodynamic study of depsipeptide (FK228) in chronic lymphocytic leukemia and acute myeloid leukemia. Blood 2004;105:959–67. https://doi.org/10.1182/blood-2004-05-1693.
[163] Berg SL, Stone J, Xiao JJ, Chan KK, Nuchtern J, Dauser R, et al. Plasma and cerebrospinal fluid pharmacokinetics of depsipeptide (FR901228) in nonhuman primates. Cancer Chemother Pharmacol 2004;54:85–8. https://doi.org/10.1007/s00280-004-0766-5.
[164] Zhao H, Wang Y, Yang C, Zhou J, Wang L, Yi K, et al. EGFR-vIII downregulated H2AZK4/7AC though the PI3K/AKT-HDAC2 axis to regulate cell cycle progression. Clin Transl Med 2020;9:e10. https://doi.org/10.1186/s40169-020-0260-7.
[165] Rothbart SB, Strahl BD. Interpreting the language of histone and DNA modifications. Biochim et Biophys Acta (BBA) Gene Regul Mech 2014;1839:627–43. https://doi.org/10.1016/j.bbagrm.2014.03.001.
[166] Copeland RA, Solomon ME, Richon VM. Protein methyltransferases as a target class for drug discovery. Nat Rev Drug Discov 2009;8:724–32. https://doi.org/10.1038/nrd2974.
[167] Morera L, Lübbert M, Jung M. Targeting histone methyltransferases and demethylases in clinical trials for cancer therapy. Clin Epigenet 2016;8:57. https://doi.org/10.1186/s13148-016-0223-4.
[168] Wiles ET, Selker EU. H3K27 methylation: a promiscuous repressive chromatin mark. Curr Opin Genet Dev 2017;43:31–7. https://doi.org/10.1016/j.gde.2016.11.001.
[169] Zhang J, Chen L, Han L, Shi Z, Zhang J, Pu P, et al. EZH2 is a negative prognostic factor and exhibits pro-oncogenic activity in glioblastoma. Cancer Lett 2015;356:929–36. https://doi.org/10.1016/j.canlet.2014.11.003.
[170] Fan T-Y, Wang H, Xiang P, Liu Y-W, Li H-Z, Lei B-X, et al. Inhibition of EZH2 reverses chemotherapeutic drug TMZ chemosensitivity in glioblastoma. Int J Clin Exp Pathol 2014;7:6662–70.
[171] Harutyunyan AS, Krug B, Chen H, Papillon-Cavanagh S, Zeinieh M, De Jay N, et al. H3K27M induces defective chromatin spread of PRC2-mediated repressive H3K27me2/me3 and is essential for glioma tumorigenesis. Nat Commun 2019;10:1262. https://doi.org/10.1038/s41467-019-09140-x.
[172] Justin N, Zhang Y, Tarricone C, Martin SR, Chen S, Underwood E, et al. Structural basis of oncogenic histone H3K27M inhibition of human polycomb repressive complex 2. Nat Commun 2016;7:11316. https://doi.org/10.1038/ncomms11316.
[173] Mohammad F, Weissmann S, Leblanc B, Pandey DP, Højfeldt JW, Comet I, et al. EZH2 is a potential therapeutic target for H3K27M-mutant pediatric gliomas. Nat Med 2017;23:483–92. https://doi.org/10.1038/nm.4293.

[174] Stazi G, Taglieri L, Nicolai A, Romanelli A, Fioravanti R, Morrone S, et al. Dissecting the role of novel EZH2 inhibitors in primary glioblastoma cell cultures: effects on proliferation, epithelial-mesenchymal transition, migration, and on the pro-inflammatory phenotype. Clin Epigenet 2019;11:173. https://doi.org/10.1186/s13148-019-0763-5.

[175] Italiano A, Soria J-C, Toulmonde M, Michot J-M, Lucchesi C, Varga A, et al. Tazemetostat, an EZH2 inhibitor, in relapsed or refractory B-cell non-Hodgkin lymphoma and advanced solid tumours: a first-in-human, open-label, phase 1 study. Lancet Oncol 2018;19:649–59. https://doi.org/10.1016/S1470-2045(18)30145-1.

[176] Zhang P, de Gooijer MC, Buil LCM, Beijnen JH, Li G, van Tellingen O. ABCB1 and ABCG2 restrict the brain penetration of a panel of novel EZH2-Inhibitors. Int J Cancer 2015;137:2007–18. https://doi.org/10.1002/ijc.29566.

[177] Zhang J, Stevens MFG, Bradshaw TD. Temozolomide: mechanisms of action, repair and resistance. Curr Mol Pharmacol n.d.;5:102–114.

[178] Gil J, Peters G. Regulation of the INK4b–ARF–INK4a tumour suppressor locus: all for one or one for all. Nat Rev Mol Cell Biol 2006;7:667–77. https://doi.org/10.1038/nrm1987.

[179] Bronner SM, Merrick KA, Murray J, Salphati L, Moffat JG, Pang J, et al. Design of a brain-penetrant CDK4/6 inhibitor for glioblastoma. Bioorg Med Chem Lett 2019;29:2294–301. https://doi.org/10.1016/j.bmcl.2019.06.021.

[180] Michaud K, Solomon DA, Oermann E, Kim JS, Zhong WZ, Prados MD, et al. Pharmacologic inhibition of cyclin-dependent kinases 4 and 6 arrests the growth of glioblastoma multiforme intracranial xenografts. Cancer Res 2010;70:3228–38. https://doi.org/10.1158/0008-5472.CAN-09-4559.

[181] Cen L, Carlson BL, Schroeder MA, Ostrem JL, Kitange GJ, Mladek AC, et al. p16-Cdk4-Rb axis controls sensitivity to a cyclin-dependent kinase inhibitor PD0332991 in glioblastoma xenograft cells. Neuro Oncol 2012;14:870–81. https://doi.org/10.1093/neuonc/nos114.

[182] Yin L, Li H, Liu W, Yao Z, Cheng Z, Zhang H, et al. A highly potent CDK4/6 inhibitor was rationally designed to overcome blood brain barrier in gliobastoma therapy. Eur J Med Chem 2018;144:1–28. https://doi.org/10.1016/j.ejmech.2017.12.003.

[183] Patel YT, Davis A, Baker SJ, Campagne O, Stewart CF. CNS penetration of the CDK4/6 inhibitor ribociclib in non-tumor bearing mice and mice bearing pediatric brain tumors. Cancer Chemother Pharmacol 2019;84:447–52. https://doi.org/10.1007/s00280-019-03864-9.

[184] Miller TW, Traphagen NA, Li J, Lewis LD, Lopes B, Asthagiri A, et al. Tumor pharmacokinetics and pharmacodynamics of the CDK4/6 inhibitor ribociclib in patients with recurrent glioblastoma. J Neurooncol 2019;144:563–72. https://doi.org/10.1007/s11060-019-03258-0.

[185] Tien A-C, Li J, Bao X, Derogatis A, Kim S, Mehta S, et al. A phase 0 trial of ribociclib in recurrent glioblastoma patients incorporating a tumor pharmacodynamic- and pharmacokinetic-guided expansion cohort. Clin Cancer Res 2019;25:5777–86. https://doi.org/10.1158/1078-0432.CCR-19-0133.

[186] Liao X, Hong Y, Mao Y, Chen N, Wang Q, Wang Z, et al. SPH3643: a novel cyclin-dependent kinase 4/6 inhibitor with good anticancer efficacy and strong blood-brain barrier permeability. Cancer Sci 2020;111:1761–73. https://doi.org/10.1111/cas.14367.

[187] Taylor JW, Parikh M, Phillips JJ, James CD, Molinaro AM, Butowski NA, et al. Phase-2 trial of palbociclib in adult patients with recurrent RB1-positive glioblastoma. J Neurooncol 2018;140:477–83. https://doi.org/10.1007/s11060-018-2977-3.

[188] Wick W, Dettmer S, Berberich A, Kessler T, Karapanagiotou-Schenkel I, Wick A, et al. N2M2 (NOA-20) phase I/II trial of molecularly matched targeted therapies plus radiotherapy in patients with newly diagnosed non-MGMT hypermethylated glioblastoma. Neuro Oncol 2019;21:95–105. https://doi.org/10.1093/neuonc/noy161.

[189] Raub TJ, Wishart GN, Kulanthaivel P, Staton BA, Ajamie RT, Sawada GA, et al. Brain exposure of two selective dual CDK4 and CDK6 inhibitors and the antitumor activity of CDK4 and CDK6 inhibition in combination with temozolomide in an intracranial glioblastoma xenograft. Drug Metab Dispos 2015;43:1360–71. https://doi.org/10.1124/dmd.114.062745.

[190] Packer RJ, Gajjar A, Vezina G, Rorke-Adams L, Burger PC, Robertson PL, et al. Phase III study of craniospinal radiation therapy followed by adjuvant chemotherapy for newly diagnosed average-risk medulloblastoma. JCO 2006;24:4202–8. https://doi.org/10.1200/JCO.2006.06.4980.

[191] Cairncross G, Wang M, Shaw E, Jenkins R, Brachman D, Buckner J, et al. Phase III trial of chemoradiotherapy for anaplastic oligodendroglioma: long-term results of RTOG 9402. J Clin Oncol 2013;31:337–43. https://doi.org/10.1200/JCO.2012.43.2674.

[192] van den Bent MJ, Brandes AA, Taphoorn MJB, Kros JM, Kouwenhoven MCM, Delattre J-Y, et al. Adjuvant procarbazine, lomustine, and vincristine chemotherapy in newly diagnosed anaplastic oligodendroglioma: long-

term follow-up of EORTC brain tumor group study 26951. J Clin Oncol 2013;31:344–50. https://doi.org/10.1200/JCO.2012.43.2229.

[193] Nabors LB, Portnow J, Ahluwalia M, Baehring J, Brem H, Brem S, et al. NCCN clinical practice guidelines in oncology. In: Central nervous system cancers; 2020. p. 2020. Version 1.

[194] De Witt M, Gamble A, Hanson D, Markowitz D, Powell C, Al Dimassi S, et al. Repurposing mebendazole as a replacement for vincristine for the treatment of brain tumors. Mol Med 2017;23:50–6. https://doi.org/10.2119/molmed.2017.00011.

[195] Wang F, Zhou F, Kruh GD, Gallo JM. Influence of blood-brain barrier efflux pumps on the distribution of vincristine in brain and brain tumors. Neuro Oncol 2010;12:1043–9. https://doi.org/10.1093/neuonc/noq056.

[196] Bai R-Y, Staedtke V, Aprhys CM, Gallia GL, Riggins GJ. Antiparasitic mebendazole shows survival benefit in 2 preclinical models of glioblastoma multiforme. Neuro Oncol 2011;13:974–82. https://doi.org/10.1093/neuonc/nor077.

[197] Kipper FC, Silva AO, Marc AL, Confortin G, Junqueira AV, Neto EP, et al. Vinblastine and antihelmintic mebendazole potentiate temozolomide in resistant gliomas. Invest New Drugs 2018;36:323–31. https://doi.org/10.1007/s10637-017-0503-7.

[198] Markowitz D, Ha G, Ruggieri R, Symons M. Microtubule-targeting agents can sensitize cancer cells to ionizing radiation by an interphase-based mechanism. OTT 2017;10:5633–42. https://doi.org/10.2147/OTT.S143096.

[199] Bai R-Y, Staedtke V, Wanjiku T, Rudek MA, Joshi A, Gallia GL, et al. Brain penetration and efficacy of different mebendazole polymorphs in a mouse brain tumor model. Clin Cancer Res 2015;21:3462–70. https://doi.org/10.1158/1078-0432.CCR-14-2681.

[200] Patil VM, Bhelekar A, Menon N, Bhattacharjee A, Simha V, Abhinav R, et al. Reverse swing-M, phase 1 study of repurposing mebendazole in recurrent high-grade glioma. Cancer Med 2020;9:4676–85. https://doi.org/10.1002/cam4.3094.

[201] Braithwaite PA, Roberts MS, Allan RJ, Watson TR. Clinical pharmacokinetics of high dose mebendazole in patients treated for cystic hydatid disease. Eur J Clin Pharmacol 1982;22:161–9. https://doi.org/10.1007/BF00542462.

[202] Prota AE, Danel F, Bachmann F, Bargsten K, Buey RM, Pohlmann J, et al. The novel microtubule-destabilizing drug BAL27862 binds to the colchicine site of tubulin with distinct effects on microtubule organization. J Mol Biol 2014;426:1848–60. https://doi.org/10.1016/j.jmb.2014.02.005.

[203] Berges R, Tchoghandjian A, Honore S, Esteve M-A, Figarella-Branger D, Bachmann F, et al. The novel tubulin-binding checkpoint activator BAL101553 inhibits EB1-dependent migration and invasion and promotes differentiation of glioblastoma stem-like cells. Mol Cancer Ther 2016;15:2740–9. https://doi.org/10.1158/1535-7163.MCT-16-0252.

[204] Berges R, Baeza-Kallee N, Tabouret E, Chinot O, Petit M, Kruczynski A, et al. End-binding 1 protein overexpression correlates with glioblastoma progression and sensitizes to Vinca-alkaloids in vitro and in vivo. Oncotarget 2014;5:12769–87. https://doi.org/10.18632/oncotarget.2646.

[205] Schmitt-Hoffmann A, Klauer D, Gebhardt K, Fullhardt P, Brendle A, Hardgreaves P, et al. Abstract C233: BAL27862: a unique microtubule-targeted agent with a potential for the treatment of human brain tumors. Tubulin-interacting agents. American Association for Cancer Research; 2009. p. C233. https://doi.org/10.1158/1535-7163.TARG-09-C233.

[206] Joerger M, Stathis A, Metaxas Y, Hess D, Mantiero M, Mark M, et al. A Phase 1 study of BAL101553, a novel tumor checkpoint controller targeting microtubules, administered as 48-h infusion in adult patients with advanced solid tumors. Invest New Drugs 2020;38:1067–76. https://doi.org/10.1007/s10637-019-00850-z.

[207] Kristeleit R, Evans J, Molife LR, Tunariu N, Shaw H, Slater S, et al. Phase 1/2a trial of intravenous BAL101553, a novel controller of the spindle assembly checkpoint, in advanced solid tumours. Br J Cancer 2020;123:1360–9. https://doi.org/10.1038/s41416-020-1010-8.

[208] Ingles Garces AH, Plummer ER, Lopez JS, Kristeleit RS, MacDonald J, Sweeting L, et al. A phase I study to assess the safety, pharmacokinetics (PK), pharmacodynamics (PD) and antitumor activities of daily oral BAL101553, a novel tumor checkpoint controller (TCC) in adult patients with progressive or recurrent glioblastoma (GBM) or high-grade glioma. JCO 2017;35:TPS2601. https://doi.org/10.1200/JCO.2017.35.15_suppl.TPS2601.

[209] Lopez JS, Kristeleit RS, Rulach R, Haris NM, Scaranti M, Mulholland PJ, et al. Phase 1/2a study of once daily oral BAL101553, a novel tumor checkpoint controller (TCC), in adult patients with progressive or recurrent glioblastoma (GBM) or high-grade glioma. JCO 2019;37:2025. https://doi.org/10.1200/JCO.2019.37.15_suppl.2025.

[210] Zhou H, Kuang J, Zhong L, Kuo W, Gray J, Sahin A, et al. Tumour amplified kinase STK15/BTAK induces centrosome amplification, aneuploidy and transformation. Nat Genet 1998;20:189–93. https://doi.org/10.1038/2496.

[211] Marumoto T, Honda S, Hara T, Nitta M, Hirota T, Kohmura E, et al. Aurora-A kinase maintains the fidelity of early and late mitotic events in HeLa cells. J Biol Chem 2003;278:51786–95. https://doi.org/10.1074/jbc.M306275200.

[212] Lehman NL, O'Donnell JP, Whiteley LJ, Stapp RT, Lehman TD, Roszka KM, et al. Aurora A is differentially expressed in gliomas, is associated with patient survival in glioblastoma and is a potential chemotherapeutic target in gliomas. Cell Cycle 2012;11:489–502. https://doi.org/10.4161/cc.11.3.18996.

[213] Kogiso M, Qi L, Braun FK, Injac SG, Zhang L, Du Y, et al. Concurrent inhibition of neurosphere and monolayer cells of pediatric glioblastoma by Aurora A inhibitor MLN8237 predicted survival extension in PDOX models. Clin Cancer Res 2018;24:2159–70. https://doi.org/10.1158/1078-0432.CCR-17-2256.

[214] Dees EC, Cohen RB, von Mehren M, Stinchcombe TE, Liu H, Venkatakrishnan K, et al. Phase I study of aurora A kinase inhibitor MLN8237 in advanced solid tumors: safety, pharmacokinetics, pharmacodynamics, and bioavailability of two oral formulations. Clin Cancer Res 2012;18:4775–84. https://doi.org/10.1158/1078-0432.CCR-12-0589.

[215] Song A, Andrews DW, Werner-Wasik M, Kim L, Glass J, Bar-Ad V, et al. Phase I trial of alisertib with concurrent fractionated stereotactic re-irradiation for recurrent high grade gliomas. Radiother Oncol 2019;132:135–41. https://doi.org/10.1016/j.radonc.2018.12.019.

[216] Xu Y, Hao S-Y, Zhang X-J, Li W-B, Qiao X-P, Wang Z-X, et al. Discovery of novel 2,4-disubstituted pyrimidines as Aurora kinase inhibitors. Bioorg Med Chem Lett 2020;30:126885. https://doi.org/10.1016/j.bmcl.2019.126885.

[217] Marchetti S, Pluim D, van Eijndhoven M, van Tellingen O, Mazzanti R, Beijnen JH, et al. Effect of the drug transporters ABCG2, Abcg2, ABCB1 and ABCC2 on the disposition, brain accumulation and myelotoxicity of the aurora kinase B inhibitor barasertib and its more active form barasertib-hydroxy-QPA. Invest New Drugs 2013;31:1125–35. https://doi.org/10.1007/s10637-013-9923-1.

[218] Francoz S, Froment P, Bogaerts S, De Clercq S, Maetens M, Doumont G, et al. Mdm4 and Mdm2 cooperate to inhibit p53 activity in proliferating and quiescent cells in vivo. Proc Natl Acad Sci U S A 2006;103:3232–7. https://doi.org/10.1073/pnas.0508476103.

[219] Sun D, Li Z, Rew Y, Gribble M, Bartberger MD, Beck HP, et al. Discovery of AMG 232, a potent, selective, and orally bioavailable MDM2–p53 inhibitor in clinical development. J Med Chem 2014;57:1454–72. https://doi.org/10.1021/jm401753e.

[220] Her N-G, Oh J-W, Oh YJ, Han S, Cho HJ, Lee Y, et al. Potent effect of the MDM2 inhibitor AMG232 on suppression of glioblastoma stem cells. Cell Death Dis 2018;9:792. https://doi.org/10.1038/s41419-018-0825-1.

[221] Erba HP, Becker PS, Shami PJ, Grunwald MR, Flesher DL, Zhu M, et al. Phase 1b study of the MDM2 inhibitor AMG 232 with or without trametinib in relapsed/refractory acute myeloid leukemia. Blood Adv 2019;3:1939–49. https://doi.org/10.1182/bloodadvances.2019030916.

[222] Gluck WL, Gounder MM, Frank R, Eskens F, Blay JY, Cassier PA, et al. Phase 1 study of the MDM2 inhibitor AMG 232 in patients with advanced P53 wild-type solid tumors or multiple myeloma. Invest New Drugs 2020;38:831–43. https://doi.org/10.1007/s10637-019-00840-1.

[223] Gupta SK, Mladek AC, Telele S, Mohammad AS, He L, Hu Z, et al. Abstract 6276: brain penetrant MDM2 inhibitor RG7388 extends survival benefit of radiation treatment in select glioblastoma patient-derived xenograft models. In: Experimental and molecular therapeutics. American Association for Cancer Research; 2020. p. 6276. https://doi.org/10.1158/1538-7445.AM2020-6276.

[224] Mai WX, Gosa L, Daniels VW, Ta L, Tsang JE, Higgins B, et al. Cytoplasmic p53 couples oncogene-driven glucose metabolism to apoptosis and is a therapeutic target in glioblastoma. Nat Med 2017;23:1342–51. https://doi.org/10.1038/nm.4418.

[225] Berberich A, Kessler T, Thomé CM, Pusch S, Hielscher T, Sahm F, Oezen I, Schmitt LM, Ciprut S, Hucke N, Ruebmann P, Fischer M, Lemke D, Breckwoldt MO, von Deimling A, Bendszus M, Platten M, Wick W. Targeting resistance against the MDM2 inhibitor RG7388 in glioblastoma cells by the MEK inhibitor trametinib. Clin Cancer Res 2019;25(1):253–65. https://doi.org/10.1158/1078-0432.CCR-18-1580.

[226] Kawabe T. G2 checkpoint abrogators as anticancer drugs. Mol Cancer Ther 2004;3:513–9.

[227] Russell P, Nurse P. Negative regulation of mitosis by wee1+, a gene encoding a protein kinase homolog. Cell 1987;49:559–67. https://doi.org/10.1016/0092-8674(87)90458-2.

[228] Mir SE, De Witt Hamer PC, Krawczyk PM, Balaj L, Claes A, Niers JM, et al. In silico analysis of kinase expression identifies WEE1 as a gatekeeper against mitotic catastrophe in glioblastoma. Cancer Cell 2010;18:244–57. https://doi.org/10.1016/j.ccr.2010.08.011.
[229] Santamaría D, Barrière C, Cerqueira A, Hunt S, Tardy C, Newton K, et al. Cdk1 is sufficient to drive the mammalian cell cycle. Nature 2007;448:811–5. https://doi.org/10.1038/nature06046.
[230] Sarcar B, Kahali S, Prabhu AH, Shumway SD, Xu Y, Demuth T, et al. Targeting radiation-induced G2 checkpoint activation with the Wee-1 inhibitor MK-1775 in glioblastoma cell lines. Mol Cancer Ther 2011;10:2405–14. https://doi.org/10.1158/1535-7163.MCT-11-0469.
[231] Pokorny JL, Calligaris D, Gupta SK, Iyekegbe DO, Mueller D, Bakken KK, et al. The efficacy of the Wee1 inhibitor MK-1775 combined with temozolomide is limited by heterogeneous distribution across the blood-brain barrier in glioblastoma. Clin Cancer Res 2015;21:1916–24. https://doi.org/10.1158/1078-0432.CCR-14-2588.
[232] Li J, Wu J, Bao X, Honea N, Xie Y, Kim S, et al. Quantitative and mechanistic understanding of AZD1775 penetration across human blood–brain barrier in glioblastoma patients using an IVIVE–PBPK modeling approach. Clin Cancer Res 2017;23:7454–66. https://doi.org/10.1158/1078-0432.CCR-17-0983.
[233] Sanai N, Li J, Boerner J, Stark K, Wu J, Kim S, et al. Phase 0 Trial of AZD1775 in first-recurrence glioblastoma patients. Clin Cancer Res 2018;24:3820–8. https://doi.org/10.1158/1078-0432.CCR-17-3348.
[234] Romo CG, Alexander BM, Agar N, Ahluwalia MS, Desai AS, Dietrich J, et al. Intratumoral drug distribution of adavosertib in patients with glioblastoma: interim results of phase I study. JCO 2020;38:2568. https://doi.org/10.1200/JCO.2020.38.15_suppl.2568.
[235] Lee TH, Turck C, Kirschner MW. Inhibition of cdc2 activation by INH/PP2A. MBoC 1994;5:323–38. https://doi.org/10.1091/mbc.5.3.323.
[236] Perrotti D, Neviani P. Protein phosphatase 2A (PP2A), a drugable tumor suppressor in Ph1(+) leukemias. Cancer Metastasis Rev 2008;27:159–68. https://doi.org/10.1007/s10555-008-9119-x.
[237] Hong CS, Ho W, Zhang C, Yang C, Elder JB, Zhuang Z. LB100, a small molecule inhibitor of PP2A with potent chemo- and radio-sensitizing potential. Cancer Biol Ther 2015;16:821–33. https://doi.org/10.1080/15384047.2015.1040961.
[238] Lu J, Kovach JS, Johnson F, Chiang J, Hodes R, Lonser R, et al. Inhibition of serine/threonine phosphatase PP2A enhances cancer chemotherapy by blocking DNA damage induced defense mechanisms. PNAS 2009;106:11697–702. https://doi.org/10.1073/pnas.0905930106.
[239] Ho WS, Wang H, Maggio D, Kovach JS, Zhang Q, Song Q, et al. Pharmacologic inhibition of protein phosphatase-2A achieves durable immune-mediated antitumor activity when combined with PD-1 blockade. Nat Commun 2018;9:2126. https://doi.org/10.1038/s41467-018-04425-z.
[240] Cui J, Wang H, Medina R, Zhang Q, Xu C, Indig IH, et al. Inhibition of PP2A with LB-100 enhances efficacy of CAR-T cell therapy against glioblastoma. Cancers 2020;12:139. https://doi.org/10.3390/cancers12010139.
[241] Chung V, Mansfield AS, Braiteh F, Richards D, Durivage H, Ungerleider RS, et al. Safety, tolerability, and preliminary activity of LB-100, an inhibitor of protein phosphatase 2A, in patients with relapsed solid tumors: an open-label, dose escalation, first-in-human, phase I trial. Clin Cancer Res 2017;23:3277–84.
[242] Maggio D, Ho WS, Breese R, Walbridge S, Wang H, Cui J, et al. Inhibition of protein phosphatase-2A with LB-100 enhances antitumor immunity against glioblastoma. J Neurooncol 2020;148:231–44. https://doi.org/10.1007/s11060-020-03517-5.
[243] Marechal A, Zou L. DNA damage sensing by the ATM and ATR kinases. Cold Spring Harb Perspect Biol 2013;5:a012716. https://doi.org/10.1101/cshperspect.a012716.
[244] Karnitz LM, Zou L. Molecular pathways: targeting ATR in cancer therapy. Clin Cancer Res 2015;21:4780–5. https://doi.org/10.1158/1078-0432.CCR-15-0479.
[245] Fròsina G, Profumo A, Marubbi D, Marcello D, Ravetti JL, Daga A. ATR kinase inhibitors NVP-BEZ235 and AZD6738 effectively penetrate the brain after systemic administration. Radiat Oncol 2018;13:76. https://doi.org/10.1186/s13014-018-1020-3.
[246] Vendetti FP, Lau A, Schamus S, Conrads TP, O'Connor MJ, Bakkenist CJ. The orally active and bioavailable ATR kinase inhibitor AZD6738 potentiates the anti-tumor effects of cisplatin to resolve ATM-deficient non-small cell lung cancer in vivo. Oncotarget 2015;6:44289–305. https://doi.org/10.18632/oncotarget.6247.
[247] Lau AY, Yates J, Wilson Z, Young LA, Hughes AM, Berges A, et al. Abstract 2494: ATR inhibitor AZD6738 as monotherapy and in combination with olaparib or chemotherapy: defining pre-clinical dose-schedules and efficacy modelling. In: Molecular and cellular biology, genetics. American Association for Cancer Research; 2017. p. 2494. https://doi.org/10.1158/1538-7445.AM2017-2494.

[248] Adams JM, Cory S. The Bcl-2 apoptotic switch in cancer development and therapy. Oncogene 2007;26:1324–37. https://doi.org/10.1038/sj.onc.1210220.
[249] Tirapelli LF, Bolini PHNA, Tirapelli DP da C, Peria FM, Becker ANP, Saggioro FP, et al. Caspase-3 and Bcl-2 expression in glioblastoma: an immunohistochemical study. Arq Neuropsiquiatr 2010;68:603–7. https://doi.org/10.1590/S0004-282X2010000400023.
[250] Walker KL, Rinella SP, Hess NJ, Turicek DP, Kabakov SA, Zhu F, Bouchlaka MN, Olson SL, Cho MM, Quamine AE, Feils AS, Gavcovich TB, Rui L, Capitini CM. CXCR4 allows T cell acute lymphoblastic leukemia to escape from JAK1/2 and BCL2 inhibition through CNS infiltration. Leuk Lymphoma 2021. https://doi.org/10.1080/10428194.2021.1910684.
[251] West DC, Qin Y, Peterson QP, Thomas DL, Palchaudhuri R, Morrison KC, et al. Differential effects of procaspase-3 activating compounds in the induction of cancer cell death. Mol Pharmaceutics 2012;9:1425–34. https://doi.org/10.1021/mp200673n.
[252] Ray SK, Patel SJ, Welsh CT, Wilford GG, Hogan EL, Banik NL. Molecular evidence of apoptotic death in malignant brain tumors including glioblastoma multiforme: upregulation of calpain and caspase-3. J Neurosci Res 2002;69:197–206. https://doi.org/10.1002/jnr.10265.
[253] Murphy ÁC, Weyhenmeyer B, Schmid J, Kilbride SM, Rehm M, Huber HJ, et al. Activation of executioner caspases is a predictor of progression-free survival in glioblastoma patients: a systems medicine approach. Cell Death Dis 2013;4:e629. https://doi.org/10.1038/cddis.2013.157.
[254] Joshi AD, Botham RC, Schlein LJ, Roth HS, Mangraviti A, Borodovsky A, et al. Synergistic and targeted therapy with a procaspase-3 activator and temozolomide extends survival in glioma rodent models and is feasible for the treatment of canine malignant glioma patients. Oncotarget 2017;8:80124–38. https://doi.org/10.18632/oncotarget.19085.
[255] Peterson QP, Hsu DC, Goode DR, Novotny CJ, Totten RK, Hergenrother PJ. Procaspase-3 activation as an anti-cancer strategy: structure – activity relationship of procaspase-activating compound 1 (PAC-1) and its cellular co-localization with caspase-3. J Med Chem 2009;52:5721–31. https://doi.org/10.1021/jm900722z.
[256] Peterson QP, Hsu DC, Novotny CJ, West DC, Kim D, Schmit JM, et al. Discovery and canine preclinical assessment of a nontoxic procaspase-3-activating compound. Cancer Res 2010;70:7232–41. https://doi.org/10.1158/0008-5472.CAN-10-0766.
[257] Schlein LJ, Fadl-Alla B, Pondenis HC, Lezmi S, Eberhart CG, LeBlanc AK, et al. Immunohistochemical Characterization of Procaspase-3 Overexpression as a Druggable Target With PAC-1, a Procaspase-3 Activator, in Canine and Human Brain Cancers. Front Oncol 2019;9:96. https://doi.org/10.3389/fonc.2019.00096.
[258] Danciu OC, Nicholas MK, Holdhoff M, Peterson RA, Venepalli N, Chowdhery R, et al. Abstract CT027: phase I study of procaspase activating compound-1 (PAC-1) in the treatment of advanced malignancies and in combination with temozolomide in refractory high-grade astrocytomas. Clinical Trials: American Association for Cancer Research; 2019. p. CT027. https://doi.org/10.1158/1538-7445.AM2019-CT027.
[259] Wang M, Medeiros BC, Erba HP, DeAngelo DJ, Giles FJ, Swords RT. Targeting protein neddylation: a novel therapeutic strategy for the treatment of cancer. Expert Opin Ther Targets 2011;15:253–64. https://doi.org/10.1517/14728222.2011.550877.
[260] Thibaudeau TA, Smith DM. A Practical Review of Proteasome Pharmacology. Pharmacol Rev 2019;71:170–97. https://doi.org/10.1124/pr.117.015370.
[261] Roth P, Mason WP, Richardson PG, Weller M. Proteasome inhibition for the treatment of glioblastoma. Expert Opin Investig Drugs 2020;29:1133–41. https://doi.org/10.1080/13543784.2020.1803827.
[262] Britton M, Lucas MM, Downey SL, Screen M, Pletnev AA, Verdoes M, et al. Selective inhibitor of proteasome's caspase-like sites sensitizes cells to specific inhibition of chymotrypsin-like sites. Chem Biol 2009;16:1278–89. https://doi.org/10.1016/j.chembiol.2009.11.015.
[263] Richardson PG, Sonneveld P, Schuster MW, Irwin D, Stadtmauer EA, Facon T, et al. Bortezomib or high-dose dexamethasone for relapsed multiple myeloma. N Engl J Med 2005;352:2487–98. https://doi.org/10.1056/NEJMoa043445.
[264] San Miguel JF, Schlag R, Khuageva NK, Dimopoulos MA, Shpilberg O, Kropff M, et al. Bortezomib plus melphalan and prednisone for initial treatment of multiple myeloma. N Engl J Med 2008;359:906–17. https://doi.org/10.1056/NEJMoa0801479.
[265] Kane RC, Dagher R, Farrell A, Ko C-W, Sridhara R, Justice R, et al. Bortezomib for the treatment of mantle cell lymphoma. Clin Cancer Res 2007;13:5291–4. https://doi.org/10.1158/1078-0432.CCR-07-0871.
[266] Hemeryck A, Geerts R, Monbaliu J, Hassler S, Verhaeghe T, Diels L, et al. Tissue distribution and depletion kinetics of bortezomib and bortezomib-related radioactivity in male rats after single and repeated intravenous

injection of 14C-bortezomib. Cancer Chemother Pharmacol 2007;60:777–87. https://doi.org/10.1007/s00280-007-0424-9.

[267] Raizer JJ, Chandler JP, Ferrarese R, Grimm SA, Levy RM, Muro K, et al. A phase II trial evaluating the effects and intra-tumoral penetration of bortezomib in patients with recurrent malignant gliomas. J Neurooncol 2016;129:139–46. https://doi.org/10.1007/s11060-016-2156-3.

[268] Odia Y, Kreisl TN, Aregawi D, Innis EK, Fine HA. A phase II trial of tamoxifen and bortezomib in patients with recurrent malignant gliomas. J Neurooncol 2015;125:191–5. https://doi.org/10.1007/s11060-015-1894-y.

[269] Morin A, Soane C, Pierce A, Sanford B, Jones KL, Crespo M, et al. Proteasome inhibition as a therapeutic approach in atypical teratoid/rhabdoid tumors. Neuro Oncol Adv 2020;2:vdaa051. https://doi.org/10.1093/noajnl/vdaa051.

[270] Quillin J, Patel R, Herzberg E, Alton D, Bikzhanova G, Geisler L, et al. A phase 0 analysis of ixazomib in patients with glioblastoma. Mol Clin Oncol 2020;13:1. https://doi.org/10.3892/mco.2020.2114.

[271] Potts BC, Albitar MX, Anderson KC, Baritaki S, Berkers C, Bonavida B, et al. Marizomib, a proteasome inhibitor for all seasons: preclinical profile and a framework for clinical trials. Curr Cancer Drug Targets 2011;11:254–84. https://doi.org/10.2174/156800911794519716.

[272] Di K, Lloyd GK, Abraham V, MacLaren A, Burrows FJ, Desjardins A, et al. Marizomib activity as a single agent in malignant gliomas: ability to cross the blood-brain barrier. Neuro Oncol 2016;18:840–8. https://doi.org/10.1093/neuonc/nov299.

[273] Manton CA, Johnson B, Singh M, Bailey CP, Bouchier-Hayes L, Chandra J. Induction of cell death by the novel proteasome inhibitor marizomib in glioblastoma in vitro and in vivo. Sci Rep 2016;6:18953. https://doi.org/10.1038/srep18953.

[274] Harrison SJ, Mainwaring P, Price T, Millward MJ, Padrik P, Underhill CR, et al. Phase I clinical trial of marizomib (NPI-0052) in patients with advanced malignancies including multiple myeloma: study NPI-0052-102 final results. Clin Cancer Res 2016;22:4559–66. https://doi.org/10.1158/1078-0432.CCR-15-2616.

[275] Roth P, Reijneveld JC, Gorlia T, Dhermain F, De Vos FYFL, Vanlancker M, et al. EORTC 1709/CCTG CE.8: a phase III trial of marizomib in combination with standard temozolomide-based radiochemotherapy versus standard temozolomide-based radiochemotherapy alone in patients with newly diagnosed glioblastoma. JCO 2019;37:TPS2072. https://doi.org/10.1200/JCO.2019.37.15_suppl.TPS2072.

[276] Bota D, Desjardins A, Mason W, Kesari S, Magge R, Winograd B, et al. ACTR-71. full enrollment results from the phase 1/2, multicenter, open-label study of marizomib (MRZ) ± bevacizumab (BEV) in recurrent who grade IV malignant glioma (glioblastoma, RGBM). Neuro Oncol 2017;19:vi16. https://doi.org/10.1093/neuonc/nox168.058.

[277] Lin GL, Wilson KM, Ceribelli M, Stanton BZ, Woo PJ, Kreimer S, et al. Therapeutic strategies for diffuse midline glioma from high-throughput combination drug screening. Sci Transl Med 2019;11:eaaw0064. https://doi.org/10.1126/scitranslmed.aaw0064.

[278] Xirodimas DP, Saville MK, Bourdon J-C, Hay RT, Lane DP. Mdm2-mediated NEDD8 conjugation of p53 inhibits its transcriptional activity. Cell 2004;118:83–97. https://doi.org/10.1016/j.cell.2004.06.016.

[279] Hua W, Li C, Yang Z, Li L, Jiang Y, Yu G, et al. Suppression of glioblastoma by targeting the overactivated protein neddylation pathway. Neuro Oncol 2015;17:1333–43. https://doi.org/10.1093/neuonc/nov066.

[280] Smith MA, Maris JM, Gorlick R, Kolb EA, Lock R, Carol H, et al. Initial testing of the investigational NEDD8-activating enzyme inhibitor MLN4924 by the pediatric preclinical testing program: MLN4924 Pediatric Preclinical Testing. Pediatr Blood Cancer 2012;59:246–53. https://doi.org/10.1002/pbc.23357.

[281] Han S, Shin H, Oh J-W, Oh YJ, Her N-G, Nam D-H. The protein neddylation inhibitor MLN4924 suppresses patient-derived glioblastoma cells via inhibition of ERK and AKT signaling. Cancers 2019;11:1849. https://doi.org/10.3390/cancers11121849.

[282] Ferdosi SR, Taylor B, Lee M, Tang N, Peng S, Bybee R, et al. Dynamic multi-OMICs of glioblastoma reveal sensitivity to neddylation inhibition dependent on nuclear PTEN and DNA replication pathways: nuclear PTEN mediates MLN4924 sensitivity in GBM. bioRxiv 2020.07.21.212571. https://doi.org/10.1101/2020.07.21.212571.

[283] Swords RT, Erba HP, DeAngelo DJ, Bixby DL, Altman JK, Maris M, et al. Pevonedistat (MLN4924), a first-in-class NEDD8-activating enzyme inhibitor, in patients with acute myeloid leukaemia and myelodysplastic syndromes: a phase 1 study. Br J Haematol 2015;169:534–43. https://doi.org/10.1111/bjh.13323.

[284] Cully M, You H, Levine AJ, Mak TW. Beyond PTEN mutations: the PI3K pathway as an integrator of multiple inputs during tumorigenesis. Nat Rev Cancer 2006;6:184–92. https://doi.org/10.1038/nrc1819.

[285] Avraham R, Yarden Y. Feedback regulation of EGFR signalling: decision making by early and delayed loops. Nat Rev Mol Cell Biol 2011;12:104–17. https://doi.org/10.1038/nrm3048.

[286] Latora V, Marchiori M. Efficient behavior of small-world networks. Phys Rev Lett 2001;87:198701. https://doi.org/10.1103/PhysRevLett.87.198701.

[287] Verhaak RGW, Hoadley KA, Purdom E, Wang V, Qi Y, Wilkerson MD, et al. Integrated genomic analysis identifies clinically relevant subtypes of glioblastoma characterized by abnormalities in PDGFRA, IDH1, EGFR, and NF1. Cancer Cell 2010;17:98–110. https://doi.org/10.1016/j.ccr.2009.12.020.

[288] Westphal M, Maire CL, Lamszus K. EGFR as a target for glioblastoma treatment: an unfulfilled promise. CNS Drugs 2017;31:723–35. https://doi.org/10.1007/s40263-017-0456-6.

[289] Neyns B, Sadones J, Joosens E, Bouttens F, Verbeke L, Baurain J-F, et al. Stratified phase II trial of cetuximab in patients with recurrent high-grade glioma. Ann Oncol 2009;20:1596–603. https://doi.org/10.1093/annonc/mdp032.

[290] Hasselbalch B, Lassen U, Hansen S, Holmberg M, Sørensen M, Kosteljanetz M, et al. Cetuximab, bevacizumab, and irinotecan for patients with primary glioblastoma and progression after radiation therapy and temozolomide: a phase II trial. Neuro Oncol 2010;12:508–16. https://doi.org/10.1093/neuonc/nop063.

[291] Westphal M, Heese O, Steinbach JP, Schnell O, Schackert G, Mehdorn M, et al. A randomised, open label phase III trial with nimotuzumab, an anti-epidermal growth factor receptor monoclonal antibody in the treatment of newly diagnosed adult glioblastoma. Eur J Cancer 2015;51:522–32. https://doi.org/10.1016/j.ejca.2014.12.019.

[292] de Groot JF, Gilbert MR, Aldape K, Hess KR, Hanna TA, Ictech S, et al. Phase II study of carboplatin and erlotinib (Tarceva, OSI-774) in patients with recurrent glioblastoma. J Neuro Oncol 2008;90:89–97. https://doi.org/10.1007/s11060-008-9637-y.

[293] van den Bent MJ, Brandes AA, Rampling R, Kouwenhoven MCM, Kros JM, Carpentier AF, et al. Randomized phase II trial of erlotinib versus temozolomide or carmustine in recurrent glioblastoma: EORTC brain tumor group study 26034. J Clin Oncol 2009;27:1268–74. https://doi.org/10.1200/JCO.2008.17.5984.

[294] Peereboom DM, Shepard DR, Ahluwalia MS, Brewer CJ, Agarwal N, Stevens GHJ, et al. Phase II trial of erlotinib with temozolomide and radiation in patients with newly diagnosed glioblastoma multiforme. J Neuro Oncol 2010;98:93–9. https://doi.org/10.1007/s11060-009-0067-2.

[295] Kesavabhotla K, Schlaff CD, Shin B, Mubita L, Kaplan R, Tsiouris AJ, et al. Phase I/II study of oral erlotinib for treatment of relapsed/refractory glioblastoma multiforme and anaplastic astrocytoma. J Exp Ther Oncol 2012;10:71–81.

[296] Clarke JL, Molinaro AM, Phillips JJ, Butowski NA, Chang SM, Perry A, et al. A single-institution phase II trial of radiation, temozolomide, erlotinib, and bevacizumab for initial treatment of glioblastoma. Neuro Oncol 2014;16:984–90. https://doi.org/10.1093/neuonc/nou029.

[297] Raizer JJ, Giglio P, Hu J, Groves M, Merrell R, Conrad C, et al. A phase II study of bevacizumab and erlotinib after radiation and temozolomide in MGMT unmethylated GBM patients. J Neuro Oncol 2016;126:185–92. https://doi.org/10.1007/s11060-015-1958-z.

[298] Rich JN, Reardon DA, Peery T, Dowell JM, Quinn JA, Penne KL, et al. Phase II trial of gefitinib in recurrent glioblastoma. J Clin Oncol 2004;22:133–42. https://doi.org/10.1200/JCO.2004.08.110.

[299] Franceschi E, Cavallo G, Lonardi S, Magrini E, Tosoni A, Grosso D, et al. Gefitinib in patients with progressive high-grade gliomas: a multicentre phase II study by Gruppo Italiano Cooperativo di Neuro-Oncologia (GICNO). Br J Cancer 2007;96:1047–51. https://doi.org/10.1038/sj.bjc.6603669.

[300] Uhm JH, Ballman KV, Wu W, Giannini C, Krauss JC, Buckner JC, et al. Phase II evaluation of gefitinib in patients with newly diagnosed Grade 4 astrocytoma: Mayo/North Central Cancer Treatment Group Study N0074. Int J Radiat Oncol Biol Phys 2011;80:347–53. https://doi.org/10.1016/j.ijrobp.2010.01.070.

[301] Chakravarti A, Wang M, Robins HI, Lautenschlaeger T, Curran WJ, Brachman DG, et al. RTOG 0211: a phase 1/2 study of radiation therapy with concurrent gefitinib for newly diagnosed glioblastoma patients. Int J Radiat Oncol Biol Phys 2013;85:1206–11. https://doi.org/10.1016/j.ijrobp.2012.10.008.

[302] Thiessen B, Stewart C, Tsao M, Kamel-Reid S, Schaiquevich P, Mason W, et al. A phase I/II trial of GW572016 (lapatinib) in recurrent glioblastoma multiforme: clinical outcomes, pharmacokinetics and molecular correlation. Cancer Chemother Pharmacol 2010;65:353–61. https://doi.org/10.1007/s00280-009-1041-6.

[303] Reardon DA, Groves MD, Wen PY, Nabors L, Mikkelsen T, Rosenfeld S, et al. A phase I/II trial of pazopanib in combination with lapatinib in adult patients with relapsed malignant glioma. Clin Cancer Res 2013;19:900–8. https://doi.org/10.1158/1078-0432.CCR-12-1707.

[304] Kim M, Laramy JK, Mohammad AS, Talele S, Fisher J, Sarkaria JN, et al. Brain distribution of a panel of epidermal growth factor receptor inhibitors using cassette dosing in wild-type and *Abcb1/Abcg2*-deficient mice. Drug Metab Dispos 2019;47:393–404. https://doi.org/10.1124/dmd.118.084210.

[305] Chagoya G, Kwatra SG, Nanni CW, Roberts CM, Phillips SM, Nullmeyergh S, et al. Efficacy of osimertinib against EGFRvIII+ glioblastoma. Oncotarget 2020;11:2074–82. https://doi.org/10.18632/oncotarget.27599.
[306] Frattini V, Trifonov V, Chan JM, Castano A, Lia M, Abate F, et al. The integrated landscape of driver genomic alterations in glioblastoma. Nat Genet 2013;45:1141–9. https://doi.org/10.1038/ng.2734.
[307] Vivanco I, Robins HI, Rohle D, Campos C, Grommes C, Nghiemphu PL, et al. Differential sensitivity of glioma- versus lung cancer-specific EGFR mutations to EGFR kinase inhibitors. Cancer Discov 2012;2:458–71. https://doi.org/10.1158/2159-8290.CD-11-0284.
[308] Liao H-W, Hsu J-M, Xia W, Wang H-L, Wang Y-N, Chang W-C, et al. PRMT1-mediated methylation of the EGF receptor regulates signaling and cetuximab response. J Clin Investig 2015;125:4529–43. https://doi.org/10.1172/JCI82826.
[309] Nathanson DA, Gini B, Mottahedeh J, Visnyei K, Koga T, Gomez G, et al. Targeted therapy resistance mediated by dynamic regulation of extrachromosomal mutant EGFR DNA. Science 2014;343:72–6. https://doi.org/10.1126/science.1241328.
[310] Wu Y-L, Ahn M-J, Garassino MC, Han J-Y, Katakami N, Kim HR, et al. CNS efficacy of osimertinib in patients with T790M-positive advanced non-small-cell lung cancer: data from a randomized phase III trial (AURA3). JCO 2018;36:2702–9. https://doi.org/10.1200/JCO.2018.77.9363.
[311] Ramalingam SS, Vansteenkiste J, Planchard D, Cho BC, Gray JE, Ohe Y, et al. Overall survival with osimertinib in untreated, *EGFR*-mutated advanced NSCLC. N Engl J Med 2020;382:41–50. https://doi.org/10.1056/NEJMoa1913662.
[312] Greig SL. Osimertinib: first global approval. Drugs 2016;76:263–73. https://doi.org/10.1007/s40265-015-0533-4.
[313] Liu X, Chen X, Shi L, Shan Q, Cao Q, Yue C, et al. The third-generation EGFR inhibitor AZD9291 overcomes primary resistance by continuously blocking ERK signaling in glioblastoma. J Exp Clin Cancer Res 2019;38:219. https://doi.org/10.1186/s13046-019-1235-7.
[314] Ballard P, Yates JWT, Yang Z, Kim D-W, Yang JC-H, Cantarini M, et al. Preclinical comparison of osimertinib with other EGFR-TKIs in EGFR-mutant NSCLC brain metastases models, and early evidence of clinical brain metastases activity. Clin Cancer Res 2016;22:5130–40. https://doi.org/10.1158/1078-0432.CCR-16-0399.
[315] Varrone A, Varnäs K, Jucaite A, Cselényi Z, Johnström P, Schou M, et al. A PET study in healthy subjects of brain exposure of ^{11}C-labelled osimertinib—a drug intended for treatment of brain metastases in non-small cell lung cancer. J Cereb Blood Flow Metab 2020;40:799–807. https://doi.org/10.1177/0271678X19843776.
[316] Makhlin I, Salinas RD, Zhang D, Jacob F, Ming G, Song H, et al. Clinical activity of the *EGFR* tyrosine kinase inhibitor osimertinib in *EGFR*-mutant glioblastoma. CNS Oncol 2019;8:CNS43. https://doi.org/10.2217/cns-2019-0014.
[317] Janku F, Yap TA, Meric-Bernstam F. Targeting the PI3K pathway in cancer: are we making headway? Nat Rev Clin Oncol 2018;15:273–91. https://doi.org/10.1038/nrclinonc.2018.28.
[318] Maynard J, Ricketts S-A, Gendrin C, Dudley P, Davies BR. 2-Deoxy-2-[18F]fluoro-d-glucose positron emission tomography demonstrates target inhibition with the potential to predict anti-tumour activity following treatment with the AKT inhibitor AZD5363. Mol Imaging Biol 2013;15:476–85. https://doi.org/10.1007/s11307-013-0613-3.
[319] Keen HG, Ricketts S-A, Maynard J, Logie A, Odedra R, Shannon AM, et al. Examining changes in [18 F]FDG and [18 F]FLT uptake in U87-MG glioma xenografts as early response biomarkers to treatment with the dual mTOR1/2 inhibitor AZD8055. Mol Imaging Biol 2014;16:421–30. https://doi.org/10.1007/s11307-013-0705-0.
[320] Maynard J, Emmas S-A, Blé F-X, Barjat H, Lawrie E, Hancox U, et al. The use of 18F-fluorodeoxyglucose positron emission tomography (18F-FDG PET) as a pathway-specific biomarker with AZD8186, a PI3Kβ/δ inhibitor. EJNMMI Res 2016;6:62. https://doi.org/10.1186/s13550-016-0220-9.
[321] Burger MT, Pecchi S, Wagman A, Ni Z-J, Knapp M, Hendrickson T, et al. Identification of NVP-BKM120 as a potent, selective, orally bioavailable class I PI3 kinase inhibitor for treating cancer. ACS Med Chem Lett 2011;2:774–9. https://doi.org/10.1021/ml200156t.
[322] Raynaud FI, Eccles SA, Patel S, Alix S, Box G, Chuckowree I, et al. Biological properties of potent inhibitors of class I phosphatidylinositide 3-kinases: from PI-103 through PI-540, PI-620 to the oral agent GDC-0941. Mol Cancer Ther 2009;8:1725–38. https://doi.org/10.1158/1535-7163.MCT-08-1200.
[323] Salphati L, Pang J, Plise EG, Chou B, Halladay JS, Olivero AG, et al. Preclinical pharmacokinetics of the novel PI3K inhibitor GDC-0941 and prediction of its pharmacokinetics and efficacy in human. Xenobiotica 2011;41:1088–99. https://doi.org/10.3109/00498254.2011.603386.
[324] Salphati L, Shahidi-Latham S, Quiason C, Barck K, Nishimura M, Alicke B, et al. Distribution of the phosphatidylinositol 3-kinase inhibitors pictilisib (GDC-0941) and GNE-317 in U87 and GS2 intracranial glioblastoma

models—assessment by matrix-assisted laser desorption ionization imaging. Drug Metab Dispos 2014;42:1110–6. https://doi.org/10.1124/dmd.114.057513.

[325] Sarker D, Ang JE, Baird R, Kristeleit R, Shah K, Moreno V, et al. First-in-human phase I study of pictilisib (GDC-0941), a potent pan–class I phosphatidylinositol-3-kinase (PI3K) inhibitor, in patients with advanced solid tumors. Clin Cancer Res 2015;21:77–86. https://doi.org/10.1158/1078-0432.CCR-14-0947.

[326] Pitz MW, Eisenhauer EA, MacNeil MV, Thiessen B, Easaw JC, Macdonald DR, et al. Phase II study of PX-866 in recurrent glioblastoma. Neuro Oncol 2015;17:1270–4. https://doi.org/10.1093/neuonc/nou365.

[327] Koul D, Fu J, Shen R, LaFortune TA, Wang S, Tiao N, et al. Antitumor activity of NVP-BKM120—a selective pan class I PI3 kinase inhibitor showed differential forms of cell death based on p53 status of glioma cells. Clin Cancer Res 2012;18:184–95. https://doi.org/10.1158/1078-0432.CCR-11-1558.

[328] de Gooijer MC, Zhang P, Buil LCM, Çitirikkaya CH, Thota N, Beijnen JH, et al. Buparlisib is a brain penetrable pan-PI3K inhibitor. Sci Rep 2018;8:10784. https://doi.org/10.1038/s41598-018-29062-w.

[329] Speranza M-C, Nowicki MO, Behera P, Cho C-F, Chiocca EA, Lawler SE. BKM-120 (Buparlisib): a phosphatidylinositol-3 kinase inhibitor with anti-invasive properties in glioblastoma. Sci Rep 2016;6:20189. https://doi.org/10.1038/srep20189.

[330] Netland IA, Førde HE, Sleire L, Leiss L, Rahman MA, Skeie BS, et al. Treatment with the PI3K inhibitor buparlisib (NVP-BKM120) suppresses the growth of established patient-derived GBM xenografts and prolongs survival in nude rats. J Neurooncol 2016;129:57–66. https://doi.org/10.1007/s11060-016-2158-1.

[331] Wen PY, Touat M, Alexander BM, Mellinghoff IK, Ramkissoon S, McCluskey CS, et al. Buparlisib in patients with recurrent glioblastoma harboring phosphatidylinositol 3-kinase pathway activation: an open-label, multicenter, multi-arm, phase II trial. JCO 2019;37:741–50. https://doi.org/10.1200/JCO.18.01207.

[332] Wen PY, Rodon JA, Mason W, Beck JT, DeGroot J, Donnet V, et al. Phase I, open-label, multicentre study of buparlisib in combination with temozolomide or with concomitant radiation therapy and temozolomide in patients with newly diagnosed glioblastoma. ESMO Open 2020;5, e000673. https://doi.org/10.1136/esmoopen-2020-000673.

[333] Rosenthal M, Clement PM, Campone M, Gil-Gil MJ, DeGroot J, Chinot O, et al. Buparlisib plus carboplatin or lomustine in patients with recurrent glioblastoma: a phase Ib/II, open-label, multicentre, randomised study. ESMO Open 2020;5:e000672. https://doi.org/10.1136/esmoopen-2020-000672.

[334] Lin F, de Gooijer MC, Hanekamp D, Chandrasekaran G, Buil LCM, Thota N, et al. PI3K–mTOR pathway inhibition exhibits efficacy against high-grade glioma in clinically relevant mouse models. Clin Cancer Res 2017;23:1286–98. https://doi.org/10.1158/1078-0432.CCR-16-1276.

[335] Rewcastle GW, Gamage SA, Flanagan JU, Frederick R, Denny WA, Baguley BC, et al. Synthesis and biological evaluation of novel analogues of the pan class I phosphatidylinositol 3-kinase (pi3k) inhibitor 2-(difluoromethyl)-1-[4,6-di(4-morpholinyl)-1,3,5-triazin-2-yl]-1 *H* -benzimidazole (ZSTK474). J Med Chem 2011;54:7105–26. https://doi.org/10.1021/jm200688y.

[336] Kelter G, Metz T, Korrat A, Giesemann T, Fiebig H, Beckers T. Abstract B271: *in vitro* profiling of traditional tubulin targeting and new antimitotic compounds by using a 40 cancer cell line panel in combination with compare analysis. In: Therapeutic agents—small molecule kinase inhibitors. American Association for Cancer Research; 2009. p. B271. https://doi.org/10.1158/1535-7163.TARG-09-B271.

[337] Barnett SF, Defeo-Jones D, Fu S, Hancock PJ, Haskell KM, Jones RE, et al. Identification and characterization of pleckstrin-homology-domain-dependent and isoenzyme-specific Akt inhibitors. Biochem J 2005;385:399–408. https://doi.org/10.1042/BJ20041140.

[338] DeFeo-Jones D, Barnett SF, Fu S, Hancock PJ, Haskell KM, Leander KR, et al. Tumor cell sensitization to apoptotic stimuli by selective inhibition of specific Akt/PKB family members. Mol Cancer Ther 2005;4:271–9.

[339] Yap TA, Yan L, Patnaik A, Fearen I, Olmos D, Papadopoulos K, et al. First-in-man clinical trial of the oral pan-AKT inhibitor MK-2206 in patients with advanced solid tumors. JCO 2011;29:4688–95. https://doi.org/10.1200/JCO.2011.35.5263.

[340] Cheng Y, Zhang Y, Zhang L, Ren X, Huber-Keener KJ, Liu X, et al. MK-2206, a novel allosteric inhibitor of Akt, synergizes with gefitinib against malignant glioma via modulating both autophagy and apoptosis. Mol Cancer Ther 2012;11:154–64. https://doi.org/10.1158/1535-7163.MCT-11-0606.

[341] Jin R, Nakada M, Teng L, Furuta T, Sabit H, Hayashi Y, et al. Combination therapy using Notch and Akt inhibitors is effective for suppressing invasion but not proliferation in glioma cells. Neurosci Lett 2013;534:316–21. https://doi.org/10.1016/j.neulet.2012.12.008.

[342] Yap TA, Yan L, Patnaik A, Tunariu N, Biondo A, Fearen I, et al. Interrogating two schedules of the AKT inhibitor MK-2206 in patients with advanced solid tumors incorporating novel pharmacodynamic and functional imaging biomarkers. Clin Cancer Res 2014;20:5672–85. https://doi.org/10.1158/1078-0432.CCR-14-0868.

[343] Machl A, Wilker EW, Tian H, Liu X, Schroeder P, Clark A, et al. M2698 is a potent dual-inhibitor of p70S6K and Akt that affects tumor growth in mouse models of cancer and crosses the blood-brain barrier. Am J Cancer Res 2016;6:806–18.

[344] Tsimberidou AM, Verschraegen CF, Weise AM, Sarantopoulos J, Lopes G, Nemunaitis JJ, et al. Precision oncology: results of a phase I study of M2698, a p70S6K/AKT targeted agent in patients with advanced cancer and tumor PI3K/AKT/mTOR (PAM) pathway abnormalities. JCO 2018;36:2584. https://doi.org/10.1200/JCO.2018.36.15_suppl.2584.

[345] Ma DJ, Galanis E, Anderson SK, Schiff D, Kaufmann TJ, Peller PJ, et al. A phase II trial of everolimus, temozolomide, and radiotherapy in patients with newly diagnosed glioblastoma: NCCTG N057K. Neuro Oncol 2015;17:1261–9. https://doi.org/10.1093/neuonc/nou328.

[346] Chinnaiyan P, Won M, Wen PY, Rojiani AM, Werner-Wasik M, Shih HA, et al. A randomized phase II study of everolimus in combination with chemoradiation in newly diagnosed glioblastoma: results of NRG Oncology RTOG 0913. Neuro Oncol 2018;20:666–73. https://doi.org/10.1093/neuonc/nox209.

[347] Wen PY, Chang SM, Lamborn KR, Kuhn JG, Norden AD, Cloughesy TF, et al. Phase I/II study of erlotinib and temsirolimus for patients with recurrent malignant gliomas: North American Brain Tumor Consortium trial 04-02. Neuro Oncol 2014;16:567–78. https://doi.org/10.1093/neuonc/not247.

[348] Wick W, Gorlia T, Bady P, Platten M, van den Bent MJ, Taphoorn MJB, et al. Phase II study of radiotherapy and temsirolimus versus radiochemotherapy with temozolomide in patients with newly diagnosed glioblastoma without MGMT promoter hypermethylation (EORTC 26082). Clin Cancer Res 2016;22:4797–806. https://doi.org/10.1158/1078-0432.CCR-15-3153.

[349] Galanis E, Buckner JC, Maurer MJ, Kreisberg JI, Ballman K, Boni J, et al. Phase II trial of temsirolimus (CCI-779) in recurrent glioblastoma multiforme: a North Central Cancer Treatment Group Study. J Clin Oncol 2005;23:5294–304. https://doi.org/10.1200/JCO.2005.23.622.

[350] Chang SM, Wen P, Cloughesy T, Greenberg H, Schiff D, Conrad C, et al. Phase II study of CCI-779 in patients with recurrent glioblastoma multiforme. Invest New Drugs 2005;23:357–61. https://doi.org/10.1007/s10637-005-1444-0.

[351] Flannery P, DeSisto J, Amani V, Venkataraman S, Lemma R, Prince E, et al. Preclinical analysis of MTOR complex 1/2 inhibition in diffuse intrinsic pontine glioma. Oncol Rep 2017;39:455–64. https://doi.org/10.3892/or.2017.6122.

[352] Kahn J, Hayman TJ, Jamal M, Rath BH, Kramp T, Camphausen K, et al. The mTORC1/mTORC2 inhibitor AZD2014 enhances the radiosensitivity of glioblastoma stem-like cells. Neuro Oncol 2014;16:29–37. https://doi.org/10.1093/neuonc/not139.

[353] Basu B, Dean E, Puglisi M, Greystoke A, Ong M, Burke W, et al. First-in-human pharmacokinetic and pharmacodynamic study of the dual m-TORC 1/2 inhibitor AZD2014. Clin Cancer Res 2015;21:3412–9. https://doi.org/10.1158/1078-0432.CCR-14-2422.

[354] Lapointe S, Mason W, MacNeil M, Harlos C, Tsang R, Sederias J, et al. A phase I study of vistusertib (dual mTORC1/2 inhibitor) in patients with previously treated glioblastoma multiforme: a CCTG study. Invest New Drugs 2020;38:1137–44. https://doi.org/10.1007/s10637-019-00875-4.

[355] Maira S-M, Stauffer F, Brueggen J, Furet P, Schnell C, Fritsch C, et al. Identification and characterization of NVP-BEZ235, a new orally available dual phosphatidylinositol 3-kinase/mammalian target of rapamycin inhibitor with potent in vivo antitumor activity. Mol Cancer Ther 2008;7:1851–63. https://doi.org/10.1158/1535-7163.MCT-08-0017.

[356] Wang W, Long L, Yang N, Zhang Q, Ji W, Zhao J, et al. NVP-BEZ235, a novel dual PI3K/mTOR inhibitor, enhances the radiosensitivity of human glioma stem cells in vitro. Acta Pharmacol Sin 2013;34:681–90. https://doi.org/10.1038/aps.2013.22.

[357] Yu Z, Xie G, Zhou G, Cheng Y, Zhang G, Yao G, et al. NVP-BEZ235, a novel dual PI3K–mTOR inhibitor displays anti-glioma activity and reduces chemoresistance to temozolomide in human glioma cells. Cancer Lett 2015;367:58–68. https://doi.org/10.1016/j.canlet.2015.07.007.

[358] Liu T-J, Koul D, LaFortune T, Tiao N, Shen RJ, Maira S-M, et al. NVP-BEZ235, a novel dual phosphatidylinositol 3-kinase/mammalian target of rapamycin inhibitor, elicits multifaceted antitumor activities in human gliomas. Mol Cancer Ther 2009;8:2204–10. https://doi.org/10.1158/1535-7163.MCT-09-0160.

[359] Netland IA, Førde HE, Sleire L, Leiss L, Rahman MA, Skeie BS, et al. Dactolisib (NVP-BEZ235) toxicity in murine brain tumour models. BMC Cancer 2016;16:657. https://doi.org/10.1186/s12885-016-2712-4.

[360] Gil del Alcazar CR, Hardebeck MC, Mukherjee B, Tomimatsu N, Gao X, Yan J, et al. Inhibition of DNA double-strand break repair by the dual PI3K/mTOR inhibitor NVP-BEZ235 as a strategy for radiosensitization of glioblastoma. Clin Cancer Res 2014;20:1235–48. https://doi.org/10.1158/1078-0432.CCR-13-1607.

[361] Bendell JC, Kurkjian C, Infante JR, Bauer TM, Burris HA, Greco FA, et al. A phase 1 study of the sachet formulation of the oral dual PI3K/mTOR inhibitor BEZ235 given twice daily (BID) in patients with advanced solid tumors. Invest New Drugs 2015;33:463–71. https://doi.org/10.1007/s10637-015-0218-6.

[362] Wise-Draper TM, Moorthy G, Salkeni MA, Karim NA, Thomas HE, Mercer CA, et al. A phase Ib study of the dual PI3K/mTOR inhibitor dactolisib (BEZ235) combined with everolimus in patients with advanced solid malignancies. Target Oncol 2017;12:323–32. https://doi.org/10.1007/s11523-017-0482-9.

[363] Carlo MI, Molina AM, Lakhman Y, Patil S, Woo K, DeLuca J, et al. A phase Ib study of BEZ235, a dual inhibitor of phosphatidylinositol 3-kinase (PI3K) and mammalian target of rapamycin (mTOR), in patients with advanced renal cell carcinoma. Oncologist 2016;21:787–8. https://doi.org/10.1634/theoncologist.2016-0145.

[364] Fazio N, Buzzoni R, Baudin E, Antonuzzo L, Hubner RA, Lahner H, et al. A phase II study of BEZ235 in patients with everolimus-resistant, advanced pancreatic neuroendocrine tumours. Anticancer Res 2016;36:713–9.

[365] Salazar R, Garcia-Carbonero R, Libutti SK, Hendifar AE, Custodio A, Guimbaud R, et al. Phase II study of BEZ235 versus everolimus in patients with mammalian target of rapamycin inhibitor-naïve advanced pancreatic neuroendocrine tumors. Oncol 2018;23:766–e90. https://doi.org/10.1634/theoncologist.2017-0144.

[366] Hargrave DR, Bouffet E, Tabori U, Broniscer A, Cohen KJ, Hansford JR, et al. Efficacy and safety of dabrafenib in pediatric patients with BRAF V600 mutation-positive relapsed or refractory low-grade glioma: results from a phase I/IIa study. Clin Cancer Res 2019;25:7303–11. https://doi.org/10.1158/1078-0432.CCR-19-2177.

[367] Manoharan N, Choi J, Chordas C, Zimmerman MA, Scully J, Clymer J, et al. Trametinib for the treatment of recurrent/progressive pediatric low-grade glioma. J Neurooncol 2020;149:253–62. https://doi.org/10.1007/s11060-020-03592-8.

[368] Selt F, van Tilburg CM, Bison B, Sievers P, Harting I, Ecker J, et al. Response to trametinib treatment in progressive pediatric low-grade glioma patients. J Neurooncol 2020;149:499–510. https://doi.org/10.1007/s11060-020-03640-3.

[369] Burger MC, Ronellenfitsch MW, Lorenz NI, Wagner M, Voss M, Capper D, et al. Dabrafenib in patients with recurrent, BRAF V600E mutated malignant glioma and leptomeningeal disease. Oncol Rep 2017;38:3291–6. https://doi.org/10.3892/or.2017.6013.

[370] See WL, Tan I-L, Mukherjee J, Nicolaides T, Pieper RO. Sensitivity of glioblastomas to clinically available MEK inhibitors is defined by neurofibromin 1 deficiency. Cancer Res 2012;72:3350–9. https://doi.org/10.1158/0008-5472.CAN-12-0334.

[371] de Gooijer MC, Zhang P, Weijer R, Buil LCM, Beijnen JH, van Tellingen O. The impact of P-glycoprotein and breast cancer resistance protein on the brain pharmacokinetics and pharmacodynamics of a panel of MEK inhibitors: impact of ABC-transporters on MEK inhibitor PK/PD. Int J Cancer 2018;142:381–91. https://doi.org/10.1002/ijc.31052.

[372] Huang A, Garraway LA, Ashworth A, Weber B. Synthetic lethality as an engine for cancer drug target discovery. Nat Rev Drug Discov 2020;19:23–38. https://doi.org/10.1038/s41573-019-0046-z.

[373] Hartwell LH, Szankasi P, Roberts CJ, Murray AW, Friend SH. Integrating genetic approaches into the discovery of anticancer drugs. Science 1997;278:1064–8.

CHAPTER

4

Molecular mechanisms in temozolomide-resistant glioblastoma

Sang Y. Lee[a] *and Gaspar J. Kitange*[b]

[a]Department of Neurology, The Johns Hopkins University School of Medicine, Baltimore, MD, United States [b]Department of Radiation Oncology, Mayo Clinic, Rochester, MN, United States

Abstract

Glioblastoma (GBM), the most common and aggressive brain tumor, can be treated using a multimodal treatment. Temozolomide (TMZ) has been the gold standard chemotherapeutic agent for newly diagnosed GBM for over a decade. However, almost all GBM patients eventually become resistant to chemotherapy including TMZ and develop recurrent tumors that are more aggressive than primary GBM. This chapter reviews the molecular mechanisms of innate and acquired TMZ resistance in GBM. It also summarizes potential treatment options and new target identification methods for TMZ-resistant GBMs.

Abbreviations

AGT	alkylguanine alkyltransferase
AKR1C	aldo-keto reductase 1C
ALDH1A3	aldehyde dehydrogenase 1 family, member A3
ANGPTL4	angiopoietin-like 4 protein
-aR	acquired TMZ resistant
ATRX	ATP-dependent helicase ATRX, X-linked helicase II, or X-linked nuclear protein
Avastin	bevacizumab
BBB	blood-brain-barrier
BCNU	carmustine wafer
BER	base excision repair
BMI1	B-cell-specific Moloney murine leukemia virus integration site 1
BMP2	bone morphogenetic protein 2
BTICs	brain tumor-initiating cells
CBTRU	Central Brain Tumor Registry of the United States
CBX5	chromobox protein homologue 5, heterochromatin protein 1 homologue alpha
CCNU	lomustine
CD133	prominin-1, a glycoprotein

Glioblastoma Resistance to Chemotherapy: Molecular Mechanisms and Innovative Reversal Strategies
https://doi.org/10.1016/B978-0-12-821567-8.00029-4

ChIP	chromatin precipitation
CREB	cAMP response element-binding protein
CRISPR	clustered regularly interspaced short palindromic repeat DNA sequences
CSCs	cancer stem cells
CTGF	connective tissue growth factor
CXCL	chemokine (C-X-C motif) ligand
DCA	dichloroacetate
DPG	dipotassium glycyrrhizinate
DYNC2H1	cytoplasmic dynein 2 heavy chain 1
EMT	epithelial to mesenchymal transition
EGCG	epigallocatechin gallate
EGFR	epidermal growth factor receptor
ER	endoplasmic reticulum
ERBB2	Erb-B2 receptor tyrosine kinase 2
EZH2	enhancer of zeste homolog 2
FADD	Fas-associated protein with death domain
FDA	Food and Drug Administration
Fn14	fibroblast growth factor-inducible 14
FUR	furanodienone
GA	glycyrrhetic acid
GBM	glioblastoma multiforme
GFAP	glial fibrillary acidic protein
GLUT-3	glucose transporter 3
GSCs	glioma stem cells
GSK3β	glycogen synthase kinase 3β
H3K9m2	dimethyl-histone-3-lysine-9 methylation
HDAC1	histone deacetylase 1
HPβCD	2-hydroxypropyl-beta-cyclodextrin
IC_{50}	50% inhibitory concentration
IFNs	interferons
ING4	inhibitor of growth 4
JNK	c-Jun terminal kinase
LGA	liver-type glutaminase
lncRNA	noncoding RNA
LSD1	lysine-specific demethylase 1A
MALAT1	metastasis-associated lung adenocarcinoma transcript 1
MAPK	mitogen-activated protein kinase
MGMT	O^6-methylguanine DNA methyltransferase
MIF	migration inhibitory factor
miRNA	microRNA
MMP-2	matrix metalloproteinase-2
MMP-14	matrix metalloproteinase-14
MMR	mismatch repair
MPG	*N*-methylpurine DNA glycosylase
MSH2	MutS homolog 2
MSH6	MutS homolog 6
MSP	methylation specific PCR
N^3-mA	N^3-methyladenine
N^7-mG	N^7-methylguanine
NCDDG	National Cooperative Drug Discovery/Development Groups
NF1	neurofibromatosis type 1
NFIA	nuclear factor I A
NKG2DL	natural killer Group 2 member D, a ligand for NKG2D

O^6-BG	O^6-benzylguanine
O^6-mG	O^6-methylguanine
OL	oleuropein
OLE	Olea Europaea leaf extract
P4HB	prolyl 4-hydroxylase, beta polypeptide
PARP1	poly [ADP-ribose] polymerase 1
PCAF	p300/CBP-associated factor
PDK1	pyruvate dehydrogenase kinase 1
PDX	patient-derived xenograft
PI3K	phosphoinositide 3-kinases, also called phosphatidylinositol 3-kinases
PIK3R1	phosphoinositide-3-kinase regulatory subunit 1
PomGnT1	*O*-mannose-*b*-1,2-*N*-acetylglucosaminyltransferase
PRC2	polycomb repressive complex-2
PTCH1	patched 1
RBBP4	retinoblastoma-binding protein 4
RORA	RAR-related orphan receptor A
RTK	receptor tyrosine kinase
SCD1	stearoyl-coenzyme A desaturase 1
sgRNAs	single-guided RNAs
SHH	sonic hedgehog
siRNA	small interfering RNA
SNAP	*S*-nitroso-*N*-acetylpenicillamine
SNHG15	small nucleolar RNA host gene 15
SNVs	single-nucleotide variations
STAT3	signal transducer and activator of transcription 3
TCGA	the Cancer Genome Atlas
TGF-β	tissue growth factor-β
TMZ	temozolomide
TNFAIP3	tumor necrosis factor-alpha-induced protein 3
TP53	tumor protein p53
TRAF6	targeting tumor necrosis factor receptor-associated factor 6
WA	Withaferin A
XPC	xeroderma pigmentosum group C-complementing protein

Conflict of interest

No potential conflicts of interest were disclosed.

Introduction

What is glioblastoma?

Glioblastoma (glioblastoma, grade IV astrocytoma, GBM) is the most common and aggressive human brain tumor. The American Cancer Society estimated that about 24,530 malignant tumors of the brain or spinal cord (13,840 in males and 10,690 in females) would be diagnosed in the United States (US) in 2021. GBMs account for about 15% of all brain tumors [1]. GBMs can develop from star-shaped glial cells but contain a mixture of different cell types. GBMs can develop from either lower-grade astrocytic tumors or de novo. Although GBMs are generally found in the cerebral hemispheres (mainly frontal and temporal lobes) of the brain, GBM can be found anywhere in the brain. GBMs can occur at any age. Sturm et al. summarized the comparison of pediatric and adult GBM at the genome and epigenome level [2].

Males are 60% more likely to develop GBM overall than females at all ages; however, female GBM patients tend to respond better than male GBM patients to standard chemotherapy drug temozolomide (Temodar, TMZ); i.e., TMZ showed a significant decline in tumor growth in female GBM patients than in male GBM patients [3, 4].

Treatments for GBM

Although the treatment methods for GBM depend on the tumor type, tumor location, tumor size, the basic procedures for GBM treatment are complete removal of tumors by surgery, followed by radiation and chemotherapy. Recently, other new approaches have been developed such as gene therapy, gamma radiation therapy, immunotherapy, cancer vaccine therapy, and a combination of chemotherapy with other therapy. At present, the Food and Drug Administration (FDA) approved the following treatments for GBM. (i) TMZ—for newly diagnosed GBM, (ii) Carmustine/Lomustine—for recurrent GBM patients, (iii) Bevacizumab/Avastin—conditionally approved for recurrent GBM, and (iv) Optune (NovoTTF-100A Device)—for both newly diagnosed and recurrent GBM.

Chemotherapy for GBM

At present, there are four FDA approved chemotherapy drugs for GBM. Carmustine wafer (BCNU, Gliadel), developed by National Cooperative Drug Discovery/Development Groups (NCDDG), first approved for recurrent GBM in 1996 and later for primary GBM in 2003. Carmustine was used as an adjunct to surgery or radiation. TMZ, developed by Schering-Plough as an oral alkylating agent, was FDA approved for the treatment of adult patients with refractory anaplastic astrocytoma in 1999, and adult patients with newly diagnosed GBM in 2005. Bevacizumab (Avastin), developed by Genentech, was granted accelerated approval in 2009 for GBM patients who had disease progression on prior therapy, later approved for recurrent GBM in 2017. Lomustine (CCNU, Ceenu), developed by Genentech, was approved for recurrent brain cancer with or without other drugs in 2017.

Among the FDA approved agents, TMZ is the standard treatment option for GBM. Bevacizumab, Lomustine, and Camustine wafer are the second line of treatment. TMZ ($75\,mg/m^2$) is administered every day during focal radiotherapy for 42 days, then a 28-day cycle of TMZ ($150\,mg/m^2$) for 6–12 cycles for newly diagnosed GBM. In each 28-day cycle, TMZ is given for the first 5 days, followed by 23 days of rest.

It is known that the mechanism of action of TMZ is based on O^6-methylguanine (O^6-mG), N^7-methylguanine (N^7-mG), and N^3-methyladenine (N^3-mA) formation during DNA replication via transporting a methyl group at the O^6 and N^7 of guanine, and N^3 of adenine [5, 6]. The methyl-guanine or methyl-adenine cause a mismatched base pair, result in DNA breaks and finally induce cell cycle arrest at G2/M and cancer cell death. The majority of TMZ-mediated altered DNA base can be repaired by the base excision repair (BER) pathway. *N*-Methylpurine DNA glycosylase (MPG) initiates BER by removing TMZ-induced N^3-methyladenine and N^7-methylguanine base lesions, leaving basic sites (AP sites) in DNA for further processing by BER. A recent review paper summarized the function of TMZ and its action mechanism [7].

Beyond DNA alkylation, TMZ can impact phenotypic changes of GBM cells via genomic and molecular signaling pathway events [8]. Therefore, the TMZ resistance in GBM is associated with not a single signal pathway or molecule but complex signaling pathways or networks. There are several review papers for the TMZ-resistant mechanisms in GBM [9–11].

In this chapter, we focus on the molecular mechanisms of innate (inherently and primary) and acquired (secondary) TMZ resistance in GBM and potential treatment options for TMZ-resistant GBMs. We also include current methods to identify new targets for TMZ-resistant GBMs.

Molecular events in GBM cells after TMZ treatment

GBM is a complex human cancer due to multiple gene mutations and protein expression alterations that result in the altered cell signaling pathway for key cellular functions such as cell proliferation, division, metabolism, migration/metastasis/invasion and angiogenesis [12, 13]. The Cancer Genome Atlas (TCGA) analysis showed the roles of ERBB2 (Erb-B2 receptor tyrosine kinase 2), NF1 (neurofibromatosis type 1), TP53 (tumor protein p53), PIK3R1 (phosphoinositide-3-kinase regulatory subunit 1), MGMT (O^6-methylguanine DNA methyltransferase) promoter methylation, and MMR (mismatch repair) deficiency in GBM. The TCGA genome analysis also identified frequent genetic alterations in three critical signaling pathways (RTK/RAS/PI3K, p53, RB signaling pathway). In addition to the abnormal activation of RTK (receptor tyrosine kinase), GBMs had constitutively activated STAT3 (signal transducer and activator of transcription 3) at Tyrosine705 via activation or overexpression of downstream transcription factors [14]. However, the impact of STAT3 activation at a different position on clinical outcomes is not consistent. For example, the phosphorylated STAT3 profile analysis of 68 GBM patients receiving alkylating agent therapy identified p-STAT3 (Tyrosine705) as an independent unfavorable prognostic factor for progression-free and overall survival [15]. Another study also found that activation of STAT3 (phospho-Tyrosine705) is associated with poor clinical outcomes of newly diagnosed supratentorial GBM [16]. In contrast, a study using 61 GBM clinical samples found that activation of STAT3 (not phospho-Tyrosine705 but phospho-Serine727) was associated with poor clinical outcome in GBM [17].

The hallmark of molecular events and signaling pathways of GBMs can be affected by TMZ treatment. TMZ-treated GBMs induce hypermutation in DNA and a subset of GBMs recur with an increased mutation rate [18]. In addition to gene mutations, TMZ can affect the expression of molecules of multiple signaling pathways in GBMs. For example, the transcriptome analysis in TMZ-treated human GBM U87-MG cells demonstrated significant enrichment of Wnt/β-catenin signaling pathway, independent of the ATM/Chk2 pathway and other pathways in the TMZ-treated cells [19]. The study also found that PI3K (phosphoinositide 3-kinases, also called phosphatidylinositol 3-kinases) inhibitor (LY294002) treatment repressed Akt activation and abolished the TMZ-mediated induction of Wnt/β-catenin pathway in TMZ-treated U87-MG cells [19].

TMZ can affect the expression of growth factors and microRNA (miRNA) molecules in GBMs. For example, CTGF (connective tissue growth factor) was significantly upregulated in the TMZ-treated human GBM U87-MG and U251 cell lines (both cells are known as TMZ-sensitive cell lines) through stem-like properties acquisition and CD44 interference

[20]. TMZ-induced CTGF upregulation was TGF-β (tissue growth factor-β)-dependent and regulated by TGF-β1 activation through Smad and ERK1/2 signaling. TMZ upregulated miR-524-5p in GBM cells (U87-MG, U251) and significantly reduced the viability of the cells [21].

TMZ can affect the expression of numerous GBM proteins. For example, TMZ treatment increased extracellular matrix breakdown enzyme MMP14 (matrix metalloproteinase-14) protein expression, especially in the stem cell condition but not in the differentiating condition, in TMZ-sensitive GBM cells (U87-MG, U251) and PDX (patient-derived xenograft) cells compared to control cells [22].

There is a limited metabolism alteration study in TMZ-treated GBMs. In the metabolomics analysis, histidine, phosphorylcholine, and choline were increased in the TMZ-treated U373 cells compared to vehicle-treated cells [23].

Molecular events in TMZ-resistant GBM cells after TMZ treatment

Innate TMZ-resistant GBM cells

The hallmark of GBM characteristics and signaling pathways are affected by TMZ treatment; however, innate TMZ-resistant GBM cells may have different transcriptome and signaling pathways to TMZ treatment than that of TMZ-sensitive GBM cells. Over 50% of primary GBM patients do not respond to TMZ at initial treatment; this phenomenon is called innate TMZ resistance. Innate drug resistance such as TMZ resistance in GBM cells is owing to multiple reasons such as heterogeneous characteristics of tumor, glioma stem cells (GSCs), intrinsic drug resistance mechanisms, epigenetic plasticity of GBM, and poor penetration of therapeutic agents into the human BBB (blood-brain barrier) [24–26]. Similar to other drug resistance mechanisms, innate TMZ resistance can occur by intertumoral and/or intratumoral heterogeneous characteristics in the GBM tumors [24, 25, 27–30]. The existence of GSCs in GBM also contributes to innate TMZ resistance [31–34]. In addition, innate TMZ resistance may occur via its inherent molecular functional characteristics such as not being responsive to TMZ or disruption of TMZ action in the GBM cells. Poor penetration of therapeutic agents into the BBB can cause innate TMZ resistance; however, TMZ has good permeability for BBB [35, 36]. Therefore, the innate TMZ resistance is not owing to BBB permeability but other reasons, such as the hypoxic microenvironment. TMZ sensitizes the boundary areas of GBMs, leaving the hypoxic environment in the core area of GBM cells without a response to TMZ owing to HIF-1α-mediated miR-26a upregulation [37]. This HIF-1α/miR-26a axis prevents Bax and Bad in mitochondria dysfunction in GBM and GBM progression. A recent report identified lipid peroxidation plays an important role in TMZ resistance in GBM [38].

In this section, we will review the innate TMZ resistance on GBMs, focusing on the molecular events and signaling pathways.

Research models (in vitro, in vivo) for innate TMZ-resistant GBM

There are multiple established human GBM cell lines known as TMZ-resistant GBM cell lines (e.g., CCF-STTG1, LN-18, LN229, LN382, SWB77, T98G, U138, U343-MG, VU-28,

VU-110, etc.). Among them, T98G is identified as the most TMZ-resistant GBM cell line by researchers. The half-maximal inhibitory concentration (IC_{50}) to TMZ in T98G cells is from 500 to 1600 μM. However, established GBM cell lines do not represent cancer heterogeneity, thus TMZ-resistant PDX lines were generated from human GBM tumor tissues by several groups. For example, Dr Jann Sarkaria's group (Mayo Clinic) reported following PDX lines as innate TMZ-resistant PDX line cells (e.g., GBM5, GBM8, GBM15, GBM39, etc.) [39]. The PDX G55 cell line developed by Dr. C. David James (UCSF) is a human GBM-derived PDX line that was passaged through nude mice and reestablished as a stable xenograft cell line [40]. The list of innate TMZ-resistant GBM cell lines and PDX lines has been previously summarized [41].

Molecular events of innate TMZ-resistant GBMs/PDXs

(i) Akt/mTOR or EGFR/MAPK/STAT or JAK/STAT pathway in innate TMZ resistance

GBM cancer stem cells (CSCs) and several established GBM cell lines (e.g., T98G, LN-18) are known as innate TMZ-resistant cells. T98G and LN-18 are human GBM cell lines generated from male Caucasian GBM tumor tissues. GBM patient specimens showed that elevated p-STAT-3 expression (= STAT3 activation) correlated with resistance to alkylation therapy such as TMZ treatment [15]. The study suggests JAK and/or EGFR/MAPK (epidermal growth factor receptor/mitogen-activated protein kinase) pathways associated with STATs may be involved in innate TMZ resistance. TMZ-resistant GBM cells, especially core cancer stem cells, have high HIF-1α and MGMT expression [42]. This data suggest that Akt/mTOR linked HIF-1α may be involved in innate TMZ resistance. Although there are several studies for the role of EGFR/STAT3 and ERK/HIF-1α pathway, the ways in which p-STAT-3 and HIF-1α influence MGMT expression remains unclear.

(ii) Hedgehog pathway in innate TMZ resistance

Activation of hedgehog/Gli1 signaling led to increased MGMT expression and enhanced resistance to TMZ treatment in primary GBM tissues [43].

(iii) Wnt/β-catenin signaling pathway in innate TMZ resistance

There is a significant correlation between MGMT expression and Wnt/β-catenin signaling in cancers such as colon carcinoma (SW480), GBM (T98G), medulloblastoma (Daoy), and neuroblastoma (SK-N-AS), etc. [44]. For example, β-catenin shRNA (short hairpin RNA or small hairpin RNA) study demonstrated that Wnt/β-catenin regulated MGMT expression in LS1747 colon carcinoma cells. These data suggest that Wnt/β-catenin may be associated with TMZ resistance in GBM.

(iv) DNA repair and DNA damage response enzymes in innate TMZ resistance

MGMT is overexpressed in a high percentage of innate TMZ-resistant GBMs. In addition, lost MMR (DNA mismatch repair) or p53 mutation status is associated with innate TMZ resistance in GBM [45–48].

MGMT is a DNA repair enzyme for damaged DNA by TMZ. Thus, increased MGMT expression is a well-known innate and acquired TMZ resistance mechanism. When MGMT expression level is high in GBMs, the cells overcome TMZ-mediated cell death and result in

TMZ resistance. In general, about 60% of human brain tumors express significantly higher MGMT than that of the normal brain [49]. Therefore, many MGMT expressing brain tumors may have a chance to be TMZ resistant.

The effect of TMZ on GBM cancer stem cells (CSCs) is controversial. It has been reported that TMZ can be cytotoxic on GBM CSCs via dose- and time-dependent manners [50]; however, it is clear that hypoxic and immature GBM CSCs are resistant to TMZ in vitro owing to overexpression of MGMT, the most well-known TMZ resistance mechanism in GBM [51].

Although high MGMT expression level is a major TMZ resistance mechanism in GBM, some MGMT deficient GBMs showed TMZ resistance. The TCGA GBM database analysis identified that *DYNC2H1* (cytoplasmic dynein 2 heavy chain 1, also called DHC2) expression correlates with poor progression-free survival in GBM and recurrent GBM patients (who did not receive TMZ treatment) with *MGMT* promoter methylation (= MGMT-deficient) [52]. The RNA expression of *DYNC2H1* was negatively correlated to *MGMT* expression. The overcome mechanism for TMZ in MGMT deficient GBMs is via the interaction between DHC2 with XPC (xeroderma pigmentosum group C-complementing protein, DNA repair protein complementing XP-C cells) and CBX5 (chromobox protein homologue 5, heterochromatin protein 1 homologue alpha) and by the regulation of their nuclear localization [52]. XPC and CBX5 mediate repair of TMZ-induced DNA damage in MGMT deficient GBMs.

Owing to the action of TMZ on DNA methylation, a defect in the DNA BER and MMR pathway contributes innate TMZ resistance in GBM. MMR proteins are nuclear enzymes, which are involved in the repair of base-base mismatch occurring during DNA replication. There are multiple DNA repair enzymes/pathways related to different types of DNA damage. Innate TMZ resistance in GBM is by MMR deficiency through the development of mutations in critical MMR genes (e.g., MSH2, MSH6, MLH1, PMS2) [12].

(v) Ligand and receptor in innate TMZ resistance

The expression level of ligands and their receptors correlated with TMZ response. For example, EGFR (epidermal growth factor receptor) amplification and mutations are found in 40% of primary GBM patients. U87-MG cells with EGFRvIII expression are more sensitive than wild-type EGFR-expressing tumor cells to treatment with TMZ alone and TMZ in combination with cetuximab [53]. GSCs tend to show higher expression of NKG2DL (natural killer group 2 member D, a ligand for NKG2D) compared to differentiated tumor cells, and result in resistance to TMZ therapy [54]. NKG2D is a molecular sensor of stressed cells and a potent activator of the immune system. Innate TMZ-resistant T98G cells had significantly higher NKG2DL (e.g., MICB and ULBP1, but not MICA and ULBP2) compared to TMZ-sensitive U251 cells [54]. These results indicate that NKG2DL may be involved in TMZ resistance.

(vi) Metabolism in innate TMZ resistance

The expression level of metabolism genes correlates with TMZ response. For example, the expression level of ALDH1A3 (aldehyde dehydrogenase 1 family, member A3), which is involved in alcohol metabolism and lipid peroxidation, was correlated with TMZ resistance in GBM cells (LN229, T98G, U87-MG), recurrent GBMs, and primary GBM tumor samples [38]. Thus, ALDH1A3 knockout cells were more sensitive to TMZ compared to wild-type cells.

(vii) miRNA and noncoding RNA (lncRNA) in innate TMZ resistance

A miRNA is a small noncoding RNA molecule that functions in RNA silencing and post-transcriptional regulation of gene expression. The expression level of a miRNA is associated with innate and acquired TMZ resistance. For example, the expression of oncogenic miRNA-21, but not tumor suppressor miRNA-16, was increased in TMZ and radiation-treated neurospheres compared to attached cells [55]. In the primary GBM specimen study, four miRNAs (miR-423-5p, miR-938, miR-1280, miR-1238) showed overexpression in TMZ-resistant primary GBM patient samples compared to TMZ-sensitive samples [56]. Meanwhile, three miRNAs (miR-93, miR-151-3p, let-7i) had decreased expression in TMZ-resistant primary GBM patient samples compared to TMZ-sensitive samples [56]. A review paper summarized the roles of miRNA in chemoresistant and recurrent brain tumors [57].

Comparative qPCR analysis between TMZ-sensitive clinical samples and TMZ-resistant clinical samples showed a significantly higher expression of long noncoding RNA (lncRNA) *SNHG15* (small nucleolar RNA host gene 15) and other genes (*Sox-2*, *β-catenin*, *EGFR*, *CDK6*) in TMZ-resistant PDX cells [58]. Overexpression of *SNHG15* inhibited tumor suppressor miR-627-5p and results in CDK6 activation.

(viii) Transcription factor in innate TMZ resistance

Several transcription factors such as STAT3 are altered in innate TMZ-resistant GBMs. STAT3 protein is an oncogenic downstream mediator of the JAK/STAT pathway. EGFR-mediated STAT3 activation is found in innate TMZ-resistant GBM LN229 cells [14].

(ix) Genes and proteins related to innate TMZ resistance

Owing to the key role of TMZ-induced DNA damage on TMZ resistance, a molecule that influences the expression of MGMT, BER, or MMR genes is involved in innate TMZ resistance. For example, RBBP4 (retinoblastoma-binding protein 4), which interacts with CBP/p300 to form a chromatin-modifying complex and binds within the promoter of MGMT and other genes, is associated with TMZ resistance in the GBM cells of GBM patient-derived xenograft model [59].

Several proteins are known for innate TMZ resistance in GBMs by altered protein expression or mutant form of protein. For example, innate TMZ-resistant T98G cells have a higher expression of a plasma membrane transporter Na+/K+-ATPase α2 and α3 subunits than TMZ-sensitive cells and normal human astrocytes [60]. Increased CD44, a cell-surface glycoprotein, is involved in TMZ resistance in GSCs. CD44 is a marker of GSCs and attenuates activation of the hippo-signaling pathway [61]. CD133 is a glycoprotein and cell surface antigen as a marker for the GBM cancer stem cell marker. CD133 positive cells express higher levels of MGMT mRNA and are significantly more resistant to chemotherapeutic agents including TMZ than CD133 negative cells [62].

TMZ resistance in GBMs depends on p53 status because functional p53 carrying GBM cells respond to DNA damage and lead to sensitivity to TMZ. However, GBM cells (SWB39, SWB77) with mutant p53 and a weak cell cycle response to DNA damage showed innate TMZ resistance [63]. In addition, Sato et al. reported that neither the status of p53 (wild type vs mutated) nor the expression of MGMT significantly affected the GBM cells (A172, T98G, YKG-1) response to TMZ [64]. Although innate TMZ resistance is not owing to a single gene mutation, we reported that *HFE* gene (one of the homeostatic iron regulator gene) mutation carrying GBM cells (e.g., CCF-STTG1, T98G, U343-MG) had innate TMZ resistance compared to wild-type *HFE* carrying GBM cells (e.g., U87-MG, U251) [65].

GSCs had overexpression and secretion of ANGPTL4 (angiopoietin-like 4 protein). Overexpression of ANGPTL4-induced GSC enrichment results in TMZ resistance in GBM through EGFR phosphorylation-mediated 4E-BP1 phosphorylation that was required for ANGPTL4-induced GSC enrichment [66].

Acquired TMZ-resistant GBM cells

Sometimes, GBM patients do not respond to TMZ during TMZ-treatment cycles; and, over 90% of relapsing GBM patients were nonresponsive to TMZ [67, 68]. We defined these types of cells as acquired TMZ-resistant GBM cells when the cells derived from recurrent GBM patients do not respond to TMZ or GBM cells acquire TMZ resistance by continuous TMZ exposure at the same concentration or gradual increment for several weeks or months. The acquired TMZ resistance can occur by the epigenetic plasticity of GBMs because GBMs have an adaptation ability to therapy by transcriptional/translational regulation or post-transcriptional/posttranslational regulation [26, 69]. Previously, we summarized the list of acquired TMZ-resistant GBM cell lines and PDX lines [41]. In this section, we will review the study models and molecular mechanisms for acquired TMZ-resistant GBMs. We summarized molecular events in innate and acquired TMZ-resistant GBM cells in Fig. 1 and Table 1.

Research models (in vitro, in vivo) for acquired TMZ-resistant GBMs/PDXs

Two approaches are used to generate acquired TMZ-resistant GBMs. Firstly, the acquired TMZ-resistant (called as -aR) GBM cell lines (e.g., A172-aR, U87-aR, U251-aR, U373EGFRvIII-aR) were developed by a low concentration of TMZ for several weeks ~6 months using TMZ-sensitive GBM cell lines. For example, U87-MG cells were cultured with a low dose of TMZ for 3 weeks. The generated TMZ-resistant U87-MG cells (U87-aR) have a higher concentration of IC_{50} to TMZ than that of parent U87-MG cells (150 μM vs <40 μM).

Secondary, acquired TMZ-resistant GBM cell lines (e.g., SF188-aR, SNB-19-aR, U251-aR, SNB19-aR, T98G-aR, U343-aR, U373-aR, LN18-aR, LN229-aR, LN308-aR) were generated using incremental concentrations of TMZ for several months with TMZ-sensitive or -resistant GBM cell lines. For example, U251 cells were cultured by twofold stepwise increases of TMZ for 10 months to generate acquired TMZ-resistant U251 cells (U251-aR). The identified U251-aR cells have a higher concentration of IC_{50} to TMZ than parent U251 cells (271 μM vs 58 μM). Also, acquired TMZ-resistant PDX cell lines (e.g., GBM3-aR, GBM5-aR) were generated using the method described earlier.

The acquired TMZ-resistant PDX xenograft animal models (e.g., GBM12-aR, GBM14-aR, GBM22-aR, GBM39-aR, GBM43-aR, GBM44-aR, GBM46-aR, GBM76-aR, GBM102-aR, GBM143-aR, HG7-aR, HG9-aR) were developed using high doses of TMZ on the established flank tumors in mice [39, 97].

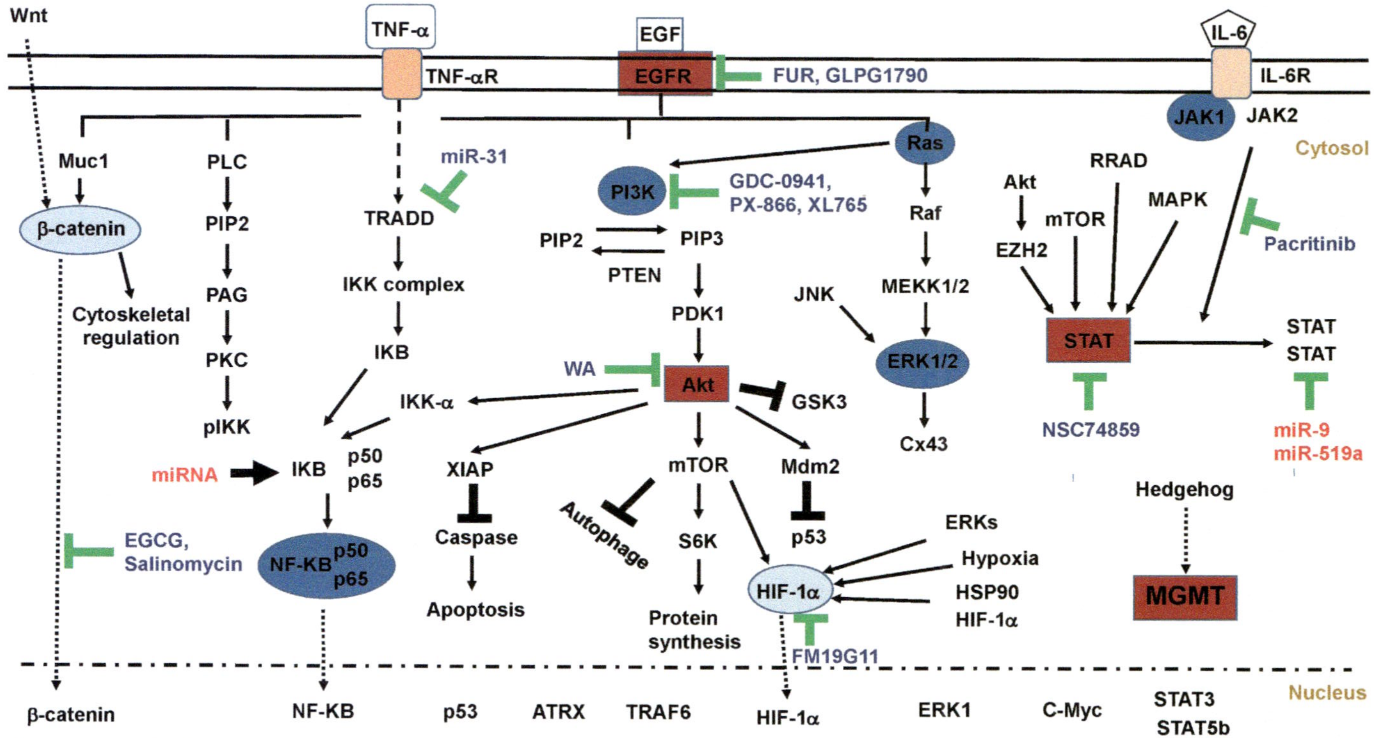

FIG. 1 A schematic diagram for critical molecular events in innate and acquired TMZ-resistant GBM/PDX cells and some blocking agents/miRNAs targeting the molecular events. While GBMs have activated Ras/PI3K pathway and STAT3 activation, TMZ induces Wnt/β-catenin pathway and TGF-β/Smad/ERK pathway in GBM cells. However, innate TMZ-resistant GBM cells have different signaling pathways to TMZ compared to TMZ-sensitive GBM cells. In addition to the activation of the Wnt/β-catenin/MGMT pathway, the innate TMZ-resistant GBM/PDX cells have activation of the EGFR/MAPK/STAT pathway, Akt/mTOR/HIF-1α pathway, and hedgehog/MGMT pathway. Acquired TMZ-resistant GBM/PDX cells have more diverse channels such as activated EGFR/JNK/ERK pathway, hedgehog pathway, NF-kB pathway, PI3K/Akt pathway, Ras pathway, and MAPK/ERK pathway. The signaling pathways in acquired TMZ-resistant GBM cells are similar to known pathways in GBM signaling pathways. Numerous transcription factors regulate the expression of genes and miRNAs, which are involved in innate and acquired TMZ resistance in GBM/PDX cells, i.e., nucleus transcription factors NF-kB and p53 are associated with MGMT expression. Nucleus HIF-1α is involved in numerous miRNA expression. All these different aberrant signaling pathways and transcription factors result in TMZ resistance in GBM/PDX cells. *Light blue circle* molecules are primarily involved in innate TMZ resistance. *Dark blue circle* molecules are predominantly engaged in acquired TMZ resistance. *Brown rectangle* molecules are associated with both inherent and acquired TMZ resistance. The *green T symbol* indicates inhibition by an agent or miRNA. The *black T symbol* indicates the inhibition molecule. The *black arrow symbol* indicates the activation pathway or signaling pathway. EGCG (epigallocatechin gallate); FUR (furanodienone); GLPG1790 (Eph receptor kinase inhibitor); IKB (inhibitory proteins of kB family); RRAD (Ras-related glycolysis inhibitor and calcium channel regulator); WA (Withaferin A); XIAP (X-linked inhibitor of apoptosis protein, an inhibitor of apoptosis protein 3 (IAP3) and baculoviral IAP repeat-containing protein 4 (BIRC4)).

TABLE 1 The summary of molecular events in the TMZ resistant GBMs.

Molecules/ pathways	GBM tissue	Innate TMZ-resistant cells		Acquired TMZ-resistant cells	
		GBM	GSC	GBM	Recurrent GBM
Cytokine				↑ CXCL2, CXCL3, CXCL8 [70]	
DNA repair, DNA damage		↑ DYNC2H1 (in low MGMT cells), MGMT, XPC/CBX5 [52] ↓ MLH1, MSH2, MSH6, PMS2 [12]		↑ BER gene *NTL1*, MGMT [71] ↓ MSH2, MSH6 [71, 72]	↑ DYNC2H1 (in low MGMT cells), XPC/CBX5 [52]
EGFR/JNK		↑ JAK, EGFR/MAPK [15]	↑ HIF-1α [42]	↑ EGFR, JNK-ERK1/2-AP1, Cx43 [73, 74]	
Hedgehog	↑ HH/Gli1 [43]			↑ PTCH [75]	
Ligand, receptor		↑ EGFR wild-type [53]	↑ NKG2DL (MICB, ULBP1) [54]		
lncRNA		↑ SNHG15 (PDX) [58]		↑ ADAMTS9-AS2 [76]	
Metabolism	↑ Adenine, ALDH1A3, isoleucine [23, 38] ↓ Adenosine, glucose, glutathione [23]	↑ ALDH1A3 [38]		↑ Arginine, leucine, myo-inositol, taurine, valine [23, 38] ↓ Alanine, choline, creatine, phosphorylcholine [23]	↑ ALDH1A3 [38]

TABLE 1 The summary of molecular events in the TMZ resistant GBMs—cont'd

Molecules/ pathways	GBM tissue	Innate TMZ-resistant cells		Acquired TMZ-resistant cells	
		GBM	GSC	GBM	Recurrent GBM
miRNA	↑ miR-9, miR-423-5p, miR-938, miR-1280, miR-1238 [56, 75] ↓ miR-93, miR-151-3p, let-7i [56]		↑ miR-21 (neurosphere) [77, 78]	↑ miR-9, miR-10a, miR-138, miR-195, miR-455-3p, miR-1238 (GBM/PDX) [75, 79, 80] ↓ miR-101, miR-126-3p, miR-146b-5p, miR-519a [81–84]	↑ miR-9, miR-138, miR-1238 [75, 79, 80]
NF-kB				↑ NFIA, NF-kB p65 ser536 [85] ↓ NFKBIA (PDX), TNFAIP3 [86]	
PI3K/AKT				↑ Actin cytoskeleton, focal adhesion, PI3K-AKT, PDK1, phagosome [87, 88] ↓ Ribosome and splicesome pathway proteins [87]	↑ PI3K, MAPK/ERK [87] ↓ AKT, CREB phosphorylation [89]
Protein		Mutant p53 [63], Mutant HFE [65] ↑ CD44, CD133, Na+/K+ ATPase α2, α3 [60–62]	↑ ANGPTL4 [66]	↑ AKR1C, EZH2, Fn14 (PDX), GLUT-3, K-M enhancer (PDX), PomGnT1, P4HB, SCD1 [69, 90–95]	↑ CD133, EZH2, Fn14, K-M enhancer (PDX) [62, 69, 95, 96]
Ras				↑ RRAD [14]	
Transcription factor	↑ STAT3 [15, 17]	↑ STAT3 [14, 15, 17]		↑ ATRX (GBM/PDX), FUS, PARP, STAT3, STAT5b [76, 97, 98] ↓ FADD [97]	↑ Snail [99]
Wnt/β-catenin		↑ Wnt/β-catenin [44]			

The number in the blanket is the reference information.

TMZ-resistance mechanisms in acquired TMZ-resistant GBM/PDX cells

(i) EGFR/JNK pathway in acquired TMZ resistance

Acquired TMZ-resistant GBM cells (U87-aR, T98G-aR) had activated EGFR and activated JNK-ERK1/2-AP-1 axis to induce Cx43 which is involved in gap junctional intercellular communication among the acquired TMZ-resistant GBM cells [73]. Moreover, activation (= increased phosphorylation of JNK and higher activation of its downstream signaling) of the JNK (c-Jun terminal kinase) signaling pathway was found in acquired TMZ-resistant GBM cells (U343-aR) compared to parent cells [74].

(ii) Hedgehog pathway in acquired TMZ resistance

PTCH, the Sonic Hedgehog (SHH) receptor, and mRNA were significantly increased in the acquired TMZ-resistant GBM cells (U87-aR, T98G-aR) whereas its corresponding protein was decreased [75]. These results suggest that TMZ could induce posttranscriptional regulation of *PTCH1* (Patched 1) in acquired TMZ-resistant GBM cells. The study identified miR-9 as the target of PTCH1 in acquired TMZ-resistant GBM cells with concomitant activation of SHH signaling. miR-9 was increased in the tissues from GBM patients and in early passage GBM cells from a patient with recurrent GBM but not from a naïve patient. These data suggest that miR-9 targets PTCH1 via an SHH-independent method in GBM cells for TMZ resistance.

(iii) NF-kB pathway in acquired TMZ resistance

The upregulation of NF-κB oncogene pathway is associated with acquired TMZ resistance in GBM cells. For example, nuclear factor I A (NFIA) contributed to acquired TMZ resistance in GBM cells (U87-aR and U251-aR) [85]. In addition, phosphorylation of NF-kB p65 at serine 536 induced TMZ resistance in U87-aR and U251-aR GBM cells [85]. A potent NF-kB signaling pathway inhibitor TNFAIP3 (tumor necrosis factor-alpha-induced protein 3) and NF-kB inhibiting IkB family member NFKBIA were significantly underexpressed in acquired O^6-alkylating agent resistant GBM PDX cells (LX-aR, ME-aR) [86]. The expression of the TNFAIP3 transcript was inversely correlated with NF-kB activation status in the acquired O^6-alkylating agent resistant GBM PDX cells.

(iv) PI3K/Akt or Ras pathway in acquired TMZ resistance

Increased protein expression related to the actin cytoskeleton, focal adhesion, and phagosome; and, the PI3K-AKT signaling pathway were identified by proteomic analysis in acquired TMZ-resistant U87-MG cells (U87-aR). In U87-aR cells, ribosome and spliceosome pathway proteins were decreased [87]. Recurrent GBM tumor tissues had activation of the PI3K and MAPK/ERK (also known as the Ras-Raf-MEK-ERK) pathways. However, AKT and CREB (cAMP response element-binding protein) phosphorylation were reduced in biopsies of resistant tumors while ERK phosphorylation remained unchanged [89]. Increased PDK1 (pyruvate dehydrogenase kinase 1) in EGFRvIII overexpressing and acquired TMZ-resistant GBM cells (U373vIII-aR) support the role of PI3K/PDK1/Akt in acquired TMZ-resistant GBM cells [88].

A Ras-related GTP-binding protein RRAD (RAS associated with diabetes, RAD) expression was correlated with acquired TMZ-resistant LN229 cells (LN229-aR) [14]. LN229-aR cells showed fast tumorsphere formation and increased stemness-regulating molecules compared to parent LN229 cells. RRAD enhanced EGFR-induced STAT3 activation; the RRAD expression was correlated with malignancy-associated properties. The results imply that RRAD expression is critical for stem cell maintaining and TMZ resistance in acquired TMZ-resistant GBM cells.

(v) Cytokine for acquired TMZ resistance

Acquired TMZ-resistant GBM cells (U373-aR, T98G-aR), generated from 0.1 μM to 1 mM gradual increments for almost 1 year, injected mice demonstrated TMZ resistance through correlation with the upregulation of CXCL (chemokine (C-X-C motif) ligand) chemokines (CXCL2, CXCL3, and CXCL8) [70]. This is not observed in oligodendroglioma cells-injected mice. The data indicate that TMZ modifies the CXC chemokine network in GBMs.

(vi) DNA repair and DNA damage response enzymes for acquired TMZ resistance

The expression level of MGMT in acquired TMZ-resistant GBM cells is controversial. MGMT is overexpressed in the acquired TMZ-resistant GBM cells (U373-aR) compared to parent cells [23]. Another group also found increased expression of MGMT in U373-aR cells [71]. Increased AGT (alkylguanine alkyltransferase) activity is the TMZ-resistant mechanism in acquired TMZ-resistant GBM cells (SF188-aR) [100]. Moreover, overexpressed MGMT is observed in acquired TMZ-resistant PDX lines (LX-aR, ME-aR) [86]. In contrast, MGMT expression is not involved in acquired TMZ resistance in GBM cells (U251-aR) [79]. Oliva et al. found that MGMT expression is downregulated in acquired TMZ-resistant GBM cells (U251-aR) [101]. Another study also found that MGMT expression is not involved in acquired TMZ resistance in GBM cells (U251-aR, U373-aR) [102]. Although the above acquired TMZ-resistant GBM cells were generated by different methods, the results suggest that MGMT expression level is not associated with TMZ response, at least in U251-aR cells, but may be associated with U373-aR cells.

In acquired TMZ-resistant SNB19-aR GBM cells, upregulation of BER gene *NTL1* and downregulation of *MSH6* (MutS Homolog 6) gene and protein expression was associated with TMZ tolerance [71]. In another study, acquired TMZ resistance is not owing to increased repair of O^6-methylguanine lesions but owing to decreased MMR activity such as MSH2 (MutS homolog 2) and MSH6 in acquired TMZ-resistant GBM cells (U87-aR, LN229-aR, A172-aR) [72].

(vii) Metabolism for acquired TMZ resistance

The metabolism comparison between acquired TMZ-resistant GBM cells and sensitive cells (U373-aR vs U373) identified the upregulation of arginine, leucine, myo-inositol, taurine, and valine in U373-aR cells than U373 cells [23]. The study also found decreased metabolites such as alanine, choline, creatine, and phosphorylcholine in U373-aR cells compared to U373 cells. In a metabolism comparison among primary GBM samples, TMZ-resistant primary GBM samples showed increased levels of adenine and isoleucine but decreased levels in adenosine, glucose, and glutathione compared to TMZ-sensitive primary GBM samples. These results suggest that there is a discrepancy in the metabolomics difference between primary GBM samples and established GBM cell lines for acquired TMZ resistance.

(viii) miRNA and lncRNA for acquired TMZ resistance

The expression of several miRNAs was increased in acquired TMZ-resistant GBM cells. For example, upregulated miRNA such as miR-10a, miR-195, and miR-455-3p was found in acquired TMZ-resistant GBM cells (U251-aR, U87-aR, M059-aR) [79]. miR-138 was upregulated in acquired TMZ-resistant GBM cells (LN18-aR, LN308-aR) compared to parental cells [80]. The expression of miR-138 also increased in recurrent GBM tissue samples after switching from TMZ/RT treatment to TMZ treatment [80]. The expression of miR-1238 was higher in acquired TMZ-resistant GBM cells (U251-aR) and acquired TMZ-resistant PDX cells (N3-aR) and the cell exosomes than in sensitive cells [103]. A higher level of miR-1238 was found in recurrent tumors compared to primary tumors [103].

In contrast, much of miRNA's expression was decreased in acquired TMZ-resistant GBMs. miR-101 was significantly downregulated in acquired TMZ-resistant GBM cells (A172-aR, U251-aR) and human specimens [81]. miR-126-3p expression was decreased in acquired TMZ-resistant GBM tissues and cells (U87-aR, U251-aR) [82]. miR-146b-5p was downregulated in acquired TMZ-resistant GBM cells (U87-aR, U251-aR) [83]. The expression of miR-519a and baseline autophagy level was lower in acquired TMZ-resistant U87-MG cells (U87-aR) and a PDX cell (B131212-aR) compared to parent cells [84]. miR-519a downregulates STAT3 and Bcl-2 expression and increases autophagy and apoptosis; therefore, decreased miR-519a may induce acquired TMZ resistance via activation of STAT3. The above miRNA expression study of the same acquired TMZ-resistant GBMs (e.g., U251-aR), generated from different investigators, indicated that different miRNAs can be identified from an acquired TMZ-resistant GBM cells.

A long noncoding RNA (lncRNA) gene *MALAT1* (metastasis-associated lung adenocarcinoma transcript 1) was significantly upregulated in acquired TMZ-resistant GBM cells (U251-aR) [104]. MALAT1 promoted chemoresistance by suppressing the miR-101 signaling pathway via directly binding it in U251-aR cells. This result is supported by the finding that miR-101 was significantly downregulated in acquired TMZ-resistant GBM cells (A172-aR, U251-aR) and human specimens [81]. These data indicate the importance of the MALAT1/miR-101 pathway in acquired TMZ resistance in GBM. Recently, ADAMTS9-AS2 lncRNA was identified as significant and overexpressed in acquired TMZ-resistant GBM cells (T98G-aR, U118-aR) [76]. ADAMTS9-AS2 expression was correlated with FUS expression via direct binding to the RRM and Znf_RanBP2 domains of FUS, and consequently increasing FUS protein expression. ADAMTS9-AS2 has also decreased the interaction between FUS and MDM2 by FUS K48 ubiquitination [76].

(ix) Transcription factor for acquired TMZ resistance

The expression levels of MGMT and STAT3 were overexpressed in acquired TMZ-resistant GBM cells (U87-aR) [98]. The expression levels of STAT3 and phosphorylated STAT3 (pSTAT3-Ser727) were highly expressed in acquired TMZ-resistant GBM cells (U251-aR, U373-aR); however, phosphorylated STAT3 (pSTAT3-Tyr705) was decreased in acquired TMZ-resistant GBM cells [102].

Snail, a transcription factor and zinc finger protein, is upregulated in recurrent GBM tumors [99]. In addition to the repression of the adhesion molecule E-cadherin to regulate epithelial to mesenchymal transition (EMT) during embryonic development, Snail promotes GBM resistance to TMZ-induced apoptosis. The expression of Snail is modulated by STAT3 activation. These results suggest that STAT3 activation is a key contributor to acquired TMZ resistance in GBMs.

Transcription factor STAT5b (signal transducer and activator of transcription 5B) and ATRX (ATP-dependent helicase ATRX, X-linked helicase II, or X-linked nuclear protein) are increased in acquired TMZ-resistant GBM/PDX cells [97]. ATRX (a transcriptional regulator) and STAT5b (a transcription factor) had higher expression levels in acquired TMZ-resistant GBM/PDX cells (LN229-aR, HG7-aR, HG9-aR) than parental cells. Increased ATRX-mediated acquired TMZ resistance occurs via strengthening DNA damage repair by higher stabilizing PARP1 (poly [ADP-ribose] polymerase 1) protein and downregulation of FADD (Fas-associated protein with death domain) expression in acquired TMZ-resistant GBM/PDX cells. The results suggest that there is a novel mechanism of the ATRX/PARP1 axis for acquired TMZ resistance in GBMs.

(x) Genes and proteins related to acquired TMZ resistance

Increased expression of EZH2 (enhancer of zeste homolog 2), which plays an important role in cancer characteristics such as invasion and drug resistance, was found in acquired TMZ-resistant GBM cells (U87-aR, U251-aR) [90] and recurrent GBMs [96]. EZH2 (histone-lysine *N*-methyltransferase) is a histone methyltransferase that constitutes the catalytic component of the polycomb repressive complex-2 (PRC2); however, EZH2 can also methylate nonhistone proteins such as the transcription factor GATA4 (GATA binding protein 4) and the nuclear receptor RORA (RAR-related orphan receptor A).

Glucose transporter and reductase are associated with acquired TMZ resistance. Acquired TMZ-resistant GBM cells (T98G-aR, U373-aR) generated by long-term treatment of human GBM cells with up to 1 mM TMZ induces increased expression of facilitative glucose transporter/solute carrier GLUT/SLC2A family members, mainly GLUT-3 (glucose transporter 3), and of the AKR1C (aldo-keto reductase 1C) family of proteins [91].

Multiple proteins are identified as markers for TMZ response in the acquired TMZ-resistant GBM cells. SCD1 (stearoyl-coenzyme A desaturase 1) is a key rate-limiting enzyme for the synthesis of monounsaturated fatty acids and was significantly upregulated in acquired TMZ-resistant GBM cells (U87-aR, T98G-aR) compared to their parental cells, through activation of Akt/GSK3β/β-catenin signaling in TMZ-resistant GBM cells [92]. *O*-Mannose-*b*-1,2-*N*-acetylglucosaminyltransferase (PomGnT1), a glycosyltransferase that participates in the formation of GlcNAc-β1,2-Man glycan, was significantly upregulated in acquired TMZ-resistant GBM cells (U87-aR, U251-aR) compared to the TMZ-sensitive cells [93]. Increased P4HB (prolyl 4-hydroxylase, beta polypeptide) expression was observed in acquired TMZ-resistant GBM cells (D54-aR, U87-aR) [94]. Fibroblast growth factor-inducible 14 (Fn14) protein was highly expressed in recurrent GBM and TMZ-resistant PDX cells derived from GBM patients in vivo [95]. Fn14 protein was highly expressed in GBM patient-derived xenografts with acquired TMZ resistance [95]. CD133 (also known as prominin-1, a glycoprotein, the most frequently used cell surface marker to detect CSCs) expression was significantly higher in recurrent GBM tissue than in their respective newly diagnosed tumors [62].

Enhancers are regulatory elements that can promote gene expression. An enhancer (K-M enhancer), located between the promoters of the marker of proliferation Ki67 (*MKI67*) and *MGMT* genes, is activated in acquired TMZ-resistant PDX lines (GBM12-aR) and recurrent tumor samples [69]. Activation of the enhancer correlated with increased MGMT expression, a well-known mechanism for TMZ resistance.

The kinomic profiling using acquired TMZ-resistant xenolines identified significantly altered kinase activity in the TMZ-resistant xenolines (two of the five tumors) [105]. The network modeling suggested that potential TMZ resistance is mediated by PKC, JAK1, PI3K, CDK2, and VEGFR.

Therapeutic approaches

Therapeutic approaches for GBM treatment

Owing to frequent genetic alterations in three critical signaling pathways in the TCGA GBM [12], gene-stabilizing agents are studied for GBM treatment. For example, a p53-stabilizing agent, CP-31398, leads GBMs cell death by p53-dependent and p53-independent mechanisms [106].

Targeting for DNA repair enzyme is studied for GBM treatment. For example, the MGMT inhibition approach was used for primary GBMs because the level of MGMT expression is a major predictor of TMZ sensitivity in GBMs. The MGMT inhibitor O^6-benzylguanine (O^6-BG) sensitizes MGMT-positive GBM cells (LN-18) to TMZ [45]. Inhibition of the Hedgehog/Gli1 signaling by cyclopamine (a drug targeting smoothened in the Hedgehog pathway) in U251 cells resulted in decreased MGMT expression and increased sensitivity to TMZ treatment [43].

GBM therapy uses the targeting of the three GBM signaling pathway inhibitors. Multiple compounds derived from natural products, microorganisms, or chemical synthesis on GBM signaling pathway inhibition are studied for GBM treatment. For example, BC3EE2, 9B, a derivative of naturally occurring carbazole alkaloid isolated from curry leaves, sensitize GBM PDX cells (GBM8401, GBM8901) via cell cycle arrest at G1 stage, autophagy-mediated cell death, activation of AMP-activated protein kinase, and attenuation of Akt and mTOR downstream signaling pathway [107]. Cordycepin, a major bioactive component in Cordyceps militaris (an entomopathogenic fungus), inhibited cellular proliferation, migration, and invasion as well as cellular apoptosis and cell cycle arrest in glioma cell lines in vitro [108]. Inhibition of STAT3 (a transcription factor) significantly enhanced the effect of TMZ on GBM cells (U87, U251) via an increase of Bcl-2, G0/G1 arrest as well as decreased Bax/cyclin D1 expression [15]. These results indicate that inhibition of STAT3 sensitizes GBM cells to TMZ, at least in part, by blocking the p-AKT and β-catenin pathways.

Targeting miRNA has also been assessed for GBM treatment. For example, there is a combinational miRNA therapy for GBM. Upregulation of miR-524-5p and miR-324-5p reduced glioma cell (U87-MG, U251) proliferation and tumor growth in an intracranial xenograft model (U87-MG) by targeting EZH2 (enhancer of zeste homolog 2) [21].

There are many review papers on the treatment of GBM [109–112].

Therapeutic approaches for TMZ-resistant GBM cells to overcome TMZ resistance: Single-agent approach

In general, TMZ resistance in GBM is owing to elevated expression of MGMT or lack of expression for MMR pathway gene/protein or p53 mutation status. Therefore, numerous genes/proteins have been tested as potential therapeutic targets based on the expression level that is associated with the innate or acquired TMZ-resistant GBMs/PDXs. We summarized tested agents for innate or acquired TMZ-resistant GBM/PDX cells in Table 2.

TABLE 2 The summary of therapeutic options for TMZ-resistant GBMs.

Group	Agent	Primary target	Pathway	TMZ sensitive (S) or Innate TMZ-resistant (R) cells	Acquired TMZ-resistant cells (-aR)	References[a]
Analog	TMZ analog (C8-imidazolyl, C8-methylimidazole tetrazines)	DNA	DNA double-strand break damage, cell cycle arrest at G2/M, apoptosis	R_T98G		[113] In vitro
Antiinflammation	DPG (dipotassium glycyrrhizinate)	Corticosteroids metabolism	NF-kB pathway ↑ *miR-16*, *miR-146a* ↓ *TRAF6*	S_U87 R_T98G		[114] In vitro
Antioxidant	Sodium selenite	Antioxidant	S phase arrest, modulate H3K9m2 level, Cas3-mediated apoptosis ↓ Histone deacetylase activity	3D cultured cells (U87, LN229, T98G)		[115] In vitro
Gene	GAB (GRB2-associated binding protein)		Protect H_2O_2-mediated cell death by decreased PI3K/AKT pathway	S_U87 R_LN229		[116] In vitro
Inhibitor	ADDA 5 (1-[2-(1-adamantyl) ethoxy]-3-(3,4-dihydro-2 (1*H*)-isoquinolinyl)-2-propanol hydrochloride)	Cytochrome *c* oxidase		S_U251	U251-aR	[117] In vitro, **in vivo**
	DCA (dichloroacetate)	Pyruvate dehydrogenase kinase 1	↑ Mitochondrial membrane potential, apoptosis		U373vIII-aR	[88] In vitro, **in vivo**
	Didox, trimidox	Ribonucleotide reductase		R_T98G		[118] In vitro
	EGCG (epigallocatechin gallate)	MGMT/Wnt/-catenin	Transcription factors (TCF1, LEF1)	R_T98G, GBM-XD		[119] In vitro

Continued

TABLE 2 The summary of therapeutic options for TMZ-resistant GBMs—cont'd

Group	Agent	Primary target	Pathway	TMZ sensitive (S) or Innate TMZ-resistant (R) cells	Acquired TMZ-resistant cells (-aR)	References[a]
	Etoposide	Topoisomerase II	Cell cycle arrest at S & G2/M, apoptosis	R_T98G		[64] In vitro
	FM19G11	HIF-1α	↓ MGMT ↓ NF-kB pathway	R_T98G		[120] In vitro
	FUR (furanodienone)	EGFR	↓ EGFR, CSPG4 signaling		U87-aR, U251-aR	[121] In vitro
	HET0016	Cyt P450		Syngenic GL261 & U251 in vivo		[122] In vitro, **in vivo**
	Irinotecan	Topoisomerase I	↓ TGFβ signaling	S_U87, U251, SJ-GBM2 R_T98G		[123, 124] **In vivo**
	GDC-0941	PI3K		A172, SHG44, T98G		[125] In vitro
	GLPG1790	Eph receptor kinase	↑ GFAP, proneural/neuronal marker ↓ Mesenchymal markers	S_A172, SW1783, U87, U251, PDXs (BT25M, BT50EF, BT53M) R_LN229, T98G, PDXs (BT12M, BT48EF) GSC (GSC-5, CSC-7)		[126] **In vivo**
	O^6-BG (O^6-benzylguanine)	MGMT	Autophagy	R_T98G		[45, 127] In vitro
	Pacritinib	JAK2/STAT3		R_BITC		[128] **In vivo**

	PF403	Hedgehog signaling	↓ MGMT	R_T98G	U251-aR	[129] **In vivo**
	RGFP109	HDAC1, HDAC3	↑ NF-kB/p65 acetylation ↓ NF-kB regulated pro-survival genes		A172-aR, U87-aR, U118-aR, U251-aR	[130] In vitro
	Statin	HMG-CoA reductase	↓ TGFβ signaling	S_U87, U251, SJ-GBM2 R_T98G		[123, 124] **In vivo**
	WA (Withaferin A)	AKT	G2/M arrest ↑ AKT/mTOR/MAPK pathway, ERK activation ↓ MGMT		U87-aR, U251-aR	[131] In vitro
	XL765	PI3K/mTOR	↑ CHOP/DR5 pathway ↓ ER stress	S_A172, U87 R_T98G		[132] **In vivo**
KD (knock-down)	Ezh2 KD	Ezh2	G1/S phase cell cycle arrest, apoptosis		U87-aR, U251-aR	[90] In vitro
	RBBP4 KD	RBBP4	↑ DNA damage	R_T98G		[59] **In vivo**
KO (knock-out)	ATRX KO	ATRX	↑ PARP1 cleavage, γ-H2AX, caspase 3/7		HG7-aR, HG9-aR	[97] **In vivo**
miRNA	Anti-miR-155	miR-155	Apoptosis	R_T98G		[133] In vitro
	Anti-miR-221	miR-221	Apoptosis	R_T98G		[133] In vitro
	Anti-miR-1238	miR-1238	CAV1/EGFR pathway		U251-aR	[103] **In vivo**
	Anti-miR-221	miR-221	Apoptosis	R_T98G		[134] In vitro
	Anti-miR-222	miR-222	Apoptosis	R_T98G		[134] In vitro

Continued

TABLE 2 The summary of therapeutic options for TMZ-resistant GBMs—cont'd

Group	Agent	Primary target	Pathway	TMZ sensitive (S) or Innate TMZ-resistant (R) cells	Acquired TMZ-resistant cells (-aR)	References[a]
	Anti-miR-124	miR-124, EZH2		GSC (G34)		[135] **In vivo**
	Anti-miR-128	miR-128, BMI1		GSC (G34)		[135] **In vivo**
	Anti-miR-137	miR-137, LSD1		GSC (G34)		[135] **In vivo**
	miR-101		↓ GSK3β		A172-aR, U251-aR	[81] **In vivo**
	miR-146b-5p		↓ TRAF6		U87-aR, U251-aR	[83] **In vivo**
	miR-126-3p		↑ SOX2 ↓ Wnt/β-catenin		U87-aR, U251-aR	[82] In vitro
	miR-519a		↑ Apoptosis, autophagy		U87-aR	[84] **In vivo**
Protein	BMP2 (bone morphogenetic protein 2)		↓ MGMT	GBM stem-like cells		[42] In vitro
siRNA	APE1 (DNA-(apurinic or apyrimidinic site) lyase) siRNA	APE1	↑ DNA damage, S phase arrest, H2AX phosphorylation, apoptosis	R_T98G		[136] In vitro
	MGMT (O^6-methylguanine-DNA methyltransferase) siRNA	MGMT		R_T98G, U251nu/nu GICs, GSC		[137] **In vivo**
	SCD1 (stearoyl-coenzyme A desaturase 1) siRNA	SCD1			U87-aR, T98G-aR	[92] In vitro

Bold indicates in vivo study.
[a] *The number in the blanket is the reference information.*

Targeting DNA repair and DNA damage response enzymes

Many DNA repair genes and proteins are overexpressed or downexpressed in the innate and acquired TMZ-resistant GBMs. MGMT is a DNA repair enzyme and responsible for resistance to O^6-alkylating agent-based chemotherapy. When innate TMZ-resistant T98G cells are exposed to MGMT inhibitor O^6-benzylguanine (O^6-BG) before TMZ treatment, the T98G cells showed a significant decrease in cell viability not by apoptosis but autophagy [127]. A natural small molecule epigallocatechin gallate (EGCG, a type of catechin) was cytotoxic on MGMT positive cells (T98G, GBM-XD) by downregulated MGMT mRNA and protein expression, via WNT/β-catenin pathway, prevent β-catenin translocation into the nucleus and inhibit the transcription factors (TCF1, LEF1) [119]. Downregulation of MGMT by small interfering RNA (siRNA) enhanced the chemosensitivity of malignant gliomas (T98G, U251nu/nu) against TMZ. TMZ-resistant glioma initiating cells are sensitized to TMZ by liposome-mediated MGMT siRNA in both in vitro and in vivo tumor models [137].

APE1 (a base excision repair pathway enzyme) knockdown by siRNA inhibited proliferation of T98G cells via increased DNA damage, S-phase arrest, and H2AX phosphorylation resulted in apoptosis induction. TMZ-treatment and APE1 knockdown significantly reduced cell invasion in TMZ-sensitive U87-MG cells and TMZ-resistant T98G cells [136].

Downregulation of Ezh2 expression altered the MDR, MRP, BCRP mRNA and protein levels, cell cycle arrest in the G1/S phase and resulted in apoptosis in acquired TMZ-resistant GBM cells (U87-aR, U251-aR) [90].

Several recurrent GBMs had mutations in the MMR genes (MSH2, MSH6, MLH1, PMS2) [12]. Thus, MSH2 attenuation is a potent mediator of TMZ resistance [72].

TMZ analogs were tested for TMZ-resistant GBM cells. Two TMZ analogs, C8-imidazolyl and C8-methylimidazole tetrazines, inhibited cell viability of innate TMZ-resistant T98G cells (MGMT-overexpressing) and MMR deficient colorectal carcinoma HCT116 cells through cell cycle arrest at G2/M, DNA double-strand break damage and apoptosis, and the same DNA incorporation form (N^7-mG, O^6-mG, N^3-mA, N^3-mT, N^3-mC) as TMZ [113].

Etoposide (DNA topoisomerase II inhibitor)-induced apoptotic cell death in innate TMZ-resistant T98G cells through S and G2/M cell cycle arrest [64].

Signaling pathway inhibitors

(i) JAK/STAT/HIF-1α pathway

For innate TMZ-resistant GBM and GSC treatment, a certain key enzyme inhibitor has been studied as a therapeutic agent. For example, Pacritinib, a JAK2/STAT3 inhibitor, improved the response to TMZ in TMZ-resistant patient-derived brain tumor-initiating cells (BTICs) [128]. BMP2 (bone morphogenetic protein 2; a protein that belongs to the TGF-β superfamily)-induced downregulation of HIF-1α and MGMT in GBM stem-like cells [42]. HIF-1α inhibitor FM19G11 decreased MGMT expression via the NF-kB pathway in T98G cells at both normoxic and hypoxic conditions [120]. FM19G11 alone was not cytotoxic to GBM cells at the HIF-1α inhibition concentration; however, it enhanced the cytotoxic effect of TMZ in hypoxic T98G cells, hypoxic GBM-XD cells, and normoxic T98G cells [120].

(ii) PI3K/AKT/mTOR & Ras/ERK pathway

The PI3K specific inhibitor GDC-0941 is cytotoxic to three GBM cells (A172, T98G, SHG44) including innate TMZ-resistant GBM cells (T98G) [125]. XL765, a dual inhibitor of the PI3K/mTOR pathway, inhibits GBM (A172, U87-MG, T98G) growth via induced ER (endoplasmic reticulum) stress and activation of CHOP/DR5 pathway-dependent apoptosis [132].

For the treatment of acquired TMZ-resistant GBMs, inhibitors for identified essential signaling pathway molecules are studied as a therapeutic agent for acquired TMZ-resistant GBMs. For example, many Akt pathway targeting compounds are tested for acquired TMZ-resistant GBMs. Withaferin A (WA), a steroidal lactone derived from several genera of the Solanaceae plant family, inhibited acquired TMZ-resistant GBM cells (U87-aR, U251-aR) via MGMT depletion, G2/M cell cycle arrest, depletion of principle proteins of the Akt/mTOR and MAPK pathways, and activation of ERK1/2 [131].

EGFR is an upstream signaling molecule of Akt. Thus, targeting EGFR is essential for TMZ-resistant GBMs. Furanodienone (FUR), a diene-type sesquiterpene isolated from the rhizomes of *Rhizoma curcumae*, exhibited a potential cytotoxic effect on acquired TMZ-resistant GBM cells (U87-aR, U251-aR) in vitro by inhibition of EGR1-mediated transcription and suppressing CSPG4-related signaling pathways [121].

Connective tissue growth factor (CTGF) promoted TMZ resistance in GBM (U87-MG, U251) through TGF-β1-dependent activation of Smad/ERK signaling [138]. Therefore, TGF-β signaling inhibitors (irinotecan, statin) are cytotoxic to MMR-deficient GBM and GBM initiating cells (SJ-GBM2 has undetectable MGMT; T98G, U87-MG, U251) [123, 124].

(iii) Hedgehog pathway

Hedgehog signaling pathway inhibitors are assessed for innate and acquired TMZ-resistant GBMs. Silencing GLI1 (zinc finger protein, glioma-associated oncogene, effectors of Hedgehog signaling) does not induce apoptosis in GBM cells (U87-MG, T98G) without TMZ cotreatment [139]. The cotreatment of GLI inhibition and TMZ enhances apoptosis in neurospheres and impairs neurosphere growth [139].

PF403, the active metabolite of CAT_3, blocked the hedgehog-signaling pathway, downregulated the expression of MGMT, and inhibited the proliferation of T98G cells. PF403 also inhibited proliferation of acquired TMZ-resistant GBM cells (U251-aR) by repression of the Hedgehog signaling pathway, MGMT expression, and abolished the effects of the Hedgehog pathway. Moreover, CAT_3 suppressed growth in the T98G subcutaneous xenograft tumor model and U251-aR orthotopic subcutaneous xenograft tumor model in vivo [129]. These data support the role of the hedgehog signaling pathway for innate TMZ resistance.

(iv) Eph receptor signaling pathway

Eph receptor kinase inhibitor GLPG1790 inhibits glioma growth in vitro and antitumor efficacy in vivo via downregulated mesenchymal markers, upregulated GFAP (glial fibrillary acidic protein), and pro-neural/neuronal markers [126].

(v) NF-kB pathway

NF-kB is a downstream signaling molecule of Akt. Thus, targeting NF-kB is also essential for treating TMZ-resistant GBMs. RGFP109, a selective histone deacetylase (HDAC1 and HDAC3) inhibitor, overcomes TMZ resistance and downregulates the expression of NF-κ

B-regulated pro-survival genes in acquired TMZ-resistant GBM cells (A172-aR, U87-aR, U118-aR, U251-aR). RGFP109 induces the hyperacetylation of NF-κB/p65 and histones and blocks interactions between NF-κB/p65 and its coactivators, p300 and p300/CBP-associated factor (PCAF). RGFP109 reduces NF-κB/p65 binding to κB-DNA and decreases the transcriptional level of κB-mediated genes. RGFP109 elevates the expression of inhibitor of growth 4 (ING4), which is typically downregulated in GBM cells. RGFP109 enhances ING4 recognition and binding to NF-κB/p65, which may be positively correlated with reduced interactions between NF-κB/p65 and p300/PCAF, thereby effecting transcription of the κB gene [130].

Dipotassium glycyrrhizinate (DPG), a dipotassium salt of glycyrrhetic acid (GA)—a compound isolated from *Glycyrrhiza glabra*, decreased cell viability of GBM cells (U87-MG, T98G) through the suppression of NF-kB signaling pathway, overexpression of *miR-16* and *miR-146a* expression, and decreased its target gene *TRAF6* mRNA expression in GBM cells (U87-MG, T98G) [114].

Targeting transcription factor and transcriptional regulator

Among the transcription factors, STAT3 is the most associated transcription factor with not only GBMs but also innate/acquired TMZ-resistant GBMs. Knockdown of STAT3 turns down the expression of Snail in mRNA and protein, thereby sensitizing the resistant GBM cells to TMZ treatment. Depletion of Snail (a family of a transcription factor) by siRNA induced the recurrent GBM cells sensitive to TMZ [99].

Similar to siRNA/RNAi, gene knockout is used to inhibit gene expression. For example, ATRX (a transcriptional regulator) knockout cells showed more cleaved PARP1, increased γ-H2AX level and CASP3/7 activity induced by TMZ, compared with control cells. ATRX elicits PARP1 stabilization through downregulating FADD expression in acquired TMZ-resistant PDX cells (HG7-aR, HG9-aR). The study provides that PARP inhibitor might be a potential adjuvant agent in overcoming ATRX-medicated TMZ resistance in glioma [97].

DNA synthesis inhibition

Many inhibitors for key enzymes such as DNA synthesis related are studied as a therapeutic target in the innate and acquired TMZ-resistant GBMs. Ribonucleotide reductase is a key enzyme that mediates the DNA synthesis in the cells. Ribonucleotide reductase inhibitors (didox, trimidox) were cytotoxic to innate TMZ-resistant GBM cells (T98G) [118].

Gene therapy

GLS2-encoded glutaminase displays tumor-suppressive properties. The GAB gene, main *GLS2* isoform, transfected innate TMZ-resistant GBM cells (T98G) decreased cell viability and migration and increased TMZ sensitivity and oxidative stress [140, 141]. The GAB transfected GBM cells (TMZ-sensitive U87-MG, TMZ-resistant LN229) protect H_2O_2-mediated cell death by less induced PI3K/AKT pathway compared to control cells [116].

Ligand and receptor

A certain ligand or receptor can be found in GBM cells. Recent studies showed that GBM cells uniquely express ERβ18, which has tumor suppressor functions [142, 143]. In vivo tumor model, ERβ sensitizes GBM to TMZ therapy via induced DNA damage response and apoptosis [144].

Metabolism inhibition

There is mitochondrial dysfunction and mitochondrial heterogeneity among GBMs. Mitochondrial heterogeneity decreased activities of mitochondrial complexes I and V and upregulated cytochrome c oxidase activity [117]. Therefore, mitochondrial complex IV or cytochrome oxidase is a potential target for GBM. The classic mitochondria inhibitors (e.g., rotenone, antimycin A, oligomycin meclizine, and cetylpyridinium chloride) targeting mitochondrial aberrant signaling cascades are considered as a potential therapeutic for GBMs [145]. For example, ADDA5, a cytochrome *c* oxidase inhibitor, showed dose-dependent inhibition of cell proliferation in TMZ-sensitive U251 cells and acquired TMZ-resistant U251 cells (U251-aR) [117].

Dichloroacetate (DCA), a metabolic inhibitor of pyruvate dehydrogenase kinase 1 (PDK1), reversed the Warburg effect, induced mitochondrial membrane potential change, and induced apoptosis in acquired TMZ-resistant U373EGFRvIII cells (U373vIII-aR) [88].

Fatty acid synthesis protein SCD1 downregulation by siRNA could resensitize acquired TMZ-resistant cells (U87-aR, T98G-aR) in vitro [92].

Sodium selenite (inorganic antioxidant compound) was more cytotoxic than TMZ in 3D cultured GBM cells (LN229, U87-MG, T98G) through S phase cell cycle arrest, decreased histone deacetylase activity and modulated H3K9m2 (dimethyl-histone-3-lysine-9 methylation) levels, and caspase 3-mediated apoptosis [115].

A selective cytochromes P450 inhibitor, HET0016, conjugated HPβCD (2-hydroxypropyl-beta-cyclodextrin) injected intravenously into the mouse brain tumors (syngeneic GL261 GBM, U251 GBM model in rat) decreased tumor growth [122].

miRNA modulation

miRNA is a subclass of noncoding RNAs (ncRNAs) of 20–22 nucleotides in length and function by targeting cellular messenger RNAs, thus preventing protein translation. Therefore, miRNAs are studied as a target for anticancer drug development [146]. Because several miRNAs are increased in innate and/or acquired TMZ-resistant GBM/PDX cells, inhibiting increased miRNAs are studied as a therapeutic approach. For example, coadministration of anti-miR-155 and anti-miR-221 peptides-induced apoptosis in innate TMZ-resistant T98G cells [133]. The loss of miR-1238 sensitized acquired TMZ-resistant GBM cells (U251-aR) by directly targeting the CAV1/EGFR pathway [103]. Cotreatment of anti-miR-221 and anti-miR-222 nanoparticle-induced apoptosis in innate TMZ-resistant T98G cells [134]. Multi-miRNAs targeting transcriptional regulators involved in GBM recurrence after therapeutic chemoradiation, including EZH2 (enhancer of zeste 2 polycomb repressive complex 2 subunit), BMI1 (B cell-specific Moloney murine leukemia virus integration site 1), and LSD1 (lysine-specific demethylase 1A), were studied. Synchronized inhibition of three miRNAs (miR-128 for BMI1 target, miR-124 for EZH2 target, miR-137 for LSD1 target) by gene therapy showed significant synergism of anticancer effect and results in a fivefold increase in survival when combined with TMZ in murine intracranial GBM model using G34 glioma stem cells [135].

In addition, restoring the decreased miRNAs are studied as a therapeutic method because many tumor suppressor miRNAs are downregulated in innate and/or acquired TMZ-resistant GBM/PDX cells. For example, overexpression of miR-101 could sensitize acquired TMZ-resistant GBM cells (A172-aR and U251-aR) through downregulation of glycogen synthase kinase 3β (GSK3β). GSK3β inhibition could enhance TMZ effect through repression of MGMT via promoter methylation [81]. Overexpression of miR-146b-5p restored sensitivity of acquired TMZ-resistant GBM cells (U87-aR, U251-aR) to TMZ by decreasing TRAF6 (targeting tumor necrosis factor receptor-associated factor 6) [83]. High levels of miR-126-3p enhanced TMZ sensitivity in GBM tissues and acquired TMZ-resistant GBM cells (U87-aR, U251-aR) by inhibiting cell viability, reducing colony-forming potential and inducing apoptosis. Furthermore, miR-126-3p-induced inactivation of Wnt/β-catenin signaling was greatly abrogated by SOX2 upregulation [82]. miR-519a dramatically enhanced TMZ-induced autophagy and apoptotic cell death in acquired TMZ-resistant GBM cells (U87-aR) [147]. These results support the feasibility of miRNA strategies for anticancer therapies in acquired TMZ-resistant GBMs.

Therapeutic approaches for TMZ-resistant GBM cells to overcome TMZ resistance: Multitarget approach with TMZ

Many combination therapies with or without TMZ have been tested in innate or acquired TMZ-resistant GBMs/PDXs. We summarized tested combination agents for innate or acquired TMZ-resistant GBM/PDX cells in Table 3.

(i) Targeting for DNA repair or DNA damage response enzymes with TMZ

In addition to a single agent, either a single or multiple agents with TMZ are studied for innate TMZ-resistant GBMs. Dual targeting of two pathways (DNA repair pathway and another pathway) is studied for GBM. The administration of O^6-BG and TMZ to T98G cells (MGMT proficient, p53 mutated) inhibited cell viability by reduced expression of MGMT mRNA and protein, and a decrease in NF-kB and mutant p53 level [153]. However, upon withdrawing drugs, the cells had increased cell viability indicating the role of high expression of NF-kB and mutant p53 for TMZ resistance. The T98G cells exposed to O^6-BG and TMZ had cell cycle arrest in the G2-M phase. Therefore, a series of MGMT inhibitors (25 O^6-alkyl, O^6-aryl, O^6-substituted-aryl guanine analogs) was synthesized/tested and found that three compounds decreased T98G cell viability when combined with TMZ [158]. The combination of multiple drugs (three or more) was investigated to sensitize glioma cells. For example, the combination of TMZ with BER inhibitor (methoxyamine hydrochloride) and NAD(+) biosynthesis inhibitor (FK866) significantly sensitized TMZ-resistant human GBM LN428/MPG/MGMT cells, which are MPG (*N*-methylpurine DNA glycosylase) and MGMT stably overexpressed cells, by increased expression of MGMT but deficient MMR expression [157].

The combination of *DYNC2H1* (cytoplasmic dynein 2 heavy chain 1, also called DHC2) knockdown by shRNA or knockout and TMZ showed a significant reduction of cell viability in MGMT negative GBM/PDX cells (U87-MG, GBM1) via increased H2AX phosphorylation (γH2AX), a marker for DNA damage status. The combination effect among *DYNC2H1* and TMZ is not observed in the MGMT positive GBM/PDX cells (T98G, GBM2). Knockdown or knockout of *DYNC2H1* had significant effects on tumorigenic potential and survival without TMZ treatment in MGMT deficient GBMs [52].

TABLE 3 The summary of combination therapy with TMZ or without TMZ for GBMs.

Group	Agent	Primary target	Pathway	TMZ sensitive (S) or innate TMZ-resistant (R) cells[a]	Acquired TMZ-resistant cells (-aR)	References[b]
Agonist	Dexamethasone + TMZ	Glucocorticoid receptor	No cell death	R_T98G		[148, 149] In vitro
Analog	Cordycepin + TMZ	Adenosine	↓ p-mTOR, p-p70S6K, MMP-2, MMP-9	S_U87, U251 R_LN18, T98G		[108] **In vivo**
Antioxidant	Luteolin + silibinin	Antioxidants	↓ PKCα, miR-7-1-3p ↑ apoptosis	S_U87 R_T98G		[150] **In vivo**
	Resveratrol + TMZ	Cyclooxygenases, lipooxygenases, kinases, sirtuins	Nuclear translocation of NF-kB ↓ MGMT, wnt/β-catenin ↑ Caspase3 cleavage	R_T98G, U138		[151] In vitro [152] **In vivo**
Inhibitor	Bortezomib + TMZ	20S proteasome	Nuclear accumulation of NF-kB, p53 stabilization, attenuation of p44/42 MAPK, AKT & STAT3 activation, inactive HIF-1α ↓ MGMT ↑ AKT/STAT3	R_T98G		[153] In vitro
	O^6-BG + TMZ	MGMT	Cell cycle arrest at G2/M ↓ MGMT, NF-kB	R_T98G		[153] In vitro
	Celecoxib + TMZ	Wnt, COX-2		R_T98G		[44] **In vivo**
	Didox + TMZ Trimidox + TMZ	Ribonucleotide reductase		R_T98G		[118] In vitro
	GANT-61 + TMZ	GLI1	G2/M arrest	R_T98G		[154] In vitro
	HET0016 + TMZ	Cyt P450		GBM811		[122] **In vivo**
	GDC-0941 + TMZ	PI3K α/δ	Autophagy ↓ MGMT, p-AKT	S_A172 R_T98G SHG44 (PDX)		[125] In vitro

Ibudilast + TMZ	Phosphodiesterase	Cell cycle arrest, apoptosis	PEX		[155] **In vivo**
KU0063794 + TMZ	mTORC1 and mTORC2	No cell death ↑ MGMT	R_T98G		[156] In vitro
Methoxyamine + TMZ	BER	↑ MPG	LN428 R_T98G		[157] In vitro
Methoxyamine hydrochloride + FK866 + TMZ	BER Nicotinamide phosphoribosyltransferase		LN428/MPG/MGMT cells		[157] In vitro
MGMT inhibitors (25 O^6-alkyl or O^6-aryl or O^6-substituted-aryl guanine analogs) + TMZ	MGMT		R_T98G		[158] In vitro
NSC 74859 + TMZ	STAT3		R_T98G	U87-aR	[98] In vitro
OKN-007 + TMZ	SULF2	IDO-1, SUMO2, PFN1	R_LN18, T98G, U138, G55 (PDX)		[159] **In vivo**
2-Propylpentanoic acid + TMZ	Histone deacetylase	G2 cell cycle arrest ↑ Apoptosis	S_SF295 R_T98G		[160] In vitro
PX-866 + TMZ	PI3K	Autophagy (treatment order is important)	R_T98G, U373vIII		[161] In vitro
Salinomycin + TMZ	Wnt/β-catenin		R_T98G		[44] In vitro
SCD1 inhibitor + TMZ	SCD (stearoyl-CoA desaturase)			U87-aR, T98G-aR	[92] In vitro
Sulfasalazine + TMZ	XCT, antiinflammatory agent	↓ Glutathione	S_A172 R_T98G		[162] In vitro
Veliparib + TMZ	PARP	Hypermethylated MGMT	GBM12 (PDX)		[39] **In vivo**
XL765 + TMZ	PI3K/mTOR		S_A172, U87 R_T98G		[132] **In vivo**

Continued

TABLE 3 The summary of combination therapy with TMZ or without TMZ for GBMs—cont'd

Group	Agent	Primary target	Pathway	TMZ sensitive (S) or innate TMZ-resistant (R) cells[a]	Acquired TMZ-resistant cells (-aR)	References[b]
Fatty acid derivative	Valproic acid + TMZ	Histone deacetylase	↓ MGMT	R_T98G, U138		[163] **In vivo**
KD (knock-down)	ATG7 KD + EGFR KD	ATG7 EGFR			T98G-aR	[164] In vitro
	DYNC2H1 KD + TMZ	DYNC2H1	↑ H2AX phosphorylation	S_U87, GBM1 (PDX)		[52] **In vivo**
	EGFR KD + TNF KD	EGFR TNF			MGMT unmethylated GBM	Recurrent GBM [165] **In vivo**
	P4HB KD + bacitracin inhibition	P4HB			D54-aR, U87-aR	[94] **In vivo**
Metabolic	Metformin + TMZ	Hepatic gluconeogenesis	↓ ROS, CD90 ↑ Bax/Bcl-2 ratio	S_U251 R_T98G	U87-aR, U251-aR	[166] **In vivo**
Mixture	OLE (Olea Europaea leaf extract) + TMZ		↑ miR-137, miR-145, miR-153, miR-1816, let-7d (in T98G) ↑ miR-137, miR-145, miR-153, (in GSCs)	S_U87 R_T98G, U138		[167] In vitro
Nucleic acid/peptide	Anti-miR-21 nucleic acid + TMZ	miR-21		R_T98G		[77] In vitro
	Anti-miR-221 nucleic acid + TMZ	miR-221		R_T98G		[78] In vitro
	Anti-miR-155 nucleic acid + anti miR-221 nucleic acid + TMZ	miR-155 miR-221		R_T98G		[133] In vitro
	Chlorotoxin + TMZ	NO donor	↓ MGMT Altered p53	S_U87 R_T98G		[168] In vitro

	SNAP (*S*-nitroso-*N*-acetylpenicillamine) + TMZ	NO donor	↓ MGMT ↑ Sub-G1 cell cycle arrest, apoptosis. intracellular HIF-1 accumulation Posttranslational modification of p53, phosphorylated N-myc downstream regulated gene 1, and MGMT protein stability	R_LN229, GBM8401	LN229-aR, GBM8401-aR	[169] **In vivo**
	tFNA + TMZ		↓ MGMT ↑ Apoptosis and autophagy	S_A172, U87 R_LN18, T98G		[170] In vitro
siRNA or shRNA	AKT3 siRNA + TMZ	AKT3	G1 cell cycle arrest, apoptosis, necrosis, autophagy	R_T98G		[171] In vitro
	PI3KCA siRNA + TMZ	PI3KCA	G1 cell cycle arrest, apoptosis, necrosis, autophagy	R_T98G		[171] In vitro
	RBBP4 shRNA + TMZ	RBBP4	↑ DNA damage	R_T98G		[59] **In vivo**
	SNHG15 siRAN + TMZ	SNHG15		PDX		[58] **In vivo**
Pathway	Sulforaphane + TMZ	KEAP-1/NRF2 signaling (Nrf2 inducer)	↓ MGMT NF-kB-dependent pathway		U87-aR, U373-aR	[172] **In vivo**
Protein	IFN + TMZ		↓ MGMT	R_T98G, U251nu/nu		[173, 174] In vitro
Virus	Oncolytic virus Delta-24-RGD + TMZ		Prevent the recruitment of p300 to the MGMT promoter by Delta-24-RGD & p300 interaction ↓ MGMT	S_U87 R_T98G In vivo		[175] **In vivo**

Bold indicates in vivo study.
[a] *A certain cell's response to TMZ is not clear whether the cells are sensitive or resistant to TMZ.*
[b] *The number in the blanket is the reference information.*

(ii) Nucleic acid with TMZ

Combination of cordycepin (an adenosine analog) and TMZ reduced the expression levels of p-mTOR, p-p70S6K, matrix metalloproteinase-2 (MMP-2), and MMP-9, compared to those in the groups treated with either cordycepin or TMZ alone in TMZ-sensitive and TMZ-resistant GBM cells (U87-MG, U251, LN18, T98G) [108].

A tetrahedral framework nucleic acid (tFNA) nanoparticle-containing TMZ enhanced the cytotoxicity on TMZ-sensitive GBM cells (A172, U87) and innate TMZ-resistant GBM cells (LN-18, T98G) via MGMT down expression, activating the cell apoptosis and autophagy pathway [170].

(iii) PI3K/Akt/mTOR pathway inhibitor with TMZ

Combination treatment of AKT3 or PI3KCA siRNA and TMZ in T98G cells significantly reduced cell viability, accumulation of subG1-phase cells, reduction of S and G2/M phases, induction of apoptosis or necrosis, and regulation of autophagy [171]. The combination of GDC-0941 and TMZ inhibited cell viability enhanced proapoptotic effect, augmented autophagy response, and attenuated migration/invasion in GBM cells (A172, T98G, SHG44) by decreased MGMT and p-AKT [125]. However, TMZ and mTOR inhibitor KU0063794 in T98G cells resulted in increased MGMT protein levels indicating that mTOR inhibition may not be a useful therapy for TMZ-resistant GBM cells [156]. A dual inhibitor of PI3K/mTOR pathway, XL765, and TMZ enhanced in vitro cell inhibition of GBM cells (A172, U87-MG, T98G) [132]. The combination of TMZ and PX-866, a BBB permeable PI3K inhibitor and an early-stage inhibitor of autophagic flux, inhibited colony formation compared to TMZ alone in T98G and U373vIII cells while maintaining its ability to inhibit PI3K/AKT signaling in glioma cells [161]. They found the importance of treatment order, i.e., the inhibition of colony formation was observed via the sequential treatment of TMZ followed by PX-866 rather than cotreatment.

(iv) Hedgehog pathway inhibitor with TMZ

TMZ and the Sonic hedgehog (SHH) pathway inhibitor GANT-61 increased DNA fragmentation and G2/M arrest in T98G cells [154].

(v) Wnt/β-catenin signaling pathway inhibitor with TMZ

Two Wnt inhibitors (e.g., celecoxib, salinomycin) augmented the cytotoxic effects of TMZ in the T98G cells [44].

(vi) Transcription factor inhibitor with TMZ

The combination treatment of STAT3 inhibitor VI (NSC 74859) or STAT3 knockdown and TMZ potentiated TMZ efficacy in innate TMZ-resistant GBM cells (T98G) and acquired TMZ-resistant GBM cells (U87-aR) [98].

(vii) Protein inhibitor with TMZ

A combination between an enzyme inhibitor and TMZ is assessed in innate TMZ-resistant GBMs. For example, the combination treatment of ribonucleotide reductase inhibitors (didox, trimidox) and TMZ were cytotoxic to innate TMZ-resistant GBM cells (T98G) [118].

Interestingly, the combination effect depends on the ratio of two drugs (didox or trimidox) and TMZ. BER inhibitor (methoxyamine) enhanced the TMZ effect in GBM cells (LN428, T98G) via overexpression of the BER initiating enzyme MPG [176]. Further, increased gene/protein expression with TMZ has been tested for innate TMZ-resistant GBMs. Stably LGA (liver-type glutaminase) transfected T98G GBM cells inhibited glioma growth and had increased sensitivity to TMZ and BCNU combination treatment [177]. Pretreatment of GBM cells (U87-MG, T98G) with a glioma-specific and nitric oxide-donating chlorotoxin derivative increased the efficacy of TMZ or BCNU through decreased MGMT level and altered p53 activity [168]. Another nitric oxide donor *S*-nitroso-*N*-acetylpenicillamine (SNAP) also sensitized acquired TMZ-resistant GBM cells (LN229-aR, GBM8401-aR) to TMZ through downregulation of MGMT expression, posttranslational modification of p53, phosphorylated N-myc downstream-regulated gene 1, and MGMT protein stability [169]. Treatment of proteasome inhibitor bortezomib followed by TMZ treatment overcame TMZ resistance in T98G cells through downregulation of MGMT, nuclear accumulation of NF-kB, attenuation of p44/42 MAPK, AKT and STAT3 activation, p53 stabilization, and inactive HIF-1α [178].

Inhibitors of several enzymes (histone deacetylase, PARP, stearoyl-coenzyme A desaturase 1, etc.) have been evaluated with TMZ in acquired TMZ-resistant GBMs. A combination of Valproic acid (a short-chain fatty acid that inhibits histone deacetylases) and TMZ had a significantly enhanced antitumor effect in TMZ-resistant glioma cells (T98G, U138) through reduced MGMT expression [163]. The combination showed enhanced apoptotic and autophagy cell death in vitro cell culture. Another histone deacetylase inhibitor, 2-propylpentanoic acid, and TMZ significantly inhibited GBM (SF295, T98G) growth via a decrease in the G2 phase cells and increased apoptotic cell death [160]. This result suggests that 2-propylpentanoic acid may enhance the effect of TMZ on GBM cells through cell cycle blocking and promote autophagy. The combination of veliparib (a PARP inhibitor) and TMZ is significantly cytotoxic in innate and acquired TMZ-resistant cells as well as extends mice survival of mouse GBM models (e.g., GBM12) which have hypermethylated MGMT, but there is no effect on MGMT overexpressed mouse GBM models [39, 179]. The combinational treatment of TMZ and an SCD1 (stearoyl-coenzyme A desaturase 1) specific inhibitor showed a combined inhibitory effect on acquired TMZ-resistant GBM cells (U87-aR, T98G-aR) [92]. SCD1 catalyzes the rate-limiting step in the production of monounsaturated fatty acids that are major components of tissue lipids. The combination of PARP inhibitor and TMZ inhibited glioma growth in ATRX wild-type TMZ-resistant intracranial xenograft models [97].

Although TMZ alone did not alter cell viability, the combination of sulfasalazine (a cysteine/glutamate antiporter xCT inhibitor and an antiinflammatory agent) and TMZ is cytotoxic to GBM cells (A172, T98G) via depletion of glutathione level [162].

TMZ and HET0016 (a selective cytochromes P450 inhibitor) showed prolonged survival in xenografts of GBM811 [122].

In human GBM specimens with MGMT methylation, macrophage migration inhibitory factor (MIF), also known as a glycosylation-inhibiting factor (GIF), was identified as overexpression and the increased MIF is associated with poor survival. When a specific MIF inhibitor (ibudilast) and TMZ combined significant synergism in the antiproliferation effect by cell cycle arrest and apoptosis was observed in vitro and in vivo PDX model [155].

RBBP4 (retinoblastoma binding protein 4) knockdown by shRNA enhanced TMZ sensitivity in innate TMZ-resistant GBM cells (T98G) and increased DNA damage signaling in response to TMZ. Moreover, RBBP4 disruption suppresses tumor growth and TMZ sensitivity in vivo [59].

(viii) Cytokine with TMZ

IFNs (interferons) are a family of cytokines and recognized as effective in cancer therapy. The combination of IFN-β and TMZ synergistically inhibits GBM growth and downregulated MGMT mRNA levels compared to treatment with IFN-beta or TMZ alone in innate TMZ-resistant T98G cells [173]. The combination of IFN-β and TMZ sensitizes TMZ-resistant glioma cells (T98G, U251nu/nu) that harbor the unmethylated MGMT promoter via p53-mediated MGMT expression downregulation [174].

(ix) miRNA with TMZ

Several studies used TMZ and miRNA (miR-21, miR-221) targeting nucleic acid-loaded nanoparticles for innate TMZ-resistant GBM cells (T98G) [77, 78]. In addition, TMZ further increased the effect of cotreatment between anti-miR-155 and anti-miR-221 peptides-mediated apoptosis in T98G cells [133].

(x) Long noncoding RNA with TMZ

The combination of SNHG15 silencing and TMZ makes TMZ-resistant PDX sensitive to TMZ [58].

(xi) Oncolytic virus with TMZ

Combination treatment of oncolytic adenovirus Delta-24-RGD and TMZ inhibited both U87-MG (TMZ-sensitive GBM) and T98G (TMZ-resistant GBM) cell proliferation [175]. The adenovirus Delta-24-RGD overcomes TMZ-mediated G2-M cell cycle arrest and downregulate MGMT mRNA expression via interaction between Delta-24-RGD and p300, resulting in the prevention of recruitment of p300 to the MGMT promoter.

(xii) Photodynamic therapy with TMZ

There is a synergism between photodynamic therapy and chemotherapy with TMZ, especially at low TMZ concentrations and higher light doses enhances the inhibition of GBM cells (U87-MG, T98G, U343), and overcomes cross-resistance mechanisms [180].

(xiii) Other compounds with TMZ

Furthermore, combination therapy between a compound and TMZ has been assessed in innate and acquired TMZ-resistant GBMs. For example, resveratrol, a natural polyphenol, and TMZ can reverse TMZ resistance of innate TMZ-resistant T98G cells in vitro and in vivo tumor model through decreased intracellular content and nuclear translocation of NF-kB and increased cleavage of caspase-3 [151]. The resveratrol sensitized TMZ-induced GBM (T98G, U138) cell apoptosis by inhibiting activation of the Wnt/β-catenin signaling pathway and downregulating MGMT expression [152]. Instead of a pure compound, a plant leaf extract is studied in innate TMZ-resistant GBM cells. For example, Olea Europaea leaf extract (OLE) causes apoptotic and necrotic cell death in GBM cells (T98G, U87-MG, U-138MG) [181]. OLE and TMZ had an antiproliferative effect on T98G cells via the upregulation of miRNAs such as miR-181b, miR-153, miR-145, miR-137, and let-7d [167].

In GSCs, OLE significantly induced the expression of miR-137, miR-145, and miR-153, and resulted in decreased TMZ target gene expression of these miRNAs [181]. Oleuropein (OL), the main compound in OLE, alone or with TMZ inhibited cell viability of T98G cells mainly via regulation of Let-7d expression [182].

The treatment of WA (Withaferin A, a steroidal lactone derived from several genera of the Solanaceae plant family, used in ayurvedic medicine) and TMZ resulted in resensitization of MGMT-mediated TMZ resistance but not resistance through mismatch repair mutations in acquired TMZ-resistant GBM cells (U87-aR, U251-aR) [131]. Metformin (an oral diabetes medicine that helps control blood sugar levels) enhances TMZ effect on GBM cells (U251, T98G) via increased Bax//Bcl-2 ratio, reduced ROS (reactive oxygen species), and decreased glioma stem cell marker CD90 [166]. Most importantly, metformin and TMZ decreased tumor growth of TMZ-resistant tumors in in vivo study; however, this effect is not observed in TMZ-sensitive tumor-bearing mice [166]. Metformin acts synergistically with TMZ to inhibit acquired TMZ-resistant GBM cells (U87-aR, U251-aR) by downregulating SOX2 expression, which reduces neurosphere formation capacity of GBM cells and inhibits GBM xenograft growth in vivo [183]. Sulforaphane (a natural compound within the isothiocyanate group of organosulfur compounds that improve heart health and digestion) significantly suppressed proliferation of TMZ-resistant GBM cells (U87-aR, U373-aR) by NF-κB-dependent pathway, then downregulating MGMT expression to reverse the chemoresistance to TMZ in T98G, U87-aR, and U373-aR cell lines [172].

A combination of OKN-007 (a nitrogen-based agent) and TMZ-treated mice had significantly better survival than TMZ alone treated G55 (TMZ-resistant cells) glioma bearing orthotopic mice [159]. The combination effect for cell proliferation was also found in TMZ-resistant GBM cells (T98G, LN18, U138, G55); however, the combination treatment showed a different direction for HIF-1α gene expression, MGMT gene expression, and MPG gene expression. The cytotoxic effect of combination treatment on TMZ-resistant GBM cells possibly via IDO-1, SUMO2, and PFN1 based on the microarray.

Not all drugs enhance the effect of TMZ in GBM. A pretreatment of dexamethasone (a steroid), blocked TMZ-induced apoptosis in innate TMZ-resistant GBM cells (T98G) [148, 149]. This result indicates that dexamethasone works as an antagonistic agent in combination with TMZ. However, acetazolomide which has the same effect as dexamethasone to alleviate vasogenic edema and inflammation pretreatment enhanced TMZ-mediated apoptosis in T98G cells [149] indicating the combination with TMZ can be drug dependent.

Therapeutic approaches for TMZ-resistant GBM cells to overcome TMZ resistance: Multitarget approach without TMZ

There are studies of the combination effect with two agents without TMZ for acquired innate or TMZ-resistant GBMs. For example, the combination of two natural flavonoids luteolin and silibinin synergistically inhibit GBM cells (U87-MG, T98G) due to suppression of PKCα and promotes apoptosis through downregulation of iNOS and increases tumor suppressor miR-7-1-3p, inhibits invasion and migration, and inhibited rapamycin-induced autophagy [150]. P4HB (prolyl 4-hydroxylase, beta polypeptide) inhibition by RNAi knockdown and bacitracin inhibition could sensitize acquired TMZ-resistant cells (D54-aR, U87-aR) to TMZ in vitro and in vivo [94]. In addition, silencing both EGFR and autophagy gene ATG7 resulted in an increase of TMZ cytotoxicity in acquired TMZ-resistant GBM cells (T98G-aR) [164]. The combination of EGFR inhibition and TNF inhibition is effective in MGMT unmethylated and recurrent EGFR-expressing GBMs orthotopic mouse model of GBM [165].

Clinical trial

According to clinicaltrials.gov, accessed on March 1, 2020, there are 352 completed and 236 ongoing clinical trials for primary GBMs. For recurrent GBMs, there are 239 completed and 128 ongoing clinical trials. Among them, there are three completed clinical trials for TMZ-resistant GBMs: (i) imatinib mesylate and hydroxyurea vs hydroxyurea alone for TMZ-resistant progressive GBM (Novartis), (ii) TMZ and O^6-BG for TMZ-resistant malignant glioma (Duke University), and (iii) VEGF trap for TMZ-resistant recurrent malignant gliomas (UCLA). However, there are no ongoing clinical trials for TMZ-resistant GBMs. Instead, we found a Bevacizumab-resistant recurrent GBM study (Beth Israel Deaconess Medical Center).

Methods for identification of novel targets for sensitizing TMZ

As mentioned in prior sections of this chapter, MGMT is the only protein considered in clinical practice as an important determinant of outcome following TMZ therapy in GBM [47, 184–187]. However, GBM patients with MGMT negative tumors due to promoter hypermethylation are initially responsive to TMZ therapy but also they eventually fail this therapy leading to recurrence [47, 188]. Therefore, as discussed in the prior sections, there may be other mechanisms beside MGMT that modulate TMZ responsiveness in GBM. As a result, there is a tremendous ongoing effort aimed at identification of these modulators of TMZ sensitivity, mainly because they may provide targets for novel TMZ-sensitizing therapies in GBM. Several studies have been conducted and many are ongoing both in vitro and in vivo to identify modulators of TMZ response in GBM [71, 138, 189, 190]. In this section, we will describe the development of TMZ-resistance models and the approaches that are deployed to discover novel targets for sensitizing TMZ in GBM.

Models for studying modulators of TMZ response in GBM

In vitro models of TMZ resistance

As previously described, the in vitro models of TMZ resistance can either be inherent (primary resistance) or acquired (secondary resistance). The in vitro primary resistant models are used to identify genes that may be driving primary resistance in GBM cells. Although primary TMZ resistance is largely owing to MGMT expression [42, 43, 49], there are several studies that have used this approach to identify other genes modulating TMZ resistance in GBM. For example, using this approach the MIF receptor CD74 was identified as a modulator of TMZ resistance in GBM [191]. Similar studies were conducted and demonstrated several proteins modulating TMZ sensitivity in GBM [85, 192–198]. Because of tumor heterogeneity, the findings from in vitro primary TMZ-resistant models are weakened by the fact that they are not developed from isogenic cells. Therefore, there is a possibility of missing the important drivers of TMZ resistance that may be dominant in other GBM cells but have not been evaluated. This weakness can be, at least in part, addressed by developing secondary TMZ-resistance models. In these models, isogenic GBM cells are treated with escalating concentrations of TMZ until cells are able to survive even when treated with concentrations similar or higher than levels that can be achieved in patients. This approach has been widely used to discover novel genes controlling TMZ resistance in GBM cells [71, 188, 199, 200]. Therefore, the in vitro models have significantly contributed to the current understanding of the mechanisms of TMZ resistance. Nonetheless, the main challenge is whether the mechanisms identified in vitro are representative of the mechanisms operating in vivo either in patients or PDXs, raising a need for generating in vivo models of resistance. This is even more relevant, in part because the in vivo models can be treated with exact dosing schedules given to patients. Thus, in the next section we describe the in vivo models of TMZ resistance and describe how they are developed and have been used to study mechanisms of TMZ resistance in vivo.

In vivo models of TMZ resistance

In vivo models of TMZ resistance are usually developed in mice and these models may be developed by implanting established or primary cultured cells from PDXs; however, the recommendation is to use PDXs because these models faithfully represent the situation in patient tumors [201]. As mentioned in the section above, one of the advantages of in vivo models is that the animals can be treated with TMZ regimen equivalent to those given to patients. In general, there are three approaches for developing in vivo models of TMZ resistance. First, in vivo models can be developed by implanting cells already known to be sensitive and resistant either primarily or secondarily induced, as described earlier. These models are specifically useful in evaluating TMZ response in vivo after biochemical manipulation of the driver resistance genes in the resistant group. Therefore, this type of model is mainly used for in vivo validation of studies conducted using in vitro models. This method has been and continues to be useful for evaluating mechanistic hypotheses related to TMZ sensitivity, especially in knockdown and knockout studies. For example, this model was used to demonstrate that knockdown of RBBP4 gene in resistant GBM cells sensitizes xenografts developed from the knockdown cells regardless of the MGMT status [59]. Several reports have reported the use of this model to confirm in vivo sensitization after manipulation of different genes, including Cx43, YKL-40, SOX9 [195, 199, 202].

The second in vivo model of TMZ resistance was developed by implanting GBM cells (primary or established) into the flanks of the mice; and, when the established tumors reached a predetermined volume mice are treated with the standard concentration of TMZ and then monitor tumor volumes. Initially, tumors will regress and when they regrow back to a predetermined volume, mice will receive another dose of TMZ but slightly higher than the first one. This tumor regrowth followed by high concentrations of TMZ treatment will be continued until the tumors fail to regress, even in response to the maximum TMZ concentration tolerable in mice. The control mice are parallel treated with a placebo. The placebo and resistant tumors can now be used to evaluate a resistance mechanism, preferably using high-throughput approaches described in the next section. Interestingly, a recent study using isogenic flank models developed by engrafting cells from GBM PDXs; and, the mice harbored tumors that regrew after the completion of escalating doses of TMZ identified multiple mechanisms of acquired TMZ resistance in MGMT methylated cells, broadly classified as MGMT-dependent and -independent mechanisms [203, 204]. The only downfall of this method is that the resistance was established using a TMZ regimen not similar to what is been used in the clinic for patients. Therefore, the resistance mechanisms induced by escalating TMZ dosing may possibly be different from the mechanisms induced in response to the three cycle TMZ regimen given to patients. Therefore, a model that will faithfully represent TMZ resistance in patients will likely be a regimen currently administered to patients and is detailed later.

The third in vivo model developed using a TMZ regimen is equivalent to the current regimen given to patients. Newly diagnosed GBM patients receive three cycles of TMZ daily for 5 days and are given after every 21 days [205–207]. Therefore, to mimic the resistance mechanisms that may be driven by each cycle of TMZ dosing, a research group at Mayo Clinic established in vivo models by treating mice harboring flank tumors with a single cycle of TMZ (50 mg/kg daily for 5 days); they harvested the recurrent tumors to evaluate the resistance mechanisms [204]. Similar to model 2 above, tumors that regrew from each mouse showed different resistant mechanisms: some showed reexpression of MGMT, though they were originally methylated, and other tumors were resistant due to unknown mechanisms.

Collectively, in vivo models especially those developed using PDX cells are feasible, in particular, those generated using a TMZ regimen similar to that given to patients, are more likely to provide mechanisms similar or close to those driving resistance in patients. These resistance models will only be meaningful if there are unbiased methods to identify the resistance mechanisms. In the next sections, we describe the methods for identifying modulators of TMZ sensitivity.

Methods used for identifying modulators of TMZ response

Methods for evaluation of a single modulator of TMZ resistance

(i) Protein expression

As discussed earlier, in studying models of TMZ resistance, the in vitro models often focus on evaluating single or few targets. In this case, there are several methods that can be used to evaluate the expression of target gene(s) comparing sensitive to corresponding resistant models. Western blotting is perhaps the most frequently used method for evaluating the expression of a limited number of proteins to establish whether they are differentially expressed between resistant and sensitive models. For example, evaluation of the MGMT protein by western blot stratified TMZ resistant from sensitive cells [43, 71, 208]. Moreover, western blotting analysis revealed expression of CD74 protein was overrepresented in the resistant compared to sensitive [191], while an inverse association was established for MMR proteins [72, 192, 209–211]. Western blotting can also be used to evaluate protein complexes that may be preferentially functional between TMZ-sensitive and -resistant cells.

(ii) RNA expression

Evaluation of mRNA is an alternative method to western blotting in a situation wherein no antibodies are available. Similar to western blotting, quantitative RT-PCR is used to validate TMZ-resistant targets identified by other high-throughput methods as outlined below.

Because TMZ-induced DNA lesions O^6-meG recognized by polymerase as thymine, tumors that regrew after failing TMZ therapy have enriched C > T mutations [212, 213]. These mutations contribute to TMZ resistance through the inactivation of MMR genes [212, 213]. While most of these C > T mutations, and perhaps other mutations induced by TMZ treatment, may not be damaging, the standard PCR followed by Sanger's sequencing is the method of choice for validating the damaging mutations that are found in resistant but not sensitive cells.

(iii) Promoter hypermethylation

Expression of genes may be suppressed without any evidence of mutations in the TMZ-sensitive vs -resistant cells. This is mostly owing to epigenetic changes that may include DNA or histone methylation. As mentioned previously, MGMT is the TMZ-resistant gene most frequently silenced by promoter methylation [47, 185–187]. Silencing by DNA methylation of a single or a cohort of genes has been linked to TMZ sensitivity [214]. DNA methylation is evaluated by first modifying the DNA by bisulfite followed by methylation-specific PCR (MSP) or Sanger's- or pyro-sequencing; the use of either approach have revealed methylated genes important for TMZ sensitivity, besides MGMT [198, 215–220]. In general, methylation of DNA recruits chromatin complexes that include histone methyltransferases [221]. Thus, the failure of these complexes to assemble upon methylated CpG nucleotides would lead to the expression of genes despite promoter methylation. This was found to be a key mechanism leading to the MGMT reexpression leading to TMZ resistance in some methylated GBM PDXs; therefore, chromatin precipitation (ChIP) assay is employed to identify specific histone changes within promoters of genes with an impacted expression on becoming resistant to TMZ [204]. In summary, we have described methods for evaluating single or few candidates that are differentially expressed among TMZ-sensitive and -resistant GBM cells. These candidates are validated after being identified using the high-throughput whole-genome screening approaches described later.

Methods for genome-wide screening modulators of TMZ resistance

(i) Proteomics

Activation of protein is mediated through posttranslational modifications, including phosphorylation, acetylation, and dephosphorylation [222]. Since DNA damage signaling is mediated mainly through protein phosphorylation TMZ-induced DSBs activate phosphorylation of proteins within not only DNA damage repair pathways but also other pro-survival proteins such as AKT [223]. Therefore, proteomics targeting protein phosphorylation has been employed as a global way of identifying proteins that are preferentially phosphorylated between TMZ-sensitive and -resistant cells [52, 87, 224]. Using proteomics, the proteome related to TMZ resistance was defined and the DHX9, HNRNPR, and RPL3 proteins were shown to be deferentially activated between sensitive and resistant GBM cells [87]. Since protein modifications are likely druggable targets, defining proteome mediating TMZ resistance would likely identify a novel target for sensitizing TMZ in GBM.

(ii) Microarray

Microarray technology, which is a DNA chip-based expression assay, was initially conceived in the 1980s but many papers using this technology were published in the late 1990s and onward [225]. This technology subsequently revolutionized the ability to evaluate the expression of multiple genes in a single sample, particularly in cancer research. As a result, microarray became an important tool for the global identification of gene expression across multiple samples [225]. For many years microarray has been used as a tool to evaluate differential gene expression between TMZ-sensitive and -resistant GBM specimens, including cell lines. For example, using microarray, the protein encoded by CD74 gene was described to be the driver of TMZ resistance in GBM PDXs [191]. Similarly, several other studies have employed the microarray technique to identify TMZ-resistant genes [190, 199, 200, 226–228]. Importantly, microarray has also been used to identify miRNA and noncoding RNA that may potentially contribute to TMZ resistance [80, 147, 228–231]. This assay is not without limitations, including reliance on prior knowledge of the genome sequence, high background, and saturation signals causing limited dynamic range of detection and last but not least, sometimes it is difficult to compare results across many samples which may require complicated normalization. These limitations increased the need for developing other genomic high-throughput techniques including RNA sequencing described later.

(iii) Transcriptome (RNA) sequencing

As described in the prior sections, TMZ is a potent mutagenic drug; therefore, cells that have been treated with this drug have been shown to accumulate DNA variations particularly single-nucleotide changes involving C>T alterations [212, 213]. Further, unlike microarray, the RNA-Seq approach can not only identify mutations but also provide information as to whether these alterations cause differential gene expression between TMZ-sensitive and -resistant cells. Accordingly, RNA-Seq is the preferred approach for identifying gene targets modulating TMZ resistance in GBM. RNA-Seq is especially useful in characterizing transcriptome changes in both in vitro and in vivo TMZ-resistance models. For example, using RNA-Seq, Kitange et al. demonstrated multiple mechanisms of TMZ resistance emerging in the same PDX tumor after failing in vivo TMZ treatment, which were stratified as MGMT dependent and MGMT independent mechanisms [204]. In addition, through RNA-Seq several other modulators of TMZ resistance have been identified in GBM [147, 232, 233]. Interestingly, reexpression of MGMT in methylated TMZ-resistant model was not linked with loss of promoter methylation but instead with the loss of suppressive histone marks coupled with increased chromatin relaxing histone marks [204]. These findings provided the rationality for conducting RNA-Seq coupled with ChIPseq, which involves chromatin precipitation followed by sequencing. Establishing the transcriptome and epigenome provides additional information as to whether the differential gene alterations observed between sensitive and resistant cells is owing to TMZ-induced chromatin changes. This view is supported by the observation that TMZ-treated cells display heterochromatin reorganization [234]. Importantly, MGMT reexpression in TMZ-resistant GBM cells was shown to be coupled with increased H3K9ac decoration within the promoter region [204], a mark for open chromatin. Collectively, RNA-Seq is the preferred assay for whole-genome screening for modulators of TMZ mainly through its potential to identify the expression and single-nucleotide variations (SNVs); however, more comprehensive data will need to be obtained when it is coupled with the epigenome evaluation using ChIPseq.

(iv) Metabolomics profiling

Altered metabolism in cancer cells is necessary for meeting energy requirements for tumor progression, a phenomenon which is called the Warburg effect, characterized by a metabolic shift from oxidative phosphorylation to glycolysis or lactate fermentation both in the presence and absence of oxygen [235]. The metabolic derangement in cancer cells may render survival advantages against therapy, including survival against TMZ therapy in GBM cells. In line with this view, several studies have demonstrated that inhibition of mitochondrial functions sensitizes TMZ in GBM cells [68, 101, 202, 226, 236–238]. Accordingly, whole-genome characterization of metabolic profiles comparing TMZ sensitive and resistant may identify modulators of TMZ sensitivity in GBM. The metabolic profiling can be done using ^{1}H NMR spectra but also standard microarray and/or RNA-Seq may identify TMZ-resistant candidates involved in cell metabolism [239]. Understanding the metabolic targets modulating TMZ sensitivity is highly relevant clinically, in part because almost all metabolism proteins are druggable targets.

(v) shRNA library screening

The methods described in the prior sections are based on the evaluation focusing on the intrinsic expression of genes and corresponding proteins. The alternative approach is to identify TMZ-resistance genes by targeting the whole genome or specific pathways using shRNA, which is referred to as shRNA library screening [240, 241]. In this approach, a library of shRNA targeting the whole genome is made and infected into cells frequently using lentivirus. Then cells expressing the shRNA library are treated with TMZ or the vehicle and allowed to grow for at least 5 days. Then RNA will be extracted from the cells and shRNA amplification is performed followed by evaluation of shRNA enrichment using microarray or RNA-Seq methods comparing vehicle with TMZ-treated group. The expectation is that genes with a positive influence on TMZ response when suppressed by shRNA will become resistant to TMZ; and, the shRNA enrichment will be similar or increased in the TMZ-treated group vs vehicle group. In contrast, for genes that drive resistance to TMZ, the shRNA suppression will enhance TMZ sensitivity and increase cell death. Since dying cells will not be available during shRNA amplification, the enrichment in TMZ will be lower than in the vehicle-treated group. The shRNA library screening method has been used by several investigators to identify modulators of TMZ therapy. For example, the RBBP4 protein was recently identified using shRNA library screening as modulator of TMZ resistance through the regulation of multiple DNA repair proteins [59]. A similar approach was used to identify BRAC1 and DDX6 as a modulator of TMZ resistance [242, 243]. An advantage of this method is that the screening can be conducted in vitro and in vivo. Nevertheless, the reproducibility may be influenced by the fact that cells are being transfected randomly. Also, off-target effects may contribute to high false discovery rate. Recently, shRNA screening has increasingly been replaced by CRISPR screening, considered to have less off-target effects and hence less false discovery rate.

(vi) Genome editing by CRISPR-Cas9 library

Clustered regularly interspaced short palindromic repeat DNA sequences (CRISPR)-Cas9 system is a genome editing technology based on DNA repeat elements found in bacteria that were initially reported in *Escherichia coli* [244]. The principles of CRISPR-Cas9 whole-genome screening are fundamentally similar to the shRNA screening. Nonetheless, shRNA reduces (knockdown) RNA expression whereas CRISPR-Cas9 screening had a complete knockdown of the gene at DNA level. Thus, CRISPR-Cas9 screening may identify genes that the complete knockdown is needed to show the desired phenotypes. The detailed description of the principles of CRISPR-Cas9 genome editing is beyond the scope of this chapter [244, 245]. Briefly, in high-throughput screening, the single-guided RNAs (sgRNAs) are designed to target genes within a pathway or are pooled to a library targeting the whole genome. The sgRNA is linked to common adapter sequences to allow amplification using PCR followed by next-generation sequencing. Enriched sgRNA in the control untreated relative to the treated cells are targeting the genes driving resistance to TMZ. Even though CRISPR-Cas9 is a relatively new technology, several studies have already used this technology to identify modulators of TMZ sensitivity in GBM cells. For example, a recent study using genome-wide CRISPR-Cas9 screens exposed genetic vulnerabilities and mechanisms of TMZ sensitivity in GBM stem cells [246]. Several similar studies have identified interesting modulators of TMZ resistance, including the gene encoding the pro-survival NF-kB protein [233].

Conclusions and future directions

We have better knowledge of the mechanisms of TMZ resistance in GBMs from the outcome of innate TMZ-resistant GBMs, acquired TMZ-resistant GBMs, PDX lines, and GBM tumor specimens study. However, we still need more information for the TMZ-resistance mechanisms, especially on MGMT deficient GBMs and recurrent GBMs, until we have better therapies utilizing TMZ. In addition to the role of known TMZ-resistance mechanisms in TMZ-resistant GBMs, studies are needed for the role of gene mutations, transcription factors, and tumor microenvironment for innate/acquired TMZ resistance using PDX lines.

Moreover, the current data indicate that we should focus on the gene and molecular level differences between innate TMZ resistance and acquired TMZ resistance in in vivo tumor models than in vitro cell culture models. We also need more studies for new combination therapies without TMZ and consider the importance of compound/therapy treatment to find novel therapeutic strategies for innate and/or acquired TMZ-resistant GBMs. Using new techniques and increasing numbers of in vivo tumor model studies, we can identify novel targets for innate and acquired TMZ-resistant GBMs. With the advance of novel target identification methods and continuous ongoing clinical trials, we will have new therapeutic options (agents or strategies) to treat devastating and malignant GBMs, especially for TMZ-resistant GBMs.

Acknowledgment

We thank Mrs Sara Langan for her editing of the book chapter.

References

[1] Ostrom QT, Cioffi G, Gittleman H, Patil N, Waite K, Kruchko C, Barnholtz-Sloan JS. CBTRUS Statistical Report: primary brain and other central nervous system tumors diagnosed in the United States in 2012–2016. Neuro Oncol 2019;21(Supplement_5):v1–v100.

[2] Sturm D, Bender S, Jones DTW, Lichter P, Grill J, Becher O, Hawkins C, Majewski J, Jones C, Costello JF, Iavarone A, Aldape K, Brennan CW, Jabado N, Pfister SM. Paediatric and adult glioblastoma: multiform (epi)genomic culprits emerge. Nat Rev Cancer 2014;14(2):92–107.

[3] Kfoury N, Sun T, Yu K, Rockwell N, Tinkum KL, Qi Z, Warrington NM, McDonald P, Roy A, Weir SJ, Mohila CA, Deneen B, Rubin JB. Cooperative p16 and p21 action protects female astrocytes from transformation. Acta Neuropathol Commun 2018;6(1):12.

[4] Yang W, Warrington NM, Taylor SJ, Whitmire P, Carrasco E, Singleton KW, Wu N, Lathia JD, Berens ME, Kim AH, Barnholtz-Sloan JS, Swanson KR, Luo J, Rubin JB. Sex differences in GBM revealed by analysis of patient imaging, transcriptome, and survival data. Sci Transl Med 2019;11(473). Pii: eaao5253.

[5] Denny BJ, Wheelhouse RT, Stevens MF, Tsang LLH, Slack JA. NMR and molecular modeling investigation of the mechanism of activation of the antitumor drug temozolomide and its interaction with DNA. Biochemistry 1994;33:9045–51.

[6] Barciszewska AM, Gurda D, Glodowicz P, Nowak S, Naskręt-Barciszewska M. A new epigenetic mechanism of temozolomide action in glioma cells. PLoS One 2015;10(8), e0136669.

[7] Strobel H, Baisch T, Fitzel R, Schilberg K, Siegelin MD, Karpel-Massler G, Debatin KM, Westhoff MA. Temozolomide and other alkylating agents in glioblastoma therapy. Biomedicines 2019;7:69.

[8] Stepanenko AA, Andreieva SV, Korets KV, Mykytenko DO, Baklaushev VP, Huleyuk NL, Kovalova OA, Kotsarenko KV, Chekhonin VP, Vassetzky YS, Avdieiev SS, Dmitrenko VV. Temozolomide promotes genomic and phenotypic changes in glioblastoma cells. Cancer Cell Int 2016;16:36.

[9] Rekers NH, Sminia P, Peters GJ. Towards tailored therapy of glioblastoma multiforme. J Chemother 2011; 23(4):187–99.

[10] Messaoudi K, Clavreul A, Lagarce F. Toward an effective strategy in glioblastoma treatment. Part I: resistance mechanisms and strategies to overcome resistance of glioblastoma to temozolomide. Drug Discov Today 2015; 20(7):899–905.

[11] Jiapaer S, Furuta T, Tanaka S, Kitabayashi T, Nakada M. Potential strategies overcoming the temozolomide resistance for glioblastoma. Neurol Med Chir (Tokyo) 2018;58(10):405–21.

[12] Cancer Genome Atlas Research Network. Comprehensive genomic characterization defines human glioblastoma genes and core pathways. Nature 2008;455(7216):1061–8.

[13] Pearson JRD, Regad T. Targeting cellular pathways in glioblastoma multiforme. Signal Transduct Target Ther 2017;2:17040.

[14] Yeom SY, Nam DH, Park C. RRAD promotes EGFR-mediated STAT3 activation and induces temozolomide resistance of malignant glioblastoma. Mol Cancer Ther 2014;13(12):3049–61.

[15] Wang Y, Chen L, Bao Z, Li S, You G, Yan W, Shi Z, Liu Y, Yang P, Zhang W, Han L, Kang C, Jiang T. Inhibition of STAT3 reverses alkylator resistance through modulation of the AKT and beta-catenin signaling pathways. Oncol Rep 2011;26(5):1173–80.

[16] Lin GS, Yang LJ, Wang XF, Chen YP, Tang WL, Chen L, Lin ZX. STAT3 Tyr705 phosphorylation affects clinical outcome in patients with newly diagnosed supratentorial glioblastoma. Med Oncol 2014;31(4):924.

[17] Masliantsev K, Pinel B, Balbous A, Guichet PO, Tachon G, Milin S, Godet J, Duchesne M, Berger A, Petropoulos C, Wager M, Karayan-Tapon L. Impact of STAT3 phosphorylation in glioblastoma stem cells radiosensitization and patient outcome. Oncotarget 2018;9(3):3968–79.

[18] Daniel P, Sabri S, Chaddad A, Meehan B, Jean-Claude B, Rak J, Abdulkarim BS. Temozolomide induced hypermutation in glioma: evolutionary mechanisms and therapeutic opportunities. Front Oncol 2019;9:41.

[19] Tomar VS, Patil V, Somasundaram K. Temozolomide induces activation of Wnt/β-catenin signaling in glioma cells via PI3K/Akt pathway: implications in glioma therapy. Cell Biol Toxicol 2020;36(3):273–8.

[20] Zeng H, Yang Z, Xu N, Liu B, Fu Z, Lian C, Guo H. Connective tissue growth factor promotes temozolomide resistance in glioblastoma through TGF-β1-dependent activation of Smad/ERK signaling. Cell Death Dis 2017;8, e2885.

[21] Zhi T, Yu T, Pan M, Nie E, Wu W, Wang X, Liu N, You Y, Wang Y, Zhang J. EZH2 alteration driven by microRNA-524-5p and microRNA-324-5p promotes cell proliferation and temozolomide resistance in glioma. Oncotarget 2017;8(56):96239–48.
[22] Ulasov IV, Mijanovic O, Savchuk S, Gonzalez-Buendia E, Sonabend A, Xiao T, Timashev P, Lesniak MS. TMZ regulates GBM stemness via MMP14-DLL4-Notch3 pathway. Int J Cancer 2020;146(8):2218–28.
[23] St-Coeur PD, Poitras JJ, Cuperlovic-Culf M, Touaibia M, Morin Jr P. Investigating a signature of temozolomide resistance in GBM cell lines using metabolomics. J Neuro-Oncol 2015;125:91–102.
[24] Weinstein IB. Cancer. Addiction to oncogenes—the achilles heal of cancer. Science 2002;297(5578):63–4.
[25] Stupp R, Mason WP, van den Bent MJ, Weller M, Fisher B, Taphoorn MJ, Belanger K, Brandes AA, Marosi C, Bogdahn U, Curschmann J, Janzer RC, Ludwin SK, Gorlia T, Allgeier A, Lacombe D, Cairncross JG, Eisenhauer E, Mirimanoff RO, European Organisation for Research and Treatment of Cancer Brain Tumor and Radiotherapy Groups, National Cancer Institute of Canada Clinical Trials Group. Radiotherapy plus concomitant and adjuvant temozolomide for glioblastoma. N Engl J Med 2005;352:987–96.
[26] Liau BB, Sievers C, Donohue LK, Gillespie SM, Flavahan WA, Miller TE, Venteicher AS, Hebert CH, Carey CD, Rodig SJ, Shareef SJ, Najm FJ, van Galen P, Wakimoto H, Cahill DP, Rich JN, Aster JC, Suva ML, Patel AP, Bernstein BE. Adaptive chromatin remodeling drives glioblastoma stem cell plasticity and drug tolerance. Cell Stem Cell 2017;20(2):233–46.
[27] Osuka S, Van Meir EG. Overcoming therapeutic resistance in glioblastoma: the way forward. J Clin Invest 2017;127(2):415–26.
[28] Mansoori B, Mohammadi A, Davudian S, Shirjang S, Baradaran B. The different mechanisms of cancer drug resistance: a brief review. Adv Pharm Bull 2017;7(3):339–48.
[29] Wang X, Zhang H, Chen X. Drug resistance and combating drug resistance in cancer. Cancer Drug Resist 2019;2:141–60.
[30] Dirkse A, Golebiewska A, Buder T, Nazarov PV, Muller A, Poovathingal S, Brons NHC, Leite S, Sauvageot N, Sarkisjan D, Seyfrid M, Fritah S, Stieber D, Michelucci A, Hertel F, Herold-Mende C, Azuaje F, Skupin A, Bjerkvig R, Deutsch A, Voss-Bohme A, Niclou AP. Stem cell-associated heterogeneity in glioblastoma results from intrinsic tumor plasticity shaped by the microenvironment. Nat Commun 2019;10(1):1787.
[31] Bao S, Wu O, McLendon RE, Hao Y, Shi Q, Hjelmeland AB, Dewhirst MW, Bigner DD, Rich JN. Glioma stem cells promote radioresistance by preferential activation of the DNA damage response. Nature 2006;444 (7120):756–60.
[32] Lathia JD, Mack SC, Mulkearns-Hubert EE, Valentim CLL, Rich JN. Cancer stem cells in glioblastoma. Genes Dev 2015;29(12):1203–17.
[33] Auffinger B, Spencer D, Pytel P, Ahmed AU, Lesniak MS. The role of flioma stem cells in chemotherapy resistance and glioblastoma multiforme recurrence. Expert Rev Neurother 2015;15(7):741–52.
[34] Garnier D, Meehan B, Kislinger T, Daniel P, Sinha A, Abdulkarim B, Nakano I, Rak J. Divergent evolution of temozolomide resistance in glioblastoma stem cells is reflected in extracellular vesicles and coupled with radiosensitization. Neuro-Oncology 2018;20(2):236–48.
[35] Pera MF, Köberle B, Masters JR. Exceptional sensitivity of testicular germ cell tumour cell lines to the new anticancer agent, temozolomide. Br J Cancer 1995;71(5):904–6.
[36] Serwer LP, James CD. Challenges in drug delivery to tumors of the central nervous system: an overview of pharmacological and surgical considerations. Adv Drug Deliv Rev 2012;64:590–7.
[37] Ge X, Pan MH, Wang L, Li W, Jiang C, He J, Abouzid K, Liu LZ, Shi Z, Jiang BH. Hypoxia-mediated mitochondria apoptosis inhibition induces temozolomide treatment resistance through miR-26a/Bad/Bax axis. Cell Death Dis 2018;9(11):1128.
[38] Wu W, Wu Y, Mayer K, von Rosenstiel C, Schecker J, Baur S, Würstle S, Liesche-Starnecker F, Gempt J, Schlegel J. Lipid peroxidation plays an important role in chemotherapeutic effects of temozolomide and the development of therapy resistance in human glioblastoma. Transl Oncol 2020;13(3):100748.
[39] Gupta SK, Kizilbash SH, Carlson BL, Mladek AC, Boakye-Agyeman F, Bakken KK, Pokorny JL, Schroeder MA, Decker PA, Cen L, Eckel-Passow JE, Sarkar G, Ballman KV, Reid JM, Jenkins RB, Verhaak RG, Sulman EP, Kitange GJ, Sarkaria JN. Delineation of MGMT hypermethylation as a biomarker for veliparib-mediated temozolomide-sensitizing therapy of glioblastoma. J Natl Cancer Inst 2015;108(5). pii: djv369.
[40] Sughrue ME, Rutkowski MJ, Kane AJ, Parsa AT. Human glioma demonstrates cell line specific results with ATP-based chemiluminescent cellular proliferation assays. J Clin Neurosci 2010;17(12):1573–7.

[41] Lee SY. Temozolomide resistance in glioblastoma multiforme. Genes Dis 2016;3(3):198–210.
[42] Persano L, Pistollato F, Rampazzo E, Della Puppa A, Abbadi S, Frasson C, Volpin F, Indraccolo S, Scienza R, Basso G. BMP2 sensitizes glioblastoma stem-like cells to temozolomide by affecting HIF-1α stability and MGMT expression. Cell Death Dis 2012;3(10), e412.
[43] Wang K, Chen D, Qian Z, Cui D, Gao L, Lou M. Hedgehog/Gli1 signaling pathway regulates MGMT expression and chemoresistance to temozolomide in human glioblastoma. Cancer Cell Int 2017;17:117.
[44] Wickström M, Dyberg C, Milosevic J, Einvik C, Calero R, Sveinbjörnsson B, Sandén E, Darabi A, Siesjö P, Kool M, Kogner P, Baryawno N, Johnsen JI. Wnt/-catenin pathway regulates MGMT gene expression in cancer and inhibition of Wnt signaling prevents chemoresistance. Nat Commun 2015;6:8904.
[45] Hermisson M, Klumpp A, Wick W, Wischhusen J, Nagel G, Roos W, Kaina B, Weller M. O6-methylguanine DNA methyltransferase and p53 status predict temozolomide sensitivity in human malignant glioma cells. J Neurochem 2006;96:766–76.
[46] Pollack IF, Hamilton RL, Sobol RW, Bumham J, Yates AJ, Holmes EJ, Zhou T, Finlay JL. MGMT expression strongly correlates with outcome in childhood malignant gliomas: results from the CCG-945 cohort. J Clin Oncol 2006;24:3431–7.
[47] Hegi ME, Liu L, Herman JG, Stupp R, Wick W, Weller M, Mehta MP, Gilbert MR. Correlation of O6-methylguanine methyltransferase (MGMT) promoter methylation with clinical outcomes in glioblastoma and clinical strategies to modulate MGMT activity. J Clin Oncol 2008;26(25):4189–99.
[48] Sarkaria JN, Kitange GJ, James CD, Plummer R, Calvert H, Weller M, Wick W. Mechanisms of chemoresistance to alkylating agents in malignant glioma. Clin Cancer Res 2008;14(10):2900–8.
[49] Silber JR, Mueller BA, Ewers TG, Berger MS. Comparison of O6-methylguanine-DNA methyltransferase activity in brain tumors and adjacent normal brain. Cancer Res 1993;53:3416–20.
[50] Beier D, Rohrl S, Pillai DR, Schwarz S, Kunz-Schughart LA, Leukel P, Proescholdt M, Brawanski A, Bogdahn U, Trampe-Kieslich A, Giebel B, Wischhusen J, Reifenberger G, Hau P, Beier CP. Temozolomide preferentially depletes cancer stem cells in glioblastoma. Cancer Res 2008;68(14):5706–15.
[51] Pistollato F, Abbadi S, Rampazzo E, Persano L, Puppa AD, Frasson C, Sarto E, Scienza R, D'avella D, Basso G. Intratumoral hypoxic gradient drives stem cells distribution and MGMT expression in glioblastoma. Stem Cells 2010;28(5):851–62.
[52] Yi GZ, Huang G, Guo M, Zhang X, Wang H, Deng S, Li Y, Xiang W, Chen Z, Pan J, Li Z, Yu L, Lei B, Liu Y, Qi S. Acquired temozolomide resistance in MGMT-deficient glioblastoma cells is associated with regulation of DNA repair by DHC2. Brain 2019;142(8):2352–66.
[53] Wachsberger PR, Lawrence RY, Liu Y, Rice B, Daskalakis C, Dicker AP. Epidermal growth factor receptor mutation status and rad51 determine the response of glioblastoma to multimodality therapy with Cetuximab, Temozolomide, and radiation. Front Oncol 2013;3:13.
[54] Flüh C, Chitadze G, Adamski V, Hattermann K, Synowitz M, Kabelitz D, Held-Feindt J. NKG2D ligands in glioma stem-like cells: expression in situ and in vitro. Histochem Cell Biol 2018;149(3):219–33.
[55] Rodrigues AR, Neto FSL, Lourenco LG, Trevisan FA, Cirino MLA, Nery B, Peria FM, Pereira-da-Silva G, Tazima MFGS, Tirapelli LF, Tiezzi DG, Carlotti Junior CG, Tirapelli DPC. Expression of oncogenic microRNA-21 in neurospheres and attached cells of a glioblastoma cell line increased after treatment with temozolomide and ionizing radiation. Genet Mol Res 2019;18(2), GMR18095.
[56] Yan W, Liu Y, Yang P, Wang Z, You Y, Jiang T. MicroRNA profiling of Chinese primary glioblastoma reveals a temozolomide-chemoresistant subtype. Oncotarget 2015;6(13):11676–82.
[57] Low SYY, Ho YK, Too HP, Yap CT, Ng WH. MicroRNA as potential modulators in chemoresistant high-grade gliomas. J Clin Neurosci 2014;21(3):395–400.
[58] Li Z, Zhang J, Zheng H, Li C, Xiong J, Wang W, Bao H, Jin H, Liang P. Modulating lncRNA SNHG15/CDK6/miR-627 circuit by Palbociclib, overcomes temozolomide resistance and reduces M2-polarization of glioma associated microglia in glioblastoma multiforme. J Exp Clin Cancer Res 2019;38(1):380.
[59] Kitange GJ, Mladek AC, Schroeder MA, Pokorny JC, Carlson BL, Zhang Y, Nair AA, Lee JH, Yan H, Decker PA, Zhang Z, Sarkaria JN. Retinoblastoma binding protein 4 modulates temozolomide sensitivity in glioblastoma by regulating DNA repair proteins. Cell Rep 2016;14(11):2587–98.
[60] Chen D, Song M, Mohamad O, Yu SP. Inhibition of Na+/K+-ATPase induces hybrid cell death and enhanced sensitivity to chemotherapy in human glioblastoma cells. BMC Cancer 2014;14:716.
[61] Xu Y, Stamenkovic I, Yu Q. CD44 attenuates activation of the hippo signaling pathway and is a prime therapeutic target for glioblastoma. Cancer Res 2010;70(6):2455–64.

[62] Liu G, Yuan X, Zeng Z, Tunici P, Ng H, Abdulkadir IR, Lu L, Irvin D, Black KL, Yu JS. Analysis of gene expression and chemoresistance of CD133+ cancer stem cells in glioblastoma. Mol Cancer 2006;5:67.
[63] Bocangel DB, Finkelstein S, Schold SC, Bhakat KK, Mitra S, Kokkinakis DM. Multifaceted resistance of gliomas to temozolomide. Clin Cancer Res 2002;8:2725–34.
[64] Sato Y, Kurose A, Ogawa A, Ogasawara K, Traganos F, Darzynkiewicz D, Sawai T. Diversity of DNA damage response of astrocytes and glioblastoma cell lines with various p53 status to treatment with etoposide and temozolomide. Cancer Biol Ther 2009;8(5):452–7.
[65] Lee SY, Liu S, Mitchell RM, Slagle-Webb B, Hong YS, Sheehan JM, Connor JR. HFE polymorphisms influence the response to chemotherapeutic agents via induction of p16INK4A. Int J Cancer 2011;129:2104–14.
[66] Tsai YT, Wu AC, Yang WB, Kao TJ, Chuang JY, Chang WC, Hsu TI. ANGPTL4 induces TMZ resistance of glioblastoma by promoting cancer stemness enrichment via the EGFR/AKT/4E-BP1 cascade. Int J Mol Sci 2019;20 (22). pii:E5625.
[67] Eramo A, Ricci-Vitiani L, Zeuner A, Pallini R, Lotti F, Sette G, Pilozzi E, Larocca LM, Peschle C, de Maria R. Chemotherapy resistance of glioblastoma stem cells. Cell Death Differ 2006;13(7):1238–41.
[68] Oliva CR, Moellering DR, Gillespie GY, Griguer CE. Acquisition of chemoresistance in gliomas is associated with increased mitochondrial coupling and decreased ROS production. PLoS One 2011;6(9), e24665.
[69] Chen X, Zhang M, Gan H, Wang H, Lee JH, Fang D, Kitange GJ, He L, Hu Z, Parney IF, Meyer FB, Giannini C, Sakaria JN, Zhang Z. A novel enhancer regulates MGMT expression and promotes temozolomide resistance in glioblastoma. Nat Commun 2018;9:2949.
[70] Bruyère C, Mijatovic T, Lonez C, Spiegl-Kreinecker S, Berger W, Kast RE, Ruysschaert JM, Kiss R, Lefranc F. Temozolomide-induced modification of the CXC chemokine network in experimental gliomas. Int J Oncol 2011;38(5):1453–64.
[71] Zhang J, Stevens MF, Laughton CA, Madhusudan S, Bradshaw TD. Acquired resistance to temozolomide in glioma cell lines: molecular mechanisms and potential translational applications. Oncology 2010;78(2):103–14.
[72] McFaline-Figueroa JL, Braun CJ, Stanciu M, Nagel ZD, Mazzucato P, Sangaraju D, Cerniauskas E, Barford K, Vargas A, Chen Y, Tretyakova N, Lees JA, Hemann MT, White FM, Samson LD. Minor changes in expression of the mismatch repair protein MSH2 exert a major impact on glioblastoma response to temozolomide. Cancer Res 2015;75(15):3127–38.
[73] Munoz JL, Rodriguez-Cruz V, Greco SJ, Ramkissoon SH, Ligon KL, Rameshwar P. Temozolomide resistance in glioblastoma cells occurs partly through epidermal growth factor receptor-mediated induction of connexin 43. Cell Death Dis 2014;5, e1145.
[74] Ueno H, Tomiyama A, Yamaguchi H, Uekita T, Shirakihara T, Nakashima K, Otani N, Wada K, Sakai R, Arai H, Mori K. Augmentation of invadopodia formation in temozolomide-resistant or adopted glioma is regulated by c-Jun terminal kinase-paxillin axis. Biochem Biophys Res Commun 2015;468:240–7.
[75] Munoz JL, Rodriguez-Cruz V, Ramkissoon SH, Ligon KL, Greco SJ, Rameshwar P. Temozolomide resistance in glioblastoma occurs by miRNA-9-targeted PTCH1, independent of sonic hedgehog level. Oncotarget 2015; 6(2):1190–201.
[76] Yan Y, Xu Z, Chen X, Wang X, Zeng S, Zhao Z, Qian L, Li Z, Wei J, Huo L, Gong Z, Sun L. Novel function of lncRNA ADAMTS9-AS2 in promoting temozolomide resistance in glioblastoma via upregulating the FUS/MDM2 ubiquitination axis. Front Cell Dev Biol 2019;7:217.
[77] Ananta JS, Paulmurugan R, Massoud TF. Nanoparticle-delivered antisense microRNA-21 enhances the effects of temozolomide on glioma cells. Mol Pharm 2015;12(12):4509–17.
[78] Bertucci A, Prasetyanto EA, Septiadi D, Manicardi A, Brognara E, Gambari R, Corradini R, Cola LD. Combined delivery of temozolomide and anti-miR221 PNA using mesoporous silica nanoparticles induces apoptosis in resistant glioma cells. Small 2015;11(42):5687–95.
[79] Ujifuku K, Mitsutake N, Takakura S, Matsuse M, Saenko V, Suzuki K, Hayashi K, Matsuo T, Kamada K, Nagata I, Yamashita S. miR-195, miR-455-3p and miR-10a(*) are implicated in acquired temozolomide resistance in glioblastoma multiforme cells. Cancer Lett 2010;296:241–8.
[80] Stojcheva N, Schechtmann G, Sass S, Roth P, Florea AM, Stefanski A, Stuhler K, Wolter M, Muller NS, Theis FJ, Weller M, Reifenberger G, Happold C. MicroRNA-138 promotes acquired alkylator resistance in glioblastoma by targeting the Bcl-2-interacting mediator BIM. Oncotarget 2016;7(11):12937–50.
[81] Tian T, Mingyi M, Qiu X, Qiu Y. MicroRNA-101 reverses temozolomide resistance by inhibition of GSK3β in glioblastoma. Oncotarget 2016;7(48):79584–95.

[82] Luo W, Yan D, Song Z, Zhu X, Liu X, Li X, Zhao S. miR-126-3p sensitizes glioblastoma cells to temozolomide by inactivating Wnt/β-catenin signaling via targeting SOX2. Life Sci 2019;226:98–106.
[83] Qian Z, Zhou S, Zhou Z, Yang X, Que S, Lan J, Qiu Y, Lin Y. miR-146b-5p suppresses glioblastoma cell resistance to temozolomide through targeting TRAF6. Oncol Rep 2017;38(5):2941–50.
[84] Li H, Chen L, Li JJ, Zhou Q, Huang A, Liu WW, Wang K, Gao L, Qi ST, Lu YT. miR-519a enhances chemosensitivity and promotes autophagy in glioblastoma by targeting STAT3/Bcl2 signaling pathway. J Hematol Oncol 2018;11(1):70.
[85] Yu X, Wang M, Zuo J, Wahafu A, Mao P, Li R, Wu W, Xie W, Wang J. Nuclear Factor I A promotes temozolomide resistance in glioblastoma via activation of nuclear factor kB pathway. Life Sci 2019;236:116917.
[86] Bredel M, Bredel C, Juric D, Duran GE, Yu RX, Harsh GR, Vogel H, Recht LD, Scheck AC, Sikic BI. Tumor necrosis factor-alpha-induced protein 3 as a putative regulator of nuclear factor-kappaB-mediated resistance to O6-alkylating agents in human glioblastomas. J Clin Oncol 2006;24:274–87.
[87] Yi GZ, Xiang W, Feng WY, Chen ZY, Li YM, Deng SZ, Guo ML, Zhao L, Sun XG, He MY, Qi ST, Liu YW. Identification of key candidate proteins and pathways associated with temozolomide resistance in glioblastoma based on subcellular proteomics and bioinformatical analysis. Biomed Res Int 2018;2018:5238760.
[88] Velpula KK, Guda MR, Sahu K, Tuszynski J, Asuthkar S, Bach SE, Lathia JD, Tsung AJ. Metabolic targeting of EGFRvIII/PDK1 axis in temozolomide resistant glioblastoma. Oncotarget 2017;8:35639–55.
[89] Haas B, Klinger V, Keksel C, Bonigut V, Kiefer D, Caspers J, Walther J, Wos-Maganga M, Weickhardt S, Röhn G, Timmer M, Frötschl R, Eckstein N. Inhibition of the PI3K but not the MEK/ERK pathway sensitizes human glioma cells to alkylating drugs. Cancer Cell Int 2018;18:69.
[90] Fan TY, Wang H, Xiang P, Xiang P, Liu YW, Li HZ, Lei BX, Yu M, Qi ST. Inhibition of EZH2 reverses chemotherapeutic drug TMZ chemosensitivity in glioblastoma. Int J Clin Exp Pathol 2014;7(10):6662–70.
[91] Le Calvé B, Rynkowski M, Mercier ML, Bruyere C, Lonez C, Gras T, Haibe-Kains B, Bontempi G, Decaestecker C, Ruysschaert JM, Kiss R, Lefranc F. Long-term in vitro treatment of human glioblastoma cells with temozolomide increases resistance in vivo through up-regulation of GLUT transporter and aldo-keto reductase enzyme AKR1C expression. Neoplasia 2010;12(9):727–39.
[92] Dai S, Yan Y, Xu Z, Zeng S, Qian L, Huo L, Li X, Sun L, Gong Z. SCD1 confers temozolomide resistance to human glioma cells via the Akt/GSK3β/β-Catenin signaling axis. Front Pharmacol 2018;8:960.
[93] Liu Q, Xue Y, Chen Q, Chen H, Zhang X, Wang L, Han C, Que S, Lou M, Lan J. PomGnT1 enhances temozolomide resistance by activating epithelial-mesenchymal transition signaling in glioblastoma. Oncol Rep 2017;38(5):2911–8.
[94] Sun S, Lee D, Ho ASW, Pu JKS, Zhang PXQ, Lee NP, Day PJR, Lui WM, Fung CF, Leung GKK. Inhibition of prolyl 4-hydroxylase, beta polypeptide (P4HB) attenuates temozolomide resistance in malignant glioma via the endoplasmic reticulum stress response (ERSR) pathways. Neuro-Oncology 2013;15(5):562–77.
[95] Hersh DS, Harder BG, Roos A, Peng S, Heath JE, Legesse T, Kim AJ, Woodworth GF, Tran NL, Winkles JA. The TNF receptor family member Fn14 is highly expressed in recurrent glioblastoma and in GBM patient-derived xenografts with acquired temozolomide resistance. Neuro-Oncology 2018;20(10):1321–30.
[96] Kuser-Abali G, Gong L, Yan J, Liu Q, Zeng W, Williamson A, Lim CB, Molloy ME, Little JB, Huang L, Yuan ZM. An EZH2-mediated epigenetic mechanism behind p53-dependent tissue sensitivity to DNA damage. Proc Natl Acad Sci U S A 2018;115(13):3452–7.
[97] Han B, Meng X, Wu P, Li Z, Li S, Zhang Y, Zha C, Ye Q, Jiang C, Cai J, Jiang T. ATRX/EZH2 complex epigenetically regulates FADD/PARP1 axis, contributing to TMZ resistance in glioma. Theranostics 2020; 10(7):3351–65.
[98] Kohsaka S, Wang L, Yachi K, Mahabir R, Narita T, Itoh T, Tanino M, Kimura T, Nishihara H, Tanaka S. STAT3 inhibition overcomes temozolomide resistance in glioblastoma by downregulating MGMT expression. Mol Cancer Ther 2012;11:1289–99.
[99] Liang H, Chen G, Li J, Yang F. Snail expression contributes to temozolomide resistance in glioblastoma. Am J Transl Res 2019;11(7):4277–89.
[100] Ma J, Murphy M, O'Dwyer PJ, Berman E, Reed K, Gallo JM. Biochemical changes associated with a multidrug-resistant phenotype of a human glioma cell line with temozolomide-acquired resistance. Biochem Pharmacol 2002;63:1219–28.
[101] Oliva CR, Nozell SE, Diers A, McClugage 3rd SG, Sarkaria JN, Markert JM, Darley-Usmar VM, Bailey SM, Gillespie GY, Landar A, Griguer CE. Acquisition of temozolomide chemoresistance in gliomas leads to remodeling of mitochondrial electron transport chain. J Biol Chem 2010;285(51):39759–67.

[102] Lee ES, Ko KK, Joe YA, Kang SG, Hong YK. Inhibition of STAT3 reverses drug resistance acquired in temozolomide-resistant human glioma cells. Oncol Lett 2011;2:115–21.
[103] Yin J, Zeng A, Zhang Z, Shi Z, Yan W, You Y. Exosomal transfer of miR-1238 contributes to temozolomide-resistance in glioblastoma. EBioMedicine 2019;42:238–51.
[104] Cai T, Liu Y, Xiao J. Long noncoding RNA MALAT1 knockdown reverses chemoresistance to temozolomide via promoting microRNA-101 in Glioblastoma. Cancer Med 2018;7(4):1404–15.
[105] Anderson JC, Duarte CW, Welaya K, Rohrbach TD, Bredel M, Yang ES, Choradia NV, Thottassery JV, Gillespie GY, Bonner JA, Willey CD. Kinomic exploration of temozolomide and radiation resistance in glioblastoma multiforme xenolines. Radiother Oncol 2014;111(3):468–74.
[106] Wischhusen J, Naumann U, Ohgaki H, Rastinejad F, Weller M. CP-31398, a novel p53-stabilizing agent, induces p53-dependent and p53-independent glioma cell death. Oncogene 2003;22:8233–45.
[107] Chen CM, Syu JP, Way TD, Huang LJ, Kuo SC, Lin CT, Lin CL. BC3EE2,9B, a synthetic carbazole derivative, upregulates autophagy and synergistically sensitizes human GBM8901 glioblastoma cells to temozolomide. Int J Mol Med 2015;36(5):1244–52.
[108] Bi Y, Li H, Yi D, Sun Y, Bai Y, Zhong S, Song Y, Zhao G, Chen Y. Cordycepin augments the chemosensitivity of human glioma cells to temozolomide by activating AMPK and inhibiting the AKT signaling pathway. Mol Pharm 2018;15(11):4912–25.
[109] Shergalis A, Bankhead A, Luesakul U, Muangsin N, Neamati N. Current challenges and opportunities in treating glioblastoma. Pharmacol Rev 2018;70(3):412–45.
[110] Jain KK. A critical overview of targeted therapies for glioblastoma. Front Oncol 2018;8:419.
[111] Bahadur S, Sahu AK, Baghel P, Saha S. Current promising treatment strategy for glioblastoma multiform: a review. Oncol Rev 2019;13(2):417.
[112] Taylor OG, Brzozowski JS, Skelding KA. Glioblastoma multiforme: an overview of emerging therapeutic targets. Front Oncol 2019;9:963.
[113] Yang Z, Wei D, Dai X, Stevens MFG, Bradshaw TD, Luo Y, Zhang J. C8-substituted imidazotetrazine analogs overcome temozolomide resistance by inducing DNA adducts and DNA damage. Front Oncol 2019;9:485.
[114] Bonafé GA, Dos Santos JS, Ziegler JV, Umezawa K, Ribeiro ML, Rocha T, Ortega MM. Growth inhibitory effects of dipotassium glycyrrhizinate in glioblastoma cell lines by targeting microRNAs through the NF-kB signaling pathway. Front Cell Neurosci 2019;13:216.
[115] Berthier S, Arnaud J, Champelovier P, Col E, Garrel C, Cottet C, Boutonnat J, Laporte F, Faure P, Hazane-Puch F. Anticancer properties of sodium selenite in human glioblastoma cell cluster spheroids. J Trace Elem Med Biol 2017;44:161–76.
[116] Majewska E, Márquez J, Albrecht J, Szeliga M. Transfection with GLS2 glutaminase (GAB) sensitizes human glioblastoma cell lines to oxidative stress by a common mechanism involving suppression of the PI3K/AKT pathway. Cancers (Basel) 2019;11(1):115.
[117] Oliva CR, Markert T, Ross LJ, White EL, Rasmussen L, Zhang W, Everts M, Moellering DR, Bailey SM, Suto MJ, Griguer CE. Identification of small molecule inhibitors of human cytochrome c oxidase that target chemoresistant glioma cells. J Biol Chem 2016;291(46):24188–99.
[118] Figul M, Söling A, Dong HJ, Chou TC, Rainov NG. Combined effects of temozolomide and the ribonucleotide reductase inhibitors didox and trimidox in malignant brain tumor cells. Cancer Chemother Pharmacol 2003;52(1):41–6.
[119] Xie CR, You CG, Zhang N, Sheng HS, Zheng XS. Epigallocatechin gallate preferentially inhibits O6-methylguanine DNA-methyltransferase expression in glioblastoma cells rather than in nontumor glial cells. Nutr Cancer 2018;70(8):1339–47.
[120] You CG, Sheng HS, Xie CR, Zhang N, Zheng XS. FM19G11 inhibits O6-methylguanine DNA-methyltransferase expression under both hypoxic and normoxic conditions. Cancer Med 2018;7(7):3292–300.
[121] Chen L, Liu YC, Zheng YY, Xu J, Zhang Y, Liu WL, Li ZY, Huang GD. Furanodienone overcomes temozolomide resistance in glioblastoma through the downregulation of CSPG4-Akt-ERK signaling by inhibiting ERG1-dependent transcription. Phytother Res 2019;33(6):1736–47.
[122] Jain M, Hamage NDH, Alsulami M, Shankar A, Achyut BR, Angara K, Rashid MH, Iskander A, Borin TF, Wenbo Z, Ara R, Ali MM, Lebedyeva I, Chwang WB, Guo A, Bagher-Ebadian H, Arbab AS. Intravenous formulation of HET0016 decreased human gliobastoma growth and implicated survival benefit in rat xenograft models. Sci Rep 2017;7:41809.

[123] Houghton PJ, Stewart CF, Cheshire PJ, Richmond LB, Kirstein MN, Poquette CA, Tan M, Friedman HS, Brent TP. Antitumor activity of temozolomide combined with irinotecan is partly independent of O6-methylguanine-DNA methyltransferase and mismatch repair phenotypes in xenograft models. Clin Cancer Res 2000;6 (10):4110–8.
[124] Xiao A, Brenneman B, Floyd D, Comeau L, Spurio K, Olmez I, Lee J, Nakano I, Godlewski J, Bronisz A, Kagaya N, Shin-ya K, Purow B. Statins affect human glioblastoma and other cancers through TGF-β. Oncotarget 2019; 10(18):1716–28.
[125] Shi F, Guo H, Zhang R, Liu H, Wu L, Wu Q, Liu J, Liu T, Zhang Q. The PI3K inhibitor GDC-0941 enhances radiosensnitization and reduces chemorsistance to temozolomide in GBM cell lines. Neuroscience 2017;346:298–308.
[126] Gravina GL, Mancini A, Colapietro A, Monache SD, Sferra R, Vitale F, Cristiano L, Martellucci S, Marampon F, Mattei V, Beirinckx F, Pujuguet P, Saniere L, Lorenzon G, van der Aar E, Festuccia C. The small molecule ephrin receptor inhibitor, GLPG1790, reduces renewal capabilities of cancer stem cells, showing anti-tumor efficacy on preclinical glioblastoma models. Cancers (Basel) 2019;11(3). Pii: E359.
[127] Kanzawa T, Bedwell J, Kondo Y, Kondo S, Germano IM. Inhibition of DNA repair for sensitizing resistant glioma cells to temozolomide. J Neurosurg 2003;99:1047–52.
[128] Jensen KV, Cseh O, Aman A, Weiss S, Luchman HA. The JAK2/STAT3 inhibitor pacritinib effectively inhibits patient-derived GBM brain tumor initiating cells in vitro and when used in combination with temozolomide increases survival in an orthotopic xenograft model. PLoS One 2017;12(12), e0189670.
[129] Ji M, Wang L, Chen J, Xue N, Wang C, Lai F, Wang R, Yu S, Jin J, Chen X. CAT_3, a Prodrug of 13a(S)-3-hydroxyl-6,7-dimethoxyphenanthro[9,10-b]-indolizidine, circumvents temozolomide-resistant glioblastoma via the hedgehog signaling pathway, independently of O^6-methylguanine DNA methyltransferase expression. Onco Targets Ther 2018;11:3671–84.
[130] Li ZY, Li QZ, Chen L, Chen BD, Wang B, Zhang XJ, Li WP. Histone deacetylase inhibitor RGFP109 overcomes temozolomide resistance by blocking NF-kB-dependent transcription in glioblastoma cell lines. Neurochem Res 2016;41(12):3192–205.
[131] Grogan PT, Sarkaria JN, Timmermann BN, Cohen MS. Oxidative cytotoxic agent Withaferin A resensitizes temozolomide-resistant glioblastomas via MGMT depletion and induces apoptosis through Akt/mTOR pathway inhibitory modulation. Investig New Drugs 2014;32(4):604–17.
[132] Zhao H, Chen G, Liang H. Dual PI3K/mTOR inhibitor, XL765, suppresses glioblastoma growth by inducing ER stress-dependent apoptosis. Onco Targets Ther 2019;12:5415–24.
[133] Milani R, Brognara E, Fabbri E, Manicardi A, Corradini R, Finotti A, Gasparello J, Borgatti M, Cosenza LC, Lampronti I, Dechecchi MC, Cabrini G, Gambari R. Targeting miR-155-5p and miR-221-3p by peptide nucleic acids induces caspase-3 activation and apoptosis in temozolomide-resistant T98G glioma cells. Int J Oncol 2019;55(1):59–68.
[134] Brognara E, Fabbri E, Montagner G, Gasparello J, Manicardi A, Corradini R, Bianchi N, Finotti A, Breveglieri G, Borgatti M, Lampronti I, Milani R, Dechecchi MC, Cabrini G, Gambari R. High levels of apoptosis are induced in human glioma cell lines by co-administration of peptide nucleic acids targeting miR-221 and miR-222. Int J Oncol 2016;48(3):1029–38.
[135] Bhaskaran V, Nowicki MO, Idriss M, Jimenez MA, Lugli G, Hayes JL, Mahmoud AB, Zane RE, Passaro C, Ligon KL, Haas-Kogan D, Bronisz A, Godlewski J, Lawler SE, Chiocca EA, Peruzzi P. The functional synergism of microRNA clustering provides therapeutically relevant epigenetic interference in glioblastoma. Nat Commun 2019;10(1):442.
[136] Montaldi AP, Godoy PR, Sakamoto-Hojo ET. APE1/REF-1 down-regulation enhances the cytotoxic effects of temozolomide in a resistant glioblastoma cell line. Mutat Res Genet Toxicol Environ Mutagen 2015;793:19–29.
[137] Kato T, Natsume A, Toda H, Iwamizu H, Sugita T, Hachisu R, Watanabe R, Yuki K, Motomura K, Bankiewicz K, Wakabayashi T. Efficient delivery of liposome-mediated MGMT-siRNA reinforces the cytotoxicity of temozolomide in GBM-initiating cells. Gene Ther 2010;17(11):1363–71.
[138] Zeng H, Xu N, Liu Y, Liu B, Yang Z, Fu Z, Lian C, Guo H. Genomic profiling of long non-coding RNA and mRNA expression associated with acquired temozolomide resistance in glioblastoma cells. Int J Oncol 2017;51(2):445–55.
[139] Melamed JR, Morgan JT, Ioele SA, Gleghorn JP, Sims-Mourtada J, Day ES. Investigating the role of hedgehog/GLI1 signaling in glioblastoma cell response to temozolomide. Oncotarget 2018;9(43):27000–15.

[140] Szeliga M, Obara-Michlewska M, Matyja E, Łazarczyk M, Lobo C, Hilgier W, Alonso FJ, Márquez J, Albrecht J. Transfection with liver-type glutaminase cDNA alters gene expression and reduces survival, migration and proliferation of T98G glioma cells. Glia 2009;57:1014–23.

[141] Martín-Rufián M, Nascimento-Gomes R, Higuero A, Crisma AR, Campos-Sandoval JA, Gómez-García MC, Cardona C, Cheng T, Lobo C, Segura JA, Alonso FJ, Szeliga M, Albrecht J, Curi R, Márquez J, Colquhoun A, DeBerardinis RJ, Matés JM. Both GLS silencing and GLS2 overexpression synergize with oxidative stress against proliferation of glioma cells. J Mol Med 2014;92:277–90.

[142] Sareddy GR, Nair BC, Gonugunta VK, Zhang QG, Brenner A, Brann DW, Tekmal RR, Vadlamudi RK. Therapeutic significance of estrogen receptor beta agonists in gliomas. Mol Cancer Ther 2012;11(5):1174–82.

[143] Liu J, Sareddy GR, Zhou M, Viswanadhapalli S, Li X, Lai Z, Tekmal RR, Brenner A, Vadlamudi RK. Differential effects of estrogen receptor β isoforms on glioblastoma progression. Cancer Res 2018;78(12):3176–89.

[144] Zhou M, Sareddy GR, Li M, Liu J, Luo Y, Venkata PP, Viswanadhapalli S, Tekmal RR, Brenner A, Vadlamudi RK. Estrogen receptor beta enhances chemotherapy response of GBM cells by down regulating DNA damage response pathways. Sci Rep 2019;9(1):6124.

[145] Guntuku L, Naidu VGM, Yerra VG. Mitochondrial dysfunction in gliomas: pharmacotherapeutic potential of natural compounds. Curr Neuropharmacol 2016;14(6):567–83.

[146] Ling H, Fabbri M, Calin GA. MicroRNAs and other non-coding RNAs as targets for anticancer drug development. Nat Rev Drug Discov 2013;12(11):847–65.

[147] Li Y, Liu Y, Ren J, Deng S, Yi G, Guo M, Shu S, Zhao L, Peng Y, Qi S. miR-1268a regulates ABCC1 expression to mediate temozolomide resistance in glioblastoma. J Neuro-Oncol 2018;138(3):499–508.

[148] Sur P, Sribnick EA, Patel SJ, Ray SK, Banik NL. Dexamethasone decreases temozolomide-induced apoptosis in human glioblastoma T98G cells. Glia 2005;50(2):160–7.

[149] Das A, Banik NL, Ray SK. Modulatory effects of acetazolamide and dexamethasone on temozolomide-mediated apoptosis in human glioblastoma T98G and U87MG cells. Cancer Investig 2008;26(4):352–8.

[150] Chakrabarti M, Ray SK. Anti-tumor activities of luteolin and silibinin in glioblastoma cells: overexpression of miR-7-1-3p augmented luteolin and silibinin to inhibit autophagy and induce apoptosis in glioblastoma in vivo. Apoptosis 2016;21(3):312–28.

[151] Huang H, Lin H, Zhang X, Li J. Resveratrol reverses temozolomide resistance by downregulation of MGMT in T98G glioblastoma cells by the NF-kB-dependent pathway. Oncol Rep 2012;27(6):2050–6.

[152] Yang HC, Wang JY, Bu XY, Yang B, Wang BQ, Hu S, Yan ZY, Gao YS, Han SY, Qu MQ. Resveratrol restores sensitivity of glioma cells to temozolomide through inhibiting the activation of Wnt signaling pathway. J Cell Physiol 2019;234(5):6783–800.

[153] Vlachostergios PJ, Hatzidaki E, Papandreou CN. MGMT repletion after treatment of glioblastoma cells with temozolomide and O6-benzylguanine implicates NFkB and mutant p53. Neurol Res 2013;35(8):879–82.

[154] Honorato JR, Hauser-Davis RA, Saggioro EM, Correia FV, Sales-Junior SF, Soares LOS, Lima LDR, Moura-Neto V, Lopes GPF, Spohr TCLS. Role of sonic hedgehog signaling in cell cycle, oxidative stress, and autophagy of temozolomide resistant glioblastoma. J Cell Physiol 2020;235(4):3798–814.

[155] Ha W, Sevim-Nalkiran H, Zaman AM, Matsuda K, Khasraw M, Nowak A, Chung L, Baxter RC, McDonald KL. Ibudilast sensitizes glioblastoma to temozolomide by targeting macrophage migration inhibitory factor (MIF). Sci Rep 2019;9(1):2905.

[156] Smalley S, Chalmers AJ, Morley SJ. mTOR inhibition and levels of the DNA repair protein MGMT in T98G glioblastoma cells. Mol Cancer 2014;13:144.

[157] Goellner EM, Grimme B, Brown AR, Lin YC, Wang XH, Sugrue KF, Mitchell L, Trivedi RN, Tang JB, Sobol RW. Overcoming temozolomide resistance in glioblastoma via dual inhibition of NAD+ biosynthesis and base excision repair. Cancer Res 2011;71:2308–17.

[158] St-Coeur PD, Cormier M, LeBlanc VC, Morin PJ, Touaibia M. Effect of O6-substituted guanine analogs on O6-methylguanine DNA-methyltransferase expression and glioblastoma cells viability. Med Chem 2016;13(1):28–39.

[159] Towner RA, Smith N, Saunders D, Brown CA, Cai X, Ziegler J, Mallory S, Dozmorov MG, Coutinho De Souza P, Wiley G, Kim K, Kang S, Kong DS, Kim YT, Fung KM, Wren JD, Battiste J. OKN-007 increases temozolomide (TMZ) sensitivity and suppresses TMZ-resistant glioblastoma (GBM) tumor growth. Transl Oncol 2019;12(2):320–35.

[160] Shao CJ, Wu MW, Chen FR, Li C, Xia YF, Chen ZP. Histone deacetylase inhibitor, 2-propylpentanoic acid, increases the chemosensitivity and radiosensitivity of human glioma cell lines in vitro. Chin Med J 2012;125(24):4338–43.

[161] Harder BG, Peng S, Sereduk CP, Sodoma AM, Kitange GJ, Loftus JC, Sarkaria JN, Tran NL. Inhibition of phosphatidylinositol 3-kinase by PX-866 suppresses temozolomide-induced autophagy and promotes apoptosis in glioblastoma cells. Mol Med 2019;25(1):49.
[162] Ignarro RS, Facchini G, Vieira AS, de Melo DR, Lopes-Cendes I, Castilho RF, Rogerio F. Sulfasalazine intensifies temozolomide cytotoxicity in human glioblastoma cells. Mol Cell Biochem 2016;418(1–2):167–78.
[163] Ryu CH, Yoon WS, Park KY, Kim SM, Lim JY, Woo JS, Jeong CH, Hou Y, Jeun SS. Valproic acid downregulates the expression of MGMT and sensitizes temozolomide-resistant glioma cells. J Biomed Biotechnol 2012;2012:987495.
[164] Palumbo S, Tini P, Toscano M, Allavena G, Angeletti F, Manai F, Miracco C, Comincini S, Pirtoli L. Combined EGFR and autophagy modulation impairs cell migration and enhances radiosensitivity in human glioblastoma cells. J Cell Physiol 2014;229(11):1863–73.
[165] Guo G, Gong K, Puliyappadamba VT, Panchani N, Pan E, Mukherjee B, Damanwalla Z, Bharia S, Hatanpaa KJ, Gerber DE, Mickey BE, Patel TR, Sarkaria JN, Zhao D, Burma S, Habib AA. Efficacy of EGFR plus TNF inhibition in a preclinical model of temozolomide-resistant glioblastoma. Neuro-Oncology 2019;21(12):1529–39.
[166] Valtorta S, Dico AL, Raccagni I, Gaglio D, Belloli S, Politi LS, Martelli C, Diceglie C, Bonanomi M, Ercoli G, Vaira V, Ottobrini L, Moresco RM. Metformin and temozolomide, a synergic option to overcome resistance in glioblastoma multiforme models. Oncotarget 2017;8(68):113090–104.
[167] Tunca B, Tezcan G, Cecener G, Egeli U, Ak S, Malyer H, Tumen G, Bilir A. Olea Europaea leaf extract alters microRNA expression in human glioblastoma cells. J Cancer Res Clin Oncol 2012;138(11):1831–44.
[168] Safdar S, Payne CA, Tu NH, Taite LJ. Targeted nitric oxide delivery preferentially induces glioma cell chemosensitivity via altered p53 and O(6)-methylguanine-DNA methyltransferase activity. Biotechnol Bioeng 2013;110(4):1211–20.
[169] Tsai CK, Huang LC, Wu YP, Kan IY, Hueng DY. SNAP reverses temozolomide resistance in human glioblastoma multiforme cells through down-regulation of MGMT. FASEB J 2019;33(12):14171–84.
[170] Fu W, You C, Ma L, Li H, Ju Y, Guo X, Shi S, Zhang T, Zhou R, Lin Y. Enhanced efficacy of temozolomide loaded by a tetrahedral framework DNA nanoparticle in the therapy for glioblastoma. ACS Appl Mater Interfaces 2019;11(43):39525–33.
[171] Paul-Samojedny M, Pudelko A, Kowalczyk M, Fila-Danilow A, Suchanek-Raif R, Borkowska P, Kowalski J. Combination therapy with AKT3 and PI3KCA siRNA enhances the antitumor effect of temozolomide and carmustine in T98G glioblastoma multiforme cells. BioDrugs 2016;30(2):129–44.
[172] Lan F, Yang Y, Han J, Wu Q, Yu H, Yue X. Sulforaphane reverses chemo-resistance to temozolomide in glioblastoma cells by NF-kB-dependent pathway downregulating MGMT expression. Int J Oncol 2016;48(2):559–68.
[173] Yoshino A, Ogino A, Yachi K, Ohta T, Fukushima T, Watanabe T, Katayama Y, Okamoto Y, Naruse N, Sano E. Effect of IFN-beta on human glioma cell lines with temozolomide resistance. Int J Oncol 2009;35(1):139–48.
[174] Natsume A, Ishii D, Wakabayashi T, Tsuno T, Hatano H, Mizuno M, Yoshida J. IFN-beta down-regulates the expression of DNA repair gene MGMT and sensitizes resistant glioma cells to temozolomide. Cancer Res 2005;65:7573–9.
[175] Alonso MM, Gomez-Manzano C, Bekele BN, Yung WK, Fueyo J. Adenovirus-based strategies overcome temozolomide resistance by silencing the O6-methylguanine-DNA methyltransferase promoter. Cancer Res 2007;67:11499–504.
[176] Tang JB, Svilar D, Trivedi RN, Wang XH, Goellner EM, Moore B, Hamilton RL, Banze LA, Brown AR, Sobol RW. N-Methylpurine DNA glycosylase and DNA polymerase beta modulate BER inhibitor potentiation of glioma cells to temozolomide. Neuro-Oncology 2011;13(5):471–86.
[177] Szeliga M, Zgrzywa A, Obara-Michlewska M, Albrecht J. Transfection of a human glioblastoma cell line with liver-type glutaminase (LGA) down-regulates the expression of DNA-repair gene MGMT and sensitizes the cells to alkylating agents. J Neurochem 2012;123(3):428–36.
[178] Vlachostergios PJ, Hatzidaki E, Befani CD, Liakos P, Papandreou CN. Bortezomib overcomes MGMT-related resistance of glioblastoma cell lines to temozolomide in a schedule-dependent manner. Investig New Drugs 2013;31(5):1169–81.
[179] Gupta SK, Mladek AC, Carlson BL, Boakye-Agyeman F, Bakken KK, Kizilbash SH, Schroeder MA, Reid J, Sarkaria JN. Discordant in vitro and in vivo chemopotentiating effects of the PARP inhibitor veliparib in temozolomide-sensitive versus –resistant glioblastoma multiforme xenografts. Clin Cancer Res 2014;20 (14):3730–41.
[180] Pellosi DS, Paula LB, de Melo MT, Tedesco AC. Targeted and synergic glioblastoma treatment: multifunctional nanoparticles delivering verteporfin as adjuvant therapy for temozolomide. Mol Pharm 2019;16(3):1009–24.

[181] Tezcan G, Tunca B, Bekar A, Budak F, Sahin S, Cecener G, Egeli U, Taskapilioglu MO, Kocaeli H, Tolunay S, Malyer H, Demir C, Tumen G. Olea Europaea leaf extract improves the treatment response of GBM stem cells by modulating miRNA expression. Am J Cancer Res 2014;4(5):572–90.
[182] Tezcan G, Aksoy SA, Tunca B, Bekar A, Mutlu M, Cecener G, Egeli U, Kocaeli H, Demirci H, Taskapilioglu MO. Oleuropein modulates glioblastoma miRNA pattern different from Olea europaea leaf extract. Hum Exp Toxicol 2019;38(9):1102–10.
[183] Yang SH, Li S, Lu G, Xue H, Kim DH, Zhu JJ, Liu Y. Metformin treatment reduces temozolomide resistance of glioblastoma cells. Oncotarget 2016;7(48):78787–803.
[184] Feldheim J, Kessler AF, Monoranu CM, Ernestus RI, Lohr M, Hagemann C. Changes of O(6)-methylguanine DNA methyltransferase (MGMT) promoter methylation in glioblastoma relapse—a meta-analysis type literature review. Cancers (Basel) 2019;11(12):1837.
[185] Hegi ME, Diserens AC, Godard S, Dietrich PY, Regli L, Ostermann S, Otten P, Van Melle G, de Tribolet N, Stupp R. Clinical trial substantiates the predictive value of O-6-methylguanine-DNA methyltransferase promoter methylation in glioblastoma patients treated with temozolomide. Clin Cancer Res 2004;10(6):1871–4.
[186] Kitange GJ, Carlson BL, Mladek AC, Decker PA, Schroeder MA, Wu W, Grogan PT, Giannini C, Ballman KV, Buckner JC, James CD, Sarkaria JN. Evaluation of MGMT promoter methylation status and correlation with temozolomide response in orthotopic glioblastoma xenograft model. J Neuro-Oncol 2009;92(1):23–31.
[187] Kreth S, Thon N, Eigenbrod S, Lutz J, Ledderose C, Egensperger R, Tonn JC, Kretzschmar HA, Hinske LC, Kreth FW. O-Methylguanine-DNA methyltransferase (MGMT) mRNA expression predicts outcome in malignant glioma independent of MGMT promoter methylation. PLoS One 2011;6(2), e17156.
[188] Happold C, Roth P, Wick W, Schmidt N, Florea AM, Silginer M, Reifenberger G, Weller M. Distinct molecular mechanisms of acquired resistance to temozolomide in glioblastoma cells. J Neurochem 2012;122(2):444–55.
[189] Xiao S, Yang Z, Qiu X, Lv R, Liu J, Wu M, Liao Y, Liu Q. miR-29c contribute to glioma cells temozolomide sensitivity by targeting O6-methylguanine-DNA methyltransferases indirectely. Oncotarget 2016;7(31):50229–38.
[190] Yoshino A, Ogino A, Yachi K, Ohta T, Fukushima T, Watanabe T, Katayama Y, Okamoto Y, Naruse N, Sano E, Tsumoto K. Gene expression profiling predicts response to temozolomide in malignant gliomas. Int J Oncol 2010;36(6):1367–77.
[191] Kitange GJ, Carlson BL, Schroeder MA, Decker PA, Morlan BW, Wu W, Ballman KV, Giannini C, Sarkaria JN. Expression of CD74 in high grade gliomas: a potential role in temozolomide resistance. J Neuro-Oncol 2010; 100(2):177–86.
[192] Perazzoli G, Prados J, Ortiz R, Caba O, Cabeza L, Berdasco M, Gonzalez B, Melguizo C. Temozolomide resistance in glioblastoma cell lines: implication of MGMT, MMR, P-glycoprotein and CD133 expression. PLoS One 2015;10(10), e0140131.
[193] Se YB, Kim SH, Kim JY, Kim JE, Dho YS, Kim JW, Kim YH, Woo HG, Kim SH, Kang SH, Kim HJ, Kim TM, Lee ST, Choi SH, Park SH, Kim IH, Kim DG, Park CK. Underexpression of HOXA11 is associated with treatment resistance and poor prognosis in glioblastoma. Cancer Res Treat 2017;49(2):387–98.
[194] Wang H, Song X, Huang Q, Xu T, Yun D, Wang Y, Hu L, Yan Y, Chen H, Lu D, Chen J. LGALS3 promotes treatment resistance in glioblastoma and is associated with tumor risk and prognosis. Cancer Epidemiol Biomark Prev 2019;28(4):760–9.
[195] Wang Z, Xu X, Liu N, Cheng Y, Jin W, Zhang P, Wang X, Yang H, Liu H, Tu Y. SOX9-PDK1 axis is essential for glioma stem cell self-renewal and temozolomide resistance. Oncotarget 2018;9(1):192–204.
[196] Wu Q, Berglund AE, Wang D, MacAulay RJ, Mule JJ, Etame AB. Paradoxical epigenetic regulation of XAF1 mediates plasticity towards adaptive resistance evolution in MGMT-methylated glioblastoma. Sci Rep 2019;9(1):14072.
[197] Yu H, Park J, Lee J, Choi K, Choi C. Constitutive expression of MAP kinase phosphatase-1 confers multi-drug resistance in human glioblastoma cells. Cancer Res Treat 2012;44(3):195–201.
[198] Zhou D, Wan Y, Xie D, Wang Y, Wei J, Yan Q, Lu P, Mo L, Xie J, Yang S, Qi X. DNMT1 mediates chemosensitivity by reducing methylation of miRNA-20a promoter in glioma cells. Exp Mol Med 2015;47, e182.
[199] Akiyama Y, Ashizawa T, Komiyama M, Miyata H, Oshita C, Omiya M, Iizuka A, Kume A, Sugino T, Hayashi N, Mitsuya K, Nakasu Y, Yamaguchi K. YKL-40 downregulation is a key factor to overcome temozolomide resistance in a glioblastoma cell line. Oncol Rep 2014;32(1):159–66.
[200] Auger N, Thillet J, Wanherdrick K, Idbaih A, Legrier ME, Dutrillaux B, Sanson M, Poupon MF. Genetic alterations associated with acquired temozolomide resistance in SNB-19, a human glioma cell line. Mol Cancer Ther 2006;5(9):2182–92.

[201] Vaubel RA, Tian S, Remonde D, Schroeder MA, Mladek AC, Kitange GJ, Caron A, Kollmeyer TM, Grove R, Peng S, Carlson BL, Ma DJ, Sarkar G, Evers L, Decker PA, Yan H, Dhruv HD, Berens ME, Wang Q, Marin BM, Klee EW, Califano A, LaChance DH, Eckel-Passow JE, Verhaak RG, Sulman EP, Burns TC, Meyer FB, O'Neill BP, Tran NL, Giannini C, Jenkins RB, Parney IF, Sarkaria JN. Genomic and phenotypic characterization of a broad panel of patient-derived xenografts reflects the diversity of glioblastoma. Clin Cancer Res 2020; 26(5):1094–104.

[202] Gielen PR, Aftab Q, Ma N, Chen VC, Hong X, Lozinsky S, Naus CC, Sin WC. Connexin43 confers temozolomide resistance in human glioma cells by modulating the mitochondrial apoptosis pathway. Neuropharmacology 2013;75:539–48.

[203] Kitange GJ, Carlson BL, Schroeder MA, Grogan PT, Lamont JD, Decker PA, Wu W, James CD, Sarkaria JN. Induction of MGMT expression is associated with temozolomide resistance in glioblastoma xenografts. Neuro-Oncology 2009;11(3):281–91.

[204] Kitange GJ, Mladek AC, Carlson BL, Schroeder MA, Pokorny JL, Cen L, Decker PA, Wu W, Lomberk GA, Gupta SK, Urrutia RA, Sarkaria JN. Inhibition of histone deacetylation potentiates the evolution of acquired temozolomide resistance linked to MGMT upregulation in glioblastoma xenografts. Clin Cancer Res 2012; 18(15):4070–9.

[205] Clarke JL, Iwamoto FM, Sul J, Panageas K, Lassman AB, DeAngelis LM, Hormigo A, Nolan CP, Gavrilovic I, Karimi S, Abrey LE. Randomized phase II trial of chemoradiotherapy followed by either dose-dense or metronomic temozolomide for newly diagnosed glioblastoma. J Clin Oncol 2009;27(23):3861–7.

[206] Dresemann G. Temozolomide in malignant glioma. Onco Targets Ther 2010;3:139–46.

[207] Houy N, Le Grand F. Administration of temozolomide: comparison of conventional and metronomic chemotherapy regimens. J Theor Biol 2018;446:71–8.

[208] van Nifterik KA, van den Berg J, van der Meide WF, Ameziane N, Wedekind LE, Steenbergen RD, Leenstra S, Lafleur MV, Slotman BJ, Stalpers LJ, Sminia P. Absence of the MGMT protein as well as methylation of the MGMT promoter predict the sensitivity for temozolomide. Br J Cancer 2010;103(1):29–35.

[209] Shinsato Y, Furukawa T, Yunoue S, Yonezawa H, Minami K, Nishizawa Y, Ikeda R, Kawahara K, Yamamoto M, Hirano H, Tokimura H, Arita K. Reduction of MLH1 and PMS2 confers temozolomide resistance and is associated with recurrence of glioblastoma. Oncotarget 2013;4(12):2261–70.

[210] Stark AM, Doukas A, Hugo HH, Hedderich J, Hattermann K, Maximilian Mehdorn H, Held-Feindt J. Expression of DNA mismatch repair proteins MLH1, MSH2, and MSH6 in recurrent glioblastoma. Neurol Res 2015;37(2):95–105.

[211] Struve N, Binder ZA, Stead LF, Brend T, Bagley SJ, Faulkner C, Ott L, Muller-Goebel J, Weik AS, Hoffer K, Krug L, Rieckmann T, Bussmann L, Henze M, Morrissette JJD, Kurian KM, Schuller U, Petersen C, Rothkamm K, Rourke DMO, Short SC, Kriegs M. EGFRvIII upregulates DNA mismatch repair resulting in increased temozolomide sensitivity of MGMT promoter methylated glioblastoma. Oncogene 2020. Online ahead of print.

[212] Hunter C, Smith R, Cahill DP, Stephens P, Stevens C, Teague J, Greenman C, Edkins S, Bignell G, Davies H, O'Meara S, Parker A, Avis T, Barthorpe S, Brackenbury L, Buck G, Butler A, Clements J, Cole J, Dicks E, Forbes S, Gorton M, Gray K, Halliday K, Harrison R, Hills K, Hinton J, Jenkinson A, Jones D, Kosmidou V, Laman R, Lugg R, Menzies A, Perry J, Petty R, Raine K, Richardson D, Shepherd R, Small A, Solomon H, Tofts C, Varian J, West S, Widaa S, Yates A, Easton DF, Riggins G, Roy JE, Levine KK, Mueller W, Batchelor TT, Louis DN, Stratton MR, Futreal PA, Wooster R. A hypermutation phenotype and somatic MSH6 mutations in recurrent human malignant gliomas after alkylator chemotherapy. Cancer Res 2006;66(8):3987–91.

[213] Yip S, Miao J, Cahill DP, Iafrate AJ, Aldape K, Nutt CL, Louis DN. MSH6 mutations arise in glioblastomas during temozolomide therapy and mediate temozolomide resistance. Clin Cancer Res 2009;15(14):4622–9.

[214] Yin AA, He YL, Etcheverry A, Liu YH, Aubry M, Barnholtz-Sloan J, Liu BL, Mosser J, Lu ZF, Zhang X. Novel predictive epigenetic signature for temozolomide in non-G-CIMP glioblastomas. Clin Epigenetics 2019; 11(1):76.

[215] Asuthkar S, Velpula KK, Chetty C, Gorantla B, Rao JS. Epigenetic regulation of miRNA-211 by MMP-9 governs glioma cell apoptosis, chemosensitivity and radiosensitivity. Oncotarget 2012;3(11):1439–54.

[216] Chen H, Li X, Li W, Zheng H. miR-130a can predict response to temozolomide in patients with glioblastoma multiforme, independently of O6-methylguanine-DNA methyltransferase. J Transl Med 2015;13:69.

[217] Lu C, Wei Y, Wang X, Zhang Z, Yin J, Li W, Chen L, Lyu X, Shi Z, Yan W, You Y. DNA-methylation-mediated activating of lncRNA SNHG12 promotes temozolomide resistance in glioblastoma. Mol Cancer 2020;19(1):28.

[218] Quintavalle C, Mangani D, Roscigno G, Romano G, Diaz-Lagares A, Iaboni M, Donnarumma E, Fiore D, De Marinis P, Soini Y, Esteller M, Condorelli G. MiR-221/222 target the DNA methyltransferase MGMT in glioma cells. PLoS One 2013;8(9), e74466.
[219] Schliesser MG, Claus R, Hielscher T, Grimm C, Weichenhan D, Blaes J, Wiestler B, Hau P, Schramm J, Sahm F, Weiss EK, Weiler M, Baer C, Schmidt-Graf F, Schackert G, Westphal M, Hertenstein A, Roth P, Galldiks N, Hartmann C, Pietsch T, Felsberg J, Reifenberger G, Sabel MC, Winkler F, von Deimling A, Meisner C, Vajkoczy P, Platten M, Weller M, Plass C, Wick W. Prognostic relevance of miRNA-155 methylation in anaplastic glioma. Oncotarget 2016;7(50):82028–45.
[220] Switzeny OJ, Christmann M, Renovanz M, Giese A, Sommer C, Kaina B. MGMT promoter methylation determined by HRM in comparison to MSP and pyrosequencing for predicting high-grade glioma response. Clin Epigenetics 2016;8:49.
[221] Cedar H, Bergman Y. Linking DNA methylation and histone modification: patterns and paradigms. Nat Rev Genet 2009;10(5):295–304.
[222] Wang YC, Peterson SE, Loring JF. Protein post-translational modifications and regulation of pluripotency in human stem cells. Cell Res 2014;24(2):143–60.
[223] Karimian A, Mir SM, Parsian H, Refieyan S, Mirza-Aghazadeh-Attari M, Yousefi B, Majidinia M. Crosstalk between phosphoinositide 3-kinase/Akt signaling pathway with DNA damage response and oxidative stress in cancer. J Cell Biochem 2019;120(6):10248–72.
[224] Wang H, Feng W, Lu Y, Li H, Xiang W, Chen Z, He M, Zhao L, Sun X, Lei B, Qi S, Liu Y. Expression of dynein, cytoplasmic 2, heavy chain 1 (DHC2) associated with glioblastoma cell resistance to temozolomide. Sci Rep 2016;6:28948.
[225] Lenoir T, Giannella E. The emergence and diffusion of DNA microarray technology. J Biomed Discov Collab 2006;1:11.
[226] Chen PH, Shen WL, Shih CM, Ho KH, Cheng CH, Lin CW, Lee CC, Liu AJ, Chen KC. The CHAC1-inhibited Notch3 pathway is involved in temozolomide-induced glioma cytotoxicity. Neuropharmacology 2017;116:300–14.
[227] Kaur S, Ramdzan ZM, Guiot MC, Li L, Leduy L, Ramotar D, Sabri S, Abdulkarim B, Nepveu A. CUX1 stimulates APE1 enzymatic activity and increases the resistance of glioblastoma cells to the mono-alkylating agent temozolomide. Neuro-Oncology 2018;20(4):484–93.
[228] Zhao C, Gao Y, Guo R, Li H, Yang B. Microarray expression profiles and bioinformatics analysis of mRNAs, lncRNAs, and circRNAs in the secondary temozolomide-resistant glioblastoma. Invest New Drugs 2019. December 10 Online ahead of print.
[229] Gao YT, Chen XB, Liu HL. Up-regulation of miR-370-3p restores glioblastoma multiforme sensitivity to temozolomide by influencing MGMT expression. Sci Rep 2016;6:32972.
[230] Liu Y, Xu N, Liu B, Huang Y, Zeng H, Yang Z, He Z, Guo H. Long noncoding RNA RP11-838N2.4 enhances the cytotoxic effects of temozolomide by inhibiting the functions of miR-10a in glioblastoma cell lines. Oncotarget 2016;7(28):43835–51.
[231] Zeng A, Wei Z, Yan W, Yin J, Huang X, Zhou X, Li R, Shen F, Wu W, Wang X, You Y. Exosomal transfer of miR-151a enhances chemosensitivity to temozolomide in drug-resistant glioblastoma. Cancer Lett 2018;436:10–21.
[232] Atkins RJ, Stylli SS, Kurganovs N, Mangiola S, Nowell CJ, Ware TM, Corcoran NM, Brown DV, Kaye AH, Morokoff A, Luwor RB, Hovens CM, Mantamadiotis T. Cell quiescence correlates with enhanced glioblastoma cell invasion and cytotoxic resistance. Exp Cell Res 2019;374(2):353–64.
[233] Huang K, Liu X, Li Y, Wang Q, Zhou J, Wang Y, Dong F, Yang C, Sun Z, Fang C, Liu C, Tan Y, Wu X, Jiang T, Kang C. Genome-wide CRISPR-Cas9 screening identifies NF-kappaB/E2F6 responsible for EGFRvIII-associated temozolomide resistance in glioblastoma. Adv Sci (Weinh) 2019;6(17):1900782.
[234] Papait R, Magrassi L, Rigamonti D, Cattaneo E. Temozolomide and carmustine cause large-scale heterochromatin reorganization in glioma cells. Biochem Biophys Res Commun 2009;379(2):434–9.
[235] Bayley JP, Devilee P. The Warburg effect in 2012. Curr Opin Oncol 2012;24(1):62–7.
[236] Chien CH, Chuang JY, Yang ST, Yang WB, Chen PY, Hsu TI, Huang CY, Lo WL, Yang KY, Liu MS, Chu JM, Chung PH, Liu JJ, Chou SW, Chen SH, Chang KY. Enrichment of superoxide dismutase 2 in glioblastoma confers to acquisition of temozolomide resistance that is associated with tumor-initiating cell subsets. J Biomed Sci 2019;26(1):77.
[237] Desai V, Jain A, Shaghaghi H, Summer R, Lai JCK, Bhushan A. Combination of biochanin A and temozolomide impairs tumor growth by modulating cell metabolism in glioblastoma multiforme. Anticancer Res 2019; 39(1):57–66.

[238] Frontinan-Rubio J, Santiago-Mora RM, Nieva-Velasco CM, Ferrin G, Martinez-Gonzalez A, Gomez MV, Moreno M, Ariza J, Lozano E, Arjona-Gutierrez J, Gil-Agudo A, De la Mata M, Pesic M, Peinado JR, Villalba JM, Perez-Romasanta L, Perez-Garcia VM, Alcain FJ, Duran-Prado M. Regulation of the oxidative balance with coenzyme Q10 sensitizes human glioblastoma cells to radiation and temozolomide. Radiother Oncol 2018; 128(2):236–44.
[239] Cuperlovic-Culf M, Ferguson D, Culf A, Morin Jr P, Touaibia M. 1H NMR metabolomics analysis of glioblastoma subtypes: correlation between metabolomics and gene expression characteristics. J Biol Chem 2012; 287(24):20164–75.
[240] Krausz E. High-content siRNA screening. Mol BioSyst 2007;3(4):232–40.
[241] Nencioni A, Sandy P, Dillon C, Kissler S, Blume-Jensen P, Van Parijs L. RNA interference for the identification of disease-associated genes. Curr Opin Mol Ther 2004;6(2):136–40.
[242] Cho YJ, Kang W, Kim SH, Sa JK, Kim N, Paddison PJ, Kim M, Joo KM, Hwang YI, Nam DH. Involvement of DDX6 gene in radio- and chemoresistance in glioblastoma. Int J Oncol 2016;48(3):1053–62.
[243] Ding J, Wu S, Zhang C, Garyali A, Martinez-Ledesma E, Gao F, Pokkulandra A, Li X, Bristow C, Carugo A, Koul D, Yung WA. BRCA1 identified as a modulator of temozolomide resistance in P53 wild-type GBM using a high-throughput shRNA-based synthetic lethality screening. Am J Cancer Res 2019;9(11):2428–41.
[244] Ishino Y, Krupovic M, Forterre P. History of CRISPR-Cas from encounter with a mysterious repeated sequence to genome editing technology. J Bacteriol 2018;200(7). e00580-17.
[245] Javed MR, Sadaf M, Ahmed T, Jamil A, Nawaz M, Abbas H, Ijaz A. CRISPR-Cas system: history and prospects as a genome editing tool in microorganisms. Curr Microbiol 2018;75(12):1675–83.
[246] MacLeod G, Bozek DA, Rajakulendran N, Monteiro V, Ahmadi M, Steinhart Z, Kushida MM, Yu H, Coutinho FJ, Cavalli FMG, Restall I, Hao X, Hart T, Luchman HA, Weiss S, Dirks PB, Angers S. Genome-wide CRISPR-Cas9 screens expose genetic vulnerabilities and mechanisms of temozolomide sensitivity in glioblastoma stem cells. Cell Rep 2019;27(3):971–86. e9.

CHAPTER

5

Glioma stem cells and associated molecular mechanisms in Glioblastoma Chemoresistance

Georgios Solomou[a,c] *and Colin Watts*[a,b]

[a]Institute of Cancer and Genome Sciences, University of Birmingham, Birmingham, United Kingdom [b]Department of Neurosurgery, University Hospitals Birmingham, Birmingham, United Kingdom [c]School of Medicine, Keele University, Stoke on Trent, United Kingdom

Abstract

GBM is one of the most devastating cancers known. Despite optimal therapy, reoccurrence is seen. GSCs can show proliferative, self-renewal, and tumor-propagating potential and thus are accused to be responsible for tumor reoccurrence. GSCs can show significant survival potential as well, in terms of enabling blood vessels formation, resisting hypoxia, and avoiding the immune's system detection mechanisms. Notably, GSC function is not dichotomous. They can switch across a spectrum of functional states making their identification challenging. TMZ is a gold-standard chemotherapeutic agent. However, GSC has "learned" to resist its action by pumping the drug out of the cells as well as detecting and correcting DNA damages induced by the drug.

Abbreviations

5-ALA	5-aminolevulinic acid
ABC	ATP-binding cassette
APE1	apurinic/apyrimidinic endonuclease 1
BBB	blood–brain barrier
BER	base excision repair
CET	contrast-enhancing tumor
CRET	complete resection of enhancing tumor
CSF	cerebrospinal fluid
EANO	European Association of Neuroncology
EGFR	epidermal growth factor receptor
FDA	Food and Drug Administration
GB	glioblastoma
GSC	glioma stem-like cell

Glioblastoma Resistance to Chemotherapy: Molecular Mechanisms and Innovative Reversal Strategies
https://doi.org/10.1016/B978-0-12-821567-8.00005-1

HIF	hypoxia-inducible factor
ID	isocitrate dehydrogenase
IL	interleukin
IoMRI	intraoperative MRI
IoUS	intraoperative ultrasonography
MGMT	O6-methylguianine-DNA-methyltransferase
MMR	mismatch repair
MRI	magnetic resonance imaging
NF	neurofibromatosis
NICE	National Institute of Clinical Excellence
NK	natural killer
OS	overall survival
PARP-1	poly ADP-ribose polymerase 1
PDGFR-a	platelet derived growth factor receptor alpha
PFS	progression free survival
PG	prostaglandin
PTEN	phosphatase and tensin homolog
STAT3	signal transducer and activator of transcription 3
TAM	tissue-associated macrophages
TCGA	the Cancer Genome Atlas
TGF-ß	tissue growth factor beta
TMZ	temozolomide
VEGF-A	vascular endothelial growth factor A

Conflict of interest

No potential conflicts of interest were disclosed.

Introduction

Glioblastoma

Brain tumors are one of the most devastating cancers known. They can be classified into benign and malignant. Malignant tumors can be further categorized into primary and secondary. Out of all primary malignant brain tumors in adults, glioblastoma (GBM) is the most common, accounting for 57.3% of the cases in the United States, with incidence rate of 3.2/100,000 [1]. Further, the incidence of GBM has doubled in the United Kingdom, from 2.4 to 5.0 per 100,000 between 1995 and 2015 [2], while in the United States it is predicted to rise in the next 30 years [3]. GBM, a WHO grade IV glioma [4], is the most aggressive, with median survival rate of 14–16 months [5]. The 2- and 5-year survival rate of GBM is between 26%–33% and 6.8%, respectively [1, 6]. The current management of GBM is based on three pillars: surgery, radiotherapy, and chemotherapy [6]. Despite optimal surgical and adjuvant therapy, GBM evolves resistance to conventional therapies leading to local reoccurrence via rapid growth and local invasion, ultimately resulting in the poor prognosis.

Several clinical trials have been conducted to date, testing the efficacy of new drugs. The vast majority were unable to progress from phase II to phase III, and/or meeting their primary end points, failing to show superior outcomes versus conventional therapy in terms of overall survival (OS) and/or progression-free survival (PFS) [7–14]. Currently, a number of new trials are ongoing. Cancer immunotherapy, targeted therapy, viral and nonviral gene therapy, and biomarker-based trials are a few important examples of current experimental strategies under investigation [15–18]. Despite substantial progress in revealing the underpinnings of GBM [19], few effective therapies remain. New therapeutic strategies are needed but in order to develop these, we first need to understand why and how current therapies fail.

Glioma stem-like cells

Malignant recurrence is the hallmark of poor prognosis. For a new tumor to be formed, there has to be a cell or cell types/s able to survive the treatment, proliferate, self-renew, invade, and possibly differentiate, leading to GBM regenesis. It has been previously described that GBM formation is a result of a complex interplay of dynamic microenvironmental alterations, and new genetic/epigenetic mutations leading to heterogeneous group of cell types, across time and space, within the same macroscopically seen tumor [20–25]. Further, this occurs through evolution and differentiation, whereby genetically and phenotypically heterogeneous population of cells, the clones, are formed, thereafter giving rise to small subpopulation of cells within the same tumor [20, 26, 27]. The cancer stem cell model suggests that for a tumor to form, in this case GBM, there has to be a subpopulation of cells, which preserve greater stem-like properties than the rest.

Indeed, more recently, the notion of functional cellular heterogeneity implies a hierarchical model of tumorigenesis, whereby a cell type, the glioma stem-like cell (GSC), can demonstrate higher self-renewal capacity and multipotency than the rest [21, 28–30]. Therefore, being the most tumorigenic and hence the founder clone [31–34]. GSCs can either proliferate and slowly differentiate or multiply and self-renew, making them a perfect candidate to be 'accused' for therapeutic resistance and malignant relapse [35, 36]. In this chapter, we collate information and outline how GSCs can survive and evade the current gold-standard therapy, primarily focusing on chemoresistance, shining some light to the intricacies of GBM aggressive reoccurrence.

Current therapies and the role of GSCs

Surgical resection

Survival benefit was shown with as little as 78% resection of contrast-enhanced tumor on T1-weighted magnetic resonance imaging (MRI), with exponential increase in survival (12 months) to near-complete resection (16 months) [37, 38]. The philosophy of management must be patient centric. Although the complete or near-complete resection of contrast-enhanced tumor leads to maximum possible survival, the operating surgeon is simultaneously challenged to preserve 'normal' neurological tissue, keeping the risk of neurological morbidity as low as possible. To overcome the challenge of fine balance between resecting more of the tumor without compromising neurological function, adjuncts in neurosurgery have been developed, trialed, and recommend by the European Association of Neuroncology (EANO) [39] and the National Institute of Clinical Excellence (NICE) in the United Kingdom [40]. This includes intraoperative MRI (IoMRI) [41–43], intraoperative ultrasonography (IoUS) [44], fluorescence-guided surgery, predominantly with 5-aminolevulinic acid (5-ALA)

[45–47], and electrophysiological stimulation with or without awake craniotomy [48, 49]. Further, in 2017, the US Food and Drug Administration (FDA) has approved 5-ALA to be used in the United States [50].

The current paradigm implies that adjuncts should be used aiding complete (gross macroscopic) resection of enhancing tumor (CRET) [51]. It is widely agreed that one of the main reasons explaining poor prognosis despite optimal therapy is the existence of GSC. GSCs are held responsible for tumor invasion, progression, and reoccurrence [32, 52], since they are capable of self-renewal and glioma genesis. After surgical and adjuvant therapy, reoccurrence is seen at the periphery of the tumor cavity [52, 53]. Studies have shown that GSCs expressing stem-cell markers exist in distinct "niches" at the periphery of the tumor [54, 55] and not solely to the perivascular and hypoxic/necrotic "niche" [56, 57]. A recent large retrospective comparative study highlighted the need to reconsider surgical strategies, by revealing that resection of GBM beyond contrast-enhancing tumor (CET), can prolong survival, regardless of molecular subgroup [58]. Since tumor reoccurrence requires cells capable of tumorigenesis, there is speculation that GSCs may reside beyond the CET hence remain unresected leading to tumor reoccurrence. Indeed, there is some evidence to suggest that infiltrating tumor cells reside beyond CET [59], supplemented by studies showing that resection beyond CET is correlated to better survival [60–63].

Bioavailability of temozolomide

Temozolomide (TMZ) is the main chemotherapeutic adjunct for GBM therapy and its usage is considered standard of care [6]. It is systemically administered, needing to cross the blood–brain barrier (BBB) to reach the tumor. The BBB is highly selective preventing a significant amount of the drug to reach the bulk of the tumor [64]. TMZ does not exhibit first-pass metabolism. This leads to high plasma to cerebrospinal fluid (CSF) ratio [65–67]. The CSF TMZ concentration is found to range from 17.8% to 20% compared to plasma [65–67]. In addition, 50 μM TMZ was shown to only deplete half of the GBM cells in vitro, whereas 5 μM (which was recorded in plasma) did not affect cell death [68, 69]. Studies have suggested that CET, which is surrounded by a disrupted BBB might allow exposure to 50 μM of TMZ. However, infiltrative glioma cells at the periphery of the tumor and plausibly outside the CET, where BBB is not disrupted, are subjected to 5 μM of TMZ [70]. Since GSCs are shown to reside at the periphery of the tumor [55, 56], they might be subjected to much lower TMZ concentrations than anticipated. Therefore, understanding whether the TMZ tumor bioavailability is sufficient to eradicate the remaining of tumor cells postsurgery and thus GSCs, which may sit outside the CET, should be prioritized. This might also partly explain the high rates of reoccurrence.

Glioma stem-like cell biology aiding survival

GSC identification

Much has been written regarding the role of GSCs in relation to tumor reoccurrence and chemoresistance. Isolating and identifying GSCs is the first step enabling their study. The first isolation of human neural stem progenitor cells was performed using CD133 (prominin-1) [31, 32]. Thereafter, a burst of studies emerged, using CD133 as a surface marker to describe brain GSCs or cancer stem cells. However, further studies have shown that human-derived fresh CD133-negative cells can still give rise to tumors [71], and can demonstrate equal tumorigenic potential as CD133-positive cells [72]. More recently, Chen and colleagues have shown CD133-negative GSCs can give rise to both CD133-positive and -negative progeny [73]. Further, they showed that a primordial CD133-negative cell gives rise to CD133-positive cells in tumors. Notably, the biological function of CD133 remains unknown. Consequently, the scientific rationale of CD133 as a GSC marker has declined.

A number of reasons exist to explain the challenge to identify and characterize what has been hypothetically defined as a GSC. Heterogeneity of tumor, low number of hypothetical GSCs as progenitor cells within the tumor, and inability to biopsy adequate amount of those GSCs are but a few examples. More recently, both the Suva and Watts groups have shown that GSC is not "there" all the time [29, 30, 74–79]. By definition, GSC needs to proliferate, self-renew, and propagate tumor formation. However, GSC was found to be in a quiescence state and able to switch from a nonactive state to a proliferative and an invasive state depending on the environment. Thus, heterogeneity not only exists in the expression of genes but also in their functional state. The current thinking is that there is no definite stem-cell marker for GSCs, but rather a plasticity of functional states, of cells able to switch from one to the other, influenced by the microenvironment around them. This concept is pertinent to the inability to adequately characterize them using a stem-cell marker such as CD133.

GSC role in resistance to hypoxia

Surgical resection and chemo- and radiotherapy can certainly create unfavorable conditions for any cell type to grow. Poor vasculature and aberrant tumor growth create a low-tissue oxygen tension. At the extremes, where oxygen concentration is very low, ATP and thus energy for the cells will be inadequate. However, GSCs are frequently found in regions of mild to moderate hypoxia [80]. GSCs were found to express hypoxia-inducible factors (HIFs), both at baseline and in recognition to low oxygen tension. Hypoxia was also found to promote expansion of CD133-positive stem cells through inactivation of HIF-1a [81], as well as regulating its tumorigenic potential [33]. Further, GSCs are able to utilize the HIF signaling pathway for survival and proliferation. It was also shown that hypoxia increases the expression of stem-cell markers and induces clonogenicity [82]. Through upregulation of downstream effectors and induction of GSC-associated genes, HIFs are able to enhance the GSC potential. Notably, the GSC hypoxia niche has a direct effect on the chemotherapeutic resistance as will be discussed later in this chapter [83]. Therefore, resistance to hypoxia plays a key role on survival of GSCs.

GSC role in angiogenesis

For a tumor to grow, a growing network of blood vessels has to be formed, allowing the exchange of nutrients. GSCs are known to be found in perivascular niches, having a close interaction with blood vessels [14]. It is established that vascular endothelial cells are

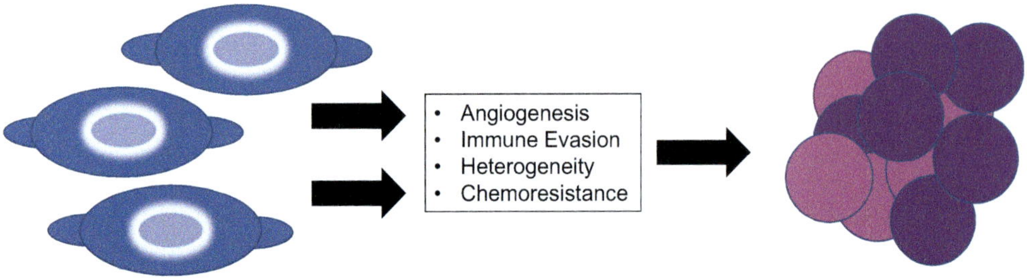

FIG. 1 Illustration of GSCs demonstrating capabilities aiding survival.

supervising the fate and survival of GSCs [84–86]. However, the reciprocal is also been suggested in the literature [87, 88]. GSCs might also interact with adjacent vascular endothelial cells embracing vascular bed formation within the growing tumor. Skog and colleagues shown that isolated tumor cells and GBM bulk are able to secrete extracellular vesicles containing angiogenic factors [89]. A recent paper found that vascular endothelial growth factor A (VEGF-A) is also released in extracellular vesicles by the perivascular GSCs [90]. Indeed, a scoping review on the "cross talk between glioma stem cells and their microenvironment," concluded that GSCs play a dominant role in tumor angiogenesis via: (i) releasing extracellular proangiogenic vesicles, (ii) recruitment of endothelial cells, and (iii) direct differentiation into endothelial cells [91]. Therefore, the pertinent role of GSC in tumor reoccurrence is better understood if we take into consideration their role in regulating blood vessel formation (Fig. 1).

GSC role in immune evasion

Currently, there are a number of new therapeutic strategies primarily focusing on harnessing and/or enhancing the immune response against tumor formation. For the tumor cells to survive even after therapy, there has to be an element of immune evasion and suppression. GSCs are known to the manipulate the tumor microenvironment [92]. Studies have shown that GSCs are evading the cancer-immunity cycle at various stages. GSCs can utilize the signal transducer and activator of transcription 3 (STAT3) pathway, aiding suppression of macrophage activation [93, 94] as well as the cytotoxicity potential of neutrophils and natural killer cells (NK) [94]. Further, they can manipulate expression of the cell-surface receptors of dendritic cells (DT) such as MHC II, CD80, and interleukin (IL)-12, preventing the DT-T-cell interaction, which can generate antitumor immunity [95]. Interestingly, the phosphorylated STAT3 pathway is held responsible for the release of a number of antiinflammatory mediators such as IL-10, prostaglandin (PG) E_2, and tissue growth factor beta (TGF-ß), which have a unique role in: (i) downregulating the Th1-mediated cytotoxicity, (ii) inducing Th17 cells activation which is immunosuppressive, and (iii) inducing T-reg and inhibit T-cell proliferation, respectively [96–99]. Further, the extracellular immunosuppressive cytokines convert the tissue-associated macrophages (TAM), into an M2 phenotype, which is vastly immunosuppressive [100]. A recent review has summarized and discussed the immune system evasion in

GBM and provided an outline of the ongoing novel immune therapies to date [101]. Herein, the complex dynamic relationship between the GBM cells, primarily GSCs, and the rest of the immune and supporting cells, aids survival and tumor recurrence.

The impact of GSC heterogeneity and plasticity on chemoresistance

The Cancer Genome Atlas (TCGA) has allowed a number of research groups to use the database and inspired others, publishing results, which have expanded and vastly transformed our knowledge on GBM and other high-grade gliomas [19, 25, 102, 103]. It is now recognized that within a high-grade glioma (WHO III–IV), three main phenotypes of cells exist: proneural, mesenchymal, and classical. The proneural phenotype is associated with favorable outcomes and isocitrate dehydrogenase gene 1 (IDH1) mutation or platelet-derived growth factor receptor a (PDGFRa) amplification, plus p53 mutation [25]. On the contrary, the classical phenotype is associated with amplified epidermal growth factor receptor (EGFR) signaling pathway, while the mesenchymal type exhibits no mutation on IDH1 but rather a neurofibromatosis 1 (NF1) deregulated gene expression. Both phenotypes are associated with worse prognosis [103]. It is recently shown that tumor reoccurrence is associated with a phenotypic shift from proneural to mesenchymal, suggesting that tumor reoccurrence could be linked to worse prognosis than the initial primary tumor [104]. Further, two different phenotypes could coexist within the same tumor, proneural and mesenchymal [104]. Interestingly, it has also been documented that increasing phenotypic heterogeneity within the same tumor group is linked to worse survival [22]. Moreover, the same study showed that cells can simultaneously "score" highly for two subtypes, meaning they came from two different phenotypic cells; herein the existence of "hybrid" states is possible, reflecting aberrant developmental programs. Therefore, within the same tumor, there might be a number of different GSC subpopulations. Thus, the deletion of one could mean the enrichment of another.

During therapy, a number of GSC subpopulations would be extinct. This applies a selective pressure to the rest of the subpopulations, which may remain. Meaning a new set of subpopulations will arise and proliferate. These new population could have survived due to two main reasons: (i) drug induce mutations making them resistance to treatment and (ii) existence of resistant clones owing to 'natural' tumor progression even before therapeutic intervention. In fact, GB-resistant subclones were found to acquire the cellular information pertinent to tumor reoccurrence at the early stages of their development, rather than accumulation of mutations over time [105]. The subclones will persist after treatment and repopulate over time, leading to tumor reoccurrence.

The cancer stem cell hypothesis proposes that the complete elimination, if at all possible, of GSCs would be able to stop reoccurrence. However, the recent discoveries denote that non-GSCs can acquire proliferative potential [83, 89]. CD133 cells were also shown to acquire GSC CD133+ properties in animal models [106]. Dahan et al. have discussed the importance of cellular plasticity in chemoresistance [107]. They showed that non-GSCs were able to dedifferentiate into GSCs following radiation. Worth noting that GSCs were proposed to interconvert to non-GSCs induced by TMZ therapy also [108]. TMZ may create cellular stressors such as hypoxia and acidity. Furthermore, HIF2a as well as loss of von Hippel–Lindau and/or phosphatase and tensin homolog (PTEN) have been proposed to induce GSC conversion

[109, 110]. It can be concluded that even if all GSCs subpopulations are eliminated, the surrounding cells can be induced to acquire oncogenic characteristics, which may lead to tumor formation (Fig. 1).

Specific molecular pathways of GSC resistance to chemotherapy

GBM has been incurable to date. GSCs remain at the center of attention, held responsible for current treatment failures and therapeutic resistance. GSCs are shown to circumvent chemotherapy in a plethora of cellular mechanisms. Preventing high accumulation of TMZ within the cell, increasing expression of specific enzymes counteracting cytotoxic effects of TMZ, and increasing or decreasing the expression of DNA repair mechanisms are but a few examples (Fig. 2).

ATP-binding cassette transporter channel

TMZ's tumor bioavailability has been questionable owing to previously mentioned drug delivery challenges presented by the BBB and local vascularity. Cellular bioavailability is further hindered by a unique cell-surface transporter, the ATP-binding cassette (ABC) transporter channels. The ABC transporter was found to be increasingly expressed in GSCs [111, 112]. Its role is to actively transport xenobiotics and metabolites, molecules such as TMZ, out of the cell as well as the BBB [113]. ABCG2 is the main transporter of reference, which has been associated with small stem-like subpopulation showing multidrug resistance [114]. ABC has also been accused to be a cancer driver, denoting functions in cancer progression [115].

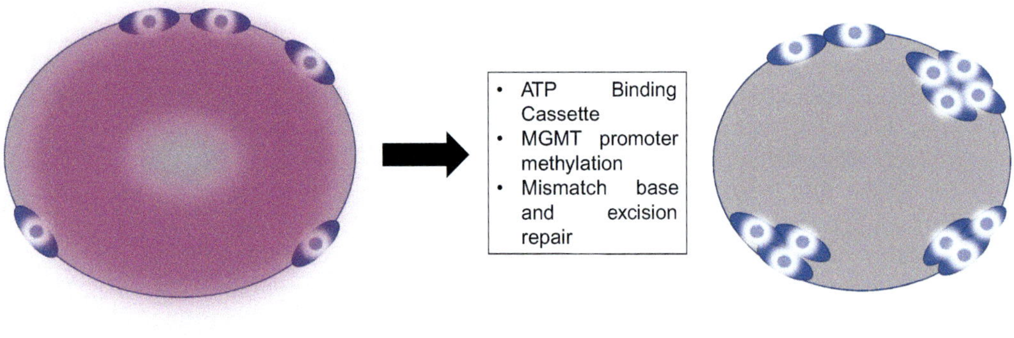

FIG. 2 Illustration of GSCs surrounding the tumor cavity able to resist TMZ therapy, leading to proliferation of cells.

Methylguanine-DNA-methyltransferase

TMZ is able to induce its cytotoxic effects in GBM cells by methylating the O6-position of guanine base in DNA. This creates a change in DNA structure, which can be detected by other cellular mechanisms, leading to cell cycle arrest and inducing apoptosis if the damage cannot be repaired. However, the methyl residue that is added on the DNA molecule by TMZ can be removed by O6-methylguianine-DNA-methyltransferase (MGMT), a repair enzyme. MGMT's promoter region can also be randomly methylated leading to inactivation of the MGMT enzyme and thus inability to repair the methylation gain in DNA molecule. Methylated DNA can be detected by cell cycle checkpoint inhibitors and if not repaired, cell cycle arrest can be induced [116–118]. Whereas if methylation of the MGMT promoter is low, an increased MGMT ability to remove methyl residue form DNA will reduce cytotoxicity, leading to poor prognosis. GSCs have been linked to increase expression of the MGMT repair enzyme, leading to TMZ resistance [119]. Notably, TMZ was shown unable to block the renewal properties of GSCs, expressing MGMT [69]. On the contrary, TMZ was found to be effective in eradicating MGMT-negative GSCs [120].

Mismatch and base excision repair mechanisms

DNA repair aberrations have been previously described and accounted for treatment resistance and recurrence. TMZ leads to methylation of a guanine base. This base-pair change is a common target for intercellular DNA repair mechanisms [121]. The mismatch repair (MMR) mechanism is responsible for recognizing the base-pair mismatches and deletions including the O6-methylguanine produced by TMZ [122]. Thereafter, if base change is irreparable, the DNA breaks, followed by initiation of apoptotic pathways for self-destruction [121]. In GSCs, MMR is found to be inactivated or at least dysfunctional [123]. TMZ main mechanism of action is thus bypassed. This allows for cell cycle and replication to continue without correction, leading to accumulation of mutations.

Via the poly(ADP-ribose) polymerase (PARP-1) pathway, another way of recognizing DNA damage, cells can activate the DNA repair systems [123]. PARP-1 was also found to be upregulated in glioma [124]. A base excision repair (BER) mechanism aids the correction of the base pairs detected [125]. BER is able to excise the common chemo/radiotherapeutic-induced DNA changes detected [125]. Subsequently, damaged DNA base pairs can be removed via a DNA glycosylate and/or apurinic/apyrimidinic endonuclease (APE1) enzyme. Following this, a DNA polymerase will add the correct bases and a ligase with enable collect closure of the DNA molecule [126]. PARP-1 affects both the MMR and BER systems. These systems are hyperactive in malignancy, herein GBM cells, leading to a cumulative effect of drastic increase in repair of DNA damaged induced by therapy.

Cell cycle checkpoints

The role of checkpoint inhibitors in cancer biology has been well known [127]. DNA checkpoints specifically ChK1 and Chk2 kinases are known to detect possible DNA alterations, prevent cycle progression, and activate genes aiding correction of DNA changes [128]. Chk1 is active, in the S and G2 cycle phase, without the detection of a DNA change. On the contrary,

Chk2 protein is activated upon detection of DNA damage at any point in the cell cycle. Chk1 has found to be overexpressed in a number of tumors including breast and colon as well as being correlated with tumor grade and recurrence [129–133]. Enhanced Chk1 has led to chemo/radiotherapeutic resistance of cancel cells, not only in GB but also in prostate and lung [134–137]. Increased expression of Chk1 has also been specifically linked to resistance to chemotherapy [137]. Chk1 is known to be essential for cell survival and viability. Increased expression may lead to increased survival because the higher the Chk1 protein concentration is, the higher the ability to detect and handle DNA damage induced by therapies [138]. Cell cycle checkpoint proteins have an overwhelming role in therapeutic resistance, and thus are extensively studied, hoping to provide future targets for therapy.

Combining a variety of intracellular counteracting chemotherapeutic mechanisms, specifically to TMZ, the GSC antitherapeutic abilities are an area of research and discovery. Despite the finding of myriad chemoresistant mechanisms, no clinical translation improving overall prognosis has been made. New drug targets combined with precision medicine protocols may offer hope to one of the most feared malignancies known to mankind.

Conclusion

GSCs exist in quiescence and active states, able to switch between functional states, influenced by the microenvironment. They play a key role in tumor reoccurrence. They are able to release key proangiogenic mediators and induce blood vessel formation. Further, they can downregulate the expression of pro-inflammatory mediators and suppress cell induce-induced cytotoxicity and antibody production. Their properties allow for new tumor to be formed. TMZ is the gold-standard chemotherapeutic agent for GBM. However, adequacy of its bioavailability within tumor is debatable. GSCs are not only able to survive but also can resist the action of TMZ. First, they can increase the expression of ABC transporters, effectively pumping out TMZ of the cells. Second, they show increased expression of MGMT, an enzyme that can efficiently correct TMZ-methylation induced changes to the DNA. Third, they demonstrate an alteration of MMR mechanism allowing for accumulation of TMZ-induced DNA changes without arresting the cell cycle. However, for the DNA changes that are detrimental to the cell, they exhibit increase in ChK1 and Chk2 kinase expression, and thus an enhanced ability to handle pertinent DNA chemotherapeutic-induced damage.

References

[1] Ostrom QT, Cioffi G, Gittleman H, Patil N, Waite K, Kruchko C, Barnholtz-Sloan JS. CBTRUS statistical report: primary brain and other central nervous system tumors diagnosed in the United States in 2012–2016. Neuro Oncol 2019;21(Suppl. 5). v1–00.

[2] Philips A, Henshaw DL, Lamburn G, O'Carroll MJ. Brain tumours: rise in glioblastoma multiforme incidence in England 1995–2015 suggests an adverse environmental or lifestyle factor. J Environ Public Health 2018;2018:7910754.

[3] Johnson DR. Rising incidence of glioblastoma and meningioma in the United States: Projections through 2050. J Clin Oncol 2012;30(Suppl 15):2065.

[4] Louis DN, Perry A, Reifenberger G, Von Deimling A, Figarella-Branger D, Cavenee WK, Ohgaki H, Wiestler OD, Kleihues P, Ellison DW. The 2016 World Health Organization classification of tumors of the central nervous system: a summary. Acta Neuropathol 2016;131(6):803–20.
[5] Gilbert MR, Wang M, Aldape KD, Stupp R, Hegi ME, Jaeckle KA, Armstrong TS, Wefel JS, Won M, Blumenthal DT, Mahajan A. Dose-dense temozolomide for newly diagnosed glioblastoma: a randomized phase III clinical trial. J Clin Oncol 2013;31(32):4085.
[6] Stupp R, Mason WP, van den Bent MJ, Weller M, Fisher B, Taphoorn MJ, et al. Radiotherapy plus concomitant and adjuvant temozolomide for glioblastoma. N Engl J Med 2005;352(10):987.
[7] Taal W, Oosterkamp HM, Walenkamp AM, Dubbink HJ, Beerepoot LV, Hanse MC, Buter J, Honkoop AH, Boerman D, de Vos FY, Dinjens WN. Single-agent bevacizumab or lomustine versus a combination of bevacizumab plus lomustine in patients with recurrent glioblastoma (BELOB trial): a randomised controlled phase 2 trial. Lancet Oncol 2014;15(9):943–53.
[8] Wick W, Gorlia T, Bendszus M, Taphoorn M, Sahm F, Harting I, Brandes AA, Taal W, Domont J, Idbaih A, Campone M. Lomustine and bevacizumab in progressive glioblastoma. N Engl J Med 2017;377(20):1954–63.
[9] Herrlinger U, Schäfer N, Steinbach JP, Weyerbrock A, Hau P, Goldbrunner R, Friedrich F, Rohde V, Ringel F, Schlegel U, Sabel M. Bevacizumab plus irinotecan versus Temozolomide in newly diagnosed O6-Methylguanine–DNA methyltransferase nonmethylated glioblastoma: the randomized GLARIUS trial. J Clin Oncol 2016;34(14):1611–9.
[10] Weller M, Butowski N, Tran DD, Recht LD, Lim M, Hirte H, Ashby L, Mechtler L, Goldlust SA, Iwamoto F, Drappatz J. Rindopepimut with temozolomide for patients with newly diagnosed, EGFRvIII-expressing glioblastoma (ACT IV): a randomised, double-blind, international phase 3 trial. Lancet Oncol 2017;18(10):1373–85.
[11] Batchelor TT, Mulholland P, Neyns B, Nabors LB, Campone M, Wick A, Mason W, Mikkelsen T, Phuphanich S, Ashby LS, DeGroot J. Phase III randomized trial comparing the efficacy of cediranib as monotherapy, and in combination with lomustine, versus lomustine alone in patients with recurrent glioblastoma. J Clin Oncol 2013;31(26):3212.
[12] Wick W, Puduvalli VK, Chamberlain MC, Van Den Bent MJ, Carpentier AF, Cher LM, Mason W, Weller M, Hong S, Musib L, Liepa AM. Phase III study of enzastaurin compared with lomustine in the treatment of recurrent intracranial glioblastoma. J Clin Oncol 2010;28(7):1168.
[13] de Groot JF, Lamborn KR, Chang SM, Gilbert MR, Cloughesy TF, Aldape K, Yao J, Jackson EF, Lieberman F, Robins HI, Mehta MP. Phase II study of aflibercept in recurrent malignant glioma: a north American brain tumor consortium study. J Clin Oncol 2011;29(19):2689.
[14] Chinot OL, Wick W, Mason W, Henriksson R, Saran F, Nishikawa R, Carpentier AF, Hoang-Xuan K, Kavan P, Cernea D, Brandes AA. Bevacizumab plus radiotherapy–temozolomide for newly diagnosed glioblastoma. N Engl J Med 2014;370(8):709–22.
[15] Filley AC, Dey M. Dendritic cell based vaccination strategy: an evolving paradigm. J Neurooncol 2017; 133(2):223–35.
[16] Weller M, Roth P, Preusser M, Wick W, Reardon DA, Platten M, Sampson JH. Vaccine-based immunotherapeutic approaches to gliomas and beyond. Nat Rev Neurol 2017;13(6):363.
[17] Bagley SJ, Desai AS, Linette GP, June CH, O'Rourke DM. CAR T-cell therapy for glioblastoma: recent clinical advances and future challenges. Neuro Oncol 2018;20(11):1429–38.
[18] Lang FF, Conrad C, Gomez-Manzano C, Yung WA, Sawaya R, Weinberg JS, Prabhu SS, Rao G, Fuller GN, Aldape KD, Gumin J. Phase I study of DNX-2401 (Delta-24-RGD) oncolytic adenovirus: replication and immunotherapeutic effects in recurrent malignant glioma. J Clin Oncol 2018;36(14):1419.
[19] Phillips HS, Kharbanda S, Chen R, Forrest WF, Soriano RH, Wu TD, Misra A, Nigro JM, Colman H, Soroceanu L, Williams PM. Molecular subclasses of high-grade glioma predict prognosis, delineate a pattern of disease progression, and resemble stages in neurogenesis. Cancer Cell 2006;9(3):157–73.
[20] Aparicio S, Caldas C. The implications of clonal genome evolution for cancer medicine. N Engl J Med 2013; 368(9):842–51.
[21] Sottoriva A, Spiteri I, Piccirillo SG, Touloumis A, Collins VP, Marioni JC, Curtis C, Watts C, Tavaré S. Intratumor heterogeneity in human glioblastoma reflects cancer evolutionary dynamics. Proc Natl Acad Sci 2013;110(10):4009–14.
[22] Patel AP, Tirosh I, Trombetta JJ, Shalek AK, Gillespie SM, Wakimoto H, Cahill DP, Nahed BV, Curry WT, Martuza RL, Louis DN. Single-cell RNA-seq highlights intratumoral heterogeneity in primary glioblastoma. Science 2014;344(6190):1396–401.

[23] Liebelt BD, Shingu T, Zhou X, Ren J, Shin SA, Hu J. Glioma stem cells: signaling, microenvironment, and therapy. Stem Cells Int 2016;2016:7849890.
[24] Szerlip NJ, Pedraza A, Chakravarty D, Azim M, McGuire J, Fang Y, Ozawa T, Holland EC, Huse JT, Jhanwar S, Leversha MA. Intratumoral heterogeneity of receptor tyrosine kinases EGFR and PDGFRA amplification in glioblastoma defines subpopulations with distinct growth factor response. Proc Natl Acad Sci 2012;109(8):3041–6.
[25] Brennan CW, Verhaak RG, McKenna A, Campos B, Noushmehr H, Salama SR, Zheng S, Chakravarty D, Sanborn JZ, Berman SH, Beroukhim R. The somatic genomic landscape of glioblastoma. Cell 2013;155(2):462–77.
[26] Greaves M, Maley CC. Clonal evolution in cancer. Nature 2012;481(7381):306.
[27] Wang J, Cazzato E, Ladewig E, Frattini V, Rosenbloom DI, Zairis S, Abate F, Liu Z, Elliott O, Shin YJ, Lee JK. Clonal evolution of glioblastoma under therapy. Nat Genet 2016;48(7):768.
[28] Meyer M, Reimand J, Lan X, Head R, Zhu X, Kushida M, Bayani J, Pressey JC, Lionel AC, Clarke ID, Cusimano M. Single cell-derived clonal analysis of human glioblastoma links functional and genomic heterogeneity. Proc Natl Acad Sci 2015;112(3):851–6.
[29] Tirosh I, Venteicher AS, Hebert C, Escalante LE, Patel AP, Yizhak K, Fisher JM, Rodman C, Mount C, Filbin MG, Neftel C. Single-cell RNA-seq supports a developmental hierarchy in human oligodendroglioma. Nature 2016;539(7628):309–13.
[30] Venteicher AS, Tirosh I, Hebert C, Yizhak K, Neftel C, Filbin MG, Hovestadt V, Escalante LE, Shaw ML, Rodman C, Gillespie SM. Decoupling genetics, lineages, and microenvironment in IDH-mutant gliomas by single-cell RNA-seq. Science 2017;355(6332):eaai8478.
[31] Singh SK, Clarke ID, Terasaki M, Bonn VE, Hawkins C, Squire J, Dirks PB. Identification of a cancer stem cell in human brain tumors. Cancer Res 2003;63(18):5821–8.
[32] Singh SK, Hawkins C, Clarke ID, Squire JA, Bayani J, Hide T, Henkelman RM, Cusimano MD, Dirks PB. Identification of human brain tumour initiating cells. Nature 2004;432(7015):396–401.
[33] Yuan X, Curtin J, Xiong Y, Liu G, Waschsmann-Hogiu S, Farkas DL, Black KL, John SY. Isolation of cancer stem cells from adult glioblastoma multiforme. Oncogene 2004;23(58):9392–400.
[34] Galli R, Binda E, Orfanelli U, Cipelletti B, Gritti A, De Vitis S, Fiocco R, Foroni C, Dimeco F, Vescovi A. Isolation and characterization of tumorigenic, stem-like neural precursors from human glioblastoma. Cancer Res 2004; 64(19):7011–21.
[35] Huang Z, Cheng L, Guryanova OA, Wu Q, Bao S. Cancer stem cells in glioblastoma—molecular signaling and therapeutic targeting. Protein Cell 2010;1(7):638–55.
[36] Ahmed AU, Auffinger B, Lesniak MS. Understanding glioma stem cells: rationale, clinical relevance and therapeutic strategies. Expert Rev Neurother 2013;13(5):545–55.
[37] Bloch O, Han SJ, Cha S, Sun MZ, Aghi MK, McDermott MW, Berger MS, Parsa AT. Impact of extent of resection for recurrent glioblastoma on overall survival. J Neurosurg 2012;117(6):1032–8.
[38] Sanai N, Polley MY, McDermott MW, Parsa AT, Berger MS. An extent of resection threshold for newly diagnosed glioblastomas. J Neurosurg 2011;115(1):3–8.
[39] Weller M, Van Den Bent M, Tonn JC, Stupp R, Preusser M, Cohen-Jonathan-Moyal E, Henriksson R, Le Rhun E, Balana C, Chinot O, Bendszus M. European Association for Neuro-Oncology (EANO) guideline on the diagnosis and treatment of adult astrocytic and oligodendroglial gliomas. Lancet Oncol 2017;18(6):e315–29.
[40] NICE. Context | Brain tumours (primary) and brain metastases in adults | Guidance | NICE. [cited 2020 April 05]; Available from: https://www.nice.org.uk/guidance/ng99/chapter/Context.
[41] Senft C, Bink A, Franz K, Vatter H, Gasser T, Seifert V. Intraoperative MRI guidance and extent of resection in glioma surgery: a randomised, controlled trial. Lancet Oncol 2011;12(11):997–1003.
[42] Kubben PL, ter Meulen KJ, Schijns OE, ter Laak-Poort MP, van Overbeeke JJ, van Santbrink H. Intraoperative MRI-guided resection of glioblastoma multiforme: a systematic review. Lancet Oncol 2011;12(11):1062–70.
[43] Kuhnt D, Becker A, Ganslandt O, Bauer M, Buchfelder M, Nimsky C. Correlation of the extent of tumor volume resection and patient survival in surgery of glioblastoma multiforme with high-field intraoperative MRI guidance. Neuro Oncol 2011;13(12):1339–48.
[44] Mahboob S, McPhillips R, Qiu Z, Jiang Y, Meggs C, Schiavone G, Button T, Desmulliez M, Demore C, Cochran S, Eljamel S. Intraoperative ultrasound-guided resection of gliomas: a meta-analysis and review of the literature. World Neurosurg 2016;92:255–63.
[45] Stummer W, Pichlmeier U, Meinel T, Wiestler OD, Zanella F, Reulen HJ, ALA-Glioma Study Group. Fluorescence-guided surgery with 5-aminolevulinic acid for resection of malignant glioma: a randomised controlled multicentre phase III trial. Lancet Oncol 2006;7(5):392–401.

[46] Zhao S, Wu J, Wang C, Liu H, Dong X, Shi C, Shi C, Liu Y, Teng L, Han D, Chen X. Intraoperative fluorescence-guided resection of high-grade malignant gliomas using 5-aminolevulinic acid–induced porphyrins: A systematic review and meta-analysis of prospective studies. PLoS One 2013;8(5):e63682.
[47] Eljamel MS, Mahboob SO. The effectiveness and cost-effectiveness of intraoperative imaging in high-grade glioma resection; a comparative review of intraoperative ALA, fluorescein, ultrasound and MRI. Photodiagnosis Photodyn Ther 2016;16:35–43.
[48] Meyer FB, Bates LM, Goerss SJ, Friedman JA, Windschitl WL, Duffy JR, Perkins WJ, O'Neill BP. Awake craniotomy for aggressive resection of primary gliomas located in eloquent brain. In: Mayo Clinic Proceeding. Elsevier; 2001. p. 677–87. Vol. 76, No. 7.
[49] Hervey-Jumper SL, Li J, Lau D, Molinaro AM, Perry DW, Meng L, Berger MS. Awake craniotomy to maximize glioma resection: methods and technical nuances over a 27-year period. J Neurosurg 2015;123(2):325–39.
[50] Hadjipanayis CG, Stummer W. 5-ALA and FDA approval for glioma surgery. J Neurooncol 2019;141(3):479–86.
[51] Suchorska B, Weller M, Tabatabai G, Senft C, Hau P, Sabel MC, Herrlinger U, Ketter R, Schlegel U, Marosi C, Reifenberger G. Complete resection of contrast-enhancing tumor volume is associated with improved survival in recurrent glioblastoma—results from the DIRECTOR trial. Neuro Oncol 2016;18(4):549–56.
[52] Oh J, Sahgal A, Sanghera P, Tsao MN, Davey P, Lam K, Symons S, Aviv R, Perry JR. Glioblastoma: patterns of recurrence and efficacy of salvage treatments. Can J Neurol Sci 2011;38(4):621–5.
[53] Giese A, Kucinski T, Knopp U, Goldbrunner R, Hamel W, Mehdorn HM, Tonn JC, Hilt D, Westphal M. Pattern of recurrence following local chemotherapy with biodegradable carmustine (BCNU) implants in patients with glioblastoma. J Neurooncol 2004;66(3):351–60.
[54] Munthe S, Petterson SA, Dahlrot RH, Poulsen FR, Hansen S, Kristensen BW. Glioma cells in the tumor periphery have a stem cell phenotype. PLoS One 2016;11(5):e0155106.
[55] Nishikawa M, Inoue A, Ohnishi T, Kohno S, Ohue S, Matsumoto S, Suehiro S, Yamashita D, Ozaki S, Watanabe H, Yano H. Significance of glioma stem-like cells in the tumor periphery that express high levels of CD44 in tumor invasion, early progression, and poor prognosis in glioblastoma. Stem Cells Int 2018;2018:53870412018.
[56] Calabrese C, Poppleton H, Kocak M, Hogg TL, Fuller C, Hamner B, Oh EY, Gaber MW, Finklestein D, Allen M, Frank A. A perivascular niche for brain tumor stem cells. Cancer Cell 2007;11(1):69–82.
[57] Li Z, Bao S, Wu Q, Wang H, Eyler C, Sathornsumetee S, Shi Q, Cao Y, Lathia J, McLendon RE, Hjelmeland AB. Hypoxia-inducible factors regulate tumorigenic capacity of glioma stem cells. Cancer Cell 2009;15(6):501–13.
[58] Molinaro AM, Hervey-Jumper S, Morshed RA, Young J, Han SJ, Chunduru P, Zhang Y, Phillips JJ, Shai A, Lafontaine M, Crane J. Association of maximal extent of resection of contrast-enhanced and non–contrast-enhanced tumor with survival within molecular subgroups of patients with newly diagnosed glioblastoma. JAMA Oncol 2020;6(4):495–503.
[59] Chamberlain MC. Radiographic patterns of relapse in glioblastoma. J Neurooncol 2011;101(2):319–23.
[60] Yan JL, van der Hoorn A, Larkin TJ, Boonzaier NR, Matys T, Price SJ. Extent of resection of peritumoral diffusion tensor imaging–detected abnormality as a predictor of survival in adult glioblastoma patients. J Neurosurg 2017;126(1):234–41.
[61] Eyüpoglu IY, Hore N, Merkel A, Buslei R, Buchfelder M, Savaskan N. Supra-complete surgery via dual intraoperative visualization approach (DiVA) prolongs patient survival in glioblastoma. Oncotarget 2016; 7(18):25755.
[62] Pessina F, Navarria P, Cozzi L, Ascolese AM, Simonelli M, Santoro A, Clerici E, Rossi M, Scorsetti M, Bello L. Maximize surgical resection beyond contrast-enhancing boundaries in newly diagnosed glioblastoma multiforme: is it useful and safe? A single institution retrospective experience. J Neurooncol 2017;135(1):129–39.
[63] Kotrotsou A, Elakkad A, Sun J, Thomas GA, Yang D, Abrol S, Wei W, Weinberg JS, Bakhtiari AS, Kircher MF, Luedi MM. Multi-center study finds postoperative residual non-enhancing component of glioblastoma as a new determinant of patient outcome. J Neurooncol 2018;139(1):125–33.
[64] Pardridge WM. Drug transport across the blood–brain barrier. J Cereb Blood Flow Metab 2012;32(11):1959–72.
[65] Brada M, Judson I, Beale P, Moore S, Reidenberg P, Statkevich P, Dugan M, Batra V, Cutler D. Phase I dose-escalation and pharmacokinetic study of temozolomide (SCH 52365) for refractory or relapsing malignancies. Br J Cancer 1999;81(6):1022–30.
[66] Ostermann S, Csajka C, Buclin T, Leyvraz S, Lejeune F, Decosterd LA, Stupp R. Plasma and cerebrospinal fluid population pharmacokinetics of temozolomide in malignant glioma patients. Clin Cancer Res 2004;10(11):3728–36.
[67] Portnow J, Badie B, Chen M, Liu A, Blanchard S, Synold TW. The neuropharmacokinetics of temozolomide in patients with resectable brain tumors: potential implications for the current approach to chemoradiation. Clin Cancer Res 2009;15(22):7092–8.

[68] Hermisson M, Klumpp A, Wick W, Wischhusen J, Nagel G, Roos W, Kaina B, Weller M. O6-methylguanine DNA methyltransferase and p53 status predict temozolomide sensitivity in human malignant glioma cells. J Neurochem 2006;96(3):766–76.
[69] Beier D, Röhrl S, Pillai DR, Schwarz S, Kunz-Schughart LA, Leukel P, Proescholdt M, Brawanski A, Bogdahn U, Trampe-Kieslich A, Giebel B. Temozolomide preferentially depletes cancer stem cells in glioblastoma. Cancer Res 2008;68(14):5706–15.
[70] Blough MD, Westgate MR, Beauchamp D, Kelly JJ, Stechishin O, Ramirez AL, Weiss S, Cairncross JG. Sensitivity to temozolomide in brain tumor initiating cells. Neuro Oncol 2010;12(7):756–60.
[71] Son MJ, Woolard K, Nam DH, Lee J, Fine HA. SSEA-1 is an enrichment marker for tumor-initiating cells in human glioblastoma. Cell Stem Cell 2009;4(5):440–52.
[72] Lottaz C, Beier D, Meyer K, Kumar P, Hermann A, Schwarz J, Junker M, Oefner PJ, Bogdahn U, Wischhusen J, Spang R. Transcriptional profiles of CD133+ and CD133− glioblastoma-derived cancer stem cell lines suggest different cells of origin. Cancer Res 2010;70(5):2030–40.
[73] Chen R, Nishimura MC, Bumbaca SM, Kharbanda S, Forrest WF, Kasman IM, Greve JM, Soriano RH, Gilmour LL, Rivers CS, Modrusan Z. A hierarchy of self-renewing tumor-initiating cell types in glioblastoma. Cancer Cell 2010;17(4):362–75.
[74] Al-Mayhani TF, Heywood RM, Vemireddy V, Lathia JD, Piccirillo SG, Watts C. A non-hierarchical organization of tumorigenic NG2 cells in glioblastoma promoted by EGFR. Neuro Oncol 2019;21(6):719–29.
[75] Neftel C, Laffy J, Filbin MG, Hara T, Shore ME, Rahme GJ, Richman AR, Silverbush D, Shaw ML, Hebert CM, Dewitt J. An integrative model of cellular states, plasticity, and genetics for glioblastoma. Cell 2019;178(4):835–49.
[76] Tirosh I, Suvà ML. Tackling the many facets of glioblastoma heterogeneity. Cell Stem Cell 2020;26(3):303–4.
[77] Finocchiaro G, Pellegatta S. NG2/CSPG4 in glioblastoma: about flexibility. Neuro Oncol 2019;21(6):697–8.
[78] Suvà ML, Rheinbay E, Gillespie SM, Patel AP, Wakimoto H, Rabkin SD, Riggi N, Chi AS, Cahill DP, Nahed BV, Curry WT. Reconstructing and reprogramming the tumor-propagating potential of glioblastoma stem-like cells. Cell 2014;157(3):580–94.
[79] Rheinbay E, Suvà ML, Gillespie SM, Wakimoto H, Patel AP, Shahid M, Oksuz O, Rabkin SD, Martuza RL, Rivera MN, Louis DN. An aberrant transcription factor network essential for Wnt signaling and stem cell maintenance in glioblastoma. Cell Rep 2013;3(5):1567–79.
[80] Seidel S, Garvalov BK, Wirta V, von Stechow L, Schänzer A, Meletis K, Wolter M, Sommerlad D, Henze AT, Nister M, Reifenberger G. A hypoxic niche regulates glioblastoma stem cells through hypoxia inducible factor 2α. Brain 2010;133(4):983–95.
[81] Soeda A, Park M, Lee D, Mintz A, Androutsellis-Theotokis A, McKay RD, Engh J, Iwama T, Kunisada T, Kassam AB, Pollack IF. Hypoxia promotes expansion of the CD133-positive glioma stem cells through activation of HIF-1α. Oncogene 2009;28(45):3949–59.
[82] Bar EE, Lin A, Mahairaki V, Matsui W, Eberhart CG. Hypoxia increases the expression of stem-cell markers and promotes clonogenicity in glioblastoma neurospheres. Am J Pathol 2010;177(3):1491–502.
[83] Pistollato F, Abbadi S, Rampazzo E, Persano L, Della Puppa A, Frasson C, Sarto E, Scienza R, D'avella D, Basso G. Intratumoral hypoxic gradient drives stem cells distribution and MGMT expression in glioblastoma. Stem Cells 2010;28(5):851–62.
[84] Galan-Moya EM, Le Guelte A, Lima-Fernandes E, Thirant C, Dwyer J, Bidere N, Couraud PO, Scott MG, Junier MP, Chneiweiss H, Gavard J. Secreted factors from brain endothelial cells maintain glioblastoma stem-like cell expansion through the mTOR pathway. EMBO Rep 2011;12(5):470–6.
[85] Galan-Moya EM, Treps L, Oliver L, Chneiweiss H, Vallette FM, Bidère N, Gavard J. Endothelial secreted factors suppress mitogen deprivation-induced autophagy and apoptosis in glioblastoma stem-like cells. PLoS One 2014;9(3):e93505.
[86] Fessler E, Borovski T, Medema JP. Endothelial cells induce cancer stem cell features in differentiated glioblastoma cells via bFGF. Mol Cancer 2015;14(1):157.
[87] Bao S, Wu Q, Sathornsumetee S, Hao Y, Li Z, Hjelmeland AB, Shi Q, McLendon RE, Bigner DD, Rich JN. Stem cell–like glioma cells promote tumor angiogenesis through vascular endothelial growth factor. Cancer Res 2006;66(16):7843–8.
[88] Folkins C, Shaked Y, Man S, Tang T, Lee CR, Zhu Z, Hoffman RM, Kerbel RS. Glioma tumor stem-like cells promote tumor angiogenesis and vasculogenesis via vascular endothelial growth factor and stromal-derived factor 1. Cancer Res 2009;69(18):7243–51.

[89] Skog J, Würdinger T, Van Rijn S, Meijer DH, Gainche L, Curry WT, Carter BS, Krichevsky AM, Breakefield XO. Glioblastoma microvesicles transport RNA and proteins that promote tumour growth and provide diagnostic biomarkers. Nat Cell Biol 2008;10(12):1470–6.

[90] Treps L, Perret R, Edmond S, Ricard D, Gavard J. Glioblastoma stem-like cells secrete the pro-angiogenic VEGF-A factor in extracellular vesicles. J Extracell Vesicles 2017;6(1):1359479.

[91] Filatova A, Acker T, Garvalov BK. The cancer stem cell niche (s): the crosstalk between glioma stem cells and their microenvironment. Biochim Biophys Acta Gen Subj 2013;1830(2):2496–508.

[92] Wei J, Barr J, Kong LY, Wang Y, Wu A, Sharma AK, Gumin J, Henry V, Colman H, Sawaya R, Lang FF. Glioma-associated cancer-initiating cells induce immunosuppression. Clin Cancer Res 2010;16(2):461–73.

[93] Lang R, Patel D, Morris JJ, Rutschman RL, Murray PJ. Shaping gene expression in activated and resting primary macrophages by IL-10. J Immunol 2002;169(5):2253–63.

[94] O'Farrell AM, Liu Y, Moore KW, Mui AL. IL-10 inhibits macrophage activation and proliferation by distinct signaling mechanisms: evidence for Stat3-dependent and-independent pathways. EMBO J 1998;17(4):1006–18.

[95] Kortylewski M, Kujawski M, Wang T, Wei S, Zhang S, Pilon-Thomas S, Niu G, Kay H, Mulé J, Kerr WG, Jove R. Inhibiting Stat3 signaling in the hematopoietic system elicits multicomponent antitumor immunity. Nat Med 2005;11(12):1314–21.

[96] Steinbrink K, Wölfl M, Jonuleit H, Knop J, Enk AH. Induction of tolerance by IL-10-treated dendritic cells. J Immunol 1997;159(10):4772–80.

[97] Williams L, Bradley L, Smith A, Foxwell B. Signal transducer and activator of transcription 3 is the dominant mediator of the anti-inflammatory effects of IL-10 in human macrophages. J Immunol 2004;172(1):567–76.

[98] Boniface K, Bak-Jensen KS, Li Y, Blumenschein WM, McGeachy MJ, McClanahan TK, McKenzie BS, Kastelein RA, Cua DJ, de Waal Malefyt R. Prostaglandin E2 regulates Th17 cell differentiation and function through cyclic AMP and EP2/EP4 receptor signaling. J Exp Med 2009;206(3):535–48.

[99] Kinjyo I, Inoue H, Hamano S, Fukuyama S, Yoshimura T, Koga K, Takaki H, Himeno K, Takaesu G, Kobayashi T, Yoshimura A. Loss of SOCS3 in T helper cells resulted in reduced immune responses and hyperproduction of interleukin 10 and transforming growth factor–β1. J Exp Med 2006;203(4):1021–31.

[100] Razavi SM, Lee KE, Jin BE, Aujla PS, Gholamin S, Li G. Immune evasion strategies of glioblastoma. Front Surg 2016;3:11.

[101] Brown NF, Carter TJ, Ottaviani D, Mulholland P. Harnessing the immune system in glioblastoma. Br J Cancer 2018;119(10):1171–81.

[102] Sturm D, Witt H, Hovestadt V, Khuong-Quang DA, Jones DT, Konermann C, Pfaff E, Tönjes M, Sill M, Bender S, Kool M. Hotspot mutations in H3F3A and IDH1 define distinct epigenetic and biological subgroups of glioblastoma. Cancer Cell 2012;22(4):425–37.

[103] Verhaak RG, Hoadley KA, Purdom E, Wang V, Qi Y, Wilkerson MD, Miller CR, Ding L, Golub T, Mesirov JP, Alexe G. Integrated genomic analysis identifies clinically relevant subtypes of glioblastoma characterized by abnormalities in PDGFRA, IDH1, EGFR, and NF1. Cancer Cell 2010;17(1):98–110.

[104] Mao P, Joshi K, Li J, Kim SH, Li P, Santana-Santos L, Luthra S, Chandran UR, Benos PV, Smith L, Wang M. Mesenchymal glioma stem cells are maintained by activated glycolytic metabolism involving aldehyde dehydrogenase 1A3. Proc Natl Acad Sci 2013;110(21):8644–9.

[105] Orzan F, De Bacco F, Crisafulli G, Pellegatta S, Mussolin B, Siravegna G, D'Ambrosio A, Comoglio PM, Finocchiaro G, Boccaccio C. Genetic evolution of glioblastoma stem-like cells from primary to recurrent tumor. Stem Cells 2017;35(11):2218–28.

[106] Wang J, Sakariassen PØ, Tsinkalovsky O, Immervoll H, Bøe SO, Svendsen A, Prestegarden L, Røsland G, Thorsen F, Stuhr L, Molven A. CD133 negative glioma cells form tumors in nude rats and give rise to CD133 positive cells. Int J Cancer 2008;122(4):761–8.

[107] Dahan P, Gala JM, Delmas C, Monferran S, Malric L, Zentkowski D, Lubrano V, Toulas C, Moyal EC, Lemarie A. Ionizing radiations sustain glioblastoma cell dedifferentiation to a stem-like phenotype through survivin: possible involvement in radioresistance. Cell Death Dis 2014;5(11):e1543.

[108] Auffinger B, Tobias AL, Han Y, Lee G, Guo D, Dey M, Lesniak MS, Ahmed AU. Conversion of differentiated cancer cells into cancer stem-like cells in a glioblastoma model after primary chemotherapy. Cell Death Differ 2014;21(7):1119.

[109] Kanno H, Sato H, Yokoyama TA, Yoshizumi T, Yamada S. The VHL tumor suppressor protein regulates tumorigenicity of U87-derived glioma stem-like cells by inhibiting the JAK/STAT signaling pathway. Int J Oncol 2013;42(3):881–6.

[110] Korkaya H, Paulson A, Charafe-Jauffret E, Ginestier C, Brown M, Dutcher J, Clouthier SG, Wicha MS. Regulation of mammary stem/progenitor cells by PTEN/Akt/β-catenin signaling. PLoS Biol 2009;7(6):e1000121.
[111] Xu ZY, Wang K, Li XQ, Chen S, Deng JM, Cheng Y, Wang ZG. The ABCG2 transporter is a key molecular determinant of the efficacy of sonodynamic therapy with Photofrin in glioma stem-like cells. Ultrasonics 2013;53(1):232–8.
[112] Dean M. ABC transporters, drug resistance, and cancer stem cells. J Mammary Gland Biol Neoplasia 2009;14(1):3–9.
[113] Fletcher JI, Williams RT, Henderson MJ, Norris MD, Haber M. ABC transporters as mediators of drug resistance and contributors to cancer cell biology. Drug Resist Updat 2016;26:1–9.
[114] Li WQ, Li YM, Tao BB, Lu YC, Hu GH, Liu HM, He J, Xu CDY, Yu HY. Downregulation of ABCG2 expression in glioblastoma cancer stem cells with miRNA-328 may decrease their chemoresistance. Med Sci Monit 2010;16(10):30.
[115] Domenichini A, Adamska A, Falasca M. ABC transporters as cancer drivers: potential functions in cancer development. Biochim Biophys Acta Gen Subj 2019;1863(1):52–60.
[116] Hegi ME, Diserens AC, Gorlia T, Hamou MF, De Tribolet N, Weller M, Kros JM, Hainfellner JA, Mason W, Mariani L, Bromberg JE. MGMT gene silencing and benefit from temozolomide in glioblastoma. N Engl J Med 2005;352(10):997–1003.
[117] Beier D, Schulz JB, Beier CP. Chemoresistance of glioblastoma cancer stem cells-much more complex than expected. Mol Cancer 2011;10(1):128.
[118] Sørensen MD, Fosmark S, Hellwege S, Beier D, Kristensen BW, Beier CP. Chemoresistance and chemotherapy targeting stem-like cells in malignant glioma. In: Stem cell biology in neoplasms of the central nervous system. Cham: Springer; 2015. p. 111–38.
[119] Liu G, Yuan X, Zeng Z, Tunici P, Ng H, Abdulkadir IR, Lu L, Irvin D, Black KL, John SY. Analysis of gene expression and chemoresistance of CD133+ cancer stem cells in glioblastoma. Mol Cancer 2006;5(1):67.
[120] Clement V, Sanchez P, De Tribolet N, Radovanovic I, i Altaba AR. HEDGEHOG-GLI1 signaling regulates human glioma growth, cancer stem cell self-renewal, and tumorigenicity. Curr Biol 2007;17(2):165–72.
[121] Bartkova J, Hamerlik P, Stockhausen MT, Ehrmann J, Hlobilkova A, Laursen H, Kalita O, Kolar Z, Poulsen HS, Broholm H, Lukas J. Replication stress and oxidative damage contribute to aberrant constitutive activation of DNA damage signalling in human gliomas. Oncogene 2010;29(36):5095–102.
[122] Yip S, Miao J, Cahill DP, Iafrate AJ, Aldape K, Nutt CL, Louis DN. MSH6 mutations arise in glioblastomas during temozolomide therapy and mediate temozolomide resistance. Clin Cancer Res 2009;15(14):4622–9.
[123] Ropolo M, Daga A, Griffero F, Foresta M, Casartelli G, Zunino A, Poggi A, Cappelli E, Zona G, Spaziante R, Corte G. Comparative analysis of DNA repair in stem and nonstem glioma cell cultures. Mol Cancer Res 2009;7(3):383–92.
[124] Wharton SB, McNelis U, Bell HS, Whittle IR. Expression of poly (ADP-ribose) polymerase and distribution of poly (ADP-ribosyl) ation in glioblastoma and in a glioma multicellular tumour spheroid model. Neuropathol Appl Neurobiol 2000;26(6):528–35.
[125] Johannessen TC, Bjerkvig R, Tysnes BB. DNA repair and cancer stem-like cells–potential partners in glioma drug resistance? Cancer Treat Rev 2008;34(6):558–67.
[126] Bobola MS, Blank A, Berger MS, Stevens BA, Silber JR. Apurinic/apyrimidinic endonuclease activity is elevated in human adult gliomas. Clin Cancer Res 2001;7(11):3510–8.
[127] Bartek J, Lukas J. Chk1 and Chk2 kinases in checkpoint control and cancer. Cancer Cell 2003;3(5):421–9.
[128] Zhang Y, Hunter T. Roles of Chk1 in cell biology and cancer therapy. Int J Cancer 2014;134(5):1013–23.
[129] Cho SH, Toouli CD, Fujii GH, Crain C, Parry D. Chk1 is essential for tumor cell viability following activation of the replication checkpoint. Cell Cycle 2005;4(1):131–9.
[130] Madoz-Gúrpide J, Cañamero M, Sanchez L, Solano J, Alfonso P, Casal JI. A proteomics analysis of cell signaling alterations in colorectal cancer. Mol Cell Proteomics 2007;6(12):2150–64.
[131] Verlinden L, Bempt IV, Eelen G, Drijkoningen M, Verlinden I, Marchal K, De Wolf-Peeters C, Christiaens MR, Michiels L, Bouillon R, Verstuyf A. The E2F-regulated gene Chk1 is highly expressed in triple-negative estrogen receptor−/progesterone receptor−/HER-2− breast carcinomas. Cancer Res 2007;67(14):6574–81.
[132] Yao H, Yang Z, Li Y. Expression of checkpoint kinase 1 and polo-like kinase 1 and its clinicopathological significance in benign and malignant lesions of the stomach. Zhong nan da xue xue bao. Yi xue ban= journal of Central South University. Med Sci 2010;35(10):1080–4.

[133] Lundgren K, Holm K, Nordenskjöld B, Borg Å, Landberg G. Gene products of chromosome 11q and their association with CCND1 gene amplification and tamoxifen resistance in premenopausal breast cancer. Breast Cancer Res 2008;10(5):R81.

[134] Bao S, Wu Q, McLendon RE, Hao Y, Shi Q, Hjelmeland AB, Dewhirst MW, Bigner DD, Rich JN. Glioma stem cells promote radioresistance by preferential activation of the DNA damage response. Nature 2006; 444(7120):756–60.

[135] Bartucci M, Svensson S, Romania P, Dattilo R, Patrizii M, Signore M, Navarra S, Lotti F, Biffoni M, Pilozzi E, Duranti E. Therapeutic targeting of Chk1 in NSCLC stem cells during chemotherapy. Cell Death Differ 2012; 19(5):768–78.

[136] Wang X, Ma Z, Xiao Z, Liu H, Dou Z, Feng X, Shi H. Chk1 knockdown confers radiosensitization in prostate cancer stem cells. Oncol Rep 2012;28(6):2247–54.

[137] Røe OD, Szulkin A, Anderssen E, Flatberg A, Sandeck H, Amundsen T, Erlandsen SE, Dobra K, Sundstrøm SH. Molecular resistance fingerprint of pemetrexed and platinum in a long-term survivor of mesothelioma. PLoS One 2012;7(8):e40521.

[138] López-Contreras AJ, Gutierrez-Martinez P, Specks J, Rodrigo-Perez S, Fernandez-Capetillo O. An extra allele of Chk1 limits oncogene-induced replicative stress and promotes transformation. J Exp Med 2012;209(3):455–61.

CHAPTER

6

Signaling in the tumor microenvironment of therapy-resistant glioblastoma

Marina Nikolopoulos[a] *and Sunit Das*[a,b]

[a]Institute for Medical Science, University of Toronto, Toronto, ON, Canada [b]Division of Neurosurgery, St. Michael's Hospital, University of Toronto, Toronto, ON, Canada

Abstract

We review how signals from glioblastoma (GBM) extracellular matrix and from tumor-stromal cell interactions activate downstream cascades to create a pro-tumorigenic environment resistant to therapy.

Abbreviations

BBB	blood-brain barrier
(CAR)T	chimeric antigen receptor
CPTEN	phosphatase and tensin homolog
CSF-1	colony-stimulating factor -1
CTL	cytotoxic T cell
CXCR4	chemokine receptor type 4
ECM	extracellular matrix
EGF	epidermal growth factor
EMT	epithelial to mesenchymal transition
GAG	glycosaminoglycan
GBM	glioblastoma
GSCs	glioblastoma stem cells
GSIs	inhibitors of γ-secretase
HA	hyaluronic acid
HIF	hypoxia-induced factors
IL-1	interleukin-1
IL-6	interleukin-6
JAK	Janus Kinase
MAPK	mitogen-activated protein kinase

Glioblastoma Resistance to Chemotherapy: Molecular Mechanisms and Innovative Reversal Strategies
https://doi.org/10.1016/B978-0-12-821567-8.00004-X

MDR-1 multidrug resistance-1
MDSCs myeloid-derived suppressor cells
MMP matrix metalloproteinase
mTOR mechanistic target of rapamycin
MSCs glioma-associated mesenchymal stem cells
NF-κB nuclear factor-κB
P-gp P-glycoprotein
PDGF platelet-derived growth factor
PDL-1 programmed cell death ligand -1
PKC protein kinase C
PLC phospholipase
RTK receptor tyrosine kinase
SDF-1 stromal cell-derived factor 1
Sox2 sex-determining region Y box
STAT3 signal transducer and activator of transcription 3
TAAs tumor-associated astrocytes
TAMs tumour-associated macrophages
TGF-β1 transforming growth factor-β1
TGF-b transforming growth factor-beta
TLR-4 toll-like receptor 4
TME tumour microenvironment
TMZ temozolomide
TNF-α tumor necrosis factor α
Tregs tegulatory T cells
uPA urokinase-type plasminogen activator
VEGF bascular endothelial growth factor
WNT Wingless and Int-1

Conflict of interest

No potential conflicts of interest were disclosed.

Introduction

Glioblastoma (GBM) is the most aggressive primary central nervous system tumor in adults, with a median survival rate of approximately 15 months following diagnosis [1]. Despite multimodal treatment consisting of maximal surgical resection of the tumor followed by concomitant radiotherapy and chemotherapy, the prognosis for GBM patients remains poor [2]. Resistance to the most commonly administered chemotherapy drug, temozolomide (TMZ), as well as to radiotherapy, is inherent, inevitably leading to tumor recurrence [3].

Cells in GBM diffusively infiltrate healthy tissue beyond the tumor margin, making complete surgical resection of the tumor difficult [4–6]. Multiple studies have suggested that a subpopulation of these invasive residual tumor cells, termed as GBM stem cells (GSCs), exhibits the ability to self-renew and repopulate the heterogeneity of GBM cell phenotypes [7–9]. These GSCs are often refractory to chemotherapy, resulting in chemoresistance and tumor recurrence [10, 11]. Accordingly, understanding the molecular mechanisms of treatment resistance can yield better management and appropriate triage of therapies.

The tumor mass is composed of a variety of cell types, extracellular matrix proteins, and secreted factors that collectively form the tumor microenvironment (TME). Tumor progression and resistance are profoundly influenced by the interactions of tumor cells with their environment, ultimately determining whether the primary tumor is eradicated or continues to grow [12]. GBM tumors actively reconfigure their environment to create a pro-tumorigenic niche, co-opting signaling pathways to increase angiogenesis, inhibit immune activity and create a pro-invasive extracellular matrix (ECM). Cell-secreted soluble factors, including transforming growth factor-β (TGF-β), IL-6, Notch, platelet-derived growth factor (PDGF), epidermal growth factor (EGF), and vascular endothelial growth factor (VEGF) serve as signaling molecules by binding to receptors on target cells and triggering pathways promoting the invasion, proliferation, and self-renewal of GSCs [13].

TME

The tumor microenvironment (TME) refers to the extracellular matrix (ECM), interstitial fluid, and various stromal cells (e.g., astrocytes, macrophages, and endothelial cells) surrounding the tumor. The TME is a regulator of tumor growth and progression, ultimately determining therapeutic response and chemoresistance [14]. The role of the TME in regulating progression in other cancers, such as cancers of the breast and prostate, has been well established [15–17].

The presence of the blood-brain barrier (BBB), myelinated and interconnected axon tracts, and a modified ECM composition all contribute to the unique challenges of targeting the TME in GBM. The BBB, despite losing its integrity throughout tumor progression, remains intact in actively invading tumor regions, where it cannot be crossed by certain chemotherapeutics [18, 19]. Axon tracts provide directional cues for cellular invasion and thus are a barrier to complete surgical resection—tumor cells invading through the parenchyma often follow and remodel the surface of myelinated tracts, as well as haptotactic cues in basement membranes [20]. GSCs that migrate along remodeled or deteriorating white-matter tracts gain access to the Notch ligand Jagged1 on exposed nerve fibers, which further promotes invasive growth [21]. Similarly, GBM cells remodel myelin to enable their invasion along myelinated tracts [22]. Microvasculature in the TME is modified in GBM, where a combination of hypervascularity, loss of BBB integrity, and change in the composition of the basement membrane allows tumor cells to rapidly invade along the vasculature. Moreover, as tumor cells migrate through the basement membrane, they disrupt astrocytic end feet contacts with endothelial cells and weaken the BBB [23].

GSCs are able to self-renew and repopulate the heterogeneity of cells typically seen in GBM in large part owing to their environment [24]. GSCs grow within specific niches that regulate their stemness and proliferation. These niches shield GSCs from therapies by anatomically blocking them from therapy exposure [25]. Recent research has pointed to the existence of four distinct zones within these GSC niches, each with its own transcriptional and epigenetic signature: subarachnoid, perineuronal, perivascular, and perinecrotic [26]. Of these, the perivascular niche is particularly susceptible to infiltration, therapeutic resistance, and tumor progression [28–32]. Here, a combination of atypical signaling and neurovascularity contribute to GSC stemness and maintenance. The leaky vasculature provides nutrient access, while its endothelium maintains GSC stemness through Notch-dependent signaling pathways [33]. This effect is further amplified in a positive feedback loop as GSCs secrete vasculogenic factors such as VEGF to form new blood vessels [34].

The need for therapies that target these residual GSCs has raised the question of how the TME promotes the survival, invasion, and proliferation of diffusely infiltrating cells.

Signals from the ECM

The ECM is a three-dimensional network of extracellular macromolecules, such as collagen and glycoproteins that provides structural support to surrounding cells [35]. In GBM, the composition of the ECM is altered, such that the level of glycosaminoglycans (GAGs) is increased by three- to fourfold [36]. GBM cells produce their own pro-invasive matrix [22] and induce stromal cells in the TME to express specific ECM components.

There is strong evidence demonstrating the role of signals from the ECM, either mechanical or chemical, in promoting a supportive environment for GSCs. The changes in matrix composition seen in GBM support cell invasion and proliferation. Stromal cells begin to produce excess laminin and collagen IV, which are mainly localized in vascular basement membranes [37, 38]. Laminins provide a supportive environment for GSC stemness, invasion, and proliferation, with evidence in human cell culture models pointing to its role in GSC growth, as GSCs are often propagated on laminin-coated cultures [39]. Laminin-binding integrin α6 is necessary for GSC renewal, proliferation, and tumor formation [40]. Fibronectin assembly reduces GBM cell migration, while its depletion increases migration and sensitizes tumors to chemotherapy in orthotopic mouse models [41–43]. Other nonstructural proteins in the ECM, such as tenascin-C, are overexpressed in aggressive gliomas [44]. Tenascin-C increases matrix stiffness, cell invasion, and proliferation [45, 46]. Several studies have shown that tenascin-C suppresses the activity of T lymphocytes by interacting with integrins to reduce mTOR signaling [47]. Similarly, tenascin-C in collagen I matrices leads to an increase in GBM invasion mediated by matrix metalloproteinase (MMP)-12 [48]. Osteopontin, another matricellular (i.e., nonstructural) protein in the ECM, has been shown to promote therapeutic resistance in GBM [49]. In the perivascular niche, osteopontin increases TME permeability and maintains GSC stemness through CD44 signaling [50]. Osteopontin also recruits tumor-associated macrophages (TAMs) and maintains their phenotype [51]. Hyaluronic acid (HA), the most abundant component of brain ECM, is a glycosaminoglycan that, when expressed in tumor and stromal cells in its low-molecular-weight form, accumulates and promotes GBM cell invasion, GSC stemness, and GSC resistance [52–54]. HA regulates GSC stemness by engaging the HA-specific cell receptor RHAMM and CD44 and by activating the transcription of stemness modulators [55, 56]. HA acts in conjunction with the TLR-4 receptors to increase the activity of the nuclear factor-kappaB (NF-κB) pathway to suppress terminal differentiation in GSCs [57]. Finally, modulation of the HA content in TME models (using brain-derived human GSC cells) recreates resistance of GBM cells to the EGFR inhibitor erlotinib, although this effect is dependent on the mutant status of EGFR [58]. These findings highlight the TME's role in maintaining GSC stemness and reduce therapeutic efficacy [59].

Matrix mechanics, topography, and interstitial fluid pressure/flow have also been shown to promote invasion and proliferation. An increase in matrix production by

GBM cells compared to nontumor cells results in a higher density matrix, which reduces perfusion and increases ECM compaction and cell damage [60, 61]. In vitro studies have shown that compaction of the matrix induces VEGF, an angiogenic signaling protein, and promotes proliferation through the amplification of EGFR signaling [62]. Although there is an increase in blood vessel formation in GBM, this new vasculature is often leaky, leading to high interstitial pressure and volume [63]. High interstitial fluid pressure is a physical barrier to effective chemotherapy delivery, as small molecules struggle to migrate to the tumor core [64]. Composition of this fluid varies by tumor region—in the core, fluid has a lower pH and is hypoxic, driving tumor cells toward invasive and pro-angiogenic phenotypes [65]. Despite its importance in GBM, there are limited therapies targeting interstitial fluid pressure.

Finally, as the ECM serves as a reservoir for several growth factors including VEGF, EGF, PDGF, and TGF-β, secretion of these factors and their interaction with their receptors leads to the activation of pro-invasive and angiogenic signaling pathways, resulting in uncontrolled cell behaviors in tumor growth, angiogenesis, and invasion [66].

Hypoxia

Hypoxia is a prominent feature of GBM, as defective blood vessels are commonly obstructed or collapse, leading to a reduction in oxygenation [67]. Hypoxia levels vary throughout the tumor region, with single-cell RNA and genome sequencing analyses demonstrating a gradient of hypoxia markers, indicating that cells in the tumor and surrounding TME respond to a range of oxygen levels [68, 69].

Low oxygen levels trigger transcription factors known as hypoxia-induced factors (HIF), which are involved in the maintenance of stemness in multiple cancers [65, 67]. HIFs play a key role in maintaining the "stem-cell" like phenotype in GSCs. For example, GSC viability is highly associated with levels of HIF-α in GBM cell lines exposed to gradients of oxygen tension [70]. HIF-α also upregulates CXCR4, a chemokine receptor, to promote the expression of pro-invasive proteins [70]. HIF-α has also been shown to physically interact with the Notch intracellular domain (NICD) to prevent differentiation in GSCs [71]. Similarly, hypoxia can stimulate expression of transcription factors such as NANOG to produce TGF-β1, a cytokine that regulates stemness-associated factors such as Sox2 and Sox4 [72, 73]. Similarly, HIF-2α promotes the expression of Oct4, another marker strongly associated with stemness [74]. Targeted knockdown of either HIF-1α or HIF-2α impairs the self-renewal abilities of GSCs [75].

Importantly, activation of the HIF transcriptional cascade triggers the secretion of VEGF and angiopoietin to stimulate angiogenesis. Despite this apparent increase in activity, these newly formed blood vessels lack the structural support of pericytes and are often leaky, thus impeding blood flow and further decreasing perfusion [76].

Hypoxia shapes the immune microenvironment in GBM by recruiting immunosuppressive cells, such as tumor-associated macrophages (TAMs), regulatory T cells (T-regs), and myeloid-derived suppressor cells (MDSCs), helping tumor cells evade immune surveillance [77]. Specifically, HIF-α can directly bind to a hypoxia response element promoter in the T cell inhibitor programmed cell death ligand -1 (PDL-1) on MDSCs, macrophages, and tumor cells. In a hypoxic environment, upregulated PDL-1 binds to PD-1 on activated T cells to inhibit their activity, demonstrating a clear immunosuppressive role for the TME [78]. Downstream effectors of HIFs overlap with the NF-κB pathway in cells of the innate immune system to regulate pro-inflammatory cytokines such as IL-1 and TNF-α [79].

Hypoxia can also activate the signal transducer and activator of transcription 3 (STAT3) pathway in solid tumors such as GBM. Research has shown that STAT3 promotes GSC survival and suppresses cells of the immune system [80]. Targeted inhibition of STAT3 can increase tumor susceptibility to cytotoxic T cell (CTL)-mediated lysis-separate depletion of HIF-α produces the same effect, suggesting that hypoxia-related pathways interfere with CTL activity targeting tumor cells [81, 82]. Similarly, angiogenesis and cell migration are diminished in GBM cell lines with small-RNA mediated STAT3 knockdown [83].

HIF signaling promotes some of the metabolic changes seen in GBM, as the stabilization of HIF-1α activates the transcription of glucose transporters, promotes aerobic glycolysis by amplifying glucose uptake in tumor cells, and promotes the conversion of pyruvate to lactate rather than acetyl- CoA [84, 85].

In addition to serving as an environmental cue for the self-renewal, invasion, and proliferation of GSCs, hypoxia also contributes to therapeutic resistance and enhances genetic instability of tumor cells [86]. Low oxygen levels impede the creation of free radicals responsible for the "antitumor" effect generated by radiotherapy [87]. The development of certain imaging methods to quantify oxygen content in tumors may help identify regional differences in hypoxia, potentially allowing for adjustments to radiation dose as a result [88]. Hypoxia may further contribute to tumor pathogenesis by attenuating DNA mismatch repair genes, subsequently promoting further genetic mutations [89]. Moreover, HIF-1 activates the multidrug resistance-1 (MDR-1) gene, which encodes P-glycoprotein (P-gp) ATP binding cassette transporters. Activated P-gp acts as a drug efflux pump to remove intracellular concentrations of chemotherapeutic drugs, reducing the efficacy of treatment [90, 91].

In sum, the hypoxic environment upregulates key signaling pathways to promote a proliferative niche for GSCs.

Tumor/stroma interactions

Stromal cells are key regulators of tumor progression. In normal tissue, local stromal cells convert to a reactive phenotype during the wound healing process, mediating blood vessel formation, immune cell, and fibroblast infiltration as well as remodeling the ECM. In GBM, tumor cells can promote the secretion of MMPs and pro-tumorigenic factors, such as VEGF and tenascin-C in these reactive stromal cells [12, 92].

Cross talk between stromal and tumor cells is essential for tumor growth and progression. A variety of cell types, spanning from monocytes, endothelial cells, and glioma-associated mesenchymal stem cells (MSCs), interact with tumor cells to foster a pro-tumorigenic environment [93]. Crucially, these nontumor cells secrete signals to modulate tumor cell survival, proliferation, and migration. For example, MSCs secrete cytokines such as IL-6 to promote proliferation and stemness in GSCs [94, 95]. Similarly, neurons secrete neuroligin-3 to promote proliferation in GBM tumor cells [96]. Vascular endothelial cells secrete IL-8 to promote

GSC stemness and migration, while GSCs upregulate IL-8 in response [97]. To meet the increased demand for blood vessel formation, tumor cells align with vascular endothelial cells [98].

Finally, tumor-associated astrocytes (TAAs) secrete factors and activate matrix-remodeling enzymes, such as MMPs, to promote tumor cell survival, proliferation, and mediate the immune cell response. As a result, immune activation within the tumor region decreases, while tumor cells become more invasive and chemoresistant [99].

The following sections cover the effects of pro-tumorigenic secretions from stromal/tumor cell interactions, and how they play key regulatory roles in response to therapy and tumor progression.

Immune infiltrates-modulators-microglia and macrophages

Many studies have identified inflammatory cells in the GBM tumor core, particularly tumor-associated macrophages (TAMs) and microglia. In fact, TAMs are the predominant immune cell infiltrating GBM tumor tissue, accounting for nearly a third of the tumor cell population [66]. The concentration of TAMs correlates with malignancy in tumor grade and their overrepresentation in hypoxic regions as well as near tumor microvasculature suggests crosstalk with GSCs [100]. TAMs are recruited to tumor tissue through the release of chemoattractants from GSCs [101]. Soluble factors secreted from GSCs, such as VEGF, matrix metalloproteinase 2 (MMP2), TGF-β, and SDF-1 have been shown to mediate this chemoattraction [102]. In general, macrophages can adopt either a pro-inflammatory phenotype known as M1, or an antiinflammatory phenotype known as M2 [103]. The presence of GSC-secreted factors, such as colony-stimulating factor 1 (CSF-1), TGF-β1, and macrophage inhibitory cytokine 1, have been shown to influence the inflammatory phenotype in TAMs [104]. For example, CSF-1 promotes the M2 phenotype, contributing to invasion. In fact, inhibition of CSF-1 reverts TAMs to the M1 phenotype, with promising preclinical utility [105]. TAMs with an M2 phenotype actively promote a GBM microenvironment with increased angiogenesis, remodeling of the ECM, and enhanced tumor cell invasion [106]. GSCs interact with TAMs to increase the production of immunosuppressive cytokines. GSCs secrete a chemoattractant, periostin, which specifically recruits immunosuppressive M2-polarized TAMs [107]. GSCs can upregulate the synthesis of arginiase-1 to promote an immunosuppressive environment via MDSCs [108]. Targeting NANOG, a stemness-associated transcription factor, replaces TAMs and Tregs with antitumor CD8+ effector T cells by modulating TGF-β1 production [109].

TAMs secrete a complex array of cytokines and growth factors affecting the proliferation, survival, and migration of glioma cells. For example, upon exposure to GBM cells, TAMs upregulate the expression of membrane-type 1-matrix metalloproteinase (MT1-MMP) to facilitate degradation of the ECM and promote GBM invasion [110]. Importantly, TAMs also secrete TGF-β, enhancing GBM cell migration through the upregulation of integrin expression [111].

Following exposure to radiation, TAMs can infiltrate the tumor region via a permeable BBB [112]. Agents targeting macrophages, such as small-molecule inhibitors of the colony-stimulating factor-1 receptor (CSF-1R) have been used to inhibit TAMs in glioma mouse models and are under clinical evaluation in combination with either standard-of-care therapy

(NCT01790503) or immune checkpoint inhibitors (NCT02829723) [13, 113]. The abundance and phenotypes of TAM populations change during the course of radiotherapy, emergence of resistance, and recurrence. Inhibition of both TAMs and monocyte-derived macrophages (MDMs) using a BBB-permeable CSF-1R tyrosine kinase inhibitor (BLZ945) enhances the initial response to radiotherapy, and combined radiation with acute CSF-1R inhibition delays glioma relapse in a mouse model of GBM. Moreover, transcriptional changes in TAMs and MDMs observed after radiotherapy are reversed with CSF-1R inhibition and correlated with increased cell apoptosis, DNA damage sensitization and tumor debulking. Interestingly, long-term CSF-1R inhibition in combination with radiotherapy inhibits glioma recurrence in preclinical trials [114].

Self-renewal pathways: WNT

The WNT/β-catenin pathway, along with the Sonic Hedgehog (SHH) and Notch pathways, is essential for self-renewal in GSCs. WNT proteins are a family of highly conserved secreted molecules, involved in the regulation of self-renewal and differentiation during the development of the central nervous system. Aberrant WNT signaling is a feature of many cancers [115].

All three WNT pathways (canonical, noncanonical cell polarity, and noncanonical WNT/calcium) are activated by the binding of a WNT protein ligand to a Frizzled (FZD) family receptor on the cell surface. In the canonical pathway, the signal is then transduced to β-catenin, which then enters the nucleus to form a complex and activate transcription of WNT target genes [116]. Noncanonical WNT pathways are β-catenin independent and are involved in cell polarity and cytoskeleton-mediated processes [117].

In normal cells, WNT regulates stemness and stem cell niches. Similarly, numerous studies have identified a similar role for WNT signaling in the maintenance of GSC stemness [118]. WNT signaling components, such as WNT6, FZD9, and FZD2, are upregulated in primary GBM specimens and GBM cell lines as a result of overexpressed PLAGL2 [119]. Similarly, a regulator of WNT signaling, FoxM1, is overexpressed in GBM patient specimens as well as GSCs and promotes self-renewal by binding to the promoter of Sox2, a regulator of GSC-self renewal [120, 121]. Interestingly, FoxM1 promotes chemoresistance in GBM via upregulation of Rad51, a critical component of the DNA damage repair process [122]. A comparative analysis of chromatin state in GSCs compared to the entire tumor has identified a set of developmental transcription factors unique to GSCs, whereby ASCL1 activates WNT signaling by repressing the negative regulator DKK1.79 [123]. This suggests WNT activation in GBM is mediated by epigenetic regulation rather than genetic mutations. A recent study has shown that a subset of primary GBM-derived stem cells exhibiting high ASCL1 requires WNT proteins for self-renewal, and this subset relies on WNT/ β-catenin signaling for enhanced tumor burden in xenograft models. In a WNT reporter model, cells that exhibit high levels of β-catenin signaling, termed Wnt^{hi} GBM cells, are faster-cycling and highly self-renewing, while cells with low levels of β-catenin signaling have decreased self-renewing capacity. Inhibition of WNT/ β-catenin as well as Notch signaling in GSCs expressing high levels of ASCL1 leads to neuronal differentiation [124]. Further, multiple studies have indicated that certain components of the WNT pathway, namely Frizzled and Dishevelled are overexpressed in GBM and result in GSC growth and stemness [125–128].

WNT signaling is also involved in tumor invasion and the epithelial to mesenchymal transition (EMT) in GBM. Overexpression of WNT regulators increases the expression of genes associated with the EMT, such as ZEB1, SNAIL, TWIST, SLUG, and N-cadherin and results in enhanced migration and invasion of GBM cells [129–134]. For example, one study showed that β-catenin expression induced the expression of ZEB1 in GBM cells and increased cell motility. Inhibition of β-catenin suppresses cellular invasion in U87MG and LN229 GBM cells [135]. Similarly, expression of WNT5A induces migration in GBM cells, while a knockdown inhibits the migratory capacity of these cells [136]. Similar observations have been made using other WNT regulators such as WNT2 and FZD2 [137]. In another study, the mRNA expression profiles of highly invasive GBM cells identified FZD4, a positive WNT regulator, as a causative effector for the invasive phenotype in GBM cells [128].

Aberrant WNT signaling contributes to resistance to both chemotherapy (TMZ) and radiotherapy. TMZ-resistant subclones expressed high levels of FZD2 [138]. Gene signatures from irradiated orthotopic xenograft tumor models showed high expression levels of WNT signaling-related genes, such as WISP1, FZD1, LEF1, TCF4, WNT9B, and AXIN2, in radioresistant GBM cells. Similarly, inhibition of the WNT signaling pathway using the inhibitor XAV939 sensitized GBM cells to irradiation [139].

These findings point to the important role of WNT signaling in self-renewal and invasion of GBM cells and provide a rationale for further investigation into potential therapeutic approaches.

Self-renewal pathways: Notch

Similar to WNT, the Notch pathway is critical for cell fate determination in cancer, especially in GBM. Notch mediates short-range communication between cells via interactions with ligands on neighboring cells. In normal stem cells, Notch is involved in the regulation of self-renewal pathways [140, 141]. In GBM, endothelial cells create a stem cell niche by providing NOTCH ligands needed for self-renewal [33]. Similarly, GSCs often adhere to vascular structures to maximize the physical proximity to endothelial cells, which promote GSC survival through the secretion of soluble pro-tumorigenic factors such as TGF-β and PDGF [142].

Notch signals through four Notch receptors (Notch 1–4) and five ligands (Jagged-1, -2, and Delta-like-1, -3, and -4), which are all type I transmembrane proteins [143]. Notch signaling can be activated through the secretion of soluble factors from endothelial cells as well as hypoxia. Aside from secreting growth factors, endothelial cells also release nitric oxide (NO), which has been found to diffuse into GSCs and activate Notch signaling via paracrine signaling and promote tumorigenicity [144]. Hypoxia can activate Notch signaling via HIFs, which stabilize the Notch intracellular domain [145, 146]. Conversely, GSCs increase the expression of several Notch genes (NOTCH1, NOTCH3, DLL1, JAGGED1, JAGGED2, HES1, HEY1, and HEY2), as well as hypoxia-related genes (HIF and VEGF) [147–149].

Activation of Notch involves sequential proteolytic cleavages that eventually lead to the release and nuclear translocation of the intracellular domains of Notch receptors (NICDs), and subsequent activation of Notch-dependent transcription [150, 151]. The γ-secretase complex, which mediates the last proteolytic step for release of the NICDs, is required for Notch activation. Notch signaling has successfully been blocked with γ-secretase (GSIs) both in vitro and in vivo [152].

Several studies have confirmed the prevalence of aberrant Notch signaling in GBM. For example, higher expression of ASCL1, Dll1, Notch1, -3, -4, and Hey1 correlates with higher glioma grade and poorer prognosis in GBM, suggesting that increased activation of Notch promotes more aggressive tumor phenotypes [149, 153–155]. Gene set enrichment analysis has shown that Notch gene sets, in conjunction with the proneural, classical, and Wnt-canonical gene sets, are more enriched in Wnt-dependent GSCs [124]. Similarly, both mRNA and protein levels of Notch gene sets such as Notch1, Notch4, Dll1, Dll4, Jagged1, CBF1, Hey1, Hey2, and Hes1 are overexpressed in GBM cells and associated with higher levels of VEGF and pAKT as well as reduced PTEN expression [149, 156–159].

Overexpression of the Notch pathway has a number of effects on angiogenesis, stemness, and cell migration. Endothelial cells in the perivascular niche upregulate DLL-4, resulting in tip cell formation [160]. In terms of effects on stemness and self-renewal, inhibition of Notch in ALSC1^{hi} GSCs results in terminal neuronal differentiation. Similarly, RNAi-mediated knock-down of Notch ligands abrogates their ability to induce self-renewal in GSCs, both in vivo and in vitro [124]. The capacity for self-renewal is impaired in GSCs when DLL1 is silenced, as are markers of stemness [161]. Another study showed that the convergence of the PDGF-Notch-NO axis drove angiogenesis and maintenance of the GSC phenotype in the perivascular niche [162]. Overexpression of Notch1 has been shown to activate AKT, thus inducing the nuclear translocation of β-catenin and NF-κB to promote cell invasion and migration via the Notch-mediated overexpression of Snail, Zeb1, and vimentin [159]. Other studies have identified a similar role for Notch1, showing that neurosphere cultures with cells expressing high levels of Notch1 exhibit a much more infiltrative phenotype than cells with low Notch1 [163]. Moreover, inhibition of the Notch-signaling-induced AKT, NF-κB, and ERK pathways via uPA/uPAR suppresses cell migration, invasion, and angiogenesis [164].

Aberrant Notch signaling has also been implicated in therapeutic resistance in GBM. Components of the Notch pathway, namely Jag1 and Hey1, are transiently activated in endothelial cells following radiation [165]. Inhibition of the Notch pathway with γ-secretase inhibitors sensitizes GSCs to clinically relevant doses of radiation but fails to do so in nonstem glioma cells. Moreover, knockdown of Notch1 or Notch2 increases radiosensitivity in GSCs. In this study, Notch inhibition did not affect the DNA damage response following radiation, instead impairing radiation-induced AKT activation and upregulating levels of the truncated apoptotic isoform of Mcl-1 [166]. Another study examining the effectiveness of bevacizumab, a humanized monoclonal antibody against VEGF, identified two subsets of GBM cells, one sensitive (Dll4$^+$/Jagged1$^+$) to bevacizumab, the other resistant (Dll4$^+$/Jagged1$^-$) [167].

Self-renewal pathways: Sonic Hedgehog

Similar to the WNT and Notch pathways, the Hedgehog pathway plays a crucial role in the development of the central nervous system. In many cancers, including glioma, the Hedgehog pathway is upregulated and plays a role in tumorigenesis and tumor progression [168]. Sonic hedgehog ligands can activate the Hedgehog pathway by binding to the transmembrane protein Patched (PTCH1). Activation of the Hedgehog pathway leads to the upregulation of the GLI1, PTCH1, cyclin D2 (CCND2), Bcl-2, and VEGF [169]. In addition, the Hedgehog pathway modulates the expression of stemness genes, such as NANOG, OCT4, and SOX, through its effector GLI1 [170–173].

Receptor tyrosine kinases

Receptor tyrosine kinases (RTKs) are a family of cell surface receptors for growth factors, hormones, neurotrophic factors, and other extracellular signaling molecules. All RTKs share a similar structure composed of an extracellular ligand-binding domain, a transmembrane domain, and an intracellular tyrosine kinase domain, which is autophosphorylated upon ligand binding [174]. RTKs activate a number of major downstream pathways involved in the regulation of cell proliferation, survival, differentiation, and angiogenesis and are thus promising targets for therapeutic agents in GBM [175].

EGFR

The epidermal growth factor receptor (EGFR) is a transmembrane receptor for the epidermal growth factor family of extracellular protein ligands. Mutations affecting EGFR expression or activity have been linked to disruptions in the regulation of cancer cell proliferation, migration, and survival [176, 177]. In GBM, amplification of the EGFR gene is found in greater than 50% of primary GBM patients and is commonly enriched in the "classical" subtype [178].

EGFRvIII is the most common mutant variant of EGFR, resulting in a constitutively active ligand-independent receptor [179]. Expression of this variant has only been found in tumors and not in healthy tissue, suggesting potential for targeted therapy [180, 181]. However, efforts to target this variant have proved challenging. The recent failure of a phase III clinical trial (ACT IV), in which an anti-EGFRvIII vaccine was administered to newly diagnosed EGFRvIII-positive patients, showcases the difficulties in translating promising preclinical work to the clinic [182]. A study from 2014 suggests a possible mechanism for resistance to anti-EGFR inhibitors such as erlotinib is the ability of EGFRvIII amplicons to "hide" during therapy and reemerge following drug withdrawal [183]. This suggests a highly specific and dynamic mechanism of therapy evasion.

EGFR dimerizes upon binding to ligands that include EGF, transforming growth factor-alpha (TGF-α), amphiregulin, betacellulin, heparin-binding EGF (HB-EGF), epigen, or epiregulin [184, 185]. Following dimerization, the receptor then autophosphorylates its C-terminal tail, which serves as the docking site for SRC homology 2 (SH2) domain-containing signaling proteins, including phosphoinositide 3-kinase (PI3K), SRC homology 2 domain-containing transforming protein 1 (SHC1), and signal transducer and activator of transcription (STAT) proteins [186, 187].

Pathways mediated by EGFR signaling from the plasma membrane include the RAS/mitogen-activated protein kinase (MAPK)/extracellular signal-regulated kinase (ERK) pathway, PI3K/AKT/mTOR) pathway, Janus Kinase (JAK)/STAT pathway, and protein kinase C (PKC) pathway [176, 177, 188]. Although both EGFR and EGFRvIII can transduce signals through classic RTK pathways, evidence suggests that EGFRvIII signaling is different both quantitatively (signal strength) and qualitatively (difference in signaling molecules) [189–192].

RAS/MAPK/ERK

Upon activation, EGFR activates RAS, a small GTPase, subsequently triggering the MAPK/ERK kinase signaling cascade, which modulates the activity of a range of proteins and transcription factors through ERK-1/2 [193]. ERK1/2 plays critical roles in regulating proliferation, survival, and metabolism. ERK1/2 can also function in tumor-suppressing pathways depending on activation strength [194]. Pathways downstream of EGFR, such as

RAS and PI3K/AKT, have been shown to mediate tumor cell resistance to agents causing DNA damage [195].

PI3K/AKT/MTOR

Initiation of EGFR/EGFRvIII signaling by ligand binding, gene amplification, or mutation can also activate the PI3K/AKT/mTOR pathway [196]. PI3K, an important downstream effector of EGFR, regulates apoptosis through its downstream effector AKT, which acts on a number of targets [197]. Activated AKT phosphorylates the FOXO subfamily of transcription factors, thereby inhibiting the transcription of several proapoptotic proteins, such as BCL2-associated death-protein (BAD) and GSK3 [198–202]. Conversely, AKT phosphorylates and degrades the inhibitor of κB (IκB), which results in increased nuclear factor kappa B (NF-κB) activity and transcriptional stimulation of pro-survival genes [203, 204]. AKT also indirectly inhibits P53, an activator of cell-cycle arrest, via its modulation of MDM2 [205]. AKT directly and indirectly activates mTOR, which has been shown to regulate HIFα, leading to the downstream activation of VEGF and increased angiogenesis [206]. P13K/AKT signaling is also negatively regulated by PTEN, a tumor suppressor frequently mutated in GBM [199]. In fact, loss of chromosome 10q (on which PTEN is found) precedes EGFR amplification in GBM [207]. Loss of PTEN occurs in more than 50% of GBMs and is the third mechanism of constitutive PI3K pathway activation [208–211]. GBM patients with an activated PI3K/Akt/mTOR pathway have a poorer prognosis than patients without oncogenic activation of the pathway [212, 213].

JAK/STAT

Activation of JAK, a receptor tyrosine kinase, results in phosphorylation of the STAT proteins, which then dimerize and translocate to the nucleus [214]. Activated JAK, another RTK, phosphorylates nuclear STAT proteins, where they regulate the transcription of a number of genes involved in inflammation-associated tumorigenesis, stemness, and migration. Alternatively, STAT3 can also be directly phosphorylated by EGFR and its downstream effector, AKT [215–217]. The role of STAT3 in GBM varies depending on the mutational status of other genes. For example, STAT3 suppresses the malignant transformation of astrocytes induced by the loss of PTEN [218]. However, in cases where EGFRvIII is expressed, STAT3 can drive malignant transformation of high-grade gliomas and promote GSC survival by both suppressing the antitumor response as well as inducing angiogenesis [219–223]. By inhibiting JAK2 receptors (and by extension, STAT), U87 EGFR and EGFRVIII cells are sensitized to the EGFR inhibitor gefitinib. Interestingly, the interaction of JAK2 with EGFR confers resistance to EGFR inhibitors [224]. In response to astrocyte-secreted factors, STAT3 signaling is activated in GSCs and increases resistance to radiation in a coculture of GSCs and astrocytes [225]. In sum, the complexity of the role of STAT in regulating GBM cell phenotype showcases the highly dynamic and interdependent nature of signaling in the TME.

Phospholipase C (PLC/PKC)

Activated EGFR is capable of recruiting and activating PLC, triggering the activation of PKC, which in turn activates a large group of effector molecules capable of regulating tumor proliferation, angiogenesis, infiltration, and survival [226]. For example, downstream effectors of PKC include regulators of the cell cycle (p53, p21), cell growth and proliferation (RAS-RAF1, GSK3), cell survival (BLC2, BAD), and inflammation (NF-κB) [227]. Critically, inhibition of PKC has been shown to decrease the viability of GBM cells, indicating that PKC plays an important role in this signaling cascade [228].

Therapeutic resistance to EGFR inhibition

Current anti-EGFR-based therapies for GBM include small-molecule tyrosine kinase inhibitors, antibodies, vaccines, chimeric antigen receptor (CAR) T cells, and RNA-based therapies. Despite showing promising results in cell-based assays and preclinical models, they have largely been ineffective in clinical trials [177].

There are several potential factors contributing to resistance with EGFR/EGFRvIII inhibition in GBM, namely, inefficient BBB penetration, intratumoral heterogeneity, compensatory signaling pathways, and secondary mutations [126, 177]. Many therapeutic options targeting EGFR do not cross the BBB effectively, and thus are limited in efficacy. While the BBB in GBM is compromised, the magnitude is unlikely sufficient to allow efficient inhibition [229]. In the biopsied posttreatment tumors of patients treated with lapatinib, a potent EGFR inhibitor, p-EGFR was only modestly inhibited [230]. Second, mutations or deregulation of the pathway may confer the ability to bypass EGFR/EGFRvIII inhibition. For example, patients with EGFR amplification but with an intact PTEN gene demonstrated a radiographic response, but ultimately did now show improved survival when treated with EGFR inhibitors. Conversely, patients whose tumors showed a loss of PTEN failed to respond radiographically to EGFR inhibitors [231]. Constitutively active mutant EGFRvIII expression sensitizes tumors to EGFR inhibitors in vitro and clinically, but only if the PTEN tumor suppressor protein is intact [232]. Similarly, EGFR-dependent phosphorylation of PTEN has also been shown to confer resistance [233]. Moreover, EGFR inhibition by erlotinib increased nuclear expression of the promyelocytic leukemia tumor suppressor, prevented cell death, and promoted tumor cell survival [234].

DNA double-stranded breaks (DSB) are the most important DNA lesions leading to cell death following radiotherapy. A complex involved in the repair process following exposure to ionizing radiation is the DNA-dependent protein kinase catalytic subunit (DNA-PKcs), which can be activated by nuclear EGFR [235, 236]. Several studies suggest that EGFR-mediated repair of DNA damage might play a role in the mechanism of radioresistance. However, further in vivo preclinical as well as clinical research is required to determine the extent of its importance, as well as to identify predictors of response to EGFR inhibitors.

Additionally, compensatory regulation of other RTK pathways adds to the complexity of targeted therapies. Inhibition of a single RTK can lead to upregulation of other RTKs. For example, inhibition of EGFR-upregulated transcription of PDGFRβ promoted resistance to EGFR inhibitor treatment [237]. Tumor cell subpopulations harboring distinct EGFR mutations or RTK amplifications add to the complexity of GBM and contribute to resistance to EGFR/EGFRvIII-directed therapies [238, 239].

PDGFR

Growth factors and growth factor receptors play a multifaceted role in regulating GBM tumorigenicity. Certain growth factors such as EGFR, PDGF, and HGF are thought to activate transcription factors related to the epithelial to mesenchymal transition of GBM cells [66].

Platelet-derived growth factor receptor (PDGFR) is involved in CNS development and its dysregulation in GBM contributes to gliomagenesis. GBMs regularly exhibit a PDGF autocrine loop that is absent in normal brain tissues [240, 241]. Data from the TCGA show that PDGRFα is amplified in 10%–13% of the cases studied and suggest that it is the second most frequently amplified RTK behind EGFR [210, 211, 242]. In GBM, all PDGF ligands (PDGF-A, PDGF-B, PDGF-C, and PDGF-D) as well as its two cell surface receptors PDGFR-α and PDGFR-β.21 are expressed. The TME plays an active role in its upregulation, as nitric oxide from the endothelium increases the tumor initiating capacity of PDGFR [144].

VEGF

Vascular endothelial growth factor (VEGF) is a signaling protein that promotes the formation of new blood vessels. The VEGF gene is activated by the nuclear translocation of HIF-1α and HIF-1β to counteract the hypoxic conditions in GBM [243]. To meet this demand for oxygen, GSCs secrete a perivascular gradient of VEGF that consequently gives rise to new capillaries by binding to receptors on endothelial cells and promoting tube formation [34]. By binding to the VEGR receptor-2 (VEGFR2), VEGF stimulates an autocrine loop that maintains stemness in GSCs and confers survival benefits. Reflecting its function, VEGF activity is mainly localized to the vascular endothelium in the perivascular niche of the tumor region. GBM tumor tissue shows high levels of VEGF expression, associated with an upregulation of VEGFR2 [244, 245]. Overexpression of VEGF is linked to poorer clinical outcomes [246, 247]. Conversely, lower VEGF expression, measured through immunohistochemistry, was found to be predictive of long-term progression-free survival [248].

In addition to its effects on angiogenesis, VEGF suppresses the antitumor response of the TME by regulating immune modulators such as Tregs, TAMs, and MDSCs [243]. VEGF inhibits the maturation of dendritic cells, resulting in the inactivation of cytotoxic T lymphocytes (CTLs), which facilitates targeted cell death. Similarly, VEGF and its receptor VEGFR2 induce the expression of programmed death ligand-1 (PD-L1) on CTLs and Tregs [78, 249, 250]. CTLs, but not Tregs, are also targeted through the secretion of VEGF, IL-10 and prostaglandin-3, which collectively induce the expression of the Fas ligand in endothelial cells to suppress the immune response [251].

The use of antiangiogenic drugs, specifically those that inhibit VEGF, such as bevacizumab, has been extensively studied and remains controversial in their effects. The two latest trials on bevacizumab, a humanized monoclonal antibody against VEGF, found that this treatment increased progression-free survival, but not overall survival [252–255].

Secreted factors: TGF-β

Transforming growth factor-beta (TGF-β) is a family of growth factors that includes proteins such as TGF-β and bone morphogenic proteins (BMP). Many cancers, including GBM, exhibit atypical and disrupted TGF-β signaling that gives rise to tumor initiation and progression.

As a multifunctional cytokine secreted from immune, stromal, and tumor cells, TGF-β binds to receptors on the cell surface, triggering signaling cascades regulating cell proliferation, differentiation, and tissue homeostasis [256].

All of the ligands of the TGF-β family are initially synthesized and secreted as latent precursors that require processing by extracellular pro-protein convertases to produce biologically active ligands [257]. Active TGF-β signals via a complex of type I and type II transmembrane serine/threonine kinase receptors and activates canonical or noncanonical pathways. Canonical TGF-β signaling involves ligand binding to TGF-β receptor II and the subsequent downstream phosphorylation of SMAD2 and SMAD3. Noncanonical TGF-β signaling stimulates SMAD-independent pathways such as PI3K/AKT, MAPK, c-Src, and NF-κB. These pathways can also respond to signals from Wnt/β-catenin, Notch, Hedgehog, TNF-α, or EGF-dependent pathways [258]. Interestingly, in endothelial cells, TGF-β signals via the type I receptor activin receptor-like kinase-1 (ALK), which activates SMAD1 and SMAD5. This alternative pathway may counteract the SMAD2 pathway and thereby balance TGF-β signaling [259].

Downstream effectors of TGF-β bind to the specific promoter regions of target genes to regulate their activity. The transcription of SMAD-dependent gene targets is amplified via DNA binding cofactors such as p300, CBP, and FOXH1 [260].

The TGF-β pathways are heavily regulated at multiple levels, be it the level of ligand binding, the expression of inhibitory SMAD proteins (SMAD-6, SMAD-7) or the stabilization of R-SMAD via phosphorylation by MAPKs, glycogen synthase kinase 3β (GSK3β), and cyclin-dependent kinases (CDKs) [261]. Such regulatory mechanisms are the most frequent defects developed by cancer cells.

There is robust evidence demonstrating elevated levels of TGF-β in GBM tissue compared to normal tissue. Several studies have demonstrated that TGF-β is overexpressed in malignant glioma tissues but undetectable in normal brain tissues, suggesting that aberrant TGF-β signaling contributes to glioma development [262–266]. More specifically, GSCs express higher amounts of TGF-β-2 than differentiated glioma cells, and the secreted TGF-β-2 appears to correlate with the pathological grade of the glioma [267]. Studies in GBM cell lines have shown that high TGFs/Smad signaling induces platelet-derived growth factors-B (PDGF-B), resulting in tumor cell proliferation. In contrast, in nonaggressive gliomas with low proliferative indices, SMAD-dependent TGF-β signaling is inactive, keeping PDGF-B unexpressed as its promoter remains hypermethylated [268]. The proliferative role of TGF-β is thought to stem from a shift from "tumor suppressive" pSmad3C signaling to tumorigenic pSmad3L/C signaling (dual phosphorylation), leading to a more invasive and metastatic tumor phenotype [269].

High levels of TGF-β have also been shown to mediate angiogenesis and immunosuppression in later stages of tumor progression [270]. TGF-β can induce GSCs expressing CXCR4 to generate vascular pericytes for blood vessel formation [271, 272]. For its role in immune evasion, TGF-β decreases MHCII expression on GBM tumor cells to prevent antigen presentation [109]. Finally, TGF-β promotes migration in GSCs, as it directly stimulates an epithelial to mesenchymal (EMT) transition, leading to a metastatic phenotype [273, 274].

TGF-β also plays a role in maintaining stemness in GSCs, as it induces the expression of Sox2 and Sox4, a process that is amplified in hypoxic conditions. Conversely, inhibition of TGF-β leads to downregulated Sox2 expression and induces GSC differentiation in orthotopic immune-compromised mice [73].

Inhibition of TGF-β also decreases the "CD44hi/ID1hi" GSC phenotype associated with poorer clinical outcomes, by reducing ID1 and ID3 transcription factor levels to prevent tumorigenesis [275].

TGF-β produced by GSCs enhances effective DNA damage response and self-renewal capacity, leading to microenvironment-mediated resistance to ionizing radiation. These effects are reversed by LY364947, a TGF-βR1 inhibitor [276]. In a study that stratified mutations in GBM tumors on the basis of: recurrent tumors without TMZ treatment, untreated tumors, TMZ-treated but nonhypermutated and TMZ-treated hypermutated tumors, the LTBP4 gene, which encodes latent TGF-β–binding protein 4 and is considered an activator of the TGF-β pathway, harbored more mutations in recurrent compared to untreated GBM, and was associated with worse prognosis in patients with primary GBM and wild-type IDH1 status. In addition, silencing of LTBP4 leads to reduced expression of genes positively regulated by TGF-β (ID1 and ID2) while also leading to upregulation of genes repressed by TGF-β in GBM (RHOB and GADD45A) [277].

In sum, there is strong evidence to support the role of TGF-β signaling in glioma invasion, migration, proliferation, angiogenesis, tumor-mediated immunosuppression, and maintenance of stemness in (GSCs).

Secreted factors: BMP

Bone morphogenic protein (BMP) is a member of the TGF-β superfamily and is activated through a similar mechanism as described for TGF-β, where the ligand-bound type II receptor recruits and phosphorylates the type I receptor, which then recruits and phosphorylates Smad family proteins. The Smad1, Smad5, and Smad8 I proteins are activated in the BMP subfamily specifically [278]. Historically, BMP was thought to guide GSCs toward a terminally differentiated astroglial fate [264, 279–281]. Recent research has shown that BMP signaling in the GBM microenvironment guides GSCs to a quiescent state [282]. For example, pSmad1(BMP) and pSmad2 (TGF-β) staining revealed that both populations of cells expressed markers of stemness, such as Sox2 and nestin, contrary to only pSmad2 (TGF-β) as expected in a proliferative cell population. SMAD2+ glioma cells were highly enriched for a DNA polymerase factor, suggesting this cell population housed the majority of dividing cells. Treatment of GSCs with recombinant BMP4 or TGF-β1 resulted in a highly reproducible expression profile, whereby BMP4-treated GSCs showed a greater than twofold decrease in cell proliferation, while exposure of GSCs to TGF-β1 resulted in an increase in cell proliferation. This suggests that GSCs in proliferative culture secrete TGF-β, with analysis of the culture media showing high levels of TGF-β1 and TGF-β2. ID1 expression in GSCs was induced by treatment with BMP4 rather than TGF-β, in contrast to previous indirect evidence suggesting ID1 as a downstream effector of TGF-β1 in GBM. Using a neurosphere assay system, it was shown that TGF-β had a positive effect on self-renewal and proliferation, while BMP4 exposure diminished but did not eliminate sphere formation. With serial passaging, BMP4-treated GSCs continued to form spheres, indicating these cells remained capable of self-renewal. Mice transplanted with BMP4-treated GSCs developed high-grade gliomas, though with prolonged latencies to tumor formation, compared to control cells. These findings are consistent with a previous report using a transgenic mouse model that ablation of ID1 does not result in loss of tumorigenicity, but actually leads to shortened tumor latency following transplantation [283]. To determine whether BMP directed cells toward a terminal astroglial fate or to a dynamic quiescent state, two long-term label-retaining cell assay studies were conducted in a GBM xenograft model. The tag will only be found in cells that have undergone DNA replication during a period overlapping that of delivery of the synthetic nucleoside. Further, the signal will diminish as that cell undergoes further cell divisions. Using ID1 as a proxy for BMP activation and pSmad2 as a marker of TGF-β, results showed rapid decay of signal in pSmad2+ cells and no loss of signal from ID1+ cells, confirming a divergence in cell cycle entry between BMP and TGF-β activated cells. Further experiments showed that pSmad1 identifies a pool of treatment resistance glioma cells. To determine whether activation of the BMP pathway could account for treatment resistance in GBM cells, 3 GSC lines were exposed to a 5-day course of TMZ and concomitant radiation following treatment with BMP or TGF-β agonists. Pretreatment with BMP4 significantly reduced the cytotoxic effect of TMZ and radiation, resulting in increased expression of the DNA repair protein, O6-methylguanine DNA methyltransferase (MGMT). Pretreatment with TGF-β increased cell death as a result of cytotoxic therapy. Taken together, these results suggest that activation of the BMP signaling pathway protects glioma cells from cytotoxic injury, conferring relative resistance to radiation and temozolomide chemotherapy. To determine the mechanism through which GSCs transition from a quiescent to an activated state, expression profiling of these two cell populations was performed in vitro. Among the genes that showed a change in expression in the BMP- and TGF-β-treated groups was p21, where expression increased within 4 hours of exposure to BMP4 and peaked by 24 hours. Coimmunoprecipitation assays showed evidence of p21-Stat3 complex formation in U251 cells treated with BMP4, but not in U251 cells in resting-state culture [282]. STAT3 is a known target of TGF-β in GBM and is a driver of GSC self-renewal and proliferation [284, 285]. These results demonstrate that p21 binds to STAT3 in U251 cells following activation of the BMP signaling pathway. By generating IC50 curves for temozolomide in U251 cells treated with BMP and scrambled siRNA or BMP and p21 siRNA, this study showed that p21 mediates the effect of BMP4 as an inhibitor of cell proliferation and a mediator of temozolomide resistance in GBM.

Another study supported these findings using a 3D organoid approach to recreate intratumoral heterogeneity [286]. They demonstrated that quiescent GBM cells exhibit higher therapy resistance than their proliferative counterparts and these quiescent cells exhibit self-renewal capacity, based on results from a neurosphere assay. Quiescent cells were more resistant to therapy compared to proliferative cells, with a gene expression profile revealing a mesenchymal shift, compared to proliferative. Previous studies have shown that neural stem cell niches were maintained by the BMP receptor, BMPR1a, and that inhibition of BMP signaling in an increase in progenitor proliferation and a subsequent decrease in neurogenesis due to exhaustion of the stem cell pool [287]. In an animal model of GBM, treatment with TMZ lead to an increase in quiescent stem-like cells, which were able to drive tumor recurrence [288].

Transcription factors: NF-κB

NF-κB refers to a highly regulated family of transcription factors contributing to multiple aspects of cancer formation, growth, and treatment response. In GBM, NF-κB activity is upregulated and contributes to the survival of migratory tumor cells. NF-κB has been shown to play a key role in signaling pathways triggered by growth factor receptors such as EGFR and PDGFR. A number of stimuli trigger activation of the NF-κB pathway, such as HA, TLR4, differentiation of GSCs, surrounding parenchymal cells as well as cytokines released by infiltrating monocytes/macrophages [289].

In the nucleus, NF-κB binds to gene regions involved in the regulation of inflammation, cell survival, and apoptosis. Cells with the mesenchymal phenotype in GBM demonstrate higher levels of NF-κB pathways genes, such as TRADD, RELB, and TNFRSF1A [242]. Similarly, activation of NF-κB can promote a mesenchymal transition in GBM cells by modulating downstream transcriptional signaling. Mutations or amplifications in NF-κB are rare in GBM, suggesting that the increased activity is due to upstream pathways mediating the NF-κB response.

The downstream effects of upregulated NF-κB span cell migration, invasion, and chemoresistance. Activation of NF-κB through TLR4 maintains GSCs in a proliferative astrocytic precursor state [57]. The mesenchymal subtype in GBM is considered the most aggressive due to its highly invasive and radioresistant phenotype contributing to poor patient outcomes. Activation of NF-κB signaling promotes a transition from a less aggressive GBM phenotype (proneural subtype) to a more aggressive mesenchymal phenotype [290]. Moreover, NF-κB activation in mesenchymal GBM cells mediates cell migration and tumor invasion through upregulation of NF-κB target genes, including cell chemoattractants (IL-8, MCP-1) and matrix metalloproteinases (MMP-9) [291]. When RANKL, a member of the TNF family, is upregulated in GBM cells, it activates neighboring astrocytes via NF-κB signaling, leading to the secretion of cytokines such as TGF-β and ultimately promoting GBM cell invasion [292]. In an inflammatory environment, Caspase 8, a protein implicated in apoptosis, promotes NF-κB transcription factor activation with consequent increase of VEGF, IL-6, IL-8, IL-1β, and MCP-1 secretion, enhancing neovascularization, and resistance to (TMZ) [293].

Effects of therapy on the TME

Radiotherapy

Despite its therapeutic benefits in GBM, radiotherapy can trigger remodeling of the TME to promote more aggressive tumors at recurrence [294]. Repeated exposure to radiation has been shown to increase angiogenesis (through VEGF), and, over time, resistance to radiation through increased secretion of insulin-like growth factor 1 [295]. In particular, GSCs, as opposed to tumor cells, can successfully survive the effects of radiotherapy by engaging DNA damage checkpoints to repair DNA damage [296].

GSCs can thrive despite the effects of radiotherapy in large part due to their pro-tumorigenic environment. For example, astrocytes can activate STAT3 signaling to mediate radiation resistance in GSCs [297]. Radiotherapy can also induce GSCs to produce HA and HAS2 to promote invasion [298].

Radiotherapy can also give rise to a "senescent" GBM cell phenotype by modulating the ECM and upregulating pro-inflammatory signaling [299]. In turn, GBM cells counteract senescence in endothelial cells by transdifferentiating into endothelial cells and forming new blood vessels [98].

Chemotherapy

Although concomitant administration of TMZ improves overall patient survival, it can remodel the TME to create a resistant, pro-invasive tumor phenotype. A recent study characterized the extent of TMZ-induced mutations and their effects on survival [285]. Tumor samples were stratified as follows: mutations in recurrent tumors without TMZ treatment, untreated tumors, TMZ-treated but nonhypermutated cases and TMZ-treated hypermutated cases. In patients treated with TMZ, 17% relapsed with hypermutated tumors (17/100), while there was no incidence of hypermutation in non-TMZ-treated patients (0/14). In cultured GBM cells treated with radiation, TMZ induced an increase in MMP-2 secretion as well as matrix-degrading invadopodia [300]. Combined administration of TMZ and dexamethasone modifies the proteoglycan and GAG content in the ECM to promote a pro-invasive matrix [301].

Targeting the TME

Targeting the tumor and surrounding TME for therapy has the potential to improve patient survival and quality of life. Although several therapies targeting various components of the TME have been developed, very few have shown promising efficacy. Strategies targeting overexpressed EGFR (erlotinib) or CSF1R to modulate TAMs (PLX3397) have been unsuccessful [302, 303]. To date, strategies targeting the hypervascularity of GBM have shown the most promise. Bevacizumab, a VEGF-A inhibitor, is currently the only drug approved to target the TME in GBM [252–254].

Unfortunately, while treatment with bevacizumab reduces tumor volume and vascularity, tumors exhibit more invasive and vascular phenotypes at recurrence [255]. Other inhibitors undergoing clinical evaluation include tivozanib, a potent pan-VEGF inhibitor targeting proliferation and invasion, as well as cediranib and sudinitib, which reduce angiogenesis [304–307]. Non-VEGF inhibitors targeting vascularity and angiogenesis include AMG 386, an angiopoietin inhibitor that reduces vascular permeability [308].

In addition to the focus on novel therapeutic development, other groups are veering toward improving drug delivery methods by leveraging the influence of the TME on tumor mechanics. For example, one approach involves inserting a catheter directly into the tumor core to ensure continuous delivery of chemotherapy and avoid the effects of interstitial pressure. Preclinical and clinical results demonstrate that convection-enhanced delivery can effectively deliver therapeutics to brain tumors, although limitations in catheter technology and imaging of delivery prevent this technique from being reliable and reproducible [309]. Another method makes use of poliovirus-based immunotherapy to engage oncolytic T cells and improve overall survival in GBM [310, 311]. As many therapeutic agents fail owing to limited bioavailability/specificity, developing methods to improve drug delivery offers potential for improved outcomes.

Engineered GBM models of the TME offer promising potential in the preclinical and clinical pipeline to better understand the complexity of molecular interactions. For example, platforms incorporating patient-specific tumor samples can aid in predicting response to therapy and developing a personalized treatment plan. Response to chemotherapeutics is currently more robustly predicted by molecular subtype, DNA methylation status, and patient age than by in vitro testing, demonstrating the urgent need for experimental systems that reflect the genetics and behavior of GBM patients. Advances in 3D printing now allow for engineered models to reflect patient-specific anatomy and pathology with the inclusion of patient-derived cells [312]. These models are helpful in studying the mechanistic challenges of therapeutic delivery, such as accounting for the effects of interstitial pressure on drug release from Gliadel wafers, for example [313]. Such models are already being developed, but require additional validation and standardization [92].

Predictive information pertaining to changes in the TME can be extracted from MR images using machine learning approaches. For example, machine learning-based parametrization of contrast enhancement in MR images expression of markers of hypoxia, matrix remodeling, and endothelial permeability [314]. Chemical exchange saturation transfer (CEST) MRI can be used to assess tumor response to therapy and can even identify early progressors prior to the start of treatment [315]. Moreover, image feature extraction can be correlated with tumor subtype and patient survival [316, 317]. Patient-specific MRI data can then be combined with other characteristics to improve diagnosis prior to surgical resection and therapy selection. Refinement of these techniques can lead to individually tailored treatment plans and allow for adaptive therapy.

Conclusion

Signals from the TME, whether from the ECM or the various cell types surrounding the tumor mass, actively mold a pro-tumorigenic environment supportive of angiogenesis, cell migration, invasion, and proliferation in GBM. Conversely, the TME is highly dynamic and susceptible to changes incurred as a result of therapy. Key survival and self-renewal pathways are coopted to promote the survival of GSCs, which have been shown to be drivers of therapeutic resistance and recurrence. Although several therapies targeting the TME are in the preclinical and clinical development stage, only therapies targeting hypervascularity are currently approved for use, with eventual revascularization and invasion. Further efforts should be directed toward the development of personalized therapies, integrating patient radiographic and genomic data to allow for evaluation of response and adaptive treatments.

References

[1] Ostrom QT, et al. CBTRUS statistical report: primary brain and other central nervous system tumors diagnosed in the united states in 2011–2015. Neuro Oncol 2018;20. iv1–iv86.
[2] Koshy M, et al. Improved survival time trends for glioblastoma using the SEER 17 population-based registries. J Neurooncol 2012;107:207–12.
[3] Stupp R, et al. Effects of radiotherapy with concomitant and adjuvant temozolomide versus radiotherapy alone on survival in glioblastoma in a randomised phase III study: 5-year analysis of the EORTC-NCIC trial. Lancet Oncol 2009;10:459–66.
[4] Watanabe M, Tanaka R, Takeda N. Magnetic resonance imaging and histopathology of cerebral gliomas. Neuroradiology 1992;34:463–9.
[5] Young RM, Jamshidi A, Davis G, Sherman JH. Current trends in the surgical management and treatment of adult glioblastoma. Ann Transl Med 2015;3:121.
[6] Sherriff J, et al. Patterns of relapse in glioblastoma multiforme following concomitant chemoradiotherapy with temozolomide. Br J Radiol 2013;86:20120414.

[7] Suva ML, Rheinbay E, Gillespie SM, et al. Reconstructing and reprogramming the tumor-propagating potential of glioblastoma stem-like cells. Cell 2014;157:580–94.

[8] Eyler CE, Rich JN. Survival of the fittest: cancer stem cells in therapeutic resistance and angiogenesis. J Clin Oncol 2008;26:2839–45.

[9] Franceschi E, et al. Treatment options for recurrent glioblastoma: pitfalls and future trends. Expert Rev Anticancer Ther 2009;9:613–9.

[10] Seymour T, Nowak A, Kakulas F. Targeting aggressive cancer stem cells in glioblastoma. Front Oncol 2015;5:159.

[11] Wang J, Ma Y, Cooper MK. Cancer stem cells in glioma: challenges and opportunities. Transl Cancer Res 2013;2:429–41.

[12] Broekman ML, Maas SLN, Abels ER, Mempel TR, Krichevsky AM, Breakefield XO. Multidimensional communication in the microenvirons of glioblastoma. Nat Rev Neurol 2018;14(8):482–95.

[13] Quail DF, Joyce JA. The Microenvironmental Landscape of Brain Tumors. Cancer Cell 2017;31(3):326–41.

[14] De Vleeschouwer S, Bergers G. Glioblastoma: To Target the Tumor Cell or the Microenvironment? In: De Vleeschouwer S, editor. Glioblastoma. Brisbane (AU): Codon Publications; 2017.

[15] Ghajar CM, et al. The perivascular niche regulates breast tumour dormancy. Nat Cell Biol 2019;15:807–17.

[16] Provenzano PP, Inman DR, Eliceiri KW, Keely PJ. Matrix density-induced mechanoregulation of breast cell phenotype, signaling and gene expression through a FAK–ERK linkage. Oncogene 2009;28:4326–43.

[17] Elahi-Gedwillo KY, Carlson M, Zettervall J, Provenzano PP. Antifibrotic therapy disrupts stromal barriers and modulates the immune landscape in pancreatic ductal adenocarcinoma. Cancer Res 2019;79:372–86.

[18] van Tellingen O, et al. Overcoming the blood–brain tumor barrier for effective glioblastoma treatment. Drug Resist Updat 2015;19:1–12.

[19] de Vries NA, Beijnen JH, Boogerd W, van Tellingen O. Blood–brain barrier and chemotherapeutic treatment of brain tumors. Expert Rev Neurother 2006;6:1199–209.

[20] Gritsenko PG, Ilina O, Friedl P. Interstitial guidance of cancer invasion. J Pathol 2012;226:185–99.

[21] Wang J, et al. Invasion of white matter tracts by glioma stem cells is regulated by a NOTCH1–SOX2 positive-feedback loop. Nat Neurosci 2019;22:91–105.

[22] Koh I, Cha J, Park J, Choi J, Kang SG, Kim P. The mode and dynamics of glioblastoma cell invasion into a decellularized tissue-derived extracellular matrix-based three-dimensional tumor model. Sci Rep 2018;8(1):4608.

[23] Watkins S, et al. Disruption of astrocyte–vascular coupling and the blood–brain barrier by invading glioma cells. Nat Commun 2014;5:4196.

[24] Neftel C, Laffy J, Filbin MG, et al. An integrative model of cellular states, plasticity, and genetics for Glioblastoma. Cell 2019;178(4):835–849.e21.

[25] Brooks MD, Sengupta R, Snyder SC, Rubin JB. Hitting them where they live: targeting the glioblastoma perivascular stem cell niche. Curr Pathobiol Rep 2013;1:101–10.

[26] Borovski T, De Sousa E, Melo F, Vermeulen L, Medema JP. Cancer stem cell niche: the place to be. Cancer Res 2011;71:634–9.

[27] Silver DJ, Lathia JD. Revealing the glioma cancer stem cell interactome, one niche at a time. J Pathol 2018; 244:260–4.

[28] Charles N, Holland EC. The perivascular niche microenvironment in brain tumor progression. Cell Cycle 2010;9:3012–21.

[29] Calabrese C, et al. A perivascular niche for brain tumor stem cells. Cancer Cell 2007;11:69–82.

[30] Shiraki Y, et al. Significance of perivascular tumour cells defined by CD109 expression in progression of glioma. J Pathol 2017;243:468–80.

[31] Wolf KJ, Lee S, Kumar SA. 3D topographical model of parenchymal infiltration and perivascular invasion in glioblastoma. APL Bioeng 2018;2, 031903.

[32] Ngo MT, Harley BAC. Perivascular signals alter global gene expression profile of glioblastoma and response to temozolomide in a gelatin hydrogel. Biomaterials 2019;198:122–34.

[33] Zhu TS, et al. Endothelial cells create a stem cell niche in glioblastoma by providing NOTCH ligands that nurture self-renewal of cancer stem-like cells. Cancer Res 2011;71:6061–72.

[34] Bao S, et al. Stem cell-like glioma cells promote tumor angiogenesis through vascular endothelial growth factor. Cancer Res 2006;66:7843–8.

[35] Novak U, Kaye AH. Extracellular matrix and the brain: components and function. J Clin Neurosci 2000;7:280–90.

[36] Kim Y, Kang H, Powathil G, et al. Role of extracellular matrix and microenvironment in regulation of tumor growth and LAR-mediated invasion in glioblastoma. PLoS One 2018;13.

[37] Mahesparan R, et al. Expression of extracellular matrix components in a highly infiltrative in vivo glioma model. Acta Neuropathol 2003;105:49–57.
[38] Bellail AC, Hunter SB, Brat DJ, Tan C, Van Meir EG. Microregional extracellular matrix heterogeneity in brain modulates glioma cell invasion. Int J Biochem Cell Biol 2004;36:1046–69.
[39] Lathia JD, et al. Laminin alpha 2 enables glioblastoma stem cell growth. Ann Neurol 2012;72:766–78.
[40] Lathia JD, et al. Integrin alpha 6 regulates glioblastoma stem cells. Cell Stem Cell 2010;6:421–32.
[41] Serres E, et al. Fibronectin expression in glioblastomas promotes cell cohesion, collective invasion of basement membrane in vitro and orthotopic tumor growth in mice. Oncogene 2014;33:3451–62.
[42] Sabari J, et al. Fibronectin matrix assembly suppresses dispersal of glioblastoma cells. PLoS One 2011;6, e24810.
[43] Yuan L, et al. Transglutaminase 2 inhibitor, KCC009, disrupts fibronectin assembly in the extracellular matrix and sensitizes orthotopic glioblastomas to chemotherapy. Oncogene 2007;26:2563–73.
[44] Miroshnikova YA, et al. Tissue mechanics promote IDH1-dependent HIF1α–tenascin C feedback to regulate glioblastoma aggression. Nat Cell Biol 2016;18:1336–45.
[45] Rauch U. Brain matrix: structure, turnover and necessity. Biochem Soc Trans 2007;35:656–60.
[46] Lundell A, et al. Structural basis for interactions between tenascins and lectican C-type lectin domains: evidence for a crosslinking role for tenascins. Structure 2004;12:1495–506.
[47] Mirzaei R, et al. Brain tumor-initiating cells export tenascin-C associated with exosomes to suppress T cell activity. Oncoimmunology 2018;7, e1478647.
[48] Sarkar S, Nuttall RK, Liu S, Edwards DR, Yong VW. Tenascin-C stimulates glioma cell invasion through matrix metalloproteinase-12. Cancer Res 2006;66:11771–80.
[49] Lamour V, et al. Targeting osteopontin suppresses glioblastoma stem-like cell character and tumorigenicity in vivo. Int J Cancer 2015;137:1047–57.
[50] Pietras A, et al. Osteopontin-CD44 signaling in the glioma perivascular niche enhances cancer stem cell phenotypes and promotes aggressive tumor growth. Cell Stem Cell 2014;14:357–69.
[51] Wei J, et al. Osteopontin mediates glioblastoma-associated macrophage infiltration and is a potential therapeutic target. J Clin Invest 2018;129:137–49.
[52] Tian X, et al. High-molecular-mass hyaluronan mediates the cancer resistance of the naked mole rat. Nature 2013;499:346–9.
[53] Chanmee T, Ontong P, Itano N. Mini-review. Hyaluronan: a modulator of the tumor microenvironment. Cancer Lett 2016;375:20–30.
[54] Chen JWE, et al. Influence of hyaluronic acid transitions in tumor microenvironment on glioblastoma malignancy and invasive behavior. Front Mater 2018;5:39.
[55] Tilghman J, et al. HMMR maintains the stemness and tumorigenicity of glioblastoma stem-like cells. Cancer Res 2014;74:3168–79.
[56] Chanmee T, Ontong P, Kimata K, Itano N. Key roles of hyaluronan and its CD44 receptor in the stemness and survival of cancer stem cells. Front Oncol 2015;5:180.
[57] Ferrandez E, Gutierrez O, Segundo DS, Fernandez-Luna JL. NFκB activation in differentiating glioblastoma stem-like cells is promoted by hyaluronic acid signaling through TLR4. Sci Rep 2018;8:6341.
[58] Pedron S, et al. Hyaluronic acid-functionalized gelatin hydrogels reveal extracellular matrix signals temper the efficacy of erlotinib against patient-derived glioblastoma specimens. Biomaterials 2019;219:119371.
[59] Xiao W, et al. Brain-mimetic 3D culture platforms allow investigation of cooperative effects of extracellular matrix features on therapeutic resistance in glioblastoma. Cancer Res 2018;78:1358–70.
[60] Mammoto T, et al. Role of collagen matrix in tumor angiogenesis and glioblastoma multiforme progression. Am J Pathol 2013;183:1293–305.
[61] Seano G, et al. Solid stress in brain tumours causes neuronal loss and neurological dysfunction and can be reversed by lithium. Nat Biomed Eng 2019;3:230–45.
[62] Umesh V, Rape AD, Ulrich TA, Kumar S. Microenvironmental stiffness enhances glioma cell proliferation by stimulating epidermal growth factor receptor signaling. PLoS One 2014;9, e101771.
[63] Heldin CH, Rubin K, Pietras K, Ostman A, Östman A. High interstitial fluid pressure—an obstacle in cancer therapy. Nat Rev Cancer 2004;4:806–13.
[64] Swartz MA, Fleury ME. Interstitial flow and its effects in soft tissues. Annu Rev Biomed Eng 2007;9:229–56.
[65] Monteiro A, Hill R, Pilkington G, Madureira P. The role of hypoxia in glioblastoma invasion. Cell 2017;6:45.
[66] Velásquez C, Mansouri S, Mora C, et al. Molecular and clinical insights into the invasive capacity of Glioblastoma cells. J Oncol 2019;19, 1740763.

[67] Colwell N, et al. Hypoxia in the glioblastoma microenvironment: shaping the phenotype of cancer stem-like cells. Neuro Oncol 2017;19:887–96.

[68] Patel AP, Tirosh I, Trombetta JJ, et al. Single-cell RNA-seq highlights intratumoral heterogeneity in primary glioblastoma. Science 2014;344(6190):1396–401.

[69] Meyer M, Reimand J, Lan X, et al. Single cell-derived clonal analysis orontof human glioblastoma links functional and genomic heterogeneity. Proc Natl Acad Sci U S A 2015;112(3):851–6.

[70] Soeda A, Park M, Lee D, et al. Hypoxia promotes expansion of the CD133-positive glioma stem cells through activation of HIF-1alpha. Oncogene 2009;28(45):3949–59.

[71] Qiang L, Wu T, Zhang H, et al. HIF-1α is critical for hypoxia-mediated maintenance of glioblastoma stem cells by activating Notch signaling pathway. Cell Death Differ 2012;19:284–94.

[72] Niu CS, Li DX, Liu YH, et al. Expression of NANOG in human gliomas and its relationship with undifferentiated glioma cells. Oncol Rep 2011;26(3):593–601.

[73] Ikushima H, Todo T, Ino Y, et al. Autocrine TGF-beta signaling maintains tumorigenicity of glioma-initiating cells through Sry-related HMG-box factors. Cell Stem Cell 2009;5(5):504–14.

[74] Heddleston JM, Li Z, McLendon RE, Hjelmeland AB, Rich JN. The hypoxic microenvironment maintains glioblastoma stem cells and promotes reprogramming towards a cancer stem cell phenotype. Cell Cycle 2009;8: 3274–84.

[75] Li Z, Bao S, Wu Q, et al. Hypoxia-inducible factors regulate tumorigenic capacity of glioma stem cells. Cancer Cell 2009;15(6):501–13.

[76] Rong Y, Durden DL, Van Meir EG, et al. 'Pseudopalisading' necrosis in glioblastoma: a familiar morphologic feature that links vascular pathology, hypoxia, and angiogenesis. J Neuropathol Exp Neurol 2006;65(6):529–39.

[77] Nizet V, Johnson RS. Interdependence of hypoxic and innate immune responses. Nat Rev Immunol 2009;9 (9):609–17.

[78] Noman MZ, Desantis G, Janji B, et al. PD-L1 is a novel direct target of HIF-1α, and its blockade under hypoxia enhanced MDSC-mediated T cell activation. J Exp Med 2014;211(5):781–90.

[79] Taylor CT, Cummins EP. The role of NF-kappaB in hypoxia-induced gene expression. Ann N Y Acad Sci 2009;1177:178–84.

[80] Pawlus MR, Wang L, Hu CJ. STAT3 and HIF1α cooperatively activate HIF1 target genes in MDA-MB-231 and RCC4 cells. Oncogene 2014;33(13):1670–9.

[81] Noman MZ, Buart S, Van Pelt J, et al. The cooperative induction of hypoxia-inducible factor-1 alpha and STAT3 during hypoxia induced an impairment of tumor susceptibility to CTL-mediated cell lysis. J Immunol 2009; 182(6):3510–21.

[82] Hasmim M, Noman MZ, Lauriol J, et al. Hypoxia-dependent inhibition of tumor cell susceptibility to CTL-mediated lysis involves NANOG induction in target cells. J Immunol 2011;187(8):4031–9.

[83] Kang SH, Yu MO, Park KJ, Chi SG, Park DH, Chung YG. Activated STAT3 regulates hypoxia-induced angiogenesis and cell migration in human glioblastoma. Neurosurgery 2010;67(5):1386–95.

[84] Martinez-Outschoorn UE, Pavlides S, Howell A, et al. Stromal epithelial metabolic coupling in cancer: integrating autophagy and metabolism in the tumor microenvironment. Int J Biochem Cell Biol 2011;43(7):1045–51.

[85] Oldham WM, Clish CB, Yang Y, Loscalzo J. Hypoxia-mediated increases in L-2-hydroxyglutarate coordinate the metabolic response to reductive stress. Cell Metab 2015;22(2):291–303.

[86] Pistollato F, Chen HL, Rood BR, et al. Hypoxia and HIF1alpha repress the differentiative effects of BMPs in high-grade glioma. Stem Cells 2009;27(1):7–17.

[87] Sheehan JP, Shaffrey ME, Gupta B, et al. Improving the radiosensitivity of radioresistant and hypoxic glioblastoma. Future Oncol 2010;6(10):1591–601.

[88] O'Connor JP, Boult JK, Jamin Y, et al. Oxygen-enhanced MRI accurately identifies, quantifies, and maps tumor hypoxia in preclinical cancer models. Cancer Res 2016;76(4):787–95.

[89] Mihaylova VT, Bindra RS, Yuan J, et al. Decreased expression of the DNA mismatch repair gene Mlh1 under hypoxic stress in mammalian cells. Mol Cell Biol 2003;23(9):3265–73.

[90] Binello E, Germano IM. Targeting glioma stem cells: a novel framework for brain tumors. Cancer Sci 2011;102:1958–66.

[91] Ho IAW, Shim WSN. Contribution of the Microenvironmental Niche to Glioblastoma Heterogeneity. Biomed Res Int 2017;2017:9634172.

[92] Wolf KJ, Chen J, Coombes JD, et al. Dissecting and rebuilding the glioblastoma microenvironment with engineered materials. Nat Rev Mater 2019;4:651–68.

[93] Osswald M, et al. Brain tumour cells interconnect to a functional and resistant network. Nature 2015;528:93–8.
[94] Figueroa J, et al. Exosomes from glioma-associated mesenchymal stem cells increase the tumorigenicity of glioma stem-like cells via transfer of miR-1587. Cancer Res 2017;77:5808–19.
[95] Hossain A, et al. Mesenchymal stem cells isolated from human gliomas increase proliferation and maintain stemness of glioma stem cells through the IL-6/gp130/STAT3 pathway. Stem Cells 2015;33:2400–15.
[96] Venkatesh HS, et al. Neuronal activity promotes glioma growth through neuroligin-3 secretion. Cell 2015;161:803–16.
[97] Infanger DW, et al. Glioblastoma stem cells are regulated by interleukin-8 signaling in a tumoral perivascular niche. Cancer Res 2013;73:7079–89.
[98] Soda Y, et al. Transdifferentiation of glioblastoma cells into vascular endothelial cells. Proc Natl Acad Sci U S A 2011;108:4274–80.
[99] Brandao M, Simon T, Critchley G, Giamas G. Astrocytes, the rising stars of the glioblastoma microenvironment. Glia 2019;67:779–90.
[100] Yi L, Xiao H, Xu M, et al. Glioma-initiating cells: a predominant role in microglia/macrophages tropism to glioma. J Neuroimmunol 2011;232(1-2):75–82.
[101] Hambardzumyan D, Gutmann DH, Kettenmann H. The role of microglia and macrophages in glioma maintenance and progression. Nat Neurosci 2016;19(1):20–7.
[102] Wang SC, Hong JH, Hsueh C, et al. Tumor-secreted SDF-1 promotes glioma invasiveness and TAM tropism toward hypoxia in a murine astrocytoma model. Lab Invest 2012;92(1):151–62.
[103] Chen Z, Hambardzumyan D. Immune microenvironment in Glioblastoma subtypes. Front Immunol 2018;9:1004.
[104] Wu A, Wei J, Kong LY, et al. Glioma cancer stem cells induce immunosuppressive macrophages/microglia. Neuro Oncol 2010;12(11):1113–25.
[105] Pyonteck SM, Akkari L, Schuhmacher AJ, et al. CSF-1R inhibition alters macrophage polarization and blocks glioma progression. Nat Med 2013;19(10):1264–72.
[106] Mantovani A, Sica A. Macrophages, innate immunity and cancer: balance, tolerance, and diversity. Curr Opin Immunol 2010;22(2):231–7.
[107] Zhou W, Ke SQ, Huang Z, et al. Periostin secreted by glioblastoma stem cells recruits M2 tumour-associated macrophages and promotes malignant growth. Nat Cell Biol 2015;17(2):170–82.
[108] Otvos B, Silver DJ, Mulkearns-Hubert EE, et al. Cancer stem cell secreted macrophage migration inhibitory factor stimulates myeloid derived suppressor cell function and facilitates glioblastoma immune evasion. Stem Cells 2016;34(8):2026–39.
[109] Hasmim M, Noman MZ, Messai Y, et al. Cutting edge: hypoxia induced Nanog favors the intratumoral infiltration of regulatory T cells and macrophages via direct regulation of TGF-β1. J Immunol 2013;191(12):5802–6.
[110] Markovic DS, Vinnakota K, Chirasani S, et al. Gliomas induce and exploit microglial MT1-MMP expression for tumor expansion. Proc Natl Acad Sci USA 2009;106(30):12530–5.
[111] Wesolowska A, Kwiatkowska A, Slomnicki L, et al. Microglia-derived TGF-β as an important regulator of glioblastoma invasion—an inhibition of TGF-β-dependent effects by shRNA against human TGF-β type II receptor. Oncogene 2008;27(7):918–30. https://doi.org/10.1038/sj.onc.1210683.
[112] Gupta K, Burns TC. Radiation-induced alterations in the recurrent glioblastoma microenvironment: therapeutic implications. Front Oncol 2018;8:503.
[113] Akkari L, Bowman RL, Tessier J, et al. Dynamic changes in glioma macrophage populations after radiotherapy reveal CSF-1R inhibition as a strategy to overcome resistance. Sci Transl Med 2020;12(552), eaaw7843.
[114] Quail DF, Bowman RL, Akkari L, et al. The tumor microenvironment underlies acquired resistance to CSF-1R inhibition in gliomas. Science 2016;352(6288), aad3018.
[115] Lee Y, Lee JK, Ahn SH, Lee J, Nam DH. WNT signaling in glioblastoma and therapeutic opportunities. Lab Invest 2016;96(2):137–50. https://doi.org/10.1038/labinvest.2015.140.
[116] Anastas JN, Moon RT. Wnt signalling pathways as therapeutic targets in cancer. Nat Rev Cancer 2013;13:11–26.
[117] Semenov MV, Habas R, Macdonald BT, et al. Snapshot: noncanonical Wnt signaling pathways. Cell 2007;131:1378.
[118] Liu C, Tu Y, Sun X, et al. Wnt/Beta-catenin pathway in human glioma: expression pattern and clinical/prognostic correlations. Clin Exp Med 2011;11:105–12.
[119] Zheng H, Ying H, Wiedemeyer R, et al. Plagl2 regulates Wnt signaling to impede differentiation in neural stem cells and gliomas. Cancer Cell 2010;17:497–509.
[120] Zhang N, Wei P, Gong A, et al. Foxm1 promotes beta-catenin nuclear localization and controls Wnt target-gene expression and glioma tumorigenesis. Cancer Cell 2011;20:427–42.

[121] Joshi K, Banasavadi-Siddegowda Y, Mo X, et al. Melk-dependent foxm1 phosphorylation is essential for proliferation of glioma stem cells. Stem Cells 2013;31:1051–63.
[122] Lee Y, Kim KH, Kim DG, et al. Foxm1 promotes stemness and radioresistance of glioblastoma by regulating the master stem cell regulator Sox2. PLoS One 2015;10, e0137703.
[123] Rheinbay E, Suva ML, Gillespie SM, et al. An aberrant transcription factor network essential for wnt signaling and stem cell maintenance in glioblastoma. Cell Rep 2013;3:1567–79.
[124] Rajakulendran N, Rowland KJ, Selvadurai HJ, et al. Wnt and Notch signaling govern self-renewal and differentiation in a subset of human glioblastoma stem cells. Genes Dev 2019;33(9-10):498–510. https://doi.org/10.1101/gad.321968.118.
[125] Pulvirenti T, Van Der Heijden M, Droms LA, et al. Dishevelled 2 signaling promotes self-renewal and tumorigenicity in human gliomas. Cancer Res 2011;71:7280–90.
[126] Roth W, Wild-Bode C, Platten M, et al. Secreted frizzled-related proteins inhibit motility and promote growth of human malignant glioma cells. Oncogene 2000;19:4210–20.
[127] Schiefer L, Visweswaran M, Perumal V, et al. Epigenetic regulation of the secreted frizzled-related protein family in human glioblastoma multiforme. Cancer Gene Ther 2014;21:297–303.
[128] Jin X, Jeon HY, Joo KM, et al. Frizzled 4 regulates stemness and invasiveness of migrating glioma cells established by serial intracranial transplantation. Cancer Res 2011;71:3066–75.
[129] Smit MA, Peeper DS. Zeb1 is required for trkb-induced epithelial mesenchymal transition, anoikis resistance and metastasis. Oncogene 2011;30:3735–44.
[130] Han SP, Kim JH, Han ME, et al. Snai1 is involved in the proliferation and migration of glioblastoma cells. Cell Mol Neurobiol 2011;31:489–96.
[131] Wellner U, Schubert J, Burk UC, et al. The Emt-activator Zeb1 promotes tumorigenicity by repressing stemness-inhibiting micro-RNAs. Nat Cell Biol 2009;11:1487–95.
[132] Mikheeva SA, Mikheev AM, Petit A, et al. Twist1 promotes invasion through mesenchymal change in human glioblastoma. Mol Cancer 2010;9:194.
[133] Yang HW, Menon LG, Black PM, et al. Snai2/Slug promotes growth and invasion in human gliomas. BMC Cancer 2010;10:301.
[134] Howe LR, Watanabe O, Leonard J, et al. Twist is up-regulated in response to Wnt1 and inhibits mouse mammary cell differentiation. Cancer Res 2003;63:1906–13.
[135] Kahlert UD, Maciaczyk D, Doostkam S, et al. Activation of canonical Wnt/Beta-catenin signaling enhances in vitro motility of glioblastoma cells by activation of Zeb1 and other activators of epithelial-tomesenchymal transition. Cancer Lett 2012;325:42–53.
[136] Kamino M, Kishida M, Kibe T, et al. Wnt-5a signaling is correlated with infiltrative activity in human glioma by inducing cellular migration and Mmp-2. Cancer Sci 2011;102:540–8.
[137] Pu P, Zhang Z, Kang C, et al. Downregulation of Wnt2 and Beta catenin by Sirna suppresses malignant glioma cell growth. Cancer Gene Ther 2009;16:351–61.
[138] Auger N, Thillet J, Wanherdrick K, et al. Genetic alterations associated with acquired temozolomide resistance in Snb-19, a human glioma cell line. Mol Cancer Ther 2006;5:2182–92.
[139] Kim Y, Kim KH, Lee J, et al. Wnt activation is implicated in glioblastoma radioresistance. Lab Invest 2012;92:466–73.
[140] Lai EC. Notch signaling: control of cell communication and cell fate. Development 2004;131:965–73.
[141] Chiba S. Notch signaling in stem cell systems. Stem Cells 2006;24:2437–47.
[142] Carmeliet P, Jain RK. Molecular mechanisms and clinical applications of angiogenesis. Nature 2011;473(7347):298–307.
[143] Bazzoni R, Bentivegna A. role of notch signaling pathway in Glioblastoma pathogenesis. Cancers (Basel) 2019;11(3):292.
[144] Charles N, et al. Perivascular nitric oxide activates notch signaling and promotes stem-like character in PDGF-induced glioma cells. Cell Stem Cell 2010;6:41–152.
[145] Qiang L, Wu T, Zhang HW, et al. HIF-1α is critical for hypoxia-mediated maintenance of glioblastoma stem cells by activating Notch signaling pathway. Cell Death Differ. 2012;19(2).
[146] Hu YY, Fu LA, Li SZ, Chen Y, Li JC, Han J, Liang L, Li L, Ji CC, Zheng MH, et al. Hif-1_and Hif-2_ differentially regulate Notch signaling through competitive interaction with the intracellular domain of Notch receptors in glioma stem cells. Cancer Lett 2014;349:67–76.
[147] Louvi A, Artavanis-Tsakonas S. Notch signalling in vertebrate neural development. Nat Rev Neurosci 2006;7:93–102.

[148] Taylor MK, Yeager K, Morrison SJ. Physiological Notch signaling promotes gliogenesis in the developing peripheral and central nervous systems. Development 2007;134:2435–47.
[149] Kanamori M, Kawaguchi T, Nigro JM, Feuerstein BG, Berger MS, et al. Contribution of Notch signaling activation to human glioblastoma multiforme. J Neurosurg 2007;106:417–27.
[150] Logeat F, Bessia C, Brou C, LeBail O, Jarriault S, Seidah NG, Israël A. The Notch1 receptor is cleaved constitutively by a furin-like convertase. Proc Natl Acad Sci U S A 1998;95:8108–12.
[151] Kopan R, Ilagan MX. The canonical Notch signaling pathway: Unfolding the activation mechanism. Cell 2009;137:216–33.
[152] Mumm JS, Schroeter EH, Saxena MT, Griesemer A, Tian X, Pan DJ, Ray WJ, Kopan RA. Ligand-induced extracellular cleavage regulates gamma-secretase-like proteolytic activation of Notch1. Mol Cell 2000;5:197–206.
[153] Somasundaram K, Reddy SP, Vinnakota K, Britto R, Subbarayan M, Nambiar S, Hebbar A, Samuel C, Shetty M, Sreepathi HK, et al. Upregulation of ASCL1 and inhibition of Notch signaling pathway characterize progressive astrocytoma. Oncogene 2005;24:7073–83.
[154] Zhang XP, Zheng G, Zou L, Liu HL, Hou LH, Zhou P, Yin DD, Zheng QJ, Liang L, Zhang SZ, et al. Notch activation promotes cell proliferation and the formation of neural stem cell-like colonies in human glioma cells. Mol Cell Biochem 2008;307:101–8.
[155] Phillips HS, Kharbanda S, Chen R, Forrest WF, Soriano RH, Wu TD, Misra A, Nigro JM, Colman H, Soroceanu L, et al. Molecular subclasses of high-grade glioma predict prognosis, delineate a pattern of disease progression, and resemble stages in neurogenesis. Cancer Cell 2006;9:157–73.
[156] Hulleman E, Quarto M, Vernell R, Masserdotti G, Colli E, Kros JM, Levi D, Gaetani P, Tunici P, Finocchiaro G, et al. A role for the transcription factor HEY1 in glioblastoma. J Cell Mol Med 2009;13:136–46.
[157] Brennan C, Momota H, Hambardzumyan D, Ozawa T, Tandon A, Pedraza A, Holland E. Glioblastoma subclasses can be defined by activity among signal transduction pathways and associated genomic alterations. PLoS One 2009;4, e7752.
[158] El Hindy N, Keyvani K, Pagenstecher A, Dammann P, Sandalcioglu IE, Sure U, Zhu Y. Implications of Dll4-Notch signaling activation in primary glioblastoma multiforme. Neuro Oncol 2013;15:1366–78.
[159] Zhang X, Chen T, Zhang J, Mao Q, Li S, Xiong W, Qiu Y, Xie Q, Ge J. Notch1 promotes glioma cell migration and invasion by stimulating b-catenin and NF-_B signaling via AKT activation. Cancer Sci 2011;103:181–90.
[160] Gersey Z, Osiason AD, Bloom L, et al. Therapeutic targeting of the notch pathway in Glioblastoma multiforme. World Neurosurg 2019;131:252–263.e2.
[161] Zhang C, Hai L, Zhu M, Yu S, Li T, Lin Y, Liu B, Zhou X, Chen L, Zhao P, et al. Actin cytoskeleton regulator Arp2/3 complex is required for DLL1 activating Notch1 signaling to maintain the stem cell phenotype of glioma initiating cells. Oncotarget 2017;8:33353–64.
[162] Jeon HM, Kim SH, Jin X, Park JB, Kim SH, Joshi K, Nakano I, Kim H. Crosstalk between glioma-initiating cells and endothelial cells drives tumor progression. Cancer Res 2014;74:4482–92.
[163] Kristoffersen K, Nedergaard MK, Villingshoj M, Borup R, Broholm H, Kjaer A, Poulsen HS, Stockhausen MT. Inhibition of Notch signaling alters the phenotype of orthotopic tumors formed from glioblastoma multiforme neurosphere cells but does not hamper intracranial tumor growth regardless of endogene Notch pathway signature. Cancer Biol Ther 2014;15:862–77.
[164] Raghu H, Gondi CS, Dinh DH, Gujrati M, Rao JS. Specific knockdown of uPA/uPAR attenuates invasion in glioblastoma cells and xenografts by inhibition of cleavage and trafficking of Notch -1 receptor. Mol Cancer 2011;10:130.
[165] Wang Z, Li Y, Banerjee S, Sarkar FH. Emerging role of Notch in stem cells and cancer. Cancer Lett 2009;279:8–12.
[166] Wang J, Wakeman TP, Lathia JD, et al. Notch promotes radioresistance of glioma stem cells. Stem Cells 2010; 28(1):17–28. https://doi.org/10.1002/stem.261.
[167] De Pascalis I, et al. Endothelial trans-differentiation in glioblastoma recurring after radiotherapy. Mod Pathol 2018;31:1361–6.
[168] Allen BK, Stathias V, Maloof ME, et al. Epigenetic pathways and glioblastoma treatment: insights from signaling cascades. J Cell Biochem 2015;116(3):351–63.
[169] Robbins DJ, Fei DL, Riobo NA. The Hedgehog signal transduction network. Sci Signal 2012;5(246):6.
[170] Carpenter RL, Paw I, Zhu H, et al. The gain-of-function GLI1 transcription factorTGLI1 enhances expression of VEGF-C and TEM7 to promote glioblastoma angiogenesis. Oncotarget 2015;6(26):22653–65.
[171] Wang KL, Pan X, Che D, Cui K, Li C. Sonic Hedgehog/ GLI1 signaling pathway inhibition restricts cell migration and invasion in human gliomas. Neurol Res 2010;32(9):975–80.

[172] Uchida H, Arita K, Yunoue S, et al. Role of sonic hedgehog signaling in migration of cell lines established from CD133- positive malignant glioma cells. J Neurooncol 2011;104(3):697–704.
[173] Honorato JR, Hauser-Davis RA, Saggioro EM, et al. Role of Sonic hedgehog signaling in cell cycle, oxidative stress, and autophagy of temozolomide resistant glioblastoma. J Cell Physiol 2020;235(4):3798–814.
[174] Carrasco-García E, Saceda M, Martínez-Lacaci I. Role of receptor tyrosine kinases and their ligands in glioblastoma. Cell 2014;3(2):199–235.
[175] Pearson J, Regad T. Targeting cellular pathways in glioblastoma multiforme. Signal Transduct Target Ther 2017;2:17040.
[176] Thorne AH, Zanca C, Furnari F. Epidermal growth factor receptor targeting and challenges in glioblastoma. Neuro Oncol 2016;18(7):914–8.
[177] An Z, Aksoy O, Zheng T, Fan QW, Weiss WA. Epidermal growth factor receptor and EGFRvIII in glioblastoma: signaling pathways and targeted therapies. Oncogene 2018;37(12):1561–75.
[178] Eskilsson E, Røsland GV, Solecki G, et al. EGFR heterogeneity and implications for therapeutic intervention in glioblastoma. Neuro Oncol 2018;20(6):743–52.
[179] Mimeault M, Batra SK. Complex oncogenic signaling networks regulate brain tumor-initiating cells and their progenies: pivotal roles of wild-type EGFR, EGFRvIII mutant and hedgehog cascades and novel multitargeted therapies. Brain Pathol 2011;21(5):479–500.
[180] Shinojima N, Tada K, Shiraishi S, Kamiryo T, Kochi M, Nakamura H, et al. Prognostic value of epidermal growth factor receptor in patients with glioblastoma multiforme. Cancer Res 2003;63:6962–70.
[181] Congdon KL, Gedeon PC, Suryadevara CM, Caruso HG, Cooper LJ, Heimberger AB, et al. Epidermal growth factor receptor and variant III targeted immunotherapy. Neuro Oncol 2014;16(Suppl 8). viii20–25.
[182] Malkki H. Trial watch: glioblastoma vaccine therapy disappointment in Phase III trial. Nat Rev Neurol 2016;12:190.
[183] Nathanson DA, Gini B, Mottahedeh J, Visnyei K, Koga T, Gomez G, et al. Targeted therapy resistance mediated by dynamic regulation of extrachromosomal mutant EGFR DNA. Science 2014;343:72–6.
[184] Azuaje F, Tiemann K, Niclou SP. Therapeutic control and resistance of the EGFR-driven signaling network in glioblastoma. Cell Commun Signal 2015;13:23.
[185] Singh B, Coffey RJ. Trafficking of epidermal growth factor receptor ligands in polarized epithelial cells. Annu Rev Physiol 2014;76:275–300.
[186] Mayer BJ. The discovery of modular binding domains: building blocks of cell signalling. Nat Rev Mol Cell Biol 2015;16:691–8.
[187] Thorpe LM, Yuzugullu H, Zhao JJ. PI3K in cancer: divergent roles of isoforms, modes of activation and therapeutic targeting. Nat Rev Cancer 2015;15:7–24.
[188] Roth P, Weller M. Challenges to targeting epidermal growth factor receptor in glioblastoma: escape mechanisms and combinatorial treatment strategies. Neuro Oncol 2014;16(suppl 8). viii14–19.
[189] Grandal MV, Zandi R, Pedersen MW, Willumsen BM, van Deurs B, Poulsen HS. EGFRvIII escapes down-regulation due to impaired internalization and sorting to lysosomes. Carcinogenesis 2007;28:1408–17.
[190] Liu F, Hon GC, Villa GR, Turner KM, Ikegami S, Yang H, et al. EGFR mutation promotes glioblastoma through epigenome and transcription factor network remodeling. Mol Cell 2015;60:307–18.
[191] Feng H, Hu B, Jarzynka MJ, Li Y, Keezer S, Johns TG, et al. Phosphorylation of dedicator of cytokinesis 1 (Dock180) at tyrosine residue Y722 by Src family kinases mediates EGFRvIIIdriven glioblastoma tumorigenesis. Proc Natl Acad Sci U S A 2012;109:3018–23.
[192] Jahani-Asl A, Yin H, Soleimani VD, Haque T, Luchman HA, Chang NC, et al. Control of glioblastoma tumorigenesis by feedforward cytokine signaling. Nat Neurosci 2016;19:798–806.
[193] Kolch W, Halasz M, Granovskaya M, Kholodenko BN. The dynamic control of signal transduction networks in cancer cells. Nat Rev Cancer 2015;15:515–27.
[194] Deschenes-Simard X, Kottakis F, Meloche S, Ferbeyre G. ERKs in cancer: friends or foes? Cancer Res 2014; 74:412–9.
[195] Nijkamp MM, Span PN, Bussink J, Kaanders JH. Interaction of EGFR with the tumour microenvironment: implications for radiation treatment. Radiother Oncol 2013;108(1):17–23.
[196] Li X, Wu C, Chen N, et al. PI3K/Akt/mTOR signaling pathway and targeted therapy for glioblastoma. Oncotarget 2016;7(22):33440–50.
[197] Klempner SJ, Myers AP, Cantley LC. What a tangled web we weave: emerging resistance mechanisms to inhibition of the phosphoinositide 3-kinase pathway. Cancer Discov 2013;3:1345–54.

[198] Hay N. Interplay between FOXO, TOR, and Akt. Biochim Biophys Acta 1813;2011:1965–70.
[199] Carracedo A, Pandolfi PP. The PTEN-PI3K pathway: of feedbacks and cross-talks. Oncogene 2008;27:5527–41.
[200] Datta SR, Dudek H, Tao X, Masters S, Fu H, Gotoh Y, et al. Akt phosphorylation of BAD couples survival signals to the cell intrinsic death machinery. Cell 1997;91:231–41.
[201] Majewska E, Szeliga M. AKT/GSK3β Signaling in Glioblastoma. Neurochem Res 2017;42(3):918–24.
[202] Fang X, Yu SX, Lu Y, Bast Jr RC, Woodgett JR, Mills GB. Phosphorylation and inactivation of glycogen synthase kinase 3 by protein kinase A. Proc Natl Acad Sci U S A 2000;97:11960–5.
[203] Ozes ON, Mayo LD, Gustin JA, Pfeffer SR, Pfeffer LM, Donner DB. NF-kappaB activation by tumour necrosis factor requires the Akt serine-threonine kinase. Nature 1999;401:82–5.
[204] Bai D, Ueno L, Vogt PK. Akt-mediated regulation of NFkappaB and the essentialness of NFkappaB for the oncogenicity of PI3K and Akt. Int J Cancer 2009;125:2863–70.
[205] Burris HA. Overcoming acquired resistance to anticancer therapy: focus on the PI3K/AKT/mTOR pathway. Cancer Chemother Pharmacol 2013;71:829–42.
[206] Land SC, Tee AR. Hypoxia-inducible factor 1alpha is regulated by the mammalian target of rapamycin (mTOR) via an mTOR signaling motif. J Biol Chem 2007;282:20534–43.
[207] Ozawa T, Riester M, Cheng YK, Huse JT, Squatrito M, Helmy K, et al. Most human non-GCIMP glioblastoma subtypes evolve from a common proneural-like precursor glioma. Cancer Cell 2014;26:288–300.
[208] Choe G, Horvath S, Cloughesy TF, et al. Analysis of the phosphatidylinositol 3′-kinase signaling pathway in glioblastoma patients in vivo. Cancer Res 2003;63:2742–6.
[209] Ermoian RP, Furniss CS, Lamborn KR, et al. Dysregulation of PTEN and protein kinase B is associated with glioma histology and patient survival. Clin Cancer Res 2002;8:1100–6.
[210] Cancer Genome Atlas Research Network. Comprehensive genomic characterization defines human glioblastoma genes and core pathways. Nature 2008;455:1061–8.
[211] Brennan CW, Verhaak RG, McKenna A, Campos B, Noushmehr H, Salama SR, et al. The somatic genomic landscape of glioblastoma. Cell 2013;155:462–77.
[212] Courtney KD, Corcoran RB, Engelman JA. The PI3K pathway as drug target in human cancer. J Clin Oncol 2010;28:1075–83.
[213] Huang PH, Xu AM, White FM. Oncogenic EGFR signaling networks in glioma. Sci Signal 2009;2(87). re6.
[214] O'Shea JJ, Schwartz DM, Villarino AV, Gadina M, McInnes IB, Laurence A. The JAK-STAT pathway: impact on human disease and therapeutic intervention. Annu Rev Med 2015;66:311–28.
[215] Park OK, Schaefer TS, Nathans D. In vitro activation of Stat3 by epidermal growth factor receptor kinase. Proc Natl Acad Sci U S A 1996;93:13704–8.
[216] Wen Z, Zhong Z, Darnell Jr JE. Maximal activation of transcription by Stat1 and Stat3 requires both tyrosine and serine phosphorylation. Cell 1995;82:241–50.
[217] Kim E, Kim M, Woo DH, Shin Y, Shin J, Chang N, et al. Phosphorylation of EZH2 activates STAT3 signaling via STAT3 methylation and promotes tumorigenicity of glioblastoma stem-like cells. Cancer Cell 2013;23:839–52.
[218] de la Iglesia N, Konopka G, Puram SV, Chan JA, Bachoo RM, You MJ, et al. Identification of a PTEN-regulated STAT3 brain tumor suppressor pathway. Genes Dev 2008;22:449–62.
[219] Fan QW, Cheng CK, Gustafson WC, Charron E, Zipper P, Wong RA, et al. EGFR phosphorylates tumor-derived EGFRvIII driving STAT3/5 and progression in glioblastoma. Cancer Cell 2013;24:438–49.
[220] See AP, Han JE, Phallen J, Binder Z, Gallia G, Pan F, et al. The role of STAT3 activation in modulating the immune microenvironment of GBM. J Neurooncol 2012;110:359–68.
[221] Gong AH, Wei P, Zhang S, Yao J, Yuan Y, Zhou AD, et al. FoxM1 drives a feed-forward STAT3-activation signaling loop that promotes the self-renewal and tumorigenicity of glioblastoma stem-like cells. Cancer Res 2015;75:2337–48.
[222] Garner JM, Fan M, Yang CH, Du Z, Sims M, Davidoff AM, et al. Constitutive activation of signal transducer and activator of transcription 3 (STAT3) and nuclear factor kappaB signaling in glioblastoma cancer stem cells regulates the Notch pathway. J Biol Chem 2013;288:26167–76.
[223] Lo HW, Cao X, Zhu H, Ali-Osman F. Constitutively activated STAT3 frequently coexpresses with epidermal growth factor receptor in high-grade gliomas and targeting STAT3 sensitizes them to Iressa and alkylators. Clin Cancer Res 2008;14:6042–54.
[224] He K, Qi Q, Chan CB, Xiao G, Liu X, Tucker-Burden C, et al. Blockade of glioma proliferation through allosteric inhibition of JAK2. Sci Signal 2013;6. ra55.

[225] Rath BH, Wahba A, Camphausen K, Tofilon PJ. Co-culture with astrocytes reduces the radiosensitivity of glioblastoma stem-like cells and identifies additional targets for radiosensitization. Cancer Med 2015;4:1705–16.
[226] do Carmo A, Balça-Silva J, Matias D, Lopes MC. PKC signaling in glioblastoma. Cancer Biol Ther 2013;14(4):287–94.
[227] Garg R, Benedetti LG, Abera MB, Wang H, Abba M, Kazanietz MG. Protein kinase C and cancer: what we know and what we do not. Oncogene 2014;33:5225–37.
[228] Fan QW, Cheng C, Knight ZA, Haas-Kogan D, Stokoe D, James CD, et al. EGFR signals to mTOR through PKC and independently of Akt in glioma. Sci Signal 2009;2:4.
[229] Papademetriou IT, Porter T. Promising approaches to circumvent the blood-brain barrier: Progress, pitfalls and clinical prospects in brain cancer. Ther Deliv 2015;6:989–1016.
[230] Wood ER, Truesdale AT, McDonald OB, et al. A unique structure for epidermal growth factor receptor bound to GW572016 (Lapatinib): relationships among protein conformation, inhibitor off-rate, and receptor activity in tumor cells. Cancer Res 2004;64(18):6652–9.
[231] Mellinghoff IK, Wang MY, Vivanco I, Haas-Kogan DA, Zhu S, Dia EQ, et al. Molecular determinants of the response of glioblastomas to EGFR kinase inhibitors. N Engl J Med 2005;353:2012–24.
[232] Gini B, Zanca C, Guo D, et al. The mTOR kinase inhibitors, CC214-1 and CC214-2, preferentially block the growth of EGFRvIII-activated glioblastomas. Clin Cancer Res 2013;19(20):5722–32.
[233] Fenton TR, Nathanson D, Ponte de Albuquerque C, Kuga D, Iwanami A, Dang J, et al. Resistance to EGF receptor inhibitors in glioblastoma mediated by phosphorylation of the PTEN tumor suppressor at tyrosine 240. Proc Natl Acad Sci U S A 2012;109:14164–9.
[234] Iwanami A, Gini B, Zanca C, Matsutani T, Assuncao A, Nael A, et al. PML mediates glioblastoma resistance to mammalian target of rapamycin (mTOR)-targeted therapies. Proc Natl Acad Sci U S A 2013;110:4339–44.
[235] Dittmann K, Mayer C, Rodemann HP. Nuclear EGFR as novel therapeutic target: insights into nuclear translocation and function. Strahlenther Onkol 2010;186:1–6.
[236] Dittmann K, Mayer C, Rodemann HP. Inhibition of radiation-induced EGFR nuclear import by C225 (Cetuximab) suppresses DNA-PK activity. Radiother Oncol 2005;76:157–61.
[237] Akhavan D, Pourzia AL, Nourian AA, Williams KJ, Nathanson D, Babic I, et al. De-repression of PDGFRbeta transcriptionpromotes acquired resistance to EGFR tyrosine kinase inhibitorsin glioblastoma patients. Cancer Discov 2013;3:534–47.
[238] Francis JM, Zhang CZ, Maire CL, Jung J, Manzo VE, Adalsteinsson VA, et al. EGFR variant heterogeneity in glioblastoma resolved through single-nucleus sequencing. Cancer Discov 2014;4:956–71.
[239] Szerlip NJ, Pedraza A, Chakravarty D, Azim M, McGuire J, Fang Y, et al. Intratumoral heterogeneity of receptor tyrosine kinases EGFR and PDGFRA amplification in glioblastoma defines subpopulations with distinct growth factor response. Proc Natl Acad Sci U S A 2012;109:3041–6.
[240] Lokker NA, Sullivan CM, Hollenbach SJ, Israel MA, Giese NA. Platelet-derived growth factor (PDGF) autocrine signaling regulates survival and mitogenic pathways in glioblastoma cells: evidence that the novel PDGF-C and PDGF-D ligands may play a role in the development of brain tumors. Cancer Res 2002;62:3729–35.
[241] Hermanson M, Funa K, Hartman M, Claesson-Welsh L, Heldin CH, Westermark B, et al. Platelet-derived growth factor and its receptors in human glioma tissue: expression of messenger RNA and protein suggests the presence of autocrine and paracrine loops. Cancer Res 1992;52:3213–9.
[242] Verhaak RG, Hoadley KA, Purdom E, et al. Cancer Genome Atlas Research Network. Integrated genomic analysis identifies clinically relevant subtypes of glioblastoma characterized by abnormalities in PDGFRA, IDH1, EGFR, and NF1. Cancer Cell 2010;17(1):98–110.
[243] Tamura R, Tanaka T, Akasaki Y, Murayama Y, Yoshida K, Sasaki H. The role of vascular endothelial growth factor in the hypoxic and immunosuppressive tumor microenvironment: perspectives for therapeutic implications. Med Oncol 2019;37(1):2.
[244] Steiner HH, Karcher S, Mueller MM, Nalbantis E, Kunze S, Herold-Mende C. Autocrine pathways of the vascular endothelial growth factor (VEGF) in glioblastoma multiforme: clinical relevance of radiation-induced increase of VEGF levels. J Neurooncol 2004;66:129–38.
[245] Joensuu H, Puputti M, Sihto H, Tynninen O, Nupponen NN. Amplification of genes encoding KIT, PDGFRalpha and VEGFR2 receptor tyrosine kinases is frequent in glioblastoma multiforme. J Pathol 2005;207:224–31.
[246] Chaudhry IH, O'Donovan DG, Brenchley PE, Reid H, Roberts IS. Vascular endothelial growth factor expression correlates with tumour grade and vascularity in gliomas. Histopathology 2001;39:409–15.

[247] Hicklin DJ, Ellis LM. Role of the vascular endothelial growth factor pathway in tumor growth and angiogenesis. J Clin Oncol 2005;23:1011–27.
[248] Lemée JM, Clavreul A, Menei P. Intratumoral heterogeneity in glioblastoma: don't forget the peritumoral brain zone. Neuro Oncol 2015;17(10):1322–32. https://doi.org/10.1093/neuonc/nov119.
[249] Voron T, Colussi O, Marcheteau E, Pernot S, Nizard M, Pointet AL, Latreche S, Bergaya S, Benhamouda N, Tanchot C, Stockmann C, Combe P, Berger A, Zinzindohoue F, Yagita H, Tartour E, Taieb J, Terme M. VEGF-A modulates expression of inhibitory checkpoints on CD8+T cells in tumors. J Exp Med 2015;212:139–48.
[250] Xue S, Hu M, Li P, Ma J, Xie L, Teng F, Zhu Y, Fan B, Mu D, Yu J. Relationship between expression of PD-L1 and tumor angiogenesis, proliferation, and invasion in glioma. Oncotarget 2017;8:49702–12.
[251] Motz GT, Santoro SP, Wang LP, Garrabrant T, Lastra RR, Hagemann IS, Lal P, Feldman MD, Benencia F, Coukos G. Tumor endothelium FasL establishes a selective immune barrier promoting tolerance in tumors. Nat Med 2014;20:607–15.
[252] Friedman HS, et al. Bevacizumab alone and in combination with irinotecan in recurrent glioblastoma. J Clin Oncol 2009;27:4733–40.
[253] Kreisl TN, et al. Phase II trial of single-agent bevacizumab followed by bevacizumab plus irinotecan at tumor progression in recurrent glioblastoma. J Clin Oncol 2009;27:740–5.
[254] Wenger KJ, et al. Bevacizumab as a last-line treatment for glioblastoma following failure of radiotherapy, temozolomide and lomustine. Oncol Lett 2017;14:1141–6.
[255] Pàez-Ribes M, et al. Antiangiogenic therapy elicits malignant progression of tumors to increased local invasion and distant metastasis. Cancer Cell 2009;15:220–31.
[256] Han J, Alvarez-Breckenridge CA, Wang QE, Yu J. TGF-β signaling and its targeting for glioma treatment. Am J Cancer Res 2015;5(3):945–55.
[257] Bellomo C, Caja L, Moustakas A. Transforming growth factor β as regulator of cancer stemness and metastasis. Br J Cancer 2016;115(7):761–9.
[258] Wakefield LM, Hill CS. Beyond TGF beta: roles of other TGFbeta superfamily members in cancer. Nat Rev Cancer 2013;13:328–41.
[259] Watabe T, Miyazono K. Roles of TGF-beta family signaling in stem cell renewal and differentiation. Cell Res 2009;19:103–15.
[260] Neuzillet C, Tijeras-Raballand A, Cohen R, et al. Targeting the TGFβ pathway for cancer therapy. Pharmacol Ther 2015;147:22–31.
[261] Kaminska B, Kocyk M, Kijewska M. TGF β signaling and its role in glioma pathogenesis. Adv Exp Med Biol 2013;986:171–87.
[262] Zhang C, Zhang X, Xu R, et al. TGF-β2 initiates autophagy via Smad and non-Smad pathway to promote glioma cells' invasion. J Exp Clin Cancer Res 2017;36:162.
[263] Bryukhovetskiy I, Shevchenko V. Molecular mechanisms of the effect of TGF-β1 on U87 human glioblastoma cells. Oncol Lett 2016;12:1581–90.
[264] Bruna A, Darken RS, Rojo F, Ocaña A, Peñuelas S, Arias A, Paris R, Tortosa A, Mora J, Baselga J. High TGFβ-Smad activity confers poor prognosis in glioma patients and promotes cell proliferation depending on the methylation of the PDGF-B gene. Cancer Cell 2007;11:147–60.
[265] Kjellman C, Olofsson SP, Hansson O, Von Schantz T, Lindvall M, Nilsson I, Salford LG, Sjögren HO, Widegren B. Expression of TGF-β isoforms, TGF-β receptors, and SMAD molecules at different stages of human glioma. Int J Cancer 2000;89:251–8.
[266] Maxwell M, Galanopoulos T, Neville-Golden J, Antoniades HN. Effect of the expression of transforming growth factor-β2 in primary human glioblastomas on immunosuppression and loss of immune surveillance. J Neurosurg 1992;76:799–804.
[267] Qiu B, Zhang D, Wang C, Tao J, Tie X, Qiao Y, Xu K, Wang Y, Wu A. IL-10 and TGF-beta2 are overexpressed in tumor spheres cultured from human gliomas. Mol Biol Rep 2011;38:3585–91.
[268] Bruna A, Darken RS, Rojo F, Ocana A, Penuelas S, Arias A, Paris R, Tortosa A, Mora J, Baselga J, Seoane J. High TGFbeta-Smad activity confers poor prognosis in glioma patients and promotes cell proliferation depending on the methylation of the PDGF-B gene. Cancer Cell 2007;11:147–60.
[269] Matsuzaki K. Smad phosphoisoform signaling specificity: the right place at the right time. Carcinogenesis 2011;32:1578–88.
[270] Huber MA, Kraut N, Beug H. Molecular requirements for epithelial-mesenchymal transition during tumor progression. Curr Opin Cell Biol 2005;17:548–58.

[271] Cheng L, Huang Z, Zhou W, Wu Q, Donnola S, Liu JK, Fang X, Sloan AE, Mao Y, Lathia JD, Min W, McLendon RE, Rich JN, Bao S. Glioblastoma stem cells generate vascular pericytes to support vessel function and tumor growth. Cell 2013;153:139–52.
[272] Cornelison RC, Brennan CE, Kingsmore KM, Munson JM. Convective forces increase CXCR4- dependent glioblastoma cell invasion in GL261 murine model. Sci Rep 2018;8:17057.
[273] Bierie B, Moses HL. Tumour microenvironment: TGFbeta: the molecular Jekyll and Hyde of cancer. Nat Rev Cancer 2006;6(7):506–20.
[274] Iwadate Y. Epithelial-mesenchymal transition in glioblastoma progression. Oncol Lett 2016;11(3):1615–20.
[275] Anido J, Saez-Borderias A, Gonzalez-Junca A, Rodon L, Folch G, Carmona MA, Prieto-Sanchez RM, Barba I, Martinez-Saez E, Prudkin L, Cuartas I, Raventos C, Martinez-Ricarte F, Poca MA, Garcia-Dorado D, Lahn MM, Yingling JM, Rodon J, Sahuquillo J, Baselga J, Seoane J. TGF-beta receptor inhibitors target the CD44(high)/Id1(high) Glioma-initiating cell population in human Glioblastoma. Cancer Cell 2010;18: 655–68.
[276] Hardee ME, Marciscano AE, Medina-Ramirez CM, Zagzag D, Narayana A, Lonning SM, Barcellos-Hoff MH. Resistance of glioblastoma-initiating cells to radiation mediated by the tumor microenvironment can be abolished by inhibiting transforming growth factor-beta. Cancer Res 2012;72:4119–29.
[277] Wang J, Cazzato E, Ladewig E, et al. Clonal evolution of glioblastoma under therapy. Nat Genet 2016; 48(7):768–76.
[278] Caja L, Bellomo C, Moustakas A. Transforming growth factor β and bone morphogenetic protein actions in brain tumors. FEBS Lett 2015;589(14):1588–97.
[279] Piccirillo SGM, et al. Bone morphogenetic proteins inhibit the tumorigenic potential of human brain tumour-initiating cells. Nature 2006;444:761–5.
[280] Peñuelas S, et al. TGF-& beta; increases glioma-initiating cell self-renewal through the induction of LIF in human glioblastoma. Cancer Cell 2009;15:315–27.
[281] Caja L, Tzavlaki K, Dadras MS, et al. Snail regulates BMP and TGFβ pathways to control the differentiation status of glioma-initiating cells. Oncogene 2018;37(19):2515–31.
[282] Sachdeva R, Wu M, Johnson K, et al. BMP signaling mediates glioma stem cell quiescence and confers treatment resistance in glioblastoma. Sci Rep 2019;9(1):14569.
[283] Barrett LE, et al. Self-renewal does not predict tumor growth potential in mouse models of high-grade glioma. Cancer Cell 2012;21:11–24.
[284] Coqueret O, Gascan H. Functional interaction of STAT3 transcription factor with the cell cycle inhibitor p21WAF1/CIP1/SDI1. J Biol Chem 2000;275:18794–800.
[285] Rahaman SO, et al. Inhibition of constitutively active Stat3 suppresses proliferation and induces apoptosis in glioblastoma multiforme cells. Oncogene 2002;21:8404–13.
[286] Mira H, et al. Signaling through BMPR-IA regulates quiescence and long-term activity of neural stem cells in the adult hippocampus. Cell Stem Cell 2010;7:78–89.
[287] Tejero R, Huang Y, Katsyv I, et al. Gene signatures of quiescent glioblastoma cells reveal mesenchymal shift and interactions with niche microenvironment. EBioMedicine 2019;42:252–69.
[288] Chen J, et al. A restricted cell population propagates glioblastoma growth after chemotherapy. Nature 2012;488:522–6.
[289] Cahill KE, Morshed RA, Yamini B. Nuclear factor-κB in glioblastoma: insights into regulators and targeted therapy. Neuro Oncol 2016;18(3):329–39.
[290] Lee DW, Ramakrishnan D, Valenta J, Parney IF, Bayless KJ, Sitcheran R. The NF-kappaB RelB protein is an oncogenic driver of mesenchymal glioma. PLoS One 2013;8(2), e57489.
[291] Tchoghandjian A, Jennewein C, Eckhardt I, Rajalingam K, Fulda S. Identification of non-canonical NF-κB signaling as a critical mediator of Smac mimetic-stimulated migration and invasion of glioblastoma cells. Cell Death Dis 2013;4(3): e564–e564.
[292] Kim J, Jin X, Sohn Y, et al. Tumoral RANKL activates astrocytes that promote glioma cell invasion through cytokine signaling. Cancer Lett 2014;353(2):194–200.
[293] Fianco G, Mongiardi MP, Levi A, De Luca T, Desideri M, Trisciuoglio D, Del Bufalo D, Cinà I, Di Benedetto A Mottolese M, et al. Caspase-8 contributes to angiogenesis and chemotherapy resistance in glioblastoma. Elife 2017;6.
[294] Barker HE, Paget JT, Khan AA, Harrington KJ. The tumour microenvironment after radiotherapy: mechanisms of resistance and recurrence [published correction appears in Nat Rev Cancer. 2015 Aug;15(8):509]. Nat Rev Cancer 2015;15(7):409–25.

[295] Osuka S, Sampetrean O, Shimizu T, et al. IGF1 receptor signaling regulates adaptive radioprotection in glioma stem cells. Stem Cells 2013;31(4):627–40.
[296] Osuka S, Van Meir EG. Overcoming therapeutic resistance in glioblastoma: the way forward. J Clin Invest 2017;127(2):415–26.
[297] Rath BH, Wahba A, Camphausen K, Tofilon PJ. Coculture with astrocytes reduces the radiosensitivity of glioblastoma stem-like cells and identifies additional targets for radiosensitization. Cancer Med 2015;4:1705–16.
[298] Yoo K-C, et al. Proinvasive extracellular matrix remodeling in tumor microenvironment in response to radiation. Oncogene 2018;37:3317–28.
[299] Jeon HY, Kim JK, Ham SW, Oh SY, Kim J, Park JB, et al. Irradiation induces glioblastoma cell senescence and senescence-associated secretory phenotype. Tumour Biol 2016;37:5857–67.
[300] Mao L, et al. Enhancement of invadopodia activity in glioma cells by sublethal doses of irradiation and temozolomide. J Neurosurg 2018;129:598–610.
[301] Tsidulko AY, et al. Conventional anti-glioblastoma chemotherapy affects proteoglycan composition of brain extracellular matrix in rat experimental model in vivo. Front Pharmacol 2018;9:1104.
[302] Westphal M, Maire CL, Lamszus K. EGFR as a target for glioblastoma treatment: an unfulfilled promise. CNS Drugs 2017;31:723–35.
[303] Butowski N, et al. Orally administered colony stimulating factor 1 receptor inhibitor PLX3397 in recurrent glioblastoma: an ivy foundation early phase clinical trials consortium phase II study. Neuro Oncol 2016;18:557–64.
[304] Momeny M, et al. Blockade of vascular endothelial growth factor receptors by tivozanib has potential anti-tumour effects on human glioblastoma cells. Sci Rep 2017;7:44075.
[305] Batchelor TT, et al. Phase II study of cediranib, an oral pan-vascular endothelial growth factor receptor tyrosine kinase inhibitor, in patients with recurrent glioblastoma. J Clin Oncol 2010;28:2817–23.
[306] Batcheor TT, et al. Phase III randomized trial comparing the efficacy of cediranib as monotherapy, and in combination with lomustine, versus lomustine alone in patients with recurrent glioblastoma. J Clin Oncol 2013;31:3212–8.
[307] Kreisl TN, et al. Continuous daily sunitinib for recurrent glioblastoma. J Neurooncol 2013;111:41–8.
[308] Neal J, Wakelee H. AMG-386, a selective angiopoietin-1/-2-neutralizing peptibody for the potential treatment of cancer. Curr Opin Mol Ther 2010;12:487–95.
[309] Vogelbaum MA, Aghi MK. Convection-enhanced delivery for the treatment of glioblastoma. Neuro Oncol 2015;17(Suppl 2). ii3–ii8.
[310] Brown MC, et al. Cancer immunotherapy with recombinant poliovirus induces IFN-dominant activation of dendritic cells and tumor antigen-specific CTLs. Sci Transl Med 2017;9. eaan4220.
[311] Desjardins A, et al. Recurrent glioblastoma treated with recombinant poliovirus. N Engl J Med 2018;379:150–61.
[312] Dai X, Ma C, Lan Q, Xu T. 3D bioprinted glioma stem cells for brain tumor model and applications of drug susceptibility. Biofabrication 2016;8, 045005.
[313] Bregy A, et al. The role of Gliadel wafers in the treatment of high-grade gliomas. Expert Rev Anticancer Ther 2013;13:1453–61.
[314] Treiber JM, et al. Molecular physiology of contrast enhancement in glioblastomas: an analysis of The Cancer Imaging Archive (TCIA). J Clin Neurosci 2018;55:86–92.
[315] Gevaert O, et al. Glioblastoma multiforme: exploratory radiogenomic analysis by using quantitative image features. Radiology 2014;273:168–74.
[316] Chow D, et al. Imaging genetic heterogeneity in glioblastoma and other glial tumors: review of current methods and future directions. Am J Roentgenol 2018;210:30–8.
[317] Lao J, et al. A deep learning-based radiomics model for prediction of survival in glioblastoma multiforme. Sci Rep 2017;7:10353.

CHAPTER

7

Intratumoral heterogeneity associated with glioblastoma drug response and resistance

Sabra K. Salim[a], *Neil Savage*[a], *William T. Maich*[a], *Chitra Venugopal*[b], *and Sheila K. Singh*[a,b]

[a]Department of Biochemistry, McMaster University, Hamilton, ON, Canada [b]Department of Surgery, McMaster University, Hamilton, ON, Canada

Abstract

Glioblastoma (GBM) is the most common primary malignant brain tumor in adults. Despite aggressive multimodal treatment, relapse is inevitable with patients succumbing to disease 15 months after diagnosis. Poor prognosis may be attributed to significant intratumoral heterogeneity at the genomic, cellular, and functional levels of a single tumor. However, this heterogeneity highlights the need to understand mechanisms of therapeutic resistance to give insight into the underpinnings of treatment failure. Thus, we review some of the various tumor-intrinsic molecular mechanisms, as well as tumor-extrinsic factors that support GBM progression and therapeutic resistance.

Abbreviations

ABCG2	ATP-binding cassette superfamily Gmember 2
AKT	protein kinase B
Apaf1	apoptotic protease activating factor 1
ATM	ataxia telangiectasia mutated
ATR	ataxia telangiectasia and Rad3 related
BBB	blood–brain barrier
BCL-2	B-cell lymphoma 2
BCL-XL	B-cell lymphoma-extra Large
BCL2	B-cell lymphoma 2
BER	base excision repair
Chk	checkpoint kinase
CpG	cytosine-phosphate-guanine
CSC	cancer stem cell
EMT	endothelial-to-mesenchymal transition

Glioblastoma Resistance to Chemotherapy: Molecular Mechanisms and Innovative Reversal Strategies
https://doi.org/10.1016/B978-0-12-821567-8.00033-6

GBM glioblastoma
GSC glioma stem cell
IDH isocitrate dehydrogenase
IL-6 interleukin 6
ITH Intratumoral Heterogeneity
Mcl-1 induced myeloid leukemia cell differentiation protein
MDR1 multidrug resistance protein 1
MDSC myeloid derived supressor cells
MGMT O6-methylguanine-DNA methyltransferase
MMR mismatch repair
MRP1 multidrug resistance-associated protein 1
mTOR mammalian target of Rapamycin
NF-κB nuclear factor kappa-light-chain-enhancer of activated B cells
pGBM primary GBM
PI3K phosphoinositide 3-kinases
PMT proneural-to-mesenchymal transition
PTEN phosphatase and tensin homolog
rGBM recurrent GBM
SoC standard of care
STAT3 signal transducer and activator of transcription-3
TAM tumor-associated macrophage
TCGA The Cancer Genome Atlas
TME tumor microenvironment environment
TMZ temozolomide
VEGF vascular endothelial growth factor

Conflict of interest

No potential conflicts of interest were disclosed.

Introduction

Glioblastoma (GBM) is an aggressive primary malignant brain tumor in adults (WHO Grade IV) [1]. With a North American incidence rate of 3.20 per 100,000 persons per year, GBM represents the most diagnosed tumor of neuroepithelial origin [2]. Upon diagnosis, most patients undergo standard of care (SoC) treatment that includes maximally safe surgical resection, chemotherapy, and radiotherapy. Despite SoC, almost all patients relapse 7–9 months after diagnosis, with median overall survival rarely extending beyond 15 months [3, 4]. Over the past decade, little has changed in patient prognosis, underscoring the need to understand mechanisms of treatment failure and disease relapse.

Prior to the current therapeutic regimen for patients with GBM, standard therapy consisted of surgical resection where feasible, and postoperative external beam radiotherapy. In later randomized studies of nitrosourea-containing adjuvant chemotherapies, compounds such as temozolomide (TMZ) showed efficacy in the management of GBM. In 2005, the Stupp protocol, which combined fractionated radiotherapy and adjuvant TMZ along with surgical debulking, became the standard therapeutic regimen which extended median survival by 3 months. Despite this breakthrough, GBM remains an incurable disease [3, 5]. Treatment failure and disease relapse are believed to be attributed to extensive cellular and genetic heterogeneity existing not only between patients, but also within a single tumor through space and time in a phenomenon known as intratumoral heterogeneity (ITH).

Genetic and transcriptomic ITH

At the genetic level, gene expression profiling in the mid-2000s started to elucidate genetic signatures associated with patient survival. When assessing mutational landscape, GBM exhibits aberrant copy number alterations, gene expression, DNA methylation profile, and pathway dysregulation [6–8]. These studies showed GBM represents multiple molecular subtypes, thus providing evidence for ITH. Later in 2010, The Cancer Genome Atlas (TCGA) stratified GBMs into four molecular subgroups at both the genomic and transcriptomic levels, which were termed as classical, mesenchymal, proneural, and neural [9]. Although GBM specimens could be assigned to a dominant subgroup, a later study by Patel et al. showed using single-cell RNA sequencing that a single tumor consisted of a heterogeneous mixture of cells representing all the different GBM subgroups. There was no impact on clinical prognosis except for in the case of the proneural subgroup, where the authors showed that patients with proneural tumors that also displayed markers of other subgroups had poorer survival, especially if the relative representation of the alternative subgroups was high in the tumor [10]. This provided significant evidence for ITH in that a single tumor could consist of cells with different molecular signatures. In 2013, Sottoriva et al. attempted to further understand ITH through a comprehensive genomic analysis. In this study, tumors showed spatial heterogeneity through objective surgical multisampling in which multiple specimens from the same neoplasm displayed different molecular subtypes. Even more so, this study confirmed temporal heterogeneity through molecular clock analysis, indicating tumoral evolution [11]. Recent studies using single-cell RNA-seq also corroborate earlier work on ITH by revealing four GBM cell states that dynamically populate each GBM. The authors also show the ability of cellular GBM subpopulations from a single state to give rise to tumors comprising all four cell states, emphasizing the highly plastic nature of GBM cells [12].

Temporal ITH is also observed over the course of disease progression between newly diagnosed (de novo) or primary GBM (pGBM) and recurrent (rGBM) GBM. Recent studies on posttreatment specimens have shown that rGBM is a genomically and transcriptomically distinct entity compared with its primary counterpart. In a recent study by Wang et al., which comprises the largest longitudinal analysis of both genomic and transcriptomic data from GBM patients through therapy, the authors show that there was a change in subtype in 63% of patients [13]. In addition, they identified mutational landscapes that corresponded exclusively to the pGBM or the rGBM, as well as mutations shared between the two. By determining how both genetic and epigenetic events are clonally selected during GBM progression and constructing phylogenetic and phylo-epigenetic trees of GBM patients at diagnosis and recurrence, these studies documented both linear and branched divergent subclonal evolution, suggesting that targeted monotherapies based on the tumor genome at diagnosis are doomed to fail [14]. A study by Johnson et al. showed that low-grade gliomas and their paired recurrences only shared a few early mutations and were highly divergent [15]. They also found that in 43% of profiled GBM cases, at least half of the mutations in the initial tumor were undetected at recurrence, suggesting that therapy acts as a selection pressure or bottleneck for tumor evolution from minority cell populations present at time of initial diagnosis. Moreover, they also discovered that therapy might drive the emergence of treatment-resistant subclones, as secondary GBMs from low-grade gliomas were found to be hypermutated and bore a TMZ-induced mutagenesis signature [15].

Cellular and functional ITH

ITH has long been known to be associated with poor clinical outcome and a key determinant in therapy failure. While genetic and transcriptomic profiling has deconvoluted and defined ITH in GBM, linking molecular heterogeneity to functional heterogeneity may explain why GBMs respond poorly to therapy.

At the cellular level, treatment resistance can be explained by two frameworks. One such mode, known as the cancer stem cell (CSC) model, may explain and comprise some of the inter- and intratumoral heterogeneity observed in GBM. In this model of cancer, CSCs are tumor-initiating cells that grow and maintain tumors in a hierarchical system. These cells most notably host stem-like properties including self-renewal and multilineage differentiation giving rise to fast-dividing, differentiated cells that comprise the heterogeneous bulk tumor [16, 17].

In human brain tumors, a distinct population of CSCs termed as brain tumor-initiating cells (BTICs) was identified as presenting with increased self-renewal, proliferation and differentiation potential, as well as the ability to recapitulate human tumor biology when injected into immunocompromised mice [17]. The BTIC model is ultimately believed to recapitulate the ITH seen in GBM by giving rise to various subpopulations within a tumor. These cells are most notably marked by the expression of the transmembrane glycoprotein CD133 [17]. In 2006, Bao et al. demonstrated that these CD133-positive BTICs were more resistant to radiotherapy compared to CD133-negative cells by activating DNA damage checkpoint responses [18]. Meanwhile, Liu et al. showed that the CD133-positive cells were more resistant to chemotherapy treatments, such as temozolomide, carboplatin, paclitaxel, and etoposide, as these cells expressed higher levels of ABCG2 and MGMT [19]. Ultimately, these cells are believed to seed tumor recurrence as they were again found in recurrent GBMs that had been exposed to chemoradiotherapy [20].

CD133 expression also correlates with disease outcomes in GBM. While CD133 is one of the most recognizable and most utilized markers, others have been used to isolate glioma CSCs (GSCs) such as CD15, CD44, and A2B5 that also confer treatment resistance [21–23]. Thus, numerous stem-like populations may exist within a single tumor and comprise and contribute to the observed ITH.

The clonal evolution or stochastic model of cancer is another framework that may explain cellular and subsequent functional ITH. First proposed by Peter Nowell in 1976, this theory posits that cancer develops in a "Darwinian" manner, in which various selective pressures differentially affect tumor cells [24]. Pressures such as chemo- and radiotherapy are capable of ablating vulnerable cells, while selecting for treatment-resistant clones with genetically distinct identities. Thus, cells within a tumor can acquire resistance in a therapy-driven event. A study by Meyer et al. demonstrated that genetically distinct clonal populations derived from single cells have variable responses to TMZ as well other drugs, linking genomic heterogeneity to functional heterogeneity [25]. Reinartz et al. later built on these concepts by isolating single-cell–derived GBM subclones with distinct genetic identity and showing that they maintained differential drug resistance profile [26].

Molecular mechanisms of treatment resistance

While ITH is a prognosticator of overall prognosis and treatment failure, understanding the molecular mechanisms of resistance gives direct insight into the underpinnings of treatment failure. ITH in GBM ultimately gives rise to a phenomenon known as polyclonal resistance in which an individual tumor can use multiple mechanisms of drug resistance to evade therapy.

O6-methylguanine-DNA methyltransferase (MGMT) status

TMZ is the most used chemotherapeutic agent to treat GBM and astrocytomas. TMZ is a derivative of dacarbazine, an alkylating chemotherapy given to melanoma and Hodgkin's lymphoma. TMZ is an orally administered prodrug that is readily absorbed in the small intestine, with good penetration of the blood–brain barrier (BBB) and into brain and tumor cells owing to its lipophilic nature [27]. TMZ is stable at a pH less than 5, but at physiologic pH within cells (i.e., greater than 7.0) it rapidly hydrolyzes into various metabolites which methylate primarily guanine nucleobases. Alkylation can occur at the N7 position of guanine but methylating the O6 site of guanine leads to mismatching nucleobases during DNA replication. This causes cells to initiate DNA mismatch repair pathways or undergo cell death. However, various groups have elucidated mechanisms of resistance in GBM to TMZ. On a broader scale, one such mechanism includes heterogeneous patterns of DNA methylation profiles in GBM samples [28]. In particular, the gene encoding O6-methylguanine-DNA methyltransferase (MGMT) has shown differential methylation status and has been linked to TMZ resistance.

MGMT is a highly evolutionarily conserved and ubiquitously expressed enzyme involved in DNA repair [29]. Multiple spatial samplings of the GBM tissue specimen have shown variations in MGMT methylation status in up to 14% of the samples, independent of the transcriptional subgrouping each sample belonged to. Located on chromosome 10, MGMT functions by removing alkyl groups (such as methyl) from the O6 position on guanine nucleobases and during this process, the alkyl group of the O6-methylguanine adduct is transferred to the MGMT protein, causing irreversible inhibition. Because MGMT can only catalyze one reaction per enzyme before it is inhibited, it is referred to as a "suicide enzyme" [30].

As previously mentioned, alkylating DNA is the mode of action for TMZ and, therefore, MGMT may be a valuable prognostic marker for how patients may respond to SoC [3]. The presence of O6-methylguanine adducts cause the mismatch pairings of guanine with thymine to trigger DNA mismatch repair responses, and if a patient's MGMT enzymes are nonfunctional, the cell eventually arrests and may undergo apoptosis. Epigenetic silencing of the MGMT gene has the same effect due to a lack of MGMT enzymes available in the cell, and this commonly happens via methylation of the cytosine-phosphate-guanine (CpG) islands within the MGMT gene promoter. MGMT methylation is observed in approximately 50% of GBMs, and as monosomy of chromosome 10 is a frequent event in GBMs, methylation of the remaining allele eliminates MGMT-mediated DNA repair. MGMT methylation is a favorable prognosis, as median survival was found to be 21.2 months compared to 14.0 months in unmethylated patients [31]. Various studies [31–35] with conflicting reports of how well patients respond to TMZ, radiation or both according to the MGMT status confounds the ability of it being a routine clinical decision maker, and therefore all patients are given the Stupp protocol when appropriate.

Drug efflux pumps

The BBB is reinforced by important drug efflux pumps to protect the brain and tumor from cytotoxic compounds. These include the multidrug resistance protein 1 (MDR1), multidrug resistance-associated protein 1 (MRP1; encoded by *ABCC1*), and ATP-binding cassette superfamily G member 2 (ABCG2). MDR1 has been investigated in GBM, as it protects GSCs and is regulated by GSCs pathways to multiple drug compounds including TMZ [36–38]. A drug compound GNE-317 was designed to overcome this efflux effect and inhibit the PI3K/mTOR kinase inhibitor [39] but this has not been taken to clinical trial.

Like MDR1, MRP1 has been investigated to elucidate its role in resistance to chemotherapy. Inhibition of MRP1 was shown to improve TMZ response in both primary and recurrent GBMs [40], and its ability to protect against other alkylating agents has also been shown [41]. Tacrolimus, an immunosuppressive drug used after allogeneic organ transplants, was shown to impose chemosensitivity on differentiated GBM cells [42] and GSCs [43]; however, this compound has also not been tested in GBM clinical trials.

ABCG2 is the most heavily described drug efflux pump in GBMs and frequently used to identify side populations in tumors [44]. Side populations as a characteristic of stemness have been a point of contention [45–48], but overall, the overexpression of ABCG2 is deleterious to GBM patients [49]. A study examining the relationship between MGMT and drug efflux pumps MDR1 and ABCG2, found no correlation between any of their expression and patient survival time, further clouding the diagnostic purposes for patients and whether a single marker, or even combined markers, could be used in a way to predict patient responses to such a heterogeneous tumor-like GBM [50]. For now, molecular and functional roles of these genes may play a role in drug resistance and contribute to tumor progression, but their role in GBMs will require further elucidation before their selective inhibition will be applied widely in the clinic.

DNA checkpoints

Dysregulation of the DNA damage pathways such as the mismatch repair (MMR) and base excision repair (BER) pathways can also promote therapy-resistance independent of the MGMT promoter methylation status. Tumor suppressor TP53 and its many isoforms are the most famous of tumor suppressor genes as it orchestrates fundamental cellular responses to a variety of stress signals including DNA damage of many cancers [51]. *TP53* mutations are frequently observed in GBM [9], as well as the amplification or overexpression of its negative regulators, MDM2 and MDM4 [52]. Together regulation of the cell cycle and maintenance of the genome is lost. Some mutations of *TP53* cause gain-of-function and increased activity leading to sequestration of the Sp1 transcription factor, thereby impeding MGMT transcription [53]. This process has been shown to sensitize GBM cells to TMZ in vitro [53]. The wild-type status of *TP53* has been reported as fundamental to initiate cell cycle arrest and GBM cells to enter the apoptotic process [54].

DNA damage also activates sensor kinases such as ataxia telangiectasia mutated (ATM) and ataxia telangiectasia and Rad3-related (ATR) [55]. In response to chemotherapy damage, cells lacking MGMT trigger an MMR cycle by phosphorylating ATM/ATR-Chk1/Chk2, leading to G2/M-cell cycle arrest [56]. ATM targets TP53 and Chk2 to regulate the G1/S and intra-S DNA checkpoints, while ATR targets primarily Chk1 to control the intra-S and G2/M DNA checkpoints, with existing functional overlaps [57]. Squatrito et al. used genetically engineered Chk2 −/− mice to exhibit the ATM/Chk2/p53 pathway as a tumor suppressor in the brain in response to radiation damage [58]. The Chk2 −/− mice displayed increased resistance to radiotherapy by failing to initiate apoptosis via DNA damage-induced cell cycle checkpoints [58]. Authors went on to show how crucial DNA damage response elements were constitutively activated in primary GBM biopsies, from patients not previously exposed to chemotherapy or radiation. Later studies went on to show that these are some of the molecular pathways GBM cells undergo in response to TMZ and other chemotherapies [59].

Apoptosis

Apoptosis is a healthy cellular program used to kill cells beyond a reparable threshold, and is essential in maintaining a balance between cell proliferation and death. Both extrinsic and intrinsic apoptotic pathways lead to activation of proteolytic caspases to induce cellular changes including chromatin condensation, DNA fragmentations, membrane blebbing, and cell shrinkage [60, 61]. The intrinsic pathway is activated from inside the cell by severe cell stress, such as DNA damage, by promoting permeabilization of the mitochondrial outer membrane and activation of BH3-only proapoptotic B-cell lymphoma 2 (BCL-2) family proteins. This releases proapoptotic proteins, such as cytochrome *c* and second mitochondria-derived activator of caspase (Smac) from the mitochondria into the cytosol. Cytochrome *c* forms a complex with apoptotic protease-activating factor-1 (Apaf-1) to activate caspase-9. This caspase, in turn, activates the effector caspase-3, -6, and -7, effectively executing apoptosis [60, 61].

GBMs innately deregulate the intrinsic apoptotic pathway, but also activate pro-survival mechanisms in response to cytotoxic compounds and radiotherapy. Research into the possible enhancement of apoptosis of GBM is a primary goal in the development of new and more effective treatments. The PI3K/AKT/mTOR pathway plays a central role in signaling pathways controlling cell survival, proliferation, motility, angiogenesis, and apoptosis and is estimated to be deregulated in 80% of all GBMs [9, 62]. Phosphorylated-AKT overexpresses MDM2, a proto-oncogene, by negatively regulating p53 [63] and prevents the apoptosis-inducing protein Bad from functioning [64]. Phosphatase and tensin homolog (PTEN) is a tumor suppressor protein that directly counteracts the PI3K/AKT/mTOR pathway but is one of the most frequently mutated genes in GBM [9]. Because this pathway is so widely spread among GBM patients, several compounds have been brought to clinical trials to inhibit this pathway [65–67].

Because GSCs are biologically distinct to GBM bulk cells, proteins that are differentially expressed in GSCs can regulate their survival, but by targeting those researchers may selectively induce apoptosis. For example, signal transducer and activator of transcription-3 (STAT3) is a transcription factor whose activation plays a pivotal role in proliferation, apoptosis induction, and differentiation of GSCs. Knockout of STAT3 reduces CD133 expression, neurosphere formation, and tumor formation in vivo [68, 69]. STX-0119 is a compound designed to prevent dimerization of STAT3, reduce BCL-XL and survivin expression, and cause caspase-3 activation ultimately triggering apoptosis in GSCs [70]. A natural compound, cardamonin, was also discovered to suppress STAT3 downstream gene targets including BCL-XL, survivin, BCL-2, Mcl-1, and VEGF [71].

Cell-intrinsic and -extrinsic mechanisms of resistance

Cell state and the proneural-to-mesenchymal transition

A variety of different cell states are seen in GBM, many of which confer resistance to chemotherapy, radiotherapy, and other potential therapeutic modalities. GBM was initially broken down into four specific subtypes—classical, mesenchymal, neural, and proneural—by Verhaak et al. [9], however, more recently, single-cell RNA sequencing by Wang et al. [72] suggested that the neural subtype should be removed, after accounting for IDH-mutant status' favorable prognosis within the proneural subtype. Different studies claim that certain subtypes confer worse prognosis, however, there is currently no definitive conclusion to this debate. When using only samples with low transcriptional heterogeneity, mesenchymal subtypes tend to be indicative of worse prognosis.

However, at the clinical level, the use of subtyping in GBM has been somewhat limited, as GBM cells demonstrate a significant degree of plasticity, and it has been proposed that cancer stem cells themselves may be plastic, resulting in the ability for nonstem-like cancer cells to become cancer stem cells, and vice versa [73, 74]. While this plasticity extends to cancer subtypes, Neftel et al. demonstrated that these subtypes exist along an axis, rather than in binary states [12].

Given that GBM subtypes are plastic, tumor cells can alter their phenotype to enhance survival benefits. In GBM, a hallmark of worsening disease upon tumor recurrence is the proneural-to-mesenchymal transition (PMT). This transition is believed to be a mechanism by which GBM is capable of evading chemotherapeutic agents, and potentially other therapeutic modalities, during which the loss of the proneural gene expression profile is observed, along with an increase in mesenchymal genes. Similar to the epithelial-to-mesenchymal transition (EMT) which plays a role in natural development, as well as tumor cell migration, metastatic invasion, and therapeutic resistance in other cancers, the PMT may mediate these processes in brain cancer as well. Transcription factors that are responsible for the EMT in other cancers have been identified as being active in the PMT in GBM including STAT3, Zeb1 and Zeb2, Twist1 and Twist2, and Snail1–3 [75–78]. While the PMT is thought to be primarily driven by activation of therapy evasion pathways, others have shown that macrophages in the tumor microenvironment of GBM may also play a role in regulating this transition [79]. Typical of the heterogeneity of GBM, in 2016 Segerman et al. demonstrated that induction of the PMT by chemoradiotherapy is partly conferred by and associated with changes in DNA methylation [80].

GSCs can enter quiescent, or noncycling, cell states, in order to evade therapy, as quiescent GSCs are less susceptible to DNA damage by TMZ and ionizing radiation. Sachdeva et al. previously showed that BMP signaling defines GSCs in a quiescent state, and that when induction of BMP signaling is relieved, their tumorigenic and self-renewal capacity is not abrogated, indicating that this quiescent state can be induced as a protective mechanism [81]. Glycerol-3-phosphate dehydrogenase 1 (GPD1) was recently shown to mark dormant GSCs, which actively proliferate during early tumor development, before situating themselves on the leading edge of tumors and entering dormancy, exiting dormancy after chemotherapy in order to contribute to tumor relapse [82]. This indicates that dormant GSCs may express unique markers that may be leveraged therapeutically.

Immune cell and niche interactions

Beyond tumor-intrinsic heterogeneity, GBMs also host a complex immunosuppressive microenvironment that has been shown to contribute to its tumorigenicity. This tumor microenvironment (TME) is a highly heterogeneous and dynamic network composed of neoplastic, as well as resident and infiltrating nonneoplastic cells. At a molecular level, the TME contains hypoxic, vascular, arteriolar, and extracellular matrix niches that are driven by secretion of various proteins and molecules such as hypoxia-inducible factors 1 and 2 and vascular endothelial growth factor [83]. However, the most predominant form of ITH in the TME is derived from its cellular composition. At the cellular level, the TME has been shown to contain various immune cells including T-cells and macrophages, as well as cancer-associated fibroblasts, endothelial cells, astrocytes, microglia, and myeloid-derived suppressor cells (MDSCs) [84]. Despite some antitumor activity by immune cells, recruitment of nonneoplastic cells in the TME has been shown to promote tumor progression by creating an immunosuppressive niche. Both nonneoplastic and tumor cells are capable of secreting immunosuppressive factors such as cytokines and chemokines to suppress antitumor immunity [85].

Tumor-associated macrophages (TAMs) represent the largest fraction of nonneoplastic cells in GBM, comprising as many as 30%–50% of all cells [86]. Many studies have implicated TAMs as mediators of tumor maintenance and progression in GBM [87]. Interestingly, TAMs have also been suggested to directly promote drug resistance of tumor cells in other cancers. In breast cancers, chemoprotective effects of TAMs were observed in a breast cancer model by chemo-induced upregulation of cathepsin proteases to abrogate tumor cell death [88]. In GBM, radioprotective effects of cathepsin-expressing TAMs have been well documented, however, further study into enhancing chemoresistance must be performed [89]. TAMs may be a therapeutically tractable strategy, heterogeneity in phenotype complicates this therapeutic window. TAMs may polarize into a pro-inflammatory M1 or antiinflammatory M2 phenotype [90]. Studies in other cancers such as osteosarcomas have revealed that some TAMs may amplify cytotoxic effects providing further evidence for heterogeneity in TAM phenotype and activity [91]. Thus, TAMs may cumulatively exhibit a combination of tumoricidal and tumor-promoting function, further contributing to the functional ITH observed in GBM.

Astrocytes, a type of stromal cell in the TME, have also been shown to elicit a chemoprotective effect on GBM cells through a contact-dependent manner. In one study, glioma cells cocultured with astrocytes showed resistance to TMZ and vincristine, another chemotherapeutic agent [92]. This effect was primarily mediated through gap junctions in which astrocytes were suggested to send antiapoptotic signals through gap junctions to regulate sensitivity of glioma cells to chemotherapeutic agents [92].

Owing to the extensive vascularity seen in primary GBM, researchers believed that antiangiogenic therapies, such as bevacizumab (anti-VEGF) would effectively abrogate tumor growth and progression. However, GBMs are not only capable of thriving in a hypoxic environment, but also the tumor microenvironment may contribute to limiting the effectiveness of such therapies. Kloepper et al. demonstrated that a bispecific therapy targeting both angiopoietin-2 and VEGF prolonged survival in murine models, and reprogrammed protumor M2 macrophages in the tumor microenvironment to antitumor M1 macrophages, implicating M2 macrophages in antivasculature therapy resistance [93]. In addition, Piao et al. demonstrated that while antiangiogenic therapy initially decreases TAM counts, hypoxia quickly results in re-infiltration of TAMs, and a significant increase in $CD11b^{+}Gr1^{+}$ MDSCs was observed [94].

Endothelial cells, which make up the vasculature within GBMs, greatly contribute to therapy resistance. Owing to the aberrant vasculature seen in GBM, there are inherent complications with delivery of drugs, including chemotherapy. These endothelial cells also secrete factors such as nitric oxide, serpins, and angiopoietins, which induce activation of tumor-promoting pathways such as sonic hedgehog. Fianco et al. demonstrated in 2017 that increases in VEGF, IL-6, and other factors were the result of caspase-8 secreted from the TME activating NF-kB resulting in vascularization and TMZ resistance [95].

Lastly, the extracellular matrix (ECM) in GBM is also capable of conferring therapy resistance by promoting a stiffened matrix, which is able to reduce recruitment of immune cells, and limit the accessibility of compounds such as chemotherapeutics to the tumor site [96].

The tumor immune microenvironment of GBM is nearly as formidable as the solid tumor itself. However, given the dependence of GBM on its microenvironment in promoting resistance to not only chemoradiotherapy but also various other therapeutic modalities as well, it opens up a potential weak spot that requires further research, with the potential to inhibit the TME, and perhaps leverage it against the tumor itself.

Conclusions

Poor prognosis and treatment failure in GBM necessitate the need to understand mechanisms of drug resistance. One such contributor to therapy resistance is the extensive ITH observed at the genomic, transcriptomic, and cellular levels of the tumor. GBMs can employ a vast array of cellular mechanisms in order to evade therapies. This includes the employment of various plastic subtypes, CSCs capable of adorning various cell states, and altering phenotype and acquiring resistance to therapeutics via clonal selection. On a molecular level, this resistance can be acquired through epigenetic mechanisms, dysregulation of DNA damage and apoptotic pathways, and expression of drug efflux pumps. Tumors may adapt and change cell states that provide protection from drug-induced DNA damage. Beyond tumor-intrinsic mechanisms, GBMs can also recruit niche-associated factors to promote therapy evasion and subsequent survival. GBMs may ultimately achieve resistance via multiple mechanisms in a phenomenon known as polyclonal resistance. While the heterogeneity and complexity of GBM may seem overwhelming, advances in both experimental and therapeutic modalities are continually improving the depth at which we can study this disease to ultimately generate new and improved therapeutic modalities.

References

[1] Louis DN, Perry A, Reifenberger G, von Deimling A, Figarella-Branger D, Cavenee WK, Ohgaki H, Wiestler OD, Kleihues P, Ellison DW. The 2016 World Health Organization classification of tumors of the central nervous system: a summary. Acta Neuropathol 2016;131(6):803–20.

[2] Ostrom QT, Gittleman H, Xu J, Kromer C, Wolinsky Y, Kruchko C, Barnholtz-Sloan JS. CBTRUS statistical report: primary brain and other central nervous system tumors diagnosed in the United States in 2009–2013. Neuro Oncol 2016;18(suppl. 5):v1–v75.

[3] Stupp R, Mason WP, van den Bent MJ, Weller M, Fisher B, Taphoorn MJ, Belanger K, Brandes AA, Marosi C, Bogdahn U, Curschmann J, Janzer RC, Ludwin SK, Gorlia T, Allgeier A, Lacombe D, Cairncross JG, Eisenhauer E, Mirimanoff RO. European organisation for R, treatment of Cancer brain T, radiotherapy G, National Cancer Institute of Canada clinical trials G. radiotherapy plus concomitant and adjuvant temozolomide for glioblastoma. N Engl J Med 2005;352(10):987–96.

[4] Walker EV, Ross J, Yuan Y, Smith TR, Davis FG. Brain cancer survival in Canada 1996–2008: effects of sociodemographic characteristics. Curr Oncol 2019;26(3):e292–9.

[5] Bower M, Newlands ES, Bleehen NM, Brada M, Begent RJ, Calvert H, Colquhoun I, Lewis P, Brampton MH. Multicentre CRC phase II trial of temozolomide in recurrent or progressive high-grade glioma. Cancer Chemother Pharmacol 1997;40(6):484–8.

[6] Mischel PS, Nelson SF, Cloughesy TF. Molecular analysis of glioblastoma: pathway profiling and its implications for patient therapy. Cancer Biol Ther 2003;2(3):242–7.

[7] Nutt CL, Mani DR, Betensky RA, Tamayo P, Cairncross JG, Ladd C, Pohl U, Hartmann C, McLaughlin ME, Batchelor TT, Black PM, von Deimling A, Pomeroy SL, Golub TR, Louis DN. Gene expression-based classification of malignant gliomas correlates better with survival than histological classification. Cancer Res 2003; 63(7):1602–7.

[8] Freije WA, Castro-Vargas FE, Fang Z, Horvath S, Cloughesy T, Liau LM, Mischel PS, Nelson SF. Gene expression profiling of gliomas strongly predicts survival. Cancer Res 2004;64(18):6503–10.

[9] Verhaak RG, Hoadley KA, Purdom E, Wang V, Qi Y, Wilkerson MD, Miller CR, Ding L, Golub T, Mesirov JP, Alexe G, Lawrence M, O'Kelly M, Tamayo P, Weir BA, Gabriel S, Winckler W, Gupta S, Jakkula L, Feiler HS, Hodgson JG, James CD, Sarkaria JN, Brennan C, Kahn A, Spellman PT, Wilson RK, Speed TP, Gray JW, Meyerson M, Getz G, Perou CM, Hayes DN, Cancer Genome Atlas Research N. Integrated genomic analysis identifies clinically relevant subtypes of glioblastoma characterized by abnormalities in PDGFRA, IDH1, EGFR, and NF1. Cancer Cell 2010;17(1):98–110.

[10] Patel AP, Tirosh I, Trombetta JJ, Shalek AK, Gillespie SM, Wakimoto H, Cahill DP, Nahed BV, Curry WT, Martuza RL, Louis DN, Rozenblatt-Rosen O, Suva ML, Regev A, Bernstein BE. Single-cell RNA-seq highlights intratumoral heterogeneity in primary glioblastoma. Science 2014;344(6190):1396–401.

[11] Sottoriva A, Spiteri I, Piccirillo SG, Touloumis A, Collins VP, Marioni JC, Curtis C, Watts C, Tavare S. Intratumor heterogeneity in human glioblastoma reflects cancer evolutionary dynamics. Proc Natl Acad Sci U S A 2013; 110(10):4009–14.

[12] Neftel C, Laffy J, Filbin MG, Hara T, Shore ME, Rahme GJ, Richman AR, Silverbush D, Shaw ML, Hebert CM, Dewitt J, Gritsch S, Perez EM, Gonzalez Castro LN, Lan X, Druck N, Rodman C, Dionne D, Kaplan A, Bertalan MS, Small J, Pelton K, Becker S, Bonal D, Nguyen QD, Servis RL, Fung JM, Mylvaganam R, Mayr L, Gojo J, Haberler C, Geyeregger R, Czech T, Slavc I, Nahed BV, Curry WT, Carter BS, Wakimoto H, Brastianos PK, Batchelor TT, Stemmer-Rachamimov A, Martinez-Lage M, Frosch MP, Stamenkovic I, Riggi N, Rheinbay E, Monje M, Rozenblatt-Rosen O, Cahill DP, Patel AP, Hunter T, Verma IM, Ligon KL, Louis DN, Regev A, Bernstein BE, Tirosh I, Suva ML. An integrative model of cellular states, plasticity, and genetics for glioblastoma. Cell 2019;178(4):835–49 e21.

[13] Wang J, Cazzato E, Ladewig E, Frattini V, Rosenbloom DI, Zairis S, Abate F, Liu Z, Elliott O, Shin YJ, Lee JK, Lee IH, Park WY, Eoli M, Blumberg AJ, Lasorella A, Nam DH, Finocchiaro G, Iavarone A, Rabadan R. Clonal evolution of glioblastoma under therapy. Nat Genet 2016;48(7):768–76.

[14] Ramaswamy V, Taylor MD. The amazing and deadly Glioma race. Cancer Cell 2015;28(3):275–7.

[15] Johnson BE, Mazor T, Hong C, Barnes M, Aihara K, McLean CY, Fouse SD, Yamamoto S, Ueda H, Tatsuno K, Asthana S, Jalbert LE, Nelson SJ, Bollen AW, Gustafson WC, Charron E, Weiss WA, Smirnov IV, Song JS, Olshen AB, Cha S, Zhao Y, Moore RA, Mungall AJ, Jones SJM, Hirst M, Marra MA, Saito N, Aburatani H, Mukasa A,

Berger MS, Chang SM, Taylor BS, Costello JF. Mutational analysis reveals the origin and therapy-driven evolution of recurrent glioma. Science 2014;343(6167):189–93.
[16] Bonnet D, Dick JE. Human acute myeloid leukemia is organized as a hierarchy that originates from a primitive hematopoietic cell. Nat Med 1997;3(7):730–7.
[17] Singh SK, Hawkins C, Clarke ID, Squire JA, Bayani J, Hide T, Henkelman RM, Cusimano MD, Dirks PB. Identification of human brain tumour initiating cells. Nature 2004;432(7015):396–401.
[18] Bao S, Wu Q, McLendon RE, Hao Y, Shi Q, Hjelmeland AB, Dewhirst MW, Bigner DD, Rich JN. Glioma stem cells promote radioresistance by preferential activation of the DNA damage response. Nature 2006;444(7120):756–60.
[19] Liu G, Yuan X, Zeng Z, Tunici P, Ng H, Abdulkadir IR, Lu L, Irvin D, Black KL, Yu JS. Analysis of gene expression and chemoresistance of CD133+ cancer stem cells in glioblastoma. Mol Cancer 2006;5:67.
[20] Liu Q, Nguyen DH, Dong Q, Shitaku P, Chung K, Liu OY, Tso JL, Liu JY, Konkankit V, Cloughesy TF, Mischel PS, Lane TF, Liau LM, Nelson SF, Tso CL. Molecular properties of CD133+ glioblastoma stem cells derived from treatment-refractory recurrent brain tumors. J Neurooncol 2009;94(1):1–19.
[21] Son MJ, Woolard K, Nam DH, Lee J, Fine HA. SSEA-1 is an enrichment marker for tumor-initiating cells in human glioblastoma. Cell Stem Cell 2009;4(5):440–52.
[22] Ogden AT, Waziri AE, Lochhead RA, Fusco D, Lopez K, Ellis JA, Kang J, Assanah M, McKhann GM, Sisti MB, McCormick PC, Canoll P, Bruce JN. Identification of A2B5+CD133- tumor-initiating cells in adult human gliomas. Neurosurgery 2008;62(2):505–14 [discussion 14-5].
[23] Anido J, Saez-Borderias A, Gonzalez-Junca A, Rodon L, Folch G, Carmona MA, Prieto-Sanchez RM, Barba I, Martinez-Saez E, Prudkin L, Cuartas I, Raventos C, Martinez-Ricarte F, Poca MA, Garcia-Dorado D, Lahn MM, Yingling JM, Rodon J, Sahuquillo J, Baselga J, Seoane J. TGF-beta receptor inhibitors target the CD44 (high)/Id1(high) Glioma-initiating cell population in human Glioblastoma. Cancer Cell 2010;18(6):655–68.
[24] Nowell PC. The clonal evolution of tumor cell populations. Science 1976;194(4260):23–8.
[25] Meyer M, Reimand J, Lan X, Head R, Zhu X, Kushida M, Bayani J, Pressey JC, Lionel AC, Clarke ID, Cusimano M, Squire JA, Scherer SW, Bernstein M, Woodin MA, Bader GD, Dirks PB. Single cell-derived clonal analysis of human glioblastoma links functional and genomic heterogeneity. Proc Natl Acad Sci U S A 2015;112(3):851–6.
[26] Reinartz R, Wang S, Kebir S, Silver DJ, Wieland A, Zheng T, Kupper M, Rauschenbach L, Fimmers R, Shepherd TM, Trageser D, Till A, Schafer N, Glas M, Hillmer AM, Cichon S, Smith AA, Pietsch T, Liu Y, Reynolds BA, Yachnis A, Pincus DW, Simon M, Brustle O, Steindler DA, Scheffler B. Functional subclone profiling for prediction of treatment-induced Intratumor population shifts and discovery of rational drug combinations in human Glioblastoma. Clin Cancer Res 2017;23(2):562–74.
[27] Wesolowski JR, Rajdev P, Mukherji SK. Temozolomide (Temodar). AJNR Am J Neuroradiol 2010;31(8):1383–4.
[28] Klughammer J, Kiesel B, Roetzer T, Fortelny N, Nemc A, Nenning KH, Furtner J, Sheffield NC, Datlinger P, Peter N, Nowosielski M, Augustin M, Mischkulnig M, Strobel T, Alpar D, Erguner B, Senekowitsch M, Moser P, Freyschlag CF, Kerschbaumer J, Thome C, Grams AE, Stockhammer G, Kitzwoegerer M, Oberndorfer S, Marhold F, Weis S, Trenkler J, Buchroithner J, Pichler J, Haybaeck J, Krassnig S, Mahdy Ali K, von Campe G, Payer F, Sherif C, Preiser J, Hauser T, Winkler PA, Kleindienst W, Wurtz F, Brandner-Kokalj T, Stultschnig M, Schweiger S, Dieckmann K, Preusser M, Langs G, Baumann B, Knosp E, Widhalm G, Marosi C, Hainfellner JA, Woehrer A, Bock C. The DNA methylation landscape of glioblastoma disease progression shows extensive heterogeneity in time and space. Nat Med 2018;24(10):1611–24.
[29] Wick W, Platten M. Understanding and targeting alkylator resistance in glioblastoma. Cancer Discov 2014;4(10):1120–2.
[30] Sabharwal A, Middleton MR. Exploiting the role of O6-methylguanine-DNA-methyltransferase (MGMT) in cancer therapy. Curr Opin Pharmacol 2006;6(4):355–63.
[31] Gilbert MR, Wang M, Aldape KD, Stupp R, Hegi ME, Jaeckle KA, Armstrong TS, Wefel JS, Won M, Blumenthal DT, Mahajan A, Schultz CJ, Erridge S, Baumert B, Hopkins KI, Tzuk-Shina T, Brown PD, Chakravarti A, Curran Jr WJ, Mehta MP. Dose-dense temozolomide for newly diagnosed glioblastoma: a randomized phase III clinical trial. J Clin Oncol 2013;31(32):4085–91.
[32] Hegi ME, Diserens AC, Gorlia T, Hamou MF, de Tribolet N, Weller M, Kros JM, Hainfellner JA, Mason W, Mariani L, Bromberg JE, Hau P, Mirimanoff RO, Cairncross JG, Janzer RC, Stupp R. MGMT gene silencing and benefit from temozolomide in glioblastoma. N Engl J Med 2005;352(10):997–1003.
[33] Stupp R, Hegi ME, Mason WP, van den Bent MJ, Taphoorn MJ, Janzer RC, Ludwin SK, Allgeier A, Fisher B, Belanger K, Hau P, Brandes AA, Gijtenbeek J, Marosi C, Vecht CJ, Mokhtari K, Wesseling P, Villa S, Eisenhauer E, Gorlia T, Weller M, Lacombe D, Cairncross JG, Mirimanoff RO, European Organisation for Research

and Treatment of Cancer Brain Tumour and Radiation Oncology Groups, National Cancer Institute of Canada Clinical Trials Group. Effects of radiotherapy with concomitant and adjuvant temozolomide versus radiotherapy alone on survival in glioblastoma in a randomised phase III study: 5-year analysis of the EORTC-NCIC trial. Lancet Oncol 2009;10(5):459–66.

[34] Wick W, Platten M, Meisner C, Felsberg J, Tabatabai G, Simon M, Nikkhah G, Papsdorf K, Steinbach JP, Sabel M, Combs SE, Vesper J, Braun C, Meixensberger J, Ketter R, Mayer-Steinacker R, Reifenberger G, Weller M, Society NOASGoN-oWGoGC. Temozolomide chemotherapy alone versus radiotherapy alone for malignant astrocytoma in the elderly: the NOA-08 randomised, phase 3 trial. Lancet Oncol 2012;13(7):707–15.

[35] Malmstrom A, Gronberg BH, Marosi C, Stupp R, Frappaz D, Schultz H, Abacioglu U, Tavelin B, Lhermitte B, Hegi ME, Rosell J, Henriksson R. Nordic clinical brain tumour study G. Temozolomide versus standard 6-week radiotherapy versus hypofractionated radiotherapy in patients older than 60 years with glioblastoma: the Nordic randomised, phase 3 trial. Lancet Oncol 2012;13(9):916–26.

[36] Nakai E, Park K, Yawata T, Chihara T, Kumazawa A, Nakabayashi H, Shimizu K. Enhanced MDR1 expression and chemoresistance of cancer stem cells derived from glioblastoma. Cancer Invest 2009;27(9):901–8.

[37] Munoz JL, Rodriguez-Cruz V, Greco SJ, Nagula V, Scotto KW, Rameshwar P. Temozolomide induces the production of epidermal growth factor to regulate MDR1 expression in glioblastoma cells. Mol Cancer Ther 2014; 13(10):2399–411.

[38] Xi G, Hayes E, Lewis R, Ichi S, Mania-Farnell B, Shim K, Takao T, Allender E, Mayanil CS, Tomita T. CD133 and DNA-PK regulate MDR1 via the PI3K- or Akt-NF-kappaB pathway in multidrug-resistant glioblastoma cells in vitro. Oncogene 2016;35(42):5576.

[39] Becker CM, Oberoi RK, McFarren SJ, Muldoon DM, Pafundi DH, Pokorny JL, Brinkmann DH, Ohlfest JR, Sarkaria JN, Largaespada DA, Elmquist WF. Decreased affinity for efflux transporters increases brain penetrance and molecular targeting of a PI3K/mTOR inhibitor in a mouse model of glioblastoma. Neuro Oncol 2015; 17(9):1210–9.

[40] Tivnan A, Zakaria Z, O'Leary C, Kogel D, Pokorny JL, Sarkaria JN, Prehn JH. Inhibition of multidrug resistance protein 1 (MRP1) improves chemotherapy drug response in primary and recurrent glioblastoma multiforme. Front Neurosci 2015;9:218.

[41] Benyahia B, Huguet S, Decleves X, Mokhtari K, Criniere E, Bernaudin JF, Scherrmann JM, Delattre JY. Multidrug resistance-associated protein MRP1 expression in human gliomas: chemosensitization to vincristine and etoposide by indomethacin in human glioma cell lines overexpressing MRP1. J Neurooncol 2004;66(1–2):65–70.

[42] Garrido W, Munoz M, San Martin R, Quezada C. FK506 confers chemosensitivity to anticancer drugs in glioblastoma multiforme cells by decreasing the expression of the multiple resistance-associated protein-1. Biochem Biophys Res Commun 2011;411(1):62–8.

[43] Torres A, Arriagada V, Erices JI, Toro MLA, Rocha JD, Niechi I, Carrasco C, Oyarzun C, Quezada C. FK506 Attenuates the MRP1-Mediated Chemoresistant Phenotype in Glioblastoma Stem-Like Cells. Int J Mol Sci 2018; 19(9):2697.

[44] Weber K, Paulus W, Senner V. The side population of gliomas exhibits decreased cell migration. J Neuropathol Exp Neurol 2010;69(6):623–31.

[45] Bleau AM, Hambardzumyan D, Ozawa T, Fomchenko EI, Huse JT, Brennan CW, Holland EC. PTEN/PI3K/Akt pathway regulates the side population phenotype and ABCG2 activity in glioma tumor stem-like cells. Cell Stem Cell 2009;4(3):226–35.

[46] Wee B, Pietras A, Ozawa T, Bazzoli E, Podlaha O, Antczak C, Westermark B, Nelander S, Uhrbom L, Forsberg-Nilsson K, Djaballah H, Michor F, Holland EC. ABCG2 regulates self-renewal and stem cell marker expression but not tumorigenicity or radiation resistance of glioma cells. Sci Rep 2016;6:25956.

[47] Golebiewska A, Bougnaud S, Stieber D, Brons NH, Vallar L, Hertel F, Klink B, Schrock E, Bjerkvig R, Niclou SP. Side population in human glioblastoma is non-tumorigenic and characterizes brain endothelial cells. Brain 2013;136(Pt 5):1462–75.

[48] Broadley KW, Hunn MK, Farrand KJ, Price KM, Grasso C, Miller RJ, Hermans IF, McConnell MJ. Side population is not necessary or sufficient for a cancer stem cell phenotype in glioblastoma multiforme. Stem Cells 2011; 29(3):452–61.

[49] Emery IF, Gopalan A, Wood S, Chow KH, Battelli C, George J, et al. Expression and function of ABCG2 and XIAP in glioblastomas. J Neurooncol 2017;133(1):47–57.

[50] Oberstadt MC, Bien-Moller S, Weitmann K, Herzog S, Hentschel K, Rimmbach C, et al. Epigenetic modulation of the drug resistance genes MGMT, ABCB1 and ABCG2 in glioblastoma multiforme. BMC Cancer 2013;13:617.

[51] Bieging KT, Mello SS, Attardi LD. Unravelling mechanisms of p53-mediated tumour suppression. Nat Rev Cancer 2014;14(5):359–70.
[52] Rao SK, Edwards J, Joshi AD, Siu IM, Riggins GJ. A survey of glioblastoma genomic amplifications and deletions. J Neurooncol 2010;96(2):169–79.
[53] Bocangel D, Sengupta S, Mitra S, Bhakat KK. p53-mediated down-regulation of the human DNA repair gene O6-methylguanine-DNA methyltransferase (MGMT) via interaction with Sp1 transcription factor. Anticancer Res 2009;29(10):3741–50.
[54] Hermisson M, Klumpp A, Wick W, Wischhusen J, Nagel G, Roos W, et al. O6-methylguanine DNA methyltransferase and p53 status predict temozolomide sensitivity in human malignant glioma cells. J Neurochem 2006;96(3):766–76.
[55] Matsuoka S, Rotman G, Ogawa A, Shiloh Y, Tamai K, Elledge SJ. Ataxia telangiectasia-mutated phosphorylates Chk2 in vivo and in vitro. Proc Natl Acad Sci U S A 2000;97(19):10389–94.
[56] Caporali S, Falcinelli S, Starace G, Russo MT, Bonmassar E, Jiricny J, et al. DNA damage induced by temozolomide signals to both ATM and ATR: role of the mismatch repair system. Mol Pharmacol 2004;66(3):478–91.
[57] Weber AM, Ryan AJ. ATM and ATR as therapeutic targets in cancer. Pharmacol Ther 2015;149:124–38.
[58] Squatrito M, Brennan CW, Helmy K, Huse JT, Petrini JH, Holland EC. Loss of ATM/Chk2/p53 pathway components accelerates tumor development and contributes to radiation resistance in gliomas. Cancer Cell 2010;18(6):619–29.
[59] Annovazzi L, Caldera V, Mellai M, Riganti C, Battaglia L, Chirio D, et al. The DNA damage/repair cascade in glioblastoma cell lines after chemotherapeutic agent treatment. Int J Oncol 2015;46(6):2299–308.
[60] García MA, Ramírez A, López-Ruiz E, et al. Apoptosis as a therapeutic target in cancer and cancer stem cells: Novel strategies and futures perspectives. Citeseer; 2012.
[61] Ghobrial IM, Witzig TE, Adjei AA. Targeting apoptosis pathways in cancer therapy. CA Cancer J Clin 2005;55(3):178–94.
[62] Mao H, Lebrun DG, Yang J, Zhu VF, Li M. Deregulated signaling pathways in glioblastoma multiforme: molecular mechanisms and therapeutic targets. Cancer Invest 2012;30(1):48–56.
[63] Daniele S, Costa B, Zappelli E, Da Pozzo E, Sestito S, Nesi G, et al. Combined inhibition of AKT/mTOR and MDM2 enhances Glioblastoma Multiforme cell apoptosis and differentiation of cancer stem cells. Sci Rep 2015;5:9956.
[64] Datta SR, Dudek H, Tao X, Masters S, Fu H, Gotoh Y, et al. Akt phosphorylation of BAD couples survival signals to the cell-intrinsic death machinery. Cell 1997;91(2):231–41.
[65] Wang WJ, Long LM, Yang N, Zhang QQ, Ji WJ, Zhao JH, et al. NVP-BEZ235, a novel dual PI3K/mTOR inhibitor, enhances the radiosensitivity of human glioma stem cells in vitro. Acta Pharmacol Sin 2013;34(5):681–90.
[66] Galanis E, Buckner JC, Maurer MJ, Kreisberg JI, Ballman K, Boni J, et al. Phase II trial of temsirolimus (CCI-779) in recurrent glioblastoma multiforme: a north central Cancer treatment group study. J Clin Oncol 2005;23(23):5294–304.
[67] Kubicek GJ, Werner-Wasik M, Machtay M, Mallon G, Myers T, Ramirez M, et al. Phase I trial using proteasome inhibitor bortezomib and concurrent temozolomide and radiotherapy for central nervous system malignancies. Int J Radiat Oncol Biol Phys 2009;74(2):433–9.
[68] Sherry MM, Reeves A, Wu JK, Cochran BH. STAT3 is required for proliferation and maintenance of multipotency in glioblastoma stem cells. Stem Cells 2009;27(10):2383–92.
[69] Li GH, Wei H, Lv SQ, Ji H, Wang DL. Knockdown of STAT3 expression by RNAi suppresses growth and induces apoptosis and differentiation in glioblastoma stem cells. Int J Oncol 2010;37(1):103–10.
[70] Ashizawa T, Miyata H, Iizuka A, Komiyama M, Oshita C, Kume A, et al. Effect of the STAT3 inhibitor STX-0119 on the proliferation of cancer stem-like cells derived from recurrent glioblastoma. Int J Oncol 2013;43(1):219–27.
[71] Wu N, Liu J, Zhao X, Yan Z, Jiang B, Wang L, et al. Cardamonin induces apoptosis by suppressing STAT3 signaling pathway in glioblastoma stem cells. Tumour Biol 2015;36(12):9667–76.
[72] Wang Q, Hu B, Hu X, Kim H, Squatrito M, Scarpace L, et al. Tumor evolution of Glioma-intrinsic gene expression subtypes associates with immunological changes in the microenvironment. Cancer Cell 2017;32(1):42–56 e6.
[73] Dirkse A, Golebiewska A, Buder T, Nazarov PV, Muller A, Poovathingal S, et al. Stem cell-associated heterogeneity in Glioblastoma results from intrinsic tumor plasticity shaped by the microenvironment. Nat Commun 2019;10(1):1787.

[74] Chaffer CL, Marjanovic ND, Lee T, Bell G, Kleer CG, Reinhardt F, et al. Poised chromatin at the ZEB1 promoter enables breast cancer cell plasticity and enhances tumorigenicity. Cell 2013;154(1):61–74.

[75] Lau J, Ilkhanizadeh S, Wang S, Miroshnikova YA, Salvatierra NA, Wong RA, et al. STAT3 blockade inhibits radiation-induced malignant progression in Glioma. Cancer Res 2015;75(20):4302–11.

[76] Mikheeva SA, Mikheev AM, Petit A, Beyer R, Oxford RG, Khorasani L, et al. TWIST1 promotes invasion through mesenchymal change in human glioblastoma. Mol Cancer 2010;9:194.

[77] Siebzehnrubl FA, Silver DJ, Tugertimur B, Deleyrolle LP, Siebzehnrubl D, Sarkisian MR, et al. The ZEB1 pathway links glioblastoma initiation, invasion and chemoresistance. EMBO Mol Med 2013;5(8):1196–212.

[78] Savary K, Caglayan D, Caja L, Tzavlaki K, Bin Nayeem S, Bergstrom T, et al. Snail depletes the tumorigenic potential of glioblastoma. Oncogene 2013;32(47):5409–20.

[79] Bhat KPL, Balasubramaniyan V, Vaillant B, Ezhilarasan R, Hummelink K, Hollingsworth F, et al. Mesenchymal differentiation mediated by NF-kappaB promotes radiation resistance in glioblastoma. Cancer Cell 2013; 24(3):331–46.

[80] Segerman A, Niklasson M, Haglund C, Bergstrom T, Jarvius M, Xie Y, et al. Clonal variation in drug and radiation response among Glioma-initiating cells is linked to proneural-Mesenchymal transition. Cell Rep 2016;17(11):2994–3009.

[81] Sachdeva R, Wu M, Johnson K, Kim H, Celebre A, Shahzad U, et al. BMP signaling mediates glioma stem cell quiescence and confers treatment resistance in glioblastoma. Sci Rep 2019;9(1):14569.

[82] Rusu P, Shao C, Neuerburg A, Acikgoz AA, Wu Y, Zou P, et al. GPD1 specifically Marks dormant Glioma stem cells with a distinct metabolic profile. Cell Stem Cell 2019;25(2):241–57 [e8].

[83] Aderetti DA, Hira VVV, Molenaar RJ, van Noorden CJF. The hypoxic peri-arteriolar glioma stem cell niche, an integrated concept of five types of niches in human glioblastoma. Biochim Biophys Acta Rev Cancer 2018; 1869(2):346–54.

[84] Won WJ, Deshane JS, Leavenworth JW, Oliva CR, Griguer CE. Metabolic and functional reprogramming of myeloid-derived suppressor cells and their therapeutic control in glioblastoma. Cell Stress 2019;3(2):47–65.

[85] Schiffer D, Annovazzi L, Casalone C, Corona C, Mellai M. Glioblastoma: microenvironment and niche concept. Cancers (Basel) 2018;11(1):5.

[86] Charles NA, Holland EC, Gilbertson R, Glass R, Kettenmann H. The brain tumor microenvironment. Glia 2012;60(3):502–14.

[87] Hambardzumyan D, Gutmann DH, Kettenmann H. The role of microglia and macrophages in glioma maintenance and progression. Nat Neurosci 2016;19(1):20–7.

[88] Shree T, Olson OC, Elie BT, Kester JC, Garfall AL, Simpson K, et al. Macrophages and cathepsin proteases blunt chemotherapeutic response in breast cancer. Genes Dev 2011;25(23):2465–79.

[89] Malla RR, Gopinath S, Alapati K, Gorantla B, Gondi CS, Rao JS. uPAR and cathepsin B inhibition enhanced radiation-induced apoptosis in gliomainitiating cells. Neuro Oncol 2012;14(6):745–60.

[90] Jayasingam SD, Citartan M, Thang TH, Mat Zin AA, Ang KC, Ch'ng ES. Evaluating the polarization of tumor-associated macrophages into M1 and M2 phenotypes in human Cancer tissue: technicalities and challenges in routine clinical practice. Front Oncol 2019;9:1512.

[91] Nardin A, Lefebvre ML, Labroquere K, Faure O, Abastado JP. Liposomal muramyl tripeptide phosphatidylethanolamine: targeting and activating macrophages for adjuvant treatment of osteosarcoma. Curr Cancer Drug Targets 2006;6(2):123–33.

[92] Chen W, Wang D, Du X, He Y, Chen S, Shao Q, et al. Glioma cells escaped from cytotoxicity of temozolomide and vincristine by communicating with human astrocytes. Med Oncol 2015;32(3):43.

[93] Kloepper J, Riedemann L, Amoozgar Z, Seano G, Susek K, Yu V, et al. Ang-2/VEGF bispecific antibody reprograms macrophages and resident microglia to anti-tumor phenotype and prolongs glioblastoma survival. Proc Natl Acad Sci U S A 2016;113(16):4476–81.

[94] Piao Y, Park SY, Henry V, Smith BD, Tiao N, Flynn DL, et al. Novel MET/TIE2/VEGFR2 inhibitor altiratinib inhibits tumor growth and invasiveness in bevacizumab-resistant glioblastoma mouse models. Neuro Oncol 2016;18(9):1230–41.

[95] Seano G. Targeting the perivascular niche in brain tumors. Curr Opin Oncol 2018;30(1):54–60.

[96] Zhang X, Ding K, Wang J, Li X, Zhao P. Chemoresistance caused by the microenvironment of glioblastoma and the corresponding solutions. Biomed Pharmacother 2019;109:39–46.

CHAPTER

8

Role of apoptosis, autophagy, and the unfolded protein response in glioblastoma chemoresistance

Sanaz Dastghaib[a,*], *Sima Hajiahmadi*[b,*], *Amir Seyfoori*[c,d,*], *Meitham Amereh*[c], *Mozhdeh Zamani*[e], *Zahra Shahsavari*[b], *Shahla Shojaei*[f], *Mohsen Akbari*[c,g,h,†], *Pooneh Mokarram*[e,i,†], *and Saeid Ghavami*[e,f,j,k,l,†]

[a]Endocrine and Metabolism Research Center, Shiraz University of Medical Sciences, Shiraz, Iran [b]Department of Clinical Biochemistry, Faculty of Medicine, Shahid Beheshti University of Medical Sciences, Tehran, Iran [c]Laboratory for Innovations in Microengineering (LiME), Department of Mechanical Engineering, University of Victoria, Victoria, BC, Canada [d]Biomaterials and Tissue Engineering Department, Breast Cancer Research Center, Motamed Cancer Institute, ACECR, Tehran, Iran [e]Autophagy Research Center, Shiraz University of Medical Sciences, Shiraz, Iran [f]Department of Human Anatomy and Cell Science, Max Rady College of Medicine, Rady Faculty of Health Sciences, University of Manitoba, Winnipeg, MB, Canada [g]Centre for Biomedical Research (CBR), University of Victoria, Victoria, BC, Canada [h]Center for Advanced Materials and Related Technologies, University of Victoria, Victoria, BC, Canada [i]Department of Biochemistry, Shiraz University of Medical sciences, Shiraz, Iran [j]Research Institute in Oncology and Hematology, CancerCare Manitoba, University of Manitoba, Winnipeg, MB, Canada [k]Faculty of Medicine, Katowice School of Technology, Katowice, Poland [l]Children's Hospital Research Institute of Manitoba, Winnipeg, MB, Canada

*These authors contributed equally to this work.

†These authors have equal senior authorship.

Glioblastoma Resistance to Chemotherapy: Molecular Mechanisms and Innovative Reversal Strategies
https://doi.org/10.1016/B978-0-12-821567-8.00016-6

Abstract

Glioblastoma (GBM) is a rapidly progressive form of brain tumor with high mortality rates. Current treatment modalities are moderately effective and chemoresistance has become a major concern in GBM treatment. Evidence suggests that cross talks between several pathways, including apoptosis, autophagy, and unfolded protein response (UPR), are involved in the induction of resistance to chemotherapeutic agents such as temozolomide (TMZ). Apoptosis is crucial for organism survival through the elimination of unwanted cells. Various conditions such as hypoxia, defective DNA mismatch repair (MMR) system, p53 loss of function, insufficient O6-methylguanine-DNA-methyltransferase (MGMT) enzyme, and dysregulation of miRNAs expression negatively influence apoptosis and lead to chemoresistance. A dual role has been reported for autophagy in GBM chemoresistance and chemosensitivity. Similarly, the UPR adaptive pathway has both cytotoxic and cytoprotective effects on GBM cells in response to chemotherapeutic agents. Considering the complexity of this malignancy, the development of comprehensive in vitro models is required to predict the effects of different therapeutic strategies by recapitulating the intricate in vivo GBM microenvironment. Three-dimensional (3D) tissue engineering using different biomaterials is a promising approach to evaluate the effectiveness of chemotherapeutics by developing bioengineered GBM models. The functional contribution of signaling pathways such as autophagy, apoptosis, and UPR can be investigated in detail through 3D tissue engineering. In this chapter, we briefly review the role of autophagy, apoptosis, and unfolded protein response in chemoresistance in GBM and later discuss the application of a 3D bioengineer model of GBM and its application in the evaluation of chemotherapy response and chemoresistance in GBM.

Abbreviations

ACD	autophagy cell death
ALDH1A3	aldehyde dehydrogenase 1 family member A3
AMPK	AMP-activated protein kinase
ASK1	apoptosis signal-regulated kinase 1
ATF4	activated transcription factors
ATF6α	activated transcription factor 6
ATG	autophagy-related gene
ATM	ataxia telangiectasia mutated
Bad	Bcl-2-associated death promotor
BafA1	bafilomycin A1
Bak	Bcl-2 homologous antagonist killer
Bax	Bcl-2-associated X protein
BBB	blood-brain barrier
Bcl-2	B-cell lymphoma 2
BCL-XL	BCL extra-large
BFP	2,4-Bis(4-fluoro-o-phenylacetyl) phloroglucinol
BID	BH3 interacting domain death agonist
Bim	Bcl-2-like protein 11
CA	carnosic acid
CASC2	cancer susceptibility 2
Cav-1	caveolin-1
CHOP	C/EBP homologous protein
cIAP	cellular inhibitor of apoptosis protein
CMA	chaperon-mediated autophagy
CNS	central nervous system
CPT-11	camptothecin-11
CQ	chloroquine
DCC	deleted-in-colorectal-carcinoma
DISC	death inducing signal complex
EphA3	ephrin receptor A

DRAM	damage regulated autophagy modulator
ECM	extracellular matrix
EGFR	epidermal growth factor receptor
eIF2α	eukaryotic initiation factor 2
EMAP-II	endothelial monocyte-activating polypeptide-II
ER	endoplasmic reticulum
ERAD	endoplasmic reticulum-associated degradation
ERK	extracellular signal-regulated kinase
ERO1α	endoplasmic reticulum redox protein 1
ETC	electron transport chain; 2-DG, 2-deoxy-D-glucose
ETM	epithelial-to-mesenchymal transition
FADD	Fas-associated death domain
FAM134B	family with sequence similarity 134 member B
Fas	cell surface death receptor
Fas-L	Fas ligand
5-FU	5-fluorouracil
GABARAP-L1	gamma-aminobutyric acid receptor-associated protein-like 2
GADD34	DNA damage-inducible protein
GBM	glioblastoma multiform
GRP	glucose-regulate protein
GSC	glioblastoma stem cell
GSCs	glioblastoma stem cells
HCQ	hydroxychloroquine
HIF-1 α	hypoxia-inducible factor-1α
HSC70	heat shock cognate 70
IAPs	IAP-like protein2
IDH	isocitrate dehydrogenase
ILP-2	inhibitor of apoptotic proteins
IRE-1α	inositol requires enzyme 1
JNK	c-Jun NH2-terminal kinase
L3Cβ I	microtubule-associated protein 1 light chain
LAMP-2a	lysosomal-associated membrane protein 2
LIR	light-chain 3-interacting regions
LOH	loss of heterozygosity
lncRNA	long noncoding RNA
MAGE-E1	MAGE family member E1
MCL-1	myeloid cell leukemia-1
3-MA	3-methyladenine
Mdm-2	mouse double minute 2 homolog
MET	methoprene-tolerant
MGMT	O-6-methylguanine-DNA methyltransferase
MHC-I	major histocompatibility complexes I
miR	microRNA
MMR	DNA mismatch repair
mTOR	mammalian target of rapamycin
NRP/B	nuclear-restricted protein/brain
NAIP	neuronal apoptosis inhibitory protein
NDP52	nuclear dot protein 52
NF1	neurofibromin 1
NF-κB	nuclear factor kappa-light-chain-enhancer of activated B cells
NIX	NIP3-like protein X
NOXA	a proapoptotic BH3-only member of the Bcl-2 family
p53	tumor protein p53

PARPs	poly ADP-ribosyl proteins
PCD	programmed cell death
PDGFR	platelet derived growth factor receptor
PDI	protein disulfide bond isomerization
PE	phosphatidylethanolamine
PERK	protein kinase RNA-like ER kinase
p16INK4A	inhibitor of cyclin-dependent kinase 4 A
PI3K	phosphoinositide 3-kinase
PI3KCA	phosphoinositide-3-kinase
PLD1	phospholipase D1
PTEN	phosphatase and tensin homolog
PUMA	P53 upregulated modulator of apoptosis
RB	retinoblastoma
RIP	regulated intramembrane proteolysis
ROS	reactive oxygen species
RTK	receptor tyrosine kinase
SAHA	suberoylanilide hydroxamic acid
STAT3	signal transducer and activator of transcription 3
TERT	human telomerase reverse transcriptase
TMZ	temozolomide
TNF	tumor necrosis factor
TP53	tumor protein 53
TRADD	TNF receptor-associated death domain
TRAF2	tumor necrosis factor receptor-related factor 2
TRAIL	TNF-related apoptosis inducing ligand
TRB3	tribbles-related protein 3
UBD	ubiquitin-binding domain
ULK1	Unc-51-like autophagy activating kinase
V-ATPase	vascular adenosine triphosphatase
VEGF	vascular endothelial growth factor
VEGFR	vascular endothelial growth factor receptor
WHO	World Health Organization
XBP-1	X-box-binding protein
XIAP	X-linked IAP

Conflict of interest

No potential conflicts of interest were disclosed.

Introduction

Approximately 60% of adult brain tumors are gliomas. Glioblastoma (GBM) is one of the most malignant primary brain tumors in the adult central nervous system (CNS) with a poor prognosis. The World Health Organization (WHO) has classified the GBM as a grade IV glioma based on the biological behavior of the tumor, response to treatment, chemotherapy outcome, and prognosis. The incidence of GBM is less than 0.01% and has increased over the past 10 years. It comprises almost 75% of all high-grade gliomas with an incidence of 0.003% in the United States [1–5].

A systematic review on the incidence of GBM in Iran during 2000–09 showed that astrocytoma and glioblastoma constituted 60% of the primary malignant tumors with the estimated incidence of 0.006% of the general population [6].

Glioblastomas are categorized into distinct disease subtypes including primary and secondary. The majority of GBM cases are primary that mostly affect elderly patients (adults over 50 years), whereas the secondary type typically develops in younger individuals (under the age of 45). The tumor progression time varies considerably between primary and secondary GBMs. Histopathological evidence of primary GBM usually manifests in less than 3 months, whereas for secondary GBM this occurs over the course of 4–5 years on average, demonstrating much slower neoplastic progression. The nature of the two disease subtypes also differs with respect to the involvement of specific genetic pathways, different RNA and protein expression profiles, and the response to existing treatments [1, 5, 7, 8].

The genetic abnormalities often observed in primary GBM include loss of heterozygosity (LOH), $p16^{INK4A}$ deletions, mutations in p53 or PTEN, altered epidermal growth factor receptor (EGFR) or platelet-derived growth factor receptor (PDGFR) gene expression and amplification or overexpression of mouse double minute 2 homolog (MDM2) or EGFR, and other common alterations such as $p16^{INK4A}$ and retinoblastoma (RB) gene/protein alterations. On the other hand, genetic abnormalities in secondary GBM are the result of LOH at chromosome arm 19q and 10q, alterations in RB protein, PTEN mutations, and loss of expression in deleted-in-colorectal-carcinoma (DCC) gene. In aggressive malignant GBMs, mutations in PTEN (also called MMAC1 and TEP1) gene, melanoma-associated antigen gene family (MAGE-E1), and nuclear-restricted protein/brain (NRP/B) genes also occur, albeit with less frequency [1, 4, 5, 8–10].

Gliomas are categorized into three groups according to the biomarkers: tumors with mutations in the telomerase reverse transcriptase (TERT) promoter, mutations in isocitrate dehydrogenase (IDH), and those characterized by codeletion of chromosome arms 1p and 19q. Other classifications of gliomas are based on the mutation status of IDH1 and IDH2 as well as NOS expression. Extremely poor survival rates are seen in GBM patients with mutations in both TERT and IDH [11, 12].

Despite the different classifications, most patients with GBM are treated similarly. Targeted therapies for these patients are limited and existing strategies hardly increase the survival rate. The blood-brain barrier (BBB) represents a significant challenge in the treatment of brain tumors by limiting the drug access on the tumor site. The BBB contains capillaries, astrocytes, and pericytes that are fixed in the capillary basal lamina. The passage of therapeutic agents through the BBB depends on factors such as molecular weight, lipophilicity, and molecular charge. Large molecules (>400 Da) cannot pass the BBB and ~98% of small molecules cannot reach the CNS [1, 4]. Current treatment consists of surgical removal of the tumor and subsequent radio- and chemotherapies. TMZ is one of the common chemotherapeutic agents used in the treatment of GBM. It is administered as a daily dose of $75\,mg/m^2$ body surface area in combination with radiotherapy for 6 weeks followed by six 5-day cycles of TMZ alone at a dose of $150–200\,mg/m^2$ every 28 days. Unfortunately, GBM patients frequently develop chemoresistance to TMZ. As compared to patients treated with radiotherapy alone, TMZ in combination with radiotherapy increases the median survival rate from 12 to about 14.6 months [1, 13]. The low survival rate, poor prognosis for elderly patients, and development of resistance to TMZ necessitate strategies to establish more effective therapeutic approaches. For instance, some promising therapeutics have been suggested that allow the drugs to pass the BBB with positive preclinical outcomes. Therefore, novel therapeutic approaches need to be developed, which involves identifying and targeting key molecular intermediates that give rise to the malignant phenotypes of GBM.

The expression of several potential drug targets, including receptor tyrosine kinases (RTKs) such as EGFR, VEGF, EphA3, PDGFR, and MET, which varies according to the genetic profiling of GBM biopsies, has been widely studied [14]. However, only limited clinical success in inhibition of these targets by conventional single-targeted therapy has been reported. Therefore, multitargeted therapies have raised considerable interest in the treatment of GBM tumors by increasing the sensitivity and reducing the chemoresistance of GBMs and enhancing drug efficiency against the BBB [10, 13]. Endoplasmic reticulum (ER) stress and associated downstream pathways, such as autophagy and apoptosis, are the most common therapeutic targets and might be susceptible to TMZ. Autophagy is a cellular homeostasis mechanism that is intricately connected to ER stress-induced apoptosis. Under mild ER stress condition, autophagy prevents the apoptosis-related caspase activation to diminish cell damage but promotes apoptosis under intense stress condition. Although some protein interactions between autophagy and apoptosis have been identified, the exact mechanisms driving the interplay between these pathways remain uncertain [9, 13].

This chapter summarizes the possible mechanisms of ER stress, autophagy, and apoptosis, and their function in regulating cell fate. In addition, it will discuss their role in current GBM treatment (TMZ) and evaluate their potential as therapeutic targets and chemoresistance. Finally, we will provide an overview of bioengineered 3D model of GBM and its application in measuring chemotherapy and chemoresistance in GBM.

Apoptosis

The main type of programmed cell death (PCD), apoptosis, was first introduced in 1972 [15]. It was labeled as type 1 programmed cell death and characterized by biochemical and morphological features that are different from necrosis [16]. Apoptosis can be initiated by various triggers that can induce two cascades: the intrinsic and extrinsic apoptotic pathways [1], as shown in Fig. 1.

Intrinsic pathway of apoptosis

A multitude of intracellular stressors such as DNA damage, cell cycle disruption, and hypoxia induce intrinsic or mitochondrial apoptotic pathways where the outer membrane of mitochondria is permeabilized by these stress signals. Mitochondrial outer membrane permeabilization (MOMP) activates proapoptotic B-cell lymphoma 2 (BCL-2) superfamily members and facilitates the subsequent release of cytochrome c and other proapoptotic factors from the intermembrane space of mitochondria to the cytoplasm [17–20].

Extrinsic pathway of apoptosis

Various cytokines, including tumor necrosis factor (TNF) receptor superfamily members, Fas ligand (Fas-L), and TNF-related apoptosis-inducing ligands (TRAIL), are responsible for the activation of death receptors, which are principal initiators of the extrinsic pathway [21]. Upon activation, death receptors interact with adaptor proteins, such as Fas-associated death domain (FADD) or TNF receptor-associated death domain (TRADD), via their intracellular

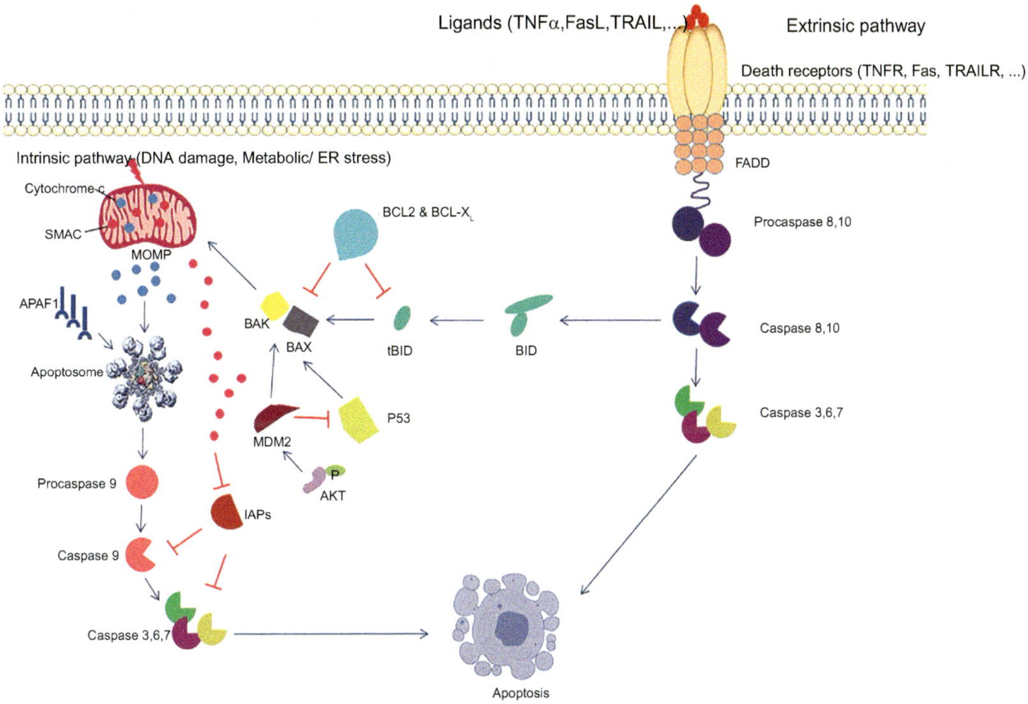

FIG. 1 Intrinsic and extrinsic pathways of apoptosis. In the intrinsic pathway of apoptosis some different intracellular stimuli such as DNA damage, ER stress and hypoxia induce the release of mitochondria cytochrome C, which results in APAF-1 activation and formation of apoptosome that converts procaspase-9 to caspase-9 and finally leads to activation of caspase-3 and apoptosis. Proapoptotic proteins such as Bax and Bak are triggered by p-53 and induce the release of mitochondria cytochrome C. On the other hand, Mdm-2 by blocking p53 and IAPs via inhibition of caspase proteins leads to apoptosis inactivation. In the extrinsic pathway of apoptosis some external ligands such as TNFα, FasL and TRAIL and also death receptors interact with FADD, which leads to procaspase-8 autoactivation and then caspase-3 activation that finally results in apoptosis initiation. Caspase-8 also converts Bid to tBid which triggers apoptosis via activation of Bax and Bak and release of cytochrome-C from mitochondria.

domains. These proteins binds to procaspase 8 and death-inducing complex (DISC) transduce a downstream signal cascade resulting in autoactivation of procaspase 8 and subsequent activation of caspases -3, -6, and/-7 [22].

Apoptosis mediators

Different apoptotic-associated proteins contribute to the apoptotic response to various targets [23]. These mediators are as follows:

Poly (ADP-ribose) polymerases (PARPs) family protein

PARPs include about 17 nuclear protein members with poly ADP-ribosyl transferase activity [24]. PARP-1 performs a substantial role in some DNA repair pathways and various biological procedures such as gene expression [25, 26]. PARPs are considered as a depleting factor for ATP in the cell, and the proteolytic inactivation of PARP-1 by caspases-3 and -7 is a warranty for ATP demand. Therefore, cellular ATP depletion is predictable owing to the inhibition of caspases [27, 28].

Bcl-2 family proteins

Bcl-2 family proteins are other important proteins in apoptosis signaling pathways that include pro- and antiapoptotic proteins. Bcl-2 family proteins have a central role in balancing the mitochondrial membrane permeability [1]. Protection of mitochondrial outer membrane integrity is the main role of antiapoptotic proteins, such as Bcl-2 and Bcl-X_L. These proteins can bind to proapoptotic proteins, such as Bax and Bak, and inhibit the cytochrome c release, leading to caspases inactivation, and ultimately cell death block [29].

Inhibitors of apoptosis proteins (IAPs)

IAPs are antiapoptotic proteins that inactivate caspases via covering the catalytic active pocket of caspase-3, -7 and -9 or breaking down of their structures [29]. The human IAP family includes XIAP, NAIP, cIAP1, cIAP2, ILP2, survivin, Bruce, and livin [18, 29].

P53

P53, known as the guardian of the genome, has a crucial role in managing cell activities in response to stress conditions [30]. This vital protein mediates apoptosis by finding DNA damage and promotes signaling pathways involved in apoptosis [31]. For instance, p53 can activate Bax as a proapoptotic protein and provoke the intrinsic pathway of apoptosis [29].

Apoptosis and chemoresistance

In general, apoptosis plays a significant role in the survival of organisms by regulating apoptotic pathways and eliminating unwanted cells, a process so-called programmed cell death. Evading apoptosis is one of the well-known hallmarks of cancer cells where they stimulate various mechanisms to evade apoptosis [32]. Studies have shown that apoptosis disorders may lead to drug resistance in cancer cells [33]. Therefore, having a clear understanding of the mechanisms involved in chemoresistance is a basic step to considerably improve the treatment. These mechanisms manipulate apoptosis signals by attenuating the proapoptotic and augmenting the antiapoptotic signals or disturbing the hemostasis of apoptosis and autophagy [32].

Chemoresistance in glioblastoma

Glioblastoma stem cells disrupt the state of equilibrium among apoptosis and survival pathways. This may lead cells to possess resistance to the chemotherapy compounds containing alkylating agents [34].

Role of Bcl-2 family proteins in chemoresistance of glioma cells

The release of cytochrome c to the cytosol is blocked in human cancer cells by downregulation of antiapoptotic proteins (like Bcl-2 and Akt) and upregulation of proapoptotic proteins (such as Bax and Bak), which inhibits the formation of mitochondrial apoptosis-induced channel (MAC). This adaptation results in chemoresistance and immortal feature of the tumor cells [32]. As an example, glioma cells downregulate the prosurvival proteins such as Bcl-2 family to acquire apoptotic-induced chemoresistance [35, 36].

Hypoxia and chemoresistance in glioma cells

There are some conditions that underlie chemotherapy drug resistance. Researches have demonstrated that solid tumors have lower oxygen concentration compared to normal tissues and therefore develop hypoxia. Studies have shown that three types of IAPs are overexpressed in hypoxic glioma cells [29]. For example, a high level of livin as a member of IAPs is observed in the hypoxic microenvironment of GBMs [33]. The function of Bad as a proapoptotic protein is also affected in hypoxia [37, 38]. It is also confirmed that blocking IAPs such as XIAP or survivin increases chemosensitivity in glioma cells [29, 39].

Role of p53 in alkylating agent chemoresistance

Mitochondrial proteins are considered as potential targets in cancer treatment. Evidence indicates that dysregulation of apoptosis genes may lead to chemoresistance in gliomas [1]. A study on glioma cells reported that TMZ-induced intrinsic apoptosis pathway depends on the cells' p53 status [40]. Loss of function of this protein during the DNA damage leads to the overexpression of antiapoptotic proteins such as Bcl-2 and Bcl-X_L and disrupts the apoptosis process. Based on the available reports, p53 is inactivated in 50% of GBM patients, which is related to high levels of Bcl-2 [31]. Therefore, chemoresistance can be affected by Bcl-2 upregulation in glioma patients [31]. Also, phosphorylated RAC-α serine/threonine-protein kinase (pAKT) induces the overexpression of MdM2 protooncogene in GBM, which negatively regulates p53. This downregulation eliminates the stimulatory effect of p53 on Bax proapoptotic protein, which subsequently inhibits the permeability of mitochondria and apoptosis [29]. However, it is difficult to identify the sensitivity or resistance of cells to chemotherapy agents, e.g., TMZ, owing to the presence of different p53 mutation subgroups in heterogeneous GBM tumors [40].

Role of miRNAs in alkylating agent chemoresistance

Cancer cells are able to escape from chemotherapy by disrupting apoptosis pathways [41]. It has been shown that the improper expression of miRNAs has a significant effect on the chemosensitivity of GBM. The correlation of TMZ resistance with several miRNAs such as miR-21, miR-195, miR-455-3p, and miR-10a are verified [41]. Generally, Bax/Bcl-2 ratio is an indicator of apoptosis level in glioma cells, i.e., higher ratio indicates a high level of apoptosis [41]. For example, an elevated ratio of Bax/Bcl-2 is observed in TMZ-treated U87MG cell line compared to untreated cells. Downregulation of Bax and upregulation of Bcl-2, owing to the overexpression of miR-21, inhibit apoptosis in treated cells [41]. Therefore, miR-21 is introduced as a chemoresistant factor in U87MG cells. Significant overexpression of miR-195, miR-455-3p and miR-10a is also reported in TMZ-chemoresistant U251R cells and their suppression has shown to be effective in increasing U251R cells sensitivity to TMZ [41].

Role of defective DNA repair systems in alkylating agent chemoresistance

Alkylating agents such as TMZ and decarbazine methylates induce DNA damage and DNA mismatch repair pathway [31]. A DNA repair enzyme, O-6Methyl guanine methyl transferase (MGMT), can directly reverse O6-methylguanine alkylation, induced by alkylating agents such as TMZ, and seems to have a substantial role in cytotoxicity resistance of alkylating drugs [31]. The chemoresistance function of MGMT has been shown in different human tumor cells and xenografts [31]. Although remarkable sensitivity to alkylating agents owing to MGMT insufficiency has been reported in severe glioma, there are still significant GBM tumors resistant to alkylating drugs. This implies the contribution of other active mechanisms in chemoresistance tumors [31]. A functional MMR system can detect different types of mutations such as mismatches in base pairs, insertion or deletion and O6-methylguanine. These damages extend double-stranded DNA breaks and induce apoptosis through MMR system. Although MMR system has a loss of function or is extremely defective in some tumor cells, such as glioblastoma stem cells (GSCs), it can enforce the cells to either endure mutant p53 or increase the level of Bcl-2 and Bcl-X_L [42]. Such abnormal activity of MMR allows cells to not only maintain the mechanism of defect replication but also to develop mutations. Cells undergoing the alkylating therapy with abnormal MMR suffer from the chemoresistance and progression of a faulty cell cycle, simultaneously [42].

As discussed above, different conditions such as hypoxia, defective DNA mismatch repair system, insufficient MGMT enzyme, and dysregulation of miRNAs expression can influence apoptosis. Also, abnormal expression of apoptosis-associated proteins such as p53, Bcl-2 family proteins, and IAPs can disrupt their efficient functionality. Therefore, further studies on the correlation between apoptosis and mechanisms such as autophagy and ER stress are necessary.

Autophagy pathway

Autophagy is a self-eating mechanism for lysosomal degradation and recycling of nonfunctional proteins and damaged organelles [43]. Various physiological and pathological stress conditions such as hypoxia, nutrient deprivation, DNA damage, and pathogen infection induce autophagy. This catabolic pathway contributes to maintaining cellular and tissues homeostasis, development, metabolism, and immunity and pathogen clearance [44]. Autophagy can be classified based on cargo contents and cargo delivery mechanisms. The contents of cargo are either nonselective or selective [45]. Recycling of the bulk cytoplasmic contents during the starvation stress is called nonselective autophagy, in which molecular building blocks and energy are provided for cell survival [45]. In contrast, selective autophagy specifically targets damaged cellular structures, such as chromatin (chromatophagy), aggregate-prone proteins (aggrephagy), mitochondria (mitophagy), endoplasmic reticulum (ER) (reticulophagy), lysosomes (lysophagy), peroxisomes (pexophagy), ferritin (ferritinophagy), and microorganisms (xenophagy) [45, 46]. The exact mechanisms of selective autophagy are still under investigation. However, it is clear that ubiquitin-independent and ubiquitin-dependent pathways can recognize and deliver specific cargoes to the lysosome [1]. The former pathway contains direct binding of the specific receptors to substrates and interacting with the autophagosome through their light-chain 3 (LC3)-interacting regions (LIR). NIP3-like protein X (NIX) and family with sequence similarity 134 Member B (FAM134B) are among the receptors for mitophagy and reticulophagy, respectively [1, 47]. In the latter pathway, mono- or poly-ubiquitinated cargos are recognized by receptors such as nuclear dot protein 52 (NDP52) for mitophagy and p62 for aggrephagy containing ubiquitin-binding domain (UBD) and also LIR motif for interaction with autophagosome [48]. Macroautophagy, microautophagy, and chaperon-mediated autophagy (CMA) are three known types of autophagias distinguished based on cargo delivery mechanisms [45]. The first two types are conserved in all eukaryotes, however, CMA is exclusive to mammalian cells [1]. Microautophagy consists of selective or nonselective transport of cytoplasmic materials into lysosomes through direct invagination of the lysosomal membrane. CMA is responsible for selective transport of the proteins, containing CMA-specific KFERQ motif, into lysosomes through interaction with heat shock cognate 70 (HSC70) and lysosomal-associated membrane protein 2 (LAMP-2a) [49]. In contrast to the microautophagy and CMA, macroautophagy includes the formation of a double-membrane organelle named autophagosome, which engulfs the cytosolic materials in a selective or nonselective manner and then fuses to the lysosomes for content degradation [50]. According to the autophagosome size (~0.5–2 μm), macroautophagy (hereafter called autophagy) is able to degrade materials too large for microautophagy and CMA, such as damaged mitochondria, ER, lysosomes, bacteria, and aggregated proteins [50]. A brief overview of autophagy is shown in Fig. 2.

Autophagy and apoptosis cross talk: self-eating and self-killing processes

Despite different mechanisms of autophagy and apoptosis, emerging evidence proposes common upstream signals which may trigger both processes [51]. These pathways have a complex interaction. Generally, autophagy contributes to stress adaptation and blocks apoptosis. However, in other cellular conditions autophagy may induce apoptosis or necrosis and excessive degradation of cytoplasm, which results in autophagic cell death (ACD) [52]. Recently, several molecular mechanisms of autophagy-apoptosis cross talk have been deciphered. One of the principal factors connecting autophagy and apoptosis is B-cell lymphoma 2 (Bcl-2)-Beclin-1. Under normal conditions, antiapoptotic Bcl-2 protein binds to

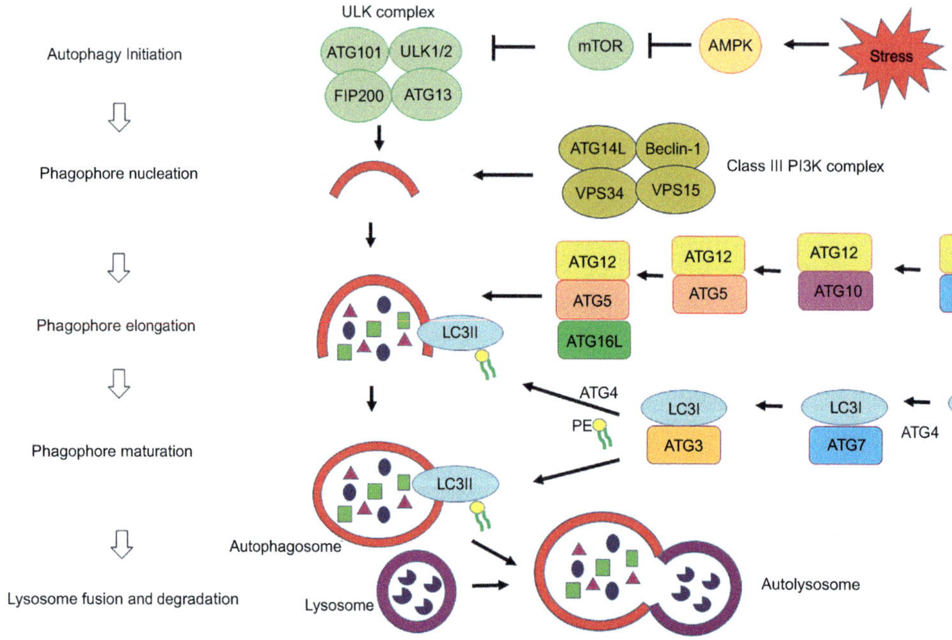

FIG. 2 A brief description of the autophagy pathway comprises initiation, nucleation, elongation, maturation, fusion, and degradation stages. mTOR is a critical regulator of autophagy. Stress conditions, such as nutrient deprivation, activate AMPK, which inhibits mTOR and activates ULK1 complex for autophagy initiation. Phagophore nucleation is mediated by class III PI3K complex. ATG12 and LC3 conjugation systems mediate phagophore elongation. Phagophore maturation is mediated by LC3 conjugation system. Mature autophagosome is fused to the lysosome and forms autolysosome to degrade the cargos.

Beclin-1 and inhibits autophagy and apoptosis [53]. During starvation, C-Jun N-terminal protein kinase 1 (JNK1) phosphorylates Bcl-2, which leads to Bcl-2-Beclin-1 complex disruption and autophagy activation [54]. Continued starvation results in binding of the phosphorylated Bcl-2 to Bcl-2-associated X protein (Bax), which preserves the mitochondrial membrane integrity and prevents the release of proapoptotic proteins into the cytoplasm. However, in excessive starvation, Bcl-2 is hyperphosphorylated by JNK1 and dissociates from Bax. This separation induces apoptosis by the caspase-3-dependent pathway [51]. Autophagy-related gene (ATG) proteins are other factors involved in autophagy/apoptosis regulation. ATG4 plays role in covalent binding of ATG8 to phosphatidylethanolamine (PE) and its delipidation at the lysosomal fusion step [51]. In apoptotic cells, the N-terminal fragment of a human ATG4 member, ATG4D, is cleaved by caspase-3, which delipidates gamma-aminobutyric acid receptor-associated protein-like 2 (GABARAP-L1) and eventually decreases autophagosome formation and promotes mitochondria-mediated apoptosis [55]. An autophagy promoting factor ATG5 is cleaved by calpains in apoptotic stimuli. Then, truncated ATG5 interacts with BCL extra-large (Bcl-X_{L}) and triggers cytochrome C release and caspase activation [56]. Another autophagy promoting factor, ATG12, binds to and inhibits myeloid cell leukemia-1 (Mcl-1) and Bcl-2, which promotes mitochondrial apoptosis during apoptotic stimuli [51].

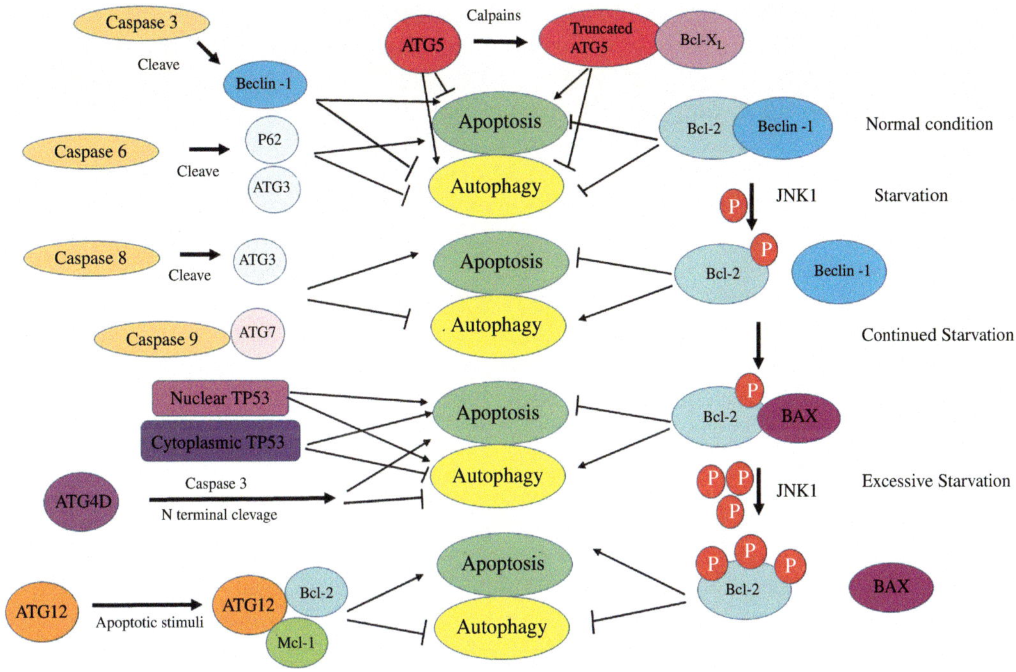

FIG. 3 Cross-regulation of autophagy and apoptosis by interactions of their related factors.

Nuclear Tp53 induces an intrinsic apoptotic pathway by promoting proapoptotic proteins such as BH3 interacting domain death agonist (BID), Bax, Tp53 upregulated modulator of apoptosis (PUMA), and a proapoptotic BH3-only member of the Bcl-2 family (NOXA) and inhibiting the Bcl-2 expression. Cytoplasmic Tp53 triggers the extrinsic apoptotic pathway through promoting the TNF-related apoptosis inducing ligand (TRAIL) receptor and Fas receptor. TP53 also has a dual role in autophagy regulation based on cellular localization [51]. Nuclear TP53 transcriptionally regulates damage regulated autophagy modulator (DRAM) and promotes autophagy, whereas cytoplasmic Tp53 inactivates AMP-activated protein kinase (AMPK), induces mammalian target of rapamycin (mTOR) signaling and subsequently suppresses autophagy [57]. In apoptotic stimuli, caspase-3, caspase-6, caspase-8, and caspase-9 contribute to autophagy suppression by cleaving Beclin-1, P62 andATG3, and interacting with ATG7, respectively [51, 58]. Therefore, proteins involved in every stage of apoptosis play critical roles in autophagy regulation, emphasizing the importance of rigorous regulation between these pathways. Interactions between autophagic and apoptotic factors are shown in Fig 3.

The dual roles of autophagy in glioblastoma promotion and suppression

Autophagy has special importance for tissues such as the brain through preventing the accumulation of damaged mitochondria and protein aggregates and maintaining iron hemostasis by recycling iron complexed with ferritin (ferritinophagy) [1, 59]. Therefore, autophagy deficiency in the brain may contribute to GBM development [1]. However, sufficient evidence supports the dual roles of autophagy in GBM-promoting and GBM-suppressing effects, depending on the environmental and intracellular stresses [60]. On one hand, autophagy induces the uncontrolled proliferation of glioma cells by facilitating the regulating receptor tyrosine kinase (RTK) oncogenic signaling, endocytic trafficking, and enabling metabolic rewiring [61]. Reactive oxygen species (ROS) enhancement induces autophagy, which promotes tumor cell survival by glycolysis activation and increase of lactate, acetoacetate and 3-hydroxybutyrate [62, 63]. Activated hypoxia-inducible factor-1α (HIF-1α) and nuclear factor kappa-light-chain-enhancer of activated B cells (NF-κB) induce autophagy during oxidative stress, which results in tumor suppressor caveolin-1 (Cav-1) degradation [64]. Autophagy also contributes to the detachment of cancer cells from the extracellular matrix (ECM), which facilitates tumor cell metastasis [65]. Tumor cell immune evasion is mediated by autophagy through the degradation of major histocompatibility complex I (MHC-I) [61, 66]. Autophagy is also upregulated in response to the treatment and promotes glioma stem cell (GSC) survival and treatment resistance [61].

On the other hand, autophagy could be implicated in suppressing tumor progression. Immunologic cell death can be stimulated by autophagy through inflammatory signals release. Autophagy overactivation also induces cell death [61]. Evidence suggests that autophagy could inhibit tumor invasion by impairing the epithelial-to-mesenchymal transition (EMT)-like process and downregulating the oncogenic C-Met (hepatocyte growth factor receptor, HGFR) signaling [61, 67]. Autophagy also limits inflammatory responses and induces glioma cell senescence [64, 68]. The autophagy dual role in cancers reveals the importance of tumor cell type and the context in autophagy function [1].

Genomic classification of GBM consists of four molecular subtypes (proneural, neural, classical, and mesenchymal) according to the frequently observed mutations in epidermal growth factor receptor (EGFR), platelet-derived growth factor receptor (PDGFR) alpha, vascular endothelial growth factor receptor (VEGFR), phosphatase and tensin homolog (PTEN), tumor protein 53 (TP53), and neurofibromin 1 (NF1). Different expression levels of LC3 protein in these subtypes indicate a possible correlation between genetic alterations and autophagy regulation in GBM [45]. Table 1 shows the dual functions of autophagy in GBM promotion and suppression.

Autophagy involvement in the initiation and overcome chemoresistance in GBM

Finding the exact role of autophagy in tumors progression and treatment has attracted great attention in the field of cancer research. The present hypothesis proposes the inhibitory role of autophagy in tumor initiation vs its promotional role in tumor progression and chemoresistance [69].

GBM is lethal owing to its chemo- and radioresistance. Strategies such as autophagy genes knockdown or a combination of pharmaceutical inhibitors of autophagy could improve the efficacy of chemotherapeutic drugs [1]. Cisplatin alone induces autophagy in glioma cells, whereas retinoblastoma (RB) knockdown results in autophagy inhibition and sensitization of GBM cells to apoptosis [70]. Some GBM cells become resistant to bevacizumab,

TABLE 1 Different GBM-promoting and GBM-suppressing functions of autophagy.

GBM-promoting effects of autophagy	GBM-suppressing effects of autophagy
Induction of uncontrolled cell proliferation Positive regulation of • Oncogenic RTK signaling • Endocytic trafficking • Metabolic rewiring	Inhibition of the tumor invasion • Inhibition of EMT-like processes • Downregulation of oncogenic Met signaling
Induction of tumor cell survival • Activation of glycolysis	Stimulation of the immunologic cell death • Release of inflammatory signals
Degradation of tumor suppressor gene Cav-1	Induction of senescence • Limits the inflammatory responses
Induction of tumor metastasis • Detachment of GBM cells from ECM	
Induction of tumor cell immune evasion • Degradation of MHC-I	
Promotion of GSCs survival and treatment resistance	

an antivascular endothelial growth factor (VEGF) antibody, by inducing hypoxia-associated autophagy and promoting cell survival. However, treatment with autophagy inhibitor hydroxychloroquine (HCQ) in combination with bevacizumab decreases GBM cells survival [71]. Mutations and amplification of EGFR are other major contributors to GBM chemoresistance. Cotreatment of GBM cells with CQ and ZD6474 (vandetanib), a dual EGFR/VEGFR inhibitor, induces GBM cell death by autophagy blocking [72]. The phosphoinositide 3-kinase (PI3K)/mTOR pathway induces cell survival signaling. Some GBM cells are resistant to PI3K/mTOR pathway inhibitors, e.g., NVP-BEZ235, which promotes autophagy and GBM cells survival. However, a combination of NVP-BEZ235 with CQ inhibits autophagy and survival in GBM cells [73]. Another study indicates that although the histone deacetylase inhibitor vorinostat induces protective autophagy in glioma cells, the coadministration with phospholipase D1 (PLD1) inhibitor leads to apoptosis [74]. These results highlight the importance of autophagy in targeted therapy and its potential role in improving cell sensitivity to drugs. Table 2 summarizes the list of combination therapies for autophagy manipulation and their effects on the chemosensitivity of GBM.

Autophagy involvement in mechanisms of anticancer activity of alkylating agents

A DNA alkylating agent, temozolomide (TMZ), is known as an integral part of GBM treatments [1]. In the genomic DNA, guanine is methylated by TMZ at the O^6 position, which leads to G-T mismatching in the next cycle of replication [89]. The exact mechanism of GBM treatment by TMZ is still under discussion and has not universally been accepted. Different

TABLE 2 Combination therapies for increasing the chemosensitivity in GB by autophagy induction or inhibition.

Glioblastoma model	Treatments	Mechanisms of chemosensitivity enhancement	References
GBM cells GSCs	TMZ/EMAP-II	Induction of autophagy	[75]
GBM cells	TMZ/RAD001	Induction of autophagy	[76]
Glioma cells	TMZ/cannabinoid THC	Induction of autophagy	[77, 78]
GBM cells	TMZ/pan-Bcl-2 inhibitor (−)-gossypol	Induction of autophagy	[79]
Glioma cells	TMZ/carnosic acid (CA)	Induction of autophagy	[80]
GBM cells	TMZ/BCNU/AK3 and PI3KCA siRNAs	Induction of autophagy	[81]
Glioma cells	TMZ/bafilomycin A1 (BafA1)	Inhibition of autophagy	[82]
Human Clinical trial phase IB	TMZ/CQ/radiotherapy	Inhibition of autophagy	[83]
Glioma cells	TMZ/vorinostat (SAHA)/CQ	Inhibition of autophagy	[84]
GBM xenografts Glioma cells	TMZ/Sirolimus (Rapamycin)/CQ	Inhibition of autophagy	[85]
Glioma cells	TMZ/2-deoxy-D-glucose (2-DG)	Inhibition of autophagy	[86]
Glioma cells GBM xenografts	TMZ/Resveratrol	Inhibition of autophagy	[87]
Glioma cells	TMZ/Simvastatin (Simva)	Inhibition of autophagy	[88]
Glioma cells	Cisplatin/RB knockdown	Inhibition of autophagy	[70]
Glioma cells	Bevacizumab/HCQ	Inhibition of autophagy	[71]
GBM xenografts	ZD6474 (Vandetanib)/CQ	Inhibition of autophagy	[73]
GBM xenografts	NVP-BEZ235/CQ	Inhibition of autophagy	[73]
Glioma cells	Vorinostat (SAHA)/PLD1 inhibitor	Inhibition of autophagy	[74]

mechanisms have been proposed for TMZ cytotoxicity including autophagy, senescence, and apoptosis pathways [90]. Failure in activation of apoptosis and autophagy may induce other cell death pathways such as necroptosis and ferroptosis [91]. Autophagy is upregulated as a stress response upon TMZ treatment [1]. TMZ concentration, DNA damage degree, metabolic, and hypoxic condition and clonal variations are among primary factors that affect the induction levels of autophagy by TMZ. Hypo-or hyperactivity of autophagy promotes cell death, while the intermediate activation may induce cell survival [92]. Therefore, it is still a matter of debate whether the induction or the inhibition of autophagy is more beneficial for GBM treatment (Table 2) [93].

Aldehyde dehydrogenase 1 family member A3 (ALDH1A3) which is involved in the oxidation of aldehydes is highly expressed in human GBM and correlates with poor prognosis and therapy resistance. One study indicated that TMZ-induced autophagy leads to ALDH1A3 degradation and improves TMZ chemosensitivity [94]. Cotreatment of U87MG glioblastoma cells and human glioblastoma stem cells (GSCs) with TMZ and endothelial monocyte-activating polypeptide-II (EMAP-II) as a proinflammatory cytokine, could increase the cytotoxicity compared with TMZ-only treatment. The combination of TMZ and EMAP-II improves the cytotoxic effect of TMZ and induces uncontrolled autophagy [75]. Another study showed that the combined treatment by TMZ and mTOR inhibitor RAD001 along with γ-radiation intensely decreased the GBM cells viability. Autophagy was reported as the main mechanism of cell death, which also effectively enhanced the radiosensitivity of cells [76]. The combination therapy of TMZ and cannabinoids promotes antitumor efficiency in a glioma cells treatment and xenograft model by autophagy induction [77, 78].

The pan-Bcl2 inhibitor gossypol is also suggested as a promising drug for increasing TMZ cytotoxicity through the induction of apoptotic and autophagic cell death in apoptosis-resistant GBM cells [79]. Carnosic acid (CA) has anticancer, antioxidant, and antiinfectious properties that enhances the TMZ cytotoxicity by inducing both autophagy and apoptosis. TMZ-induced autophagy is prompted by CA through inhibition of PI3k/AKT pathway, p62 downregulation, and LC3-I to LC3-II transition [80].

The abovementioned reports indicate that although TMZ treatment alone is unable to induce enough autophagic cell death, in combination with other drugs it can induce autophagy as an alternative trigger of cell death in TMZ-resistant glioblastoma cells. Therefore, autophagy induction may be utilized to increase tumor regression and eventually to improve patient survival [93].

Autophagy involvement in chemoresistance to alkylating drugs in GBM

Autophagy is proposed as the main mechanism of chemoresistance to alkylating drugs. TMZ treatment at different doses induces different autophagy levels based on the glioma cell types. As shown in Fig.4, different molecular mechanisms seem to be involved in autophagy activation by alkylating agents [1]. Cotreatment of GMB8901 cells with TMZ and carbazole alkaloid BC3EE2, 9B induces autophagy through adenosine monophosphate-activated protein kinase (AMPK) activation and AKT/mTOR inhibition [95]. Downregulation of a long noncoding RNA (lncRNA), cancer susceptibility 2 (CASC2) results in enhancing the miR-193a-5p levels, decreasing mTOR expression, and inducing protective autophagy [96]. Knockdown of AKT3 and phosphoinositide-3-kinase (PI3KCA) in T98G cells combined with treatment by TMZ and carmustine (BCNU) prevents the proliferation and induces autophagy and apoptosis [81]. Activation of ataxia telangiectasia mutated (ATM)/AMPK/Unc-51 like autophagy activating kinase (ULK1) pathway in GBM cells is another mechanism of TMZ-induced autophagy [97]. Treatment of U87MG and GBM 8401 cells by TMZ promotes ER stress and autophagy induction through activation of c-Jun N-terminal kinases (JNK) and $Ca2+$ signaling pathways [93]. Moreover, ROS generation leads to activation of the extracellular signal-regulated kinase (ERK) pathway, which induces autophagy in GBM cells [93]. Therefore, based on the available evidence, autophagy inhibition can be proposed as an efficient strategy for the treatment of chemoresistant GBMs.

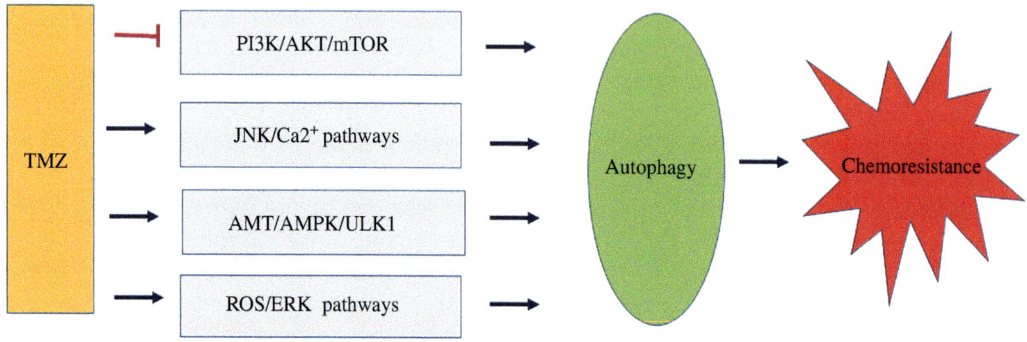

FIG. 4 Schematic description of the proposed mechanisms of TMZ-induced autophagy and chemoresistance.

3-Methyladenine (3-MA), a class III PI3-K inhibitor, suppresses autophagy at the early stages. Simultaneous administration of 3-MA and TMZ promotes tumor cell survival, while the administration of 3-MA after TMZ treatment increases TMZ cytotoxicity. These contradictory results emphasize the importance of 3-MA administration time in TMZ treatment of glioma cells [1]. The stage of autophagy inhibition is another key factor in GBM treatment. Bafilomycin A1 (BafA1), vascular adenosine triphosphatase (V-ATPase) inhibitor, suppresses autophagy at the last stage by blocking autophagosome-lysosome fusion and improves TMZ cytotoxicity via inducing apoptosis [82]. CQ and HCQ inhibit autophagy completion, reduce tumor hypoxia, and improve radiosensitivity (1). Based on preclinical data, the most beneficial response to CQ is observed in *EGFRvIII*-expressing glioblastoma owing to autophagy dependency. The combination of CQ, daily TMZ and radiotherapy is in the phase IB trial for newly diagnosed glioblastoma [83]. Cotreatment with TMZ and a specific histone deacetylase inhibitor, suberoylanilide hydroxamic acid (SAHA), could also increase TMZ cytotoxicity. Autophagic flux is enhanced as a survival response by both TMZ and SAHA, which impairs the efficacy of these drugs. CQ improves antiglioma effects of TMZ and SAHA by inhibiting the autophagy termination [84]. One study suggested that combined triple therapy with autophagy inducer [sirolimus (rapamycin)], autophagy inhibitor (CQ), and alkylating agent (TMZ) can synergize the antiglioma effects. This triple therapy strongly decreases the mitochondrial function and promotes lysosome-dependent apoptosis cell death [85]. A combination of an alkylating agent with 2-deoxy-D-glucose (2-DG) enhances chemosensitivity in GBM cells by decreasing autophagy [86]. Previous reports indicated the improvement in synergistic effects of antioxidant resveratrol combined with TMZ against glioma cells by downregulation of the overexpressed O-6-methylguanine-DNA methyltransferase (MGMT), caused by TMZ, and inhibition of the signal transducer and activator of transcription 3 (STAT3)/Bcl-2/survivin signaling pathway [87]. The combination of simvastatin (Simva) with TMZ also increases TMZ cytotoxicity by blocking autophagosome-lysosome fusion [88]. The effects of combination therapies on autophagy and chemosensitivity of GBM are summarized in Table 2. In conclusion, autophagy plays a critical dual role, by induction/inhibition of chemosensitivity, in the treatment of GBM with chemotherapeutic agents. However, further studies on the ER stress are required to clarify the precise mechanism of chemoresistance in GBM.

Endoplasmic reticulum stress, and unfolded protein response (UPR)

The endoplasmic reticulum (ER) (rough and smooth) consists of a complex network of closely related cisternae, sacs, and tubules, and a continuous network connects it into the nuclear membrane [98–100]. This major organelle is involved in the synthesis, folding, post-translational modification, secretion, and transfer of approximately one-third of all proteins to a defined destination [101]. As a quality control gate, ER responds to the cellular stress by ER chaperones including GRP78, GRP94, and GRP18. The glucose-regulated protein (GRP) detects the presence of unfolded or misfolded proteins [102, 103]. The constant interaction of ER with the nucleus, mitochondria, and the Golgi apparatus makes it a key regulator in normal cell function [104]. Various internal or external stressors, including hypoxia, starvation, infection, calcium homeostasis, and/or disruption of cytotoxic compounds, directly or indirectly lead to the instability of the ER microenvironment and abnormal protein accumulation in the ER lumen commonly referred to as ER stress [105, 106]. In this condition, the normal balance between protein biosynthesis and normal protein folding in the ER is perturbed. Depending on the intensity and exposure time of the stress, adaptive defense systems are triggered by various compensation mechanisms such as the unfolded protein response (UPR), endoplasmic reticulum-associated degradation (ERAD), and autophagy [107, 108]. Cell death occurs when the adaptive pathway fails to compensate for the ER normal function [109, 110]. Activation of UPR in endoplasmic reticulum stress enhances normal protein folding by expression of the essential genes in protein folding, secretion, and quality control. Not only UPR regulates ER function, but also modulates the mitochondrial biology, lipid metabolism, and intracellular apoptotic pathways [111]. Nowadays, UPR is not only considered as a protein folding pathway [1] but new functions have been recognized, including lipid metabolism, energy control, and inflammation [112].

The UPR is an important target in the treatment of diabetes, cancers, immune diseases, neurodegenerative disorders, cataracts, pulmonary fibrosis, cardiomyopathy, atherosclerosis, ischemia, nephrotic syndrome, nonalcoholic steatosis, and psoriasis [113]. Nowadays, ER stress, UPR and other cell homeostasis pathways have attracted great attention in the field of cancer therapy. The ER stress response is regularly monitored by three specific UPR sensors: inositol requires enzyme 1 (IRE-1α and IRE-1β), protein kinase RNA-like ER kinase (PERK), and activated transcription factor 6 (ATF6α and ATF6β subtypes). These sensors are located on the outer membrane of the ER but are connected to GRP78 during an inactive state. Depending on the cell type, the stimuli, duration, and intensity of the stress, UPR may cause cell adaptation or cell death [105, 114].

GRP78 as a main regulator of the ER stress promotes the assembly, folding, and translocation of newly synthesized nascent proteins. It can inhibit protein accumulation and remove unfolded/misfolded proteins through the ER-related degradation (ERAD) process [114, 115]. GRP78 also regulates calcium homeostasis by ER-mitochondrial calcium cross talk [116].

In ER stress status, activated PERK dissociates from GRP78 and phosphorylates eukaryotic initiation factor 2 (eIF2α) to reduce the pressure on the ER. Dissociated PERK then suppresses the translation of excess cytoplasmic mRNA and prevents the newly synthesized proteins from entering the ER [115, 117, 118]. Phosphorylated eIF2α interferes with the translation of intracellular proteins, except for some important transcription factors such as activated transcription factors (ATF4) [119, 120]. Moreover, activated IRE-1 affects the X-box-binding protein (XBP-1) mRNA transcription factor through the endonuclease activity. Spliced and activated XBP-1 develops the protein folding capacity of ER by regulating the expression of several genes involved in ER quality control, ERAD, autophagy, and apoptosis [121]. Finally, ATF6 is separated from GRP78 and is transported to the Golgi organelle. The transported ATF6 is then activated by regulated intramembrane proteolysis (RIP) by site 1 and site 2 (SP) proteases. The cytoplasmic domain (50 kDa) of the activated ATF6 translocates to the nucleus [122, 123] and increases the levels of ER chaperones, XBP-1, and protein folding capacity. The entire cascade of UPR pathway is presented in Fig.5.

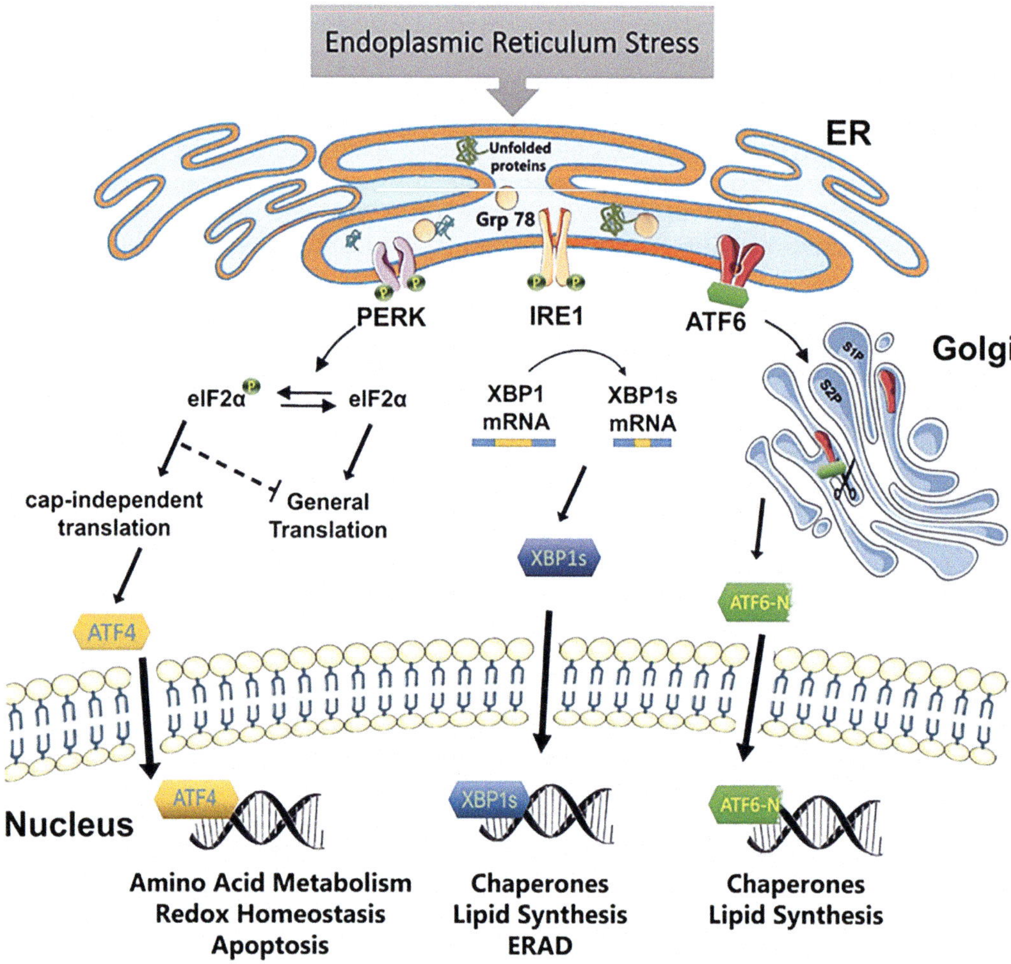

FIG. 5 Schematic overview of the UPR pathway. Under ER stress, UPR sensors (e.g., PERK, IRE1 and, ATF6), which are located in outer-membrane of ER, are dissociated from GRP78 and initiate UPR cascade. Activated of the IRE1 arm of UPR induces splicing of XBP mRNA, subsequently leading to activation of UPR target genes. Activated PERK promotes ATF4 activation via phosphorylation of eIF2α. ATF6 arm is activated through cleavage in the Golgi with proteolysis, which is later translocated to the nucleus and induces expression of UPR-responsive genes.

Cross talk between UPR, autophagy, and apoptosis

ER stress and UPR can be activated by several cellular stimuli such as changes in extracellular factors (nutrition or oxygen) and intracellular damages (carcinogens or cancer-causing mutations). UPR is involved in several signal transduction pathways such as hypoxia, autophagy, mitochondrial biogenesis (oxidative stress), reactive oxygen species (ROS), and apoptosis [112, 124]. Importantly, autophagy and ER stress determines the cell fate, however, the exact molecular mechanisms have remained elusive [115]. The interaction of UPR with autophagy, especially with the PERK-eIF2a arm, is essential in autophagy induction [125]. Transcription of ATG genes (3, 5, 7, 10, 12) is activated by ATF4 which is the main transcription factor in PERK signaling pathway. ATF4 also contributes to cell growth suppression by provoking C/EBP homologous protein (CHOP), a DNA damage-inducible protein (GADD34), and proapoptotic BCL-2 family proteins [126]. On the other hand, IRE1 also induces autophagy through tumor necrosis factor (TNF) receptor-related factor 2 (TRAF2) and c-Jun N-terminal kinase (JNK) [127]. It has been reported that JNK causes phosphorylation of Bcl-2 and dissociation of Beclin-1 (autophagy regulatory protein), and leads to activation of phosphoinositide 3-kinase (PI3K) complex and autophagy induction [128]. The spliced XBP-1 (the downstream element of the IRE-1 arm) initiates autophagy, matures it through the Bcl-2/Beclin-1 complex, and converts microtubule-associated protein 1 light chain L3Cβ I to L3Cβ II [1, 108]. UPR also interacts with apoptosis through different intermediates including CHOP, JNK, proapoptotic BH3-containing protein (NOXA), and Caspase-4 and 12 [127]. ATF6, PERK and IRE1 arms trigger the proapoptotic process by activation of the downstream molecules (such as CHOP or JNK) and inducing cell death. Generally, three UPR arms induce transcription of CHOP, however, the PERK-eIF2α-ATF4 signaling branch is considerably more essential [129]. P38 MAPK, as a substrate of apoptosis signal-regulated kinase 1 (ASK1), is also recruited to IRE1-TRAF2 and activates CHOP through phosphorylation of serine residues 78 and 8. Thus, during uncontrolled stress, the IRE-1 and the PERK pathways may mediate synergistic effects via the CHOP molecule [130, 131]. CHOP intermediate apoptotic molecules are BCL2, GADD34, endoplasmic reticulum redox protein 1 (ERO1α) and tribbles-related protein 3 (TRB3) [132, 133] which consequently activates Bax and Bak, and caspase activations are detected. In summary, UPR arms change the cell fate from cell protection to proapoptotic cell death status [134]. The cross talk between UPR, autophagy, and apoptosis is presented in Fig. 6.

Role of UPR in glioblastoma chemoresistance

Three UPR arms may profit in several functions including tumor initiation, development, and progression, cell transformation, patient survival, antitumor immune responses, angiogenesis, metastasis and invasion, and chemoresistance properties [100, 135–138]. Several studies showed that knockdown of GRP78, ATF6, ATF4, and XBP1s is associated with the re-sensitization of chemoresistant cells [139, 140]. Generally, ER stress and UPR activation promote the development and progression of prostate cancer, breast cancer, brain cancer, and colon cancer [137, 141, 142]. Glioblastoma is an invasive, incurable, poorly prognosed,

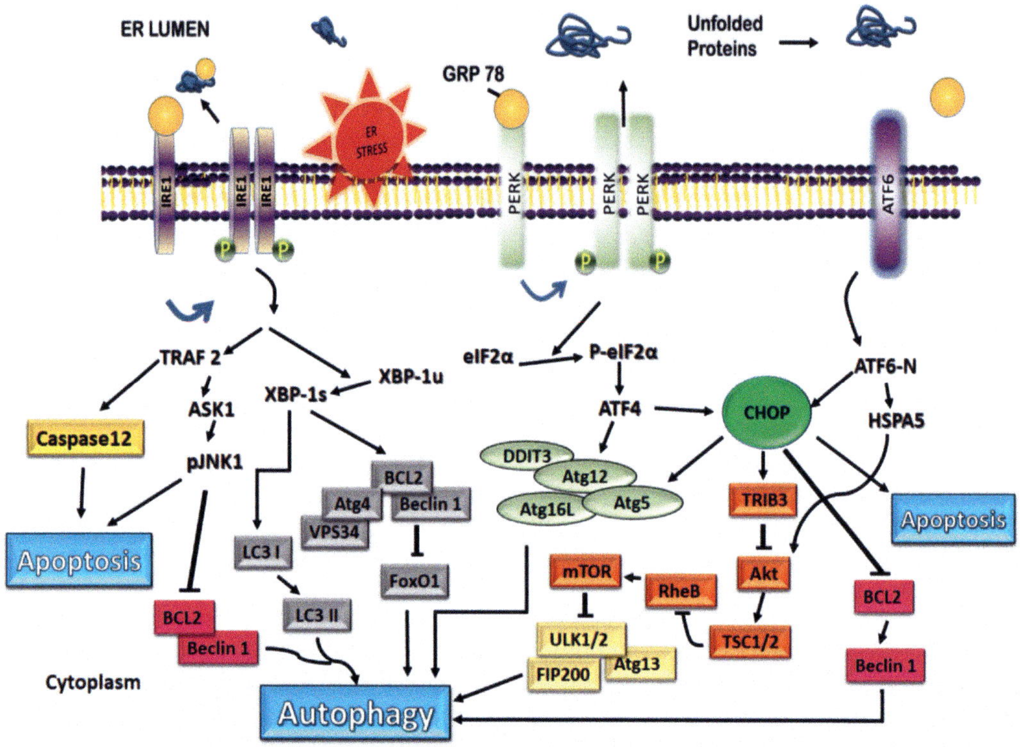

FIG. 6 Overview of close connection between ER stress, UPR, autophagy and apoptosis. The ER stress can activate autophagy and apoptosis through UPR arms (e.g., IRE1α, PERK, and the ATF6α) through activated UPR sensors. Activated PERK affects the expression of ATG12, ATG5, DDIT3 and ATG16L by ATF4 transcription factor; similarly CHOP activates TRIB3 that suppress the activity of Akt/mTOR pathway induced autophagy. The ATF6α can also induce autophagy by inhibiting phosphorylation at Akt and mTOR pathway. IRE1α induces autophagy through TRAF/ASK/JNK1 branch and mediates phosphorylation of Bcl2, which leads to Beclin-1 dissociation. Moreover, spliced XBP1 directly enhances the formation of LC3-I and LC3-II, which triggers Beclin-1 via decrease of FoxO1 activity. ER stress and UPR can induce apoptosis through an intrinsic pathway involving cytochrome c release from mitochondria and caspase activation. Activated IRE1α via TRAF2 directly or TRAF/ASK/JNK1 pathway induces caspase 12 activation and leads to apoptosis. Activated PERK and ATF6 can induce apoptosis through activation of CHOP.

and inevitably recurrence type of cancer [1]. Unfortunately, resistance to the only available chemotherapeutic drug, TMZ, rapidly occurs in glioblastoma and there are limited effective therapeutic approaches [143]. Nowadays, several combination therapies are considered for glioblastoma remedy [144]. Recent studies on molecular mechanisms involved in the cancer cells escape from chemotherapy-induced cell death have provided promising findings. For instance, researchers showed that cyclosporin A is cytotoxic to glioma cells and induces autophagy through the activation of PERK and IRE1 branches [112]. Similar to UPR, autophagy is primarily cells adaptive response against damages. Suppression of UPR or autophagy sensitizes glioma cells to cyclosporin A treatment, suggesting that UPR-dependent autophagy may enhance the treatment of chemoresistance [1, 145].

During tumor development and metastasis, the folding of proteins and intracellular transports are accelerated in the ER. Uncontrolled tumor cell growth leads to hypoxia, starvation, oxidative stress, decreased cellular calcium, ER stress, and eventually UPR initiation [138, 146]. It has been shown that UPR branches via GRP78, X-box-binding protein 1 (XBP1) with activated ATF6 promote GB chemoresistance [1, 147]. The role of UPR signaling pathway involved in chemoresistance and tumor survival is presented in Fig. 7.

Role of UPR branches in GBM-induced chemoresistant to TMZ

GRP78

In the normal adult brain, GRP78 is usually expressed at very low levels [148]. However, this level is elevated in glioma cell lines [149]. There is a direct association between the expression of GRP78 and the grade of tumors [1, 147]. Unfortunately, it is observed that the overexpression of BIP/GRP78, in response to TMZ treatment, activates UPR and autophagy, inhibits apoptosis, and consequently leads to drug resistance [150]. This overexpression in glioma cells increases their resistance to etoposide and cisplatin-induced apoptosis. However, the silencing of GRP78 increases their sensitivity to etoposide, cisplatin, radiation,

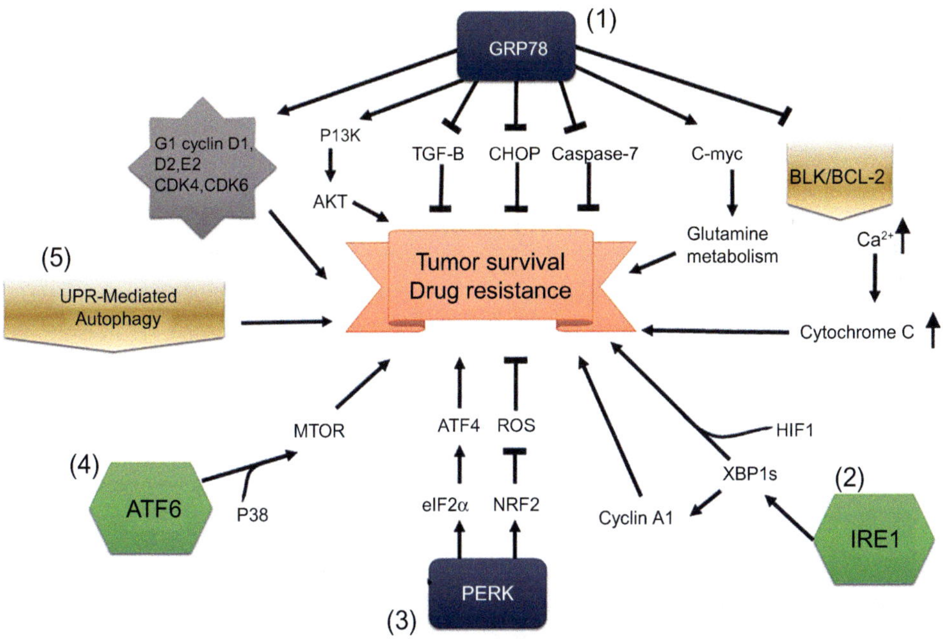

FIG. 7 UPR effects tumor survival and drug resistance via crosstalk with various pathways. The roles of (1) GRP78, (2) IRE-1/XBP1, (3) PERK, (4) ATF-6, and (5) autophagy activated under ER stress in tumor survival. CKD, cyclin-dependent kinase; HIF1α, hypoxia inducible factor-1; ROS, reactive oxygen species; PI3K, phosphatidylinositol 3-kinase; TGF-β, transforming growth factor-β; PDIA5, protein disulfide isomerase 5; NRF2, nuclear factor erythroid-2-related factor 2.

TMZ, 5-fluorouracil (5-FU), and camptothecin-11 (CPT-11) [149, 151]. Targeting the ATP-binding domain of GRP78 with epigallocatechin gallate inhibits its protective function and makes U251 glioma cells sensitive to TMZ [149]. Since TMZ activates UPR, the inhibition of UPR signaling may reduce the adaptive pro-survival effect of UPR and makes cells sensitive to TMZ. Alternatively, ER stress overload used in combination with TMZ and UPR activators may trigger fatal ER stress [1, 152].

IRE1

The IRE1/XBP-1s demonstrates the prosurvival effects in adaptive response. Inhibition of IRE1 or XBP1 enhances the toxicity of celecoxib derivative, OSU-03012, in GB cell lines, indicating that IRE1/XBP1 signaling promotes chemoresistance [153]. Evidence shows that XBP1 is abundantly expressed in gliomas and the knockdown of its expression leads to cell death [154]. Moreover, XBP1 silencing reduces glioma cell viability under hypoxic conditions, suggesting that XBP1 is involved in glioma cell survival [70, 154]. On the other hand, IRE1-JNK signaling increases the expression of epidermal growth factor receptor (EGFR) ligand, epiregulin, in human glioma cells [155]. Therefore, EGFR signaling is known to contribute to classical glial chemoresistance [156]. Glioma cells that express dominant-negative IRE1α have remarkably less growth rate and angiogenic signaling pathway activity [157].

PERK

PERK overexpression is observed in human glioma tissues. Higher levels of phosphorylated PERK correspond to tumors with higher grades. Moreover, the phosphorylation of eIF2α and induction of ATF4 are reported in GBM tumors [158]. Under low glucose conditions, silencing PERK decreases the survival rate of glioma cells and tumor formation capacity, making them susceptible to ER-stress-induced cell death [158, 159]. However, the exact role of PERK in the chemoresistance of GBM remains to be determined.

ATF6

The ATF6 pathway has been recently shown to enhance the viability of irradiated GB cells [160]. The knockdown of ATF6 promotes apoptotic cell death and decreases clone formation [161]. Although ATF6 is involved in tumor cell viability, further studies are needed to determine the contribution of ATF6 to chemoresistance in GBM cells.

UPR affects chemotherapy-induced cell death in GBM

Induction of UPR signaling in glioma cells is not always associated with chemoresistance-related responses. According to Johnson, White, and Grimaldi, (2011), classic endoplasmic reticulum stress inducers such as thapsigargin, tunicamycin, and brefeldin A, or celecoxib and bortezomib, increase ER stress responses and lead to glioma cell death [162]. Martin et al. [163] showed that the fatty acid 2-hydroxyoleic acid also induces endoplasmic reticulum stress, autophagy, and cell death in human glioma cell lines. Another study revealed that the novel phloroglucinol derivative 2,4-bis(4-fluoro-o-phenylacetyl) phloroglucinol (BFP) disrupts autophagy and apoptosis bridge connection by GRP78 or CHOP knockdown (PERK target and cross-promoting factor) and induces cell death in human glioma cell lines [164]. According to Meng et al. (2009), ruthenium derivative compound 11 (RDC11) elevates UPR markers including GRP78, XBP1, apoptosis protein 1 (PDI), and CHOP [1]. Similarly, siRNA-directed CHOP knockdown reduces cell death in U87MG glioma cells treated with RDC11, indicating the main role of CHOP in U87MG cell death [165]. It was also shown that PERK signaling is associated with the induction of human GBM cell death via melanoma differentiation-related gene 7 (mda-7) that encodes the interleukin 24 (IL-24), where ceramide synthase 6 and thioredoxin and autophagy are involved [1]. Another group showed that celecoxib derivative, OSU-03012, increases autophagy-induced cell death in U251 GBM cells [166]. Indeed, prolonged activation of PERK/eIF2α signaling by OSU-03012 regulates autophagy-induced tumor cell death [166, 167]. Combination of a novel ER stress inducer agent, JLK1486 with TMZ increases cell death in GBM cell lines, which is shown to be mediated by elevated levels of GRP78, ATF4, and CHOP. This implicates the role of the PERK arm through which the ATF4 and CHOP are induced [150, 168]. More studies are required to determine whether inhibition or activation of the UPR is more beneficial in GBM treatment. Therefore, it is possible to promote cancer cell death by either inhibiting the adaptive UPR or increasing the level of ER stress beyond the capacity of the cells [1, 143].

How simvastatin/TMZ affect GBM chemoresistance

As mentioned in previous sections, UPR can be exploited to target cancerous cells in the following two methods: (1) targeting each arm of UPR individually to directly induce cell death, (2) exacerbating the whole cascade to increase cell susceptibility to chemotherapies [1, 152]. However, more investigations are needed to determine which arm(s) of the UPR should be targeted to decrease chemoresistance and increase the patient survival rate. Finding the best combination of UPR inhibitors and chemotherapy to achieve better treatment for different types of brain tumors will be an important future challenge. It has been shown that statins, as HMG-CoA reductase inhibitors, suppress invasion and induce apoptosis in GBM cells. Statins decrease cancer cell proliferation, migration, and invasion by targeting cancer cell cycle, apoptosis and regulation of small Rho GTPase protein activity [169]. Recent investigations revealed that a combination of mevalonate cascade inhibitors including simvastatin, fluvastatin, atorvastatin, and lovastatin enhanced the efficacy of chemotherapy agents, especially TMZ [88, 170, 171]. The exact mechanism involved in sensitizing GBM cells to TMZ-induced apoptosis is not still clear. Our team is currently working in this field to identify such mechanisms involved in GBM sensitivity to TMZ-induced apoptosis. Shojaei et al. showed that simvastatin (simva) increases TMZ-induced apoptosis by targeting autophagosome/lysosome fusions and accumulation of autophagosome in the cytoplasm, leading to activation of caspases in GBM cell lines (U87, U251) and primary cells [88]. Our team has recently demonstrated that cotreatment of simvastatin and TMZ increases the sensitivity of GBM cells to the TMZ-induced cell death. This occurred simultaneously with UPR induction and autophagy flux inhibition, independent of HMG-CoA inhibition, and mevalonate cascade (unpublished data). Cotreatment with simvastatin and TMZ can induce UPR signaling, especially IRE-1 and PERK branches, and autophagic flux in U87 and U251 cells. It also increases the expression of GRP78, IRE-1, XBP-1s, cleaved caspase-3, Beclin-1, p62 and LC3β-II, and reduces the ratio of p-eIF2α/eIF2α as a marker of PERK activity. It seems that simvastatin modulates autophagy flux inhibition through the induction of IRE-1 in vitro GBM models. It is concluded that simvastatin, as a UPR inducer, manipulates the proapoptotic signaling arms and enhances the efficacy of TMZ treatment in GBM. These findings must be evaluated by GBM animal models and tissue engineered-tumor models to pinpoint the precise mechanisms in tumor initiation and progression.

Tissue-engineered tumor models

Tissue engineering is a multidisciplinary enterprise that combines the fundamentals of engineering, cellular biology, and biomaterials to develop biomimetic disease models for drug screening applications [172–174]. Tissue-engineered tumor models are able to imitate the complexities of tumors in vivo and overcome the limitations of conventional two-dimensional (2D) cell cultures [175–177]. Tissue engineering strategies open up a broad range of fundamental studies and allow for the characterization of mechanical and physical properties of the tumor extracellular matrix (ECM) [178, 179] and studying the underlying mechanisms of tumor vascularization, invasion, and migration [180–183]. An appropriate tumor model should embrace all components of natural tissue including both stromal cells and stromal ECM [184]. Cancer cells experience various mechanical and chemical signals from both their microenvironment and surrounding cells. Therefore, it is essential to imitate these signals within engineered matrices to mimic the functionality of the cells [180].

A standard approach for making in vitro models is to use the tissues harvested from animals or patients [185, 186], which are then explanted and cultured in vitro. Although these models retain the original 3D architecture in culture, they are only useful in relatively short-term cultures. Another popular approach is to use stem cells or tumor cells from malignant cell lines or fragments of human tumors and form cell aggregates, known as "tumor spheroid/organoids" [187, 188]. The transport of bioactive molecules and oxygen in these models allows researchers to reproduce the complexity of the multicellular cytostructure of in vivo tumors and facilitates extracellular matrix deposition [189]. Therefore, spheroid-based models are more accurate for drug screening compared to conventional 2D cultures. Despite the advantages, in vitro spheroid models are limited in providing physiological relevance, interactions between tumor cells and stroma cells, heterogeneous microenvironment, etc.

Biomaterials with properties similar to the physicochemical properties of the brain tumor are central to the successful development of various bioengineered models [175]. In addition to biocompatibility and biodegradability, ideal tumor matrices must be able to mimic the nurturing environment of tumors and recapitulate the mechanical properties of the tumor microenvironment. Naturally derived materials including collagen type I [190], hyaluronic acid (HA) [180, 191], chitosan [191], and gelatin [192] are among the most popular biomaterials that have been extensively used for the development of 3D models. To modify the degradation and mechanical properties of these matrices, synthetic materials such as polyethylene glycol (PEG) can be added to the naturally derived polymers [192]. In this section, we have highlighted the recent advances in the development of bioengineered brain tumor models and their application in drug studies, with particular attention to glioblastoma and neuroblastoma.

Bioengineered glioblastoma and neuroblastoma in vitro models

Primary brain tumors, such as GBM, are rapidly progressive and deadly forms of brain tumors with high mortality rates [193]. They arise from either astrocytes—the star-shaped cells that make up the "glue-like," or supportive tissue of the brain [194]. These tumors are usually highly malignant and proliferate and form a supportive network of blood vessels. They also have a complex microenvironment composed of various ECM proteins, glycosaminoglycans (GAGs) [195, 196] and vascular-related cytokines which promote cancer cell proliferation and migration [197]. Predicting the effects of different therapeutic pathways requires comprehensive in vitro models to recapitulate the complexity of the GBM in-vivo microenvironment. In a pioneering study, Lee et al. developed a physiological glioma-vascular niche model using 3D bioprinting technology for investigating the interaction of glioma and endothelial cells [198]. The vascular channels were fabricated by 3D printing gelatin fibers, as sacrificial layers, covered by collagen type I and loaded with patient-derived glioma cells. In another study, chitosan and HA composite hydrogel (C-HA), as the most abundant GAG composition of the ECM of the natural brain tumor, was used as scaffolding material [191]. Results showed that the composite scaffold enhances the tumor spheroid formation and promotes the stem-like properties of glioma cells compared to 2D cultures (Fig. 8A). Additionally, the invasiveness of GBM cells grown in C-HA scaffolds was significantly enhanced (Fig. 8B). To study the impact of stromal cells on tumor cell progression, the neuroblastoma tumor microenvironment was modeled through co-encapsulation of cancer cells combined with different ratios of mesenchymal stem cells (MSCs) in a collagen matrix [199]. The expression of signaling factors including vascular endothelial growth factor (VEGF), transforming growth factors (TGF-β), and collagen type I was studied. Results showed that the rate of tumor cell proliferation was increased by changing the percentages of MSCs leading to irregularly shaped outgrowth. Recently, several bioengineered 3D in vitro models have been developed to study the role of the microenvironment and particularly the matrix stiffness, structure, and density on the invasiveness and migration of tumor cells [200, 201]. The cell-ECM interactions play a critical role in GBM invasion. Therefore, most of in vitro models study the effect of biomaterials and their mechanical properties on cellular behavior. Harley's group has used HA/PEG/gelatin hydrogel with tunable stiffness and degradability to produce a 3D in vitro model of GBM [192]. They reported a significant change in cell morphology by increasing HA content in the matrix (Fig. 8C). Heffernan et al. developed a high-throughput approach for the measurement of GBM proliferation and invasion in HA/PEG/gelatin hydrogels [195]. They showed that by altering the scaffold stiffness from 5 to 2 kPa the rate and depth of invasion increased and the morphology of the cells was shifted from rounded to elongated. They also showed that HA hydrogel had dose-dependent effects on the behavior of U87MG encapsulated cells; the greater HA content the lower invasion and

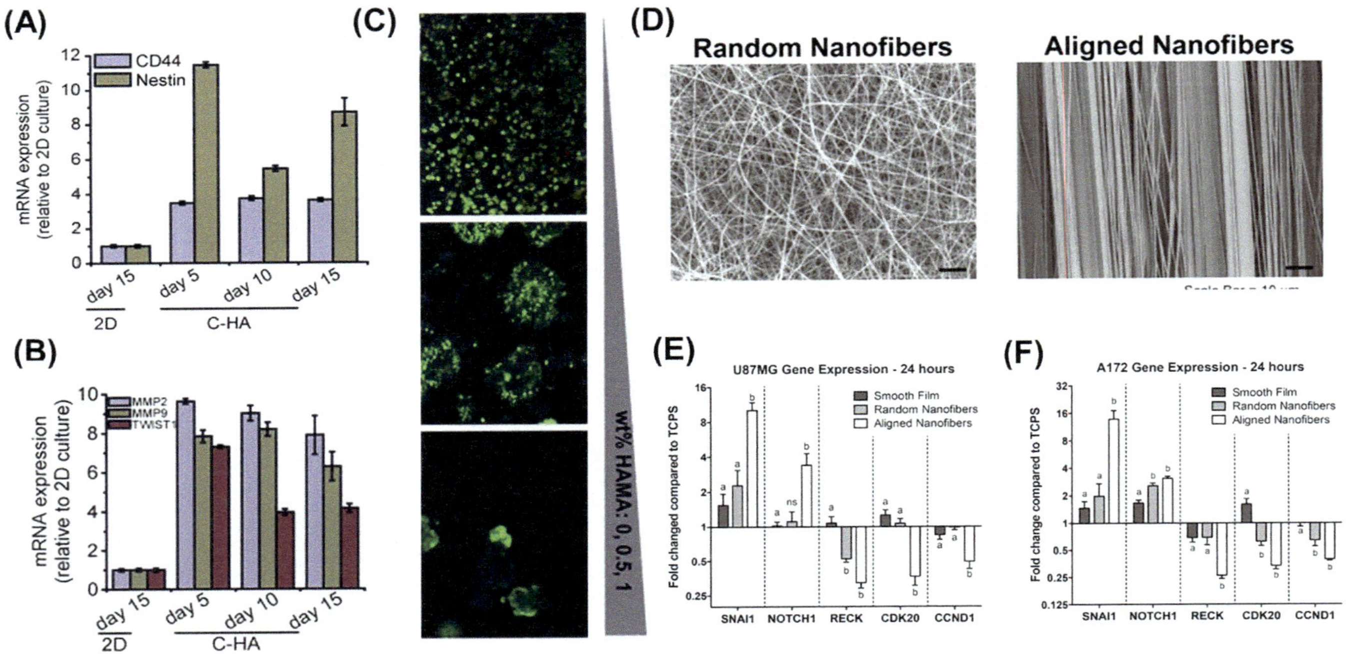

FIG. 8 (A) Expression of CD44 and Nestin on day 15, which were normalized with the values from 2D culture. In immunofluorescence images, CD44 and Nestin are stained *green* and *red*, respectively; and nuclei are counterstained blue. (B) C-HA scaffolds promote invasive phenotype. Expression of MMP-2, MMP-9, and TWIST1 by cells cultured on C-HA scaffolds over a 15-day period as related to cells cultured on 2D at day 15. (C) U87MG cells aggregated in the presence of HA in PEG. Scale bar 100 μm. (D) Scanning electron microscopy images of scaffolds with a smooth surface, random, and aligned nanofibers. (E) Quantitative real-time PCR results of gene expression in U87MG and (F) A172 cells when grown on smooth film, random, and aligned nanofibers for 24 h. *Panel (B) Reprinted with permission from Katt ME, Placone AL, Wong AD, Xu ZS, Searson PC. In vitro tumor models: advantages, disadvantages, variables, and selecting the right platform. Front Bioeng Biotechnol 2016;4. Copyright 2013, Elsevier. Panel (C) Reprinted with permission from Doke SK, Dhawale SC. Alternatives to animal testing: a review. Saudi Pharm J 2015;23(3):223–229. Copyright 2013, Elsevier. Panel (E and F) Reprinted with permission from Lee VK, Dai G, Zou H, Yoo S-S, editors. Generation of 3-D glioblastoma-vascular niche using 3-D bioprinting. 2015 41st Annual Northeast Biomedical Engineering Conference (NEBEC); 2015: IEEE.*

cell proliferation rate were observed owing to the downregulation of MMP-2-related genes. These results are aligned with the results of another research in which the collagen matrix with agarose as stiffening agents were used. It was demonstrated that increasing matrix stiffness by adding agarose reduced glioma spheroid invasion through inhibition of the traction-mediated remodeling of collagen fibers. However, in single collagen hydrogel, increasing collagen density promotes cell invasion through the fibers available for cell-ECM adhesions and traction [202].

ECM and stromal cells of the brain tumor microenvironment have a crucial role in tuning cancer cell motility and eventually cell migration and metastasis. Kumar's group has reviewed a variety of tissue engineering platforms designed to investigate the impact of signaling factors and different stromal cells such as endothelial, fibroblasts, immune cells, and astrocytes in GBMs [196]. A 3D bioprinted mini brain was developed to investigate the role of glioblastoma-associated macrophages (GAMs) in the progression and invasiveness of GBMs. It was observed that GBM progression was induced by infiltration and phenotype adjustment of macrophages through their interaction with glioblastoma cells (GL261). Moreover, overexpression of vimentin (Vim) and nestin (Nes) and significant loss of e-cadherin (Cdh1) were observed on the cocultured GBM cells, indicating the migratory phenotype of the glioblastoma cells [203]. Interstitial flow and ECM nano features are also important parameters in glioma cell invasion [204, 205]. A study on the invasion of U87MG cells, cultured on the HA/collagen matrix in the presence of CXCL12-driven autologous chemotaxis, showed that glioma cells mostly migrate through the floe direction [206]. To mimic ECM nano topographies, aligned polycaprolactone (PCL) nanofibers were coated with collagen and seeded with glioma cells (Fig. 8D) [207]. Overexpression of SNAI1 and NOTCH1, upregulation of epithelial to mesenchymal transition, and downregulation of CDK20 and CCND1 proteins were observed, indicating that the invasion and metastatic of GBM cells were increased (Fig. 8E and F). Overall, bioengineered in vitro models are promising platforms to study the role of cell-cell and cell-matrix interactions in tumor progression and invasion. Recapitulating the physiochemical properties of the native tumor and controlling the microenvironment at the cellular level are key factors in reproducing tumor-like pathophysiology and studding cell migration in bioengineered models [208].

Brain tumor tissue-engineered models for chemotherapeutic evaluations

Bioengineered models can be used to study the chemotherapeutic resistance in brain tumors by inducing hypoxia or inhibiting apoptosis as the primary mechanisms of chemosensitivity or resistance [209, 210]. Bioengineered in vitro models enable adjusting the specific parameters of the GBM microenvironment to induce the intracellular signaling pathways involved in drug resistance [211, 212]. Moreover, these models can clarify the role of heterogeneity in tumor cell populations in chemoresistance [208, 213, 214]. Compared to 2D cell culture, multicellular tumor spheroids (MCTS) are widely considered as suitable candidates for high-throughput analysis of chemotherapeutic drugs [213, 215]. These spheroids were used in a microfluidic platform to screen the combinatorial treatment of pitavastatin and irinotecan drugs on GBM (Fig. 9A) [216]. Testing these drugs on U87 tumor spheroids cultured in microwells enabled the simultaneous analysis of both induced apoptosis and

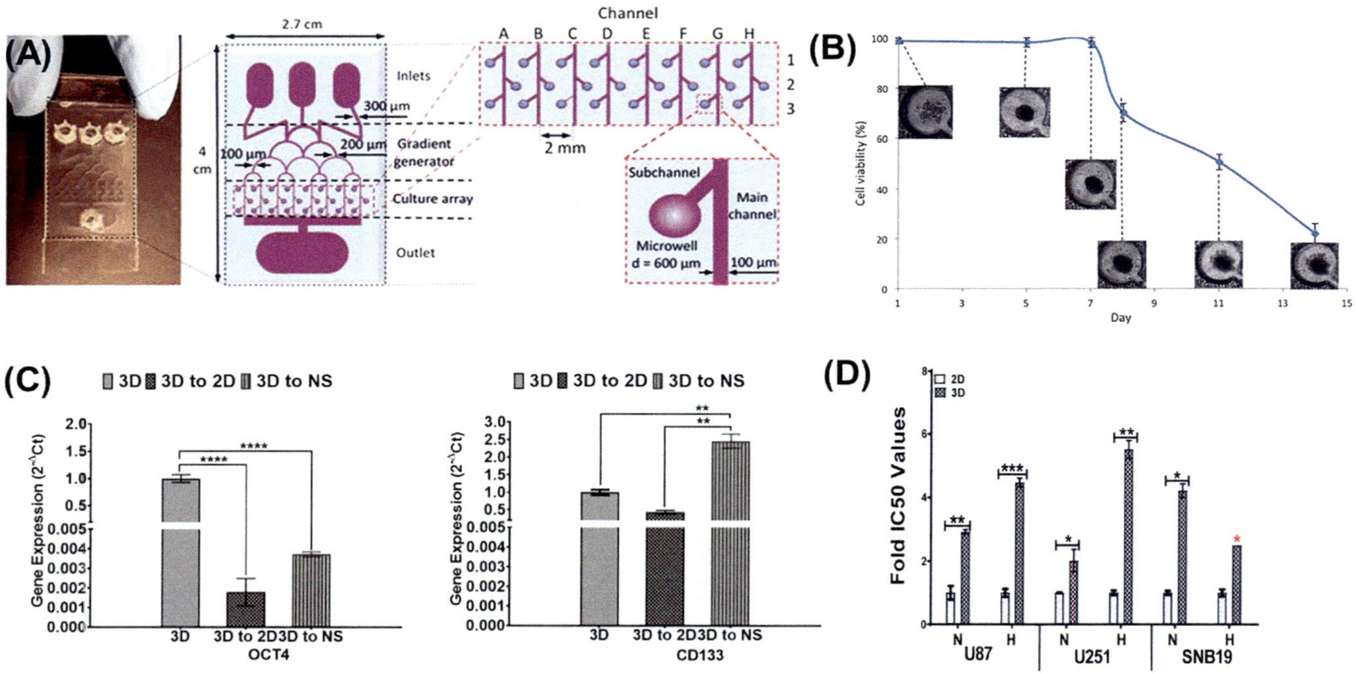

FIG. 9 (A) A microfluidic device with microchannels and microwells after adding inlet and outlet reservoirs on top of the inlet and outlet orifices. (B) Time-lapse images of U87 cells cultured in microwells after exposing them to a single dose of drug on day 7. Drugs were continuously released from the hydrogel layer to microwell and affected cancer spheroid. (C) Differences in the mRNA expression of CD133 and OCT4 in the 2D, 3D and neurosphere assays of U251 brain tumor cell culture. (D) Sensitivity of GBM cells to temozolomide in the 2D, 3D in terms of IC50 value. It shows significant resistance to TMZ in 3D cultures cells compared to 2D cultures ones. *Panel (B) Reprinted with permission from Ananthanarayanan B, Kim Y, Kumar S. Elucidating the mechanobiology of malignant brain tumors using a brain matrix-mimetic hyaluronic acid hydrogel platform. Biomaterials 2011;32(31):7913–7923. Panel (D) Reprinted with permission from Wartenberg et al. Wartenberg M, Ling FC, Müschen M, Klein F, Acker H, Gassmann M, Petrat K, Pütz V, Hescheler J, Sauer H. Regulation of the multidrug resistance transporter P-glycoprotein in multicellular tumor spheroids by hypoxia-inducible factor-1 and reactive oxygen species. FASEB J 2003;17(3):1–22.*

inhibited cell growth in treated spheroids. The results showed that at a certain gradient of the combinatorial drug concentrations, cancer cells are gradually dissociated from the tumor body and the spheroids became loose owing to autophagic cell death (Fig. 9B) [216]. To study 3D dynamic tumor recapitulation, GBM spheroids were used by Wang and coworkers to generate a glioblastoma model by manipulating the human glioma cells (U251) [217]. They conducted high-throughput monitoring of cellular responses in different fluidic microenvironments with various anticancer drug treatments (VCR and BLM). High-throughput analysis of drug concentration-related responses showed that a high concentration of VCR and BLM resulted in a significant decrease in tumor size. They also showed that the number of apoptotic cells was increased indicating a greater total resistance in lower concentration compared to the 2D condition. This integrated microfluidic design facilitated the mitochondrial depolarization assay and confirmed the apoptotic mechanisms of drug treatment on U251 cell spheroids. Moreover, the activation of the caspase-3 signaling pathway that leads to the apoptosis mechanism of the cell death was validated by observing as an increment in the number of caspase-3^+ cells with the treatment time [217]. To investigate the patient-specific response to the clinically observed chemotherapy resistance in a combination of TMZ and radiation therapies, Yi et al. developed a bioprinted model of human GBM which mimics all the pathophysiological complexities of the native human GBM [218]. In their study, the brain decellularized ECM ink generated a heterogeneous tumor cell environment surrounded by the microvessels where a hypoxic gradient condition occurred from the tumor core to the peripheral. They showed that under the hypoxic condition, $SOX2^+$-resistant cells emerged in the necrotic core of the tumor while they were surrounded by the $SHMT2^+$ pseudopalisading forming cells. This model was successfully used to observe different responses of GBMs to [218]. Cancer stem cells (CSC), hypoxia, and proinflammatory responses play important roles in the occurrence of chemoresistance in primary brain tumors [219, 220]. Hypoxia-inducible factor (HIF-2α) induces the expression of stem cell-associated genes in neuroblastoma [221]. Therefore, microfabricated tumor spheroids with hypoxic core can be used as potential in vitro 3D models to investigate the role of neuroblastoma stem cells in the chemoresistance process and therefore to assess the CSC targeted therapeutic compounds [222]. Hypoxia-initiated chemoresistance to TMZ was also studied on the 3D cultured glioblastoma cells using the Cultrex-based three-dimensional tumor growth assay (3D-TGA). The 3D cultured cells formed neurospheres in a hypoxia chamber with around 1% oxygen concentration. As demonstrated in Fig. 9C, tumor microenvironmental conditions can significantly affect the expression of various stem cell markers. It was shown that CD133 and OCT4 expressions were higher in the neurosphere and 3D culture assays, respectively [223]. Compared to 2D cultures, cells in 3D structures are normally more exposed to hypoxia conditions which lead to the in vivo like TMZ resistant owing to the presence of the stem cell populations within the specific tumor matrix, shown in Fig. 9D. Since apoptosis is highly sensitive to cell-ECM adhesion ligands, apoptosis resistance in GBM can be modeled by bioengineered ECM models. These models mimic 3D structure of biopolymers such as collagen and fibronectin and are able to tune the extent of integrin protein adhesion to ECM-based ligands. Therefore, they can change the level of induced resistance in cancer cells [224, 225]. Autophagy is another important intracellular mechanism of inducing chemoresistance in primary brain tumors, which has been studied using 3D tissue engineering concepts. Microenvironmental stresses applied to GBM cells activate the cell survival process through the autophagy mechanism, therefore one can tune the applied stresses on cancer cells in 3D developed structure of ECM and obtain the feedback of autophagy signaling pathways [226]. The same experiment was conducted on a 3D platform of laminin-rich ECM for breast cancer cell lines. It was shown that autophagy, which was induced in 3D culture of breast cancer cells by rapamycin treatment, increased the resistance to this drug. However, in 2D monolayer culture the autophagy mechanism was inhibited and alleviated drug resistance was observed [227].

Conclusion

GBM is a fatal disease and existing treatment modalities are moderately effective. To improve survival rates for GBM, evidence suggests that a novel and personalized therapeutic approach for the treatment of GBM patients is necessary. Differences in the genetic properties of GMB imply that extra off targets should be determined for an effective treatment. Indeed, several hopeful preclinical effects of targeted inhibitors have been observed. However, the identification and translation into the clinic of novel therapeutic agents that can cross the BBB remain a challenge. Although UPR or autophagy targeted therapies have less and limited efficacy when used as single agents, the combination of several targeted therapies may be more effective. The effect of signaling pathways such as autophagy and apoptosis can be investigated more precisely through 3D tissue engineering concepts. Therefore, further studies are certainly needed to design novel therapeutic policies for the treatment of GBM and GBM chemoresistance.

References

[1] Hombach-Klonisch S, Mehrpour M, Shojaei S, Harlos C, Pitz M, Hamai A, Siemianowicz K, Likus W, Wiechec E, Toyota BD. Glioblastoma and chemoresistance to alkylating agents: involvement of apoptosis, autophagy, and unfolded protein response. Pharmacol Ther 2018;184:13–41.

[2] Khosla D. Concurrent therapy to enhance radiotherapeutic outcomes in glioblastoma. Ann Trans Med 2016; 4(3):54–62.

[3] Ostrom QT, Gittleman H, Liao P, Rouse C, Chen Y, Dowling J, Wolinsky Y, Kruchko C, Barnholtz-Sloan J. CBTRUS statistical report: primary brain and central nervous system tumors diagnosed in the United States in 2007-2011. Neuro Oncol 2014;16(Suppl 4):iv1–63.

[4] Desai R, Suryadevara CM, Batich KA, Farber SH, Sanchez-Perez L, Sampson JH. Emerging immunotherapies for glioblastoma. Expert Opin Emerg Drugs 2016;21(2):133–45.

[5] Hanif F, Muzaffar K, Perveen K, Malhi SM, Simjee SU. Glioblastoma multiforme: a review of its epidemiology and pathogenesis through clinical presentation and treatment. Asian Pac J Cancer Prev 2017;18(1):3–9.

[6] Jazayeri SB, Rahimi-Movaghar V, Shokraneh F, Saadat S, Ramezani R. Epidemiology of primary CNS tumors in Iran: a systematic review. Asian Pac J Cancer Prev 2013;14(6):3979–85.

[7] Taylor OG, Brzozowski JS, Skelding KA. Glioblastoma multiforme: an overview of emerging therapeutic targets. Front Oncol 2019;9:963.

[8] Ohgaki H, Kleihues P. Genetic pathways to primary and secondary glioblastoma. Am J Pathol 2007; 170(5):1445–53.

[9] Warren KE, Gururangan S, Geyer JR, McLendon RE, Poussaint TY, Wallace D, Balis FM, Berg SL, Packer RJ, Goldman S, Minturn JE, Pollack IF, Boyett JM, Kun LE. A phase II study of O6-benzylguanine and temozolomide in pediatric patients with recurrent or progressive high-grade gliomas and brainstem gliomas: a Pediatric Brain Tumor Consortium study. J Neurooncol 2012;106(3):643–9.

[10] Waugh MG. Chromosomal instability and phosphoinositide pathway gene signatures in glioblastoma multiforme. Mol Neurobiol 2016;53(1):621–30.
[11] Cohen AL, Holmen SL, Colman H. IDH1 and IDH2 mutations in gliomas. Curr Neurol Neurosci Rep 2013; 13(5):345.
[12] Tran AN, Boyd NH, Walker K, Hjelmeland AB. NOS expression and NO function in glioma and implications for patient therapies. Antioxid Redox Signal 2017;26(17):986–99.
[13] He Y, Su J, Lan B, Gao Y, Zhao J. Targeting off-target effects: endoplasmic reticulum stress and autophagy as effective strategies to enhance temozolomide treatment. Onco Targets Ther 2019;12:1857–65.
[14] Alexandru O, Horescu C, Sevastre A-S, Cioc CE, Baloi C, Oprita A, Dricu A. Receptor tyrosine kinase targeting in glioblastoma: performance, limitations and future approaches. Contemp Oncol 2020;24(1):55.
[15] Kerr E, Holohan C, McLaughlin K, Majkut J, Dolan S, Redmond K, Riley J, McLaughlin K, Stasik I, Crudden M. Identification of an acetylation-dependant Ku70/FLIP complex that regulates FLIP expression and HDAC inhibitor-induced apoptosis. Cell Death Differ 2012;19(8):1317–27.
[16] Alizadeh J, Zeki AA, Mirzaei N, Tewary S, Moghadam AR, Glogowska A, Nagakannan P, Eftekharpour E, Wiechec E, Gordon JW. Mevalonate cascade inhibition by simvastatin induces the intrinsic apoptosis pathway via depletion of isoprenoids in tumor cells. Sci Rep 2017;7(1):1–14.
[17] Goldar S, Khaniani MS, Derakhshan SM, Baradaran B. Molecular mechanisms of apoptosis and roles in cancer development and treatment. Asian Pac J Cancer Prev 2015;16(6):2129–44.
[18] Khan KH, Blanco-Codesido M, Molife LR. Cancer therapeutics: targeting the apoptotic pathway. Crit Rev Oncol Hematol 2014;90(3):200–19.
[19] Shakeri R, Hosseinkhani S, Los MJ, Davoodi J, Jain MV, Cieślar-Pobuda A, Rafat M, Ardestani SK. Role of the salt bridge between glutamate 546 and arginine 907 in preservation of autoinhibited form of Apaf-1. Int J Biol Macromol 2015;81:370–4.
[20] Ghavami S, Eshraghi M, Kadkhoda K, Mutawe MM, Maddika S, Bay GH, Wesselborg S, Halayko AJ, Klonisch T, Los M. Role of BNIP3 in TNF-induced cell death—TNF upregulates BNIP3 expression. Biochim Biophys Acta Mol Cell Res 2009;1793(3):546–60.
[21] Ghavami S, Hashemi M, Ande SR, Yeganeh B, Xiao W, Eshraghi M, Bus CJ, Kadkhoda K, Wiechec E, Halayko AJ. Apoptosis and cancer: mutations within caspase genes. J Med Genet 2009;46(8):497–510.
[22] Shahsavari Z, Karami-Tehrani F, Salami S, Ghasemzadeh M. RIP1K and RIP3K provoked by shikonin induce cell cycle arrest in the triple negative breast cancer cell line, MDA-MB-468: necroptosis as a desperate programmed suicide pathway. Tumor Biol 2016;37(4):4479–91.
[23] Vidal S, Rodriguez-Bravo V, Galsky M, Cordon-Cardo C, Domingo-Domenech J. Targeting cancer stem cells to suppress acquired chemotherapy resistance. Oncogene 2014;33(36):4451–63.
[24] Wang D, Berglund A, Kenchappa RS, Forsyth PA, Mulé JJ, Etame AB. BIRC3 is a novel driver of therapeutic resistance in glioblastoma. Sci Rep 2016;6:21710.
[25] Gibson BA, Zhang Y, Jiang H, Hussey KM, Shrimp JH, Lin H, Schwede F, Yu Y, Kraus WL. Chemical genetic discovery of PARP targets reveals a role for PARP-1 in transcription elongation. Science 2016;353(6294):45–50.
[26] Majuelos-Melguizo J, Rodríguez MI, López-Jiménez L, Rodríguez-Vargas JM, Martín-Consuegra JMM, Serrano-Sáenz S, Gavard J, de Almodóvar JMR, Oliver FJ. PARP targeting counteracts gliomagenesis through induction of mitotic catastrophe and aggravation of deficiency in homologous recombination in PTEN-mutant glioma. Oncotarget 2015;6(7):4790.
[27] Barazzuol L, Jena R, Burnet NG, Meira LB, Jeynes JC, Kirkby KJ, Kirkby NF. Evaluation of poly (ADP-ribose) polymerase inhibitor ABT-888 combined with radiotherapy and temozolomide in glioblastoma. Radiat Oncol 2013;8(1):1–11.
[28] Liu X, Li Y, Hua C, Li S, Zhao G, Song H, Yu M, Huang Q. Oxidative stress inhibits growth and induces apoptotic cell death in human U251 glioma cells via the caspase-3-dependent pathway. Eur Rev Med Pharmacol Sci 2015;19(21):4068–75.
[29] Valdés-Rives SA, Casique-Aguirre D, Germán-Castelán L, Velasco-Velázquez MA, González-Arenas A. Apoptotic signaling pathways in glioblastoma and therapeutic implications. Biomed Res Int 2017;2017: 7403747.
[30] Nam G-H, Jo K-J, Park Y-S, Kawk HW, Kim S-Y, Kim Y-M. In vitro and in vivo induction of p53-dependent apoptosis by extract of euryale ferox salisb in A549 human caucasian lung carcinoma cancer cells is mediated through Akt signaling pathway. Front Oncol 2019;9:406.
[31] Sarkaria JN, Kitange GJ, James CD, Plummer R, Calvert H, Weller M, Wick W. Mechanisms of chemoresistance to alkylating agents in malignant glioma. Clin Cancer Res 2008;14(10):2900–8.

[32] Chen L, Zeng Y, Zhou S-F. Role of apoptosis in cancer resistance to chemotherapy. In: Current understanding of apoptosis-programmed cell death. Intech Open 2018.
[33] Zhang X, Ding K, Wang J, Li X, Zhao P. Chemoresistance caused by the microenvironment of glioblastoma and the corresponding solutions. Biomed Pharmacother 2019;109:39–46.
[34] Daniele S, Pietrobono D, Costa B, Giustiniano M, La Pietra V, Giacomelli C, La Regina G, Silvestri R, Taliani S, Trincavelli ML. Bax activation blocks self-renewal and induces apoptosis of human glioblastoma stem cells. ACS Chem Nerosci 2018;9(1):85–99.
[35] Pont LMB, Spoor JK, Venkatesan S, Swagemakers S, Kloezeman JJ, Dirven CM, van der Spek PJ, Lamfers ML, Leenstra S. The Bcl-2 inhibitor Obatoclax overcomes resistance to histone deacetylase inhibitors SAHA and LBH589 as radiosensitizers in patient-derived glioblastoma stem-like cells. Genes Cancer 2014;5(11–12):445.
[36] Foster KA, Jane EP, Premkumar DR, Morales A, Pollack IF. Co-administration of ABT-737 and SAHA induces apoptosis, mediated by Noxa upregulation, Bax activation and mitochondrial dysfunction in PTEN-intact malignant human glioma cell lines. J Neurooncol 2014;120(3):459–72.
[37] Uribe D, Torres Á, Rocha JD, Niechi I, Oyarzún C, Sobrevia L, San Martín R, Quezada C. Multidrug resistance in glioblastoma stem-like cells: Role of the hypoxic microenvironment and adenosine signaling. Mol Aspects Med 2017;55:140–51.
[38] Hsieh C-H, Lin Y-J, Wu C-P, Lee H-T, Shyu W-C, Wang C-C. Livin contributes to tumor hypoxia–induced resistance to cytotoxic therapies in glioblastoma multiforme. Clin Cancer Res 2015;21(2):460–70.
[39] Lopez PL, Filippi-Chiela EC, Silva AO, Cordero EA, Garcia-Santos D, Pelegrini AL, Reder GM, Barbieri NL, Lenz G. Sensitization of glioma cells by X-linked inhibitor of apoptosis protein knockdown. Oncology 2012;83(2):75–82.
[40] Beier D, Schulz JB, Beier CP. Chemoresistance of glioblastoma cancer stem cells-much more complex than expected. Mol Cancer 2011;10(1):1–11.
[41] Ma Z, Zhou H, Chang Q, Xue X. MicroRNA-based chemoresistance in glioblastoma. Glioma 2019;2(2):83.
[42] Auffinger B, Spencer D, Pytel P, Ahmed AU, Lesniak MS. The role of glioma stem cells in chemotherapy resistance and glioblastoma multiforme recurrence. Expert Rev Neurother 2015;15(7):741–52.
[43] Noguchi M, Hirata N, Tanaka T, Suizu F, Nakajima H, Chiorini JA. Autophagy as a modulator of cell death machinery. Cell Death Dis 2020;11(7):1–12.
[44] Saha S, Panigrahi DP, Patil S, Bhutia SK. Autophagy in health and disease: A comprehensive review. Biomed Pharmacother 2018;104:485–95.
[45] Zhang X, Deibert CP, Jaman E, Rao AV, Lotze MT, Amankulor NM. Autophagy inhibition is the next step in the treatment of glioblastoma patients following the Stupp era. Cancer Gene Ther 2020;8:1–13.
[46] Johansen T, Lamark T. Selective autophagy: ATG8 family proteins, LIR motifs and cargo receptors. J Mol Biol 2020;432(1):80–103.
[47] Mizushima N. A brief history of autophagy from cell biology to physiology and disease. Nat Cell Biol 2018;20 (5):521–7.
[48] Khaminets A, Behl C, Dikic I. Ubiquitin-dependent and independent signals in selective autophagy. Trends Cell Biol 2016;26(1):6–16.
[49] Escamilla-Ramírez A, Castillo-Rodríguez RA, Zavala-Vega S, Jimenez-Farfan D, Anaya-Rubio I, Briseño E, Palencia G, Guevara P, Cruz-Salgado A, Sotelo J. Autophagy as a potential therapy for malignant glioma. Pharmaceuticals 2020;13(7):156.
[50] Yim WW-Y, Mizushima N. Lysosome biology in autophagy. Cell Dis 2020;6(1):1–12.
[51] Li M, Gao P, Zhang J. Crosstalk between autophagy and apoptosis: potential and emerging therapeutic targets for cardiac diseases. Int J Mol Sci 2016;17(3):332.
[52] Denton D, Kumar S. Autophagy-dependent cell death. Cell Death Differ 2019;26(4):605–16.
[53] Decuypere J-P, Parys JB, Bultynck G. Regulation of the autophagic bcl-2/beclin 1 interaction. Cell 2012; 1(3):284–312.
[54] Xu H-D, Qin Z-H. Beclin 1, Bcl-2 and autophagy. In: Autophagy: biology and diseases. Springer; 2019. p. 109–26.
[55] Fu Y, Huang Z, Hong L, Lu J-H, Feng D, Yin X-M, Li M. Targeting ATG4 in cancer therapy. Cancer 2019;11(5):649.
[56] Kessel D, Reiners JJ. Photodynamic therapy: autophagy and mitophagy, apoptosis and paraptosis. Autophagy 2020;16(11):2098–101.
[57] Mrakovcic M, Fröhlich LF. p53-mediated molecular control of autophagy in tumor cells. Biomolecules 2018;8 (2):14.

[58] Lamb HM. Double agents of cell death: novel emerging functions of apoptotic regulators. FEBS J 2020; 287:2647–63.
[59] Gerónimo-Olvera C, Massieu L. Autophagy as a homeostatic mechanism in response to stress conditions in the central nervous system. Mol Neurobiol 2019;56(9):6594–608.
[60] Feng F, Zhang M, Yang C, Heng X, Wu X. The dual roles of autophagy in gliomagenesis and clinical therapy strategies based on autophagic regulation mechanisms. Biomed Pharmacother 2019;120:109441.
[61] Simpson JE, Gammoh N. The impact of autophagy during the development and survival of glioblastoma. Open Biol 2020;10(9):200184.
[62] Pavlides S, Vera I, Gandara R, Sneddon S, Pestell RG, Mercier I, Martinez-Outschoorn UE, Whitaker-Menezes D, Howell A, Sotgia F. Warburg meets autophagy: cancer-associated fibroblasts accelerate tumor growth and metastasis via oxidative stress, mitophagy, and aerobic glycolysis. Antioxid Redox Signal 2012;16(11):1264–84.
[63] Han W, Shi J, Cao J, Dong B, Guan W. Emerging roles and therapeutic interventions of aerobic glycolysis in glioma. Onco Targets Ther 2020;13:6937.
[64] Yang K, Niu L, Bai Y, Le W. Glioblastoma: Targeting the autophagy in tumorigenesis. Brain Res Bull 2019;153:334–40.
[65] Mowers EE, Sharifi MN, Macleod KF. Autophagy in cancer metastasis. Oncogene 2017;36(12):1619–30.
[66] Yamamoto K, Venida A, Yano J, Biancur DE, Kakiuchi M, Gupta S, Sohn AS, Mukhopadhyay S, Lin EY, Parker SJ. Autophagy promotes immune evasion of pancreatic cancer by degrading MHC-I. Nature 2020; 581(7806):100–5.
[67] Colella B, Faienza F, Di Bartolomeo S. EMT regulation by autophagy: a new perspective in glioblastoma biology. Cancer 2019;11(3):312.
[68] Kang C, Elledge SJ. How autophagy both activates and inhibits cellular senescence. Autophagy 2016;12(5):898–9.
[69] Wojton J, Meisen WH, Kaur B. How to train glioma cells to die: molecular challenges in cell death. J Neurooncol 2016;126(3):377–84.
[70] Liu X, Sun K, Wang H, Dai Y. Knockdown of retinoblastoma protein may sensitize glioma cells to cisplatin through inhibition of autophagy. Neurosci Lett 2016;620:137–42.
[71] Liu L-Q, Wang S-B, Shao Y-F, Shi J-N, Wang W, Chen W-Y, Ye Z-Y, Fang Q-X, Zhang G-B. Hydroxychloroquine potentiates the anti-cancer effect of bevacizumab on glioblastoma via the inhibition of autophagy. Biomed Pharmacother 2019;118:109339.
[72] Shen J, Zheng H, Ruan J, Fang W, Li A, Tian G, Niu X, Luo S, Zhao P. Autophagy inhibition induces enhanced proapoptotic effects of ZD6474 in glioblastoma. Br J Cancer 2013;109(1):164–71.
[73] Yu Z, Xie G, Zhou G, Cheng Y, Zhang G, Yao G, Chen Y, Li Y, Zhao G. NVP-BEZ235, a novel dual PI3K–mTOR inhibitor displays anti-glioma activity and reduces chemoresistance to temozolomide in human glioma cells. Cancer Lett 2015;367(1):58–68.
[74] Kang DW, Hwang WC, Noh YN, Kang Y, Jang Y, Kim JA, Min DS. Phospholipase D1 is upregulated by vorinostat and confers resistance to vorinostat in glioblastoma. J Cell Physiol 2020;236(1):549–60.
[75] Yu Q, Liu L, Wang P, Yao Y, Xue Y, Liu Y. EMAP-II sensitize U87MG and glioma stem-like cells to temozolomide via induction of autophagy-mediated cell death and G2/M arrest. Cell Cycle 2017;16(11):1085–92.
[76] Josset E, Burckel H, Noel G, Bischoff P. The mTOR inhibitor RAD001 potentiates autophagic cell death induced by temozolomide in a glioblastoma cell line. Anticancer Res 2013;33(5):1845–51.
[77] López-Valero I, Saiz-Ladera C, Torres S, Hernández-Tiedra S, García-Taboada E, Rodríguez-Fornés F, Barba M, Dávila D, Salvador-Tormo N, Guzmán M. Targeting glioma initiating cells with a combined therapy of cannabinoids and temozolomide. Biochem Pharmacol 2018;157:266–74.
[78] Luís A, Marcelino H, Rosa C, Domingues F, Pereira L, Cascalheira JF. The effects of cannabinoids on glioblastoma growth: a systematic review with meta-analysis of animal model studies. Eur J Pharmacol 2020;876:173055.
[79] Kim HY, Lee BI, Jeon JH, Kim DK, Kang S-G, Shim J-K, Kim SY, Kang SW, Jang H. Gossypol suppresses growth of temozolomide-resistant glioblastoma tumor spheres. Biomolecules 2019;9(10):595.
[80] Shao N, Mao J, Xue L, Wang R, Zhi F, Lan Q. Carnosic acid potentiates the anticancer effect of temozolomide by inducing apoptosis and autophagy in glioma. J Neurooncol 2019;141(2):277–88.
[81] Paul-Samojedny M, Pudełko A, Kowalczyk M, Fila-Daniłow A, Suchanek-Raif R, Borkowska P, Kowalski J. Combination therapy with AKT3 and PI3KCA siRNA enhances the antitumor effect of temozolomide and carmustine in T98G glioblastoma multiforme cells. BioDrugs 2016;30(2):129–44.

[82] Kanzawa T, Germano I, Komata T, Ito H, Kondo Y, Kondo S. Role of autophagy in temozolomide-induced cytotoxicity for malignant glioma cells. Cell Death Differ 2004;11(4):448–57.
[83] Compter I, Eekers DB, Hoeben A, Rouschop KM, Reymen B, Ackermans L, Beckervordersantforth J, Bauer NJ, Anten MM, Wesseling P. Chloroquine combined with concurrent radiotherapy and temozolomide for newly diagnosed glioblastoma: a phase IB trial. Autophagy 2020;1–9. https://doi.org/10.1080/15548627.2020.1816343.
[84] Gonçalves RM, Agnes JP, Delgobo M, de Souza PO, Thomé MP, Heimfarth L, Lenz G, Moreira JCF, Zanotto-Filho A. Late autophagy inhibitor chloroquine improves efficacy of the histone deacetylase inhibitor SAHA and temozolomide in gliomas. Biochem Pharmacol 2019;163:440–50.
[85] Hsu SP, Kuo JS, Chiang H-C, Wang H-E, Wang Y-S, Huang C-C, Huang Y-C, Chi M-S, Mehta MP, Chi K-H. Temozolomide, sirolimus and chloroquine is a new therapeutic combination that synergizes to disrupt lysosomal function and cholesterol homeostasis in GBM cells. Oncotarget 2018;9(6):6883.
[86] Jalota A, Kumar M, Das BC, Yadav AK, Chosdol K, Sinha S. Synergistic increase in efficacy of a combination of 2-deoxy-D-glucose and cisplatin in normoxia and hypoxia: switch from autophagy to apoptosis. Tumor Biol 2016;37(9):12347–58.
[87] Liu Y, Song X, Wu M, Wu J, Liu J. Synergistic effects of resveratrol and temozolomide against glioblastoma cells: underlying mechanism and therapeutic implications. Cancer Manag Res 2020;12:8341–54.
[88] Shojaei S, Koleini N, Samiei E, Aghaei M, Cole LK, Alizadeh J, Islam MI, Ar V, Albokashy M, Butterfield Y. Simvastatin increases temozolomide-induced cell death by targeting the fusion of autophagosomes and lysosomes. FEBS J 2020;287(5):1005–34.
[89] Pawlowska E, Szczepanska J, Szatkowska M, Blasiak J. An interplay between senescence, apoptosis and autophagy in glioblastoma multiforme—role in pathogenesis and therapeutic perspective. Int J Mol Sci 2018;19(3):889.
[90] Oliver L, Lalier L, Salaud C, Heymann D, Cartron PF, Vallette F. Drug resistance in glioblastoma: are persisters the key to therapy? Cancer Drug Res 2020;3:287–301.
[91] Buccarelli M, Marconi M, Pacioni S, De Pascalis I, D'Alessandris QG, Martini M, Ascione B, Malorni W, Larocca LM, Pallini R. Inhibition of autophagy increases susceptibility of glioblastoma stem cells to temozolomide by igniting ferroptosis. Cell Death Dis 2018;9(8):1–17.
[92] Ulasov I, Lenz G, Lesniak MS. Autophagy in glioma cells: an identity crisis with a clinical perspective. Cancer Lett 2018;428:139–46.
[93] Taylor MA, Das BC, Ray SK. Targeting autophagy for combating chemoresistance and radioresistance in glioblastoma. Apoptosis 2018;23(11–12):563–75.
[94] Wu W, Schecker J, Würstle S, Schneider F, Schönfelder M, Schlegel J. Aldehyde dehydrogenase 1A3 (ALDH1A3) is regulated by autophagy in human glioblastoma cells. Cancer Lett 2018;417:112–23.
[95] Chen C-M, Syu J-P, Way T-D, Huang L-J, Kuo S-C, Lin C-T, Lin C-L. BC3EE2, 9B, a synthetic carbazole derivative, upregulates autophagy and synergistically sensitizes human GBM8901 glioblastoma cells to temozolomide. Int J Mol Med 2015;36(5):1244–52.
[96] Jiang C, Shen F, Du J, Fang X, Li X, Su J, Wang X, Huang X, Liu Z. Upregulation of CASC2 sensitized glioma to temozolomide cytotoxicity through autophagy inhibition by sponging miR-193a-5p and regulating mTOR expression. Biomed Pharmacother 2018;97:844–50.
[97] Jiapaer S, Furuta T, Tanaka S, Kitabayashi T, Nakada M. Potential strategies overcoming the temozolomide resistance for glioblastoma. Neurol Med Chir 2018;58(10):405.
[98] Saveljeva S, Mc Laughlin S, Vandenabeele P, Samali A, Bertrand M. Endoplasmic reticulum stress induces ligand-independent TNFR1-mediated necroptosis in L929 cells. Cell Death Dis 2015;6(1). e1587-e.
[99] Sovolyova N, Healy S, Samali A, Logue SE. Stressed to death–mechanisms of ER stress-induced cell death. J Biol Chem 2014;395(1):1–13.
[100] Oakes SA. Endoplasmic reticulum stress signaling in cancer cells. Am J Pathol 2020;190(5):934–46.
[101] King AP, Wilson JJ. Endoplasmic reticulum stress: an arising target for metal-based anticancer agents. Chem Soc Rev 2020;49(22):8113–36.
[102] Hosseinzadeh S, Daemi A, Bolhassani A. Heat shock proteins as the efficient vehicle in cancer. J Solid Tumors 2012;2(3):47.
[103] Mohan S, Brown L, Ayyappan P. Endoplasmic reticulum stress: A master regulator of metabolic syndrome. Eur J Pharmacol 2019;860:172553.
[104] Venditti R, Masone MC, De Matteis MA. ER-Golgi membrane contact sites. Biochem Soc Trans 2020; 48(1):187–97.

[105] Dastghaib S, Kumar PS, Aftabi S, Damera G, Dalvand A, Sepanjnia A, Kiumarsi M, Aghanoori MR, Sohal SS, Ande SR, Alizadeh J, Mokarram P, Ghavami S, Sharma P, Zeki AA. Mechanisms targeting the unfolded protein response in asthma. Am J Respir Cell Mol Biol 2020;64(1):29–38.
[106] Di Conza G, Ho P-C. ER stress responses: an emerging modulator for innate immunity. Cell 2020;9(3):695.
[107] Chevet E, Hetz C, Samali A. Endoplasmic reticulum stress-activated cell reprogramming in oncogenesis. Cancer Discov 2015;5(6):586–97.
[108] Ghavami S, Yeganeh B, Zeki AA, Shojaei S, Kenyon NJ, Ott S, Samali A, Patterson J, Alizadeh J, Moghadam AR, Dixon I, Unruh H, Knight D, Post M, Klonisch T, Halayko A. Autophagy and the unfolded protein response promote profibrotic effects of TGF-β1 in human lung fibroblasts. Am J Physiol Lung Cell Mol Physiol 2018;314 (3):L493–504.
[109] Qi Z, Chen L. Endoplasmic reticulum stress and autophagy. In: Autophagy: biology and diseases. Springer; 2019. p. 167–77.
[110] Lindner P, Christensen SB, Nissen P, Moller JV, Engedal N. Cell death induced by the ER stressor thapsigargin involves death receptor 5, a non-autophagic function of MAP1LC3B, and distinct contributions from unfolded protein response components. Cell Commun Signal 2020;18(1):12.
[111] Poplawski T, Pytel D, Dziadek J, Majsterek I. Interplay between redox signaling, oxidative stress, and unfolded protein response (UPR) in pathogenesis of human diseases. Oxid Med Cell Longev 2019;2019:6949347.
[112] Senft D, Ze'ev AR. UPR, autophagy, and mitochondria crosstalk underlies the ER stress response. Trends Biochem Sci 2015;40(3):141–8.
[113] Lindholm D, Korhonen L, Eriksson O, SJFic K. Recent insights into the role of unfolded protein response in ER stress in health and disease. Front Cell Dev Biol 2017;5:48.
[114] Hwang J, Qi L. Quality control in the endoplasmic reticulum: crosstalk between ERAD and UPR pathways. Trends Biochem Sci 2018;43(8):593–605.
[115] Lu G, Luo H, Zhu X. Targeting the GRP78 pathway for cancer therapy. Front Med 2020;7:351.
[116] Gopal U, Pizzo SV. The endoplasmic reticulum chaperone GRP78 also functions as a cell surface signaling receptor. In: Cell surface GRP78, a new paradigm in signal transduction biology. Elsevier; 2018. p. 9–40.
[117] Mandula J, Mohamed E, Sierra-Mondragon R, Cao Y, Trillo-Tinoco J, Rodriguez P. Expression of cancer cell-associated Protein kinase R (PKR)-like endoplasmic reticulum kinase (PERK) regulates protective anti-tumor immunity in melanoma. J Immunol 2020;204. 162.13.
[118] Yeganeh B, Jäger R, Gorman AM, Samali A, Ghavami S. Induction of autophagy: Role of endoplasmic reticulum stress and unfolded protein response. In: Autophagy: cancer, other pathologies, inflammation, immunity, infection, and aging. Elsevier; 2015. p. 91–101.
[119] Ghavami S, Mutawe MM, Unruh H, Halayko AJ. The mevalonate cascade modulates endoplasmic reticulum (ER) stress during human airway mesenchymal cell apoptosis. In: A65 gene expression, phenotype, and function in airway smooth muscle: new insights. American Thoracic Society; 2010. p. A2124-A.
[120] Ghavami S, Yeganeh B, Stelmack G, Kashani H, Sharma P, Cunnington R, Rattan S, Bathe K, Klonisch T, Dixon I, Feed D, Halayko AJ. Apoptosis, autophagy and ER stress in mevalonate cascade inhibition-induced cell death of human atrial fibroblasts. Cell Death Dis 2012;3(6), e330.
[121] Moosavi MA, Rahmati M, Ashtari N, Alizadeh J, Hashemi M, Bathaei SZ, Ghavami S. Apoptosis, autophagy, and unfolded protein response and cerebellar development. In: Development of the cerebellum from molecular aspects to diseases. Springer; 2017. p. 153–78.
[122] Jin Y, Saatcioglu F. Targeting the unfolded protein response in hormone-regulated cancers. Trends Cancer 2020;6(2):160–71.
[123] Kim HS, Kim Y, Lim MJ, Park Y-G, Park SI, Sohn J. The p38-activated ER stress-ATF6α axis mediates cellular senescence. FASEB J 2019;33(2):2422–34.
[124] Cook KL, Shajahan AN, Wärri A, Jin L, Hilakivi-Clarke LA, Clarke R. Glucose-regulated protein 78 controls cross-talk between apoptosis and autophagy to determine antiestrogen responsiveness. Cancer Res 2012;72 (13):3337–49.
[125] Zheng W, Xie W, Yin D, Luo R, Liu M, Guo F. ATG5 and ATG7 induced autophagy interplays with UPR via PERK signaling. Cell Commun Signal 2019;17(1):1–16.
[126] Liu Y, Sun L, Ma Y, Wei B, Gao M, Shang L. High glucose and bupivacaine-induced cytotoxicity is mediated by enhanced apoptosis and impaired autophagy via the PERK-ATF4-CHOP and IRE1-TRAF2 signaling pathways. Mol Med Rep 2019;20(3):2832–42.
[127] Song S, Tan J, Miao Y, Li M, Zhang Q. Crosstalk of autophagy and apoptosis: Involvement of the dual role of autophagy under ER stress. J Cell Physiol 2017;232(11):2977–84.

[128] Zhang C, Jia X, Wang K, Bao J, Li P, Chen M, Wan J-B, Su H, Mei Z, He C. Polyphyllin VII induces an autophagic cell death by activation of the JNK pathway and inhibition of PI3K/AKT/mTOR pathway in HepG2 cells. PLoS One 2016;11(1), e0147405.
[129] Ge H, Li Z, Jiang L, Li Q, Geng C, Yao X, Shi X, Liu Y, Cao J. Cr (VI) induces crosstalk between apoptosis and autophagy through endoplasmic reticulum stress in A549 cells. Chem Biol Interact 2019;298:35–42.
[130] Moustafa EM, Rashed LA, El-Sebaie MM, Thabet NM, Abdel-Rafei MK. Crosstalk between ER-stress and apoptosis in irradiated HepG2 cells with gemcitabine: implication of PI3K/AKT and IκB/NF-κB signaling pathways. J Radiat Res Appl Sci 2020;13(1):144–54.
[131] Chen Q, Men Y, Wang D, Xu D, Liu S, Xiao S, Fang L. Porcine reproductive and respiratory syndrome virus infection induces endoplasmic reticulum stress, facilitates virus replication, and contributes to autophagy and apoptosis. Sci Rep 2020;10(1):1–13.
[132] Szegezdi E, Logue SE, Gorman AM, Samali A. Mediators of endoplasmic reticulum stress-induced apoptosis. EMBO Rep 2006;7(9):880–5.
[133] Tabas I, Ron D. Integrating the mechanisms of apoptosis induced by endoplasmic reticulum stress. Nat Cell Biol 2011;13(3):184–90.
[134] Li Y, Guo Y, Tang J, Jiang J, Chen Z. New insights into the roles of CHOP-induced apoptosis in ER stress. Acta Biochim Biophys Sin 2014;46(8):629–40.
[135] Li X, Zhang K, Li Z. Unfolded protein response in cancer: the physician's perspective. J Hematol Oncol 2011; 4(1):8.
[136] Hsu S-K, Chiu C-C, Dahms H-U, Chou C-K, Cheng C-M, Chang W-T, Cheng K-C, Wang H-MD, Lin I. Unfolded protein response (UPR) in survival, dormancy, immunosuppression, metastasis, and treatments of cancer cells. Int J Mol Sci 2019;20(10):2518.
[137] Hazari YM, Bashir A, ul Haq E, Fazili KM. Emerging tale of UPR and cancer: an essentiality for malignancy. Tumor Biol 2016;37(11):14381–90.
[138] Urra H, Dufey E, Avril T, Chevet E, Hetz C. Endoplasmic reticulum stress and the hallmarks of cancer. Trends Cancer 2016;2(5):252–62.
[139] Jing X, Shi Q, Bi W, Zeng Z, Liang Y, Wu X, Xiao S, Liu J, Yang L, Tao E. Rifampicin protects PC12 cells from rotenone-induced cytotoxicity by activating GRP78 via PERK-eIF2α-ATF4 pathway. PLoS One 2014;9(3), e92110.
[140] Gupta A, Hossain MM, Miller N, Kerin M, Callagy G, Gupta S. NCOA3 coactivator is a transcriptional target of XBP1 and regulates PERK–eIF2α–ATF4 signalling in breast cancer. Oncogene 2016;35(45):5860–71.
[141] Liu J, Xiao M, Li J, Wang D, He Y, He J, Gao F, Mai L, Li Y, Liang Y. Activation of UPR signaling pathway is associated with the malignant progression and poor prognosis in prostate cancer. Prostate 2017;77(3):274–81.
[142] Scriven P, Coulson S, Haines R, Balasubramanian S, Cross S, Wyld L. Activation and clinical significance of the unfolded protein response in breast cancer. Br J Cancer 2009;101(10):1692–8.
[143] Markouli M, Strepkos D, Papavassiliou AG, Piperi C. Targeting of endoplasmic reticulum (er) stress in gliomas. Pharmacol Res 2020;157:104823.
[144] Fajardo NMP, Meijer C, Kruyt FA. The endoplasmic reticulum stress/unfolded protein response in gliomagenesis, tumor progression and as a therapeutic target in glioblastoma. Biochem Pharmacol 2016;118:1–8.
[145] Groenendyk J, Paskevicius T, Urra H, Viricel C, Wang K, Barakat K, Hetz C, Kurgan L, Agellon LB, Michalak M. Cyclosporine A binding to COX-2 reveals a novel signaling pathway that activates the IRE1α unfolded protein response sensor. Sci Rep 2018;8(1):1–15.
[146] Giampietri C, Petrungaro S, Conti S, Facchiano A, Filippini A, Ziparo E. Cancer microenvironment and endoplasmic reticulum stress response. Mediators Inflamm 2015;2015:417281.
[147] Epple LM, Dodd RD, Merz AL, Dechkovskaia AM, Herring M, Winston BA, Lencioni AM, Russell RL, Madsen H, Nega M, Dusto NL, White J, Bigner DD, Nicchitta CV, Serkova NJ, Graner MW. Induction of the unfolded protein response drives enhanced metabolism and chemoresistance in glioma cells. PLoS One 2013;8(8), e73267.
[148] Zhu G, Lee AS. Role of the unfolded protein response, GRP78 and GRP94 in organ homeostasis. J Cell Physiol 2015;230(7):1413–20.
[149] Pyrko P, Schönthal AH, Hofman FM, Chen TC, Lee AS. The unfolded protein response regulator GRP78/BiP as a novel target for increasing chemosensitivity in malignant gliomas. Cancer Res 2007;67(20):9809–16.
[150] Weatherbee JL, Kraus J-L, Ross AH. ER stress in temozolomide-treated glioblastomas interferes with DNA repair and induces apoptosis. Oncotarget 2016;7(28):43820.

[151] Lee HK, Xiang C, Cazacu S, Finniss S, Kazimirsky G, Lemke N, Lehman NL, Rempel SA, Mikkelsen T, Brodie C. GRP78 is overexpressed in glioblastomas and regulates glioma cell growth and apoptosis. Neuro Oncol 2008;10(3):236–43.
[152] Healy SJ, Gorman AM, Mousavi-Shafaei P, Gupta S, Samali A. Targeting the endoplasmic reticulum-stress response as an anticancer strategy. Eur J Pharmacol 2009;625(1–3):234–46.
[153] Booth L, Roberts JL, Cruickshanks N, Grant S, Poklepovic A, Dent P. Regulation of osu-03012 toxicity by ER stress proteins and ER stress–inducing drugs. Mol Cancer Ther 2014;13(10):2384–98.
[154] Liu Y, Zhang X, Liang Y, Yu H, Chen X, Zheng T, Zheng B, Wang L, Zhao L, Shi CJN. neurobiology a. Targeting X box-binding protein-1 (XBP1) enhances sensitivity of glioma cells to oxidative stress. Neuropathol Appl Neurobiol 2011;37(4):395–405.
[155] Auf G, Jabouille A, Delugin M, Guerit S, Pineau R, North S, Platonova N, Maitre M, Favereaux A, Vajkoczy P, Seno M, Bikfalvi A, Minchenko D, Minchenko O, Moenner M. High epiregulin expression in human U87 glioma cells relies on IRE1α and promotes autocrine growth through EGF receptor. BMC Cancer 2013;13(1):1–12.
[156] Huang PH, Xu AM, White FM. Oncogenic EGFR signaling networks in glioma. Sci Signal 2009;2(87). re6-re.
[157] Auf G, Jabouille A, Guerit S, Pineau R, Delugin M, Bouchecareilh M, Magnin N, Favereaux A, Maitre M, Gaiser T, von Deimling A, Czabanka M, Vajkoczy P, Chevet E, Bikfalvi A, Moenner M. Inositol-requiring enzyme 1α is a key regulator of angiogenesis and invasion in malignant glioma. Proc Natl Acad Sci U S A 2010;107(35):15553–8.
[158] Hou X, Liu Y, Liu H, Chen X, Liu M, Che H, Guo F, Wang C, Zhang D, Wu J, Chen X, Shen C, Li C, Peng F, Bi Y, Yang Z, Yang G, Ai J, Gao X, Zhao S. PERK silence inhibits glioma cell growth under low glucose stress by blockage of p-AKT and subsequent HK2's mitochondria translocation. Sci Rep 2015;5:9065.
[159] Dadey DY, Kapoor V, Khudanyan A, Thotala D, Hallahan DE. PERK regulates glioblastoma sensitivity to ER stress although promoting radiation resistance. Mol Cancer Res 2018;16(10):1447–53.
[160] Dadey DY, Kapoor V, Khudanyan A, Urano F, Kim AH, Thotala D, Hallahan DE. The ATF6 pathway of the ER stress response contributes to enhanced viability in glioblastoma. Oncotarget 2016;7(2):2080.
[161] Shah A, Kumar A. Methamphetamine-mediated endoplasmic reticulum (ER) stress induces type-1 programmed cell death in astrocytes via ATF6, IRE1α and PERK pathways. Oncotarget 2016;7(29):46100.
[162] Johnson GG, White MC, Grimaldi M. Stressed to death: targeting endoplasmic reticulum stress response induced apoptosis in gliomas. Curr Pharm Des 2011;17(3):284–92.
[163] Martin ML, Barceló-Coblijn G, de Almeida RF, Noguera-Salvà MA, Teres S, Higuera M, Liebisch G, Schmitz G, Busquets X, Escriba PV. The role of membrane fatty acid remodeling in the antitumor mechanism of action of 2-hydroxyoleic acid. Biochim Biophys Acta Biomemb 2013;1828(5):1405–13.
[164] Lu D-Y, Chang C-S, Yeh W-L, Tang C-H, Cheung C-W, Leung Y-M, Liu J-F, Wong K-L. The novel phloroglucinol derivative BFP induces apoptosis of glioma cancer through reactive oxygen species and endoplasmic reticulum stress pathways. Phytomedicine 2012;19(12):1093–100.
[165] Meng X, Leyva ML, Jenny M, Gross I, Benosman S, Fricker B, Harlepp S, Hébraud P, Boos A, Wlosik P. A ruthenium-containing organometallic compound reduces tumor growth through induction of the endoplasmic reticulum stress gene CHOP. Cancer Res 2009;69(13):5458–66.
[166] Park MA, Yacoub A, Rahmani M, Zhang G, Hart L, Hagan MP, Calderwood SK, Sherman MY, Koumenis C, Spiegel S. OSU-03012 stimulates PKR-like endoplasmic reticulum-dependent increases in 70-kDa heat shock protein expression, attenuating its lethal actions in transformed cells. Mol Pharmacol 2008;73(4):1168–84.
[167] Shen K, Johnson DW, Vesey DA, McGuckin MA, Gobe GCJCS. Chaperones. Role of the unfolded protein response in determining the fate of tumor cells and the promise of multi-targeted therapies. J Cell Stress Chaperones 2018;23(3):317–34.
[168] Weatherbee JL, Kraus J-L, Moser RP, Ross AH. ER stress in temozolomide-treated glioblastomas interferes with DNA repair and induces apoptosis. Oncotarget 2013;12(28):43820–34.
[169] Ahmadi M, Amiri S, Pecic S, Machaj F, Rosik J, Łos MJ, Alizadeh J, Mahdian R, da Silva Rosa SC, Schaafsma D, Madrakian T, Zeki AA, Ghavami S. Pleiotropic effects of statins: a focus on cancer. Biochim Biophys Acta Mol Basis Dis 2020;165968. https://doi.org/10.1016/j.bbadis.2020;1866.
[170] Shojaei S, Alizadeh J, Thliveris J, Koleini N, Kardami E, Hatch GM, Xu F, Hombach-Klonisch S, Klonisch T, Ghavami S. Statins: a new approach to combat temozolomide chemoresistance in glioblastoma. J Invest Med 2018;66(8):1083–7.
[171] Pirmoradi L, Seyfizadeh N, Ghavami S, Zeki AA, Shojaei S. Targeting cholesterol metabolism in glioblastoma: a new therapeutic approach in cancer therapy. J Invest Med 2019;67(4):715–9.

[172] Akbari M, Tamayol A, Bagherifard S, Serex L, Mostafalu P, Faramarzi N, Mohammadi MH, Khademhosseini A. Textile technologies and tissue engineering: a path toward organ weaving. Adv Healthcare Mater 2016;5(7):751–66.
[173] Khademhosseini A, Langer R. Microengineered hydrogels for tissue engineering. Biomaterials 2007;28(34):5087–92.
[174] Khademhosseini A, Vacanti JP, Langer R. Progress in tissue engineering. Sci Am 2009;300(5):64–71.
[175] Annabi N, Tamayol A, Uquillas JA, Akbari M, Bertassoni LE, Cha C, Camci-Unal G, Dokmeci MR, Peppas NA, Khademhosseini A. 25th anniversary article: rational design and applications of hydrogels in regenerative medicine. Adv Mater 2014;26(1):85–124.
[176] Kharaziha M, Memic A, Akbari M, Brafman DA, Nikkhah M. Nano-enabled approaches for stem cell-based cardiac tissue engineering. Adv Healthcare Mater 2016;5(13):1533–53.
[177] Shin SR, Zihlmann C, Akbari M, Assawes P, Cheung L, Zhang K, Manoharan V, Zhang YS, Yüksekkaya M, Kt W. Reduced graphene oxide-GelMA hybrid hydrogels as scaffolds for cardiac tissue engineering. Small 2016;12(27):3677–89.
[178] Grinnell F, Petroll WM. Cell motility and mechanics in three-dimensional collagen matrices. Annu Rev Cell Dev Biol 2010;26:335–61.
[179] Herrmann D, Conway JR, Vennin C, Magenau A, Hughes WE, Morton JP, Timpson P. Three-dimensional cancer models mimic cell–matrix interactions in the tumour microenvironment. Carcinogenesis 2014;35(8):1671–9.
[180] Song H-HG, Park KM, Gerecht S. Hydrogels to model 3D in vitro microenvironment of tumor vascularization. Adv Drug Deliv Rev 2014;79:19–29.
[181] Szot CS, Buchanan CF, Freeman JW, Rylander MN. 3D in vitro bioengineered tumors based on collagen I hydrogels. Biomaterials 2011;32(31):7905–12.
[182] Talukdar S, Mandal M, Hutmacher DW, Russell PJ, Soekmadji C, Kundu SC. Engineered silk fibroin protein 3D matrices for in vitro tumor model. Biomaterials 2011;32(8):2149–59.
[183] Pathak A, Kumar S. Independent regulation of tumor cell migration by matrix stiffness and confinement. Proc Natl Acad Sci U S A 2012;109(26):10334–9.
[184] Katt ME, Placone AL, Wong AD, Xu ZS, Searson PC. In vitro tumor models: advantages, disadvantages, variables, and selecting the right platform. Front Bioeng Biotechnol 2016;4.
[185] Doke SK, Dhawale SC. Alternatives to animal testing: a review. Saudi Pharm J 2015;23(3):223–9.
[186] Elliott NT, Yuan F. A review of three-dimensional in vitro tissue models for drug discovery and transport studies. J Pharm Sci 2011;100(1):59–74.
[187] Kunz-Schughart LA, Freyer JP, Hofstaedter F, Ebner R. The use of 3-D cultures for high-throughput screening: the multicellular spheroid model. J Biomol Screen 2004;9(4):273–85.
[188] Friedrich J, Seidel C, Ebner R, Kunz-Schughart LA. Spheroid-based drug screen: considerations and practical approach. Nat Protoc 2009;4(3):309–24.
[189] Mehta G, Hsiao AY, Ingram M, Luker GD, Takayama S. Opportunities and challenges for use of tumor spheroids as models to test drug delivery and efficacy. J Control Release 2012;164(2):192–204.
[190] Chung S, Sudo R, Mack PJ, Wan C-R, Vickerman V, Kamm RD. Cell migration into scaffolds under co-culture conditions in a microfluidic platform. Lab Chip 2009;9(2):269–75.
[191] Florczyk SJ, Wang K, Jana S, Wood DL, Sytsma SK, Sham JG, Kievit FM, Zhang M. Porous chitosan-hyaluronic acid scaffolds as a mimic of glioblastoma microenvironment ECM. Biomaterials 2013;34(38):10143–50.
[192] Pedron S, Becka E, Harley BA. Regulation of glioma cell phenotype in 3D matrices by hyaluronic acid. Biomaterials 2013;34(30):7408–17.
[193] Gao Y, Fotovati A, Lee C, Wang M, Cote G, Guns E, Toyota B, Faury D, Jabado N, Dunn SE. Inhibition of Y-box binding protein-1 slows the growth of glioblastoma multiforme and sensitizes to temozolomide independent O6-methylguanine-DNA methyltransferase. Mol Cancer Ther 2009;8(12):3276–84.
[194] Panigrahi M, Das P, Parikh P. Brain tumor and Gliadel wafer treatment. Indian J Cancer 2011;48(1):11.
[195] Heffernan JM, Overstreet DJ, Le LD, Vernon BL, Sirianni RW. Bioengineered scaffolds for 3D analysis of glioblastoma proliferation and invasion. Ann Biomed Eng 2015;43(8):1965–77.
[196] Rape A, Ananthanarayanan B, Kumar S. Engineering strategies to mimic the glioblastoma microenvironment. Adv Drug Deliv Rev 2014;79:172–83.
[197] Knott JC, Mahesparan R, Garcia-Cabrera I, Bølge Tysnes B, Edvardsen K, Ness GO, Mørk S, Lund-Johansen M, Bjerkvig R. Stimulation of extracellular matrix components in the normal brain by invading glioma cells. Int J Cancer 1998;75(6):864–72.

[198] Lee VK, Dai G, Zou H, Yoo S-S, editors. Generation of 3-D glioblastoma-vascular niche using 3-D bioprinting. 2015 41st Annual Northeast Biomedical Engineering Conference (NEBEC). IEEE; 2015.
[199] Yeung P, Sin HS, Chan S, Chan GCF, Chan BP. Microencapsulation of neuroblastoma cells and mesenchymal stromal cells in collagen microspheres: a 3D model for cancer cell niche study. PLoS One 2015;10(12), e0144139.
[200] Lu P, Weaver VM, Werb Z. The extracellular matrix: a dynamic niche in cancer progression. J Cell Biol 2012;196 (4):395–406.
[201] Wang C, Tong X, Yang F. Bioengineered 3D brain tumor model to elucidate the effects of matrix stiffness on glioblastoma cell behavior using PEG-based hydrogels. Mol Pharm 2014;11(7):2115–25.
[202] Ulrich TA, de Juan Pardo EM, Kumar S. The mechanical rigidity of the extracellular matrix regulates the structure, motility, and proliferation of glioma cells. Cancer Res 2009;69(10):4167–74.
[203] Heinrich MA, Bansal R, Lammers T, Zhang YS, Michel Schiffelers R, Prakash J. 3D-bioprinted mini-brain: a glioblastoma model to study cellular interactions and therapeutics. Adv Mater 2019;31(14):1806590.
[204] Nagano M, Hoshino D, Koshikawa N, Akizawa T, Seiki M. Turnover of focal adhesions and cancer cell migration. Int J Cell Biol 2012;2012:310616.
[205] Ananthanarayanan B, Kim Y, Kumar S. Elucidating the mechanobiology of malignant brain tumors using a brain matrix-mimetic hyaluronic acid hydrogel platform. Biomaterials 2011;32(31):7913–23.
[206] Munson JM, Bellamkonda RV, Swartz MA. Interstitial flow in a 3D microenvironment increases glioma invasion by a CXCR4-dependent mechanism. Cancer Res 2013;73(5):1536–46.
[207] Beliveau A, Thomas G, Gong J, Wen Q, Jain A. Aligned nanotopography promotes a migratory state in glioblastoma multiforme tumor cells. Sci Rep 2016;66(1):26143.
[208] Wolf KJ, Chen J, Coombes JD, Aghi MK, Kumar S. Dissecting and rebuilding the glioblastoma microenvironment with engineered materials. Nat Rev Mater 2019;4(10):651–68.
[209] Patra B, Peng C-C, Liao W-H, Lee C-H, Tung Y-C. Drug testing and flow cytometry analysis on a large number of uniform sized tumor spheroids using a microfluidic device. Sci Rep 2016;6.
[210] Wartenberg M, Ling FC, Müschen M, Klein F, Acker H, Gassmann M, Petrat K, Pütz V, Hescheler J, Sauer H. Regulation of the multidrug resistance transporter P-glycoprotein in multicellular tumor spheroids by hypoxia-inducible factor-1 and reactive oxygen species. FASEB J 2003;17(3):1–22.
[211] Misra S, Hascall VC, Markwald RR, Ghatak S. Interactions between hyaluronan and its receptors (CD44, RHAMM) regulate the activities of inflammation and cancer. Front Immunol 2015;6:201.
[212] Petrie RJ, Doyle AD, Yamada KM. Random versus directionally persistent cell migration. Nat Rev Mol Cell Biol 2009;10(8):538–49.
[213] Pradhan S, Hassani I, Clary JM, Lipke EA. Polymeric biomaterials for in vitro cancer tissue engineering and drug testing applications. Tissue Eng 2016;22(6):470–84.
[214] Ferreira LP, Gaspar VM, Mano JF. Decellularized extracellular matrix for bioengineering physiomimetic 3D in vitro tumor models. Trends Biotechnol 2020;38(12):1397–414.
[215] Seyfoori A, Samiei E, Jalili N, Godau B, Rahmanian M, Farahmand L, Majidzadeh-A K, Akbari M. Self-filling microwell arrays (SFMAs) for tumor spheroid formation. Lab Chip 2018;18(22):3516–28.
[216] Fan Y, Nguyen DT, Akay Y, Xu F, Akay M. Engineering a brain cancer chip for high-throughput drug screening. Sci Rep 2016;6(1):1–12.
[217] Liu W, Xu J, Li T, Zhao L, Ma C, Shen S, Wang J. Monitoring tumor response to anticancer drugs using stable three-dimensional culture in a recyclable microfluidic platform. Anal Chem 2015;87(19):9752–60.
[218] Yi H-G, Jeong YH, Kim Y, Choi Y-J, Moon HE, Park SH, Kang KS, Bae M, Jang J, Youn H. A bioprinted human-glioblastoma-on-a-chip for the identification of patient-specific responses to chemoradiotherapy. Nat Biomed Eng 2019;3(7):509–19.
[219] Buhagiar A, Ayers D. Chemoresistance, cancer stem cells, and miRNA influences: the case for neuroblastoma. Anal Cell Pathol 2015;2015:150634.
[220] Dean M, Fojo T, Bates S. Tumour stem cells and drug resistance. Nat Rev Cancer 2005;5(4):275–84.
[221] Garner EF, Beierle EA. Cancer stem cells and their interaction with the tumor microenvironment in neuroblastoma. Cancer 2015;8(1):5.
[222] Galli R. The neurosphere assay applied to neural stem cells and cancer stem cells. In: Target identification and validation in drug discovery. Springer; 2013. p. 267–77.
[223] Musah-Eroje A, Watson S. A novel 3D in vitro model of glioblastoma reveals resistance to temozolomide which was potentiated by hypoxia. J Neurooncol 2019;142(2):231–40.

[224] Aoudjit F, Vuori K. Integrin signaling inhibits paclitaxel-induced apoptosis in breast cancer cells. Oncogene 2001;20(36):4995–5004.
[225] Weaver VM, Lelièvre S, Lakins JN, Chrenek MA, Jones JC, Giancotti F, Werb Z, Bissell MJ. β4 integrin-dependent formation of polarized three-dimensional architecture confers resistance to apoptosis in normal and malignant mammary epithelium. Cancer Cell 2002;2(3):205–16.
[226] Sui X, Chen R, Wang Z, Huang Z, Kong N, Zhang M, Han W, Lou F, Yang J, Zhang Q. Autophagy and chemotherapy resistance: a promising therapeutic target for cancer treatment. Cell Death Dis 2013;4(10), e838.
[227] Gomes L, Vessoni A, Menck C. Three-dimensional microenvironment confers enhanced sensitivity to doxorubicin by reducing p53-dependent induction of autophagy. Oncogene 2015;34(42):5329–40.

CHAPTER

9

Role of multidrug resistance in glioblastoma chemoresistance: Focus on ABC transporters

Mohammed Ahmed[a,b], *Maite Verreault*[a], *Xavier Declèves*[c,d], *and Ahmed Idbaih*[e]

[a]Sorbonne Université, Institut du Cerveau - Paris Brain Institute - ICM, Inserm, CNRS, APHP, Hôpital de la Pitié Salpêtrière, Paris, France [b]Faculté de Médecine Paris-Sud - Université Paris-Saclay, Saint-Aubin, France [c]Inserm, UMR-S 1144, Université de Paris, Therapeutic Optimization in Neuropsychopharmacology, Paris, France [d]Biologie du Médicament et Toxicologie, Hôpital Cochin, Assistance Publique-Hôpitaux de Paris, Paris, France [e]Sorbonne Université, Inserm, CNRS, UMR S 1127, Institut du Cerveau, ICM, AP-HP, Hôpitaux Universitaires La Pitié Salpêtrière - Charles Foix, Service de Neurologie 2-Mazarin, Paris, France

Abstract

Glioblastoma (GBM) is the most frequent and aggressive primary cancer of the brain in adults. Despite aggressive therapeutic interventions, the median overall survival is below 18 months after initial diagnosis. The current standard of care for patients with newly diagnosed GBM includes concurrent administration of temozolomide (TMZ) and radiotherapy followed by adjuvant TMZ. Since 2005, TMZ remained the first-line chemotherapy in the treatment of GBM patients with its ability to cross the blood–brain barrier. However, after initial efficacy, GBM cells acquire resistance to TMZ and other chemotherapeutic agents via multiple mechanisms, including the expression of ATP-binding cassette (ABC) efflux proteins. These transporters not only are involved in normal physiological functions, i.e., physiological cholesterol transport and elimination of toxins, but also play a role in pathological conditions, i.e., chemotherapies drug resistance. In humans, each ABC protein has specific tissue locations and specific functions. In this review, we highlight the role of ABC protein superfamily members ABCB1, ABCC1, and ABCG2 in the resistance of GBM cells to chemotherapy.

Abbreviations

ABC	ATP binding cassette
ABT-414	depatuxizumab mafodotin
ADC	antibody-drug conjugates

Glioblastoma Resistance to Chemotherapy: Molecular Mechanisms and Innovative Reversal Strategies
https://doi.org/10.1016/B978-0-12-821567-8.00025-7

ALD	adrenoleukodystrophy
ATP	adenosine triphosphate
BBB	blood–brain barrier
BCRP	breast cancer resistance protein
BMP7	bone morphogenetic protein 7
CAXII	carbonic anhydrase XII
CNS	central nervous system
EC	endothelial cells
GBM	glioblastoma
GSC	GBM stem cells
MDR	multidrug resistance
MK571	MRP1 and MRP4 inhibitor
MRP	multidrug resistance associated protein
NBD	nucleotide binding domains
PDCL	patient-derived cell line
PET	positron emission tomography
P-gly	P-glycoprotein
SNP	single-nucleotide polymorphism
TMD	trans membranous domain
TMZ	temozolomide
UMBO	ultrasound-mediated blood–brain barrier opening

Conflict of interest

Ahmed Idbaih reports grants and travel funding from Carthera; research grants from Transgene, Sanofi, Air Liquide, and Nutritheragene; travel funding from Leo Pharma; and advisory board from Novocure and Leo Pharma.

Introduction

Glioblastoma (GBM) is the most common and the deadliest primary brain cancer in adults with a median overall survival below 18 months after initial diagnosis [1]. The current standard of care for newly diagnosed GBM patients, which was established in 2005, relies on concurrent administration of temozolomide (TMZ) and radiotherapy regimen followed by adjuvant TMZ alone [2]. Despite remarkable efforts in the neuro-oncology field to develop new treatments, TMZ remains today the standard first-line chemotherapy in the treatment of GBM patients [1, 2]. TMZ is an alkylating agent with a small molecular weight (194.15 g/mol) that readily crosses the blood–brain barrier (BBB) [3].

The essential and complex organ which is the human brain is separated from the blood by the BBB. The BBB is specific to the central nervous system (CNS) blood vessels. The BBB isolates the brain from the blood for protection purposes [4]. Indeed, it prevents potentially toxic molecules circulating in the bloodstream to access brain cells, while ensuring the supply of nutrients to maintain homeostasis [5]. Highly specialized brain capillary endothelial cells (ECs) form an important part of the BBB. In addition to ECs, various cells contribute to the biophysical structure of BBB described in Fig. 1. To date, five mechanisms are known to regulate the exchanges of molecules from the blood to the brain and vice versa. Passive diffusion of molecules through the BBB can occur paracellularly for very low molecular weight molecules (e.g., inorganic ions, water, gases) and transcellularly for lipophilic compounds. An active transport can also occur either by: (i) transcytosis for some proteins (e.g., leptin, insulin,

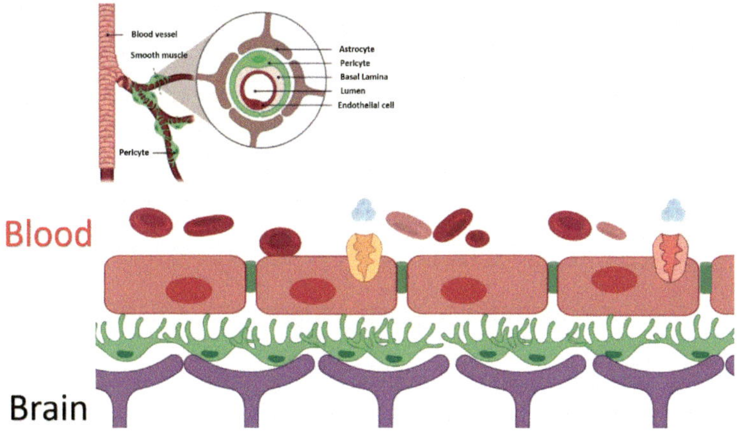

FIG. 1 The blood–brain barrier (BBB) is formed of different types of cells tightly knit together. Highly specialized endothelial cells (ECs) surround blood vessels and form part of the BBB. In addition to brain ECs, various cells contribute to the structure of BBB. Pericytes (represented in *green*) are attached to endothelium cells via gap junctions while astrocytes end feet (represented in *purple*) surround endothelial cells of the BBB, providing structural and functional support to these cells. Five mechanisms are known to regulate the entry of molecules to the brain. The efflux pumps pathway is considered a mechanism of active transport through the BBB [6].

transferrin) or (ii) carrier-mediated proteins belonging to two major transporter superfamilies for small molecules. The solute carrier (SLC) superfamily contains more than 400 transporters in humans that allow exchanges of small molecules through the BBB while the transporters of the ATP-binding cassette (ABC) superfamily limit CNS penetration of small molecules by effluxing substrates from the brain ECs directly into the bloodstream (Fig. 1) [4, 6, 7].

GBM cells acquire resistance to anticancer drugs via multiple mechanisms not limited to the following: (i) acquisition of mutation in DNA repair genes, (ii) activation of alternative signaling pathways, (iii) immune escape, (iv) invasive switch from angiogenic growth and, (v) multidrug resistance mechanism (MDR) [8–10]. The MDR phenotype is observed in many types of cancers and induces: (i) increased efflux of drugs outside tumor cells and, (ii) reduced influx of drugs inside tumor cells [11]. ABC transporters are efflux pump proteins involved in MDR. To date, 49 members of ABC proteins have been identified to be involved in different biological mechanisms within the human body and are classified in 7 subfamilies: ABCA (12 proteins), ABCB (11 proteins), ABCC (13 proteins), ABCD (4 proteins), ABCE (1 protein), ABCF (3 proteins), and ABCG (5 proteins) [12].

ABC transporters are expressed in various tissues such as the liver and the intestine and have a distinct role in the absorption, distribution, and excretion of drugs. Some ABC transporters are predominantly expressed in ECs of the BBB [13]. Indeed, ABCB1, ABCC4–5, and ABCG2 are expressed in ECs of the human BBB, while others (ABCCs, ABCG2) can also be found in other cells such as astrocytes and neurons [5, 13, 14]. In humans, each ABC protein is expressed in specific locations and exhibits specific functions. ABCA subfamily members are expressed mainly in the CNS while ABCB subfamily members are mainly expressed in the BBB and liver [15]. ABC proteins were studied in several types of cancers [16]. In GBM, three proteins (ABCB1, ABCC1, and ABCG2) were extensively studied and were shown to have an impact on GBM cell biology. In this review, we are highlighting the role of the ABC protein superfamily, mainly (ABCB1, ABCC1, and ABCG2) in resistance of GBM cells to chemotherapy.

Structure and functions of ABC proteins

Structure of ABC proteins

The BBB was described for the first time in the 20th century when an intravenous injection of Evans blue significantly stained all tissue except the CNS, while a direct intrathecal injection was able to stain only the brain tissue. This staining pattern highlighted the possibility of a barrier, termed the BBB, that prevents the dye from reaching the brain tissue [4]. The integrity of the BBB is preserved throughout life to maintain homeostasis and to regulate the influx and efflux of nutrients/metabolites between the blood and the brain [17]. ABC transporters within the BBB play a pivotal role in brain protection by eliminating harmful agents.

The primary function of the ABC transporters is to actively transport their substrates, ranging from low-molecular-weight molecules to polypeptides, outside of the cells. Despite the large number of ABC transporters, they share structural similarities. In general, a typical ABC protein includes two functional units called transmembrane domain (TMD) and nucleotide-binding domains (NBD) (Fig. 2). NBDs are the ATP-binding units and they contain the Walker A motif—a phosphate-binding structure, Walker B motif, and Walker C motif. The C motif is specific for ABC proteins while Walker A and B are present in all ATP dependent proteins. The TMDs include 6 or 10 transmembrane helices depending on the ABC superfamily members. There are two types of transmembrane helices: the inward part (open to the cytoplasm) and the outward part (open to the extracellular environment). These helices determine the direction of transport of the ABC transporters, i.e., importer or exporter [14, 19].

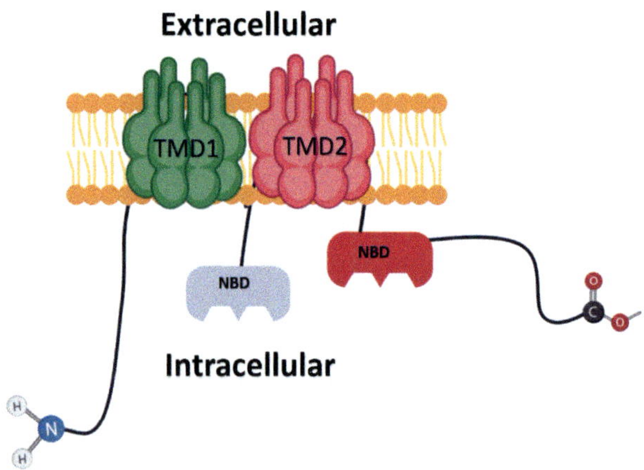

FIG. 2 A full ABC transporter consists of four building units. The first two building units, called TMDs, are formed by six transmembrane segments. TMD1 and TMD2 are colored in *green* and *pale red*, respectively. The two other building units are called NBDs, NBD1 (*pale blue*), and NBD2 (*red*). ABC half transporters have only one TMD and one NBD and need to dimerize to become a functional protein. Additionally, some other ABC transporters have an additional TMD unit that is conjugated to the N-terminus of the protein and called the "Long" ABC transporter [18].

Functions of ABC proteins

Several physiological functions are reported for each ABC subfamily. However, the main function of ABC proteins is to actively transport cytotoxic xenobiotics and endobiotics against their concentration gradient [19]. According to their location, they transport many substrates including anionic and cationic charged molecules, inorganic ions, i.e., chloride, peptides, and lipophilic compounds [13]. The ABCA subfamily is mainly responsible for lipids and cholesterol transport while the ABCB, ABCC, and ABCG subfamily members are mainly associated with drug resistance and elimination of xenobiotics. Genetic variants in ABC proteins are linked to genetic disorders, e.g., a pathogenic variant in the ABCD2 is responsible for 95% of cases of X-linked adrenoleukodystrophy (ALD) disorder [20]. Table 1 summarizes the physiological functions of each ABC subfamily.

Mechanisms of action of ABC transporters

ABC proteins hydrolyze ATP to efflux chemical agents against their concentration gradient. The active transport cycle starts with the binding of a substrate, e.g., xenobiotic to a high-affinity structure formed by the TMDs and two ATP molecule-binding NBDs.

TABLE 1 Identified subfamilies of ABC transporters and their physiological functions.

ABC subfamily	ABC proteins	Physiological functions	Reference
ABCA	**12**	• Lipid and cholesterol transport, ABCA2 is involved in drug resistance	[21]
ABCB	**11**	• Elimination of toxins • Inhibition of apoptosis • Volume dependent Cl^- channel regulator • Phospholipid translocation (can translocate short-chain phospholipids) • Maintenance of plasma membrane cholesterol esterification • Drug resistance	[21, 22]
ABCC	**13**	• Anion efflux • Drug resistance • Nucleoside transport	[21]
ABCD	**4**	• Mainly expressed in peroxisomes • ABCD2 fatty acid transport and a major modifier locus for clinical diversity in X-linked ALD	[21]
ABCE/ ABCF	**1 ABCE** **3 ABCF**	• Along with ABCE1, ABCF members have ATP-binding domains, but no transmembrane domains, making transporter function unlikely • Mainly regulate protein synthesis or expression	[21]
ABCG	**5**	• Transport of diverse drug substrates, sterols and lipids • ABCG4 is expressed in macrophages • Drug resistance	[21]

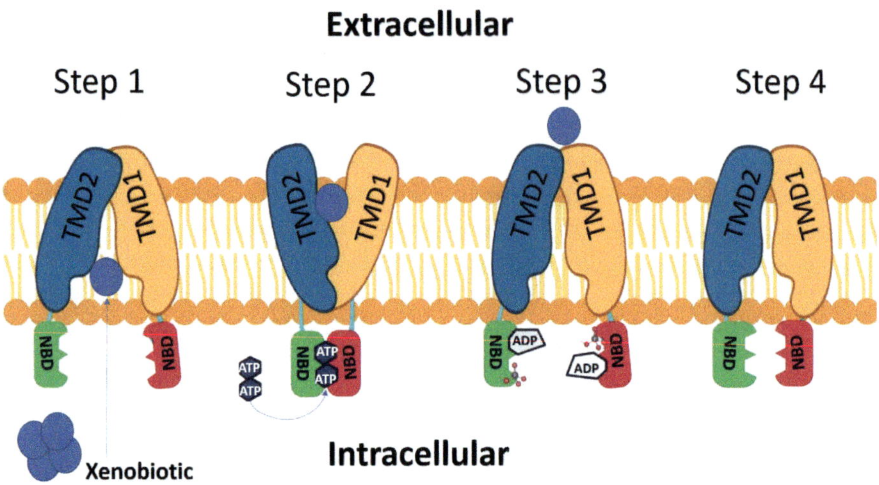

FIG. 3 ABC transporters are transmembrane proteins capable of actively transporting a xenobiotic from the intracellular to the extracellular compartments. This active transport requires the hydrolysis of ATP to provide the energy necessary for the transport. The active transport cycle starts with the binding of the xenobiotic to a high-affinity structure formed by the TMDs (step 1). As a result, a conformational change makes the NBD units more exposed to ATP binding. The NBD dimer induces a major conformational change on the TMDs allowing the xenobiotic to be translocated (step 2). The hydrolysis of ATP allows the NBD dimers to be dissociated and again induces a TMD conformational change (step 3). A final step of restoration of the open NBD-dimer conformation then takes place (step 4) [14, 19].

The ATP-binding sites at the NBD residues are formed by a conformational contribution from each NBD monomer. As a result, a conformational change in TMDs takes place from either outward to inward (importer) or vice versa (exporter), allowing the NBD units to form a dimer. The NBD dimer induces a major conformational change on the TMDs allowing the xenobiotic to be translocated across the plasma membrane. The hydrolysis of ATP allows the NBD dimers to be dissociated and again inducing the TMD conformational change, resulting in the xenobiotic to be released. A final step of restoration of the open NBD-dimer conformation then takes place (Fig. 3) [13, 14, 19, 21, 23].

ABC transporters in glioblastoma

We reviewed public data obtained from PubMed and Google Scholar. Original published articles were obtained using the keywords (ABC transporters, glioblastoma, TMZ, and chemotherapy). 151 articles appeared in the results from the search engines. Another step was carried out to exclude the duplicated and review articles. Following the removal of duplicated and review articles, 91 abstracts were then reviewed (abstract review) and only articles that studied ABC transporters in glioblastoma were selected. Full texts were obtained for all 36 selected articles from https://universiteparissud.focus.universite-parissaclay.fr/ and https://insermbiblio.inist.fr/ using an institutional access. Fig. 4 summarizes the methodology used in the review processes.

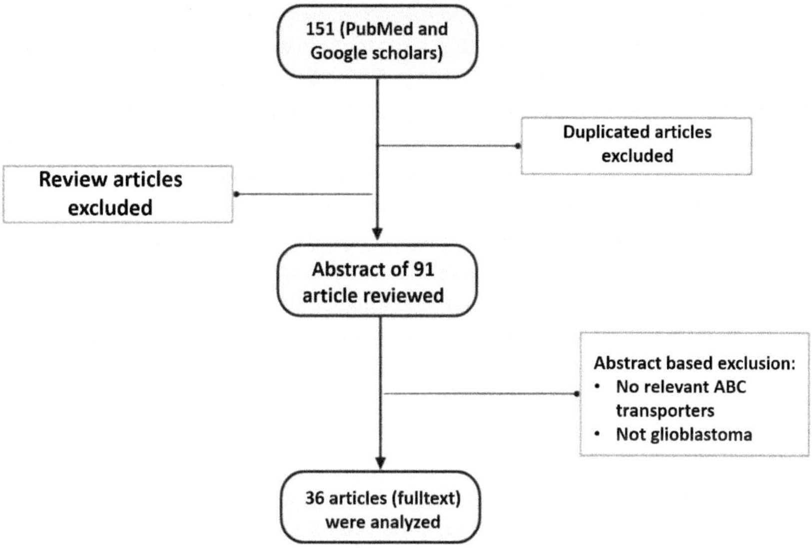

FIG. 4 Represents the methodology of this literature review. The key words (ABC transporters, glioblastoma, TMZ, and chemotherapy) were used in PubMed and Google Scholar search engines. One hundred fifty-one articles appeared in the results, and then a few steps were carried out to exclude the duplicated and review articles. Ninety-one abstracts were then reviewed, and from the abstract, only articles that studied ABC transporters or GBM were selected. Full texts were obtained for all 36 selected articles.

ABCB1 (MDR1, P-glycoprotein)

ABCB1, which is also known as MDR1 and P-glycoprotein (P-gp), was identified by Ling in 1976 [24], making it the first studied protein among all ABC proteins. ABCB1 protein is a 170 kDa glycoprotein highly expressed in the endocrine tissues, liver, gallbladder, and brain, and it is usually coexpressed with the ABCG2 protein [25]. The physiological impact of the ABCB1 protein was accidentally identified in 1994 by Schinkel, Smit [26], who found that a homozygous knockout of *Abcb1a/b* in laboratory mice induced a 100-fold increase in susceptibility to antiparasitic medications [27].

In humans, the ABCB1 protein is encoded by the *ABCB1* gene. An update was published in 2011 to illustrate the role of *ABCB1* genetic polymorphisms, which accounts for more than 65 exon related single-nucleotide polymorphisms (SNPs) [28]. These SNPs could be responsible for the differences in drug response and toxicity in several types of cancers [15]. In the brain, ABCB1 is localized in the luminal membrane of ECs of the BBB [13]. It has an essential role in protecting the brain from a possible brain uptake of toxic molecules or metabolic substances with a wide range of known substrates including TMZ [29].

Schaich et al. investigated the role of three different SNPs of ABCB1 in GBM patients treated with TMZ. He showed that the rs1128503 SNP in *MDR1* exon 12 is an independent predictive biomarker of response to TMZ. Patients with GBM exhibiting the homozygous allele (C/C allele) have better survival compared to their heterozygous variant counterparts [30]. However, more recently, another large clinical cohort analyzed the clinical impact of four SNPs variants (rs2229109, rs1128503, rs2032582, and rs1045642) in patients with newly diagnosed GBM patients treated with the standard of care. They did not find any clinical value of the SNPs investigated in a large Swedish cohort and hence could not validate the results obtained from the Schaich study [31]. One pilot clinical trial tried to evaluate the ABCB1 protein among glioma patients. They measured the uptake of (^{11}C) *N*-desmethyl-loperamide ((^{11}C)dLop) using positron emission tomography (PET) imaging as a marker of ABCB1 activity. The aim of the clinical study was to recruit 10 patients; however, only 2 registered patients were available in the clinical trial database, suggesting that the reason for early termination of the trial was the lack of patients who fit the inclusion criteria of the study [32].

Several studies have investigated the role of ABCB1 in the context of TMZ treatment in GBM preclinical models. ABCB1 downregulation was associated with increased efficacy of TMZ in U87 cell lines [33, 34]. Two recent studies showed that the downregulation of ABCB1 also increases the efficacy of TMZ *in vitro* and *in vivo* in GBM preclinical models [35, 36]. Furthermore, an *in vivo* study reported a higher concentration of irinotecan in the brain of *Abcb1a/Mdr1a* (−/−) mice versus wild-type mice when both were exposed to the same dose of irinotecan [37]. The antitumor efficacy of TMZ against three intracranial tumor GBM models was significantly enhanced when *Abcb1a/b* and *Abcg2* were genetically deficient or pharmacologically inhibited [29, 38]. ABCB1 expression can be altered by several compounds including carbonic anhydrase XII (CAXII), Bone morphogenetic protein 7 (BMP7), and TMZ [35, 39, 40]. Tso et al. found that BMP7 sensitizes GBM stem cells to clinically relevant doses of TMZ [35], while Riganti and Salaroglio have found that GBM exposure to TMZ downregulates the expression of ABCB1 [39]. A recent study showed that CAXII could also reduce ABCB1 activity and sensitize GBM cells to TMZ [40]. Table 2 summarizes the xenobiotics that alter the function of ABCB1, ABCG2, and ABCC1 transporters in GBM.

ABCC1 (MRP1)

The multidrug resistance-associated protein 1 (MRP1) is encoded by the *ABCC1* gene. It was described for the first time by Cole et al. 1992 [66]. The ABCC1 protein is a 180–190 kDa protein and is ubiquitously expressed in many tissues in humans. It is highly expressed in the intestine, kidney and testis, while a lower expression is detected in the lung, colon, and brain [67]. The ABCC1 protein has a wide range of substrates including anticancer drugs tested in GBM cell lines, i.e., vinca alkaloids (vincristine and vinblastine) and topoisomerase inhibitors (mitoxantrone) [64, 68]. Many genetic alterations were detected in the *ABCC1* gene, and most of them are SNPs in noncoding sequences and introns. A complete list of all *ABCC1* SNPs can be obtained from the available public database accessible from https://www.ncbi.nlm.nih.gov/snp/?term=ABCC1.

In the literature, ABCC1 inhibitors including KIAP—an antiapoptotic protein—reduce ABCC1 activity and sensitize the U251 GBM cell line to TMZ [54]. Two *in vitro* studies found that ABCC1 inhibition sensitizes cells to vincristine and etoposide but not to TMZ [64]. Furthermore, a study showed that MK571, an inhibitor of ABCC1 and ABCC4, increased the antitumor efficacy of vincristine and etoposide in primary glioblastoma cell lines [23]. On the other hand, the overexpression of both ABCB1 and ABCG1 in GBM cell lines (U87 and U251) is associated with resistance to TMZ [60, 69].

TABLE 2 Analysis of published studies that show the effects of ABC transporters on chemotherapeutic agents used in GBM.

Effect	ABC transporter involved	Drug involved	Reference
High dose of cyclosporine[a] doubles the plasma concentration of etoposide among glioma patients	**ABCB1**	Cyclosporine	[41]
Nimodipine[b] enhances sensitivity to procarbazine in viability tests *in vitro* using PDCLs obtained from glioblastoma patients	**ABCB1 and ABCG2**	Nimodipine	[42]
Paclitaxel in combination with valspodar[c] significantly decreases the tumor volume of U-118 MG tumors compared to the control and paclitaxel groups in mice	**ABCB1**	Valspodar	[43]
Vincristine exposure induces an elevated expression of ABCG1 in rat brains. This effect could lead to the assumption that ABCB1 is partially responsible for the observed resistance of a relapsed tumors	**ABCB1**	Vincristine	[44]
Overexpression of ABCG2 in human GBM cell lines is associated with mitoxanthone resistance	**ABCG2**	Mitoxanthone	[45]
GBM cell lines overexpressing ABCB1 exhibit high resistance to carmustine, carboplatin and etoposide	**ABCB1**	Carmustine, carboplatin, and etoposide	[46]
Elacridar[d] sensitizes GBM cell lines to dasatinib. Homozygous knockout of ABCG2 in mice results in a better overall survival compared to the wild type when treated with dasatinib	**ABCB1 and ABCG2**	Elacridar	[47]
Similar brain-to-plasma concentration was observed for sunitinib in both ABCB1 and ABCG2 knockout mice model and with elacridar treatment in mice. However, mild effect was observed with the zosuquidar[e] and no effect with KO143[f]	**ABCB1 and ABCG2**	Elacridar, KO143 and Zosuquidar	[48]
Inhibition of ABCB1 and ABCG2 with ABT-888 improves the efficacy of TMZ therapy in GBM patients	**ABCB1 and ABCG2**	ABT-888	[49]
Mdr1 −/− mice show a higher concentration of irinotecan compared to mdr1a+/+ mice when both exposed to the same dose of irinotecan	**ABCB1**	Irinotecan	[37]
MRP1 inhibition enhanced Vincristine and Etoposide but not TMZ chemotherapeutic effect however the combined inhibition of MRP1 and P glycoprotein (P-gp) using Reversan[g] increased TMZ response in GBM PDCLs	**ABCC1**	MK571 and Etoposide	[23]
Inhibition of ABCB1 and ABCG2 with verapamil and KO143 increases the efficacy of TMZ when combined with MGMT inhibitors	**ABCB1 and ABCG2**	Verapamil and KO143	[50]

Continued

TABLE 2 Analysis of published studies that show the effects of ABC transporters on chemotherapeutic agents used in GBM—cont'd

Effect	ABC transporter involved	Drug involved	Reference
Melatonin enhances ABCG2 promoter methylation hence sensitizes GBM cell lines to mitoxantrone, doxorubicin, temozolomide	**ABCG2**	Melatonin	[51]
Limited drug delivery into brain tumors may significantly limit the efficacy of rucaparib[h] combined with TMZ in GBM	**ABCB1 and ABCG2**	Rucaparib	[52]
AZD2461[h] has a limited brain permeability in vivo owing to its efflux by ABCB1 protein	**ABCB1**	AZD2461	[38]
Downregulation of ABCB1 and ABCG2 by bone morphogenetic protein 7 sanitize the glioblastoma stem cells to the clinically relevant dose of TMZ	**ABCB1 and ABCG2**	BMP7	[35]
ABCE1 downregulation enhance the efficacy of TMZ in GBM cells (U87 and A172)	**ABCE1**	TMZ	[36]
ABCG2 knockdown in several GBM cell lines resulted in upregulation of other drug transporters ABCB1 and ABCC3 when treated with TMZ	**ABCG2**	TMZ.	[53]
KIAP-anti apoptotic protein-sensitizes U251 cells to TMZ through reduction of the ABCC1 expression	**ABCC1**	TMZ	[54]
The study was not conclusive. Only 7% of the 125 cases studied showed detectable MDR1 expression, suggesting that ABCB1 was not a major contributor to drug resistance in the selected cohort	**ABCB1**	TMZ	[55]
Histone-lysine *N*-methyltransferase (EZH2) enzyme silencing decreases the ABCB1, ABCC1 and ABCG2 mRNA and protein levels, which would lead to reduce efflux pump activity	**ABCC1, ABCB1, and ABCG2**	EZH2	[56]
TMZ treatment upregulate the expression of ABCC3 compared to control mice. ABCC3 protect natural killers from TMZ. A GL261 syngeneic mouse model was used in this study	**ABCC3**	TMZ	[57]
LRIG1, human EGFR inhibitor, reversed MDR in GBM cell lines (U87 and U251) by the inhibiting EGFR and a secondary inhibition of ABCB1 and ABCG2	**ABCB1 and ABCG2**	TMZ	[58]
CDK6 knockdown in glioblastoma cell line (U251) resulted in significant downregulation of MDR1, MRP which enhanced the TMZ response	**ABCB1, ABCC1**	TMZ	[59]
The antitumor efficacy of TMZ against three different intracranial tumor models was significantly enhanced by a homozygous knockout of Abcb1a/b and Abcg2 genes	**ABCB1 and ABCG2**	TMZ	[29]

TABLE 2 Analysis of published studies that show the effects of ABC transporters on chemotherapeutic agents used in GBM—cont'd

Effect	ABC transporter involved	Drug involved	Reference
The single-nucleotide polymorphism (SNPs) in the MDR1 gene exon12 C1236T is an independent predictive factor for prediction of the temozolomide treatment in glioblastoma patients	**ABCB1**	TMZ	[30]
Overexpression of MDR and MRP in GBM cells (U87, U251, U373) is associated with a high resistance to TMZ	**ABCB1 and ABCC1**	TMZ	[60]
Activated EGFR kinase enhanced the ability of GBM cells (U87 and T98G) to resist TMZ through the upregulation of MDR1	**ABCB1**	TMZ	[61]
Loss and gain of function for MDR1 showed an enhanced and reduced efficacy of TMZ in GBM cell lines (U87 and T98G)	**ABCB1**	TMZ	[34]
Inhibition of ABCG2 enhanced the efficacy of TMZ and is considered a promising therapeutic target in GBMs	**ABCG2**	TMZ	[62]
TMZ downregulate the expression of ABCB1 in glioblastoma stem cells	**ABCB1**	TMZ	[39]
MDR1 and ABCG2 are responsible for the resistant of recurrent GBM tumors to TMZ. Following a TMZ exposure in U87 and T89G cells eightfolds expression MDR1 and fourfolds for ABCG2 in the cells compared to naive cells was recorded using real-time PCR	**ABCB1 and ABCG2**	TMZ	[63]
Multiple inhibition of the MDR1 protein showed no enhanced associated with an enhanced sensitivity of TMZ in GBM cell line (T98G)	**ABCB1, ABCB1**	TMZ	[64]
Downregulation of p-glycoprotein enhance the efficacy of TMZ in glioblastoma U87 cell line	**ABCB1**	TMZ	[33]
ABCA13 overexpression is associated with a decreased progression free survival in univariate and multivariate analyses in GBM patients	**ABCA13**	TMZ	[65]
Carbonic anhydrase XII (CAXII) sensitize primary GBM cells to TMZ by reduction of ABCB1 protein activity	**ABCB1**	TMZ	[40]

[a] *Immunosuppressant medication with ABCB1 blocking activity.*
[b] *Calcium channel blocker with ABCB1 blocking activity.*
[c] *An experimental cancer drug with ABCB1 inhibition properties. It is a derivative of cyclosporine.*
[d] *An experimental small molecule that has a dual ABCB1 and ABCG2 inhibition.*
[e] *A potent ABCB1 inhibitor has reached clinical trials.*
[f] *Experimental drug with ABCG2 inhibition activity.*
[g] *It is an experimental drug with ABCC1 and ABCB1 inhibition.*
[h] *Poly (ADP ribose) polymerase (PARP) inhibitors.*

ABCG2, breast cancer resistance protein (BCRP)

The ABCG2 protein was the first MRP-associated protein to be discovered. This 72 kDa protein was first identified in 1998 after being cloned from a human breast cancer cell line, which led to its alias, BCRP [70]. In humans, it is highly expressed in the small intestine, colon, rectum, placenta, and smooth muscles while a lower expression is detected in the adrenal and thyroid glands, lung, and cerebral cortex [25]. In isolated brain microvessels and cortex biopsies from 12 patients with epilepsy or glioma, the expression of the ABCG2 protein was 1.6-folds higher than the expression of ABCB1 [71]. The ABCG2 protein is a ABC half transporter, and therefore it requires the dimerization of two NBDs to function as a drug efflux pump [72]. Many SNPs were identified in the *ABCG2* gene. The frequency of SNPs in the *ABCG2* gene is highly variable among ethnic groups, potentially associated with heterogeneous drug responses among these groups [73].

In 2017, a study enrolling 50 Caucasian GBM patients found a correlation between expression of eight different proteins (ABCG2, XIAP, MGMT, MSH2, pATM, pTp53, pAKT, Nestin) including ABCG2 and they reported a correlation between ABCG2 and the poor prognosis among GBM patients treated with the TMZ [62]. To study the role of ABCG2 protein *in vitro*, they used GBM stem cells (GSC) and noticed an enhanced efficacy of TMZ following the inhibition of ABCG2. Therefore, they considered ABCG2 a promising therapeutic target in GBMs [62]. Another study reported, however, that ABCG2 knockdown results in the upregulation of other drug transporters (ABCB1 and ABCC3) when treated with TMZ [53], suggesting that there might be compensating mechanisms between transporters.

Several studies reported the importance of ABCG2 in drug response in GBM. An *in vivo* study reported that *Abcg2* knockout in mice is associated with better overall survival compared to wild-type mice when treated with dasatinib, an Src inhibitor [47]. Another study showed that melatonin enhanced ABCG2 promoter methylation and sensitized GBM cell lines to mitoxantrone, doxorubicin, and TMZ [74]. Consistent with this study, the overexpression of ABCG2 in human GBM cell lines is found to be associated with mitoxantrone resistance [45]. It was also reported that triple knockout of Abcb1a/b and Abcg2 improves efficacy of TMZ therapy in spontaneous GBM mouse models [49]. Finally, TMZ exposure in U87 and T89G cells was found to increase ABCB1 and ABCG2 mRNA expression. Therefore, exposure to TMZ itself could modulate the levels of ABC proteins and induce TMZ resistance among patients [61].

Clinical value

To date, several strategies have been developed to overcome the ABC transporter-mediated MDR. These strategies are summarized in Fig. 5 and they vary from partial/complete inhibition to bypass approaches: (i) nanocarrier technologies, (ii) antibody-drug conjugates (ADC), and (iii) ultrasound-mediated BBB opening (UMBO).

Partial or complete inhibitors of ABC transporters can be combined with their substrates to enhance their CNS penetration and anticancer activity. A variety of modulators were tested to suppress activity of efflux proteins, mainly (ABCB1 and ABCG2), and few were successful enough to reach clinical trials. In GBM, a limited number of clinical trials were initiated to

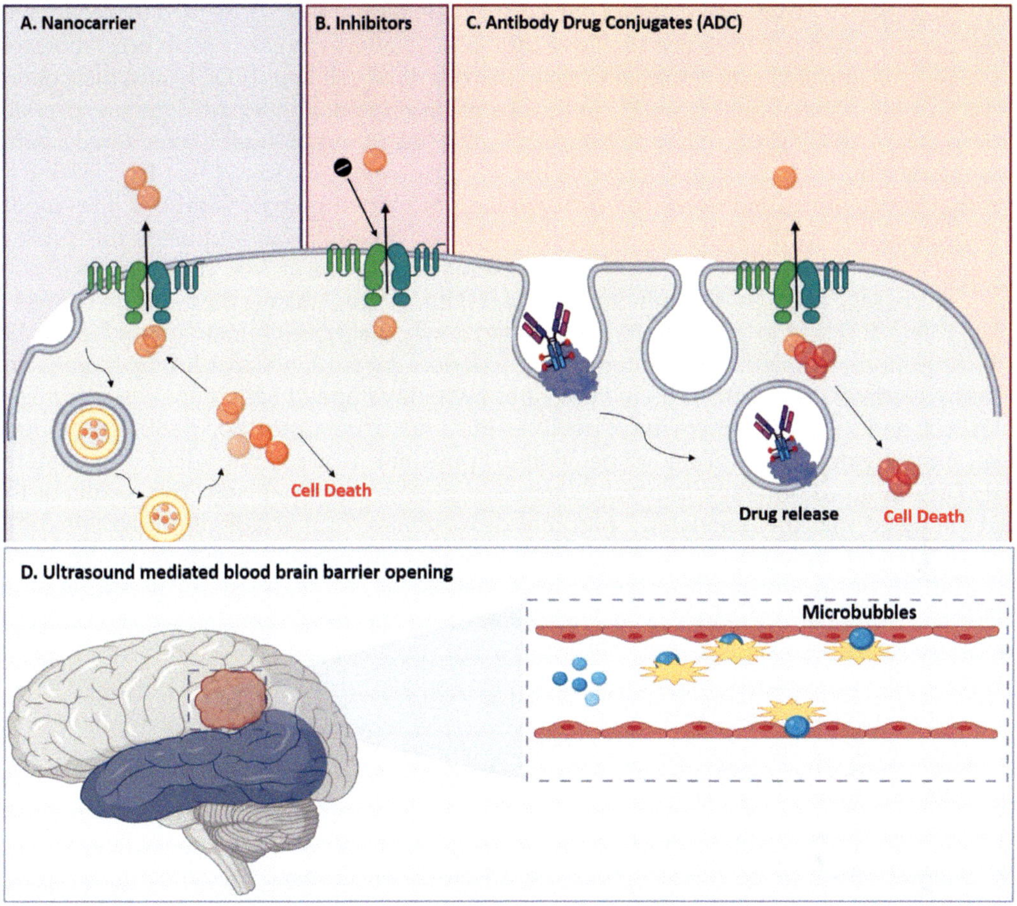

FIG. 5 Summary of methods that are being developed to overcome ABC transporters. *Panel A* represents the development of nanocarriers that allows the drug to enter via endocytosis and bypass ABC transporters. *Panel B* shows another method by using a partial or complete antagonist that can be administered in combination with the efflux pumps substrates and, as a result, enhances the activity of the substrates. *Panel C* shows the antibody-drug conjugates approach that relies on an antibody to carry the substrates when binding to its ligand. *Panel D* represents the combined effect of using microbubbles and low intensity ultrasound to open the BBB. These four methods have been used *in vitro* and *in vivo* to develop strategies to overcome ABC efflux pumps [75].

modulate ABC proteins in GBM patients. To date, these trials failed to show significant clinical benefit, which could be related to a few reasons. First, the study design was not optimal, i.e., in the early clinical trials, the patients were not stratified based on high expression of ABC proteins, and a precise evaluation of the role of ABCB1 and ABCG2 transporters in patients could not be conducted. Second, modulators of ABC proteins could change the pharmacokinetics of other drugs reducing their antitumor properties, and safety issues have been in evidence in patients during clinical trials with ABC inhibitors. Third, the dose that was selected to inhibit the ABC protein was not sufficient or a higher dose could not be applied safely in patients.

The selection of cancer cell lines is a crucial step in *in vitro* studies dedicated to GBM. Many commercial human cell lines are available for GBM. However, a study from our laboratory has tested ABC protein expression in GBM patient-derived cell lines (PDCL) and their parental tumors [65]. The study showed that PDCLs recapitulated a better ABC gene expression pattern of human GBM compared to commercial cell lines and can thus be considered a better model to test the biology of ABC proteins in GBM. Moreover, we found that fetal bovine serum that is usually added to the cell culture medium for commercial GBM cell lines modulates resistance to TMZ. Recently, a study highlighted the importance of using low passage number PDCL and serum-free medium when studying the role of ABC transporters *in vitro*. The high passage number of commercial GBM cell lines could change the expression level of ABC protein and lead to conclusions irrelevant to newly diagnosed human tumor [65, 76, 77].

Furthermore, sophisticated strategies using nanocarriers (nanocapsules, liposomes, micelles, dendrimers) for ABC protein substrates were developed to allow these substrates to enter via endocytosis, improving drug half-life and drug protection [78]. One clinical study investigated the efficacy of pegylated liposomal doxorubicin when administered with TMZ and radiotherapy in newly diagnosed glioblastoma patients. The results of this study showed that the pegylated liposomal doxorubicin form is safe and tolerable; however, no meaningful efficacy was observed from either the prolongation of TMZ therapy or the addition of liposomal doxorubicin (NCT00944801) [79]. Another study evaluated the safety and the pharmacokinetics of a liposomal form of irinotecan; the study confirmed the safety of the liposomal form of irinotecan but the efficacy is still under investigation [80]. Therefore, the utilization of these or similar nanocarrier forms could be effective tools in future clinical studies [78].

Another strategy consists in the use of antibody-drug conjugates (ADCs) against a variety of targets in GBM, and this approach is currently under investigation [81]. ADCs are newly developed biopharmaceutical compounds that allow the targeting of tumor cells while sparing healthy cells. This method is based on the use of an antibody to carry the substrate when binding to its ligand. In 2017, a published clinical trial showed a promising efficacy of depatuxizumab mafodotin (ABT-414), an ADC specific for the activated form of EGFR to selectively deliver a cytotoxin, in epidermal growth factor receptor (*EGFR*)-amplified recurrent GBM patients with manageable adverse effects [82]. Another clinical trial was followed to evaluate its efficacy in newly diagnosed GBM patients (NCT02573324).

Finally, ultrasound-mediated BBB opening (UMBO) was used in preclinical models to bypass BBB efflux transporters and to increase the penetration in the brain of a wide variety of therapeutics. Low intensity pulsed ultrasound can be delivered to the brain to induce a safe oscillation of intravenously injected microbubbles within blood vessels [83]. Oscillation of these microbubbles opens the BBB by reversibly disrupting the tight junctions between ECs. A range of drugs have been tested for use with UMBO for treating gliomas and include temozolomide, carmustine, irinotecan, carboplatin, doxorubicin, and drug-loaded liposomes [84, 85]. A new study showed that UMBO could decrease the expression of the ABCB1 protein in cerebral vessels without affecting the integrity of other proteins [86]. UMBO has recently moved to clinical trials where its clinical safety was confirmed: Sonocloud-induced UMBO was found to be safe and tolerable among GBM patients [87]. Two other phase 1 and phase 2 clinical trials are currently in progress to evaluate the efficacy of UMBO in combination with carboplatin in patients with recurrent GBM (NCT03744026).

Conclusion

GBM is an aggressive primary brain tumor with dismal prognosis. Over the last 15 years, no new drugs were found to be superior to TMZ. The long-term limited efficacy of TMZ is explained, at least partly, by the effect of MDR proteins (ABCB, ABCC1, and ABCG2). Accumulating evidence is rising to connect the effect of chemotherapies and ABC proteins. The clinical role of ABC proteins is still under investigation and the failure of previous clinical trials raised several questions regarding the strategies to overcome MDR in GBM. A few clinical recommendations are currently being reported in the literature regarding future clinical trials. First, all drugs that are going to be used in the trials should be tested against the major ABC proteins (ABCB, ABCC1, and ABCG2). A wide range of *in vitro* and *in vivo* models could allow a precise testing of the novel drugs [76]. Second, newly available noninvasive diagnostic imaging approaches, i.e., PET scanning, have the potential to determine whether ABCB1 or other transporters are functioning to reduce drug accumulation and whether inhibition can change drug uptake in solid tumors [88, 89]. Furthermore, dual ABC inhibitors with high specificity could be developed. Indeed, for example, the ABCB1 specific inhibitor zosuquidar enhanced sunitinib brain concentration in mice, but not to the same level as the dual inhibitions of ABCB1 and ABCG2 [48]. Following these recommendations could lead to the design of clinical trials that might successfully demonstrate the therapeutic potential of ABC protein inhibition in GBM treatment.

Acknowledgments

This work was supported by the European Union's Horizon 2020 research and innovation program under the Marie Skłodowska-Curie grant agreement #766069 (GLIO-TRAIN). We would like to acknowledge the online BioRender platform that provided all the tools and icons to draw the manuscript's graphs.

The research leading to these results has received funding from the program Investissements d'avenir" ANR-10-IAIHU-06. Institut Universitaire de Cancérologie. INCA-DGOS-Inserm_12560 (SiRIC CURAMUS) benefits from support from Institut National du Cancer, Ministère des Solidarités et de la Santé and Inserm.

Authors' contributions

M.A. performed the literature analysis, drawing of the figures, writing of the manuscript, and final approval of the final version. A.I. participated in the original concept of the article, reviewing and editing of the figures, reviewing the manuscript, and final approval of the manuscript. M.V. and X.D. participated in reviewing the figures and manuscript and final approval of the manuscript.

References

[1] Ostrom QT, Bauchet L, Davis FG, Deltour I, Fisher JL, Langer CE, et al. The epidemiology of glioma in adults: a "state of the science" review. Neuro Oncol 2014;16(7):896–913.

[2] Pace A, Dirven L, Koekkoek JAF, Golla H, Fleming J, Ruda R, et al. European Association for Neuro-Oncology (EANO) guidelines for palliative care in adults with glioma. Lancet Oncol 2017;18(6). e330-e40.

[3] National Center for Biotechnology Information. PubChem Compound Summary for CID 5394, Temozolomide; 2021. Available from https://pubchem.ncbi.nlm.nih.gov/compound/Temozolomide [Retrieved 21 April 2021].

[4] Saunders NR, Dreifuss JJ, Dziegielewska KM, Johansson PA, Habgood MD, Møllgård K, et al. The rights and wrongs of blood-brain barrier permeability studies: a walk through 100 years of history. Front Neurosci 2014;8:404.
[5] Shen S, Zhang W. ABC transporters and drug efflux at the blood-brain barrier. Rev Neurosci 2010;21(1):29–53.
[6] Drean A, Goldwirt L, Verreault M, Canney M, Schmitt C, Guehennec J, et al. Blood-brain barrier, cytotoxic chemotherapies and glioblastoma. Expert Rev Neurother 2016;16(11):1285–300.
[7] Chaves C, Shawahna R, Jacob A, Scherrmann JM, Decleves X. Human ABC transporters at blood-CNS interfaces as determinants of CNS drug penetration. Curr Pharm Des 2014;20(10):1450–62.
[8] Gillet JP, Gottesman MM. Mechanisms of multidrug resistance in cancer. Methods Mol Biol 2010;596:47–76.
[9] Ramirez YP, Weatherbee JL, Wheelhouse RT, Ross AH. Glioblastoma multiforme therapy and mechanisms of resistance. Pharmaceuticals (Basel, Switzerland) 2013;6(12):1475–506.
[10] Franceschi E, Lamberti G, Visani M, Paccapelo A, Mura A, Tallini G, et al. Temozolomide rechallenge in recurrent glioblastoma: when is it useful? Future Oncol 2018;14(11):1063–9.
[11] Turk D, Hall MD, Chu BF, Ludwig JA, Fales HM, Gottesman MM, et al. Identification of compounds selectively killing multidrug-resistant cancer cells. Cancer Res 2009;69(21):8293–301.
[12] Drean A, Rosenberg S, Lejeune FX, Goli L, Nadaradjane AA, Guehennec J, et al. Correction to: ATP binding cassette (ABC) transporters: expression and clinical value in glioblastoma. J Neurooncol 2018;138(3):487.
[13] Mahringer A, Fricker G. ABC transporters at the blood-brain barrier. Expert Opin Drug Metab Toxicol 2016;12(5):499–508.
[14] Linton KJ, Higgins CF. Structure and function of ABC transporters: the ATP switch provides flexible control. Archiv Eur J Physiol 2007;453(5):555–67.
[15] Lockhart AC, Tirona RG, Kim RB. Pharmacogenetics of ATP-binding cassette transporters in cancer and chemotherapy. Mol Cancer Ther 2003;2(7):685–98.
[16] Schinkel AH, Jonker JW. Mammalian drug efflux transporters of the ATP binding cassette (ABC) family: an overview. Adv Drug Deliv Rev 2003;55(1):3–29.
[17] Pulgar VM. Transcytosis to cross the blood brain barrier, new advancements and challenges. Front Neurosci 2018;12:1019.
[18] Deeley RG, Westlake C, Cole SP. Transmembrane transport of *endo*- and xenobiotics by mammalian ATP-binding cassette multidrug resistance proteins. Physiol Rev 2006;86(3):849–99.
[19] Zolnerciks JK, Andress EJ, Nicolaou M, Linton KJ. Structure of ABC transporters. Essays Biochem 2011;50(1):43–61.
[20] Raymond GV, Moser AB, Fatemi A. X-linked adrenoleukodystrophy. In: Adam MP, Ardinger HH, Pagon RA, et al., editors. GeneReviews® [internet]. Seattle (WA): University of Washington, Seattle; 1999. [Updated 2018 Feb 15]. 1993-2020. Bookshelf URL https://www.ncbi.nlm.nih.gov/books/.
[21] Vasiliou V, Vasiliou K, Nebert DW. Human ATP-binding cassette (ABC) transporter family. Hum Genomics 2009;3(3):281–90.
[22] Johnstone RW, Ruefli AA, Smyth MJ. Multiple physiological functions for multidrug transporter P-glycoprotein? Trends Biochem Sci 2000;25(1):1–6.
[23] Tivnan A, Zakaria Z, O'Leary C, Kogel D, Pokorny JL, Sarkaria JN, et al. Inhibition of multidrug resistance protein 1 (MRP1) improves chemotherapy drug response in primary and recurrent glioblastoma multiforme. Front Neurosci 2015;9:218.
[24] Juliano RL, Ling V. A surface glycoprotein modulating drug permeability in Chinese hamster ovary cell mutants. Biochim Biophys Acta 1976;455(1):152–62.
[25] The Human Protein Atlas. 2020 [Updated Feb, 6]. Available from: https://www.proteinatlas.org/ENSG00000085563-ABCB1/tissue.
[26] Schinkel AH, Smit JJ, van Tellingen O, Beijnen JH, Wagenaar E, van Deemter L, et al. Disruption of the mouse mdr1a P-glycoprotein gene leads to a deficiency in the blood-brain barrier and to increased sensitivity to drugs. Cell 1994;77(4):491–502.
[27] Borst P, Schinkel AH. P-glycoprotein ABCB1: a major player in drug handling by mammals. J Clin Invest 2013;123(10):4131 3.
[28] Wolf SJ, Bachtiar M, Wang J, Sim TS, Chong SS, Lee CG. An update on ABCB1 pharmacogenetics: insights from a 3D model into the location and evolutionary conservation of residues corresponding to SNPs associated with drug pharmacokinetics. Pharmacogenomics J 2011;11(5):315–25.

[29] de Gooijer MC, de Vries NA, Buckle T, Buil LCM, Beijnen JH, Boogerd W, et al. Improved brain penetration and antitumor efficacy of Temozolomide by inhibition of ABCB1 and ABCG2. Neoplasia 2018;20(7):710–20.
[30] Schaich M, Kestel L, Pfirrmann M, Robel K, Illmer T, Kramer M, et al. A MDR1 (ABCB1) gene single nucleotide polymorphism predicts outcome of temozolomide treatment in glioblastoma patients. Ann Oncol 2009;20(1): 175–81.
[31] Malmström A, Łysiak M, Åkesson L, et al. ABCB1 single-nucleotide variants and survival in patients with glioblastoma treated with radiotherapy concomitant with temozolomide. Pharmacogenomics J 2020;20:213–9. https://doi.org/10.1038/s41397-019-0107-z.
[32] ICH GCP. (11C)dLop as a marker of P-glycoprotein function in patients with gliomas. Available from https://clinicaltrials.gov/ct2/show/NCT01281982?id=NCT01281982&draw=2&rank=1&load=cart; 2012.
[33] Zhang Y, Wang SX, Ma JW, Li HY, Ye JC, Xie SM, et al. EGCG inhibits properties of glioma stem-like cells and synergizes with temozolomide through downregulation of P-glycoprotein inhibition. J Neurooncol 2015;121(1): 41–52.
[34] Munoz JL, Walker ND, Scotto KW, Rameshwar P. Temozolomide competes for P-glycoprotein and contributes to chemoresistance in glioblastoma cells. Cancer Lett 2015;367(1):69–75.
[35] Tso JL, Yang S, Menjivar JC, Yamada K, Zhang Y, Hong I, et al. Bone morphogenetic protein 7 sensitizes O6-methylguanine methyltransferase expressing-glioblastoma stem cells to clinically relevant dose of temozolomide. Mol Cancer 2015;14:189.
[36] Zhang P, Chen XB, Ding BQ, Liu HL, He T. Down-regulation of ABCE1 inhibits temozolomide resistance in glioma through the PI3K/Akt/NF-κB signaling pathway. Biosci Rep 2018;38(6):BSR20181711. https://doi.org/10.1042/BSR20181711.
[37] Goldwirt L, Beccaria K, Carpentier A, Farinotti R, Fernandez C. Irinotecan and temozolomide brain distribution: a focus on ABCB1. Cancer Chemother Pharmacol 2014;74(1):185–93.
[38] de Gooijer MC, Buil LCM, Citirikkaya CH, Hermans J, Beijnen JH, van Tellingen O. ABCB1 attenuates the brain penetration of the PARP inhibitor AZD2461. Mol Pharm 2018;15(11):5236–43.
[39] Riganti C, Salaroglio IC, Caldera V, Campia I, Kopecka J, Mellai M, et al. Temozolomide downregulates P-glycoprotein expression in glioblastoma stem cells by interfering with the Wnt3a/glycogen synthase-3 kinase/beta-catenin pathway. Neuro Oncol 2013;15(11):1502–17.
[40] Salaroglio IC, Mujumdar P, Annovazzi L, Kopecka J, Mellai M, Schiffer D, et al. Carbonic anhydrase XII inhibitors overcome P-glycoprotein-mediated resistance to Temozolomide in Glioblastoma. Mol Cancer Ther 2018;17 (12):2598–609.
[41] Lum BL, Kaubisch S, Yahanda AM, Adler KM, Jew L, Ehsan MN, et al. Alteration of etoposide pharmacokinetics and pharmacodynamics by cyclosporine in a phase I trial to modulate multidrug resistance. J Clin Oncol 1992;10(10):1635–42.
[42] Durmaz R, Deliorman S, Uyar R, Isiksoy S, Erol K, Tel E. The effects of anticancer drugs in combination with nimodipine and verapamil on cultured cells of glioblastoma multiforme. Clin Neurol Neurosurg 1999;101(4): 238–44.
[43] Fellner S, Bauer B, Miller DS, Schaffrik M, Fankhanel M, Spruss T, et al. Transport of paclitaxel (Taxol) across the blood–brain barrier in vitro and in vivo. J Clin Investig 2002;110(9):1309–18.
[44] Balayssac D, Cayre A, Authier N, Bourdu S, Penault-Llorca F, Gillet JP, et al. Patterns of P-glycoprotein activity in the nervous system during vincristine-induced neuropathy in rats. J Peripher Nerv Syst 2005;10(3):301–10.
[45] Rao VK, Wangsa D, Robey RW, Huff L, Honjo Y, Hung J, et al. Characterization of ABCG2 gene amplification manifesting as extrachromosomal DNA in mitoxantrone-selected SF295 human glioblastoma cells. Cancer Genet Cytogenet 2005;160(2):126–33.
[46] Nakai E, Park K, Yawata T, Chihara T, Kumazawa A, Nakabayashi H, et al. Enhanced MDR1 expression and chemoresistance of cancer stem cells derived from glioblastoma. Cancer Investig 2009;27(9):901–8.
[47] Agarwal S, Mittapalli RK, Zellmer DM, Gallardo JL, Donelson R, Seiler C, et al. Active efflux of Dasatinib from the brain limits efficacy against murine glioblastoma: broad implications for the clinical use of molecularly targeted agents. Mol Cancer Ther 2012;11(10):2183–92.
[48] Oberoi RK, Mittapalli RK, Elmquist WF. Pharmacokinetic assessment of efflux transport in sunitinib distribution to the brain. J Pharmacol Exp Ther 2013;347(3):755–64.
[49] Lin F, de Gooijer MC, Roig EM, Buil LC, Christner SM, Beumer JH, et al. ABCB1, ABCG2, and PTEN determine the response of glioblastoma to temozolomide and ABT-888 therapy. Clin Cancer Res 2014;20(10):2703–13.

[50] Tomaszowski KH, Schirrmacher R, Kaina B. Multidrug efflux pumps attenuate the effect of MGMT inhibitors. Mol Pharm 2015;12(11):3924–34.
[51] Martin V, Sanchez-Sanchez AM, Herrera F, Gomez-Manzano C, Fueyo J, Alvarez-Vega MA, et al. Melatonin-induced methylation of the ABCG2/BCRP promoter as a novel mechanism to overcome multidrug resistance in brain tumor stem cells. Br J Cancer 2013;108(10):2005–12.
[52] Parrish KE, Cen L, Murray J, Calligaris D, Kizilbash S, Mittapalli RK, et al. Efficacy of PARP inhibitor rucaparib in orthotopic glioblastoma xenografts is limited by ineffective drug penetration into the central nervous system. Mol Cancer Ther 2015;14(12):2735–43.
[53] Chua C, Zaiden N, Chong KH, See SJ, Wong MC, Ang BT, et al. Characterization of a side population of astrocytoma cells in response to temozolomide. J Neurosurg 2008;109(5):856–66.
[54] Liu Y, Guo Q, Zhang H, Li GH, Feng S, Yu XZ, et al. Effect of siRNA-Livin on drug resistance to chemotherapy in glioma U251 cells and CD133(+) stem cells. Exp Ther Med 2015;10(4):1317–23.
[55] Fruehauf JP, Brem H, Brem S, Sloan A, Barger G, Huang W, et al. In vitro drug response and molecular markers associated with drug resistance in malignant gliomas. Clinical Cancer Res 2006;12(15):4523–32.
[56] Fan TY, Wang H, Xiang P, Liu YW, Li HZ, Lei BX, et al. Inhibition of EZH2 reverses chemotherapeutic drug TMZ chemosensitivity in glioblastoma. Int J Clin Exp Pathol 2014;7(10):6662–70.
[57] Pessina S, Cantini G, Kapetis D, Cazzato E, Di Ianni N, Finocchiaro G, et al. The multidrug-resistance transporter Abcc3 protects NK cells from chemotherapy in a murine model of malignant glioma. Oncoimmunology 2016; 5(5), e1108513.
[58] Liu B, Guo Z, Dong H, Daofeng T, Cai Q, Ji B, et al. LRIG1, human EGFR inhibitor, reverses multidrug resistance through modulation of ABCB1 and ABCG2. Brain Res 1611;2015:93–100.
[59] Li B, He H, Tao BB, Zhao ZY, Hu GH, Luo C, et al. Knockdown of CDK6 enhances glioma sensitivity to chemotherapy. Oncol Rep 2012;28(3):909–14.
[60] Liang F, Wang B, Bao L, Zhao YS, Zhang SM, Zhang SQ. Overexpression of ILK promotes temozolomide resistance in glioma cells. Mol Med Rep 2017;15(3):1297–304.
[61] Munoz JL, Rodriguez-Cruz V, Greco SJ, Nagula V, Scotto KW, Rameshwar P. Temozolomide induces the production of epidermal growth factor to regulate MDR1 expression in glioblastoma cells. Mol Cancer Ther 2014; 13(10):2399–411.
[62] Emery IF, Gopalan A, Wood S, Chow KH, Battelli C, George J, et al. Expression and function of ABCG2 and XIAP in glioblastomas. J Neurooncol 2017;133(1):47–57.
[63] Munoz JL, Rodriguez-Cruz V, Ramkissoon SH, Ligon KL, Greco SJ, Rameshwar P. Temozolomide resistance in glioblastoma occurs by miRNA-9-targeted PTCH1, independent of sonic hedgehog level. Oncotarget 2015;6(2):1190–201.
[64] Peignan L, Garrido W, Segura R, Melo R, Rojas D, Carcamo JG, et al. Combined use of anticancer drugs and an inhibitor of multiple drug resistance-associated protein-1 increases sensitivity and decreases survival of glioblastoma multiforme cells in vitro. Neurochem Res 2011;36(8):1397–406.
[65] Drean A, Rosenberg S, Lejeune FX, Goli L, Nadaradjane AA, Guehennec J, et al. ATP binding cassette (ABC) transporters: expression and clinical value in glioblastoma. J Neurooncol 2018;138(3):479–86.
[66] Cole SP, Bhardwaj G, Gerlach JH, Mackie JE, Grant CE, Almquist KC, et al. Overexpression of a transporter gene in a multidrug-resistant human lung cancer cell line. Science 1992;258(5088):1650–4.
[67] Uhlén M, Fagerberg L, Hallström BM, Lindskog C, Oksvold P, Mardinoglu A, et al. Tissue-based map of the human proteome. Science 2015;347(6220):1260419.
[68] Yin J, Zhang J. Multidrug resistance-associated protein 1 (MRP1/ABCC1) polymorphism: from discovery to clinical application. J Cent S Univ Med Sci 2011;36(10):927–38.
[69] Benyahia B, Huguet S, Declèves X, Mokhtari K, Crinière E, Bernaudin JF, et al. Multidrug resistance-associated protein MRP1 expression in human gliomas: chemosensitization to vincristine and etoposide by indomethacin in human glioma cell lines overexpressing MRP1. J Neurooncol 2004;66(1–2):65–70.
[70] Doyle LA, Yang W, Abruzzo LV, Krogmann T, Gao Y, Rishi AK, et al. A multidrug resistance transporter from human MCF-7 breast cancer cells. Proc Natl Acad Sci U S A 1998;95(26):15665–70.
[71] Shawahna R, Uchida Y, Declèves X, Ohtsuki S, Yousif S, Dauchy S, et al. Transcriptomic and quantitative proteomic analysis of transporters and drug metabolizing enzymes in freshly isolated human brain microvessels. Mol Pharm 2011;8(4):1332–41.

[72] Mao Q, Unadkat JD. Role of the breast cancer resistance protein (BCRP/ABCG2) in drug transport—an update. AAPS J 2015;17(1):65–82.
[73] Hira D, Terada T. BCRP/ABCG2 and high-alert medications: biochemical, pharmacokinetic, pharmacogenetic, and clinical implications. Biochem Pharmacol 2018;147:201–10.
[74] Martin V, Sanchez-Sanchez AM, Herrera F, Gomez-Manzano C, Fueyo J, Alvarez-Vega MA, et al. Melatonin-induced methylation of the ABCG2/BCRP promoter as a novel mechanism to overcome multidrug resistance in brain tumour stem cells. Br J Cancer 2013;108(10):2005–12.
[75] Li W, Zhang H, Assaraf YG, Zhao K, Xu X, Xie J, et al. Overcoming ABC transporter-mediated multidrug resistance: molecular mechanisms and novel therapeutic drug strategies. Drug Resist Updat 2016;27:14–29.
[76] Tamaki A, Ierano C, Szakacs G, Robey RW, Bates SE. The controversial role of ABC transporters in clinical oncology. Essays Biochem 2011;50(1):209–32.
[77] Leonard GD, Fojo T, Bates SE. The role of ABC transporters in clinical practice. Oncologist 2003;8(5):411–24.
[78] Zhao M, van Straten D, Broekman MLD, Préat V, Schiffelers RM. Nanocarrier-based drug combination therapy for glioblastoma. Theranostics 2020;10(3):1355–72.
[79] Beier CP, Schmid C, Gorlia T, Kleinletzenberger C, Beier D, Grauer O, et al. RNOP-09: pegylated liposomal doxorubicine and prolonged temozolomide in addition to radiotherapy in newly diagnosed glioblastoma—a phase II study. BMC Cancer 2009;9:308.
[80] Clarke JL, Molinaro AM, DeSilva AA, Rabbitt JE, Drummond DC, Chang SM, et al. A phase I trial of intravenous liposomal irinotecan in patients with recurrent high-grade gliomas. J Clin Oncol 2015;33(15_suppl):2029.
[81] Gan HK, van den Bent M, Lassman AB, Reardon DA, Scott AM. Antibody-drug conjugates in glioblastoma therapy: the right drugs to the right cells. Nat Rev Clin Oncol 2017;14(11):695–707.
[82] Reardon DA, Lassman AB, van den Bent M, Kumthekar P, Merrell R, Scott AM, et al. Efficacy and safety results of ABT-414 in combination with radiation and temozolomide in newly diagnosed glioblastoma. Neuro Oncol 2017;19(7):965–75.
[83] Hynynen K, McDannold N, Vykhodtseva N, Jolesz FA. Noninvasive MR imaging-guided focal opening of the blood-brain barrier in rabbits. Radiology 2001;220(3):640–6.
[84] Beccaria K, Canney M, Goldwirt L, Fernandez C, Piquet J, Perier MC, et al. Ultrasound-induced opening of the blood-brain barrier to enhance temozolomide and irinotecan delivery: an experimental study in rabbits. J Neurosurg 2016;124(6):1602–10.
[85] Goldwirt L, Canney M, Horodyckid C, Poupon J, Mourah S, Vignot A, et al. Enhanced brain distribution of carboplatin in a primate model after blood-brain barrier disruption using an implantable ultrasound device. Cancer Chemother Pharmacol 2016;77(1):211–6.
[86] Choi H, Lee EH, Han M, An SH, Park J. Diminished expression of P-glycoprotein using focused ultrasound is associated with JNK-dependent signaling pathway in cerebral blood vessels. Front Neurosci 2019;13:1350.
[87] Carpentier A, Canney M, Vignot A, Reina V, Beccaria K, Horodyckid C, Karachi C, Leclercq D, Lafon C, Chapelon JY, Capelle L, Cornu P, Sanson M, Hoang-Xuan K, Delattre JY, Idbaih A. Clinical trial of blood-brain barrier disruption by pulsed ultrasound. Sci Transl Med 2016;8(343):343re2. https://doi.org/10.1126/scitranslmed.aaf6086.
[88] Crews KR, Hicks JK, Pui CH, Relling MV, Evans WE. Pharmacogenomics and individualized medicine: translating science into practice. Clin Pharmacol Ther 2012;92(4):467–75. https://doi.org/10.1038/clpt.2012.120.
[89] Bauer M, Romermann K, Karch R, Wulkersdorfer B, Stanek J, Philippe C, et al. Pilot PET study to assess the functional interplay between ABCB1 and ABCG2 at the human blood-brain barrier. Clin Pharmacol Ther 2016;100(2):131–41.

CHAPTER

10

Role of mitochondrial reactive oxygen species in glioblastoma drug resistance and strategies for therapeutic intervention

Alessia Lo Dico[a,*], *Daniela Salvatore*[a,*], *Cecilia Diceglie*[a], *Cristina Martelli*[a], *and Luisa Ottobrini*[a,b]

[a]Department of Pathophysiology and Transplantation, University of Milan, Segrate, MI, Italy
[b]Department of Molecular Bioimaging and Physiology (IBFM), CNR, Segrate, MI, Italy

Abstract

Glioblastoma (GBM) is the most aggressive tumor of CNS characterized by a poor prognosis mainly due to the resistance to the available treatments. Temozolomide (TMZ) is the gold standard for GBM treatment although this drug even in association with radiotherapy provides a median survival of 14.6 months. There are evidences that the increase in intracellular ROS is able to sustain gliomagenesis, tumor proliferation rate, and resistance to treatments by supporting proliferative pathways and metabolism as well as by generating genetic instability causing the establishment of protumorigenic mutants. At the same time, an increase in intracellular ROS, being involved in the oxidative damage of DNA, lipids, and proteins, is involved in the impairment of cell function, triggering also cell death.

Reactive oxygen species (ROS) are highly reactive molecules originating from different enzymatic and nonenzymatic reactions occurring in mitochondria and other different intracellular organelles. They can have also an exogenous origin related to the effects of xenobiotics or radiations. Under basal conditions, ROS generation and their scavenging are tightly controlled, and redox homeostasis is maintained, but in case of an imbalance in ROS turnover, oxidative stress can occur producing cell damage. For these reasons, ROS modulation has been pursued by different treatment plans to increase cytotoxicity and overcome GBM resistance to treatment. Since the double-edged sword role covered by ROS, which mediates cell survival signaling, both a positive and a negative ROS modulation have been assessed for therapeutic purposes.

[*]These authors have contributed equally to this work.

Glioblastoma Resistance to Chemotherapy: Molecular Mechanisms and Innovative Reversal Strategies
https://doi.org/10.1016/B978-0-12-821567-8.00026-9

In this chapter, the mechanisms involved in the relation between ROS and cell response to treatment are described with particular attention focused on the induction of chaperone-mediated autophagy (CMA) whose activity together with the reduction of hypoxia-inducible factor-1α (HIF-1) activity after treatment have been related to GBM cell responsiveness to TMZ.

Abbreviations

ALDH1A3	aldehyde dehydrogenase 1A3
ALDH4A1	aldehyde dehydrogenase 4 family member A1
APNG	*alkylpurine–DNA-N*-glycosylase
AMPK	AMP-activated protein kinase
AP-1	activator protein-1
AQPs	aquaporins
ARE	antioxidant response elements
Bcl-2	B-cell lymphoma 2
BER	base-excision repair
BSO	buthionine sulfoximine
CDC	cell division cycle 25
Chk	checkpoint kinase 1 and 2
CMA	chaperone mediated autophagy
COXs	cyclooxygenases
DDR	DNA damage response
DSB	double-strand breaks
EF-1α	elongation factor
EGF	epidermal growth factor
EGFR	epidermal growth factor receptor
EpRE	electrophile response elements
ER	endoplasmic reticulum
FDA	Food and Drug Administration
FoxO	forkhead homeobox type O
GAPDH	glyceraldehyde-3-phosphate dehydrogenase
GBM	glioblastoma multiforme
GFAP	glial fibrillary acidic protein
GLUT	glucose transporter type
GLS2	glutaminase 2
GSH	glutathione
GOX	glucose oxidase
GPx	glutathione peroxidase
GSR	glutathione reductase
GSS	glutathione synthetase
HG	hydroxy-glutarate
HIF-1α	hypoxia inducible factor-1α
HK	hexokinase
HO-1	heme oxygenase
HRE	hypoxia response elements
HSC70	heat shock cognate 70 kDa protein
Hsp90	heat shock proteins
IDH1	isocitrate dehydrogenase
KEAP	kelch-like erythroid cell-derived protein
KFERQ	lysine-phenylalanine-glutamic acid-arginine-glutamine
LOX	lipoxygenase
LAMP	lysosome-associated membrane protein
MA	mithramycin A
MAPK	mitogen-activated protein kinase

MDM2	mouse double minute 2 homolog
MDR1	multidrug resistance 1
MMR	mismatch repair
MnSOD	manganese superoxide dismutase
MPTP	mitochondrial permeability transition pore
mTOR	mammalian target of rapamycin
mTORC1	mTOR complex 1
NADPH	nicotinamide adenine dinucleotide phosphate
NF-κB	nuclear factor kappa B
NOX	nicotinamide adenine dinucleotide phosphate (NADPH) oxidase
Nrf2	nuclear factor E2-related factor 2
O6MG	O6-methylguanin
MGMT	O6-methylguanine-DNA methyltransferase
ODD	oxygen-dependent degradation
OXPHOS	oxidative phosphorylation
OS	overall survival
PAS	Per-Arnt-Sim
PDG	platelet-derived growth factor
PDK	pyruvate dehydrogenase kinase
PFS	progression-free survival
PHD	prolylhydroxylases
PI3K	phosphatidylinositol 3-kinase
PPP	pentose phosphate pathway
PRDXs	peroxiredoxins
PTEN	phosphatase and tensin homolog
ROS	reactive oxygen species
SOD	superoxide dismutase
Sp1	specificity protein 1
SSB	single-strand breaks
STUB-1	STIP1 homology and U-box containing protein 1
TCA	tricarboxylic acid cycle
TMZ	temozolomide
TKIs	tyrosine kinase inhibitors
Tp53	tumor suppressor protein
UPR	unfolded protein response
VDACs	voltage-dependent anion channels
VEGF	vascular-endothelial growth factor
VHL	von Hippel-Lindau

Conflict of interest

No potential conflicts of interest were disclosed.

Introduction

Glioblastoma

Glioblastoma (GBM) is the most common astrocytic-derived brain tumor in adults, characterized by a poor prognosis mainly owing to the resistance to the available treatments. Temozolomide (TMZ) is the gold standard for GBM treatment (Temodar, Temodal, TMZ). TMZ was first approved by the US Food and Drug Administration (FDA) for the treatment of refractory anaplastic astrocytoma and then in 2005 for newly diagnosed adult GBM [1]. Then, Roger Stupp proposed a combined treatment, the so-called Stupp protocol, based on the use of both TMZ and radiotherapy. The Stupp protocol showed a clinically meaningful and statistically significant survival benefit without adding additional toxicity. The overall survival of this combined treatment offers an improvement in the median survival from 12.1 months to 14.6. Despite this improvement, tumor recurrence occurs with a frequency of almost 90%. The difficulty to treat this kind of cancer depends on several reasons based on intrinsic features such as invasiveness and staminal potential as well as its ability to develop drug resistance [2–5].

There is evidence that the increase in intracellular reactive oxygen species (ROS) is able to sustain gliomagenesis, tumor proliferation rate, and resistance to treatments. Inflammation is one of the main causes of ROS increase in the microenvironment being involved in the induction of oxidative stress-provoking genetic instability, mitochondrial dysfunction, and abnormal metabolism, leading to glioma development, since ROS sustain the activity of several transduction pathways. At the same time, an increase of intracellular ROS is involved in the oxidative damage of DNA, lipids, and proteins impairing cell function and triggering apoptosis. These phenomena are particularly known as biological effects induced by different antineoplastic drugs and radiotherapy. For these reasons, new therapeutics have been aimed at modulating the microenvironment and cellular redox status [6].

TMZ: Mechanism of action and resistance

TMZ (4-methyl-5-oxo-2,3,4,6,8-pentazabicyclo [4.3.0] nona-2,7,9-triene-9-carboxamide) is an imidazotetrazine that derives from the alkylating agent dacarbazine. It exists as a prodrug and is stable at acidic pH (<5), whereas at neutral or alkaline pH (>7), it undergoes hydroxylation forming the (3-methyltriazen-1-yl) imidazole-4-carboxamide (MTIC) that can be released in the tumor tissue as methyldiazonium ion. This exerts its cytotoxic effect by transferring methyl groups to DNA bases [7, 8]. TMZ as an alkylating agent methylates DNA for almost 90% at N7 position of guanine and at the N3 position of adenine [9]. Methylated complexes are repaired by the base-excision repair (BER) mechanism counteracting the TMZ-mediated cytotoxic effect. The abrogation or the downmodulation of this system of repair allows the drug to exert its cytotoxicity. The expression of the nuclear alkylpurine-DNA-N-glycosylase (APNG), an enzyme involved in removing alkylated DNA bases and in repairing the N7-meG and N3-meA, has been positively correlated with worse survival compared to patients with no or low APNG expression [10, 11].

Beyond the N-related adducts, TMZ cytotoxic effect is mainly owing to the O6-methylguanine-(O6MG)-induced DNA damage although its amount is very limited (about 5%–10%) [12]. During DNA synthesis, polymerase inserts a thymine opposite in the pairs with O6MGs. O6MG-thymine mispairs are recognized by the DNA mismatch repair (MMR) system that can remove the thymine paired with the O6MG, but not the O6MG itself. Then, a futile cycle occurs mediating the formation of DNA single-strand breaks (SSB) and double-strand breaks (DSB) roughly 2–3 cell cycles after TMZ administration [13, 14]. This condition finally causes stalled replication forks with damage to DNA, phosphorylation of the checkpoint kinase 1 and 2 (Chk1 and Chk2), inactivation of the cell division cycle 25 (cdc25), cell cycle arrest, and accumulation of cells at the G2/M boundary. The result is cell cycle arrest aimed at the recovery of TMZ-mediated damage or if the damage cannot be recovered, the induction of cell death through apoptosis, senescence, or mitotic catastrophe [15].

Methylated nucleotides are often removed in a single-step reaction, mediated by the enzyme O6-methylguanine-DNA methyltransferase (MGMT). The amount of functional MGMT enzyme is proportional to drug resistance. Indeed, resistance to TMZ could be addressed to the expression of this protein or more specifically to the percentage of MGMT promoter methylation status. MGMT expression, in fact, is usually suppressed owing to promoter CpG methylation. Tumors that do not express MGMT protein are often sensitive to TMZ, although other several mechanisms can be involved in GBM resistance to the drug. An improved progression-free survival (PFS) and overall survival (OS) after TMZ treatment has been correlated with MGMT epigenetic silencing in newly diagnosed patients [16].

Furthermore, MGMT expression regulation depends also upon the activation status of its enhancer. Indeed, a distal enhancer able to promote MGMT expression has been identified. This enhancer, called K-M, is located 560 kb away from the MGMT promoter between the gene encoding the well-known marker of proliferation Ki67 (MKI67) and MGMT promoter and it is able to switch on MGMT gene expression even when hypermethylation occurs. GBMs in which the enhancer is deleted show lower MGMT expression, are TMZ responders, and possess an impaired proliferation [17].

Additional TMZ mechanism of action

Nowadays, MGMT promoter methylation has been considered an important prognostic and predictive biomarker. Actually, there is a percentage of patients in which MGMT expression level or MGMT promoter methylation do not correlate with GBM responsiveness to TMZ [18]. Moreover, its value is limited to the analysis of damages provoked by O6-methylguanine-induced toxicity, while the contribution of other mechanisms of action of the drug is not considered by using this biomarker.

TMZ treatment, beyond its alkylating activity, causes multiple effects into the cell, which might promote death or survival, by regulating GBM responsiveness to chemo- and radio-treatment [19]. For example, the expression of an efflux pump (P-glycoprotein, multidrug resistance 1, MDR1), has been described as allowing the drug to be excreted from the cells, thus counteracting its cytotoxicity. This mechanism is well known as MDR and has become the target of several therapeutic approaches aimed at overcoming resistance in GBM [20].

Moreover, TMZ-mediated cell cycle G_2 arrest is strictly related to the phosphorylation/activation of the Chk1, which is known to interact with the heat shock cognate 70 kDa protein (HSC70) [21, 22]. In detail, Chk1 represents the first protein reported to undergo chaperone-mediated autophagy (CMA)-mediated degradation and, what is more, it has been described that also etoposide (topoisomerase II inhibitor)-mediated DNA damage is able to trigger phosphorylated-Chk1 degradation through CMA. Moreover, the authors have observed that after etoposide treatment, HSC70 was present in the nucleus, with the hypothesis that this chaperone might target phosphorylated Chk1 and deliver it from the nucleus to the lysosome for the consecutive degradation [22].

Other TMZ mechanisms of action have been related to autophagy and apoptosis induction and have been also referred to as the release in the cytoplasm of ROS. Under stressors,

GBM cells activate survival pathways based on autophagy or, on the contrary, react switching on the apoptosis pathway when the cell is not able to sustain stress conditions and vital functions [23, 24].

It has been reported that TMZ induces a concerted mechanism involving endoplasmic reticulum (ER) stress and mitochondrial membrane depolarization, resulting in a cytoplasmic ROS release that contributes to the autophagic pathway activation [19, 25]. These TMZ-induced concerted steps are owing to the fact that mitochondria and ER are physically interacting [26, 27]. ER is an important cellular organelle which is strictly bound to the mitochondria and is involved in the synthesis of transmembrane proteins and lipids, in the cellular redox homeostasis (being the source of 25% of ROS within the cell), and in the modulation of intracellular calcium rates [28]. ER stress is intended as an accumulation of misfolded proteins in the ER itself and a release of Ca^{2+} from ER to the cytoplasm, in turn provoking the opening of the mitochondrial permeability transition pore (MPTP), its depolarization, the increase of ROS into the cytoplasm, and the induction of an unfolded protein response (UPR) for reestablishing the homeostasis [28, 29]. Folding proteins in ER is an energy-demanding process, thus metabolic alterations regarding both glycolysis and oxidative phosphorylation (OXPHOS) can induce an impairment of protein folding, ER stress, and UPR [30].

Recent evidence showed that TMZ-induced ROS release from mitochondria is able to induce autophagy and in detail a specific kind of autophagy, the CMA. CMA is able to degrade key proteins associated with cell survival, cell metabolism, proliferation, stemness, angiogenesis, and invasiveness, fulfilling a crucial role in tumor development, progression, and resistance to treatments [31]. Among CMA targets, it is possible to find enzymes involved in glucose metabolism, proteins regulating cell cycle, translation, and transcription factors able to reprogram cell expression profile. One of these factors is the hypoxia-inducible factor-1α, (HIF-1α) that beyond the well-known role in angiogenesis and cell response to hypoxia, has been described to mediate cell resistance to therapy, by inhibiting cell apoptosis and promoting all mechanisms involved in the cell survival and proliferation [26, 32–34].

HIF-1α degradation mediated by CMA has been related to GBM sensitivity to TMZ-based treatments [34]. In detail, HIF-1α can interact with the key CMA players, HSC70 and lysosome-associated membrane protein (LAMP)-2A, as assessed in a study where the modulation of these proteins resulted in the inverse HIF-1α expression level and activity [35]. Of note, it has been demonstrated that ubiquitin-ligase STIP1 homology and U-box containing protein 1 (STUB-1) is able to interact with CMA key molecules (as HSC70 and the heat shock proteins Hsp90) and CMA substrates, by ubiquitinating them for their consequent CMA degradation. STUB1-mediated HIF-1α ubiquitination on lysine 63 (K63) has been recently confirmed to be pivotal for HIF-1α degradation through CMA even in hypoxic conditions [36] (Fig. 1).

Responsiveness to TMZ is mediated by a complex series of events that, based on the DNA repair mechanisms, apoptosis, and autophagy induction and CMA interplay, make the study of GBM resistance a clinical dilemma. To complicate this scenario, extrinsic factors such as hypoxia, nutrient availability, and redox homeostasis have to be considered.

Drug resistance is defined as the result of different phenomena which help cancer cells to become tolerant to pharmacological treatment. In detail, cancer cells acquire a specific

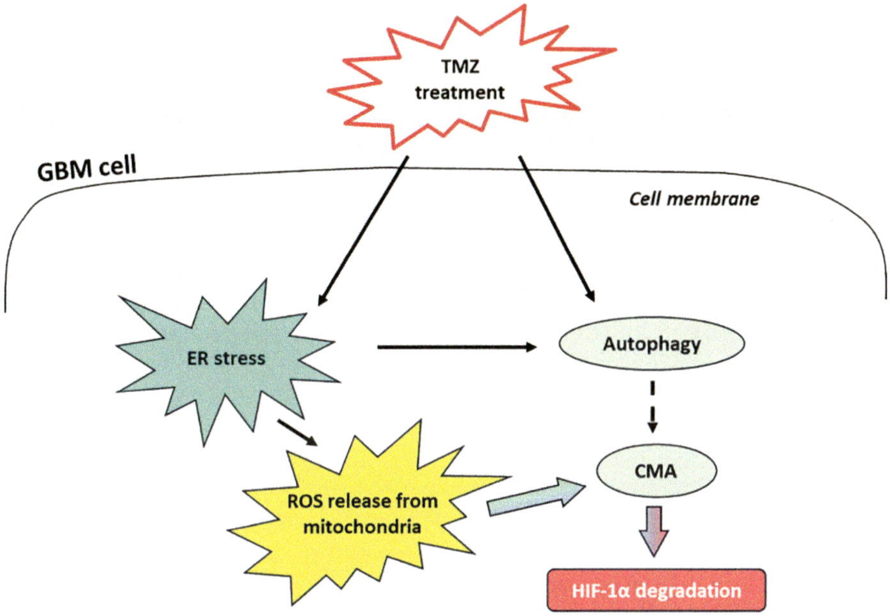

FIG. 1 TMZ mechanism of action in GBM cells. In addition to the well-known mechanism involving base methylation, TMZ produces other important effects within the cells. Indeed, it provokes an ER stress allowing ROS release by the mitochondria and finally inducing CMA. This cascade finally result in the downregulation of HIF-1 activity that becomes biomarker of TMZ responsiveness.

behavior becoming refractory to the used drug. Drug resistance can be classified as intrinsic or acquired. Intrinsic resistance is a condition in which cancer cells show an innate drug resistance already at the onset of the tumor. Cancer cells or patients before the start of the treatment are indeed refractory to the drug. This condition could be owing to genetic features protecting cancer cells from the cytotoxic effect of the drug. In this view, tumor heterogeneity can account for escaping from the cytotoxic effect and can contribute to tumor relapse. On the other hand, acquired resistance occurs in a two-phase process: at first tumor cells are responsive to the treatment, and then they develop resistance to the drug. It is an acquired feature. This process could be owing to the activation of an oncogene, to the downregulation of an oncosuppressor gene, or to the involvement of the microenvironment. Indeed, it is an ability of the cells to develop defense mechanisms against the treatment.

Within this acquired resistance, in GBM, genetic background covers the main role—in fact, it is crucial for tumor development and progression. In GBM, mutations are responsible for specific features related to chemoresistance. Moreover, genetic mutations and their accumulation, account for the progression from the lower to the higher GBM grade [37, 38]. The main players involved in cancer progression and resistance to treatments in GBM are:

Tumor Suppressor Protein (Tp53): The first genetic alterations identified in astrocytic brain tumors and the most recurring mutations have been detected in the *Tp53* gene on chromosome 17p. Alterations of the *Tp53* gene have been found in about 30% of benign and anaplastic astrocytomas as well as GBMs. Anaplastic astrocytomas (grade III), found in preexistent low-grade astrocytomas or detected *de novo*, have a similar incidence of *Tp53* mutations but, in addition, show a loss of heterozygosity on chromosome 19q in more than 40% of the cases [39]. Tp53 regulates cell survival or apoptosis in relation to cell damages but it is involved also in the modulation of cell metabolism, and of transcriptional regulation of superoxide dismutase (SOD)-2, glutathione peroxidase (GPx), and aldehyde dehydrogenase four family member A1 (ALDH4A1), all genes involved in redox regulation. Mutations can induce a gain of function profile that has been related to a number of mechanisms involved in tumor survival, proliferation, and resistance to treatment. In fact, mutant Tp53 can increase MGMT expression, modulate apoptosis and autophagy, and in general it can reprogram cell gene expression and cell metabolism by also increasing ROS levels [40–43]. Tp53 is also able to induce the expression of Tp53-induced glycolysis and apoptosis regulator (TIGAR) leading to a decrease of glycolysis and an increase of the pentose phosphate pathway (PPP) involved in the production of nicotinamide adenine dinucleotide phosphate (NADPH), an important antioxidant molecule. Finally, Tp53 induces the expression of glutaminase 2 (GLS2) providing glutamate for glutathione (GSH) synthesis and concurring in the detoxification from ROS [44].

MDM2 (Mouse Double Minute 2 homolog): *MDM2* amplification or overexpression constitutes an alternative mechanism of escaping from p53-regulated control of cell growth by binding to Tp53 and blunting its activity. Overexpression of *MDM2* is the second most common gene mutation in GBMs and is observed in 10%–15% of patients. Some studies show that this mutation has been associated with a poor prognosis. It is involved in GBM resistance because of its role in reducing Tp53-mediated response after treatment [43].

PTEN (Phosphatase and Tensin homolog): *PTEN* mutations are found in ~70% of advanced human GBM. PTEN overexpression has been related to resistance because of its ability to modulate cancer stem cell features [45]. Overexpression of recombinant functional PTEN in GBMs demonstrated the involvement of its axis in the regulation of HIF-1α levels and activity. In fact, the expression of the recombinant functional form leads to a marked reduction in HIF-1α expression. Normal PTEN is a negative regulator of the phosphatidylinositol 3-kinase PI3K/Akt pathway (by dephosphorylating phosphatidylinositol 3,4,5-triphosphate). Its mutation or inactivation by oxidation results in hyperactivation of the PI3K/Akt pathway that may contribute to the enhancement of HIF-1α stability, nuclear translocation, dimerization, and transcriptional activity, even though this event is context- and cell type-dependent. The ROS increase has been associated with the hyper stimulation of this pathway through the activation of tyrosine (Tyr) kinase receptors, supporting thus the recruitment of the PI3Ks. Furthermore, the switch on of AKT by means of PI3K pathway promotes also the release of ROS through the activation of the NADPH oxidase (NOX) [46].

Isocitrate Dehydrogenase (IDH1/IDH2): In GBM, mutations in the genes encoding the cytosolic and the mitochondrial forms of isocitrate dehydrogenase (*IDH1*, mutation occurring in the arginine, R132 and *IDH2*, the mitochondrial homolog in the R172, respectively) are frequent especially in low-grade GBM progressing to higher grades. These enzymes catalyze the conversion of isocitrate and NADP+ to 2-oxo-glutarate and NADPH within tricarboxylic acid cycle (TCA). *IDH1/2* mutations are neomorphic in that mutant IDH leads to the accumulation of the D-2-hydroxy-glutarate (D-2HG), that as an oncometabolite, can also be used as a biomarker in GBM (since it occurs in ~85% of adult secondary GBMs). D2HG molecular activities can be referred to as resistance. For example, IDH mutation triggers the decrease of the prolylhydroxylases (PHD) activity with the consequent increase of HIF-1α that has been involved in resistance to TMZ [47].

Regarding the clinical outcome associated with IDH mutations, controversial theories have been proposed since they have been associated with longer overall survival in both *de novo* and secondary GBM, even if IDH1/IDH2 mutations have been proposed to lead gliomagenesis by epigenetic, metabolic, and transcriptional modulations [48–52]. The mechanism of action could involve a reduction in NADPH and GSH antioxidants and an increase of intracellular ROS [53]. Different models have been proposed also relating IDH1/2 mutations to Tp53 mutation and glutamate availability, showing that their gained function could be different in distinct molecular contexts [54].

Epidermal Growth Factor Receptor (EGFR): The amplification or the activating mutations occurring in *EGFR* are common genetic alterations in GBM (occurring in 50% of GBM) and they are associated with a malignant phenotype and with resistance to therapy. Indeed, EGFR, by triggering tyrosine kinase-mediated signaling, promotes the activation of growth-related pathways, such as those involving PI3K/AKT and the mitogen-activated protein kinases (MAPK). Several clinical trials based on the use of EGFR tyrosine kinase inhibitors (TKIs) (gefinitib, erlotinib, lepatinib) or monoclonal antibodies that block this receptor (cetuximab, panitumumab), have been considered to counteract this molecular axis [55]. Furthermore, the amplification of the EGFR pathway is also connected to ROS amount in the cell and to the tumor progression, being critical also for the resistance to treatment [56, 57]. EGF binding to EGFR induces intracellular ROS increase and, in turn, high hydrogen peroxide levels (200 mM) are able to sustain Tyr autophosphorylation in EGFR supporting further ROS increment. This ROS boost can modulate downstream effectors inducing cell survival and proliferation.

Reactive oxygen species in GBM

GBM and ROS

In GBM, ROS assume different roles since they are involved in carcinogenesis, in tumor progression as well as in tumor cell damage and cell death. In these tumors, cell redox imbalance has been described. In this context, it has been confirmed that ROS are implicated in several signaling pathways related to cellular growth and are strongly linked to resistance to treatment. In fact, ROS possess the ability to interact with pyrimidines, purines, and chromatin proteins provoking genomic instability, favoring the onset of gene expression alterations, and finally inducing carcinogenesis [58]. High ROS content can also sustain cell signaling contributing to resistance to treatments by leading the activity of kinases and small G proteins inducing cell proliferation and cytoskeletal remodeling [59]. On the other hand, oxidative stress has long been used with the aim of inducing cell damage and finally tumor cell death. A number of pharmacological treatments as well as radiotherapy have been described to induce cell oxidative stress aimed at provoking oxidative damage in cell macromolecules and organelles.

In GBM, several molecular pathways involved in tumor onset and progression can be modulated by ROS and their fluctuations. Some of those pathways driven by ROS and linked to GBM progression are related to metabolic activity, transcription factor modulation (HIF-1, nuclear factor-κB, NF-κB), oncosuppressor activity (p53, PTEN), growth factors (vascular-endothelial growth factor [VEGF]; epidermal growth factor [EGF]; platelet-derived growth factor, PDG, PI3K, MAPK), proliferation (RAS, PI3K), telomerase activity, and autophagy. At the same time, it has been described that an increase of antioxidant molecules such as intracellular GSH and the activation of GSH-related enzymes can play a key role in mediating resistance to TMZ treatment, demonstrating the therapeutic role of ROS [60, 61]. In a work of Zhu et al., it has been shown that glutathione reductase (GSR) has favored resistance to cisplatin treatment in GBM cell models and that GSR silencing has been able to augment the overall survival in murine GBM xenografts [62].

In the oncological field, tumor cell metabolic reprogramming is one of the main pathways activated by ROS and involved in cancer cell survival [63]. Both glucose deprivation and ROS, for instance, are able to induce 5′ AMP-activated protein kinase (AMPK) activity mediating a metabolic reprogramming favoring catabolism. Moreover, by inhibiting the mTOR complex 1 (mTORC1) activity and increasing FoxO3 phosphorylation, AMPK is also able to induce autophagy [64].

As mentioned earlier, ROS are also involved in HIF-1α stabilization. In fact, the OXPHOS damage provokes a huge ROS production, which is correlated with the inhibition of PHD activity, the abrogation of HIF-1α degradation through the UPS system, and the increase of its transcriptional activity. HIF-1α acts as a transcription factor, inducing the expression of a number of genes involved in GBM metabolic reprogramming [65]. At the same time, UPS inhibition favors also the increase of the autophagic degradation pathway activity that is also involved in HIF-1α degradation through chaperone-mediated autophagy [66]. This scenario demonstrates how ROS can influence metabolic reprogramming, but also how cells possess a buffering potential to counteract any potentially harmful change for cell homeostasis. For these reasons, it is so complex to understand the molecular net regulating cell functions and to pharmacologically interact with it.

ROS

Reactive oxygen species (ROS) are chemical entities containing highly reactive oxygen atoms. ROS are distinguished in free-radical species and in nonradical ones. The most encountered radical species are the hydroxyl and superoxide ones ($OH^\cdot$, O_2^-) characterized by the presence of unpaired electrons in the outer shell of distinct orbits, which lead to oxygen instability, with consequent reactivity owing to its partial reduction. The nonradical species are mostly represented by the hydrogen peroxide (H_2O_2) and singlet oxygen ($^1O_2$); although they do not contain unpaired electrons, they can drive the production of radical forms [67,68]. The main ROS, from the less reactive species to the most reactive ones, are hydrogen peroxide (H_2O_2), superoxide anion (O_2^-), singlet oxygen ($^1O_2$, which is the triplet molecular oxygen, $^3O_2$, excited form by UV and light), hydroxyl radical ($OH^\cdot$), alkoxyl radical ($RO^\cdot$), and peroxyl radical ($ROO^\cdot$) [69–76].

Oxidative stress

Under basal conditions, in the cells there is a balance between generated and scavenged ROS [76]. This phenomenon is called "redox homeostasis." However, when an imbalance between generation and scavenging processes occurs, we refer to a condition called "oxidative stress," which could drive both physiological and pathological conditions, being involved with the evolution of neurodegenerative diseases, cardiovascular disorders, atherosclerosis, diabetes, and neoplastic transformation [77–84]. A confirmation of the important role played by ROS in these diseases is related to the age of appearance of these pathological conditions having a higher likelihood of appearing with ageing, when redox balance results usually impaired [85]. ROS are involved in intracellular signaling and intercellular communication [86,87]. ROS are produced in different compartments: mitochondria, cytosol, peroxisomes, endoplasmic reticulum-derived vesicles, endosomes, phagosomes, cell membrane-released exosomes, and extracellular fluids, including plasma [76]. The factors that lead to endogenous and exogenous ROS production are different. Endogenous ROS in tumors are produced by the enhanced metabolic activity, by mitochondrial dysfunctions (included ER stress), by oncogenes, by different enzymes such as oxidoreductases, NADPH oxidase, xanthine oxidase, myeloperoxidase, nitric oxide synthase, and by the oxidations catalyzed from metals. Moreover, cytochrome P450 enzymes, lipoxygenases (LOXs), cyclooxygenases (COXs), heme oxygenase (HO), glucose oxidase (GOX) play a crucial role in producing ROS [88]. On the other hand, exogenous ROS are produced by the presence of xenobiotics, drugs, ionizing radiations, variation of nutrient availability, and by the interplay between cancer cells and the immune system recruited at the tumor site [85].

Metabolism and mitochondrial dysfunction modulate ROS production

ROS release from mitochondria has been the object of several studies, for assessing their role in GBM malignant phenotype and in relation to proliferation rate. Mitochondria are the powerhouse of cells; they are indeed involved in cellular respiration and energy production. All their modifications or instability that could cause partial or total inactivity could impair cell integrity and survival. Since cancer cells require a high amount of energy, alterations in mitochondria become pivotal and this is particularly evident in GBM [67].

ROS can be generated within the mitochondrial compartment as an alternative product deriving by OXPHOS and by fatty acid metabolism [68, 69]. In particular, it has been reported that superoxide anion, one of the most reactive and unstable species, is produced within the mitochondrial matrix by complexes I and II of the mitochondrial respiratory chain and within the intermembrane space by complex III owing to electron leakage. Glioma cells need high energy availability deriving mainly by glucose metabolism (glycolysis and OXPHOS) [70, 71]. ROS content increases proportionally with the metabolic rate. Actually, like many other tumors, glioma cells prefer to switch toward the so-called Warburg effect or aerobic glycolysis [72]. Indeed, although the standard metabolic flow in which glycolysis produces pyruvate entering then in the TCA cycle and oxidative phosphorylation, cancer cells prefer to perform aerobic glycolysis, in which pyruvate is converted into lactate even if it is less efficient in producing ATP from glucose.

In this view, the pyruvate dehydrogenase kinase (PDK), the mitochondrial enzyme that inhibits the pyruvate dehydrogenase implicated in the conversion of cytosolic pyruvate into mitochondrial acetyl-CoA, is able to counteract the Warburg effect promoting the shift from glycolysis to mitochondrial oxidation and showing cytotoxic activity in several kinds of tumors [73].

To sustain the inefficient energy production characterizing the Warburg effect and since glioma cells need a high amount of energy, other biological substrates have to be oxidized. This is, for example, the case of fatty acids. Indeed, although glioma cells mainly use the glycolytic pathway to produce energy, other metabolic pathways have been recently taken into account. Among these pathways, lipid metabolism often supports cell proliferation as suggested by the high lipogenesis and the concomitant use of the exogenous sources of fatty acids observed in GBM [74]. In particular, glioma cells produce a large amount of fatty acids by fatty acid synthase (FAS) activity both from glucose (providing almost 60% of the carbon used for fatty acid synthesis in mitochondria) and from TCA intermediates (producing long-chain fatty acids, such as palmitate). Fatty acid availability and metabolism are strictly linked to ROS levels. Indeed, GBM cells oxidize fatty acids to sustain their proliferation, producing radical species (O_2^- and H_2O_2) as a result of mitochondrial β-oxidation [75–78].

A consistent hypothesis is that the oxidation of fatty acids, by producing NADPH concomitantly to ROS, allows the cells to protect themselves from oxidative stress, counteracting cell death [76]. Regarding the interplay between ROS and fatty acids in GBM, it is important to know that the increase of the FAS enzyme is usually correlated with glioma malignancy. The reduction of FAS and consequently of fatty acids, leads both to a downregulation of NADPH and a reduced GSH content, increasing finally ROS levels thus contributing to the apoptosis switch-on and to the cell death [76]. It is thus evident how ROS levels and mitochondrial activity are so interconnected features and are so crucial for cell survival. Since FAS is overexpressed in many kinds of tumors, its inhibition has been recently proposed as a new therapeutic option also in GBM [79].

Other processes that have to be taken into account, being connected with ROS production, are the abnormalities in mitochondria morphology (crystolysis, swelling, changes upon the dynamics such as fission and fusion cycles) and the modulation of the apoptotic process. Modulation of the antiapoptotic gene (i.e., B-cell lymphoma 2, *Bcl-2*) has a significant role in GBM chemoradioresistance [34]. Indeed it has been described that Bcl-2 is able to reduce ROS release inducing an antioxidant effect that sustains cell survival [80].

Once it has been produced, superoxide anion can follow different fates. It can be converted through the manganese superoxide dismutase (MnSOD, or termed SOD2 as well), within the mitochondrial matrix, in hydrogen peroxide, which can be transformed therein in water and molecular oxygen by means of GPx or catalase, or which can diffuse toward mitochondrial intermembrane space and then it can be released within the cytosolic compartment through the mitochondrial aquaporins (also defined peroxyporins, AQPs) [77, 81, 82]. Superoxide anion can also be delivered toward the cytosol through the opening of the voltage-dependent anion channels (VDACs, also known as mitochondrial porins, whose isoform 1 is the most studied for its potential engagement in therapeutic strategies) after a change in mitochondrial membrane potential (mtΔψ) [83].

Anion superoxide, once transferred from the mitochondrial compartment to the cytosolic one, is immediately converted by the cytosolic SOD1 to hydrogen peroxide, which, in turn, is transformed by cytosolic catalase to water and molecular oxygen. Beyond catalase and peroxidase-mediated pathways, hydrogen peroxide can be reduced also by the Fenton reaction, in which metal ions, such as iron or copper, are able to convert hydrogen peroxide into the harmful and highly unstable hydroxyl radical $OH^{\cdot}$, which could react with other biomolecules [84].

An important source of exogenous radicals in the oncological field, as expected, is radiotherapy. In fact, superoxide anion, hydrogen peroxide, and hydroxyl radical can be produced both in the extracellular and in the intracellular compartment as a result of physical and chemical events owing to exposure to ionizing radiation. These radicals can alter mitochondrial OXPHOS function, can directly damage membranes by peroxidation, and induce protein misfolding and DNA damages.

Another exogenous trigger for the increase of intracellular ROS consists in antineoplastic treatments. Several drugs, tested on GBM cell lines, demonstrated the ability to increase intracellular ROS levels or decrease the antioxidant machinery favoring oxidative damage stabilization and favoring response to treatment. These drugs are able to control important pathways such as those related to NF-κB, Nfr2, JNK/STAT, MAPs modulating their activity, and the consequent ROS production, release, and intracellular persistence [6, 63]. Tumors are able to develop different strategies to counteract the potential damage produced by ROS, as by increasing antioxidants, by altering cell metabolism, and by inducing a hypoxic response [85].

Direct and indirect ROS effects

ROS exhibit direct or indirect effects on their targets, depending on the type of interaction that they establish with the biomolecules. In fact, they directly interact with proteins, lipids, and DNA, by causing oxidative stress and by generally inducing cell death. Furthermore, a ROS overhead can provoke indirect detrimental consequences on mitochondria, by inducing apoptotic cell death and increasing proinflammatory signals through the positive modulation of NF-κB [86].

The direct ROS effects consist in the oxidation of biological structures such as proteins, DNA, and lipids. Generally, protein oxidation may occur both in physiological and pathological conditions [87]. One of the most deleterious ROS-mediated effects on proteins is their fragmentation in peptide chains, by modifying the electric charge, causing protein cross-linking, and making unfolded proteins prone to proteolytic degradation [88]. Furthermore, oxidative stress, by promoting protein aggregation and proteolysis, could indirectly modify and/or abrogate enzyme activities and protein binding ability (for instance, for kinases and phosphatases). In fact, several transcription factors have been described, whose level and activity resulted deeply affected by ROS presence. These molecules are involved in different regulatory mechanisms affecting genetic instability, kinase activities, proliferation, mitochondrial integrity, and function [89]. Some of these factors, also influencing GBM features, are forkhead homeobox type O (FoxO), nuclear factor E2-related factor 2 (Nrf2), tumor suppressor protein 53 (p53), nuclear factor-kappa B NF-κB, activator protein-1 (AP-1), and HIF-1α (as shown in Fig. 2) [86, 90–94].

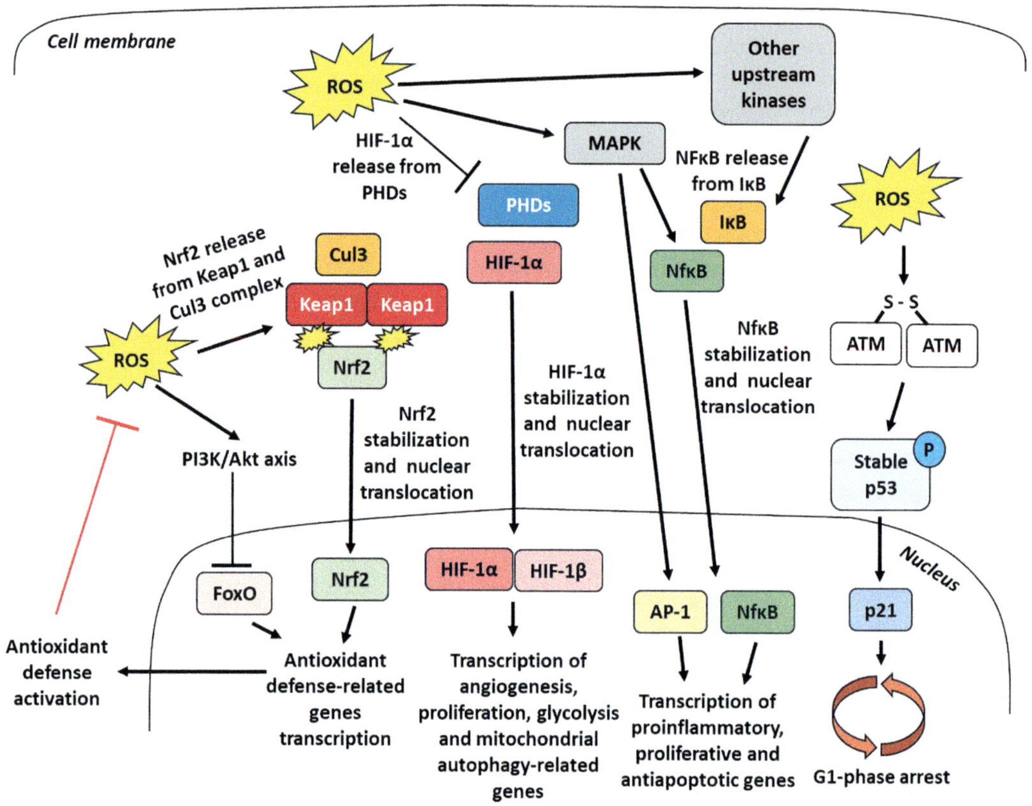

FIG. 2 Schematic representation of ROS influence on the modulation of several transcription factors: FoxO, Nrf2, p53, NF-κB, AP-1, and HIF-1α. The presented signaling pathways emphasize the ROS importance in the supporting of tumor cells.

ROS-mediated oxidative stress is also able to produce direct DNA damage by provoking both single- and double-strand breaks (SSBs and DSBs) and contributing to cell death. DNA damage leads to a typical cellular reaction, defined DNA damage response (DDR), in which different upstream kinases related to repair mechanisms are activated [95].

ROS can also damage membranes impairing their integrity and the correct functionality, by abrogating membrane-receptors activity [88]. In fact, ROS are attracted by the C—C double bonds which characterize lipids, in particular the polyunsaturated fatty acids (PUFAs) of the cell membrane. This interaction gives also rise to the formation of toxic lipidic aldehyde intermediates, whose presence provokes cytotoxicity and cell death [96]. Lipid peroxidation covers a crucial role in GBM cells: in fact, in a recent work by Wu and colleagues, it has been reported that the overexpression of the aldehyde dehydrogenase 1A3 (ALDH1A3) confers resistance to TMZ-based treatment, both *in vitro* and *in vivo* in GBM cell lines. This phenomenon is owing to the detoxifying activity exerted by the enzyme ALDH1A3, able to eliminate toxic aldehyde molecules produced following lipid peroxidation due to the TMZ-mediated oxidative stress [97].

Moreover, ROS and especially hydrogen peroxide are able to regulate different targets, such as kinases, phosphatases, transcription factors, ion channels, antioxidants, and metabolic enzymes. In physiological conditions, hydrogen peroxide stimulates endothelial cell proliferation in different cell tissues and it has been shown that this reactive species increases blood-brain barrier permeability in pathological conditions [98]. Furthermore, ROS are able to influence calcium signaling, autophagy, destruction of pathogens, inflammatory processes, blood pressure control, and response to physical exercise [99].

Modulation of ROS scavenging

Oxidative stress possesses an important role in sustaining tumorigenesis and cell survival and proliferation, however, ROS can be detrimental for cell structures and functions. For this reason, cells have developed strategies to balance the increase in these reactive species inducing a concurrent increase of detoxifying mechanisms to allow the maintenance of regulated redox homeostasis. Detox machinery is composed of different antioxidant systems [100, 101]. This apparatus is composed of enzymes, as superoxide dismutases, catalases, peroxidases and reductases, and molecules, such as GSH, peroxiredoxins, thioredoxins, and vitamins (C, A, and E). Thus, since ROS are mostly generated by intracellular organelles such as mitochondria, peroxisomes, and endoplasmic reticulum, antioxidant machinery is generally specialized for the intracellular environment [99].

Superoxide anion (O_2^-) generated from molecular oxygen or from mitochondrial OXPHOS can be converted into hydrogen peroxide (H_2O_2) and into molecular oxygen through the action of different superoxide dismutases, thus limiting cell damages owing to this very high reactive radical. About the SOD is important to know that three different isoforms are expressed being located in the specific cellular compartment: SOD1 is cytosolic, SOD2 is in the mitochondria, and SOD3 within the extracellular milieu, demonstrating the importance of superoxide anion detoxification ability [102].

Hydrogen peroxide, in turn, can be transformed into H_2O by the catalase enzyme. Reducing power is provided by the reduced nicotinamide adenine dinucleotide phosphate (NADPH). NADPH possesses a fundamental role in yielding electrons to reduce several substrates and molecules, sustaining detox systems [103]. It is produced during the first oxidative phase of the pentose phosphate pathway. Hydrogen peroxide can be also converted into water by peroxiredoxins (PRDXs). These antioxidant enzymes are induced by Nrf2 and can play a differential role in different tumors limiting or sustaining cancer growth by blocking c-Myc transcriptional activity or activating NF-κB, respectively. Moreover, some members of their family are also involved in tumor resistance to treatment and have been considered as potential therapeutic targets to overcome this resistance [104].

Nrf2 which belongs to the bZIP protein family is a transcription factor that "senses" cell redox state. In the absence of oxidative stress, Nrf2-induced transcription is abrogated, since this factor is bound to its inhibitor the Cul3 ubiquitin ligase complex adaptor, Kelch-like erythroid cell-derived protein (Keap1), which favors Nrf2 transport to the proteasome for its consecutive degradation [105]. On the other hand, under oxidative stress, Nrf2 is no longer degraded, it is stabilized and it accumulates into the nucleus, where it binds to specific electrophile response elements/antioxidant response elements

(EpRE/ARE) activating the expression of all those genes aimed at balancing the cellular redox state [106].

GSH is another key player in mediating ROS scavenging. Three amino acids (cysteine, glutamate, and glycine) compose this peptide, which is concentrated in cells at millimolar rates. Availability of nonessential cysteine represents the highest limit for the GSH production and, consequently, for its antioxidative stress activity [107]. For ROS neutralization, the oxidized glutathione (GSSG) is converted in the reduced form (GSH) through the conversion of NADPH in NADP+ by GSR [108]. The study of GSH/GSSG ratio helps to understand the redox state of the cells. In the mitochondrial compartment, a different GSH pool is present, ready to scavenge ROS due to OXPHOS activity. GSH can also interact with different enzymes inducing protein glutathionylation that similar to other protein modifications (phosphorylation or sumoylation), is able to modulate protein activity. GSH metabolism master regulator is the ubiquitous transcription factor Nrf2; in fact the expression of different regulatory subunits of the enzymes (GSH, GPx2, GR, and GCL) is transcriptionally modulated by the activity of this transcription factor [101].

Another enzyme, important in maintaining redox cell homeostasis, is heme oxygenase-1 (HO-1). Its overexpression has been described in relation to cell metastatic potential, angiogenesis, poor prognosis, and activation of infiltrating macrophages. HO-1 protects tumor cells avoiding ROS increase after treatment. Its expression is related again to Nrf2 activity and it has been proposed as the therapeutic target as well [109].

ROS therapeutic activity in GBM

The discrimination between cell survival and cell death driven by ROS levels is dependent on the amount and persistence of ROS exposure. It is thus evident that the correlation between ROS amount and the cellular consequence is not obvious. In the view of developing new therapeutic strategies, different approaches based on the modulation of ROS levels have been taken into account. The therapeutic strategy based on the increase of ROS cell content is of immediate comprehension since their ability to produce DNA and whole-cell damage. These damages first sustain genomic instability and DNA mutations that are accumulated in the cells without an efficient mechanism of repair, and finally are able to promote cell death.

Most aggressive tumor cells often present high ROS levels concomitant to a strong antioxidant system. Main signaling pathways modulated by ROS are those able to promote and sustain cell growth and proliferation (Nrf2, PI3H/AKT, HIF1, etc.). In this context, to exert their toxic effect, ROS have to overcome a threshold beyond which ROS from protumorigenic become toxic [100].

Some treatments have been thought to achieve this aim by increasing intracellular ROS levels or by decreasing antioxidant cell activity. For example, cannabidiol is able to induce an increase of ROS in GBM cell cultures and to prolong the survival of GBM-bearing mice [110]. Another treatment tested for its efficacy in GBM cells is dichloroacetate, a pyruvate dehydrogenase inhibitor. This has been proposed for its potential in reverting the Warburg effect and its ability in inducing ROS release because of the increase of mitochondrial pyruvate oxidation. Radiotherapy is another classical strategy to increase intracellular ROS. Its association with dichloroacetate improved treatment efficacy. In a work by Shen et al. dichloroacetate was also able to increase the efficacy of TMZ [6, 111]. Muscarinic M2 receptor activation by arecaidine can also induce ROS and apoptosis increase [112]. In the same way, some flavonoids such as apigenin and quercetin showed the ability to limit GBM cell line growth by increasing ROS also in combination with TMZ [113–115].

Nevertheless, the persistent increase of ROS content often is not efficacious in glioma therapy. Many cells become resistant to exogenous oxidative stress and become adapted to this condition. This phenomenon can be reported to the so-called threshold concept for cancer therapy [116]. In this mechanism, a key role is played by transcription factors such as NF-κB and HIF-1α, which in turn promote antioxidant molecules making these cells refractory to other oxidative stressors. Indeed, this mechanism can induce an increase in the oxidative tolerability threshold.

On the other hand, since ROS importance in signaling, also ROS-depletion strategy has been evaluated for therapeutic purposes. This strategy has been pursued by improving ROS scavenger potential, for example, by increasing antioxidant (Catalase, GSH, SOD2) molecules or by inhibiting NOX (NADPH oxidase) activity using molecules such as minodronate and histamine. Buthionine sulfoximine (BSO) has been also evaluated for its ability to enhance the efficacy of other drugs in that it is a blocker of GSH synthesis [117, 118]. Improving ROS scavenger potential or inhibiting ROS production would induce a decrease in intracellular ROS abrogating their signaling, thus counteracting cell growth [119–123].

ROS therapeutic activity in GBM has been already confirmed as one of the mechanisms pursued by the most used drug, TMZ. In fact, it has been widely reported that even TMZ-induced ER stress is able to provoke ROS release from mitochondria and that this phenomenon is tightly bound to responsiveness to TMZ treatment [19, 124]. In the Stupp protocol, radiotherapy treatment further increases oxidative stress, improving the efficacy [5].

In a study performed by Chang et al., co-treatment with TMZ and mithramycin A (MA, a specificity protein 1, Sp1, inhibitor) has been capable to restore sensitivity to TMZ treatment, since Sp1 is a transcription factor which activates SOD2 expression and which normally favors resistance to TMZ treatment. The use of the Sp1 inhibitor helped in reducing the detox machinery of the cells favoring ROS-mediated toxicity [125].

ROS production has also been reported to activate the autophagic pathway. A ROS overproduction could sequestrate iron, which, being a cofactor for activating PHDs, could result in an inactivation of these enzymes. This step inhibits the proteasome-dependent degradation of different proteins, which will shift to the autophagic pathway [66, 84]. Recently, it has been demonstrated the CMA engagement in mediating responsiveness to TMZ treatment in TMZ-responsive cells, which is strictly related to the transitory ROS release from mitochondria. Furthermore, since HIF-1α has been described as both a PHD and a CMA-target, its modulation owing to the drug has been proposed also as a biomarker of responsiveness to TMZ [26].

GBM and HIF-1α activity

Historically, hypoxia has been one of the main features taken into account in GBM treatment planning since it drives several mechanisms correlated with the resistance to chemotherapeutic agents and radiation therapy [126]. In particular, hypoxia reduces chemotherapeutic efficacy since it positively selects cells with reduced susceptibility for apoptosis. At the same time, since radio sensitivity is generally owing to the presence of molecular oxygen that induces DNA damage through the formation of oxygen free radicals, which occurs primarily after the interaction of radiation with intracellular water, tumor sensitivity to radiation therapy is progressively limited when the O_2 partial pressure in a tumor is less than 25–30 mmHg leading to a significant radio resistance in the presence of hypoxia [127–129].

HIF-1 is the master regulator of cell response to hypoxia, first identified by Semenza and colleagues as a regulator of hypoxia-induced erythropoietin expression and of the expression of more than 30 genes. HIF-1 is a basic helix-loop-helix PAS (Per-Arnt-Sim) transcription factor composed of two subunits, a constitutively expressed subunit called HIF-1β (or ARNT) and the catalytic and oxygen-dependent subunit, HIF-1α, whose expression is tightly regulated by multiple mechanisms at different levels (transcriptional, translational, and post translational). In normoxic conditions, the HIF-1α subunit undergoes rapid decay via ubiquitin-proteasome degradation pathway involving the von Hippel-Lindau (VHL) tumor suppressor gene product pVHL. The binding of pVHL to HIF-1α requires the posttranslational hydroxylation of proline residues (Pro402 and Pro564) within the HIF-1α oxygen-dependent degradation (ODD) domain by soluble PHD. These enzymes are 2-oxoglutarate-dependent dioxygenases that require molecular oxygen and iron as co-substrates. In hypoxic conditions, this modification is prevented, thus allowing HIF-1α to escape proteolysis, to dimerize with HIF-1β, and to translocate into the nucleus where this transcription factor sustains the expression of specific genes whose promoters contain hypoxia response elements (HRE) and are related to angiogenesis, metabolism, proliferation, stemness, and invasiveness. A separately controlled, O_2-dependent hydroxylation of asparagine 803 in the HIF-1α C-terminal transactivation domain inhibits HIF-1α interaction with the p300/CBP coactivator, thereby blocking HIF-1α transcriptional activity in the presence of oxygen [130].

Overexpression of HIF-1α protein has been reported in several human cancers, where it has been associated with tumor progression, treatment failure, and thus poor survival. GBMs are characterized by several hypoxic areas correlated with insufficient or aberrant angiogenesis. In this condition, HIF-1α is often highly expressed being correlated to several GBM features (prominent angiogenesis, high proliferation rate, high ranking of invasiveness, apoptosis modulation) and to resistance to treatment. HIF-1α is linked to tumoral progression since its overexpression is involved in the angiogenetic switch, in the invasiveness, in the metabolic reprogramming, and in the modulation of the apoptosis in addition to regulation of different genes related to stemness. To name a few, HIF-1α induces the VEGF, promoting the formation of new vessels and triggers cell invasiveness and migration by stimulating epithelial-mesenchymal transition (EMT). Furthermore, HIF-1α acts in the metabolic reprogramming being involved in the overexpression of genes such as the glucose transporter type (GLUT)-1, hexokinase (HK)-2, and glyceraldehyde-3-phosphate dehydrogenase (GAPDH)

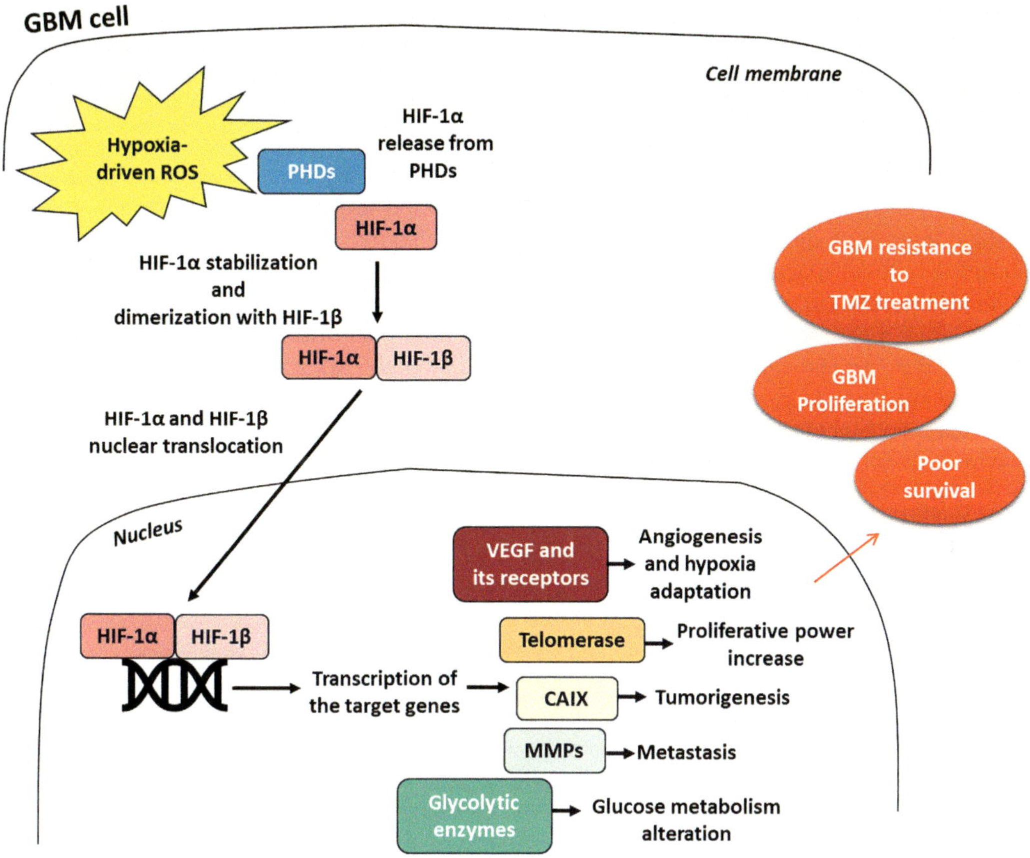

FIG. 3 Schematic representation of HIF-1 activity. Impairment in HIF-1α degradation induces its dimerization with HIF-1β. The complex translocates into the nucleus where it drives the expression of genes involved in tumor cell proliferation and survival contributing to cell resistance to treatments.

that among all are engaged in the Warburg effect typical of GBM. It is possible to affirm that there is a self-sustaining vicious circle in which the increase of ROS can induce HIF-1α stabilization producing a metabolic reprogramming that can, in turn, sustain ROS production. The breakage of this circle could imply the collapse of cell homeostasis inducing potentially cell death. This series of concepts support the assumption that HIF-1α plays a pivotal role in determining GBM features and response to treatment [131] (Fig. 3).

Reactive oxygen species and chaperone-mediated autophagy

The close interaction between cytoplasmic ROS fluctuation and cell signaling as well as cell processes has been documented. GBM response to treatment in relation to ROS fluctuation has been recently related to a crucial process involved in the manifestation of the TMZ-dependent cytotoxicity, CMA [26].

CMA represents a selective type of autophagy which, under basal conditions entails the degradation and recycling of those proteins which might affect cell physiological conditions by converting themselves into intermediates capable to produce aggregates and/or inclusions. In particular, differently to macro-autophagy, which degrades protein by exploiting autophagosome-autolysosome formation, CMA is a specific form of autophagy able to recognize consensus sequences named "KFERQ" (lysine-phenylalanine-glutamic acid-arginine-glutamine)-motif within proteins and drive KFERQ-containing proteins to specific lysosomes for the consecutive degradation. In fact, the HSC70 (also named HspA8, is a 70 kDa protein belonging to the heat shock protein family), in association with other chaperones, Heat Shock Protein 90 (Hsp90), is involved in the recognition of this pentapeptide motif, the binding of the recognized protein and its shipping to the lysosomal membrane. At this site, the complex interacts with the cytosolic tail of the homotrimer LAMP-2A allowing the lysosomal HSC70-mediated substrate translocation within the lysosomal lumen for its degradation by means of lysosomal proteases. CMA is a tightly regulated process. Its positive or negative modulation is owing to a delicate balance between phosphorylation and dephosphorylation mechanisms. LAMP-2A activation (which corresponds to its multimerization to a homotrimer) is regulated by the indirect action of the glial fibrillary acidic protein (GFAP), which is able to stabilize the translocation complex at the lysosomal membrane level. This step strictly depends on the elongation factor (EF-1α). It is indeed the limiting factor since it inhibits the binding of GFAP with its phosphorylated form (GFAP-P) allowing GFAP itself to bind and stabilize LAMP-2A multimer. Moreover, a fine intracellular amino acid sensing owing to the serine-threonine kinase mammalian target of rapamycin (mTOR), through mTORC2 complex, tightly regulates CMA activation. In detail, mTORC2 complex phosphorylates lysosomal Akt1, which in turn, phosphorylates GFAP favoring the destabilization of the translocation complex and CMA inhibition. On the other hand, the GTPase Rac1 ensures the availability at the lysosomal membrane of the pleckstrin homology domain and leucine-rich repeat protein phosphatase 1 (PHLPP1), which dephosphorylates Akt1 by activating the CMA pathway [132]. Moreover, it has been widely reported that CMA activation is owing also to LAMP-2A availability on the lysosomal membrane in relation to its degradation rate and its relocalization from the lumen to the lysosomal membrane more than its *de novo* protein synthesis [124, 133] (Fig. 4). CMA upregulation can occur under different stimuli, such as starvation-nutrient deprivation, oxidative stress, and DNA damage.

Among the aforementioned stimuli, oxidative stress represents an important CMA inducer. In fact, Kiffin and colleagues have reported that a paraquat-induced mild oxidative stress was able to cause the oxidation-mediated unfolding protein response by shortening the oxidized substrate recognition time and by accelerating its consecutive translocation to the lysosome membrane for degradation. This should be owing to the fact that the KFERQ-like motif could be more exposed and subjected to CMA-specific chaperone recognition in this oxidant milieu [124, 132]. As shown in a recent publication, it has been also demonstrated that the oxidative stress sensor Nrf2 transcription factor is able to bind and activate a specific transcription profile by binding antioxidant response element (ARE) sequences, placed, for example, within LAMP-2A promoter, both under physiological and under stress conditions, favoring CMA activation [134]. Among CMA targets, there is a number of proteins involved in GBM important features as reported in Table 1 [135, 136].

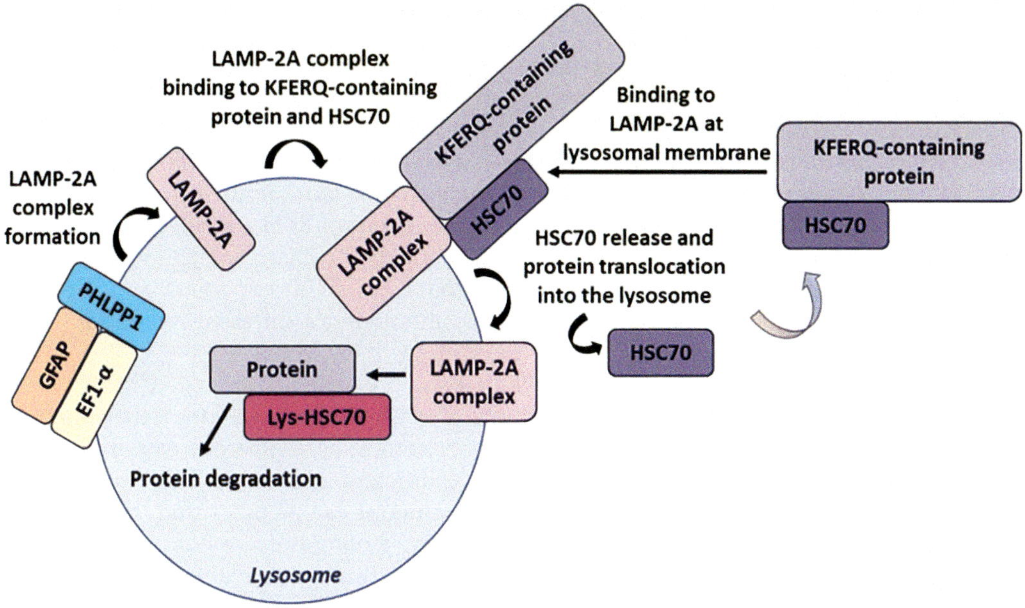

FIG. 4 Graphical representation of CMA machinery. Different CMA key players (PHLPP1, HSC70, and LAMP-2A) are responsible for CMA switch on. The depletion of even one of them impairs the machinery counteracting GBM cell responsiveness to TMZ.

TABLE 1 CMA targets involved in the modulation of oxidative stress.

Protein	Function in the cell
Mutant p53	Oncogene acting as oncometabolite
Cell cycle checkpoint kinase (Chk)	Cell cycle checkpoint modulator
Alpha-synuclein	Modulator of oxidative stress and of mitochondrial morphology [135]
Ubiquitin	Protein degradation
Glutathione transferase	Detoxification from xenobiotics
EIF4A1, EIF4H, DDX3X	Translation initiation factors [136]
Pyruvate kinase M2 (PKM2), glyceraldehyde-3 phosphate dehydrogenase (GAPDH), hexokinase 2 (HK2), glucose-6 phosphate dehydrogenase (G6PD), aldolase B	Enzymes involved in glucose metabolism
Myocyte enhancer factor 2D (MEF2D) and HIF-1α	Transcription factor involved in oxidative stress

ROS-CMA axis as a therapeutic target for GBM

In this chapter, ROS content has been described as a key player in tumor growth as well as in apoptosis and in metabolic reprogramming. It has been described that an increase of the oxidative stress can be used for therapeutic purposes since ROS are involved in the cytotoxic effect responsible for cell death, and the balance between ROS and their scavenger factors (i.e., antioxidant molecules) is crucial for the responsiveness to chemoradiotherapy. Besides, it has been described that cancer cells present different basal levels of ROS, together with a specific threshold of tolerance for oxidative stress. A recent paper by Lo Dico et al. [26] has supported this evidence in human glioma cell lines, showing that U251 TMZ-responsive and T98 TMZ-resistant cells show different ROS basal levels accompanying a different molecular profile that characterizes each cell line.

It is interesting that one of the mechanisms that could be used to activate CMA is a temporary ROS increase in the cytoplasm and that as a result of CMA activation there can be a decrease in HIF-1α activity leading a metabolic reprogramming with a decrease in ROS production. In fact, the U251 human glioma cell line is characterized by a low ROS level whereas after TMZ treatment a fluctuation of their cytosolic amount can be measured. In these cells, ROS fluctuation is able to induce CMA activity reducing among the number of CMA targets, HIF-1α expression and activity, and, at the same time, leading the activation of the apoptotic process and reducing cell viability in culture, as expected, since these cells are responsive to TMZ.

On the other hand, T98 TMZ resistant cells are characterized by an eightfold higher cytoplasmic ROS level compared to U251. In these cells, high ROS concur in sustaining all the processes involved in cell proliferation, metabolic reprogramming, and mitochondrial modification that together lead to chemoresistance. What is more, these cells are refractory to induce the ROS fluctuation after TMZ treatment and no CMA activation can be observed. HIF-1α activity is maintained and no apoptotic death is induced, confirming the resistant behavior.

Results by Lo Dico et al. showed also that by inducing an exogenous ROS fluctuation in previously resistant cells, it is possible to overcome their tolerance threshold obtaining a "responsive-like" cell behavior and molecular profile. All these results suggest that, although cells can present different ROS content, the comprehension of their "tolerance-threshold" could be useful for planning therapeutic strategies to overcome the intrinsic resistance to standard therapy [26].

The identification of biomarkers that sustain diagnosis, prognosis, and therapeutic efficacy is of fundamental importance for a precision medicine approach in the improvement of clinical management of cancer patients. Biomarkers are "characteristics that are objectively measured and evaluated as indicators of normal biological processes, pathogenic processes, or pharmacologic responses to a therapeutic intervention" as reported by the working group of the National Institutes of Health Director's Initiative on Biomarkers and Surrogate Endpoints [137]. In all the oncological diseases, but even more in GBM for which only a few therapeutic options are available, and time plays a crucial role, it is pivotal to know the molecular profile of the tumor for optimal treatment planning. Lo Dico et al. in a recent paper demonstrated that each GBM cell line is characterized by a specific molecular profile after TMZ treatment that can be examined soon after 24h of treatment and that is in relation to ROS fluctuation and cell responsiveness to the drug. This molecular signature is able to discriminate between TMZ-responsive and TMZ-resistant cells and it takes into account:

- Differential HIF-1α/LAMP-2A expression
- CMA-related gene expression (PHLPP1, HSC70 in addition to LAMP-2A)
- Genes involved in detox machinery
- Structure and function of respiratory chain

It is of note that HIF-1α and LAMP-2A show a differential pattern of expression after treatment in responsive and resistant cells confirming an interplay between CMA and HIF-1α in determining cell responsiveness to TMZ [26]. In a recent paper, Lo Dico et al. showed a reciprocal expression of HIF-1α and LAMP-2A in TMZ-responsive cells after 24h TMZ treatment. These data are in line with a TMZ-mediated CMA activation and the consequent HIF-1α activity reduction producing a responsive-expression signature [34]. Moreover, all these considerations elect HIF-1α not only as an early biomarker of responsiveness to TMZ but also a biomarker of CMA activation, a necessary condition to obtain a response to the drug [138].

For what concerns CMA-related gene expression (PHLPP1, HSC70 in addition to LAMP-2A), the investigations have shown that by monitoring CMA-related gene expression (HIF-1α) and its target genes (for instance, VEGF), it is possible to estimate CMA activation. Within this pathway, a key role is played by PHLLP1 both as a positive prognostic marker and as a molecular supporter of LAMP-2A-driven CMA. Indeed, TMZ-responsive cells are characterized by a CMA-network in which LAMP-2A, HSC70, and PHLPP1 are significantly upregulated by TMZ treatment, while an opposite molecular profile is observed in resistant cells.

In the same paper, it has been described that GBM cells can be divided between responsive and resistant to TMZ also for their expression of the genes involved in the detox machinery after TMZ treatment, in fact, responsive cells showed a negative modulation of GPx, glutathione synthetase (GSS), SOD-2, and CATALASE whereas GSS and catalase genes were upregulated in T98 TMZ-resistant cells.

Finally, responsive and resistant cells, showing different ROS amounts, were also differentiated for the activity of some complexes of the respiratory chain. TMZ-resistant cells showed higher activity of complexes, both at basal level and after TMZ treatment [26].

These observations support the crucial role played by HIF-1α and the fundamental involvement of CMA activation in mediating GBM cell responsiveness to TMZ as well as the importance of ROS fluctuation in the activation of this degradative pathway.

Conclusions

ROS have been described as a double-edge sword in that they sustain gliomagenesis by favoring genetic instability, metabolic reprogramming, and finally proliferation and resistance. At the same time, an increase of intracellular ROS has been demonstrated to be involved in the oxidative damage of DNA, lipids, and proteins and thus be implicated in the impairment of cell function and finally in the triggering of cell death.

ROS can be considered as second messengers fundamental for the transduction of signals through different pathways such as those related to NF-κB, Nrf2-ARE, PI3K/AKT, and

MAPs. For these reasons, cells have developed a ROS scavenging machinery able to maintain a ROS homeostasis ensuring transduction signaling but avoiding oxidative damage.

Looking at this scenario, two options can be considered. The first could consist in the attempt to reduce intracellular ROS with the aim of decreasing ROS-mediated signaling pathways and, in turn, cell proliferation rate, metabolism, and invasiveness. The other option is aimed at increasing intracellular ROS levels to exploit their oxidative potential to damage cells and trigger their death. This second therapeutic option is pursued by several antineoplastic treatments. Among them, TMZ has been described to induce a transitory increase of ROS in the cytoplasm. If this ROS fluctuation overcomes the cell threshold of oxidative tolerability, it is able to induce CMA activation. Cells can be characterized by the different thresholds of oxidative tolerability. This feature should be taken into account to improve therapeutic results.

Recently, CMA has been proposed as one of the main mechanisms responsible for GBM sensitivity to TMZ. In fact, the simple silencing of even one of the genes encoding the proteins involved in this machinery results in GBM cell resistance to the drug. Among the proteins degraded by CMA, HIF-1α plays a crucial role in GBM being involved in the regulation of metabolism, proliferation, angiogenesis, and invasiveness. Indeed, a correlation has been described between HIF-1 activity decrease, after TMZ administration, and sensitivity to this treatment in GBM cell lines. In this scenario, treatments directly aimed at increasing intracellular ROS or at decreasing the activity of the scavenging machinery could help in overcoming the threshold of oxidative tolerability of the cells resulting in CMA activation and consequently to HIF-1 activity reduction, resulting in an antineoplastic effect.

These results pave the way for further analyses to be conducted to elucidate the relation between ROS increase and cell death and to unravel all the molecular mechanisms linking oxidative damage to the overcoming of GBM resistance to treatment.

Acknowledgments

The authors thank Professor Giovanni Lucignani for his fundamental and continuous mentoring, his brilliant mind, his insights, his ability to see ever beyond, and his extraordinary ability to teach that allowed us to grow in a motivating and stimulating atmosphere. We all recognize that his teachings will be the foundation that will allow us to go on in the research, in the academic, and in our self-improvement.

References

[1] Newlands ES, Blackledge GR, Slack JA, Rustin GJ, Smith DB, Stuart NS, Quarterman CP, Hoffman R, Stevens MF, Brampton MH. Phase I trial of temozolomide (CCRG 81045: M&B 39831: NSC 362856). Br J Cancer 1992 Feb;65(2):287–91.

[2] Ramirez YP, Weatherbee JL, Wheelhouse RT, Ross AH. Glioblastoma multiforme therapy and mechanisms of resistance. Pharmaceuticals (Basel) 2013 Nov 25;6(12):1475–506.

[3] Wang X, Zhang H, Chen X. Drug resistance and combating drug resistance in cancer. Cancer Drug Res 2019 Jun 19;2(2):141–60.

[4] Perazzoli G, Prados J, Ortiz R, Caba O, Cabeza L, Berdasco M, Gónzalez B, Melguizo C. Temozolomide resistance in glioblastoma cell lines: implication of MGMT, MMR, P-glycoprotein and CD133 expression. PLoS ONE 2015;10(10), e0140131.

[5] Stupp R, Mason WP, Van Den Bent MJ, Weller M, Fisher B, Taphoorn MJB, Belanger K, Brandes AA, Marosi C, Bogdahn U, Curschmann J, Janzer RC, Ludwin SK, Gorlia T, Allgeier A, Lacombe D, Cairncross JG, Eisenhauer

E, Mirimanoff RO. Radiotherapy plus concomitant and adjuvant temozolomide for glioblastoma. N Engl J Nucl Med 2005;352(10):987–96.

[6] Salazar-Ramiro A, Ramírez-Ortega D, Pérez de la Cruz V, Hérnandez-Pedro NY, González-Esquivel DF, Sotelo J, Pineda B. Role of redox status in development of glioblastoma. Front Immunol 2016;7:156. https://doi.org/10.3389/fimmu.2016.00156.

[7] Denny BJ, Wheelhouse RT, Stevens MF, Tsang LL, Slack JA. NMR and molecular modeling investigation of the mechanism of activation of the antitumor drug temozolomide and its interaction with DNA. Biochemistry 1994 Aug 9;33(31):9045–51.

[8] Koukourakis GV, Kouloulias V, Zacharias G, Papadimitriou C, Pantelakos P, Maravelis G, Fotineas A, Beli I, Chaldeopoulos D, Kouvaris J. Temozolomide with radiation therapy in high grade brain gliomas: pharmaceuticals considerations and efficacy; a review article. Molecules 2009 Apr 16;14(4):1561–77.

[9] Newlands ES, Stevens MF, Wedge SR, Wheelhouse RT, Brock C. Temozolomide: a review of its discovery, chemical properties, pre-clinical development and clinical trials. Cancer Treat Rev 1997 Jan;23(1):35–61.

[10] Boysen G, Pachkowski BF, Nakamura J, Swenberg JA. The formation and biological significance of N7-guanine adducts. Mutat Res 2009 Aug 2;678(2):76–94.

[11] Agnihotri S, Gajadhar AS, Ternamian C, Gorlia T, Diefes KL, Mischel PS, Kelly J, McGown G, Thorncroft M, Carlson BL, Sarkaria JN, Margison GP, Aldape K, Hawkins C, Hegi M, Guha A. Alkylpurine-DNA-N-glycosylase confers resistance to temozolomide in xenograft models of glioblastoma multiforme and is associated with poor survival in patients. J Clin Invest 2012 Jan;122(1):253–66.

[12] Esteller M, Garcia-Foncillas J, Andion E, Goodman SN, Hidalgo OF, Vanaclocha V, Baylin SB, Herman JG. Inactivation of the DNA-repair gene MGMT and the clinical response of gliomas to alkylating agents. N Engl J Med 2000 Nov 9;343(19):1350–4.

[13] Hirose Y, Berger MS, Pieper RO. p53 effects both the duration of G2/M arrest and the fate of temozolomide-treated human glioblastoma cells. Cancer Res 2001 Mar 1;61(5):1957–63.

[14] Günther W, Pawlak E, Damasceno R, Arnold H, Terzis AJ. Temozolomide induces apoptosis and senescence in glioma cells cultured as multicellular spheroids. Br J Cancer 2003 Feb 10;88(3):463–9.

[15] von Bueren AO, Bacolod MD, Hagel C, Heinimann K, Fedier A, Kordes U, Pietsch T, Koster J, Grotzer MA, Friedman HS, Marra G, Kool M, Rutkowski S. Mismatch repair deficiency: a temozolomide resistance factor in medulloblastoma cell lines that is uncommon in primary medulloblastoma tumours. Br J Cancer 2012 Oct 9;107(8):1399–408.

[16] Minniti G, Salvati M, Arcella A, Buttarelli F, D'Elia A, Lanzetta G, Esposito V, Scarpino S, Maurizi Enrici R, Giangaspero F. Correlation between O6-methylguanine-DNA methyltransferase and survival in elderly patients with glioblastoma treated with radiotherapy plus concomitant and adjuvant temozolomide. J Neurooncol 2011 Apr;102(2):311–6.

[17] Chen X, Zhang M, Gan H, Wang H, Lee J-H, Fang D, Kitange GJ, He L, Hu Z, Parney IF, Meyer FB, Giannini C, Sarkaria JN, Zhang Z. A novel enhancer regulates MGMT expression and promotes temozolomide resistance in glioblastoma. Nat Commun 2018 Jul 27;9(1):2949.

[18] Kitange GJ, Carlson BL, Schroeder MA, Grogan PT, Lamont JD, Decker PA, Wu W, James CD, Sarkaria JN. Induction of MGMT expression is associated with temozolomide resistance in glioblastoma xenografts. Neuro Oncol 2009 Jun;11(3):281–91.

[19] He Y, Su J, Lan B, Gao Y, Zhao J. Targeting off-target effects: endoplasmic reticulum stress and autophagy as effective strategies to enhance temozolomide treatment. Onco Targets Ther 2019 Mar 7;12:1857–65.

[20] SCITECH. TMZ resistance in glioblastoma: role of P-glycoprotein. J Neurosurg Imag Tech 2020;2473-1943. Available from: https://www.scitcentral.com/article/38/170/TMZ-Resistance-in-Glioblastoma:-Role-of-P-Glycoprotein#tabs2.

[21] Hirose Y, Katayama M, Berger MS, Pieper RO. Cooperative function of Chk1 and p38 pathways in activating G2 arrest following exposure to temozolomide. J Neurosurg 2004 Jun;100(6):1060–5.

[22] Park C, Suh Y, Cuervo AM. Regulated degradation of Chk1 by chaperone-mediated autophagy in response to DNA damage. Nat Commun 2015 Apr 16;6(1):6823.

[23] Yan Y, Xu Z, Dai S, Qian L, Sun L, Gong Z. Targeting autophagy to sensitive glioma to temozolomide treatment. J Exp Clin Cancer Res 2016 Feb 2;35(1):23.

[24] Knizhnik AV, Roos WP, Nikolova T, Quiros S, Tomaszowski K-H, Christmann M, Kaina B. Survival and death strategies in glioma cells: autophagy, senescence and apoptosis triggered by a single type of temozolomide-induced DNA damage. PLoS ONE 2013;8(1), e55665.

[25] Lin C-J, Lee C-C, Shih Y-L, Lin C-H, Wang S-H, Chen T-H, Shih C-M. Inhibition of mitochondria- and endoplasmic reticulum stress-mediated autophagy augments temozolomide-induced apoptosis in glioma cells. PLoS ONE 2012;7(6), e38706.

[26] Lo Dico A, Salvatore D, Martelli C, Ronchi D, Diceglie C, Lucignani G, Ottobrini L. Intracellular redox-balance involvement in temozolomide resistance-related molecular mechanisms in glioblastoma. Cells 2019;8(11):1315. https://doi.org/10.3390/cells8111315.

[27] Valtorta S, Dico AL, Raccagni I, Gaglio D, Belloli S, Politi LS, Martelli C, Diceglie C, Bonanomi M, Ercoli G, Vaira V, Ottobrini L, Moresco RM. Metformin and temozolomide, a synergic option to overcome resistance in glioblastoma multiforme models. Oncotarget 2017 Dec 7;8(68):113090–104.

[28] Mekahli D, Bultynck G, Parys JB, De Smedt H, Missiaen L. Endoplasmic-reticulum calcium depletion and disease. Cold Spring Harb Perspect Biol 2011;3(6):a004317.

[29] Bahar E, Kim H, Yoon H. ER stress-mediated signaling: Action potential and Ca(2+) as key players. Int J Mol Sci 2016;17(9):1558. https://doi.org/10.3390/ijms17091558.

[30] Kim JK, Kang KA, Ryu YS, Piao MJ, Han X, Oh MC, Boo SJ, Jeong SU, Jeong YJ, Chae S, Na S-Y, Hyun JW. Induction of endoplasmic reticulum stress via reactive oxygen species mediated by luteolin in melanoma cells. Anticancer Res 2016 May;36(5):2281–9.

[31] Thorburn A, Debnath J. Targeting chaperone-mediated autophagy in cancer. Sci Transl Med 2011 Nov 16;3 (109):109ps45.

[32] Köhler T, Schill C, Deininger MW, Krahl R, Borchert S, Hasenclever D, Leiblein S, Wagner O, Niederwieser D. High Bad and Bax mRNA expression correlate with negative outcome in acute myeloid leukemia (AML). Leukemia 2002 Jan;16(1):22–9.

[33] Pattingre S, Tassa A, Qu X, Garuti R, Liang XH, Mizushima N, Packer M, Schneider MD, Levine B. Bcl-2 antiapoptotic proteins inhibit Beclin 1-dependent autophagy. Cell 2005 Sep 23;122(6):927–39.

[34] Lo Dico A, Martelli C, Diceglie C, Lucignani G, Ottobrini L. Hypoxia-inducible factor-1α activity as a switch for glioblastoma responsiveness to temozolomide. Front Oncol 2018;8:249. https://doi.org/10.3389/fonc.2018.00249.

[35] Hubbi ME, Hu H, Kshitiz G, Ahmed I, Levchenko A, Semenza GL. Chaperone-mediated autophagy targets hypoxia-inducible factor-1α (HIF-1α) for lysosomal degradation. J Biol Chem 2013 Apr 12;288(15):10703–14.

[36] Vasco Ferreira J, Rosa Soares A, Silva Ramalho J, Pereira P, Girao H. K63 linked ubiquitin chain formation is a signal for HIF1A degradation by chaperone-mediated autophagy. Sci Rep 2015 May 11;5(1):10210.

[37] Foo J, Michor F. Evolution of acquired resistance to anti-cancer therapy. J Theor Biol 2014 Aug 21;355:10–20.

[38] Mansoori B, Mohammadi A, Davudian S, Shirjang S, Baradaran B. The different mechanisms of cancer drug resistance: a brief review. Adv Pharm Bull 2017 Sep;7(3):339–48.

[39] Alderson LM, Castleberg RL, Harsh GR, Louis DN, Henson JW. Human gliomas with wild-type p53 express bcl-2. Cancer Res 1995 Mar 1;55(5):999–1001.

[40] Wang X, Chen J-X, Liu J-P, You C, Liu Y-H, Mao Q. Gain of function of mutant TP53 in glioblastoma: prognosis and response to temozolomide. Ann Surg Oncol 2014 Apr;21(4):1337–44.

[41] Muller PAJ, Vousden KH. Mutant p53 in cancer: new functions and therapeutic opportunities. Cancer Cell 2014 Mar 17;25(3):304–17.

[42] Zhang C, Liu J, Liang Y, Wu R, Zhao Y, Hong X, Lin M, Yu H, Liu L, Levine AJ, Hu W, Feng Z. Tumour-associated mutant p53 drives the Warburg effect. Nat Commun 2013 Dec 17;4(1):2935.

[43] Zhang Y, Dube C, Gibert M Jr, Cruickshanks N, Wang B, Coughlan M, Yang Y, Setiady I, Deveau C, Saoud K, Grello C, Oxford M, Yuan F, Abounader R. The p53 pathway in glioblastoma. Cancers (Basel) 2018;10(9):297. https://doi.org/10.3390/cancers10090297.

[44] Hu W, Zhang C, Wu R, Sun Y, Levine A, Feng Z. Glutaminase 2, a novel p53 target gene regulating energy metabolism and antioxidant function. Proc Natl Acad Sci U S A 2010 Apr 20;107(16):7455–60.

[45] Bleau A-M, Hambardzumyan D, Ozawa T, Fomchenko EI, Huse JT, Brennan CW, Holland EC. PTEN/PI3K/Akt pathway regulates the side population phenotype and ABCG2 activity in glioma tumor stem-like cells. Cell Stem Cell 2009 Mar 6;4(3):226–35.

[46] Koundouros N, Poulogiannis G. Phosphoinositide 3-kinase/Akt signaling and redox metabolism in cancer. Front Oncol 2018;8:160. https://doi.org/10.3389/fonc.2018.00160.

[47] Zhao S, Lin Y, Xu W, Jiang W, Zha Z, Wang P, Yu W, Li Z, Gong L, Peng Y, Ding J, Lei Q, Guan K-L, Xiong Y. Glioma-derived mutations in IDH1 dominantly inhibit IDH1 catalytic activity and induce HIF-1alpha. Science 2009 Apr 10;324(5924):261–5.
[48] Dang L, Jin S, Su SM. IDH mutations in glioma and acute myeloid leukemia. Trends Mol Med 2010 Sep;16 (9):387–97.
[49] Cairns RA, Mak TW. Oncogenic isocitrate dehydrogenase mutations: mechanisms, models, and clinical opportunities. Cancer Discov 2013 Jul;3(7):730–41.
[50] Thompson CB. Metabolic enzymes as oncogenes or tumor suppressors. N Engl J Med 2009 Feb 19;360(8):813–5.
[51] Losman J-A, Kaelin WG. What a difference a hydroxyl makes: mutant IDH, (R)-2-hydroxyglutarate, and cancer. Genes Dev 2013 Apr 15;27(8):836–52.
[52] Ma S, Jiang B, Deng W, Gu Z-K, Wu F-Z, Li T, Xia Y, Yang H, Ye D, Xiong Y, Guan K-L. D-2-hydroxyglutarate is essential for maintaining oncogenic property of mutant IDH-containing cancer cells but dispensable for cell growth. Oncotarget 2015 Apr 1;6(11):8606–20.
[53] Shi J, Sun B, Shi W, Zuo H, Cui D, Ni L, Chen J. Decreasing GSH and increasing ROS in chemosensitivity gliomas with IDH1 mutation. Tumour Biol 2015 Feb;36(2):655–62.
[54] Huang LE. Friend or foe—IDH1 mutations in glioma 10 years on. Carcinogenesis 2019 Nov 25;40(11):1299–307.
[55] Taylor TE, Furnari FB, Cavenee WK. Targeting EGFR for treatment of glioblastoma: molecular basis to overcome resistance. Curr Cancer Drug Targets 2012 Mar;12(3):197–209.
[56] Weng M-S, Chang J-H, Hung W-Y, Yang Y-C, Chien M-H. The interplay of reactive oxygen species and the epidermal growth factor receptor in tumor progression and drug resistance. J Exp Clin Cancer Res 2018 Mar 16;37(1):61.
[57] Bae YS, Kang SW, Seo MS, Baines IC, Tekle E, Chock PB, Rhee SG. Epidermal growth factor (EGF)-induced generation of hydrogen peroxide. Role in EGF receptor-mediated tyrosine phosphorylation. J Biol Chem 1997 Jan 3;272(1):217–21.
[58] Tudek B, Winczura A, Janik J, Siomek A, Foksinski M, Oliński R. Involvement of oxidatively damaged DNA and repair in cancer development and aging. Am J Transl Res 2010 May 15;2(3):254–84.
[59] Ghosh S, Tewari R, Dixit D, Sen E. TNFalpha induced oxidative stress dependent Akt signaling affects actin cytoskeletal organization in glioma cells. Neurochem Int 2010 Jan;56(1):194–201.
[60] Kohsaka S, Takahashi K, Wang L, Tanino M, Kimura T, Nishihara H, Tanaka S. Inhibition of GSH synthesis potentiates temozolomide-induced bystander effect in glioblastoma. Cancer Lett 2013 Apr 30;331(1):68–75.
[61] Rocha CRR, Kajitani GS, Quinet A, Fortunato RS, Menck CFM. NRF2 and glutathione are key resistance mediators to temozolomide in glioma and melanoma cells. Oncotarget 2016 Jul 26;7(30):48081.
[62] Zhu Z, Du S, Du Y, Ren J, Ying G, Yan Z. Glutathione reductase mediates drug resistance in glioblastoma cells by regulating redox homeostasis. J Neurochem 2018;144. [cited 2020 Sep 29] Available from: https://pubmed.ncbi.nlm.nih.gov/29105080/.
[63] Marengo B, Nitti M, Furfaro AL, Colla R, Ciucis CD, Marinari UM, Pronzato MA, Traverso N, Domenicotti C. Redox homeostasis and cellular antioxidant systems: crucial players in cancer growth and therapy. Oxid Med Cell Longev 2016;2016:6235641.
[64] Zhao Y, Hu X, Liu Y, Dong S, Wen Z, He W, Zhang S, Huang Q, Shi M. ROS signaling under metabolic stress: cross-talk between AMPK and AKT pathway. Mol Cancer 2017;16(1):79. https://doi.org/10.1186/s12943-017-0648-1.
[65] Lennicke C, Rahn J, Lichtenfels R, Wessjohann LA, Seliger B. Hydrogen peroxide—production, fate and role in redox signaling of tumor cells. Cell Commun Signal 2015;13:39. https://doi.org/10.1186/s12964-015-0118-6.
[66] Qutub AA, Popel AS. Reactive oxygen species regulate hypoxia-inducible factor 1alpha differentially in cancer and ischemia. Mol Cell Biol 2008 Aug;28(16):5106–19.
[67] Guerra F, Arbini AA, Moro L. Mitochondria and cancer chemoresistance. Biochim Biophys Acta Bioenerg 2017 Aug 1;1858(8):686–99.
[68] Bouchez C, Devin A. Mitochondrial biogenesis and mitochondrial reactive oxygen species (ROS): a complex relationship regulated by the cAMP/PKA signaling pathway. Cells 2019;8(4):287.
[69] Rosca MG, Vazquez EJ, Chen Q, Kerner J, Kern TS, Hoppel CL. Oxidation of fatty acids is the source of increased mitochondrial reactive oxygen species production in kidney cortical tubules in early diabetes. Diabetes 2012 Aug;61(8):2074–83.

[70] Boveris A, Cadenas E. Mitochondrial production of hydrogen peroxide regulation by nitric oxide and the role of ubisemiquinone. IUBMB Life 2000 Nov;50(4–5):245–50.
[71] Sies H. Role of metabolic H2O2 generation redox signaling and oxidative stress. J Biol Chem 2014 Mar 28;289 (13):8735–41.
[72] Guntuku L, Naidu VGM, Yerra VG. Mitochondrial dysfunction in gliomas: pharmacotherapeutic potential of natural compounds. Curr Neuropharmacol 2016;14(6):567–83.
[73] Sutendra G, Michelakis ED. Pyruvate dehydrogenase kinase as a novel therapeutic target in oncology. Front Oncol 2013;3:38. https://doi.org/10.3389/fonc.2013.00038.
[74] Taïb B, Aboussalah AM, Moniruzzaman M, Chen S, Haughey NJ, Kim SF, Ahima RS. Lipid accumulation and oxidation in glioblastoma multiforme. Sci Rep 2019;9(1):19593.
[75] Sperry J, Belle JEL, Condro MC, Guo L, Braas D, Vanderveer-Harris N, Kim KKO, Pope WB, Divakaruni AS, Lai A, Christofk H, Kornblum HI. Metabolism of fatty acids and ketone bodies for glioblastoma growth: implications for ketogenic diet therapy. iScience 2020;23(9):101453.
[76] Pike LS, Smift AL, Croteau NJ, Ferrick DA, Wu M. Inhibition of fatty acid oxidation by etomoxir impairs NADPH production and increases reactive oxygen species resulting in ATP depletion and cell death in human glioblastoma cells. Biochim Biophys Acta 2011 Jun;1807(6):726–34.
[77] Brand MD. The sites and topology of mitochondrial superoxide production. Exp Gerontol 2010 Aug;45 (7–8):466–72.
[78] Quijano C, Trujillo M, Castro L, Trostchansky A. Interplay between oxidant species and energy metabolism. Redox Biol 2015 Nov 30;8:28–42.
[79] Zhao W, Kridel S, Thorburn A, Kooshki M, Little J, Hebbar S, Robbins M. Fatty acid synthase: a novel target for antiglioma therapy. Br J Cancer 2006 Oct 9;95(7):869–78.
[80] Susnow N, Zhang L, Margineantu D, Hockenbery DM. Bcl-2 family proteins as regulators of oxidative stress. Semin Cancer Biol 2009 Feb;19(1):42–9.
[81] Tamma G, Valenti G, Grossini E, Donnini S, Marino A, Marinelli RA, Calamita G. Aquaporin membrane channels in oxidative stress, cell signaling, and aging: recent advances and research trends. Oxid Med Cell Longev 2018;2018:1501847. https://doi.org/10.1155/2018/1501847.
[82] Georgieva E, Ivanova D, Zhelev Z, Bakalova R, Gulubova M, Aoki I. Mitochondrial dysfunction and redox imbalance as a diagnostic marker of "free radical diseases". Anticancer Res 2017;37(10):5373–81.
[83] Camara AKS, Zhou Y, Wen P-C, Tajkhorshid E, Kwok W-M. Mitochondrial VDAC1: a key gatekeeper as potential therapeutic target. Front Physiol 2017;8:460.
[84] Redza-Dutordoir M, Averill-Bates DA. Activation of apoptosis signalling pathways by reactive oxygen species. Biochim Biophys Acta 2016;1863(12):2977–92.
[85] Kim W, Lee S, Seo D, Kim D, Kim K, Kim E, Kang J, Seong KM, Youn H, Youn B. Cellular stress responses in radiotherapy. Cells 2019;8(9):1105. https://doi.org/10.3390/cells8091105.
[86] Rudziński W, Swiat M, Tomaszewski M, Krejza J. Cerebral hemodynamics and investigations of cerebral blood flow regulation. Nucl Med Rev Cent East Eur 2007;10(1):29–42.
[87] Reeg S, Grune T. Protein oxidation in aging: does it play a role in aging progression? Antioxid Redox Signal 2015 Jul 20;23(3):239–55.
[88] Tarafdar A, Pula G. The role of NADPH oxidases and oxidative stress in neurodegenerative disorders. Int J Mol Sci 2018 Nov 30;19(12). [cited 2020 Oct 1] Available from: https://www.ncbi.nlm.nih.gov/pmc/articles/PMC6321244/.
[89] Zhang J, Wang X, Vikash V, Ye Q, Wu D, Liu Y, Dong W. ROS and ROS-mediated cellular signaling. Oxid Med Cell Longev 2016;2016:4350965.
[90] Klotz L-O, Sánchez-Ramos C, Prieto-Arroyo I, Urbánek P, Steinbrenner H, Monsalve M. Redox regulation of FoxO transcription factors. Redox Biol 2015 Dec 1;6:51–72.
[91] Tonelli C, Chio IIC, Tuveson DA. Transcriptional Regulation by Nrf2. Antioxid Redox Signal 2018;29 (17):1727–45.
[92] Yang H, Xiang S, Kazi A, Sebti SM. The GTPase KRAS suppresses the p53 tumor suppressor by activating the NRF2-regulated antioxidant defense system in cancer cells. J Biol Chem 2020 Mar 6;295(10):3055–63.
[93] Surh Y-J, Kundu JK, Na H-K, Lee J-S. Redox-sensitive transcription factors as prime targets for chemoprevention with anti-inflammatory and antioxidative phytochemicals. J Nutr 2005;135(12 Suppl):2993S–3001S.

[94] Chen R, Lai UH, Zhu L, Singh A, Ahmed M, Forsyth NR. Reactive oxygen species formation in the brain at different oxygen levels: the role of hypoxia inducible factors. Front Cell Dev Biol 2018;6. [cited 2020 Oct 1] Available from: https://www.frontiersin.org/articles/10.3389/fcell.2018.00132/full.

[95] Sharma V, Collins LB, Chen T, Herr N, Takeda S, Sun W, Swenberg JA, Nakamura J. Oxidative stress at low levels can induce clustered DNA lesions leading to NHEJ mediated mutations. Oncotarget 2016 Mar 23;7 (18):25377–90.

[96] Perluigi M, Coccia R, Butterfield DA. 4-Hydroxy-2-Nonenal, a reactive product of lipid peroxidation, and neurodegenerative diseases: a toxic combination illuminated by redox proteomics studies. Antioxid Redox Signal 2012 Dec 1;17(11):1590–609.

[97] Wu W, Wu Y, Mayer K, von Rosenstiel C, Schecker J, Baur S, Würstle S, Liesche-Starnecker F, Gempt J, Schlegel J. Lipid peroxidation plays an important role in chemotherapeutic effects of temozolomide and the development of therapy resistance in human glioblastoma. Trans Oncol 2020 Mar 1;13(3):100748.

[98] Bretón-Romero R, Lamas S. Hydrogen peroxide signaling in vascular endothelial cells. Redox Biol 2014 Jan 1;2:529–34.

[99] Bodega G, Alique M, Puebla L, Carracedo J, Ramírez RM. Microvesicles: ROS scavengers and ROS producers. J Extracell Vesicles 2019;8(1):1626654.

[100] Rinaldi M, Caffo M, Minutoli L, Marini H, Abbritti RV, Squadrito F, Trichilo V, Valenti A, Barresi V, Altavilla D, Passalacqua M, Caruso G. ROS and brain gliomas: an overview of potential and innovative therapeutic strategies. Int J Mol Sci 2016 Jun;17(6):984.

[101] Lewerenz J, Hewett SJ, Huang Y, Lambros M, Gout PW, Kalivas PW, Massie A, Smolders I, Methner A, Pergande M, Smith SB, Ganapathy V, Maher P. The cystine/glutamate antiporter system xc− in health and disease: from molecular mechanisms to novel therapeutic opportunities. Antioxid Redox Signal 2013 Feb 10;18(5):522–55.

[102] Fukai T, Ushio-Fukai M. Superoxide dismutases: role in redox signaling, vascular function, and diseases. Antioxid Redox Signal 2011 Sep 15;15(6):1583–606.

[103] Xiao W, Wang R-S, Handy DE, Loscalzo J. NAD(H) and NADP(H) redox couples and cellular energy metabolism. Antioxid Redox Signal 2018 Jan 20;28(3):251–72.

[104] Wardyn JD, Ponsford AH, Sanderson CM. Dissecting molecular cross-talk between Nrf2 and NF-κB response pathways. Biochem Soc Trans 2015 Aug 1;43(4):621–6.

[105] Silva MM, Rocha CRR, Kinker GS, Pelegrini AL, Menck CFM. The balance between NRF2/GSH antioxidant mediated pathway and DNA repair modulates cisplatin resistance in lung cancer cells. Sci Rep 2019 Nov 27;9(1):17639.

[106] Shelton P, Jaiswal AK. The transcription factor NF-E2-related Factor 2 (Nrf2): a protooncogene? FASEB J 2013 Feb;27(2):414–23.

[107] Combs JA, DeNicola GM. The non-essential amino acid cysteine becomes essential for tumor proliferation and survival. Cancers (Basel) 2019;11(5):678. https://doi.org/10.3390/cancers11050678.

[108] Gordillo GM, Biswas A, Khanna S, Spieldenner JM, Pan X, Sen CK. Multidrug resistance-associated protein-1 (MRP-1)-dependent glutathione disulfide (GSSG) efflux as a critical survival factor for oxidant-enriched tumorigenic endothelial cells. J Biol Chem 2016 May 6;291(19):10089–103.

[109] Li VG. Role of heme oxygenase-1 (HSP32) and HSP90 in glioblastoma. EuroMediterranean Biomed J 2017;12:99–107.

[110] Singer E, Judkins J, Salomonis N, Matlaf L, Soteropoulos P, McAllister S, Soroceanu L. Reactive oxygen species-mediated therapeutic response and resistance in glioblastoma. Cell Death Dis 2015 Jan 15;6, e1601.

[111] Shen H, Decollogne S, Dilda PJ, Hau E, Chung SA, Luk PP, Hogg PJ, McDonald KL. Dual-targeting of aberrant glucose metabolism in glioblastoma. J Exp Clin Cancer Res 2015 Feb 5;34:14.

[112] Ferretti M, Fabbiano C, Di Bari M, Ponti D, Calogero A, Tata AM. M2 muscarinic receptors inhibit cell proliferation in human glioblastoma cell lines. Life Sci 2012 Nov 27;91(21–22):1134–7.

[113] Das A, Banik NL, Ray SK. Flavonoids activated caspases for apoptosis in human glioblastoma T98G and U87MG cells but not in human normal astrocytes. Cancer 2010 Jan 1;116(1):164–76.

[114] Kim H, Moon JY, Ahn KS, Cho SK. Quercetin induces mitochondrial mediated apoptosis and protective autophagy in human glioblastoma U373MG cells. Oxid Med Cell Longev 2013;2013:596496.

[115] Bądziul D, Jakubowicz-Gil J, Langner E, Rzeski W, Głowniak K, Gawron A. The effect of quercetin and imperatorin on programmed cell death induction in T98G cells in vitro. Pharmacol Rep 2014 Mar 1;66 (2):292–300.

[116] Kong Q, Beel JA, Lillehei KO. A threshold concept for cancer therapy. Med Hypotheses 2000 Jul;55(1):29–35.

[117] Iida M, Doi H, Asamoto S, Sugiyama H, Sakagami H, Kuribayashi N, Takeda M, Okamura Y, Matsumoto K. Effect of glutathione-modulating compounds on platinum compounds-induced cytotoxicity in human glioma cell lines. Anticancer Res 1999 Dec;19(6B):5383–4.

[118] Iida M, Sunaga S, Hirota N, Kuribayashi N, Sakagami H, Takeda M, Matsumoto K. Effect of glutathione-modulating compounds on hydrogen-peroxide-induced cytotoxicity in human glioblastoma and glioma cell lines. J Cancer Res Clin Oncol 1997;123(11–12):619–22.

[119] Khromova NV, Kopnin PB, Stepanova EV, Agapova LS, Kopnin BP. p53 hot-spot mutants increase tumor vascularization via ROS-mediated activation of the HIF1/VEGF-A pathway. Cancer Lett 2009 Apr 18;276 (2):143–51.

[120] Xia C, Meng Q, Liu L-Z, Rojanasakul Y, Wang X-R, Jiang B-H. Reactive oxygen species regulate angiogenesis and tumor growth through vascular endothelial growth factor. Cancer Res 2007 Nov 15;67(22):10823–30.

[121] Martín V, Herrera F, García-Santos G, Antolín I, Rodriguez-Blanco J, Rodriguez C. Signaling pathways involved in antioxidant control of glioma cell proliferation. Free Radic Biol Med 2007 Jun 1;42(11):1715–22.

[122] Son Y, Cheong Y-K, Kim N-H, Chung H-T, Kang DG, Pae H-O. Mitogen-activated protein kinases and reactive oxygen species: how can ROS activate MAPK pathways? J Signal Trans 2011;2011:e792639. Hindawi. [cited 2020 Oct 1] Available from: https://www.hindawi.com/journals/jst/2011/792639/.

[123] Guo Y-J, Pan W-W, Liu S-B, Shen Z-F, Xu Y, Hu L-L. ERK/MAPK signalling pathway and tumorigenesis. Exp Ther Med 2020 Mar;19(3):1997–2007.

[124] Kiffin R, Christian C, Knecht E, Cuervo AM. Activation of chaperone-mediated autophagy during oxidative stress. Mol Biol Cell 2004 Nov;15(11):4829–40.

[125] Chang K-Y, Hsu T-I, Hsu C-C, Tsai S-Y, Liu J-J, Chou S-W, Liu M-S, Liou J-P, Ko C-Y, Chen K-Y, Hung J-J, Chang W-C, Chuang C-K, Kao T-J, Chuang J-Y. Specificity protein 1-modulated superoxide dismutase 2 enhances temozolomide resistance in glioblastoma, which is independent of O6-methylguanine-DNA methyltransferase. Redox Biol 2017 Oct 1;13:655–64.

[126] Kaur B, Tan C, Brat DJ, Post DE, Van Meir EG. Genetic and hypoxic regulation of angiogenesis in gliomas. J Neurooncol 2004 Nov;70(2):229–43.

[127] Wilson WR, Hay MP. Targeting hypoxia in cancer therapy. Nat Rev Cancer 2011 Jun;11(6):393–410.

[128] Jensen RL. Hypoxia in the tumorigenesis of gliomas and as a potential target for therapeutic measures. Neurosurg Focus 2006 Apr 15;20(4), E24.

[129] Osinsky S, Zavelevich M, Vaupel P. Tumor hypoxia and malignant progression. Exp Oncol 2009 Jun;31(2):80–6.

[130] Liu Z-J, Semenza GL, Zhang H-F. Hypoxia-inducible factor 1 and breast cancer metastasis. J Zhejiang Univ Sci B 2015 Jan;16(1):32–43.

[131] Keith B, Simon MC. Hypoxia-inducible factors, stem cells, and cancer. Cell 2007 May 4;129(3):465–72.

[132] Alfaro IE, Albornoz A, Molina A, Moreno J, Cordero K, Criollo A, Budini M. Chaperone mediated autophagy in the crosstalk of neurodegenerative diseases and metabolic disorders. Front Endocrinol (Lausanne) 2019;9:778. https://doi.org/10.3389/fendo.2018.00778.

[133] Valdor R, García-Bernal D, Riquelme D, Martinez CM, Moraleda JM, Cuervo AM, Macian F, Martinez S. Glioblastoma ablates pericytes antitumor immune function through aberrant up-regulation of chaperone-mediated autophagy. PNAS 2019 Oct 8;116(41):20655–65.

[134] Pajares M, Rojo AI, Arias E, Díaz-Carretero A, Cuervo AM, Cuadrado A. Transcription factor NFE2L2/NRF2 modulates chaperone-mediated autophagy through the regulation of LAMP2A. Autophagy 2018;14 (8):1310–22.

[135] Duan J, Ying Z, Su Y, Lin F, Deng Y. α-Synuclein binds to cytoplasmic vesicles in U251 glioblastoma cells. Neurosci Lett 2017 Mar 6;642:148–52.

[136] Hao Y, Kacal M, Ouchida AT, Zhang B, Norberg E, Vakifahmetoglu-Norberg H. Targetome analysis of chaperone-mediated autophagy in cancer cells. Autophagy 2019;15(9):1558–71.

[137] Biomarkers Definitions Working Group. Biomarkers and surrogate endpoints: preferred definitions and conceptual framework. Clin Pharmacol Ther 2001 Mar;69(3):89–95.

[138] Lo Dico A, Martelli C, Valtorta S, Raccagni I, Diceglie C, Belloli S, Gianelli U, Vaira V, Politi LS, Bosari S, Lucignani G, Moresco RM, Ottobrini L. Identification of imaging biomarkers for the assessment of tumour response to different treatments in a preclinical glioma model. Eur J Nucl Med Mol Imaging 2015 Jun;42(7):1093–105.

CHAPTER

11

Glioblastoma cell-induced immunosuppression causing chemoresistance

Peter P. Sordillo[a] *and Laura A. Sordillo*[b]

[a]Department of Hematology-Oncology, Lenox Hill Hospital, New York, NY, United States
[b]Departments of Physics and Electrical Engineering, The City College of New York, New York, NY, United States

Abstract

Glioblastomas (GBMs) secrete hundreds of molecules that facilitate their growth and invasiveness. GBMs also release numerous pro- and antiinflammatory cytokines, checkpoint molecules and other molecules in order to cause profound, local and systemic, immunosuppression, a mechanism by which they can resist chemotherapy, radiation therapy, vaccines, and other treatments. This is thought to be a major factor responsible for the extremely poor prognosis of patients with GBM. In this chapter, we look closely at some of the most important of these immunosuppressive mechanisms, as well as at potential ways of increasing GBM patient survival by reversing this immunosuppression.

Abbreviations

ADAM	a disintegrin and metalloproteinase
AKT	protein kinase B
CAR-T	chimeric antigen receptor T cells
CCL	chemokine (C-C motif) ligand
CD	cluster of differentiation
CTLA4	cytotoxic T-lymphocyte-associated protein 4
CXCL	chemokine (C-X-C motif) ligand
EGFRvIII	epidermal growth factor receptor variant III
ERK	extracellular-signal-regulated kinase
FOXP3	forkhead box P3
GBM	glioblastoma
G-CSF	granulocyte colony-stimulating factor
GM-CSF	granulocyte-macrophage colony-stimulating factor

Glioblastoma Resistance to Chemotherapy: Molecular Mechanisms and Innovative Reversal Strategies
https://doi.org/10.1016/B978-0-12-821567-8.00008-7

IDO	indoleamine 2,3-dioxygenase
IL	interleukin
JAK	Janus kinase
JNK	C-Jun N-terminal kinase
LAG3	lymphocyte-activation gene 3
MAPK	mitogen-activated protein kinase
MEK	mitogen-activated protein kinase kinase
miR-124	microRNA-124
MMP	matrix metalloproteinase
mTOR	mammalian target of rapamycin
NF-κB	nuclear factor kappa-light-chain-enhancer of activated B cells
NK	natural killer
PD-L1	programmed death ligand 1
PFS	progression-free survival
PI3K	phosphatidylinositol 3-kinase
RANTES	regulated on activation, normal T cell expressed and secreted
S1P	sphingosine 1-phosphate
SK	sphingosine kinase
STAT3	signal transducer and activator of transcription 3
TDO	tryptophan 2,3-dioxygenase
TGF-β	transforming growth factor-β
TIM-3	T-cell immunoglobulin and mucin domain-containing protein 3
TNF-α	tumor necrosis factor-α
Tregs	T regulator cells

Conflict of interest

No potential conflicts of interest were disclosed.

Introduction

Glioblastoma (GBM) is an extremely aggressive brain cancer with less than 5% of patients living for 5 years after diagnosis [1, 2]. Surgery is rarely 100% successful, and radiation provides a definite, but only a limited, relief of symptoms and increase in survival. Chemotherapy is largely ineffective. The alkylating agent, temozolomide, extends survival a few months when given as adjuvant therapy after surgery and radiation therapy [3]. Salvage agents, such as lomustine, carmustine, and bevacizumab, may produce short-term responses, but, in general, do not have a major impact on survival [4, 5]. The use of alternating electromagnetic field therapy in combination with radiation therapy and temozolomide has given some early, encouraging, results; however, the overwhelming number of patients who do respond still relapse quickly [6, 7]. Numerous new chemotherapy drugs, vaccines, monoclonal antibodies, and targeted therapies are now under investigation [8–11]. Owing to the role that GBM-induced immunosuppression plays in its resistance to treatment, agents that might block immunosuppressive molecules or pathways in the GBM microenvironment have been studied extensively. GBMs have multiple mechanisms by which they can grow and spread throughout the brain, and cause local immunosuppression. These include expression on the tumor cell surface of immune checkpoints and secretion of immunosuppressive molecules, which alter the local microenvironment around the cancer in order to suppress the patient's immune response. In this chapter, we discuss some of these important immunosuppressive mechanisms, as well as outline possible strategies to overcome them.

The glioblastoma secretome

Cancers secrete thousands of proteins to stimulate their growth, to help them invade locally and metastasize, to resist therapy, and recur after treatment is initially effective. The kinds and amounts of the proteins released vary greatly between types of cancers and between tumors of single histology. Also released in some cancers are proteins that can cause severe immunosuppression. Analysis of the secretome has been used to find markers for diagnosis of an individual type of cancer, to determine subtypes, to understand the nature of a cancer, and to inform treatment strategies. Robinson et al. recently conducted a pan-cancer analysis of the secretomes from 32 cancer histologies, as well as from 30 normal tissues [12]. Especially important across histologic types were matrix metalloproteinases (MMPs), zinc-based enzymes that digest extracellular matrix, and ADAMs (composed of a disintegrin and an MMP) [12].

The GBM secretome is the group of molecules released by the GBM stem cells, mature GBM cells and GBM support cells at various time points in the neoplastic process. The secretome includes proteins such as pro- and antiinflammatory cytokines, growth factors, adhesion molecules, and other molecules. The secretome varies depending on the individual GBM, and partially determines the malignant potential of the cancer. Thus, much work has been devoted to analyzing the GBM secretome in attempts to look for GBM markers and markers of poor prognosis, as well as to understand more fully the nature of this cancer. Formolo et al. looked at the secretome profiles of four GBM cell lines (U87, T98, U118, and LN18), by measuring the 1100 protein signatures of each [13]. MMPs and ADAMs, which digest extracellular matrix and facilitate GBM spread, as well as cathepsins, which degrade proteins, were especially prominent in the U87 line, the most malignant of the GBM cell lines [13]. As we will see in the next section, this is one mechanism by which GBM-induced cytokine secretion enables GBM spread, as these enzymes not only break barriers that might slow tumor spread, they directly stimulate cytokines such as IL-1β [14]. Kast et al. note that many GBMs secrete large amounts of granulocyte colony-stimulating factor (G-CSF) and granulocyte-macrophage colony-stimulating factor (GM-CSF). This causes immunosuppression by shifting hematopoiesis away from lymphocyte production [15]. Other molecules are secreted by cancers specifically to cause immunosuppression in the microenvironment by suppressing T effector and natural killer (NK) cells and increasing the production of immunosuppressive T regulatory cells (Tregs), myeloid-derived suppressor cells and M2 macrophages [16]. Particularly important immunosuppressive molecules secreted are the cytokines, interleukin-1β (IL-1β), and transforming growth factor-β (TGF-β), as well as growth factors, fibroblast growth factor, epidermal growth factor, and platelet-derived growth factor [16]. Similar results were reported by Liu et al. who noted the importance of cancer-secreted MMP-2, interleukin-6 (IL-6), TGF-β, and CXCL-12 [17]. In GBM, other secreted cytokines, such as interleukin-10 (IL-10) and interleukin-8 (IL-8), are also extremely important [18].

Cytokine-induced immunosuppression in glioblastoma

As noted, GBMs secrete numerous cytokines to cause local immunosuppression and can increase these secretions in response to chemotherapy or other treatments (see Table 1)

TABLE 1 Important cytokines increased in glioblastoma.

Cytokine	Function
CCL2 (monocyte chemoattractant protein 2)	Chemokine, attracts lymphocytes and monocytes to inflammation sites, induces IL-1, IL-6, IL-12 secretion, stimulates Tregs and immunosuppressive macrophages
CCL5 (RANTES)	Chemokine, T-cell chemotaxis, also attracts eosinophils and basophils, activates NK cells, associated with development and progression of cancer, enables entry of viruses, including HIV, into lymphocytes
CXCL1 (neutrophil-activating protein 3)	Chemokine, angiogenesis, tumorigenesis, neuroprotective activity, wound healing
CXCL12 (stromal cell-derived factor)	Chemokine, neurogenesis, neuroinflammation, tumor angiogenesis
IL-1α	Proinflammatory cytokine, stimulates TNF-α release from endothelial cells, hematopoietic activity
IL-1β	Major proinflammatory cytokine, regulates production of IL-2, IL-6, IL-8 and interferon γ, increases sphingosine kinase activity, stimulation of phagocytosis and programmed cell death, early CNS development
IL-2	Proinflammatory cytokine, T-cell lymphocyte differentiation, once used in treatment for malignant melanoma, renal carcinoma and other cancers
IL-4	Antiinflammatory cytokine, also stimulates B-cell proliferation
IL-5	Antiinflammatory cytokine, eosinophilia, allergies
IL-6	Major proinflammatory cytokine, upregulates STAT3 in GBM, B-cell differentiation, neurogenesis, myokine, also has antiinflammatory functions (suppression of IL-1 and TNF-α)
IL-8 (CXCL8)	Major proinflammatory cytokine (chemokine), neutrophil chemotaxis
IL-10	Major antiinflammatory cytokine, important suppressor of TNF-α, interferon γ, IL-2 and IL-1β, also has T-cell stimulatory functions
IL-12	Proinflammatory cytokine, T-cell differentiation
IL-13	Antiinflammatory cytokine, pro-allergic, induces IgE secretion, induces matrix metalloproteinases
IL-17	Important proinflammatory cytokine, produced by helper 17 cells, stimulates immunosuppressive mesenchymal stem cells
Interferon-γ	Proinflammatory cytokine, T- and B- cell lymphocyte and macrophage activation, stimulates IDO
MCP-1α	Proinflammatory cytokine, neutrophil activation, monocyte and basophil chemotaxis
TGF-β	Antiinflammatory cytokine, also has proinflammatory functions, causes T effector cell paralysis in GBM through suppression of interferon γ
TNF-α	Major proinflammatory cytokine, insulin resistance, stimulates apoptosis, may also have neuroprotective functions, stimulates IDO and PD-L1

[19–21]. Prominently elevated among these are the proinflammatory cytokines, interleukin-1β (IL-1β), IL-6, tumor necrosis factor-α (TNF-α) and interferon-gamma (interferon γ), as well as the antiinflammatory cytokines, IL-10 and transforming growth factor-β (TGF-β) [19, 22–24]. After initial secretion by GBM cells, these cytokines stimulate Tregs and other cells to increase further their concentrations in the tumor microenvironment. These cytokines can cause immunosuppression by numerous other mechanisms, including causing the release of immune checkpoints. They can also stimulate the growth of GBM stem cells.

Interleukin-6 (IL-6) and STAT3 signaling

IL-6 is an extremely important cytokine that can be massively stimulated and can accumulate at the inflammation site after infection or serious injury. It has both pro- and anti-inflammatory activities [25–27]. An increase in IL-6 is seen in and plays a role in the causation of numerous diseases as diverse as renal fibrosis [28], aortic aneurysm [29]; viral infection [30], and ischemic stroke [31]. IL-6 is released by numerous cancers, including GBM, and is associated with rapid progression and a poor prognosis [32–35]. In GBM, in particular, increased levels of IL-6 are associated with higher glioma grade and shorter survival. IL-6 is known to promote GBM growth because of its effects on multiple cancer-promoting pathways. IL-6 forms a complex with the IL-6 receptor and the common signal-transducing receptor, glycoprotein 130 (gp130) [32]. Through its activation of Janus kinases, it stimulates the mitogen-activated protein kinase (MAPK), phosphatidylinositol 3-kinase/protein kinase B (PI3K/AKT), and signal transducer and activator of transcription 3 (STAT3) signaling pathways [36–39]. Particularly important in GBM is STAT3 (Fig. 1), which has long been known to be a major prognostic factor in these patients [40]. IL-6-induced STAT3 plays a key role in establishing an immunosuppressive environment around a cancer, and STAT3 expression is associated with an increased chance of recurrence and reduced survival [40–42]. Activated STAT3 is found preferably in GBM compared to lower-grade gliomas [43]. It is through STAT3 signaling that IL-6 primarily works to activate and increase immunosuppressive myeloid-derived suppressor cells. STAT3 also inhibits natural killer (NK) and cytotoxic T cells, and activates T regulatory cells (Tregs). IL-6-induced STAT3 increases the release of other cytokines by GBM cells, particularly IL-1β, IL-10, and TGF-β, and also stimulates GBM secretion of prostaglandin E2 and the immunosuppressive molecule, fibrinogen-like protein-2 [18]. Heiland et al. have shown STAT3 also signals the secretion of IL10, TGF-β, and G-CSF from microglia and astrocytes in the cancer microenvironment [44]. Inhibition of STAT3 has become a major strategy for reversing GBM immunosuppression. Numerous agents have been shown to be STAT3 inhibitors, including peptide inhibitors, small molecules, natural products, antisense oligonucleotides, G-quartet oligodeoxynucleotides and other molecules, reflecting the importance of this mechanism in cancer immunosuppression. These agents have been discussed in detail recently by Lau et al. [45]. See et al. showed that STAT3 inhibition increased antitumor responsiveness when tested in three human GBM cell lines which express high levels of phospho-STAT3 (HS683, U57, and U251) [46]. Senft et al. showed that inhibition of the STAT3 activator, Janus kinase 2, reduced the invasive and migratory potential of five GBM cell lines [47]. Hussain et al. reported on a small molecule that can cross the blood-brain barrier and reverse the immune tolerance caused by GBM. It inhibits STAT3 and induces

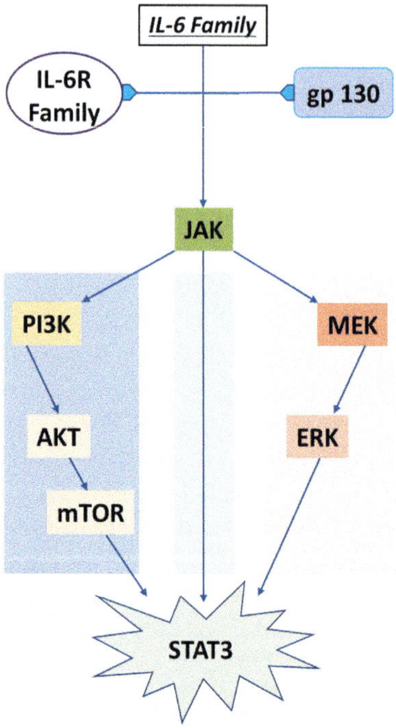

FIG. 1 A simplified model of the IL-6/JAK/STAT3 pathway. gp130: glycoprotein 130; JAK: Janus kinase; PI3K: phosphatidylinositol 3-kinase; AKT: protein kinase B; mTOR: mammalian target of rapamycin; MEK: mitogen-activated protein kinase kinase; ERK: extracellular-signal-regulated kinase; STAT3: signal transducer and activator of transcription 3.

proliferation of effector T cells [48]. Similarly, Wei et al. demonstrated that miR-124, which is absent in gliomas, could, after upregulation in GBM stem cells, inhibit STAT3 and reverse immunosuppression [49]. Likewise, the natural product, quercetin, has been shown to reduce IL-6 induction of STAT3 signaling and to inhibit GBM growth and migration in two GBM cell lines (T986 and U87) [50]. Bi et al. reported that honokiol, derived from magnolia bark, reduced STAT3 phosphorylation, caused apoptosis of U251 and U-87MG human GBM cells in culture, and eliminated GBM stem cells [51]. We have reported on the multiple mechanisms whereby curcumin may be useful in the treatment of GBM, including STAT3 inhibition [52]. A list of natural products that may reduce the immunosuppressive actions of STAT3 is given in Table 2. The Janus kinase inhibitor, pacritinib, currently being studied clinically as a potential treatment for myelofibrosis, has been shown to overcome the resistance to the chemotherapy drug, temozolomide, in GBM cells through STAT3 inhibition [96].

IL-6 has other mechanisms by which it allows GBM to resist treatment. IL-6 can convert mature cancer cells to cancer stem cells, which are resistant to both chemotherapy and radiation therapy [97, 98]. IL-6 stimulates myeloid cells to secrete PD-L1, an important immunosuppressive mechanism [99]. IL-6 can also cause profound immunosuppression by stimulating the release of the enzyme, indoleamine 2,3-dioxygenase (IDO), thereby causing excess tryptophan metabolism along the kynurenine pathway [100] (PD-L1 and IDO will be discussed extensively later in this chapter).

TABLE 2 Natural products reported to suppress STAT3.

Product	Source	Reference
Quercetin	Onions, capers, kale, berries	[50, 53–60]
Resveratrol	Grapes, berries, peanuts	[61–65]
Curcumin	Turmeric	[66–75]
Honokiol	Bark of magnolia tree	[76, 77]
Epigallocatechin gallate	Green tea	[71, 78, 79]
Luteolin	Broccoli, celery, various plant leaves	[57, 80–82]
Betulinic acid	Bark from white birch and other trees	[83, 84]
Oleanic acid	Olive oil, garlic, other plants	[84]
Bruceantinol	Fruit of Brucei javanica tree	[85]
Caffeic acid	Coffee, spearmint, sage, thyme, Ceylon cinnamon	[86–88]
Cinnamic acid	Cinnamon	[89]
Ursolic acid	Apple peels	[90–93]
Tocotrienol	Vegetable oils, seeds, wheat germ, barley	[94, 95]

Interleukin-1β (IL-1β) and the sphingolipid rheostat

The IL-1 family is a group of cytokines that play a crucial role in the inflammatory response to infection or severe injury [25]. IL-1 is also required for tumor angiogenesis and invasion [101]. Most important in this group is IL-1β. Like IL-6, IL-1β also has both pro- and antiinflammatory functions. However, after chronic inflammation, IL-1β drives tumor growth and spread [102]. Increased levels of IL-1β have been detected in numerous GBM models [103]. IL-1β stimulates multiple tumor-related signaling pathways, including JNK and NF-κβ. IL-1β drives GBM cells to increase their secretion of IL-6, further stimulating STAT3 [24, 104, 105]. Tarassishin et al. showed that exogenous IL-1-stimulated U251 cells could cause the upregulation of IL-8, monocyte chemoattractant protein 1 (CCL2), galectin-1 and tenascin-C, as well as could increase markedly the level of IL-1 itself in the GBM secretome [106]. CCL2, long known for its ability to attract myeloid cells to sites of inflammation, has been found to have both tumoricidal and immunosuppressive effects, including stimulation of other cytokines such as IL-6 and IL-12, as well as stimulation of Tregs and immunosuppressive macrophages [107]. Galectin-1, highly expressed in GBM and other solid tumors and associated with poor patient survival, is cytotoxic to T effector cells and has recently been found to maintain immunosuppression by also preventing T effector cells from entering the tumor microenvironment [108]. Tenascin-C protects cancer stem cells by preventing T effector cell activation, proliferation, and cytokine production [109]. Mirzaei et al. have shown that α5β1 tenascin-C specifically protects GBM stem cells by binding to integrin α5β1 and integrin αvβ6 on T cells [110].

Importantly, IL-1β upregulates the expression of sphingosine kinase, a key enzyme that shifts the balance from cancer-destructive ceramides toward sphingosine 1-phosphate (S1P), which favors GBM survival [111, 112] (Fig 2). This "sphingolipid rheostat" is a basic mechanism that determines the fate of a cell and is governed by the extent of sphingosine kinase (SK), which, thus, plays a key role in numerous diseases [113–116]. SK stimulates cancer growth and invasion, as well as resistance to chemotherapy [117–121]. Overexpression of SK has been detected in numerous cancers [119, 120, 122, 123], but is particularly important in GBM [124]. The increase in S1P produced by GBM cells high in SK drives GBM growth and

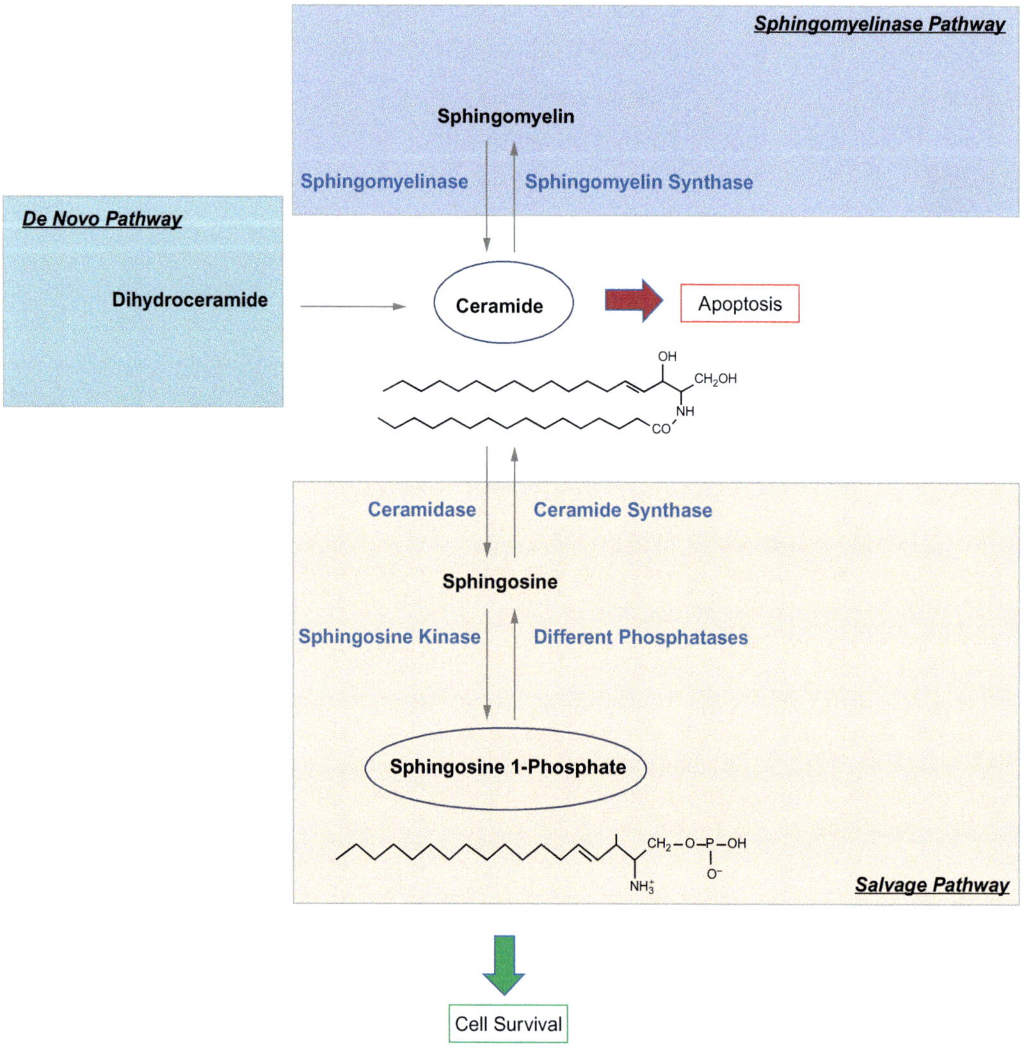

FIG. 2 The sphingolipid rheostat. Sphingosine kinase governs metabolism between destructive ceramide and cell-preserving sphingosine 1-phosphate.

invasion [125, 126], and correlates with poor patient survival [127, 128]. Increased S1P also correlates with increased GBM tumor grade [129]. Both SK1 and SK2 also stimulate the cytokine IL-17, which dramatically increases the immunosuppressive effects of mesenchymal stem cells [130, 131]. A recent paper by Riboni et al. summarizes the extensive evidence of the essential role of SK in GBM survival through its effect of converting cancer-destroying ceramides into tumor-protecting S1P [132]. A number of commonly used medications have SK inhibitor activity, such as fingolimod [133], tyrosine kinase inhibitors [134], calcium channel blockers [135], antimalarials [136], and antihistamines [137]. Numerous new SK inhibitors have been designed and are showing promise in multiple cancers [138–140].

Transforming growth factor-β (TGF-β) and T cell paralysis

TGF-β is a pleotropic cytokine that has both tumor-suppressant and tumor-stimulating actions. Tumor-suppressant actions appear to occur early in carcinogenesis, but, later, tumor-stimulating actions predominate [141, 142]. Three isoforms of TGF-β have been found in humans. Increased levels of this cytokine have been detected in many cancers, including malignant melanoma, glioblastoma, and carcinomas of the breast, lung, stomach, ovary, and uterus [143], and signify a poor prognosis [144]. It is primarily through STAT3 signaling that GBM cells are stimulated to secrete TGF-β [18], which, in turn, has multiple mechanisms by which it can promote tumor growth. In GBM, TGF-β stimulates invasion, promotes angiogenesis, and predisposes to multidrug resistance [142, 145]. TGF-β stimulates GBM stem cells, which are resistant to both chemotherapy and radiation therapy [146]. GBM-secreted TGF-β also causes profound immunosuppression. In an important paper published a few months ago, Dimiloe et al. demonstrated that, after phosphorylation of Smad proteins, TGF-β markedly reduced interferon γ production by $CD4^+$ T cells, resulting in immunosuppression through T cell metabolic paralysis [147]. Further immunosuppression occurs because of additional TGF-β that is secreted by Tregs and by macrophages that have been stimulated by the initial GBM cell secretion [18]. Because of an increasing appreciation of the key role that TGF-β plays in GBM progression and resistance to therapy, there has been much work over the last few years on finding and designing new TGF-β inhibitors, many of which are now being tested against this cancer [141, 148–150]. Many medications, including statins [151], estrogens [152], and androgens [153], and natural products such as quercetin [154], curcumin [155], rutin [156], resveratrol [157], and many other molecules also suppress TGF-β. Statins, as well as many of these natural products, demonstrate anticancer efficacy in many GBM models.

Interleukin-8 (IL-8) and other cytokines

IL-8 (CXCL8, neutrophil chemotactic factor), a chemokine, is released in response to infection, acidosis, hypoxia, or local trauma. IL-8 is involved in neutrophil chemotaxis and angiogenesis [158, 159]. Like other cytokines, elevated levels of IL-8 are seen with multiple cancers and are associated with a poor prognosis [160, 161]. GBM cells are stimulated to secrete IL-8 by other cytokines, particularly by TNF-α and IL-1β [24, 162]. Levels of IL-8 are particularly high in the GBM secretome. IL-8 promotes GBM growth and invasiveness [163]. In GBM, very elevated levels of IL-8 are detected at the cancer margins, where invasive and angiogenic activity is strongest [164], IL-8 plays a major role in maintaining local immunosuppression. Just as IL-8 can induce neutrophils to travel to sites of infection or tissue injury, it can cause myeloid-derived suppressor cells to infiltrate the tumor microenvironment [165]. Like IL-6, IL-8 induces STAT3 and increases the activity of checkpoints LAG3, PD-1, and CTLA4 on T effector cells [166].

Long regarded as the most important antiinflammatory cytokine and a tumor-suppressant, it is now apparent that IL-10 has major tumor-promoting actions as well. Levels of IL-10 have been found to be markedly elevated in numerous cancers and these level increases are associated with markedly reduced patient survival [167]. Elevated levels are also associated with higher tumor grade [168]. Like other cytokines, IL-10 upregulates STAT3 [169, 170]. Major increases in IL-10 occur from GBM cells themselves after IL-6 STAT3 stimulation, and again, secondarily, from Tregs and M2 macrophages [18], thus, markedly facilitating GBM growth and spread.

CCL5 (also known as RANTES) potently attracts T cells, eosinophils, and basophils to sites of inflammation. With its receptor, CCR5 is associated with the development and progression of numerous cancers [171–173]. Although Fujimoto et al. reported that high CCL5 blood levels were associated initially with a better prognosis in patients with breast carcinoma [174], Walens et al. have shown that this cytokine promotes recurrence in breast cancer patients through macrophage recruitment [175]. CCL5 and its receptor CCR5 are highly expressed in GBMs, can direct interactions with macrophages, and stimulate proliferation of GBM stem cells [176–178]. Numerous inhibitors of CCL5/CCR5 are now under investigation. Leronlimab, a monoclonal antibody that inhibits CCR5, has recently shown encouraging results in patients with breast cancer, and it is expected that a trial of this drug will begin shortly in patients with other solid tumors, including those with GBM [179, 180].

TNF-α is a major proinflammatory cytokine involved in GBM initiation, progression and resistance to treatment, particularly driving the production of an immunosuppressive environment [181, 182]. TNF-α can also have tumor destructive effects, and an increased number of cases of GBM have been found to occur in patients taking TNF-α inhibitors for autoimmune disorders [182–184]. Despite this, TNF-α inhibition has had some success in the treatment of GBM in experimental models [185]. It should be noted that TNF-α stimulates the secretion of two extremely important, immunosuppressive, checkpoint molecules, indoleamine 2,3-dioxygenase [186], and programmed death-ligand 1 [187].

Indoleamine 2,3-dioxygenase and the kynurenine pathway

One of the major mechanisms by which GBM causes immunosuppression is by inducing abnormal or unbalanced tryptophan metabolism. GBMs express high levels of the enzyme indoleamine 2,3-dioxygenase (IDO). This causes increased tryptophan metabolism down the kynurenine pathway, resulting in high levels of immunotoxic and neurotoxic tryptophan metabolites (Fig 3). These metabolites cause apoptosis of T effector cells and recruitment of Tregs. Increased metabolism along this pathway can also result in reduced activity along the serotonin-melatonin pathway and local deficiencies in tryptophan, serotonin, and melatonin (Fig 4). Normal activity along the kynurenine pathway is actually physiologic since a normal

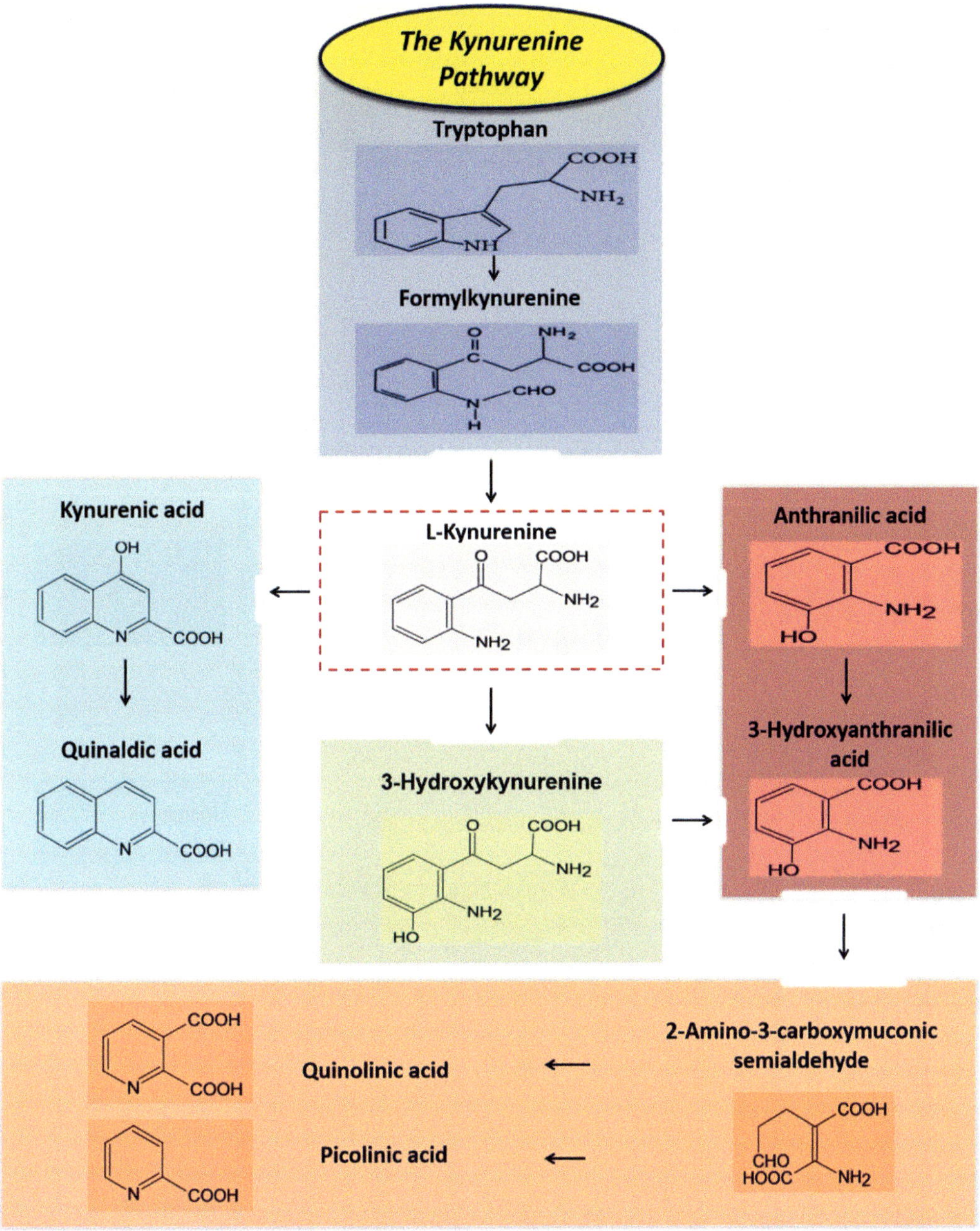

FIG. 3 Tryptophan metabolism through the kynurenine pathway. *From Sordillo LA, Sordillo PP. Optical spectroscopy of tryptophan metabolites in neurodegenerative disease. In: Neurophotonics and biomedical spectroscopy. Elsevier; 2019, p. 137–157.*

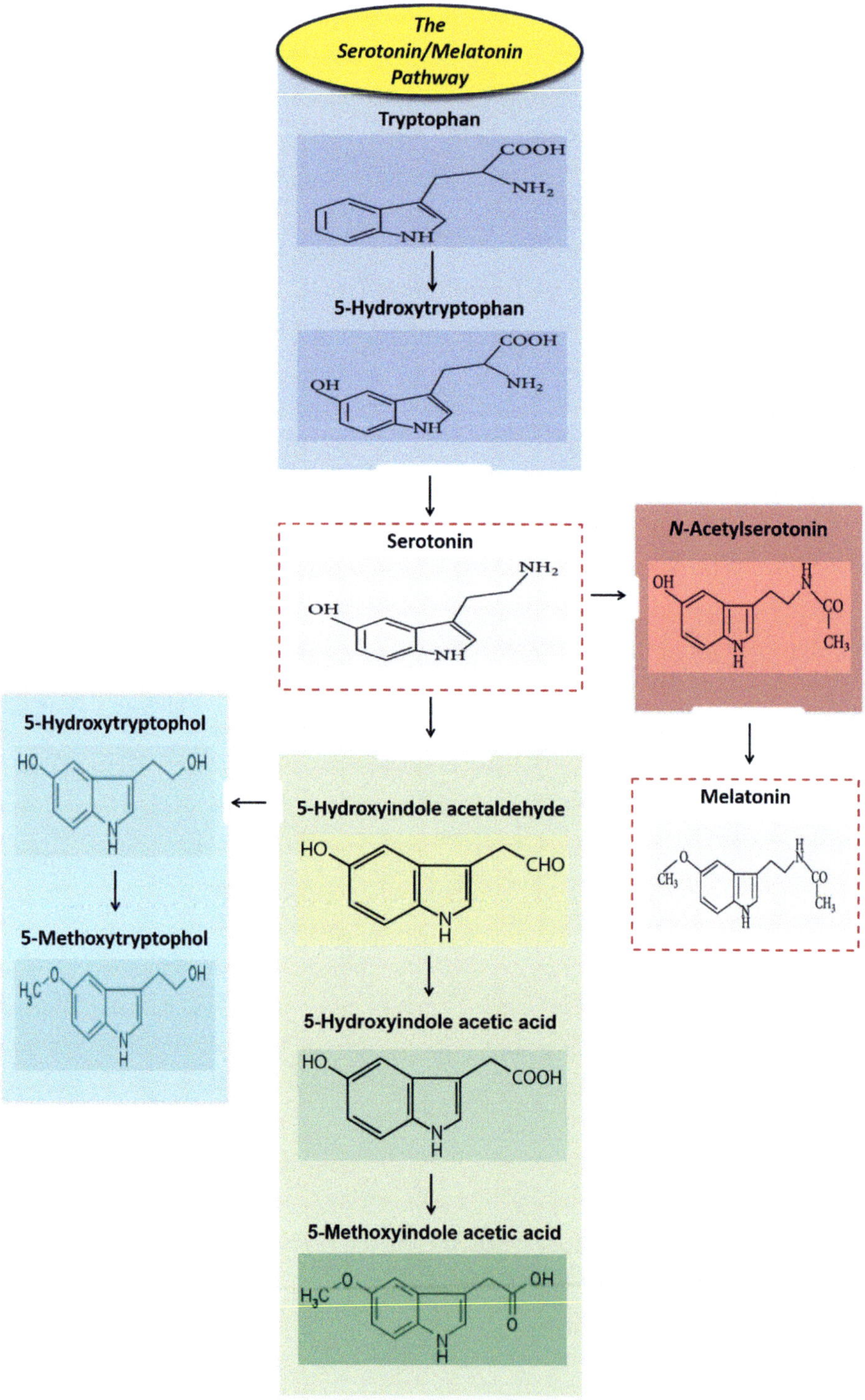

FIG. 4 Tryptophan metabolism through the serotonin/melatonin pathway. *From Sordillo LA, Sordillo PP. Optical spectroscopy of tryptophan metabolites in neurodegenerative disease. In: Neurophotonics and biomedical spectroscopy. Elsevier; 2019, p. 137–157.*

degree of T cell immunosuppression is necessary for self-tolerance and to protect the fetus from the T cells of the mother. GBM uses this immunosuppressive mechanism for its own survival. Yu et al. have shown, in a meta-analysis of 2706 cancer patients from 24 studies, that higher IDO expression strongly correlated with higher clinical stage and with reduced survival [188]. Mitsuka et al. reported similar results in samples from GBM patients [189].

There are actually two forms of IDO, IDO1 and IDO2. IDO1 is expressed in numerous tissues, including macrophages, mesenchymal stromal cells, dendritic cells, and neurons, as well as in cancer cells, while IDO2 is primarily expressed in the liver, brain, colon, placenta, and kidney [190]. A third, distinct enzyme, tryptophan 2,3-dioxygenase (TDO), also causes tryptophan to be metabolized along the kynurenine pathway. Particularly toxic to both T cells and neurons are products of this pathway, 3-hydroxyanthranilic acid, 3-hydroxykynurenine, and quinolinic acid [191]. These and other toxic kynurenines can cause immunosuppression through multiple mechanisms. IDO can suppress mechanistic target of rapamycin (mTOR) activity, resulting in T-cell anergy [192]. Not only can IDO activate mature Tregs, but it can also induce CD4(+) T cells to differentiate into FOX3+ Treg cells, and prevent the conversion of Tregs back to T helper cells [193]. In addition, local deficiencies in tryptophan stimulate the general control nondepressable-2 kinase pathway. This causes altered protein translation and interferes with the activation of T cells [194]. Because the brain is an "immune-privileged" site, kynurenine pathway metabolites play even greater role in the immunosuppression associated with GBM [195].

O'Rourke et al. have recently reported their phase 1 study on the use of EGFRvIII chimeric antigen receptor T cells (CAR-T cells) against GBM. No major responses were seen, and the authors attributed this to the profound immunosuppression associated with GBM [196], suggesting future trials should include a checkpoint inhibitor arm [197]. As part of their protocol, these scientists were able to compare resected GBM samples from before CAR-T therapy with those done at various time points after surgery, in order to study the changes in checkpoints and other immunosuppressive molecules that were induced by treatment. Strikingly, major increases in both IDO and FoxP3 were seen on days 6 and 7 after treatment. In some cases, increases in TGF-β, PD-L1 and IL-10 were also noted. The authors concluded that their CAR-T therapy had some initial activity, based on decreased levels of EGFRvIII activity, but the GBMs adapted to therapy by increasing their immunosuppressive activity, primarily by increasing secretion of IDO and PD-L1, and by recruitment of FoxP3 Tregs [196].

Numerous medications have been found in experimental models to be suppressors of IDO, including cyclooxygenase (COX_2) inhibitors, antiestrogens, statins, salicylates, chloroquine, and acyclovir, as well as natural products, such as resveratrol, quercetin, curcumin, and genistein [198]. Because of this, there has been intense interest over the last 5 years in IDO inhibitors as a third major class of immune checkpoint inhibitors, along with PD-1/PD-L1 and CTLA4 blockers [199–202]. This excitement has cooled significantly with the failure of Incyte's epacadostat in the ECHO-301 trial. However, as Muller et al. note, this failure may be secondary to epacadostat itself, rather than to the idea of IDO suppression as a treatment modality. The ineffectiveness of this drug as an IDO inhibitor, including, possibly, its inability to cause sufficient drug exposure or to act within the tumor, and the failure of the investigators to consider IDO2 and TDO, may have also played a role in epacadostat's ineffectiveness. Both IDO and TDO contribute to the malignancy in GBM and other cancers [203, 204].

Programmed death-ligand 1 (PD-L1) (cluster of differentiation, CD 274)

Expression of PD-L1 on the tumor cell membrane is a major mechanism by which cancers can suppress the patient immune system, and the use of PD-1 or PD-L1 inhibitors has been a successful strategy for treating some cancers (Table 3). These immune checkpoint inhibitors block the binding of PD-1 on the T cell to PD-L1 on the cancer cell membrane and prevent the inhibitory signal that PD-L1/PD-1 binding would send to depress T cell function. PD-L1 and PD-1 inhibitors are now being used widely (Table 3).

For example, the use of these agents either alone or in combination with the CTLA-4 inhibitor, ipilimumab, has been successful in increasing the survival of patients with malignant melanoma [205], and great strides have been made in the treatment of other cancers. Treatment of patients with GBM has not been successful, however, partly because of the multiple mechanisms by which GBM can suppress the immune system. Omuro et al. studied treatment with nivolumab, with and without ipilimumab, against recurrent GBM in the Checkmate 143 study. Among 40 patients, there were three partial responses and eight patients had stable disease for greater than 12 weeks. Diarrhea was a significant toxicity, especially for patients on both drugs [206]. In another phase 2 trial, of pembrolizumab in 14 evaluable patients, there were no complete responses, but three patients had partial responses and one had stable disease. Median overall survival was 20 months and progression-free survival was 4.5 months [207]. Of importance was the finding of both a marked reduction of T cells and a marked increase of immunosuppressive CD68+ macrophages in the tumor microenvironment [207]. It is known that high expression of CD68-related genes correlates with increased immunosuppression, poor response to treatment, and poor survival [208]. The authors concluded that PD-1 inhibition is insufficient alone to reverse the immunosuppression caused by GBM. Recent results with these agents in GBM have continued to be disappointing. An updated report (Sept 2019) on the checkmate 548 study comparing nivolumab and standard of care

TABLE 3 Currently approved immune checkpoint inhibitors.

Inhibitor	Target	Approved for
Pembrolizumab	PD-1	Melanoma, nonsmall cell lung, gastric, head and neck, renal, liver, esophageal, cervical and bladder carcinomas, Hodgkin's disease, cancers with high microsatellite instability/mismatched repair
Nivolumab	PD-1	Melanoma, renal, hepatic, bladder, head and neck, and nonsmall cell lung carcinoma, Hodgkin's disease, cancers with high microsatellite instability/ mismatched repair
Durvalumab	PD-L1	Urothelial and nonsmall cell lung carcinoma
Avelumab	PD-L1	Renal and bladder carcinoma, metastatic Merkel cell carcinoma with microsatellite insufficiency
Atezolizumab	PD-L1	Urothelial, nonsmall lung, small-cell lung and triple-negative breast carcinoma
Cemiplimab	PD-1	Cutaneous squamous cell carcinoma
Ipilimumab	CTLA-4	Melanoma, renal and nonsmall cell lung carcinoma, colorectal carcinoma with microsatellite instability

(temozolomide and RT) versus standard of care alone in patients with O6-methylguanine-DNA methyltransferase (MGMT) methylated GBM showed no increase in progression-free survival (PFS) with the addition of nivolumab. Likewise, the final report of checkmate 498 study comparing nivolumab + RT with temozolomide + RT in MGMT unmethylated GBM showed no increase in overall survival with nivolumab [209]. It is likely that obtaining a significant number of responses in patients with GBM will require blocking of the immunosuppression caused by multiple of the mechanisms discussed in this chapter. Trials combining PD-1/PD-L1 inhibitors with inhibitors of CTLA4, IDO, TIM-3, or LAG3 against GBM are ongoing or planned [210–216].

References

[1] Cantrell JN, Waddle MR, Rotman M, Peterson JL, Ruiz-Garcia H, Heckman MG, Quiñones-Hinojosa A, Rosenfeld SS, Brown PD, Trifiletti DM. Progress toward long-term survivors of glioblastoma. In: Mayo Clinic Proceedings. Elsevier; 2019. Jun 20.

[2] Norden AD, Korytowsky B, You M, Kim Le T, Dastani H, Bobiak S, Singh P. A real-world claims analysis of costs and patterns of care in treated patients with glioblastoma multiforme in the United States. J Manag Care Spec Pharm 2019;25(4):428–36.

[3] Lukas RV, Wainwright DA, Ladomersky E, Sachdev S, Sonabend AM, Stupp R. Newly diagnosed glioblastoma: a review on clinical management. Brain 2019;33(3):91–100.

[4] Zhang H, Wang R, Yu Y, Liu J, Luo T, Fan F. Glioblastoma treatment modalities besides surgery. J Cancer 2019;10(20):4793.

[5] Lyon KA, Huang JH. Bevacizumab as an adjuvant therapy for glioblastoma in elderly patients: the facts. Transl Cancer Res 2018;7(Suppl. 7):S802.

[6] Ornelas AS, Porter AB, Sharma A, Knox MG, Marks LA, Wingerchuk DM, O'Carroll CB. What is the role of tumor-treating fields in newly diagnosed glioblastoma? Neurologist 2019;24(2):71–3.

[7] Ballo MT, Urman N, Lavy-Shahaf G, Grewal J, Bomzon ZE, Toms S. Correlation of tumor treating fields dosimetry to survival outcomes in newly diagnosed glioblastoma: a large-scale numerical simulation-based analysis of data from the phase 3 EF-14 randomized trial. Int J Radiat Oncol Biol Phys 2019;104(5):1106–13.

[8] Taylor OG, Brzozowski JS, Skelding KA. Glioblastoma multiforme: an overview of emerging therapeutic targets. Front Oncol 2019;9:963.

[9] Jahan N, Talat H, Alonso A, Saha D, Curry WT. Triple combination immunotherapy with GVAX, anti-PD-1 monoclonal antibody, and agonist anti-OX40 monoclonal antibody is highly effective against murine intracranial glioma. Oncoimmunology 2019;8(5), e1577108.

[10] Sadahiro H, Kang KD, Gibson JT, Minata M, Yu H, Shi J, Chhipa R, Chen Z, Lu S, Simoni Y, Furuta T. Activation of the receptor tyrosine kinase AXL regulates the immune microenvironment in glioblastoma. Cancer Res 2018;78(11):3002–13.

[11] McGranahan T, Therkelsen KE, Ahmad S, Nagpal S. Current state of immunotherapy for treatment of glioblastoma. Curr Treat Options Oncol 2019;20(3):24.

[12] Robinson JL, Feizi A, Uhlén M, Nielsen J. A systematic investigation of the malignant functions and diagnostic potential of the cancer secretome. Cell Rep 2019;26(10):2622–35.

[13] Formolo CA, Williams R, Gordish-Dressman H, MacDonald TJ, Lee NH, Hathout Y. Secretome signature of invasive glioblastoma multiforme. J Proteome Res 2011;10(7):3149–59.

[14] Rodríguez D, Morrison CJ, Overall CM. Matrix metalloproteinases: what do they not do? New substrates and biological roles identified by murine models and proteomics. Biochim Biophys Acta Mol Cell Res 2010;1803 (1):39–54.

[15] Kast RE, Hill QA, Wion D, Mellstedt H, Focosi D, Karpel-Massler G, Heiland T, Halatsch ME. Glioblastoma-synthesized G-CSF and GM-CSF contribute to growth and immunosuppression: potential therapeutic benefit from dapsone, fenofibrate, and ribavirin. Tumor Biol 2017;39(5). 1010428317699797.

[16] Ziani L, Chouaib S, Thiery J. Alteration of the antitumor immune response by cancer-associated fibroblasts. Front Immunol 2018;9:414.

[17] Liu T, Han C, Wang S, Fang P, Ma Z, Xu L, Yin R. Cancer-associated fibroblasts: an emerging target of anti-cancer immunotherapy. J Hematol Oncol 2019;12(1):1–5.
[18] Nduom EK, Weller M, Heimberger AB. Immunosuppressive mechanisms in glioblastoma. Neuro Oncol 2015;17(Suppl. 7):vii9–14.
[19] Albulescu R, Codrici E, Popescu ID, Mihai S, Necula LG, Petrescu D, Teodoru M, Tanase CP. Cytokine patterns in brain tumour progression. Mediat Inflamm 2013;2013:979748.
[20] Nijaguna MB, Patil V, Hegde AS, Chandramouli BA, Arivazhagan A, Santosh V, Somasundaram K. An eighteen serum cytokine signature for discriminating glioma from normal healthy individuals. PLoS One 2015; 10(9), e0137524.
[21] Schwartzbaum J, Wang M, Root E, Pietrzak M, Rempala GA, Huang RP, Johannesen TB, Grimsrud TK. A nested case-control study of 277 prediagnostic serum cytokines and glioma. Plos One 2017;12(6), e0178705.
[22] Gieryng A, Pszczolkowska D, Walentynowicz KA, Rajan WD, Kaminska B. Immune microenvironment of gliomas. Lab Invest 2017;97(5):498–518.
[23] Zhenjiang L, Rao M, Luo X, Valentini D, von Landenberg A, Meng Q, Sinclair G, Hoffmann N, Karbach J, Altmannsberger HM, Jäger E. Cytokine networks and survivin peptide-specific cellular immune responses predict improved survival in patients with glioblastoma multiforme. EBioMedicine 2018;33:49–56.
[24] Yeung YT, McDonald KL, Grewal T, Munoz L. Interleukins in glioblastoma pathophysiology: implications for therapy. Br J Pharmacol 2013;168(3):591–606.
[25] Sordillo PP, Sordillo LA, Helson L. Bifunctional role of pro-inflammatory cytokines after traumatic brain injury. Brain Injury 2016;30(9):1043–53.
[26] Coughlin BA, Trombley BT, Mohr S. Interleukin-6 (IL-6) mediates protection against glucose toxicity in human Müller cells via activation of VEGF-A signaling. Biochem Biophys Res Commun 2019;517(2):227–32.
[27] Schett G. Physiological effects of modulating the interleukin-6 axis. Rheumatology 2018;57(Suppl. 2):ii43–50.
[28] Chen W, Yuan H, Cao W, Wang T, Chen W, Yu H, Fu Y, Jiang B, Zhou H, Guo H, Zhao X. Blocking interleukin-6 trans-signaling protects against renal fibrosis by suppressing STAT3 activation. Theranostics 2019;9(14):3980.
[29] Paige E, Clément M, Lareyre F, Sweeting M, Raffort J, Grenier C, Finigan A, Harrison J, Peters JE, Sun BB, Butterworth AS. Interleukin-6 receptor signaling and abdominal aortic aneurysm growth rates. Circ Genom Precis Med 2019;12(2), e002413.
[30] Velazquez-Salinas L, Verdugo-Rodriguez A, Rodriguez LL, Borca MV. The role of interleukin 6 during viral infections. Front Microbiol 2019;10:1057.
[31] Georgakis MK, Malik R, Gill DK, Franceschini N, Sudlow CL, Dichgans M, NVENT Consortium, CHARGE Inflammation Working Group. Interleukin-6 signaling effects on ischemic stroke and other cardiovascular outcomes: a Mendelian randomization study. Circ Genom Precis Med 2020;13(3):e002872.
[32] Waldner MJ, Foersch S, Neurath MF. Interleukin-6-A key regulator of colorectal cancer development. Int J Biol Sci 2012;8(9):1248.
[33] Ham IH, Oh HJ, Jin H, Bae CA, Jeon SM, Choi KS, Son SY, Han SU, Brekken RA, Lee D, Hur H. Targeting Interleukin-6 as a strategy to overcome stroma-induced resistance to chemotherapy in gastric cancer. Mol Cancer 2019;18(1):68.
[34] Kumari N, Dwarakanath BS, Das A, Bhatt AN. Role of interleukin-6 in cancer progression and therapeutic resistance. Tumor Biol 2016;37(9):11553–72.
[35] Masjedi A, Hashemi V, Hojjat-Farsangi M, Ghalamfarsa G, Azizi G, Yousefi M, Jadidi-Niaragh F. The significant role of interleukin-6 and its signaling pathway in the immunopathogenesis and treatment of breast cancer. Biomed Pharmacother 2018;108:1415–24.
[36] Kumari N, Das A, Bhatt AN. Interleukin-6 confers radio-resistance by inducing Akt-mediated glycolysis and reducing mitochondrial damage in cells. J Biochem 2020;167(3):303–14.
[37] Zegeye MM, Lindkvist M, Fälker K, Kumawat AK, Paramel G, Grenegård M, Sirsjö A, Ljungberg LU. Activation of the JAK/STAT3 And PI3K/AKT pathways are crucial for IL-6 trans-signaling-mediated pro-inflammatory response in human vascular endothelial cells. Cell Commun Signal 2018;16(1):55.
[38] Fajgenbaum DC, Langan RA, Japp AS, Partridge HL, Pierson SK, Singh A, Arenas DJ, Ruth JR, Nabel CS, Stone K, Okumura M. Identifying and targeting pathogenic PI3K/AKT/mTOR signaling in IL-6 blockade–refractory idiopathic multicentric castleman disease. J Clin Investig 2019;129(10):4451–63.
[39] Shen TN, Kanazawa S, Kado M, Okada K, Luo L, Hayashi A, Mizuno H, Tanaka R. Interleukin-6 stimulates AKT and P38 MAPK phosphorylation and fibroblast migration in non-diabetic but not diabetic mice. Plos One 2017;12(5), e0178232.

[40] Piperi C, Papavassiliou KA, Papavassiliou AG. Pivotal role of STAT3 in shaping glioblastoma immune microenvironment. Cells 2019;8(11):1398.
[41] Jones LM, Broz ML, Ranger JJ, Ozcelik J, Ahn R, Zuo D, Ursini-Siegel J, Hallett MT, Krummel M, Muller WJ. STAT3 establishes an immunosuppressive microenvironment during the early stages of breast carcinogenesis to promote tumor growth and metastasis. Cancer Res 2016;76(6):1416–28.
[42] Ferguson SD, Srinivasan VM, Heimberger AB. The role of STAT3 in tumor-mediated immune suppression. J Neuro Oncol 2015;123(3):385–94.
[43] Lo HW, Cao X, Zhu H, Ali-Osman F. Constitutively activated STAT3 frequently coexpresses with epidermal growth factor receptor in high-grade gliomas and targeting STAT3 sensitizes them to Iressa and alkylators. Clin Cancer Res 2008;14(19):6042–54.
[44] Heiland DH, Ravi VM, Behringer SP, Frenking JH, Wurm J, Joseph K, Garrelfs NW, Strähle J, Heynckes S, Grauvogel J, Franco P. Tumor-associated reactive astrocytes aid the evolution of immunosuppressive environment in glioblastoma. Nat Commun 2019;10(1):2541.
[45] YTK L, Ramaiyer M, Johnson DE, Grandis JR. Targeting STAT3 in cancer with nucleotide therapeutics. Cancers 2019;11(11):1681.
[46] See AP, Han JE, Phallen J, Binder Z, Gallia G, Pan F, Jinasena D, Jackson C, Belcaid Z, Jeong SJ, Gottschalk C. The role of STAT3 activation in modulating the immune microenvironment of GBM. J Neuro Oncol 2012;110(3): 359–68.
[47] Senft C, Priester M, Polacin M, Schröder K, Seifert V, Kögel D, Weissenberger J. Inhibition of the JAK-2/STAT3 signaling pathway impedes the migratory and invasive potential of human glioblastoma cells. J Neuro Oncol 2011;101(3):393–403.
[48] Hussain SF, Kong LY, Jordan J, Conrad C, Madden T, Fokt I, Priebe W, Heimberger AB. A novel small molecule inhibitor of signal transducers and activators of transcription 3 reverses immune tolerance in malignant glioma patients. Cancer Res 2007;67(20):9630–6.
[49] Wei J, Wang F, Kong LY, Xu S, Doucette T, Ferguson SD, Yang Y, McEnery K, Jethwa K, Gjyshi O, Qiao W. Mir-124 inhibits STAT3 signaling to enhance t cell–mediated immune clearance of glioma. Cancer Res 2013; 73(13):3913–26.
[50] Michaud-Levesque J, Bousquet-Gagnon N, Béliveau R. Quercetin abrogates IL-6/STAT3 signaling and inhibits glioblastoma cell line growth and migration. Exp Cell Res 2012;318(8):925–35.
[51] Bi L, Yu Z, Wu J, Yu K, Hong G, Lu Z, Gao S. Honokiol inhibits constitutive and inducible STAT3 signaling via PU.1-induced SHP1 expression in acute myeloid leukemia cells. Tohoku J Exp Med 2015;237(3):163–72.
[52] Sordillo LA, Sordillo PP, Helson L. Curcumin for the treatment of glioblastoma. Anticancer Res 2015; 35(12):6373–8.
[53] Shafabakhsh R, Asemi Z. Quercetin: a natural compound for ovarian cancer treatment. J Ovarian Res 2019; 12(1):55–64.
[54] Granato M, Rizzello C, Montani MS, Cuomo L, Vitillo M, Santarelli R, Gonnella R, D'Orazi G, Faggioni A, Cirone M. Quercetin induces apoptosis and autophagy in primary effusion lymphoma cells by inhibiting PI3K/AKT/MTOR and STAT3 signaling pathways. J Nutr Biochem 2017;41:124–36.
[55] Yu D, Ye T, Xiang Y, Shi Z, Zhang J, Lou B, Zhang F, Chen B, Zhou M. Quercetin inhibits epithelial–mesenchymal transition, decreases invasiveness and metastasis, and reverses IL-6 induced epithelial–mesenchymal transition, expression of MMP by inhibiting STAT3 signaling in pancreatic cancer cells. Oncotargets Ther 2017;10:4719–29.
[56] Wang G, Wang JJ, Chen XL, Du SM, Li DS, Pei ZJ, Lan H, Wu LB. The JAK2/STAT3 and mitochondrial pathways are essential for quercetin nanoliposome-induced C6 glioma cell death. Cell Death Dis 2013;4(8), e746.
[57] Wu L, Li J, Liu T, Li S, Feng J, Yu Q, Zhang J, Chen J, Zhou Y, Ji J, Chen K. Quercetin shows anti-tumor effect in hepatocellular carcinoma LM3 cells by abrogating JAK2/STAT3 signaling pathway. Cancer Med 2019; 8(10):4806–20.
[58] Cao HH, Tse AK, Kwan HY, Yu H, Cheng CY, Su T, Fong WF, Yu ZL. Quercetin exerts anti-melanoma activities and inhibits STAT3 signaling. Biochem Pharmacol 2014;87(3):424–34.
[59] Ma Z, Xia Y, Hu C, Yu M, Yi H. Quercetin promotes the survival of granulocytic myeloid-derived suppressor cells via the ESR2/STAT3 signaling pathway. Biomed Pharmacother 2020;125:109922.
[60] Seo HS, Ku JM, Choi HS, Choi YK, Woo JK, Kim M, Kim I, Na CH, HuR HA, Jang BH, Shin YC. Quercetin induces caspase-dependent extrinsic apoptosis through inhibition of signal transducer and activator of transcription 3 signaling in HER2-overexpressing BT-474 breast cancer cells. Oncol Rep 2016;36(1):31–42.

[61] Suzuki Y, Ito S, Sasaki R, Asahi M, Ishida Y. Resveratrol suppresses cell proliferation via inhibition of STAT3 phosphorylation and Mcl-1 and cIAP-2 expression in HTLV-1-infected T cells. Leuk Res 2013;37(12):1674–9.

[62] Kim C, Baek SH, Um JY, Shim BS, Ahn KS. Resveratrol attenuates constitutive STAT3 and STAT5 activation through induction of PTPε and SHP-2 tyrosine phosphatases and potentiates sorafenib-induced apoptosis in renal cell carcinoma. BMC Nephrol 2016;17(1):19.

[63] Kotha A, Sekharam M, Cilenti L, Siddiquee K, Khaled A, Zervos AS, Carter B, Turkson J, Jove R. Resveratrol inhibits Src and Stat3 signaling and induces the apoptosis of malignant cells containing activated Stat3 protein. Mol Cancer Ther 2006;5(3):621–9.

[64] Li D, Wang G, Jin G, Yao K, Zhao Z, Bie L, Guo Y, Li N, Deng W, Chen X, Chen B. Resveratrol suppresses colon cancer growth by targeting the AKT/STAT3 signaling pathway. Int J Mol Med 2019;43(1):630–40.

[65] Baek SH, Ko JH, Lee H, Jung J, Kong M, Lee JW, Lee J, Chinnathambi A, Zayed ME, Alharbi SA, Lee SG. Resveratrol inhibits STAT3 signaling pathway through the induction of SOCS-1: role in apoptosis induction and radiosensitization in head and neck tumor cells. Phytomedicine 2016;23(5):566–77.

[66] Hahn YI, Kim SJ, Choi BY, Cho KC, Bandu R, Kim KP, Kim DH, Kim W, Park JS, Han BW, Lee J. Curcumin interacts directly with the cysteine 259 residue of STAT3 and induces apoptosis in H-Ras transformed human mammary epithelial cells. Sci Rep 2018;8(1):6409.

[67] Alexandrow MG, Song LJ, Altiok S, Gray J, Haura EB, Kumar NB. Curcumin: a novel Stat 3 pathway inhibitor for chemoprevention of lung cancer. Eur J Cancer Prevent 2012;21(5):407.

[68] Glienke W, Maute L, Wicht J, Bergmann L. Curcumin inhibits constitutive STAT3 phosphorylation in human pancreatic cancer cell lines and downregulation of survivin/BIRC5 gene expression. Cancer Investig 2009;28 (2):166–71.

[69] Lin L, Hutzen B, Zuo M, Ball S, Deangelis S, Foust E, Pandit B, Ihnat MA, Shenoy SS, Kulp S, Li PK. Novel STAT3 phosphorylation inhibitors exhibit potent growth-suppressive activity in pancreatic and breast cancer cells. Cancer Res 2010;70(6):2445–54.

[70] Liu Y, Wang X, Zeng S, Zhang X, Zhao J, Zhang X, Chen X, Yang W, Yang Y, Dong Z, Zhu J. The natural polyphenol curcumin induces apoptosis by suppressing STAT3 signaling in esophageal squamous cell carcinoma. J Exp Clin Cancer Res 2018;37(1):303.

[71] Chung SS, Vadgama JV. Curcumin and epigallocatechin gallate inhibit the cancer stem cell phenotype via down-regulation of STAT3–NFκB signaling. Anticancer Res 2015;35(1):39–46.

[72] Chiablaem K, Lirdprapamongkol K, Keeratichamroen S, Surarit R, Svasti J. Curcumin suppresses vasculogenic mimicry capacity of hepatocellular carcinoma cells through STAT3 and PI3K/AKT inhibition. Anticancer Res 2014;34(4):1857–64.

[73] Bill MA, Nicholas C, Mace TA, Etter JP, Li C, Schwartz EB, Fuchs JR, Young GS, Lin L, Lin J, He L. Structurally modified curcumin analogs inhibit STAT3 phosphorylation and promote apoptosis of human renal cell carcinoma and melanoma cell lines. Plos One 2012;7(8), e40724.

[74] Bharti AC, Donato N, Aggarwal BB. Curcumin (diferuloylmethane) inhibits constitutive and IL-6-inducible STAT3 phosphorylation in human multiple myeloma cells. J Immunol 2003;171(7):3863–71.

[75] Wong KE, Ngai SC, Chan KG, Lee LH, Goh BH, Chuah LH. Curcumin nanoformulations for colorectal cancer: a review. Front Pharmacol 2019;10:152.

[76] Fan Y, Xue W, Schachner M, Zhao W. Honokiol eliminates glioma/glioblastoma stem cell-like cells via JAK-STAT3 signaling and inhibits tumor progression by targeting epidermal growth factor receptor. Cancers 2019;11(1):22.

[77] Pan J, Lee Y, Zhang Q, Xiong D, Wan TC, Wang Y, You M. Honokiol decreases lung cancer metastasis through inhibition of the STAT3 signaling pathway. Cancer Prevent Res 2017;10(2):133–41.

[78] Xiao X, Jiang K, Xu Y, Peng H, Wang Z, Liu S, Zhang G. (−)-Epigallocatechin-3-gallate induces cell apoptosis in chronic myeloid leukaemia by regulating Bcr/Abl-mediated P38-MAPK/JNK and JAK 2/STAT 3/AKT signalling pathways. Clin Exp Pharmacol Physiol 2019;46(2):126–36.

[79] Lin HY, Hou SC, Chen SC, Kao MC, Yu CC, Funayama S, Ho CT, Way TD. (−)-Epigallocatechin gallate induces Fas/CD95-mediated apoptosis through inhibiting constitutive and IL-6-induced JAK/STAT3 signaling in head and neck squamous cell carcinoma cells. J Agric Food Chem 2012;60(10):2480–9.

[80] Song S, Su Z, Xu H, Niu M, Chen X, Min H, Zhang B, Sun G, Xie S, Wang H, Gao Q. Luteolin selectively kills STAT3 highly activated gastric cancer cells through enhancing the binding of STAT3 to SHP-1. Cell Death Dis 2017;8(2), e2612.

[81] Palombo R, Savini I, Avigliano L, Madonna S, Cavani A, Albanesi C, Mauriello A, Melino G, Terrinoni A. Luteolin-7-glucoside inhibits IL-22/STAT3 pathway, reducing proliferation, acanthosis, and inflammation in keratinocytes and in mouse psoriatic model. Cell Death Dis 2016;7(8), e2344.

[82] Xia N, Chen G, Liu M, Ye X, Pan Y, Ge J, Mao Y, Wang H, Wang J, Xie S. Anti-inflammatory effects of luteolin on experimental autoimmune thyroiditis in mice. Exp Ther Med 2016;12(6):4049–54.

[83] Pandey MK, Sung B, Aggarwal BB. Betulinic acid suppresses STAT3 activation pathway through induction of protein tyrosine phosphatase SHP-1 in human multiple myeloma cells. Int J Cancer 2010;127(2):282–92.

[84] Su D, Gao YQ, Dai WB, Hu Y, Wu YF, Mei QX. Helicteric acid, oleanic acid, and betulinic acid, three triterpenes from Helicteres Angustifolia L., inhibit proliferation and induce apoptosis in HT-29 colorectal cancer cells via suppressing NF-κB and STAT3 signaling. Evid Based Complement Alternat Med 2017;2017:5180707.

[85] Wei N, Li J, Fang C, Chang J, Xirou V, Syrigos NK, Marks BJ, Chu E, Schmitz JC. Targeting colon cancer with the novel STAT3 inhibitor bruceantinol. Oncogene 2019;38(10):1676.

[86] Jung JE, Kim HS, Lee CS, Park DH, Kim YN, Lee MJ, Lee JW, Park JW, Kim MS, Ye SK, Chung MH. Caffeic acid and its synthetic derivative CADPE suppress tumor angiogenesis by blocking STAT3-mediated VEGF expression in human renal carcinoma cells. Carcinogenesis 2007;28(8):1780–7.

[87] Agilan B, Rajendra Prasad N, Kanimozhi G, Karthikeyan R, Ganesan M, Mohana S, Velmurugan D, Ananthakrishnan D. Caffeic acid inhibits chronic UVB-induced cellular proliferation through JAK-STAT 3 signaling in mouse skin. Photochem Photobiol 2016;92(3):467–74.

[88] Ren X, Liu J, Hu L, Liu Q, Wang D, Ning X. Caffeic acid phenethyl ester inhibits the proliferation of HEp2 cells by regulating Stat3/Plk1 pathway and inducing S phase arrest. Biol Pharm Bull 2019;42(10):1689–93.

[89] Gabriele E, Brambilla D, Ricci C, Regazzoni L, Taguchi K, Ferri N, Asai A, Sparatore A. New sulfurated derivatives of cinnamic acids and rosmaricine as inhibitors of STAT3 and NF-κB transcription factors. J Enzyme Inhib Med Chem 2017;32(1):1012–28.

[90] Pathak AK, Bhutani M, Nair AS, Ahn KS, Chakraborty A, Kadara H, Guha S, Sethi G, Aggarwal BB. Ursolic acid inhibits STAT3 activation pathway leading to suppression of proliferation and chemosensitization of human multiple myeloma cells. Mol Cancer Res 2007;5(9):943–55.

[91] Wang W, Zhao C, Jou D, LÜ J, Zhang C, Lin L, Lin J. Ursolic acid inhibits the growth of colon cancer-initiating cells by targeting STAT3. Anticancer Res 2013;33(10):4279–84.

[92] Shanmugam MK, Rajendran P, Li F, Nema T, Vali S, Abbasi T, Kapoor S, Sharma A, Kumar AP, Ho PC, Hui KM. Ursolic acid inhibits multiple cell survival pathways leading to suppression of growth of prostate cancer xenograft in nude mice. J Mol Med 2011;89(7):713.

[93] Liu T, Ma H, Shi W, Duan J, Wang Y, Zhang C, Li C, Lin J, Li S, Lv J, Lin L. Inhibition of STAT3 signaling pathway by ursolic acid suppresses growth of hepatocellular carcinoma. Int J Oncol 2017;51(2):555–62.

[94] Ye C, Zhao W, Li M, Zhuang J, Yan X, Lu Q, Chang C, Huang X, Zhou J, Xie B, Zhang Z. δ-tocotrienol induces human bladder cancer cell growth arrest, apoptosis and chemosensitization through inhibition of STAT3 pathway. Plos One 2015;10(4), e0122712.

[95] Rajendran P, Li F, Manu KA, Shanmugam MK, Loo SY, Kumar AP, Sethi G. γ-Tocotrienol is a novel inhibitor of constitutive and inducible STAT3 signalling pathway in human hepatocellular carcinoma: potential role as an antiproliferative, pro-apoptotic and chemosensitizing agent. Br J Pharmacol 2011;163(2):283–98.

[96] Chuang HY, Su YK, Liu HW, Chen CH, Chiu SC, Cho DY, Lin SZ, Chen YS, Lin CM. Preclinical evidence of STAT3 inhibitor, pacritinib, overcomes temozolomide resistance via down-regulating Mir-21-enriched exosomes from M2 glioblastoma-associated macrophages. J Clin Med 2019;8:959.

[97] Sordillo PP, Helson L. Curcumin and cancer stem cells: curcumin has asymmetrical effects on cancer and normal stem cells. Anticancer Res 2015;35(2):599–614.

[98] Iliopoulos D, Hirsch HA, Wang G, Struhl K. Inducible formation of breast cancer stem cells and their dynamic equilibrium with non-stem cancer cells via IL6 secretion. Proc Natl Acad Sci U S A 2011;108(4):1397–402.

[99] Lamano JB, Lamano JB, Li YD, DiDomenico JD, Choy W, Veliceasa D, Oyon DE, Fakurnejad S, Ampie L, Kesavabhotla K, Kaur R. Glioblastoma-derived IL6 induces immunosuppressive peripheral myeloid cell PD-L1 and promotes tumor growth. Clin Cancer Res 2019;25(12):3643–57.

[100] Li F, Wei L, Li S, Liu J. Indoleamine-2, 3-dioxygenase and interleukin-6 associated with tumor response to neoadjuvant chemotherapy in breast cancer. Oncotarget 2017;8(64):107844.

[101] Fahey E, Doyle SL. IL-1 family cytokine regulation of vascular permeability and angiogenesis. Front Immunol 2019;10:1426.

[102] Bent R, Moll L, Grabbe S, Bros M. Interleukin-1 beta—a friend or foe in malignancies? Int J Mol Sci 2018; 19(8):2155.
[103] Litmanovich A, Khazim K, Cohen I. The role of interleukin-1 in the pathogenesis of cancer and its potential as a therapeutic target in clinical practice. Oncol Ther 2018;6(2):109–27.
[104] Tanabe K, Kozawa O, Iida H. Midazolam suppresses interleukin-1β-induced interleukin-6 release from rat glial cells. J Neuroinflamm 2011;8(1):68.
[105] Spooren A, Mestdagh P, Rondou P, Kolmus K, Haegeman G, Gerlo S. IL-1β potently stabilizes IL-6 Mrna in human astrocytes. Biochem Pharmacol 2011;81(8):1004–15.
[106] Tarassishin L, Lim J, Weatherly DB, Angeletti RH, Lee SC. Interleukin-1-Induced changes in the glioblastoma secretome suggest its role in tumor progression. J Proteomics 2014;99:152–68.
[107] Gschwandtner M, Derler R, Midwood KS. More than just attractive: how CCL2 influences myeloid cell behavior beyond chemotaxis. Front Immunol 2019;10:2759.
[108] Nambiar DK, Aguilera T, Cao H, Kwok S, Kong C, Bloomstein J, Wang Z, Rangan VS, Jiang D, Von Eyben R, Liang R. Galectin-1-driven T cell exclusion in the tumor endothelium promotes immunotherapy resistance. J Clin Investig 2019;129(12):5553–67.
[109] Jachetti E, Caputo S, Mazzoleni S, Brambillasca CS, Parigi SM, Grioni M, Piras IS, Restuccia U, Calcinotto A, Freschi M, Bachi A. Tenascin-C protects cancer stem–like cells from immune surveillance by arresting T-cell activation. Cancer Res 2015;75(10):2095–108.
[110] Mirzaei R, Sarkar S, Dzikowski L, Rawji KS, Khan L, Faissner A, Bose P, Yong VW. Brain tumor-initiating cells export tenascin-C associated with exosomes to suppress T cell activity. Onco Immunol 2018;7(10), e1478647.
[111] Paugh BS, Bryan L, Paugh SW, Wilczynska KM, Alvarez SM, Singh SK, Kapitonov D, Rokita H, Wright S, Griswold-Prenner I, Milstien S. Interleukin-1 regulates the expression of sphingosine kinase 1 in glioblastoma cells. J Biol Chem 2009;284(6):3408–17.
[112] Sordillo LA, Sordillo PP, Helson L. Sphingosine kinase inhibitors as maintenance therapy of glioblastoma after ceramide-induced response. Anticancer Res 2016;36(5):2085–95.
[113] El Jamal A, Bougault C, Mebarek S, Magne D, Cuvillier O, Brizuela L. The role of sphingosine 1-phosphate metabolism in bone and joint pathologies and ectopic calcification. Bone 2019;130:115087.
[114] Schulz ME, Katunaric B, Hockenberry JC, Gutterman DD, Freed JK. Manipulation of the sphingolipid rheostat influences the mediator of flow-induced dilation in the human microvasculature. J Am Heart Assoc 2019;8(17), e013153.
[115] Chakinala RC, Khatri A, Gupta K, Koike K, Epelbaum O. Sphingolipids in COPD. Eur Respir Rev 2019; 28(154):190047.
[116] Sharma L, Nadella V, Kumar P, Gupta P, Gupta U, Tripathy SP, Potani S, Qadri SS, Prakash H. Sphingosine-1 phosphate (S-1P) promotes differentiation of naïve macrophages and enhances protective immunity against M. Tuberculosis. Front Immunol 2019;10:3085.
[117] Napolitano F, Rosa R, D'Amato V, Orsini RC, Ciciola P, Di Mauro C, Santaniello A, Marciano R, De Placido S, Bianco R. 60P investigating the role of nuclear sphingosine kinase 1 (Sphk1) in lung cancer. Ann Oncol 2018; 29(suppl_3):mdy047–009.
[118] Lewis CS, Voelkel-Johnson C, Smith CD. Targeting sphingosine kinases for the treatment of cancer. In: Advances in cancer research, vol. 140. Academic Press; 2018. p. 295–325.
[119] Acharya S, Yao J, Li P, Zhang C, Lowery FJ, Zhang Q, Guo H, Qu J, Yang F, Wistuba II, Piwnica-Worms H. Sphingosine kinase 1 signaling promotes metastasis of triple-negative breast cancer. Cancer Res 2019;79 (16):4211–26.
[120] Sibley CD, Morris EA, Kharel Y, Brown AM, Bevan DR, Lynch KR, Santos WL. Discovery of a Small side cavity in sphingosine kinase 2 that enhances inhibitor potency and selectivity. J Med Chem 2020;63(3):1178–98.
[121] Bandara G, Muñoz-Cano R, Tobío A, Yin Y, Komarow HD, Desai A, Metcalfe DD, Olivera A. Targeting Sphingosine kinase isoforms effectively reduces growth and survival of neoplastic mast cells with D816V-kit. Front Immunol 2018;9:631.
[122] Wang P, Yuan Y, Lin W, Zhong H, Xu K, Qi X. Roles of sphingosine-1-phosphate signaling in cancer. Cancer Cell Int 2019;19(1):295.
[123] Imbert C, Montfort A, Fraisse M, Marcheteau E, Gilhodes J, Martin E, Bertrand F, Marcellin M, Burlet-Schiltz O, de Peredo AG, Garcia V. Resistance of melanoma to immune checkpoint inhibitors is overcome by targeting the sphingosine kinase-1. Nat Commun 2020;11(1):1–4.

[124] Tea MN, Poonnoose SI, Pitson SM. Targeting the sphingolipid system as a therapeutic direction for glioblastoma. Cancers 2020;12(1):111.
[125] Vann LR, Payne SG, Edsall LC, Twitty S, Spiegel S, Milstien S. Involvement of sphingosine kinase in TNF-α-stimulated tetrahydrobiopterin biosynthesis in C6 glioma cells. J Biol Chem 2002;277(15):12649–56.
[126] Young N, Van Brocklyn JR. Roles of sphingosine-1-phosphate (S1P) receptors in malignant behavior of glioma cells. Differential effects of S1P2 on cell migration and invasiveness. Exp Cell Res 2007;313(8):1615–27.
[127] Van Brocklyn JR, Jackson CA, Pearl DK, Kotur MS, Snyder PJ, Prior TW. Sphingosine kinase-1 expression correlates with poor survival of patients with glioblastoma multiforme: roles of sphingosine kinase isoforms in growth of glioblastoma cell lines. J Neuropathol Exp Neurol 2005;64(8):695–705.
[128] Yoshida Y, Nakada M, Sugimoto N, Harada T, Hayashi Y, Kita D, Uchiyama N, Hayashi Y, Yachie A, Takuwa Y, Hamada JI. Sphingosine-1-phosphate receptor type 1 regulates glioma cell proliferation and correlates with patient survival. Int J Cancer 2010;126(10):2341–52.
[129] Abuhusain HJ, Matin A, Qiao Q, Shen H, Kain N, Day BW, Stringer BW, Daniels B, Laaksonen MA, Teo C, McDonald KL. A metabolic shift favoring sphingosine 1-phosphate at the expense of ceramide controls glioblastoma angiogenesis. J Biol Chem 2013;288(52):37355–64.
[130] Barra G, Lepore A, Gagliardi M, Somma D, Matarazzo MR, Costabile F, Pasquale G, Mazzoni A, Gallo C, Nuzzo G, Annunziato F. Sphingosine kinases promote IL-17 expression in human T lymphocytes. Sci Rep 2018; 8(1):13233.
[131] Han X, Yang Q, Lin L, Xu C, Zheng C, Chen X, Han Y, Li M, Cao W, Cao K, Chen Q. Interleukin-17 enhances immunosuppression by mesenchymal stem cells. Cell Death Diff 2014;21(11):1758.
[132] Riboni L, Abdel Hadi L, Navone SE, Guarnaccia L, Campanella R, Marfia G. Sphingosine-1-phosphate in the tumor microenvironment: a signaling hub regulating cancer hallmarks. Cells 2020;9(2):337.
[133] Brunkhorst R, Vutukuri R, Pfeilschifter W. Fingolimod for the Treatment of neurological diseases—state of play and future perspectives. Front Cell Neurosci 2014;8:283.
[134] Bonhoure E, Lauret A, Barnes DJ, Martin C, Malavaud B, Kohama T, Melo JV, Cuvillier O. Sphingosine kinase-1 is a downstream regulator of imatinib-induced apoptosis in chronic myeloid leukemia cells. Leukemia 2008; 22(5):971.
[135] Alemany R, Kleuser B, Ruwisch L, Danneberg K, Lass H, Hashemi R, Spiegel S, Jakobs KH, Meyer zu Heringdorf D. Depolarisation induces rapid and transient formation of intracellular sphingosine-1-phosphate. FEBS Lett 2001;509(2):239–44.
[136] Pankova-Kholmyansky I, Flescher E. Potential new antimalarial chemotherapeutics based on sphingolipid metabolism. Chemotherapy 2006;52(4):205–9.
[137] Sun WY, Abeynaike LD, Escarbe S, Smith CD, Pitson SM, Hickey MJ, Bonder CS. Rapid histamine-induced neutrophil recruitment is sphingosine kinase-1 dependent. Am J Pathol 2012;180(4):1740–50.
[138] Magli E, Corvino A, Fiorino F, Frecentese F, Perissutti E, Saccone I, Santagada V, Caliendo G, Severino B. Design of sphingosine kinases inhibitors: challenges and recent developments. Curr Pharm Des 2019;25(9):956–68.
[139] Cao M, Ji C, Zhou Y, Huang W, Ni W, Tong X, Wei JF. Sphingosine kinase inhibitors: a patent review. Int J Mol Med 2018;41(5):2450–60.
[140] Khan FI, Lai D, Anwer R, Azim I, Khan MK. Identifying novel sphingosine kinase 1 inhibitors as therapeutics against breast cancer. J Enzyme Inhib Med Chem 2020;35(1):172–86.
[141] Yeh HW, Lee SS, Chang CY, Lang YD, Jou YS. A new switch for TGFβ in cancer. Cancer Res 2019;79(15): 3797–805.
[142] Roy LO, Poirier MB, Fortin D. Differential expression and clinical significance of transforming growth factor-beta isoforms in GBM tumors. Int J Mol Sci 2018;19(4):1113.
[143] Li J, Shen C, Wang X, Lai Y, Zhou K, Li P, Liu L, Che G. Prognostic value of TGF-β in lung cancer: systematic review and meta-analysis. BMC Cancer 2019;19(1):691.
[144] Teicher BA. Malignant cells, directors of the malignant process: role of transforming growth factor-beta. Cancer Metastasis Rev 2001;20(1–2):133–43.
[145] Seystahl K, Papachristodoulou A, Burghardt I, Schneider H, Hasenbach K, Janicot M, Roth P, Weller M. Biological role and therapeutic targeting of TGF-β3 in glioblastoma. Mol Cancer Ther 2017;16(6):1177–86.
[146] Anido J, Sáez-Borderías A, Gonzàlez-Juncà A, Rodón L, Folch G, Carmona MA, Prieto-Sánchez RM, Barba I, Martínez-Sáez E, Prudkin L, Cuartas I. TGF-β receptor inhibitors target the CD44high/Id1high glioma-initiating cell population in human glioblastoma. Cancer Cell 2010;18(6):655–68.

[147] Dimeloe S, Gubser P, Loeliger J, Frick C, Develioglu L, Fischer M, Marquardsen F, Bantug GR, Thommen D, Lecoultre Y, Zippelius A. Tumor-derived TGF-β inhibits mitochondrial respiration to suppress IFN-γ production by human CD4+ T cells. Sci Signal 2019;12(599):eaav3334.
[148] Papachristodoulou A, Silginer M, Weller M, Schneider H, Hasenbach K, Janicot M, Roth P. Therapeutic targeting of TGFβ ligands in glioblastoma using novel antisense oligonucleotides reduces the growth of experimental gliomas. Clin Cancer Res 2019;25(23):7189–201.
[149] Batlle E, Massagué J. Transforming growth factor-β signaling in immunity and cancer. Immunity 2019; 50(4):924–40.
[150] Huynh LK, Hipolito CJ, ten Dijke P. A perspective on the development of TGF-β inhibitors for cancer treatment. Biomolecules 2019;9(11):743.
[151] Xiao A, Brenneman B, Floyd D, Comeau L, Spurio K, Olmez I, Lee J, Nakano I, Godlewski J, Bronisz A, Kagaya N. Statins affect human glioblastoma and other cancers through TGF-β inhibition. Oncotarget 2019;10(18):1716.
[152] Avouac J, Pezet S, Gonzalez V, Baudoin L, Cauvet A, Ruiz B, Boleto G, Brandely ML, Elmerich M, Allanore Y. Estrogens counteract the profibrotic effects of TGF-β and their inhibition exacerbates experimental dermal fibrosis. J Investig Dermatol 2020;140(3):593–601.
[153] Bruckheimer EM, Kyprianou N. Dihydrotestosterone enhances transforming growth factor-β-induced apoptosis in hormone-sensitive prostate cancer cells. Endocrinology 2001;142(6):2419–26.
[154] Liu Y, Dai E, Yang J. Quercetin suppresses glomerulosclerosis and TGF-β signaling in a rat model. Mol Med Rep 2019;19(6):4589–96.
[155] Saidi A, Kasabova M, Vanderlynden L, Wartenberg M, Kara-Ali GH, Marc D, Lecaille F, Lalmanach G. Curcumin inhibits the TGF-β1-dependent differentiation of lung fibroblasts via PPARγ-driven upregulation of cathepsins B and L. Sci Rep 2019;9(1):491.
[156] Vengoji R, Macha MA, Batra SK, Shonka NA. Natural products: a hope for glioblastoma patients. Oncotarget 2018;9(31):22194.
[157] Song Y, Chen Y, Li Y, Lyu X, Cui J, Cheng Y, Zheng T, Zhao L, Zhao G. Resveratrol suppresses epithelial-mesenchymal transition in GBM by regulating Smad-dependent signaling. Biomed Res Int 2019;2019:1321973.
[158] Moore BB, Kunkel SL. Attracting attention: discovery of IL-8/CXCL8 and the birth of the chemokine field. J Immunol 2019;202(1):3.
[159] Hasan T, Caragher SP, Shireman JM, Park CH, Atashi F, Baisiwala S, Lee G, Guo D, Wang JY, Dey M, Wu M. Interleukin-8/CXCR2 signaling regulates therapy-induced plasticity and enhances tumorigenicity in glioblastoma. Cell Death Dis 2019;10(4):1–7.
[160] Burz C, Bojan A, Balacescu L, Pop VV, Silaghi C, Lupan I, Aldea C, Sur D, Samasca G, Cainap C, Chiorean B. Interleukin 8 as predictive factor for response to chemotherapy in colorectal cancer patients. Acta Clin Belg 2021;76(2):113–8.
[161] Limpakan S, Wongsirisin P, Yodkeeree S, Chakrabandhu B, Chongruksut W, Limtrakul P. Interleukin-8 associated with chemosensitivity and poor chemotherapeutic response to gastric cancer. J Gastrointest Oncol 2019;10(6):1120.
[162] Qazi BS, Tang K, Qazi A. Recent advances in underlying pathologies provide insight into interleukin-8 expression-mediated inflammation and angiogenesis. Int J Inflamm 2011;2011:908468.
[163] Sharma I, Singh A, Siraj F, Saxena S. IL-8/CXCR1/2 signalling promotes tumor cell proliferation, invasion and vascular mimicry in glioblastoma. J Biomed Sci 2018;25(1):62.
[164] Dwyer J, Hebda JK, Le Guelte A, Galan-Moya EM, Smith SS, Azzi S, Bidere N, Gavard J. Glioblastoma cell-secreted interleukin-8 induces brain endothelial cell permeability via CXCR2. Plos One 2012;7(9), e45562.
[165] David J, Dominguez C, Hamilton D, Palena C. The IL-8/IL-8R axis: a double agent in tumor immune resistance. Vaccines 2016;4(3):22.
[166] Somasundaram A, Cillo A, Oliveri L, Herman J, Kirkwood J, Ferris RL, Bruno TC, Vignali D. IL-6, IL-8 drive LAG3/PD1 immune suppression on effector and naïve, peripheral blood CD8+ T cells in cancer patients. J Immunol 2019;202(1 Suppl):195–9.
[167] Feng L, Qi Q, Wang P, Chen H, Chen Z, Meng Z, Liu L. Serum levels of IL-6, IL-8, and IL-10 are indicators of prognosis in pancreatic cancer. J Int Med Res 2018;46(12):5228–36.
[168] Huettner C, Czub S, Kerkau S, Roggendorf W, Tonn JC. Interleukin 10 is expressed in human gliomas in vivo and increases glioma cell proliferation and motility in vitro. Anticancer Res 1997;17(5A):3217–24.
[169] Qi L, Yu H, Zhang Y, Zhao D, Lv P, Zhong Y, Xu Y. IL-10 secreted by M2 macrophage promoted tumorigenesis through interaction with JAK2 in glioma. Oncotarget 2016;7(44):71673.

[170] Marriott I, Burmeister AR. The interleukin-10 family of cytokines and their role in the CNS. Front Cell Neurosci 2018;12:458.
[171] Singh SK, Mishra MK, Eltoum IE, Bae S, Lillard JW, Singh R. CCR5/CCL5 axis interaction promotes migratory and invasiveness of pancreatic cancer cells. Sci Rep 2018;8(1):1–2.
[172] An G, Wu F, Huang S, Feng L, Bai J, Gu S, Zhao X. Effects of CCL5 on the biological behavior of breast cancer and the mechanisms of its interaction with tumor-associated macrophages. Oncol Rep 2019;42(6):2499–511.
[173] Jiao X, Nawab O, Patel T, Kossenkov AV, Halama N, Jaeger D, Pestell RG. Recent advances targeting CCR5 for cancer and its role in immuno-oncology. Cancer Res 2019;79(19):4801–7.
[174] Fujimoto Y, Inoue N, Morimoto K, Watanabe T, Hirota S, Imamura M, Matsushita Y, Katagiri T, Okamura H, Miyoshi Y. Significant association between high serum CCL5 levels and better disease-free survival of patients with early breast cancer. Cancer Sci 2020;111(1):209.
[175] Walens A, DiMarco AV, Lupo R, Kroger BR, Damrauer JS, Alvarez JV. CCL5 promotes breast cancer recurrence through macrophage recruitment in residual tumors. Elife 2019;8, e43653.
[176] Zhao L, Wang Y, Xue Y, Lv W, Zhang Y, He S. Critical roles of chemokine receptor CCR5 in regulating glioblastoma proliferation and invasion. Acta Biochim Biophys Sin 2015;47(11):890–8.
[177] Kranjc MK, Novak M, Pestell RG, Lah TT. Cytokine CCL5 and receptor CCR5 axis in glioblastoma multiforme. Radiol Oncol 2019;53(4):397–406.
[178] Wu CY, Chen CH, Lin CY, Feng LY, Lin YC, Wei KC, Huang CY, Fang JY, Chen PY. CCL5 of glioma-associated microglia/macrophages regulates glioma migration and invasion via calcium-dependent matrix metalloproteinase-2. Neuro Oncol 2020;22(2):253–66.
[179] Miao M, De Clercq E, Li G. Clinical significance of chemokine receptor antagonists. Exp Opin Drug Metab Toxicol 2020;16(1):11–30.
[180] Pestell RG, Cristofanilli M, Rui H, Jiao X, Wang M. Leronlimab, a humanized monoclonal antibody to CCR5, restrains breast cancer metastasis and enhances cell death induced by DNA damaging chemotherapies. In: Proceedings of the 2019 San Antonio breast cancer symposium. San Antonio, TX. Philadelphia (PA): AACR; 2019. Dec 10-14. Cancer Res 2020;80(4 Suppl):Abstract nr PD4-04.
[181] Ha ET, Antonios JP, Soto H, Prins RM, Yang I, Kasahara N, Liau LM, Kruse CA. Chronic inflammation drives glioma growth: cellular and molecular factors responsible for an immunosuppressive microenvironment. Neuroimmunol Neuroinflamm 2014;1:66–76.
[182] Montfort A, Colacios C, Levade T, Andrieu-Abadie N, Meyer N, Ségui B. The TNF paradox in cancer progression and immunotherapy. Front Immunol 2019;10:1818.
[183] Lu C, Chen X, Wang Q, Xu X, Xu B. TNFα promotes glioblastoma A172 cell mitochondrial apoptosis via augmenting mitochondrial fission and repression of MAPK–ERK–YAP signaling pathways. Oncotargets Ther 2018;11:7213.
[184] Beukelman T, Xie F, Chen L, Horton DB, Lewis JD, Mamtani R, Mannion MM, Saag KG, Curtis JR. Risk of malignancy associated with paediatric use of tumour necrosis factor inhibitors. Ann Rheum Dis 2018;77(7):1012–6.
[185] Guo G, Gong K, Puliyappadamba VT, Panchani N, Pan E, Mukherjee B, Damanwalla Z, Bharia S, Hatanpaa KJ, Gerber DE, Mickey BE. Efficacy of EGFR plus TNF inhibition in a preclinical model of temozolomide-resistant glioblastoma. Neuro Oncol 2019;21(12):1529–39.
[186] Sordillo LA, Sordillo PP. Optical spectroscopy of tryptophan metabolites in neurodegenerative disease. In: Neurophotonics and biomedical spectroscopy. Elsevier; 2019, p. 137–157.
[187] Wang X, Yang L, Huang F, Zhang Q, Liu S, Ma L, You Z. Inflammatory cytokines IL-17 and TNF-α up-regulate PD-L1 expression in human prostate and colon cancer cells. Immunol Lett 2017;184:7–14.
[188] Yu CP, Fu SF, Chen X, Ye J, Ye Y, Kong LD, Zhu Z. The clinicopathological and prognostic significance of IDO1 expression in human solid tumors: evidence from a systematic review and meta-analysis. Cell Physiol Biochem 2018;49(1):134–43.
[189] Mitsuka K, Kawataki T, Satoh E, Asahara T, Horikoshi T, Kinouchi H. Expression of indoleamine 2,3-dioxygenase and correlation with pathological malignancy in gliomas. Neurosurgery 2013;72(6):1031–9.
[190] Mbongue J, Nicholas D, Torrez T, Kim NS, Firek A, Langridge W. The role of indoleamine 2, 3-dioxygenase in immune suppression and autoimmunity. Vaccines 2015;3(3):703–29.
[191] Badawy AA. Hypothesis kynurenic and quinolinic acids: the main players of the kynurenine pathway and opponents in inflammatory disease. Med Hypotheses 2018;118:129–38.

[192] Moon YW, Hajjar J, Hwu P, Naing A. Targeting the indoleamine 2, 3-dioxygenase pathway in cancer. J Immunother Cancer 2015;3(1):51.
[193] Munn DH. Indoleamine 2, 3-dioxygenase, Tregs and cancer. Curr Med Chem 2011;18(15):2240–6.
[194] Godin-Ethier J, Hanafi LA, Piccirillo CA, Lapointe R. Indoleamine 2, 3-dioxygenase expression in human cancers: clinical and immunologic perspectives. Clin Cancer Res 2011;17(22):6985–91.
[195] Routy JP, Routy B, Graziani GM, Mehraj V. The kynurenine pathway is a double-edged sword in immune-privileged sites and in cancer: implications for immunotherapy. Int J Tryptophan Res 2016;9. IJTR-S38355.
[196] O'Rourke DM, Nasrallah MP, Desai A, Melenhorst JJ, Mansfield K, Morrissette JJ, Martinez-Lage M, Brem S, Maloney E, Shen A, Isaacs R. A single dose of peripherally infused EGFRvIII-directed CAR T cells mediates antigen loss and induces adaptive resistance in patients with recurrent glioblastoma. Sci Transl Med 2017;9 (399), eaaa0984.
[197] Chuntova P, Downey KM, Hegde B, Almeida ND, Okada H. Genetically engineered T-cells for malignant glioma: overcoming the barriers to effective immunotherapy. Front Immunol 2019;9:3062.
[198] Sordillo PP, Sordillo LA, Helson L. The kynurenine pathway: a primary resistance mechanism in patients with glioblastoma. Anticancer Res 2017;37(5):2159–71.
[199] Prendergast GC, Malachowski WP, DuHadaway JB, Muller AJ. Discovery of IDO1 inhibitors: from bench to bedside. Cancer Res 2017;77(24):6795–811.
[200] Sari S, Tomek P, Leung E, Reynisson J. Discovery and characterisation of dual inhibitors of tryptophan 2, 3-dioxygenase (TDO2) and indoleamine 2, 3-dioxygenase 1 (IDO1) using virtual screening. Molecules 2019;24(23):4346.
[201] Komiya T, Huang CH. Updates in clinical development of indoleamine 2,3-dioxygenase inhibitors for human cancers. Front Oncol 2018;8:423.
[202] Sordillo LA, Sordillo PP, Helson L, Inventors. Treatment for Glioblastoma. United States patent US 10,485,768. 2019 Nov 26.
[203] Du L, Xing Z, Tao B, Li T, Yang D, Li W, Zheng Y, Kuang C, Yang Q. Both IDO1 and TDO contribute to the malignancy of gliomas via the Kyn–Ahr–AQP4 signaling pathway. Signal Transduct Target Ther 2020; 5(1):1–3.
[204] Muller AJ, Manfredi MG, Zakharia Y, Prendergast GC. Inhibiting IDO pathways to treat cancer: lessons from the ECHO-301 trial and beyond. In: Seminars in immunopathology, vol. 41, No. 1. Berlin Heidelberg: Springer; 2019. p. 41–8. Jan 11.
[205] Lebbé C, Meyer N, Mortier L, Marquez-Rodas I, Robert C, Rutkowski P, Menzies AM, Eigentler T, Ascierto PA, Smylie M, Schadendorf D. Evaluation of two dosing regimens for nivolumab in combination with ipilimumab in patients with advanced melanoma: results from the phase IIIb/IV checkmate 511 trial. J Clin Oncol 2019; 37(11):867.
[206] Omuro A, Vlahovic G, Lim M, Sahebjam S, Baehring J, Cloughesy T, Voloschin A, Ramkissoon SH, Ligon KL, Latek R, Zwirtes R. Nivolumab with or without ipilimumab in patients with recurrent glioblastoma: results from exploratory phase I cohorts of CheckMate 143. Neuro Oncol 2017;20(5):674–86.
[207] de Groot J, Penas-Prado M, Alfaro-Munoz KD, Hunter K, Pei B, O'brien B, Weathers SP, Loghin M, Kamiya Matsouka C, Yung WK, Mandel J. Window-of-opportunity clinical trial of pembrolizumab in patients with recurrent glioblastoma reveals predominance of immune-suppressive macrophages. Neuro Oncol 2019; 22(4):539–49.
[208] Wang L, Zhang C, Zhang Z, Han B, Shen Z, Li L, Liu S, Zhao X, Ye F, Zhang Y. Specific clinical and immune features of CD68 in glioma via 1,024 samples. Cancer Manag Res 2018;10:6409.
[209] Wang X, Guo G, Guan H, Yu Y, Lu J, Yu J. Challenges and potential of PD-1/PD-L1 checkpoint blockade immunotherapy for glioblastoma. J Exp Clin Cancer Res 2019;38(1):87.
[210] Romani M, Pistillo MP, Carosio R, Morabito A, Banelli B. Immune checkpoints and innovative therapies in glioblastoma. Front Oncol 2018;8:464.
[211] Huang J, Liu F, Liu Z, Tang H, Wu H, Gong Q, Chen J. Immune checkpoint in glioblastoma: promising and challenging. Front Pharmacol 2017;8:242.
[212] Wainwright DA, Chang AL, Dey M, Balyasnikova IV, Kim CK, Tobias A, Cheng Y, Kim JW, Qiao J, Zhang L, Han Y. Durable therapeutic efficacy utilizing combinatorial blockade against IDO, CTLA-4, and PD-L1 in mice with brain tumors. Clin Cancer Res 2014;20(20):5290–301.

[213] Harris-Bookman S, Mathios D, Martin AM, Xia Y, Kim E, Xu H, Belcaid Z, Polanczyk M, Barberi T, Theodros D, Kim J. Expression of LAG-3 and efficacy of combination treatment with anti-LAG-3 and anti-PD-1 monoclonal antibodies in glioblastoma. Int J Cancer 2018;143(12):3201–8.
[214] Long L, Zhang X, Chen F, Pan Q, Phiphatwatchara P, Zeng Y, Chen H. The promising immune checkpoint LAG-3: from tumor microenvironment to cancer immunotherapy. Genes Cancer 2018;9(5–6):176.
[215] Puhr HC, Ilhan-Mutlu A. New emerging targets in cancer immunotherapy: the role of LAG3. ESMO Open 2019;4(2):e000482.
[216] Kim JE, Patel MA, Mangraviti A, Kim ES, Theodros D, Velarde E, Liu A, Sankey EW, Tam A, Xu H, Mathios D. Combination therapy with anti-PD-1, anti-TIM-3, and focal radiation results in regression of murine gliomas. Clin Cancer Res 2017;23(1):124–36.

CHAPTER

12

Role of cell quiescence in glioblastoma cytotoxic resistance and strategies for therapeutic intervention

Ryan J. Atkins[a], *Stanley S. Stylli*[a,b], *Niall M. Corcoran*[a], *Christopher M. Hovens*[a], *and Theo Mantamadiotis*[a,c]

[a]Department of Surgery (RMH), The University of Melbourne, Parkville, VIC, Australia
[b]Department of Neurosurgery, The Royal Melbourne Hospital, Parkville, VIC, Australia
[c]Department of Microbiology & Immunology, The University of Melbourne, Parkville, VIC, Australia

Abstract

The understanding of how cancer stem cells behave is important in understanding how tumors are initiated and how they recur following initial treatment. More specifically to understand how cancer stem cells behave, the different signaling mechanisms orchestrating their growth, cell cycle dynamics, differentiation, trans-differentiation, and survival following cytotoxic challenges need to be deciphered. Ultimately, this will advance the ability to predict how these cells will behave in individual patients and under different therapeutic conditions. Glioblastoma (GBM) stem cells, which have low proliferative capacity, exhibiting characteristics similar to quiescent stem cells are a likely source of treatment-resistant cells that contribute to tumor recurrence, which occurs in all patients. The molecular mechanisms regulating quiescent tumor cells are not well understood. Investigating the molecular mechanisms that regulate quiescence or slow cell cycling within a cell population requires that specific cells be isolated, examined, and compared to faster cycling cells. Identification of cells with different cycling rates, in vitro, has been made possible by the development of nontoxic vital fluorescent dyes, making it possible to track and select cells with different proliferation rates. In this review, we illustrate the use of vital dyes to select slow-cycling quiescent-like GBM cells to examine key biological mechanisms responsible for drug and radiation resistance.

Abbreviations

BrdU 5′-bromo-2′-deoxyuridine
CFSE carboxyfluorescein succinimidyl ester
EMT epithelial to mesenchymal transition

Glioblastoma Resistance to Chemotherapy: Molecular Mechanisms and Innovative Reversal Strategies
https://doi.org/10.1016/B978-0-12-821567-8.00003-8

FACS fluorescence-activated cell sorting
GBM glioblastoma
MGMT O^6-methylguanine DNA methyltransferase
TMZ temozolomide

Conflict of interest

No potential conflicts of interest were disclosed.

Introduction

Gliomas are the most prevalent primary central nervous system malignancy in adults [1,2]. Pathologically, gliomas are tumors that demonstrate immunohistochemical, histological, and ultrastructural signs of glial cell origin [3] and are broadly classified on the basis of the dominant neural cell type present, namely, astrocytic, ependymal, or oligodendroglial [4]. Clinically, there are four glioma grades, with the most aggressive being grade IV or glioblastoma (GBM), the most common and deadly brain tumor.

Despite years of research and advances in multimodal therapies, GBM patients still face a dismal prognosis, with an average survival of 14 months postdiagnosis and less than 5% survival at 5 years [5,6]. The lack of identifiable risk factors for developing GBM prevents any rational attempt to routinely screen groups at risk or to implement effective preventative strategies. The invasive nature of GBM indicates that cells capable of seeding new tumors migrate beyond the visible tumor margins, thus limiting surgical resection [7]. Indeed, when neurosurgeon Harvey Cushing and physician Percival Bailey first systematically described the histopathology of GBM in 1926, they recognized that a surgical cure would be difficult if not impossible [8].

Cytotoxic adjuvant GBM therapies including radiotherapy and the standard chemotherapeutic drug used in GBM, temozolomide (TMZ), elicit their therapeutic effect by damaging DNA, causing apoptosis and/or necrosis in the most susceptible cells, which are the dividing cells [9,10]. DNA repair mechanisms in these cytotoxic-sensitive cells are unable to sufficiently repair DNA damage between cell divisions [11–13]. However, the slow-cycling/quiescent subpopulation of treatment-resistant cells is largely spared by these treatments, and ultimately contributes to tumor recurrence [14].

Understanding the molecular mechanisms that regulate oncogenesis, tumor progression, and cytotoxic resistance is critical for identifying novel therapies. Recent developments in neuro-oncology have expanded our understanding of GBM pathogenesis and paved the way toward the development of precision medicine strategies for GBM [15]. Extensive characterization of GBM mutations has provided a better understanding of the genetic aberrations and signaling pathways that underlie its development, while the discovery of rare slow-cycling cancer stem cells provides a cellular drug target to afford long-term suppression of tumor growth and ultimately prevention of GBM recurrence [16]. Cancer stem cells are believed to retain many of the traits typically associated with their nonmalignant stem cell counterparts, including a reversible state of cellular quiescence [17]. Several studies have implicated a role for quiescent tumor cells in treatment resistance and tumor recurrence [18,19].

Resistance mechanisms in GBM

GBM tumor location within the central nervous system and its ability to defy therapeutic intervention renders this cancer difficult to treat. Its diffuse and infiltrative growth, high stem cell component, and location within a highly sensitive and specialized organ with limited repair capacity all come together in a veritable "perfect storm" of malignancy. However, the increasing characterization of GBM tumors has given us insights into potential weaknesses and novel ways to treat the disease. These include improved surgical imaging techniques, novel reprogramming of tumor stromal components, using established chemotherapeutics in new ways, and radical gene transfer therapy. Numerous innovative treatments have been proposed, including photodynamic therapy, but the current standard treatment with maximal safe surgery and concomitant radiotherapy and TMZ chemotherapy remains the norm [5]. One predictor of improved outcome with TMZ treatment is O^6-methylguanine DNA methyltransferase (MGMT) promoter methylation. A study by Hegi and colleagues reported that within the MGMT methylated subgroup, those treated with radiotherapy and TMZ had a significantly improved median overall survival of 21.7 months (95% CI, 17.4–30.4) compared to 15.3 months (95% CI, 13.0–20.9) for those treated with radiotherapy alone [20]. It is now likely that this is just one example of what might represent a genome-wide epigenetic phenomenon, but how the methylation status of various genes influences tumor activity remains to be fully elucidated. The correlation of MGMT promoter region methylation with improved treatment-related survival emphasizes the potential of precision-targeted therapies.

While new (effective) targets continue to be identified, it is important to optimize currently available treatments. Advances have been made in precision ionizing radiation delivery including stereotactic radiosurgery (SRS), intensity-modulated radiation therapy (IMRT), three-dimensional (3D) conformal radiation therapy, and proton beam therapy [21–24]. However, given glial tumors comprise a diffuse and often large tumor bulk, advances in this modality alone may have a limited impact. Therefore, GBM cells must be resensitized not only to radiotherapy, but also to other currently available cytotoxic therapies such as TMZ. As most cancer cells are deficient in at least one DNA repair pathway [25,26], therapeutic targeting of these pathways leads to compounding genomic damage that ultimately causes cytotoxicity in susceptible cancer cells via a process referred to as synthetic lethality [27], where cancer cells deficient in DNA repair can be selectively targeted, while sparing normal cells with intact DNA repair systems.

Resistance to radiation

Radiotherapy remains a keystone treatment for GBM despite advances in the understanding of mechanisms contributing to oncogenesis. The radiation dose delivered to a patient's tumor is limited by the sensitivity of the noncancerous cells in the adjacent tissue. Strategies to improve the efficacy of radiotherapy aim to sensitize tumor cells and cancer stem cells, so a lower radiation dose can be used to limit the effect of radiation on the surrounding healthy tissue [28,29]. Radiation therapy triggers cell death by damaging DNA via the induction of single- and double-strand breaks, both directly and as a by-product of generating oxygen-free radicals, as well as causing oxidized base damage and the generation of abasic sites. As such, the mechanisms to repair these radiation-induced insults are shared with chemotherapeutic resistance pathways [11,30]. Single-strand breaks are intermediates of the base excision repair pathway that occurs following cleavage of a damaged residue from the genome. In GBM cells, chemo- and radiotherapy repair mechanisms converge as poly(ADP-ribose) polymerase binds single-strand breaks and recruits additional proteins to resynthesize and rejoin the damaged strands [31]. Inhibition of base excision repair pathway proteins polymerase leads to enhanced formation of double-strand DNA breaks and increases radiosensitivity of glioma cells both in vitro and in vivo [32].

In addition to DNA repair pathways, the other arm of the DNA damage response consists of cell cycle checkpoint proteins, which halt cell cycle progression in recognition of DNA damage. Cancer cells are almost universally deficient in G1/S cell cycle checkpoint control, largely owing to mutations in *TP53* or proteins associated regulating p53-associated functions. Targeting additional checkpoint proteins such as checkpoint kinase-1 (Chk1) and 2 (Chk2) at S and G2/M cell cycle checkpoints denies tumor cells the opportunity to pause and repair their genome, forcing them prematurely into mitosis, causing further DNA damage, and triggering apoptosis [33,34]. In vivo models using the dual Chk1 and Chk2 inhibitor AZD7762 showed a significant inhibition of tumor growth in xenografts, and subsequent clinical trials have also shown efficacy in potentiating patient response to conventional chemo- and radiotherapy [33,35]. The selectivity of targeting DNA repair pathways and checkpoint proteins will depend on the molecular profile of the individual tumor, and treatment schedules will need to be tailored on a patient-to-patient basis.

Like many solid tumors, GBM tumors harbor many hypoxic regions, and the reduced oxygen concentration limits the efficacy of radiotherapy due to a reduced production of DNA-damaging free radicals, indirectly protecting the tumor cells. At about 1% O_2, cells exhibit enhanced resistance to radiation [36]. In vitro and in vivo experiments have shown that transient exposure of hypoxic cells to increased levels of O_2 increases tumor cell sensitivity to radiation, even if radiation is administered after cells have returned to hypoxic levels—an effect thought to be mediated by hypoxic-inducible factor-1 (HIF1) [36]. In this study, hyperoxic exposure did not sensitize human astrocytes in vitro, suggesting tumor selective efficacy.

Chemotherapy resistance

Numerous chemotherapeutics have since been tested in GBM patients, but restrictions including insufficient drug penetration into the central nervous system and dose-limiting side effects have limited their therapeutic efficacy. The critical function of the brain, combined with the presence of the blood–brain barrier, poses a unique challenge for drug delivery systems: the drugs must not only be small enough to cross the blood–brain barrier, but also must spare healthy brain cells from cytotoxicity. As such, chemotherapy for GBM has been typically limited to small cytotoxic nitrosourea-class compounds and alkylating agents that kill rapidly dividing cells, rather than specifically targeting cancer cells, which are the same strategy used for all cancers, where no targeted therapy exists. This paradigm may increase patient survival by essentially debulking the tumor, but this strategy ultimately fails to eliminate quiescent cancer stem cells which are the source of recurrent tumors.

There are a variety of mechanisms by which quiescent tumor and GBM stem cells resist chemotherapy, including upregulation of DNA repair proteins and increased activity of multidrug efflux pumps [37]. The mechanisms regulating cytotoxic resistance in tumor cells, especially cancer stem cells need to be understood so that the factors involved in these mechanisms can be targeted to improve patient response to chemotherapy by extending remission.

Cell quiescence

Adult stem cells are predominantly maintained in a quiescent, nondividing G0 state from which they may exit and begin to rapidly proliferate and differentiate in response to stress. This tiered stem cell hierarchy, with a single stem cell giving rise to a small number of progenitor cells that proliferate into transit-amplifying cells, finally giving rise to an expanded population of postmitotic differentiated cells suggests that stem cells, at least under nonmalignant conditions, are relatively quiescent. However, quiescence is not a prerequisite state of a stem cell; slow-cycling stem cells have been identified in both normal and malignant tissues [17,38,39]. However, slow division is thought to be important for the maintenance of stem cells by preventing them from exhausting their replicative capacity, inhibiting differentiation, and limiting potential errors over multiple cycles of DNA replication [40].

Quiescence is distinct from cellular senescence, which occurs once a cell has reached its Hayflick limit (typically due to shortening of telomeres) and is no longer capable of cell division [41]. Since the progeny of a senescent cell would be nonviable, senescence acts as an alternative to apoptosis. Quiescence, rather, is a reversible state, adopted by the majority of normal mammalian stem cells and a proportion of cancer stem cells [17,42]. Recent refinements of stem cell isolation methods have provided insights into the molecular regulation of quiescence, revealing it to be an actively maintained state poised for rapid reactivation given the appropriate environmental cues [38]. A more detailed understanding of the mechanisms that govern quiescence may be crucial in developing novel cancer treatments.

Pulse-chase assays that utilize the label-retaining properties of nondividing cells contrasting with the label-diluting properties of rapidly dividing cells have been used to select out quiescent populations with stem cell properties in a number of cell lines, including GBM [43,44]. Studies in the adult mouse brain demonstrated the presence of quiescent, label-retaining cells that can proliferate upon elimination of the mitotic cells, in vivo, and have the ability to produce long-term cell lines, in vitro [45]. Actively dividing cells of the subventricular zone are abolished with antimitotic treatment, while quiescent astrocytes survive treatment [46]. Codega and colleagues identified quiescent and activated neural stem cells from the subventricular zone of adult mice that are able to transition between quiescent and proliferative states [47].

As normal adult stem and progenitor cells which gain a series of oncogenic mutations are the likely origin of cancer stem cells, the cancer stem cells retain traits of their nonmalignant precursors, including quiescence. Several studies demonstrate the existence of quiescent stem cell populations across a range of human cancers, including sebaceous tumors, melanoma, pancreatic, breast, and ovarian cancer [18,19,48–50]. The dominant transcriptional signature of label-retaining, nonmalignant human mammary gland stem cells was more common to high-grade breast tumors compared to low-grade tumors, and grade 3 tumors exhibited a higher proportion of quiescent cells [19]. Breast tumor cells expressing the quiescent gene signature were also more efficient at forming spheroids in vitro and tumors in vivo, suggesting a role for quiescence in malignant progression and cell stemness. Direct evidence linking quiescence with cancer cell stemness was shown in pancreatic adenocarcinoma cells by fluorescence-activated cell sorting (FACS). Quiescent cells were resistant to chemotherapy, displayed an elongated, fibroblast-like morphology, and an increase in the expression of epithelial to mesenchymal transition (EMT) genes, consistent with an increase in stemness [18].

Importantly, the quiescent population exhibited up to 10-fold increase in clone-forming capacity in soft agar, were twice as invasive, and were 10-fold more efficient at forming tumors in vivo compared with the nonquiescent cells, indicating that quiescence correlates with stem-like properties.

Multiple lines of evidence indicate that label-retaining cells can give rise to both label- and nonlabel-retaining progeny, and vice versa, that nonlabel-retaining cells can also recapitulate a proliferative spectrum. The ability of both quiescent and mitotically active populations to give rise to progeny that encompasses the whole proliferative spectrum suggests that although label retention enriches for cancer stem cells, a dynamic plasticity exists wherein nonlabel-retaining cells are also able to adopt a cancer stem cell-like phenotype. The molecular mechanisms that underpin this plasticity remain to be fully elucidated.

Quiescence and treatment resistance

Rapidly dividing normal cells such as those in hair follicles, the gastrointestinal tract, and blood precursors and rapidly dividing tumor cells which constitute the bulk of a tumor are similarly sensitive to the cellular toxicity elicited by chemoradiotherapy. However, nontumor stem cells are resistant to chemoradiotherapy, and through transient proliferation and subsequent differentiation can replenish the cells within a tissue/organ following administration of chemotherapy. Since tissue-specific stem cells persist for the life of the organism, they are inherently better equipped to survive insult than nonstem cells due to active or enhanced protective cellular mechanisms, including checkpoint activation, DNA repair, and expression of drug efflux pumps [51]. Cancer stem cells, including glioma stem cells retain these refractory abilities [52]. An intriguing recent study suggests that GBM cells transition through different proliferative states and gradually acquire an irreversible resistance to temozolomide. The study identified a four gene expression signature, which correlates with a transient quiescent state during exposure to temozolomide where the cells exhibit cell cycle quiescence, specifically G2/M arrest and reduced malignancy in vivo [53]. This implies that the existence of a GBM cell Achilles Heel, represented by a GBM stem cell-specific gene signature, where targeting the GBM stem cells may be possible. Identification of other GBM quiescent cell molecular targets, such as *O*-acetylated gangliosides [54] may be useful in targeting GBM stem cells. Targeting the *O*-acetylated ganglioside, *O*-acetyl-GD2, using a monoclonal antibody demonstrated efficacy in killing temozolomide-resistant GBM stem-like cells [55]. The mechanism remains to be elucidated but the authors demonstrated that the *O*-acetyl-GD2 led to enhanced temozolomide uptake by tumor cells in vitro and a decrease in quiescent stem-like tumor cells in vivo. Time and again, research opens a window to the possibility of targeting quiescent GBM cells but to date the precise biological mechanisms at play in these cells remain elusive.

Molecular drivers of quiescence

Conventional therapies such as TMZ chemotherapy and radiotherapy are largely impotent at killing quiescent cancer cells and given their importance to patient relapse and mortality, effective future treatment regimens will need to address this resistant population directly. This will require a detailed understanding of the molecular mechanisms that govern quiescence in cancer stem cells. Our current understanding of the mechanisms that govern cellular quiescence are grounded in conditional experiments that induced fibroblasts into quiescence by either contact inhibition, loss of adhesion, or mitogen deprivation. Such studies of quiescent human fibroblasts by Coller et al. identified an expression signature that included downregulated genes associated with cell cycle progression such as cyclin B1, cdc20, cul-1, and myc, and upregulated genes associated with cell cycle regulation, including p53, cyclin D2, and MXI1 [56]. Also upregulated in this signature were genes in key stem cell regulatory pathways such as the Wnt, bone morphogenic protein (BMP), and Notch pathways.

The cyclin-dependent kinase inhibitor, p21 (CDKN1A), regulates initiation and maintenance of quiescence in multiple cell types and is overexpressed in quiescent HSCs [56]. p21-null mice exhibit an increased number of stem cells that lose their ability to repopulate bone marrow upon serial transplantation, suggesting the loss of p21 leads to exhaustion of the stem cell pool [57]. Notably, p21 is an important factor in the regulation of neural stem cell quiescence [58]. While quiescence mechanisms in normal cells are well understood, information about the control of quiescent cancer stem cells is still limited. It is reasonable to assume that there is an overlap between drivers of quiescence in normal stem cells and cancer stem cells. Consistent with this, p21 has also been shown to play a role in the resistance of GBM cells to radiotherapy [59]. Furthermore, HCT116-p21-null colon cancer cells generated tumors, which were 10-fold smaller compared to wild-type cells and were also unable to form spheres, in vitro [60]. Targeting of p21 or its downstream effectors thus represents a potential way to directly deplete cancer stem cells. Gene expression analysis shows that high p21 expression in glioma correlates with poorer patient survival [61].

Hedgehog, Notch, and Wnt signaling pathways are associated with cell cycle control and are all frequently dysregulated in cancer. The Notch pathway has been shown to be important in the response of a number of cancer stem cells (including GBM stem cells) to treatment, and has also been implicated in maintenance of stem cell quiescence [62,63]. Inhibition of Notch reduces in vitro clonogenicity of GBM stem cells, and improves the survival of mouse glioma xenograft models, suggesting that therapies directed at this pathway may provide a novel way of driving cancer stem cells into proliferation and making them more susceptible to treatment [64].

Targeting quiescent GBM tumor cells

To date, there are no clinically useful drugs for targeting quiescent tumor cells for brain cancer or any other cancer. Various strategies aimed at targeting quiescent cancer cells have been proposed along several lines of logic. There are several ways that dormant cancer cells could be targeted. Quiescent tumor cells could be either (1) directly killed, (2) directly targeted to lock them in a quiescent state so they cannot contribute to tumor recurrence or (3) triggered out of dormancy to sensitize them to antiproliferative cytotoxic therapy. These approaches and the relevant cellular mechanisms are detailed and discussed in a review on the pharmacological targeting of quiescent cancer cells [65]. Although there is no consensus on which of the aforementioned strategies may be the best approach in a therapeutic setting, considering the biological mechanisms and based on experimental evidence, it is likely that a strategy triggering quiescent tumor cells to "awaken" from dormancy is an achievable goal. Shifting quiescent tumor cells into a proliferative state will allow sensitization of the cells to first line, commonly used cytotoxic therapies which cause DNA damage and trigger apoptosis. A landmark study which demonstrated that interferon-alpha stimulation caused quiescent cancer stem cells to transiently proliferate [66]. The study showed that acute interferon-alpha stimulation triggered quiescent hematopoietic stem cells to transiently proliferate, and that in a leukemia mouse model, interferon-alpha administration followed by fluorouracil (5-FU) treatment, mice led to enhanced long-term survival. In a recent study, tumor cell intrinsic interferon signal activation in prostate cancer cells, promoted tumor cell proliferation, promoted long-term antitumor immunity, and blocked metastatic cancer growth in bone [67]. These studies demonstrate that interferon signaling is one of the key targetable pathways regulating tumor cell quiescence. To date, there are no studies that have replicated or directly demonstrated that interferon signaling regulates glioma cell quiescence. However, there is evidence that interferon stimulation of quiescent glioma stem-like cells sensitizes the cells to TMZ [68]. Although, this study determined that interferon downregulated O-6-alkylguanine DNA alkyltransferase (MGMT), thus sensitizing the cells to TMZ, it is not unreasonable to postulate that interferon also sensitized quiescent cells by changing the tumor cells from a quiescent state to a proliferative state, which would also sensitize the cells to TMZ.

Methods to identify and isolate quiescent cells

The first step for identifying the fundamental differences between the quiescent and highly proliferative cell fractions is to isolate them. One of the widely used methods for identifying proliferating cells is the incorporation of the thymidine analog 5′-bromo-2′-deoxyuridine (BrdU) into the newly replicated DNA. Incubating cells with BrdU permits incorporation of BrdU into DNA, in place of the nucleic acid thymine. However, there are some limitations of this method. Identification of BrdU-incorporating cells requires cell fixation, permeabilization, before incubation with an anti-BrdU antibody. The method is relatively labor intensive and does not allow for the identification and isolation of living cells, which limits subsequent experimental investigation. BrdU has also been shown to induce global loss of DNA methylation and consequently affect the astrocytic differentiation of neural stem cells, making it problematic for studies in which the stem cell state may be important [69].

Other approaches for labeling cells based on their proliferative potential amenable to downstream live cell assays can be used. Nontoxic fluorescent dyes can be applied to cells in a pulse-chase assay, whereby cells are labeled with the dye (the pulse), and then allowed to proliferate for a period (the chase). As cells divide, the label is apportioned 50:50 to each daughter cell, with a halving of intracellular dye content per cell division. Therefore, cells that have divided following the labeling event will contain lower levels of the dye, and subsequently a less intense signal compared to quiescent cells [70,71]. There are two major classes of dye used to label cells: succinimidyl esters, which bind to the amine groups of proteins and activated to their fluorescent form only after intracellular esterases cleave acetate side chains, and lipophilic dyes that label cell membranes [70,72]. The advent of fluorescent dyes as a method of labeling cells to quantify proliferation came serendipitously in studies aimed at tracking lymphocyte migration [73]. Westein and Parish (1990) showed that cells could be fluorescently labeled and later interrogated in a pulse-chase assay that directly relied on the cells proliferation rate to determine their retention of the dye. This allowed discrimination of cells based on how quiescent or proliferative they were, by fluorescence imaging or FACS. The study determined that the dye most suitable for discriminating between quiescent and proliferating cells was carboxyfluorescein succinimidyl ester (CFSE), which could passively cross membranes and bind intracellular proteins before being converted to a fluorescent form, and stable and intense enough to track cells for about six cell divisions. Oregon Green is a fluorinated analog of CFSE, differing only in the presence of two fluorine atoms that render the dye more stable and less prone to photobleaching, making it ideal for identifying quiescent cells [74].

Entry of Oregon Green into the cell is mediated by two acetate side chains, which make the dye highly membrane permeable and allow it to rapidly cross the cell membrane (Fig. 1). Once the dye has entered the cell, intracellular esterases cleave off the two acetate groups and give rise to the fluorescent form of the dye, which, due to the loss of its acetate side chains, is less membrane permeable and, therefore, sequestered within the cell. The amino-reactive side chain then binds to intracellular proteins. Once inside the cell and bound to protein the dye is stable. With each cell division, the dye is diluted 50:50 between each daughter cell, allowing cells to be identified for migration or proliferation experiments (Fig. 2). Given that succinimidyl esters such as Oregon Green bind to proteins, it is important to determine the optimal dye concentration to ensure the dye does not interfere with cellular biochemistry which may inhibit cellular processes, killing the cell.

Vital dye retention, proliferative potential, and cytotoxic sensitivity in glioblastoma cells

CD133 has been extensively used to isolate and separate GBM stem cells into $CD133^+$ and $CD133^-$ subpopulations to study their molecular and functional differences [75–77]. Previous investigations using established GBM cell lines and patient-derived GBM stem cells showed that using CD133 alone did not allow for the purification of a functionally distinct cell subpopulation with stem cell or tumor cell characteristics, from a mixed pool of cultured cells [76]. CD133 does not distinguish cells with low proliferative potential or sensitivity to chemotherapeutic exposure (temozolomide), although $CD133^+$ cells exhibit enhanced sphere-forming capacity. A study performed in our laboratory demonstrated the utility of Oregon Green and CFSE to identify and isolate slow-cycling quiescent GBM cells from a highly proliferative bulk cell culture and application of downstream cytotoxicity assays [78]. The study also demonstrated that Oregon Green showed superior fluorescence intensity at lower concentrations, allowing for more robust identification and selection of quiescent cells. In vitro experiments show that Oregon Green fluorescence reduced over the 7 days which directly correlated with continuing cell doubling on each subsequent day (Fig. 3). Importantly, about 1% of the cells in the bulk culture retain maximum fluorescence (day zero levels) at 7 days, thus identifying a subset of slow-cycling quiescent cells, which can be purified by FACS.

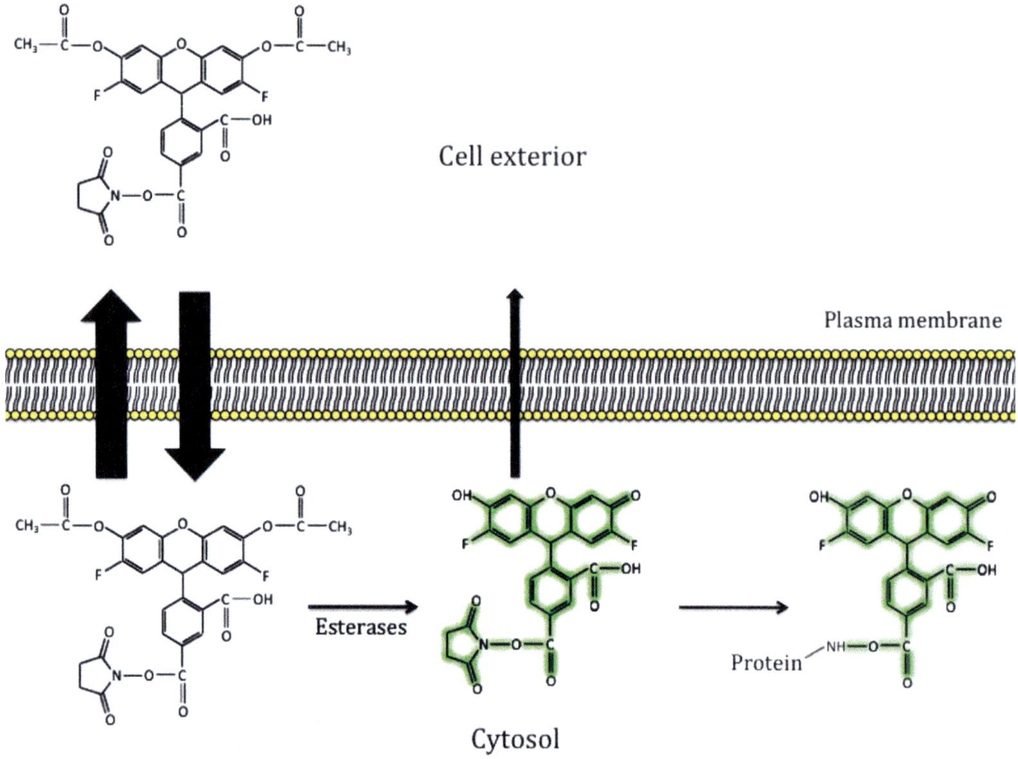

FIG. 1 Schematic representation of Oregon Green cell entry and activation. Two acetate side chains facilitate passive diffusion across the cell membrane. Once inside the cell, intracellular esterases cleave the acetate side chains simultaneously, generating the fluorescent form of the dye and reducing its ability to exit the cell. The amine-reactive side chain then binds intracellular proteins, thereby sequestering the dye in the cell.

The quiescent GBM cell subpopulation for multiple cell lines exhibited enhanced resistance to cytotoxic treatment (Fig. 4). FACS analysis revealed that the quiescent cells also showed significantly higher forward- and side-scatter profiles compared to proliferative cells, suggesting that quiescent cells are both larger and more morphologically complex compared to the proliferating cells; a state linked to cancer cells with enhanced clonogenicity [78]. These phenotypic differences at the extreme ends of the proliferative spectrum suggest that there are quiescent cell properties beyond proliferation that plays a role in treatment resistance. Moreover, plasticity between the quiescent and proliferative states appears to be dependent on the cytotoxic challenge cells experience (temozolomide vs radiation), the treatment intensity, and the underlying genomics of the cell line. Another malignant functional hallmark of aggressive brain cancer cells, cell migration (motility) and invasion (secrete extracellular matrix digesting enzymes) is enhanced in quiescent GBM cells, highlighting that GBM cell quiescence correlates with multiple oncogenic functions, and the challenges which need to be overcome if GBM therapy is to be improved [78].

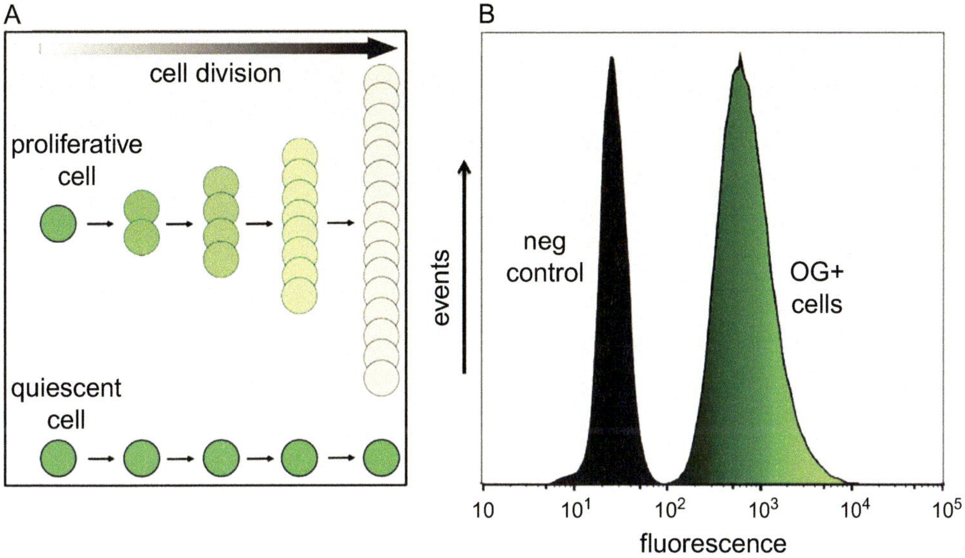

FIG. 2 Oregon Green dye fluorescence can be used to isolate distinct subpopulations of quiescent and proliferative GBM cells. (A) Schematic of Oregon Green retention in proliferative and quiescent cells. (B) FACS profile of Oregon Green-labeled GBM cell line U138MG cells 7 days after labeling vs untreated negative control.

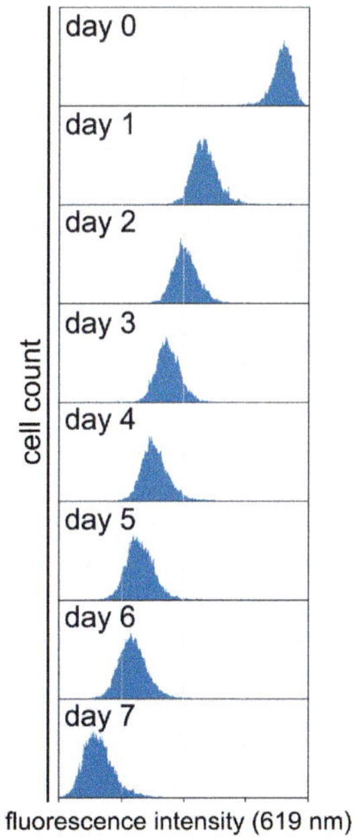

FIG. 3 Intracellular Oregon Green fluorescence over time. FACS profile of Oregon Green fluorescence in GBM U87MG cells over 7 days. Fluorescence intensity decreases over 7 days due to cell division.

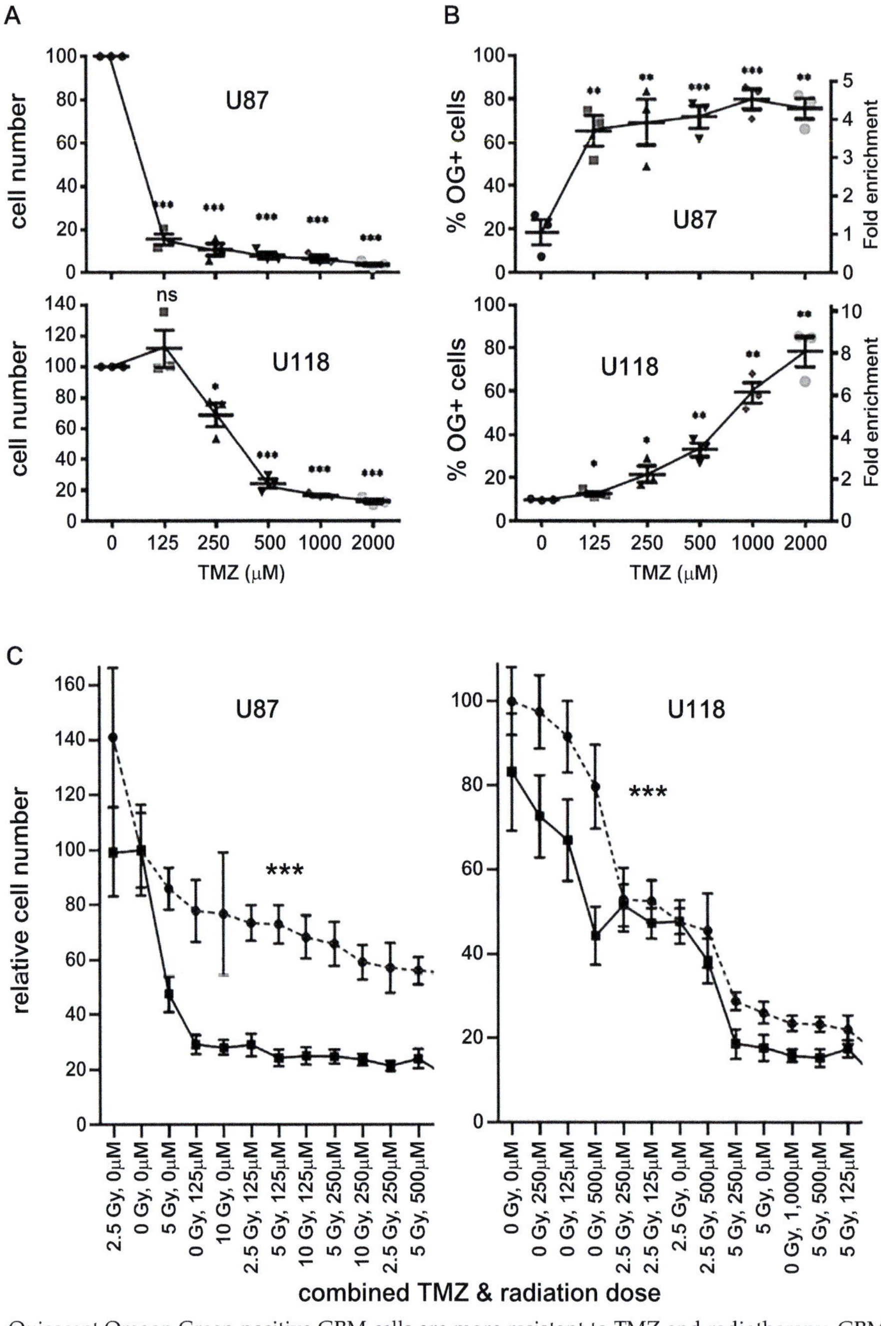

FIG. 4 Quiescent Oregon Green-positive GBM cells are more resistant to TMZ and radiotherapy. GBM cell lines U87MG and U118MG cells were labeled with Oregon Green and treated with either increasing doses of temozolomide (A), (B), or (C) both temozolomide and radiation. (A) Cell viability in the bulk cell culture (all cells) and (B) the proportion (%) of Oregon Green-positive cells showing enrichment of these cells with increasing TMZ concentration. Cells were FACS sorted to isolate cells with the most intense Oregon Green-retaining fluorescent cells (correlating with quiescence) (*dotted line*) and least fluorescent cells (correlating to the most proliferative cells) (*solid line*) and then treated with increasing doses of TMZ and/or radiation for a further 7 days. Relative cell number was determined by standardizing each treatment group to the untreated group set at 100. The percentage of OG+ cells was determined by dividing the number of OG+ cells × 100 by the total number of cells for each treatment group. Student's *t*-test was used to compare Oregon Green retention between untreated controls and TMZ or radiation treatment groups ($n=3$, error bars = SEM; *, $P<0.05$, **, $P<0.01$, ***, $P<0.001$).

Conclusions

There is a relationship between proliferative quiescence and GBM cell malignant properties, including treatment resistance. It is these properties of quiescent or slow-cycling tumor cells, in GBM, that protects them from cytotoxic challenge. The posttreatment survival and persistence of quiescent GBM tumor cells and their contribution to the almost inevitable posttherapy tumor recurrence emphasizes the importance for the better understanding the mechanisms regulating cell quiescence and designing novel drugs to selectively and effectively target these. Since cell surface biomarkers have not proven to be robust identifiers of quiescent tumor cells, the use of vital fluorescent dyes, including CFSE and Oregon Green, is a reliable way to discriminate tumor cells based on the rate of cell division. In vitro selection of these cells then allows for further experiments, including assessment of treatment response and resistance. The possibility to study quiescent cancer cells to understand the mechanisms regulating quiescence and cytotoxic resistance is a path toward novel treatments that specifically target the quiescent cell subpopulation, in combination with existing treatments to drastically improve patient outcomes not only for GBM patients, but also for patients with other cancers.

Acknowledgments

For some of the work used in this review, the authors received funding from the Royal Melbourne Hospital Neuroscience Foundation (S.S.), NHMRC project grants 1104010 (C.M.H., N.M.C.) and 1047581 (C.M.H., N.M.C.), a grant from the Australian Department of Health and Aging to the Epworth Cancer Centre, Epworth Hospital (N.M.C., C.M. H.), and a seed grant from the Department of Surgery (RMH) and School of Biomedical Sciences, the University of Melbourne.

References

[1] Ohgaki H, Kleihues P. Genetic pathways to primary and secondary glioblastoma. Am J Pathol 2007;170:1445–53.
[2] Friedmann-Morvinski D, Bushong EA, Ke E, Soda Y, Marumoto T, Singer O, et al. Dedifferentiation of neurons and astrocytes by oncogenes can induce gliomas in mice. Science 2012;338:1080–4.
[3] Liu C, Sage JC, Miller MR, Verhaak RGW, Hippenmeyer S, Vogel H, et al. Mosaic analysis with double markers reveals tumor cell of origin in glioma. Cell 2011;146:209–21.
[4] See S-J, Gilbert MR. Chemotherapy in adults with gliomas. Ann Acad Med Singapore 2007;36:364–6.
[5] Stupp R, Mason WP, van den Bent MJ, Weller M, Fisher B, Taphoorn MJB, et al. Radiotherapy plus concomitant and adjuvant temozolomide for glioblastoma. N Engl J Med 2005;352:987–96.
[6] Milano MT, Okunieff P, Donatello RS, Mohile NA, Sul J, Walter KA, et al. Patterns and timing of recurrence after temozolomide-based chemoradiation for glioblastoma. Int J Radiat Oncol Biol Phys 2010;78:1147–55.
[7] Claes A, Idema AJ, Wesseling P. Diffuse glioma growth: a guerilla war. Acta Neuropathol 2007;114:443–58.
[8] Bailey P, Cushing H. A classification of the glioma group on a histogenetic basis. Philaderphia: Lippincott; 1926.
[9] Mitchison TJ. The proliferation rate paradox in antimitotic chemotherapy. Mol Biol Cell 2012;23:1–6.
[10] Vogin G, Foray N. The law of Bergonié and Tribondeau: a nice formula for a first approximation. Int J Radiat Biol 2013;89:2–8.
[11] Atkins RJ, Ng W, Stylli SS, Hovens CM, Kaye AH. Repair mechanisms help glioblastoma resist treatment. J Clin Neurosci 2015;22:14–20.
[12] Beeharry N, Broccoli D. Telomere dynamics in response to chemotherapy. Curr Mol Med 2005;5:187–96.
[13] West CM, Barnett GC. Genetics and genomics of radiotherapy toxicity: towards prediction. Genome Med 2011;3:52.

[14] Visvader JE, Lindeman GJ. Cancer stem cells in solid tumours: accumulating evidence and unresolved questions. Nat Rev Cancer 2008;8:755–68.
[15] Cho D-Y, Lin S-Z, Yang W-K, Lee H-C, Hsu D-M, Lin H-L, et al. Targeting cancer stem cells for treatment of glioblastoma multiforme. Cell Transplant 2013;22:731–9.
[16] Cruceru ML, Neagu M, Demoulin J-B, Constantinescu SN. Therapy targets in glioblastoma and cancer stem cells: lessons from haematopoietic neoplasms. J Cell Mol Med 2013;17:1218–35.
[17] Moore N, Lyle S. Quiescent, slow-cycling stem cell populations in cancer: a review of the evidence and discussion of significance. J Oncol 2011;2011. https://doi.org/10.1155/2011/396076.
[18] Dembinski JL, Krauss S. A distinct slow-cycling cancer stem-like subpopulation of pancreatic adenocarcinoma cells is maintained in vivo. Cancers 2010;2:2011–25.
[19] Pece S, Tosoni D, Confalonieri S, Mazzarol G, Vecchi M, Ronzoni S, et al. Biological and molecular heterogeneity of breast cancers correlates with their cancer stem cell content. Cell 2010;140:62–73.
[20] Hegi ME, Diserens A-C, Gorlia T, Hamou M-F, de Tribolet N, Weller M, et al. MGMT gene silencing and benefit from temozolomide in glioblastoma. N Engl J Med 2005;352:997–1003.
[21] Combs SE, Widmer V, Thilmann C, Hof H, Debus J, Schulz-Ertner D. Stereotactic radiosurgery (SRS): treatment option for recurrent glioblastoma multiforme (GBM). Cancer 2005;104:2168–73.
[22] Saeed AM, Khairnar R, Sharma AM, Larson GL, Tsai HK, Wang CJ, et al. Clinical outcomes in patients with recurrent glioblastoma treated with proton beam therapy reirradiation: analysis of the multi-institutional proton collaborative group registry. Adv Radiat Oncol 2020;5:978–83.
[23] Tsien CI, Brown D, Normolle D, Schipper M, Piert M, Junck L, et al. Concurrent temozolomide and dose-escalated intensity-modulated radiation therapy in newly diagnosed glioblastoma. Clin Cancer Res 2012;18:273–9.
[24] Thomas RJ, Chen H, Fellows Z, Chen CC, Li H, Deville C. Comparative in silico analysis of pre-operative scanning beam proton therapy, intensity modulated photon radiation therapy, and 3D conformal photon radiation therapy in adult soft tissue sarcoma. Int J Radiat Oncol Biol Phys 2020. https://doi.org/10.1016/j.ijrobp.2020.07.1014, e11.
[25] Hanahan D, Weinberg RA. Hallmarks of cancer: the next generation. Cell 2011;144:646–74.
[26] Dianov GL. Base excision repair targets for cancer therapy. Am J Cancer Res 2011;1:845–51.
[27] Burdak-Rothkamm S, Rothkamm K. DNA damage repair deficiency and synthetic lethality for cancer treatment. Trends Mol Med 2021;27(1):91–2.
[28] Moding EJ, Kastan MB, Kirsch DG. Strategies for optimizing the response of cancer and normal tissues to radiation. Nat Rev Drug Discov 2013;12:526–42.
[29] Olivares-Urbano MA, Griñán-Lisón C, Marchal JA, Núñez MI. CSC radioresistance: a therapeutic challenge to improve radiotherapy effectiveness in cancer. Cell 2020;9(7):1651.
[30] Karanam NK, Story MD. An overview of potential novel mechanisms of action underlying tumor treating fields-induced cancer cell death and their clinical implications. Int J Radiat Biol 2020;29:1–11
[31] Barazzuol L, Jena R, Burnet NG, Meira LB, Jeynes JCG, Kirkby KJ, et al. Evaluation of poly (ADP-ribose) polymerase inhibitor ABT-888 combined with radiotherapy and temozolomide in glioblastoma. Radiat Oncol 2013;8:65.
[32] Russo AL, Kwon H-C, Burgan WE, Carter D, Beam K, Weizheng X, et al. In vitro and in vivo radiosensitization of glioblastoma cells by the poly (ADP-ribose) polymerase inhibitor E7016. Clin Cancer Res 2009;15:607–12.
[33] Bucher N, Britten CD. G2 checkpoint abrogation and checkpoint kinase-1 targeting in the treatment of cancer. Br J Cancer 2008;98:523–8.
[34] Zabludoff SD, Deng C, Grondine MR, Sheehy AM, Ashwell S, Caleb BL, et al. AZD7762, a novel checkpoint kinase inhibitor, drives checkpoint abrogation and potentiates DNA-targeted therapies. Mol Cancer Ther 2008;7:2955–66.
[35] Morgan MA, Parsels LA, Zhao L, Parsels JD, Davis MA, Hassan MC, et al. Mechanism of radiosensitization by the Chk1/2 inhibitor AZD7762 involves abrogation of the G2 checkpoint and inhibition of homologous recombinational DNA repair. Cancer Res 2010;70:4972–81.
[36] Clarke RH, Moosa S, Anzivino M, Wang Y, Floyd DII, Purow BW, et al. Sustained radiosensitization of hypoxic glioma cells after oxygen pretreatment in an animal model of glioblastoma and in vitro models of tumor hypoxia. PLoS One 2014;9, e111199.

[37] Moitra K, Lou H, Dean M. Multidrug efflux pumps and cancer stem cells: insights into multidrug resistance and therapeutic development. Clin Pharmacol Ther 2011;89:491–502.
[38] Cheung TH, Rando TA. Molecular regulation of stem cell quiescence. Nat Rev Mol Cell Biol 2013;14:329–40.
[39] Barker N, van Es JH, Kuipers J, Kujala P, van den Born M, Cozijnsen M, et al. Identification of stem cells in small intestine and colon by marker gene Lgr5. Nature 2007;449:1003–7.
[40] Sang L, Roberts JM, Coller HA. Hijacking HES1: how tumors co-opt the anti-differentiation strategies of quiescent cells. Trends Mol Med 2010;16:17–26.
[41] Hayflick L. The limited in vitro lifetime of human diploid cell strains. Exp Cell Res 1965;37:614–36.
[42] Li L, Clevers H. Coexistence of quiescent and active adult stem cells in mammals. Science 2010;327:542–5.
[43] Potten CS, Kellett M, Roberts SA, Rew DA, Wilson GD. Measurement of in vivo proliferation in human colorectal mucosa using bromodeoxyuridine. Gut 1992;33:71–8.
[44] Richichi C, Brescia P, Alberizzi V, Fornasari L, Pelicci G. Marker-independent method for isolating slow-dividing cancer stem cells in human glioblastoma. Neoplasia 2013;15:840–7.
[45] Morshead CM, Reynolds BA, Craig CG, McBurney MW, Staines WA, Morassutti D, et al. Neural stem cells in the adult mammalian forebrain: a relatively quiescent subpopulation of subependymal cells. Neuron 1994;13:1071–82.
[46] Pastrana E, Cheng L-C, Doetsch F. Simultaneous prospective purification of adult subventricular zone neural stem cells and their progeny. Proc Natl Acad Sci U S A 2009;106:6387–92.
[47] Codega P, Silva-Vargas V, Paul A, Maldonado-Soto AR, Deleo AM, Pastrana E, et al. Prospective identification and purification of quiescent adult neural stem cells from their in vivo niche. Neuron 2014;82:545–59.
[48] Gao M-Q, Choi Y-P, Kang S, Youn JH, Cho N-H. CD24+ cells from hierarchically organized ovarian cancer are enriched in cancer stem cells. Oncogene 2010;29:2672–80.
[49] Vaillant F, Asselin-Labat M-L, Shackleton M, Forrest NC, Lindeman GJ, Visvader JE. The mammary progenitor marker CD61/beta3 integrin identifies cancer stem cells in mouse models of mammary tumorigenesis. Cancer Res 2008;68:7711 7.
[50] Roesch A, Fukunaga-Kalabis M, Schmidt EC, Zabierowski SE, Brafford PA, Vultur A, et al. A temporarily distinct subpopulation of slow-cycling melanoma cells is required for continuous tumor growth. Cell 2010;141:583–94.
[51] Borst P. Cancer drug pan-resistance: pumps, cancer stem cells, quiescence, epithelial to mesenchymal transition, blocked cell death pathways, persisters or what? Open Biol 2012;2:120066.
[52] Bao S, Wu Q, McLendon RE, Hao Y, Shi Q, Hjelmeland AB, et al. Glioma stem cells promote radioresistance by preferential activation of the DNA damage response. Nature 2006;444:756–60.
[53] Rabé M, Dumont S, Álvarez-Arenas A, Janati H, Belmonte-Beitia J, Calvo GF, et al. Identification of a transient state during the acquisition of temozolomide resistance in glioblastoma. Cell Death Dis 2020;11:19.
[54] Fougeray S, Fleurence J, Faraj S, Bahri M, Cochonneau D, Terme M, et al. O-acetylated gangliosides: structure, biosynthesis, immunogenicity, functions and their potential for cancer immunotherapy. Available: https://www.hal.inserm.fr/inserm-01419857/; 2016.
[55] Fleurence J, Bahri M, Fougeray S, Faraj S, Vermeulen S, Pinault E, et al. Impairing temozolomide resistance driven by glioma stem-like cells with adjuvant immunotherapy targeting O-acetyl GD2 ganglioside. Int J Cancer 2020;146:424–38.
[56] Coller HA, Sang L, Roberts JM. A new description of cellular quiescence. PLoS Biol 2006;4, e83.
[57] Brugarolas J, Chandrasekaran C, Gordon JI, Beach D, Jacks T, Hannon GJ. Radiation-induced cell cycle arrest compromised by p21 deficiency. Nature 1995;377:552–7.
[58] Kippin TE, Martens DJ, van der Kooy D. p21 Loss compromises the relative quiescence of forebrain stem cell proliferation leading to exhaustion of their proliferation capacity. Genes Dev 2005;19:756–67.
[59] Bacco FD, De Bacco F, D'Ambrosio A, Casanova E, Orzan F, Neggia R, et al. MET inhibition overcomes radiation resistance of glioblastoma stem-like cells. EMBO Mol Med 2016;550–68. https://doi.org/10.15252/emmm.201505890.
[60] Mueller S, Cadenas E, Schönthal AH. p21WAF1 regulates anchorage-independent growth of HCT116 colon carcinoma cells via E-cadherin expression. Cancer Res 2000;60:156–63.
[61] Uhlen M, Zhang C, Lee S, Sjöstedt E, Fagerberg L, Bidkhori G, et al. A pathology atlas of the human cancer transcriptome. Science 2017;357. https://doi.org/10.1126/science.aan2507.
[62] Capaccione KM, Pine SR. The Notch signaling pathway as a mediator of tumor survival. Carcinogenesis 2013;1420–30. https://doi.org/10.1093/carcin/bgt127.
[63] Woodward WA, Chen MS, Behbod F, Rosen JM. On mammary stem cells. J Cell Sci 2005;118:3585–94.

[64] Purow BW, Haque RM, Noel MW, Su Q, Burdick MJ, Lee J, et al. Expression of Notch-1 and its ligands, Delta-like-1 and Jagged-1, is critical for glioma cell survival and proliferation. Cancer Res 2005;65:2353–63.
[65] Recasens A, Munoz L. Targeting cancer cell dormancy. Trends Pharmacol Sci 2019;40:128–41.
[66] Essers MAG, Offner S, Blanco-Bose WE, Waibler Z, Kalinke U, Duchosal MA, et al. IFNalpha activates dormant haematopoietic stem cells in vivo. Nature 2009;458:904–8.
[67] Owen KL, Gearing LJ, Zanker DJ, Brockwell NK, Khoo WH, Roden DL, et al. Prostate cancer cell-intrinsic interferon signaling regulates dormancy and metastatic outgrowth in bone. EMBO Rep 2020;21, e50162.
[68] Shen D, Guo C-C, Wang J, Qiu Z-K, Sai K, Yang Q-Y, et al. Interferon-α/β enhances temozolomide activity against MGMT-positive glioma stem-like cells. Oncol Rep 2015;34:2715–21.
[69] Schneider L, d'Adda di Fagagna F. Neural stem cells exposed to BrdU lose their global DNA methylation and undergo astrocytic differentiation. Nucleic Acids Res 2012;40:5332–42.
[70] Filby A, Begum J, Jalal M, Day W. Appraising the suitability of succinimidyl and lipophilic fluorescent dyes to track proliferation in non-quiescent cells by dye dilution. Methods 2015;82:29–37.
[71] Parish CR. Fluorescent dyes for lymphocyte migration and proliferation studies. Immunol Cell Biol 1999;77:499–508.
[72] Tario Jr JD, Gray BD, Wallace SS, Muirhead KA, Ohlsson-Wilhelm BM, Wallace PK. Novel lipophilic tracking dyes for monitoring cell proliferation. Immunol Invest 2007;36:861–85.
[73] Weston SA, Parish CR. New fluorescent dyes for lymphocyte migration studies. Analysis by flow cytometry and fluorescence microscopy. J Immunol Methods 1990;133:87–97.
[74] Quah BJC, Parish CR. New and improved methods for measuring lymphocyte proliferation in vitro and in vivo using CFSE-like fluorescent dyes. J Immunol Methods 2012;379:1–14.
[75] Beier D, Hau P, Proescholdt M, Lohmeier A. CD133+ and CD133− glioblastoma-derived cancer stem cells show differential growth characteristics and molecular profiles. Cancer Res 2007. Available: http://cancerres.aacrjournals.org/content/67/9/4010.short.
[76] Brown DV, Filiz G, Daniel PM, Hollande F, Dworkin S, Amiridis S, et al. Expression of CD133 and CD44 in glioblastoma stem cells correlates with cell proliferation, phenotype stability and intra-tumor heterogeneity. PLoS One 2017;12, e0172791.
[77] Wang J, Sakariassen PØ, Tsinkalovsky O, Immervoll H, Bøe SO, Svendsen A, et al. CD133 negative glioma cells form tumors in nude rats and give rise to CD133 positive cells. Int J Cancer 2008;122:761–8.
[78] Atkins RJ, Stylli SS, Kurganovs N, Mangiola S, Nowell CJ, Ware TM, et al. Cell quiescence correlates with enhanced glioblastoma cell invasion and cytotoxic resistance. Exp Cell Res 2019;374:353–64.

CHAPTER

13

The molecular and cellular effects of radiotherapy-induced microenvironment changes on potential chemoresistance in glioblastoma

Kshama Gupta

Departments of Cancer Biology and Radiation Oncology, Mayo Clinic, Jacksonville, FL, United States

Abstract

Glioblastoma (GBM) is uniformly fatal with a median survival of 1–2 years, despite best available treatments including, radiotherapy (RT), and chemotherapy (CHT). Despite continued advances in image-guided hypofractionated stereotactic radiosurgery, GBM remains one of the most invasive tumors with 100% recurrence rate. Multiple biological processes contribute to the development of resistance to cytotoxic chemotherapy, and perpetuation of cell molecular changes which cause treatment failure, tumor cell escape, and reestablishment of new tumor foci in the brain. Impacts of prior brain RT in GBM pathophysiology are less understood, though increasing evidence suggests RT-induced changes in the brain microenvironment contribute to recurrent GBM phenotype. The tumor microenvironment impacts malignant cells directly and indirectly through stromal cells and stroma-GBM cell interactions that support tumor progression. Changes in extracellular matrix (ECM), abnormal vasculature, hypoxia, and inflammation have been reported to promote tumor aggressiveness that could be exacerbated by prior RT. Prior radiation may have long-term impacts on microglia and brain-infiltrating monocytes, leading to lasting alterations in cell and cytokine signaling which can cause irreversible changes in ECM, thus affecting the extracellular distribution of small molecules, and chemotherapeutic drugs making them sparsely available to combat tumor growth. Since RT is vital to GBM care, but substantially alters the tumor microenvironment, we here discuss the challenges, knowledge gaps, and therapeutic opportunities relevant to targeting protumorigenic features of the GBM microenvironment. Understanding the intricate biological processes involved can allow us to incorporate new targeted interventions in current treatment approaches and during postprimary treatment regimens early enough to overcome the setting of secondary tumor growth. We suggest that insights gained from RT-induced changes in the tumor microenvironment may provide

Glioblastoma Resistance to Chemotherapy: Molecular Mechanisms and Innovative Reversal Strategies
https://doi.org/10.1016/B978-0-12-821567-8.00035-X

opportunities for development of new combinatorial therapies that could help better manage the late effects of RT, which otherwise get amplified in previously radiated brain microenvironment making its stroma permissive for recurrent tumors that are chemo-radiotherapy resistant and cause poor outcome in glioblastoma.

Abbreviations

AMPK	AMP-activated protein kinase
ASC	ascorbate
ATP/GTP	adenosine triphosphate/guanosine triphosphate
BBB	blood-brain barrier
BBTB	blood-brain tumor barrier
BDNF	brain-derived neurotrophic factor
BILs	brain infiltrating leucocytes
BL	basal lamina
CHT, C&Rx	chemotherapy
CNS	central nervous system
CS	chondroitin sulfate
CSPG	chondroitin sulfate proteoglycan
DDR1	discoidin domain receptor 1
eATP/eGTP	extracellular ATP/extracellular GTP
ECM	extracellular matrix
EGFR	epidermal growth factor receptor
EMT	epithelial to mesenchymal transition
eROS	extracellular ROS
FGF	fibroblast growth factor
GAG	glycosaminoglycans
GBM	glioblastoma multiforme
GPI	glycosylphosphatidylinositol
GSH	glutathione
HA	hyaluronan/hyaluronic acid
HIF-1	hypoxia inducible factor-1
HSPG	heparan sulfate proteoglycan
ICAM-1	intercellular adhesion molecule-1
IR	irradiation
LTP	long-term potentiation
MAPK	mitogen-activated protein kinase
MG	microglia
MMP	matrix metalloprotease
MP	macrophages
mTOR	mammalian target of rapamycin
NADPH	nicotinamide adenine dinucleotide phosphate (reduced)
NO	nitric oxide
NOS/iNOS	nitric oxide synthase/inducible-NOS
OPC	oligodendrocyte progenitor cell
PDGFR	platelet-derived growth factor receptor
PI3K	phosphatidylinositol-3-kinase
PNN	perineuronal nets
RIBE	radiation-induced bystander effects
ROS/RNS	reactive oxygen species/reactive nitrogen species
RT	radiation therapy
SASP	senescence-associated secretory phenotype
TAM	tumor-associated macrophages
TGF-β	transforming growth factor beta

TILs	tumor infiltrating leucocytes
TIMP	tissue inhibitors of metalloproteinase
TME	tumor microenvironment
TMZ	temozolomide
Trx	thioredoxin
TSP	thrombospondin
VEGFR	vascular endothelial growth factor receptor

Conflict of interest

No potential conflicts of interest were disclosed.

Introduction

Glioblastoma (GBM) is the most common brain malignancy affecting primarily adults > 65 years of age [1–3]. The highly aggressive nature of GBM is characterized by its diffuse infiltration into the brain parenchyma mediated through glioma cell interaction with the brain extracellular matrix (ECM) [4–8]. GBMs are incurable; multimodality treatment with surgical resection followed by radiation therapy (RT) and chemotherapy (CHT) represents the standard of care [9].

Surgery is generally the first step in the treatment of GBM, the goal of which is to remove as much of the tumor as possible. However, factors such as location of the tumor and consequential loss of functions postsurgery are deterministic in governing if the tumor can be completely removed. Partial tumor removal may also be performed owing to the diffuse character of the tumor.

Radiation therapy is the mainstay of care for GBM and is administered following safe surgical resection. A landmark study by Stupp et al. in 2004, involving a randomized phase-III trial reported improved, progression-free, and overall survival for patients with GBM when treated with the addition of concomitant and adjuvant temozolomide (TMZ), an oral alkylating agent, to RT, as compared to the RT treatment alone. This study led chemoradiotherapy, with RT plus TMZ, to become the new standard of care [10].

While patients who receive RT have improved survival (>6 months), 50%–90% of survivors exhibit a disabling and progressive cognitive decline [10–18]. High-dose RT effects include hemorrhage, neurodegeneration, and premature senescence, which can progress over time [19–21]. The association between cancer, aging, and therapy-associated aging is well documented [22, 23]. Recent studies have shown that radiation-induced brain injury or radiation-associated microenvironmental alterations can be protumorigenic [24–28]. The dose and time dependence of radiation exposure can significantly alter the impact of RT on the tumor microenvironment (TME) by affecting tumor or stromal cell behavior, migration, and treatment response [3, 29–39]. It is therefore reasonable to postulate that both acute events and alterations accumulated over time post-RT can have a cumulative long-term effect on the structure and composition of the brain matrix and cell extrinsic milieu. These modulations in the brain microenvironment can make it permissive for GBM cell migration and invasion, which eventually can manifest itself with an aggressive tumor recurrence [24–26, 28, 40, 41].

Thus, despite being an indispensable treatment modality for GBM, radiation therapy, owing to its cytotoxic effects on proliferative tumor cells, requires further investigation in specific areas offering scope for improvement and innovation [3, 42]. Few of these include safe radiation regimes for older or elderly patients, reirradiation options, technological advances in focal RT administration, dose-time modulation to reduce long-term debilitating effects (cognitive decline, seizers), and opportunities for control of tumor recurrence.

We discuss here the available evidence related to RT-induced brain microenvironmental alterations, with specific focus on ECM remodeling and tumor adaptation processes that can facilitate aggressiveness during tumor recurrence. These RT-induced changes in the brain stroma involving unusual matrix composition, disturbed vascular architecture, metabolic shifts, and inflammation can functionally affect the sustenance of brain homeostasis and severely impact the availability, distribution, and diffusion of chemotherapy drugs, preventing them from reaching concentrations relevant for optimal and overall therapeutic response [24, 27, 28, 43–75]. Furthermore, cellular senescence and radiation-induced bystander effects can lead to neo-transformations supporting treatment resistance and tumor recurrence [15, 17, 18, 76–83]. Considering the preponderance of tumor recurrence in GBM, insights gained from these microenvironmental disturbances occurring post-RT can provide possibilities to potentiate current therapeutic regimes, opening new avenues to advance GBM care.

Extracellular microenvironment of the brain in glioblastoma

Brain extracellular matrix

The extracellular space makes up approximately 15%–25% of the brain volume [84]. The components found within this extracellular space include various ions, metabolites, neurohormones, peptides, and ECM molecules that are secreted by neurons and glia. The ECM of the normal brain comprises high levels of space-filling carbohydrate molecules, the glycosaminoglycans (GAG), those free/unbound to proteins such as hyaluronan/hyaluronic acid (HA), and protein-bound GAGs, the proteoglycans and glycoproteins. Under a physiological state, the brain matrix has a predominance of chondroitin sulfate proteoglycans (CSPGs) and heparan sulfate proteoglycans (HSPGs) and a paucity, but not absence, of otherwise frequent ECM glycoproteins like fibronectin and collagens. The distinctive composition of ECM of the normal brain involves proteoglycans of the lectican/hyalectan family and their binding partner, hyaluronan, link proteins and glycoproteins. Lecticans are large secreted proteoglycans with chondroitin sulfate (CS) side chains and the HA-binding domain. Lecticans such as brevican, versican-V2, neurocan are specific to the central nervous system (CNS). The other class of proteoglycans, HSPG, is presently bound to the cell surface through a glycosylphosphatidylinositol (GPI) anchor (glypicans) or intercalated into the cell membranes (syndecans, NG2). Fibrous glycoproteins are mainly associated with the basement membranes of the brain vasculature, which include collagen-IV, fibronectin, and laminin [85]. Ligands of the brain matrix proteoglycans include growth factors, other ECM molecules, and cell surface molecules [86]. The biological functions of several growth factors, such as fibroblast growth factor (FGF-2) and intracellular signaling, depend strongly on the composition and interaction with the ECM [87]. Neuronal functions modulated by ECM include neurogenesis, neurite growth and inhibition, cell survival, migration, synapse formation, and synaptic transmission. ECM composition is critical to the development of brain, and in maintenance of brain health [42, 85]. Matrix alterations are a normal response to brain injury and in tumor development. Hence, modulation of ECM is central to promoting repair or ameliorating disease.

Brain extracellular matrix in development and homeostasis

ECM components are expressed in dynamic patterns in the developing and adult nervous system. ECM forms a basal lamina (BL) surrounding the blood vessels throughout the brain. Structural components of the neocortex include CSPGs and glycoproteins as reelin, tenascins, and others whose expression is spatially governed [88]. ECM is crucial for axon formation and myelination and is a component of glial scar tissue at sites of brain injury [89–91]. Oligodendrocytes are cells that myelinate the brain axons. Tenascin-C, laminin, and their cognate integrin receptors play roles at different developmental stages, including oligodendrocyte progenitor cell (OPC) migration and proliferation, immature oligodendrocyte survival, and axon myelination [92]. Additionally, several proteoglycans, tenascin-R, laminin, and dystroglycan contribute to the formation of nodal matrices. ECM surrounding nodes of Ranvier regulate the local concentration of cations and clusters of voltage-gated sodium channels, which allow for saltatory electrical conductivity [88].

Synapses are embedded in protein meshwork secreted by neurons and astrocytes [90, 93, 94]. Perineuronal nets (PNNs) are the meshwork of ECM enwrapping neuronal cell bodies and processes that affect their development and function and provide the supporting matrix for synapse formation. Hyaluronan (HA) forms the backbone of PNNs. PNNs are in a dynamic state during nervous system maturation. Thrombospondin 1 (TSP isoform-1) is a multifunctional extracellular matrix protein that is crucial in neuronal migration, axonal outgrowth, and guidance in the developing brain [95]. In addition to axonal growth and guidance, thrombospondin proteins secreted by glial cells play a critical role in synaptogenesis and maintain synapse stability [96]. Modulating the neuronal matrix molecules affects synapse morphology and function, interfering with the induction and maintenance of long-term potentiation (LTP), and synaptic plasticity [94].

Tumors are known to reactivate several developmental processes that are otherwise silenced in the adult brain. Studies have shown PNNs to play a neuroprotective role by attenuating lipid peroxidation and reactive oxygen damage [97, 98]. Thus, changes in the expression, structure, and topography of the ECM proteins in the brain can influence the fate of neurons during cancer progression and treatment response, which can have neurotoxic or neuroprotective effects.

Microglia: The guards in brain parenchyma

Owing to the extensive involvement of the ECM in sustaining normal brain functions, the ECM must be heavily guarded from extrinsic and intrinsic threats. Microglia are the sentinels for innate immunity in the brain, continually monitoring and screening the brain extracellular microenvironment, walling it off from foreign invaders and from internal damage by removing cell debris or dead cells [99, 100]. Microglia comprise about 10%–15% of the total glial population in the brain [101–103]. They are associated with regulatory processes critical for development, maintenance of the neural environment, injury, and repair. Microglia are derived primarily from mesodermal precursor cells of hematopoietic-myeloid lineage that populate the brain early in development and differentiate into resident microglia; a small and transient population of peripheral monocyte-derived microglia is found only during brain injury [104, 105]. Microglia are functionally annotated to have M1 and M2 phenotypes, though the suitability of this terminology is debated. These phenotypes allow them to actively exhibit the immunoinflammatory or immunosuppressive responses, respectively, and participate in pathological processes [106, 107]. In the healthy mature brain, microglia have a ramified morphology associated with the microglial "resting state." In this state, microglia actively scan the brain extracellular environment through immune surveillance [108–110]. A brain insult causes microglia to undergo intracellular changes and conjure "microglial activation." Surveillance, intervention, and cytotoxicity all constitute the microglia mediated support and protection system in the brain.

Neuron-microglia communication relies mostly on the release of signaling molecules from axonal varicosities. Mitogens for microglial proliferation are secreted in response to neuronal deafferentation, dysfunction, or damage and include growth factors (brain-derived neurotrophic factor, BNDF), cytokines, chemokines (Fractalkine [CX_3CL1]), and purinergic receptors (P2X7) [111–116]. Transforming growth factor beta (TGF-β) acts as an inhibitor of microglial proliferation [117]. Microglial cell migration toward sites of acute brain damage is well documented. The main chemoattractant sensed by microglia is adenosine triphosphate (ATP) [113, 117–121]. Extracellular ATP released from the damaged neurons or astrocytes regulates the microglial process dynamics and rapid response toward injury, a process which can have implications in aging and cancer [121]. Moreover, extracellular adenosine, present at high concentrations in cancer tissues, regulates cancer growth, metastatic dissemination, and phagocytosis by activated microglia and is a crucial mediator in the alteration of immune cell functions in cancer [28]. Neuropeptides are also chemoattractants for microglia. Different intracellular signaling cascades are activated downstream of these stimulants, indicating that chemoattractants control distinct microglial functions.

Microglial activation and brain matrix regulation

Microglial activation is a highly regulated process in order to limit neuronal damage. The "surveillance" phenotype is maintained, in part, through neuron-derived signals, including CX_3C-chemokine ligand 1 (CX_3CL1), which acts through their cognate receptors expressed at the microglial plasma membrane. During an acute inflammatory response, microglia rapidly shift to an activated proinflammatory state (M1 phenotype) and, soon after the immune defense function is executed, microglia shift to the "alternative activated" state (M2 phenotype) that promotes wound healing and functional recovery. Microglia can both produce and respond to cytokines, which can be either neuroprotective and/or neurotoxic [108]. Aberrant activation of microglia is highly neurotoxic. Glioma-infiltrating microglia acquire the M2-phenotype during tumor progression [109]. Hypoxia further alters microglial activation by supporting the M2 phenotype [122].

In addition to being the primary immune effector cells, microglia play an important role in orchestrating the behavior of other immune cells that enter the brain, by secreting cytokines and chemokines, and members of the matrix metalloproteinase (MMP) family, which degrade the ECM, thus facilitating inflammatory cell migration [123, 124]. The regulatory control between microglia and the ECM components is, however, bidirectional, indicating that microglial activation can not only cause ECM remodeling; conversely, it is also activated differentially by the matrix components. Fibronectin and vitronectin are stronger substrates for microglial attachment than laminins, as well as being potent activators of microglia [125, 126]. Fibronectin and vitronectin expression within the brain is increased in tumors [127, 128]. Further clinical relevance comes from the observation that serum has a higher content of circulating fibronectin and vitronectin, but not laminin [129]. Insults such as tumor and radiation damage can cause alterations in the permeability of the blood-brain barrier, thereby dramatically increasing the concentration of these glycoproteins in the brain [130]. These changes can collectively lead to aberrant microglial activation and profound alterations in the overall brain parenchyma, assisting GBM pathology. Vitronectin has been reported to also facilitate radioresistance in tumors [131].

Extracellular matrix in glioblastoma

Brain tumors craft the brain matrix to cause significant structural and functional changes in the microenvironment [4–8, 27, 54, 132–142]. These changes include both matrix biopolymer synthesis and degradation and are elicited as a consequence of tumor cell compactness, changes in local oxygen tension, and inflammatory reactions, all of which occur owing to uncontrolled proliferation and/or altered antigen presentation by tumor cells [143]. Biosynthetic changes in the tumor stroma involve the secretion of new ECM molecules and expression of new receptor adhesion molecules by GBM cells. These involve a specific increase in expression of fibrillar glycoproteins such as collagens, laminin, fibronectin, vitronectin, and tenascins and also increased expression of specific proteoglycans that contribute to neoangiogenesis (collagens, tenascins), maintaining the blood-brain barrier integrity and facilitating entry of macrophages into the brain (fibronectin, laminin, and col-IV) and microglial activation (fibronectin, vitronectin), and acting as scaffolds to promote GBM cell migration and invasion.

Elevated levels of CSPGs, such as versican and brevican, are found in GBM and support its progression. The versican isoform expressed by migratory GBM cells interacts with surface receptors, e.g., EGFR, and is involved in tumor promotion [144–146]. Several other molecules such as HA, CD44, tenascin, and transforming growth factor beta 2 (TGFβ2) also interact with versican and promote GBM cell invasion [147–149]. HSPGs are highly immune-active. They act as coreceptors for many cytokines and growth factors and contribute to neuroinflammation. HSPGs also participate in promoting axonal growth, which is one of the bystander mechanisms in tumors that compensate for tumor-associated insult, and support brain functions during tumor progression. Fibrillar proteins as collagens and fibronectin make the ECM rigid in GBM, which contributes to the GBM pathophysiology by promoting angiogenesis, or by acquiring specific topography to act as scaffolds for GBM cell migration [150]. Degradative changes in the GBM matrix are caused by secretion of matrix metalloproteases (as MMP-2, MMP-9) that are essential for cell invasion [151].

Intratumoral hypoxia is well studied in tumor ecology. Reports show a plausible role of tumor hypoxia in promoting matrix production and remodeling, which facilitates the cross talks between ECM and the inflammatory microenvironment having prosurvival effects [152–154]. Intratumoral hypoxia causes the recruitment of macrophages into the brain parenchyma through Sema3/Nrp1 signaling and supports the microglial M-2 profibrotic state [155]. Biosynthesis of ECM proteins is a major anabolic requirement during cancer progression. Cell anabolism depends on a high amount of reducing equivalents such as NADPH, which are generated by a metabolic shift to the pentose phosphate pathway [156]. NADPH plays a central role in the reductive biosynthesis of membrane lipids, maintenance of cell integrity, protein synthesis, and redox homeostasis, and its production is inevitable for tumor growth [157]. Metabolic modulators, adenosine receptor blockade, and cell migration inhibitors form an arsenal of potential therapeutic targets for GBM care (Table 1) [28, 193, 236].

TABLE 1 RT-induced alterations in the brain stroma, or glioblastoma tumor microenvironment, and relevant therapeutic opportunities.

Primary target	RT effect (+/–), consequence	Therapeutic approaches	Reference
Extracellular matrix remodeling • Poor delivery and distribution of chemotherapy drugs in the tumor • Tumor invasion to distant foci, with escape from effective chemotherapeutic gradients • Altered perineuronal nets (PNN) lead to impaired neurotransmission and neurogenesis, exacerbate damage of nontransformed cells, neuronal cell death, cognitive decline, and premature brain aging			
ECM–component biosynthesis and degradation			
Collagens and fibrillar proteins	Up (+), migration, invasion, altered PNN	Functional inhibition of enzymes: P4HA1, P4HA2, PLOD2, and LOX; HSP47 targeting	[137, 138, 152, 158–161]
ECM glycoproteins (Tenascin-C)	Up (+), proliferation, invasion, altered PNN	Inhibition of tenascin-C expression and, TNIIIA2 peptide processing, integrin α5β1 blockage, PDGF-Rβ blockage	[47, 125, 162–164]
Proteoglycans (PG): brevican, versican, hyaluronan	Up (+), survival, migration, invasion, inflammation, neo-vascularization, altered PNN	Targeting PG synthesis, xylosyltransferase 1, and others	[27, 44, 52, 144, 165–174]
MDA-9/syntenin	Up (+), metastasis, tumor progression	Inhibitors to MDA-9/syntenin	[46, 175]
LOX	Up (+), migration	Lox inhibitors	[49]
MMPs, TIMP	Up (+), invasion	MMP-13 and others	[26, 48, 53, 55, 176]
ECM-cell (ligand/receptor) interaction			
HA—CD44	Up (+), invasion	CD44 blockade	[54, 177, 178]
Sema3A—Nrp1 GDNF—Nrp1	Up (+), invasion	Anti-Nrp1 therapy	[155, 179, 180]
Fibronectin/ vitronectin—integrin	Up (+), invasion	Receptor (integrin) blocking	[91, 92, 125, 181, 182]

TABLE 1 RT-induced alterations in the brain stroma, or glioblastoma tumor microenvironment, and relevant therapeutic opportunities—cont'd

Primary target	RT effect (+/–), consequence	Therapeutic approaches	Reference
ECM receptors: NTRK4/DDR-1, ICAM-1, α5β1, αvβ3	Up (+), invasion	Receptor blocking	[24, 137, 183–186]
Tumor cell adaptation processes • Tumor progression, with development of resistance to chemotherapy • Tumor invasion to distant foci, with escape from effective chemotherapeutic gradients • Glioma cell stemness and GSC maintenance, with escape from tumor immunity and cytotoxic treatments • Metabolic shifts cause tumor progression, invasion treatment resistance, and recurrence • Altered BBTB lead to tumor progression, progressive inflammation, and tissue damage • Impaired neurotransmission and neurogenesis exacerbate damage of nontransformed cells, causing neuronal cell death, cognitive decline, and premature brain aging • RT-induced bystander effects cause progressive genomic alterations, neo-transformation, chemotherapy resistance, tumor recurrence, cognitive decline, and premature brain aging			
Cytoskeleton (myosins)	Up (+), migration, invasion	Myosin inhibitors	[187–190]
Antiapoptosis ($BCL2/BCL_XL$)	Up (+), migration, invasion	Apoptosis inducers	[24, 191, 192]
Cell signaling pathways			
PI3K/mTOR, Ras/Raf/Rock	Up (+), proliferation, migration, invasion	Inhibitors to Ras/Raf/Rock, PI3K/mTOR	[51, 193–198]
VEGF/VEGF-R PDGF/PDGF-R	Up (+), migration, invasion	Inhibitors to VEGF/VEGF-R and PDGF/PDGF-R axis (TKI)	[187, 194, 199]
SDF-1/CXCR4	Up (+), migration, invasion	Anti-CXCR4 monoclonal neutralizing antibody	[5]
Glutamate pathway, ion channels	Up/down(+/−), glia activation, invasion, glioma cell stemness and maintenance, abnormal neuronal firing (seizers), impaired neurogenesis, etc.	Receptor blocking, ion-channel (K^+) blocking, inhibitors to AETs, metabolic therapy, antioxidant supplementation, others	[43, 45, 58, 200–206]
Inflammation (cytokines, chemokines, chemokine receptors)	Up (+), neuronal damage, tumor proliferation, migration, invasion	Immunotherapy, novel strategic dose-/time-dependent RT-regimes	[61, 68, 74, 207–212]
Oxygen tension (HIF-1) leaky BBB BBTB angiogenesis (NOS, VEGF, Ang)	Up (+), Hypoxia, angiogenesis, glioma stem cell niche, inflammation, malignancy, cerebral necrosis	Targeting: HIF-1, NOS, PDGF/VEGF signaling, MMPs (MMP-2, others) YKL-40 (CHI3L1); inhibitors to AETs; novel strategic dose-/time-dependent RT-regimes	[50, 57, 63, 64, 67, 71, 73, 75, 159, 200, 213–226]

Continued

TABLE 1 RT-induced alterations in the brain stroma, or glioblastoma tumor microenvironment, and relevant therapeutic opportunities—cont'd

Primary target	RT effect (+/–), consequence	Therapeutic approaches	Reference
Metabolism			
Redox regulation	ROS/RNS, NO, NOX4: Up (+), senescence, inflammation, migration, invasion, ischemic injury, etc.	TSP-1/CD47 axes: anti-CD47 monoclonal antibodies senolytic therapy	[19, 61, 227]
	ASC, GSH: down (−), tumor proliferation, migration, invasion, recurrence	ROS scavengers/antioxidants (Vit C and E)	[28, 228]
	NAD^+: Down (–)	Nicotinamide administration	[229, 230]
		NADPH oxidation: inhibition of NADPH oxidase, 5-lipooxygenase	[153]
Energy carriers (ATP, GTP)	Up (+), inflammation, migration, invasion; DNA repair and treatment resistance	Adenosine receptor blocking; inhibition of DNA repair pathways, targeting NAD^+ synthesis	[28, 62, 65, 70, 193, 231–233]
Radiation induced bystander effects			
Bystander effects to nontargeted cells	Up (+), chronic inflammation, immune escape, therapeutic resistance, senescence, neo-transformation, tumor recurrence, cognitive decline, brain aging	Receptor blocking, ion-channel (K^+) blocking, metabolic therapy, antioxidant supplementation, senolytic therapy, angiotensin-converting enzyme inhibitors (ACEI) or angiotensin II type 1 receptor blockers (ARB); novel strategic dose- /time-dependent RT-regimes	[13, 15–17, 76, 77, 79, 80, 234, 235]

The table summarizes the biological processes (or primary targets) modulated by RT in the preradiated brain stroma or glioblastoma microenvironment (column-1, from left to right), with having them categorized under two prime phenomenon, ECM remodeling and tumor cell adaptation processes. How RT modulates these biological phenomenon (up or downregulation) and, its pathophysiological consequence is mentioned in column-2; column-3 describes the therapeutic approaches (current or futuristic), to target them or prevent them. Radiation therapy, RT; extracellular matrix, ECM; perineuronal nets, PNN; prolyl 4-hydroxylase subunit Alpha 1, P4HA1; Prolyl 4-hydroxylase subunit Alpha 2, P4HA2; procollagen-lysine 2-oxoglutarate 5-dioxygenase 2, PLOD2; lysl oxidase, LOX; hyaluronic acid, HA; MDA-9, melanoma differentiation-associated gene-9; neurotropic tyrosine kinase 4, NTRK4, discoidin domain receptor-1, DDR-1; intercellular adhesion molecule 1, ICAM-1; Integrins $\alpha5\beta1$; Integrin, $\alpha v\beta3$; matrix metalloproteinases, MMPs; tissue inhibitor of matrix metalloproteinase, TIMP; Heat shock protein 47, HSP47; hypoxia-inducible factor 1, HIF-1; active efflux transporters, ATEs; vascular endothelial growth factor, VEGF; vascular endothelial growth factor receptor, VEGF-R; platelet-derived growth factor, PDGF; platelet-derived growth factor receptor, PDGF-R; tyrosine kinase inhibitor, TKI; phosphatidylinositol-3-kinase, *PI3K*; mammalian target of rapamycin, *mTOR; Rat Sarcoma, Ras;* rapidly accelerated fibrosarcoma, Raf; rho-associated *protein* kinase, *Rock*; blood-brain barrier, *BBB; blood-brain tumor barrier, BBTB; hypoxia inducible factor-1, HIF-1;* angiotensins, Ang; NADPH oxidase 4, NOX4; reactive oxygen species, ROS; reactive nitrogen species, RNS; nitric oxide, NO; nitric oxide synthase, NOS; thrombospondin, TSP; adenosine triphosphate, ATP; guanosine triphosphate, GTP; nicotinamide adenine dinucleotide, NAD; nicotinamide adenine dinucleotide phosphate (reduced), NADPH; ascorbate, ASC; glutathione, GSH; cluster of differentiation, CD; B-cell lymphoma-2, BCL2; BCL_XL, B-cell lymphoma-extra-large; vitamin, vit (C, E); potassium, K^+; cytokines: tumor necrosis factor-α, TNF-α; transforming growth factor-β, TGF-β, IL, interleukins (IL-6, IL-8, IL-1β). Chemokines: CX3C family (CX3CL1, Fractalkine), CCL family (CCL2, CCL3, CCL7, CCL8, CCL12), CXC family (CXCL4, CXCL12/stromal cell-derived factor 1, *SDF1*). Chemokine receptors: CC-chemokine receptor family (CCR1, CCR2). RT-effects: +, positive; Up, upregulation/increased; −, negative; Down, downregulation/decreased.

Radiation-induced changes in the brain and GBM microenvironment

Multiple biological processes contribute to treatment failure in GBM, owing to the development of resistance to cytotoxic chemotherapy and perpetuation of molecular and cellular changes that allow tumor recurrence. In contrast to chemotherapy, radiation uniformly penetrates both the brain and tumor parenchyma, overcomes chemoresistant cells, and thus remains an essential therapeutic modality.

Radiation therapy

Ionizing radiation (IR) can damage cellular macromolecules, such as DNA, proteins, and membranes, primarily by generation of reactive oxygen species (ROS). RT is based on the failure of the cell repair system to mend the DNA double strand breaks that are caused in effect by IR (directly or by radicals produced from water radiolysis), which triggers apoptosis in rapidly proliferating cells. In nonreplicative cells, IR can mediate structural changes in the proteins and in plasma membrane lipids that can affect cellular interactions with the ECM molecules and alter intracellular signaling [50]. Neuronal cell membrane damage can further exert physiologically detrimental effects, such as axonal demyelination, leading to inefficient neurotransmission, and neuronal cell death. Clinical challenges in RT have been reviewed by Chao et al. (2013) and others [66, 237, 238].

Molecular and cellular adaptations to RT

Radiation therapy exerts a profound influence on the GBM microenvironment, which plays a deterministic role in tumor response to therapy and in tumor recurrence. ROS production is a prime consequence of RT administration, which can have multitudes of effects on GBM pathology [19, 61, 239].

The extracellular microenvironment is a unique feature of the tumor. It forms the matrix composition and allows ECM modulation and cell-matrix interactions that characterize aggressiveness in GBM [54, 138, 140, 158, 159, 162]. Radiation injury can significantly alter matrix composition and functions [59, 69]. Changes in stroma caused by tumor intrinsic and extrinsic factors, such as IR, can lead to reduced treatment response and development of resistance (Fig. 1) [50]. Various pathophysiological processes elicited by brain radiation include oxidative stress (caused by production of ROS/RNS or activation of NF-κB, NOX4), neuroinflammation (mediated by release of cytokines TNF-α, IL-1β, IL-6, TGF-β), and chemokine (CX_3CL1 or fractalkine, CCL3 or macrophage inhibitory factor 1, CCL2, CCL3, CCL5)) expression of cell adhesion molecules (intercellular adhesion molecule 1 or ICAM-1, P-selectin), ECM modulation (mediated by MMPs, TIMPs, LOX, and through ROS/RNS), neo-angiogenesis (by upregulation of VEGF, angiotensins), stem/progenitor cell death, and induction of cellular senescence and impaired neurogenesis [61, 68, 69, 74, 207–211, 240–247].

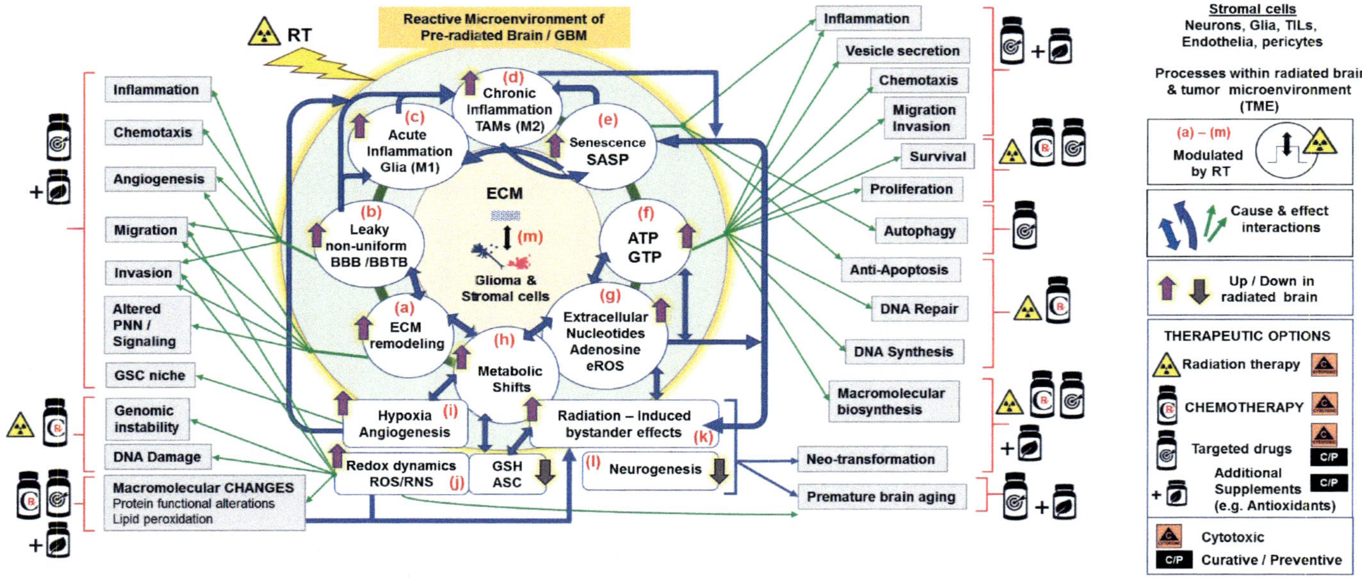

FIG. 1 Impact of RT on brain and glioblastoma (GBM) microenvironment (TME) and its therapeutic cross talks. ECM and its interaction with cellular components, glia (microglia and astrocytes), and peripherally derived tumor infiltrating leukocytes (TILs), play a central role in generation of reactive GBM TME, which contributes to tumor cell survival, proliferation, migration, and invasion. Radiation injury leads to neuronal damage and overactivation of M1 microglia, and elicits an acute inflammatory response, with high ROS production in neurons and glia cells. There is an alteration in ECM composition and ECM-cell interactions, which alters the composition of perineuronal nets (PNN), thereby affecting signaling and transmission across neurons. MMP/TIMP imbalance degrades Col-IV in basement membranes, which leads to blood-brain barrier leakage. Pronounced inflammation causes infiltration of leukocytes (monocyte-derived cell populations, macrophages), which, along with activated microglia, forms the TAMs. TME of progressive tumors favors the M2 phenotype and establishment of chronic inflammation. Redox dysregulation due to RT causes exacerbation of the SASP phenotype and tumor cell adaptive processes, such as hypoxia, metabolic shifts, redox regulation (ROS/RNS production), leading to neo-angiogenesis and furthered ECM remodeling. Exhaustion of antioxidants (GSH, ASC) and increased production of energy carriers (ATP, GTP) due to stress-induced signaling (AMPK) and proliferative cues mediate the release of extracellular ROS and extracellular nucleotides (eATP, eGTP, adenosine, guanosine), which activate signaling pathways in an autocrine and paracrine manner, causing therapeutic resistance and tumor progression. These alterations collectively make the TME permissive to glioma cell migration and invasion, thereby contributing to resistance and an aggressive recurrent tumor. Additionally, dying and degenerative cells in preradiated brain and tumor stroma secrete extracellular nucleotides, stress mediators, chronic inflammatory signals, and damaged DNA fragments. These impact the nontargeted cells, setting in radiation-induced bystander effects, which ultimately lead to genomic alterations in noncancerous cells, causing neo-transformations, cognitive decline, and premature brain aging. The green circle represents the reactive microenvironment of the preradiated brain or GBM TME. (a)–(m) state various cell molecular mechanisms modulated in effect of RT and *blue arrows* indicate their positive cross talks that form an integral part of the glioblastoma pathology. Purple small arrows represent the upregulation of these phenomena in radiated

(Continued)

Radiation effects on ECM remodeling and invasiveness

Radiation-induced oxidative stresses affect ECM biosynthesis and degradation, leading to activation of cellular processes such as inflammation, autophagy, or senescence, which can regulate GBM cell proliferation, migration, and invasion [27]. Oxygen and ROS are involved in different steps of ECM protein transcription, translation, and posttranslational modification, such as the hydroxylation of proline residues in collagen chains within the cells, collagen fibril formation, and extracellular aggregation. Secretion of lysyl oxidase, the key enzyme responsible for stabilization of collagen aggregates in the ECM, has been found to be induced by IR [49, 248]. Specific proteins involved in ECM biosynthesis (brevican, tenascin C, hyaluronan, lysyl oxidase) and degradation (matrix metalloproteinases) [44, 49, 52, 125, 137, 160–167, 249], signal transduction (melanoma differentiation-associated gene 9/Syntenin) [46, 175], and ECM-glioma cell interactions [ICAM-1, DDR-1, integrins, CD44, Neurophilin-1 (Nrp1)] [27, 54, 177–179, 183, 184], have been shown to be upregulated postradiation.

Elevated levels of ROS cause matrix dysfunction through remodeling and fragmentation of collagen and proteoglycans, pronounced protease activity, and altered cytoskeletal contractility, thereby fueling senescent phenotypes marked by irregular collagen meshworks and ECM degradation [250, 251]. The GAG chains are nonenzymatically fragmented by ROS generated in effect of IR [252]. ECM proteins maintained by a favorable balance of the MMPs/TIMPs system play a pivotal role in maintaining normal brain homeostasis. Radiation can alter this balance, thereby reducing the tension and elasticity of affected tissue. These alterations collectively increase malignancy and enhance the migration and invasion of tumor cells [24, 27, 52, 91, 138, 144, 155, 159, 168–172, 180, 181, 185, 186].

Epithelial to mesenchymal transition (EMT), a property of highly malignant and aggressive GBM, depends on the MMP mediated ECM remodeling, which in turn is a function of ROS. ECM degradative processes alter cell-matrix interactions, thereby affecting the intracellular cytoskeleton, causing cells to acquire a mesenchymal phenotype [187–189, 253]. Radiation also promotes GBM cell migration and invasiveness in a p53-independent manner involving enhanced integrin expression, altered MMP expression and activity, altered

FIG. 1, CONT'D TME and *gray small arrows* represent those downregulated. *Green arrows* indicate the specific biological (functional) processes that are impacted by each of these RT-induced alterations in cell mechanisms. The therapeutic opportunities available against each of these biological processes are illustrated respectively as radiation therapy (RT), chemotherapy, targeted therapy and their combinations, along with futuristic approaches involving additional supplements (antioxidants). The nature of these treatment modalities is indicated in the side-key, categorizing them as cytotoxic, curative, or preventive. While chemoradiotherapy combats the tumor growth by killing proliferative cancer cells having DNA damage, genomic mutations, and rearrangements, the RT-induced changes in the stromal microenvironment can cause severe redox stress and metabolic shifts that can affect DNA methylation status and develop new genomic landscapes with altered signaling, macromolecular synthesis, and DNA repair, thereby causing chemotherapeutic escape, resistance, neo-transformation, and aggressive tumor recurrence. RT, Radiation therapy; ECM, extracellular matrix; MMP, matrix metalloproteinases; TIMP, tissue inhibitors of metalloproteinases; TAM, Tumor-associated macrophages; TILs, tumor infiltrating leukocytes; BBB, blood-brain barrier; BBTB, blood-brain tumor barrier; ROS, reactive oxygen species; RNS, reactive nitrogen species; SASP, senescence-associated secretory phenotype; M1/M2, proinflammatory or immune suppressive phenotypes of TAMs, respectively; ATP, adenosine triphosphate; GTP, guanosine triphosphate; ASC, ascorbate; GSH, glutathione (reduced); PNN, perineuronal nets; GSC, glioma stem cells. Symbols used: fluorescent yellow lightening for RT administration via ionizing radiations; standard radiation symbol, for radiation therapy; drug bottles: "C&Rx," chemotherapy; "target icon," targeted therapy; "leaf icon," for additional supplements (natural antioxidants, vitamins, etc.).

membrane type 1 MMP and tissue inhibitor of MMP-2 expression, and altered BCL-2/BAX expression, resulting in apoptosis resistance, enhanced migration, [24, 191, 192] and cognitive decline [241].

Additional radiation-induced modifications in the tumor or preradiated stromal microenvironment are associated with poor outcome in GBM [254, 255]. These alterations include neurogenesis impairment [58, 244] and severe metabolic stresses (metabolic shifts, rise in energy carriers such as intracellular/extracellular ATP and GTP, adenine, guanine, with reduction in levels of coenzymes (NAD^+), TCA cycle intermediates, and antioxidants (glutathione, GSH; ascorbate, ASC)). These metabolomic changes can lead to tumor immune escape, GSC niche formation, maintenance and GSC activation, microglial chemotaxis, microglial and astrocytic activation, and inflammation, angiogenesis, migration, invasion, proliferation, and secondary tumor growth [28, 62, 65, 70, 193, 231–233]. It is therefore reasonable to postulate that the aggressive character during tumor recurrence is a consequence of radiation-induced changes in the GBM microenvironment.

Radiation-induced redox dynamics and modulation of inflammation

Reactive oxygen/nitrogen species (ROS/RNS) are by-products of metabolism. In rapidly proliferating cells, they exert cytotoxicity by damaging the cell macromolecules and induction of apoptosis. Radiotherapy exploits this principle to kill proliferating cancer cells. Tumors adapt to the redox stress through several adaptive mechanisms including metabolic shifts, increasing levels of antioxidant peptides, and by generation of intra-tumoral hypoxia [256, 257]. Radiation-induced stabilization and activation of hypoxia inducible factor 1 (HIF-1) elicit protective processes through regulation of downstream target genes, such as nitric oxide (NO), which can stimulate responses that are immunosuppressive and antiapoptotic [258, 259]. Moreover, under low oxygen conditions, DNA peroxide radicals are either not formed or do not suffice to induce apoptosis. Tumor hypoxia is therefore associated with radioresistance and recurrence [237, 260].

Altered matrix composition and vasculature are hallmarks of radiation injury, mediated by various inflammatory processes. The BBB is physiologically impermeable to immune cells. However, the cytokines and chemokines released by IR injured neurons, activated astrocytes, and microglia can lead to endothelial cell activation and a subsequent increase in BBB permeability [50, 57, 61, 63, 64, 67, 69, 71, 73, 75, 159, 176, 200, 213–226, 261, 262]. Matrix metalloproteinases (MMPs) and plasminogen activators, the two main matrix-degrading proteolytic enzymes, are induced by radiation, within and around the endothelial cells. They disrupt the basal lamina enwrapping cerebral capillaries and open the BBB [226].

Radiation effects on vasculature integrity elicit a series of events causing vasculopathy, vascular depletion, hypoxia induction, and subsequent neo-angiogenesis. ECM remodeling at the basal lamina and vasculature breakdown leads to chemotaxis and activation of resident microglia at the site of injury, recruitment of peripheral monocyte-derived populations, potentiation of intra-tumoral hypoxia, and establishment of chronic inflammation [59–61, 66, 69, 73, 75, 226, 263]. Tumor hypoxia is widely associated with radioresistance and recurrence [237, 264]. Radiation-dependent Hypoxia inducible factor-1 (HIF-1) stabilization and activation elicit multiple cell protective processes, through regulation of its downstream target genes such as VEGF and nitric oxide (NO) production, which can stimulate wound-healing responses that are angiogenic, immunosuppressive, and antiapoptotic [258]. Chronic inflammation in tumors leads to high intracellular NO in GBM cells that causes stabilization of HIF-1 and inhibition of TIMP-1, a tissue inhibitor of matrix metalloproteases-1. Both these events contribute to ECM remodeling and facilitate tumor cell invasion [176, 260]. Radiation-induced NO production in vascular cells can protect cells and tissues from radiation damage by stimulating soluble guanylate cyclase production of cGMP to promote cell survival pathways, which can be counteracted by the matricellular protein, thrombospondin-1 (TSP1). Whether NO would stimulate immunosuppressive/prosurvival signaling depends on its concentration gradient, and inhibition of NO signaling by TSP1 radiosensitizes the cells. Therefore, targeting the TSP1/CD47 axis could have radioprotective effects on nontransformed cells in the GBM microenvironment or distant from the primary tumor [219, 227, 260].

Redox signaling has a profound impact on microglial fate during tumor insult or radiation-induced brain injury. Microglia (MG) and macrophages (MPs) represent a significant component of the inflammatory response to GBMs [211, 265]. The differential expression of iNOS and arginase-1 (ARG1) in M1 compared with M2 profiles provides the basis of redox control of the MG/MP phenotypes (also called the tumor-associated macrophage, TAM phenotypes) [262, 266]. Transcription factor nuclear factor (erythroid-derived 2)-like 2 (NRF2) is the master regulator of redox dynamics and favors an M2 phenotype, which has cytoprotective effects [267]. GBM and stromal cells maintain a pool of antioxidant peptides (glutathione, GSH; thioredoxin, Trx) that protects them against redox stress induced by RT. The enzymatic processes involve glutathione reductase and thioredoxin reductase. Their expression is particularly increased during cancer. In microglia, GSH is the prime molecule for antioxidant defense, whose antioxidant effects are mediated by an additional nonenzymatic mechanism, the γ-glutamyl cycle, which requires exchange of intermediates between the cell and extracellular space. This further emphasizes the importance of stromal biochemistry in survival of activated microglia [262]. Differences in the intrinsic ability of GBM cells to modulate such redox reactions explain a mechanism that causes certain cancer cells to be more radioresistant than others and acts as seeds for tumor recurrence.

In addition to apoptosis, radiation exposure of cells can result in other cell death mechanisms, including necrosis or autophagy, or may lead to induction of accelerated cellular senescence [268, 269]. While the selection process for a specific mode of cell death or senescence is not clearly defined, studies indicate that it is affected by the radiation dose, the dose rate, and multiple cellular intrinsic and extrinsic processes, such as induction of intracellular stress responses (e.g., ER stress), and secretion of various cytokines and growth factors [268]. GBM cells harbor a variety of mutations that can prevent cells from undergoing cell death despite harboring the irreparable DNA lesions, leading to accumulation of mutations, with constitutive activation of mechanisms for DNA repair that would otherwise render normal cells unviable [270–273]. These survival mechanisms culminate in the formation of chemoradioresistant clones, which can fuel tumor recurrence.

Radiation-induced cell senescence and bystander effects

Cellular senescence occurs as a defensive mechanism in response to excessive extracellular or intracellular stress pathways. A plethora of stresses can be provoked by radiation which includes severe or irreparable DNA damage and chromatin disruption (genotoxic stress), oxidative stress (by ROS/RNS), endoplasmic reticulum stress, and mitochondrial deterioration. When oxidative damage may surpass a crucial threshold, it results in ineffective maintenance of functional mitochondria, thus fueling senescent phenotypes marked with altered cytoskeletal dynamics, and a disproportionate ECM having excess deposition of fibronectins, stiffening of ECM at local regions, irregular collagen meshworks, and ECM degradation [27, 139, 274–276].

Radiation-induced vasculature permeability, triggered by activated microglia and astrocytes, leads to infiltration of leukocytes and monocyte-derived cells/macrophages into the brain parenchyma. Chronic inflammation occurs when the rate of leukocyte infiltration during an acute inflammatory process is not sufficiently abundant to halt the process. This leads to persistent activation of microglia and astrocytes. Prolonged microglial activation leads to a vicious circle, where secretion of proinflammatory cytokines and other neurotoxic agents (ROS and RNS) leads to further neuronal damage and cell death, which maintains microglial cells in their activated status. Radiation-induced cellular senescence is characterized by chronic inflammatory processes, where there is an interplay between the acute inflammation owing to detrimental processes, such as lipid peroxidation and neuronal damage, and the immune suppression mediated by infiltrating immunosuppressive lymphocytes or altered microglial phenotypes [277–282]. This renders the brain parenchyma a disorganized meshwork of ECM that can be permissive to invasion and an aggressive recurrence. In addition, the GBM cell survival processes such as hypoxia and metabolic adaptations favor a shift from the M1 to M2 phenotype in the tumor-associated macrophages/microglia (TAMs) [211, 283]. These TAMs of M2-type initiate a wound healing-like process, angiogenesis, ECM production, remodeling, and other prosurvival processes through secretion of cytokines such as TGFβ, which collectively lead to a tumor progressive phenotype.

In addition, a continuum of mutations accumulated in GBM cells can lead to oncogene-induced senescence [284, 285]. The initiation of senescence is affected by the radiation dose and rate [78, 286, 287]. Senescent cells develop a metabolically active, proinflammatory phenotype, described as a senescence-associated secretory phenotype (SASP), with release of various soluble factors, cytokines, chemokines, and growth factors (GM-CSF, IL-6, IL-8, IL-1α, IL-1β, TGFβ) that can trigger survival signaling pathways, ECM remodeling and result in chronic inflammation in the GBM TME [78, 288, 289], making it promigratory and protumorigenic [288, 290]. SASP, through its secretory phenotype, can exhibit autocrine and paracrine effects, spreading the effects to neighboring cells, and may intensify local healthy tissue deterioration or even exacerbate the chronic state to exhibit long-term deleterious effects [27].

Radiation-induced bystander effects (RIBE) contribute to tumor promotion and recurrence [291]. Bystander effects are defined by a cell reacting to its neighboring irradiated cell. The consequence of such effects is the inevitable damage caused to the nearby healthy regions of the brain. Bystander effects are triggered by changes in the extracellular microenvironment, mediated by production of oxidized cell-free DNA, extracellular nucleotides, and reactive oxygen species (eATP, eGTP, adenine, guanine, eROS), extracellular vesicles (EVs), all of which can communicate their effect distant from the primary tumor. Oxidized cell-free DNA fragments can augment chemotherapy resistance in surviving tumor cells after chemoradiotherapy. Irradiated GBM cells have been shown to induce bystander effects including cell growth, micronucleus formation, and apoptosis in non-irradiated tumor cells [13, 15–17, 76, 77, 79, 80, 234, 235]. These bystander processes can ignite ROS production and mitochondrial dysfunction, leading to persistent or irreparable DNA damage, activation of DNA damage responses, irreversible cell cycle arrest, and culmination of cellular senescence [78].

Therapeutic implications and future directions

Understanding the RT-induced microenvironmental changes in the brain and GBM tumors is critical for optimization of current standard of care with the development of new cytotoxic, metabolic, and molecular targeted therapies to improve GBM care [27, 292]. Surgery followed by standard radiotherapy with concomitant and adjuvant chemotherapy with TMZ is the standard of care. However, treatment resistance and recurrence is observed in 100% of GBM patients. Common genetic abnormalities in GBM are associated with aberrant activation or suppression of cellular signal transduction pathways that can contribute to tumor progression and development of resistance to radiation and chemotherapy. Inhibitors to signaling cascades such as epidermal growth factor receptor (EGFR), vascular endothelial growth factor receptor (VEGFR), platelet derived growth factor receptor (PDGF), phosphatidylinositol-3kinase and mammalian target of rapamycin (PI3K/Akt, mTOR), and Ras/Raf/mitogen-activated protein-kinase (Ras/Raf/MAPK) pathways have been examined in preclinical and clinical malignant glioma trials, including bevacizumab, enzastaurin (antiangiogenic), gefitinib and erlotinib (anti-EGFR), temsirolimus, everolimus (anti-mTOR), and cilengitide (integrin targeting). Although the preliminary clinical results of these targeted agents have not been promising, trials to use combinations of multitargeted drugs with cytotoxic chemotherapy and radiotherapy are underway [213, 215, 225, 283, 293].

Resistance to chemotherapy observed in nearly all patients of GBM results from the failure of drugs reaching effectively into the brain, owing to the BBB and to upregulation of drug efflux transporters in GBM [64, 67, 294–296]. Recent studies have shown a plausible role for RT-induced changes in the tumor and its microenvironment to confer treatment resistance, via demodulation of several of the protumorigenic survival, angiogenic and antiapoptotic signaling pathways and alterations in extracellular signaling mediated by changes in metabolism, ECM, ion channels, and ionic gradients [5, 28, 51, 58, 62, 65, 159, 187, 193–199, 201–203, 216, 220, 222, 231, 297].

Despite systemic radiosensitizers in combination with RT being explored to enhance the effects of radiation [298], the highly diffusive characteristic of GBM leads to the inability of RT to control GBM. ECM forms the substratum in the brain that facilitates the integrity of perineuronal nets (PNNs) for proper neurotransmission to occur; it also permits the diffusion of ions, molecules, cytokines, chemokines for maintenance of brain homeostasis. Additionally, ECM forms the structural component of the blood-brain barrier. Radiation-induced changes in the ECM involve alterations in its structure and composition in the GBM tumor and stromal compartment and changes in the vascular architecture, making the BBB leaky [27]. These contribute to altered diffusion coefficients of molecules in the extracellular space and impact the availability and distribution of chemotherapeutic agents within the brain, preventing them from reaching therapeutically relevant concentrations at all sites of the tumor. Additionally, ECM, GBM cell, and stromal cell interactions trigger several intracellular signaling pathways that are inflammatory, promigratory, and proinvasive [27]. Thus, the functional relevance of ECM modulation after RT makes it a potential therapeutic target

(Table 1). Studies on the inhibition of ECM biosynthetic and degradative processes and receptor blockade to prevent ECM-cell interactions for cancer prevention further support this idea [299]. Targeting cytoskeletal dynamics and neo-vascularization has also been proposed as a promising therapeutic strategy [27, 71, 181, 213–215, 218, 225, 300].

Targeting specific microglial-ECM interactions that promote protumor phenotypes in GBM may also offer opportunities for therapeutic intervention. Further investigation of RT-induced injury to the BBB holds the potential for enhanced delivery of targeted chemotherapeutic agents [57, 63, 64, 67, 301]. For example, image-guided chemotherapeutic delivery following BBB permeability modulation has been previously investigated and may provide avenues to enhance glioma treatment [302]. Additionally, while nanoparticles are widely documented for their radiosensitization effects, the potential for RT to improve nanomedicine efficacy and delivery of chemotherapy and targeted therapies shows promise for combining nanoparticles and RT for GBM care [303–306].

Conclusions

The GBM microenvironment is highly dynamic, consisting of the extracellular matrix and an array of various cell types including neurons, glia, pericytes, endothelial cells, TAMs, Tumor infiltrating leucocytes (TILs), and brain infiltrating leucocytes (BILs), which interact with each other and the tumor cells leading to formation of a continually changing tissue ecosystem which is inevitable for the progression of glioblastoma pathophysiology. These microenvironmental cellular and molecular changes account for the development of chemoresistance and observed recurrence in GBM.

Understanding the radiation-induced alterations in the GBM stroma adds a new dimension to how these microenvironmental orchestrations promote tumor immune escape, treatment resistance, and tumor aggressiveness during recurrence. The dependence of the tumor for its survival and aggressive character on its stromal microenvironment, with having a relatively stable genetic composition of the stroma, as opposed to the highly heterogeneous tumor cells, makes the stromal compartment a better candidate for targeted therapies. Understanding the RT-induced changes in ECM synthesis and degradation offers opportunities to harness matrix composition/pore size and BBB/vasculature remodeling to better the availability and distribution of chemotherapeutic agents in highly malignant diffused gliomas. Additionally, investigating the radiation-induced changes in the glioblastoma microenvironment opens several new avenues for combinatorial synergistic treatments involving ECM stabilizers, migration and proliferation inhibitors with metabolic modulators to overcome the late effects of RT that are permissive to tumor recurrence by restoring brain homeostasis and health.

References

[1] Meyer MA. Malignant gliomas in adults. N Engl J Med 2008;359:1850. author reply 1850.

[2] Louis DN, Perry A, Reifenberger G, von Deimling A, Figarella-Branger D, Cavenee WK, et al. The 2016 World Health Organization classification of tumors of the central nervous system: a summary. Acta Neuropathol 2016;131:803–20.

[3] Mann J, Ramakrishna R, Magge R, Wernicke AG. Advances in radiotherapy for glioblastoma. Front Neurol 2017;8:748.
[4] Farin A, Suzuki SO, Weiker M, Goldman JE, Bruce JN, Canoll P. Transplanted glioma cells migrate and proliferate on host brain vasculature: a dynamic analysis. Glia 2006;53:799–808.
[5] Zagzag D, Esencay M, Mendez O, Yee H, Smirnova I, Huang Y, et al. Hypoxia- and vascular endothelial growth factor-induced stromal cell-derived factor-1alpha/CXCR4 expression in glioblastomas: one plausible explanation of Scherer's structures. Am J Pathol 2008;173:545–60.
[6] Seifert S, Sontheimer H. Bradykinin enhances invasion of malignant glioma into the brain parenchyma by inducing cells to undergo amoeboid migration. J Physiol 2014;592:5109–27.
[7] Xia S, Lal B, Tung B, Wang S, Goodwin CR, Laterra J. Tumor microenvironment tenascin-C promotes glioblastoma invasion and negatively regulates tumor proliferation. Neuro Oncol 2016;18:507–17.
[8] Manini I, Caponnetto F, Bartolini A, Ius T, Mariuzzi L, Di Loreto C, et al. Role of microenvironment in glioma invasion: what we learned from in vitro models. Int J Mol Sci 2018;19(1):147.
[9] Stupp R, Hegi ME, Gilbert MR, Chakravarti A. Chemoradiotherapy in malignant glioma: standard of care and future directions. J Clin Oncol 2007;25:4127–36.
[10] Stupp R, Mason WP, van den Bent MJ, Weller M, Fisher B, Taphoorn MJ, et al. Radiotherapy plus concomitant and adjuvant temozolomide for glioblastoma. N Engl J Med 2005;352:987–96.
[11] Makale MT, McDonald CR, Hattangadi-Gluth JA, Kesari S. Mechanisms of radiotherapy-associated cognitive disability in patients with brain tumours. Nat Rev Neurol 2017;13:52–64.
[12] Li MD, Burns TC, Kumar S, Morgan AA, Sloan SA, Palmer TD. Aging-like changes in the transcriptome of irradiated microglia. Glia 2015;63:754–67.
[13] Burns TC, Awad AJ, Li MD, Grant GA. Radiation-induced brain injury: low-hanging fruit for neuroregeneration. Neurosurg Focus 2016;40, E3.
[14] Wu PH, Coultrap S, Pinnix C, Davies KD, Tailor R, Ang KK, et al. Radiation induces acute alterations in neuronal function. PLoS One 2012;7, e37677.
[15] Greene-Schloesser D, Moore E, Robbins ME. Molecular pathways: radiation-induced cognitive impairment. Clin Cancer Res 2013;19:2294–300.
[16] Greene-Schloesser D, Robbins ME. Radiation-induced cognitive impairment—from bench to bedside. Neuro Oncol 2012;14(Suppl 4):iv37–44.
[17] Surma-aho O, Niemela M, Vilkki J, Kouri M, Brander A, Salonen O, et al. Adverse long-term effects of brain radiotherapy in adult low-grade glioma patients. Neurology 2001;56:1285–90.
[18] Douw L, Klein M, Fagel SS, van den Heuvel J, Taphoorn MJ, Aaronson NK, et al. Cognitive and radiological effects of radiotherapy in patients with low-grade glioma: long-term follow-up. Lancet Neurol 2009;8:810–8.
[19] Azzam EI, Jay-Gerin JP, Pain D. Ionizing radiation-induced metabolic oxidative stress and prolonged cell injury. Cancer Lett 2012;327:48–60.
[20] Greene-Schloesser D, Robbins ME, Peiffer AM, Shaw EG, Wheeler KT, Chan MD. Radiation-induced brain injury: a review. Front Oncol 2012;2:73.
[21] Sharma NK, Sharma R, Mathur D, Sharad S, Minhas G, Bhatia K, et al. Role of ionizing radiation in neurodegenerative diseases. Front Aging Neurosci 2018;10:134.
[22] Gonzalez LC, Ghadaouia S, Martinez A, Rodier F. Premature aging/senescence in cancer cells facing therapy: good or bad? Biogerontology 2016;17:71–87.
[23] Leonardi GC, Accardi G, Monastero R, Nicoletti F, Libra M. Ageing: from inflammation to cancer. Immun Ageing 2018;15:1.
[24] Wild-Bode C, Weller M, Rimner A, Dichgans J, Wick W. Sublethal irradiation promotes migration and invasiveness of glioma cells: implications for radiotherapy of human glioblastoma. Cancer Res 2001;61:2744–50.
[25] Shankar A, Kumar S, Iskander AS, Varma NR, Janic B, deCarvalho A, et al. Subcurative radiation significantly increases cell proliferation, invasion, and migration of primary glioblastoma multiforme in vivo. Chin J Cancer 2014;33:148–58.
[26] Moncharmont C, Levy A, Guy JB, Falk AT, Guilbert M, Trone JC, et al. Radiation-enhanced cell migration/invasion process: a review. Crit Rev Oncol Hematol 2014;92:133–42.
[27] Gupta K, Burns TC. Radiation-induced alterations in the recurrent glioblastoma microenvironment: therapeutic implications. Front Oncol 2018;8:503.
[28] Gupta K, Vuckovic I, Zhang S, Xiong Y, Carlson BL, Jacobs J, et al. Radiation induced metabolic alterations associate with tumor aggressiveness and poor outcome in glioblastoma. Front Oncol 2020;10:535.

[29] Sundahl N, Duprez F, Ost P, De Neve W, Mareel M. Effects of radiation on the metastatic process. Mol Med 2018;24:16.
[30] Arnold KM, Flynn NJ, Raben A, Romak L, Yu Y, Dicker AP, et al. The impact of radiation on the tumor microenvironment: effect of dose and fractionation schedules. Cancer Growth Metastasis 2018;11. 1179064418761639.
[31] Betlazar C, Middleton RJ, Banati RB, Liu GJ. The impact of high and low dose ionising radiation on the central nervous system. Redox Biol 2016;9:144–56.
[32] Chen H, Chong ZZ, De Toledo SM, Azzam EI, Elkabes S, Souayah N. Delayed activation of human microglial cells by high dose ionizing radiation. Brain Res 1646;2016:193–8.
[33] Stapleton S, Jaffray D, Milosevic M. Radiation effects on the tumor microenvironment: implications for nanomedicine delivery. Adv Drug Deliv Rev 2017;109:119–30.
[34] Leroi N, Lallemand F, Coucke P, Noel A, Martinive P. Impacts of ionizing radiation on the different compartments of the tumor microenvironment. Front Pharmacol 2016;7:78.
[35] Shibamoto Y, Miyakawa A, Otsuka S, Iwata H. Radiobiology of hypofractionated stereotactic radiotherapy: what are the optimal fractionation schedules? J Radiat Res 2016;57(Suppl 1):i76–82.
[36] Moravan MJ, Olschowka JA, Williams JP, O'Banion MK. Brain radiation injury leads to a dose- and time-dependent recruitment of peripheral myeloid cells that depends on CCR2 signaling. J Neuroinflamm 2016;13:30.
[37] Yang L, Yang J, Li G, Li Y, Wu R, Cheng J, et al. Pathophysiological responses in rat and mouse models of radiation-induced brain injury. Mol Neurobiol 2017;54:1022–32.
[38] Brown JM, Carlson DJ, Brenner DJ. The tumor radiobiology of SRS and SBRT: are more than the 5 Rs involved? Int J Radiat Oncol Biol Phys 2014;88:254–62.
[39] Kaminaga T, Shirai K. Radiation-induced brain metabolic changes in the acute and early delayed phase detected with quantitative proton magnetic resonance spectroscopy. J Comput Assist Tomogr 2005;29:293–7.
[40] Wank M, Schilling D, Schmid TE, Meyer B, Gempt J, Barz M, et al. Human glioma migration and infiltration properties as a target for personalized radiation medicine. Cancers (Basel) 2018;10(11):456.
[41] Zhang X, Wang X, Xu R, Ji J, Xu Y, Han M, et al. YM155 decreases radiation-induced invasion and reverses epithelial-mesenchymal transition by targeting STAT3 in glioblastoma. J Transl Med 2018;16(1):79.
[42] Zschaeck S, Wust P, Graf R, Misch M, Onken J, Ghadjar P, et al. Locally dose-escalated radiotherapy may improve intracranial local control and overall survival among patients with glioblastoma. Radiat Oncol 2018;13:251.
[43] Yool AJ, Ramesh S. Molecular targets for combined therapeutic strategies to limit glioblastoma cell migration and invasion. Front Pharmacol 2020;11:358.
[44] Yoo KC, Suh Y, An Y, Lee HJ, Jeong YJ, Uddin N, et al. Proinvasive extracellular matrix remodeling in tumor microenvironment in response to radiation. Oncogene 2018;37:3317–28.
[45] Strickland M, Stoll EA. Metabolic reprogramming in glioma. Front Cell Dev Biol 2017;5:43.
[46] Kegelman TP, Wu B, Das SK, Talukdar S, Beckta JM, Hu B, et al. Inhibition of radiation-induced glioblastoma invasion by genetic and pharmacological targeting of MDA-9/Syntenin. Proc Natl Acad Sci U S A 2017;114:370–5.
[47] Fidoamore A, Cristiano L, Antonosante A, d'Angelo M, Di Giacomo E, Astarita C, et al. Glioblastoma stem cells microenvironment: the paracrine roles of the niche in drug and radioresistance. Stem Cells Int 2016;2016:6809105.
[48] Pei J, Park IH, Ryu HH, Li SY, Li CH, Lim SH, et al. Sublethal dose of irradiation enhances invasion of malignant glioma cells through p53-MMP 2 pathway in U87MG mouse brain tumor model. Radiat Oncol 2015;10:164.
[49] Shen CJ, Sharma A, Vuong DV, Erler JT, Pruschy M, Broggini-Tenzer A. Ionizing radiation induces tumor cell lysyl oxidase secretion. BMC Cancer 2014;14:532.
[50] Yoshimura M, Itasaka S, Harada H, Hiraoka M. Microenvironment and radiation therapy. Biomed Res Int 2013;2013:685308.
[51] Li HF, Kim JS, Waldman T. Radiation-induced Akt activation modulates radioresistance in human glioblastoma cells. Radiat Oncol 2009;4:43.
[52] Donson AM, Erwin NS, Kleinschmidt-DeMasters BK, Madden JR, Addo-Yobo SO, Foreman NK. Unique molecular characteristics of radiation-induced glioblastoma. J Neuropathol Exp Neurol 2007;66:740–9.
[53] Park CM, Park MJ, Kwak HJ, Lee HC, Kim MS, Lee SH, et al. Ionizing radiation enhances matrix metalloproteinase-2 secretion and invasion of glioma cells through Src/epidermal growth factor receptor-mediated p38/Akt and phosphatidylinositol 3-kinase/Akt signaling pathways. Cancer Res 2006;66:8511–9.
[54] Bellail AC, Hunter SB, Brat DJ, Tan C, Van Meir EG. Microregional extracellular matrix heterogeneity in brain modulates glioma cell invasion. Int J Biochem Cell Biol 2004;36:1046–69.

[55] Nirmala C, Jasti SL, Sawaya R, Kyritsis AP, Konduri SD, Ali-Osman F, et al. Effects of radiation on the levels of MMP-2, MMP-9 and TIMP-1 during morphogenic glial-endothelial cell interactions. Int J Cancer 2000;88:766–71.

[56] Wang JL, Sun Y, Wu S. Gamma-irradiation induces matrix metalloproteinase II expression in a p53-dependent manner. Mol Carcinog 2000;27:252–8.

[57] Wang D, Wang C, Wang L, Chen Y. A comprehensive review in improving delivery of small-molecule chemotherapeutic agents overcoming the blood-brain/brain tumor barriers for glioblastoma treatment. Drug Deliv 2019;26:551–65.

[58] Tong J, Li J, Zhang QS, Yang JK, Zhang L, Liu HY, et al. Delayed cognitive deficits can be alleviated by calcium antagonist nimodipine by downregulation of apoptosis following whole brain radiotherapy. Oncol Lett 2018;16:2525–32.

[59] George N, Geller HM. Extracellular matrix and traumatic brain injury. J Neurosci Res 2018;96:573–88.

[60] Sarkaria JN, Hu LS, Parney IF, Pafundi DH, Brinkmann DH, Laack NN, et al. Is the blood-brain barrier really disrupted in all glioblastomas? A critical assessment of existing clinical data. Neuro Oncol 2018;20:184–91.

[61] Lumniczky K, Szatmari T, Safrany G. Ionizing radiation-induced immune and inflammatory reactions in the brain. Front Immunol 2017;8:517.

[62] Kojima S, Ohshima Y, Nakatsukasa H, Tsukimoto M. Role of ATP as a key signaling molecule mediating radiation-induced biological effects. Dose Response 2017;15. 1559325817690638.

[63] Appelboom G, Detappe A, LoPresti M, Kunjachan S, Mitrasinovic S, Goldman S, et al. Stereotactic modulation of blood-brain barrier permeability to enhance drug delivery. Neuro Oncol 2016;18:1601–9.

[64] Oberoi RK, Parrish KE, Sio TT, Mittapalli RK, Elmquist WF, Sarkaria JN. Strategies to improve delivery of anticancer drugs across the blood-brain barrier to treat glioblastoma. Neuro Oncol 2016;18:27–36.

[65] Tsukimoto M. Purinergic signaling is a novel mechanism of the cellular response to ionizing radiation. Biol Pharm Bull 2015;38:951–9.

[66] Bhowmik A, Khan R, Ghosh MK. Blood brain barrier: a challenge for effectual therapy of brain tumors. Biomed Res Int 2015;2015:320941.

[67] van Tellingen O, Yetkin-Arik B, de Gooijer MC, Wesseling P, Wurdinger T, de Vries HE. Overcoming the blood-brain tumor barrier for effective glioblastoma treatment. Drug Resist Updat 2015;19:1–12.

[68] Chiang CS, Fu SY, Wang SC, Yu CF, Chen FH, Lin CM, et al. Irradiation promotes an m2 macrophage phenotype in tumor hypoxia. Front Oncol 2012;2:89.

[69] Lee WH, Warrington JP, Sonntag WE, Lee YW. Irradiation alters MMP-2/TIMP-2 system and collagen type IV degradation in brain. Int J Radiat Oncol Biol Phys 2012;82:1559–66.

[70] Sanli T, Rashid A, Liu C, Harding S, Bristow RG, Cutz JC, et al. Ionizing radiation activates AMP-activated kinase (AMPK): a target for radiosensitization of human cancer cells. Int J Radiat Oncol Biol Phys 2010;78:221–9.

[71] Kioi M, Vogel H, Schultz G, Hoffman RM, Harsh GR, Brown JM. Inhibition of vasculogenesis, but not angiogenesis, prevents the recurrence of glioblastoma after irradiation in mice. J Clin Invest 2010;120:694–705.

[72] Higano S, Yun X, Kumabe T, Watanabe M, Mugikura S, Umetsu A, et al. Malignant astrocytic tumors: clinical importance of apparent diffusion coefficient in prediction of grade and prognosis. Radiology 2006;241:839–46.

[73] Yuan H, Gaber MW, Boyd K, Wilson CM, Kiani MF, Merchant TE. Effects of fractionated radiation on the brain vasculature in a murine model: blood-brain barrier permeability, astrocyte proliferation, and ultrastructural changes. Int J Radiat Oncol Biol Phys 2006;66:860–6.

[74] Tabatabai G, Frank B, Mohle R, Weller M, Wick W. Irradiation and hypoxia promote homing of haematopoietic progenitor cells towards gliomas by TGF-beta-dependent HIF-1alpha-mediated induction of CXCL12. Brain 2006;129:2426–35.

[75] van Vulpen M, Kal HB, Taphoorn MJ, El-Sharouni SY. Changes in blood-brain barrier permeability induced by radiotherapy: implications for timing of chemotherapy? (Review). Oncol Rep 2002;9:683–8.

[76] Sergeeva VA, Ershova ES, Veiko NN, Malinovskaya EM, Kalyanov AA, Kameneva LV, et al. Low-dose ionizing radiation affects mesenchymal stem cells via extracellular oxidized cell-free DNA: a possible mediator of bystander effect and adaptive response. Oxid Med Cell Longev 2017;2017:9515809.

[77] Nikitaki Z, Mavragani IV, Laskaratou DA, Gika V, Moskvin VP, Theofilatos K, et al. Systemic mechanisms and effects of ionizing radiation: A new 'old' paradigm of how the bystanders and distant can become the players. Semin Cancer Biol 2016;37–38:77–95.

[78] Jeon HY, Kim JK, Ham SW, Oh SY, Kim J, Park JB, et al. Irradiation induces glioblastoma cell senescence and senescence-associated secretory phenotype. Tumour Biol 2016;37:5857–67.

[79] Havaki S, Kotsinas A, Chronopoulos E, Kletsas D, Georgakilas A, Gorgoulis VG. The role of oxidative DNA damage in radiation induced bystander effect. Cancer Lett 2015;356:43–51.
[80] Glebova K, Veiko N, Kostyuk S, Izhevskaya V, Baranova A. Oxidized extracellular DNA as a stress signal that may modify response to anticancer therapy. Cancer Lett 2015;356:22–33.
[81] Prise KM, O'Sullivan JM. Radiation-induced bystander signalling in cancer therapy. Nat Rev Cancer 2009;9:351–60.
[82] Shao C, Stewart V, Folkard M, Michael BD, Prise KM. Nitric oxide-mediated signaling in the bystander response of individually targeted glioma cells. Cancer Res 2003;63:8437–42.
[83] Shao C, Aoki M, Furusawa Y. Bystander effect on cell growth stimulation in neoplastic HSGc cells induced by heavy-ion irradiation. Radiat Environ Biophys 2003;42:183–7.
[84] Quirico-Santos T, Fonseca CO, Lagrota-Candido J. Brain sweet brain: importance of sugars for the cerebral microenvironment and tumor development. Arq Neuropsiquiatr 2010;68:799–803.
[85] Zimmermann DR, Dours-Zimmermann MT. Extracellular matrix of the central nervous system: from neglect to challenge. Histochem Cell Biol 2008;130:635–53.
[86] Bandtlow CE, Zimmermann DR. Proteoglycans in the developing brain: new conceptual insights for old proteins. Physiol Rev 2000;80:1267–90.
[87] Vlodavsky I, Miao HQ, Medalion B, Danagher P, Ron D. Involvement of heparan sulfate and related molecules in sequestration and growth promoting activity of fibroblast growth factor. Cancer Metastasis Rev 1996;15: 177–86.
[88] Barros CS, Franco SJ, Muller U. Extracellular matrix: functions in the nervous system. Cold Spring Harb Perspect Biol 2011;3:a005108.
[89] Rolls A, Shechter R, Schwartz M. The bright side of the glial scar in CNS repair. Nat Rev Neurosci 2009;10:235–41.
[90] Colognato H, ffrench-Constant C, Feltri ML. Human diseases reveal novel roles for neural laminins. Trends Neurosci 2005;28:480–6.
[91] Baron W, Colognato H, ffrench-Constant C. Integrin-growth factor interactions as regulators of oligodendroglial development and function. Glia 2005;49:467–79.
[92] ffrench-Constant C, Colognato H. Integrins: versatile integrators of extracellular signals. Trends Cell Biol 2004;14:678–86.
[93] Bukalo O, Schachner M, Dityatev A. Modification of extracellular matrix by enzymatic removal of chondroitin sulfate and by lack of tenascin-R differentially affects several forms of synaptic plasticity in the hippocampus. Neuroscience 2001;104:359–69.
[94] Saghatelyan AK, Dityatev A, Schmidt S, Schuster T, Bartsch U, Schachner M. Reduced perisomatic inhibition, increased excitatory transmission, and impaired long-term potentiation in mice deficient for the extracellular matrix glycoprotein tenascin-R. Mol Cell Neurosci 2001;17:226–40.
[95] Moller JC, Klein MA, Haas S, Jones LL, Kreutzberg GW, Raivich G. Regulation of thrombospondin in the regenerating mouse facial motor nucleus. Glia 1996;17:121–32.
[96] Wang B, Guo W, Huang Y. Thrombospondins and synaptogenesis. Neural Regen Res 2012;7:1737–43.
[97] Morawski M, Bruckner MK, Riederer P, Bruckner G, Arendt T. Perineuronal nets potentially protect against oxidative stress. Exp Neurol 2004;188:309–15.
[98] Schuppel K, Brauer K, Hartig W, Grosche J, Earley B, Leonard BE, et al. Perineuronal nets of extracellular matrix around hippocampal interneurons resist destruction by activated microglia in trimethyltin-treated rats. Brain Res 2002;958:448–53.
[99] Block ML, Hong JS. Microglia and inflammation-mediated neurodegeneration: multiple triggers with a common mechanism. Prog Neurobiol 2005;76:77–98.
[100] Kim YS, Joh TH. Microglia, major player in the brain inflammation: their roles in the pathogenesis of Parkinson's disease. Exp Mol Med 2006;38:333–47.
[101] Greter M, Merad M. Regulation of microglia development and homeostasis. Glia 2013;61:121–7.
[102] Harry GJ, Kraft AD. Microglia in the developing brain: a potential target with lifetime effects. Neurotoxicology 2012;33:191–206.
[103] Vilhardt F. Microglia: phagocyte and glia cell. Int J Biochem Cell Biol 2005;37:17–21.
[104] Greter M, Lelios I, Croxford AL. Microglia versus myeloid cell nomenclature during brain inflammation. Front Immunol 2015;6:249.
[105] Hanisch UK, Kettenmann H. Microglia: active sensor and versatile effector cells in the normal and pathologic brain. Nat Neurosci 2007;10:1387–94.

[106] Luo XG, Chen SD. The changing phenotype of microglia from homeostasis to disease. Transl Neurodegener 2012;1:9.
[107] Kigerl KA, Gensel JC, Ankeny DP, Alexander JK, Donnelly DJ, Popovich PG. Identification of two distinct macrophage subsets with divergent effects causing either neurotoxicity or regeneration in the injured mouse spinal cord. J Neurosci 2009;29:13435–44.
[108] Mecca C, Giambanco I, Donato R, Arcuri C. Microglia and aging: the role of the TREM2-DAP12 and CX3CL1-CX3CR1 axes. Int J Mol Sci 2018;19(1):318.
[109] Arcuri C, Mecca C, Bianchi R, Giambanco I, Donato R. The pathophysiological role of microglia in dynamic surveillance, phagocytosis and structural remodeling of the developing CNS. Front Mol Neurosci 2017;10:191.
[110] Nimmerjahn A, Kirchhoff F, Helmchen F. Resting microglial cells are highly dynamic surveillants of brain parenchyma in vivo. Science 2005;308:1314–8.
[111] Yang H, Feng GD, Liang Z, Vitale A, Jiao XY, Ju G, et al. In vitro beneficial activation of microglial cells by mechanically-injured astrocytes enhances the synthesis and secretion of BDNF through p38MAPK. Neurochem Int 2012;61:175–86.
[112] Zhou LJ, Yang T, Wei X, Liu Y, Xin WJ, Chen Y, et al. Brain-derived neurotrophic factor contributes to spinal long-term potentiation and mechanical hypersensitivity by activation of spinal microglia in rat. Brain Behav Immun 2011;25:322–34.
[113] Bianco F, Ceruti S, Colombo A, Fumagalli M, Ferrari D, Pizzirani C, et al. A role for P2X7 in microglial proliferation. J Neurochem 2006;99:745–58.
[114] Verge GM, Milligan ED, Maier SF, Watkins LR, Naeve GS, Foster AC. Fractalkine (CX3CL1) and fractalkine receptor (CX3CR1) distribution in spinal cord and dorsal root ganglia under basal and neuropathic pain conditions. Eur J Neurosci 2004;20:1150–60.
[115] Zujovic V, Benavides J, Vige X, Carter C, Taupin V. Fractalkine modulates TNF-alpha secretion and neurotoxicity induced by microglial activation. Glia 2000;29:305–15.
[116] Ganter S, Northoff H, Mannel D, Gebicke-Harter PJ. Growth control of cultured microglia. J Neurosci Res 1992;33:218–30.
[117] Liu W, Tang Y, Feng J. Cross talk between activation of microglia and astrocytes in pathological conditions in the central nervous system. Life Sci 2011;89:141–6.
[118] Noda M, Ifuku M, Mori Y, Verkhratsky A. Calcium influx through reversed NCX controls migration of microglia. Adv Exp Med Biol 2013;961:289–94.
[119] Wu LJ, Vadakkan KI, Zhuo M. ATP-induced chemotaxis of microglial processes requires P2Y receptor-activated initiation of outward potassium currents. Glia 2007;55:810–21.
[120] Haynes SE, Hollopeter G, Yang G, Kurpius D, Dailey ME, Gan WB, et al. The P2Y12 receptor regulates microglial activation by extracellular nucleotides. Nat Neurosci 2006;9:1512–9.
[121] Davalos D, Grutzendler J, Yang G, Kim JV, Zuo Y, Jung S, et al. ATP mediates rapid microglial response to local brain injury in vivo. Nat Neurosci 2005;8:752–8.
[122] Block ML, Zecca L, Hong JS. Microglia-mediated neurotoxicity: uncovering the molecular mechanisms. Nat Rev Neurosci 2007;8:57–69.
[123] Raivich G, Bohatschek M, Kloss CU, Werner A, Jones LL, Kreutzberg GW. Neuroglial activation repertoire in the injured brain: graded response, molecular mechanisms and cues to physiological function. Brain Res Brain Res Rev 1999;30:77–105.
[124] Kreutzberg GW. Microglia: a sensor for pathological events in the CNS. Trends Neurosci 1996;19:312–8.
[125] Milner R, Campbell IL. The extracellular matrix and cytokines regulate microglial integrin expression and activation. J Immunol 2003;170:3850–8.
[126] Milner R, Campbell IL. Cytokines regulate microglial adhesion to laminin and astrocyte extracellular matrix via protein kinase C-dependent activation of the alpha6beta1 integrin. J Neurosci 2002;22:1562–72.
[127] Sobel RA, Chen M, Maeda A, Hinojoza JR. Vitronectin and integrin vitronectin receptor localization in multiple sclerosis lesions. J Neuropathol Exp Neurol 1995;54:202–13.
[128] Venstrom KA, Reichardt LF. Extracellular matrix. 2: role of extracellular matrix molecules and their receptors in the nervous system. FASEB J 1993;7:996–1003.
[129] Felding-Habermann B, Cheresh DA. Vitronectin and its receptors. Curr Opin Cell Biol 1993;5:864–8.
[130] Bart J, Nagengast WB, Coppes RP, Wegman TD, van der Graaf WT, Groen HJ, et al. Irradiation of rat brain reduces P-glycoprotein expression and function. Br J Cancer 2007;97:322–6.
[131] Wei F, Tang L, He Y, Wu Y, Shi L, Xiong F, et al. BPIFB1 (LPLUNC1) inhibits radioresistance in nasopharyngeal carcinoma by inhibiting VTN expression. Cell Death Dis 2018;9:432.

[132] Diao W, Tong X, Yang C, Zhang F, Bao C, Chen H, et al. Behaviors of glioblastoma cells in in vitro microenvironments. Sci Rep 2019;9:85.
[133] Gao YF, Zhu T, Chen J, Liu L, Ouyang R. Knockdown of collagen alpha-1(III) inhibits glioma cell proliferation and migration and is regulated by miR128-3p. Oncol Lett 2018;16:1917–23.
[134] Wang C, Tong X, Yang F. Bioengineered 3D brain tumor model to elucidate the effects of matrix stiffness on glioblastoma cell behavior using PEG-based hydrogels. Mol Pharm 2014;11:2115–25.
[135] Motegi H, Kamoshima Y, Terasaka S, Kobayashi H, Houkin K. Type 1 collagen as a potential niche component for CD133-positive glioblastoma cells. Neuropathology 2014;34:378–85.
[136] Rao SS, Dejesus J, Short AR, Otero JJ, Sarkar A, Winter JO. Glioblastoma behaviors in three-dimensional collagen-hyaluronan composite hydrogels. ACS Appl Mater Interfaces 2013;5:9276–84.
[137] Payne LS, Huang PH. The pathobiology of collagens in glioma. Mol Cancer Res 2013;11:1129–40.
[138] Huijbers IJ, Iravani M, Popov S, Robertson D, Al-Sarraj S, Jones C, et al. A role for fibrillar collagen deposition and the collagen internalization receptor endo180 in glioma invasion. PLoS One 2010;5, e9808.
[139] Hegedus B, Marga F, Jakab K, Sharpe-Timms KL, Forgacs G. The interplay of cell-cell and cell-matrix interactions in the invasive properties of brain tumors. Biophys J 2006;91:2708–16.
[140] Zamecnik J. The extracellular space and matrix of gliomas. Acta Neuropathol 2005;110:435–42.
[141] Kaufman LJ, Brangwynne CP, Kasza KE, Filippidi E, Gordon VD, Deisboeck TS, et al. Glioma expansion in collagen I matrices: analyzing collagen concentration-dependent growth and motility patterns. Biophys J 2005;89:635–50.
[142] Zamecnik J, Vargova L, Homola A, Kodet R, Sykova E. Extracellular matrix glycoproteins and diffusion barriers in human astrocytic tumours. Neuropathol Appl Neurobiol 2004;30:338–50.
[143] Bryan N, Ahswin H, Smart N, Bayon Y, Wohlert S, Hunt JA. Reactive oxygen species (ROS)—a family of fate deciding molecules pivotal in constructive inflammation and wound healing. Eur Cell Mater 2012;24:249–65.
[144] Ricciardelli C, Sakko AJ, Ween MP, Russell DL, Horsfall DJ. The biological role and regulation of versican levels in cancer. Cancer Metastasis Rev 2009;28:233–45.
[145] Sim H, Hu B, Viapiano MS. Reduced expression of the hyaluronan and proteoglycan link proteins in malignant gliomas. J Biol Chem 2009;284:26547–56.
[146] Viapiano MS, Matthews RT. From barriers to bridges: chondroitin sulfate proteoglycans in neuropathology. Trends Mol Med 2006;12:488–96.
[147] Xu Y, Stamenkovic I, Yu Q. CD44 attenuates activation of the hippo signaling pathway and is a prime therapeutic target for glioblastoma. Cancer Res 2010;70:2455–64.
[148] Wiranowska M, Ladd S, Moscinski LC, Hill B, Haller E, Mikecz K, et al. Modulation of hyaluronan production by CD44 positive glioma cells. Int J Cancer 2010;127:532–42.
[149] Annabi B, Bouzeghrane M, Moumdjian R, Moghrabi A, Beliveau R. Probing the infiltrating character of brain tumors: inhibition of RhoA/ROK-mediated CD44 cell surface shedding from glioma cells by the green tea catechin EGCg. J Neurochem 2005;94:906–16.
[150] Ulrich TA, de Juan Pardo EM, Kumar S. The mechanical rigidity of the extracellular matrix regulates the structure, motility, and proliferation of glioma cells. Cancer Res 2009;69:4167–74.
[151] Wang M, Wang T, Liu S, Yoshida D, Teramoto A. The expression of matrix metalloproteinase-2 and -9 in human gliomas of different pathological grades. Brain Tumor Pathol 2003;20:65–72.
[152] Gilkes DM, Semenza GL, Wirtz D. Hypoxia and the extracellular matrix: drivers of tumour metastasis. Nat Rev Cancer 2014;14:430–9.
[153] Edderkaoui M, Hong P, Vaquero EC, Lee JK, Fischer L, Friess H, et al. Extracellular matrix stimulates reactive oxygen species production and increases pancreatic cancer cell survival through 5-lipoxygenase and NADPH oxidase. Am J Physiol Gastrointest Liver Physiol 2005;289:G1137–47.
[154] Vaquero EC, Edderkaoui M, Nam KJ, Gukovsky I, Pandol SJ, Gukovskaya AS. Extracellular matrix proteins protect pancreatic cancer cells from death via mitochondrial and nonmitochondrial pathways. Gastroenterology 2003;125:1188–202.
[155] Casazza A, Laoui D, Wenes M, Rizzolio S, Bassani N, Mambretti M, et al. Impeding macrophage entry into hypoxic tumor areas by Sema3A/Nrp1 signaling blockade inhibits angiogenesis and restores antitumor immunity. Cancer Cell 2013;24:695–709.
[156] Pavlova NN, Thompson CB. The emerging hallmarks of cancer metabolism. Cell Metab 2016;23:27–47.
[157] Snyder V, Reed-Newman TC, Arnold L, Thomas SM, Anant S. Cancer stem cell metabolism and potential therapeutic targets. Front Oncol 2018;8:203.

[158] Koh I, Cha J, Park J, Choi J, Kang SG, Kim P. The mode and dynamics of glioblastoma cell invasion into a decellularized tissue-derived extracellular matrix-based three-dimensional tumor model. Sci Rep 2018;8:4608.
[159] Mammoto T, Jiang A, Jiang E, Panigrahy D, Kieran MW, Mammoto A. Role of collagen matrix in tumor angiogenesis and glioblastoma multiforme progression. Am J Pathol 2013;183:1293–305.
[160] Jiang X, Zhou T, Wang Z, Qi B, Xia H. HSP47 promotes glioblastoma stemlike cell survival by modulating tumor microenvironment extracellular matrix through TGF-beta pathway. ACS Chem Neurosci 2017;8:128–34.
[161] Pointer KB, Clark PA, Schroeder AB, Salamat MS, Eliceiri KW, Kuo JS. Association of collagen architecture with glioblastoma patient survival. J Neurosurg 2017;126:1812–21.
[162] Sarkar S, Nuttall RK, Liu S, Edwards DR, Yong VW. Tenascin-C stimulates glioma cell invasion through matrix metalloproteinase-12. Cancer Res 2006;66:11771–80.
[163] Fujita M, Yamamoto T, Iyoda T, Fujisawa T, Nagai R, Kudo C, et al. Autocrine production of PDGF stimulated by the tenascin-C-derived peptide TNIIIA2 induces hyper-proliferation in glioblastoma cells. Int J Mol Sci 2019;20.
[164] Sarkar S, Zemp FJ, Senger D, Robbins SM, Yong VW. ADAM-9 is a novel mediator of tenascin-C-stimulated invasiveness of brain tumor-initiating cells. Neuro Oncol 2015;17:1095–105.
[165] Nutt CL, Matthews RT, Hockfield S. Glial tumor invasion: a role for the upregulation and cleavage of BEHAB/brevican. Neuroscientist 2001;7:113–22.
[166] Jaworski DM, Kelly GM, Piepmeier JM, Hockfield S. BEHAB (brain enriched hyaluronan binding) is expressed in surgical samples of glioma and in intracranial grafts of invasive glioma cell lines. Cancer Res 1996;56:2293–8.
[167] Gladson CL, Wilcox JN, Sanders L, Gillespie GY, Cheresh DA. Cerebral microenvironment influences expression of the vitronectin gene in astrocytic tumors. J Cell Sci 1995;108(Pt 3):947–56.
[168] Pu A, Stephenson EL, Yong VW. The extracellular matrix: focus on oligodendrocyte biology and targeting CSPGs for remyelination therapies. Glia 2018;66:1809–25.
[169] Keasey MP, Jia C, Pimentel LF, Sante RR, Lovins C, Hagg T. Blood vitronectin is a major activator of LIF and IL-6 in the brain through integrin-FAK and uPAR signaling. J Cell Sci 2018;131.
[170] Lau LW, Cua R, Keough MB, Haylock-Jacobs S, Yong VW. Pathophysiology of the brain extracellular matrix: a new target for remyelination. Nat Rev Neurosci 2013;14:722–9.
[171] Yang YL, Sun C, Wilhelm ME, Fox LJ, Zhu J, Kaufman LJ. Influence of chondroitin sulfate and hyaluronic acid on structure, mechanical properties, and glioma invasion of collagen I gels. Biomaterials 2011;32:7932–40.
[172] Hu B, Kong LL, Matthews RT, Viapiano MS. The proteoglycan brevican binds to fibronectin after proteolytic cleavage and promotes glioma cell motility. J Biol Chem 2008;283:24848–59.
[173] Wade A, Robinson AE, Engler JR, Petritsch C, James CD, Phillips JJ. Proteoglycans and their roles in brain cancer. FEBS J 2013;280:2399–417.
[174] Rooney P, Kumar S, Ponting J, Wang M. The role of hyaluronan in tumour neovascularization (review). Int J Cancer 1995;60:632–6.
[175] Kegelman TP, Das SK, Hu B, Bacolod MD, Fuller CE, Menezes ME, et al. MDA-9/syntenin is a key regulator of glioma pathogenesis. Neuro Oncol 2014;16:50–61.
[176] Gardner J, Ghorpade A. Tissue inhibitor of metalloproteinase (TIMP)-1: the TIMPed balance of matrix metalloproteinases in the central nervous system. J Neurosci Res 2003;74:801–6.
[177] Wiranowska M, Rojiani AM, Gottschall PE, Moscinski LC, Johnson J, Saporta S. CD44 expression and MMP-2 secretion by mouse glioma cells: effect of interferon and anti-CD44 antibody. Anticancer Res 2000;20:4301–6.
[178] Sameshima T, Nabeshima K, Toole BP, Yokogami K, Okada Y, Goya T, et al. Expression of emmprin (CD147), a cell surface inducer of matrix metalloproteinases, in normal human brain and gliomas. Int J Cancer 2000;88:21–7.
[179] Sun S, Lei Y, Li Q, Wu Y, Zhang L, Mu PP, et al. Neuropilin-1 is a glial cell line-derived neurotrophic factor receptor in glioblastoma. Oncotarget 2017;8:74019–35.
[180] Hu B, Guo P, Bar-Joseph I, Imanishi Y, Jarzynka MJ, Bogler O, et al. Neuropilin-1 promotes human glioma progression through potentiating the activity of the HGF/SF autocrine pathway. Oncogene 2007;26:5577–86.
[181] Gandalovicova A, Rosel D, Fernandes M, Vesely P, Heneberg P, Cermak V, et al. Migrastatics-anti-metastatic and anti-invasion drugs: promises and challenges. Trends Cancer 2017;3:391–406.
[182] Rasheed BK, Wiltshire RN, Bigner SH, Bigner DD. Molecular pathogenesis of malignant gliomas. Curr Opin Oncol 1999;11:162–7.
[183] Kesanakurti D, Chetty C, Rajasekhar Maddirela D, Gujrati M, Rao JS. Essential role of cooperative NF-kappaB and Stat3 recruitment to ICAM-1 intronic consensus elements in the regulation of radiation-induced invasion and migration in glioma. Oncogene 2013;32:5144–55.

[184] Ongusaha PP, Kim JI, Fang L, Wong TW, Yancopoulos GD, Aaronson SA, et al. p53 induction and activation of DDR1 kinase counteract p53-mediated apoptosis and influence p53 regulation through a positive feedback loop. EMBO J 2003;22:1289–301.
[185] Leitinger B. Discoidin domain receptor functions in physiological and pathological conditions. Int Rev Cell Mol Biol 2014;310:39–87.
[186] Ram R, Lorente G, Nikolich K, Urfer R, Foehr E, Nagavarapu U. Discoidin domain receptor-1a (DDR1a) promotes glioma cell invasion and adhesion in association with matrix metalloproteinase-2. J Neurooncol 2006;76:239–48.
[187] Smith CL, Kilic O, Schiapparelli P, Guerrero-Cazares H, Kim DH, Sedora-Roman NI, et al. Migration phenotype of brain-cancer cells predicts patient outcomes. Cell Rep 2016;15:2616–24.
[188] Maddirela DR, Kesanakurti D, Gujrati M, Rao JS. MMP-2 suppression abrogates irradiation-induced microtubule formation in endothelial cells by inhibiting alphavbeta3-mediated SDF-1/CXCR4 signaling. Int J Oncol 2013;42:1279–88.
[189] Beadle C, Assanah MC, Monzo P, Vallee R, Rosenfeld SS, Canoll P. The role of myosin II in glioma invasion of the brain. Mol Biol Cell 2008;19:3357–68.
[190] Bond LM, Tumbarello DA, Kendrick-Jones J, Buss F. Small-molecule inhibitors of myosin proteins. Future Med Chem 2013;5:41–52.
[191] Lucki NC, Villa GR, Vergani N, Bollong MJ, Beyer BA, Lee JW, et al. A cell type-selective apoptosis-inducing small molecule for the treatment of brain cancer. Proc Natl Acad Sci U S A 2019;116:6435–40.
[192] Ziegler DS, Wright RD, Kesari S, Lemieux ME, Tran MA, Jain M, et al. Resistance of human glioblastoma multiforme cells to growth factor inhibitors is overcome by blockade of inhibitor of apoptosis proteins. J Clin Invest 2008;118:3109–22.
[193] Tan SK, Jermakowicz A, Mookhtiar AK, Nemeroff CB, Schurer SC, Ayad NG. Drug repositioning in glioblastoma: a pathway perspective. Front Pharmacol 2018;9:218.
[194] Taylor OG, Brzozowski JS, Skelding KA. Glioblastoma multiforme: an overview of emerging therapeutic targets. Front Oncol 2019;9:963.
[195] Pearson JRD, Regad T. Targeting cellular pathways in glioblastoma multiforme. Signal Transduct Target Ther 2017;2:17040.
[196] Zhao HF, Wang J, Shao W, Wu CP, Chen ZP, To ST, et al. Recent advances in the use of PI3K inhibitors for glioblastoma multiforme: current preclinical and clinical development. Mol Cancer 2017;16:100.
[197] Sun Y, Alberta JA, Pilarz C, Calligaris D, Chadwick EJ, Ramkissoon SH, et al. A brain-penetrant RAF dimer antagonist for the noncanonical BRAF oncoprotein of pediatric low-grade astrocytomas. Neuro Oncol 2017;19:774–85.
[198] Lo HW. Targeting Ras-RAF-ERK and its interactive pathways as a novel therapy for malignant gliomas. Curr Cancer Drug Targets 2010;10:840–8.
[199] Lokker NA, Sullivan CM, Hollenbach SJ, Israel MA, Giese NA. Platelet-derived growth factor (PDGF) autocrine signaling regulates survival and mitogenic pathways in glioblastoma cells: evidence that the novel PDGF-C and PDGF-D ligands may play a role in the development of brain tumors. Cancer Res 2002;62:3729–35.
[200] Vengoji R, Ponnusamy MP, Rachagani S, Mahapatra S, Batra SK, Shonka N, et al. Novel therapies hijack the blood-brain barrier to eradicate glioblastoma cancer stem cells. Carcinogenesis 2019;40:2–14.
[201] Obara-Michlewska M, Szeliga M. Targeting glutamine addiction in gliomas. Cancers (Basel) 2020;12.
[202] Mehta S, Lo CC. Developmentally regulated signaling pathways in glioma invasion. Cell Mol Life Sci 2018;75:385–402.
[203] Oh MC, Kim JM, Safaee M, Kaur G, Sun MZ, Kaur R, et al. Overexpression of calcium-permeable glutamate receptors in glioblastoma derived brain tumor initiating cells. PLoS One 2012;7, e47846.
[204] Corsi L, Mescola A, Alessandrini A. Glutamate receptors and glioblastoma multiforme: an old "route" for new perspectives. Int J Mol Sci 2019;20.
[205] Lefranc F, Le Rhun E, Kiss R, Weller M. Glioblastoma quo vadis: will migration and invasiveness reemerge as therapeutic targets? Cancer Treat Rev 2018;68:145–54.
[206] Dienel GA. Astrocytic energetics during excitatory neurotransmission: what are contributions of glutamate oxidation and glycolysis? Neurochem Int 2013;63:244–58.
[207] Aldape K, Brindle KM, Chesler L, Chopra R, Gajjar A, Gilbert MR, et al. Challenges to curing primary brain tumours. Nat Rev Clin Oncol 2019;16:509–20.

[208] Artene SA, Turcu-Stiolica A, Ciurea ME, Folcuti C, Tataranu LG, Alexandru O, et al. Comparative effect of immunotherapy and standard therapy in patients with high grade glioma: a meta-analysis of published clinical trials. Sci Rep 2018;8:11800.

[209] Wang Y, Deng W, Li N, Neri S, Sharma A, Jiang W, et al. Combining immunotherapy and radiotherapy for cancer treatment: current challenges and future directions. Front Pharmacol 2018;9:185.

[210] Reznik E, Smith AW, Taube S, Mann J, Yondorf MZ, Parashar B, et al. Radiation and immunotherapy in high-grade gliomas: where do we stand? Am J Clin Oncol 2018;41:197–212.

[211] Ye XZ, Xu SL, Xin YH, Yu SC, Ping YF, Chen L, et al. Tumor-associated microglia/macrophages enhance the invasion of glioma stem-like cells via TGF-beta1 signaling pathway. J Immunol 2012;189:444–53.

[212] Park B, Yee C, Lee KM. The effect of radiation on the immune response to cancers. Int J Mol Sci 2014;15:927–43.

[213] Simon JE, Prabhu VC, Barton K, Borys E, Piedras-Renteria E, Melian E. Synergistic therapies for recurrent malignant gliomas. World Neurosurg 2020;133:237–9.

[214] Ameratunga M, Pavlakis N, Wheeler H, Grant R, Simes J, Khasraw M. Anti-angiogenic therapy for high-grade glioma. Cochrane Database Syst Rev 2018;11, CD008218.

[215] Jain KK. A critical overview of targeted therapies for glioblastoma. Front Oncol 2018;8:419.

[216] Aderetti DA, Hira VVV, Molenaar RJ, van Noorden CJF. The hypoxic peri-arteriolar glioma stem cell niche, an integrated concept of five types of niches in human glioblastoma. Biochim Biophys Acta Rev Cancer 1869;2018:346–54.

[217] Hira VVV, Aderetti DA, van Noorden CJF. Glioma stem cell niches in human glioblastoma are periarteriolar. J Histochem Cytochem 2018;66:349–58.

[218] Peters KB, Lipp ES, Miller E, Herndon 2nd JE, McSherry F, Desjardins A, et al. Phase I/II trial of vorinostat, bevacizumab, and daily temozolomide for recurrent malignant gliomas. J Neurooncol 2018;137:349–56.

[219] Roberts DD, Kaur S, Isenberg JS. Regulation of cellular redox signaling by matricellular proteins in vascular biology, immunology, and cancer. Antioxid Redox Signal 2017;27:874–911.

[220] Shao R, Taylor SL, Oh DS, Schwartz LM. Vascular heterogeneity and targeting: the role of YKL-40 in glioblastoma vascularization. Oncotarget 2015;6:40507–18.

[221] Lubelski D, Abdullah KG, Weil RJ, Marko NF. Bevacizumab for radiation necrosis following treatment of high grade glioma: a systematic review of the literature. J Neurooncol 2013;115:317–22.

[222] Mongiardi MP. Angiogenesis and hypoxia in glioblastoma: a focus on cancer stem cells. CNS Neurol Disord Drug Targets 2012;11:878–83.

[223] Yang L, Lin C, Wang L, Guo H, Wang X. Hypoxia and hypoxia-inducible factors in glioblastoma multiforme progression and therapeutic implications. Exp Cell Res 2012;318:2417–26.

[224] Matuschek C, Bolke E, Nawatny J, Hoffmann TK, Peiper M, Orth K, et al. Bevacizumab as a treatment option for radiation-induced cerebral necrosis. Strahlenther Onkol 2011;187:135–9.

[225] Argyriou AA, Kalofonos HP. Molecularly targeted therapies for malignant gliomas. Mol Med 2009;15:115–22.

[226] Adair JC, Baldwin N, Kornfeld M, Rosenberg GA. Radiation-induced blood-brain barrier damage in astrocytoma: relation to elevated gelatinase B and urokinase. J Neurooncol 1999;44:283–9.

[227] Isenberg JS, Maxhimer JB, Hyodo F, Pendrak ML, Ridnour LA, DeGraff WG, et al. Thrombospondin-1 and CD47 limit cell and tissue survival of radiation injury. Am J Pathol 2008;173:1100–12.

[228] Wang H, Bouzakoura S, de Mey S, Jiang H, Law K, Dufait I, et al. Auranofin radiosensitizes tumor cells through targeting thioredoxin reductase and resulting overproduction of reactive oxygen species. Oncotarget 2017;8:35728–42.

[229] Kwak JY, Ham HJ, Kim CM, Hwang ES. Nicotinamide exerts antioxidative effects on senescent cells. Mol Cells 2015;38:229–35.

[230] Song SB, Jang SY, Kang HT, Wei B, Jeoun UW, Yoon GS, et al. Modulation of mitochondrial membrane potential and ROS generation by nicotinamide in a manner independent of SIRT1 and mitophagy. Mol Cells 2017;40:503–14.

[231] Sanli T, Steinberg GR, Singh G, Tsakiridis T. AMP-activated protein kinase (AMPK) beyond metabolism: a novel genomic stress sensor participating in the DNA damage response pathway. Cancer Biol Ther 2014;15:156–69.

[232] Bonora M, Patergnani S, Rimessi A, De Marchi E, Suski JM, Bononi A, et al. ATP synthesis and storage. Purinergic Signal 2012;8:343–57.

[233] Hertz L, Peng L, Dienel GA. Energy metabolism in astrocytes: high rate of oxidative metabolism and spatiotemporal dependence on glycolysis/glycogenolysis. J Cereb Blood Flow Metab 2007;27:219–49.

[234] Short S, Fielder E, Miwa S, von Zglinicki T. Senolytics and senostatics as adjuvant tumour therapy. EBioMedicine 2019;41:683–92.
[235] McKinley MJ, Albiston AL, Allen AM, Mathai ML, May CN, McAllen RM, et al. The brain renin-angiotensin system: location and physiological roles. Int J Biochem Cell Biol 2003;35:901–18.
[236] Vigano S, Alatzoglou D, Irving M, Menetrier-Caux C, Caux C, Romero P, et al. Targeting adenosine in cancer immunotherapy to enhance T-cell function. Front Immunol 2019;10:925.
[237] Rey S, Schito L, Koritzinsky M, Wouters BG. Molecular targeting of hypoxia in radiotherapy. Adv Drug Deliv Rev 2017;109:45–62.
[238] Chao ST, Ahluwalia MS, Barnett GH, Stevens GH, Murphy ES, Stockham AL, et al. Challenges with the diagnosis and treatment of cerebral radiation necrosis. Int J Radiat Oncol Biol Phys 2013;87:449–57.
[239] Brieger K, Schiavone S, Miller Jr FJ, Krause KH. Reactive oxygen species: from health to disease. Swiss Med Wkly 2012;142:w13659.
[240] Wang SC, Yu CF, Hong JH, Tsai CS, Chiang CS. Radiation therapy-induced tumor invasiveness is associated with SDF-1-regulated macrophage mobilization and vasculogenesis. PLoS One 2013;8, e69182.
[241] Lee YW, Cho HJ, Lee WH, Sonntag WE. Whole brain radiation-induced cognitive impairment: pathophysiological mechanisms and therapeutic targets. Biomol Ther (Seoul) 2012;20:357–70.
[242] Deng Z, Sui G, Rosa PM, Zhao W. Radiation-induced c-Jun activation depends on MEK1-ERK1/2 signaling pathway in microglial cells. PLoS One 2012;7, e36739.
[243] Wei M, Li H, Huang H, Xu D, Zhi D, Liu D, et al. Increased expression of EMMPRIN and VEGF in the rat brain after gamma irradiation. J Korean Med Sci 2012;27:291–9.
[244] Acharya MM, Lan ML, Kan VH, Patel NH, Giedzinski E, Tseng BP, et al. Consequences of ionizing radiation-induced damage in human neural stem cells. Free Radic Biol Med 2010;49:1846–55.
[245] Lee WH, Sonntag WE, Mitschelen M, Yan H, Lee YW. Irradiation induces regionally specific alterations in pro-inflammatory environments in rat brain. Int J Radiat Biol 2010;86:132–44.
[246] Collins-Underwood JR, Zhao W, Sharpe JG, Robbins ME. NADPH oxidase mediates radiation-induced oxidative stress in rat brain microvascular endothelial cells. Free Radic Biol Med 2008;45:929–38.
[247] Kyrkanides S, Moore AH, Olschowka JA, Daeschner JC, Williams JP, Hansen JT, et al. Cyclooxygenase-2 modulates brain inflammation-related gene expression in central nervous system radiation injury. Brain Res Mol Brain Res 2002;104:159–69.
[248] Le QT, Harris J, Magliocco AM, Kong CS, Diaz R, Shin B, et al. Validation of lysyl oxidase as a prognostic marker for metastasis and survival in head and neck squamous cell carcinoma: radiation therapy oncology Group trial 90-03. J Clin Oncol 2009;27:4281–6.
[249] Leins A, Riva P, Lindstedt R, Davidoff MS, Mehraein P, Weis S. Expression of tenascin-C in various human brain tumors and its relevance for survival in patients with astrocytoma. Cancer 2003;98:2430–9.
[250] Dasgupta J, Kar S, Liu R, Joseph J, Kalyanaraman B, Remington SJ, et al. Reactive oxygen species control senescence-associated matrix metalloproteinase-1 through c-Jun-N-terminal kinase. J Cell Physiol 2010;225:52–62.
[251] Rees MD, Hawkins CL, Davies MJ. Hypochlorite and superoxide radicals can act synergistically to induce fragmentation of hyaluronan and chondroitin sulphates. Biochem J 2004;381:175–84.
[252] Moseley R, Waddington RJ, Embery G. Degradation of glycosaminoglycans by reactive oxygen species derived from stimulated polymorphonuclear leukocytes. Biochim Biophys Acta 1997;1362:221–31.
[253] Iwadate Y. Epithelial-mesenchymal transition in glioblastoma progression. Oncol Lett 2016;11:1615–20.
[254] Barker HE, Paget JT, Khan AA, Harrington KJ. The tumour microenvironment after radiotherapy: mechanisms of resistance and recurrence. Nat Rev Cancer 2015;15:409–25.
[255] Thompson RF, Maity A. Radiotherapy and the tumor microenvironment: mutual influence and clinical implications. Adv Exp Med Biol 2014;772:147–65.
[256] Sabatino ME, Grondona E, Sosa LDV, Mongi Bragato B, Carreno L, Juarez V, et al. Oxidative stress and mitochondrial adaptive shift during pituitary tumoral growth. Free Radic Biol Med 2018;120:41–55.
[257] Fiaschi T, Chiarugi P. Oxidative stress, tumor microenvironment, and metabolic reprogramming: a diabolic liaison. Int J Cell Biol 2012;2012:762825.
[258] Kim YH, Yoo KC, Cui YH, Uddin N, Lim EJ, Kim MJ, et al. Radiation promotes malignant progression of glioma cells through HIF-1alpha stabilization. Cancer Lett 2014;354:132–41.
[259] Kil WJ, Tofilon PJ, Camphausen K. Post-radiation increase in VEGF enhances glioma cell motility in vitro. Radiat Oncol 2012;7:25.

[260] Thomas DD, Heinecke JL, Ridnour LA, Cheng RY, Kesarwala AH, Switzer CH, et al. Signaling and stress: the redox landscape in NOS2 biology. Free Radic Biol Med 2015;87:204–25.
[261] Chang AL, Miska J, Wainwright DA, Dey M, Rivetta CV, Yu D, et al. CCL2 produced by the glioma microenvironment is essential for the recruitment of regulatory T cells and myeloid-derived suppressor cells. Cancer Res 2016;76:5671–82.
[262] Rojo AI, McBean G, Cindric M, Egea J, Lopez MG, Rada P, et al. Redox control of microglial function: molecular mechanisms and functional significance. Antioxid Redox Signal 2014;21:1766–801.
[263] Trnovec T, Kallay Z, Bezek S. Effects of ionizing radiation on the blood brain barrier permeability to pharmacologically active substances. Int J Radiat Oncol Biol Phys 1990;19:1581–7.
[264] Hsieh CH, Wu CP, Lee HT, Liang JA, Yu CY, Lin YJ. NADPH oxidase subunit 4 mediates cycling hypoxia-promoted radiation resistance in glioblastoma multiforme. Free Radic Biol Med 2012;53:649–58.
[265] Piao Y, Liang J, Holmes L, Zurita AJ, Henry V, Heymach JV, et al. Glioblastoma resistance to anti-VEGF therapy is associated with myeloid cell infiltration, stem cell accumulation, and a mesenchymal phenotype. Neuro Oncol 2012;14:1379–92.
[266] Zhang I, Alizadeh D, Liang J, Zhang L, Gao H, Song Y, et al. Characterization of arginase expression in glioma-associated microglia and macrophages. PLoS One 2016;11, e0165118.
[267] Ryoo IG, Lee SH, Kwak MK. Redox modulating NRF2: a potential mediator of cancer stem cell resistance. Oxid Med Cell Longev 2016;2016:2428153.
[268] Maier P, Hartmann L, Wenz F, Herskind C. Cellular pathways in response to ionizing radiation and their targetability for tumor radiosensitization. Int J Mol Sci 2016;17(1):102.
[269] Panganiban RA, Snow AL, Day RM. Mechanisms of radiation toxicity in transformed and non-transformed cells. Int J Mol Sci 2013;14:15931–58.
[270] Domina EA, Philchenkov A, Dubrovska A. Individual response to ionizing radiation and personalized radiotherapy. Crit Rev Oncog 2018;23:69–92.
[271] Erasimus H, Gobin M, Niclou S, Van Dyck E. DNA repair mechanisms and their clinical impact in glioblastoma. Mutat Res Rev Mutat Res 2016;769:19–35.
[272] Mukherjee B, McEllin B, Camacho CV, Tomimatsu N, Sirasanagandala S, Nannepaga S, et al. EGFRvIII and DNA double-strand break repair: a molecular mechanism for radioresistance in glioblastoma. Cancer Res 2009;69:4252–9.
[273] Lindsay KJ, Coates PJ, Lorimore SA, Wright EG. The genetic basis of tissue responses to ionizing radiation. Br J Radiol 2007;80:S2–6. Spec No 1.
[274] Levi N, Papismadov N, Solomonov I, Sagi I, Krizhanovsky V. The ECM path of senescence in aging: components and modifiers. FEBS J 2020;287(13):2636–46.
[275] Liao YX, Zhang ZP, Zhao J, Liu JP. Effects of fibronectin 1 on cell proliferation, senescence and apoptosis of human glioma cells through the PI3K/AKT signaling pathway. Cell Physiol Biochem 2018;48:1382–96.
[276] Cescon M, Chen P, Castagnaro S, Gregorio I, Bonaldo P. Lack of collagen VI promotes neurodegeneration by impairing autophagy and inducing apoptosis during aging. Aging (Albany NY) 2016;8:1083–101.
[277] Amor S, Peferoen LA, Vogel DY, Breur M, van der Valk P, Baker D, et al. Inflammation in neurodegenerative diseases—an update. Immunology 2014;142:151–66.
[278] Kohanbash G, Okada H. Myeloid-derived suppressor cells (MDSCs) in gliomas and glioma-development. Immunol Invest 2012;41:658–79.
[279] Zhang Z, Zhang ZY, Wu Y, Schluesener HJ. Lesional accumulation of CD163+ macrophages/microglia in rat traumatic brain injury. Brain Res 2012;1461:102–10.
[280] Perry VH, Nicoll JA, Holmes C. Microglia in neurodegenerative disease. Nat Rev Neurol 2010;6:193–201.
[281] Umemura N, Saio M, Suwa T, Kitoh Y, Bai J, Nonaka K, et al. Tumor-infiltrating myeloid-derived suppressor cells are pleiotropic-inflamed monocytes/macrophages that bear M1- and M2-type characteristics. J Leukoc Biol 2008;83:1136–44.
[282] O'Connor RA, Malpass KH, Anderton SM. The inflamed central nervous system drives the activation and rapid proliferation of Foxp3+ regulatory T cells. J Immunol 2007;179:958–66.
[283] Gregoire H, Roncali L, Rousseau A, Cherel M, Delneste Y, Jeannin P, et al. Targeting tumor associated macrophages to overcome conventional treatment resistance in glioblastoma. Front Pharmacol 2020;11:368.
[284] Coppe JP, Desprez PY, Krtolica A, Campisi J. The senescence-associated secretory phenotype: the dark side of tumor suppression. Annu Rev Pathol 2010;5:99–118.

[285] Kuilman T, Michaloglou C, Vredeveld LC, Douma S, van Doorn R, Desmet CJ, et al. Oncogene-induced senescence relayed by an interleukin-dependent inflammatory network. Cell 2008;133:1019–31.
[286] Chen Z, Cao K, Xia Y, Li Y, Hou Y, Wang L, et al. Cellular senescence in ionizing radiation (Review). Oncol Rep 2019;42:883–94.
[287] Li M, You L, Xue J, Lu Y. Ionizing radiation-induced cellular senescence in normal, non-transformed cells and the involved DNA damage response: a mini review. Front Pharmacol 2018;9:522.
[288] Mikula-Pietrasik J, Niklas A, Uruski P, Tykarski A, Ksiazek K. Mechanisms and significance of therapy-induced and spontaneous senescence of cancer cells. Cell Mol Life Sci 2020;77:213–29.
[289] Ruhland MK, Loza AJ, Capietto AH, Luo X, Knolhoff BL, Flanagan KC, et al. Stromal senescence establishes an immunosuppressive microenvironment that drives tumorigenesis. Nat Commun 2016;7:11762.
[290] Faget DV, Ren Q, Stewart SA. Unmasking senescence: context-dependent effects of SASP in cancer. Nat Rev Cancer 2019;19:439–53.
[291] Marin A, Martin M, Linan O, Alvarenga F, Lopez M, Fernandez L, et al. Bystander effects and radiotherapy. Rep Pract Oncol Radiother 2015;20:12–21.
[292] Jarosz-Biej M, Smolarczyk R, Cichon T, Kulach N. Tumor microenvironment as a "game changer" in cancer radiotherapy. Int J Mol Sci 2019;20(13):3212.
[293] Minniti G, Muni R, Lanzetta G, Marchetti P, Enrici RM. Chemotherapy for glioblastoma: current treatment and future perspectives for cytotoxic and targeted agents. Anticancer Res 2009;29:5171–84.
[294] Gomez-Zepeda D, Taghi M, Scherrmann JM, Decleves X, Menet MC. ABC transporters at the blood-brain interfaces, their study models, and drug delivery implications in gliomas. Pharmaceutics 2019;12(1):20.
[295] Papademetriou IT, Porter T. Promising approaches to circumvent the blood-brain barrier: progress, pitfalls and clinical prospects in brain cancer. Ther Deliv 2015;6:989–1016.
[296] Ingram WJ, Crowther LM, Little EB, Freeman R, Harliwong I, Veleva D, et al. ABC transporter activity linked to radiation resistance and molecular subtype in pediatric medulloblastoma. Exp Hematol Oncol 2013;2:26.
[297] Desbats MA, Giacomini I, Prayer-Galetti T, Montopoli M. Metabolic plasticity in chemotherapy resistance. Front Oncol 2020;10:281.
[298] Chang JE, Khuntia D, Robins HI, Mehta MP. Radiotherapy and radiosensitizers in the treatment of glioblastoma multiforme. Clin Adv Hematol Oncol 2007;5:894–902. 907-815.
[299] Harisi R, Jeney A. Extracellular matrix as target for antitumor therapy. Onco Targets Ther 2015;8:1387–98.
[300] Hidalgo-Carcedo C, Hooper S, Chaudhry SI, Williamson P, Harrington K, Leitinger B, et al. Collective cell migration requires suppression of actomyosin at cell-cell contacts mediated by DDR1 and the cell polarity regulators Par3 and Par6. Nat Cell Biol 2011;13:49–58.
[301] Angeli E, Nguyen TT, Janin A, Bousquet G. How to make anticancer drugs cross the blood-brain barrier to treat brain metastases. Int J Mol Sci 2019;21(1):22.
[302] Gao X, Yue Q, Liu Y, Fan D, Fan K, Li S, et al. Image-guided chemotherapy with specifically tuned blood brain barrier permeability in glioma margins. Theranostics 2018;8:3126–37.
[303] Hainfeld JF, Ridwan SM, Stanishevskiy Y, Panchal R, Slatkin DN, Smilowitz HM. Iodine nanoparticles enhance radiotherapy of intracerebral human glioma in mice and increase efficacy of chemotherapy. Sci Rep 2019;9:4505.
[304] Retif P, Pinel S, Toussaint M, Frochot C, Chouikrat R, Bastogne T, et al. Nanoparticles for radiation therapy enhancement: the key parameters. Theranostics 2015;5:1030–44.
[305] Cooper DR, Bekah D, Nadeau JL. Gold nanoparticles and their alternatives for radiation therapy enhancement. Front Chem 2014;2:86.
[306] Baumann BC, Kao GD, Mahmud A, Harada T, Swift J, Chapman C, et al. Enhancing the efficacy of drug-loaded nanocarriers against brain tumors by targeted radiation therapy. Oncotarget 2013;4:64–79.

CHAPTER

14

Molecular and cellular mechanisms in recurrent glioblastoma chemoresistance

Sabrina Battista, Laura Cerchia, and Monica Fedele

Institute for Experimental Endocrinology and Oncology (IEOS), National Research Council (CNR), Naples, Italy

Abstract

Glioblastoma (GBM) is the most lethal form of brain cancer. To date, no therapy exists to save patients from this deadly disease. Even though most of the patients subjected to surgical resection followed by radiotherapy and chemotherapy undergo remission, the tumor always recurs, leading to death in 15 months from diagnosis. Besides the blood-brain barrier (BBB), which represents a first-line shield from therapeutic molecules, GBM exploits a plethora of extracellular, cellular, and molecular weapons to escape therapy, either by hindering the drug uptake or by overriding its toxic effect. GBM stem cells (GSCs) are not only the hub of tumor initiation but also the immortal fuel of chemoresistance, owing to their innate and acquired molecular equipment. Therefore, the molecular mechanisms responsible for GSC chemoresistance have attracted the interest of scientists for their potential therapeutic applications. The scenario, however, appears very complex, since chemoresistance is not only attributable to GSCs, but also to other cellular mechanisms, such as the transformation of proneural and endothelial cells into mesenchymal stem cells. Moreover, the involved molecular mechanisms foresee the expression of efflux transporters, the activation of DNA repair mechanisms (such as O^6-methylguanine-DNA methyl-transferase, base excision repair, mismatch repair), metabolic switching, autophagy, and depends on the expression of several genes. On the other hand, a pivotal role is played by extracellular mechanisms, involving the BBB, tumor microenvironment, and immunologic resistance.

Abbreviations

ABC	ATP-binding cassette
ACNU	nimustine
ALDH1A3	aldehyde dehydrogenase 1A3
ATM	ataxia telangiectasia mutated
ATR	ataxia telangiectasia and Rad3 related
ATRA	all-trans retinoic acid
AURKA	Aurora A kinase

Glioblastoma Resistance to Chemotherapy: Molecular Mechanisms and Innovative Reversal Strategies
https://doi.org/10.1016/B978-0-12-821567-8.00011-7

BCNU	carmustine
BER	base excision repair
CAFs	cancer-associated fibroblasts
CCNU	lomustine
CHKα	choline kinase alpha
Chk1	checkpoint kinase 1
Chk2	checkpoint kinase 2
CL	classical
CMA	chaperon-mediated autophagy
CNU	chloroethylating nitrosourea
DMAMCL	dimethylaminomicheliolide
DNA-PK	DNA-dependent protein kinase
DSBs	double-strand breaks
ECD	ectodomain
ECM	extracellular matrix
ECs	endothelial cells
EGFR	epidermal growth factor receptor
EGFRvIII	EGFR mutation variant III
EMT	epithelial-mesenchymal transition
GASC	glioma-associated stem cell
GBM	glioblastoma
GSC	glioma stem cell
HIF	hypoxia-inducible factors
HK2	hexokinase 2
HR	homologous recombination
Lig4	DNA ligase IV
mAbs	monoclonal antibodies
MDSCs	myeloid-derived suppressor cells
MES	mesenchymal
MGMT	O^6-methylguanine-DNA methyl-transferase
MMR	mismatch repair
MRN	Mre11/Rad50/Nbs1 complex
Msi1	Musashi-1
MTIC	5-3-(methyl)-1-(triazen-1-yl) imidazole-4-carboxamide
NBD	nucleotide binding domain
NHEJ	nonhomologous end joining
NL	neuronal
NSC	neural stem cell
PARP-1	poly ADP-ribose polymerase
PDGFR	platelet-derived growth factor receptor
PI3K	phosphatidylinositol-3-kinase
PKM2	pyruvate kinase 2
PMT	proneural-to-mesenchymal transition
PN	proneural
PTEN	phosphatase and tensin homolog
RISC	recurrent-initiating stem-like cells
ROS	reactive oxygen species
RTK	receptor tyrosine kinase
SFE	sphere-forming efficiency
TAMs	tumor-associated macrophages
TCGA	Tumor Cancer Gene Atlas
TGF-β	transforming growth factor-β
TKIs	tyrosine kinase inhibitors

TME	tumor microenvironment
TMZ	temozolomide
VEGF	vascular endothelial growth factor

Conflict of interest

No potential conflicts of interest were disclosed.

Introduction

Glioblastoma (GBM) is the most common and lethal tumor of the brain, with a 5-year survival rate of 19%, 8%, and 5% in patients 20–44, 45–54, and 55–64 years of age, respectively, and an average survival from diagnosis of only 12–15 months. Conventional therapy, consisting of tumor resection followed by radiotherapy and temozolomide (TMZ)-based chemotherapy, has limited efficacy. After a temporary remission, all GBMs recur, owing to mechanisms of preexisting and acquired resistance [1].

According to standard protocols, after surgical resection, GBM patients receive an initial treatment of radiotherapy (5 days/weak at 1–2 Gy/day), focused on the tumor mass and adjacent margin, delivered in combination with the alkylating chemotherapeutic TMZ (75 mg/m^2) over the course of 6 weeks [2–5]. Then, nearly all patients receive further adjuvant TMZ for a few months. During this phase, most patients develop radiological evidence of local recurrence around the surgical cavity. Once the tumor grows back, patients may receive further therapies, including additional tumor resection, bevacizumab (an antivascular endothelial factor (VEGF) antibody), different chemotherapy, or additional radiation therapy focused exclusively on the tumor site [6]. Yet, there is little evidence that such salvage treatments increase survival. Targeted therapies, including, among others, antiangiogenic antibodies [7], epidermal growth factor receptor (EGFR) inhibitors [8], and phosphatidylinositol-3-kinase (PI3K) inhibitors [9], have also demonstrated minimal clinical efficacy. Tumor cells resistant to multiple therapies persist in the brain parenchyma around the tumor cavity and underlie tumor repopulation, constituting a critical target to overcome tumor recurrence [6].

GBM stem cells (GSCs) play a central role in chemoresistance owing to both their innate and acquired molecular features, making them able to survive drugs either by hampering their uptake or by activating mechanisms overriding drug-induced damage. GSCs responsible for the onset of recurrent GBM are more aggressive than GSCs derived from the primary GBM [10–12] and display different markers (loss of CD133 and gain of CD15, BMI1, and SOX2) [12]. Therefore it has been proposed to call them recurrent-initiating stem-like cells (RISC) [6]. Such cells are a subset of GSCs that develop increased innate resistance to treatment through further genetic mutations and acquire further adaptive resistance through epigenetic alterations during the course of therapy. Therefore, understanding the genomic alterations and molecular architecture of RISC cells is critical for the development of successful therapies that could be deployed immediately after surgery, thereby preventing their adaptive resistance and expansion into a new tumor mass [6].

In this chapter, we review both cellular and molecular mechanisms, either intrinsic or extrinsic, associated with chemoresistance in recurrent GBM.

Recurrent glioblastoma

When a patient is newly diagnosed with GBM the standard of care is the surgical removal of the tumor followed by radiochemotherapy. However, in almost all cases, after 6–9 months of treatment, the tumor recurs and is less sensitive to therapy than the primary GBM, leading to patient death within 12–15 months from the first diagnosis. The recurrent GBM is also more aggressive than the primary tumor; it invades the surrounding brain tissue, hampering a secondary surgery. Indeed, only 20%–30% of recurrent GBMs can be surgically excised, and often the tissue quality of the removed sample is very low due to large areas of necrotic tissue, thus hindering large-scale studies on the biology of recurrent GBM [13]. As a matter of fact, most of the studies on the GBM response to chemotherapy have been paradoxically done on primary untreated tumors rather than recurrent GBM [14]. This could be somehow misleading because a number of studies that compared molecular differences of paired primary and recurrent GBMs suggested that mutation and expression of oncogenes and markers associated with therapy resistance differ between the two groups [15–21]. For example, MLH1, a mismatch repair enzyme involved in the resistance to TMZ, has been found to be significantly downregulated in recurrent versus primary GBM [16, 21].

It is believed that the origin of recurrent GBM is in tumor cells, located in the tumor periphery, which lie dormant or migrate there soon before the resection of the tumor. About two-thirds of recurrent GBM are within 2 cm from the original tumor (local recurrence). The remaining one-third, which develop far away in other brain regions (distant recurrence), including the contralateral hemisphere, has been suggested to be a new primary GBM that might have resulted from radio-chemotherapeutic treatment [22], but subsequent studies have demonstrated sufficient genetic similarities between the original GBM and its distant recurrence proving they are indeed originated from the same tumor as a result of tumor cells that diffusely invaded the brain at early stages of the disease [1, 23]. Irrespective of their origin, these cells have molecular characteristics distinct from the resected tumor. A genomic analysis of the initial and recurrent tumor specimens from 38 GBM patients revealed a substantial divergence in the landscape of driver alterations associated with distant appearance of a recurrent tumor from the initial tumor, providing an explanation of the ineffectiveness of targeted therapies based on the genomic profile of the initial GBM [1]. These differences further increase during DNA-damaging radiochemotherapy for the accumulation of new mutations, or under the pressure of targeted- and immune-therapy for the selection of resistant clones. The result is the high intratumor heterogeneity, which is a hallmark of both primary and recurrent GBM. Noteworthy, the intratumor heterogeneity of GBM is owing to the coexistence of different cell clones that show a different repertoire of molecular characteristics in function of time and location, which is a big challenge for targeted therapy [24]. In particular, it seems that TMZ induces a reduction in the number of the different subclones in recurrent GBM, subclones that were however already present in the primary tumor either as dominant or minor (Fig. 1). Top genes associated with these TMZ-resistant subclones were regulators of the transforming growth factor-β (TGF-β) and AKT pathways, which are involved in cell growth, differentiation, immunosuppression, and metabolism [25]. Moreover, since the expression of both hypoxia-inducible factors (HIF)-1α and HIF-2α increases upon TMZ therapy and knockdown of HIFs expression inhibits the interconversion between nonstem glioma

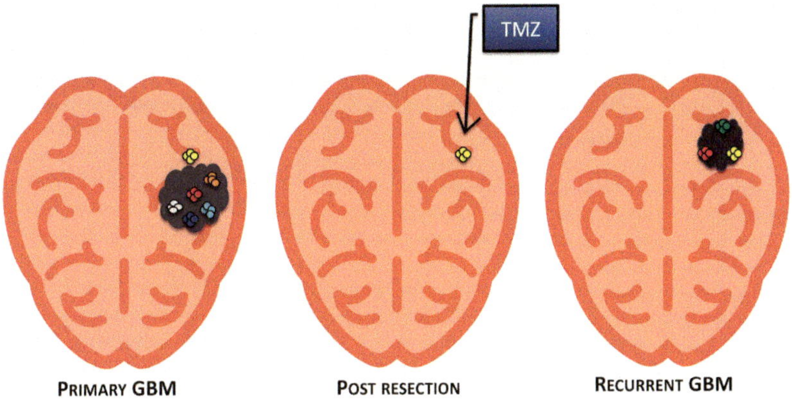

FIG. 1 Schematic representation of the origin of recurrent GBM. It is believed that recurrent GBMs originate from tumor cells located in the tumor periphery, which lay dormant or migrate there soon before the resection of the tumor. These cells have molecular characteristics that differ from the primary tumor owing to the intratumor heterogeneity given by the coexistence of different cell clones with different repertoires of molecular characteristics. This diversity further changes during therapy and recurrence. In particular, it seems that TMZ induces a reduction in the number of different subclones in recurrent GBM.

cells and GSCs posttherapy, it has been proposed that accumulation of HIFs upon GBM chemotherapy induces the formation of therapy-resistant GSCs responsible for recurrence [26]. On the other hand, Lo Dico et al. demonstrated that TMZ treatment upregulates HIF-1α expression only in TMZ-resistant cells while downregulates it in sensitive cells. Accordingly, HIF-1α silencing in sensitive cells did not induce any further modulation of apoptosis-related genes beyond that obtained after TMZ treatment, but, interestingly, HIF-1α silencing in resistant cells restored a "sensitive-like pattern" in apoptosis-related gene expression, with the upregulation of *Bax*, *Bad*, and *Trail* expression, and a reduction in *Bcl2* transcript levels [27].

From the comparison of pre- and posttreatment GBMs, recent large longitudinal studies have established that there are at least two distinct recurrent subtypes, *hypermutant* and *nonhypermutant* [28, 29]. For hypermutant recurrent tumors, the hallmark identifiers include: (i) fold increases in subclonal mutations across the whole genome; (ii) enrichment of the C:G > T:A mutational signature indicative of TMZ mutagenesis; and (iii) gain of inactivating mutations in MMR pathway components [30]. Understanding the underlying molecular mechanisms responsible for the maintenance of malignant features in these specific subtypes of recurrent GBMs will likely be essential for the identification of targeted therapies against each subtype.

Glioblastoma stem cells

GBM recurrence is inextricably linked to GSCs since they constitute the resistant seed of the tumor. Even though GSCs represent just a small subpopulation in the bulk of the tumor, they

TABLE 1 Similarities and differences between NSCs and GSCs.

	NSCs	GSCs	References
Markers	CD133, CD15, CD44	CD133, CD15, CD44	[31, 33, 34]
Stemness factors	Nestin, SOX2, HMGA1/2, BMI, MSI1, NOTCH1, EZH2, etc.	Nestin, SOX2, HMGA1/2, BMI, MSI1, NOTCH1, EZH2, etc.	[34–36]
Proliferative potential	+	+	[37]
Motility	+	+	[37]
Diversity of progeny	+	+	[37]
Asymmetric division	Yes	Reduced if compared with NSCs	[38]
Symmetric division	Yes	Increased if compared with NSCs	[38]
Nonrandom chromosome segregation	Yes	Reduced if compared with NSCs	[38, 39]
Association with vasculature	+	+	[37]
Bilateral cross talk with niche components	+	+	[37]
Tumor initiation ability in xeno-transplants	−	+	[36]

are believed to be the cells from which the tumor initiates [31]. GSCs derive from subventricular zone-residing neural stem cells (NSCs) following driver genetic mutations [32]. They share with them many common molecules such as nestin, CD133, CD15, SOX2, and other stemness markers and properties, such as self-renewal ability, a high proliferative potential, the ability to generate a differentiated progeny, the association with blood vessels, and the communication with niche components, including astrocytes and extracellular matrix (Table 1). Similar to NSCs, GSCs possess molecular efflux pumps able to exclude dyes and drugs from the cell, which make them resistant to standard therapies [33, 40].

GSCs represent not only the origin of the tumor, but also its propagation supply; most noticeably, owing to their innate and acquired resistance to therapeutic drugs, GSCs are the culprits of tumor recurrence. Standard anticancer treatments kill tumor cells, resulting in temporary tumor shrinkage, but GSCs, which are armed with molecular defences against drug entrance and/or activity, survive the treatment and shortly form a new, more aggressive tumor mass [41]. In addition, GSC plasticity makes current therapies even more ineffective, since, in the population of spared resistant cells, it allows the emergence of the best-suited cells to withstand ensuing therapies. Therefore, significant research efforts have been devoted to identifying GSC-specific molecular targets in order to develop therapies able to completely annihilate recurrent GBMs. Fig. 2 shows the strategies used by researchers to isolate and study GSCs. They can be isolated from patient tumors and cultured in suspension as neurospheres, in the absence of serum, in the presence of EGF and basic-fibroblast growth factor (b-FGF) (Fig. 2).

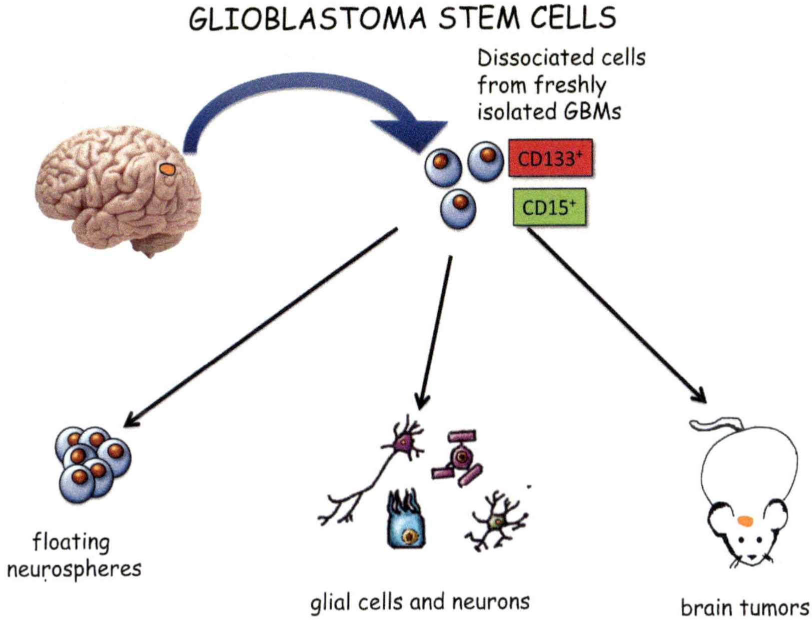

FIG. 2 Experimental approaches to GSCs. GSCs can be isolated from freshly resected GBM samples for their ability to grow in the absence of plate adhesion and serum, and cultured in the presence of EGF and b-FGF. Specific stem cell markers, such as CD133 and CD15, can be used to isolate and characterize GSCs. in low-adherence plates and in serum-free culture media, Purified CD133^{+}, but not CD133^{-} cells, generate neurospheres, which contain both GSCs and tumor GSC descendants. GSCs can be induced to differentiate into glial cells and neurons in cell differentiation conditions. Following xenotransplantation, GSCs are able to induce brain tumors resembling the parental tumor.

GSCs can be characterized based on their functional and molecular properties. Functional assays include aldehyde dehydrogenase (ALDH) activity [42] and Hoechst efflux, the latter not uniformly considered a proper assay [43]. Surface and nonsurface markers, such as CD133, CD15/SSEA1, CD44, platelet-derived growth factor receptor (PDGFR), Notch, nestin, Musashi-1, EZH2, STAT3, and Nanog have been reported to identify and characterize GSCs in vitro and in vivo, even though the validity of some of them as stem cell markers is controversial [34]. The ability of GSCs to generate GBMs can be proven by xenotransplantation into recipient mice, which give rise to tumors histologically identical to the tumor of origin. Together with the in vitro "sphere formation assay" (see section "Deregulation of GSC symmetric/asymmetric division in GBM recurrence"), xenotransplantation of patient-derived or in vitro cultured GSCs represents the functional gold standard to determine their tumor-initiation ability, which is a *bona fide* measure of their capacity to promote relapses.

Main mechanisms of chemoresistance in GSCs

GSCs are endowed with multiple intrinsic mechanisms, influenced by extrinsic factors, acting in concert to hamper the efficacy of current therapeutic treatments.

Aberrant DNA repair

Alkylating agents induce methylation adducts of several DNA nitrogen bases, generating DNA double-strand breaks (DSBs), finally leading to apoptosis. Two types of alkylating agents are used against GBM: chloroethylating nitrosoureas (CNUs) (including carmustine (BCNU), nimustine (ACNU), lomustine (CCNU)), and TMZ [8-carbamoyl-3-methylimidazo (5, 1-*d*)-1, 2, 3, 5-tetrazin-4(3*H*)-one].

CNUs alkylate the N^7-position of guanine and the N^3-position of adenine or induce chloroethylation at the O^6-position of guanine; the latter provokes an N^1-deoxyguanosinyl-N^3-deoxycytidyl cross-link, which cannot be corrected properly, giving rise to DNA strand breaks during mitosis. In cells expressing wild-type p53, CNUs-induced DNA damage leads to apoptosis, while kills cells by necrosis in cells lacking functional p53. Therefore, CNUs are effective even in the absence of an active repair system [44].

TMZ, at physiologic pH, is converted in its active compound 5-3-(methyl)-1-(triazen-1-yl) imidazole-4-carboxamide (MTIC), which induces DNA methylation at the *O*-position of guanine or at the *N*-position of both guanine and arginine, resulting in the activation of two different repair mechanisms: whereas methylation at the O^6-position of guanine generates a base mispairing with thymine in the first replication cycle (O^6G-T) and induces futile cycles of mismatch repair (MMR), methylation of N^7-position of guanine or N^3-position of adenine activate the base excision repair (BER) mechanism (Fig. 3). MMR and BER systems, in turn, are regulated by poly ADP-ribose polymerase (PARP-1), which detects the damage, recruits DNA repair proteins and leads to apoptotic cell death. This process generates either DSBs or apurinic/athymidinic site secondary lesions; the ensuing replication fork collapse induces tertiary lesions such as DSBs and sister chromatid exchange, which finally trigger the DDR response and apoptosis [44]. For this reason, TMZ therapy requires a functional DNA repair system in order to be effective. DSBs, also generated by ionizing radiation, genotoxic drugs and reactive oxygen species (ROS), are the most dramatic lesions for the mammalian genetic material. Cells have developed a huge molecular machinery to sense the damage and activate a cascade effect on cell cycle checkpoints, cell cycle arrest, and finally repair, performed either by homologous recombination (HR) or by nonhomologous end joining (NHEJ) (Fig. 4). In actively cycling cells, DSBs are repaired by HR proteins, such as Rad51, XRCC2, XRCC3, BRCA1, BRCA2, and proteins of the Mre11/Rad50/Nbs1 (MRN) complex, which resynthesize a faithful copy of DNA, once a complementary undamaged sister chromatid is available as a template. Conversely, cells in G0 and G1 phases, mostly correct DSBs by NHEJ proteins, including the DNA-dependent protein kinase (DNA-PK) (constituted by Ku70/80 and DNA-PKcs subunits), the DNA ligase IV (Lig4), XRCC4 and Artemis, which, by not necessitating a template, modify the original genetic information (Fig. 4).

The major characteristic that makes GSCs, as opposed to nonstem tumor cells, resistant to current DNA replication-targeted chemo- and radiotherapy, is their more efficient DNA repair machinery and constitutive expression of DNA repair proteins [45]. A similar machinery allows $CD133^+$ glioma cells, but not $CD133^-$ cells, to survive both ionizing radiations [40] and alkylating agents [46]. Strikingly, both radiotherapy and TMZ induce an increase in the $CD133^+$ subpopulation [40, 46], suggesting that these treatments promote the selection of resistant cancer stem cells, responsible for recurrence [40, 46]. Given their dramatic role in GBM resistance, DNA repair mechanisms are the object of intense research, aiming at identifying effective therapies.

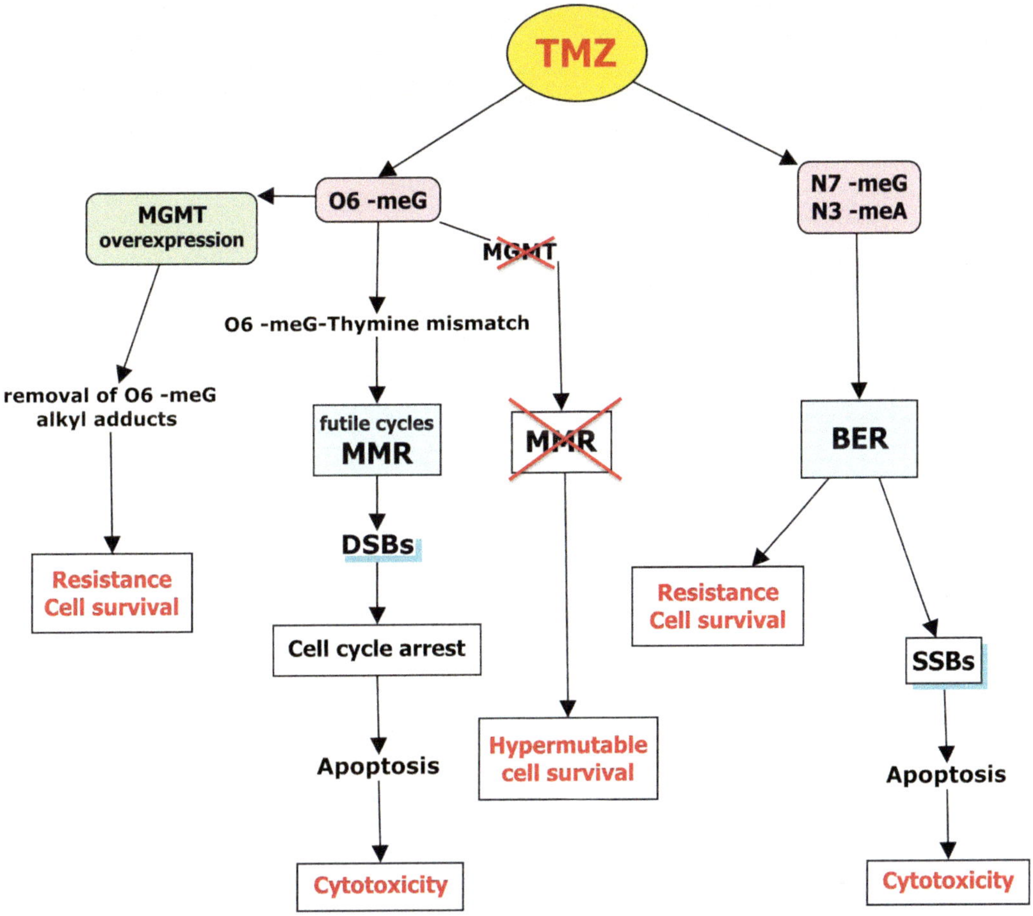

FIG. 3 TMZ-mediated cytotoxicity and mechanisms of resistance and survival in GSCs. *TMZ*, temozolomide; *MGMT*, O^6-methylguanine-DNA methyl-transferase; *O^6-meG*, O^6-methylguanine; *N7-meG*, N7-methyl-guanine; *N3-meA*, N3-methy-adenine; *MMR*, mismatch repair; *DSBs*, double-strand breaks; *BER*, base excision repair; *SSBs*, single-strand breaks. *Adapted from Annovazzi L, Mellai M, Schiffer D. Chemotherapeutic drugs: DNA damage and repair in glioblastoma. Cancers (Basel) 2017;9:57.*

Among the DNA repair tools, GSCs overexpress, in comparison to differentiated GBM cells, the O^6-methylguanine-DNA methyl-transferase (MGMT). This removes alkyl adducts from the O^6 position of guanine, compromising the cytotoxic effects of alkylating agents (Fig. 3). Whereas the BER system is overactive in GBM [34], MMR proteins are frequently lost in GSCs [47], allowing these cells to replicate in spite of the drug-induced DNA damage, therefore, not only therapy is ineffective, but it also induces additional mutations (Fig. 3) [48] resulting in a wide variation in clinical responsiveness. This raises concerns about the appropriateness of chemotherapy in this type of tumor.

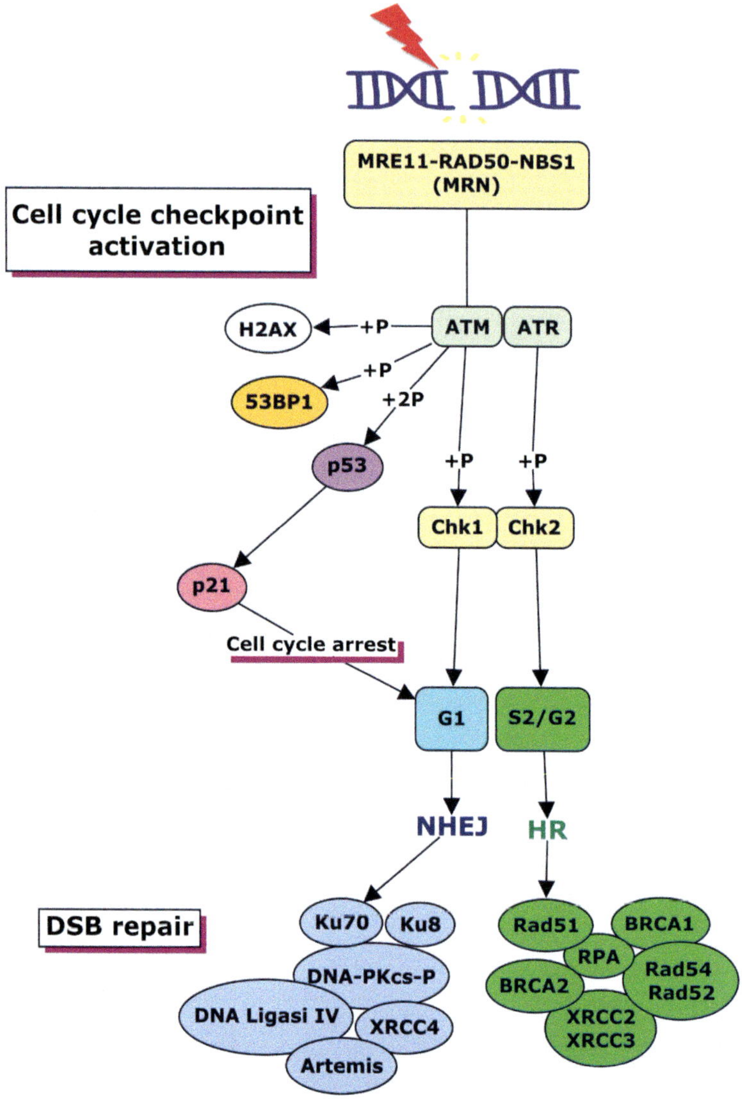

FIG. 4 Scheme of cell cycle checkpoint/DNA repair signaling downstream double-strand break (DSB) damage. *Adapted from Annovazzi L, Mellai M, Schiffer D. Chemotherapeutic drugs: DNA damage and repair in glioblastoma. Cancers (Basel) 2017;9:57.*

PARP-1 is found upregulated in GSCs [49] and, together with the other repair systems, contributes to establishing a population of cells able to overcome the burden of DNA damage induced by chemo- and radiotherapy.

Different transcription factors control GSC maintenance and activate DNA damage repair pathways, thus contributing to therapeutic resistance of GBM. They include OLIG2, c-Myc, BMI1, SOX2, NANOG, OCT4, and ID1. All of these are overexpressed in recurrent GBM as a result of extracellular signaling pathways, epigenetic regulation, and microRNA action [6]. Similarly, EGFR amplification and mutation leading to the constitutively active ligand-independent EGFR variant III (EGFRvIII) have been shown to increase DSB repair and recurrence in GBM cell cultures and in GBM patients [50]. In addition, PI3K/Akt signaling upregulation and loss of tumor suppressor phosphatase and tensin homolog (PTEN) are responsible for therapy resistance by increasing DNA repair [45].

Cell cycle checkpoints deregulation

Untreated GSCs are equipped with increased basal activation of checkpoints [47], leading to extremely elongated cell cycles and allowing GSCs to correct therapy-induced DNA damages, contributing to resistance [45]. Moreover, GSCs may aberrantly recover from DNA damage induced by radio- and chemotherapy owing to a second deregulated machinery, which is the one predisposed to check the correctness of DNA, before allowing the cell cycle to proceed; the so-called DNA damage checkpoint pathway. Following DSBs, the damage-sensing complex MRN activates the ataxia telangiectasia mutated (ATM) and Rad3-related (ATR) kinases which phosphorylate downstream targets, hence triggering cell cycle arrest (necessary for damage elaboration before the arrival of the replication fork), apoptosis and DNA repair (Fig. 4). Whereas ATM responds to DSBs, ATR is fostered by single-stranded DNA stretches, which are formed as a consequence of stalled replication forks, BER or DSBs repair. In particular, ATM-phosphorylates the H2AX histone, producing γ-H2AX (an indirect indicator of DSBs), which in turn recruits the DNA repair proteins, 53BP1, MDC1, RAD51, BRCA1, and the MRN complex, to sites of DSBs [51] and replication fork collapse [52]. ATM also phosphorylates the checkpoint kinases 1 (Chk1) and 2 (Chk2), which arrest the cell cycle in G1/S, S or G2/M phases [45]. Whereas Chk1 expression is limited to S and G2 phases of the cell cycle and is active even in the absence of DNA damage, Chk2 is constitutively expressed during the cell cycle [53], but activated only following ATM-mediated response to DSBs. Still, Chk1 is able to respond to stalled replication fork and damaged DNA [54]. These two kinases hamper cell cycle progression in the presence of DNA damage. In GSCs, their activity is constitutively increased, making cells insensitive to DNA damaging drugs [34]. Accordingly, specific inhibitors of the Chk1 and Chk2 are able to sensitize CD133-positive GSCs to therapy [40]. ATM also phosphorylates the tumor suppressor p53, which induces p21 activation, followed by a cyclin-dependent inhibition of the cell cycle [55] (Fig. 4). In p53-positive cells, TMZ induces p53 and p21-mediated G_2-M arrest, leading to cell apoptosis and senescence. In the absence of p53, TMZ is able to induce only a temporary G2-M arrest, making cells more susceptible to DNA repair alteration and survival [44].

Deregulation of GSC symmetric/asymmetric division in GBM recurrence

One of the major features common to NSCs and GSCs is that they both give rise to the complete hierarchy of cell types of the tissue. Whereas NSCs produce all the neural and glial cells of the neural system, maintaining tissue homeostasis, GSCs originate all the cell types constituting the tumor, providing an albeit aberrant tumor homeostasis. This task is accomplished in both types of stem cells by undertaking a peculiar cell division process, known as "asymmetric division," which gives rise to one self-renewing stem cell, identical to the mother cell, and one daughter precursor cell, which will undergo differentiation, and replenishing the dead cells. Whereas in a normal tissue this is vital for maintaining tissue function, in a tumor

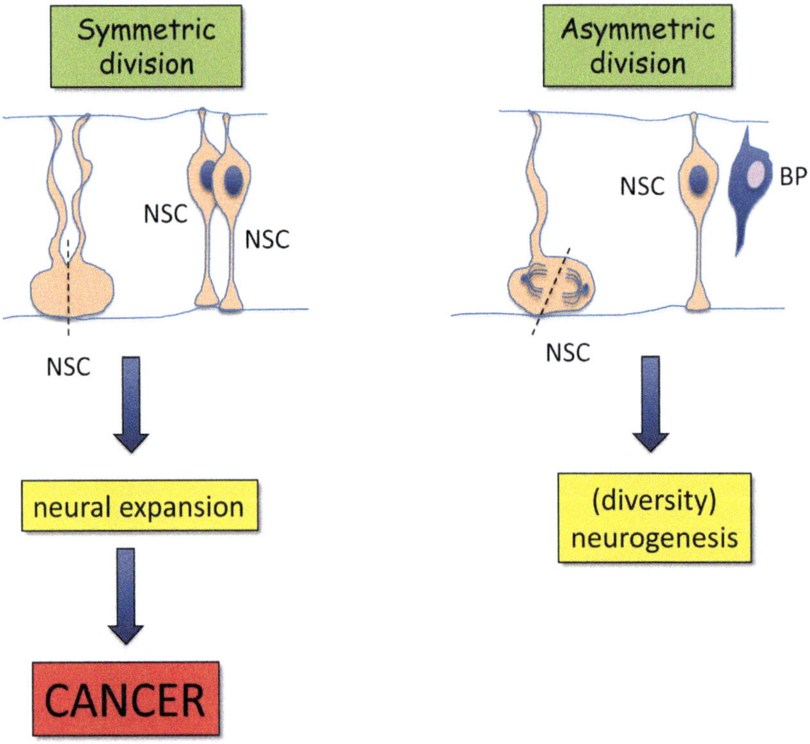

FIG. 5 Symmetric and asymmetric division in neural stem cell homeostasis, neurogenesis and cancer. Types of NSC divisions in the ventricular zone are determined by spindle orientation and the inheritance of cell fate determinants. Symmetric division generates two identical NSCs and is physiologic during neural expansion. Asymmetric divisions generate one NSC and one differentiating daughter cell, providing for tissue homeostasis and neurogenesis. The uncontrolled persistence of symmetric divisions, owing to deregulation of genes controlling cell polarity, induces an increase of the stem cell pool, hyperproliferation and, possibly, cancer.

this process fuels tumor growth, heterogeneity (for this reason, tumors in general, and GBMs in particular, are composed of cells with mixed features) and, more strikingly, recurrence. Alternatively, both NSCs and GSCs, can undergo symmetric divisions, which generates two identical stem cells, leading to the expansion of the stem cell pool. Whereas a rise in cell number is physiologic during development, preserving a certain cell number is fundamental in adulthood. NSCs, like other stem cells, finely regulate the number and type of cells by properly choosing between symmetric and asymmetric division. As expected, neural expansion during embryonic development is characterized by a higher number of symmetric divisions whereas neurogenesis entails a shift toward prevailing asymmetric divisions [37] (Fig. 5). Indeed, an undue increase in NSC symmetrical divisions, owing to the disruption of molecular regulators, is the prelude to the conversion to GSCs and tumors in invertebrates [56, 57]. In mammals, even though the direct demonstration that modification of polarity genes converting NSCs into GSCs is missing [38], it has been shown that loss of p53 or NUMB increases symmetric divisions in cancer stem cell from the mammary gland [58], whereas interference of the Snail pathway, which is correlated with cetuximab resistance in colorectal cancer patients, reduces the symmetric divisions in colorectal cancer stem cells and attenuates tumorigenicity [59], suggesting the importance of symmetric/asymmetric division balance also in human cancer stem cells. Moreover, in humans and mice, it has been shown that asymmetry-defective oligodendrocytes are glioma precursors [60], indicating that the regulation of the division mode might play a role at different levels of brain tumor transformation.

The different balance between symmetric and asymmetric division can be exploited in in vitro assays evaluating the sphere-forming efficiency (SFE) in successive passages. For NSCs, which divide asymmetrically, SFE efficiency remains stable over passages; conversely, for GSCs, which mostly divide symmetrically, the number of sphere-forming cells increases at every passage and SFE results in a rising slope. The silencing of oncogenes favoring symmetric division tends to flatten the SFE slope (Fig. 6).

Asymmetric cell division foresees the creation of a stem cell polarity prior to mitosis through the unequal distribution of cell fate determinants [38], gathered in an apical and a basal polarity complex. Signaling effectors, including NG2 [60], EGFR [61], and CD133 [31] and polarity regulators, including Aurora A kinase (AURKA), PAR3, PAR6 [62], aPKCζ, cdc42 as well as the RNA-binding protein Musashi-1 [38], preferably colocalize apically in the undifferentiated daughter cell during asymmetric cell division, whereas members such as NUMB, LLGL1, SCRIB, and DLG, become localized basolaterally and are inherited by the differentiating cell [63]. aPKC, PAR3, and PAR6 constitute the core members of the polarity complex, mediating cascade of events regulating NUMB localization. At mitosis, AURKA phosphorylates PAR6 in the Par complex, thereby activating aPKC, kept inhibited by PAR6. Activated aPKC phosphorylates LLGL1, causing its disengagement from the cell membrane. At this point, PAR3 is allowed to enter into the Par complex, and allows aPKC to phosphorylate NUMB on one pole of the cell cortex. p-NUMB is unable to localize at the cell membrane [64, 65], determining the asymmetric NUMB localization into the opposite cortical crescent [66], which in turn regulates spindle orientation [38] and cell cycle [67].

NUMB has been found to be asymmetrically segregated between daughter cells of dividing GSCs [68]. As for other polarity proteins, NUMB activity has not only to be tightly coupled to the cell cycle but also localized to a specific cortical domain, to correctly perform asymmetric cell division. NUMB is found to lose its preferential asymmetrical localization in GSCs [58] or is found downregulated [69] or alternatively spliced [70]. In GBM, it has been found downregulated in proneural tumors, owing to homozygous deletion or mRNA reduction [68].

On the other hand, the expression profiles of the Tumor Cancer Gene Atlas (TCGA) tumor samples show that polarity proteins (including LLGL1, LLGL2, CRB1, CRB2, CRB3, DLG1, DLG2, DLG3, DLG4, DLG5, INADL, PARD3, PARD6A, PARD6B, PARD6G, MPDZ, MPP5, aPKCζ, and aPKCι) significantly differ between low-grade gliomas and GBMs [63], supporting a pivotal role of neural stem cell polarity and division mode in GBM recurrence and aggressiveness. Accordingly, aPKC expression is inversely related to GBM patient survival [71]. In GSCs, PTEN loss leads to activation of aPKC and inactivation of LGL1 [72], whereas knockdown of aPKC results in neural differentiation of GSCs, supporting the role of symmetric division in GBM recurrence [73].

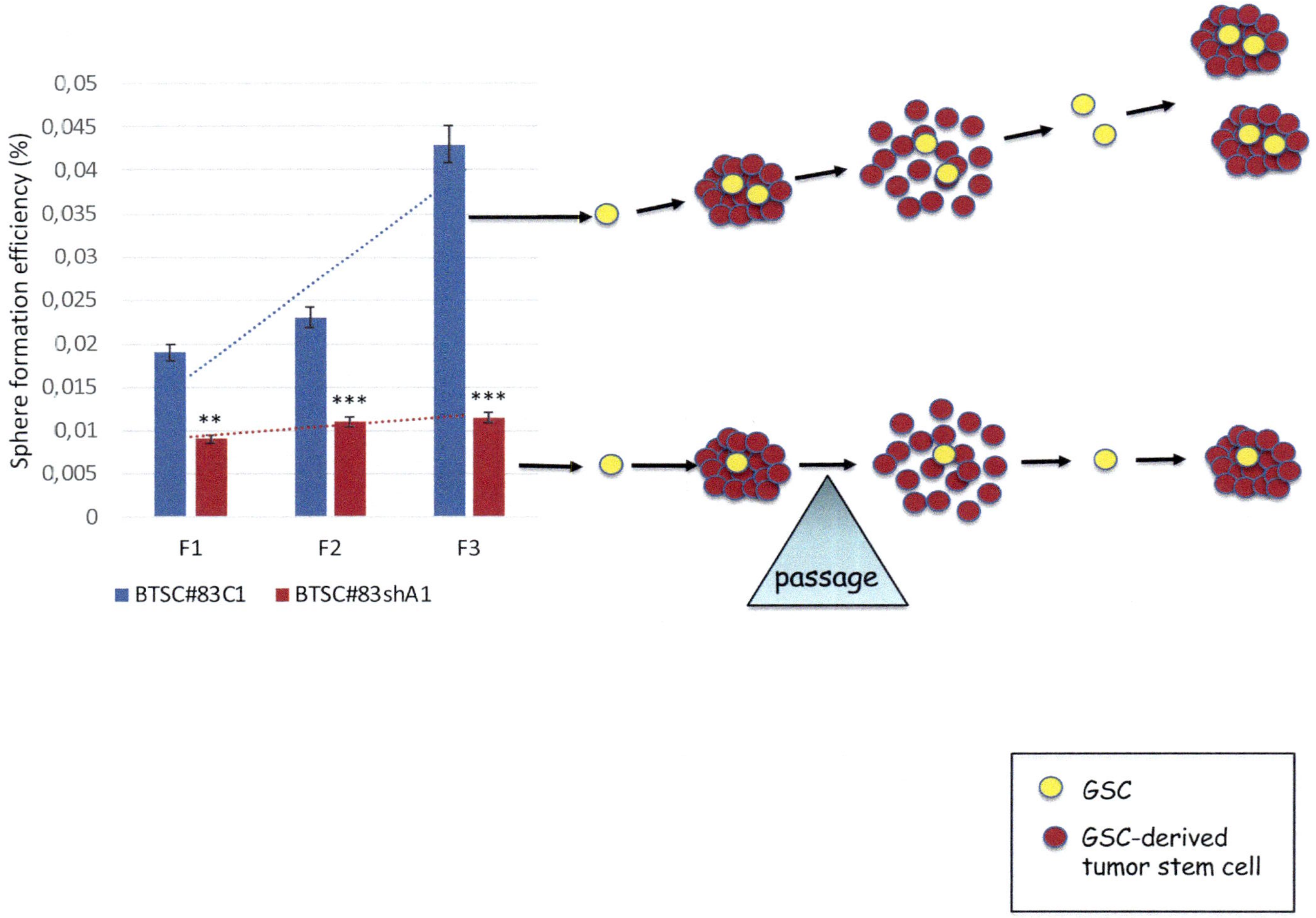

FIG. 6 Sphere-forming efficiency (SFE) in serial passages is an indirect measure of GSCs symmetric division. (Left) In this experiment, GSCs have been silenced for the architectural transcription factor HMGA1 (BTSC#83shA1) or have been transfected with a scramble construct (BTSC#83C1). HMGA1 silencing flattens the slope, compared to control (C1) cells. Adapted from Colamaio M, Tosti N, Puca F, Mari A, Gattordo R, Kuzay Y, Federico A, Pepe A, Sarnataro D, Ragozzino E, Raia M, Hirata H, Gemei M, Mimori K, Del Vecchio L, Battista S, Fusco A. HMGA1 silencing reduces stemness and temozolomide resistance in glioblastoma stem cells. Expert Opin Ther Targets 2016;20:1169–79. (Right) Cartoons depicting the correlation between the SFE slope in control and HMGA1-knockdown GSCs and symmetric/asymmetric division: since spheres are generated only by GSCs, the number of spheres increases at every passage for GSCs undergoing symmetric division; conversely, the number of spheres is kept constant for GSCs (or NSCs) undergoing asymmetric division.

In GSCs the treatment with all-trans retinoic acid (ATRA) has been shown to recover asymmetric division, asymmetric NUMB distribution, and differentiation [74]. Moreover, the silencing of the HMGA1 chromatinic architectural factor, which increases NUMB expression and favors its asymmetric partitioning [75] (S. Battista, unpublished results), has been shown to increase asymmetric divisions and to reduce TMZ resistance [35]. Conversely, the upregulation of the AURKA/PLK1 pathway has been identified as a mechanism of resistance to PI3K inhibitors [76] and the inhibition of this pathway inhibits GSCs growth in vitro and in vivo [77].

Another aspect that makes asymmetric division even more interesting is the fact that, whereas cell fate determinants are unequally segregated, old and newly synthesized DNA are coordinately separated and distributed to the daughter cells, a process known as "nonrandom chromosome segregation"; to this aim, asymmetrically dividing stem cells "mark" and recognize the newly synthesized, error-prone DNA, and, at mitosis, sequester it to the differentiating daughter cell [39]. This mechanism contributes to preserving the correctness of the genetic information of the long-lived stem cells while confining DNA replication errors to the cell which undergoes differentiation and death. Conversely, in symmetrically dividing cells, old and newly synthesized DNA molecules are segregated randomly (random chromosome segregation), which implicates a random distribution of replication-associated DNA errors to daughter stem cells, seemingly fueling tumor heterogeneity (Fig. 7). Genetic and epigenetic heterogeneity is the raw material on which the selective pressure of drugs takes action, killing the sensitive cells, but sparing any fitter heterogeneous cell. The more the population is heterogeneous, the more it will harbor resistant cells, able to reconstitute the tumor mass. Accordingly, GBM heterogeneity has been identified as one of the main causes of drug resistance and recurrence [78]. Therefore, targeting symmetric division may be beneficial to hamper GBM recurrence.

How the segregation of old and new DNA is coordinated with the uneven distribution of cell fate determinants is still a matter of intense research. In 2013, T. Rando proposed the existence of a mechanism able to "sense" the newly synthesized DNA, seemingly based on the double-strand breaks generated in the new DNA, and to deliver it to the differentiating cell through the unequal ability of mother and daughter centrosome to form the spindle [39]. Since expression of cell fate determinants is regulated by proteasomal activity, which is asymmetrically associated to mother centrosome pericentrosomal material [79], centrosomes would be the cell components able to bridge the asymmetric partitioning of chromosomes with asymmetric and local depletion of cell fate determinants, leading to their resulting unequal localization. However, whereas it is widely accepted that in GSCs the balance between the two division modes is deregulated, there are still controversies on the role each division mode plays in GBM recurrence. Strikingly, both seem to have some role in recurrent GBM. Whereas symmetric division might affect tumor relapse by increasing tumor heterogeneity and hampering stem cell differentiation, it has also been proposed that, by dividing asymmetrically, GSCs maintain a population of slow-proliferating stem cells, insensitive to cell cycle regulation [47], while providing the bulk of drug-sensitive tumor cells and regrowing the tumor mass after the treatment. A unifying appealing view has proposed that asymmetric division might be a feature of CSCs from early-stage, well-differentiated tumors, whereas symmetric division might prevail in CSCs from end-stage tumors [80].

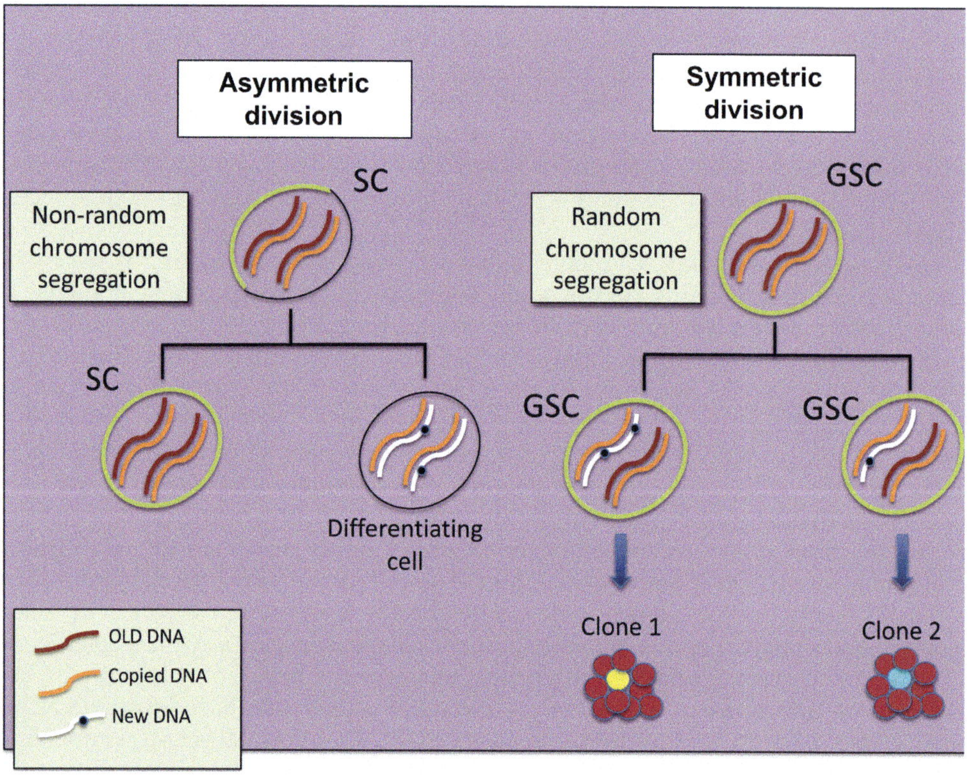

FIG. 7 Random chromosome segregation associated with symmetric division contributes to tumor heterogeneity, favoring chemoresistance. G1-phase chromosomes consist of one older strand (brown), serving as a template, and one newer strand (orange), copied on the template. In the following S-phase, both old and copied strand will in turn serve as a template for the synthesis of new error-prone DNA (dotted white). Nonrandom chromosome segregation allows cells to set aside newly synthesized chromosomes, whereas its coordination with the asymmetric distribution of cell fate determinants (such as Numb; green) allows addressing the younger chromosomes to the differentiating daughter cell, which is destined to senesce/dye. Conversely, random distribution of younger chromosomes, associated with symmetric division, represents the first step for originating different clones, carrying independent mutations, constituting the base for tumor heterogeneity, and resistance.

Efflux pumps

Efflux pumps constitute not only the initial barrier against the access of anticancer drugs in the brain, first, and subsequently into the tumor, as they are the molecular components actively protecting the brain and the tumor from the uptake of chemotherapeutic drugs at the blood-brain barrier (BBB) and blood-tumor barrier, but also a pivotal molecular weapon associated with the GSC subpopulations in recurrent GBM. They are represented by membrane pumps, which actively expel a number of anticancer drugs (hence the name of multidrug resistance proteins) against their concentration gradient, in an ATP-dependent or ATP-independent way [81]. The former group of transporters constitutes the ATP-binding cassette (ABC) transporters family, including the well-studied ABCG2 and ABCB1 (or P-gp).

These proteins bind ATP with their nucleotide-binding domain (NBD) and obtain energy from its hydrolysis to actively translocate the substrate. ABCB1 and ABCG2 overexpression, stimulated by Raf activation, p53 mutations, SHH, Bcl-2, Akt, Survivin, or DNA-damaging anticancer drugs, has been detected in GSCs and recognized as responsible for both intrinsic and acquired resistance [81, 82]; moreover, mutations in the NBD have been found in GBM and not in low-grade gliomas [82].

Other intrinsic and extrinsic mechanisms involved in the recurrence of GBM

Besides GSC mechanisms of resistance, a number of cellular processes and molecular mechanisms, either intrinsic to the tumor cells or derived by paracrine signals from the tumor microenvironment (TME), cooperate to determine chemoresistance in recurrent GBMs.

Phenotypic plasticity and trans-differentiation

Phenotypic plasticity is a quite general term referring to diverse processes involving different types of cells in the tumor. The capacity of dispersing GBM cells, having escaped from therapy, to quickly generate a new tumor in the complex brain environment implies that these cells can adapt to conditions and stress that they are not used to handle by changing their phenotype.

GSCs phenotypic plasticity is a process linked to the physiological property of stem cells to undergo differentiation into different cell types and may be altered and/or influenced by the particular TME. In this way, GSCs adopt the phenotype more suited for that particular environment. Such plasticity has also been demonstrated in differentiated tumor cells where chronic treatment with TMZ has been shown to both increase the $CD133^+$ stem cells population and induce stemness in differentiated tumor cells, leading to increased tumor aggressiveness, grafting and invasion [78, 83]. Even though this increase may be the result of "natural selection," where the few resistant GSCs are selected and expanded because of the drug, active trans-differentiation or dedifferentiation has also been implicated [78].

Recently, it has been proposed that the adaptation of GBM to new challenges occurs via exploratory changes in gene regulation. In particular, a theoretical model of exploratory adaptation has been developed showing that small random perturbations to the gene-regulatory network can result in changes in cellular phenotype [84]. Moreover, the Ivy Glioblastoma Atlas Project has constructed a transcriptional atlas of GBMs that aligns tumor anatomical regions with specific histopathological features [85]. Based on this, spatial-phenotypic trajectories have been proposed along with phenotypic transitions toward adaptation. In other words, spreading GBM cells could employ stochastic exploration of the regulatory network to adapt themselves to a new microenvironment, but they would be constrained to pass through specific spatial-phenotypic ranges. Therefore, it is possible to predict phenotypic variation over spatial trajectories by translating each location's transcriptomic profile to a phenotype [86]. Interestingly, cells can stop their phenotypic transition to an intermediary state, which is just sufficient for the local adaptation, leaving open the possibility to quickly change upon new challenges and thus survive. Blocking these intermediate semistable phenotypes could be the key to a targeted therapy that disrupts the cells' ability to survive [86].

The high plasticity of GBM is however also an intrinsic feature of the tumor. It is owing to the high heterogeneity either inside the same GBM or among different GBMs. The latter intertumor heterogeneity is exemplified by the presence of the different gene expression profiles described by the TCGA consortium that classified GBMs into four molecular subtypes, proneural (PN), mesenchymal (MES), neuronal (NL), and classical (CL) with different prognostic values [87]. Among them, the most prevalent subtypes are PN and MES [88], and PN displays a prognosis better than MES, which has the worst clinical outcome [87]. However, a further intra-subtype heterogeneity has been described. Within the PN subtype, for example, expression of the transcriptional regulator PATZ1 and the chemokine receptor CXCR4 can stratify patients into two prognostic groups, where the worse is characterized by lower levels of PATZ1 and higher levels of CXCR4 [89]. Interestingly, PATZ1 has been shown to downregulate the expression of CXCR4 [89], and CXCR4 can induce the expression of epithelial-mesenchymal transition (EMT) related genes in GBM cells [90]. Indeed, these different subtypes can coexist in the same tumor, and transitions from a subtype to another can occur, especially after treatment and/or disease progression [24, 91]. Particularly, recurrent GBMs tend to shift from PN toward a MES phenotype, a phenomenon known as proneural-to-mesenchymal transition (PMT), reminiscent of EMT that confers a more aggressive phenotype to carcinomas [92, 93]. The PMT may imply a reprogramming process by which differentiated cells acquire stem-like features including the capacity to differentiate into different cell types. Functional niches (prominently the perivascular space and hypoxic regions) provide instructive cues to induction of cellular plasticity toward a stem-like phenotype [94]. According to the clonal variation theory, a single GBM can give rise to a range of stem-like clones with different degrees of chemoresistance. Upon multimodal treatment, most of them transit from the PN to the MES phenotype [93, 95]. An emerging important player in such transition, associated with GBM cell reprogramming, genotoxic resistance, and tumor phenotype plasticity, is OLIG2 [96]. The absence of OLIG2 leads to PMT in distinct GSC cell lines [96, 97]. A strong inductor of PMT in recurrent GBM is also the TME, i.e., the proinflammatory response of macrophages/microglia consequent to the therapy [98]. Consistently, MES GBM has a high degree of necrosis and macrophage/microglia infiltration [99, 100]. Owing to PMT potential reversibility, as for EMT, elucidating its regulatory mechanisms can open a way to new therapeutic options to overcome resistance and recurrence. Consistently, it has been recently shown that silencing of LncRNA-H19 is able to decrease the resistance of glioma cells to TMZ by suppressing PMT via the Wnt/β-catenin pathway [101].

Another phenomenon linked to the high plasticity of GBM is the ability of GBM cells to trans-differentiate into endothelial cells (ECs) to couple neovascularization with tumor growth. Chemotherapeutic stress seems to induce this process. Consistently, recent studies observed increased tumor-derived vessels in TMZ-recurrent GBMs [102].

Apoptosis escape

The key role of most chemotherapeutic agents is to induce cancer cells to suicide through apoptosis. Apoptosis is one of the most extensively studied forms of programmed cell death.

At the molecular level, two major signaling pathways (extrinsic, mediated by death receptors, and intrinsic, mediated by mitochondria) have been delineated that are involved in the regulation of apoptotic cell death and eventually lead to the activation of caspases as a common effector mechanism. The caspases are a family of cysteine proteases that are present as inactive proenzymes in the cytosol and become activated during apoptosis by their proteolytic cleavage at specific aspartate residues and activate a wide range of proteins directly involved in the apoptotic cell death. Caspase action is counteracted by the IAP (inhibitor of apoptosis) molecules, which are in turn inhibited by Smac/Diablo proteins [103]. The tumor suppressor p53, which is activated by chemotherapeutic agents, is upstream of both intrinsic and extrinsic apoptotic pathways, inducing apoptosis upon aggressive treatment.

Recurrent GBMs, like other cancers, have evolved different mechanisms to escape apoptosis, thus allowing their resistance to both radio- and chemotherapy. Indeed, the efficacy of most current anticancer therapies critically depends on intact signal transduction pathways to cell death, and reactivation of defective cell death programs is currently considered as a promising approach for the treatment of cancer [104]. As already mentioned, DNA damage checkpoint proteins can render cancer cells more resistant to therapy-induced apoptosis through increased efficiency in the repair of damaged DNA. In particular, MGMT removes mutagenic alkyl adducts, thereby protecting DNA from the damage induced by TMZ. Therefore, loss of MGMT expression by its promoter hypermethylation enhances sensitivity to TMZ treatment, while its overexpression, frequently observed in recurrent GBM owing to epigenetic regulation, can be responsible for GSC resistance to TMZ [105]. It has been recently shown that the inhibitor of NF-kB kinase subunit epsilon enhances the resistance of GBM cells to TMZ by activating the AKT/NF-kB signaling pathway to regulate the expression of MGMT [106]. However, even in GBM patients with MGMT methylation, recurrent GBMs have activated alternative pathways to escape apoptosis. In a recent study, by profiling proteins expressed in GBMs from 36 MGMT methylated patients, overexpression of macrophage migration inhibitory factor was identified in those associated with poor survival outcomes [107].

As a further mechanism, it has been reported that recurrent GBMs show increased levels of the ubiquitin-specific protease USP4, a negative regulator of p53, thus contributing to escaping apoptosis after TMZ treatment. In fact, when USP4 knockdown cells were treated with TMZ, the cell viability was decreased significantly in a p53-dependent manner [108].

Autophagy

Autophagy, including macroautophagy, microautophagy, and chaperon-mediated autophagy (CMA), is a catabolic process essential in maintaining cellular homeostasis since it acts as a quality control mechanism for proteins and organelles after cellular or environmental stress, such as starvation, hypoxia, and chemo/radiotherapy. During this process, damaged organelles and/or unfolded proteins are conveyed to the lysosome where they are degraded. Therefore, it has an important tumor suppressing function by avoiding the accumulation of intracellular stress and consequent DNA damage. However, at the end of this catabolic process, the cell gains energy and nutrients from itself, thus finding a way to survive in nutrient deprivation conditions, such as the insufficient nutrient supply from the lining vasculature in a rapidly growing tumor mass. For these reasons, autophagy is considered a dual tumor-suppressing/tumor-promoting process depending on the tumor stage. At an early stage, it acts as tumor suppressor by avoiding accumulation of mutations; at a late stage, it acts as tumor promoter by allowing survival in stress conditions, such as hypoxia and nutrient deprivation, while inhibiting apoptosis. This dual role is also present in response to therapy, depending on the intensity of its induction. If moderately induced, autophagy blocks apoptosis and acts as a pro-survival mechanism by allowing adaptation to the therapeutic regimen; conversely, if strongly induced, it causes cell death [109].

In GBM, pro-survival autophagy is induced upon TMZ treatment. Consistently, its inhibition ameliorates the level of apoptosis following TMZ-based therapy. In particular, autophagy inhibitors, such as chloroquine and its analogs, have been used in clinical trials preceding TMZ treatment [110]. Chloroquine and its analog natural compounds, including chrysin, resveratrol and others, have been found to inhibit TMZ-induced cytoprotective autophagy and sensitize glioma cells to apoptotic cell death [111, 112]. On the other hand, some other plant-derived compounds, like the *Cannabis sativa*, can enhance TMZ sensitivity by inducing autophagic cell death [113, 114]. Moreover, it has been recently suggested that the TMZ-dependent reduction of HIF-1α activity in TMZ-sensitive cells (see section "Recurrent glioblastoma"), which in turn reduces MGMT expression in GSCs and sensitizes glioma cells to TMZ treatment [115], is mediated by CMA [116]. Therefore, inhibiting CMA not only inhibits HIF-1α downmodulation in sensitive GBM cells but also correlates with their acquisition of a TMZ-resistant behavior [116]. Subsequent studies by the same group established that fluctuations of cytoplasmic ROS levels were accountable for CMA induction and cytotoxic effects observed in TMZ-sensitive cells after treatment, while in resistant cells, TMZ failed in producing an increase in cytoplasmic ROS levels and CMA activation, preventing GBM cell toxicity. Then, by increasing oxidative stress, CMA activation was recovered, as also cell cytotoxicity, especially in combination with TMZ treatment [116].

So, what should we do with autophagy during TMZ therapy? It is likely that with low doses and short-term TMZ treatment autophagy is a survival mechanism and therefore it should be inhibited, whereas, upon persistent TMZ treatment, autophagy becomes a mechanism out of control that we should induce to enhance cell death [117].

Importantly, autophagy plays a crucial role also in GSC biology, promoting their differentiation [118]. This implies that when autophagy is inhibited, GSCs undergo uncontrolled self-renewal and rapid proliferation in the absence of differentiation. Moreover, GSCs infiltrate within the healthy brain parenchyma, thus resulting in tumor progression toward a more invasive and therapy-resistant GBM [119]. Consistently, autophagy induction has been shown to impair GSC sphere formation [120]. Conversely, autophagy is a hypoxia-induced adaptive mechanism promoting survival also for the GSCs, thus enhancing resistance to chemotherapeutic agents. The NIX-mediated mitochondrial autophagy (mitophagy) regulates tumor survival in the hypoxic niche of GBM TME [121].

However, it is worth noting that autophagy regulates specific immune functions such as antigen presentation, cell activation, and differentiation [122], and could be activated in GBM as a way to suppress the immune system. Indeed, it has been recently shown that GBM ablates pericyte antitumor immune function through aberrant upregulation of CMA [123]. On the other hand, another study showed that a doxorubicin-polyglycerol-nanodiamond composite stimulates GBM cell immunogenicity through activation of autophagy [124]. Therefore, the use of autophagy inhibitors may either strengthen or counteract the effectiveness of the current immunotherapies.

In conclusion, autophagy is a "double-edged sword" in both GBM development and its chemoresistance, depending on cellular context and tumor type, stage of the tumor, cell viability, and intracellular environment; it is a very complex cellular mechanism that should be deeply studied for each specific case before using it as a target for any adjuvant therapy for GBM treatments.

Cell metabolism

Compared to normal cells, cancer cells strongly upregulate glucose uptake and glycolysis to give rise to increased levels of intermediate glycolytic metabolites and the end product pyruvate. Moreover, glycolysis is uncoupled from the mitochondrial tricarboxylic acid cycle and oxidative phosphorylation in cancer cells. Consequently, the majority of glycolysis-derived pyruvate is converted to lactate and kept away from mitochondrial oxidative metabolism. This metabolic phenotype is known as the Warburg effect and avoids mitocondrial release of ROS that could be toxic to the cell. In this context, mitochondria play a central role in cellular energy production, apoptosis, and free radical generation. Glioma cells, which rely on glycolytic metabolism, readily adapt to bioenergetic stress by engaging their mitochondrial pathway in order to survive and grow [125].

Oxidative stress is an important and clinically relevant component of TMZ-induced therapeutic effects in GBMs. However, as previously mentioned, differently from TMZ-sensitive cells, in resistant GBM cells ROS release is not increased after TMZ treatment, accounting for the absence in these cells of CMA activation and proapoptotic genes upregulation [116]. It has been recently shown that ROS cytotoxicity is mediated by aldehydes resulting from lipid peroxidation and that aldehyde dehydrogenase activity is able to reduce the number of toxic aldehydes. Recurrent GBMs show significantly higher aldehyde dehydrogenase 1A3 (ALDH1A3) expression than the respective samples from the primary tumor, thus showing resistance to TMZ. Consistently, patients suffering from GBM with high ALDH1A3 expression showed a shorter median survival time [126]. Another recent study revealed that hexokinase 2 (HK2)-mediated glycolysis is crucial for chemoresistance in GBM. In contrast to the healthy brain, HK2 expression is much higher in human GBM, especially in those patients with GBM recurrence. Moreover, high HK2 expression is negatively related to overall survival, while HK2 depletion in GBM cells increases sensitivity to TMZ-induced apoptosis. The abnormal expression of HK2 is modulated by the long noncoding RNA HOTAIR by targeting miR-125, suggesting that the impairment of the HOTAIR/miR-125 axis could sensitize GBM cells to TMZ treatment [127]. Pyruvate kinase 2 (PKM2), the last rate-limiting enzyme of glycolysis, is known to regulate aerobic glycolysis and considered a novel cancer therapeutic target. Recently, the active form of the small molecular compound dimethylaminomicheliolide (DMAMCL), which has been used in clinical trials for recurrent GBM in Australia, has been shown to bind to monomeric PKM2, thus promoting its tetramerization and improving its pyruvate kinase activity in GBM cells. Following the treatment with DMAMCL many metabolites involved in glycolysis, including lactate and glucose-6-phosphate, are downregulated, suggesting a rewiring of the aerobic glycolysis, which in turn suppresses

GBM proliferation [128]. Using an in vitro TMZ-resistant model expressing EGFRvIII (see section "Extracellular signaling pathways" for further details on this variant), Velpula et al. demonstrated that dichloroacetate (DCA), a metabolic inhibitor of pyruvate dehydrogenase kinase 1 (PDK1), reverses the Warburg effect and TMZ resistance by binding to EGFRvIII and PDK1 at high affinity, suggesting implications of cellular tyrosine kinase signaling in cell metabolism of recurrent GBM [129].

Apart from glucose metabolism, other metabolic pathways appear involved in the recurrence of GBM. Recent evidences in GBM cells cultures showed that expression of EMT markers, a characteristic of the recurrent phenotype (see section "Phenotypic plasticity and trans-differentiation"), is associated with aberrant choline metabolism, resulting in high intracellular levels of phosphocholine and total choline derivatives. Importantly, inhibition of the pivotal choline metabolism regulator choline kinase alpha (CHKα) significantly reduces the cell viability, invasiveness, clonogenicity, and expression of EMT-associated genes in GBM cells, suggesting a way to target chemoresistant GSCs. Consistently, the combination of TMZ and CHKα inhibitors results in a more efficacious cytotoxic effect [130].

Extracellular signaling pathways

The GBM TME consists of different cellular populations, including tumor cells, ECs, pericytes of blood vessels, innate immune cells (microglia, monocytes, macrophages, neutrophils, and mast cells), T cells, neurons, astrocytes, oligodendrocytes, and GSCs, other than extracellular matrix (ECM) components. Growth factors and cytokines as well as tumor cell contacts and tumor-stroma interactions involving tumor cells, immune cells and ECs, activate different molecular pathways in tumor cells contributing to chemoresistance in recurrent GBM.

Exosome secretion

GBM cells use multiple ways of communication with the TME. Among these, extracellular vesicles, also named exosomes, have been deeply studied in the last few years. They are circular bilayer membrane vesicles released by cells, containing soluble proteins, DNA, mRNA, and noncoding RNAs by which they can modulate the phenotype of recipient cells. On the surface, they express adhesion molecules characteristic of their cell of origin. Recipient cells can internalize exosomes in different ways, including phagocytosis, endocytosis, or simple membrane fusion. Once the vesicle is internalized its cargo is released into the cytosol and eventually transported to the nucleus, inducing functional changes. In addition, the interaction between exosome membrane proteins and receptors on the surface of the recipient cell can stimulate intracellular signaling pathways [131].

Through exosome secretion, GBM cells can modulate their TME, with consequences on tumor progression, invasion, and recurrence. Indeed, it has been recently shown that exosomes released by GBM are able to modify the phenotype of tumor-associated macrophages (TAMs), which include monocyte, macrophages, and microglia, particularly enriched in the GBM TME, in the immune suppressive and tumor-supportive M2-like phenotype [36, 132]. It has been shown that the release of these vesicles in the tumor, by both cancer and stromal cells, promote tumor-induced immune suppression, angiogenesis, and metastasis [133]. Regarding gliomas, it has been shown that glioma-associated stem cells (GASCs) increase the aggressiveness of GSCs through the release of exosomes [134]. Semaphorin7A (SEMA7A), a secreted and membrane-bound glycoprotein belonging to a family of proteins known as important guidance signals in the development of the central nervous system [135], is exposed on GASC-exosomes' surface and signals to GSC through integrin β1. This interaction activates focal adhesion kinase into GSCs and increases their motility [136]. Moreover, other members of the semaphorin family were identified in exosomes of the GBM TME. SEMA3A released in GSC exosomes disrupts the endothelial barrier, thus promoting the vascular permeability and invasion in the surrounding brain parenchyma [137].

Another recent study showed that GBM cells remodel the TME to promote tumor chemotherapy-resistance by secreting the oncogenic long noncoding RNA SBF2 antisense RNA 1 (lncRNA SBF2-AS1) through exosomes. This implies that exosomal lncSBF2-AS1 in human serum may serve as a possible diagnostic marker for recurrent therapy-refractory GBM [138].

Receptor tyrosine kinase

Transmembrane receptor tyrosine kinases (RTKs), interacting with a huge variety of extracellular signaling molecules, play crucial roles in cell proliferation, survival, migration, differentiation and angiogenesis in human cancers [139]. Areas of intense research are dedicated to finding next-generation strategies to efficiently target major RTK-dependent signaling pathways that are altered in GBM cells by using combination drugs and novel experimental drugs including oligonucleotide aptamers, vaccines, chimeric antigen receptors, decoy molecules, and synthetic small compounds, with the final aim of overcoming the occurrence of parallel signaling and drug resistance.

EGFR gene amplification, associated with high protein expression, is a signature pathogenic event in GBM, occurring in about 57.4% of primary GBMs and enriched in the classical subtype [140]. Half of the EGFR-amplified tumors, as well as a significant percentage of recurrent GBMs, harbor the EGFRvIII mutant oncogene (also known as EGFR type III, de2-7, ΔEGFR), an intragenic rearrangement generated by in-frame deletion of exons 2–7 from EGFR, yielding a constitutively active variant receptor that cannot bind the ligand [141]. Interestingly, in addition to the deletion in the ectodomain (ECD) of EGFR, leading to the EGFRvIII mutant, some point mutations residing in the ECD of EGFR have been identified in GBM [142], most of which cause constitutive receptor activation [143, 144]. The clinical importance of EGFRvIII is evident as its expression is closely linked to increased tumor aggressiveness, invasion, and poor prognosis for survival. Further, EGFRvIII appears to be relatively resistant to treatment with conventional anti-EGFR agents such as ligand-blocking monoclonal antibodies (mAbs) or EGFR tyrosine kinase inhibitors (TKIs), and its expression contributes to chemo/radioresistance in various cancers, especially in GBM [145]. One way by which EGFR contribute to GBM resistance is by activating both HR and NHEJ mechanisms of DSB repair [50]. Most cells in GBM with amplification of both EGFR and EGFRvIII express predominantly EGFR or EGFRvIII but the coexpression of both receptors within a tumor increases malignancy and confers a worse prognosis than the expression of a single receptor alone [146–148]. Paracrine interactions exist among cells expressing the two receptors separately. Indeed, cells expressing EGFRvIII alone secrete elevated levels of interleukin-6 family cytokines, which promote the proliferation of adjacent cells expressing the wild-type receptor, thus maintaining the tumor cell heterogeneity [149]. Also, it has been shown that GBM cells can share EGFRvIII RNA and protein, which are transferred among cells *via* microvesicles that increase tumor malignancy [150, 151]. In addition, several sophisticated mechanisms of crosstalk between coexpressed EGFR and EGFRvIII, which may include [152, 153] or not [154] the physical association of the receptors, have been reported that ultimately enhance the fitness of the tumor.

A variety of therapies targeting EGFR or its mutant EGFRvIII, including TKIs and mAbs, have only shown limited efficacy for GBM in the clinic [140, 155, 156], in part owing to their scarce bioavailability and severe toxicity but also because of the occurrence of multiple mechanisms of acquired drug resistance. First, the coactivation of multiple RTKs often occurs in the same tumor [157]. For example, GBM tumors with amplification of PDGFR-alpha frequently show amplification of MET or EGFR. Further, EGFRvIII is tumorigenic, in part, through its transactivation of AXL, PDGFR, VEGF receptor 3, and fibroblast growth factor receptor and other RTKs [158] and different mechanisms of cooperation exist among EGFRvIII and MET [159, 160]. Therefore, targeting a single RTK may not be sufficient to inhibit GBM. Moreover, inhibition of an RTK can cause the upregulation of other receptors. A strong evidence for this last mechanism has been provided by Akhavan et al. [161], showing that EGFRvIII represses PDGFRβ expression. EGFR TKIs cause the derepression of PDGFRβ allowing the tumors to become addicted to its expression and signaling for survival. Accordingly, we proved that two nuclease-resistant RNA-aptamers, which specifically target and neutralize EGFR/EGFRvIII [162] and PDGFRβ [163], strongly cooperate in inhibiting GBM proliferation and migration.

As mentioned earlier, GBM is a highly heterogeneous tumor and its heterogeneity also extends to the TME composition, whose complexity contributes to resistance to conventional treatments [164]. One of the hottest areas of GBM research indeed aims to clarify the mechanisms by which GBM interacts with the different components of its TME in order to create opportunities for better treatment options. Transmembrane RTKs play essential roles in the interaction with the extracellular TME in several human cancers. It has been shown that in hypoxic and vitronectin-enriched microenvironments, integrin β3 interacts with EGFRvIII and activates an SRC/FAK/EGFRvIII signaling axis to promote GBM cell invasion [165]. Noteworthy, we proved that the binding of EGFR aptamer to the extracellular domain of the receptor hampers the formation of integrin αvβ3-EGFR complex on the surface of TNBC cells thus preventing the integrin attachment to the ECM and, consequently, inhibiting vasculogenic mimicry and tumor growth [166]. Interrupting the EGFR/integrin complex formation could be a valuable therapeutic strategy also in GBM.

Changes in the TME

The TME in recurrent GBM encounters substantial changes that contribute to drug resistance. First of all, it is characterized by immune-suppressive cells, recruited by the GSCs through the secretion of cytokines and chemokines, which allow evasion from host immune surveillance. Among them, the main types observed in GBM are TAMs and myeloid-derived suppressor cells (MDSCs). As already mentioned, most of the macrophages recruited into TME become M2 subtype (see section "Exosome secretion") and induce T-cell anergy, secrete ECM components, stimulate angiogenesis and enhance tumorigenicity and drug resistance [167].

Also abundant in the GBM TME, the MDSCs are involved in the inhibition of T-cell proliferation and activity through the secretion of several substances, including inducible nitric oxide synthase, ROS, cyclooxygenase-2 and TGF-β [168].

Other components of the GBM TME involved in chemoresistance are represented by cancer-associated fibroblasts (CAFs) and neutrophils. CAFs contribute to the maintenance of GSCs and stimulate the growth and migration of glioma cells, promoting tumor aggressiveness and resistance to therapeutic treatments [169]. Neutrophils also are involved in the regulation of GSCs and appear to be a marker of poor prognosis in GBM [168, 170].

A peculiar characteristic of GBM, relative to the ECM, another essential component of the TME, is its stiffness that could be a significant obstacle to the recruitment of inflammatory cells and drug delivery of macromolecules and influences GSC plasticity [170]. A soluble component of the ECM, galectin-1, has been recently described to be involved in therapy resistance, especially to TMZ [171].

Furthermore, as already mentioned, an essential aspect of the TME changes in recurrent GBM is hypoxia, which enhances GSC maintenance and chemoresistance through the regulation of genes, such as HIFs, that influence different tumor mechanisms of aggressiveness, including resistance to therapies, genomic instability, and metabolic changes. Under hypoxic condition, paclitaxel-induced mechanisms of apoptosis appear impaired in GBM, thus limiting the effect of this chemotherapy [172]. The effect of hypoxia on GBM chemoresistance is also known to be mediated by ABC transporters. In particular, two different studies showed that HIF-1α plays a crucial role in the induction of ABC transporters [173, 174], leading to acquired resistance to doxorubicin and etoposide in T98G glioma cells [174]. On the other hand, another study has shown that hypoxia regulates calcium-activated potassium channels that in turn enhance GBM resistance to cisplatin [175], and appears to be involved in resistance to alkylating compounds [176].

Finally, another critical stress factor of TME that promotes chemoresistance is the acidosis induced by the high levels of lactate, reducing the uptake and the efficacy of different drugs, such as anthracyclines, anthraquinones, and vinca alkaloids [177, 178]. Furthermore, in human glioma cells, lactate appears to be involved in promoting a GSC phenotype [179].

The perivascular niche

In GBMs, leaky blood vessels infiltrate the tumor mass and create a microenvironment with high interstitial pressure and poor delivery of oxygen and nutrients, which limits penetration of the drugs [34]. In addition, this particular TME contains certain chemoattractive molecules, such as bradykinin or CXCL12 that provide chemotactic signals to glioma cells in the perivascular niche and facilitate invasion [180, 181]. Often, GSCs that are in the perivascular niches are observed to physically interact with ECs and this interaction is a key driver of tumor progression [132, 182]. ECs promote GSCs survival through secretion of TGF-β and PDGF, which in turn increase expression of stemness-related genes such as Sox-2, Olig-2, Bmi-1, and CD133 in GSCs [183]. Direct interaction between ECs and GSCs also activates key stemness pathways such as nitric oxide- (NO-) cyclic GMP pathway [184] and Notch signaling [185], promoting malignancy and poor patient survival [183].

The blood-brain barrier

The blood-brain barrier (BBB) composed of ECs, astrocytes, and pericytes, isolating the brain microenvironment from the blood, is strongly linked to the drug-resistant phenotype of GBM, being a major physical and physiological obstacle for the delivery of chemotherapeutic agents into the brain. In fact, poor permeability of anticancer drugs through the BBB and their inability to reach the tumor mass at a therapeutic concentration decrease the effectiveness, thus contributing to treatment failure. It has been reported that all GBM patients present intact regions of BBB sufficient to prevent drug distribution to the tumor cells. In particular, it has been estimated that 100% of large-molecule and 98% of small-molecule drugs cannot access the brain parenchyma from the bloodstream owing to the tight junctions between capillary ECs and the efflux activity of ABC transporters [170]. Moreover, ABC transporters erect a further barrier at the blood-tumor boundary, making the access of anticancer drugs even more challenging [82].

To overcome this obstacle and increase drug delivery, different methods have been explored. Among them, noninvasive approaches, such as nanotechnologies, osmotic agents, and efflux pump inhibitors, are preferred because of their limited side effects [170]. In this regard, we have demonstrated that the already mentioned PDGFRβ aptamer (see section "Receptor tyrosine kinase") is able to transcytose the BBB and actively deliver drug-loaded nanoparticles to brain tumors [186]. The same aptamer was conjugated to another aptamer targeting GBM cells and used to assemble nanocarriers able to cross an in vitro BBB model [187], highlighting the aptamer potential for targeted drug delivery into GBM [188].

Conclusion

The failure of chemoradiation, which culminates with the inevitability of recurrence and acquisition of a chemoresistant phenotype, has induced investigations to find novel therapeutic protocols effective for recurrent GBM.

GBM is a dynamic tumor that, under the selective pressure of a specific therapy, can alter both its genetic composition and physiologic processes. Favorite clones, i.e., those that have greater adaptation to the changed TME, such as those endowed with a mutator phenotype, are selected to generate recurrent GBM. Central in the acquisition of chemoresistance in recurrent GBM are the GSCs, which are intrinsically equipped with enhanced DNA repair abilities to recover after drug-induced DNA damages. Furthermore, they can acquire resistance as a way to adapt and survive in a new TME. Notably, they can be stimulated by chemotherapeutic agents to increase their ratio of symmetric/asymmetric division, which leads to increased heterogeneity and, consequently, higher resistance. Moreover, upon different types of treatment, GSCs are forced to go through a PMT process, reaching a MES phenotype that confers a more aggressive and resistant behavior to the recurrent GBM.

The definition of the cellular and molecular mechanisms involved in the enhanced chemoresistance of recurrent GBM, also taking into account the high degree of heterogeneity and molecular complexity of the tumor, will help to design targeted therapies necessary to resensitize it toward cytotoxicity when using TMZ and possibly other chemotherapeutic agents.

Acknowledgment

We acknowledge Fondazione AIRC per la Ricerca sul Cancro for its support to our research.

References

[1] Kim J, Lee I-H, Cho HJ, Park C-K, Jung Y-S, Kim Y, Nam SH, Kim BS, Johnson MD, Kong D-S, Seol HJ, Lee J-I, Joo KM, Yoon Y, Park W-Y, Lee J, Park PJ, Nam D-H. Spatiotemporal evolution of the primary glioblastoma genome. Cancer Cell 2015;28:318–28.

[2] Stupp R, Hegi ME, Mason WP, van den Bent MJ, Taphoorn MJ, Janzer RC, Ludwin SK, Allgeier A, Fisher B, Belanger K, Hau P, Brandes AA, Gijtenbeek J, Marosi C, Vecht CJ, Mokhtari K, Wesseling P, Villa S, Eisenhauer E, Gorlia T, Weller M, Lacombe D, Cairncross JG, Mirimanoff RO, European Organisation for Research and Treatment of Cancer Brain Tumour and Radiation Oncology Groups, National Cancer Institute of Canada Clinical Trials Group. Effects of radiotherapy with concomitant and adjuvant temozolomide versus radiotherapy alone on survival in glioblastoma in a randomised phase III study: 5-year analysis of the EORTC-NCIC trial. Lancet Oncol 2009;10:459–66.

[3] Weller M, van den Bent M, Hopkins K, Tonn JC, Stupp R, Falini A, Cohen-Jonathan-Moyal E, Frappaz D, Henriksson R, Balana C, Chinot O, Ram Z, Reifenberger G, Soffietti R, Wick W, European Association for Neuro-Oncology (EANO) Task Force on Malignant Glioma. EANO guideline for the diagnosis and treatment of anaplastic gliomas and glioblastoma. Lancet Oncol 2014;15:e395–403.

[4] Van Meir EG, Hadjipanayis CG, Norden AD, Shu HK, Wen PY, Olson JJ. Exciting new advances in neuro-oncology: the avenue to a cure for malignant glioma. CA Cancer J Clin 2010;60:166–93.

[5] Ryken TC, Kalkanis SN, Buatti JM, Olson JJ, AANS/CNS Joint Guidelines Committee. The role of cytoreductive surgery in the management of progressive glioblastoma: a systematic review and evidence-based clinical practice guideline. J Neuro-Oncol 2014;118:479–88.

[6] Osuka S, Van Meir EG. Overcoming therapeutic resistance in glioblastoma: the way forward. J Clin Invest 2017;127:415–26.

[7] Taal W, Oosterkamp HM, Walenkamp AM, Dubbink HJ, Beerepoot LV, Hanse MC, Buter J, Honkoop AH, Boerman D, de Vos FY, Dinjens WN, Enting RH, Taphoorn MJ, van den Berkmortel FW, Jansen RL, Brandsma D, Bromberg JE, van Heuvel I, Vernhout RM, van der Holt B, van den Bent MJ. Single-agent bevacizumab or lomustine versus a combination of bevacizumab plus lomustine in patients with recurrent glioblastoma (BELOB trial): a randomised controlled phase 2 trial. Lancet Oncol 2014;15:943–53.

[8] Van den Bent MJ, Brandes AA, Rampling R, Kouwenhoven MCM, Kros JM, Carpentier AF, Clement PM, Frenay M, Campone M, Baurain J-F, Armand J-P, Taphoorn MJP, Tosoni A, Kletzl H, Klughammer B, Lacombe D, Gorlia T. Randomized phase II trial of erlotinib versus temozolomide or carmustine in recurrent glioblastoma: EORTC brain tumor group study 26034. J Clin Oncol 2009;27:1268–74.

[9] Pitz MW, Eisenhauer EA, MacNeil MV, Thiessen B, Easaw JC, Macdonald DR, Eisenstat DD, Kakumanu AS, Salim M, Chalchal H, Squire J, Tsao MS, Kamel-Reid S, Banerji S, Tu D, Powers J, Hausman DF, Mason WP. Phase II study of PX-866 in recurrent glioblastoma. Neuro-Oncology 2015;17:1270–4.

[10] Meyer M, Reiman J, Lan X, Head R, Zhu X, Bayani J, Pressey JC, Lionel AC, Clarke ID, Cusimano M, Squire JA, Scherer SW, Bernstein M, Woodin MA, Bader GD, Dirks PB. Single cell-derived clonal analysis of human glioblastoma links functional and genomic heterogeneity. Proc Natl Acad Sci U S A 2015;112:851–6.

[11] Qazi MA, Vora P, Venugopal C, McFarlane N, Subapanditha MK, Murty NK, Hassell JA, Hallett RM, Singh SK. A novel stem cell culture model of recurrent glioblastoma. J Neuro-Oncol 2016;126:57–67.

[12] Huang Q, Zhang QB, Dong J, Wu YY, Shen YT, Zhao YD, Zhu YD, Diao Y, Wang AD, Lan Q. Glioma stem cells are more aggressive in recurrent tumors with malignant progression than in the primary tumor, and both can be maintained long-term in vitro. BMC Cancer 2008;8:304.

[13] Weller M, Cloughesy T, Perry JR, Wick W. Standards of care for treatment of recurrent glioblastoma—are we there yet? Neuro-Oncology 2013;15:4–27.

[14] Campos B, Olsen LR, Urup T, Poulsen HS. A comprehensive profile of recurrent glioblastoma. Oncogene 2016;35:5819–25.

[15] Stark AM, Witzel P, Strege RJ, Hugo H-H, Mehdorn HM. p53, mdm2, EGFR, and msh2 expression in paired initial and recurrent glioblastoma multiforme. J Neurol Neurosurg Psychiatry 2003;74:779–83.

[16] Stark AM, Doukas A, Hugo H-H, Mehdorn HM. The expression of mismatch repair proteins MLH1, MSH2 and MSH6 correlates with the Ki67 proliferation index and survival in patients with recurrent glioblastoma. Neurol Res 2010;32:816–20.
[17] Martinez R, Setien F, Voelter C, Casado S, Quesada MP, Schackert G, et al. CpG island promoter hypermethylation of the pro-apoptotic gene caspase-8 is a common hallmark of relapsed glioblastoma multiforme. Carcinogenesis 2007;28:1264–8.
[18] Martinez R, Rohde V, Schackert G. Different molecular patterns in glioblastoma multiforme subtypes upon recurrence. J Neuro-Oncol 2010;96:321–9.
[19] Bilzer T, Stavrou D, Wechsler W, Wöhler B, Keiditsch E. Antigen variation in a human glioblastoma: from the primary tumor to the second recurrence, permanent cell line and xenotransplantation tumors. Anticancer Res 1991;11:547–53.
[20] Ilhan-Mutlu A, Wöhrer A, Berghoff AS, Widhalm G, Marosi C, Wagner L, Preusser M. Comparison of microRNA expression levels between initial and recurrent glioblastoma specimens. J Neuro-Oncol 2013;112:347–54.
[21] Shinsato Y, Furukawa T, Yunoue S, Yonezawa H, Minami K, Nishizawa Y, Ikeda R, Kawahara K, Yamamoto M, Hirano H, Tokimura H, Arita K. Reduction of MLH1 and PMS2 confers temozolomide resistance and is associated with recurrence of glioblastoma. Oncotarget 2013;4:2261–70.
[22] Reis RM, Herva R, Brandner S, Koivukangas J, Mironov N, Bär W, Kleihues P, Ohgaki H. Second primary glioblastoma. J Neuropathol Exp Neurol 2001;60:208–15.
[23] Van Nifterik KA, Elkhuizen PHM, van Andel RJ, Stalpers LJA, Leenstra S, Lafleur MVM, Vandertop WP, Slotman BJ, Hulsebos TJM, Sminia P. Genetic profiling of a distant second glioblastoma multiforme after radio-therapy: recurrence or second primary tumor? J Neurosurg 2006;105:739–44.
[24] Sottoriva A, Spiteri I, Piccirillo SGM, Touloumis A, Collins VP, Marioni JC, Curtis C, Watts C, Tavaré S. Intratumor heterogeneity in human glioblastoma reflects cancer evolutionary dynamics. Proc Natl Acad Sci U S A 2013;110:4009–14.
[25] Andor N, Harness JV, Müller S, Mewes HW, Petritsch C. EXPANDS: expanding ploidy and allele frequency on nested subpopulations. Bioinformatics 2014;30:50–60.
[26] Lee G, Auffinger B, Guo D, Hasan T, Deheeger M, Tobias AL, Kim JY, Atashi F, Zhang L, Lesniak MS, James CD, Ahmed AU. Dedifferentiation of glioma cells to glioma stem-like cells by therapeutic stress-induced HIF signaling in the recurrent GBM model. Mol Cancer Ther 2016;15:3064–76.
[27] Lo Dico A, Martelli C, Diceglie C, Lucignani G, Ottobrini L. Hypoxia-inducible factor-1α activity as a switch for glioblastoma responsiveness to temozolomide. Front Oncol 2018;8:249.
[28] Wang J, Cazzato E, Ladewig E, Frattini V, Rosenbloom DI, Zairis S, Abate F, Liu Z, Elliott O, Shin YJ, Lee JK, Lee IH, Park WY, Eoli M, Blumberg AJ, Lasorella A, Nam DH, Finocchiaro G, Iavarone A, Rabadan R. Clonal evolution of glioblastoma under therapy. Nat Genet 2016;48:768–76.
[29] Indraccolo S, Lombardi G, Fassan M, Pasqualini L, Giunco S, Marcato R, Gasparini A, Candiotto C, Nalio S, Fiduccia P, Fanelli GN, Pambuku A, Della Puppa A, D'Avella D, Bonaldi L, Gardiman MP, Bertorelle R, De Rossi A, Zagonel V. Genetic, epigenetic and immunologic profiling of MMR-deficient relapsed glioblastoma. Clin Cancer Res 2019;25:1828–37.
[30] Daniel P, Sabri S, Chaddad A, Meehan B, Bertrand J-C, Rak J, Abdulkarim BS. Temozolomide induced hypermutation in glioma: evolutionary mechanisms and therapeutic opportunities. Front Oncol 2019;9:41.
[31] Lathia JD, Hitomi M, Gallagher J, Gadani SP, Adkins J, Vasanji A, Liu L, Eyler CE, Heddleston JM, Wu Q, Minhas S, Soeda A, Hoeppner DJ, Ravin R, McKay RDJ, McLendon RE, Corbeil D, Chenn A, Hjelmeland AB, Park DM, Rich JN. Distribution of CD133 reveals glioma stem cells self-renew through symmetric and asymmetric cell divisions. Cell Death Dis 2011;2, e200.
[32] Lee JH, Lee JH. The origin-of-cell harboring cancer-driving mutations in human glioblastoma. BMB Rep 2018;51:481–3.
[33] Singh SK, Hawkins C, Clarke ID, Squire JA, Bayani J, Hide T, Henkelman RM, Cusimano MD, Dirks PB. Identification of human brain tumour initiating cells. Nature 2004;432:396–401.
[34] Auffinger B, Spencer D, Pytel P, Ahmed AU, Lesniak MS. The role of glioma stem cells in chemotherapy resistance and glioblastoma multiforme recurrence. Expert Rev Neurother 2015;15:741–52.
[35] Colamaio M, Tosti N, Puca F, Mari A, Gattordo R, Kuzay Y, Federico A, Pepe A, Sarnataro D, Ragozzino E, Raia M, Hirata H, Gemei M, Mimori K, Del Vecchio L, Battista S, Fusco A. HMGA1 silencing reduces stemness and temozolomide resistance in glioblastoma stem cells. Expert Opin Ther Targets 2016;20:1169–79.

[36] Lathia JD, Heddleston JM, Venere M, Rich JN. Deadly teamwork: neural cancer stem cells and the tumor microenvironment. Cell Stem Cell 2011;8:482–5.
[37] Matarredona ER, Pastor AM. Neural stem cells of the subventricular zone as the origin of human glioblastoma stem cells. Therapeutic implications. Front Oncol 2019;9:779. https://doi.org/10.3389/fonc.2019.00779.
[38] Lewis KM, Petritsch C. Asymmetric cell division: implications for glioma development and treatment. Transl Neurosci 2013;4:484–503.
[39] Charville GW, Rando TA. The mortal strand hypothesis: non-random chromosome inheritance and the biased segregation of damaged DNA. Semin Cell Dev Biol 2013;24:653–60.
[40] Bao S, Wu Q, McLendon RE, Hao Y, Shi Q, Hjelmeland AB, Dewhirst MW, Bigner DD, Rich JN. Glioma stem cells promote radioresistance by preferential activation of the DNA damage response. Nature 2006; 444:756–60.
[41] Chen J, Li Y, Yu TS, McKay RM, Burns DK, Kernie SG, Parada LF. A restricted cell population propagates glioblastoma growth after chemotherapy. Nature 2012;488:522–6.
[42] Toledo-Guzmán ME, Hernández MI, Gómez-Gallegos ÁA, Ortiz-Sánchez E. ALDH as a stem cell marker in solid tumors. Curr Stem Cell Res Ther 2019;14:375–88.
[43] Broadley KWR, Hunn MK, Farrand KJ, Kylie M, Price KM, Grasso C, Miller RJ, Hermans IF, McConnell MJ. Side population is not necessary or sufficient for a cancer stem cell phenotype in glioblastoma multiforme. Stem Cells 2011;29:452–61.
[44] Beier D, Schulz JB, Beier CP. Chemoresistance of glioblastoma cancer stem cells—much more complex than expected. Mol Cancer 2011;10:128.
[45] Annovazzi L, Mellai M, Schiffer D. Chemotherapeutic drugs: DNA damage and repair in glioblastoma. Cancers (Basel) 2017;9:57.
[46] Perazzoli G, Prados J, Ortiz R, Caba O, Cabeza L, Berdasco M, Gónzalez B, Melguizo C. Temozolomide resistance in glioblastoma cell lines: implication of MGMT, MMR, P-glycoprotein and CD133 expression. PLoS One 2015;10, e0140131.
[47] Ropolo M, Daga A, Griffero F, Foresta M, Casartelli G, Zunino A, Poggi A, Cappelli E, Zona G, Spaziante R, Corte G, Frosina G. Comparative analysis of DNA repair in stem and nonstem glioma cell cultures. Mol Cancer Res 2009;7:383–92.
[48] Fink D, Aebi S, Howell SB. The role of DNA mismatch repair in drug resistance. Clin Cancer Res 1998;4:1–6.
[49] Johannessen TC, Bjerkvig R, Tysnes BB. DNA repair and cancer stem-like cells—potential partners in glioma drug resistance? Cancer Treat Rev 2008;34:558–67.
[50] Mukherjee B, McEllin B, Camacho CV, Tomimatsu N, Sirasanagandala S, Nannepaga S, Hatanpaa KJ, Mickey B, Madden C, Maher E, Boothman DA, Furnari F, Cavenee WK, Bachoo RM, Burma S. EGFRvIII and DNA double-strand break repair: a molecular mechanism for radioresistance in glioblastoma. Cancer Res 2009;69:4252–9.
[51] Fillingham J, Keogh MC, Krogan NJ. GammaH2AX and its role in DNA double-strand break repair. Biochem Cell Biol 2006;84:568–77.
[52] Furuta T, Takemura H, Liao ZY, Aune GJ, Redon C, Sedelnikova OA, Pilch DR, Rogakou EP, Celeste A, Chen HT, Nussenzweig A, Aladjem MI, Bonner WM, Pommier Y. Phosphorylation of histone H2AX and activation of Mre11, Rad50, and Nbs1 in response to replication-dependent DNA double-strand breaks induced by mammalian DNA topoisomerase I cleavage complexes. J Biol Chem 2003;278:20303–12.
[53] Lukas C, Bartkova J, Latella L, Falck J, Mailand N, Schroeder T, Sehested M, Lukas J, Bartek J. DNA damage-activated kinase Chk2 is independent of proliferation or differentiation yet correlates with tissue biology. Cancer Res 2001;61:4990–3.
[54] Bartek J, Lukas J. 2003. Chk1 and Chk2 kinases in checkpoint control and cancer. Cancer Cell 2003;3:421–9.
[55] Eich M, Roos WP, Nikolova T, Kaina B. Contribution of ATM and ATR to the resistance of glioblastoma and malignant melanoma cells to the methylating anticancer drug temozolomide. Mol Cancer Ther 2013;12:2529–40.
[56] Caussinus E, Gonzalez C. Induction of tumor growth by altered stem-cell asymmetric division in Drosophila melanogaster. Nat Genet 2005;37:1125–9.
[57] Paglia S, Sollazzo M, Di Giacomo S, de Biase D, Pession A, Grifoni D. Failure of the PTEN/aPKC/Lgl axis primes formation of adult brain tumours in Drosophila. Biomed Res Int 2017;2017:2690187.
[58] Cicalese A, Bonizzi G, Pasi CE, Faretta M, Ronzoni S, Giulini B, Brisken C, Minucci S, Di Fiore PP, Pelicci PG. The tumor suppressor p53 regulates polarity of self-renewing divisions in mammary stem cells. Cell 2009;138:1083–95.

[59] Hwang W-L, Jiang J-K, Yang S-H, Huang T-S, Lan H-Y, Teng H-W, Yang C-Y, Tsai Y-P, Lin C-H, Wang H-W, Yang M-H. MicroRNA-146a directs the symmetric division of Snail-dominant colorectal cancer stem cells. Nat Cell Biol 2014;16:268–80.
[60] Sugiarto S, Persson AI, Munoz EG, Waldhuber M, Lamagna C, Andor N, Hanecker P, Ayers-Ringler J, Phillips J, Siu J, Lim DA, Vandenberg S, Stallcup W, Berger MS, Bergers G, Weiss WA, Petritsch C. Asymmetry-defective oligodendrocyte progenitors are glioma precursors. Cancer Cell 2011;20:328–40.
[61] Sun Y, Goderie SK, Temple S. Asymmetric distribution of EGFR receptor during mitosis generates diverse CNS progenitor cells. Neuron 2005;45:873–86.
[62] Gómez-López S, Lerner RG, Petritsch C. Asymmetric cell division of stem and progenitor cells during homeostasis and cancer. Cell Mol Life Sci 2014;71:575–97.
[63] Lin W-H, Asmann YW, Anastasiadis PZ. Expression of polarity genes in human cancer. Cancer Informat 2015;14:15–28.
[64] Nishimura T, Kaibuchi K. Numb controls integrin endocytosis for directional cell migration with aPKC and PAR-3. Dev Cell 2007;13:15–28.
[65] Smith CA, Lau KM, Rahmani Z, Dho SE, Brothers G, She YM, Berry DM, Bonneil E, Thibault P, Schweisguth F, Le Borgne R, McGlade CJ. aPKC-mediated phosphorylation regulates asymmetric membrane localization of the cell fate determinant Numb. EMBO J 2007;26:468–80.
[66] Wirtz-Peitz F, Nishimura T, Knoblich JA. Linking cell cycle to asymmetric division: Aurora-A phosphorylates the Par complex to regulate Numb localization. Cell 2008;135:161–73.
[67] Hirai M, Arita Y, McGlade CJ, Lee KF, Chen J, Evans SM. Adaptor proteins NUMB and NUMBL promote cell cycle withdrawal by targeting ERBB2 for degradation. J Clin Invest 2017;127:569–82.
[68] Jiang X, Xing H, Kim T-M, Jung Y, Huang W, Yang HW, Song S, Park PJ, Carroll RS, Johnson MD. Numb regulates glioma stem cell fate and growth by altering epidermal growth factor receptor and Skp1-Cullin-F-box ubiquitin ligase activity. Stem Cells 2012;30:1313–26.
[69] Colaluca IN, Tosoni D, Nuciforo P, Senic-Matuglia F, Galimberti V, Viale G, Pece S, Di Fiore PP. NUMB controls p53 tumour suppressor activity. Nature 2008;2008451:76–80.
[70] Lu Y, Xu W, Ji J, Feng D, Sourbier C, Yang Y, Qu J, Zeng Z, Wang C, Chang X, Chen Y, Mishra A, Xu M, Lee MJ, Lee S, Trepel J, Linehan WM, Wang X, Yang Y, Neckers L. Alternative splicing of the cell fate determinant Numb in hepatocellular carcinoma. Hepatology 2015;62:1122–31.
[71] Kusne Y, Carrera-Silva EA, Perry AS, Rushing EJ, Mandell EK, Dietrich JD, Errasti AE, Gibbs D, Berens ME, Loftus JC, Hulme C, Yang W, Lu Z, Aldape K, Sanai N, Rothlin CV, Ghosh S. Targeting aPKC disables oncogenic signaling by both the EGFR and the proinflammatory cytokine TNFα in glioblastoma. Sci Signal 2014;7:ra75.
[72] Gont A, Hanson JEL, Lavictoire SJ, Daneshmand M, Nicholas G, Woulfe J, Kassam A, Da Silva VF, Lorimer IAJ. Inhibition of glioblastoma malignancy by Lgl1. Oncotarget 2014;5:11541–51.
[73] Gont A, Hanson JEL, Lavictoire SJ, Parolin DA, Daneshmand M, Restall IJ, Soucie M, Nicholas G, Woulfe J, Kassam A, Da Silva VF, Lorimer IAJ. PTEN loss represses glioblastoma tumor initiating cell differentiation via inactivation of Lgl1. Oncotarget 2013;4:1266–79.
[74] Wang R, Liu C. All-trans retinoic acid therapy induces asymmetric division of glioma stem cells from the U87MG cell line. Oncol Lett 2019;18:3646–54.
[75] Puca F, Tosti N, Federico A, Kuzay Y, Pepe A, Morlando S, Savarese T, D'Alessio F, Colamaio M, Sarnataro D, Ziberi S, De Martino M, Fusco A, Battista S. HMGA1 negatively regulates NUMB expression at transcriptional and post transcriptional level in glioblastoma stem cells. Cell Cycle 2019;18:1446–57.
[76] Li X, Martinez-Ledesma E, Zhang C, Gao F, Zheng S, Ding J, Wu S, Nguyen N, Clifford SC, Wen PY, Ligon KL, Yung WKA, Koul D. Cancer Res 2019;79:5088–101.
[77] Van Brocklyn JR, Wojton J, Meisen WH, Kellough DA, Ecsedy JA, Kaur B, Lehman NL. Aurora-A inhibition offers a novel therapy effective against intracranial glioblastoma. Cancer Res 2014;74:5364–70.
[78] Auffinger B, Tobias AL, Han Y, Lee G, Guo D, Dey M, Lesniak MS, Ahmed AU. Conversion of differentiated cancer cells into cancer stem-like cells in a glioblastoma model after primary chemotherapy. Cell Death Differ 2014;21:1119–31.
[79] Vora SM, Phillips BT. The benefits of local depletion: the centrosome as a scaffold for ubiquitin-proteasome-mediated degradation. Cell Cycle 2016;15:2124–34.
[80] Bu P, Chen KY, Lipkin SM, Shen X. Asymmetric division: a marker for cancer stem cells in early stage tumors? Oncotarget 2013;4:950–1.

[81] Haar CP, Hebbar P, Wallace 4th GC, Das A, Vandergrift 3rd WA, Smith JA, Giglio P, Patel SJ, Ray SK, Banik NL. Drug resistance in glioblastoma: a mini review. Neurochem Res 2012;37:1192–200.
[82] Wijaya J, Fukuda Y, Schuetz JD. Obstacles to brain tumor therapy: key ABC transporters. Int J Mol Sci 2017;18:2544.
[83] Bleau AM, Holland EC. Chemotherapeutic treatment of gliomas increases the amount of cancer stem-like cells. Med Sci 2009;25:775–7.
[84] Shreier HI, Soen Y, Brenner N. Exploratory adaptation in large random networks. Nat Commun 2017;21 (8):14826.
[85] Puchalsky RB, Shah N, Miller J, Dalley R, Nomura SR, Yoon JG, Smith KA, Lankerovich M, Bertagnolli D, Bickley K, Boe AF, Brouner K, Butler S, Caldejon S, Chapin M, Datta S, Dee N, Desta T, Dolbeare T, Dotson N, Ebbert A, Feng D, Feng X, Fisher M, Gee G, Goldy J, Gourley L, Gregor BW, Gu G, Hejazinia N, Hohmann J, Hothi P, Howard R, Joines K, Kriedberg A, Kuan L, Lau C, Lee F, Lee H, Lemon T, Long F, Mastan N, Mott E, Murthy C, Ngo K, Olson E, Reding M, Riley Z, Rosen D, Sandman D, Shapovalova N, Slaughterbeck CR, Sodt A, Stockdale G, Szafer A, Wakeman W, Wohnoutka PE, White SJ, Marsh D, Rostomily RC, Ng L, Dang C, Jones A, Keogh B, Gittleman HR, Barnholtz-Sloan JS, Cimino PJ, Uppin MS, Keene CD, Farrokhi FR, Lathia JD, Berens ME, Iavarone A, Bernard A, Lein E, Phillips JW, Rostad SW, Cobbs C, Hawrylycz MJ, Foltz GD. An anatomical transcriptional atlas of human glioblastoma. Science 2018;360:660–3.
[86] Celiku O, Gilbert MR, Lavi O. Computational modeling demonstrates that glioblastoma cells can survive spatial environmental challenges through exploratory adaptation. Nat Commun 2019;10:5704.
[87] Verhaak RG, Hoadley KA, Purdom E, Wang V, Qi Y, Wilkerson MD, Miller CR, Ding L, Golub T, Mesirov JP, Alexe G, Lawrence M, O'Kelly M, Tamayo P, Weir BA, Gabriel S, Winckler W, Gupta S, Jakkula L, Feiler HS, Hodgson JG, James CD, Sarkaria JN, Brennan C, Kahn A, Spellman PT, Wilson RK, Speed TP, Gray JW, Meyerson M, Getz G, Perou CM, Hayes DN, Cancer Genome Atlas Research Network. Integrated genomic analysis identifies clinically relevant subtypes of glioblastoma characterized by abnormalities in PDGFRA, IDH1, EGFR, and NF1. Cancer Cell 2010;17:98–110.
[88] Lin N, Yan W, Gao K, Wang Y, Zhang J, You Y. Prevalence and clinicopathologic characteristics of the molecular subtypes in malignant glioma: a multi-institutional analysis of 941 cases. PLoS One 2014;9, e94871.
[89] Guadagno E, Vitiello M, Francesca P, Calì G, Caponnetto F, Cesselli D, Camorani S, Borrelli G, Califano M, Cappabianca P, Arra C, Crescenzi E, Cerchia L, De Caro MLDB, Fedele M. Page 2PATZ1 is a new prognostic marker of glioblastoma associated with the stem-like phenotype and enriched in the proneural subtype. Oncotarget 2017;8:59282–300.
[90] Zhu Y, Yang P, Zhang X, Zhang L, Cui G, Wang Q, Lv L, Zhang Y, Xin X, Yan T, Zhao M, Zhang N. The effect and mechanism of CXCR4 silencing on metastasis suppression of human glioma U87 cell line. Anat Rec 2013;296:1857–64.
[91] Patel AP, Tirosh I, Trombetta JJ, Shalek AK, Gillespie SM, Wakimoto H, Cahill DP, Nahed BV, Curry WT, Martuza RL, Louis DN, Rozenblatt-Rosen O, Suvà ML, Regev A, Bernstein BE. Single-cell RNA-seq highlights intratumoral heterogeneity in primary glioblastoma. Science 2014;344:1396–401.
[92] Phillips HS, Kharbanda S, Chen R, Forrest WF, Soriano RH, Wu TD, Misra A, Nigro JM, Colman H, Soroceanu L, Williams PM, Modrusan Z, Feuerstein BG, Aldape K. Molecular subclasses of high-grade glioma predict prognosis, delineate a pattern of disease progression, and resemble stages in neurogenesis. Cancer Cell 2006;9:157–73.
[93] Fedele M, Cerchia L, Pegoraro S, Sgarra R, Manfioletti G. Proneural-mesenchymal transition: phenotypic plasticity to acquire multitherapy resistance in glioblastoma. Int J Mol Sci 2019;20:2746.
[94] Heddleston JM, Hitomi M, Venere M, Flavahan WA, Yang K, Kim Y, Minhas S, Rich JN, Hjelmeland AB. 5 Glioma stem cell maintenance: the role of the microenvironment. Curr Pharmacol Des 2011;17: 2386–401.
[95] Segerman A, Niklasson M, Haglund C, Bergström T, Jarvius M, Xie Y, Westermark A, Sönmez D, Hermansson A, Kastemar M, Naimaie-Ali Z, Nyberg F, Berglund M, Sundström M, Hesselager G, Uhrbom L, Gustafsson M, Larsson R, Fryknäs M, Segerman B, Westermark B. Clonal variation in drug and radiation response among glioma-initiating cells is linked to proneural-mesenchymal transition. Cell Rep 2016;17:2994–3009.
[96] Kosty J, Lu F, Kupp R, Mehta S, Lu QR. Harnessing OLIG2 function in tumorigenicity and plasticity to target malignant gliomas. Cell Cycle 2017;16:1654–60.
[97] Kupp R, Shtayer L, Tien AC, Szeto E, Sanai N, Rowitch DH, Mehta S. Lineage-restricted OLIG2-RTK signaling governs the molecular subtype of glioma stem-like cells. Cell Rep 2016;16:2838–45.

[98] Bhat KPL, Balasubramaniyan V, Vaillant B, Ezhilarasan R, Hummelink K, Hollingsworth F, Wani K, Heathcock L, James JD, Goodman LD, Conroy S, Long L, Lelic N, Wang S, Gumin J, Raj D, Kodama Y, Raghunathan A, Olar A, Joshi K, Pelloski CE, Heimberger A, Kim SH, Cahill DP, Rao G, Den Dunnen WFA, Boddeke HWGM, Phillips HS, Nakano I, Lang FF, Colman H, Sulman EP, Aldape K. Mesenchymal differentiation mediated by NF-κB promotes radiation resistance in glioblastoma. Cancer Cell 2013;24:331–46.
[99] Cooper LA, Gutman DA, Chisolm C, Appin C, Kong J, Rong Y, Kurc T, Van Meir EG, Saltz JH, Moreno CS, Brat DJ. The tumor microenvironment strongly impacts master transcriptional regulators and gene expression class of glioblastoma. Am J Pathol 2012;180:2108–19.
[100] Engler JR, Robinson AE, Smirnov I, Hodgson JG, Berger MS, Gupta N, James CD, Molinaro A, Phillips JJ. Increased microglia/macrophage gene expression in a subset of adult and pediatric astrocytomas. PLoS One 2012;7, e43339.
[101] Jia L, Tian Y, Chen Y, Zhang G. The silencing of LncRNA-H19 decreases chemoresistance of human glioma cells to temozolomide by suppressing epithelial-mesenchymal transition via the Wnt/β-Catenin pathway. Onco Targets Ther 2018;11:313–21.
[102] Baisiwala S, Auffinger B, Caragher SP, Shireman JM, Ahsan R, Lee G, Hasan T, Park C, Saathoff MR, Christensen AC, Ahmed AU. Chemotherapeutic stress induces transdifferentiation of glioblastoma cells to endothelial cells and promotes vascular mimicry. Stem Cells Int 2019;2019:6107456.
[103] Taylor RC, Cullen SP, Martin SJ. Apoptosis: controlled demolition at the cellular level. Nat Rev Mol Cell Biol 2008;9:231–41.
[104] Fulda S. Tumor resistance to apoptosis. Int J Cancer 2009;124:511–5.
[105] Liu G, Yuan X, Zeng Z, Tunici P, Ng H, Abdulkadir IR, Lu L, Irvin D, Black KL, Yu JS. Analysis of gene expression and chemoresistance of CD133+ cancer stem cells in glioblastoma. Mol Cancer 2006;5:67.
[106] Guo G, Sun Y, Hong R, Xiong J, Lu Y, Liu Y, Lu J, Zhang Z, Guo C, Nan Y, Huang Q. IKBKE enhances TMZ-chemoresistance through upregulation of MGMT expression in glioblastoma. Clin Transl Oncol 2019. https://doi.org/10.1007/s12094-019-02251-3.
[107] Ha W, Sevim-Nalkiran H, Zaman AM, Matsuda K, Khasraw M, Nowak AK, Chung L, Baxter RC, McDonald KL. Ibudilast sensitizes glioblastoma to temozolomide by targeting Macrophage Migration Inhibitory Factor (MIF). Sci Rep 2019;9:2905.
[108] Qin N, Han F, Li L, Ge Y, Lin W, Wang J, Wu L, Zhao G, Deng Y, Zhang J. Deubiquitinating enzyme 4 facilitates chemoresistance in glioblastoma by inhibiting P53 activity. Oncol Lett 2019;17:958–64.
[109] Kaur J, Debnath J. Autophagy at the crossroads of catabolism and anabolism. Nat Rev Mol Cell Biol 2015;16:461–72.
[110] Knizhnik AV, Roos WP, Nikolova T, Quiros S, Tomaszowski KH, Christmann M, Kaina B. Survival and death strategies in glioma cells: autophagy, senescence and apoptosis triggered by a single type of temozolomide-induced DNA damage. PLoS One 2013;8, e55665.
[111] Liao CL, Chen CM, Chang YZ, Liu GY, Hung HC, Hsieh TY, Lin CL. Pine (Pinus morrisonicola Hayata) needle extracts sensitize GBM8901 human glioblastoma cells to temozolomide by downregulating autophagy and O (6)-methylguanine-DNA methyltransferase expression. J Agric Food Chem 2014;62:10458–67.
[112] Lin CJ, Lee CC, Shih YL, Lin TY, Wang SH, Lin YF, Shih CM. Resveratrol enhances the therapeutic effect of temozolomide against malignant glioma in vitro and in vivo by inhibiting autophagy. Free Radic Biol Med 2012;52:377–91.
[113] Salazar M, Carracedo A, Salanueva IJ, Hernández-Tiedra S, Lorente M, Egia A, Vázquez P, Blázquez C, Torres S, García S, Nowak J, Fimia GM, Piacentini M, Cecconi F, Pandolfi PP, González-Feria L, Iovanna JL, Guzmán M, Boya P, Velasco G. Cannabinoid action induces autophagy-mediated cell death through stimulation of ER stress in human glioma cells. J Clin Invest 2009;119:1359–72.
[114] Torres S, Lorente M, Rodríguez-Fornés F, Hernández-Tiedra S, Salazar M, García-Taboada E, Barcia J, Guzmán M, Velasco G. A combined preclinical therapy of cannabinoids and temozolomide against glioma. Mol Cancer Ther 2011;10:90–103.
[115] Tang JH, Ma ZX, Huang GH, Xu QF, Xiang Y, Li N, Sidlauskas K, Zhang EE, Sheng-Lv Q. Downregulation of HIF-1α sensitizes U251 glioma cells to the temozolomide (TMZ) treatment. Exp Cell Res 2016;343:148–58.
[116] Lo Dico A, Salvatore D, Martelli C, Ronchi D, Diceglie C, Lucignani G, Ottobrini L. Intracellular redox-balance involvement in temozolomide resistance-related molecular mechanisms in glioblastoma. Cells 2019;8:1315.

[117] Yan Y, Xu Z, Dai S, Qian L, Sun L, Gong Z. Targeting autophagy to sensitive glioma to temozolomide treatment. J Exp Clin Cancer Res 2016;35:23.
[118] Tao Z, Li T, Ma H, Yang Y, Zhang C, Hai L, Liu P, Yuan F, Li J, Yi L, Tong L, Wang Y, Xie Y, Ming H, Yu S, Yang X. Autophagy suppresses self-renewal ability and tumorigenicity of glioma-initiating cells and promotes Notch1 degradation. Cell Death Dis 2018;9:1063.
[119] Ryskalin L, Gaglione A, Limanaqi F, Biagioni F, Familiari P, Frati A, Esposito V, Fornai F. The autophagy status of cancer stem cells in gliobastoma multiforme: from cancer promotion to therapeutic strategies. Int J Mol Sci 2019;20:3824.
[120] Zhuang W, Long L, Zheng B, Ji W, Yang N, Zhang Q, Liang Z. Curcumin promotes differentiation of glioma-initiating cells by inducing autophagy. Cancer Sci 2012;103:684–90.
[121] Jung J, Zhang Y, Celiku O, Zhang W, Song H, Williams BJ, Giles AJ, Rich JN, Abounader R, Gilbert MR, Park DM. Mitochondrial NIX promotes tumor survival in the hypoxic niche of glioblastoma. Cancer Res 2019;79:5218–32.
[122] Clarke AJ, Simon AK. Autophagy in the renewal, differentiation and homeostasis of immune cells. Nat Rev Immunol 2019;19:170–83.
[123] Valdor R, García-Bernal D, Riquelme D, Martinez CM, Moraleda JM, Cuervo AM, Macian F, Martinez S. Glioblastoma ablates pericytes antitumor immune function through aberrant up-regulation of chaperone-mediated autophagy. Proc Natl Acad Sci U S A 2019;116:20655–65.
[124] Li T-F, Xu Y-X, Li K, Wang C, Liu X, Yue Y, Chen Z, Yuan S-J, Wen Y, Zhang Q, Han M, Komatsu N, Zhao L, Chen X. Doxorubicin-polyglycerol-nanodiamond composites stimulate glioblastoma cell immunogenicity through activation of autophagy. Acta Biomater 2019;86:381–94.
[125] Criguer CE, Oliva CR. Bioenergetics pathways and therapeutic resistance in gliomas: emerging role of mitochondria. Curr Pharm Des 2011;17:2421–7.
[126] Wu W, Wu Y, Mayer K, von Rosenstiel C, Schecker J, Baur S, Würstle S, Liesche-Starnecker F, Gempt J, Schlegel J. Lipid peroxidation plays an important role in chemotherapeutic effects of temozolomide and the development of therapy resistance in human glioblastoma. Transl Oncol 2020;13:100748.
[127] Zhang J, Chen G, Gao Y, Liang H. HOTAIR/miR-125 axis-mediated Hexokinase 2 expression promotes chemoresistance in human glioblastoma. J Cell Mol Med 2020;24:5707–17.
[128] Guo J, Xue Q, Liu K, Ge W, Liu W, Wang J, Zhang M, Li QY, Cai D, Shan C, Zhang C, Liu X, Li J. Dimethylaminomicheliolide (DMAMCL) suppresses the proliferation of glioblastoma cells via targeting pyruvate kinase 2 (PKM2) and rewiring aerobic glycolysis. Front Oncol 2019;9:993.
[129] Velpula KK, Guda MR, Sahu K, Tuszynski J, Asuthkar S, Bach SE, Lathia JD, Tsung AJ. Metabolic targeting of EGFRvIII/PDK1 axis in temozolomide resistant glioblastoma. Oncotarget 2017;8:35639–55.
[130] Koch K, Hartmann R, Schröter F, Suwala AK, Maciaczyk D, Krüger AC, Willbold D, Kahlert UD, Maciaczyk J. Reciprocal regulation of the cholinic phenotype and epithelial-mesenchymal transition in glioblastoma cells. Oncotarget 2016;7:73414–31.
[131] Mathieu M, Martin-Jaular L, Lavieu G, Théry C. Specificities of secretion and uptake of exosomes and other extracellular vesicles for cell-to-cell communication. Nat Cell Biol 2019;21:9–17.
[132] Calabrese C, Poppleton H, Kocak M, Hogg TL, Fuller C, Hamner B, Oh EY, Gaber MW, Finklestein D, Allen M, Frank A, Bayazitov IT, Zakharenko SS, Gajjar A, Davidoff A, Gilbertson RJ. A perivascular niche for brain tumor stem cells. Cancer Cell 2007;11:69–82.
[133] Quail DF, Joyce JA. Microenvironmental regulation of tumor progression and metastasis. Nat Med 2013;19:1423–37.
[134] Bourkoula E, Mangoni D, Ius T, Pucer A, Isola M, Musiello D, Marzinotto S, Toffoletto B, Sorrentino M, Palma A, Caponnetto F, Gregoraci G, Vindigni M, Pizzolitto S, Falconieri G, De Maglio G, Pecile V, Ruaro ME, Gri G, Parisse P, Casalis L, Scoles G, Skrap M, Beltrami CA, Beltrami AP, Cesselli D. Glioma-associated stem cells: a novel class of tumor-supporting cells able to predict prognosis of human low-grade gliomas. Stem Cells 2014;32:1239–53.
[135] Kolodkin AL, Matthes DJ, Goodman CS. The semaphorin genes encode a family of transmembrane and secreted growth cone guidance molecules. Cell 1993;75:1389–99.
[136] Manini I, Ruaro ME, Sgarra R, Bartolini A, Caponnetto F, Ius T, Skrap M, Di Loreto C, Beltrami AP, Manfioletti G, Cesselli D. Semaphorin-7A on exosomes: a promigratory signal in the glioma microenvironment. Cancers (Basel) 2019;11:758.

[137] Treps L, Edmond S, Harford-Wright E, Galan-Moya EM, Schmitt A, Azzi S, Citerne A, Bidère N, Ricard D, Gavard J. Extracellular vesicle-transported Semaphorin3A promotes vascular permeability in glioblastoma. Oncogene 2016;35:2615–23.
[138] Zhang Z, Yin J, Lu C, Wei Y, Zeng A, You Y. Exosomal transfer of long non-coding RNA SBF2-AS1 enhances chemoresistance to temozolomide in glioblastoma. J Exp Clin Cancer Res 2019;38:166.
[139] Lemmon MA, Schlessinger J. Cell signaling by receptor-tyrosine kinases. Cell 2010;141:1117–34.
[140] An Z, Aksoy O, Zheng T, Fan Q-W, Weiss WA. Epidermal growth factor receptor and EGFRvIII in glioblastoma: signaling pathways and targeted therapies. Oncogene 2018;37:1561–75.
[141] Guo G, Gong K, Wohlfeld B, Hatanpaa KJ, Zhao D, Habib AA. Ligand independent EGFR signaling. Cancer Res 2018;75:3436–41.
[142] Brennan CW, Verhaak RGW, McKenna A, Campos B, Noushmehr H, Salama SR, Zheng S, Chakravarty D, Sanborn JZ, Berman SH, Beroukhim R, Bernard B, Wu CJ, Genovese G, Shmulevich I, Barnholtz-Sloan J, Zou L, Vegesna R, Shukla SA, Ciriello G, Yung WK, Zhang W, Sougnez C, Mikkelsen T, Aldape K, Bigner DD, Van Meir EG, Prados M, Sloan A, Black KL, Eschbacher J, Finocchiaro G, Friedman W, Andrews DW, Guha A, Iacocca M, O'Neill BP, Foltz G, Myers J, Weisenberger DJ, Penny R, Kucherlapati R, Perou CM, Hayes DN, Gibbs R, Marra M, Mills GB, Lander E, Spellman P, Wilson R, Sander C, Weinstein J, Meyerson M, Gabriel S, Laird PW, Haussler D, Getz G, Chin L, TCGA Research Network. The somatic genomic landscape of glioblastoma. Cell 2013;155:462–77.
[143] Lee JC. Epidermal growth factor receptor activation in glioblastoma through novel missense mutations in the extracellular domain. PLoS Med 2006;3:e485.
[144] Orellana L, Thorne AH, Lema R, Gustavsson J, Parisian AD, Hospital A, Cordeiro TN, Bernadó P, Scott AM, Brun-Heath I, Lindahl E, Cavenee WK, Furnari FB, Orozco M. Oncogenic mutations at the EGFR ectodomain structurally converge to remove a steric hindrance on a kinase-coupled cryptic epitope. Proc Natl Acad Sci 2019;116:10009–18.
[145] Huang K, Liu X, Li Y, Wang Q, Zhou J, Wang Y, Dong F, Yang C, Sun Z, Fang C, Liu C, Tan Y, Wu X, Jiang T, Kang C. Genome-wide CRISPR-Cas9 screening identifies NF-κB/E2F6 responsible for EGFRvIII-associated temozolomide resistance in glioblastoma. Oncogene 2018;37:1561–75.
[146] Nishikawa R, Ji XD, Harmon RC, Lazar CS, Gill GN, Cavenee WK, Huang HJ. A mutant epidermal growth factor receptor common in human glioma confers enhanced tumorigenicity. Proc Natl Acad Sci U S A 1994;91:7727–31.
[147] Nagane M, Coufal F, Lin H, Bögler O, Cavenee WK, Huang HJ. A common mutant epidermal growth factor receptor confers enhanced tumorigenicity on human glioblastoma cells by increasing proliferation and reducing apoptosis. Cancer Res 1996;56:5079–86.
[148] Shinojima N, Tada K, Shiraishi S, Kamiryo T, Kochi M, Nakamura H, Makino K, Saya H, Hirano H, Kuratsu J, Oka K, Ishimaru Y, Ushio Y. Prognostic value of epidermal growth factor receptor in patients with glioblastoma multiforme. Cancer Res 2003;63:6962–70.
[149] Inda MM, Bonavia R, Mukasa A, Narita Y, Sah DW, Vandenberg S, Brennan C, Johns TG, Bachoo R, Hadwiger P, Tan P, Depinho RA, Cavenee W, Furnari F. Tumor heterogeneity is an active process maintained by a mutant EGFR-induced cytokine circuit in glioblastoma. Genes Dev 2010;24:1731–45.
[150] Al-Nedawi K, Meehan B, Micallef J, Lhotak V, May L, Guha A, Rak J. Intercellular transfer of the oncogenic receptor EGFRvIII by microvesicles derived from tumour cells. Nat Cell Biol 2008;10:619–24.
[151] Skog J, Wurdinger T, van Rijn S, Meijer DH, Gainche L, Sena-Esteves M, Curry Jr WT, Carter BS, Krichevsky AM, Breakefield XO. Glioblastoma microvesicles transport RNA and proteins that promote tumour growth and provide diagnostic biomarkers. Nat Cell Biol 2008;10:1470–6.
[152] Luwor RB, Zhu HJ, Walker F, Vitali AA, Perera RM, Burgess AW, Scott AM, Johns TG. The tumor-specific de2-7 epidermal growth factor receptor (EGFR) promotes cells survival and heterodimerizes with the wild-type EGFR. Oncogene 2004;23:6095–104.
[153] Li L, Chakraborty S, Yang CR, Hatanpaa KJ, Cipher DJ, Puliyappadamba VT, Rehman A, Jiwani AJ, Mickey B, Madden C, Raisanen J, Burma S, Saha D, Wang Z, Pingle SC, Kesari S, Boothman DA, Habib AA. An EGFR wild type-EGFRvIII-HB-EGF feed-forward loop regulates the activation of EGFRvIII. Oncogene 2014;33:4253–64.
[154] Fan QW, Cheng CK, Gustafson WC, Charron E, Zipper P, Wong RA, Chen J, Lau J, Knobbe-Thomsen C, Weller M, Jura N, Reifenberger G, Shokat KM, Weiss WA. EGFR phosphorylates tumor-derived EGFRvIII driving STAT3/5 and progression in glioblastoma. Cancer Cell 2013;24:438–49.

[155] Cloughesy TF, Cavenee WK, Mischel PS. Glioblastoma: from molecular pathology to targeted treatment. Annu Rev Pathol 2014;9:1–25.

[156] Li B, Yuan M, Kim IA, Chang CM, Bernhard EJ, Shu HK. Mutant epidermal growth factor receptor displays increased signaling through the phosphatidylinositol-3 kinase/AKT pathway and promotes radioresistance in cells of astrocytic origin. Oncogene 2004;23:4594–602.

[157] Stommel JM, Kimmelman AC, Ying H, Nabioullin R, Ponugoti AH, Wiedemeyer R, Stegh AH, Bradner JE, Ligon KL, Brennan C, Chin L, DePinho RA. Coactivation of receptor tyrosine kinases affects the response of tumor cells to targeted therapies. Science 2007;318:287–90.

[158] Greenall SA, Donoghue JF, Van Sinderen M, Dubljevic V, Budiman S, Devlin M, Street I, Adams TE, Johns TG. EGFRvIII-mediated transactivation of receptor tyrosine kinases in glioma: mechanism and therapeutic implications. Oncogene 2015;34:5277–87.

[159] Huang PH, Mukasa A, Bonavia R, Flynn RA, Brewer ZE, Cavenee WK, Furnari FB, White FM. Quantitative analysis of EGFRvIII cellular signaling networks reveals a combinatorial therapeutic strategy for glioblastoma. Proc Natl Acad Sci U S A 2007;104:12867–72.

[160] Li L, Puliyappadamba VT, Chakraborty S, Rehman A, Vemireddy V, Saha D, Souza RF, Hatanpaa KJ, Koduru P, Burma S, Boothman DA, Habib AA. EGFR wild type antagonizes EGFRvIII-mediated activation of Met in glioblastoma. Oncogene 2015;34:129–34.

[161] Akhavan D, Pourzia AL, Nourian AA, Williams KJ, Nathanson D, Babic I, Villa GR, Tanaka K, Nael A, Yang H, Dang J, Vinters HV, Yong WH, Flagg M, Tamanoi F, Sasayama T, James CD, Kornblum HI, Cloughesy TF, Cavenee WK, Bensinger SJ, Mischel PS. De-repression of PDGFRβ transcription promotes acquired resistance to EGFR tyrosine kinase inhibitors in glioblastoma patients. Cancer Discov 2013;3:534–47.

[162] Camorani S, Crescenzi E, Colecchia D, Carpentieri A, Amoresano A, Fedele M, Chiariello M, Cerchia L. Aptamer targeting EGFRvIII mutant hampers its constitutive autophosphorylation and affects migration, invasion and proliferation of glioblastoma cells. Oncotarget 2015;6:37570–87.

[163] Camorani S, Esposito CL, Rienzo A, Catuogno S, Iaboni M, Condorelli G, de Franciscis V, Cerchia L. Inhibition of receptor signaling and of glioblastoma-derived tumor growth by a novel PDGFRβ aptamer. Mol Ther 2014;22:828–41.

[164] Tomaszewski W, Sanchez-Perez L, Gajewski TF, Sampson JH. Brain tumor microenvironment and host state: implications for immunotherapy. Clin Cancer Res 2019;25:4202–10.

[165] Liu Z, Han L, Dong Y, Tan Y, Li Y, Zhao M, Xie H, Ju H, Wang H, Zhao Y, Zheng Q, Wang Q, Su J, Fang C, Fu S, Jiang T, Liu J, Li X, Kang C, Ren H. EGFRvIII/integrin β3 interaction in hypoxic and vitronectin enriching microenvironment promote GBM progression and metastasis. Oncotarget 2016;7:4680–94.

[166] Camorani S, Crescenzi E, Gramanzini M, Fedele M, Zannetti A, Cerchia L. Aptamer-mediated impairment of EGFR-integrin αvβ3 complex inhibits vasculogenic mimicry and growth of triple-negative breast cancers. Sci Rep 2017;7:46659.

[167] Jinushi M, Chiba S, Yoshiyama H, Masutomi K, Kinoshita I, Dosaka-Akita H, Yagita H, Takaoka A, Tahara H. Tumor-associated macrophages regulate tumorigenicity and anticancer drug responses of cancer stem/initiating cells. Proc Natl Acad Sci U S A 2011;108:12425–30.

[168] Audia A, Conroy S, Glass R, Bhat KPL. The impact of the tumor microenvironment on the properties of glioma stem-like cells. Front Oncol 2017;7:143.

[169] Trylcova J, Busek P, Smetana K, Balaziova E, Dvorankova B, Mifkova A, Sedo A. Effect of cancer-associated fibroblasts on the migration of glioma cells in vitro. Tumour Biol 2015;36:5873–9.

[170] Da Ros M, De Gregorio V, Iorio AL, Giunti L, Guidi M, de Martino M, Genitori L, Sardi I. Glioblastoma chemoresistance: the double play by microenvironment and blood-brain barrier. Int J Mol Sci 2018;19:2879.

[171] De Vleeschouwer S, Bergers G. Glioblastoma: to target the tumor cell or the microenvironment? In: De Vleeschouwer S, editor. Glioblastoma. Brisbane, Australia: Codon Publications; 2017. p. 315–40. Chapter 16.

[172] Merighi S, Benini A, Mirandola P, Gessi S, Varani K, Leung E, Maclennan S, Baraldi PG, Borea PA. Hypoxia inhibits paclitaxel-induced apoptosis through adenosine-mediated phosphorylation of bad in glioblastoma cells. Mol Pharmacol 2007;72:162–72.

[173] Nardinocchi L, Pantisano V, Puca R, Porru M, Aiello A, Grasselli A, Leonetti C, Safran M, Rechavi G, Givol D, Farsetti A, D'Orazi G. Zinc downregulates hif-1α and inhibits its activity in tumor cells in vitro and in vivo. PLoS One 2010;5, e15048.

[174] Chen L, Feng P, Li S, Long D, Cheng J, Lu Y, Zhou D. Effect of hypoxia-inducible factor-1α silencing on the sensitivity of human brain glioma cells to doxorubicin and etoposide. Neurochem Res 2009;34:984–90.

[175] Rosa P, Catacuzzeno L, Sforna L, Mangino G, Carlomagno S, Mincione G, Petrozza V, Ragona G, Franciolini F, Calogero A. Bk channels blockage inhibits hypoxia-induced migration and chemoresistance to cisplatin in human glioblastoma cells. J Cell Physiol 2018;233:6866–77.

[176] Weiler M, Blaes J, Pusch S, Sahm F, Czabanka M, Luger S, Bunse L, Solecki G, Eichwald V, Jugold M, Hodecker S, Osswald M, Meisner C, Hielscher T, Rübmann P, Pfenning PN, Ronellenfitsch M, Kempf T, Schnölzer M, Abdollahi A, Lang F, Bendszus M, von Deimling A, Winkler F, Weller M, Vajkoczy P, Platten M, Wick W. mTOR target NDRG1 confers MGMT-dependent resistance to alkylating chemotherapy. Proc Natl Acad Sci U S A 2014;111:409–14.

[177] Kahlon AS, Alexander M, Kahlon A, Wright J. Lactate levels with glioblastoma multiforme. Proceedings (Baylor University Medical Center) 2016;29:313–4.

[178] Peppicelli S, Andreucci E, Ruzzolini J, Laurenzana A, Margheri F, Fibbi G, Del Rosso M, Bianchini F, Calorini L. The acidic microenvironment as a possible niche of dormant tumor cells. Cell Mol Life Sci 2017;74:2761–71.

[179] Hjelmeland AB, Wu Q, Heddleston JM, Choudhary GS, MacSwords J, Lathia JD, McLendon R, Lindner D, Sloan A, Rich JN. Acidic stress promotes a glioma stem cell phenotype. Cell Death Differ 2011;18:829–40.

[180] Montana V, Sontheimer H. Bradykinin promotes the chemotactic invasion of primary brain tumors. J Neurosci 2011;31:4858–67.

[181] Würth R, Bajetto A, Harrison JK, Barbieri F, Florio T. CXCL12 modulation of CXCR4 and CXCR7 activity in human glioblastoma stem-like cells and regulation of the tumor microenvironment. Front Cell Neurosci 2014;8:1–19.

[182] Caspani EM, Crossley PH, Redondo-Garcia C, Martinez S. Glioblastoma: a pathogenic crosstalk between tumor cells and pericytes. PLoS One 2014;9, e101402.

[183] Ho IAW, Shim WSN. Contribution of the microenvironmental niche to glioblastoma heterogeneity. Biomed Res Int 2017;2017:9634172.

[184] Mujoo K, Krumenacker JS, Murad F. Nitric oxide-cyclic GMP signaling in stem cell differentiation. Free Radic Biol Med 2011;51:2150–7.

[185] Zhu TS, Costello MA, Talsma CE, Flack CG, Crowley JG, Hamm LL, He X, Hervey-Jumper SL, Heth JA, Muraszko KM, DiMeco F, Vescovi AL, Fan X. Endothelial cells create a stem cell niche in glioblastoma by providing NOTCH ligands that nurture self-renewal of cancer stem-like cells. Cancer Res 2011;71:6061–72.

[186] Monaco I, Camorani S, Colecchia D, Locatelli E, Calandro P, Oudin A, Niclou S, Arra C, Chiariello M, Cerchia L, Comes FM. Aptamer functionalization of nanosystems for glioblastoma targeting through the blood-brain barrier. J Med Chem 2017;60:4510–6.

[187] Shi S, Fu W, Lin S, Tian T, Li S, Shao X, Zhang Y, Zhang T, Tang Z, Zhou Y, Lin Y, Cai X. Targeted and effective glioblastoma therapy via aptamer-modified tetrahedral framework nucleic acid-paclitaxel nanoconjugates that can pass the blood brain barrier. Nanomed Nanotechnol Biol Med 2019;21:102061.

[188] Camorani S, Crescenzi E, Fedele M, Cerchia L. Oligonucleotide aptamers against tyrosine kinase receptors: prospect for anticancer applications. Biochim Biophys Act Rev Cancer 1869;2018:263–77.

CHAPTER

15

Role of long noncoding RNAs in temozolomide-resistant glioblastoma

M. Janaki Ramaiah[a], *Shaik Mohammad Naushad*[b], *Rajasekhar Reddy Manyam*[c], *and Vijay Kumar Kutala*[d]

[a]Functional Genomics and Disease Biology Laboratory, School of Chemical and Biotechnology, SASTRA Deemed University, Thanjavur, Tamil Nadu, India [b]Sandor Speciality Diagnostics Private Limited, Hyderabad, Telangana, India [c]Department of Computer Science and Engineering, Koneru Lakshmaiah Education Foundation, Vaddeswaram, Andhra Pradesh, India [d]Department of Clinical Pharmacology and Therapeutics, Nizam's Institute of Medical Sciences, Hyderabad, Telangana, India

Abstract

Chemoresistance is considered as one of the major obstacles in the postgenomic era of cancer treatment. Recent studies envisaged the crucial regulatory role of long noncoding RNA (lncRNA)/microRNA (miRNA) and their interaction in cancer chemotherapy and personalized medicine. Glioblastoma (GBM) is one of the devastating brain cancers and is highly malignant and resistant to therapy. LncRNAs are long noncoding RNAs with nucleotide length of more than 200bp, lack protein-coding potential, and were found to regulate cellular processes such as chromatin epigenetics, posttranscriptional mRNA processing, gene regulation, glioma cancer stem cell (GCSC) self-renewal, drug metabolism, and chemoresistance. In GBM, the survival period of most of the patients is less than 15 months. The most common drug utilized for treating GBM is temozolomide as a part of first-line therapy. Resistance to temozolomide results in tumor recurrence, epithelial to mesenchymal transition (EMT), and therapy failure. This chapter describes how various lncRNAs contribute to drug resistance to temozolomide in GBM patients and interaction of lncRNA with miRNAs, modulation of chromatin epigenetics by lncRNA, the role of RNA modifications such as m6A and m5C in mRNA in GBM cancer progression and TMZ-drug resistance.

Abbreviations

ALKBH5	AlkB homolog 5, RNA demethylase
ANRIL	antisense noncoding RNA in the INK4 locus

Glioblastoma Resistance to Chemotherapy: Molecular Mechanisms and Innovative Reversal Strategies
https://doi.org/10.1016/B978-0-12-821567-8.00001-4

BBB	blood-brain barrier
Bcl2	B-cell lymphoma 2
BRCP1	breast cancer resistance protein
CACS2	cancer susceptibility candidate 2
CCNA2	cyclin A2
CDC20	cell division cycle 20 (CDC20)
CDC45	cell division control protein 45
CDK6	cyclin-dependent kinase 6
CDKN1A	cyclin-dependent kinase inhibitor 1A
EGFR	epidermal growth factor receptor
EIF3a/b/h	eukaryotic translation initiation factor 3a/b/h
EMT	epithelial-mesenchymal transition
EPIC	epigenetically induced lncRNA I
FOXM1-AS	forkheadboxm1-antisense RNA
FTO	fat mass and obesity-associated protein
GBM	glioblastoma multiforme
GSC	glioblastoma stem cells
HCC	hepatocellular carcinoma
HMGB1	high mobility group Box1 protein
hnRNPA2/B1	heterogeneous nuclear ribonucleoprotein A2/B1
hnRNPC	heterogeneous nuclear ribonucleoprotein C1/C2
HOTAIR	HOX transcript antisense RNA
HOTTIP	Hox distal transcript antisense RNA
iASPP	inhibitor of apoptosis stimulating protein of p53
IDH	isocitrate dehydrogenase
IGF2BP1/2/3	insulin-like growth factor 2 mRNA-binding protein 1/2/3
lncRNA	long noncoding RNA
LPS	lipopolysaccharides
m5C	5-methylcytosine
m6A	N6-methyl adenosine
m6Am	N6 2-*O*-dimethyl adenosine
MALAT1	metastasis-associated lung adenocarcinoma transcript 1
MDR	multidrug resistance protein
METTL3	methyltransferase like 3
miRNA	microRNA
MMP16	matrix metalloproteinase 16
MRP1	multidrug resistance-associated protein 1
NEAT1	nuclear-enriched abundant transcript 1
NF-kB	nuclear factor-κB
PARP	poly (ADP-ribose) polymerase
PDAC	pancreatic ductal adenocarcinoma
PRC	polycomb repressive complex
PVT1	plasma-cytoma variant translocation 1
RG4	RNA G-quadraplexes
RMP	RNA modifying proteins
RPPH1	ribonuclease P RNA component H1
SNHG12	small nucleolar host gene 12
SNHG15	small nucleolar RNA host gene 15
SOX2	sex determining region Y-box 2
SOX2OT	SOX2 overlapping transcript
SP1	specificity protein 1
STAT3	signal transducer and activator of transcription 3
TALC	temozolomide-associated lncRNA

TMZ	temozolomide
TMZ-R	temozolomide resistance
TP73-AS1	TP73 antisense RNA 1
TS	thymidylate synthase
TUG1	taurine upregulated 1
UCA1	urothelial cancer-associated 1
UTR	untranslated regions
XIST X	inactive specific transcript
YTHDC1/2/3	YTH N6-methyladenosine RNA-binding protein 1/2/3
ZEB1	zinc finger *E*-box-binding homeobox 1
ψ	pseudouridine

Conflict of interest

No potential conflicts of interest were disclosed.

Introduction

The most common therapeutic option for cancer followed in clinics is chemotherapy, radiotherapy, and surgical resection [1, 2]. The development of novel technologies and bioinformatics approaches have enabled the study and understanding of the role of long noncoding RNAs and other genes related to drug resistance against various cancers [3].

GBM cancer

Gliomas are brain cancers classified into two groups: low-grade gliomas (low-grade astrocytoma, low-grade oligodendrogliomas, and mixed glioma) and high-grade gliomas such as anaplastic astrocytoma and anaplastic oligodendrogliomas and glioblastoma (GBM) (World health organization) [4]. WHO 2016 classification on diffuse gliomas includes low-grade glioma (WHO grade II), WHO grade III (anaplastic), and WHO grade IV (glioblastoma) [5]. Gliomas are also further classified based on gene mutations as biomarkers, including WHO II (low grade) and WHO grade III (i.e., anaplastic). Histologically, the isocitrate dehydrogenase (IDH) wild type and the IDH-mutant, 1p/19q-non-co-deleted diffuse gliomas belong to astrocytic, and the IDH-mutant, 1p/19q-co-deleted tumors are grouped into oligodendroglial phenotypes.

GBM is the most common primary brain tumor. Patients with GBM have poor survival rates of up to 15 months. Temozolomide (TMZ) is an oral alkylating agent that can pass through the blood-brain barrier (BBB) and is used for the treatment of GBM patients as a part of conventional chemotherapy, but its use is limited by acquired drug resistance. Studies have shown that resistance to TMZ occurs during the last stages of cancer chemotherapy in up to 50% of GBM patients. In GBM, deregulated expression of microRNAs and long noncoding RNAs were found to be implicated in the initiation, invasion, migration, metastasis, and tumor recurrence. In this review, we emphasize the role of lncRNAs and their effects on miRNA in regulating TMZ resistance in aggressive GBMs.

TMZ is a prodrug that undergoes pH-dependent degradation into 4-amino-5-imidazole-carboxamide (AIC) and a methyl diazonium cation [6]. TMZ causes methylation of DNA particularly at O6-methyl guanine (O6-meG), N7-methyl guanine, N3-methyl guanine, and N3-methyl adenine. As a part of DNA repair, the methyl guanine methyltransferase (MGMT) enzyme repairs the O6-meG lesion and results in chemoresistance. TMZ alkylates the DNA at several positions including O6meG. The modified guanine pairs with thymidine during DNA replication and results in the activation of mismatch repair pathway (MMR) leading to double-strand breaks and apoptosis [7, 8]. The key biomarkers utilized to date to diagnose various types of gliomas include isocitrate dehydrogenase 1 (IDH1) and isocitrate dehydrogenase II (IDH-II) mutations, 1p/19q deletion, ATRX mutation, mutations I p53 gene, histone K27M, BRAF V600 E, and telomerase reverse transcriptase (TERT) promoter methylation.

Cancer drug resistance occurs owing to intrinsic factors such as deregulated gene expression, as well as extrinsic factors such as molecular perturbations. Previous studies have identified methylation of methyl guanine methyltransferase (MGMT) [9], aberrant expression of ABC transporters [10], isocitrate dehydrogenase (IDHI) gene mutations (i.e., R132 H in case of IDH 1 and R172K in case of IDH2) and p53 gene mutations [11], dysregulated DNA repair events, microRNAs [12], changes in LncRNA expression, cancer stem cells, expression of multidrug-resistance-related proteins such as breast cancer resistance protein (Bcrp1/Abcg2), multidrug resistance mutation (MDR1) [13] as responsible for TMZ resistance. The inhibitors of poly (ADP-ribose) polymerase-1 and -2 (PARP) were found to suppress the emergence of resistance in GBM cells and enhance the effectiveness of TMZ [14].

Studies have indicated that TMZ resistance is owing to GBM cancer stem cells (CSC) that have effective DNA repair capability [15]. In CSC, nuclear factor-kappa-B (NF-kB) was found to mediate the transcriptional activation of O6-methyl guanine DNA methyltransferase (MGMT) when compared to normal cells [16] (Fig. 1). In addition, GBM CSC cells with CD133+ interfere with progenitor cellular programming and acquire the effective DNA repair capacity. The function of MGMT in normal cells is to remove the methyl groups from O6-guanine. MGMT expression is regulated mainly at the level of transcription by two mechanisms such as methylation of promoter and transcriptional activation [17]. Various DNA elements such as glucocorticoid response elements (GRE) and many transcription factors such as activator protein (AP-1), nuclear factor-kappa-light chain enhancer of activated B cells (NF-kB), p53 (tumor suppressor), hypoxia-inducible factor 1α (HIF-1α), protein kinase C (PKC) were found to regulate the GBM cancer stem cell growth and fate [18].

In addition, altered metabolic reprogramming in mitochondria also contributes to chemoresistance. During mitochondrial metabolism, superoxide dismutase (SOD) acts as a radical scavenger and plays a crucial role in the clonogenic activity and various other cellular processes. The transcription factor Sp1 regulates the superoxide dismutase 2 (SOD2) or MnSOD expression and regulates oxidative stress and cancer cell energy metabolism [19, 20].

Studies by Jiaper et al. [21] have suggested the combination of Wnt/β-catenin inhibitor such as celecoxib, salinomycin, MEK/ERK inhibitor such as Sl327, U0126, Aurora A kinase inhibitors such as MLN8237, and Janus kinase (JNK), bone morphogenic protein (BMP), sox2 vaccine might regulate TMZ-mediated cytotoxicity in GBM.

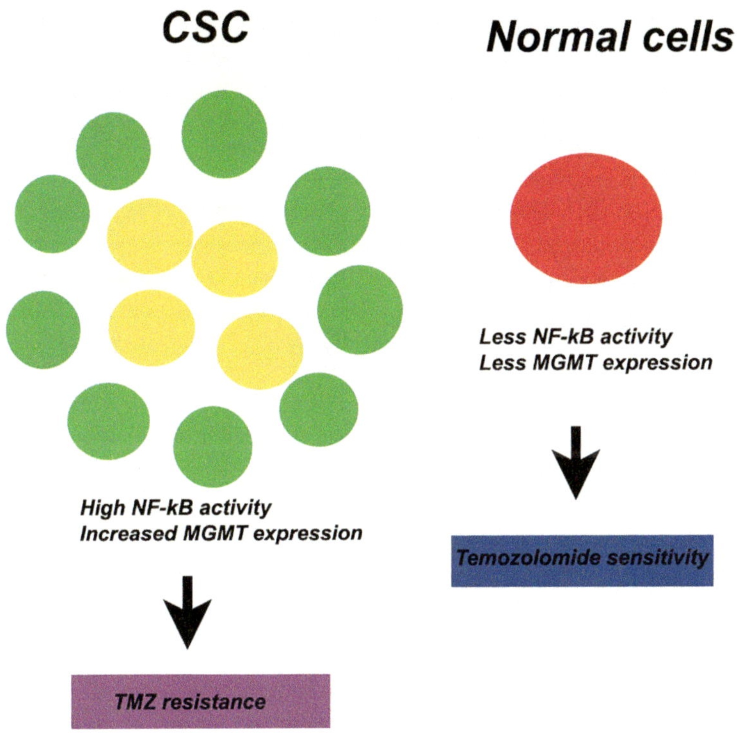

FIG. 1 Cancer stem cells and temozolomide (TMZ) resistance. GBM cancer stem cells express high levels of nuclear factor kappa B (NF-kB). The high expression of NF-kB transactivates the MGMT promoter leading to therapeutic resistance. In normal cells, less NF-kB expression leads to low MGMT promoter activity.

Recent evidence has emphasized the importance of the regulatory role of noncoding RNAs, such as lncRNAs and miRNAs, in the chemoresistance of cancers [12]. Thus, identifying novel mechanisms regulated by lncRNAs may provide useful clues in combating the therapeutic resistance against GBM.

Long noncoding RNA

Lon noncoding RNAs are competing for endogenous RNAs that bind to microRNA (miRNA) and regulate posttranscriptional gene silencing (PTGS). The LncRNAs play an important regulatory role in cancer formation and tissue physiology [22]. LncRNAs are transcribed from conserved promoters and are regulated by various transcription factors [23]. In the human genome, it was established that more than 50,000 lncRNA genes are present [22]. These molecules are both oncogenic and tumor suppressive [24]. LncRNAs are involved in the regulation of various cellular processes such as cell-cycle regulation, alternate splicing, RNA splicing, transcription, translation, epigenetic regulation, cell physiology, nuclear, and cytoplasmic transport, drug resistance, and tumorigenesis [22, 25, 26].

Long noncoding RNAs are processed in a similar manner as mRNA and are found in nucleocytoplasmic fractions [27]. Noncoding RNAs are divided into short (<200nt) or long (>200nt). These are classified into natural antisense transcripts (NATs) (i.e., *cis*-NAT or *trans*-NAT) and these regulate the expression of sense partner transcript, long intergenic noncoding RNAs (lincRNAs) that are encoded within introns or between protein-coding genes. lncRNAs can function as tumor suppressors or oncogenes and their levels are up- or downregulated during cancer formation [24]. These RNA molecules can function as diagnostic and prognostic markers as well as potential target molecules for anticancer drug discovery [28].

The biological databases used for studying lncRNA is NONCODE containing a total of 487,164 lncRNA transcripts belonging to 16 different species including humans and mice [29]. LncRNAMap contains 23,355 lncRNA for humans; LNCipedia contains 127,802 human lncRNAs [30], and Noncode v3.0 database contains 73,370 lncRNA from 1239 organisms [31]. The number has increased up to 548,640 lncRNA transcripts in the Noncode v5.0 database [32]. In human cells, only 2% of the genome codes for protein-coding genes (PCGs) and up 70%–80% of the genome is transcribed into noncoding RNAs [33] and thus anticancer drug response might not be only based on predicted coding genes [34]. Recent studies indicate that lncRNAs are better predictors of drug response when compared with protein-coding genes (PcG) [35]. Most studies have shown that lncRNA genes are less conserved compared to protein-coding genes [36]. Conservation analysis of lncRNAs with a window size of 50nt has shown that the majority of lncRNAs possess higher conservation patches and might be involved in interactions with RNA-binding proteins [37]. Aberrant expression of lncRNA results in cancer. Interestingly, many studies have indicated lncRNAs as biomarkers for the clinical outcome in GBM patients. It is very interesting to note that lncRNA molecules participate in driving GBM and other cancer phenotypes via their interactions with other key biological molecules such as DNA, protein, and RNA [38]. The long noncoding RNAs that are involved in chemoresistance toward TMZ are TP73-AS1, SOX2OT, SNHG12, MALAT1, TALC, HOTAIR, and XIST.

TP73-AS1

TP 73-AS1 (TP 73-antisense RNA 1) is a lncRNA gene that is highly expressed in glioma and is located in 1p36.32 (chromosome 1). It contains six exons, and the total nucleotide sequence is 4690bp in length. lncRNA was found to overlap with the p73 gene which is a p53 family member [39] and regulates p53-dependent apoptotic genes. The knockdown of TP73-AS1lncRNA suppressed the aggressive glioma. The lncRNA sponges miR-124 and induce glioblastoma. The downregulation of TP73-AS1 causes a reduction in expression levels of mitochondrial as well as nucleotide metabolism-related genes in G7 and G26 GBM cells [40]. Surprisingly, the restoration of miR-124 in GBM cell lines resulted in aggressiveness and therapeutic resistance in GBM [41, 42]. The epigenetic changes such as DNA methylation at the TP73AS1 promoter resulted in increased chemosensitivity [43] whereas increased histone acetylation at K27 [40] was observed in GBM cancer stem cells and therapy resistance. This indicates the epigenetic regulation of noncoding RNA by DNA methylation and histone acetylation at lncRNA promoter dictates chemoresistance and cancer stemness (Fig. 2; Table 1).

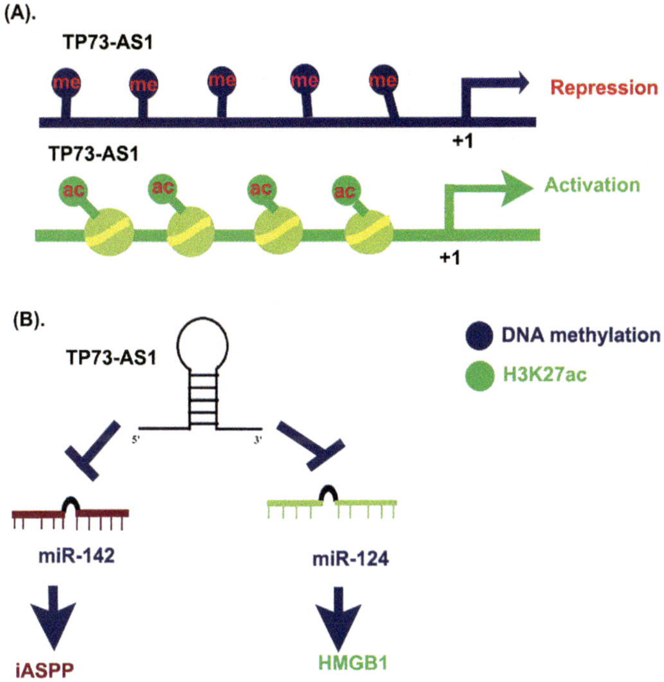

FIG. 2 TP73-AS1 epigenetics and chemo-resistance. (A) Methylation of TP 73-AS1 (TP 73-antisense RNA 1) promoter results in low expression of TP73-AS1 and leads to enhanced TMZ sensitivity in GBM cells. Histone acetylation on TP73-AS1 promoter in glioma cancer stem cells leads to transcriptional activation of the TP73AS1 promoter and result in chemoresistance. (B) TP73-AS1 acts as a competing endogenous RNA and sponges miR-124 leading to the activation of HMGB1 and GBM progression. Also, TP73-AS1 sponges miR-142 and regulates inhibitor of apoptosis stimulating protein (iASPP) leading to GBM.

TABLE 1 Role of TP 73-AS1 in various cancers.

S. no.	lncRNA function	Cancer type	References
1.	TP73 AS1 sponges miR-494-3p and upregulate Eukaryotic translation initiation factor (E1F) 5A2 which located at 3q25.27. EIF5A2 gene target gene of miR-494-3p	Medulloblastoma	[39]
2.	TP73-AS1 sponges miR-141-3p and cause NSCLC	Nonsmall cell lung cancer (NSCLC)	[44]
3.	TP73AS1 is highly expressed in hepatocellular carcinoma (HCC) and downregulation of the lncRNA result in reduced radioresistance in HCC and act via PTEN/AKT signaling	Hepato-cellular carcinoma (HCC)	[45]
4.	TP73 AS1 functions as microRNA sponge for miR-142 that targets HMGB1 gene. Silencing of lncRNA resulted in reduced glioma proliferation and invasion	Glioma	[41]
5.	TP73-AS1 sponges miR-124 and regulate inhibitor of apoptosis stimulating protein of p53 (iASPP)	Glioma	[46]
6.	TP73-AS1 downregulates miR-19-3p and stimulates cancer cell proliferation	Retinoblastoma	[47]
7.	TP73-AS1 can inhibit the invasion and migration properties in gastric cancer. Downregulation of the lncRNA was observed in gastric cancer. The miR-223-5p target TP-73AS1 in order to promote invasion as well as migration events	Gastric cancer	[42, 48]

FIG. 3 PAXIP1-AS1 regulation of GBM. LncRNA PAXIP1-AS1 facilitates the recruitment of the ETS1 transcription factor on the KLF14 promoter leading to promoter activation. The increased KLF14 gene expression results in invasion, migration, and aggressiveness in GBM.

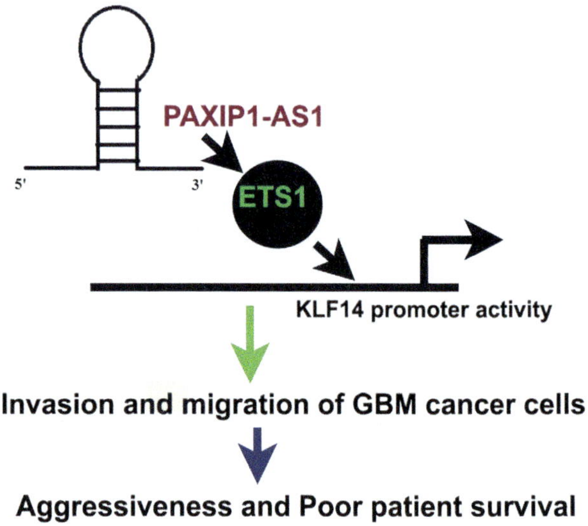

lncRNA PAXIP1-AS1

Systematic analysis of LncRNA Modulator Atlas in Pan-cancer (LncMAP) has predicted that the lncPaxIP-AS1 interacts with KLF-14 LncRNA-Pax-interacting protein 1-antisense RNA 1 that is more expressed in glioma patients and regulates glioma cell migration, invasion, and angiogenesis processes. lncRNA recruits the ETS1 transcription factor (i.e., acts as an oncogene) and thus regulates the expression of KLF-14 promoter activity [49, 50]. The increased expression of KLF-4 is correlated with tumor aggressiveness and patient survival in GBM. High expression of ETS1 is involved in angiogenesis and helps in tumor invasion between endothelial cells and epithelial cells, which ultimately results in epithelial to mesenchymal transition (EMT) [51] (Fig. 3).

LncRNA SOX2OT

SOX2 overlapping transcript (SOX2OT) occurs as six transcript variants and is highly expressed in embryonic stem cells. Recent studies have indicated the increased expression level of SOX2OT is correlated with increased GBM cancer risk [52] (Table 2).

The SOX2OT gene is composed of 10 exons with more than two transcriptions start sites and is located in human chromosome 3q26.3 (Chr3q26.3) [58] and is transcribed in the same direction of SOX2 [59]. Dysregulation of SOX2OT expression was observed in lung cancer, gastric cancer, esophageal cancer, breast cancer, hepatocellular carcinoma, ovarian cancer, and GBM. SOX2OT functions as an oncogene and is involved in growth, cancer cell proliferation, invasion, and migration,

SOX2OT lncRNA is known to regulate chromatin architecture [60] and is located 12 kb upstream of SOX2. The expression of SOX2OT is very high in TMZ-resistant cells and it was found that knockdown of SOX2OT resulted in enhanced TMZ sensitivity in TMZ-resistant GBM cancer cells (Fig. 4). Besides, the increased SOX2OT expression levels also induce

TABLE 2 SOX2OT in various cancers.

S. no.	lncRNA function	Cancer type	References
1.	Ectopic expression of SOX2OT result in several folds (i.e., up to 20-folds) increases in the expression SOX2 and cause anchorage-independent growth in breast cancer	Breast cancer	[53]
2.	SOX2OT overexpression induces migration and invasion phenotype in hepatocellular carcinoma (HCC)	HCC	[54]
3.	Knockdown of SOX2OT causes reduction in the expression ofEZH2, Cyclin B1 and Cdc2 proteins involved in regulation of cell-cycle leading to G2/M arrest in lung cancer cells High expressions of SOX2OT were observed in stage III and IV lung cancers	Lung cancer	[55]
4.	Knockdown of SOX2OT causes cell-cycle arrest and results in decreased Epithelial to mesenchymal transition (EMT) markers such as N-Cadherin and increased expression of *E*-Cadherin. Thus, SOX2OT knockdown causes EMT to MET (mesenchymal to epithelial transition) transition	Ovarian cancer	[56]
5.	Pancreatic ductal adenocarcinoma (PDAC) patients with low SOX2OT can survive longer time period than patients with higher SOX2OT level In PDAC cells SOX2OT sponges miR-200 and this lead to increased expression of SOX2	PDAC	[57]

abnormal cell proliferation and inhibition of apoptosis. SOX2 is located in the intronic region of SOX2OT and is a transcription factor that belongs to high mobility group domain-containing (HMG) proteins. It plays an important role in pluripotency and maintenance of stem cells and embryonic development [61]. RNA immunoprecipitation experiments have revealed that α-ketoglutarate-dependent dioxygenase Alk B homolog 5 (ALKBH5) protein interacts with SOX2OT (ENST 0000004855035) and regulates the transcription of SOX2 promoter. Also, GBM patients with high levels of SOX2OT expression have exhibited increased expression of drug resistance proteins BCRP1, MDR, MRP1 as well as Wnt pathway proteins Wnt 5a, β-catenin proteins, and stem cell factor SOX2. These proteins cause abnormal GBM cancer cell proliferation and chemoresistance (Fig. 3).

MIR-155 HG

MiR-155 HG is a highly expressed lncRNA in glioma patients. In GBM patients, polypyrimidine-tract-binding proteins (PTBP1) interact with miR-155 HG and regulate TMZ resistance [62].

LINC00021

Long noncoding RNA LINC00021 overexpression is highly correlated with TMZ resistance and poor prognosis and is tightly associated with the NOTCH pathway [63, 64].

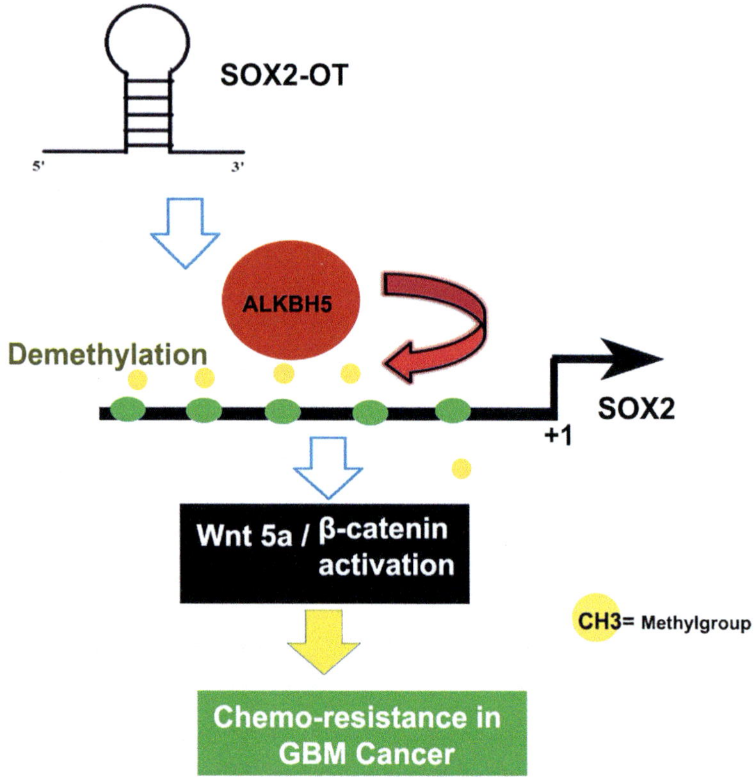

FIG. 4 SOX2OT regulate chemoresistance. SOX2 overlapping transcript is transcribed in the same direction of its target gene sex-determining region Y box 2 (SOX2) at chromosome location 3q26.3. SOX2OT interacts and recruits ALKBH5 (i.e., RNA demethylase and an eraser of methylation mark) to the SOX2OT promoter resulting in abnormal cancer growth and chemoresistance via activation of the Wnt pathway.

SNHG12

Small nuclear RNA host gene 12 (SNHG12) is a competitive endogenous RNA (ceRNA) with miRNA-binding sites in the 5′-untranslated region (5′-UTR), regulating target genes by sponging miRNAs. SNHG12 is responsible for acquired resistance toward TMZ and has a higher density of miRNA response elements (MREs) for target miRNAs, and thus prevents miRNA from binding to other transcripts [65]. SNHG12 is highly expressed in TMZ-resistant GBM. The knockdown of SNHG12 lncRNA results in TMZ sensitivity. The promoter activity of SNHG12 is regulated by specificity proteins (Sp1) proteins. The regions on SNHG12 promoter regulated by Sp1 factors include −231 to −241 region, −843 to −853 region, and −1587 to −1597bp region. Thus, the high expression of SNGH2 is implicated in TMZ resistance [66]. Zeng et al. [67] have indicated that SNHG12 lncRNA acts as a miRNA sponge (for miR-129-5p) and plays a crucial role in chemoresistance. Interestingly studies have also indicated that SNHG12 interacts with Argonaute 2 protein (AGO2) and might regulate TMZ resistance. Thus, RISC complex proteins also participate in TMZ resistance (Fig. 5).

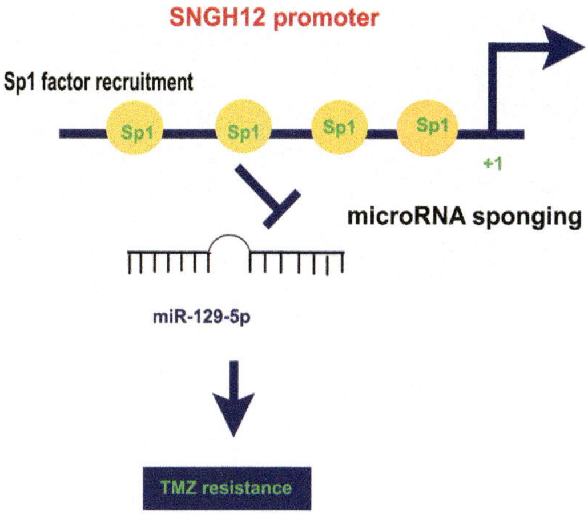

FIG. 5 SNGH12 transcription and regulation of drug resistance. The binding of specificity protein 1 (Sp1) was found to activate the SNGH12 promoter and result in the high expression of SNGH12. The lncRNA sponges miR-129-5p and results in TMZ resistance.

SNHG15

Small nucleolar RNA host gene 15 (SNHG15) is a short-lived long noncoding RNA dysregulated in many cancers. Glioma-associated microglial cells (GAMs) are functionally similar to tumor-associated macrophages; they interact and communicate with cancer cells via cytokines and regulate GBM progression [68, 69]. Increased expression of SNHG15 results in enhanced expression of epidermal growth factor receptor (EGFR), CDK6 (cyclin-dependent kinase 6), stemness marker SOX2, EMT markers such as β-catenin, and M2-polarization of GAM cells. All these events result in TMZ drug resistance in GBM cells [70]. Overexpression of miRNA-627-5p, the target for SNHG15, results in the suppression of GBM. Also, SNHG15 silencing induces sensitivity in TMZ-resistant (TMZ-R) cells. The studies by Li et al. has found that the treatment of GBM cells with palbociclib, the CDK6 inhibitor, caused a decrease in SNHG15 and an increase in miR-627-5p. Thus, the drug can be effectively used as a modulator of lncRNA and miRNA for future therapy against GBM (Table 3).

TABLE 3 SNHG15 involvement in other cancer types.

S. no.	lncRNA function	Cancer type	References
1.	Overexpression of SNHG15 in osteosarcoma sponges miR-141-3p and causes proliferation and invasion	Osteosarcoma	[71, 72]
2.	SNHG15 effectively binds and inhibits miR-338-3p and causes upregulation of FKBP propyl isomerase1A (FKBP1A) SNHG15 sponges miR-338-3p	Prostate cancer	[73]

EPIC1

The overexpression of epigenetically induced lncRNA I (EPIC1) located in chromosome 22q13.31 regulates various processes such as cell cycle, migration, and drug resistance in GBM patients. The CDC 20 overexpression reverses the EPIC1-mediated tumor progression in glioma [74]. EPIC1 (129–283 nt) interacts with MYC and promotes the cell cycle in luminal B breast cancer. It is interesting to note that EPIC 1 regulates the MYC binding ability to its target genes such as CDKN1A (p21), cyclin A2 (CCNA2), cell division cycle 20 (CDC20), and cell division control protein 45 (CDC45) via *E*-box (CACGTG). Thus, Myc RNA acts as a guide RNA and regulates the cell-cycle in cancer [35, 75].

MALAT1

Metastasis-associated lung adenocarcinoma transcript1 (MALAT1), also known as nuclear-enriched transcript 2, is 2.7 kb in length and is located on chromosome 11q13. It localizes to nuclear speckles and participates in alternate splicing of pre-mRNA [76]. MALAT1 promoter activity is regulated by NF-kB transcription factor p50 (i.e., p105) and P53 to MGMT promoter. During TMZ treatment, phosphorylation of p50 occurs and the release of p50 takes place, which facilitates binding p53 and ultimately resulting in increased expression of MALAT1. Further, MALAT1 transcriptional activation is found to be regulated by lipopolysaccharides (LPS) and cell type. Studies have indicated that NF-kB-dependent factors alter the response of GBM cells to TMZ treatment [77].

In general, p53 is a key tumor suppressor protein that can modulate the NF-kB response, and deregulation of the p53 pathway was observed in many GBM cases. Studies have indicated that lncRNA is regulated by p53 activity [78].

MALAT1 regulates many miRNAs that are 21–22 nt in length and promote chemoresistance and epithelial to mesenchymal transition (EMT). MALAT1 binds to miR-101 and promotes chemoresistance. The knockdown of MALAT1 was found to sensitize TMZ-resistant GBM cells toward TMZ. MicroRNA-101 was found to downregulate glycogen synthase kinase 3β (GSK3β) and increase chemosensitivity [79–82]. Also, chromatin modifiers interact with MALAT1 and regulate cellular growth. Thus, MALAT1 regulates epigenetic factors such as miRNA, and chromatin modulating factors regulate MALAT1. MALAT1 was found to enhance TMZ chemoresistance by sponging miR-203 and thereby increasing the expression of thymidylate synthase (TS) and Zinc finger *E*-box-binding homeobox 1 (ZEB1) [83].

NEAT1

lncRNA nuclear-enriched abundant transcript 1 (NEAT1) overexpressed in GBM is responsible for TMZ resistance [84, 85]. The knockdown of NEAT1 inhibits the expression of MGMT. It was observed that the overexpression of MGMT enhances TMZ resistance in NEAT1 silenced GBM cell lines. Also, NEAT1 levels are positively correlated with MGMT expression levels and TMZ resistance. Some studies have identified well-established lncRNAs, such as Hox distal transcript antisense RNA (HOTTIP) that are involved in gemcitabine resistance in pancreatic cancer, and urothelial cancer-associated 1 (UCA1) is

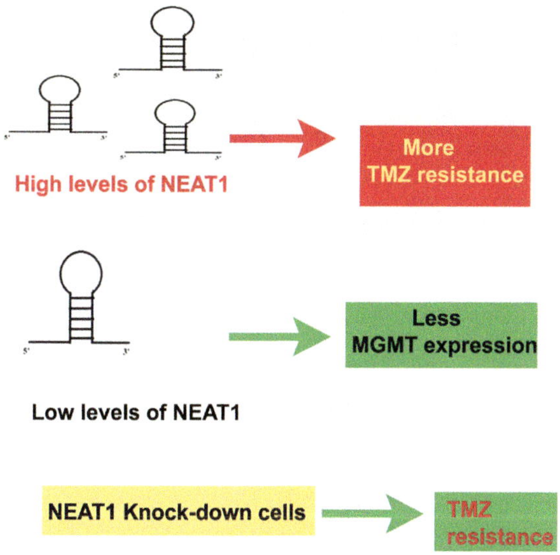

FIG. 6 Nuclear-enriched abundant transcript 1 (NEAT1) and modulation of TMZ resistance. The balance of NEAT1 was found to regulate TMZ effectively in GBM cells. High levels of NEAT1 result in an increased TMZ resistance. Lower levels of NEAT1 lead to lesser expression of MGMT expression. The cells with knockdown of NEAT1 lncRNA exhibit TMZ resistance. This indicates the amount of NEAT1 expression dictates TMZ resistance.

involved in cisplatin resistance in bladder cancer and plasmacytoma, whereas variant translocation 1 (PVT1) is involved in multidrug resistance in gastric cancer [86–88] (Fig. 6).

Cancer susceptibility candidate 2

Cancer susceptibility candidate 2 (CASC2) is a long-noncoding RNA located on chromosomal location 10q26. CASC2 has five exons and owing to alternate splicing results in the production of three variants such as CASC2a, CASC2b, and CASC2c [89]. Interestingly, in astrocytoma, CASC2 was found to be involved in the processing of miRNA-101 (i.e., pre-miRNA to mature miRNA). Thus, lncRNA has a role in miRNA processing also [71, 72]. CASC2 is downregulated in most cancer types, and functions as a tumor-suppressive protein. It also serves as a therapeutic target and a prognostic and diagnostic marker. The overexpression of the lncRNA can sensitize TMZ-resistant GBM cells. High expression of CASC2 suppresses GBM cell growth, proliferation, and invasion by inhibition of Wnt downstream target genes such as C-Myc, Cyclin D1, and β-Catenin [67, 90]. CASC2 inhibits the miRNA-181a expression and causes upregulation of phosphatase and tensin homolog (PTEN) as well as reduction of phosphorylated Akt and is involved in cell-cycle regulation and apoptosis [91, 92]. Thus, CASC2 and PTEN play a crucial regulatory role in cancer chemoresistance in GBM.

Lnc-TALC

MiRNA miR-20b-3p binds to the TMZ-associated lncRNA in GBM recurrence (lnc-TALC, TMZ-associated lncRNA in glioblastoma recurrence) and to receptor tyrosine kinase MET mRNA at the 3′UTR regions. Lnc-TALC was found to regulate the c-Met signaling pathway

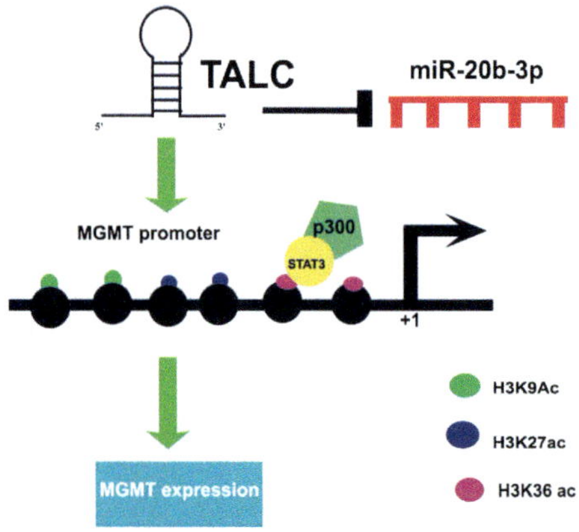

FIG. 7 LncTALC (TMZ associated lncRNA in GBM recurrence) effect on drug-resistance. LncTALC is known to bind miRNA miR-20b-3p and inhibits its activity. Also, lncTALC enhances MGMT promoter activity via the c-met/Stat-3/p300 axis. The high level of MGMT expression results in TMZ resistance.

by binding to miR-20b-3p and was found to be involved in the activation of the signal transducer and activator of transcription 3 (STAT3)/p300 complex to promote MGMT expression. Also, TALC has involved in the activation of c-Met in TMZ-resistant cells. Furthermore, TALC regulates histone acetylation at H3K36, H3 K27, and H3 K9 on MGMT promoter via C-met/STAT-3/p300 axis and thus plays a crucial role in MGMT expression [93]. LncRNA did not change the methylation status of the MGMT promoter. Thus, lncTALC activation in TMZ-resistant cells results in the activation of MGMT and c-Met (Fig. 7).

HOTAIR

HOTAIR lncRNA is involved in the development of breast cancer, lung cancer, pancreatic cancer, and GBM. Studies have demonstrated that long noncoding RNA homeobox transcript antisense intergenic RNA "HOTAIR" is known to modulate hexokinase 2 (HK2) gene expression. In GBM, hexokinase contributes to chemoresistance by regulating the glycolysis process during cancer cell metabolism. The loss or deletion of HK2 sensitizes the GBM cells to chemotherapeutic drugs. The high levels of Hexokinase 2 are correlated with TMZ resistance in GBM. In the absence of HOTAIR, the HK2 expression is increased both at the mRNA level and protein level. The lncRNA sponges miR-125 and thereby inhibits hexokinase gene expression [94]. Apart from glycolysis, HK2 also regulates the mitochondrial permeability transition pore opening (MPTP) and is involved in chemoresistance [63, 64]. Studies by Sa et al. [95] have reported that knockdown of HOTAIR, which sponges miR-148-3p in glioma microvascular endothelial cells (GECs), promotes drug delivery at the blood-tumor barrier. Thus, knockdown of HOTAIR could serve to enhance TMZ sensitivity (Table 4).

XIST

The lncRNA X inactive specific transcript (XIST) acts as an oncogene and is considered as one of the causative factors for chemoresistance in GBM [99]. XIST was found to inhibit miRNA miR-29c. Studies by Du et al. [100] have indicated that Sp1 and MGMT expression was correlated with XIST expression in GBM patients. Targeting XIST and restoration can lead to chemosensitization and improving other chemotherapies used against GBM [100].

TABLE 4 HOTAIR involvement in other cancer types.

S. no.	lncRNA function	Cancer type	References
1.	HOTAIR	Regulates DNMT1/MEG3/p53 by sponging miR-148b in hepatocellular carcinoma (HCC)	[96]
2.	HOTAIR	Sponges miR-331-3p in miRNP complex and regulates the expression of human epithelial growth factor receptor 2 (HER2) in gastric cancer	[97]
3.	HOTAIR	Inhibits the expression of miR-7 and targets the SETDB1 oncogene in cancer stem cells and reverse the cancer metastasis	[98]

LncRNA 003092.1

LncRNA AC003092.1 was decreased in TMZ-resistant GBM cells. The lncRNA regulates TFPI-2 expression by sponging miR-195 and is involved in apoptosis [84, 85].

Key cellular components in therapy resistance

RNA G-quadruplexes (RG4) are four-stranded structures (i.e., RG4) known to regulate the translation process involved in tumor formation [101]. RG4s are extremely stable structures formed by stacking of two or more G-quarters, each of which is composed of four guanine nucleotides interacting via Hoogsteen binding. These structures are formed at 5′-untranslated regions (5′-UTR) or 3′-untranslated regions (3′-UTR) and deregulate gene expression [102]. The RG4s bind to 5′-UTR and suppress translation without affecting transcription as observed in the case of genes such as Bcl2, NRAS, matrix metalloproteinase 16 (MMP16), and ARPC2. Interestingly, these G-quadruplexes interact with many proteins such as heterogeneous nuclear ribonucleoproteins (hnRNPs) and result in the stalling of ribosomes. Thus, protein G-quadruplexes interactions tightly regulate transcription and translation processes and could vary among various cancer types [103, 104].

Yang et al. [105] have indicated that methylation or demethylation of polycomb2 (PC2) protein) regulates the location of growth control genes between polycomb bodies (PcGs) and interchromatin granules (ICGs). Noncoding RNAs such as TUG1 and MALAT1/NEAT2 mediate the assembly of corepressors/coactivators and can serve to switch mark recognition by histone code [94].

Long noncoding RNA (lncRNA)-mediated gene silencing through polycomb mechanisms

Polycomb group of proteins form two complexes namely polycomb repressive complex 1 (PRC1) and polycomb repressive complex 2 (PRC2). PRC1 contributes to chromatin compaction and PRC2 catalyzes di- and tri methylation of histone H3 at Lys 27 and are involved in cell proliferation, differentiation, cellular density, and plasticity of stem cells [106] (Fig. 8). Studies

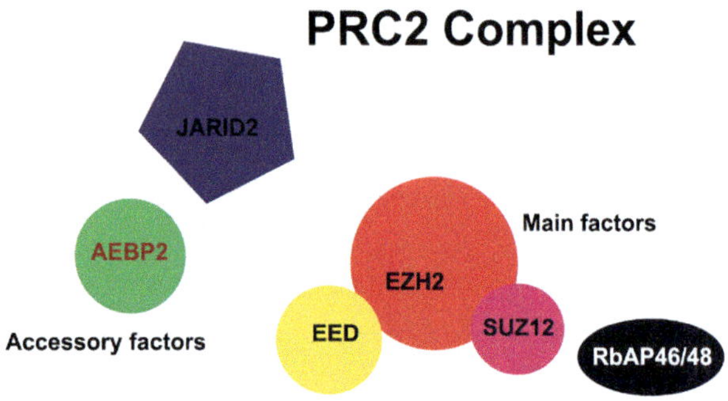

FIG. 8 PRC2 complex involved in lncRNA interaction. Polycomb repressive complex2 is divided into two complexes such as major factors (EED, EZH12, and SUZ12) as well as accessory factors such as AEBP2 and JARID2. These complexes of proteins regulate chromatin epigenetics mediated by lncRNA.

have indicated that noncoding RNA regulates PRC2 binding to chromatin either directly or indirectly with other factors [107]. Also, these lncRNAs are involved in recruiting the epigenetic factors to a particular locus, and in turn, these molecules are targeted by polycomb repressive complex.

Noncoding RNA epigenetics

The most common DNA modifications in eukaryotes include 5-methyl cytosine (m5c) [108] and N6-methyl adenosine (m6A) in the case of prokaryotes. But in unicellular eukaryotes, the modifications include m6A and m5C [109]. Several noncoding RNAs undergo epigenetic modifications such as m6A, m5c, and pseudo uridine (ψ), and these epigenetic changes were found to regulate the stability and binding ability of lncRNAs with other proteins. Thus, noncoding RNA epigenetics or epi-transcriptomics play a crucial role in the regulation of various functions of lncRNAs. Some of the lncRNAs with various epigenetic modifications include steroid receptor RNA activator (SRA1) at m6A, ψ; growth arrest special 5′-antisense (GAS5-AS1) (m5c); human telomerase RNA component (TERC) (m5c, ψ); taurine upregulated 1 (TUG1) (m6A); antisense noncoding RNA in the INK4 locus antisense to CDKN2B locus (ANRIL) (m6A, m5c); small nuclear RNA host gene 1 (SNGH1) (ψ); SNGH 7 (ψ); SNGH12 (ψ); HOTAIR (m6A, m5c); plasmacytoma variant translocation 1 PVT1 (m6A, m5c, ψ), metastasis-associated lung adenocarcinoma transcript 1 MALAT1 (m6A, m5c, ψ); XIST (m6A, m5c, ψ); NEAT1 (m6A, m5c, ψ); SRA (m6A, m5c, ψ), and ribonuclease P RNA component H1 (RPPH1) (m5c) [109]. Recent studies by Liu et al. [3, 28] have reported that MALAT1 modified at A2515 and A2577 resulted in destabilization of hairpin structure and facilitated the heterogeneous nuclear ribonucleoprotein C (HNRNPC) protein binding [3, 28] (Fig. 9).

Using next-generation sequencing technology, researchers have identified 170 RNA modifications in mRNA, rRNA, and tRNA in almost all living organisms. Thus, epitranscriptomics has emerged as an important subject to be considered in understanding cancer growth, proliferation, cancer stemness, and drug resistance. Recent studies have identified the key role of RNA modifications during various stages of cancer progression and regulation of cellular processes. These modifications include N6-methyladenosine (m6A), N1-methyl adenosine (m1A), N6 2-*O*-dimethyl adenosine (m6Am), 5-methylcytosine (m5C), inosine (I), and pseudouridine (ψ). RNA modifying proteins (RMPs) include

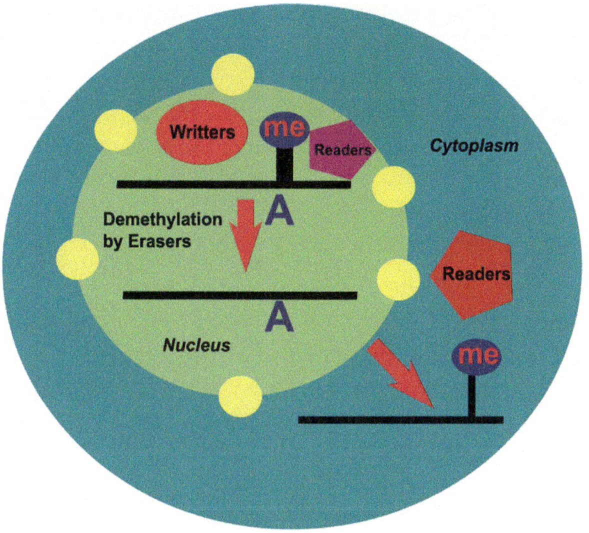

FIG. 9 Role of writers, erasers, and readers of methylation on lncRNA and TMZ resistance. Methyltransferase like 3 (METTL3) is a writer of methylation mark at m6A and α-keto glutarate-dependent dioxygenase AlkB homolog 5 (ALKBH5), and fat mass obesity associated-protein (FTO) is an eraser of methylation mark in the nucleus. Readers of methylation mark in the nucleus is YTHDC1, and in the cytoplasm, the readers of RNA methylation are YTHDF1/2/3 and YTHDC2.

methyltransferase like 3 (METTL3), and α-ketoglutarate-dependent dioxygenase AlkB homolog 5 (ALKBH5). These proteins act as writers and erasers and play an important role in cancer stemness and malignant phenotype in GBM (Fig. 9). METTL3 expression causes m6A modification in 3′-UTR regions of SRY-box transcription factor 2 (SOX2) mRNA, and this leads to recruitment of human antigen R (HuR) that results in stabilization of SOX2 mRNA and GSC maintenance [110]. N6-Methyl adenosine (m6A) modifications occur most commonly in the 3′-UTR stop codon, and long exons of mRNA occur with a consensus RRACH site (i.e., R=A/G; H=A.C or U) [111]. This modification plays a pivotal role in cancer progression and drug resistance. The m6A erasers include fat mass and obesity-associated protein (FTO) and ALKBH5 overexpressed in patient-derived GBM stem cells (GSC) cells. ALKBH5 functions as demethylase and enhances the expression of ALKBH5. Interestingly, long noncoding RNA forkheadboxM1-antisense RNA (FOXM1-AS) provides interactions between FOXM1 and ALKBH5 [112].

Several m6A-binding proteins act as readers of mRNA modification such as YTH N6-methyladenosine RNA-binding protein 1/2/3 (YTHDF1/2/3) and YTH domain-containing 1/2/3 (YTHDC1/2/3). These proteins contain the YTH domain and they recognize RNA modification, binding to these marks and playing a crucial role in protein stability of writers involved in the degradation of m^6A-labeled RNAs [113, 114]. The other readers include heterogeneous nuclear ribonucleoprotein A2/B1 (hnRNPA2/B1) and heterogeneous nuclear ribonucleoprotein C1/C2 (hnRNPC), insulin-like growth factor 2 mRNA-binding protein 1/2/3 (IGF2BP1/2/3), also known as IMP1/2/3, and eukaryotic translation initiation factor 3a/b/h (eIF3a/b/h). The readers such as YTHDC1, hnRNPC, and hnRNPA3B1 function in the nucleus and participate in RNA stability, RNA export, miRNA maturation, alternate splicing, whereas YTHDF1/2/3 and YTHDC2 act as readers in the cytoplasm and are involved in the translation process. In GBM, the protein levels of writers (METTL3), erasers (FTO, ALKBH5), and readers (YTHDC1) of m6A modifications were found to be upregulated and are involved in cancer cell metabolism. The functional implications of these proteins in the TMZ resistance are yet to be elucidated.

lncRNA classification

LncRNAs do not exhibit sequence conservation as in protein-coding genes but exhibit evolutionary conservation [115]. These are potential diagnostic and prognostic markers for GBM [67]. These molecules were found to regulate various cellular processes in the both cytoplasm and nucleus [116]. Interestingly, lncRNAs interact with other lncRNAs, miRNAs, other small RNA species, and protein factors to regulate cellular processes.

LncRNAs are broadly classified based on how they relate to nearby coding genes [117, 118].

1. **Antisense lncRNAs:** These are lncRNAs that originate inside the coding gene of a particular gene or 3′- of protein-coding genes, and transcription of these RNAs occur in the reverse direction to the protein-coding genes with at least one exon overlap.
2. **Intronic lncRNAs:** These lncRNAs originate within the intron of a particular gene and are present in both the forward and reverse direction with no overlapping of exons.
3. **Bidirectional lncRNAs:** These lncRNA transcripts initiate in a divergent fashion from the promoter of the coding genes and with few bases cutoff (1000 bp).
4. **Intergenic lncRNA or lincRNAs:** These RNAs have independent transcriptional units from coding genes and originate from intergene loci of coding genes.
5. **Sense-overlapping lncRNAs:** These lncRNAs are transcribed in the direction of the sense strand of a protein-coding gene (Fig. 10).

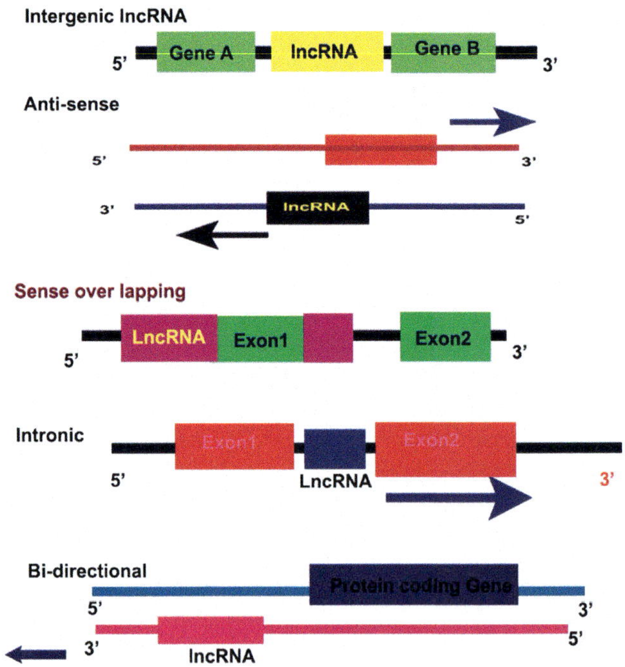

FIG. 10 Long noncoding RNA classification based on relativeness. The long noncoding RNAs are classified into intergenic, antisense, sense, intronic, and bi-directional lncRNAs. Antisense lncRNAs originate in the reverse direction to protein-coding genes, intrinsic lncRNA do not overlap with exons; sense lncRNA are transcribed in the sense direction as protein-coding genes. Intergenic lncRNAs originate from the intergenic regions and are known as lincRNAs. In bi-directional lncRNAs, the transcriptional initiation occurs in a divergent fashion.

Functions of LncRNA in the nucleus

(i) LncRNAs modulate gene expression by recruiting chromatin modification complex to the sites of transcription and cause epigenetic modification that enhances or inhibits the binding of other protein factors.
(ii) Long noncoding RNAs can bind at the sites of transcription and regulate the activation or repression of transcription. Thus, locus-specific transcription can be achieved.
(iii) Long noncoding RNAs act in trans and recruit various protein factors away from their sites of transcription and regulate gene expression.
(iv) Long noncoding RNAs sequester the transcription factors and prevent transcription factors association or interaction with the target gene (i.e., chromatin).

Functions of LncRNAs in regulating gene expression in the cytoplasm

(i) **LncRNAs identify and dictate the half-life and coding of mRNAs.** The best example is terminal differentiation-induced lncRNA (TINCR). These lncRNAs interact with staufess 1 protein (STAU1) and target mRNA containing the TINCR box motif and regulate their stability [119].
(ii) **lncRNA represses the translation of mRNAs.** Some lncRNAs regulate the fate of mRNAs in the cytoplasm. HUR, the ubiquitous RNA-binding protein (RBP) that regulates cell proliferation, cancer formation, and immune functions, modulates mRNA stability and translation processes as well as interacting with mRNA decay-promoting proteins such as AUF1 and TTP [120–123]. In the absence of HUR, long noncoding RNA LincRNA-p21 complex (i.e., a transcriptional target of p53 and HIF-1 α) interact with β-catenin (CTNNB1) and JUN-B mRNAs and a translational repressor RCK to repress the translation of targeted mRNAs [124].
(iii) **LncRNAs promote the translation of mRNAs.** The expression of lncRNA such as antisense Uchl I leads to an increase in translation owing to the binding of a large number of ribosomes without affecting mRNA stability [125].
(iv) **Certain lncRNAs such as STAU1-binding site RNAs.** 1/2-SbsRNAs have Alu sites bind to Alu elements in the 3′-UTR of actively transcribing RNAs. This leads to the binding of STAU1 with the lncRNA and these mRNAs undergo STAU1-mediated messenger RNA decay (SMD) [125].

Mechanisms of actions of lncRNA

LncRNA can function mainly in four different ways. At first lncRNA can acts as a signal for recruiting various transcription factors to regulate gene expression. The second way is lncRNAs can titrate various proteins that regulate transcription of a particular gene or they act as decoys for miRNA target sites and inhibit miRNA function. The third type of action includes the lncRNAs recruit chromatin-modifying enzymes near the target genes. The fourth type of action involves lncRNA as a scaffold that brings various transcription factors and induces chromatin epigenetic modifications and regulates gene expression [25].

LncRNA classification based on region implicated by the lncRNA

***cis*-lncRNAs:** These lncRNAs regulate the expression of genes in the nearby genomic proximity. These lncRNAs are transcribed from promoter regions and interfere with the transcriptional activation of the neighboring gene and act via transcription factor recruitment (i.e., chromatin remodeling complex) or DNA-RNA triple helix; e.g., X-inactive specific transcripts (XIST).

***trans*-lncRNAs:** These lncRNAs function in trans-mode and regulate the distant gene loci. These function by chromatin remodeling complex or by binding to RNA polymerase II or transcription elongation factors.

This lncRNA does not require sequence complementarily against the target gene. HOTAIR is the best example of Trans-lncRNA. HOTAIR from HOX-C (homeobox cluster c) locus chromosome 12q13.13 is transported to the HOXD genomic loci on chromosome 2q31 by polycomb repressive complex 2 (PRC2). The 5′ end of HOTAIR interacts with PRC2 (1–300) and 3′ end interacts with LSD1. PRC2 causes histone lysine 27 trimethylation and LSD1 is a lysine demethylase that removes H3K4 me2. Thus, HOTAIR acts as a bifunctional scaffold RNA [126].

Conclusions

GBM has a poor prognosis. TMZ is one of the most important drugs used in the treatment of GBM. Unfortunately, many GBM patients demonstrate resistance to TMZ. In this chapter, we discussed the role of long noncoding RNAs and miRNAs in the regulation of various proteins and cell signaling pathways that contribute to TMZ resistance (Fig. 11). Further, the link between lncRNA expression, chromatin modifications, and cellular signaling pathways in cancer stem cells in GBM was discussed. Long noncoding RNAs regulate various cellular processes such as cell cycle, growth, metabolism, drug resistance, and transcription. Interestingly, few lncRNAs can function as oncogenes and several others act as tumor-suppressive genes. Although the role played by lncRNAs, such as HOTAIR, MALAT1, and TP73-AS1 in cancer progression, invasion, metastasis, and TMZ resistance have been explored mechanistically in GBM, a huge number of lncRNAs still remain unexplored against GBM with respect to cancer progression and drug response.

Recent studies have focussed on combinations of various drugs that can sensitize GBM cells to TMZ. In general, cancer metabolism is a vital process in cancer progression and drug resistance. Thus, understanding the role of lncRNAs that participate in cancer metabolism is of urgent importance for future therapy against GBM. It is evident that cancer stem cells (CSCs) present in GBM cancer contribute toward uncontrolled growth and drug resistance. In this chapter, we described how NF-kB is a transcription factor that regulates cancer stemness and drug resistance via regulation of O6-methyl guanine DNA methyl transferase (MGMT) promoter activity. In GBM, MGMT is a vital gene whose expression regulates the

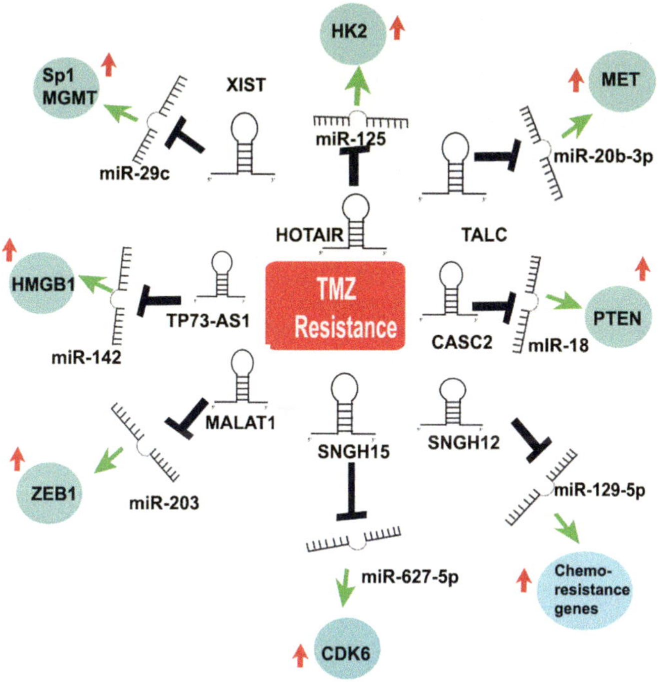

FIG. 11 LncRNAs involved in TMZ resistance. A graphical picture representing various lncRNAs and their target miRNA and respective mRNA expression. HOTAIR targets miR-125 and regulates HK2 gene expression. TALC regulates miR-20b-3p and regulates the expression of the MET gene. CASC2 sponges miR-18 and regulates PTEN expression. SNGH12 sponges miR-129-5p and regulates genes involved in chemoresistance. SNGH15 interacts with miR-627-5p and regulates CDK6 gene expression. MALAT1 sponges miR-203 and regulates ZEB1 expression. TP73-AS1 sponges miR-142 and regulates the expression of HMGB1. XIST1 sponges miR-29c and regulates the MGMT gene expression.

sensitivity of cancer cells to TMZ. Recent studies have identified the pivotal role of lncRNAs, such as NEAT1 and TALC, in the regulation of the MGMT expression and drugresistance.

Many lncRNAs, for example, HOTAIR involved in GBM, are regulated by epigenetic factors such as polycomb group of genes (PcG) (i.e., EZH2, suz12, EED). This indicates the interaction of epigenetic factors (i.e., lncRNA and PcG) dictates the outcome of drug resistance. Emerging studies have demonstrated that in-depth study on epigenetic modifications such as methylation at single base level, i.e., m6A, m5C, and several other base modifications also regulate the lncRNA expression and its ability to bind to miRNA. Thus, it is highly important to fully understand the epitranscriptomic changes and epigenetic changes on lncRNA and miRNA promoters as well as various genes involved in CSC maintenance to definitely provide useful information on how to sensitize GBM cells toward TMZ.

Acknowledgments

The work was partially supported by the Science and Engineering Research Board (SERB), Govt. of INDIA with grant number EMR/2017/001201. The corresponding author thanks the management of SASTRA Deemed University for their constant support and encouragement.

References

[1] Bray F, Ferlay J, Soerjomataram I, Siegel RL, Torre LA, Jemal A. Global cancer statistics 2018: GLOBOCAN estimates of incidence and mortality worldwide for 36 cancers in 185 countries. CA Cancer J Clin 2018;68:394–424.

[2] Ethun CG, Bilen MA, Jani AB, Maithel SK, Ogan K, Master VA. Frailty and cancer: implications for oncology surgery, medical oncology, and radiation oncology. CA Cancer J Clin 2017;67:362–77.

[3] Liu X, Gao Y, Lu Y, Zhang J, Li L, Yin F. Oncogenes associated with drug-resistance in ovarian cancer. J Cancer Res Clin Oncol 2015;141:381–95.
[4] Louis DN, Ohgaki H, Wiestler OD, Cavenee WK, Burger PC, Jouvet A, Scheithauer BW, Kleihues P. The 2007 WHO classification of tumours of the central nervous system. Acta Neuropathol 2007;114(2):97–109.
[5] Wesseling P, Capper D. WHO 2016 classification of gliomas. Neuropathol Appl Neurobiol 2018;44:139–50.
[6] Wang T, Pickard AJ, Gallo JM. Histone methylation by temozolomide: a classic DNA methylating anticancer drug. Anticancer Res 2016;36(7):3289–99.
[7] Knizhnik AV, Roos WP, Nikolova T, Quiros S, Tomaszowski KH, Christmann M, Kaina B. Survival and death strategies in glioma cells: autophagy, senescence and apoptosis triggered by a single type of temozolomide-induced DNA damage. PLoS One 2013;8(1), e55665.
[8] Bobola MS, Varadarajan S, Smith NW, Goff RD, Kolstoe DD, Blank A, Gold B, Silber JR. Human glioma cell sensitivity to the sequence-specific alkylating agent methyl-lexitropsin. Clin Cancer Res 2007;13(2):612–20.
[9] Fan C-H, Liu W-L, Cao H, Wen C, Chen L, Jiang G. O6-Methylguanine DNA methyltransferase as a promising target for the treatment of temozolomide-resistant gliomas. Cell Death Dis 2013;4(10):e876.
[10] Schaich M, Kestel L, Pfirrmann M, Robel K, Illmer T, Kramer M, Dill C, Ehninger G, Schackert G, Krex D. A MDR1 (ABCB1) gene single nucleotide polymorphism predicts outcome of temozolomide treatment in glioblastoma patients. Ann Oncol 2009;20(1):175–81.
[11] Hartmann C, Meyer J, Balss J, Capper D, Mueller W, Christians A, Felsberg J, Wolter M, Mawrin C, Wick W, Weller M, Herold-Mende C, Unterberg A, Jeuken JWM, Wesseling P, Reifenberger G, Deimling A. Type and frequency of IDH1 and IDH2 mutations are related to astrocytic and oligodendroglial differentiation and age: a study of 1,010 diffuse gliomas. Acta Neuropathol 2009;118(4):469–74.
[12] Ujifuku K, Mitsutake N, Takakura S, Matsuse M, Saeno V, Suzuki K, Hayashi K, Matsuo T, Kamada K, Nagata I, Yamashita S. miR-195, miR-455-3p and miR-10a (*) are implicated in acquired Temozolomide resistance in glioblastoma multiforme cells. Cancer Lett 2010;296(2):241–8.
[13] Hombach-klonisch S, Mehrpour M, Shojaei S, Harlos C, Pitz M, Hamai A, Siemianowicz K, Likus W, Wiechec E, Toyota BD, Hoshyar R, Seyfoori A, Sepehri Z, Ande SR, Khadem F, Akbari M, Gorman AM, Samali A, Klonisch T, Ghavami S. Glioblastoma and chemoresistance to alkylating agents: involvement of apoptosis, autophagy, and unfolded protein response. Pharmacol Ther 2018;184:13–41.
[14] Yuan AL, Ricks CB, Bohm AK, Lun X, Maxwell L, Safdar S, Bukhari S, Gerber A, Sayeed W, Bering EA, Pedersen H, Chan JA, Shen Y, Marra M, Kaplan DR, Mason W, Goodman LD, Ezhilarasan R, Kaufmann AB, Cabral M, Robbins SM, Senger DL, Cahill DP, Sulman EP, Cairncross JG, Blough M. ABT-888 restores sensitivity in temozolomide resistant glioma cells and xenografts. PLoS One 2018;13(8):e0202860.
[15] Happold C, Stojcheva N, Silginer M, Roth P, Reifenberger G, Heinrich H, Weller M. Transcriptional control of O6-methyl guanine DNA methyl transferase expression and temozolomide resistance in glioblastoma. J Neurochem 2017;144:780–90.
[16] Chumakova A, Lathia JD. Outlining involvement of stem cell program in regulation of O6-methylguanine DNA methyltransferase and development of temozolomide resistance in glioblastoma. J Neurochem 2018; 144(6):688–90.
[17] Brell M, Tortosa A, Verger F, Gill JM, Viñolas N, Villá S, Acwbes JJ, Caral L, Pujol T, Ferrer I, Ribalta T, Graus F. Prognostic significance of O6-methylguanine-DNA methyltransferase determined by promoter hypermethylation and immunohistochemical expression in anaplastic gliomas. Clin Cancer Res 2005;11:5167–74.
[18] Persano L, Pistollato F, Rampazzo E, Puppa DA, Abbadi S, Frasson C, Volpin F, Indraccolo S, Scienza R, Basso C. BMP2 sensitizes glioblastoma stem-like cells to Temozolomide by affecting HIF-1α stability and MGMT expression. Cell Death Dis 2012;3, e412.
[19] Xu Y, Porntadavity S, St Clair DK. Transcriptional regulation of the human manganese superoxide dismutase gene: the role of specificity protein 1 (Sp1) and activating protein-2 (AP-2). Biochem J 2002;362(2):401–12.
[20] Holley AK, Dhar SK, St Clair DK. Curbing cancer's sweet tooth: is there a role for MnSOD in regulation of the Warburg effect? Mitochondrion 2013;13(3):170–88.
[21] Jiaper S, Furuta T, Tanaka S, Kitabayashi T, Nakada M. Potential strategies overcoming the temozolomide resistance for glioblastoma. Neuro Med Chir (Tokyo) 2018;58(10):405–21.
[22] Iyer MK, Niknafs YS, Malik R, Singhal U, Sahu A, Hosono Y, Barette TR, Prensner JR, Evans JR, Zhao S, Poliakov A, Cao X, Dhanasekaran SM, Wu Y-M, Robinson DR, Beer DG, Feng FY, Iyer HK, Chinnaiyan AM. The landscape of long noncoding RNAs in the human transcriptome. Nat Genet 2015;47(3):199–208.

[23] Guttman M, Amit I, Garber M, French C, Lin MF, Feldser D, Huarte M, Zuk O, Carey BW, Cassady JP, Cabili MN, Jaenisch R, Mikkelsen TS, Jacks T, Hacohen N, Bernstein BE, Kellis M, Regev A, Rinn JL, Lander ES. Chromatin signature reveals over a thousand highly conserved large non-coding RNAs in mammals. Nature 2009;458(7235):223–7.

[24] Prensner JR, Chinnaiyan AM. The emergence of lncRNAs in cancer biology. Cancer Discov 2011;1(5):391–407.

[25] Wang KC, Chang HY. Molecular mechanisms of long noncoding RNAs. Mol Cell 2011;43(6):904–14.

[26] Kopp F, Mendell JT. Functional classification and experimental dissection of long noncoding RNAs. Cell 2018;172(3):393–407.

[27] Caeninci P, Kasukawa T, Katayama S, Gough J, Frith MC, Maeda N, Oyama R, Ravasi T, Lenhard B, Wells C, Kodzius R, Shimokawa K, Bajic VB, Brenner SE, Bataloy S, Forrest ARR, Zavolan M, Davis MJ, Wilming LG, Aidinis V, Allen JE, Ambesi-Impiombato A, Apweiler R, Aturaliya RN, Bailey TL, Bansal M, Baxter L, Beisel KW, Bersano T, Bono H, Chalk AM, Chiu KP, Choudhary V, Christoffels A, Clutterbuck DR, Crowe ML, Dalla E, Dalrymple BP, de Bono B, Gatta GD, di Bernardo D, Down T, Engstrom P, Fagiolini M, Faulkner G, Fletcher CF, Fukushima T, Furuno M, Futaki S, Gariboldi M, Georgii-Hemming P, Gingeras TR, Gojobori T, Green RE, Gustincich S, Harbers M, Hayashi Y, Hensch TK, Horokawa N, Hill D, Huminiecki L, Iacono M, Ikeo K, Iwama A, Ishikawa T, Jakt M, Kanapin A, Katoh M, Kawasawa Y, Kelso J, Kitamura H, Kitano H, Kollias G, Krishnan SPT, Kruger A, Kummerfeld SK, Kurochkin IV, Lareau LF, Lazarevic D, Lipovich L, Liu J, Liuni S, Mc William S, Madan Babu M, Madera M, Marchionni L, Matsuda H, Matsuzawa S, Miki H, Mignone F, Miyake S, Morris K, Mottagui-Tabar S, Mulder N, Nakano N, Nakauchi H, Ng P, Nilsson R, Nishiguchi S, Nishikawa S, Nori F, Ohara O, Okazaki Y, Orlando V, Pang KC, Pavan WJ, Pavesi G, Pesole G, Petrovsky N, Piazza S, Reed J, Reid JF, Ring BZ, Ringwald M, Rost B, Ruan Y, Salzberg SL, Sandelin A, Schneider C, Schönbach C, Seki guchi K, Semple CAM, Seno S, Sessa L, Sheng Y, Shibata Y, Shimada H, Shimada K, Silva D, Sinclair B, Sperling S, Stupka E, Sugiura K, Takenaka Y, Taki K, Tammoja K, Tan SL, Tang S, Taylor MS, Tegner J, Teichmann SA, Ueda HR, van Nimwegen E, Verardo R, Wei CL, Yagi K, Yamanishi HY, Zabarovsky E, Zhu S, Zimmer A, Hide W, Bult C, Grimmond SM, Teasdale RD, Liu ET, Brusic V, Quackenbush J, Wahlestedt C, Mattick JS, Huma DA, Kai C, Sasaki D, Tomaru Y, Fukuda S, Kanamori-Katayama M, Suzuki M, Aoki J, Arakawa T, Iida J, Imamura K, Itoh M, Kato T, Kawaji H, Kawagashira N, Kawashima T, Kojima M, Kondo S, Konno H, Nakano K, Ninomiya N, Nishio T, Okada M, Plessy C, Shibata K, Shiraki T, Suzuki S, Tagami M, Waki K, Watahiki A, Okamura-Oho Y, Suzuki H, Kawai J, Hayashizaki Y, FANTOM Consortium; RIKEN Genome Exploration Research Group and Genome Science Group (Genome Network Project Core Group). The transcriptional landscape of the mammalian genome. Science 2005;309(5740):1559–63.

[28] Liu Y-W, Sun M, Xia R, Zhang E-b, Liu X-h, Zhang Z, Xu T-p, De W, Liu B-r, Wang Z-x. Linc HOTAIR epigenetically silences miR34a by binding to PRC2 to promote the epithelial-to-mesenchymal transition in human gastric cancer. Cell Death and Disease 2015;6:1802.

[29] Liu C, Bai B, Skogerbø G, Cai L, Deng W, Zhang Y, Bu D, Zhao Y, Chen R. NONCODE: an integrated knowledge database of non-coding RNAs. Nucleic Acids Res 2005;33:D112–5.

[30] Grillone K, Riillo C, Scionti F, Rocca R, Tradigo G, Guzzi P-H, Alcaro S, Di Martino MT, Tagliaferri P, Tassone P. Non-coding RNAs in cancer: platforms and strategies for investigating the genomic "dark matter". J Exp Clin Cancer Res 2020;39:117.

[31] Bu D, Yu K, Sun S, Xie C, Skogerbø G, Miao R, Xiao H, Liao Q, Luo H, Zhao G, Zhao H, Liu Z, Liu C, Chen R, Zhao Y. NONCODE v3.0: integrative annotation of long noncoding RNAs. Nucleic Acids Res 2012;40(Database issue):D210–5.

[32] Fang SS, Zhang LL, Guo JC, Niu YW, Wu Y, Li H, Zhao LH, Li XY, Teng XY, Sun XU, Sun L, Zhang MQ, Chen RS, Zhao Y. NONCODEV5: a comprehensive annotation database for long non-coding RNAs. Nucleic Acids Res 2018;46(Database issue):D308–14.

[33] Derrien T, Johnson R, Bussotti G, Tanzer A, Djebali S, Tilgner H, Guernec G, Martin D, Merkel A, Knowles DG, Lagarde J, Veeravalli L, Ruan X, Ruan Y, Lassmann T, Carninci P, Brown JB, Lipovich L, Gonzalez JM, Thomas M, Davis CA, Shiekhattar R, Gingeras TR, Hubbard TJ, Nogtredame C, Harrow J, Guigó R. The GENCODE v7 catalog of human long noncoding RNAs: analysis of their gene structure, evolution, and expression. Genome Res 2012;22(9):1775–89.

[34] Abdel Malek MAY, Jagannthan S, Malek E, Sayed DM, Elgammal SA, Abd El-Azeem HG, Thabet NM, Driscoll JJ. Molecular chaperone GRP78 enhances aggresome delivery to autophagosomes to promote drug resistance in multiple myeloma. Oncotarget 2015;6:3098–110.

[35] Wang Y, Wang Z, Xu J, Li J, Li S, Zhang M, Yang D. Systematic identification of non-coding pharmacogenomic landscape in cancer. Nat Commun 2018;9(3192):1–15.

[36] Mercer TR, Dinger ME, Mattick JS. Long non-coding RNAs: insights into functions. Nat Rev Genet 2009; 10(3):155–9.

[37] Pang KC, Frith MC, Mattick JS. Rapid evolution of noncoding RNAs: lack of conservation does not mean lack of function. Trends Genet 2006;22(1):1–5.

[38] Schmitt AM, Chang HY. Long noncoding RNAs in cancer pathways. Cancer Cell 2016;29(4):452–63.

[39] Li B, Shen M, Yao H, Chen X, Xiao Z. Long noncoding RNA TP73-AS1 modulates medulloblastoma progression in vitro and in vivo by sponging miR-494-3p and targeting EIF5A2. Onco Targets Ther 2019;12:9873–85.

[40] Mazor G, Levin L, Picard D, Ahmadov U, Caren H, Borkhardt A, Reifenberger G, Leprivier G, Remke M, Rotblat B. The lncRNA TP73-AS1 is linked to aggressiveness in glioblastoma and promotes temozolomide resistance in glioblastoma cancer stem cells. Cell Death Dis 2019;10(246). https://doi.org/10.1038/s41419-019-1477-5.

[41] Zhang R, Jin H, Lou F. The long non-coding RNA TP73-AS1 interacted with miR-142 to modulate brain glioma growth through HMGB1/RAGE pathway. J Cell Bio Chem 2018;119(4):3007–16.

[42] Zhang W, Zhai Y, Wang W, Cao M, Ma C. Enhanced expression of lncRNA TP73-AS1 predicts unfavorable prognosis for gastric cancer and promotes cell migration and invasion by induction of EMT. Gene 2018;678:377–83.

[43] Pang JC-S, Li KK-W, Lau KM, Ng YL, Wong J, Chung NYF, Li H-M, Chui Y-L, Lui VWY, Chen Z-p, Chan DTM, Poon WS, Wang Y, Mao Y, Zhou L, Ng H-K. KIAA0495/PDAM is frequently downregulated in oligodendroglial tumors and its knockdown by siRNA induces cisplatin resistance in glioma cells. Brain Pathol 2010;20: 1021–32.

[44] Liu X, Wang M, Cui Y. LncRNA TP73-AS1 interacted with miR-141-3p to promote the proliferation of non-small cell lung cancer. Arch Med Sci 2019;15(6):1547–54.

[45] Song W, Zhang J, Xia Q, Sun M. Down-regulated lncRNA TP73-AS1 reduces radio-resistance in hepatocellular carcinoma via the PTEN/Akt signaling pathway. Cell Cycle 2019;18(22):3177–88.

[46] Xiao S, Wang R, Wu X, Liu W, Ma S. The long noncoding RNA TP73-AS1 interacted with miR-124 to modulate glioma growth by targeting inhibitor of apoptosis-stimulating protein of p53. DNA Cell Biol 2018;37(2):117–25.

[47] Xia Z, Yang X, Wu S, Feng Z, Qu L, Chen X, Liu L, Ma Y. LncRNA TP73-AS1 down-regulates miR-139-3p to promote retinoblastoma cell proliferation. Biosci Rep 2019;39(5). BSR 20190475.

[48] Liu J, Wang H, Liao X. MicroRNA-223-5p targets long non-coding RNA TP73 antisense RNA1 to promote the invasion of gastric cancer. Hum Cell 2020;33(3):676–82.

[49] Xu H, Zhao G, Zhang Y, Jiang H, Wang W, Zhao D, Yu H, Qi L. Long non-coding RNA PAXIP1-AS1 facilitates cell invasion and angiogenesis of glioma by recruiting transcription factor ETS1 to up-regulate KLF14 expression. J Exp Clin Cancer Res 2019;38:486.

[50] Weirick T, Militello G, Ponomareva Y, John D, Döring C, Dimmeler S, Uchida S. Logic programming to infer complex RNA expression patterns from RNA-seq data. Brief Bioinform 2018;19(2):199–209.

[51] Dittmer J. The biology of the Ets1 proto-oncogene. Mol Cancer 2003;2(1):1–21.

[52] Jing X, Lin J, Wang H, Tian L, Tian R, Zhang Y, Chen X, Zhang J. Prognostic value of lncRNA SOX2OT for Chinese cancer patients: a meta-analysis. PLoS One 2017;12(5), e0176889.

[53] Askarian-Amiri ME, Seyfoddin V, Smart CE, Wang J, Kim JE, Hansji H, Baguley BC, Finlay GJ, Leung EY. Emerging role of long non-coding RNA SOX2OT in SOX2 regulation in breast cancer. PLoS One 2014;9(7), e102140.

[54] Shi X-M, Teng F. Up-regulation of long non-coding RNA Sox2ot promotes hepatocellular carcinoma cell metastasis and correlates with poor prognosis. Int J Clin Exp Pathol 2015;8(4):4008–14.

[55] Kamel LM, Atef DM, Mackawy AMH, Shalaby SM, Abdelraheim N. Circulating long non-coding RNA GAS5 and SOX2OT as potential biomarkers for diagnosis and prognosis of non-small cell lung cancer. Biotechnol Appl Biochem 2019. https://doi.org/10.1002/bab.1764.

[56] Han L, Zhang W, Zhang B, Zhan L. Long non-coding RNA SOX2OT promotes cell proliferation and motility in human ovarian cancer. Exp Ther Med 2018;15(2):2182–8.

[57] Li Z, Jiang P, Li J, Peng M, Zhao X, Zhang X, Chen K, Zhang Y, Liu H, Gan L, Bi H, Zhen P, Zhu J, Li X. Tumor-derived exosomal lnc-Sox2ot promotes EMT and stemness by acting as a ceRNA in pancreatic ductal adenocarcinoma. Oncogene 2018;37:3822–38.

[58] Shahryari A, Jazi MS, Samaei NM, Mowla SJ. Long non-coding RNA SOX2OT: expression signature, splicing patterns, and emerging roles in pluripotency and tumorigenesis. Front Genet 2015;6:196.
[59] Fantes J, Ragge NK, Lynch S-A, McGill NI, Collin JRO, Peebles PNH, Hayward C, Vivian AJ, Williamson K, Heyningen VV, Fitzpatrick DR. Mutations in SOX2 cause anophthalmia. Nat Genet 2003;33:462–3.
[60] Amaral PP, Neyt C, Wilkins SJ, Askarian-Amiri ME, Sunkin SM, Perkins AC, Mattick JS. Complex architecture and regulated expression of the Sox2ot locus during vertebrate development. RNA 2009;15(11):2013–27.
[61] Bowles J, Schepers G, Koopman P. Phylogeny of the SOX family of developmental transcription factors based on sequence and structural indicators. Dev Biol 2000;227:239–55.
[62] He X, Sheng J, Yu W, Wang K, Zhu S, Liu Q. LncRNA MIR155HG promotes temozolomide resistance by activating the Wnt/β-catenin pathway via binding to PTBP1 in glioma. Cell Mol Neurol 2020. https://doi.org/10.1007/s10571-020-00898-z.
[63] Zhang S, Guo S, Liang C, Lian M. Long intergenic noncoding RNA 00021 promotes glioblastoma temozolomide resistance by epigenetically silencing p21 through Notch pathway. IUBMB Life 2020. https://doi.org/10.1002/iub.2301.
[64] Zhang J, Chen G, Gao Y, Lian H. HOTAIR/miR-125 axis-mediated Hexokinase 2 expression promotes chemoresistance in human glioblastoma. J Cell Mol Med 2020;1–11.
[65] Tan JY, Marques A. miRNA-mediated crosstalk between transcripts: the missing "linc"? Bioessays 2016;38(3):295–301.
[66] Lu C, Wei Y, Wang X, Zhang Z, Yin J, Li W, Chen L, Lyu X, Shi Z, Yan W, You Y. DNA-methylation-mediated activating of lncRNA SNHG12 promotes temozolomide resistance in glioblastoma. Mol Cancer 2020;19(1):28.
[67] Zeng T, Li L, Zhou Y, Gao L. Exploring long non-coding RNAs in glioblastoma regulatory mechanism and clinical potentials. Int J Genomics 2018. Article ID 2895958.
[68] Hussain SF, Yang D, Suki D, Aldape K, Grimm E, Heimberger AB. The role of human glioma-infiltrating microglia/macrophages in mediating antitumor immune responses. Neuro Oncol 2006;8(3):261–79.
[69] Matias D, Predes D, Filho PN, Lopes MC, Abreu JG, Lima FRS, Neto VM. Microglia-glioblastoma interactions: new role for Wnt signalling. Biochim Biophys Acta Rev Cancer 2017;1868:333–40.
[70] Li Z, Zhang J, Zheng H, Li C, Xiong J, Wang W, Bao H, Jin H, Liang P. Modulating lncRNA SNGH15/CDK6/miR-625 circuit by palbociclib, overcomes temozolomide resistance and reduces M2-polarization of glioma associated microglia in glioblastoma multiforme. J Exp Clin Care Res 2019;38(380). https://doi.org/10.1186/s13046-019-1371-0.
[71] Liu K, Hou Y, Liu Y, Zheng J. LncRNA SNHG15 contributes to proliferation, invasion, and autophagy in osteosarcoma cells by sponging miR-141. J Biomed Sci 2017;24:46.
[72] Liu C, Sun Y, She X, Tu C, Cheng X, Wang L, Yu Z, Li P, Liu Q, Yang H, Li G, Wu M. CASC2C as an unfavourable prognosis factor interacts with miR-101 to mediate astrocytoma tumorigenesis. Cell Death Dis 2017;8, e2639.
[73] Zhang Y, Zhang D, Lv J, Wang S, Zhang Q. LncRNA SNHG15 acts as an oncogene in prostate cancer by regulating miR-338-3p/FKBP1A axis. Gene 2019;705:44–50.
[74] Wang JJ, Yang S, Ji Q, Li Q, Zhou F, Li Y, Yuan F, Liu J, Tian Y, Zhao Y, Zheng Y. Long non-coding RNA EPIC1 promotes cell proliferation and motility and drug-resistance in glioma. Mol Ther Oncolytics 2020;17:130–7.
[75] Wang Z, Yang B, Zhang M, Guo WW, Wu Z, Wang Y, Jia L, Li S. lncRNA epigenetic landscape analysis identifies EPIC1as an oncogenic lncRNA that interacts with MYC and promotes cell-cycle progression in cancer. Cancer Cell 2018;33:706–20.
[76] Tripathi V, Ellis JD, Shen Z, Song DY, Pan Q, Watt AT, Freier SM, Bennett FC, Sharma A, Bubulya PA, Blencowe BJ, Prasanth SG, Prasanth KV. The nuclear –retained non-coding RNA MALAT1 regulates alternative splicing by modulating SR splicing factor phosphorylation. Mol Cell 2010;39(6):925–38.
[77] Voce DJ, Bernal GM, Wu L, Crawley CD, Zhang W, Mansour NM, Cahill KE, Szymura SJ, Uppal A, Raleigh DR, Spretz R, Nunez L, Larsen G, Khodarey NN, Weichselbaum RR, Yamini B. Temozolomide treatment induces lncRNA MALAT1 in an NF-kB and p53 co-dependent manner in glioblastoma. Cancer Res 2019;79(10):2536–48.
[78] Chaudhary R, Lal A. Long noncoding RNAs in the p53 network. Wiley Interdiscip Rev RNA 2017;8(3). https://doi.org/10.1002/wrna.1410.
[79] Zhou X, Liu S, Cai G, Kong L, Zhang T, Ren Y, Wu Y, Mei M, Zhang L, Wang X. Long non-coding RNA MALAT1 promotes tumor growth and metastasis by inducing epithelial-mesenchymal transition in oral squamous cell carcinoma. Sci Rep 2015;5. Article number 15972.

[80] Li H, Yuan X, Yan D, Li D, Guan F, Dong Y, Wang H, Liu X, Yang B. Long non-coding RNA MALAT1 decreases the sensitivity of resistant glioblastoma cell lines to temozolomide. Cell Physiol Biochem 2017;42:1192–201.
[81] Cai T, Liu Y, Xiao J. Long noncoding RNA MALAT1 knockdown reverses chemoresistance to temozolomide via promoting microRNA-101 in glioblastoma. Cancer Med 2018;7(4):1404–15.
[82] Tian T, Mingyi M, Qiu X, Qiu Y. MicroRNA-101 reverses temozolomide resistance by inhibition of GSK3β in glioblastoma. Oncotarget 2016;7:79584.
[83] Chen Y, Xiao Z, Hu M, Luo X, Cui Z. Diagnostic efficacy of long non-coding RNA MALAT1 in human cancers: a meta-analysis study. Oncotarget 2017;8:102291–300.
[84] Xu K-II, Xu Q-S, Wu Z-X, Xu S-J, Shena J. LncRNA NEAT1 is involved in temozolomide resistance by regulating MGMT in glioblastoma multiforme. Clin Surg Res Commun 2018;1:24–30.
[85] Xu N, Liu B, Lian C, Doycheva DM, Fu Z, Liu Y, Zhou J, He Z, Yang Z, Huang Q, Zeng H, Guo H. Long non-coding RNA AC003092.1 promotes temozolomide chemo-sensitivity through miR-195/TFPI-2 signaling modulation in glioblastoma. Cell Death Dis 2018;9(12):1139.
[86] Li Z, Zhao X, Zhou Y, Liu Y, Zhou Q, Ye H, Wang YX, Zeng J, Song Y, Gao W, Zheng SY, Zhuang B, Chen H, Li W, Li H, Li H, Fu Z, Chen R. The long non-coding RNA HOTTIP promotes progression and gemcitabine resistance by regulating HOXA13 in pancreatic cancer. J Transl Med 2015;13:84.
[87] Fan Y, Shen B, Tan M, Mu X, Qin Y, Zhang F, Liu Y. Long non-coding RNA UCA1 increases chemoresistance of bladder cancer cells by regulating Wnt signaling. FEBS J 2014;281(7):1750–8.
[88] Zhang XW, Bu P, Liu L, Zhang X-Z, Li J. Overexpression of long non-coding RNA PVT1 in gastric cancer cells promotes the development of multiple drug resistance. Biochem Biophys Res Commun 2015;462(3):227–32.
[89] Baldinu P, Cossu A, Manca A, Satta MP, Sini MC, Rozzo C, Dessole S, Cherchi P, Gianfrancesco F, Pintus A, Carboni A, Deiana A, Tanda F. Identification of a novel candidate gene, CASC2, in a region of common allelic loss at chromosome 10q26 in human endometrial cancer. Hum Mutat 2004;23:318–26.
[90] Wang R, Li Y, Zhu G, Tian B, Zeng W, Yang Y, Li Z. Long non-coding RNA CASC2 predicts the prognosis of glioma patients and functions as a suppressor for glioma by suppressing Wnt/β-catenin pathway. Neuropsychiatr Dis Treat 2017;13:1805–13.
[91] Liao Y, Shen LF, Zhao H, Liu Q, Fu J, Guo Y, Peng R, Cheng L. LncRNA CASC2 interacts with miR-181a to modulate glioma growth and resistance to TMZ through PTEN pathway. J Cell Biochem 2017;118(7):1889–99.
[92] Song L, Li D, Gu Y, Wen Z-M, Jie J, Zhao D, Peng L-P. MicroRNA-126 targeting PIK3R2 inhibits NSCLC A549 cell proliferation, migration, and invasion by regulation of PTEN/PI3K/AKT pathway. Clin Lung Cancer 2016;17(5):e65–75.
[93] Wu P, Cai J, Chen Q, Han B, Meng X, Li Y, Li Z, Wang R, Lin L, Duan C, Kang C, Jiang C. Lnc-TALC promotes O6-methylguanine DNA methyl transferase expression via regulating the c-Met pathway by competitively binding with miR-20b-3p. Nat Commun 2019;10. Article number 2045.
[94] Ayub ALP, Papaiz DD, Soares RdS, Jasiulionis MG. In: Tutar L, Aras S, Tutar E, editors. The function of lncRNAs as epigenetic regulators. Non-coding RNAs. IntechOpen; 2019. https://doi.org/10.5772/intechopen.88071.
[95] Sa L, Li Y, Zhao L, Liu Y, Wang P, Liu L, Li Z, Ma J, Cai H, Xue Y. The role of HOTAIR/miR-148b-3p/USF1 on regulating the permeability of BTB. Front Mol Neurosci 2017;10:194.
[96] Bian E-B, Wang Y-Y, Yang Y, Wu B-M, Xu T, Meng X-M, Huang C, Zhang L, Lv XW, Xiong Z-G, Li J. HOTAIR facilitates hepatic stellate cells activation and fibrogenesis in the liver. Biochim Biophys Acta Mol Basis Dis 2017;1863(3):674–86.
[97] Liu X-h, Sun M, Nie F-g, Ge Y-b, Zhang E-b, Yin D-d, Kong R, Xia R, Lu K-h, Li J-h, De W, Wang K-m, Wang Z-x. LncRNA HOTAIR functions as a competing endogenous RNA to regulate HER2 expression sponging miR-331p in gastric cancer. Mol Cancer 2014;13:92.
[98] Zhang H, Cai K, Wang J, Wang X, Cheng K, Shi F, Jiang L, Zhang Y, Dou J. MiR-7 inhibited indirectly by lincRNA HOTAIR, directly inhibits SETDB1 and reverses the EMT of breast cancer stem cells by down-regulating the STAT3 pathway. Stem Cells 2014;32(11):2858–68.
[99] Yao Y, Ma J, Xue Y, Wang P, Li Z, Liu J, Chen L, Xi Z, Teng H, Wang Z, Li Z, Liu Y. Knockdown of long non-coding RNA XIST exerts tumor-suppressive functions in human glioblastoma stem cells by up-regulating miR-152. Cancer Lett 2015;359(1):75–86.

[100] Du P, Zhao H, Peng R, Liu Q, Yuan J, Peng G, Liao Y. LncRNA-XIST interacts with miR-29c to modulate the chemo-resistance of glioma cell to TMZ through DNA mismatch repair pathway. Biosci Rep 2017;37(5). BSR20170696.
[101] Herviou P, Le Bras M, Dumas L, Hieblot C, Gilhodes J, Cioci G, Hugnot JP, Ameadan A, Guillonneau F, Dassi E, Cammas A, Millevoi S. HnRNP H/F drive RNA G-quadruplex-mediated translation linked to genomic instability and therapy resistance in glioblastoma. Nat Commun 2020;11:2661.
[102] Huppert JL, Bugaut A, Kumari S, Balasubramanian S. G-quadruplexes: the beginning and end of UTRs. Nucleic Acids Res 2008;36(19):6260–8.
[103] Bugaut A, Balasubramanian S. 5′-UTR RNA G-quadruplexes translation regulation and targeting. Nucleic Acids Res 2012;40(11):4727–41.
[104] Serikawa T, Spanos C, von Hacht A, Budisa N, Rappsilber J, Kurreck J. Comprehensive identification of proteins binding to RNA G-quadruplex motifs in the 5′UTR of tumor associated mRNAs. Biochimie 2018;144: 169–84.
[105] Yang L, Lin C, Liu W, Zhang J, Ohgi KA, Grinstein JD, Dorrestein PC, Rosenfeld MG. ncRNA- and Pc2 methylation-dependent gene relocation between nuclear structures mediates gene activation programs. Cell 2011;147(4):773–88.
[106] Margueron P, Reinberg D. The polycomb complex PRC2 and its mark in life. Nature 2011;469:343–9.
[107] Rinn JL, Kertesz M, Wang JK, Squazzo SL, Xu X, Brugmann SA, Goodnough H, Helms JA, Famham PJ, Segal F, Chang HY. Functional demarcation of active and silent chromatin domains in human HOX loci by non-coding RNAs. Cell 2007;129(7):1311–23.
[108] Vanyushin BF, Tkacheva SG, Belozersky AN. Rare bases in animal DNA. Nature 1970;225:948–9.
[109] Luo G-Z, Blanco MA, Greer EL, He C, Shi Y. DNA N6-methyl adenine: a new epigenetic mark in eukaryotes? Nat Rev Mol Cell Biol 2015;16(12):705–10.
[110] Visvanathan A, Patil V, Arora A, Hegde AS, Arivazhagan A, Santosh V, Somasundaram K. Essential role of METTL3-mediated m6A modification in glioma stem-like cells maintenance and radio-resistance. Oncogene 2018;37(4):522–33.
[111] Zhou Z, Lv J, Yu H, Han J, Yang X, Feng D, Wu Q, Yuan B, Lu Q, Yang H. Mechanism of RNA modification N6-methyl adenosine in human cancer. Mol Cancer 2020;19:104.
[112] Zhang S, Zhao BS, Zhou A, Lin K, Zheng S, Lu Z, Chen Y, Sulman EP, Xie K, Bögler O, Majumder S, He C, Huang S. The m6A demethylase ALKBH5 maintains tumorigenicity of glioblastoma stem-like cells by sustaining FOXM1 expression and cell proliferation. Cancer Cell 2017;31(4):591–606.
[113] Stoilov P, Rafalska I, Stamm S. YTH: a new domain in nuclear proteins. Trends Biochem Sci 2002;27: 495–7.
[114] Luo S, Tong L. Molecular basis for the recognition of methylated adenines in RNA by the eukaryotic YTH domain. Proc Natl Acad Sci 2014;111(38):13834–9.
[115] Johnsson P, Lipovich L, Grander D, Morris KV. Evolutionary conservation of long non-coding RNAs: sequence, structure, function. Biochim Biophys Acta 2014;1840(3):1063–71.
[116] Miao H, Wang L, Zhan H, Dai J, Chang Y, Wu F, Liu Z, Gao C, Li L, Song X. A long non-coding RNA distributed in both nucleus and cytoplasm operates in the PYCARD-regulated apoptosis by coordinating the epigenetic & translational regulation. PLoS Genet 2019;15(5), e1008144.
[117] Rinn JL, Chang HY. Genome regulation by long non-coding RNAs. Annu Rev Biochem 2012;81:145–66.
[118] Bhan A, Soleimani M, Mandal SS. Long non-coding RNA and cancer: a new paradigm. Cancer Res 2017; 77(15):3965–81.
[119] Kretz M. TINCR, staufen1, and cellular differentiation. RNA Biol 2013;10(10):1597–601.
[120] Hinman MN, Lou H. Diverse molecular functions of Hu proteins. Cell Mol Life Sci 2008;65:3168–81.
[121] Abdelmohsen K, Gorospe M. Posttranscriptional regulation of cancer traits by HuR. Wiley Interdiscip Rev RNA 2010;1:214–29.
[122] Lal A, Mazan-Mamczarz K, Kawai T, Yang X, Martindale JL, Gorospe M. Concurrent versus individual binding of HuR and AUF1 to common labile target mRNAs. EMBO J 2004;23:3092–102.
[123] Young LE, Sanduja S, Bemis-Standoli K, Pena EA, Price RL, Dixon DA. The mRNA binding proteins HuR and tristetraprolin regulate cyclooxygenase 2 expression during colon carcinogenesis. Gastroenterology 2009; 136:1669–79.

[124] Yoon J-H, Abdelmohsen K, Srikantan S, Yang X, Martindale JL, De S, Huarte M, Zhan M, Becker KG, Gorospe M. LincRNA-p21 suppresses target mRNA translation. Mol Cell 2012;47(4):648–55.
[125] Gong C, Maquat L. "Alu" strious long ncRNAs and their roles in shortening mRNA half-lives. Cell Cycle 2011;10(12):1882–3.
[126] Tsai M-C, Manor O, Wan Y, Mosammaparast N, Wang JK, Lan F, Shi Y, Segal E, Chang HY. Long non-coding RNA as modulator scaffold of histone modification complexes. Science 2010;329(5992):689–93.

CHAPTER

16

Mechanisms of glioblastoma resistance to antiangiogenic agents and reversal approaches

Kartik Angara[a] *and Ali S. Arbab*[b]

[a]Caris Life Sciences, Phoenix, AZ, United States [b]Georgia Cancer Center, Augusta University, Augusta, GA, United States

Abstract

Glioblastoma (GBM) is the most prevalent malignant tumor in the central nervous system. Owing to its hypervascular state and inevitable recurrence following standard therapies of surgical resection, radiation, and chemotherapies, clinicians as well as scientists have instituted antiangiogenic agents to counter the development of blood vessels in the recurrent and rapidly growing GBM. However, investigations from our group and from other laboratories, as well as from clinical data, indicate the development of therapy resistance and activation of alternate neovascularization in GBM following antiangiogenic therapies (AATs). Alternate neovascularization occurs in three distinct pathways, such as angiogenesis, vasculogenesis, and vascular mimicry. Beside angiogenesis, different mechanisms are postulated for the development of alternative neovascularization and therapy resistance in GBM following AAT including bone marrow-derived cell-mediated vasculogenesis, transdifferentiation of glioma stem cells to form vascular mimicry, and vasculogenic as well as angiogenic myeloid cell-mediated neovascularization. Specific agents that target bone marrow-derived cells, myeloid cells, and transdifferentiation of glioma stem cells could be added to current therapeutic strategies in GBM management to overcome the resistance.

Abbreviations

AAT	antiangiogenic therapies
ADAMTS1	a disintegrin and metalloproteinase with thrombospondin motifs 1
BBB	blood-brain barrier
BMDCs	bone marrow-derived cells
C-MET	also called tyrosine-protein kinase Met or hepatocyte growth factor
CNS	central nervous system
CPCs	circulating endothelial and hematopoietic progenitor cells

Glioblastoma Resistance to Chemotherapy: Molecular Mechanisms and Innovative Reversal Strategies
https://doi.org/10.1016/B978-0-12-821567-8.00006-3

DCE-MRI	dynamic contrast-enhanced magnetic resonance imaging
Dll4	delta like-4
ECs	endothelial cells
ECM	extracellular matrix
EGFR	epidermal growth factor receptor
EPCs	endothelial progenitor cells
EZH2	enhancer of zeste homolog 2
FDA	Food and Drug Administration
FGF2	fibroblast growth factor
GBM	glioblastoma
GPCRs	guanine-protein-coupled-receptors
GSCs	glioma stem cells
HIF-1α	hypoxia-inducible factor-1α
ICAM-1	intercellular adhesion molecule-1
IL	interleukin
KRAS	Kirsten rat sarcoma viral oncogene homolog
LOX	lysine 6-oxidase
MMP	matrix metalloproteinases
NF-κB	nuclear factor kappa B
PARP	poly (ADP-ribose) polymerase
PAS	periodic acid-Schiff
PDAC	pancreatic ductal adenocarcinoma
PDGF	platelet-derived growth factor
PFS	progression free survival
PI3K	phosphatidyl-inositol-3-kinase
PlGF	placental growth factor
PLK1	polo-like kinase 1
RT	radiotherapy
RTKs	receptor tyrosine kinases
SDF-1	stromal cell-derived factor-1
shRNA	short hairpin RNA
STAT3	signal transducer and activator of transcription 3
TAMs	tumor-associated macrophages
TDECs	tumor-derived endothelial cells
TKI	tyrosine kinase inhibitor
TME	tumor microenvironment
TMZ	temozolomide
TSP1	thrombospondin-1
VCAM-1	vascular cell adhesion molecule-1
VEGF	vascular endothelial growth factor
VM	vascular mimicry
WHO	World Health Organization

Conflict of interest

No potential conflicts of interest were disclosed.

Introduction

Glioma refers to the primary brain tumors of glial cell origin, which predominantly include astrocytic tumors (astrocytoma, anaplastic astrocytoma, and glioblastoma (GBM)), oligodendrogliomas, ependymomas, and mixed gliomas [1–3]. GBM is the most hypervascular, hyperproliferative, invasive, undifferentiated, and highly aggressive diffuse astrocytic glioma. The current international standard for the nomenclature and diagnosis of gliomas is the WHO (World Health Organization) classification, which classifies gliomas into grade I–IV based on the level of malignancy, proliferative potential, and invasiveness. Grade I gliomas have low proliferative potential and can be cured by surgical resection. However, grade II–IV gliomas are highly malignant and invasive. GBM is categorized as a grade IV astrocytoma based on WHO classification [4, 5]. GBM is the most common brain and central nervous system (CNS) malignancy, accounting for 80% of malignant primary brain tumors, 54% of all gliomas, and 16% of all primary brain and CNS tumors [1, 3]. GBM remains a highly lethal and incurable disease, with a median survival of 15 months, despite the current standard of care and a variety of modern therapies [6, 7]. Current treatment consists of maximal safe surgical resection followed by radiotherapy (RT) with concurrent temozolomide (TMZ) chemotherapy, which has significantly improved survival of patients with GBM [8]. Surgical intervention has decompressive, cytoreductive, and antiseizure effects that offer a significant survival advantage and improved quality of life [9–11]. If left untreated, patients diagnosed with GBM have a median survival time of 3 months [12, 13].

Tumor neovascularization through angiogenesis, vasculogenesis, and vascular mimicry

The hypervascular nature of GBM can be singularly attributed to the pathological neovascularization that irrigates the tumor. Tumor neovasculature differs radically from the normal host blood vessels, both morphologically, and physiologically. The host endothelial vascular structures are comprised of endothelial cells (ECs) and supporting cells such as pericytes and astrocytes to form an intact highly regulated membranous blood-brain barrier (BBB). On the other hand, GBM vessels are tortuous, disorganized, highly permeable, destabilized structures with abnormal endothelial and pericyte coverage [14, 15]. Different forms of tumor neovascularization are schematically represented in Fig. 1.

Angiogenesis and vasculogenesis

Tumor development was previously attributed classically to two main mechanisms of vascularization, the classical sprouting angiogenesis, in which the preexisting ECs proliferate and migrate to form neovessels, and vasculogenesis, in which the cells from the bone marrow are recruited to the tumor sites to contribute to the formation of neovascular structures [16, 17]. Several studies have also reported the contributions of bone marrow-derived cells (BMDCs) to tumor vascularization [18], but contradictory findings have also been reported [19]. Moreover, transplanted Tie2-expressing mononuclear cells have been shown to home to the angiogenic sites in the tumors where they interacted closely with the ECs [20]. Furthermore, vascular endothelial growth factor (VEGF) and placental growth factor (PlGF) enhanced the recruitment of endothelial progenitor cells (EPCs) to the tumor, thereby facilitating vasculogenesis at the tumor site [21]. Angiopoietin-1 and 2 are also differentially involved in the recruitment of BMDCs, with a net gain of Ang-2 activity being positively

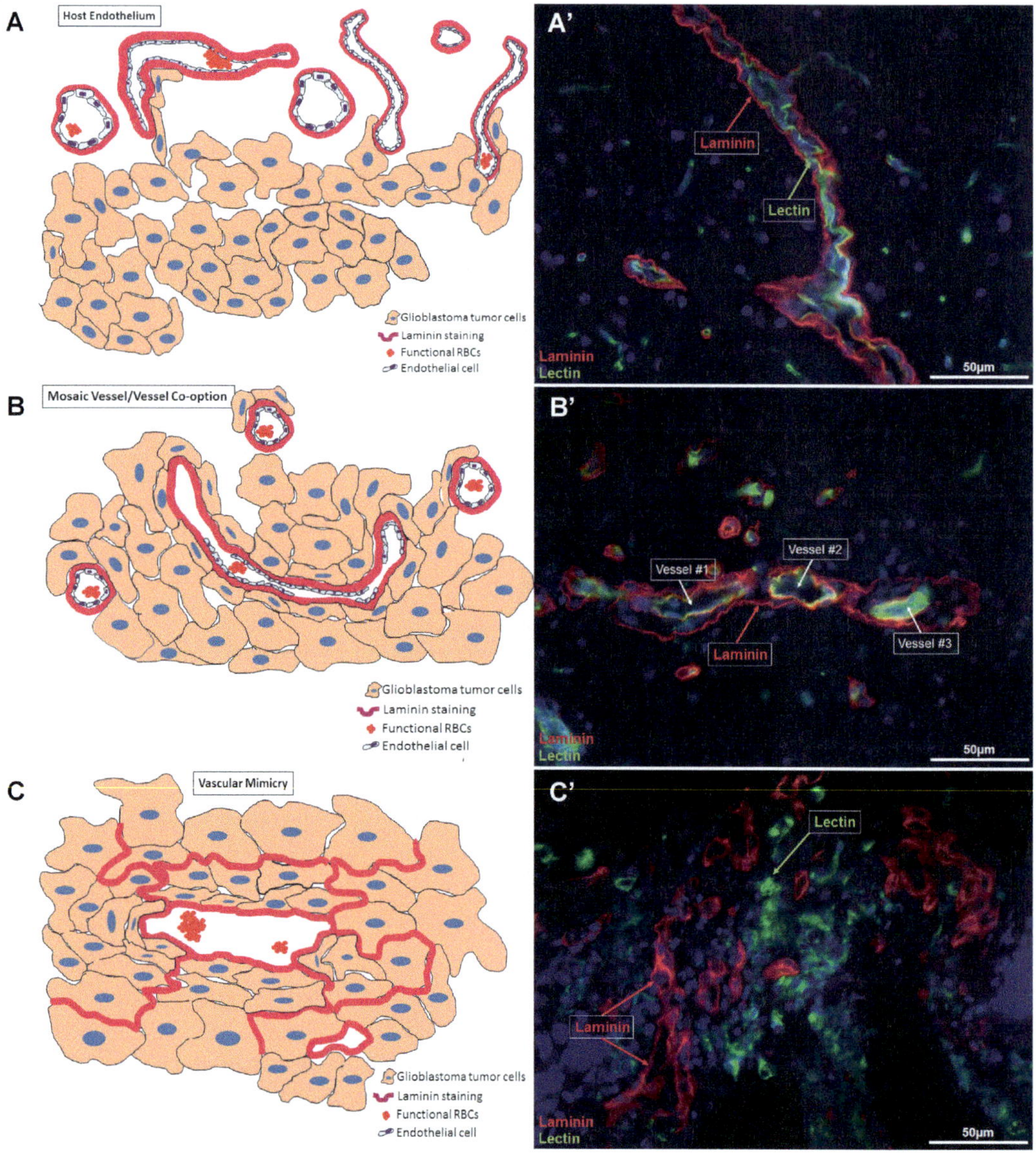

FIG. 1 Schematic of vascular mimicry development driving antiangiogenic therapy (AAT) resistance in glioblastoma. (A and A′) Host endothelium, (B and B′) vessel cooption of the endothelial cells by the tumor cells, and (C and C′) vascular mimicry.

correlated with the creation of a proangiogenic niche in the tumors to recruit the BMDCs [22]. As tumors grow in size, they need an increased supply of blood vessels to keep them viable. Avascular tumors that are unable to acquire new blood vessels to meet their metabolic and nutrient demands eventually regress [23–25]. During the early formative stages of tumor development, Ang-2 is expressed on the coopted tumor vasculature, and as the tumors grow in size.

There is a widespread increase in the expression of Ang-2 on both the endothelial cells and the tumor cells. The resulting destabilization of blood vessels caused by the disengagement of pericytes and loss of endothelial barrier integrity promotes tumor hypoxia and necrosis [15, 16, 26]. The initiation of hypoxic conditions in the tumor prompt the translocation of hypoxia-inducible factor-1α (HIF-1α) into the nuclei of cells and thereby upregulating the expression of HIF-1α-dependent expression of proangiogenic factors such as matrix metalloproteinases (MMP-2 and 9), VEGF, stromal cell-derived factor-1α (SDF-1α) and interleukin-8 (IL-8). The combined action of these factors causes the degradation of the extracellular matrix (ECM) scaffold and basement membrane to facilitate the creation of new areas in and around tumors, where neovessels can be established. In addition, these alterations of the ECM scaffolds in the tumor microenvironment (TME) bolster the metastatic capability of tumor cells by synergizing the migration and extravasation of tumor cells [27–29]. VEGF can subsequently upregulate the expression of VEGFR2 on the ECs and facilitate sprouting angiogenesis and vascular remodeling in conjunction with Ang-2. Many studies have already established the role of Ang-2 in the initiation of vessel sprouting and pericyte loss to promote neovessel formation in the tumors [16, 30–34]. Delta like-4 (Dll4)-Notch-1 signaling and Ephrin B2 regulates the sprouting and branching processes in the tip cells in response to VEGF. Several proangiogenic growth factors and chemokines such as VEGF, SDF-1α, platelet-derived growth factor (PDGF), IL8, fibroblast growth factor (FGF2), and Ang-1 create a conducive environment for EC recruitment, proliferation, and migration [15]. Integrin complexes such as αvβ3, αvβ5, and α5β1 are crucial to tumor angiogenesis [15] with αvβ3 integrin mediating FGF2-dependent angiogenesis and αvβ5 integrin being responsible for VEGF-dependent angiogenesis [35].

Vascular mimicry

Tumor growth and metastasis have long been associated with the processes of angiogenesis and vasculogenesis assuming a pathological form. This central dogma of classical angiogenesis and vasculogenesis-dependent perfusion of tumors was rendered obsolete when Maniotis et al. introduced a novel nonangiogenic-dependent pathway called vascular mimicry (VM) in 1999. VM was described as an unprecedented vascular system feeding tumor cells in malignant melanoma, which utilizes the uncanny ability of aggressive melanoma cells to transdifferentiate into a stem cell-like state and to subsequently assume an endothelial-like phenotype [36]. Following its initial discovery and detailed report in melanomas, VM was also reported in many other nonmelanoma neoplastic malignancies such as breast [37], ovarian [38, 39], prostate [39], lung [40], and also in GBM [41]. However, the phenomenon of VM has been a subject of intense controversy and many studies have questioned the validity of VM [42, 43].

VM is described as the uncanny ability of genetically dysregulated tumor cells to transdifferentiate and acquire endothelial-like phenotypes through an intermediate stem-cell-like state. VM enables the tumor cells to form functional matrix-embedded fluid conducting vascular structures devoid of endothelial lining containing blood cells and plasma to promote their growth and invasion while concurrently meeting with the increased metabolic and nutrient demands. Neovascular structures so formed are capable of irrigating the tumors in the event of therapeutic insult through AAT. The formation of neovasculatures through VM is host endothelial-cell independent, classifying it as a form of vasculogenesis to mimic the function of the blood vessels and thereby aptly defining the phenomenon of VM [36, 44–46]. VM vessels are extracellular component-rich patterned matrices that enclose small nests or lobules of tumor cells that eventually pave the way to the formation of a lumen by the movement of tumor cells. VM vessels are matrix-rich structures rich in vascular-specific laminin, positive for periodic acid-Schiff (PAS) staining for ECM components, and are often found encapsulating nests or lobules of tumor cells as closed loops. The formation of extracellular patterned matrices rich in laminin (see Fig. 2), proteoglycans, heparan sulfate, and collagen IV and VI as a part of their basement membranes visualized by the aforementioned PAS staining have already been established as crucial histopathological evidence of VM in hypervascular tumors [44, 45, 47–50].

Antiangiogenic therapies and development of resistance

GBM is characterized by increased invasiveness, hypervascularity, and pathological hyperproliferation, which induces rapid hypoxia at the core of the tumor while the peripheral migrating front retains its invasive potential. GBM has a high cellular and genetic heterogeneity in its makeup, which facilitates the rapid onset of pathological neovascularization and refractory therapy resistance. The mean survival period of patients after initial diagnosis is 1–3 years with the best of treatments [51, 52].

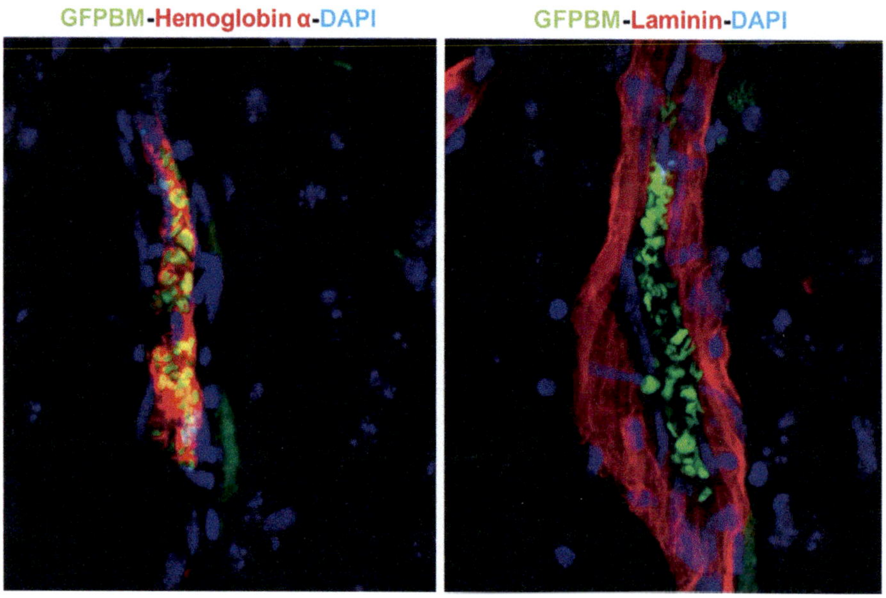

FIG. 2 GFP+ bone marrow (BM) cells enclosed in the laminin-positive VM structures (GBM tumor cells implanted into athymic nude mice with GFP+ BM engrafted cells) are also positive for hemoglobin indicating that these BM-derived cells are RBCs and that these VM structures are functional in nature.

Ide et al. in 1939 made landmark observations about the formation of new blood vessels, angiogenesis, juxtaposing the tumor growth [1, 2]. Folkman et al. demonstrated the survival of tumor cells following transplantation into syngeneic mice and proved that the growth of these tumors was angiogenesis dependent. These pioneering observations made angiogenesis inhibition a valuable therapeutic target, so much so that tumor angiogenesis has become one of the "hallmarks of cancer." [23, 53, 54]. The hypervascular nature of the GBM tumors prompted the inclusion of antiangiogenic therapies (AAT) including vatalanib, sunitinib, and cediranib primarily for two purposes: (i) to attenuate the abnormal angiogenesis and (ii) to simultaneously facilitate tumor vasculature normalization for improving the bioavailability and therapeutic efficacy of other combinatorial chemotherapies [55–60]. The AATs predominantly target the endothelial cells, which rely on the VEGF-VEGFR pathway to initiate and sustain angiogenesis. Since GBM tumor cells have significantly higher genetic instability compared to the endothelial cells, targeting the VEGF-VEGFR pathway on the endothelial cells is a more viable option to counter angiogenesis without imposing drug resistance. GBM vessels are structurally different from the normal blood vessels in that they are more tortuous, disorganized, highly permeable, destabilized structures with abnormal endothelial and pericyte coverage, making them highly leaky and more irrigational [14, 15]. Regrettably, the benefits offered by AAT were transitory and the tumors relapsed faster with an additional therapy-resistance characteristic of potent refractoriness [61].

Receptor tyrosine kinase inhibitors and anti-VEGF antibodies

The targeting of VEGFR2 using vatalanib (PTK787) has proved to be a failure because of evidence of refractory tumor growth from studies in our lab using DCE-MRI (dynamic contrast-enhanced magnetic resonance imaging) [62]. The preclinical trial results of cediranib and sunitinib, inhibitors of VEGFR, PDGFR, and c-kit receptors, were initially promising; however, they were marred with major setbacks owing to the limited therapeutic efficacy and transient benefits with high toxicity rates [63–67]. A similar disappointment was encountered in the clinical trials of imatinib and vatalanib owing to the lack of any considerable benefits in therapeutic efficacy [68, 69].

Furthermore, another study has reported that there was an increase in the plasma concentration of FGF-2 in colon cancer and GBM patients following AAT [70]. The utility of vatalanib (PTK787, Novartis Pharmaceuticals Corporation, East Hanover, NJ), a pan-VEGFR, PDGFR, and c-Kit tyrosine kinase inhibitor (TKI) has been studied both, as monotherapy [71], or in combination with temozolomide and lomustine [72] in recurrent GBM. Nintedanib (BIBF 1120), a small, orally available, triple angiokinase inhibitor in phase III development targeting VEGFR 1–3, FGFR 1–3, and PDGFR-α/β demonstrated limited but clinically nonrelevant antitumor activity in patients with recurrent GBM who had failed RT plus TMZ as first-line and subsequent bevacizumab-based second-line therapy [73]. In another clinical trial ascertaining clinical efficacy, nintedanib was not effective against recurrent high-grade gliomas, regardless of prior intervention with bevacizumab [74]. Dovitinib, an oral, potent inhibitor of FGFR and VEGFR, also failed to prolong the progression-free survival of patients with recurrent GBM, irrespective of prior AAT intervention with bevacizumab [75]. Another therapeutic agent, tivozanib, a pan-inhibitor of VEGF receptors, inhibited proliferation of GBM cells through a G2/M cell cycle arrest via inhibition of polo-like kinase 1 (PLK1) signaling pathway and downmodulation of Aurora kinases A and B, cyclin B1, and CDC25C. In the same study, the authors also reported that tivozanib decreased the adhesive potential of these cells through the reduction of intercellular adhesion molecule-1 (ICAM-1) and vascular cell adhesion molecule-1 (VCAM-1) and that a combination of tivozanib and EGFR small molecule-inhibitor gefitinib synergistically increased sensitivity to gefitinib [76].

Bevacizumab, a humanized monoclonal antibody against VEGF-165, the predominant isoform of VEGF-A was the first commercially successful antiangiogenic drug marketed [77] that received the approval of the Food and Drug Administration (FDA) in 2004. The combination of bevacizumab and cytotoxic chemotherapy was an effective antitumor regimen that prolonged progression-free survival (PFS) and reduced the need for corticosteroids [78]. Since targeting VEGF alone with bevacizumab failed to improve the overall survival, other combinatorial therapeutic regimens were employed in the treatment of GBM. A phase 2, double-blinded, placebo-controlled trial of bevacizumab plus trebananib (a novel Fc fusion protein that sequesters Ang1/Ang2) failed to improve 6-month PFS, PFS, and OS over bevacizumab alone in recurrent GBM [79]. Similarly, treatment with lomustine and bevacizumab prolonged PFS but did not confer OS advantage over monotherapies in patients with progressive GBM [80]. Combination therapy of vorinostat, bevacizumab, and TMZ also failed to improve the 6-month PFS (PFS6) beyond historical controls [81]. In patients with newly diagnosed GBM, a combination of bevacizumab with RT and TMZ is safe and feasible but yielded a low overall response, reflected in the unchanged median overall survival [82]. Rindopepimut, a vaccine targeting the tumor-specific EGF driver mutation, EGFRvIII, can be an effective addition to standard bevacizumab treatment to improve the outcome for patients with relapsed, EGFRvIII-positive GBM [83]. Despite the initial success, prolonged use of bevacizumab caused paradoxical resistance which led to deteriorated clinical outcomes and worse prognosis [84–86].

Mechanisms of AAT resistance and possible intervention

AAT works by disrupting the tumor neovasculatures inducing pathologically marked hypoxia characterized by HIF-1α-mediated upregulation of chemokine and growth factors [87–89]. Hypoxia induction produces protumorigenic and proangiogenic exosomal contents such as protein-lysine 6-oxidase (LOX), thrombospondin-1 (TSP1), VEGF, and a disintegrin and metalloproteinase with thrombospondin motifs 1 (ADAMTS1), which produce marked changes in normoxic tumor and surrounding stromal cells, thus significantly contributing to tumor progression and AAT resistance [90]. Vatalanib treatment also increased the expression of other proangiogenic molecules such as VEGF, SDF-1α, HIF-1alpha, FGF-1, FGF-2, Ephrin-A1, Ephrin-A2, Angiopoietin-1, VEGFR2, VEGFR3, and EGFR at the tumor invasive front [91]. Recent studies from our lab have also shown that vatalanib significantly increased CD68+ myeloid cells, and other endothelial cell signatures such as CD133+, CD34+, and Tie2+ in a mouse model of GBM [92]. The resultant increase in tumor size and paradoxical growth of the tumor caused by vatalanib treatment was controlled by inhibiting the mobilization of BMDCs and interaction of CXCR4-SDF-1 using whole-body irradiation and AMD3100, respectively [93]. The bevacizumab-resistant glioma-cell-derived xenograft model showed upregulation of c-MET [94]. Furthermore, c-MET potentiates glioma cell migration by forming a complex with β1-integrin extensively in bevacizumab-resistant GBM [95]. Bevacizumab treatment also upregulates the critical transcription factors of SNAI2 and ZEB [95]. Trebananib, an Ang1/2 inhibitor, failed to show an improved outcome in a recent phase II study in recurrent GBM, both as monotherapy and in combination with bevacizumab [96]. Potential mechanisms of bevacizumab-mediated therapy resistance also include an upregulated FGF signaling and increased PDGF-mediated pericyte coverage [74]. The vascular remodeling induced by anti-VEGF treatment leads to a potent hypoxia-inducible factor 1α induction and coactivation of the phosphatidyl inositol-3-kinase pathway (PI3K) favoring metabolic modifications in tumor cells toward glycolysis, thereby leading to enhanced tumor cell invasiveness into the normal brain [97]. A different anti-VEGF therapy resistance mechanism included the upregulation of PDGF-C (platelet-derived growth factor C) and c-Met expression and infiltration by myeloid cells in cediranib-treated GBM. In these recurrent GBM tumors, cediranib treatment facilitated a novel escape mechanism that was characterized by a lack of rebound angiogenesis [98]. Vessel cooption has also been shown to be an important contributor to therapy resistance during antiangiogenic treatment in preclinical GBM models [99]. In a detailed review, the different alternative pathways of neovascularization involved in the development of resistance to AAT have been explored [17]. One of the novel neovascularization mechanisms, VM, has been extensively studied in our laboratory [87, 88, 100].

Endothelial cells, bone marrow-derived progenitor cells, and angiogenic myeloid and vasculogenic cells following AAT

AAT produces significant changes not only in the tumor cells but also on the surrounding stromal cells as well. Bevacizumab treatment led to a refractory increase in the expression of Ang-2 on endothelial cells in GBM, thereby facilitating endothelial cell/myeloid cell crosstalk as a potential mechanism of treatment resistance. An interesting study investigated the effects of platelet releasate on GBM-derived endothelial cells (GECs) and demonstrated for the first time that platelet-derived VEGF might exert a powerful proangiogenic effect on GECs, prompting further studies into the contribution of platelets in anti-VEGF/VEGFR therapy [101]. Tumor-associated endothelial cell-derived CXCL12 exerts a strong trophic effect on the surrounding GBM cells through the tumor CXCR4/7 receptors and blocking the CXCL12-CXCR4/7 pathway can complement the other AAT regimens utilized for controlling tumor growth and burden [102]. Moreover, GBM tumor cell-endothelial cell hybrids were formed by fusion and may participate in tumor microvascular proliferation and treatment resistance [103]. Soda et al. have proposed the formation of tumor-derived endothelial cells as potential instigators of AAT resistance in GBM [104]. The combination of VEGF and Ang-2 inhibitors promoted survival, decreased vascular permeability, depleted CD206+ tumor-associated macrophages, improved pericyte coverage, and increased the numbers of intratumoral T lymphocytes in bevacizumab-resistant GBM [105]. Dual inhibition of Ang-2 and VEGF receptors led to enhanced morphological normalization of vessels and reprogramming of tumor-associated macrophages (TAMs) to reduce the tumor burden in preclinical GBM models [106]. Moreover, Kloepper et al. reported that utilization of a specific Ang-2/VEGF (CrossMab, A2V) preferentially reprogrammed the pro-tumor M2 (alternatively activated) macrophages to antitumor M1 macrophages in GBM [107]. Another potential neovascularization mechanism is through the recruitment of the endothelial progenitor cells to the U87 intracranial tumors via the formylpeptide receptor expressed on the tumor cells [108]. Surprisingly, GBM cells recruit bone marrow-derived progenitor cells (BMDC) in a spatiotemporal fashion and regulate regional contributions by associating with tumor vasculature in a VEGF-dependent manner. In the absence of VEGF, Ang-2 regulated the recruitment of BMDCs to maintain the vulnerable central vasculature, highlighting the need to target the influx of BMDCs to prevent pathological revascularization in GBM [109]. Tumor-associated endothelial cell-derived laminin α2 juxtracrine signaling promotes the self-renewal of the glioma stem cells (GSCs), thereby serving as a prime driver of GBM malignancy [110]. The discovery of CD45(dim)/CD34+/CD133+ circulating endothelial and hematopoietic progenitor cells (CPCs) in patients with GBM has also warranted further studies to verify the role of CPC levels in patients with brain tumors [110]. Rafat et al. have also shown that circulating endothelial progenitor cells harbor the potential of being utilized as biomarkers of GBM [111]. We have also observed decreased numbers of angiogenic MDSCs and vasculogenic leukocytes in tumors following treatment with HET0016 (a 20-HETE inhibitor) and SB225002 (a CXCR2 antagonist) (unpublished data). The postradiation influx of BMDCs, mediated by the HIF-1α-dependent stromal cell-derived factor-1 (SDF-1) and its receptor CXCR4 were the primary cause of tumor relapse. Pharmacologic inhibition of HIF-1 or SDF-1/CXCR4 interaction using a small molecule-inhibitor AMD3100 successfully abrogated tumor regrowth by attenuating vasculogenesis initiated by these BMDCs [112]. Studies from our lab also established that conditional deletion of canonical NF-κB signaling (p65) in myeloid cells inhibited syngeneic GBM through decreased CD45 infiltration in tumors by reducing the influx of tumor-associated macrophages (CD206+) and myeloid-derived suppressor cells (Gr1+ CD11b+), and an increased influx of dendritic cells (CD86+) and cytotoxic T cells (CD8+). This antitumor response was absent in p65KO athymic mice with human GBM, implying the critical role of T-cells in mediating this antitumor response [113]. Achyut et al. elegantly demonstrated that AAT increased the signatures of CD68+ myeloid cells and CD133+, CD34+, and Tie2+ endothelial cells in tumors, thereby promoting tumor resistance and relapse [92]. In further experiments, the authors employed a CSF1R inhibitor GW2580 that reduced tumor growth by decreasing myeloid (Gr1+ CD11b+ and F4/80+) and angiogenic (CD202b+ and VEGFR2+) cell signatures in the tumor microenvironment [92]. In addition, pretreatment with whole-body irradiation or the CXCR4 small molecule-inhibitor AMD3100 potentiated the antitumorigenic responses of vatalanib in an orthotopic rat glioma model [93]. Similarly, a synergistic anti-VEGF and anti-CXCR4 therapeutic regimen exerted antitumorigenic effects by inducing marked changes in tumor cell proliferation and apoptosis and also reduced the numbers of Nestin+ tumor-initiating cells [114]. Combination therapy of CSF1R inhibitor and CXCR2 inhibitor demonstrated strong antitumorigenic effects by blocking the tumor-infiltrating TAMs and MDSCs. Further, the CSF1R and CXCR2 dual inhibition enhanced the effects of immunotherapeutic responses of PD-1 antibody in preclinical mouse tumor models [115].

Transdifferentiation of GSCs and evolution of neovascular channels by VM

The heterogeneity and high degree of plasticity of the neural stem and progenitor cell pools have a multifaceted effect on the differential control on the growth of GBM. Depending on the location and degree of multipotency of these progenitor pools, the tumorigenicity of GBM presents with a complicated growth trajectory indicating the involvement of different cellular hierarchies underlying their generation [116, 117]. The development of novel methods to isolate and enrich these GSCs has enabled the cellular and genetic profiling of the pools of self-renewing, tumorigenic cells in primary gliomas and other primary brain and CNS tumors [118–120]. GSCs undergo a sequential transformation from epithelial to mesenchymal to endothelial phenotypes through an intermediate state resembling embryonic stem cells to form neovascular channels, an exemplary display of multipotency [45]. GSCs drive neovascularization through the upregulation of VEGF [121]. Moreover, the GSCs juxtapose the tumor-associated endothelial cells in the perivascular tumor niche [122] facilitating the engagement of the endothelial cell Notch ligands and GSC Notch receptors to promote GSC self-renewal [123].

The transdifferentiation of tumor cells into endothelial-like cells (termed tumor-derived endothelial cells, TDECs) was first described in a series of elegant experiments [104]. These TDECs described by Soda et al. comprise a subpopulation of specialized cells that have undergone a critical switch in the first step toward initiation and subsequent sustenance of VM-like channels in the GBM microenvironment. The multipotent CD133+CD144- cells obtained upon specialized isolation and enrichment procedures from the tumor mass generate specialized GSC-endothelial progenitors (CD133+CD144+) in the TME due to the tumor-secreted angiogenic factors and further accelerate VM in GBM tumors [124]. Previous studies from our lab and others have also reported the endothelial progenitor-like characteristics of CD133+ GSCs or glioma initiating cells [125–128]. In addition, hypoxia and metabolic stress in the tumor microenvironment cause the development of adaptive therapeutic resistance mechanisms in GSCs [129–133]. The employment of AATs and the subsequent hypoxia-induced upregulation of CD144 (VE-Cadherin) has been cited as one of the key factors driving VM in GBM [134]. GSCs with such prolific multipotency acquire elevated resistance phenotypes to several chemotherapeutic drugs such as TMZ, carboplatin, VP16, and Taxol compared to the non-GSC population [135]. Besides, the remainder of GBM cells following bevacizumab treatment becomes highly invasive possibly owing to the disinhibition of c-MET [136]. The multifaceted heterogeneity characteristic of GBM tumors underlies therapy resistance and the GSCs reinitiate and maintain tumor growth following treatment [137]. GSCs also contribute to radio-resistance by activating the DNA damage response machinery [121]. The presence of evolutionarily diverse GSC populations within the tumor highlights the importance of intratumoral heterogeneity in contributing to differential therapeutic responses [138]. The alteration [139] and abnormal activation of redundant receptor tyrosine kinases (RTKs) can induce tumor refractoriness to therapies [140]. The tumor heterogeneity is further compounded by the differential gene expression profiles controlling protumorigenic signaling, proliferation, immune responses, and hypoxia within the cells of the same tumor [141], thereby resulting in poor prognosis and a negative outcome on patient survival. The evolution of improvised design for rapid DNA repair aided by an elevated Notch, NF-κB, EZH2, and PARP-mediated survival machinery confers the resistance to therapies seen in GBM [121, 142–145] and necessitates the initiation of combination therapies to attenuate GBM growth and invasion. In light of these evidences, it becomes imperative to understand the contribution of GSCs toward tumor-associated neovascularization by focusing on the unexplored molecular signaling mechanisms to develop novel therapeutics to target tumor growth and metastasis.

The emergence of IL-CXCR2 pathway as a potential therapeutic target

The potent promigratory and proangiogenic functions of IL-8 in GBM have been well established [146–148]. An IL-8-CXCR1/2 signaling pathway is crucial in maintaining GBM stemness. CXCR1 and CXCR2 belong to class A, rhodopsin-like guanine-protein-coupled-receptors (GPCRs), and the genes encoding these two IL-8 cognate receptors are located in close proximity on chromosome 2 (2q34-35) [149–151]. The CXCR1 gene has a single intronic region flanked by two exons whereas the CXCR2 gene has three exons separated by two intronic segments [152, 153]. The possibility of evolutionary gene duplication as the derivation of these two IL-8 cognate receptors draws from the evidence of these two genes sharing a 77% amino acid homology and retaining common structural motifs [154]. However, the expression of both CXCR1 and CXCR2 exhibits strong contingencies and cell-to-cell type variations in expression patterns owing to pathogenicity and tumorigenicity [146]. Apart from the traditional chemotactic functions, IL-8 also facilitates glioma invasion through NF-κB pathway [155], thereby worsening the tumor grade in astrocytic neoplasms [156]. IL-8 has also been shown to increase the apoptotic threshold of ECs and stimulate the production of matrix metalloproteinases to remodel the ECM to accelerate pathological glioma neovascularization [146, 157]. The engagement of GBM-derived IL-8 and tumoral EC CXCR2 receptors induce the potentiation of endothelial cell permeability to facilitate pathological neovascularization [158]. Recently, Zhou et al. have demonstrated the role of neurotensin-dependent activation of a key transcription factor STAT3 through an IL8-CXCR1-mediated pathway in the preservation of GSCs [159]. The siRNA-mediated knockdown of CXCR2 completely abolished the hyperproliferative neoplasia inducing effects of IL-8 in the GSCs in vivo, thus establishing the singular contribution of EC-derived IL-8 rich tumoral perivascular niche in GSC growth, migration, and stemness through CXCR2 receptor [160]. The constitutively active mutant epidermal growth factor receptor, ΔEGFR, or EGFRvIII-mediated pathway converges onto IL8 through the NF-κB-activator protein 1 (AP-1)-CCAAT/enhancer-binding protein (C/EBP)-interleukin-8 pathway and serves a proangiogenic and protumorigenic role [161]. Previous studies from our lab have shown that intratumoral IL-8 has a crucial role in the GBM tumor growth [162] and that blocking IL-8 using HET0016 significantly reduced the tumor size and VM [88]. In light of this evidence, the IL-8-CXCR2 pathway serves as a viable target to control tumor-associated neovascularization and growth. However, the pleiotropic effects of the IL8 coupled with the presence and coregulation of other cytokines such as CXCL1/GRO-α, CXCL2/GRO-β, CXCL3/GRO-γ, CXCL5 [163] and CXCL7 [164] mediating CXCR1/2 signaling limit the efficacy and clinical translation of IL8 ligand-targeted therapies. The development of therapy-induced resistance in glioma has a serious and bleak outcome on the progression free survival (PFS) and overall survival (OS) of patients with GBM. To counter these therapeutic limitations of ligand-based therapies, CXCR1/2 receptor-based interventions can be developed to have improved and targeted therapeutic utility.

Our lab has recently demonstrated the presence of increased numbers of CXCR2+ GSCs and specialized subpopulations with an endothelial-like phenotype following AAT treatment [100]. The schematic shown in Fig. 3 is adapted from Angara et al. [100]. In addition, orthotopic and subcutaneous models of GBM hone the tumorigenic and vascular potential of these CXCR2+ glioma cells to form vascular laminin-positive VM channels as shown in Fig. 4, adapted from Angara et al. [100]. It is a well-known fact now that VM in gliomas is

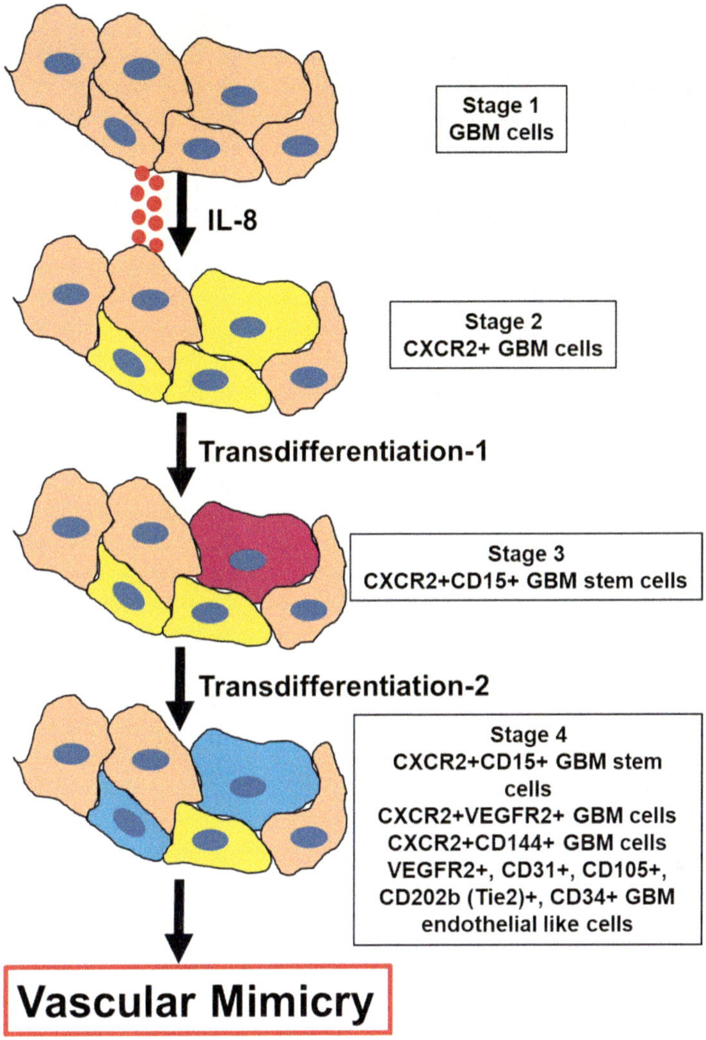

FIG. 3 Schematic of endothelial phenotype acquisition by GBM cells.

associated with extremely poor prognosis and positively correlates with high grades of tumor malignancy and invasiveness [165, 166]. Therefore, utilizing CXCR2-targeting drugs and including them in the therapeutic regimen to attenuate GBM growth in combination with other chemotherapeutic drugs might be a valid option to counter therapy resistance.

Numerous preclinical studies have highlighted that targeting CXCR2 elicits a potential antitumorigenic response in different forms of cancers. A synergistic antitumoral response was elicited using a CXCR2 inhibitor in combination with sorafenib in an ovarian cancer model [167]. Orally active small-molecule inhibitors of CXCR2, SCH-479833, or SCH-527123 demonstrated significant antitumorigenic potential in addition to attenuating chemotaxis, invasion, and microvessel density [168]. Singh et al. in subsequent studies have also

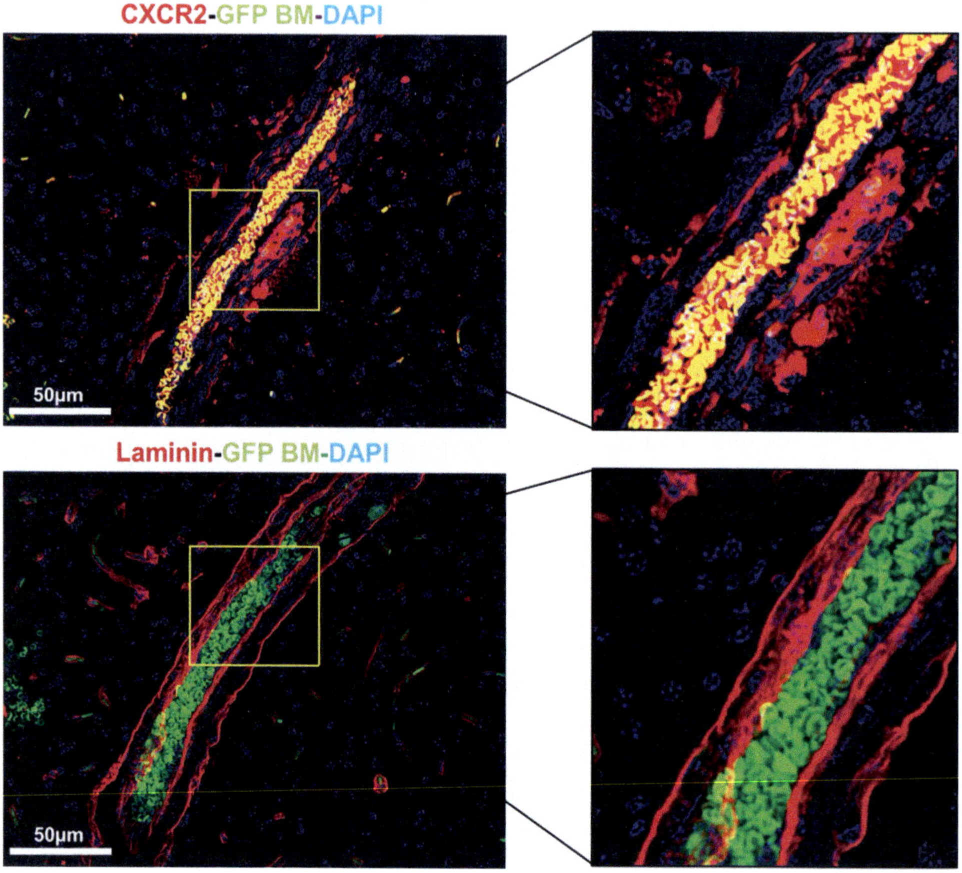

FIG. 4 CXCR2+ GBM cells form laminin-positive VM structures in vatalanib-treated tumors.

neatly demonstrated that shRNA-mediated CXCR2 knockdown in A375-SM melanoma cells attenuated cell proliferation, migration, and invasion in vitro and further decreased the tumor growth and tumor-associated vasculature in athymic nude mice [168–170]. CXCR2 mediates ELR+ CXC chemokine-dependent angiogenesis and attenuating the CXCR2 receptor inhibits tumor growth and angiogenesis in a preclinical model of lung cancer [171]. Downregulating the expression of CXCR2 in KRAS(G12D) mutation-bearing PDAC cells inhibited tumor cell proliferation and altered KRAS protein levels both in vitro and in vivo [172]. The IL-8-CXCR2 axis also plays a singular role in invasion, angiogenesis, and metastasis of liver cancer [173]. shRNA-mediated knockdown of CXCR2 in metastatic murine mammary tumor cells (C166 and 4T1) attenuated cell invasion without affecting cell proliferation. Orthotopic implantation of these CXCR2 knockdown tumor cells showed a reduction in spontaneous lung metastasis by 40% compared to control cells and displayed enhanced cytotoxicity to doxorubicin and paclitaxel in both in vitro and in vivo models [174, 175].

CXCR2 signaling is strongly activated in the GBM microenvironment, thereby promoting rebound tumor neovascularization, invasion, and tumor recurrence. The negative outcomes of overactive CXCR2 signaling in the GBM microenvironment, which include pathological tumor-associated neovascularization and subsequent tumor recurrence, have been blocked using antibodies against CXCR2 or chemical inhibitors to suppress glioma growth and cell migration [176, 177]. CXCR2 inhibition suppressed GBM growth and cell migration [176, 177]. The tube-forming capability of HUVEC cells was enhanced using coculture with these transformed tumor cells and inhibited with the CXCR2 antagonist SB225002 [178]. CXCR2-NHERF1-PLCβ3 complex regulates CXCR2 signaling activity and is crucial in tumor progression and invasion [179]. Our lab has reported some interesting findings of an improved therapeutic response following treatment with HET0016 (anti-IL8) and SB225002, a selective CXCR2 inhibitor [88, 100]. We employed both orthotopic and subcutaneous models of GBM and observed decreased tumor size, the overall reduction in the numbers of CXCR2 + GSC, and other endothelial cell-like subpopulations and decreased VM formation after anti-CXCR2 treatment (both HET0016 and SB225002). In addition, the CXCR2 inhibitor disrupted the tube-like structures formed by GBM tumor cells in an in vitro matrigel assay [100]. The results of the study have been summarized schematically in Fig. 5, adapted from Angara et al. [100].

These findings conclusively show that attenuating the IL-8-CXCR2 signaling axis by targeting the CXCR2 receptor in both tumor cells and the tumor-infiltrating myeloid cells is a compelling reason for either repurposing the already approved FDA drugs (CXCR2 inhibitors) or the development of novel, small-molecule inhibitors of CXCR2 to counter therapy resistance in tumors. Intervening on the IL-8-CXCR2 axis provides the dual targeting of both the tumor cells and the immune cells in the tumor microenvironment, which contributes to tumor refractoriness to therapies.

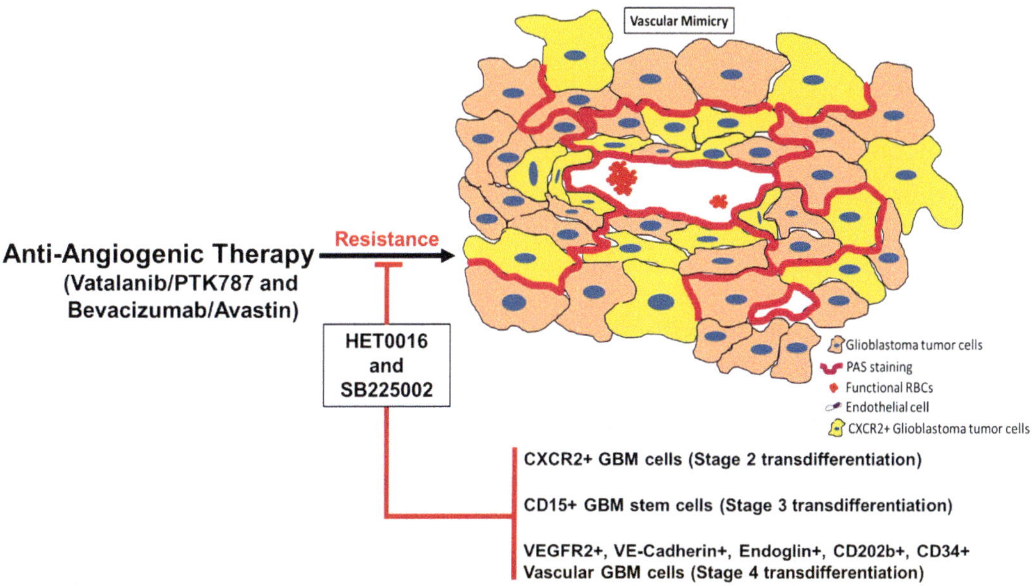

FIG. 5 Proposed schematic representation of the acquisition of VM-mediated resistance to AAT and intervention of the IL-8-CXCR2 axis using HET0016 and SB225002.

References

[1] Agnihotri S, Burrell KE, Wolf A, Jalali S, Hawkins C, Rutka JT, Zadeh G. Glioblastoma, a brief review of history, molecular genetics, animal models and novel therapeutic strategies. Arch Immunol Ther Exp 2013;61(1):25–41.

[2] Holland EC, Celestino J, Dai C, Schaefer L, Sawaya RE, Fuller GN. Combined activation of Ras and Akt in neural progenitors induces glioblastoma formation in mice. Nat Genet 2000;25(1):55–7.

[3] Schwartzbaum JA, Fisher JL, Aldape KD, Wrensch M. Epidemiology and molecular pathology of glioma. Nat Clin Pract Neurol 2006;2(9):494–503. quiz 1 p following 516.

[4] Jovčevska I, Muyldermans S. The therapeutic potential of nanobodies. BioDrugs 2020;34(1):11–26.

[5] Louis DN, Ohgaki H, Wiestler OD, Cavenee WK, Burger PC, Jouvet A, Scheithauer BW, Kleihues P. The 2007 WHO classification of tumours of the central nervous system. Acta Neuropathol 2007;114(2):97–109.

[6] Koshy M, Villano JL, Dolecek TA, Howard A, Mahmood U, Chmura SJ, Weichselbaum RR, McCarthy BJ. Improved survival time trends for glioblastoma using the SEER 17 population-based registries. J Neuro-Oncol 2012;107(1):207–12.

[7] Tran B, Rosenthal MA. Survival comparison between glioblastoma multiforme and other incurable cancers. J Clin Neurosci 2010;17(4):417–21.

[8] Stupp R, Weber DC. The role of radio- and chemotherapy in glioblastoma. Onkologie 2005;28(6–7):315–7.

[9] Lacroix M, Abi-Said D, Fourney DR, Gokaslan ZL, Shi W, DeMonte F, Lang FF, McCutcheon IE, Hassenbusch SJ, Holland E, Hess K, Michael C, Miller D, Sawaya R. A multivariate analysis of 416 patients with glioblastoma multiforme: prognosis, extent of resection, and survival. J Neurosurg 2001;95(2):190–8.

[10] Newton HB, Dalton J, Goldlust S, Pearl D. Retrospective analysis of the efficacy and tolerability of levetiracetam in patients with metastatic brain tumors. J Neuro-Oncol 2007;84(3):293–6.

[11] Stummer W, Reulen HJ, Meinel T, Pichlmeier U, Schumacher W, Tonn JC, Rohde V, Oppel F, Turowski B, Woiciechowsky C, Franz K, Pietsch T, ALA-Glioma Study Group. Extent of resection and survival in glioblastoma multiforme: identification of and adjustment for bias. Neurosurgery 2008;62(3):564–76. discussion 564–76.

[12] Malmstrom A, Gronberg BH, Marosi C, Stupp R, Frappaz D, Schultz H, Abacioglu U, Tavelin B, Lhermitte B, Hegi ME, Rosell J, Henriksson R, Nordic Clinical Brain Tumour Study Group (NCBTSG). Temozolomide versus standard 6-week radiotherapy versus hypofractionated radiotherapy in patients older than 60 years with glioblastoma: the Nordic randomised, phase 3 trial. Lancet Oncol 2012;13(9):916–26.

[13] Percy AK, Elveback LR, Okazaki H, Kurland LT. Neoplasms of the central nervous system. Epidemiologic considerations. Neurology 1972;22(1):40–8.

[14] de Vries NA, Beijnen JH, Boogerd W, van Tellingen O. Blood-brain barrier and chemotherapeutic treatment of brain tumors. Expert Rev Neurother 2006;6(8):1199–209.

[15] Jain RK, di Tomaso E, Duda DG, Loeffler JS, Sorensen AG, Batchelor TT. Angiogenesis in brain tumours. Nat Rev Neurosci 2007;8(8):610–22.

[16] Bergers G, Benjamin LE. Tumorigenesis and the angiogenic switch. Nat Rev Cancer 2003;3(6):401–10.

[17] Hardee ME, Zagzag D. Mechanisms of glioma-associated neovascularization. Am J Pathol 2012;181(4):1126–41.

[18] Peters BA, Diaz LA, Polyak K, Meszler L, Romans K, Guinan EC, Antin JH, Myerson D, Hamilton SR, Vogelstein B, Kinzler KW, Lengauer C. Contribution of bone marrow-derived endothelial cells to human tumor vasculature. Nat Med 2005;11(3):261–2.

[19] Purhonen S, Palm J, Rossi D, Kaskenpaa N, Rajantie I, Yla-Herttuala S, Alitalo K, Weissman IL, Salven P. Bone marrow-derived circulating endothelial precursors do not contribute to vascular endothelium and are not needed for tumor growth. Proc Natl Acad Sci U S A 2008;105(18):6620–5.

[20] De Palma M, Venneri MA, Roca C, Naldini L. Targeting exogenous genes to tumor angiogenesis by transplantation of genetically modified hematopoietic stem cells. Nat Med 2003;9(6):789–95.

[21] Li B, Sharpe EE, Maupin AB, Teleron AA, Pyle AL, Carmeliet P, Young PP. VEGF and PlGF promote adult vasculogenesis by enhancing EPC recruitment and vessel formation at the site of tumor neovascularization. FASEB J 2006;20(9):1495–7.

[22] Udani V, Santarelli J, Yung Y, Cheshier S, Andrews A, Kasad Z, Tse V. Differential expression of angiopoietin-1 and angiopoietin-2 may enhance recruitment of bone-marrow-derived endothelial precursor cells into brain tumors. Neurol Res 2005;27(8):801–6.

[23] Folkman J. Tumor angiogenesis: therapeutic implications. N Engl J Med 1971;285(21):1182–6.

[24] Hanahan D, Folkman J. Patterns and emerging mechanisms of the angiogenic switch during tumorigenesis. Cell 1996;86(3):353–64.

[25] Yancopoulos GD, Davis S, Gale NW, Rudge JS, Wiegand SJ, Holash J. Vascular-specific growth factors and blood vessel formation. Nature 2000;407(6801):242–8.
[26] Holash J, Maisonpierre PC, Compton D, Boland P, Alexander CR, Zagzag D, Yancopoulos GD, Wiegand SJ. Vessel cooption, regression, and growth in tumors mediated by angiopoietins and VEGF. Science 1999;284 (5422):1994–8.
[27] Brat DJ, Castellano-Sanchez AA, Hunter SB, Pecot M, Cohen C, Hammond EH, Devi SN, Kaur B, Van Meir EG. Pseudopalisades in glioblastoma are hypoxic, express extracellular matrix proteases, and are formed by an actively migrating cell population. Cancer Res 2004;64(3):920–7.
[28] Forsyth PA, Wong H, Laing TD, Rewcastle NB, Morris DG, Muzik H, Leco KJ, Johnston RN, Brasher PM, Sutherland G, Edwards DR. Gelatinase-A (MMP-2), gelatinase-B (MMP-9) and membrane type matrix metalloproteinase-1 (MT1-MMP) are involved in different aspects of the pathophysiology of malignant gliomas. Br J Cancer 1999;79(11–12):1828–35.
[29] Forsythe JA, Jiang BH, Iyer NV, Agani F, Leung SW, Koos RD, Semenza GL. Activation of vascular endothelial growth factor gene transcription by hypoxia-inducible factor 1. Mol Cell Biol 1996;16(9):4604–13.
[30] Ahmad SA, Liu W, Jung YD, Fan F, Wilson M, Reinmuth N, Shaheen RM, Bucana CD, Ellis LM. The effects of angiopoietin-1 and -2 on tumor growth and angiogenesis in human colon cancer. Cancer Res 2001;61(4):1255–9.
[31] Hu B, Guo P, Fang Q, Tao H-Q, Wang D, Nagane M, Huang H-JS, Gunji Y, Nishikawa R, Alitalo K, Cavenee WK, Cheng S-Y. Angiopoietin-2 induces human glioma invasion through the activation of matrix metalloprotease-2. Proc Natl Acad Sci U S A 2003;100(15):8904–9.
[32] Vajkoczy P, Farhadi M, Gaumann A, Heidenreich R, Erber R, Wunder A, Tonn JC, Menger MD, Breier G. Microtumor growth initiates angiogenic sprouting with simultaneous expression of VEGF, VEGF receptor-2, and angiopoietin-2. J Clin Invest 2002;109(6):777–85.
[33] Yu Q, Stamenkovic I. Angiopoietin-2 is implicated in the regulation of tumor angiogenesis. Am J Pathol 2001;158(2):563–70.
[34] Tanaka S, Mori M, Sakamoto Y, Makuuchi M, Sugimachi K, Wands JR. Biologic significance of angiopoietin-2 expression in human hepatocellular carcinoma. J Clin Invest 1999;103(3):341–5.
[35] Friedlander M, Brooks PC, Shaffer RW, Kincaid CM, Varner JA, Cheresh DA. Definition of two angiogenic pathways by distinct alpha v integrins. Science (New York, NY) 1995;270(5241):1500–2.
[36] Maniotis AJ, Folberg R, Hess A, Seftor EA, Gardner LM, Pe'er J, Trent JM, Meltzer PS, Hendrix MJ. Vascular channel formation by human melanoma cells in vivo and in vitro: vasculogenic mimicry. Am J Pathol 1999;155 (3):739–52.
[37] Shirakawa K, Tsuda H, Heike Y, Kato K, Asada R, Inomata M, Sasaki H, Kasumi F, Yoshimoto M, Iwanaga T, Konishi F, Terada M, Wakasugi H. Absence of endothelial cells, central necrosis, and fibrosis are associated with aggressive inflammatory breast cancer. Cancer Res 2001;61(2):445–51.
[38] Sood AK, Fletcher MS, Zahn CM, Gruman LM, Coffin JE, Seftor EA, Hendrix MJC. The clinical significance of tumor cell-lined vasculature in ovarian carcinoma: implications for anti-vasculogenic therapy. Cancer Biol Ther 2002;1(6):661–4.
[39] Sharma N, Seftor REB, Seftor EA, Gruman LM, Heidger PM, Cohen MB, Lubaroff DM, Hendrix MJC. Prostatic tumor cell plasticity involves cooperative interactions of distinct phenotypic subpopulations: role in vasculogenic mimicry. Prostate 2002;50(3):189–201.
[40] Passalidou E, Trivella M, Singh N, Ferguson M, Hu J, Cesario A, Granone P, Nicholson AG, Goldstraw P, Ratcliffe C, Tetlow M, Leigh I, Harris AL, Gatter KC, Pezzella F. Vascular phenotype in angiogenic and non-angiogenic lung non-small cell carcinomas. Br J Cancer 2002;86(2):244–9.
[41] Yue W-Y, Chen Z-P. Does vasculogenic mimicry exist in astrocytoma? J Histochem Cytochem 2005;53(8):997–1002.
[42] Fausto N. Vasculogenic mimicry in tumors. Fact or artifact? Am J Pathol 2000;156(2):359.
[43] McDonald DM, Munn L, Jain RK. Vasculogenic mimicry: how convincing, how novel, and how significant? Am J Pathol 2000;156(2):383–8.
[44] Clemente M, Perez-Alenza MD, Illera JC, Pena L. Histological, immunohistological, and ultrastructural description of vasculogenic mimicry in canine mammary cancer. Vet Pathol 2010;47(2):265–74.
[45] Folberg R, Maniotis AJ. Vasculogenic mimicry. APMIS 2004;112(7–8):508–25.
[46] Paulis YW, Soetekouw PM, Verheul HM, Tjan-Heijnen VC, Griffioen AW. Signalling pathways in vasculogenic mimicry. Biochim Biophys Acta 2010;1806(1):18–28.

[47] Chen X, Maniotis AJ, Majumdar D, Pe'er J, Folberg R. Uveal melanoma cell staining for CD34 and assessment of tumor vascularity. Invest Ophthalmol Vis Sci 2002;43(8):2533–9.
[48] Clarijs R, Otte-Holler I, Ruiter DJ, de Waal RM. Presence of a fluid-conducting meshwork in xenografted cutaneous and primary human uveal melanoma. Invest Ophthalmol Vis Sci 2002;43(4):912–8.
[49] Seftor RE, Seftor EA, Koshikawa N, Meltzer PS, Gardner LM, Bilban M, Stetler-Stevenson WG, Quaranta V, Hendrix MJ. Cooperative interactions of laminin 5 gamma2 chain, matrix metalloproteinase-2, and membrane type-1-matrix/metalloproteinase are required for mimicry of embryonic vasculogenesis by aggressive melanoma. Cancer Res 2001;61(17):6322–7.
[50] Folberg R, Rummelt V, Parys-Van Ginderdeuren R, Hwang T, Woolson RF, Pe'er J, Gruman LM. The prognostic value of tumor blood vessel morphology in primary uveal melanoma. Ophthalmology 1993;100(9):1389–98.
[51] Remer S, Murphy ME. The challenges of long-term treatment outcomes in adults with malignant gliomas. Clin J Oncol Nurs 2004;8(4):368–76.
[52] Dhermain F, Ducreux D, Bidault F, Bruna A, Parker F, Roujeau T, Beaudre A, Armand JP, Haie-Meder C. Use of the functional imaging modalities in radiation therapy treatment planning in patients with glioblastoma. Bull Cancer 2005;92(4):333–42.
[53] Folkman J, Long DM, Becker FF. Growth and metastasis of tumor in organ culture. Cancer 1963;16:453–67.
[54] Hanahan D, Weinberg RA. Hallmarks of cancer: the next generation. Cell 2011;144(5):646–74.
[55] Dietrich J, Norden AD, Wen PY. Emerging antiangiogenic treatments for gliomas – efficacy and safety issues. Curr Opin Neurol 2008;21(6):736–44.
[56] Rahman R, Smith S, Rahman C, Grundy R. Antiangiogenic therapy and mechanisms of tumor resistance in malignant glioma. J Oncol 2010;2010:251231.
[57] Saharinen P, Eklund L, Pulkki K, Bono P, Alitalo K. VEGF and angiopoietin signaling in tumor angiogenesis and metastasis. Trends Mol Med 2011;17(7):347–62.
[58] Samples J, Willis M, Klauber-DeMore N. Targeting angiogenesis and the tumor microenvironment. Surg Oncol Clin N Am 2013;22(4):629.
[59] Bruno A, Pagani A, Magnani E, Rossi T, Noonan DM, Cantelmo AR, Albini A. Inflammatory angiogenesis and the tumor microenvironment as targets for cancer therapy and prevention. Cancer Treat Res 2014;159:401–26.
[60] Mittal K, Ebos J, Rini B. Angiogenesis and the tumor microenvironment: vascular endothelial growth factor and beyond. Semin Oncol 2014;41(2):235–51.
[61] Miller KD, Sweeney CJ, Sledge Jr GW. Can tumor angiogenesis be inhibited without resistance? EXS 2005;94:95–112.
[62] Ali MM, Kumar S, Shankar A, Varma NRS, Iskander ASM, Janic B, Chwang WB, Jain R, Babajeni-Feremi A, Borin TF, Bagher-Ebadian H, Brown SL, Ewing JR, Arbab AS. Effects of tyrosine kinase inhibitors and CXCR4 antagonist on tumor growth and angiogenesis in rat glioma model: MRI and protein analysis study. Transl Oncol 2013;6(6):660–9.
[63] Kamoun WS, Ley CD, Farrar CT, Duyverman AM, Lahdenranta J, Lacorre DA, Batchelor TT, di Tomaso E, Duda DG, Munn LL, Fukumura D, Sorensen AG, Jain RK. Edema control by cediranib, a vascular endothelial growth factor receptor-targeted kinase inhibitor, prolongs survival despite persistent brain tumor growth in mice. J Clin Oncol Off J Am Soc Clin Oncol 2009;27(15):2542–52.
[64] Batchelor TT, Duda DG, di Tomaso E, Ancukiewicz M, Plotkin SR, Gerstner E, Eichler AF, Drappatz J, Hochberg FH, Benner T, Louis DN, Cohen KS, Chea H, Exarhopoulos A, Loeffler JS, Moses MA, Ivy P, Sorensen AG, Wen PY, Jain RK. Phase II study of cediranib, an oral pan-vascular endothelial growth factor receptor tyrosine kinase inhibitor, in patients with recurrent glioblastoma. J Clin Oncol 2010;28(17):2817–23.
[65] Gerstner ER, Chen P-J, Wen PY, Jain RK, Batchelor TT, Sorensen G. Infiltrative patterns of glioblastoma spread detected via diffusion MRI after treatment with cediranib. Neuro-Oncology 2010;12(5):466–72.
[66] Neyns B, Sadones J, Chaskis C, Dujardin M, Everaert H, Lv S, Duerinck J, Tynninen O, Nupponen N, Michotte A, De Greve J. Phase II study of sunitinib malate in patients with recurrent high-grade glioma. J Neuro-Oncol 2011;103(3):491–501.
[67] Reardon DA, Vredenburgh JJ, Coan A, Desjardins A, Peters KB, Gururangan S, Sathornsumetee S, Rich JN, Herndon JE, Friedman HS. Phase I study of sunitinib and irinotecan for patients with recurrent malignant glioma. J Neurooncol 2011;105(3):621–7. https://doi.org/10.1007/s11060-011-0631-4.
[68] Reardon DA, Dresemann G, Taillibert S, Campone M, van den Bent M, Clement P, Blomquist E, Gordower L, Schultz H, Raizer J, Hau P, Easaw J, Gil M, Tonn J, Gijtenbeek A, Schlegel U, Bergstrom P, Green S, Weir A,

Nikolova Z. Multicentre phase II studies evaluating imatinib plus hydroxyurea in patients with progressive glioblastoma. Br J Cancer 2009;101(12):1995–2004.

[69] Reardon DA, Egorin MJ, Desjardins A, Vredenburgh JJ, Beumer JH, Lagattuta TF, Gururangan S, Herndon 2nd JE, Salvado AJ, Friedman HS. Phase I pharmacokinetic study of the vascular endothelial growth factor receptor tyrosine kinase inhibitor vatalanib (PTK787) plus imatinib and hydroxyurea for malignant glioma. Cancer 2009;115(10):2188–98.

[70] Lieu C, Heymach J, Overman M, Tran H, Kopetz S. Beyond VEGF: inhibition of the fibroblast growth factor pathway and antiangiogenesis. Clin Cancer Res 2011;17(19):6130–9.

[71] Conrad C, Friedman H, Reardon D, Provenzale J, Jackson E, Serajuddin H, Laurent D, Chen B, Yung WKA. A phase I/II trial of single-agent PTK 787/ZK 222584 (PTK/ZK), a novel, oral angiogenesis inhibitor, in patients with recurrent glioblastoma multiforme (GBM). J Clin Oncol 2004;22(14_suppl):1512.

[72] Reardon DA, Friedman HS, Yung WKA, Brada M, Conrad C, Provenzale J, Jackson E, Serajuddin H, Chen B, Laurent D. A phase I/II trial of PTK787/ZK 222584 (PTK/ZK), a novel, oral angiogenesis inhibitor, in combination with either temozolomide or lomustine for patients with recurrent glioblastoma multiforme (GBM). J Clin Oncol 2004;22(14 suppl):1513.

[73] Muhic A, Poulsen HS, Sorensen M, Grunnet K, Lassen U. Phase II open-label study of nintedanib in patients with recurrent glioblastoma multiforme. J Neuro-Oncol 2013;111(2):205–12.

[74] Norden AD, Schiff D, Ahluwalia MS, Lesser GJ, Nayak L, Lee EQ, Rinne ML, Muzikansky A, Dietrich J, Purow B, Doherty LM, LaFrankie DC, Pulverenti JR, Rifenburg JA, Ruland SF, Smith KH, Gaffey SC, McCluskey C, Ligon KL, Reardon DA, Wen PY. Phase II trial of triple tyrosine kinase receptor inhibitor nintedanib in recurrent high-grade gliomas. J Neuro-Oncol 2015;121(2):297–302.

[75] Sharma M, Schilero C, Peereboom DM, Hobbs BP, Elson P, Stevens GHJ, McCrae K, Nixon AB, Ahluwalia MS. Phase II study of Dovitinib in recurrent glioblastoma. J Neuro-Oncol 2019;144(2):359–68.

[76] Momeny M, Moghaddaskho F, Gortany NK, Yousefi H, Sabourinejad Z, Zarrinrad G, Mirshahvaladi S, Eyvani H, Barghi F, Ahmadinia L, Ghazi-Khansari M, Dehpour AR, Amanpour S, Tavangar SM, Dardaei L, Emami AH, Alimoghaddam K, Ghavamzadeh A, Ghaffari SH. Blockade of vascular endothelial growth factor receptors by tivozanib has potential anti-tumour effects on human glioblastoma cells. Sci Rep 2017;7:44075.

[77] Rosen LS. Clinical experience with angiogenesis signaling inhibitors: focus on vascular endothelial growth factor (VEGF) blockers. Cancer Control 2002;9(2 Suppl):36–44.

[78] Norden AD, Young GS, Setayesh K, Muzikansky A, Klufas R, Ross GL, Ciampa AS, Ebbeling LG, Levy B, Drappatz J, Kesari S, Wen PY. Bevacizumab for recurrent malignant gliomas: efficacy, toxicity, and patterns of recurrence. Neurology 2008;70(10):779–87.

[79] Lee EQ, Zhang P, Wen PY, Gerstner ER, Reardon DA, Aldape KD, deGroot JF, Pan E, Raizer JJ, Kim LJ, Chmura SJ, Robins HI, Connelly JM, Battiste JD, Villano JL, Wagle N, Merrell RT, Wendland MM, Mehta MP. NRG/RTOG 1122: A phase 2, double-blinded, placebo-controlled study of bevacizumab with and without trebananib in patients with recurrent glioblastoma or gliosarcoma. Cancer 2020;126(12):2821–8. https://doi.org/10.1002/cncr.32811.

[80] Song J, Xue Y-Q, Zhao M-M, Xu P. Effectiveness of lomustine and bevacizumab in progressive glioblastoma: a meta-analysis. OncoTargets Ther 2018;11:3435–9.

[81] Peters KB, Lipp ES, Miller E, Herndon 2nd JE, McSherry F, Desjardins A, Reardon DA, Friedman HS. Phase I/II trial of vorinostat, bevacizumab, and daily temozolomide for recurrent malignant gliomas. J Neuro-Oncol 2018;137(2):349–56.

[82] van Linde ME, Verhoeff JJC, Richel DJ, van Furth WR, Reijneveld JC, Verheul HMW, Stalpers LJA. Bevacizumab in combination with radiotherapy and temozolomide for patients with newly diagnosed glioblastoma multiforme. Oncologist 2015;20(2):107–8.

[83] Reardon DA, Desjardins A, Vredenburgh JJ, O'Rourke DM, Tran DD, Fink KL, Nabors LB, Li G, Bota DA, Lukas RV, Ashby LS, Duic JP, Mrugala MM, Cruickshank S, Vitale L, He Y, Green JA, Yellin MJ, Turner CD, Keler T, Davis TA, Sampson JH, ReACT Trial Investigators. Rindopepimut with bevacizumab for patients with relapsed EGFRvIII-expressing glioblastoma (ReACT): results of a double-blind randomized phase II trial. Clin Cancer Res 2020;26(7):1586–94.

[84] Norden AD, Drappatz J, Wen PY. Antiangiogenic therapy in malignant gliomas. Curr Opin Oncol 2008;20(6):652–61.

[85] Kreisl TN, Zhang W, Odia Y, Shih JH, Butman JA, Hammoud D, Iwamoto FM, Sul J, Fine HA. A phase II trial of single-agent bevacizumab in patients with recurrent anaplastic glioma. Neuro-Oncology 2011;13(10):1143–50.

[86] Reardon DA, Desjardins A, Peters KB, Gururangan S, Sampson JH, McLendon RE, Herndon 2nd JE, Bulusu A, Threatt S, Friedman AH, Vredenburgh JJ, Friedman HS. Phase II study of carboplatin, irinotecan, and bevacizumab for bevacizumab naive, recurrent glioblastoma. J Neuro-Oncol 2012;107(1):155–64.
[87] Angara K, Borin TF, Arbab AS. Vascular mimicry: a novel neovascularization mechanism driving anti-angiogenic therapy (AAT) resistance in glioblastoma. Transl Oncol 2017;10(4):650–60.
[88] Angara K, Rashid MH, Shankar A, Ara R, Iskander A, Borin TF, Jain M, Achyut BR, Arbab AS. Vascular mimicry in glioblastoma following anti-angiogenic and anti-20-HETE therapies. Histol Histopathol 2017;32(9):917–28.
[89] Arbab AS, Rashid MH, Angara K, Borin TF, Lin P-C, Jain M, Achyut BR. Major challenges and potential microenvironment-targeted therapies in glioblastoma. Int J Mol Sci 2017;18:2732. https://doi.org/10.3390/ijms18122732.
[90] Kore RA, Edmondson JL, Jenkins SV, Jamshidi-Parsian A, Dings RPM, Reyna NS, Griffin RJ. Hypoxia-derived exosomes induce putative altered pathways in biosynthesis and ion regulatory channels in glioblastoma cells. Biochem Biophys Rep 2018;14:104–13.
[91] Ali MM, Janic B, Babajani-Feremi A, Varma NRS, Iskander ASM, Anagli J, Arbab AS. Changes in vascular permeability and expression of different angiogenic factors following anti-angiogenic treatment in rat glioma. PLoS One 2010;5(1), e8727.
[92] Achyut BR, Shankar A, Iskander ASM, Ara R, Angara K, Zeng P, Knight RA, Scicli AG, Arbab AS. Bone marrow derived myeloid cells orchestrate antiangiogenic resistance in glioblastoma through coordinated molecular networks. Cancer Lett 2015;369(2):416–26.
[93] Shaaban S, Alsulami M, Arbab SA, Ara R, Shankar A, Iskander A, Angara K, Jain M, Bagher-Ebadian H, Achyut BR, Arbab AS. Targeting bone marrow to potentiate the anti-tumor effect of tyrosine kinase inhibitor in preclinical rat model of human glioblastoma. Int J Cancer Res 2016;12(2):69–81.
[94] Jahangiri A, De Lay M, Miller LM, Carbonell WS, Hu Y-L, Lu K, Tom MW, Paquette J, Tokuyasu TA, Tsao S, Marshall R, Perry A, Bjorgan KM, Chaumeil MM, Ronen SM, Bergers G, Aghi MK. Gene expression profile identifies tyrosine kinase c-Met as a targetable mediator of antiangiogenic therapy resistance. Clin Cancer Res 2013;19(7):1773–83.
[95] Jahangiri A, Chen W, Yagnik G, De Lay M, Wagner J, Sidorov M, Flanigan PM, Aghi MK. 333 Obtaining the genetic fingerprint of resistance to glioblastoma through a novel multigenerational xenograft model. Neurosurgery 2016;63(CN_suppl_1):197.
[96] Reardon DA, Lassman AB, Schiff D, Yunus SA, Gerstner ER, Cloughesy TF, Lee EQ, Gaffey SC, Barrs J, Bruno J, Muzikansky A, Duda DG, Jain RK, Wen PY. Phase 2 and biomarker study of trebananib, an angiopoietin-blocking peptibody, with and without bevacizumab for patients with recurrent glioblastoma. Cancer 2018;124(7):1438–48.
[97] Keunen O, Johansson M, Oudin A, Sanzey M, Rahim SA, Fack F, Thorsen F, Taxt T, Bartos M, Jirik R, Miletic H, Wang J, Stieber D, Stuhr L, Moen I, Rygh CB, Bjerkvig R, Niclou SP. Anti-VEGF treatment reduces blood supply and increases tumor cell invasion in glioblastoma. Proc Natl Acad Sci U S A 2011;108(9):3749–54.
[98] di Tomaso E, Snuderl M, Kamoun WS, Duda DG, Auluck PK, Fazlollahi L, Andronesi OC, Frosch MP, Wen PY, Plotkin SR, Hedley-Whyte ET, Sorensen AG, Batchelor TT, Jain RK. Glioblastoma recurrence after cediranib therapy in patients: lack of "rebound" revascularization as mode of escape. Cancer Res 2011;71(1):19–28.
[99] Rubenstein JL, Kim J, Ozawa T, Zhang M, Westphal M, Deen DF, Shuman MA. Anti-VEGF antibody treatment of glioblastoma prolongs survival but results in increased vascular cooption. Neoplasia (New York, NY) 2000;2(4):306–14.
[100] Angara K, Borin TF, Rashid MH, Lebedyeva I, Ara R, Lin P-C, Iskander A, Bollag RJ, Achyut BR, Arbab AS. CXCR2-expressing tumor cells drive vascular mimicry in antiangiogenic therapy-resistant glioblastoma. Neoplasia (New York, NY) 2018;20(10):1070–82.
[101] Di Vito C, Navone SE, Marfia G, Abdel Hadi L, Mancuso ME, Pecci A, Crisa FM, Berno V, Rampini P, Campanella R, Riboni L. Platelets from glioblastoma patients promote angiogenesis of tumor endothelial cells and exhibit increased VEGF content and release. Platelets 2017;28(6):585–94.
[102] Rao S, Sengupta R, Choe EJ, Woerner BM, Jackson E, Sun T, Leonard J, Piwnica-Worms D, Rubin JB. CXCL12 mediates trophic interactions between endothelial and tumor cells in glioblastoma. PLoS One 2012;7(3), e33005.
[103] El Hallani S, Boisselier B, Peglion F, Rousseau A, Colin C, Idbaih A, Marie Y, Mokhtari K, Thomas J-L, Eichmann A, Delattre J-Y, Maniotis AJ, Sanson M. A new alternative mechanism in glioblastoma vascularization: tubular vasculogenic mimicry. Brain 2010;133(4):973–82.

[104] Soda Y, Marumoto T, Friedmann-Morvinski D, Soda M, Liu F, Michiue H, Pastorino S, Yang M, Hoffman RM, Kesari S, Verma IM. Transdifferentiation of glioblastoma cells into vascular endothelial cells. Proc Natl Acad Sci U S A 2011;108(11):4274–80.

[105] Scholz A, Harter PN, Cremer S, Yalcin BH, Gurnik S, Yamaji M, Di Tacchio M, Sommer K, Baumgarten P, Bahr O, Steinbach JP, Trojan J, Glas M, Herrlinger U, Krex D, Meinhardt M, Weyerbrock A, Timmer M, Goldbrunner R, Deckert M, Braun C, Schittenhelm J, Frueh JT, Ullrich E, Mittelbronn M, Plate KH, Reiss Y. Endothelial cell-derived angiopoietin-2 is a therapeutic target in treatment-naive and bevacizumab-resistant glioblastoma. EMBO Mol Med 2016;8(1):39–57.

[106] Peterson TE, Kirkpatrick ND, Huang Y, Farrar CT, Marijt KA, Kloepper J, Datta M, Amoozgar Z, Seano G, Jung K, Kamoun WS, Vardam T, Snuderl M, Goveia J, Chatterjee S, Batista A, Muzikansky A, Leow CC, Xu L, Batchelor TT, Duda DG, Fukumura D, Jain RK. Dual inhibition of Ang-2 and VEGF receptors normalizes tumor vasculature and prolongs survival in glioblastoma by altering macrophages. Proc Natl Acad Sci U S A 2016;113 (16):4470–5.

[107] Kloepper J, Riedemann L, Amoozgar Z, Seano G, Susek K, Yu V, Dalvie N, Amelung RL, Datta M, Song JW, Askoxylakis V, Taylor JW, Lu-Emerson C, Batista A, Kirkpatrick ND, Jung K, Snuderl M, Muzikansky A, Stubenrauch KG, Krieter O, Wakimoto H, Xu L, Munn LL, Duda DG, Fukumura D, Batchelor TT, Jain RK. Ang-2/VEGF bispecific antibody reprograms macrophages and resident microglia to anti-tumor phenotype and prolongs glioblastoma survival. Proc Natl Acad Sci U S A 2016;113(16):4476–81.

[108] Xu C-p, Zhang H-r, Chen F-l, Yao X-h, Liang Z-q, Zhang R, Cui Y, Qian C, Bian X-w. Human malignant glioma cells expressing functional formylpeptide receptor recruit endothelial progenitor cells for neovascularization. Int Immunopharmacol 2010;10(12):1602–7.

[109] Lathia JD, Li M, Hall PE, Gallagher J, Hale JS, Wu Q, Venere M, Levy E, Rani MRS, Huang P, Bae E, Selfridge J, Cheng L, Guvenc H, McLendon RE, Nakano I, Sloan AE, Phillips HS, Lai A, Gladson CL, Bredel M, Bao S, Hjelmeland AB, Rich JN. Laminin alpha 2 enables glioblastoma stem cell growth. Ann Neurol 2012;72 (5):766–78.

[110] Alexiou GA, Vartholomatos G, Karamoutsios A, Batistatou A, Kyritsis AP, Voulgaris S. Circulating progenitor cells: a comparison of patients with glioblastoma or meningioma. Acta Neurol Belg 2013;113(1):7–11.

[111] Rafat N, Beck GC, Schulte J, Tuettenberg J, Vajkoczy P. Circulating endothelial progenitor cells in malignant gliomas. J Neurosurg 2010;112(1):43–9.

[112] Kioi M, Vogel H, Schultz G, Hoffman RM, Harsh GR, Brown JM. Inhibition of vasculogenesis, but not angiogenesis, prevents the recurrence of glioblastoma after irradiation in mice. J Clin Invest 2010;120(3):694–705.

[113] Achyut BR, Angara K, Jain M, Borin TF, Rashid MH, Iskander ASM, Ara R, Kolhe R, Howard S, Venugopal N, Rodriguez PC, Bradford JW, Arbab AS. Canonical NFkappaB signaling in myeloid cells is required for the glioblastoma growth. Sci Rep 2017;7(1):13754.

[114] Barone A, Sengupta R, Warrington NM, Smith E, Wen PY, Brekken RA, Romagnoli B, Douglas G, Chevalier E, Bauer MP, Dembowsky K, Piwnica-Worms D, Rubin JB. Combined VEGF and CXCR4 antagonism targets the GBM stem cell population and synergistically improves survival in an intracranial mouse model of glioblastoma. Oncotarget 2014;5(20):9811–22.

[115] Kumar V, Donthireddy L, Marvel D, Condamine T, Wang F, Lavilla-Alonso S, Hashimoto A, Vonteddu P, Behera R, Goins MA, Mulligan C, Nam B, Hockstein N, Denstman F, Shakamuri S, Speicher DW, Weeraratna AT, Chao T, Vonderheide RH, Languino LR, Ordentlich P, Liu Q, Xu X, Lo A, Pure E, Zhang C, Loboda A, Sepulveda MA, Snyder LA, Gabrilovich DI. Cancer-associated fibroblasts neutralize the anti-tumor effect of CSF1 receptor blockade by inducing PMN-MDSC infiltration of tumors. Cancer Cell 2017;32(5):654–68. e5.

[116] Gibson P, Tong Y, Robinson G, Thompson MC, Currle DS, Eden C, Kranenburg TA, Hogg T, Poppleton H, Martin J, Finkelstein D, Pounds S, Weiss A, Patay Z, Scoggins M, Ogg R, Pei Y, Yang Z-J, Brun S, Lee Y, Zindy F, Lindsey JC, Taketo MM, Boop FA, Sanford RA, Gajjar A, Clifford SC, Roussel MF, McKinnon PJ, Gutmann DH, Ellison DW, Wechsler-Reya R, Gilbertson RJ. Subtypes of medulloblastoma have distinct developmental origins. Nature 2010;468(7327):1095–9.

[117] Lottaz C, Beier D, Meyer K, Kumar P, Hermann A, Schwarz J, Junker M, Oefner PJ, Bogdahn U, Wischhusen J, Spang R, Storch A, Beier CP. Transcriptional profiles of CD133+ and CD133- glioblastoma-derived cancer stem cell lines suggest different cells of origin. Cancer Res 2010;70(5):2030–40.

[118] Hemmati HD, Nakano I, Lazareff JA, Masterman-Smith M, Geschwind DH, Bronner-Fraser M, Kornblum HI. Cancerous stem cells can arise from pediatric brain tumors. Proc Natl Acad Sci U S A 2003;100(25):15178–83.

[119] Singh SK, Clarke ID, Terasaki M, Bonn VE, Hawkins C, Squire J, Dirks PB. Identification of a cancer stem cell in human brain tumors. Cancer Res 2003;63(18):5821–8.

[120] Galli R, Binda E, Orfanelli U, Cipelletti B, Gritti A, De Vitis S, Fiocco R, Foroni C, Dimeco F, Vescovi A. Isolation and characterization of tumorigenic, stem-like neural precursors from human glioblastoma. Cancer Res 2004;64 (19):7011–21.
[121] Bao S, Wu Q, Sathornsumetee S, Hao Y, Li Z, Hjelmeland AB, Shi Q, McLendon RE, Bigner DD, Rich JN. Stem cell-like glioma cells promote tumor angiogenesis through vascular endothelial growth factor. Cancer Res 2006;66(16):7843–8.
[122] Calabrese C, Poppleton H, Kocak M, Hogg TL, Fuller C, Hamner B, Oh EY, Gaber MW, Finklestein D, Allen M, Frank A, Bayazitov IT, Zakharenko SS, Gajjar A, Davidoff A, Gilbertson RJ. A perivascular niche for brain tumor stem cells. Cancer Cell 2007;11(1):69–82.
[123] Zhu TS, Costello MA, Talsma CE, Flack CG, Crowley JG, Hamm LL, He X, Hervey-Jumper SL, Heth JA, Muraszko KM, DiMeco F, Vescovi AL, Fan X. Endothelial cells create a stem cell niche in glioblastoma by providing NOTCH ligands that nurture self-renewal of cancer stem-like cells. Cancer Res 2011;71(18):6061–72.
[124] Wang R, Chadalavada K, Wilshire J, Kowalik U, Hovinga KE, Geber A, Fligelman B, Leversha M, Brennan C, Tabar V. Glioblastoma stem-like cells give rise to tumour endothelium. Nature 2010;468(7325):829–33.
[125] Janic B, Guo AM, Iskander ASM, Varma NRS, Scicli AG, Arbab AS. Human cord blood-derived AC133+ progenitor cells preserve endothelial progenitor characteristics after long term in vitro expansion. PLoS One 2010;5 (2), e9173.
[126] Janic B, Arbab AS. The role and therapeutic potential of endothelial progenitor cells in tumor neovascularization. Sci World J 2010;10:1088–99.
[127] Chiao M-T, Yang Y-C, Cheng W-Y, Shen C-C, Ko J-L. CD133+ glioblastoma stem-like cells induce vascular mimicry in vivo. Curr Neurovasc Res 2011;8(3):210–9.
[128] He H, Niu CS, Li MW. Correlation between glioblastoma stem-like cells and tumor vascularization. Oncol Rep 2012;27(1):45–50.
[129] Heddleston JM, Li Z, McLendon RE, Hjelmeland AB, Rich JN. The hypoxic microenvironment maintains glioblastoma stem cells and promotes reprogramming towards a cancer stem cell phenotype. Cell Cycle (Georgetown, Tex) 2009;8(20):3274–84.
[130] Li Z, Bao S, Wu Q, Wang H, Eyler C, Sathornsumetee S, Shi Q, Cao Y, Lathia J, McLendon RE, Hjelmeland AB, Rich JN. Hypoxia-inducible factors regulate tumorigenic capacity of glioma stem cells. Cancer Cell 2009;15 (6):501–13.
[131] Hjelmeland AB, Wu Q, Heddleston JM, Choudhary GS, MacSwords J, Lathia JD, McLendon R, Lindner D, Sloan A, Rich JN. Acidic stress promotes a glioma stem cell phenotype. Cell Death Differ 2011;18(5):829–40.
[132] Flavahan WA, Wu Q, Hitomi M, Rahim N, Kim Y, Sloan AE, Weil RJ, Nakano I, Sarkaria JN, Stringer BW, Day BW, Li M, Lathia JD, Rich JN, Hjelmeland AB. Brain tumor initiating cells adapt to restricted nutrition through preferential glucose uptake. Nat Neurosci 2013;16(10):1373–82.
[133] Xie Q, Wu Q, Horbinski CM, Flavahan WA, Yang K, Zhou W, Dombrowski SM, Huang Z, Fang X, Shi Y, Ferguson AN, Kashatus DF, Bao S, Rich JN. Mitochondrial control by DRP1 in brain tumor initiating cells. Nat Neurosci 2015;18(4):501–10.
[134] Mao X-G, Xue X-Y, Wang L, Zhang X, Yan M, Tu Y-Y, Lin W, Jiang X-F, Ren H-G, Zhang W, Song S-J. CDH5 is specifically activated in glioblastoma stemlike cells and contributes to vasculogenic mimicry induced by hypoxia. Neuro-Oncology 2013;15(7):865–79.
[135] Liu G, Yuan X, Zeng Z, Tunici P, Ng H, Abdulkadir IR, Lu L, Irvin D, Black KL, Yu JS. Analysis of gene expression and chemoresistance of CD133+ cancer stem cells in glioblastoma. Mol Cancer 2006;5:67.
[136] Lu C, Ward PS, Kapoor GS, Rohle D, Turcan S, Abdel-Wahab O, Edwards CR, Khanin R, Figueroa ME, Melnick A, Wellen KE, O'Rourke DM, Berger SL, Chan TA, Levine RL, Mellinghoff IK, Thompson CB. IDH mutation impairs histone demethylation and results in a block to cell differentiation. Nature 2012;483(7390):474–8.
[137] Chen J, Li Y, Yu TS, McKay RM, Burns DK, Kernie SG, Parada LF. A restricted cell population propagates glioblastoma growth after chemotherapy. Nature 2012;488(7412):522–6.
[138] Sottoriva A, Spiteri I, Piccirillo SG, Touloumis A, Collins VP, Marioni JC, Curtis C, Watts C, Tavare S. Intratumor heterogeneity in human glioblastoma reflects cancer evolutionary dynamics. Proc Natl Acad Sci U S A 2013;110(10):4009–14.
[139] Verhaak RGW, Hoadley KA, Purdom E, Wang V, Qi Y, Wilkerson MD, Miller CR, Ding L, Golub T, Mesirov JP, Alexe G, Lawrence M, O'Kelly M, Tamayo P, Weir BA, Gabriel S, Winckler W, Gupta S, Jakkula L, Feiler HS, Hodgson JG, James CD, Sarkaria JN, Brennan C, Kahn A, Spellman PT, Wilson RK, Speed TP, Gray JW,

Meyerson M, Getz G, Perou CM, Hayes DN, Cancer Genome Atlas Research Network. Integrated genomic analysis identifies clinically relevant subtypes of glioblastoma characterized by abnormalities in PDGFRA, IDH1, EGFR, and NF1. Cancer Cell 2010;17(1):98–110.
[140] Stommel JM, Kimmelman AC, Ying H, Nabioullin R, Ponugoti AH, Wiedemeyer R, Stegh AH, Bradner JE, Ligon KL, Brennan C, Chin L, DePinho RA. Coactivation of receptor tyrosine kinases affects the response of tumor cells to targeted therapies. Science (New York, NY) 2007;318(5848):287–90.
[141] Patel AP, Tirosh I, Trombetta JJ, Shalek AK, Gillespie SM, Wakimoto H, Cahill DP, Nahed BV, Curry WT, Martuza RL, Louis DN, Rozenblatt-Rosen O, Suva ML, Regev A, Bernstein BE. Single-cell RNA-seq highlights intratumoral heterogeneity in primary glioblastoma. Science 2014;344(6190):1396–401.
[142] Wang J, Wakeman TP, Lathia JD, Hjelmeland AB, Wang X-F, White RR, Rich JN, Sullenger BA. Notch promotes radioresistance of glioma stem cells. Stem Cells (Dayton, Ohio) 2010;28(1):17–28.
[143] Bhat KP, Balasubramaniyan V, Vaillant B, Ezhilarasan R, Hummelink K, Hollingsworth F, Wani K, Heathcock L, James JD, Goodman LD, Conroy S, Long L, Lelic N, Wang S, Gumin J, Raj D, Kodama Y, Raghunathan A, Olar A, Joshi K, Pelloski CE, Heimberger A, Kim SH, Cahill DP, Rao G, Den Dunnen WF, Boddeke HW, Phillips HS, Nakano I, Lang FF, Colman H, Sulman EP, Aldape K. Mesenchymal differentiation mediated by NF-kappaB promotes radiation resistance in glioblastoma. Cancer Cell 2013;24(3):331–46.
[144] Venere M, Hamerlik P, Wu Q, Rasmussen RD, Song LA, Vasanji A, Tenley N, Flavahan WA, Hjelmeland AB, Bartek J, Rich JN. Therapeutic targeting of constitutive PARP activation compromises stem cell phenotype and survival of glioblastoma-initiating cells. Cell Death Differ 2014;21(2):258–69.
[145] Kim S-H, Joshi K, Ezhilarasan R, Myers TR, Siu J, Gu C, Nakano-Okuno M, Taylor D, Minata M, Sulman EP, Lee J, Bhat KPL, Salcini AE, Nakano I. EZH2 protects glioma stem cells from radiation-induced cell death in a MELK/FOXM1-dependent manner. Stem Cell Rep 2015;4(2):226–38.
[146] Brat DJ, Bellail AC, Van Meir EG. The role of interleukin-8 and its receptors in gliomagenesis and tumoral angiogenesis. Neuro-Oncology 2005;7(2):122–33.
[147] Carlsson A, Persson O, Ingvarsson J, Widegren B, Salford L, Borrebaeck CAK, Wingren C. Plasma proteome profiling reveals biomarker patterns associated with prognosis and therapy selection in glioblastoma multiforme patients. Proteomics Clin Appl 2010;4(6-7):591–602.
[148] Zhu VF, Yang J, Lebrun DG, Li M. Understanding the role of cytokines in Glioblastoma Multiforme pathogenesis. Cancer Lett 2012;316(2):139–50.
[149] Holmes WE, Lee J, Kuang WJ, Rice GC, Wood WI. Structure and functional expression of a human interleukin-8 receptor. Science 1991;253(5025):1278–80.
[150] Murphy PM, Tiffany HL. Cloning of complementary DNA encoding a functional human interleukin-8 receptor. Science 1991;253(5025):1280–3.
[151] Lloyd A, Modi W, Sprenger H, Cevario S, Oppenheim J, Kelvin D. Assignment of genes for interleukin-8 receptors (IL8R) A and B to human chromosome band 2q35. Cytogenet Cell Genet 1993;63(4):238–40.
[152] Sprenger H, Lloyd AR, Meyer RG, Johnston JA, Kelvin DJ. Genomic structure, characterization, and identification of the promoter of the human IL-8 receptor A gene. J Immunol 1994;153(6):2524–32.
[153] Sprenger H, Lloyd AR, Lautens LL, Bonner TI, Kelvin DJ. Structure, genomic organization, and expression of the human interleukin-8 receptor B gene. J Biol Chem 1994;269(15):11065–72.
[154] Ahuja SK, Ozcelik T, Milatovitch A, Francke U, Murphy PM. Molecular evolution of the human interleukin-8 receptor gene cluster. Nat Genet 1992;2(1):31–6.
[155] Raychaudhuri B, Vogelbaum MA. IL-8 is a mediator of NF-kappaB induced invasion by gliomas. J Neuro-Oncol 2011;101(2):227–35.
[156] Van Meir E, Ceska M, Effenberger F, Walz A, Grouzmann E, Desbaillets I, Frei K, Fontana A, de Tribolet N. -Interleukin-8 is produced in neoplastic and infectious diseases of the human central nervous system. Cancer Res 1992;52(16):4297–305.
[157] Li A, Dubey S, Varney ML, Dave BJ, Singh RK. IL-8 directly enhanced endothelial cell survival, proliferation, and matrix metalloproteinases production and regulated angiogenesis. J Immunol 2003;170(6):3369–76.
[158] Dwyer J, Hebda JK, Le Guelte A, Galan-Moya E-M, Smith SS, Azzi S, Bidere N, Gavard J. Glioblastoma cell-secreted interleukin-8 induces brain endothelial cell permeability via CXCR2. PLoS One 2012;7(9), e45562.
[159] Zhou J, Yi L, Ouyang Q, Xu L, Cui H, Xu M. Neurotensin signaling regulates stem-like traits of glioblastoma stem cells through activation of IL-8/CXCR1/STAT3 pathway. Cell Signal 2014;26(12):2896–902.
[160] Infanger DW, Cho Y, Lopez BS, Mohanan S, Liu SC, Gursel D, Boockvar JA, Fischbach C. Glioblastoma stem cells are regulated by interleukin-8 signaling in a tumoral perivascular niche. Cancer Res 2013;73(23):7079–89.

[161] Bonavia R, Inda MM, Vandenberg S, Cheng S-Y, Nagane M, Hadwiger P, Tan P, Sah DWY, Cavenee WK, Furnari FB. EGFRvIII promotes glioma angiogenesis and growth through the NF-κB, interleukin-8 pathway. Oncogene 2012;31(36):4054–66.

[162] Jain M, Gamage NH, Alsulami M, Shankar A, Achyut BR, Angara K, Rashid MH, Iskander A, Borin TF, Wenbo Z, Ara R, Ali MM, Lebedyeva I, Chwang WB, Guo A, Bagher-Ebadian H, Arbab AS. Intravenous formulation of HET0016 decreased human glioblastoma growth and implicated survival benefit in rat xenograft models. Sci Rep 2017;7:41809.

[163] Bieche I, Chavey C, Andrieu C, Busson M, Vacher S, Le Corre L, Guinebretiere JM, Burlinchon S, Lidereau R, Lazennec G. CXC chemokines located in the 4q21 region are up-regulated in breast cancer. Endocr Relat Cancer 2007;14(4):1039–52.

[164] Desurmont T, Skrypek N, Duhamel A, Jonckheere N, Millet G, Leteurtre E, Gosset P, Duchene B, Ramdane N, Hebbar M, Van Seuningen I, Pruvot FR, Huet G, Truant S. Overexpression of chemokine receptor CXCR2 and ligand CXCL7 in liver metastases from colon cancer is correlated to shorter disease-free and overall survival. Cancer Sci 2015;106(3):262–9.

[165] Liu X-m, Zhang Q-p, Mu Y-g, Zhang X-h, Sai K, Pang JC-S, Ng H-K, Chen Z-p. Clinical significance of vasculogenic mimicry in human gliomas. J Neuro-Oncol 2011;105(2):173–9.

[166] Wang S-Y, Ke Y-Q, Lu G-H, Song Z-H, Yu L, Xiao S, Sun X-L, Jiang X-D, Yang Z-L, Hu C-C. Vasculogenic mimicry is a prognostic factor for postoperative survival in patients with glioblastoma. J Neuro-Oncol 2013;112(3):339–45.

[167] Devapatla B, Sharma A, Woo S. CXCR2 inhibition combined with sorafenib improved antitumor and antiangiogenic response in preclinical models of ovarian cancer. PLoS One 2015;10(9), e0139237.

[168] Singh S, Sadanandam A, Nannuru KC, Varney ML, Mayer-Ezell R, Bond R, Singh RK. Small-molecule antagonists for CXCR2 and CXCR1 inhibit human melanoma growth by decreasing tumor cell proliferation, survival, and angiogenesis. Clin Cancer Res 2009;15(7):2380–6.

[169] Singh S, Sadanandam A, Varney ML, Nannuru KC, Singh RK. Small interfering RNA-mediated CXCR1 or CXCR2 knock-down inhibits melanoma tumor growth and invasion. Int J Cancer 2010;126(2):328–36.

[170] Singh S, Nannuru KC, Sadanandam A, Varney ML, Singh RK. CXCR1 and CXCR2 enhances human melanoma tumourigenesis, growth and invasion. Br J Cancer 2009;100(10):1638–46.

[171] Keane MP, Belperio JA, Xue YY, Burdick MD, Strieter RM. Depletion of CXCR2 inhibits tumor growth and angiogenesis in a murine model of lung cancer. J Immunol 2004;172(5):2853–60.

[172] Purohit A, Varney M, Rachagani S, Ouellette MM, Batra SK, Singh RK. CXCR2 signaling regulates KRAS(G(1)(2)D)-induced autocrine growth of pancreatic cancer. Oncotarget 2016;7(6):7280–96.

[173] Li L, Khan MN, Li Q, Chen X, Wei J, Wang B, Cheng J-W, Gordon JR, Li F. G31P, CXCR1/2 inhibitor, with cisplatin inhibits the growth of mice hepatocellular carcinoma and mitigates high? Dose cisplatin-induced nephrotoxicity. Oncol Rep 2015;33(2):751–7.

[174] Nannuru KC, Sharma B, Varney ML, Singh RK. Role of chemokine receptor CXCR2 expression in mammary tumor growth, angiogenesis and metastasis. J Carcinog 2011;10:40.

[175] Sharma B, Nawandar DM, Nannuru KC, Varney ML, Singh RK. Targeting CXCR2 enhances chemotherapeutic response, inhibits mammary tumor growth, angiogenesis, and lung metastasis. Mol Cancer Ther 2013;12(5):799–808.

[176] Yang L, Liu Z, Wu R, Yao Q, Gu Z, Liu M. Correlation of C-X-C chemokine receptor 2 upregulation with poor prognosis and recurrence in human glioma. Onco Targets Ther 2015;8:3203–9.

[177] Brandenburg S, Muller A, Turkowski K, Radev YT, Rot S, Schmidt C, Bungert AD, Acker G, Schorr A, Hippe A, Miller K, Heppner FL, Homey B, Vajkoczy P. Resident microglia rather than peripheral macrophages promote vascularization in brain tumors and are source of alternative pro-angiogenic factors. Acta Neuropathol 2016;131(3):365–78.

[178] Matsuo Y, Raimondo M, Woodward TA, Wallace MB, Gill KR, Tong Z, Burdick MD, Yang Z, Strieter RM, Hoffman RM, Guha S. CXC-chemokine/CXCR2 biological axis promotes angiogenesis in vitro and in vivo in pancreatic cancer. Int J Cancer 2009;125(5):1027–37.

[179] Jiang Y, Wang S, Holcomb J, Trescott L, Guan X, Hou Y, Brunzelle J, Sirinupong N, Li C, Yang Z. Crystallographic analysis of NHERF1-PLC?3 interaction provides structural basis for CXCR2 signaling in pancreatic cancer. Biochem Biophys Res Commun 2014;446(2):638–43.

CHAPTER

17

Treatment and resistance of glioblastoma to CAR T-cell immunotherapies

Stephen J. Bagley

Division of Hematology/Oncology, Department of Medicine, Perelman School of Medicine at the University of Pennsylvania, Philadelphia, PA, United States

Abstract

Chimeric antigen receptor (CAR) T cells are a form of autologous immunotherapy that has led to unprecedented responses in refractory hematologic malignancies. CAR T-cell therapy involves apheresis for T-cell collection, ex vivo genetic modification to encode a synthetic receptor that binds a specific tumor antigen, and infusion of the T cells back into the patient. A rapidly increasing body of research is focused on translating this treatment to solid tumors. However, the clinical efficacy of CAR T cells in nonhematologic malignancies, including glioblastoma (GBM), has been elusive. Key challenges in GBM and other solid cancers include a paucity of unique tumor-target antigens, limited CAR T-cell trafficking to tumor sites, tumor heterogeneity and antigen loss, and the severely immunosuppressive tumor microenvironment. This chapter provides an overview of previous clinical studies of CAR T-cell therapies for GBM, recently identified mechanisms of resistance to CAR T-cell therapy in GBM, and current and future research directions for potentially overcoming these challenges.

Abbreviations

AE	adverse event
ALL	acute lymphoblastic leukemia
APC	antigen-presenting cell
BCMA	B-cell maturation antigen
BiTE	bispecific T-cell engager
CAIX	carobxy-anhydrase-IX
CAR	chimeric antigen receptor
CEA	carcino-embryonic antigen
CNS	central nervous system
CR	complete remission

Glioblastoma Resistance to Chemotherapy: Molecular Mechanisms and Innovative Reversal Strategies
https://doi.org/10.1016/B978-0-12-821567-8.00023-3

CMV	cytomegalovirus
CRC	colorectal cancer
CRISPR	clustered regularly interspaced short palindromic repeats
CRS	cytokine-release syndrome
CT	computed tomography
CTLA4	cytotoxic T lymphocyte-associated protein 4
DLBCL	diffuse large B-cell lymphoma
DLT	dose-limiting toxicity
EBV	Epstein-Barr virus
EGFR	endothelial growth factor receptor
FDA	Food and Drug Administration
FITC	fluorescein isothiocyanate
FR	folate receptor
GBM	glioblastoma
HER2	human epidermal growth factor 2
HIF	hypoxia inducible factor
HLA	human leukocyte antigen
HLH	hemophagocytic lymphohistiocytosis
IDO	indoleamine 2,3-dioxygenase
MAS	macrophage activation syndrome
MAV	metabolically active volume
MRI	magnetic resonance imaging
MPM	malignant pleural mesothelioma
PD-1	programmed cell death protein 1
PD-L1	programmed death-ligand 1
PDAC	pancreatic ductal adenocarcinoma
PET	positron emission tomography
PSA	prostate-specific antigen
PSMA	prostate-specific membrane antigen
PSCA	prostate stem cell antigen
qPCR	quantitative polymerase chain reaction
ROR1	receptor tyrosine kinase-like orphan receptor 1
cFV	single-chain variable fragment
TAA	tumor-associated antigen
TCR	T-cell receptor
TGF-β	transforming growth factor β

Conflict of interest

S.J.B. is a coinventor of intellectual property licensed by the University of Pennsylvania to Novartis.

Introduction to CAR T-cell therapy

Chimeric antigen receptors (CARs) are genetically engineered, artificial fusion proteins that incorporate an extracellular antigen-recognition domain, a transmembrane and hinge domain that anchors the receptor on the cell surface and projects the antigen-targeting moiety out to the extracellular space, and an intracellular T-cell signaling domain that is triggered on antigen engagement [1, 2]. Following transfection of the CAR construct into autologous or allogeneic peripheral blood T cells, the T cells are infused into the patient to target a surface-exposed tumor antigen that is specified by the CAR's antigen-binding domain, which is usually in the form of a single-chain variable fragment (scFv) [3, 4]. When the CAR binds to its associated antigen, primary T-cell activation occurs and leads to cytokine release, cytolytic degranulation, and T-cell proliferation [5]. Other effector mechanisms and memory responses also occur in a manner dependent on the mechanism of costimulation. This may include costimulation with 4-1BB or CD28 in the case of "second-generation CARs," or two or more signaling domains for "third-generation CARs" [6, 7]. Importantly, unlike vaccines and immunomodulatory agents that rely on in vivo priming of endogenous tumor-reactive cells and are therefore human leukocyte antigen (HLA) restricted, CAR T cells exert their effect in a universal, HLA-independent manner [8].

The first step in developing CAR T cells for an individual patient is leukapheresis. During this process, blood is removed and leukocytes are separated. After a sufficient number of leukocytes have been harvested, the product is enriched for T cells, which then undergo an activation process that includes incubation with the viral vector encoding the CAR. Lentiviral vectors are used most frequently, but other methods of gene transfer, such as Sleeping Beauty synthetic DNA or mRNA transposon systems, are being studied. When the cell expansion process is completed, the cells are cryopreserved in infusible medium. When the patient is ready for treatment, the frozen cells are transported to the treatment site and thawed prior to infusion. In patients with hematologic malignancies, and in certain solid tumor CAR T-cell trials, lymphodepleting chemotherapy is administered prior to the T cells. Lymphodepletion can substantially increase the in vivo expansion of the infused CAR T cells by multiple effects, including reduction of the patient's lymphoid cell pool to make "space" for the CAR T cells, increasing homeostatic cytokines, and ameliorating the tumor inhibitory microenvironment.

As of the writing of this chapter, most clinical progress for CAR T cells has occurred in the treatment of hematologic malignancies, with the United States Food and Drug Administration (FDA) having approved two therapies. Tisagenlecleucel, a CD19-targeted CAR T-cell therapy formerly known as CTL019, was first approved for the treatment of patients up to 25 years of age with B-cell precursor acute lymphoblastic leukemia (ALL) that is refractory or in second or later relapse [9]. Several key studies led to this approval, demonstrating complete remission (CR) rates between 60% and 90% in patients with heavily pretreated, relapsed/refractory B-cell ALL [10, 11]. Subsequently, axicabtagene ciloleucel, a second anti-CD19 CAR T-cell treatment, was approved for large B-cell lymphoma patients, who have failed at least two prior therapies [12]. Axicabtagene ciloleucel results in CR in approximately half of patients with refractory B-cell lymphoma, and remissions are sustained for over 3 years [12]. Most recently, tisagenlecleucel was also approved by the FDA for adult patients with relapsed or refractory large B-cell lymphoma after two or more lines of systemic therapy, including diffuse large B-cell lymphoma (DLBCL), high-grade B-cell lymphoma, and DLBCL arising from follicular lymphoma [13, 14]. In addition to ALL and DLBCL, CAR T cells have shown promise in other hematologic malignancies. In multiple myeloma, for example, an FDA-approved CAR T-cell therapy may be on the short-term horizon [15].

CAR T cells have unique toxicities that are distinct from those of cytotoxic chemotherapy, small-molecule-targeted therapies, and other forms of immunotherapy. The most commonly observed toxicity is cytokine-release syndrome (CRS), which most commonly manifests as fever, hypotension, hypoxia, and/or multiorgan toxicity [16]. CRS is triggered by the activation of T cells upon engagement of their CARs with cognate antigens expressed by tumor cells. When this occurs, both the activated T cells, as well as bystander immune cells including monocytes and macrophages, release cytokines and chemokines that lead to a systemic inflammatory state that mirrors sepsis physiology. CRS typically causes constitutional symptoms, such as fever, malaise, anorexia, and myalgias, but can ultimately impact any organ in the body. CRS is managed in accordance with the grade of this toxicity according to established guidelines [16]. Detailed discussion of the management of CRS is beyond the scope of this chapter, but includes supportive care such as intravenous fluid boluses and anti-IL-6 therapy with tocilizumab or siltuximab. In an effort to avoid hindering the activity of the CAR T cells, corticosteroids are typically reserved for cases that are refractory to anti-IL6 therapy.

In contrast to the successes observed in hematologic malignancies, consistent clinical efficacy of CAR T cells has not been demonstrated for solid tumors, including GBM [17]. The remainder of this chapter will review key clinical trials of CAR T cells in GBM that have been reported to date, obstacles to therapeutic success that have been identified thus far, and the scientific and translational efforts that are being pursued to make CAR T-cell therapy a clinical reality for this common and deadly brain tumor.

CAR T-cell therapy in GBM: Recent clinical advances

IL13Rα2 CAR T cells

One of the most extensively studied targets for CAR T-cell therapy in GBM is IL13Rα2, a membrane-bound protein expressed in over 75% of GBMs that is associated with activation of the phosphatidylinositol-3 kinase/Akt/mammalian target of rapamycin (mTOR) pathway [18–20], increased tumor invasiveness, and poor prognosis [21]. Owing to its expression on GBM tumor cells and limited expression in normal brain and other tissues [22], IL13Rα2 is an attractive target for CAR T-cell therapy [23]. In the initial trial of a first-generation IL13Rα2-specific CAR, termed "IL13 zetakine," repeated intracranial administration of the CAR T-cell product was conducted in three patients with recurrent GBM following surgical resection of the tumor [24]. One subject was also treated with direct intratumoral CAR T-cell infusions at a distant site of tumor recurrence. This first-in-human study showed that IL13Rα2-directed CAR T cells could be manufactured and delivered to patients with recurrent GBM via an implanted reservoir/catheter system. Importantly, the CAR T cells were well tolerated with only mild headaches and transient, manageable neurologic deficits. In addition, some potential preliminary signals of antiglioma activity were observed, as patients experienced an increase in necrotic tumor volume by MRI, loss of IL13Rα2 tumor cell expression, and encouraging overall survival. In a follow-up trial utilizing a second-generation 4-1BB costimulatory IL13 zetakine CAR, a 50-year-old patient with recurrent multifocal GBM, including leptomeningeal disease, received 6 weekly intracavitary infusions of the CAR T-cell product following surgical resection of three of his five progressing intracranial tumors [25]. Although the locally CAR T-cell treated site remained stable, other intracranial lesions progressed and new spinal lesions developed. The patient then received 10 additional CAR T-cell infusions delivered intraventricularly via a catheter device placed in the right lateral ventricle. Remarkably, all of the intracranial and spinal tumors regressed, and the response lasted for 7.5 months. Although the patient subsequently progressed at new locations distinct from his previous tumors, this case report underscores the therapeutic potential for CAR T cells in GBM.

EGFRvIII CAR T cells

EGFR variant III (EGFRvIII), resulting from an in-frame deletion of exons 2–7 [26], is present in about 25–30% of newly diagnosed GBM and represents another potential target for CAR T cells [27]. The amino acid sequence resulting from EGFRvIII yields a novel glycine residue at the junction of exons 1 and 8, generating a tumor-specific and immunogenic epitope within the extracellular domain of EGFR. In a first-in-human phase I trial, 10 patients with recurrent GBM received a single dose of peripherally infused EGFRvIII-directed CAR T cells [28]. No patient on this study experienced off-tumor toxicity or CRS. While the study was not designed to evaluate for efficacy, no objective tumor responses were observed (although one patient had residual stable disease lasting >18 months). All treated patients had detectable engraftment of EGFRvIII CAR T cells in the peripheral blood. Of note, however, the degree of engraftment was significantly lower than what is typically seen with CD19-specific CAR T cells bearing the same 4-1BB costimulatory domain, lentiviral backbone, and manufacturing process [29]. Seven of the 10 patients had post-CAR T-cell surgical intervention, allowing for tissue-specific analysis of CAR T-cell trafficking and assessment of the tumor microenvironment. In two of these subjects who underwent tumor resection within 2 weeks of CAR T-cell infusion, CART-EGFRvIII cells were found at higher concentrations in the brain than in the peripheral blood at the same time point. Specifically, EGFRvIII DNA sequences by qPCR were 3 times and 100 times higher, respectively, in brain specimens than in corresponding blood samples from these patients, suggesting that the CAR T cells had effectively trafficked and expanded in situ within active regions of GBM. In addition, CAR T-cell-treated tumor samples in most patients demonstrated specific loss or decreased expression of EGFRvIII. Although it is possible that some decrease in EGFRvIII expression was owing to the natural loss of EGFRvIII over time and unrelated to CAR T-cell therapy [30], the majority of EGFRvIII+ GBMs maintain EGFRvIII positivity at recurrence [27]. This suggests that antigen loss was more likely related to successful targeting of EGFRvIII+ tumor cells by CAR T cells. Findings related to CAR T-cell-induced changes in the tumor microenvironment are detailed later in this chapter.

HER2 CAR T cells

Human epidermal growth factor receptor 2 (HER2), a receptor tyrosine kinase overexpressed in numerous solid tumors, has also been evaluated clinically as a target for CAR T-cell therapy in GBM [31–33]. In a recent phase I trial, 17 patients with progressive HER2+ GBM were treated with peripheral blood infusions of HER2-specific CAR-modified virus-specific T cells [34]. Investigators in this study utilized a second-generation CAR with an FRP5-based exodomain and a CD28 signaling endodomain. No dose-limiting toxicity was observed. HER2-CAR T cells were detected by qPCR in all patients after the infusion, peaking in 15 of 17 patients at 3h after the infusion and at 1 and 2 weeks in the other 2 patients, respectively. At 6 weeks after the infusion, HER2-CAR T cells were detectable in 7 of 15 patients, with blood levels declining further every month thereafter (with two samples remaining positive up to 12 months, and none at 18 months). This suggested that the HER2-CAR T cells did not expand after infusion but could persist for up to 1 year at a low frequency. Of 16 evaluable patients, 1 had a partial response lasting for more than 9 months, and 7 had stable disease ranging between 8 weeks and 29 months (with 3 of these patients remaining free of progression during 24–29 months of follow-up). A unique and important element of this study was the use of virus-specific T cells, which can provide the expected antitumor activity through their CAR but may also receive appropriate costimulation following native T-cell receptor engagement by latent virus antigens presented by professional antigen-presenting cells [34, 35]. The investigators in this study administered CAR-modified T cells specific for adenovirus, Epstein-Barr virus (EBV), or cytomegalovirus (CMV). Among the 17 patients treated, the CAR T cells of all patients contained adenovirus- and EBV-specific T cells, and all CAR T cells from CMV seropositive patients contained CMVpp65-specific T cells as determined by interferon gamma ELISpot assays. Overall, this phase I trial demonstrated the feasibility and safety of peripherally infused virus-specific CAR T cells in GBM and, despite the lack of expansion of the CAR T cells in the blood, displayed potential efficacy signals.

Challenges and mechanisms of resistance to CAR T-cell therapy in GBM

Paucity of tumor-specific targets and tumor heterogeneity

One of the most significant challenges to CAR T-cell therapy in GBM, and for solid tumors in general, is the identification of appropriate target antigens. Unlike hematologic malignancies, most GBM tumors do not express one tumor-specific antigen (i.e., a neoantigen that is entirely absent from normal human tissues). Rather, there are numerous tumor-associated antigens that are enriched on tumors but also expressed at low levels on normal tissues [36]. When CAR T cells are designed to target an antigen that is not specific to tumor cells, patients are subjected to the potential risk of on-target off-tumor toxicity. Another problem related to antigen selection in GBM is that traditional CAR T cells require a target antigen that is expressed on the cell surface. Unfortunately, only about 1% of total cellular proteins are actually expressed on the cell surface [37], meaning that a large number of potential tumor-target antigens are not available to CAR T cells.

Even when an antigen is identified that is commonly expressed in GBM and is tumor-specific (e.g., EGFRvIII), heterogeneity of the antigen's expression can remain a significant problem [36]. Molecular heterogeneity in GBM can be both spatial and temporal in nature, making it one of the most significant hurdles to targeted therapy of any sort in this disease. Key examples of tumor heterogeneity and its detrimental impact on CAR T-cell therapy for GBM can be seen in the few trials completed to date. With both EGFRvIII and IL13Rα2 CAR T cells in GBM, patients had progressive disease with antigen-negative tumors despite successful initial antigen targeting by CAR T cells [25, 28]. Although it is possible that CAR T-cell therapy targeted against a heterogeneously expressed antigen could result in a secondary immune response against other tumor-specific neoantigens or epitopes, a phenomenon termed epitope spreading [38], this has not yet been proven to occur in humans. Efforts being undertaken to address the problem of tumor heterogeneity for CAR T-cell therapy in GBM are described later in this chapter.

Impaired trafficking to tumor

Following infusion, CAR T cells must localize to the tumor bed for effective tumor cell killing. However, T cells typically do not express the cognate receptors for the chemokines produced by tumors, and tumors produce only very small amounts of the chemokines needed for successful T-cell trafficking [17]. In addition, there are environmental barriers to CAR T-cell extravasation and tumor infiltration, including endothelial cell dysfunction, downregulation of key adhesion molecules [39, 40], and extracellular matrix barriers [41]. Given these significant detriments to effective infiltration of the tumor by CAR T cells, alternative routes of administration beyond peripheral infusion have been evaluated. Intratumoral and other modes of CAR T-cell administration are discussed in detail in the "Increased trafficking to tumor" section.

T-cell intrinsic dysfunction

If successful trafficking of CAR T cells to the tumor site occurs and cognate antigen in encountered, the cells must then undergo rapid expansion. In fact, the extent of in vivo expansion is correlated with CAR T-cell efficacy across multiple tumor types [11, 42, 43]. Unfortunately, there are significant intrinsic T-cell deficits in patients with GBM and other solid tumors that impair CAR T-cell proliferation and persistence [44]. In a study of 195 children with a variety of hematologic and solid tumor malignancies, Das et al. performed an extensive analysis of T-cell fitness and phenotyping in relation to the receipt of progressive cycles of cytotoxic chemotherapy [45]. Using single-cell techniques and a functional expansion assay that is similar to the culture process used for manufacturing CAR T cells, the authors found that the proportion of patients with enough T cells that "pass" a previously defined threshold for successful CAR T-cell manufacturing declined significantly with subsequent cycles of chemotherapy. Thus, certain patients with heavily pretreated GBM may be not be optimal candidates for CAR T-cell therapy. In patients with GBM, there are additional factors at play as well. For example, severe lymphopenia is often present in GBM even prior to treatment, which was recently demonstrated to be owing to T-cell sequestration in the bone marrow [46]. In addition, the standard of care for GBM, consisting of radiation in combination with temozolomide chemotherapy, is extremely lymphodepleting and leads to CD4 counts less than 200/mm^3 in over one-third of patients [47]. Collection of an adequate number of viable T cells for CAR T-cell manufacturing from such patients can prove challenging.

Complicating matters further, some T-cell subtypes show superior antitumor activity in vivo compared to others. In ALL and non-Hodgkin lymphoma, the presence of early lineage T cells (naïve and early memory) prior to genetic engineering has been found to correlate with better expansion [48]. Similarly, in preclinical high-grade glioma models CAR CD4+ T cells have superior antitumor activity compared to CAR CD8+ T cells, and this advantage is lost upon mixing the CAR CD4+ cells with CAR CD8+ T cells [49]. These results suggest that the differentiation status of CAR T cells is an important consideration for optimizing their expansion and persistence. To date, most clinical trials using CAR T cells have used cell products prepared from unselected "bulk" T cells [50].

A final factor to consider in optimizing CAR T-cell expansion and persistence is whether to utilize lymphodepleting chemotherapy. In hematologic malignancies, lymphodepleting chemotherapy is typically administered prior to CAR T cells and can improve their in vivo expansion by multiple proposed mechanisms, including reducing the lymphoid cell pool to make "space" for the CAR T cells, increasing homeostatic cytokines, depleting immunosuppressive regulatory T cells (Tregs), and ameliorating the tumor inhibitory microenvironment

[51–54]. In GBM and other solid tumors, the value of lymphodepletion and the optimal chemotherapy regimens to achieve it are not clearly defined. One proposed solution in patients with GBM is to use temozolomide, which is lymphodepleting and already used as standard of care for GBM [55]. However, the impact of using temozolomide for lymphodepletion in GBM CAR T-cell therapy has not been established in humans. Additional studies are needed to better delineate the necessity of lymphodepleting chemotherapy in solid tumor CAR T-cell trials, as well as the optimal regimens to be used.

Immune checkpoints and other immunosuppressive microenvironment factors

Even if CAR T cells traffic to the tumor and expand upon antigen recognition, the GBM tumor microenvironment poses significant challenges for T-cell effector function and persistence. One problem is that the immunostimulatory cytokines required for optimal killing activity are typically downregulated in the GBM microenvironment [56]. Since systemic delivery of cytokine therapy carries excessive toxicity, several transgenic-based strategies have been developed to deliver higher local levels of key cytokines in other tumors, including IL-12 [57], IL-15 [58], IL-18 [59], and IL-21 [60]. However, because these cytokines carry multiple context-dependent roles that can sometimes be immunosuppressive, it remains controversial as to whether they would help or hinder the CAR T cells [56].

An equally pressing, if not more difficult issue, is that immune checkpoints and soluble immune inhibitory factors are often upregulated in the GBM microenvironment [61]. With regard to CAR T cells specifically, the critical role of immune checkpoints and other immunosuppressive molecules was highlighted in the first-in-human study of EGFRvIII-targeted CAR T cells in GBM [28]. Immunohistochemical stains in post-CAR T-cell surgical specimens compared to matched pretreatment samples displayed consistent and significant upregulation of immune checkpoints and other soluble immunosuppressive molecules, including indoleamine 2,3-dioxygenase (IDO) 1, programmed death (PD) ligand 1 (PD-L1), transforming growth factor (TGF)-β, and IL-10. These observations suggest that CAR T-cell targeting of EGFRvIII+ tumor cells induced a compensatory immunosuppressive response in the tumor microenvironment. In addition to immune checkpoints, competition for key nutrients including glucose, amino acids, and fatty acids, as well as the hypoxic tumor microenvironment, have also emerged as critical factors that restrict the antitumor activity of CAR T cells in GBM and other solid tumors [62].

Future directions to overcome resistance

Optimizing antigen selection and addressing heterogeneity

In addition to the most commonly studied target antigens in GBM, which include IL13Rα2, EGFRvIII, and HER2, as described in the "CAR T-cell therapy in GBM: Recent clinical advances" section, numerous other promising targets are being actively explored for GBM CAR T-cell therapy. Examples include EphA2, CD70, and other EGFR targets, including wild-type EGFR and EGFR extracellular domain point mutations [63]. One monoclonal antibody being evaluated as a potential synthetic receptor for EGFR-targeted CAR T cells in

GBM is mAb806, which recognizes tumor-specific untethered EGFR, including EGFR A289V and EGFRvIII, and has shown antitumor activity in preclinical glioma models [64, 65]. When attempting to target wild-type EGFR, however, concerns remain over limiting systemic side effects due to target antigen expression by normal healthy tissues. Thus, when attempting to target wild-type EGFR or any novel antigen for solid tumors, it is critical to recognize and plan for the risk of on-target off-tumor toxicities.

One way to address uncertainty regarding on-target off-tumor toxicity is to use mRNA CAR T cells for initial human studies. This was recently accomplished in a phase I trial in patients with pancreatic ductal adenocarcinoma (PDAC). In this study, investigators used autologous T cells genetically modified with mRNA via electroporation to transiently express a CAR specific for mesothelin, which is overexpressed by PDAC cells and also found on the lining of the peritoneum, pleura, and pericardium [66]. Six patients with chemotherapy-refractory, metastatic PDAC received CAR T-cells intravenously three times weekly for 3 weeks. Because there were no dose-limiting toxicities (DLTs) and no CRS experienced by any of the patients, lentiviral-transduced CAR T cells against mesothelin can now be studied in human trials with greater confidence that they will be safe.

Another way to mitigate the risk of on-target off-tumor toxicity is to genetically engineer the T cells in an attempt to limit CAR T-cell activation to tumor sites. There are numerous elegant methods that can be utilized to achieve this. One technique is to design a T cell "circuit" in which a synthetic Notch receptor ("SynNotch" receptor), when bound by its cognate antigen, translocates to the nucleus and induces expression of a CAR for a second antigen [67]. Other methods of attempting to limit CAR T-cell activation to tumor sites include (1) using two separate CARs on the same T cell (one to provide antigen-specific zeta-signaling and one to provide antigen-specific costimulation, thus restricting full T-cell activation to tumors that express both antigens), (2) using "switchable" CAR T cells that are only activated in the presence of an exogenously administered "switch," e.g., a small molecule or a monoclonal antibody [68, 69], or (3) designing CARs to include an inhibitory peptide that reversibly blocks the scFv and keeps the T cells in an off state until they reach the tumor microenvironment, where local conditions result in cleavage of the inhibitory peptide and activation of the T cells [56]. As a recent example of this approach in GBM, CAR T cells targeted against carbonic anhydrase IX (CAIX) were studied in a patient-derived GBM xenograft model. Importantly, CAIX expression is induced by hypoxia and therefore highly expressed in the hypoxic microenvironment of GBM, theoretically minimizing off-target effects in normoxic tissues throughout the rest of the body [70]. In this study, direct intratumoral injection of the anti-CAIX CAR T cells cured ~20% of mice without detectable systemic effects. A phase I trial in humans is planned for the near future.

In addition to carefully selecting target antigens to minimize toxicity, tumor heterogeneity must also be successfully addressed in order for CAR T cells to be efficacious in GBM. In an ideal world, the target antigen would be expressed uniformly by GBM cells and not expressed at all on normal healthy tissues. Although no such antigen has been conclusively identified in human GBM, promising candidates are being explored, such as chlorotoxin, a 36-amino acid peptide isolated from the venom of scorpions [71]. Until CAR T cells for such an antigen are proven to be safe and effective in humans, a logical strategy is to design bispecific or multivalent CAR T cells, as targeting more than one antigen with CAR T cells may ultimately reduce or delay antigen escape. There are several ways to develop T-cell products for multispecificity [72]. A simple example is to simultaneously or sequentially administer T-cell products that are separately transduced for different CARs. Alternatively, one could combine vectors for two different CARs during cell production to achieve a mixed product. An even more sophisticated approach is to engineer a single CAR molecule to recognize multiple antigens. This can be accomplished by linking two different antigen binders on one molecule (i.e., a single "tandem" CAR) [73], or by using one vector to encode two or three separate CARs that can be expressed on a single T cell [74]. A unique strategy altogether has been proposed by Choi et al., who developed a bicistronic construct to drive dual expression of an EGFRvIII-specific CAR and a bispecific T-cell engager (BiTE) against wild-type EGFR in GBM [75]. The engineered cells secreted EGFR-specific BiTEs in the local tumor microenvironment that redirected CAR T cells and recruited untransduced bystander T cells against wild-type EGFR. As a result, these cells eliminated mouse tumors that heterogeneously expressed EGFRvIII. Importantly, toxicity against human skin grafts, a potential concern with targeting wild-type EGFR, was not observed in vivo. Multiple clinical trials are underway testing different variations of multispecific CAR T cells to address tumor heterogeneity.

Increased trafficking to tumor

One logical way to improve CAR T-cell recruitment to tumor cells is to target the tumor bed itself with direct injection of CAR T cells. This has been performed in previous clinical trials with hepatic artery infusions of CAR T cells for liver metastases [76], intratumoral injections of CAR T cells for breast cancer, intrapleural delivery of mesothelin-targeted CAR T cells for malignant pleural mesothelioma [77], and intratumoral/intraventricular delivery of IL-13 Rα2-targeted CAR T cells for GBM [24, 25]. Another potential way to improve CAR T-cell trafficking, even if delivered peripherally, is to augment chemokine communication. In a recent preclinical GBM model, for example, investigators used CD70-targeted CAR T cells with a modified IL-8 receptor (CXCR1/CXCR2) that allowed for improved CAR T-cell trafficking to the tumor site, where tumor cell IL-8 production is high following ionizing radiation [78]. Other methods of increasing T-cell tumor trafficking that have been explored include normalization of the vasculature by low-dose angiogenesis inhibitors [79], or engineering CAR T cells to express the enzyme heparanase, which degrades sulfate proteoglycans in the extracellular matrix and can promote tumor T-cell infiltration into stroma-rich solid tumors [80].

Improving intrinsic T-cell fitness

Ongoing efforts are under way to improve intrinsic T-cell fitness in patients with solid tumors such as GBM. One option is to use allogeneic CAR T cells from healthy donors, which are presumably better functioning than those from a patient with cancer. However, this approach has been limited by concerns regarding rejection in both the host and graft directions [44]. Another strategy involves simply shifting CAR T-cell therapy (or at least apheresis for cell collection) to the first-line setting, prior to the receipt of cytotoxic chemotherapy and resultant damage to lymphocyte populations. Clinical trials employing this approach are ongoing. Moving forward, additional focus will also be placed on the starting lymphocyte material used to generate CAR T cells. One idea is the use of a phosphoionositide 3-kinase (PI3K/0) inhibitor during manufacturing, which has been shown to preserve a less differentiated T-cell state and potentially improve T-cell function [81]. In addition, it has been shown that epigenetic reprogramming may be used to alter CAR T-cell differentiation and result in a central memory phenotype [82], raising the possibility of using drugs with epigenome-influencing drugs during the CAR T-cell manufacturing process.

CAR T cells in combination with other therapies

In an effort to augment their potential therapeutic effect, CAR T-cell therapies are also being explored in combination with viral therapies, vaccines, and other immunotherapeutic classes of treatment. Both native and genetically engineered viruses are being developed for treatment across a variety of solid tumors, including an FDA-approved herpes simplex virus (HSV-1) for melanoma [83] and numerous viruses in clinical trials for GBM [84, 85]. Because oncolytic and other viral therapies can lead to increased immune cell infiltration and proinflammatory cytokines and have the potential to lyse tumor cells and release additional tumor-associated antigens, there is strong rationale for combining these treatments with CAR T cells [86]. In addition, viral therapies can be armed with therapeutic transgenes that could further enhance the effector functions of T cells. Numerous preclinical studies have explored the combination of viral therapies with CAR T cells [87–91]; clinical trials are in development.

Owing to the limited efficacy of CAR T cells in GBM and other solid tumors to date, coupled with the known ability of therapeutic vaccination to enhance endogenous T-cell responses against cancer [92], CAR T cells are also being studied in combination with vaccination. Multiple studies have demonstrated that introducing a CAR together with a second antigen receptor specific for a target peptide (or using virus-specific endogenous T cells to prepare the CAR T cells) and then vaccinating recipients against the secondary or viral antigen can improve CAR T-cell efficacy [93–95]. In one recent study, investigators used a vaccine strategy to re-stimulate CAR T cells directly within the native lymph node microenvironment [96]. This was accomplished by vaccinating mice with an albumin-binding phospholipid polymer linked to a peptide antigen. By binding to albumin after injection, these amphiphilic polymers are directed to lymph nodes, where the polymers (including their peptide antigen cargo) are preferentially displayed on resident antigen-presenting cells (APCs). These APC can then prime CAR T cells encountered in the lymph nodes. Mice in this study were treated with CAR T cells targeted against the antigen fluorescein isothiocyanate (FITC), followed by immunization with the amphiphilic polymers linked to FITC. In vivo T cell proliferation was significantly improved by this approach. The experiment was then repeated using the tumor antigen EGFRvIII in glioma-bearing mice and demonstrated improved CAR T-cell proliferation and survival, as well as improved infiltration of activated CAR T cells into tumor sites [96].

Beyond virus and vaccine combinations, pairing CAR T cells with immune checkpoint inhibitors and/or small-molecule inhibitors (IDO, TGF-β, etc.) is also a logical next step for GBM trials. A clinical trial of CART-EGFRvIII cells in combination with the PD-1 inhibitor pembrolizumab for newly diagnosed GBM is currently ongoing (NCT03726515), as well a trial of IL13Rα2-targeted CAR T-cells with or without nivolumab and ipilimumab in recurrent GBM (NCT04003649). However, there are significant challenges that must be recognized in using these combinations, such as the possibility for additive side effects and toxicity, as well as the lack of blood-brain barrier penetration of large monoclonal antibodies. As a result, an appealing alternative to using checkpoint inhibitors and other immune modulating drugs with CAR T cells is to perform additional genetic engineering of the CAR T cells themselves to circumvent the immunosuppressive GBM microenvironment. One option is to transduce T cells not only for a tumor antigen-specific CAR, but also for a chimeric switch receptor containing the extracellular domain of an immune inhibitory signaling molecule, which, in turn, is fused to the transmembrane and cytoplasmic domains of a costimulatory molecule. This was described by Liu et al. using a PD1 switch receptor [97]. When the PD1 portion of the switch receptor engages the PDL1 ligand, it transmits an activating signal via the CD28 cytoplasmic domain instead of the inhibitory signal normally transduced by the PD1 cytoplasmic domain. This approach augmented CAR T-cell control of large, established solid tumors in multiple models of aggressive human solid tumors expressing relevant tumor antigens [97]. Another T-cell engineering strategy to overcome the immunosuppressive GBM microenvironment is to selectively knockout or knockin key endogenous genes through targeted nucleases, e.g., via CRISPR-Cas9 gene editing [98]. Several strategies based on CRISPR are being applied to develop novel CAR T cells by multiplexed genome editing, including (a) knockout of endogenous genes (TCRs, MHC, or self-antigens) to build allogenic universal CAR T-cells [99], (b) disruption of inhibitory receptors such as PD-1 or CTLA-4 [100], and (c) generation of T cells with a CAR knocked into the T-cell receptor α constant (TRAC) locus, which places the CAR expression under control of the TCR promoter and results in enhanced potency and uniform CAR expression [101]. In GBM specifically, a recent study used the CRISPR-Cas9 system to create universal CAR T cells resistant to the immune inhibitory effects of the PD-1/PD-L1 axis through multiplexed gene disruption of the endogenous T-cell receptor, beta-2 microglobulin, and PD-1. Intracerebral administration of these CAR T cells resulted in significant antitumor activity in preclinical glioma models, resulting in prolonged survival in mice bearing intracranial tumors [102].

Conclusions

CAR T-cell therapy has completely changed the treatment paradigm for patients with hematologic malignancies. Although efficacy in GBM and other solid tumors has been limited thus far, additional preclinical and clinical data to optimize CAR T-cell therapy for GBM are being generated rapidly. The main barriers to CAR T-cell therapy in GBM include impaired T-cell trafficking to the tumor, lack of expansion and persistence in a severely immunosuppressive tumor microenvironment, and marked antigen heterogeneity. It is imperative that each of these hurdles are addressed as this complex field moves forward.

References

[1] Sadelain M, Brentjens R, Riviere I. The basic principles of chimeric antigen receptor design. Cancer Discov 2013;3(4):388–98. https://doi.org/10.1158/2159-8290.Cd-12-0548.
[2] Srivastava S, Riddell SR. Engineering CAR-T cells: design concepts. Trends Immunol 2015;36(8):494–502. https://doi.org/10.1016/j.it.2015.06.004.

[3] Willemsen RA, Debets R, Hart E, Hoogenboom HR, Bolhuis RL, Chames P. A phage display selected fab fragment with MHC class I-restricted specificity for MAGE-A1 allows for retargeting of primary human T lymphocytes. Gene Ther 2001;8(21):1601–8. https://doi.org/10.1038/sj.gt.3301570.

[4] Eshhar Z, Waks T, Bendavid A, Schindler DG. Functional expression of chimeric receptor genes in human T cells. J Immunol Methods 2001;248(1–2):67–76. https://doi.org/10.1016/s0022-1759(00)00343-4.

[5] Hombach A, Wieczarkowiecz A, Marquardt T, Heuser C, Usai L, Pohl C, Seliger B, Abken H. Tumor-specific T cell activation by recombinant immunoreceptors: CD3 zeta signaling and CD28 costimulation are simultaneously required for efficient IL-2 secretion and can be integrated into one combined CD28/CD3 zeta signaling receptor molecule. J Immunol 2001;167(11):6123–31. https://doi.org/10.4049/jimmunol.167.11.6123.

[6] van der Stegen SJ, Hamieh M, Sadelain M. The pharmacology of second-generation chimeric antigen receptors. Nat Rev Drug Discov 2015;14(7):499–509. https://doi.org/10.1038/nrd4597.

[7] Jensen MC, Riddell SR. Designing chimeric antigen receptors to effectively and safely target tumors. Curr Opin Immunol 2015;33:9–15. https://doi.org/10.1016/j.coi.2015.01.002.

[8] Chen D, Yang J. Development of novel antigen receptors for CAR T-cell therapy directed toward solid malignancies. Transl Res 2017;187:11–21. https://doi.org/10.1016/j.trsl.2017.05.006.

[9] Buechner J, Grupp SA, Maude SL, Boyer M, Bittencourt H, Laetsch TW, Bader P, Verneris MR, Stefanski H, Myers GD, Qayed M, Pulsipher MA, De Moerloose B, Hiramatsu H, Schlis K, Davis K, Martin PL, Nemecek E, Peters C, Wood P, Taran T, Mueller KT, Zhang Y, Rives S. Global registration trial of efficacy and safety of CTL019 in pediatric and young adult patients with relapsed/refractory (R/R) acute lymphoblastic leukemia (ALL): update to the interim analysis. Clin Lymphoma Myeloma Leuk 2017;17:S263–4. https://doi.org/10.1016/j.clml.2017.07.030.

[10] Park JH, Riviere I, Gonen M, Wang X, Senechal B, Curran KJ, Sauter C, Wang Y, Santomasso B, Mead E, Roshal M, Maslak P, Davila M, Brentjens RJ, Sadelain M. Long-term follow-up of CD19 CAR therapy in acute lymphoblastic leukemia. N Engl J Med 2018;378(5):449–59. https://doi.org/10.1056/NEJMoa1709919.

[11] Maude SL, Frey N, Shaw PA, Aplenc R, Barrett DM, Bunin NJ, Chew A, Gonzalez VE, Zheng Z, Lacey SF, Mahnke YD, Melenhorst JJ, Rheingold SR, Shen A, Teachey DT, Levine BL, June CH, Porter DL, Grupp SA. Chimeric antigen receptor T cells for sustained remissions in leukemia. N Engl J Med 2014;371(16):1507–17. https://doi.org/10.1056/NEJMoa1407222.

[12] Neelapu SS, Locke FL, Bartlett NL, Lekakis LJ, Miklos DB, Jacobson CA, Braunschweig I, Oluwole OO, Siddiqi T, Lin Y, Timmerman JM, Stiff PJ, Friedberg JW, Flinn IW, Goy A, Hill BT, Smith MR, Deol A, Farooq U, McSweeney P, Munoz J, Avivi I, Castro JE, Westin JR, Chavez JC, Ghobadi A, Komanduri KV, Levy R, Jacobsen ED, Witzig TE, Reagan P, Bot A, Rossi J, Navale L, Jiang Y, Aycock J, Elias M, Chang D, Wiezorek J, Go WY. Axicabtagene ciloleucel CAR T-cell therapy in refractory large B-cell lymphoma. N Engl J Med 2017;377(26):2531–44. https://doi.org/10.1056/NEJMoa1707447.

[13] Schuster SJ, Svoboda J, Chong EA, Nasta SD, Mato AR, Anak O, Brogdon JL, Pruteanu-Malinici I, Bhoj V, Landsburg D, Wasik M, Levine BL, Lacey SF, Melenhorst JJ, Porter DL, June CH. Chimeric antigen receptor T cells in refractory B-cell lymphomas. N Engl J Med 2017;377(26):2545–54. https://doi.org/10.1056/NEJMoa1708566.

[14] Schuster SJ, Bishop MR, Tam CS, Waller EK, Borchmann P, McGuirk JP, Jager U, Jaglowski S, Andreadis C, Westin JR, Fleury I, Bachanova V, Foley SR, Ho PJ, Mielke S, Magenau JM, Holte H, Pantano S, Pacaud LB, Awasthi R, Chu J, Anak O, Salles G, Maziarz RT. Tisagenlecleucel in adult relapsed or refractory diffuse large B-cell lymphoma. N Engl J Med 2019;380(1):45–56. https://doi.org/10.1056/NEJMoa1804980.

[15] Raje N, Berdeja J, Lin Y, Siegel D, Jagannath S, Madduri D, Liedtke M, Rosenblatt J, Maus MV, Turka A, Lam LP, Morgan RA, Friedman K, Massaro M, Wang J, Russotti G, Yang Z, Campbell T, Hege K, Petrocca F, Quigley MT, Munshi N, Kochenderfer JN. Anti-BCMA CAR T-cell therapy bb2121 in relapsed or refractory multiple myeloma. N Engl J Med 2019;380(18):1726–37. https://doi.org/10.1056/NEJMoa1817226.

[16] Neelapu SS, Tummala S, Kebriaei P, Wierda W, Gutierrez C, Locke FL, Komanduri KV, Lin Y, Jain N, Daver N, Westin J, Gulbis AM, Loghin ME, de Groot JF, Adkins S, Davis SE, Rezvani K, Hwu P, Shpall EJ. Chimeric antigen receptor T-cell therapy—assessment and management of toxicities. Nat Rev Clin Oncol 2018;15(1):47–62. https://doi.org/10.1038/nrclinonc.2017.148.

[17] Long KB, Young RM, Boesteanu AC, Davis MM, Melenhorst JJ, Lacey SF, DeGaramo DA, Levine BL, Fraietta JA. CAR T cell therapy of non-hematopoietic malignancies: detours on the road to clinical success. Front Immunol 2018;9:2740. https://doi.org/10.3389/fimmu.2018.02740.

[18] Mintz A, Gibo DM, Slagle-Webb B, Christensen ND, Debinski W. IL-13Ralpha2 is a glioma-restricted receptor for interleukin-13. Neoplasia 2002;4(5):388–99. https://doi.org/10.1038/sj.neo.7900234.
[19] Thaci B, Brown CE, Binello E, Werbaneth K, Sampath P, Sengupta S. Significance of interleukin-13 receptor alpha 2-targeted glioblastoma therapy. Neuro Oncol 2014;16(10):1304–12. https://doi.org/10.1093/neuonc/nou045.
[20] Tu M, Wange W, Cai L, Zhu P, Gao Z, Zheng W. IL-13 receptor alpha2 stimulates human glioma cell growth and metastasis through the Src/PI3K/Akt/mTOR signaling pathway. Tumour Biol 2016;37(11):14701–9. https://doi.org/10.1007/s13277-016-5346-x.
[21] Brown CE, Warden CD, Starr R, Deng X, Badie B, Yuan YC, Forman SJ, Barish ME. Glioma IL13Ralpha2 is associated with mesenchymal signature gene expression and poor patient prognosis. PLoS One 2013;8(10). https://doi.org/10.1371/journal.pone.0077769, e77769.
[22] Debinski W, Gibo DM, Slagle B, Powers SK, Gillespie GY. Receptor for interleukin 13 is abundantly and specifically over-expressed in patients with glioblastoma multiforme. Int J Oncol 1999;15(3):481–6. https://doi.org/10.3892/ijo.15.3.481.
[23] Rodriguez A, Brown C, Badie B. Chimeric antigen receptor T-cell therapy for glioblastoma. Transl Res 2017;187:93–102. https://doi.org/10.1016/j.trsl.2017.07.003.
[24] Brown CE, Badie B, Barish ME, Weng L, Ostberg JR, Chang WC, Naranjo A, Starr R, Wagner J, Wright C, Zhai Y, Bading JR, Ressler JA, Portnow J, D'Apuzzo M, Forman SJ, Jensen MC. Bioactivity and safety of IL13Ralpha2-redirected chimeric antigen receptor CD8+ T cells in patients with recurrent glioblastoma. Clin Cancer Res 2015;21(18):4062–72. https://doi.org/10.1158/1078-0432.Ccr-15-0428.
[25] Brown CE, Alizadeh D, Starr R, Weng L, Wagner JR, Naranjo A, Ostberg JR, Blanchard MS, Kilpatrick J, Simpson J, Kurien A, Priceman SJ, Wang X, Harshbarger TL, D'Apuzzo M, Ressler JA, Jensen MC, Barish ME, Chen M, Portnow J, Forman SJ, Badie B. Regression of glioblastoma after chimeric antigen receptor T-cell therapy. N Engl J Med 2016;375(26):2561–9. https://doi.org/10.1056/NEJMoa1610497.
[26] Li G, Wong AJ. EGF receptor variant III as a target antigen for tumor immunotherapy. Expert Rev Vaccines 2008;7(7):977–85. https://doi.org/10.1586/14760584.7.7.977.
[27] Felsberg J, Hentschel B, Kaulich K, Gramatzki D, Zacher A, Malzkorn B, Kamp M, Sabel M, Simon M, Westphal M, Schackert G, Tonn JC, Pietsch T, von Deimling A, Loeffler M, Reifenberger G, Weller M. Epidermal growth factor receptor variant III (EGFRvIII) positivity in EGFR-amplified glioblastomas: prognostic role and comparison between primary and recurrent tumors. Clin Cancer Res 2017;23(22):6846–55. https://doi.org/10.1158/1078-0432.Ccr-17-0890.
[28] O'Rourke DM, Nasrallah MP, Desai A, Melenhorst JJ, Mansfield K, Morrissette JJD, Martinez-Lage M, Brem S, Maloney E, Shen A, Isaacs R, Mohan S, Plesa G, Lacey SF, Navenot JM, Zheng Z, Levine BL, Okada H, June CH, Brogdon JL, Maus MV. A single dose of peripherally infused EGFRvIII-directed CAR T cells mediates antigen loss and induces adaptive resistance in patients with recurrent glioblastoma. Sci Transl Med 2017;9(399). https://doi.org/10.1126/scitranslmed.aaa0984.
[29] Rapoport AP, Stadtmauer EA, Aqui N, Vogl D, Chew A, Fang HB, Janofsky S, Yager K, Veloso E, Zheng Z, Milliron T, Westphal S, Cotte J, Huynh H, Cannon A, Yanovich S, Akpek G, Tan M, Virts K, Ruehle K, Harris C, Philip S, Vonderheide RH, Levine BL, June CH. Rapid immune recovery and graft-versus-host disease-like engraftment syndrome following adoptive transfer of costimulated autologous T cells. Clin Cancer Res 2009; 15(13):4499–507. https://doi.org/10.1158/1078-0432.Ccr-09-0418.
[30] Del Vecchio CA, Giacomini CP, Vogel H, Jensen KC, Florio T, Merlo A, Pollack JR, Wong AJ. EGFRvIII gene rearrangement is an early event in glioblastoma tumorigenesis and expression defines a hierarchy modulated by epigenetic mechanisms. Oncogene 2013;32(21):2670–81. https://doi.org/10.1038/onc.2012.280.
[31] Liu G, Ying H, Zeng G, Wheeler CJ, Black KL, Yu JS. HER-2, gp100, and MAGE-1 are expressed in human glioblastoma and recognized by cytotoxic T cells. Cancer Res 2004;64(14):4980–6. https://doi.org/10.1158/0008-5472.Can-03-3504.
[32] Andersson U, Guo D, Malmer B, Bergenheim AT, Brannstrom T, Hedman H, Henriksson R. Epidermal growth factor receptor family (EGFR, ErbB2-4) in gliomas and meningiomas. Acta Neuropathol 2004;108(2):135–42. https://doi.org/10.1007/s00401-004-0875-6.
[33] Zhang C, Burger MC, Jennewein L, Genssler S, Schonfeld K, Zeiner P, Hattingen E, Harter PN, Mittelbronn M, Tonn T, Steinbach JP, Wels WS. ErbB2/HER2-specific NK cells for targeted therapy of glioblastoma. J Natl Cancer Inst 2016;108(5). https://doi.org/10.1093/jnci/djv375.

[34] Ahmed N, Brawley V, Hegde M, Bielamowicz K, Kalra M, Landi D, Robertson C, Gray TL, Diouf O, Wakefield A, Ghazi A, Gerken C, Yi Z, Ashoori A, Wu MF, Liu H, Rooney C, Dotti G, Gee A, Su J, Kew Y, Baskin D, Zhang YJ, New P, Grilley B, Stojakovic M, Hicks J, Powell SZ, Brenner MK, Heslop HE, Grossman R, Wels WS, Gottschalk S. HER2-specific chimeric antigen receptor-modified virus-specific T cells for progressive glioblastoma: a phase 1 dose-escalation trial. JAMA Oncol 2017;3(8):1094–101. https://doi.org/10.1001/jamaoncol.2017.0184.

[35] Pule MA, Savoldo B, Myers GD, Rossig C, Russell HV, Dotti G, Huls MH, Liu E, Gee AP, Mei Z, Yvon E, Weiss HL, Liu H, Rooney CM, Heslop HE, Brenner MK. Virus-specific T cells engineered to coexpress tumor-specific receptors: persistence and antitumor activity in individuals with neuroblastoma. Nat Med 2008;14(11):1264–70. https://doi.org/10.1038/nm.1882.

[36] Martinez M, Moon EK. CAR T cells for solid tumors: new strategies for finding, infiltrating, and surviving in the tumor microenvironment. Front Immunol 2019;10:128. https://doi.org/10.3389/fimmu.2019.00128.

[37] Walseng E, Koksal H, Sektioglu IM, Fane A, Skorstad G, Kvalheim G, Gaudernack G, Inderberg EM, Walchli S. A TCR-based chimeric antigen receptor. Sci Rep 2017;7(1):10713. https://doi.org/10.1038/s41598-017-11126-y.

[38] Gulley JL, Madan RA, Pachynski R, Mulders P, Sheikh NA, Trager J, Drake CG. Role of antigen spread and distinctive characteristics of immunotherapy in cancer treatment. J Natl Cancer Inst 2017;109(4). https://doi.org/10.1093/jnci/djw261.

[39] Dirkx AE, Oude Egbrink MG, Kuijpers MJ, van der Niet ST, Heijnen VV, Bouma-ter Steege JC, Wagstaff J, Griffioen AW. Tumor angiogenesis modulates leukocyte-vessel wall interactions *in vivo* by reducing endothelial adhesion molecule expression. Cancer Res 2003;63(9):2322–9.

[40] Gust J, Hay KA, Hanafi LA, Li D, Myerson D, Gonzalez-Cuyar LF, Yeung C, Liles WC, Wurfel M, Lopez JA, Chen J, Chung D, Harju-Baker S, Ozpolat T, Fink KR, Riddell SR, Maloney DG, Turtle CJ. Endothelial activation and blood-brain barrier disruption in neurotoxicity after adoptive immunotherapy with CD19 CAR-T cells. Cancer Discov 2017;7(12):1404–19. https://doi.org/10.1158/2159-8290.Cd-17-0698.

[41] Overstreet MG, Gaylo A, Angermann BR, Hughson A, Hyun YM, Lambert K, Acharya M, Billroth-Maclurg AC, Rosenberg AF, Topham DJ, Yagita H, Kim M, Lacy-Hulbert A, Meier-Schellersheim M, Fowell DJ. Inflammation-induced interstitial migration of effector CD4(+) T cells is dependent on integrin alphaV. Nat Immunol 2013;14(9):949–58. https://doi.org/10.1038/ni.2682.

[42] Kalos M, Levine BL, Porter DL, Katz S, Grupp SA, Bagg A, June CH. T cells with chimeric antigen receptors have potent antitumor effects and can establish memory in patients with advanced leukemia. Sci Transl Med 2011;3(95):95ra73. https://doi.org/10.1126/scitranslmed.3002842.

[43] Porter DL, Hwang WT, Frey NV, Lacey SF, Shaw PA, Loren AW, Bagg A, Marcucci KT, Shen A, Gonzalez V, Ambrose D, Grupp SA, Chew A, Zheng Z, Milone MC, Levine BL, Melenhorst JJ, June CH. Chimeric antigen receptor T cells persist and induce sustained remissions in relapsed refractory chronic lymphocytic leukemia. Sci Transl Med 2015;7(303):303ra139. https://doi.org/10.1126/scitranslmed.aac5415.

[44] Leick M, Maus MV. Wishing on a CAR: understanding the scope of intrinsic T-cell deficits in patients with cancer. Cancer Discov 2019;9(4):466–8. https://doi.org/10.1158/2159-8290.Cd-19-0073.

[45] Das RK, Vernau L, Grupp SA, Barrett DM. Naive T-cell deficits at diagnosis and after chemotherapy impair cell therapy potential in pediatric cancers. Cancer Discov 2019;9(4):492–9. https://doi.org/10.1158/2159-8290.Cd-18-1314.

[46] Chongsathidkiet P, Jackson C, Koyama S, Loebel F, Cui X, Farber SH, Woroniecka K, Elsamadicy AA, Dechant CA, Kemeny HR, Sanchez-Perez L, Cheema TA, Souders NC, Herndon JE, Coumans JV, Everitt JI, Nahed BV, Sampson JH, Gunn MD, Martuza RL, Dranoff G, Curry WT, Fecci PE. Sequestration of T cells in bone marrow in the setting of glioblastoma and other intracranial tumors. Nat Med 2018;24(9):1459–68. https://doi.org/10.1038/s41591-018-0135-2.

[47] Grossman SA, Ye X, Lesser G, Sloan A, Carraway H, Desideri S, Piantadosi S. Immunosuppression in patients with high-grade gliomas treated with radiation and temozolomide. Clin Cancer Res 2011;17(16):5473–80. https://doi.org/10.1158/1078-0432.Ccr-11-0774.

[48] Singh N, Perazzelli J, Grupp SA, Barrett DM. Early memory phenotypes drive T cell proliferation in patients with pediatric malignancies. Sci Transl Med 2016;8(320):320ra3. https://doi.org/10.1126/scitranslmed.aad5222.

[49] Wang D, Aguilar B, Starr R, Alizadeh D, Brito A, Sarkissian A, Ostberg JR, Forman SJ, Brown CE. Glioblastoma-targeted CD4+ CAR T cells mediate superior antitumor activity. JCI Insight 2018;3(10). https://doi.org/10.1172/jci.insight.99048.

[50] Guedan S, Posey Jr AD, Shaw C, Wing A, Da T, Patel PR, McGettigan SE, Casado-Medrano V, Kawalekar OU, Uribe-Herranz M, Song D, Melenhorst JJ, Lacey SF, Scholler J, Keith B, Young RM, June CH. Enhancing CAR T cell persistence through ICOS and 4-1BB costimulation. JCI Insight 2018;3(1). https://doi.org/10.1172/jci.insight.96976.
[51] Anthony SM, Rivas SC, Colpitts SL, Howard ME, Stonier SW, Schluns KS. Inflammatory signals regulate IL-15 in response to lymphodepletion. J Immunol 2016;196(11):4544–52. https://doi.org/10.4049/jimmunol.1600219.
[52] Gattinoni L, Finkelstein SE, Klebanoff CA, Antony PA, Palmer DC, Spiess PJ, Hwang LN, Yu Z, Wrzesinski C, Heimann DM, Surh CD, Rosenberg SA, Restifo NP. Removal of homeostatic cytokine sinks by lymphodepletion enhances the efficacy of adoptively transferred tumor-specific CD8+ T cells. J Exp Med 2005;202(7):907–12. https://doi.org/10.1084/jem.20050732.
[53] Heczey A, Louis CU, Savoldo B, Dakhova O, Durett A, Grilley B, Liu H, Wu MF, Mei Z, Gee A, Mehta B, Zhang H, Mahmood N, Tashiro H, Heslop HE, Dotti G, Rooney CM, Brenner MK. CAR T cells administered in combination with lymphodepletion and PD-1 inhibition to patients with neuroblastoma. Mol Ther 2017; 25(9):2214–24. https://doi.org/10.1016/j.ymthe.2017.05.012.
[54] Muranski P, Boni A, Wrzesinski C, Citrin DE, Rosenberg SA, Childs R, Restifo NP. Increased intensity lymphodepletion and adoptive immunotherapy—how far can we go? Nat Clin Pract Oncol 2006;3(12):668–81. https://doi.org/10.1038/ncponc0666.
[55] Suryadevara CM, Desai R, Abel ML, Riccione KA, Batich KA, Shen SH, Chongsathidkiet P, Gedeon PC, Elsamadicy AA, Snyder DJ, Herndon 2nd JE, Healy P, Archer GE, Choi BD, Fecci PE, Sampson JH, Sanchez-Perez L. Temozolomide lymphodepletion enhances CAR abundance and correlates with antitumor efficacy against established glioblastoma. Onco Targets Ther 2018;7(6). https://doi.org/10.1080/2162402x.2018.1434464, e1434464.
[56] Labanieh L, Majzner RG, Mackall CL. Programming CAR-T cells to kill cancer. Nat Biomed Eng 2018;2(6):377–91. https://doi.org/10.1038/s41551-018-0235-9.
[57] Zhang L, Morgan RA, Beane JD, Zheng Z, Dudley ME, Kassim SH, Nahvi AV, Ngo LT, Sherry RM, Phan GQ, Hughes MS, Kammula US, Feldman SA, Toomey MA, Kerkar SP, Restifo NP, Yang JC, Rosenberg SA. Tumor-infiltrating lymphocytes genetically engineered with an inducible gene encoding interleukin-12 for the immunotherapy of metastatic melanoma. Clin Cancer Res 2015;21(10):2278–88. https://doi.org/10.1158/1078-0432.Ccr-14-2085.
[58] Krenciute G, Prinzing BL, Yi Z, Wu MF, Liu H, Dotti G, Balyasnikova IV, Gottschalk S. Transgenic expression of IL15 improves antiglioma activity of IL13Ralpha2-CAR T cells but results in antigen loss variants. Cancer Immunol Res 2017;5(7):571–81. https://doi.org/10.1158/2326-6066.Cir-16-0376.
[59] Hu B, Ren J, Luo Y, Keith B, Young RM, Scholler J, Zhao Y, June CH. Augmentation of antitumor immunity by human and mouse CAR T cells secreting IL-18. Cell Rep 2017;20(13):3025–33. https://doi.org/10.1016/j.celrep.2017.09.002.
[60] Spolski R, Leonard WJ. Interleukin-21: a double-edged sword with therapeutic potential. Nat Rev Drug Discov 2014;13(5):379–95. https://doi.org/10.1038/nrd4296.
[61] Ngwa VM, Edwards DN, Philip M, Chen J. Microenvironmental metabolism regulates antitumor immunity. Cancer Res 2019. https://doi.org/10.1158/0008-5472.Can-19-0617.
[62] Schurich A, Magalhaes I, Mattsson J. Metabolic regulation of CAR T cell function by the hypoxic microenvironment in solid tumors. Immunotherapy 2019;11(4):335–45. https://doi.org/10.2217/imt-2018-0141.
[63] Bagley SJ, Desai AS, Linette GP, June CH, O'Rourke DM. CAR T-cell therapy for glioblastoma: recent clinical advances and future challenges. Neuro Oncol 2018;20(11):1429–38. https://doi.org/10.1093/neuonc/noy032.
[64] Ravanpay AC, Gust J, Johnson AJ, Rolczynski LS, Cecchini M, Chang CA, Hoglund VJ, Mukherjee R, Vitanza NA, Orentas RJ, Jensen MC. EGFR806-CAR T cells selectively target a tumor-restricted EGFR epitope in glioblastoma. Oncotarget 2019;10(66):7080–95. https://doi.org/10.18632/oncotarget.27389.
[65] Binder ZA, Thorne AH, Bakas S, Wileyto EP, Bilello M, Akbari H, Rathore S, Ha SM, Zhang L, Ferguson CJ, Dahiya S, Bi WL, Reardon DA, Idbaih A, Felsberg J, Hentschel B, Weller M, Bagley SJ, JJD M, Nasrallah MP, Ma J, Zanca C, Scott AM, Orellana L, Davatzikos C, Furnari FB, O'Rourke DM. Epidermal growth factor receptor extracellular domain mutations in glioblastoma present opportunities for clinical imaging and therapeutic development. Cancer Cell 2018;34(1):163–177.e7. https://doi.org/10.1016/j.ccell.2018.06.006.
[66] Beatty GL, O'Hara MH, Lacey SF, Torigian DA, Nazimuddin F, Chen F, Kulikovskaya IM, Soulen MC, McGarvey M, Nelson AM, Gladney WL, Levine BL, Melenhorst JJ, Plesa G, June CH. Activity of mesothelin-specific chimeric antigen receptor T cells against pancreatic carcinoma metastases in a phase 1 trial. Gastroenterology 2018;155(1):29–32. https://doi.org/10.1053/j.gastro.2018.03.029.

[67] Roybal KT, Rupp LJ, Morsut L, Walker WJ, McNally KA, Park JS, Lim WA. Precision tumor recognition by T cells with combinatorial antigen-sensing circuits. Cell 2016;164(4):770–9. https://doi.org/10.1016/j.cell.2016.01.011.

[68] Becerra CR, Hoof P, Paulson AS, Manji GA, Gardner O, Malankar A, Shaw J, Blass D, Ballard B, Yi X, Anumula M, Foster AE, Senesac J, Woodard P. Ligand-inducible, prostate stem cell antigen (PSCA)-directed GoCAR-T cells in advanced solid tumors: preliminary results from a dose escalation. J Clin Oncol 2019;37(4_Suppl):283. https://doi.org/10.1200/JCO.2019.37.4_suppl.283.

[69] Raj D, Yang MH, Rodgers D, Hampton EN, Begum J, Mustafa A, Lorizio D, Garces I, Propper D, Kench JG, Kocher HM, Young TS, Aicher A, Heeschen C. Switchable CAR-T cells mediate remission in metastatic pancreatic ductal adenocarcinoma. Gut 2019;68(6):1052–64. https://doi.org/10.1136/gutjnl-2018-316595.

[70] Cui J, Zhang Q, Song Q, Wang H, Dmitriev P, Sun MY, Cao X, Wang Y, Guo L, Indig IH, Rosenblum JS, Ji C, Cao D, Yang K, Gilbert MR, Yao Y, Zhuang Z. Targeting hypoxia downstream signaling protein, CAIX, for CAR T-cell therapy against glioblastoma. Neuro Oncol 2019;21(11):1436–46. https://doi.org/10.1093/neuonc/noz117.

[71] Wang D, Starr R, Chang WC, Aguilar B, Alizadeh D, Wright SL, Yang X, Brito A, Sarkissian A, Ostberg JR, Li L, Shi Y, Gutova M, Aboody K, Badie B, Forman SJ, Barish ME, Brown CE. Chlorotoxin-directed CAR T cells for specific and effective targeting of glioblastoma. Sci Transl Med 2020;12(533). https://doi.org/10.1126/scitranslmed.aaw2672.

[72] Majzner RG, Mackall CL. Tumor antigen escape from CAR T-cell therapy. Cancer Discov 2018;8(10):1219–26. https://doi.org/10.1158/2159-8290.Cd-18-0442.

[73] Hegde M, Mukherjee M, Grada Z, Pignata A, Landi D, Navai SA, Wakefield A, Fousek K, Bielamowicz K, Chow KK, Brawley VS, Byrd TT, Krebs S, Gottschalk S, Wels WS, Baker ML, Dotti G, Mamonkin M, Brenner MK, Orange JS, Ahmed N. Tandem CAR T cells targeting HER2 and IL13Ralpha2 mitigate tumor antigen escape. J Clin Invest 2016;126(8):3036–52. https://doi.org/10.1172/jci83416.

[74] Bielamowicz K, Fousek K, Byrd TT, Samaha H, Mukherjee M, Aware N, Wu MF, Orange JS, Sumazin P, Man TK, Joseph SK, Hegde M, Ahmed N. Trivalent CAR T cells overcome interpatient antigenic variability in glioblastoma. Neuro Oncol 2018;20(4):506–18. https://doi.org/10.1093/neuonc/nox182.

[75] Choi BD, Yu X, Castano AP, Bouffard AA, Schmidts A, Larson RC, Bailey SR, Boroughs AC, Frigault MJ, Leick MB, Scarfo I, Cetrulo CL, Demehri S, Nahed BV, Cahill DP, Wakimoto H, Curry WT, Carter BS, Maus MV. CAR-T cells secreting BiTEs circumvent antigen escape without detectable toxicity. Nat Biotechnol 2019. https://doi.org/10.1038/s41587-019-0192-1.

[76] Katz SC, Burga RA, McCormack E, Wang LJ, Mooring W, Point GR, Khare PD, Thorn M, Ma Q, Stainken BF, Assanah EO, Davies R, Espat NJ, Junghans RP. Phase I hepatic immunotherapy for metastases study of intra-arterial chimeric antigen receptor-modified T-cell therapy for CEA+ liver metastases. Clin Cancer Res 2015;21(14):3149–59. https://doi.org/10.1158/1078-0432.Ccr-14-1421.

[77] Adusumilli PS, Zauderer MG, Rusch VW, O'Cearbhaill R, Zhu A, Ngai D, McGee E, Chintala N, Messinger J, Cheema W, Halton E, Diamonte C, Pineda J, Vincent A, Modi S, Solomon SB, Jones DR, Brentjens RJ, Riviere I, Sadelain M. Regional delivery of mesothelin-targeted CAR T cells for pleural cancers: safety and preliminary efficacy in combination with anti-PD-1 agent. J Clin Oncol 2019;37(15_Suppl):2511. https://doi.org/10.1200/JCO.2019.37.15_suppl.2511.

[78] Jin L, Tao H, Karachi A, Long Y, Hou AY, Na M, Dyson KA, Grippin AJ, Deleyrolle LP, Zhang W, Rajon DA, Wang QJ, Yang JC, Kresak JL, Sayour EJ, Rahman M, Bova FJ, Lin Z, Mitchell DA, Huang J. CXCR1- or CXCR2-modified CAR T cells co-opt IL-8 for maximal antitumor efficacy in solid tumors. Nat Commun 2019;10(1):4016. https://doi.org/10.1038/s41467-019-11869-4.

[79] Shrimali RK, Yu Z, Theoret MR, Chinnasamy D, Restifo NP, Rosenberg SA. Antiangiogenic agents can increase lymphocyte infiltration into tumor and enhance the effectiveness of adoptive immunotherapy of cancer. Cancer Res 2010;70(15):6171–80. https://doi.org/10.1158/0008-5472.Can-10-0153.

[80] Caruana I, Savoldo B, Hoyos V, Weber G, Liu H, Kim ES, Ittmann MM, Marchetti D, Dotti G. Heparanase promotes tumor infiltration and antitumor activity of CAR-redirected T lymphocytes. Nat Med 2015;21(5):524–9. https://doi.org/10.1038/nm.3833.

[81] Zheng W, O'Hear CE, Alli R, Basham JH, Abdelsamed HA, Palmer LE, Jones LL, Youngblood B, Geiger TL. PI3K orchestration of the *in vivo* persistence of chimeric antigen receptor-modified T cells. Leukemia 2018;32(5):1157–67. https://doi.org/10.1038/s41375-017-0008-6.

[82] Fraietta JA, Nobles CL, Sammons MA, Lundh S, Carty SA, Reich TJ, Cogdill AP, Morrissette JJD, JE DN, Reddy S, Hwang Y, Gohil M, Kulikovskaya I, Nazimuddin F, Gupta M, Chen F, Everett JK, Alexander KA, Lin-Shiao E, Gee MH, Liu X, Young RM, Ambrose D, Wang Y, Xu J, Jordan MS, Marcucci KT, Levine BL, Garcia KC, Zhao Y,

Kalos M, Porter DL, Kohli RM, Lacey SF, Berger SL, Bushman FD, June CH, Melenhorst JJ. Disruption of TET2 promotes the therapeutic efficacy of CD19-targeted T cells. Nature 2018;558(7709):307–12. https://doi.org/10.1038/s41586-018-0178-z.

[83] Andtbacka RH, Kaufman HL, Collichio F, Amatruda T, Senzer N, Chesney J, Delman KA, Spitler LE, Puzanov I, Agarwala SS, Milhem M, Cranmer L, Curti B, Lewis K, Ross M, Guthrie T, Linette GP, Daniels GA, Harrington K, Middleton MR, Miller Jr WH, Zager JS, Ye Y, Yao B, Li A, Doleman S, VanderWalde A, Gansert J, Coffin RS. Talimogene laherparepvec improves durable response rate in patients with advanced melanoma. J Clin Oncol 2015;33(25):2780–8. https://doi.org/10.1200/jco.2014.58.3377.

[84] Desjardins A, Gromeier M, Herndon 2nd JE, Beaubier N, Bolognesi DP, Friedman AH, Friedman HS, McSherry F, Muscat AM, Nair S, Peters KB, Randazzo D, Sampson JH, Vlahovic G, Harrison WT, McLendon RE, Ashley D, Bigner DD. Recurrent glioblastoma treated with recombinant poliovirus. N Engl J Med 2018;379(2):150–61. https://doi.org/10.1056/NEJMoa1716435.

[85] Cloughesy TF, Landolfi J, Vogelbaum MA, Ostertag D, Elder JB, Bloomfield S, Carter B, Chen CC, Kalkanis SN, Kesari S, Lai A, Lee IY, Liau LM, Mikkelsen T, Nghiemphu P, Piccioni D, Accomando W, Diago OR, Hogan DJ, Gammon D, Kasahara N, Kheoh T, Jolly DJ, Gruber HE, Das A, Walbert T. Durable complete responses in some recurrent high-grade glioma patients treated with Toca 511 + Toca FC. Neuro Oncol 2018;20(10):1383–92. https://doi.org/10.1093/neuonc/noy075.

[86] Guedan S, Alemany R. CAR-T cells and oncolytic viruses: joining forces to overcome the solid tumor challenge. Front Immunol 2018;9:2460. https://doi.org/10.3389/fimmu.2018.02460.

[87] Nishio N, Diaconu I, Liu H, Cerullo V, Caruana I, Hoyos V, Bouchier-Hayes L, Savoldo B, Dotti G. Armed oncolytic virus enhances immune functions of chimeric antigen receptor-modified T cells in solid tumors. Cancer Res 2014;74(18):5195–205. https://doi.org/10.1158/0008-5472.Can-14-0697.

[88] Watanabe K, Luo Y, Da T, Guedan S, Ruella M, Scholler J, Keith B, Young RM, Engels B, Sorsa S, Siurala M, Havunen R, Tähtinen S, Hemminki A, June CH. Pancreatic cancer therapy with combined mesothelin-redirected chimeric antigen receptor T cells and cytokine-armed oncolytic adenoviruses. JCI Insight 2018; 3(7). https://doi.org/10.1172/jci.insight.99573.

[89] Moon EK, Wang L-CS, Bekdache K, Lynn RC, Lo A, Thorne SH, Albelda SM. Intra-tumoral delivery of CXCL11 via a vaccinia virus, but not by modified T cells, enhances the efficacy of adoptive T cell therapy and vaccines. Oncoimmunology 2018;7(3). https://doi.org/10.1080/2162402X.2017.1395997, e1395997.

[90] Tanoue K, Rosewell Shaw A, Watanabe N, Porter C, Rana B, Gottschalk S, Brenner M, Suzuki M. Armed oncolytic adenovirus-expressing PD-L1 mini-body enhances antitumor effects of chimeric antigen receptor T cells in solid tumors. Cancer Res 2017;77(8):2040–51. https://doi.org/10.1158/0008-5472.Can-16-1577.

[91] Rosewell Shaw A, Porter CE, Watanabe N, Tanoue K, Sikora A, Gottschalk S, Brenner MK, Suzuki M. Adenovirotherapy delivering cytokine and checkpoint inhibitor augments CAR T cells against metastatic head and neck cancer. Mol Ther 2017;25(11):2440–51. https://doi.org/10.1016/j.ymthe.2017.09.010.

[92] van der Burg SH, Arens R, Ossendorp F, van Hall T, Melief CJ. Vaccines for established cancer: overcoming the challenges posed by immune evasion. Nat Rev Cancer 2016;16(4):219–33. https://doi.org/10.1038/nrc.2016.16.

[93] Tanaka M, Tashiro H, Omer B, Lapteva N, Ando J, Ngo M, Mehta B, Dotti G, Kinchington PR, Leen AM, Rossig C, Rooney CM. Vaccination targeting native receptors to enhance the function and proliferation of chimeric antigen receptor (CAR)-modified T cells. Clin Cancer Res 2017;23(14):3499–509. https://doi.org/10.1158/1078-0432.Ccr-16-2138.

[94] Wang X, Wong CW, Urak R, Mardiros A, Budde LE, Chang WC, Thomas SH, Brown CE, La Rosa C, Diamond DJ, Jensen MC, Nakamura R, Zaia JA, Forman SJ. CMVpp65 vaccine enhances the antitumor efficacy of adoptively transferred CD19-redirected CMV-specific T cells. Clin Cancer Res 2015;21(13):2993–3002. https://doi.org/10.1158/1078-0432.Ccr-14-2920.

[95] Slaney CY, von Scheidt B, Davenport AJ, Beavis PA, Westwood JA, Mardiana S, Tscharke DC, Ellis S, Prince HM, Trapani JA, Johnstone RW, Smyth MJ, Teng MW, Ali A, Yu Z, Rosenberg SA, Restifo NP, Neeson P, Darcy PK, Kershaw MH. Dual-specific chimeric antigen receptor T cells and an indirect vaccine eradicate a variety of large solid tumors in an immunocompetent, self-antigen setting. Clin Cancer Res 2017;23(10):2478–90. https://doi.org/10.1158/1078-0432.Ccr-16-1860.

[96] Ma L, Dichwalkar T, Chang JYH, Cossette B, Garafola D, Zhang AQ, Fichter M, Wang C, Liang S, Silva M, Kumari S, Mehta NK, Abraham W, Thai N, Li N, Wittrup KD, Irvine DJ. Enhanced CAR-T cell activity against solid tumors by vaccine boosting through the chimeric receptor. Science 2019;365(6449):162–8. https://doi.org/10.1126/science.aav8692.

[97] Liu X, Ranganathan R, Jiang S, Fang C, Sun J, Kim S, Newick K, Lo A, June CH, Zhao Y, Moon EK. A chimeric switch-receptor targeting PD1 augments the efficacy of second-generation CAR T cells in advanced solid tumors. Cancer Res 2016;76(6):1578–90. https://doi.org/10.1158/0008-5472.Can-15-2524.

[98] Liu J, Zhou G, Zhang L, Zhao Q. Building potent chimeric antigen receptor T cells with CRISPR genome editing. Front Immunol 2019;10:456. https://doi.org/10.3389/fimmu.2019.00456.

[99] Ren J, Liu X, Fang C, Jiang S, June CH, Zhao Y. Multiplex genome editing to generate universal CAR T cells resistant to PD1 inhibition. Clin Cancer Res 2017;23(9):2255–66. https://doi.org/10.1158/1078-0432.Ccr-16-1300.

[100] Rupp LJ, Schumann K, Roybal KT, Gate RE, Ye CJ, Lim WA, Marson A. CRISPR/Cas9-mediated PD-1 disruption enhances anti-tumor efficacy of human chimeric antigen receptor T cells. Sci Rep 2017;7(1):737. https://doi.org/10.1038/s41598-017-00462-8.

[101] Eyquem J, Mansilla-Soto J, Giavridis T, van der Stegen SJ, Hamieh M, Cunanan KM, Odak A, Gonen M, Sadelain M. Targeting a CAR to the TRAC locus with CRISPR/Cas9 enhances tumour rejection. Nature 2017;543(7643):113–7. https://doi.org/10.1038/nature21405.

[102] Choi BD, Yu X, Castano AP, Darr H, Henderson DB, Bouffard AA, Larson RC, Scarfò I, Bailey SR, Gerhard GM, Frigault MJ, Leick MB, Schmidts A, Sagert JG, Curry WT, Carter BS, Maus MV. CRISPR-Cas9 disruption of PD-1 enhances activity of universal EGFRvIII CAR T cells in a preclinical model of human glioblastoma. J Immunother Cancer 2019;7(1):304. https://doi.org/10.1186/s40425-019-0806-7.

CHAPTER

18

On-chip analysis of glioblastoma cell chemoresistance

Chi Yan Wong[a,b], *Huilin Shao*[a,c,d,e,f], *and Hakho Lee*[g,h]

[a]Institute for Health Innovation and Technology, National University of Singapore, Singapore [b]Department of Medicine, Yong Loo Lin School of Medicine, National University of Singapore, Singapore [c]Department of Biomedical Engineering, Faculty of Engineering, National University of Singapore, Singapore [d]Department of Surgery, Yong Loo Lin School of Medicine, National University of Singapore, Singapore [e]Institute of Molecular and Cell Biology, Agency for Science, Technology and Research, Singapore [f]National Neuroscience Institute, Singapore [g]Center for Systems Biology, Massachusetts General Hospital, Boston, MA, United States [h]Department of Radiology, Massachusetts General Hospital/Harvard Medical School, Boston, MA, United States

Abstract

The survival of patients with glioblastoma (GBM) remains poor despite radiotherapy and temozolomide (TMZ) chemotherapy after resection. The DNA repair enzymes O^6-methylguanine-DNA methyltransferase (MGMT) and alkylpurine-DNA-*N*-glycosylase (APNG) play key roles in repairing TMZ-induced DNA damages, and resistance to TMZ is correlated to the levels of these enzymes in tissues. Despite the need to monitor treatment efficacy serially, repeat brain biopsies are invasive and complex, and can entail significant complications. Herein, we elaborate on the use of a microfluidic chip designed to measure mRNA levels of MGMT and APNG in enriched cancer exosomes from blood. We demonstrate that exosomal mRNA levels reflect these key enzyme levels in parent cells and that serial exosome analyses of MGMT and APNG levels for patients undergoing TMZ treatment can be correlated to treatment response. As such, this technology has the potential to monitor and predict chemotherapy efficacy in GBM patients in a minimally invasive manner.

Abbreviations

APNG alkylpurine DNA *N*-glycosylase
GBM glioblastoma multiforme
MGMT O^6-methylguanine-DNA methyltransferase
TMZ temozolomide

Glioblastoma Resistance to Chemotherapy: Molecular Mechanisms and Innovative Reversal Strategies
https://doi.org/10.1016/B978-0-12-821567-8.00028-2

Conflict of interest

No potential conflicts of interest were disclosed.

Introduction

Glioblastoma and chemotherapy monitoring

Glioblastoma (GBM) is the most common type of malignant brain tumor. It is very aggressive, with GBM patients having a 5-year survival rate of 6.8% [1]. The standard treatment for GBM is maximal safe surgical resection, complemented by radiotherapy and temozolomide (TMZ) chemotherapy as an adjunct [2]. Both radiotherapy and TMZ chemotherapy have been shown to improve overall survival [3, 4]. In particular, the concomitant treatment of TMZ increased overall survival in GBM patients from 12.1 months with radiotherapy alone to 14.6 months with radiotherapy and TMZ [4]. Despite the increase in survival, response to treatment remains poor. The effect of TMZ treatment differs between patients, and patients eventually acquire resistance [5, 6]. As such, there is a critical need to identify and monitor factors that play a role in chemotherapy efficacy and chemoresistance.

TMZ treatment efficacy has been linked to the DNA repair enzymes O^6-methylguanine-DNA methyltransferase (MGMT) [7] and alkylpurine-DNA-*N*-glycosylase (APNG) [8]. TMZ is a DNA alkylating agent, and MGMT and APNG play key roles in repairing the lesions produced. Specifically, MGMT repairs the most cytotoxic lesion, O^6-methylguanine, while APNG repairs N3-methyladenine and N7-methylguanine. When these lesions are not resolved, cytotoxicity and apoptosis occur [9, 10]. Resistance to TMZ is also increased with the coexpression of MGMT and APNG. This agrees with the finding that the decreased production of these proteins is associated with improvement in survival in GBM patients who receive TMZ treatment, owing to their compromised DNA repair ability. The production of these proteins is in turn controlled by the promoter methylation status of these genes. In short, increased promoter methylation leads to epigenetic silencing of the genes, resulting in decreased mRNA production and thereby less enzymes that repair DNA damage. Hence, GBM patients with methylated MGMT and APNG promoters benefit more from TMZ treatment than those with unmethylated promoters.

Despite the known link between MGMT and APNG promoter methylation and TMZ treatment efficacy, overall patient survival remains low owing to acquired TMZ resistance. A recent study has revealed that a novel enhancer, when activated, is associated with increased MGMT expression, TMZ resistance, and tumor recurrence [11]. In addition, another study found a MGMT-independent mechanism for TMZ resistance, where DHC2 is involved in the nuclear transport of the DNA repair enzymes XPC and CBX5 in MGMT-deficient GBM [12]. Therefore, it is important to monitor treatment efficacy and regulate the dosage of TMZ to prevent chemoresistance and improve treatment outcome. Monitoring chemotherapy efficacy and chemoresistance via tissue samples, however, is not ideal owing to the need for repeat biopsies. Therefore, minimally invasive biomarkers and complementary analytical approaches that can track treatment efficacy are desired. To this end, extracellular vesicles, including exosomes, are promising circulating biomarkers that can be used as a surrogate for their parental cells.

Exosomes as biomarkers for chemotherapy monitoring

Exosomes are membrane-bound nanoscale vesicles (50–200nm) that are secreted by all cells in the body, especially by actively dividing tumor cells [13, 14]. They are present in many body fluids, including blood and urine, and contain diverse molecular cargoes (e.g., proteins and nucleic acids) that are reflective of their parent cells [15–17]. These molecular cargoes allow exosomes to participate in intercellular recognition and communication [17], and be used as a surrogate for the parent cells [18–21]. Of note, the small size of exosomes allows them to pass through the blood-brain barrier, enabling the study of GBM in a minimally invasive manner through GBM-derived exosomes found in the blood [15, 18, 19, 22].

As exosomes are present in complex body fluids, conventional molecular detection methods (e.g., enzyme-linked immunosorbent assay and western blotting) require extensive purification and enrichment to obtain high concentration and purity of exosomes for analysis [23]. Exosome purification and enrichment by ultracentrifugation, sucrose-gradient centrifugation, precipitation-based techniques or size exclusion chromatography are also time-consuming and intensive [24, 25], which is impractical for clinical translation. To this end, we have developed several on-chip nanoscale biosensing technologies tailored for the detection of molecular cargoes of exosomes—proteins [19, 20, 26], nucleic acids [18], sugar modifications [27], and associated molecules [28].

To study GBM chemotherapy efficacy and chemoresistance, we have devised an integrated microfluidic platform that directly measures RNA from exosomes in body fluids [18]. Named immunomagnetic exosome RNA (iMER), the technology comprises three components: immunomagnetic enrichment of GBM-derived exosomes from blood, capture of exosomal mRNA, and subsequent real-time quantitative polymerase chain reaction (RT-qPCR) of exosomal mRNA. The iMER system has a small sample input requirement (~100 μL of blood), can generate fast readouts (~2h), and has high sensitivity. Utilizing iMER, we found that the mRNA levels of MGMT and APNG in blood exosomes were indicative of treatment response in GBM patients administered with TMZ, demonstrating the potential of exosomal mRNA in lieu of tissue biopsies for predicting and monitoring chemotherapy response.

Exosomal RNA analysis for monitoring chemoresistance in GBM

Overview of iMER

The iMER platform consists of three components. First, the body fluid undergoes immunomagnetic enrichment of GBM-derived exosomes (Fig. 1A, left). This is done by functionalizing the surface of magnetic microbeads (3 μm) with anti-EGFR (epidermal growth factor receptor) antibodies that target both the wild-type EGFR and EGFRvIII mutation. This enrichment exploits the fact that GBM-derived exosomes have a higher expression level of EGFR and a unique EGFRvIII deletion mutation [15, 19, 29–31]. The use of magnetic enrichment also overcomes the complexity of native biological samples due to their inherent lack of ferromagnetic properties [32–35]. These magnetic microbeads were shown to be able to capture exosomes derived from the GLI36vIII cell line (Fig. 1B).

After immunoenrichment, captured exosomes were lysed with guanidine-containing buffer and exosomal RNA was captured on a glass bead filter through electrostatic interactions

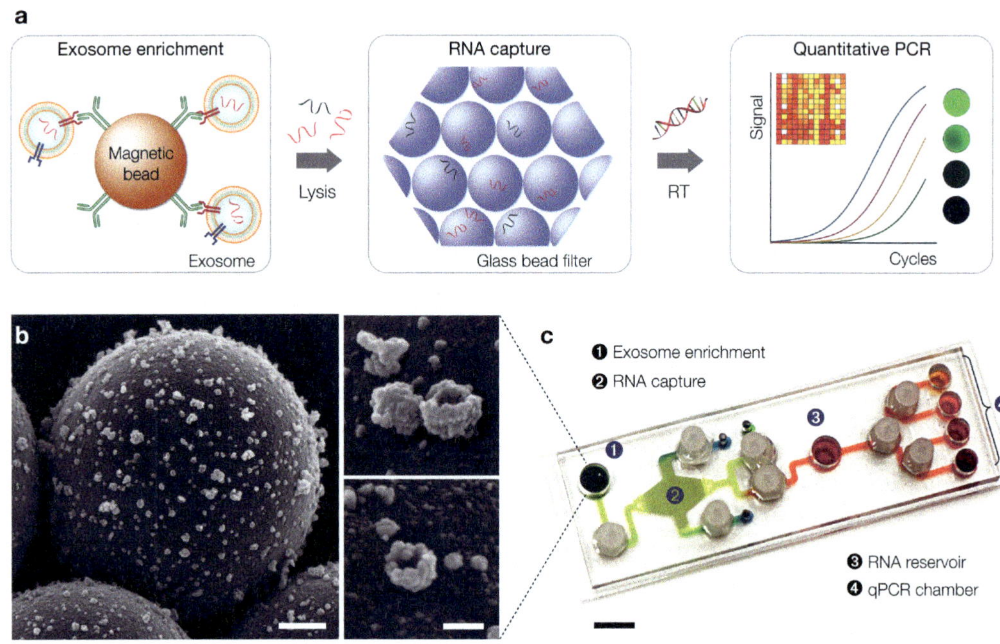

FIG. 1 Immunomagnetic exosomal RNA (iMER) platform. (A) The iMER platform comprises three modules on-chip. Magnetic beads functionalized with antibodies targeting surface proteins overexpressed in GBM-derived exosomes are used to capture the desired exosome populations. Exosomes are lysed to release their molecular contents, and RNAs are captured by adsorption onto a densely packed glass bead filter by electrostatic interactions. Captured RNA are reverse-transcribed and quantitative PCR is carried out on-chip via a custom thermal cycler and fluorescence detector. (B) Scanning electron micrographs of magnetic beads after immunoaffinity capture. Magnetic beads (left, 3 μm) functionalized with anti-EGFRvIII antibodies captured exosomes derived from GLI36vIII cells. High-magnification micrographs (right) of captured exosomes. Scale bars, 500 nm (left), 100 nm (right). (C) The iMER platform. Exosome enrichment, RNA capture, a RNA reservoir for reverse transcription, and qPCR chambers for multiplexed detection were integrated into a single microfluidic chip. Scale bar, 1 cm. *Reproduced with permission from Shao H, Chung J, Lee K, Balaj L, Min C, Carter BS, Hochberg FH, Breakefield XO, Lee H, Weissleder R. Chip-based analysis of exosomal mRNA mediating drug resistance in glioblastoma. Nat Commun 2015;6:6999. Copyright 2015 Nature Publishing Group.*

(Fig. 1A, middle). Washing was carried out to remove DNA and other contaminants; then, eluted RNA was reverse-transcribed and measured by qPCR (Fig. 1A, right). Fig. 1C shows the incorporation of the iMER components into a single microfluidic chip that takes patient serum as input and gives RNA expression levels as output via a custom-designed PCR system.

Microfluidic chip operation

We designed a microfluidic chip to facilitate all three functional steps of iMER processing (Fig. 2).

(1) Glass bead filling. With the torque-actuated valves V1 to V4 open, glass beads (20 μm) in 100% ethanol are flushed into the main chamber through the inlet. Weir-style physical

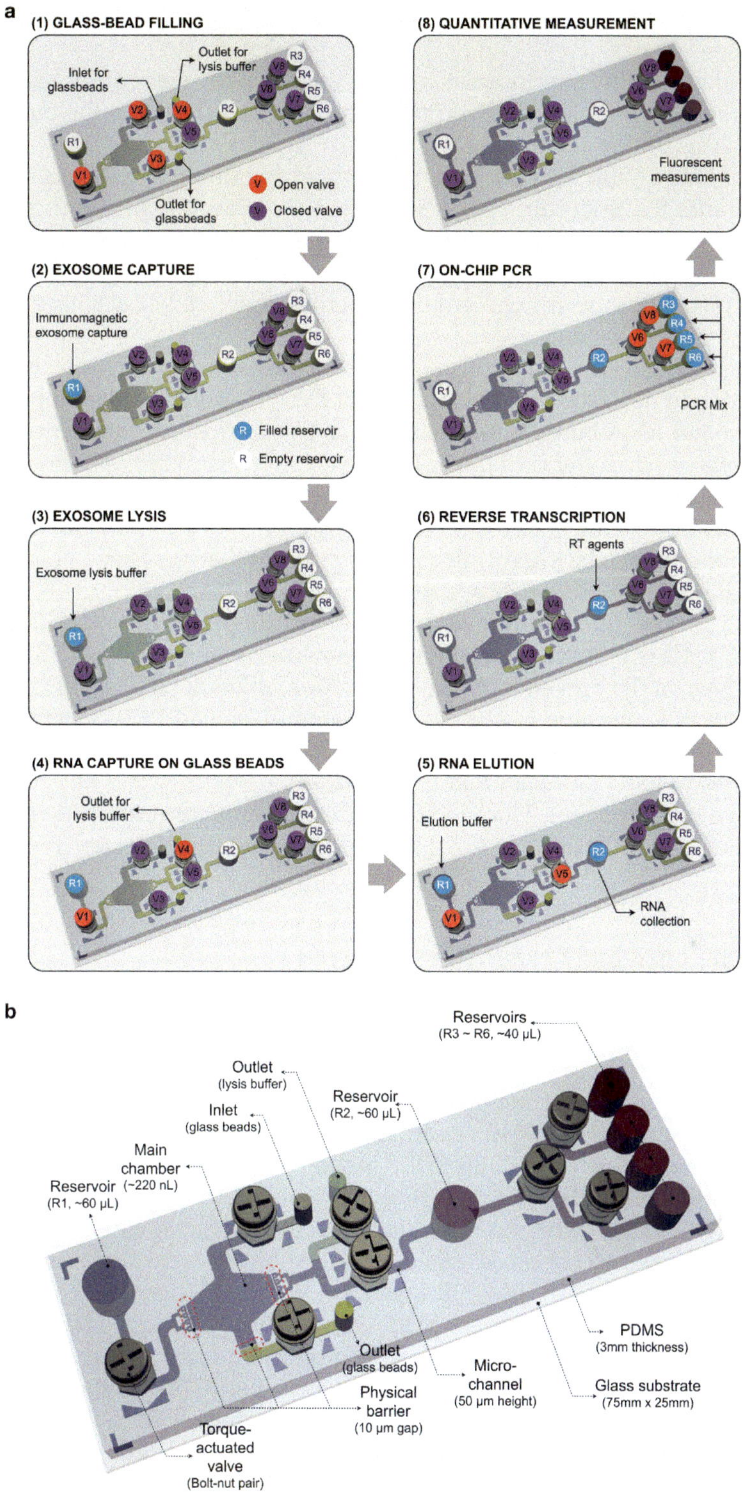

FIG. 2 The iMER microfluidic chip. (A) Steps for quantitative measurement of exosomal RNA using the iMER chip. (B) Device dimensions and maximum volumes. *Reproduced with permission from Shao H, Chung J, Lee K, Balaj L, Min C, Carter BS, Hochberg FH, Breakefield XO, Lee H, Weissleder R. Chip-based analysis of exosomal mRNA mediating drug resistance in glioblastoma. Nat Commun 2015;6:6999. Copyright 2015 Nature Publishing Group.*

barriers [36] at the three boundaries around the chamber prevent the glass beads from escaping. Excess ethanol is removed from the reservoir (R1), and the outlets for glass beads and lysis buffer are closed.

(2) Exosome capture. With V1 closed, exosomes and immunomagnetic beads (3 μm) are introduced into the reservoir R1 and incubated for 15 min. A permanent magnet is then put beneath R1 to pull down the beads and therefore capture bound exosomes. Exosomes, which are not bound to the beads, are aspirated out.

(3) Exosome lysis. After the enrichment of exosomes, they are lysed with guanidine-containing buffer (Qiagen).

(4) RNA capture on glass beads. Exosome lysate flows into the glass bead chamber for RNA collection due to negative pressure applied at the lysis buffer outlet when V1 and V4 are open. Exosomal RNA binds to the glass beads by electrostatic interactions. To remove contaminants, wash buffer containing DNase I (Qiagen) and 80% ethanol are sequentially introduced into the chamber.

(5) RNA elution. Pure water is introduced in R1 and flown to R2 via negative pressure. RNA is eluted from the beads when water passes through the glass bead chamber.

(6) Reverse transcription. Reagents (QuantiTech Reverse Transcription, Qiagen) are loaded into R2 and reverse transcription occurs by incubating the chip at 42°C.

(7) On-chip PCR. R2 is distributed into R3–R6 upon the opening of V6 to V8 by gravitational energy difference. To prevent sample evaporation, mineral oil (Sigma) is used to seal the samples. For PCR, the chip is placed on a custom-designed PCR system with a thermal cycler and a fluorescence detector. Amplification conditions are 1 cycle of 95°C for 30 s, and 40 cycles of 95°C for 3 s and 60°C for 30 s.

(8) Quantitative measurement. Fluorescence measurements for each PCR reservoir are conducted at each PCR cycle using a portable fluorescence detector (ESElog, Qiagen).

iMER exosome enrichment and RNA analysis

To confirm the feasibility of EGFR for exosome enrichment for GBM, we conducted a preliminary study with clinical samples obtained from GBM patients ($n=25$) and normal controls ($n=8$). Exosomes were isolated from the serum and the levels of EGFR, EGFRvIII, and the generic exosomal marker CD63 [16, 37] were determined. When normalized to CD63, expression levels of EGFR were found to be higher in GBM patients (Fig. 3A).

To determine the performance of the iMER exosome enrichment and RNA capture, we cultured SKMG3 and GLI36vIII GBM cell lines and isolated their exosomes. These exosomes were specifically captured by immunomagnetic beads functionalized with anti-EGFR. Exosome capture was not observed when magnetic beads were functionalized with anti-IgG isotope control antibodies (Fig. 3B and C). Capture efficiencies were >93% for both anti-EGFR and anti-CD63 enrichment (Fig. 3C). Western blotting further indicated that exosome populations expressing EGFR could be enriched from a mixture of exosomes derived from EGFR-positive and EGFR-negative cell lines (Fig. 3D). Together, these results demonstrated the ability of immunomagnetic selection to capture specific exosome populations.

To evaluate the efficacy of RNA capture, we compared the performance of RNA extraction from GLI36vIII exosomal lysate between the iMER cartridge and a commercial column filter. Exosomes harvested from GLI36vIII were first lysed on iMER. Given the same lysate

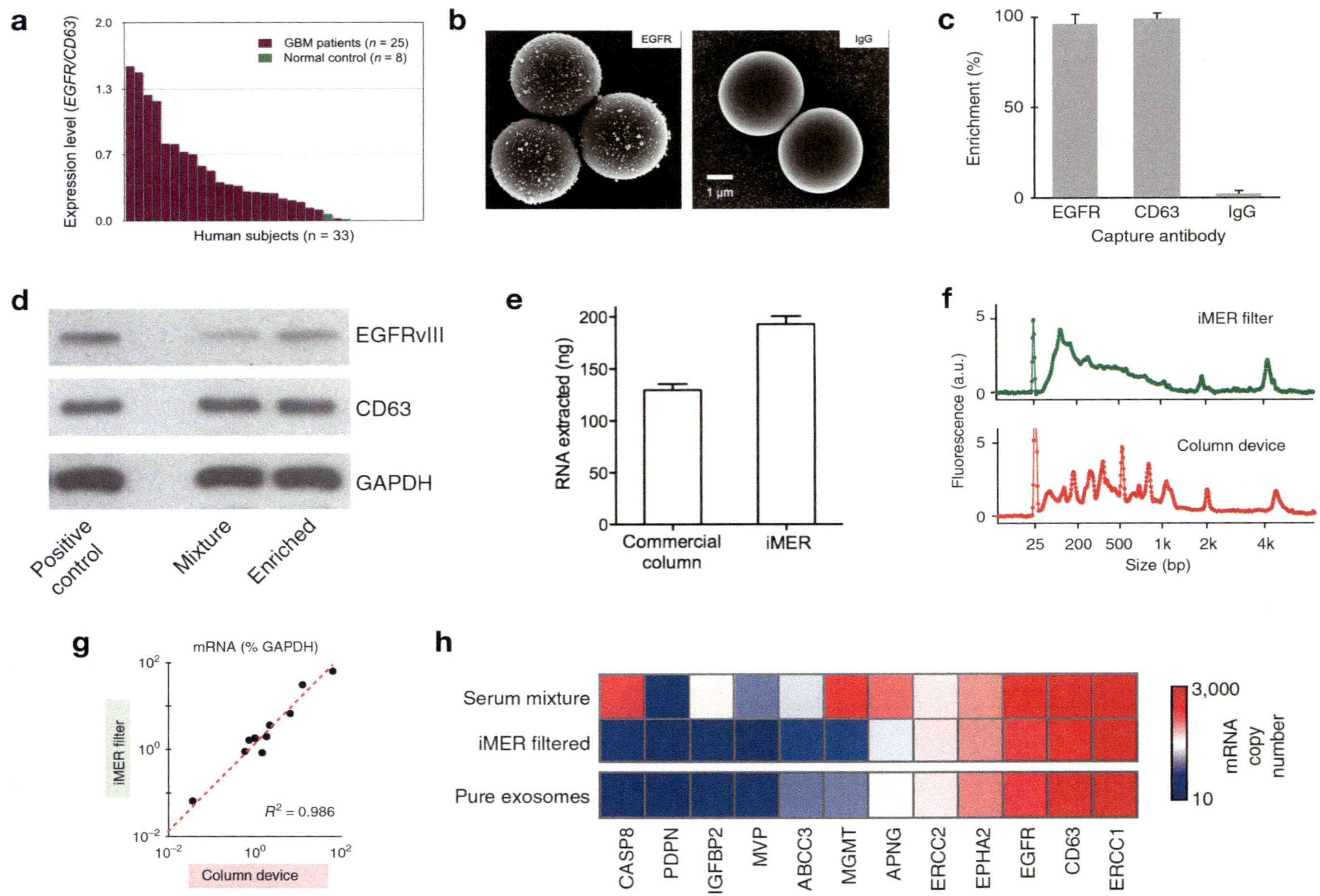

FIG. 3 See the figure legend on next page.

FIG. 3 iMER exosome enrichment and RNA analysis. (A) Expression level of EGFR in clinical samples. Clinical samples were obtained from GBM patients ($n=25$) and normal controls ($n=8$). Exosomal EGFR and EGFRvIII protein levels, normalized to the exosomal marker CD63, were measured. As exosomes from GBM patients have higher EGFR levels, EGFR was used for exosome enrichment. (B) Scanning electron micrographs of magnetic beads after immunoaffinity capture of GLI36vIII-derived exosomes. Beads functionalized with anti-EGFR (left) specifically captured and enriched for GLI36vIII exosomes, while beads with anti-IgG control did not (right). (C) Exosome enrichment with immunoaffinity capture of pure SKMG3-derived exosomes. Capture efficiencies were >93% for beads functionalized with anti-EGFR and anti-CD63 (generic exosomal marker) and <2% for the anti-IgG isotope control. (D) Exosome enrichment with immunoaffinity capture of mixed population exosomes. Equal amounts of exosomes derived from EGFRvIII-positive GLI36vIII cells and negative GBM20/3 cells were mixed ("mixture") and incubated with beads functionalized with anti-EGFR. "Enriched" shows the specific enrichment of exosomes expressing EGFR from the mixture compared against an equal amount of pure GLI36vIII exosomes. (E) RNA extraction yields for exosome lysates using a commercial column and with the iMER platform. The iMER platform had a higher yield for the same volume of exosome lysate. (F) RNA quality for exosome lysates using a commercial column and with the iMER platform. RNA extracted using the iMER platform remained intact and of high quality. (G) Correlation between RNA extracted with iMER and commercial column. Levels of RNA targets were highly correlated ($R^2=0.986$). RNA levels were normalized against intrinsic GAPDH. (H) iMER exosome enrichment and RNA analysis in mock clinical samples. Pure SKMG3 exosomes were spiked into normal human serum "serum mixture"). When analyzed using the iMER platform with beads functionalized with anti-EGFR, "iMER filtered" sample showed a similar RNA profile as pure SKMG3 exosomes, unlike the profile obtained from the serum mixture. All measurements were performed in triplicate and the data are displayed as mean ± SD. *Reproduced with permission from Shao H, Chung J, Lee K, Balaj L, Min C, Carter BS, Hochberg FH, Breakefield XO, Lee H, Weissleder R. Chip-based analysis of exosomal mRNA mediating drug resistance in glioblastoma. Nat Commun 2015;6:6999. Copyright 2015 Nature Publishing Group.*

volume, the iMER cartridge was able to extract 50% more RNA than the commercial column (Fig. 3E) and the extracted RNA remained intact and of high quality (Fig. 3F). The RNA was analyzed by RT-qPCR to quantify the expression of various mRNAs extracted by both methods. Using RT-qPCR analysis, the levels of RNA targets extracted with the iMER cartridge correlated well with column-extracted RNA ($R^2=0.986$; Fig. 3G). These validations demonstrated the use of the iMER platform for unbiased, high quantity, and quality extraction of exosomal RNA.

With a mock clinical sample, iMER profiled mRNA targets accurately from a specific exosome population. Exosomes harvested from SKMG3 cells were spiked into normal human serum. Using iMER, with the enrichment for EGFR-expressing exosomes in the serum mixture, an mRNA profile similar to that of pure SKMG3 exosomes was obtained (Fig. 3H). Therefore, the iMER platform is a fast (~2 h vs 6–8 h for standard exosomal RNA analysis) and accurate on-chip technology for multiplexed analysis of exosomal RNA, while requiring minimal sample volume (~100 μL vs >1 mL for standard exosomal RNA analysis).

Exosome mRNA profiles reflect parent cellular status

Next, we examined whether exosomal mRNA profiles are representative of their parent cell mRNA profiles. This would allow exosomal mRNA to be used as a proxy for cellular mRNA and eliminate the need for tissue biopsies. We cultured 13 GBM cell lines and a control (normal human astrocytes, NHA) and isolated the released exosomes. Analysis of cellular mRNA content with conventional RT-qPCR and exosomal mRNA using iMER was

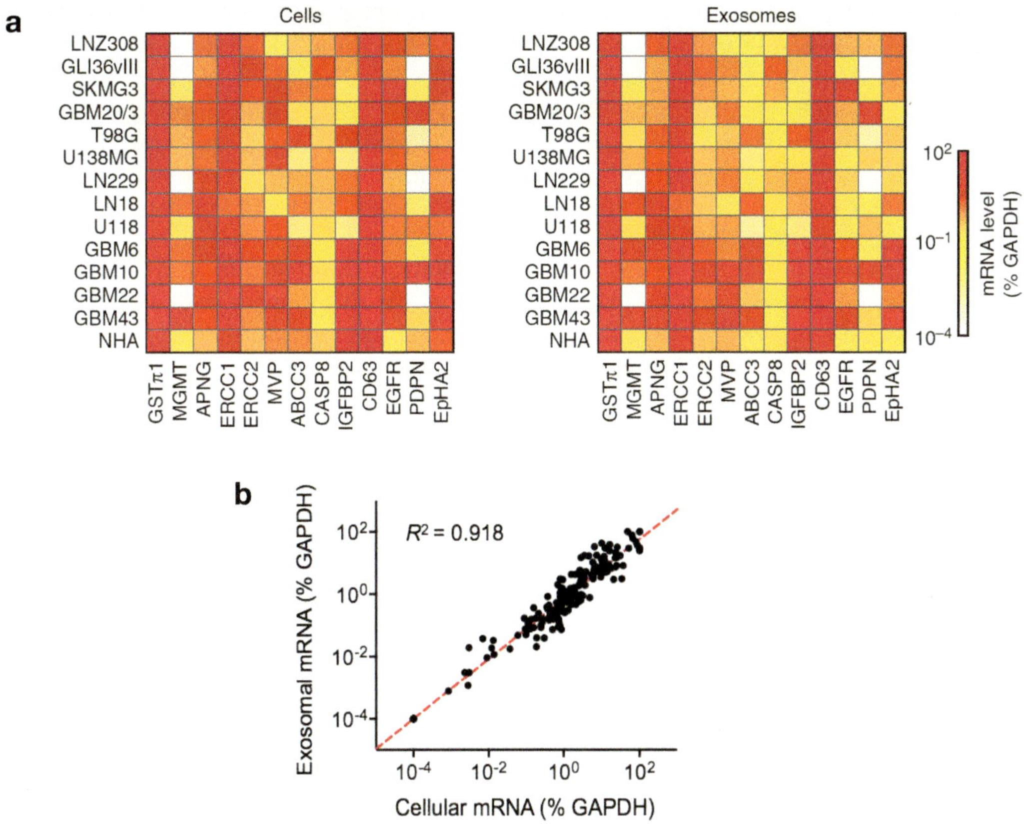

FIG. 4 Exosomal mRNA profiles reflect cellular mRNA. (A) Comparison of cellular and exosomal mRNA profiles. A total of 13 GBM cell lines and 1 control (normal human astrocytes, NHA) were cultured. Exosomes were isolated from all cell lines. The cellular mRNA levels (using conventional RT-qPCR) and exosomal mRNA levels (using an iMER platform) were determined for a panel of mRNA targets (diagnostic markers, chemoresistance markers, and a generic exosome marker). RNA levels were normalized against intrinsic GAPDH. All measurements were performed in triplicate and the data are displayed as mean. (B) Correlation between cellular and exosomal mRNA levels. The cellular and exosomal mRNA levels matched well ($R^2=0.918$). *Reproduced with permission from Shao H, Chung J, Lee K, Balaj L, Min C, Carter BS, Hochberg FH, Breakefield XO, Lee H, Weissleder R. Chip-based analysis of exosomal mRNA mediating drug resistance in glioblastoma. Nat Commun 2015;6:6999. Copyright 2015 Nature Publishing Group.*

conducted (Fig. 4A). A panel of mRNA targets were chosen based on previous studies: (i) diagnostic markers—EGFR [31], podoplanin (PDPN) [38], ephrin type-A receptor 2 (EPHA2) [39]; (ii) chemoresistance markers—MGMT [7], APNG [8], glutathione *S*-transferase (GSTp1) [40], excision repair cross-complementation (ERCC1 [41] and ERCC2 [42, 43]), major vault protein (MVP) [44, 45], ATP-binding cassette/multidrug resistance-associated protein 3 (ABCC3) [46], caspase-8 (CASP8) [47], and insulin-like growth factor binding protein 2 (IGFBP2) [48]; and (iii) a generic exosome marker—CD63 [16, 37]. We found that the mRNA expression levels in cells and corresponding exosomes correlated well ($R^2=0.918$; Fig. 4B), demonstrating that exosomal mRNA profiles reflect cellular mRNA profiles.

Subsequently, we investigated whether exosomal mRNA can be used as a surrogate for the status of their parent cells. Using various GBM cell lines, we showed that exosomal MGMT mRNA levels correlated inversely with MGMT promoter methylation in parent cells (Fig. 5A). The levels of cellular MGMT and APNG proteins also correlated well with exosomal mRNA levels (Fig. 5B). Using mouse xenograft models, the trend remained. GLI36vIII, which expresses negligible MGMT, and GBM10, which expresses high MGMT, were implanted into the flanks of mice ($n=5$ for each cell line). MGMT protein levels in extracted tumor tissues (Fig. 5C) reflected the mRNA expression levels in cell lines (Fig. 4A). Analysis of mouse serum with iMER showed that the human MGMT mRNA levels in mice implanted with GBM10 were high while those implanted with GLI36vIII had nondetectable levels (Fig. 5D). Taken together, exosomal mRNA could be used as an indicator for cellular methylation status and therefore chemoresistance status, and tumor status is reflected in serum exosomes.

Exosomal mRNA as a biomarker for treatment efficacy

The chemoresistance of various GBM cell lines was determined to investigate the relationship between chemoresistance and the mRNA levels of chemoresistance genes. GBM cell lines were treated with 0–1000 μM of TMZ, and their viability was measured after 4 days (Fig. 6A). Cell lines with a low 50% effective dose ($ED_{50} < 1000\,\mu M$) were classified as sensitive to TMZ (GLI36vIII and SKMG3) while those with a high 50% effective dose ($ED_{50} > 1000\,\mu M$) were classified as resistant (all other cell lines).

To examine the potential of exosomal mRNA levels as a biomarker for treatment efficacy, GBM cell lines with different chemoresistance levels to TMZ were cultured and their exosomes harvested. It was found that the cell lines classified as sensitive to TMZ had negligible levels of both MGMT and APNG exosomal mRNAs, while TMZ-resistant cell lines had high exosomal mRNA expression levels of MGMT, APNG, or both (Fig. 6B). Hence, the levels of MGMT and APNG exosomal mRNAs could reflect the efficacy of TMZ treatment.

Further study of exosomal mRNA levels revealed that exosomal MGMT and APNG mRNA levels changed upon cell TMZ treatment. TMZ-resistant cell lines had increased exosomal mRNA levels of MGMT, APNG, or both (Fig. 6C, left), while TMZ-sensitive cell lines had decreased levels of these mRNAs (Fig. 6C, right). These TMZ-induced changes were also reflected in the cellular mRNA levels (Fig. 6D).

Exosomal mRNA reflects TMZ treatment efficacy in clinical samples

Serum samples from GBM patients were obtained, and levels of exosomal MGMT mRNA were measured. GBM patients with negative MGMT promoter methylation in primary tissues had significantly higher MGMT exosomal levels than those with positive MGMT promoter methylation and healthy controls (Fig. 7A). GBM patients with positive MGMT promoter methylation had similar levels of exosomal MGMT mRNA as the healthy controls (Fig. 7A). This indicates that exosomal MGMT mRNA levels reflect methylation status in

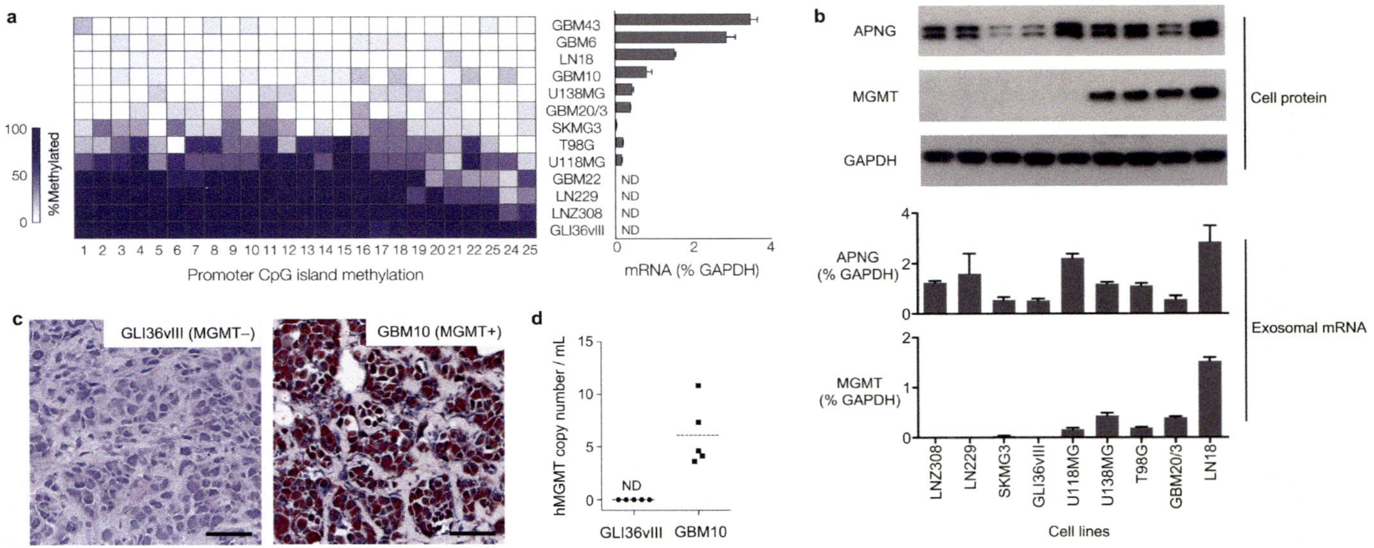

FIG. 5 Exosomal mRNA profiles reflect parent cellular status. (A) Comparison of MGMT promoter methylation levels in parent cells with MGMT mRNA levels in exosomes. Twenty-five specific CpG islands in the MGMT promoter were analyzed by bisulfite sequencing. MGMT promoter methylation levels in various GBM cell lines (left) correlated inversely with MGMT mRNA levels in derived exosomes (right). RNA levels were normalized against intrinsic GAPDH. *ND*, nondetectable. (B) Comparison of protein levels in parent cells with mRNA levels in exosomes. APNG and MGMT protein levels in various GBM cell lines (top) correlated with their respective mRNA levels in derived exosomes (bottom). All measurements were performed with GAPDH as an intrinsic control. (C) Immunohistochemistry of tumor tissues of mice implanted with GLI36vIII and GBM10 cells. MGMT protein expression in tumor tissues corresponded to MGMT mRNA levels in the implanted GBM cell lines. (D) Quantification of human MGMT mRNA in serum from the mouse models. Serum was collected from mice implanted with GLI36vIII ($n=5$) and GBM10 ($n=5$). Using iMER, human MGMT mRNA in mice serum was quantified. Human MGMT mRNA was highly expressed in mice implanted with GBM10 (MGMT positive) and nondetectable (ND) in mice implanted with GLI36vIII (MGMT negative). All measurements were performed in triplicate and the data are displayed as mean ± SD. *Reproduced with permission from Shao H, Chung J, Lee K, Balaj L, Min C, Carter BS, Hochberg FH, Breakefield XO, Lee H, Weissleder R. Chip-based analysis of exosomal mRNA mediating drug resistance in glioblastoma. Nat Commun 2015;6:6999. Copyright 2015 Nature Publishing Group.*

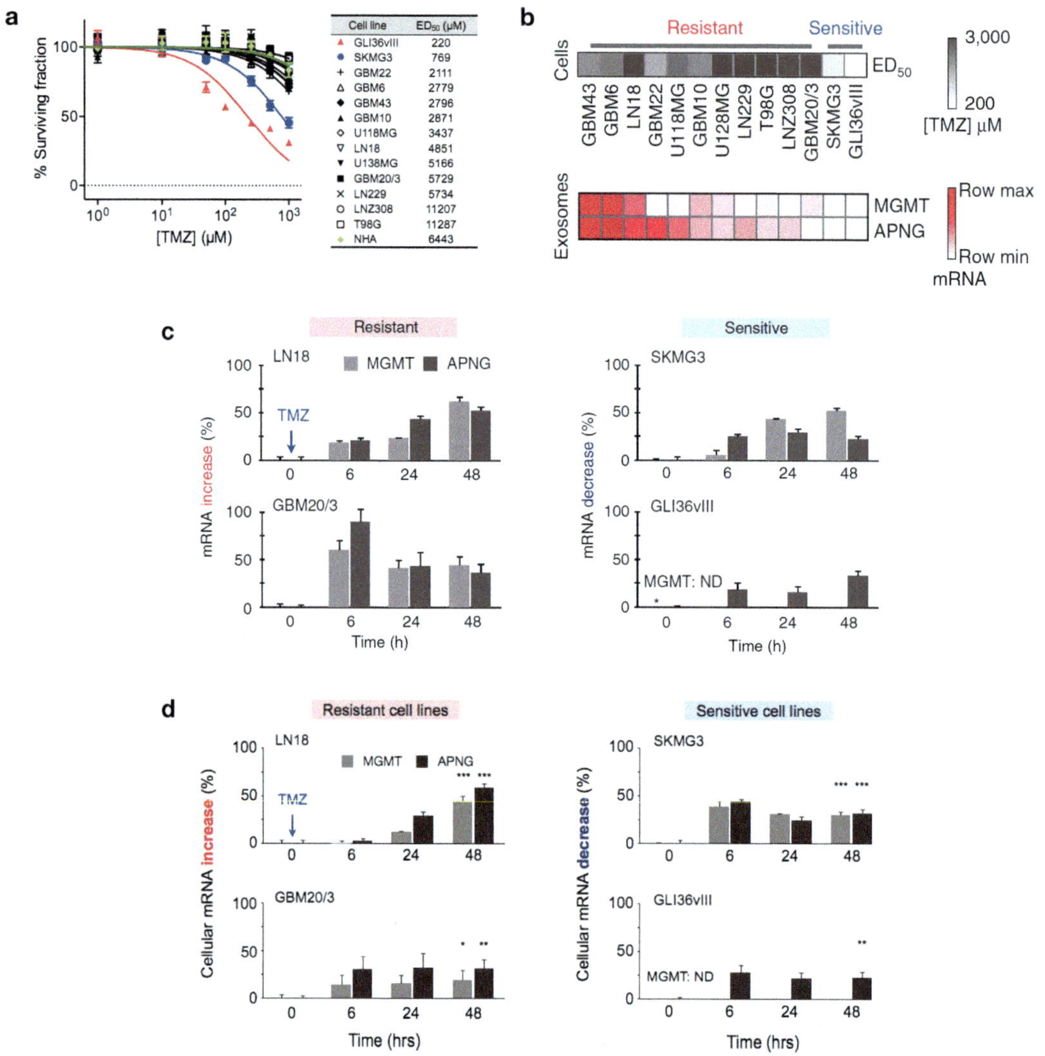

FIG. 6 Exosomal mRNA as a biomarker for treatment efficacy. (A) TMZ chemoresistance of GBM cell lines. Cell lines were treated with varying concentrations of TMZ for 4 days to determine their chemoresistance levels. ED_{50} = 50% effective dose. (B) Comparison between the chemoresistance status of GBM cell lines with exosomal MGMT and APNG mRNA levels. Cell lines sensitive to TMZ had negligible levels of both MGMT and APNG exosomal mRNAs. (C) Changes in exosomal mRNA levels of MGMT and APNG upon cell TMZ treatment. Exosomal mRNA levels of MGMT, APNG, or both increased for TMZ-resistant cells, and vice versa for TMZ-sensitive cells. *ND*, nondetectable. (D) Changes in cellular mRNA levels of MGMT and APNG upon cell TMZ treatment. Cellular mRNA levels matched well with exosomal mRNA levels. All measurements were performed in triplicate and the data are displayed as mean ± SD. *Reproduced with permission from Shao H, Chung J, Lee K, Balaj L, Min C, Carter BS, Hochberg FH, Breakefield XO, Lee H, Weissleder R. Chip-based analysis of exosomal mRNA mediating drug resistance in glioblastoma. Nat Commun 2015;6:6999. Copyright 2015 Nature Publishing Group.*

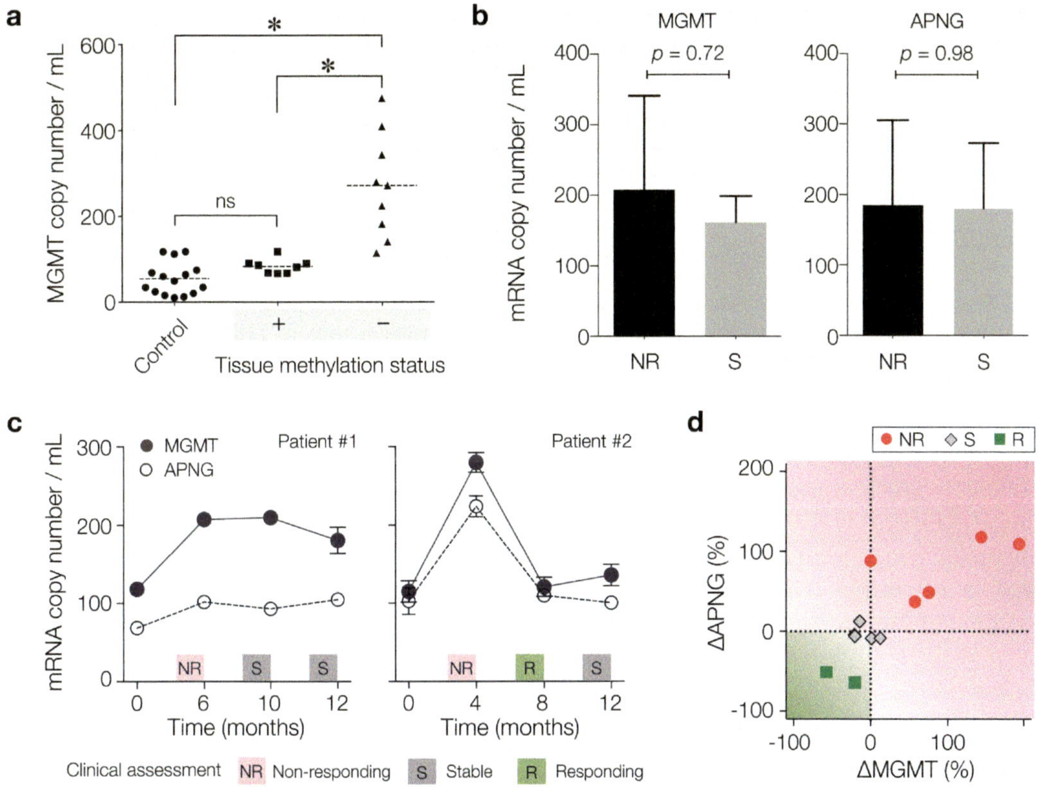

FIG. 7 Exosomal mRNA reflects TMZ treatment efficacy in clinical samples. (A) Exosomal MGMT mRNA levels in serum samples. For GBM patients, MGMT promoter DNA methylation status in primary tissues correlated inversely with exosomal MGMT mRNA levels. GBM patients with negative MGMT methylation had significantly higher levels of exosomal MGMT mRNA (*$P < .001$; Tukey's multiple comparison test). (B) Comparison of initial exosomal MGMT and APNG mRNA levels with eventual response to TMZ treatment. There was no significant difference (unpaired two-tailed *t*-test) in MGMT and APNG exosomal mRNA levels observed between nonresponding (NR, $n = 3$) and stable (S, $n = 4$) patients. (C) Longitudinal analysis of MGMT and APNG exosomal mRNA for two patients undergoing TMZ treatment. Clinical assessment at various time points corresponded to the change in MGMT and APNG levels. (D) Relationship between changes in MGMT and APNG exosomal mRNA levels and clinical assessment in GBM patients ($n = 7$) undergoing TMZ treatment. Using two adjacent time points, the changes in MGMT and APNG levels could stratify clinical assessments. All measurements were performed in triplicate and the data are displayed as mean ± SD. *Reproduced with permission from Shao H, Chung J, Lee K, Balaj L, Min C, Carter BS, Hochberg FH, Breakefield XO, Lee H, Weissleder R. Chip-based analysis of exosomal mRNA mediating drug resistance in glioblastoma. Nat Commun 2015;6:6999. Copyright 2015 Nature Publishing Group.*

primary tissues, and shows the clinical potential of exosomal MGMT mRNA for monitoring treatment efficacy in a minimally invasive manner.

To validate the clinical use, the MGMT and APNG exosomal mRNA levels of seven GBM patients undergoing treatment were profiled with iMER longitudinally. There was no correlation between the levels of MGMT and APNG initially and eventual response to treatment (Fig. 7B, NR: nonresponding, S: stable). However, when analyzed longitudinally, the change in exosomal mRNA levels reflected treatment trajectories (Fig. 7C) in agreement with the trends observed when cell lines were treated with TMZ. Specifically, the MGMT and APNG mRNA levels increased for nonresponding patients, decreased for responding patients, and remained relatively constant for stable patients. The change in MGMT and APNG levels between two adjacent time points also had the potential to predict treatment response (Fig. 7D, $n = 7$). However, a larger cohort is required to obtain statistical significance.

Outlook

The use of exosomes as a surrogate for primary tissue or circulating tumor cells allows minimally invasive real-time monitoring of the primary tissue status. This is especially so for GBM, where repeat biopsies to obtain primary tissue are laborious and dangerous. For the analysis of circulating tumor cells, large volumes of blood (~7 mL) are required to obtain sufficient numbers of the rare cells (~1–10 cells/mL). On the other hand, exosomes can pass through the blood-brain barrier, are found in high concentrations in the blood (~10^9/mL), are stable, and contain molecular cargoes from parent cells. Moreover, exosomes can be analyzed serially and report the entirety and heterogeneity of the tumor. GBM TMZ chemoresistance has been linked to the nuclear proteins MGMT and APNG, which participate in repairing TMZ-induced DNA damage. However, these proteins are unevenly partitioned into exosomes as they are primarily localized in the nucleus. As mRNAs are located in the cytoplasm during translation, the detection of exosomal mRNA instead of proteins overcomes this challenge. Therefore, exosomal mRNA is a promising biomarker for tumor status.

The iMER platform harnesses the potential of exosomes for determining tumor status longitudinally. Here, GBM-derived exosomes are enriched and the mRNA profiles are obtained in real time. Immunomagnetic enrichment of exosomes, exosome lysis and RNA capture, and RT-qPCR analysis are carried out with a single microfluidic chip. Using the iMER platform, we showed that (i) exosomal mRNA profiles reflected parent cellular status, (ii) mRNA of chemoresistance genes MGMT and APNG can be measured from patient serum, and (iii) serial exosome analyses can be correlated to the TMZ treatment response. Through optimization of the iMER microfluidic design and assay operation, high yields in exosome enrichment (>93%) and RNA extraction (150% compared to a commercial column) were obtained. This enabled the sensitive measurements of exosomal mRNA, which indicate cellular levels of proteins, including nucleus-localized proteins.

This platform presents several advantages for the analysis of GBM chemoresistance. On-chip processing allows streamlining of exosome enrichment and RNA analysis. Using a minimal sample volume of ~100 μL of serum, iMER can carry out rapid (~2 h) and sensitive multiplexed analysis of exosomal mRNA. In a preliminary study, iMER enabled the monitoring of the mRNA of chemoresistance genes (indicator of epigenetic silencing), which reflects the TMZ treatment response. This allowed a change in treatment regimen for patient #2 in Fig. 7C, where a better clinical response occurred upon the change from treatment with TMZ to bevacizumab/avastin.

Further improvements to this chip-based GBM chemoresistance platform can be made. Instead of only enriching for exosomes expressing EGFR, other surface proteins amplified in GBM (e.g., neuronal proteins) can be included to cater for tumor heterogeneity and GBM subtypes. Four distinct molecular subtypes of GBM have been identified—proneural, neural, classical, and mesenchymal [49]. Despite the generally low survival rate for GBM, patients with the mesenchymal subtype have the worst prognosis [50–52]. Hence, the inclusion of additional mRNA markers that can differentiate GBM subtypes would be useful to predict subtype-dependent prognosis and treatment response. In addition, apart from measuring mRNA, the study of exosomal proteins (proteins found within exosomes and on surface of exosomes) would provide more insight into tumor status, and aid in GBM subtyping. In terms of fluidics, more and smaller PCR chambers can be used to increase throughput for a more comprehensive understanding of the tumor.

Advances in other new analytical technologies can also be applied to study GBM chemoresistance. Biomimetic systems to study GBM in vitro can be done by capturing and culturing circulating tumor cells and glioma-propagating cells on microfluidic chips [36, 53]. These cultured cells can then be treated with TMZ on chip to investigate and predict TMZ resistance in vivo. Molecular detection technologies [54–56] have the potential to improve the sensitivity and ease of nucleic acid measurements, as well as reveal new links between cellular protein localization and TMZ resistance. This is especially relevant given the findings of recent studies, that the recruitment of DNA repair enzymes into the nucleus causes acquired TMZ resistance [12]. Improved exosome purification technologies [57], further analyses of exosomal modifications [27] and associated molecules [28], particularly at the single exosome resolution [58, 59], as well as analyses of different sizes of extracellular vesicles [60] may also enable the discovery of new clinical signatures for GBM diagnosis, prognosis, and chemoresistance.

Clinically, such technologies can complement clinical assessments made with advanced imaging. Although advanced imaging is the gold standard for brain tumors, it has unsatisfactory accuracy [2], where diagnostic specificity for GBM from the other types of brain tumor ranges from 50% to 80%. Furthermore, the use of advanced imaging for treatment monitoring is hindered by multiple factors. First, the maximum spatial resolution of millimeters is unable to detect the change in tumor dimension (micrometers) after initial resection. Next, concomitant antiangiogenic therapies prevent the exit of imaging contrast agents from blood vessels, resulting in low tumor contrast enhancement [61, 62]. Finally, imaging is unable to differentiate between radiation necrosis cells and viable tumor cells [63]. Hence, more information on tumor staging as well as subtyping, prognosis, treatment monitoring, and optimal treatment regimes are required for better clinical outcomes. To this end, the development of new technologies that enable fast, easy, and multiplexed molecular analysis of small volumes of clinical samples obtained in a minimally-invasive manner is critical. These can provide comprehensive information on the genetic, epigenetic, and transcriptomic status of tumors at various time points that may reveal new insights, as well as determine the real-time efficacy of chemotherapy, allowing clinicians to better design and rationalize treatment regimens.

Acknowledgments

Funding

This work was supported in part by funding from National University of Singapore (NUS), NUS Research Scholarship, Ministry of Education, National Medical Research Council, Institute for Health Innovation & Technology, IMCB Independent Fellowship and NUS Early Career Research Award.

Competing interests

H.L. declares that he has received consultancy payments from Exosome Diagnostics and Accure Health. All patents associated with H.L. have been assigned to and handled by Massachusetts General Hospital. C.Y.W. and H.S. declare no competing interests.

References

[1] Ostrom QT, Cioffi G, Gittleman H, Patil N, Waite K, Kruchko C, Barnholtz-Sloan JS. CBTRUS statistical report: primary brain and other central nervous system tumors diagnosed in the United States in 2012–2016. Neuro Oncol 2019;21:v1–v100.
[2] Hochberg FH, Atai NA, Gonda D, Hughes MS, Mawejje B, Balaj L, Carter RS. Glioma diagnostics and biomarkers: an ongoing challenge in the field of medicine and science. Expert Rev Mol Diagn 2014;14:439–52.
[3] Perkins A, Liu G. Primary brain tumors in adults: diagnosis and treatment. Am Fam Physician 2016;93:211–7.
[4] Stupp R, Mason WP, van den Bent MJ, Weller M, Fisher B, Taphoorn MJ, Belanger K, Brandes AA, Marosi C, Bogdahn U, Curschmann J, Janzer RC, Ludwin SK, Gorlia T, Allgeier A, Lacombe D, Cairncross JG, Eisenhauer E, Mirimanoff RO, European OFRATOCBTARG, National CIOCCTG. Radiotherapy plus concomitant and adjuvant temozolomide for glioblastoma. N Engl J Med 2005;352:987–96.
[5] Felsberg J, Rapp M, Loeser S, Fimmers R, Stummer W, Goeppert M, Steiger HJ, Friedensdorf B, Reifenberger G, Sabel MC. Prognostic significance of molecular markers and extent of resection in primary glioblastoma patients. Clin Cancer Res 2009;15:6683–93.
[6] Happold C, Roth P, Wick W, Schmidt N, Florea AM, Silginer M, Reifenberger G, Weller M. Distinct molecular mechanisms of acquired resistance to temozolomide in glioblastoma cells. J Neurochem 2012;122:444–55.
[7] Hegi ME, Diserens AC, Gorlia T, Hamou MF, de Tribolet N, Weller M, Kros JM, Hainfellner JA, Mason W, Mariani L, Bromberg JE, Hau P, Mirimanoff RO, Cairncross JG, Janzer RC, Stupp R. MGMT gene silencing and benefit from temozolomide in glioblastoma. N Engl J Med 2005;352:997–1003.
[8] Agnihotri S, Gajadhar AS, Ternamian C, Gorlia T, Diefes KL, Mischel PS, Kelly J, McGown G, Thorncroft M, Carlson BL, Sarkaria JN, Margison GP, Aldape K, Hawkins C, Hegi M, Guha A. Alkylpurine-DNA-N-glycosylase confers resistance to temozolomide in xenograft models of glioblastoma multiforme and is associated with poor survival in patients. J Clin Invest 2012;122:253–66.
[9] Liu L, Markowitz S, Gerson SL. Mismatch repair mutations override alkyltransferase in conferring resistance to temozolomide but not to 1,3-bis(2-chloroethyl)nitrosourea. Cancer Res 1996;56:5375–9.
[10] Ochs K, Kaina B. Apoptosis induced by DNA damage O6-methylguanine is Bcl-2 and caspase-9/3 regulated and Fas/caspase-8 independent. Cancer Res 2000;60:5815–24.
[11] Chen X, Zhang M, Gan H, Wang H, Lee JH, Fang D, Kitange GJ, He L, Hu Z, Parney IF, Meyer FB, Giannini C, Sarkaria JN, Zhang Z. A novel enhancer regulates MGMT expression and promotes temozolomide resistance in glioblastoma. Nat Commun 2018;9:2949.
[12] Yi GZ, Huang G, Guo M, Zhang X, Wang H, Deng S, Li Y, Xiang W, Chen Z, Pan J, Li Z, Yu L, Lei B, Liu Y, Qi S. Acquired temozolomide resistance in MGMT-deficient glioblastoma cells is associated with regulation of DNA repair by DHC2. Brain 2019;142:2352–66.
[13] Lötvall J, Hill AF, Hochberg F, Buzás EI, Di Vizio D, Gardiner C, Gho YS, Kurochkin IV, Mathivanan S, Quesenberry P, Sahoo S, Tahara H, Wauben MH, Witwer KW, Théry C. Minimal experimental requirements for definition of extracellular vesicles and their functions: a position statement from the International Society for Extracellular Vesicles. J Extracell Vesicles 2014;3:26913.
[14] Théry C, Zitvogel L, Amigorena S. Exosomes: composition, biogenesis and function. Nat Rev Immunol 2002;2:569–79.
[15] Skog J, Würdinger T, van Rijn S, Meijer DH, Gainche L, Sena-Esteves M, Curry WT, Carter BS, Krichevsky AM, Breakefield XO. Glioblastoma microvesicles transport RNA and proteins that promote tumour growth and provide diagnostic biomarkers. Nat Cell Biol 2008;10:1470–6.
[16] Théry C, Ostrowski M, Segura E. Membrane vesicles as conveyors of immune responses. Nat Rev Immunol 2009;9:581–93.
[17] Mathivanan S, Ji H, Simpson RJ. Exosomes: extracellular organelles important in intercellular communication. J Proteomics 2010;73:1907–20.

[18] Shao H, Chung J, Lee K, Balaj L, Min C, Carter BS, Hochberg FH, Breakefield XO, Lee H, Weissleder R. Chip-based analysis of exosomal mRNA mediating drug resistance in glioblastoma. Nat Commun 2015;6:6999.
[19] Shao H, Chung J, Balaj L, Charest A, Bigner DD, Carter BS, Hochberg FH, Breakefield XO, Weissleder R, Lee H. Protein typing of circulating microvesicles allows real-time monitoring of glioblastoma therapy. Nat Med 2012;18:1835–40.
[20] Im H, Shao H, Park YI, Peterson VM, Castro CM, Weissleder R, Lee H. Label-free detection and molecular profiling of exosomes with a nano-plasmonic sensor. Nat Biotechnol 2014;32:490–5.
[21] Lim CZJ, Zhang L, Zhang Y, Sundah NR, Shao H. New sensors for extracellular vesicles: insights on constituent and associated biomarkers. ACS Sens 2020;5:4–12.
[22] Simpson RJ, Lim JW, Moritz RL, Mathivanan S. Exosomes: proteomic insights and diagnostic potential. Expert Rev Proteomics 2009;6:267–83.
[23] Shao H, Chung J, Issadore D. Diagnostic technologies for circulating tumour cells and exosomes. Biosci Rep 2015;36, e00292.
[24] Witwer KW, Buzás EI, Bemis LT, Bora A, Lässer C, Lötvall J, Nolte-'t Hoen EN, Piper MG, Sivaraman S, Skog J, Théry C, Wauben MH, Hochberg F. Standardization of sample collection, isolation and analysis methods in extracellular vesicle research. J Extracell Vesicles 2013;2:20360.
[25] Shao H, Im H, Castro CM, Breakefield X, Weissleder R, Lee H. New technologies for analysis of extracellular vesicles. Chem Rev 2018;118:1917–50.
[26] Wu X, Zhao H, Natalia A, Lim CZJ, Ho NRY, Ong CJ, Teo MCC, So JBY, Shao H. Exosome-templated nanoplasmonics for multiparametric molecular profiling. Sci Adv 2020;6. eaba2556.
[27] Wang Z, Sun X, Natalia A, Tang CSL, Ang CBT, Ong C-AJ, Teo MCC, So JBY, Shao H. Dual-selective magnetic analysis of extracellular vesicle glycans. Matter 2020;2:150–66.
[28] Lim CZJ, Zhang Y, Chen Y, Zhao H, Stephenson MC, Ho NRY, Chen Y, Chung J, Reilhac A, Loh TP, Chen CLH, Shao H. Subtyping of circulating exosome-bound amyloid β reflects brain plaque deposition. Nat Commun 2019;10:1144.
[29] Sanderson MP, Keller S, Alonso A, Riedle S, Dempsey PJ, Altevogt P. Generation of novel, secreted epidermal growth factor receptor (EGFR/ErbB1) isoforms via metalloprotease-dependent ectodomain shedding and exosome secretion. J Cell Biochem 2008;103:1783–97.
[30] Chen WW, Balaj L, Liau LM, Samuels ML, Kotsopoulos SK, Maguire CA, Loguidice L, Soto H, Garrett M, Zhu LD, Sivaraman S, Chen C, Wong ET, Carter BS, Hochberg FH, Breakefield XO, Skog J. BEAMing and droplet digital PCR analysis of mutant IDH1 mRNA in glioma patient serum and cerebrospinal fluid extracellular vesicles. Mol Ther Nucleic Acids 2013;2, e109.
[31] Graner MW, Alzate O, Dechkovskaia AM, Keene JD, Sampson JH, Mitchell DA, Bigner DD. Proteomic and immunologic analyses of brain tumor exosomes. FASEB J 2009;23:1541–57.
[32] Shao H, Min C, Issadore D, Liong M, Yoon TJ, Weissleder R, Lee H. Magnetic nanoparticles and microNMR for diagnostic applications. Theranostics 2012;2:55–65.
[33] Rho J, Chung J, Im H, Liong M, Shao H, Castro CM, Weissleder R, Lee H. Magnetic nanosensor for detection and profiling of erythrocyte-derived microvesicles. ACS Nano 2013;7:11227–33.
[34] Issadore D, Park YI, Shao H, Min C, Lee K, Liong M, Weissleder R, Lee H. Magnetic sensing technology for molecular analyses. Lab Chip 2014;14:2385–97.
[35] Chen Y, Ding X, Zhang Y, Natalia A, Sun X, Wang Z, Shao H. Design and synthesis of magnetic nanoparticles for biomedical diagnostics. Quant Imaging Med Surg 2018;8:957–70.
[36] Chung J, Shao H, Reiner T, Issadore D, Weissleder R, Lee H. Microfluidic cell sorter (μFCS) for on-chip capture and analysis of single cells. Adv Healthc Mater 2012;1:432–6.
[37] Pols MS, Klumperman J. Trafficking and function of the tetraspanin CD63. Exp Cell Res 2009;315:1584–92.
[38] Mishima K, Kato Y, Kaneko MK, Nishikawa R, Hirose T, Matsutani M. Increased expression of podoplanin in malignant astrocytic tumors as a novel molecular marker of malignant progression. Acta Neuropathol 2006;111:483–8.
[39] Wykosky J, Gibo DM, Stanton C, Debinski W. EphA2 as a novel molecular marker and target in glioblastoma multiforme. Mol Cancer Res 2005;3:541–51.
[40] Townsend DM, Tew KD. The role of glutathione-S-transferase in anti-cancer drug resistance. Oncogene 2003;22:7369–75.
[41] Chen HY, Shao CJ, Chen FR, Kwan AL, Chen ZP. Role of ERCC1 promoter hypermethylation in drug resistance to cisplatin in human gliomas. Int J Cancer 2010;126:1944–54.

[42] Galluzzi L, Senovilla L, Vitale I, Michels J, Martins I, Kepp O, Castedo M, Kroemer G. Molecular mechanisms of cisplatin resistance. Oncogene 2012;31:1869–83.
[43] Hui L, Yue S, Gao G, Chang H, Li X. Association of single-nucleotide polymorphisms in ERCC1 and ERCC2 with glioma risk. Tumour Biol 2014;35:7451–7.
[44] Berger W, Spiegl-Kreinecker S, Buchroithner J, Elbling L, Pirker C, Fischer J, Micksche M. Overexpression of the human major vault protein in astrocytic brain tumor cells. Int J Cancer 2001;94:377–82.
[45] Steiner E, Holzmann K, Elbling L, Micksche M, Berger W. Cellular functions of vaults and their involvement in multidrug resistance. Curr Drug Targets 2006;7:923–34.
[46] Kuan CT, Wakiya K, Herndon JE, Lipp ES, Pegram CN, Riggins GJ, Rasheed A, Szafranski SE, McLendon RE, Wikstrand CJ, Bigner DD. MRP3: a molecular target for human glioblastoma multiforme immunotherapy. BMC Cancer 2010;10:468.
[47] Qi L, Bellail AC, Rossi MR, Zhang Z, Pang H, Hunter S, Cohen C, Moreno CS, Olson JJ, Li S, Hao C. Heterogeneity of primary glioblastoma cells in the expression of caspase-8 and the response to TRAIL-induced apoptosis. Apoptosis 2011;16:1150–64.
[48] Mehrian-Shai R, Chen CD, Shi T, Horvath S, Nelson SF, Reichardt JK, Sawyers CL. Insulin growth factor-binding protein 2 is a candidate biomarker for PTEN status and PI3K/Akt pathway activation in glioblastoma and prostate cancer. Proc Natl Acad Sci U S A 2007;104:5563–8.
[49] Verhaak RG, Hoadley KA, Purdom E, Wang V, Qi Y, Wilkerson MD, Miller CR, Ding L, Golub T, Mesirov JP, Alexe G, Lawrence M, O'Kelly M, Tamayo P, Weir BA, Gabriel S, Winckler W, Gupta S, Jakkula L, Feiler HS, Hodgson JG, James CD, Sarkaria JN, Brennan C, Kahn A, Spellman PT, Wilson RK, Speed TP, Gray JW, Meyerson M, Getz G, Perou CM, Hayes DN, Cancer Genome Atlas Research Network. Integrated genomic analysis identifies clinically relevant subtypes of glioblastoma characterized by abnormalities in PDGFRA, IDH1, EGFR, and NF1. Cancer Cell 2010;17:98–110.
[50] Huse JT, Phillips HS, Brennan CW. Molecular subclassification of diffuse gliomas: seeing order in the chaos. Glia 2011;59:1190–9.
[51] Phillips HS, Kharbanda S, Chen R, Forrest WF, Soriano RH, Wu TD, Misra A, Nigro JM, Colman H, Soroceanu L, Williams PM, Modrusan Z, Feuerstein BG, Aldape K. Molecular subclasses of high-grade glioma predict prognosis, delineate a pattern of disease progression, and resemble stages in neurogenesis. Cancer Cell 2006;9:157–73.
[52] Zheng S, Chheda MG, Verhaak RG. Studying a complex tumor: potential and pitfalls. Cancer J 2012;18:107–14.
[53] Lim GS, Hor JH, Ho NRY, Wong CY, Ng SY, Soh BS, Shao H. Microhexagon gradient array directs spatial diversification of spinal motor neurons. Theranostics 2019;9:311–23.
[54] Ho NRY, Lim GS, Sundah NR, Lim D, Loh TP, Shao H. Visual and modular detection of pathogen nucleic acids with enzyme-DNA molecular complexes. Nat Commun 2018;9:3238.
[55] Wang L, Meng Z, Martina F, Shao H, Shao F. Fabrication of circular assemblies with DNA tetrahedrons: from static structures to a dynamic rotary motor. Nucleic Acids Res 2017;45:12090–9.
[56] Sundah NR, Ho NRY, Lim GS, Natalia A, Ding X, Liu Y, Seet JE, Chan CW, Loh TP, Shao H. Barcoded DNA nanostructures for the multiplexed profiling of subcellular protein distribution. Nat Biomed Eng 2019;3:684–94.
[57] Van Deun J, Jo A, Li H, Lin HY, Weissleder R, Im H, Lee H. Integrated dual-mode chromatography to enrich extracellular vesicles from plasma. Adv Biosyst 2020;e1900310.
[58] Lee K, Fraser K, Ghaddar B, Yang K, Kim E, Balaj L, Chiocca EA, Breakefield XO, Lee H, Weissleder R. Multiplexed profiling of single extracellular vesicles. ACS Nano 2018;12:494–503.
[59] Fraser K, Jo A, Giedt J, Vinegoni C, Yang KS, Peruzzi P, Chiocca EA, Breakefield XO, Lee H, Weissleder R. Characterization of single microvesicles in plasma from glioblastoma patients. Neuro Oncol 2019;21:606–15.
[60] Yekula A, Minciacchi VR, Morello M, Shao H, Park Y, Zhang X, Muralidharan K, Freeman MR, Weissleder R, Lee H, Carter B, Breakefield XO, Di Vizio D, Balaj L. Large and small extracellular vesicles released by glioma cells in vitro and in vivo. J Extracell Vesicles 2020;9:1689784.
[61] Li Y, Ali S, Clarke J, Cha S. Bevacizumab in recurrent glioma: patterns of treatment failure and implications. Brain Tumor Res Treat 2017;5:1–9.
[62] Fischer I, Cunliffe CH, Bollo RJ, Raza S, Monoky D, Chiriboga L, Parker EC, Golfinos JG, Kelly PJ, Knopp EA, Gruber ML, Zagzag D, Narayana A. High-grade glioma before and after treatment with radiation and Avastin: initial observations. Neuro Oncol 2008;10:700–8.
[63] Rock JP, Hearshen D, Scarpace L, Croteau D, Gutierrez J, Fisher JL, Rosenblum ML, Mikkelsen T. Correlations between magnetic resonance spectroscopy and image-guided histopathology, with special attention to radiation necrosis. Neurosurgery 2002;51:912–9.

CHAPTER

19

Three-dimensional in vitro models to study potentiation of glioblastoma resistance by hypoxia

Ahmed Musah-Eroje[a,b,c]

[a]School of Life Sciences, University of Bedfordshire, Luton, United Kingdom [b]School of Science and Technology, Nottingham Trent University, Nottingham, United Kingdom [c]Department of Anatomy, Niger Delta University, Amassama, Bayelsa State, Nigeria

Abstract

Glioblastoma (GBM) is the most common brain tumor in adults with a median survival rate of between 15 and 17 months. There are heterogeneous regions in GBM, as a result of oxygen gradients that range from 0.1% to 10% in vivo. Emerging evidence suggests that tumor hypoxia leads to increased aggressiveness and chemo/radiotherapy resistance. It is therefore important that hypoxia be taken into account in the in vitro study of GBM. For instance, a subpopulation of cells called cancer stem cells (CSCs) is found within a GBM. These cells are upregulated in a hypoxic microenvironment and they are thought to mediate chemo/radiotherapy resistance. GBM is traditionally investigated in vitro by culturing cells as a monolayer (2D culture) or as neurospheres (clusters enriched in cancer stem cells), but neither system accurately reflects the complexity of the three-dimensional (3D) chemoresistant microenvironment of GBM. Hypoxia plays a pivotal role in changing the behavior of GBM cells. A 3D in vitro model to predict GBM resistance would provide a marked clinical benefit that would allow tailored treatment of GBM. This chapter presents a general background of GBM biology in hypoxia and how 3D models can be utilized to study this.

Abbreviations

2D	2-dimensional
3D	3-dimensional
BCRP1	breakpoint cluster region pseudogene 1
CD133	prominin 1
CSCs	cancer stem cells
ECM	extracellular matrix
GBM	glioblastoma
HIF-1α	hypoxia inducible factor 1-alpha

Glioblastoma Resistance to Chemotherapy: Molecular Mechanisms and Innovative Reversal Strategies
https://doi.org/10.1016/B978-0-12-821567-8.00021-X

HIF-2α	hypoxia inducible factor 2-alpha
HIFs	hypoxia inducible factors
HRE	hypoxia responsive element
MGMT	methylguanine DNA methyltransferase (MGMT)
mRNA	messenger RNA
NOD-SCID	nonobese diabetic-severe combined immunodeficiency
TFEB	transcription factor EB
WHO	World Health Organisation

Conflict of interest

No potential conflicts of interest were disclosed.

Introduction

Glioblastoma (GBM) is the most aggressive form of glioma classified by the World Health Organization (WHO) as grade IV astrocytoma [1, 2]. Despite notable achievements that have been made in the past, the clinical management of GBM patients is still a major challenge [3], which may primarily be attributed to its heterogeneous nature [4]. Although inhibition of TFEB oligomerization by cotreatment of melatonin with vorinostat has been shown to promote therapeutic sensitivity in GBM and glioma stem cells [5], temozolomide still remains the standard-of-care chemotherapy for glioblastoma [6]. Unfortunately, most patients with GBM die of the disease in less than 1 year after diagnosis [7] as a result of chemoresistance associated with a cancer stem cell population and the topographically diffuse nature of the tumor [8]. To complicate this challenge, preclinical models that faithfully recapitulate relevant aspects of GBM biology in vitro have still not been established [9].

Furthermore, tumor cells, including GBM, are traditionally cultured in vitro on a plastic substrate, with an oxygen tension of 20%. However, the oxygen tension in GBM in vivo ranges from 0.1% to 10% [8]. Many studies do not take into account the physiologically relevant oxygen tension [10] as well as the changes in the extracellular matrix, which can affect biological properties such as proliferation and motility when investigating GBM cell in vitro [11].

Numerous data on tumor biology are largely dependent on either established or primary cell lines that are cultured and maintained in vitro [12]. Traditionally, these studies have been performed in a two-dimensional (2D) assay. In the 2D environment, cells are maintained in selected medium on flat plastic plates. In this orientation, the cells make contact with the flat culture surface and with neighboring cells only at their periphery. The 2D environment is largely dissimilar from the extracellular matrix (ECM) substrate that provides natural support for cells in vivo [13]. Many scientists are beginning to appreciate the drawbacks of 2D cell culture since it fails to recapitulate in vivo signals and have opted for a more clinically relevant 3D culture that mimics the true life situation to a greater extent. Cells in the 3D environment are closer to cells in vivo; they are in contact with each other and to the matrix thereby generating some important cell-cell biological signals that are missing in 2D cultures. For instance, acinar structures were formed when mammary epithelial cells were cultured on a concentrated laminin-rich extracellular matrix [14]. This chapter explores a general background of hypoxia in the tumor microenvironment of GBM and examines how a 3D in vitro assay can be utilized to understand chemoresistance in GBM.

Hypoxia as a tumor microenvironment component in glioblastoma

The tumor microenvironment consists of a myriad of components, including extracellular cellular matrix (ECM) components [15] and variations in oxygen tension [16] that may affect tumor biology. It has also been shown that the ECM composition may affect cancer cell proliferation and motility [11]. GBMs are characterized by regions of hypoxia [17] ranging from 0.1% to 10% [8], which are thought to play important roles in the aggressiveness of the tumors [8]. As such, hypoxia has been considered to serve as a biomarker for a poor prognosis, and to be associated with radioresistance and chemoresistance, as well as tumor cell migration and invasion [18]. Adaptation to hypoxia through the activation of pro-survival pathways can affect tumor growth [19] and invasion [20], making hypoxia a suitable biomarker for predicting a malignant glioma [21]. Hypoxia has been reported to be important in the biology and aggressiveness of glial brain tumor and increases with increasing WHO grading of gliomas with grade 4 tumors associated with severe hypoxia while grade 2 tumors are associated with modest cellular hypoxia [8]. The aggressive phenotype of GBM that leads to tumor invasion and chemoresistance as well as loss of apoptosis is thought to be developed by hypoxia [22]. In addition, radioresistance is also thought to result from hypoxic regions of tumors [23]. If oxygen supply to a tumor microenvironment is reduced, as in the case of hypoxia, there will be poor delivery of therapy to the tumor cells, which will eventually reduce the efficacy of the therapy [24].

Hypoxia inducible factors (HIFs) have been postulated to be the primary mediators of hypoxic responses [25]. The HIFs (HIF-1α and HIF-2α) are known to be able to bind to the same DNA sequence, referred to as hypoxia responsive element (HRE), and to heterodimerise with HIF-1β to activate target genes. However, HIF-1α and HIF-2α may also exhibit functional differences, including different protein-protein interactions [26] and tissue specificities [27]. HIF-1α has been found to promote the survival of glioma cancer stem cells under hypoxic conditions [28], while HIF-2α is thought to be linked to the GBM stem cell phenotype within the hypoxic niche [29].

Hypoxia and cancer stem cell

The concept of cancer stem cells revolves around the fact that the cells initiating tumors are organized in a hierarchy. In this way, they give rise to mature cells that are more committed with a limited proliferative potential. The cancer stem cells reside at the top in the hierarchy and the size of their niche may be the factor that determines the disease course [30]. The properties of stem cells and their mechanism of tumorigenesis can be easily studied once they are identified and isolated [31]. Hypoxic regions in GBM have been shown to harbor GBM stem cell populations [29], which have been implicated in tumor recurrence and drug failure [32].

Hypoxia and cancer stem cell marker

Several markers have been employed to differentiate CSCs from the vast majority of cells within a tumor sample [33]. In GBM, CD133 is commonly used as a CSC marker [34]. Evidence has shown that CD133+ cells define a subpopulation of glioma stem cells. Irrespective of their volume, these cells were capable of self-renewal, proliferation, and differentiation in both in vitro and in vivo experiments [35]. Expression of CD133 has been shown to correlate with poor patient survival in glioma, and thus CD133 may also be a putative CSC marker in GBM [36]. In a study of the molecular properties of CD133+ GBM stem cells, it was reported that they have the ability to reconstitute the phenotypes of the patient's original tumor and to initiate tumor in the NOD-SCID mouse brain [37]. CD133 positive glioma cells have shown higher levels of BCRP1 and MGMT mRNA expression as well as genes that inhibit apoptosis in comparison to CD133 negative cells [37]. Furthermore, CD133 positive cells are significantly resistant to chemotherapeutic agents such as temozolomide and etoposide compared with CD133 negative cells [38].

In vitro 3D model for glioblastoma

Preclinical testing of drugs has largely relied on either two-dimensional (2D) in vitro cell models or animal studies. However, data from the 2D cultures are largely unreliable, as cells cultured in this model are not a true representative of the in vivo tumor microenvironment and, animal studies, as a result of interspecies differences, fail to fully recapitulate the response to drugs in humans with about 95% of anticancer drug candidates failing to make it to the clinic, thereby wasting significant time and resources [39]. The neurosphere assay is an acceptable 3D model for maintaining GBM in vitro [40–42]. Although this assay has undoubtedly been regarded as the gold standard assay that selects for stem cell populations, it has a major drawback of allowing cells to form their own niche, with more differentiated cells positioned at the center than on the surface as well as containing a mixed population of cells and a small number of true stem cells [43].

The 2D monolayer culture system, though spontaneous and easily manipulated, has generated scientific concerns, as some drug agents that have been successfully classed as promising in this system have become clinically irrelevant, with only about 10% making it to the clinic leading to money and time wastage [39]. For instance, it has been shown that GSCs rely on kinases in the 2D substrates but not in a 3D environment [44].

3D in vitro models are now required for GBM as cells cultured in 2D monolayers display aberrant cell-cell interactions [45]. As such, most of the conclusions from this model of culture do not accurately apply to the tumors in vivo. Therefore, 3D in vitro culture systems offer more realistic cell-cell and cell-matrix interactions that are more physiological, providing a better alternative to the 2D systems [46, 47]. Recently, a 3D model based on a polystyrene scaffold has been used to predict a drug-radiation combination for glioblastoma [48]. Furthermore, glioblastoma tumor-initiating cells have been successfully maintained in microscale alginate hydrogel tubes (or AlgTubes) that allows affordable cost for drug discovery [49]. In addition, a 3D in vitro system for GBM has been developed with a Cultrex-based three-Dimensional Tumor Growth Assay (3D-TGA) which allows routine drug testing and molecular manipulation, e.g., gene knockdown [9].

Conclusion

The fact that there are disparities between in vitro culture and in vivo milieu has given researchers the reason to develop physiologically relevant in vitro assays that better reflects the tumor-microenvironment. The 3D assay could be used to further characterize GBM cells, including potential stem cells, and the pathways through which the tumor microenvironment influences their characteristics and numbers, including their drug resistance and ability to invade and metastasize. Ongoing research hopes to develop a reproducible, predictable, and scalable 3D in vitro assay for GBM that can be utilized for preclinical drug testing.

References

[1] Zong H, Verhaak RG, Canoll P. The cellular origin for malignant glioma and prospects for clinical advancements. Expert Rev Mol Diagn 2012;12(4):383–94.

[2] Lee E, et al. Comparison of glioblastoma (GBM) molecular classification methods. In: Seminars in cancer biology. Elsevier; 2018.

[3] Mansouri A, et al. MGMT promoter methylation status testing to guide therapy for glioblastoma: refining the approach based on emerging evidence and current challenges. Neuro Oncol 2019;21(2):167–78.

[4] Liu Y, et al. Relationship between glioblastoma heterogeneity and survival time: an MR imaging texture analysis. Am J Neuroradiol 2017;38(9):1695–701.

[5] Sung GJ, et al. Inhibition of TFEB oligomerization by co-treatment of melatonin with vorinostat promotes the therapeutic sensitivity in glioblastoma and glioma stem cells. J Pineal Res 2019;66(3), e12556.

[6] Nagasawa DT, et al. Temozolomide and other potential agents for the treatment of glioblastoma multiforme. Neurosurg Clin N Am 2012;23(2):307–22.

[7] Holland EC. Glioblastoma multiforme: the terminator. Proc Natl Acad Sci USA 2000;97(12):6242–4.

[8] Evans SM, et al. Hypoxia is important in the biology and aggression of human glial brain tumors. Clin Cancer Res 2004;10(24):8177–84.

[9] Musah-Eroje A, Watson S. A novel 3D in vitro model of glioblastoma reveals resistance to temozolomide which was potentiated by hypoxia. J Neurooncol 2019;142(2):231–40.

[10] Richards R, et al. Cell cycle progression in glioblastoma cells is unaffected by pathophysiological levels of hypoxia. PeerJ 2016;4, e1755.

[11] Gilkes DM, Semenza GL, Wirtz D. Hypoxia and the extracellular matrix: drivers of tumour metastasis. Nat Rev Cancer 2014;14(6):430–9.

[12] Gillet J-P, et al. Redefining the relevance of established cancer cell lines to the study of mechanisms of clinical anti-cancer drug resistance. Proc Natl Acad Sci USA 2011;108(46):18708–13.

[13] Bissell MJ, et al. The organizing principle: microenvironmental influences in the normal and malignant breast. Differentiation 2002;70(9–10):537–46.

[14] Murtagh J, et al. Organization of mammary epithelial cells into 3D acinar structures requires glucocorticoid and JNK signaling. J Cell Biol 2004;166(1):133–43.

[15] Balkwill FR, Capasso M, Hagemann T. The tumor microenvironment at a glance. The Company of Biologists Ltd; 2012.

[16] Brown JM, Wilson WR. Exploiting tumour hypoxia in cancer treatment. Nat Rev Cancer 2004;4(6):437–47.

[17] Yang L, et al. Hypoxia and hypoxia-inducible factors in glioblastoma multiforme progression and therapeutic implications. Exp Cell Res 2012;318(19):2417–26.

[18] Chou C-W, et al. Tumor cycling hypoxia induces chemoresistance in glioblastoma multiforme by upregulating the expression and function of ABCB1. Neuro Oncol 2012;14(10):1227–38.

[19] Bartoszewska S, Collawn JF. Unfolded protein response (UPR) integrated signaling networks determine cell fate during hypoxia. Cell Mol Biol Lett 2020;25(1):1–20.

[20] Monteiro AR, et al. The role of hypoxia in glioblastoma invasion. Cell 2017;6(4):45.

[21] Kawai N, et al. Correlation between 18 F-fluoromisonidazole PET and expression of HIF-1α and VEGF in newly diagnosed and recurrent malignant gliomas. Eur J Nucl Med Mol Imaging 2014;41(10):1870–8.

[22] Jensen RL. Brain tumor hypoxia: tumorigenesis, angiogenesis, imaging, pseudoprogression, and as a therapeutic target. J Neurooncol 2009;92(3):317–35.

[23] Clarke RH, et al. Sustained radiosensitization of hypoxic glioma cells after oxygen pretreatment in an animal model of glioblastoma and in vitro models of tumor hypoxia. PLoS One 2014;9(10), e111199.

[24] Li JZ, Gao W, Chan JY-W, Ho W-K, Wong T-S. Hypoxia in head and neck squamous cell carcinoma. Int Sch Res Not 2012;2012:708974. https://doi.org/10.5402/2012/708974.

[25] Semenza GL. Hypoxia-inducible factors: mediators of cancer progression and targets for cancer therapy. Trends Pharmacol Sci 2012;33(4):207–14.

[26] Loboda A, Jozkowicz A, Dulak J. HIF-1 and HIF-2 transcription factors—similar but not identical. Mol Cells 2010;29(5):435–42.

[27] Holmquist-Mengelbier L, et al. Recruitment of HIF-1α and HIF-2α to common target genes is differentially regulated in neuroblastoma: HIF-2α promotes an aggressive phenotype. Cancer Cell 2006;10(5):413–23.

[28] Soeda A, et al. Hypoxia promotes expansion of the CD133-positive glioma stem cells through activation of HIF-1α. Oncogene 2009;28(45):3949–59.

[29] Seidel S, et al. A hypoxic niche regulates glioblastoma stem cells through hypoxia inducible factor 2α. Brain 2010;133(4):983–95.

[30] Lan X, et al. Fate mapping of human glioblastoma reveals an invariant stem cell hierarchy. Nature 2017;549 (7671):227–32.

[31] Ludwig K, Kornblum HI. Molecular markers in glioma. J Neurooncol 2017;134(3):505–12.

[32] Auffinger B, et al. The role of glioma stem cells in chemotherapy resistance and glioblastoma multiforme recurrence. Expert Rev Neurother 2015;15(7):741–52.

[33] Karsten U, Goletz S. What makes cancer stem cell markers different? Springerplus 2013;2(1):301.

[34] Pollard SM, et al. Glioma stem cell lines expanded in adherent culture have tumor-specific phenotypes and are suitable for chemical and genetic screens. Cell Stem Cell 2009;4(6):568–80.

[35] Singh SK, et al. Identification of human brain tumour initiating cells. Nature 2004;**432**(7015):396–401.

[36] Zeppernick F, et al. Stem cell marker CD133 affects clinical outcome in glioma patients. Clin Cancer Res 2008;14(1):123–9.

[37] Liu Q, et al. Molecular properties of CD133+ glioblastoma stem cells derived from treatment-refractory recurrent brain tumors. J Neurooncol 2009;94(1):1–19.

[38] Liu G, et al. Analysis of gene expression and chemoresistance of CD133+ cancer stem cells in glioblastoma. Mol Cancer 2006;5(1):67.

[39] Hutchinson L, Kirk R. High drug attrition rates—where are we going wrong? Nat Rev Clin Oncol 2011;8(4):189–90.

[40] Reynolds BA, Weiss S. Generation of neurons and astrocytes from isolated cells of the adult mammalian central nervous system. Science 1992;**255**(5052):1707–10.

[41] Quereda V, et al. A cytotoxic three-dimensional-spheroid, high-throughput assay using patient-derived glioma stem cells. SLAS Discov 2018;23(8):842–9.

[42] Kogiso M, et al. Concurrent inhibition of neurosphere and monolayer cells of pediatric glioblastoma by Aurora A inhibitor MLN8237 predicted survival extension in PDOX models. Clin Cancer Res 2018;24(9):2159–70.

[43] Bez A, et al. Neurosphere and neurosphere-forming cells: morphological and ultrastructural characterization. Brain Res 2003;993(1–2):18–29.

[44] Fernandez-Fuente G, et al. Culture dimensionality influences the resistance of glioblastoma stem-like cells to multikinase inhibitors. Mol Cancer Ther 2014;13(6):1664–72.

[45] Cheung KJ, Ewald AJ. Illuminating breast cancer invasion: diverse roles for cell–cell interactions. Curr Opin Cell Biol 2014;30:99–111.

[46] Hickman JA, et al. Three-dimensional models of cancer for pharmacology and cancer cell biology: capturing tumor complexity in vitro/ex vivo. Biotechnol J 2014;9(9):1115–28.

[47] Ahmed EM, et al. A HIF-independent, CD133-mediated mechanism of cisplatin resistance in glioblastoma cells. Cell Oncol 2018;41(3):319–28.

[48] Gomez-Roman N, et al. A novel 3D human glioblastoma cell culture system for modeling drug and radiation responses. Neuro Oncol 2017;19(2):229–41.

[49] Li Q, et al. Scalable culturing of primary human glioblastoma tumor-initiating cells with a cell-friendly culture system. Sci Rep 2018;8(1):1–13.

CHAPTER

20

Chemoresistance mechanisms in mouse models of glioblastoma

Takamasa Kinoshita[a,b], *Masafumi Miyai*[a], *Toru Iwama*[a], *Akira Hara*[b], *and Hiroyuki Tomita*[b]

[a]Department of Neurosurgery, Gifu University Graduate School of Medicine, Gifu, Japan
[b]Department of Tumor Pathology, Gifu University Graduate School of Medicine, Gifu, Japan

Abstract

Glioblastoma multiforme is the most well-known malignant brain tumor. Because of its resistance to treatment, the prognosis is extremely poor, and various studies are currently being conducted to improve the prognosis. In particular, the development of new therapeutic methods is an urgent issue. Factors related to chemotherapy resistance are being studied in detail and mouse models for studying the effectiveness of chemotherapy are being developed. Major factors related to resistance to temozolomide, which is mainly used in chemotherapy, include MGMT, MMR pathway, and apoptosis regulating factors. Mouse models have been created to study the effects of these factors and new therapeutic agents. The cell line transplantation model often used at the beginning had problems such as the need to use immunodeficient mice, the time required for cell culture, and the generation of new mutations during long-term culture. Therefore, mouse models such as a patient-derived xenograft model, which do not require much time for culturing, and a genetically engineered mouse model, which do not require the use of immunodeficient mice, have been used. Here, we review the factors involved in chemoresistance and then briefly summarize the progress of mouse models.

Abbreviations

GBM glioblastoma
GEMM genetically engineered mouse model
MGMT O^6-methylguanine-DNA methyltransferase
MMR mismatch repair
PDX patient-derived xenograft
TMZ temozolomide

Conflict of interest

No potential conflicts of interest were disclosed.

Glioblastoma Resistance to Chemotherapy: Molecular Mechanisms and Innovative Reversal Strategies
https://doi.org/10.1016/B978-0-12-821567-8.00031-2

Introduction

Glioblastoma (GBM) is the most common malignant brain tumor, accounting for 45% of all brain tumor cases [1]. The current standard of care, resection, and subsequent chemoradiation is ineffective or unsustainable owing to drug resistance and the blood-brain barrier, with the average survival being 12–15 months [2–4].

Before temozolomide (TMZ) became the standard treatment for GBM in 2005, survival had not improved for decades [5]. Even after the introduction of TMZ, the harsh reality has been that an increase in patient survival of only 2.5 months represents a significant improvement [1].

The molecular mechanisms underlying chemoresistance are only partially understood. Therefore, the development of new chemotherapy drugs for GBM is urgently needed and requires advances in preclinical research using mouse models and other approaches. Here, we review the factors involved in GBM chemoresistance and mouse models used for research in this area. We further describe how mouse models are a good means to elucidate the mechanisms.

Factors involved in GBM chemoresistance

O^6-methylguanine-DNA methyltransferase (MGMT)

TMZ, commonly used for GBM treatment, is a DNA alkylating agent that causes lethal DNA lesions in rapidly dividing cancer cells, inducing sustained O^6 methylguanine addition. This triggers an uncontrolled DNA mismatch repair cycle, eventually leading to DNA double-strand breaks and apoptosis [6]. MGMT is an enzyme that repairs the DNA damage induced by TMZ and plays an important role in chemoresistance by removing the O^6 position of guanine conferred by alkylating drugs [1]. This suggests that MGMT is a strong predictor of the treatment response in GBM [7] and, in particular, that increased MGMT activity is significantly associated with increased TMZ resistance [2].

MGMT activity is inversely correlated with *MGMT* promoter methylation in GBM cell lines. However, the degree of promoter methylation is highly variable, even in GBM lines deficient in MGMT; accordingly, promoter methylation is not necessarily a reliable measure of tumor MGMT activity [2]. Hegi et al. also found that a small proportion of patients without *MGMT* promoter methylation exhibits long-term survival after treatment with TMZ [7]. Therefore, it is possible that there are mechanisms underlying MGMT suppression other than *MGMT* promoter methylation [8]; the identification of these mechanisms has been a focus of various studies.

Mismatch repair (MMR) pathway

The MMR pathway is an alternative mechanism underlying resistance to TMZ. The MMR pathway is composed of several proteins such as hMLH1, hPMS2 and so on, and plays an important role in neutralizing the toxicity of O^6 methylguanine [6]. The MMR pathway plays an important role in maintaining genomic stability by correcting generated errors such as interbase mismatches and insertion/deletion mispairs during DNA replication [9]. The MMR pathway prevents permanent mutations in dividing cells and reduces the number of replication-related errors, resulting in preventing both short-term mutagenesis and long-term tumorigenesis [10]. When thymine and O^6 methylguanine are mismatched, thymine on the daughter strand is eliminated in an MMR-dependent manner [11]. However, because O^6 methylguanine remains on the template strand, a new thymine is introduced at the site, and repeated cycles of MMR lead to double-strand breaks [12].

MMR proteins MutSα and MutLα are required for alkylation damage signaling and induction of apoptosis, leading to the activation of numerous downstream targets such as p53, Chk1, Chk2, SMC1, and CDC25A [12]. Therefore, defects in the MMR pathway result in tolerance to O^6 methylguanine mispairing, which contributes to drug resistance [6, 10]. In fact, Nagel et al. showed that a higher MMR capacity increases sensitivity to TMZ [2].

Factors related to the regulation of apoptosis

In addition to the factors involved in DNA repair described above, the dysregulation of genes and proteins related to apoptosis also contributes to chemoresistance. For example, the loss of p53 function and overexpression of EGFR play major roles in GBM chemoresistance.

p53 is a tumor suppressor gene located on chromosome 17p13.1 [13]. *p53* plays a central role in the detection of DNA damage and regulates the signaling pathway required to induce apoptosis [6]. *p53* mutations are relatively rare in primary GBMs but common in secondary GBMs [14]. Glioma cell lines harboring *p53* mutations are significantly more sensitive to alkylating drugs than are cell lines with wild-type *p53* [15]. An examination of GBM samples collected from multiple centers in 15 countries has shown that *p53* mutations are associated with a significantly better prognosis [16]. Blough et al. reported that, in the concentration range of 10–100 μM, sensitivity to TMZ is significantly higher in cell lines harboring *p53* mutations than in cell lines with wild-type *p53* [15]. At the same time, they related that U87MG cells, which do not express MGMT, also exhibit increased sensitivity to TMZ, indicating that the effect of p53 mutations on TMZ sensitivity is likely mediated by an MGMT-independent mechanism [15].

The EGFR receptor is a member of the ErbB family; it is an oncogene that undergoes frequent amplification and mutation in malignant gliomas [17]. Amplification occurs in about 40% of GBMs, promoting the activation of the EGFR signaling pathway and thus leading to tumor growth [18]. In addition, several clinical and histopathological studies have suggested that EGFR amplification affects the response of malignant gliomas to chemotherapy [6, 19]. In studies on the effects of chemotherapy at the preclinical stage, cell cultures have been shown to rapidly lose *EGFR* gene amplification [20], hindering the development of better chemotherapy agents. It has recently been reported that *EGFR* mutations could be maintained under specific conditions [18].

Other factors

Various additional factors, such as *PTEN* and glioma stem cells, are involved in chemoresistance [2, 3, 21–24]. Various targeted therapies, including those targeting the abovementioned factors, as well as combination therapies with TMZ, are the focus of current research aimed at improving chemoresistance [25–27].

Mouse models

Transplantation models

Various mouse models have been used for transplantation with GBM cell lines to evaluate chemoresistance and novel therapeutic targets [21] (Table 1). Gupta et al. verified the effect of veliparib, a PARP inhibitor, using GBM cells with different levels of MGMT expression [29]. Sharpe et al. confirmed the effect of a monoamine oxidase B-activated pro-drug using BT111 with unmethylated MGMT [31]. Murphy et al. reported the efficacy of a connexin 43 inhibitor for MGMT-independent TMZ-resistant GBMs [1]. In addition to changes in the levels of MGMT, changes in EGFR and p53 levels have also been examined. Cho et al. concluded that EGFR-targeted monoclonal antibodies, such as erlotinib and cetuximab, are effective against GBMs with EGFR carboxyl-terminus domain deletion [28]. Wei et al. examined the effect of artesunate using cell lines with different levels of p53 expression [35]. Using allografts, Zhang et al. investigated a new treatment strategy for ErbB2-positive GBMs using GL261 [33] (Table 1).

In addition to the investigation of treatment approaches, mouse models are also useful for the identification of the underlying causes of GBM. For example, Liu et al. reported that EGFR mutations promote GBM formation [20]. The gene expression profiles of commonly used cell lines are summarized in Table 2. Particularly, common cell line transplantation models have

TABLE 1 Recent study of transplantation models.

Year	Study	Type	Cell line	Gene expression
2011	Cho et al. [28]	Xenograft	LN443	EGFRvIII or EGFR CT Del1 deletion
2014	Gupta et al. [29]	Xenograft	GBM12TMZ-mgmt high or low	MGMT high or low
2014	Yoshida et al. [30]	Xenograft	GBM6/GBM39/U87/GBM12	EGFRvIII amplified (GBM 6,39) EGFRvIII (U87) amplified wild-type EGFR(GBM12)
2015	Duan et al. [21]	Xenograft (TALEN)	Human NSC	PTEN knockout
2015	Liu et al. [20]	Xenograft	U87EGFRvIII	EGFRvIII
2015	Sharpe et al. [31]	Xenograft	BT111	Unmethylated MGMT, lack of EGFR
2015	Parrish et al. [32]	Xenograft	GBM12,GBM39	MGMT hypermethylated
2016	Murphy et al. [1]	Xenograft	LN229	MGMT deficient
2016	Zhang et al. [33]	Allograft	GL261	ErbB2 expression
2017	William et al. [18]	Xenograft	HROG 22,33,59	EGFR not amplified (HROG 22) EGFR amplified (HROG 33, 59)
2020	Koga et al. [34]	Xenograft (CRISPER)	Human NPC	PTEN/NF1 deletion or TP53/PDGFRA deletion
2020	Wei et al. [35]	Xenograft	U87, U251	p53 wild (U87) p53 mutant (U251)

TABLE 2 Gene expression of common cell lines.

MGMT [1, 15]	Positive	T98G, LN-18
	Negative	SF295, U87MG, A172, LN229, SF-268, U251, SNB-75, LNZ308
p53 [14, 15, 17, 35]	Wild type	SF767, LN443, U87MG, LNZ308, SK-N-SH
	Mutant type	U251MG, U373MG, U138, LN215, LN18, GL261

enabled rapid investigations of cancer- and metastasis-related genes and been useful for evaluations of novel drug candidates [4, 36]. The xenotransplantation of cell lines must be performed by using immunodeficient mice to prevent immunorejection [4, 36]. However, this approach is not suitable for studying the role of the immune system in tumor development and response to therapy [36]. Furthermore, when cancer cells are cultured for a long period, new mutations arise and important features are lost [36, 37]. Additionally, the transplantation of large numbers of cancer cells makes it impossible to mimic the original tumor environment [17]. For this reason, a patient-derived xenograft (PDX) model that does not require long-term culture is frequently used [37]. Parrish et al. verified that the effect of rucaparib, a PARP inhibitor, is limited owing to its poor transferability to the brain [32]. Yoshida et al. also used a PDX model with different levels of EGFR expression to examine the effect of a pan-ErbB inhibitor [30].

The PDX model addresses various limitations of in vitro analyses, including tumor gene mutations. However, it is not always possible to maintain tumors for long periods using this model, and it is difficult to identify tumors that are suitable for transplantation [4, 38]. Therefore, mutant tumor cells have been engineered using TALEN or CRISPR/Cas9. PTEN knockout neural stem cells generated using TALEN and pluripotent stem cells with PTEN/NF1 deletions or TP53/PDGFRA deletions generated using CRISPER/Cas9 have been utilized for research [21, 34] (Table 2).

Genetically engineered mouse models (GEMMs)

As an alternative to transplantation models, GEMMs that do not require the use of immunodeficient mice have been used [36]. Details of a GEMM of astrocytoma are summarized in Table 3.

Reilly et al. generated the first mouse model lacking NF1 and p53, two tumor suppressors [39]. Zhu et al. generated a p53-deficient mouse strain carrying a conditional allele of the NF1 tumor suppressor, which negatively regulates Ras signaling, and observed a 100% rate of tumor formation [42]. Zheng et al. reported that 73% of mice (42/57) with p53 and PTEN deficiencies develop high-grade gliomas at 15–40 weeks [45]. Chow et al. found that 87% (55/63) of *p53* and *PTEN*-deficient mice develop high-grade gliomas [47]. Kwon et al. reported that high-grade gliomas developed in 100% of mice with PTEN heterozygosity in an NF1/p53-inactive model [44]. According to Zuckermann et al., 100% of mice ($n=8$) with *p53, PTEN*, and *NF1* deleted by CRISPER/Cas9 developed highly aggressive tumors similar to human GBM at 6–14 weeks [49]. Lei et al. reported that the deletion of *p53* and *PTEN* with PDGF stimulation resulted in GBM in 100% of mice ($n=16$) [48].

TABLE 3 Previously reported GEMMs of astrocytomas.

Year	Study	Initiating alternations	Major grade	Incidence
2000	Reilly et al. [39]	NF1; p53	Grade 2–4	30%–75% (15–55 weeks)
2001	Ding et al. [40]	GFAP–HRasV12	Grade 2–4	95% (by 4–6 months:D7) ~ 100% (by 2 weeks:B8)
2002	Xiao et al. [41]	GFAP-T121 transgene	Grade 2–3	100% (by 25 weeks)
2005	Zhu et al. [42]	p53; NF1	Grade 2–4	100% (5–10 months)
2006	Wei et al. [43]	GFAP-HRasV12; PTEN	Grade 2–4	93% (6–15 weeks)
2008	Kwon et al. [44]	NF1; p53; PTEN	Grade 3–4	100% (10–18 weeks)
2008	Zheng et al. [45]	p53; PTEN	Grade 3–4	73% (15–40 weeks)
2009	Zhu et al. [46]	EGFRvIII/vIII; PTEN; INK4a/ARF	Grade 4	100% (by 10 weeks)
2011	Chow et al. [47]	p53; PTEN	Grade 2–4	100% (by 52 weeks)
2011	Lei et al. [48]	p53; PTEN; PDGF	Grade 4	100% (median survival: 27 days)
2013	Shchors et al. [14]	GFAP-HRasV12; p53	Grade 3–4	NR (Ave: 9 weeks)
2015	Zuckermann et al. [49]	NF1; PTEN; p53	Grade 4	100% (6–14 weeks)
2017	Pathania et al. [50]	H3.3K27M; p53; ATRX	Grade 4	100% (9 months)
2018	Oldrini et al. [51]	PDGFB+p53 or PTEN or Cdkn2a	Grade 4	Various penetrance (various latency)

NR: not reported, Ave: average.

A GFAP model has also been used. For example, Ding et al. developed the HrasV12 model [40] and Xiao et al. developed the T121 transgene model [41]. The HRasV12 model was used to show that 79% of mice with *PTEN* mutations (11/14) develop high-grade gliomas [43]. Shchors et al. further showed that tumors develop earlier in a model with additional p53 mutations than in the HRasV12 model (17 weeks vs 9 weeks) [14].

In addition, Zhu et al. reported that the addition of EGFR vIII/vIII to mice with a loss of p16Ink4a, p19Arf, and PTEN results in GBM in 100% of cases within 10 weeks [46]. Pathania et al. reported that p53/ATRX knockdown in H3.3K27M mice results in smaller proliferative lesions in the cortex and striatum developing into high-grade glioma by 9 months [50]. Oldrini et al. evaluated tumor incidence with respect to a combination of PDGFB and various gene mutations [51].

However, recent advances in the design and production of nucleases capable of introducing site-specific DNA double-strand breaks have led to reliable genomic manipulation methods, overcoming the inefficiency of homologous recombination [52]. For example, Cre/LoxP enables accurate, targeted genetic recombination, and CRISPR-cas9 enables gene knockouts and functional screening [53]. In particular, gene targeting with CRISPR/Cas9 in neural stem cells is very efficient, easy, and versatile [54].

GEMMs are particularly useful for the development of targeted therapies [17, 55] and have the ability to induce mutations at specific times and in specific tissue types using Cre-ERT [56]. GEMMs can thus be used to assess glioma development, progression, and origin [57]. However, the multistep process of carcinogenesis cannot be imitated in GEMMs [36].

Conclusion

As described above, various mouse models have been considered for the discovery of factors related to chemoresistance and new therapeutic agents, and are advancing day by day. In addition, the development of mouse model making technology such as CRISPER/Cas9 is also remarkable. On the other hand, mouse models have various limitations with respect to understanding human diseases, including interspecific differences in telomere lengths and susceptibility to antitumor drugs [21]. As a result, although a wide range of mouse models have been developed, none can be considered perfect [37]. However, preclinical mouse models are critical for investigations of chemoresistance in GBMs in the context of the development of new anticancer drugs, such as immune checkpoint-targeting drugs.

Acknowledgment

We thank all members of our laboratories.

References

[1] Murphy SF, Varghese RT, Lamouille S, Guo S, Pridham KJ, Kanabur P, Osimani AM, Sharma S, Jourdan J, Rodgers CM, Simonds GR, Gourdie RG, Sheng Z. Connexin 43 inhibition sensitizes chemoresistant glioblastoma cells to Temozolomide. Cancer Res 2016;76:139–49.
[2] Nagel ZD, Kitange GJ, Gupta SK, Joughin BA, Chaim IA, Mazzucato P, Lauffenburger DA, Sarkaria JN, Samson LD. DNA repair capacity in multiple pathways predicts chemoresistance in glioblastoma multiforme. Cancer Res 2017;77:198–206.
[3] Cloughesy TF, Cavenee WK, Mischel PS. Glioblastoma: from molecular pathology to targeted treatment. Annu Rev Pathol 2014;9:1–25.
[4] Miyai M, Tomita H, Soeda A, Yano H, Iwama T, Hara A. Current trends in mouse models of glioblastoma. J Neurooncol 2017;135:423–32.
[5] Grossman SA, Ye X, Piantadosi S, Desideri S, Nabors LB, Rosenfeld M, Fisher J, NABTT CNS Consortium. Survival of patients with newly diagnosed glioblastoma treated with radiation and temozolomide in research studies in the United States. Clin Cancer Res 2010;16:2443–9.
[6] Sarkaria JN, Kitange GJ, James CD, Plummer R, Calvert H, Weller M, Wick W. Mechanisms of chemoresistance to alkylating agents in malignant glioma. Clin Cancer Res 2008;14:2900–8.
[7] Hegi ME, Diserens AC, Gorlia T, Hamou MF, de Tribolet N, Weller M, Kros JM, Hainfeller JA, Mason W, Mariani L, Bromberg JE, Hau P, Mirimanoff RO, Cairncross JG, Janzer RC, Stupp R. MGMT gene silencing and benefit from Temozolomide in glioblastoma. N Engl J Med 2005;352:997–1003.
[8] Danam RP, Howell SR, Brent TP, Harris LC. Epigenetic regulation of O6-methylguanine-DNA methyltransferase gene expression by histone acetylation and methyl-CpG binding proteins. Mol Cancer Ther 2005;4:61–9.
[9] Sameer AS, Nissar S, Fatima K. Mismatch repair pathway: molecules, functions, and role in colorectal carcinogenesis. Eur J Cancer Prev 2014;23:246–57.
[10] Li GM. Mechanisms and functions of DNA mismatch repair. Cell Res 2008;18:85–98.

[11] Thomas A, Tanaka M, Trepel J, Reinhold WC, Rajapakse VN, Pommier Y. Temozolomide in the era of precision medicine. Cancer Res 2017;77:823–6.
[12] Hsieh P, Yamane K. DNA mismatch repair: molecular mechanism, cancer, and ageing. Mech Ageing Dev 2008;129:391–407.
[13] Wales MM, Biel MA, el Deiry W, Nelkin BD, Issa JP, Cavenee WK, Kuerbitz SJ, Baylin SB. p53 activates expression of HIC-1, a new candidate tumour suppressor gene on 17p13.3. Nat Med 1995;1:570–7.
[14] Shchors K, Persson AI, Rostker F, Tihan T, Lyubynska N, Li N, Swigart LB, Berger MS, Hanahan D, Weiss WA, Evan GI. Using a preclinical mouse model of high-grade astrocytoma to optimize p53 restoration therapy. Proc Natl Acad Sci U S A 2013;110:E1480–9.
[15] Blough MD, Beauchamp DC, Westgate MR, Kelly JJ, Cairncross JG. Effect of aberrant p53 function on Temozolomide sensitivity of glioma cell lines and brain tumor initiating cells from glioblastoma. J Neurooncol 2011;102:1–7.
[16] Ohgaki H, Dessen P, Jourde B, Horstmann S, Nishikawa T, Di Patre PL, Burkhard C, Schuler D, Probst-Hensch NM, Majorka PC, Baeza N, Pisani P, Yonekawa Y, Yasargil MG, Lutolf UM, Kleihues P. Genetic pathways to glioblastoma. A Population-Based Study. Cancer Res 2004;64:6892–9.
[17] Stylli SS, Luwor RB, Ware TM, Tan F, Kaye AH. Mouse models of glioma. J Clin Neurosci 2015;22:619–26.
[18] William D, Mokri P, Lamp N, Linnebacher M, Classen CF, Erbersdobler A, Schneider B. Amplification of the EGFR gene can be maintained and modulated by variation of EGF concentrations in in vitro models of glioblastoma multiforme. PLoS One 2017;12, e0185208.
[19] Meng X, Zhao Y, Han B, Zha C, Zhang Y, Li Z, Wu P, Qi T, Jiang C, Liu Y, Cai J. Dual functionalized brain-targeting nanoinhibitors restrain Temozolomide-resistant glioma via attenuating EGFR and MET signaling pathways. Nat Commun 2020;11:594.
[20] Liu F, Hon GC, Villa GR, Turner KM, Ikegami S, Yang H, Ye Z, Li B, Kuan S, Lee AY, Zanca C, Wei B, Lucey G, Jenkins D, Zhang W, Barr CL, Furnari FB, Cloughesy TF, Yong WH, Gahman TC, Shiau AK, Cavenee WK, Ren B, Mischel PS. EGFR mutation promotes glioblastoma through epigenome and transcription factor network remodeling. Mol Cell 2015;60:307–18.
[21] Duan S, Yuan G, Liu X, Ren R, Li J, Zhang W, Wu J, Xu X, Fu L, Li Y, Yang J, Zhang W, Bai R, Yi F, Suzuki K, Gao H, Rodriguez Esteban C, Zhang C, Izpisua Belmonte JC, Chen Z, Wang X, Jiang T, Qu J, Tang F, Liu GH. PTEN deficiency reprogrammes human neural stem cells towards a glioblastoma stem cell-like phenotype. Nat Commun 2015;6:10068.
[22] Zhang P, Chen XB, Ding BQ, Liu HL, He T. Down-regulation of ABCE1 inhibits Temozolomide resistance in glioma through the PI3K/Akt/NF-kappaB signaling pathway. Biosci Rep 2018;38.
[23] Noh H, Zhao Q, Yan J, Kong LY, Gabrusiewicz K, Hong S, Xia X, Heimberger AB, Li S. Cell surface vimentin-targeted monoclonal antibody 86C increases sensitivity to Temozolomide in glioma stem cells. Cancer Lett 2018;433:176–85.
[24] Liu T, Xu H, Huang M, Ma W, Saxena D, Lustig RA, Alonso-Basanta M, Zhang Z, O'Rourke DM, Zhang L, Gong Y, Kao GD, Dorsey JF, Fan Y. Circulating glioma cells exhibit stem cell-like properties. Cancer Res 2018;78:6632–42.
[25] Shen J, Zhang T, Cheng Z, Zhu N, Wang H, Lin L, Wang Z, Yi H, Hu M. Lycorine inhibits glioblastoma multiforme growth through EGFR suppression. J Exp Clin Cancer Res 2018;37:157.
[26] Phillips AC, Boghaert ER, Vaidya KS, Falls HD, Mitten MJ, DeVries PJ, Benatuil L, Hsieh CM, Meulbroek JA, Panchal SC, Buchanan FG, Durbin KR, Voorbach MJ, Reuter DR, Mudd SR, Li L, Ralston SL, Cao D, Gan HK, Scott AM, Reilly EB. Characterization of ABBV-221, a tumor-selective EGFR-targeting antibody drug conjugate. Mol Cancer Ther 2018;17:795–805.
[27] Li H, Chen L, Li JJ, Zhou Q, Huang A, Liu WW, Wang K, Gao L, Qi ST, Lu YT. miR-519a enhances chemosensitivity and promotes autophagy in glioblastoma by targeting STAT3/Bcl2 signaling pathway. J Hematol Oncol 2018;11:70.
[28] Cho J, Pastorino S, Zeng Q, Xu X, Johnson W, Vanderberg S, Verhaak R, Cherniack AD, Watanabe H, Dutt A, Kwon J, Chao YS, Onofrio RC, Chiang D, Yuza Y, Kesari S, Meyerson M. Glioblastoma-derived epidermal growth factor receptor carboxyl-terminal deletion mutants are transforming and are sensitive to EGFR-directed therapies. Cancer Res 2011;71:7587–96.
[29] Gupta SK, Mladek AC, Carlson BL, Boakye-Agyeman F, Bakken KK, Kizilbash SH, Schroeder MA, Reid J, Sarkaria JN. Discordant in vitro and in vivo chemopotentiating effects of the PARP inhibitor Veliparib in Temozolomide-sensitive versus -resistant glioblastoma multiforme xenografts. Clin Cancer Res 2014;20:3730–41.
[30] Yoshida Y, Ozawa T, Yao TW, Shen W, Brown D, Parsa AT, Raizer JJ, Cheng SY, Stegh AH, Mazar AP, Giles FJ, Sarkaria JN, Butowski N, Nicolaides T, James CD. NT113, a pan-ERBB inhibitor with high brain penetrance, inhibits the growth of glioblastoma xenografts with EGFR amplification. Mol Cancer Ther 2014;13:2919–29.

[31] Sharpe MA, Livingston AD, Gist TL, Ghosh P, Han J, Baskin DS. Successful treatment of intracranial glioblastoma xenografts with a monoamine oxidase B-activated pro-drug. EBioMedicine 2015;2:1122–32.
[32] Parrish KE, Cen L, Murray J, Calligaris D, Kizilbash S, Mittapalli RK, Carlson BL, Schroeder MA, Sludden J, Boddy AV, Agar NY, Curtin NJ, Elmquist WF, Sarkaria JN. Efficacy of PARP inhibitor Rucaparib in Orthotopic glioblastoma xenografts is limited by ineffective drug penetration into the central nervous system. Mol Cancer Ther 2015;14:2735–43.
[33] Zhang C, Burger MC, Jennewein L, Genbler S, Schonfeld K, Zeiner P, Hattingen E, Hater PN, Mittelbron M, Tonn T, Steinbach JP, Weis WS. ErbB2/HER2-specific NK cells for targeted therapy of glioblastoma. J Natl Cancer Inst 2016; 108(5):djv375.
[34] Koga T, Chaim IA, Benitez JA, Markmiller S, Parisian AD, Hevner RF, Turner KM, Hessenauer FM, D'Antonio M, Nguyen MD, Saberi S, Ma J, Miki S, Boyer AD, Ravitz J, Frazer KA, Bafner V, Chen CC, Mischel PS, Yeo GW, Furnari FB. Longitudinal assessment of tumor development using cancer avatars derived from genetically engineered pluripotent stem cells. Nat Commun 2020;11:550.
[35] Wei S, Liu L, Chen Z, Yin W, Liu Y, Ouyang Q, Zeng F, Nie Y, Chen T. Artesunate inhibits the mevalonate pathway and promotes glioma cell senescence. J Cell Mol Med 2020;24:276–84.
[36] Kersten K, de Visser KE, van Miltenburg MH, Jonkers J. Genetically engineered mouse models in oncology research and cancer medicine. EMBO Mol Med 2017;9:137–53.
[37] Patrizii M, Bartucci M, Pine SR, Sabaawy HE. Utility of glioblastoma patient-derived orthotopic xenografts in drug discovery and personalized therapy. Front Oncol 2018;8:23.
[38] Tuveson D, Clevers H. Cancer modeling meets human organoid technology. Science 2019;364:952–5.
[39] Reilly KM, Loisel DA, Bronson RT, McLaughlin ME, Jacks T. Nf1;Trp53 mutant mice develop glioblastoma with evidence of strain-specific effects. Nat Genet 2000;26:109–13.
[40] Ding H, Roncari L, Shannon P, Wu X, Lau N, Karaskova J, Gutmann DH, Squire JA, Nagy A, Guha A. Astrocyte-specific expression of activated p21-ras results in malignant astrocytoma formation in a transgenic mouse model of human gliomas. Cancer Res 2001;61:3826–36.
[41] Xiao A, Wu H, Pandolfi PP, Louis DN, Van Dyke T. Astrocyte inactivation of the pRb pathway predisposes mice to malignant astrocytoma development that is accelerated by PTEN mutation. Cancer Cell 2002;1:157–68.
[42] Zhu Y, Guignard F, Zhao D, Liu L, Burns DK, Mason RP, Messing A, Parada LF. Early inactivation of p53 tumor suppressor gene cooperating with NF1 loss induces malignant astrocytoma. Cancer Cell 2005;8:119–30.
[43] Wei Q, Clarke L, Scheidenhelm DK, Qian B, Tong A, Sabha N, Karim Z, Bock NA, Reti R, Swoboda R, Purev E, Lavoie JF, Bajanaru ML, Shannon P, Herlyn D, Kaplan D, Henkelman RM, Gutmann DH, Guha A. High-grade glioma formation results from postnatal pten loss or mutant epidermal growth factor receptor expression in a transgenic mouse glioma model. Cancer Res 2006;66:7429–37.
[44] Kwon CH, Zhao D, Chen J, Alcantara S, Li Y, Burns DK, Mason RP, Lee EY, Wu H, Parada LF. Pten haploinsufficiency accelerates formation of high-grade astrocytomas. Cancer Res 2008;68:3286–94.
[45] Zheng H, Ying H, Yan H, Kimmelman AC, Hiller DJ, Chen AJ, Perry SR, Tonon G, Chu GC, Ding Z, Stommel JM, Dunn KL, Wiedemeyer R, You MJ, Brennan C, Wang YA, Ligon KL, Wong WH, Chin L, DePinho RA. p53 and Pten control neural and glioma stem/progenitor cell renewal and differentiation. Nature 2008;455:1129–33.
[46] Zhu H, Acquaviva J, Ramachandran P, Boskovitz A, Woolfenden S, Pfanni R, Bronson RT, Chen JW, Weissleder R, Housman DE, Charest A. Oncogenic EGFR signaling cooperates with loss of tumor suppressor gene functions in Gliomagenesis. Proc Natl Acad Sci USA 2009;106:2712–6.
[47] Chow LM, Endersby R, Zhu X, Rankin S, Qu C, Zhang J, Broniscer A, Ellison DW, Baker SJ. Cooperativity within and among Pten, p53, and Rb pathways induces high-grade astrocytoma in adult brain. Cancer Cell 2011;19:305–16.
[48] Lei L, Sonabend AM, Guarnieri P, Soderquist C, Ludwig T, Rosenfeld S, Bruce JN, Canoll P. Glioblastoma models reveal the connection between adult glial progenitors and the proneural phenotype. PLoS One 2011;6, e20041.
[49] Zuckermann M, Hovestadt V, Knobbe-Thomsen CB, Zapatka M, Northcott PA, Schramm K, Belic J, Jones DT, Tschida B, Moriarty B, Largaespada D, Roussel MF, Korshunov A, Reifenberger G, Pfister SM, Lichter P, Kawauchi D, Gronych J. Somatic CRISPR/Cas9-mediated tumour suppressor disruption enables versatile brain tumour modelling. Nat Commun 2015;6:7391.
[50] Pathania M, De Jay N, Maestro N, Harutyunyan AS, Nitarska J, Pahlavan P, Henderson S, Mikael LG, Richard-Londt A, Zhang Y, Costa JR, Hébert S, Khazaei S, Ibrahim NS, Herrero J, Riccio A, Albrecht S, Ketteler R, Brandner S, Kleinman CL, Jabado N, Salomoni P. H3.3(K27M) cooperates with Trp53 loss and PDGFRA gain in mouse embryonic neural progenitor cells to induce invasive high-grade gliomas. Cancer Cell 2017;32:684–700.

[51] Oldrini B, Curiel-Garcia A, Marques C, Matia V, Uluckan O, Grana-Castro O, Torres-Ruiz R, Rodriguez-Perales S, Huse JT, Squatrito M. Somatic genome editing with the RCAS-TVA-CRISPR-Cas9 system for precision tumor modeling. Nat Commun 2018;9:1466.
[52] Mali P, Yang L, Esvelt KM, Aach J, Guell M, DiCarlo JE, Norville JE, Church GM. RNA-guided human genome engineering via Cas9. Science 2013;339:823–6.
[53] Noorani I. Genetically engineered mouse models of gliomas: technological developments for translational discoveries. Cancer 2019;11:1335.
[54] Bressan RB, Dewari PS, Kalantzaki M, Gangoso E, Matjusaitis M, Garcia-Diaz C, Blin C, Grant V, Bulstrode H, Gogolok S, Skarnes WC, Pollard SM. Efficient CRISPR/Cas9-assisted gene targeting enables rapid and precise genetic manipulation of mammalian neural stem cells. Development 2017;144:635–48.
[55] Simeonova I, Huillard E. In vivo models of brain tumors: roles of genetically engineered mouse models in understanding tumor biology and use in preclinical studies. Cell Mol Life Sci 2014;71:4007–26.
[56] Vooijs M, Jonkers J, Berns A. A highly efficient ligand-regulated Cre recombinase mouse line shows that LoxP recombination is position dependent. EMBO Rep 2001;2:292–7.
[57] Maire CL, Ligon KL. Glioma models: new GEMMs add "class" with genomic and expression correlations. Cancer Cell 2011;19:295–7.

CHAPTER

21

Therapeutic approaches to overcome temozolomide resistance in glioblastoma

Marina Gergues[a,b], *Vibha Harindra Savanur*[a,b], *David Prado*[a], *Adam Eljarrah*[a], *James K. Liu*[c], *Mohammed Jaloudi*[a,d], *and Pranela Rameshwar*[a]

[a]Department of Medicine, Rutgers New Jersey Medical School, Newark, NJ, United States [b]School of Graduate Studies, Rutgers New Jersey Medical School, Newark, NJ, United States [c]Department of Neurosurgery, Rutgers New Jersey Medical School, Newark, NJ, United States [d]Rutgers Cancer Institute of New Jersey, University Hospital, Newark, NJ, United States

Abstract

Glioblastoma (GBM) is a grade IV glioma, which is the most common and aggressive malignant brain tumor in adults. GBM is characterized by high infiltration into surrounding brain tissue, chemoresistance, a high rate of recurrence, all of which contribute to poor prognosis. Median survival following standard care therapy is approximately 14.5 months, with relapse commonly occurring after 6 months. Standard care therapy consists of maximal surgical resection followed by concurrent radiotherapy and chemotherapy with the alkylating agent temozolomide (TMZ). Overcoming TMZ resistance, whether intrinsic or adaptive, poses a significant clinical challenge. Here, we discuss mechanisms linked to TMZ resistance and summarize potential treatment options for patients with TMZ-resistant GBM. Despite advances in identifying novel targets and potential molecular, pharmacological, immunological, and noninvasive therapies, TMZ resistance remains a hurdle in establishing long-term treatment for GBM patients.

Abbreviations

20(*S*)-Rg3 20(*S*)-ginsenoside-Rg3
APE1 apurinic/apyrimidinic endonuclease-1
APNG alkyl purine-DNA-N-glycosylase
BBB blood–brain barrier
BER base excision repair

Glioblastoma Resistance to Chemotherapy: Molecular Mechanisms and Innovative Reversal Strategies
https://doi.org/10.1016/B978-0-12-821567-8.00012-9

bFGF	basic fibroblast growth factor
COX2	cyclooxygenase-2
CpG	cysteine-phosphate-guanine
CSC	cancer stem cell
DSB	double-strand break
EGFR	epidermal growth factor receptor
FGFR	fibroblast growth factor receptor
GBM	glioblastoma multiforme
G-CSC	glioma cancer stem cell
HAT	histone acetyltransferases
HDAC	histone deacetyltransferases
IFN-β	interferon-β
JNK	c-Jun NH2-terminal kinases
MAP	Mitogen activated-protein
MGMT	O^6-methyguanine-DNA methyltransferase
MMR	mismatch Repair
N3-A	N3-adenine
N7-G	N7-guanine
O6-BG	O6-benzylguanine
O6-BTG	O6-(4-bromoethyl)guanine
O6-MeG	O6-methylguanine
OS	overall survival
PARP1	poly (ADP-ribose) polymerase-1
PFS	progression-free survival
RITA	reactivation of p53 and induction of tumor cell apoptosis
RT	radiotherapy
RTK	receptor tyrosine kinase
SAHA	suberanilohydroxamic acid
SSB	single-strand break
TMZ	temozolomide
TTF	tumor treating fields
UA	ursolic acid
VA	valropic acid
VEGF	vascular endothelial growth factor
WA	Withaferin A

Conflict of interest

No potential conflicts of interest were disclosed.

Introduction

Prior to 1996, the only form of treating GBM was resection followed by radiotherapy (RT), which extended survival from 3 to 9 months (Fig. 1) [1]. In 1996, Gliadel wafers were approved by FDA as the first brain cancer treatment to release chemotherapy directly into the tumor site [2]. Gliadel wafers are biodegradable carmustine-loaded wafers implanted at the time of tumor resection. This offers a marginal increase in overall survival (OS) (16.2 vs 14 months) and minimal systemic toxicity. However, such treatment was associated with high rates of complication, including seizures. Hence, Gliadel wafers were not widely accepted. In 1999, temozolomide (TMZ) was approved for the treatment of anaplastic astrocytoma in adults, and in 2005, for newly diagnosed GBM [3]. TMZ is an alkylating agent that can penetrate the blood–brain barrier (BBB) and is orally administered. In 2009 bevacizumab, a

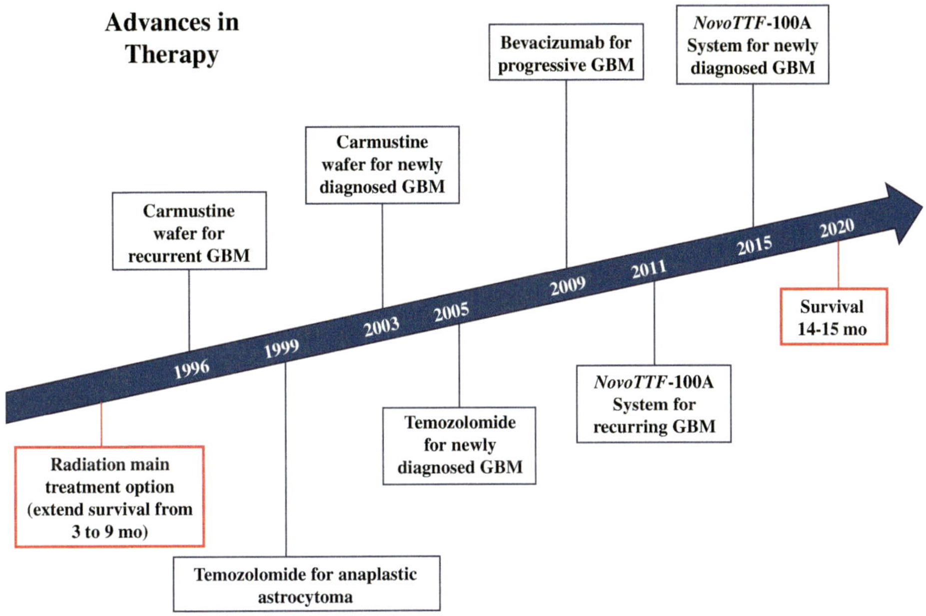

FIG. 1 Timeline advancements in GBM treatments.

VEGF antibody was approved for treating progressive GBM, despite showing no increase in overall survival (OS), as compared to current standard treatment [4]. More recently, tumor-treating fields (TTF), a noninvasive therapy with alternating electrical fields to disrupt mitosis, gained approval for refractory and newly diagnosed GBM [5]. This method extended patient survival by approximately 4 months as compared to RT and TMZ therapy. Despite a modest impact on survival, there is concern on the cost-effectiveness of TTF, which limits its clinical application [6].

Current standard therapy for GBM consists of maximal surgical resection followed by concurrent RT and TMZ [7]. Even successful surgical resection has the potential of leaving at least 10% of the bulk tumor behind owing to diffuse and ill-defined borders [8]. Therefore, despite aggressive treatment, recurrence within approximately 6 months postresection is common and survival remains 14–15 months [7]. Recurrent GBM is typically characterized as TMZ resistant, with no effective follow-up treatment [9]. Here, we discuss mechanisms linked to TMZ resistance with summaries of potential treatment options for GBM patients.

Targeting DNA repair mechanisms

MGMT-targeted therapies

Expression of the DNA repair protein, O6-methylguanine-DNA methyltransferase (MGMT), has been attributed as a significant contributor to TMZ resistance in GBM [10]. TMZ exerts its cytotoxic effect by methylating DNA at the O6-guanine, N7-guanine (N7-G), and N3-adenine (N3-A) positions. O6-methylguanine (O6-MeG) lesions mismatch with thymine during DNA replication, resulting in DNA strand breaks, which trigger cell cycle arrest and apoptosis. MGMT removes methyl groups from the O6 position of guanine, resulting in TMZ resistance [10].

Mismatch repair (MMR) recognizes and repairs mismatched bases that may arise during DNA synthesis [11]. O6-MeG improperly matches with thymine during replication and the MMR system excises the lesion creating a long gap of more than 1000 bps [12]. However, during repair, O6-MeG mispairs with thymine, which instigates a futile cycle of excision and repair, leaving the cell in S-phase. This eventually leads to DNA double-strand breaks triggering apoptosis in cells that are MGMT deficient and MMR sufficient. Therefore, depleting cells of MGMT has been a large focus in overcoming TMZ resistance [13].

O6-benzylguanine (O6-BG) is a low-molecular-weight substrate that can irreversibly inactivate MGMT through alkyl group transfer [14]. It has been widely used to deplete MGMT and sensitize glioma cells to alkylating agents. There are several studies attempting to make the best use of O6-BG in combination with currently approved therapies. Phase I trials O6-BG with carmustine wafers demonstrated that 120 mg/m^2 of O6-BG for 1 h followed by continuous infusion of 30 mg/m^2/day for days in adults with recurrent malignant glioma was identified as the dose that limits toxicity [15]. Follow-up phase II trials showed promising results in patients with anaplastic glioma. However, there was no improvement in OS for GBM patients irrespective of their MGMT methylation status [16]. Similarly, a phase III study of radiotherapy and carmustine with O6-BG vs radiotherapy and carmustine alone showed no added benefit for use of O6-BG [17]. Despite many efforts and clinical trials, O6-BG has yet to be approved for standard treatment owing to side effects and conflicting data as to its efficacy [7].

Another MGMT pseudosubstrate, O6-(4-bromothenyl) guanine (O6-BTG), has a 10-fold higher potency and has been studied in several advanced solid tumors [14]. Delivering liposomal O6-BTG by magnetic resonance image-guided microbubble-enhanced low-intensity pulsed focused ultrasound completely depleted MGMT in vivo and sensitized GBM to TMZ [18]. However, MGMT inhibitors such as O6-BG and O6-BTG are known to cause dose-limiting hematologic toxicities, such as neutropenia, leukopenia, and thrombocytopenia [14]. Therefore, identifying alternate mechanisms through administration and dosing may be necessary to reach the maximum potential of MGMT inhibitors. For example, tumor cells frequently overexpress glucose transporters that preferentially take up glucose-conjugated inhibitors to deplete MGMT in tumors [19].

Ursolic acid (UA) is a pentacyclic triterpene acid derivative existing in food herbs and other plants [20]. UA has become a major focus in cancer research because of its ability to induce apoptosis in multidrug-resistant cells while exhibiting minimal systemic toxicity. Although UA is implicated in several pathways such as JAK/STAT and JNK/Bcl-2, its mechanisms of action are broad and remain under investigation [21]. One study revealed UA has the potential of downregulating MGMT expression and attenuating TMZ resistance in GBM cells [22]. Most notably, UA can pass through the BBB, making it an attractive agent for further investigation.

Interferon-β (IFN-β) is a member of the IFN class I family. Clinical trials testing IFN-β for multiple sclerosis proved its safety and use for brain disease [23]. Additionally, IFN-β was used as a single agent or in combination with other treatments for malignant gliomas and melanomas. Studies have revealed that IFN-β sensitizes GBM cells to TMZ by inducing p53 to decrease MGMT transcription [24]. Unfortunately, when tested in phase II clinical trials, few patients benefited from IFB-β treatment [25]. However, the results proved that IFN-β facilitated TMZ sensitivity, reduced angiogenesis, and enhanced immunological surveillance. On the other hand, these studies revealed that IFN-β can induce autocrine activation of pathways involved in immune escape and minimize the benefit of immune checkpoint inhibitors [26]. With these preliminary findings, further investigation into this class of cytokines for effective translation to patient care is warranted.

Other drugs implicated in diminishing MGMT to attenuate TMZ resistance include withaferin A (WA), 20(*S*)-ginsenoside-Rg3 (20(*S*)-Rg3), cordycepin, and resveratrol. WA is a 28-carbon steroidal lactone derived from the ashwagandha plant with promising pleiotropic anticancer effects [27]. When used for GBM, WA prevented proliferation by arresting cells in G2/M phase, downregulating AKT/mTOR proteins, and depletion of MGMT, all of which potentiated TMZ-mediated apoptosis. 20(*S*)-Rg3 and cordycepin are also nontoxic natural chemicals that were independently tested in glioma cell lines. Studies revealed that these compounds might be novel agents in downregulating MGMT expression via the Wnt/β-catenin pathway and resensitizing GBM to TMZ treatment [28, 29]. Resveratrol, found in several plants such as grapes and nuts, has been used for hundreds of years in traditional Asian medicine for inflammatory conditions [30]. In vitro and in vivo studies reveal that resveratrol holds anticancer properties by promoting apoptosis, inhibiting COX activity, and regulating the cell cycle. In GBM, the mechanisms of resveratrol involve the Wnt/β-catenin and NFκB pathways to reverse TMZ resistance and this resulted in downregulated MGMT expression. Clinical trials testing resveratrol for colorectal cancers showed some success and therefore warrants further investigation in GBM [31].

MGMT expression requires several pathways such as the c-Jun NH2-terminal kinases (JNK), a member of the mitogen activated-protein (MAP) kinase family [32]. JNK inhibition sensitizes TMZ via regulation of MGMT expression. JNK integrates extracellular signals to initiate cellular processes including proliferation, apoptosis, and differentiation. Several human cancers showed high expression of JNK, especially in the self-renewing cancer stem cells (CSCs). JNK was silenced in gliomas CSCs (G-CSCs) using a JNK-specific inhibitor, SP600125, and by a JNK-shRNA. JNK inhibition reduced tumorsphere formation in GSCs, reduced stem cell markers, and increased differentiation markers such as GFAP and β-tubulin. Additional studies displayed the significance of JNK-targeting using SP600125 or JNK-shRNA for GSCs, which suppressed MGMT expression with sensitization to TMZ. Another novel JNK inhibitor, AS602801, was equally efficacious in experimental in vitro and in vivo assays [13]. However, efficacy linked to targeting JNK only occurred with MGMT expression.

Basic excision repair (BER)-targeted therapies

As discussed above, the methylations of N7-guanine and N3-adenine constitute more than 80% of TMZ activity [33]. However, these adducts are rapidly repaired by the BER system, which is a multicatalytic reaction consisting of DNA glycosylases, endonucleases, polymerases, and DNA ligases. Inhibiting one or more of these enzymes results in DNA repair deficiency, which can contribute to TMZ cytotoxicity. Components of BER that have the most therapeutic promise include poly (ADP-ribose) polymerase-1 (PARP1), apurinic/apyrimidinic endonuclease-1 (APE1), and alkyl purine-DNA-N-glycosylase (APNG).

Alkylated bases such as N7-MeG and N3-MeA cause a replication stall and breakdown of the replication fork, leading to single-strand breaks (SSBs), which eventually generate double-strand breaks (DSBs) [34, 35]. PAPR1 recognizes SSBs and uses NAD^+ as a substrate to initiate the repair process. PARP1 with NAD^+ catalyzes the covalent binding of PAR polymers on itself and other acceptor proteins. The PAR chain recruits DNA repair proteins to the damaged site and negates TMZ cytotoxicity. Thus, PARP has become a therapeutic target in modulating the BER pathway. In vitro studies testing PARP inhibitors, talazoparib and rucaparib, were studied in orthotopic GBM mouse models. However, these studies showed poor BBB penetration with no considerable TMZ sensitization [36, 37]. In contrast, veliparib (ABT-888) was able to penetrate the BBB and was effective in sensitizing GBM to TMZ [38, 39]. Reduced tolerance was noted in a follow-up phase I/II trial when veliparib was combined with standard TMZ and RT [40]. Currently, trials testing veliparib with either TMZ or RT alone are underway. Similar trials have tested veliparib with TMZ and RT in pediatric gliomas. This combination of drugs was well tolerated, yet veliparib did not improve OS [41].

APE1 is an essential enzyme of the BER pathway that repairs abasic sites and regulates transcription factors in an active reduced state. Silencing APE1 by in vitro studies was able to enhance the cytotoxicity of TMZ in resistant populations [42]. Similarly, using the small molecule inhibitor methoxyamine, APE1 expression was slightly downregulated and GBM resistance was repaired [43]. Another attractive target of the BER family is APNG. Overexpression of this enzyme increased chemotherapeutic sensitivity to alkylating agents in breast, ovarian, and glioma cell lines [44–46]. However, conflicting follow-up correlation studies suggested that APNG-positive tumors had poorer OS in GBM patients than APNG-negative tumors [47]. Furthermore, silencing APNG in GBM cells attenuated TMZ-induced apoptosis. Genes linked to BER remain appealing as a potential therapeutic option. However, in order to get this category of molecules in clinical trials, there needs further research on BER mechanisms.

Epigenetic targeting

Methylation of the promoter site in the cysteine-phosphate-guanine (CpG) MGMT gene island directly correlates to expression level [48]. Methylation of the MGMT promotor decreased its expression. Several studies indicate that GBM tumors with methylated MGMT promoters correlate to prolonged patient survival. Therefore, epigenetic modifications such as DNA methylation, chromatin remodeling, and histone modifications are considered key mechanisms in GBM development. Histone modifications can activate or repress transcription by the addition of acetyl or methyl groups respectively on lysines of H3 and H4 [49]. Histone acetyltransferases (HATs) and histone deacetyltransferases (HDACs) maintain the balance of histone acetylation levels and thereby gene transcription levels. Upregulation of HDAC expression has been linked to tumor formation, which is why HDAC inhibitors are being extensively studied for GBM therapy [50].

Several HDAC inhibitors have been approved by the FDA and have been clinically tested in GBM, including valproic acid (VA), vorinostat, romidepsin, panobinostat, and belinostat

[51]. Retrospective studies combining VA with TMZ and RT improved survival rates in patients from 13.96 to 17.35 months as compared to TMZ alone [52]. VA readily passes the BBB and has dual indications due to its effectiveness in controlling seizure. Excellent tolerance is reported for VA use in GBM clinical trials thereby holding promise for future use [53, 54]. Vorinostat tested in phase I and II trials, showed acceptable tolerance, yet did not extend OS when combined with TMZ and RT for newly diagnosed GBM [55]. Phase I and II trials testing of romidepsin and panobinostat as either single agents or in combination, were either ineffective or toxic [56, 57]. Lastly, belinostat, a pan-HDAC inhibitor that is structurally similar to suberanilohydroxamic acid (SAHA), which is a first-generation HDAC inhibitor, has greater BBB penetration as compared to SAHA and may be a potential successful agent for GBM treatment [58].

Attenuating TMZ resistance through downregulation of MGMT expression can be a useful therapeutic agent. However, the efficacy of MGMT inhibitors in treating GBM remains controversial and potentially disadvantageous to overall patient health. Furthermore, there remains controversy over the relevance and weight of targeting MGMT to overcome TMZ resistance. Several studies, for example, suggest that depleting MGMT by O6-BG does not reduce TMZ resistance nor does it improve sensitivity. Additional studies will determine the relevance of targeting DNA repair mechanisms and HDACs in this setting.

Targeting receptor tyrosine kinases and cellular pathways

Receptor tyrosine kinases (RTKs) are a family of approximately 60 cell surface receptors that integrate signals from growth factors, hormones, cytokines, and other extracellular signaling molecules [59]. RTKs regulate critical cellular processes such as cell proliferation, survival, differentiation, migration, and cell cycle control. RTKs share structures, composed of an extracellular ligand-binding domain, a single transmembrane helix, and a cytoplasmic region with tyrosine kinase activity. Ligand binding activates receptor dimerization and autophosphorylation of the tyrosine kinase domain.

Extracellular ligands activate RTKs to regulate cellular processes through major intracellular pathways such as Ras/MAPK/ERK, Ras/PI3K/AKT, p53, and NFκB. Several cancers, including GBM, arise from genetic alterations in the activity, abundance, distribution, or regulation of RTKs. Studies indicate that RTK dysregulation can lead to drug resistance, which makes them prospective therapeutic targets [60]. Moreover, several small molecules, discussed below, are more affordable and accessible than biologics. Table 1 summarizes the ongoing and completed clinical trials linked to RTK and downstream pathways.

Epidermal growth factor receptor (EGFR)

The four ErbB family receptors include ErbB1 (EGFR, Her1), ErbB2 (Her2, Neu), ErbB3 (Her3), and ErbB4 (Her4) [61]. Since the 1970s, when EGFR was first identified, hyperactivation, overexpressed, and dysregulated signaling of this gene have been demonstrated in many caners. Mutations and amplifications in the EGFR gene were characterized in approximately 50% of GBM cases and correlated with poor patient prognosis [62]. Within the

TABLE 1 Clinical trials with RTK and cellular pathway-targeted therapies.

Target	Drug	Trial ID	Trial Phase	Status	Treatment regimen
EGFR/ EGFRvIII	Gefitinib (Iressa, ZD1839)	NCT00250887	II	Completed	Pre-and postoperative use of gefitinib in recurrent GBM after standard surgery, RT and chemotherapy
		NCT00014170	II	Completed	Gefitinib in newly diagnosed GBM
		NCT00016991	II	Completed	Gefitinib for first relapse GBM patients
		NCT00025675	II	Completed	Gefitinib in treating patients who have recurrent or progressive central nervous system tumors
		NCT00052208	I/II	Completed	Gefitinib and RT for GBM
		NCT00238797	II	Completed	Gefitinib and RT for GBM
		NCT00027625	I	Completed	Gefitinib and TMZ for malignant primary gliomas
		NCT00132158	I	Completed	Gefitinib and irinotecan in treating young patients with refractory solid tumors
	Erlotinib (Tarceva)	NCT00337883	II	Completed	Erlotinib in patients with first relapse of GBM
		NCT00054496	II	Unknown	Erlotinib for recurrent or progressive GBM
		NCT00045110	I/II	Ongoing	Erlotinib for recurrent malignant glioma or recurrent or progressive meningioma
		NCT00301418	I/II	Completed	Erlotinib for recurrent/residual GBM and anaplastic astrocytoma
		NCT00124657	I/II	Completed	Erlotinib with RT in young patients with newly diagnosed glioma
		NCT01257594	I	Ongoing	Erlotinib for recurrent EGFRvIII mutant malignant gliomas
		NCT00187486	II	Completed	Erlotinib with TMZ and RT in patients diagnosed with GBM or gliosarcoma postoperative
		NCT00274833	II	Completed	Erlotinib with RT and TMZ in newly diagnosed GBM
		NCT00039494	II	Completed	Erlotinib with TMZ and RT in patients with GBM or other brain tumors
		NCT00086879	II	Completed	Erlotinib as compared to TMZ or carmustine in recurrent GBM
		NCT00671970	II	Completed	Erlotinib with bevacizumab for recurrent malignant gliomas
		NCT00720356	II	Completed	Erlotinib with bevacizumab after RT and TMZ in newly diagnosed GBM or gliosarcoma
		NCT00445588	II	Completed	Erlotinib with sorafenib in progressive or recurrent GBM
		NCT00609999	I	Completed	Erlotinib with dasatinib in recurrent GBM or gliosarcoma

	Lapatinib (GW572016)	NCT01591577	II	Active, not recruiting	Lapatinib with RT and TMZ in newly diagnosed GBM
		NCT00099060	I/II	Completed	Lapatinib for recurrent GBM
		NCT00107003	II	Completed	Lapatinib for recurring malignant brain tumors
		NCT00095940	I/II	Completed	Lapatinib for young patients with recurrent or refractory central nervous system tumors
		NCT02101905	I	Active, not recruiting	Lapatinib preoperatively in recurring high-grade glioma
		NCT00350727	II	Completed	Lapatinib with pazopanib in recurring Grade III or IV malignant gliomas
	Afatinib (BIBW2992)	NCT00977431	I	Completed	Afatinib and RT with or without TMZ in newly diagnosed GBM
		NCT00727506	II	Completed	Afatinib with or without TMZ for recurrent malignant glioma
		NCT02423525	I	Recruiting	Afatinib safety study for brain cancer
	Dacomitinib (PF-299804)	NCT01520870	II	Completed	Dacomitinib for recurring GBM with EGFR amplification or mutation
		NCT01112527	II	Completed	Dacomitinib for adults with recurrent GBM
	Osimertinib (AZD9291)	NCT03732352	II	Recruiting	Osimertinib and 18F-FDG PET in patients with EGFR activated GBM
	Cetuximab (C225)	NCT01044225	II	Terminated	Cetuximab or cilengitide with RT and TMZ for newly diagnosed MGMT-promoter unmethylated GBM
		NCT00311857	I/II	Unknown	Cetuximab with RT and TMZ in primary GBM
		NCT03603379	I	Recruiting	Cetuximab immunoliposomes loaded with doxorubicin administered intravenously in patients with relapsed or refractory high-grade gliomas
		NCT02861898	I/II	Recruiting	Cetuximab administered by superselective intra-arterial repeated infusion for newly diagnosed GBM
		NCT01238237	I	Completed	Cetuximab administered by superselective intra-arterial cerebral infusion for treatment of relapsed GBM and anaplastic astrocytoma
		NCT00463073	II	Completed	Cetuximab with irinotecan and bevacizumab compared to irinotecan and bevacizumab alone for GBM
		NCT02800486	II	Recruiting	Cetuximab administered by superselective intra-arterial repeated infusion with reirradiation for relapsed GBM, anaplastic astrocytoma and anaplastic oligoastrocytoma
		NCT01884740	I/II	Recruiting	Cetuximab and bevacizumab administered by intra-arterial infusion for refractory intracranial glioma in patients under 22

Continued

TABLE 1 Clinical trials with RTK and cellular pathway-targeted therapies—cont'd

Target	Drug	Trial ID	Trial Phase	Status	Treatment regimen
	Panitumumab	NCT01017653	II	Terminated	Panitumumab with irinotecan for malignant gliomas
	Nimotuzumab	NCT03388372	II	Completed	Nimotuzumab with RT and TMZ for GBM
		NCT00753246	III	Completed	Nimotuzumab efficacy study for GBM
		NCT00561873		Completed	Nimotuzumab efficacy study for children with high-grade glioma
	Rindopepimut (CDX-110)	NCT01498328	II	Completed	Rindopepimut with GM-CSF for patients with EGFRvIII-positive GBM
		NCT01480479	III	Completed	Rindopepimut with GM-CSF for newly diagnosed GBM
		NCT00458601	II	Completed	Rindopepimut with TMZ for GBM
	EGFRvIII-directed CAR T	NCT03283631	I	Suspended	Intracerebral EGFRvIII CAR-T cells for recurrent GBM
		NCT02844062	I	Unknown	Anti-EGFRvIII CAR-T cells for recurring GBM
		NCT02937844	I	Unknown	Truncated EGFR incorporated in chimeric switch receptor modified T cells in recurrent GBM
		NCT01454596	I/II	Completed	EGFRvIII targeted-CAR T cells for malignant gliomas expressing EGFRvIII
VEGFR	Tivozanib	NCT01846871	II	Completed	Tivozanib safety and efficacy study for recurring GBM
	Cediranib (AZD2171)	NCT01310855	II	Terminated	Cediranib with or without gefitinib in recurrent or progressive GBM
		NCT02974621	II	Active, not recruiting	Cediranib and olaparib compared to bevacizumab in recurrent GBM
		NCT00777153	III	Completed	Cediranib with lomustine compared to cediranib alone or lomustine with placebo in recurrent GBM
		NCT01062425	II	Active, not recruiting	Cediranib with or without RT and TMZ in newly diagnosed GBM
		NCT00979862	I	Completed	Cediranib and cilengitide in progressive or recurring GBM
		NCT00305656	II	Completed	Cediranib efficacy study in recurring GBM
		NCT00662506	I/II	Completed	Cediranib with RT and TMZ in newly diagnosed GBM
		NCT00503204	I	Completed	Cediranib with lomustine in recurrent malignant brain tumor
		NCT00458731	I	Completed	Cediranib with bevacizumab for metastatic or unresectable solid tumors including GBM

Lenvatinib	NCT01433991	I/II	Completed	Lenvatinib with E7050 in advanced solid tumors, unresectable melanoma and recurring GBM
	NCT03797326	II	Active, not recruiting	Lenvatinib plus pembrolizumab in patients with previously treated solid tumors
Sorafenib (BAY 439006)	NCT00544817	II	Completed	Sorafenib plus RT and TMZ for GBM
	NCT00597493	II	Completed	Sorafenib with protracted TMZ in recurrent GBM
	NCT00734526	I	Terminated	Sorafenib with RT and TMZ in newly diagnosed GBM or gliosarcoma
	NCT02599090	II	Withdrawn	Sorafenib with RT and TMZ in newly diagnosed GBM or gliosarcoma
	NCT00335764	I/II	Completed	Sorafenib with erlotinib, tipifarnib or temsirolimus in recurring GBM or gliosarcoma
	NCT00621686	II	Completed	Sorafenib and bevacizumab in recurrent GBM
	NCT00329719	I/II	Completed	Sorafenib and temsirolimus for recurrent GBM
	NCT01434602	I/II	Recruiting	Sorafenib and everolimus in recurring high-grade gliomas
	NCT01817751	II	Active, not recruiting	Sorafenib, valproic acid and sildenafil citrate for recurring high-grade gliomas
	NCT00884416	I	Completed	Sorafenib with RT and TMZ for newly diagnosed high-grade gliomas
	NCT00093613	I	Completed	Sorafenib efficacy study for recurring or progressive malignant gliomas
Sunitinib	NCT01100177	II	Completed	Sunitinib before and during RT in newly diagnosed GBM
	NCT02928575	II	Unknown	Sunitinib combined with TMZ and RT in unmethylated MGMT promoter GBM
	NCT00535379	II	Unknown	Sunitinib efficacy and safety study for recurring or progressive GBM
	NCT00864864	I	Completed	Sunitinib efficacy study for passing BBB in recurring GBM
	NCT00606008	II	Completed	Sunitinib for recurrent anaplastic astrocytoma and GBM
	NCT03025893	II/III	Recruiting	Intermittent sunitinib compared to lomustine for recurrent GBM
	NCT00611728	I	Completed	Sunitinib with irinotecan safety study for malignant gliomas
	NCT00923117	II	Terminated	Sunitinib for recurrent malignant gliomas who are naive or resistant to bevacizumab therapy
	NCT01122888	I	Terminated	Sunitinib combined with cilengitide for advanced solid tumors or GBM

Continued

TABLE 1 Clinical trials with RTK and cellular pathway-targeted therapies—cont'd

Target	Drug	Trial ID	Trial Phase	Status	Treatment regimen
		NCT00499473	II	Completed	Sunitinib efficacy study for recurring malignant gliomas
		NCT01441388	I	Withdrawn	Crizotinib in combination with VEGF inhibitors including sunitinib for advanced solid tumors
	Pazopanib	NCT02331498	I/II	Recruiting	Pazopanib combined with TMZ for newly diagnosed GBM after surgery and RT
		NCT00459381	II	Completed	Pazopanib safety and efficacy study for recurring GBM
		NCT01931098	II	Completed	Oral pazopanib combined with oral topotecan for recurrent GBM with and without bevacizumab
	Nintedanib (BIBF1120)	NCT01666600	I/II	Terminated	Nintedanib combined with RT for GBM patients qualified for reirradiation
		NCT01251484	II	Completed	Nintedanib efficacy study for recurring GBM
		NCT01380782	II	Completed	Nintedanib efficacy study for high-grade gliomas
	Vatalanib (PTK787)	NCT00385853	I	Completed	Vatalanib combined with TMZ and RT for GBM patients on EIACDs
		NCT00128700	I/II	Completed	Vatalanib combined with TMZ and RT compared to TMZ and RT alone in newly diagnosed GBM
		NCT00387933	I	Completed	Imatinib, vatalanib and hydroxyurea combined therapy for recurrent malignant gliomas
	TTAC-0001	NCT03856099	II	Recruiting	TTAC-0001 safety and efficacy study for recurrent GBM patients on bevacizumab
		NCT03722342	I	Active, not recruiting	TTAC-001 combined with pembrolizumab for recurring GBM
		NCT03033524	II	Unknown	TTAC-0001 safety and efficacy study for recurrent GBM
	VXM01	NCT03750071	I/II	Recruiting	VXM01 combined with avelumab for progressive GBM
		NCT02718443	I	Completed	VXM01 safety study for operable recurrent GBM
FGFR	BGJ398	NCT01975701	II	Completed	BGJ398 treatment for patients with recurrent GBM
	TAS-120	NCT02052778	I/II	Ongoing	TAS-120 in patients with advanced solid tumors with FGF/FGFR aberrations

p53	Adenovirus-p53	NCT00004041	I	Completed	Ad5CMV-p53 gene therapy with surgery for patients with recurrent malignant Gliomas
		NCT00004080	I	Completed	Recombinant adenovirus-p53 SCH-58500 gene therapy with surgery for treating patients with recurrent or progressive brain tumors
	SGT-53	NCT02340156	II	Terminated	Combination of TMZ and SGT-53 for treatment of recurrent GBM
		NCT03554707	I	Not yet recruiting	SGT-53 treatment with RT and chemotherapy in children with recurrent or progressive CNS malignancies
NFκB	Bortezomib	NCT03643549	I/II	Recruiting	Bortezomib and TMZ in recurrent GBM with unmethylated MGMT promoter
		NCT01435395	I	Completed	Bortezomib with TMZ and bevacizumab for recurrent GBM
		NCT00641706	II	Completed	Bortezomib and vorinostat in treating patients with progressive or recurrent GBM
		NCT00998010	II	Completed	Bortezomib, TMZ, and RT in treating patients with newly diagnosed GBM or gliosarcoma
		NCT00611325	II	Completed	Avastin plus bortezomib for patients with recurrent malignant glioma
		NCT00006773	I	Terminated	Bortezomib in treating patients with recurrent gliomas
		NCT03607643	I/II	Not yet recruiting	Efficacy of bortezomib and cannabidiol in patients with multiple myeloma, GBM, or GI malignancies
		NCT00544284	I	Completed	Bortezomib and TMZ in treating patients with brain tumors or other solid tumors resistant to treatment
	Arsenic trioxide	NCT00045565	I	Completed	Arsenic trioxide plus RT in treating patients with newly diagnosed malignant gliomas
		NCT00720564	I	Completed	RT, arsenic trioxide, and TMZ in treating patients with newly diagnosed high-grade glioma
		NCT00095771	I	Completed	Arsenic trioxide and RT in treating young patients with newly diagnosed gliomas
		NCT00275067	I/II	Ongoing	Arsenic trioxide, TMZ, and RT in treating patients with malignant glioma postoperative
	Curcumin	NCT01712542	N.A.	Completed	Curcumin to treat GBM patients
	BKM120	NCT01473901	I	Completed	BKM120 with RT and TMZ in patients with newly diagnosed GBM

Continued

TABLE 1 Clinical trials with RTK and cellular pathway-targeted therapies—cont'd

Target	Drug	Trial ID	Trial Phase	Status	Treatment regimen
		NCT01349660	I/II	Completed	Combination of BKM120 and bevacizumab in refractory solid tumors and recurrent GBM
		NCT01339052	II	Completed	BKM120 for patients with recurrent GBM
		NCT01934361	I	Completed	BKM120 plus carboplatin or lomustine in patients with recurrent GBM
		NCT01870726	I/II	Terminated	INC280 and BKM120 in patients with recurrent GBM
		NCT01576666	I	Completed	BKM120 combined with oral LDE225 in advanced solid tumors
	Aspirin	NCT00790452	II	Terminated	Aspirin prophylaxis to treat venous thromboembolism in GBM
	Sulfasalazine	NCT04205357	I	Recruiting	Sulfasalazine and stereotactic radiosurgery for recurrent GBM
pRb	Palbociclib (PD0332991)	NCT01227434	I	Terminated	Palbociclib pre- and postoperative for recurrent GBM
		NCT03158389	I/II	Recruiting	Palbociclib (with APG101, alectinib, idasanutlin, atezolizumab, vismodegib and temsirolimus) plus RT for patients with newly diagnosed GBM
		NCT02255461	I	Terminated	Palbociclib and isethionate in treating younger patients with recurrent, progressive, or refractory CNS tumors
	Ribociclib	NCT02345824	I	Unknown	Ribociclib for recurrent GBM
		NCT02933736	I	Recruiting	Ribociclib for GBM and meningioma
		NCT03834740	I	Recruiting	Ribociclib with everolimus for GBM and high-grade gliomas
		NCT03355794	I	Ongoing	Ribociclib with everolimus for GBM and high-grade gliomas
		NCT03434262	I	Recruiting	Ribociclib with gemcitabine, sonidegib, trametinib and filgrastim treatment for children and young adults with recurrent brain tumors
	Abemaciclib	NCT02981940	II	Ongoing	Abemaciclib treatment with surgery for recurrent GBM
		NCT04391595	I	Recruiting	Abemaciclib combined with LY3214996 for recurrent GBM
		NCT04074785	I	Recruiting	Abemaciclib with bevacizumab in recurrent GBM
		NCT02977780	II	Recruiting	Abemaciclib with TMZ in GBM
		NCT04238819	I	Not yet recruiting	Abemaciclib combined with TMZ and irinotecan

Autophagy	Chloroquine	NCT02378532	I	Completed	Chloroquine adjuvant therapy with RT and TMZ
		NCT00224978	III	Completed	Chloroquine treatment for GBM
		NCT02432417	II	Not yet recruiting	Chloroquine addition with TMZ and RT for GBM
		NCT04397679	I	Not yet recruiting	Chloroquine with TMZ and RT for GBM
		NCT03243461	III	Recruiting	Chloroquine and RT compared to TMZ, valproic acid and RT in children with high-grade gliomas
	Hydroxy-chloroquine	NCT00486603	I/II	Completed	Hydroxychloroquine with RT and TMZ for newly diagnosed GBM
		NCT01602588	II	Completed	Hydroxychloroquine with short course RT in older patients with high-grade gliomas
		NCT03008148	II/III	Recruiting	RT and TMZ with or with without hydroxychloroquine for newly diagnosed GBM

ErbB family, there is also an aberrant expression of the ErbB2 receptor in up to 40% of GBM. A breakdown of EGFR expression revealed that approximately 45% of patients failed to express EGFR, 25% with overexpression of wild-type EGFR (wtEGFR), and about 30% expressed a mutated EGFR (EGFRvIII, ΔEGFR). The truncated EGFRvIII is constitutively activated, ligand-independent, resulting in cancer progression. Current methods of targeting EGFR/EGFRvIII include small molecule inhibitors, antibodies, vaccines, and chimeric antigen receptor (CAR) T cells.

Gefitinib, erlotinib, and lapatinib are among the first generation of small molecules that inhibit EGFR/EGFRvIII. Despite extensive studies of these inhibitors and promising preclinical models, they showed little effect in human clinical trials. Second-generation EGFR/EGFRvIII inhibitors including afatinib and dacomitinib were designed to irreversibly bind to the tyrosine kinase domain [63]. Afatinib was barely able to penetrate the BBB, which might explain its poor outcome in phase I/II clinical trial [64]. Preclinical model testing of dacomitinib suggested that it could cross the BBB in mice and inhibit tumor growth [65]. However, phase II trials had disappointing global results with a median progression-free survival (PFS) of 2.8 months [66]. The third-generation EGFR/EGFRvIII inhibitors, rociletinib, and AZD9291, were also tested in preclinical models. AZD9291 showed excellent BBB penetration and is >10 times more potent than first-generation inhibitors in reducing GBM proliferation. Follow-up in vivo studies revealed that AZD9291 treatment significantly inhibited tumor growth and extended animal survival [67].

Alternatives to small molecules are antibodies such as cetuximab, panitumumab, and nimotuzumab [68]. These FDA-approved EGFR antibodies bind to the L2 domain of the EGFR, thereby preventing ligand binding and dimerization. Cetuximab leads to internalization of the EGFRvIII, but does not affect GBM cell viability or proliferation [69]. Nimotuzumab, used in combination with rapamycin, was able to slow down the growth of high-grade gliomas that were TMZ resistant, regardless of EGFR status [70]. Additionally, there was the development of a chimeric monoclonal antibody (mAb806) to neutralize EGFRvIII [71]. To date, there is no report on the success of anti-EGFR agents in GBM. This may be due to two apparent limitations in using anti-EGFR antibodies: (1) poor BBB penetration, therefore requiring better drug delivery technology, and (2) intra-tumoral EGFR heterogeneity might limit efficacy.

Vaccines have become another potential option for GBM. Antitumor vaccines can activate the host immune system to recognize and target existing cancer cells. Rindopepimut (CDX-110), a 13 mer peptide spanning the mutation site of EGFRvIII has been tested in a series of preclinical studies and clinical trials [72–74]. Unfortunately, a phase III trial using combined rindopepimut and TMZ treatment for newly diagnosed GBM did not increase survival with a mean OS of 20.1 months in the rindopepimut group versus 20.0 months in the control group.

CARs are engineered receptors with a single-chain variable fragment derived from monoclonal antibodies bound to the transmembrane and intracellular activation domains of surface receptors on T and NK cells. Preclinical studies have demonstrated that EGFRvIII-directed CAR T-cells can inhibit tumor growth in both in vitro and in vivo orthotopic models. A seven-subject clinical trial (NCT02664363) to test CAR-T cells targeting EGFRvIII was halted; perhaps owing to lack of funding [75]. Another CAR-T therapy to target HER2 on GBM cells reduced tumor infiltration, although there was no major improvement in overall survival [76].

Vascular endothelial growth factor receptor (VEGFR)

Recent clinical studies indicated significant radiographic responses with the VEGF monoclonal antibody, bevacizumab [62]. Despite modest improvement in OS with anti-VEGF, there was the initiation of several studies to interrogate VEGF-VEGFR activation. The VEGFR family includes VEGFR1 (Flt1), VEGFR2 (KDR), VEGFR3, neuropilin1 (NRP1), and NRP2, each with varied binding affinities to the VEGF homologs. Upregulation of VEGF expression is often associated with a concomitant increase in VEGFR2, resulting in autocrine stimulation to increase tumor vascularity.

Small molecule inhibitors targeting VEGFR include tivozanib, cediranib, lenvatinib, sorafenib, sunitinib, pazopanib, nintedanib, and vatalanib (PTK787) [68, 77, 78]. Phase II testing of tivozanib as a monotherapy for recurrent GBM patients had a median PFS of 2.3 months and median OS of 8.1 months [79]. Despite reduced vascular permeability, blood flow, and tissue oxygenation, tivozanib did not show significant antitumor activity as monotherapy [79]. Several clinical trials tested cediranib as monotherapy or in combination with other treatments [80–82]. As monotherapy or in combination with lomustine or cediranib, this drug failed to improve PFS [80]. Other therapeutic agents in clinical testing against VEGR are TTAC-001 antibody and the VXM01 vaccine [68]. In addition to receptor targets, there are developments of several ligand inhibitors. Their use as antiangiogenic therapies is further explained below.

Fibroblast growth factor receptor (FGFR)

FGF receptors constitute a family of four homologs, FGFR1–4, and are involved in processes including migration, differentiation, and survival [62]. Most FGFR gene mutations cause gain-of-function. An analysis of the TCGA database indicated approximately 3% of GBM patients with amplified FGFR. Additionally, higher FGFR1 expression in gliomas as compared to healthy brain tissue indicates a role for FGFR in oncogenesis.

There is a number of small molecule of RTK inhibitors to target FGFR1–3 isoforms, such as PD173074, AZ4547, BGJ398, JNJ-493, and TAS-120 [83]. Phase I/II trials testing of AZ4547 gliomas and FGFR1-TACC1 or FGFR3-TACC3 fusions for recurrent IDH wild type were terminated after analyses of the data from a cohort of 12 patients [83]. Preclinical studies suggest that BGJ398 can target FGFR3-TACC3 fusion protein thereby blocking MAPK pathway upregulation in many cancers including GBM [84]. Phase II trial testing BGJ398 in recurrent GBM with confirmed FGFR1-TACC1 fusion, FGFR3-TACC3 fusion, and/or activated FGFR1–3 mutations have been completed, yet data remains to be published (NCT01975701). TAS-120, an irreversible FGFR inhibitor is undergoing phase I/II clinical trial in solid tumors with FGF/FGFR aberrations, including GBM (NCT02052778). There is a need for further clinical trials to fully determine if FGFR-targeting could benefit GBM.

Tumor suppressor p53

According to the Cancer Genome Atlas Research Network, up to 87% of GBM cases have an altered p53 pathway [62]. Stress signals such as DNA damage, hypoxia, heat shock, and cold shock can initiate the p53 pathway to disrupt the fidelity of DNA replication and mitosis [85]. Upon activation, p53 can elicit cell cycle arrest, cellular senescence, and apoptosis. Mutation or alteration to p53 can compromise its tumor-suppressive functions. Thus, restoring its function can produce antitumor effects.

A phase I clinical trial sought to introduce p53 and express the functional protein by intratumoral injection using an adenovirus vector as a vehicle (Ad-p53) [86]. Tumors demonstrated apoptosis via upregulation of p21, indicating a beneficial antitumor effect. A separate ongoing phase II study tests the use of SGT-53, in combination with TMZ, for recurrent GBM (NCT02340156) [87]. SGT-53 is a complex of cationic liposomes that can efficiently cross the BBB to deliver p53 cDNA to tumor cells. This trial hopes to express wild-type p53 to GBM to downregulate MGMT activity and resensitize cells to TMZ. Indeed, the preclinical studies support this trial.

Preclinical studies indicated that chemical agents could inhibit p53 and its signaling pathway. Nutlin-3a, a cis-imidazoline analog, activates p53 by functioning as an antagonist of murine double minute (MDM2) [88]. MDM2 is overexpressed in several cancers and, as the primary inhibitor of p53, it confers increased tumor growth, survival, and chemoresistance [88, 89]. Treating glioma cells with nutlin-3a inhibited p53-MDM2 interaction thereby activating p53 and inducing apoptosis [89]. Other MDM2 inhibitors tested in GBM cell lines include RITA (reactivation of p53 and induction of tumor cell apoptosis), AMG232, and RG7112 [90–92]. Use of this MDM2 enabled activation of p53, reduced tumor growth, and increased survival in experimental in vivo models. These studies provide strong support for future clinical testing in GBM.

Tumor suppressor NFκB

NFκB is a family of inducible transcription factors that have a role in DNA transcription, cytokine production, immune response, and cell survival [93]. NFκB exists in the cytoplasm in an inactive complex bound to IκB. Several pathways, including PI3K/AKT, activate NFκB by phosphorylating IκBa to release the active NFκB complex, resulting in transcription of genes such as MGMT [94]. Strategies targeting NFκB in GBM include proteasomal inhibition, as well as direct pharmacological inhibition.

The NFκB subunit p65 (RelA) regulates the expression of MGMT by binding to two sites on the *MGMT* promoter. Proteasomal activity determines NFκB activation by catalyzing the degradation of the inactive precursor p50/p65 and the inhibitory factor, IκBa [95]. Therefore, targeting the proteasomal pathway retains dimer integrity and prevents NFκB-dependent MGMT expression.

Bortezomib (BTZ) is an FDA-approved proteasome inhibitor for multiple myeloma and mantle cell lymphoma. BTZ is under investigation for a variety of solid tumors either as a single agent or in combination with other agents [96]. GBM cell lines that are inherently chemoresistant showed increased apoptosis with BTZ when given before TMZ administration [97]. Further studies confirmed that the effects coincided with decreased IκBa proteasomal degradation, downregulated nuclear translocation of NFκB, and reduced MGMT expression [98]. Several BTZ clinical safety and efficacy trials are ongoing. A phase II study of BTZ/RT/TMZ combination therapy demonstrated that BTZ increases PFS and

OS, with more benefit in MGMT methylated tumors [99]. Taken together, this data suggests that BTZ can function as an indirect MGMT inhibitor to overcome TMZ resistance.

Several GBM cell lines exhibit activation of NFκB as compared to healthy brain tissue [100]. Treatment with NFκB pharmacological inhibitors and NFκB-p65 siRNA significantly decreased the viability of GBM as compared to MAPK, EGFR, or PKC inhibitors. Testing NFκB pathway inhibitors, BAY117082, parthenolide, arsenic trioxide, MG132, and curcumin indicated inhibitor selectivity for GBM tumor cells as compared to astrocytes [101]. These inhibitors were also able to abolish chemoresistance and induce apoptosis when administered in combination with cisplatin. The pan class 1 PI3K inhibitor, BKM120, shows similar effects when tested in glioma cells. BKM120 was able to inhibit the PI3K/AKT/NFκB signaling pathway, thereby reversing MGMT-mediated TMZ resistance [102]. Other agents that directly or indirectly target NFκB include dihydrotanshinone, lactacystin, EC-70124, triterpenoids, and NSAIDs [103].

Despite success in preclinical GBM models, clinical trials have been limited to nonspecific agents targeting NFκB. Sulfasalazine, an agent commonly used to manage rheumatoid arthritis, was able to inhibit IκB and cause growth arrest of primary brain tumors in vitro [104]. However, sulfasalazine did not significantly extend survival in GBM phase I/II trials [105]. While there is no strong report on a direct link between NFκB and GBM progression, targeting this transcription factor could be a significant contribution to treat TMZ-resistant GBM.

Tumor suppressor pRb

During normal cell cycle progression, CDK4/6 complexes with cyclin D to phosphorylate retinoblastoma protein (pRb) [106]. Hyperphosphorylated pRb is inactive and allows E2F transcription factors to initiate G1/S-phase transition. In its active form, hypophosphorylated pRb binds and inhibits transcription factors of the E2F family, thereby halting S-phase and cell cycle progression. The pRb/E2F complex also attracts HDAC proteins to chromatin and this reduces binding of S-phase promoting factors [107]. Approximately 75% of GBM tumors have altered pRb pathway, with 7.6%–11% of all cases exhibiting Rb gene deletion or mutation [62]. Palbociclib (PD0332991), ribociclib, and abemaciclib inhibit CDK4/6 and prevent hyperphosphorylation of pRb [108]. These inhibitors received FDA approval from 2015 to 2018 are commonly used in combination with tamoxifen, aromatase inhibitors, or fulvestrant in hormone receptor-positive metastatic breast cancer [108]. Despite promising in vivo experimental models, BBB permeability remains relatively low, limiting drug delivery to the central nervous system. Most recently, CDK4/6 inhibitors showed significantly greater BBB penetration with a favorable pharmacological profile [109]. The molecule, Compound 11, when tested in an orthotopic in vivo model of GBM, showed a 162% increase in median survival as compared to vehicle-treated mice.

Autophagy pathways

Autophagy is a double-edged sword in cancer therapy. It is a conserved process that maintains cellular homeostasis by discarding stressful cellular components [110]. Autophagy protects healthy cells by discarding damaged organelles triggered by environmental stress such as high heat or nutrient starvation. However, cancer cells can use the same process to their advantage when stressed by chemotherapy or radiation [111]. Therefore, autophagy not only recycles damaged proteins and organelles but also manipulates the innate process to resist therapeutic targeting.

Chloroquine (CQ) is an antimalarial drug that can efficiently inhibit autophagy by degrading the autolysosome. In vitro and in vivo models demonstrated that CQ can trigger stress and inhibit autophagy in GBM to increase TMZ sensitivity [112]. Phase I trial with CQ and TMZ showed a trend toward reduced death (PMID 16520474). However, the results of this trial were not statistically significant, and need a larger cohort of patients. A phase I/I trial-tested hydroxychloroquine (HCQ), a CQ derivative, with standard RT and chemotherapy [113]. There was no consistency in inhibiting autophagy due to the dose-limiting toxicity of HCQ, as well as the lack of improved OS. Therefore, despite efforts to repurpose CQ for cancer treatment, clinical trials have not shown a significant increase in OS, as compared to standard treatment. Furthermore, high doses of CQ may cause systemic toxicities such as cardiac disease, visual deficiency, and renal failure. Other compounds with similar molecular activity and less toxicity may have greater therapeutic potential.

Several barriers remain in achieving successful use of the aforementioned molecules such as BBB infiltration, potential bone marrow failure, and other toxicities. One such obstacle is the use of enzyme-induced anticonvulsant drugs (EIACDs) for patients experiencing epileptic symptoms [114]. EIACDs induce cytochrome P450 isoenzyme CYP3A4 expression, which rapidly metabolizes small molecules such as gefitinib, lapatinib, and AEE788. However, targeting RTKs and their associated pathways may still hold promise for overcoming TMZ resistance despite the high rate of clinical trial failures.

Antiangiogenic therapies

Angiogenesis is the proliferation of blood vessel networks from preexisting blood vessels. This process is vital for the delivery of oxygen and nutrients to healthy and healing tissue. Tumors secrete various factors to promote local angiogenesis to sustain tumor growth. Antiangiogenic drugs offer a potential avenue to disrupt the tumor supportive microenvironment. Numerous angiogenic pathways have been targets of pharmaceutical interventions. These pathways include, but are not limited to, basic fibroblast growth factor (bFGF), endostatin via cyclooxygenase-2 inhibition (COX2), vascular endothelial growth factor (VEGF), integrin, and matrix metalloproteinases (MMP).

VEGF-inhibitors and VEGF-trap

The VEGF-A ligand binds to VEGFR2, leading to the activation of various intracellular pathways such as the PLCγ/PKC/Raf, Ras/PI3K/AKT, and Src [115]. These pathways are respectively involved in the initiation of DNA synthesis and cell growth, increased endothelial-cell survival, and actin cytoskeleton changes for migration, all of which are required processes for angiogenesis. Antiangiogenic therapeutic targets include the VEGF ligand and VEGFR2 [115]. The monoclonal antibody bevacizumab, approved by the FDA for recurrent GBM in 2009, is the most utilized VEGF inhibitor for GBM [116, 117]. In newly diagnosed GBM, bevacizumab monotherapy showed no increase in overall survival in comparison to the current standard of care. Though bevacizumab increased PFS from 7.2 months in the placebo group to 10.7 months, treatment was associated with systemic toxicities such as hypertension, increased thromboembolic disease, wound dehiscence, fatigue, visceral perforation, hemorrhage, neutropenia, and neurocognitive delay. Thus, those involved in the trials deduced that despite increased PFS, such outcome fails to outweigh the diminished quality of life [118]. Bevacizumab was not recommended for newly diagnosed GBM but rather for recurrent GBM. In the recurrent setting, bevacizumab alone or in combination did provide a modest prolongation of overall survival. PFS and OS of 8.5 and 22.4 months, 16.6 and 36.2 months, and 18.2 and 39.5 months in control, monotherapy, and combination therapy groups, respectively [116]. Additionally, bevacizumab efficacy was similar when used in early or late recurrence [119].

Bevacizumab, in combination with a variety of drugs for recurrent GBM, including topoisomerase I inhibitor, irinotecan, showed 6-month PFS rates similar to bevacizumab alone [120]. Similarly, alkylating nitrosourea compound lomustine or the immune checkpoint inhibitor (Ipilimumab), in combination with bevacizumab, showed no benefit to PFS or OS [121–124]. More promising combinations have demonstrated some survival benefits such as reirradiation therapy [125]. There are ongoing studies with various drug combinations.

Other attempts to target VEGF include treatment with the soluble recombinant decoy receptor, aflibercept [126–128]. Experimental models with a GBM cell line in nude mice indicated that aflibercept is a more effective treatment when combined with radiation as compared to radiation alone [126]. However, in phase II clinical trials, aflibercept failed to prove a viable option in the adjuvant and recurrent setting, with adverse events related to VEGF inhibitors [129]. Given the above findings, bevacizumab remains the most commonly used VEGF sequestering agent.

FGF-inhibitors and combinations

Inhibiting bFGF can disrupt angiogenesis and reduce GBM cell growth [130]. Thalidomide, a bFGF inhibitor, was approved for recurrent GBM following a promising phase II clinical trial [131]. Further clinical trials showed that thalidomide, in combination with TMZ, increased patient survival to 103 weeks versus thalidomide alone with a median of 63 weeks, thus solidifying its role in combination with other drugs for GBM. The treatment was also found to be well tolerated with mild-to-moderate toxicity [132, 133]. When compared to first-line TMZ treatment, there was no clinical benefit in the adjuvant setting of GBM, though thalidomide showed marginal effects in the recurrent setting [132].

Later studies with several drugs such as thalidomide and TMZ found no survival benefit [134]. The inconsistent benefit in survival with thalidomide led to this drug deemed as a last resort. Since thalidomide is a sedative with significant adverse effects, its current use is limited to the optimization of satisfactory sleep patterns [135].

Celecoxib, a COX2 inhibitor upregulates endostatin to impart apoptosis and sensitizes cancer cells to DNA damage. Celecoxib is a second-generation NSAID that was also tested with

TMZ. This combination showed no improvement in PFS for recurrent GBM regardless of the MGMT status [134, 136]. Isotretinoin, a Vitamin A derivative, when combined with TMZ showed a decrease in PFS [134]. Many therapies look for success in newly diagnosed and recurrent GBM. However, with respect to drug-resistant GBM, many of the treatments combine DNA-damaging medications. The rationale is the premise that drug-resistant GBM is better at DNA repair following TMZ-induced DNA lesions. Thus, MGMT status is often investigated in parallel to novel drug treatments [13].

Integrase and matrix metalloproteinases inhibitors

Integrase inhibitors such as cilengitide, which selectively block activation of cell surface $\alpha\nu\beta3$ and $\alpha\nu\beta5$ integrins, halt migration, and survival of endothelial cells, show promise for MGMT-positive GBM. PFS (13.4 months) and OS (23.2 months) were significant in the cilengitide with TMZ arm for MGMT-positive GBM as compared to PFS (3.4 months) and OS (13.1 months) in MGMT-negative groups [137]. However, when repeated in the CENTRIC phase III trial with a larger cohort, PFS and OS between MGMT status groups were similar at 26.3 months. Some concern for the negative results suggested that the therapeutic level of cilengitide may be below in the CENTRIC phase III trial. This underscores the persistent challenge to deliver drugs to the CNS. Nevertheless, cilengitide was determined to have no additional benefit when compared to TMZ, a disappointment for a very promising drug [138–140]. This is more unfortunate given that integrase inhibitors are relatively safe and lack major toxicity or adverse events [141].

MMP inhibitors can decrease MMP secretion, which is associated with tumor metastasis and progression [142]. NFκB pathway, which regulates MMP, is also associated with tissue inflammation. Studies show a link between inflammation and tumor progression. Proteases, secreted by the tumor cells, trigger stromal cells to degrade the extracellular membrane to facilitate cell migration and metastatic potential of tumor cells. Clinical trials of synthetic MMPs, such as prinomastat, marimastat, metastat, tanomastat have yielded no survival benefit. However, there is abundant call for naturally occurring MMP inhibitors such as green tea extract, curcumin, and resveratrol, which showed promising results [143]. As an example, curcumin was used to treat mice xenografted with U87 GBM cells and the results indicated an increase in animal survival with reduced tumor growth [144].

Current and future therapies

Comparing the various anti-angiogenic medications, bevacizumab demonstrated better 6-month PFS and objective response rates. However, it did not differ in 1-year OS as compared to drugs within the same category [145]. The current therapy for newly diagnosed GBM is gross total resection, with subsequent RT and TMZ. For recurrent GBM, treatments may include TMZ, VEGFR inhibitors such as bevacizumab, and TTF monotherapy or in combination. Integrase inhibitors and synthetic MMP inhibitors are accessible through clinical trials if available. Lastly, thalidomide may be a last resort to warrant palliative care and quality of life issues.

Targeting cancer stem cells

CSCs are a subpopulation of cells that possess the property of self-renewal as well as differentiation into lineages [146]. They are involved in tumor initiation, growth, and progression, in addition to resisting drug treatments. Thus, CSCs have been implicated as the main reason for cancer relapse and drug resistance [147]. CSCs express stem cell genes such as Sox2, Oct4, Bmi1, and Nanog [148, 149]. Early reports of $CD133^+$ as a CSC marker later showed its $expression^+$ on normal neural cell/stem cells, including G-CSCs [150, 151]. Other reliable markers include CD44 and ABC transporters [152]. Currently, researchers focusing on this field suggest using multiple markers to identify true CSCs.

Drug efflux transporters

CSCs have overexpression of ATP-binding cassettes or ABC transporter genes [153]. This gene family comprises 49 transmembrane proteins that facilitate the transport of cytotoxic chemotherapies out of the cells, such as multiple drug resistance protein 1 (MDR1), ABCG2, and ABCC1 [152, 154]. CSCs also have a high mitochondrial ATP output, enhanced DNA activity, and increased expression of cell survival genes [152, 155]. All of these factors coupled with increased elimination of multiple drugs by efflux proteins contribute to chemoresistance or multiple drug resistance, preventing the eradication of cancer [156]. Thus, one of the treatment strategies includes targeting drug efflux mechanisms in the hopes of increasing their sensitivity to chemotherapies. Munoz *et al.* showed that blockage of mir-9 using an antagomir successfully increased GBM chemosensitivity to TMZ [157, 158]. They demonstrated that the anti-miR-9 reversed MDR1 gene expression resulting in the reduction of TMZ chemoresistance [158]. Lee *et al.* published a study where they targeted ABCB5 using a monoclonal antibody. The study reported high expression of ABCB5 in three GBM cell lines LN-229, LN-18, and U87, and tested the change in expression of $CD133^+$ cells after treatment with ABCB5-mab and an isotype. They observed a reversal of G2/M cell cycle arrest due to TMZ treatment in those that received ABCB5-mab with increased chemosensitivity of the cell lines to TMZ [159]. Thus, the strategy of blocking MDR by various methods seems most effective as an adjuvant treatment, facilitating the chemosensitization of GBM to TMZ.

Cell survival signals

CSCs, like other stem cells, depend on various signaling pathways regulated by internal and external stimuli, which play a vital role in maintaining the balance between CSC self-renewal and differentiation, thus ensuring their long-term survival. CSCs upregulate survival signals in order to evade extermination by therapies. The next section discussed some of these pathways.

Notch signaling

Notch is a transmembrane protein whose signaling is important in regulating cell survival, cell-to-cell interactions, and apoptosis [160]. The activation of the Notch ligand and the Delta receptor on neighboring cells led to Notch signaling activation to release the Notch intracellular domain (NICD). This occurs by γ-secretase exerting proteolytic cleavage of the Notch receptor. The NCID complex translocates to the nucleus and induces the transcription of HES and HEY transcription factors, resulting in tumor cell growth [161]. Several cancers, such as GBM, medulloblastoma, and neuroblastoma have increased Notch activity [162]. Thus, scientists have focused on inhibiting Notch signaling as a therapeutic approach to curtail tumor progression. Blocking of NCID cleavage with γ-secretase inhibitors can inhibit the Notch pathway. The report indicated slower tumor growth when neurospheres generated from primary human GBM tissue were treated with γ-secretase inhibitors [162]. This suggested that γ-secretase might be important to target the G-CSC. Similarly, other studies have used γ-secretase inhibitors to induce apoptosis in tumor cells and inhibit tumor growth [163–165]. Alternatively, dual targeting of ErbB1/2 and Notch receptors prevents intracellular crosstalk, resulting in apoptosis [166].

Sonic Hedgehog signaling (SHH)

SHH signaling is important for stem cell maintenance. The SHH pathway is activated when the SHH ligand binds to the transmembrane protein PTCH1. This interaction causes the release of smoothened (SMO) and activation of its downstream genes, which eventually leads to the expression of an oncogene Gli-1 which promotes cell proliferation and contributes to tumor growth. A large number of GBM highly express Gli-1 [167]. Since Gli is downstream of the SHH signaling, it is not a surprise that there are proposed methods to blunt SHH signaling to slow tumor growth. Cyclopamine, an inhibitor of SHH signaling, downregulated the expression of stem cell-associated genes to negatively affect sphere-forming ability in GBM cells. Experimental studies with xenograft models validated the ability of cyclopamine to inhibit tumor growth [168, 169]. Another SHH inhibitor, vismodegib (monotherapy or in combination), was also investigated for recurrent GBM [170]. Munoz *et al.* showed the following by vismodegib increased chemosensitivity of TMZ-resistant GBM cells by successfully reducing SHH signaling, as indicated by reduced SMO. This correlated with a significant reduction of multidrug transporters and increased apoptosis [171]. Similarly, a number of clinical trials using SHH inhibitors, either alone or in combination are underway.

WNT-PI3K-AKT signaling

WNT signaling is widely applied in several biological processes such as its requirement in neural stem cell expansion and forebrain development [172, 173]. β-Catenin, a component of the WNT pathway, is activated in gliomas, in particular those with EGFRvIII mutation [174, 175]. Poor prognosis is associated when EGFRvIII is highly coexpressed with CD133 on glioma cells [176]. AKT signaling is increased in approximately 50% GBMs with EGFR amplification, PI3K overexpression, or PTEN mutations [177]. Wnt-β-catenin activation in GBM may occur through AKT, suggesting that tandem targeting of the Wnt and AKT pathways may be required to achieve significant tumor reduction [178]. Sorafenib, an inhibitor of PI3K/AKT and MAPK pathways, reduced GBM proliferation and significantly induced apoptosis by downregulation of the cell survival factor Mcl-1 [179]. Diclofenac and celecoxib, two NSAIDs, which inhibit the WNT/β-catenin pathway, showed a significant decrease in human GBM proliferation, colony formation, and migration [180]. Thus far, there are no known clinical trials incorporating NSAIDs in GMB. It appears that the next step relevant to GJIC in GBM may be to incorporate a combinational approach, along with the standard of care treatment.

Intercellular communication among GBM cells

Gap junctional intercellular communication (GJIC) is mediated through gap junctional channels formed by the connexin (Cx) family of proteins. GJIC facilitates the exchange of small molecules such as microRNA, ions, and other metabolites within the extracellular matrix [181]. The literature reported on increased *Cx43* expression in human primary GBM cells and such increase correlates with increased TMZ resistance [182, 183]. However, Cx43 is not the only Cx in GBM. In fact, reports indicate increased Cx46 in G-CSCs, which was relevant to CSC survival, including their proliferation and self-renewal [184, 185]. Thus, it was proposed that disrupting intercellular communication among GBM cells might be therapeutic. Clofazimine, an inhibitor of Cx46, reduced the proliferation, self-renewal, and survival of G-CSCs. This treatment sensitized the cells with TMZ resulting in increased apoptosis and diminished tumor growth [185]. In other ongoing studies, a Cx43 mimetic—ACT1, disrupted GJIC, with a significant reduction in "tumorigenic potential" of G-CSCs [186]. Since Cx46 is highly expressed in GBM CSCs, this Cx may be a potential marker in identifying and isolating G-CSCs.

Tumor treating fields

Tumor treating fields (TTF), is a relatively new noninvasive technique approved for use in the adjuvant and recurrent setting of GBM. Data has shown TTF to increase PFS and OS when used as adjuvant therapy with TMZ vs TMZ monotherapy [187, 188]. TTF encompasses a series of electrodes attached to a cap, which applies varying magnetic fields via an alternating electrical current. The magnetic field disrupts mitosis, thereby killing dividing cells. This process spares quiescent, or noncancerous cells, especially in the brain [189]. However, there is no information on the response of slow dividing GBM cells such as GBM stem cells when they are in the vicinity of the magnetic field. A further discussion on preclinical studies, clinical trial outcomes, potential adverse side effects, impact on quality of life, and future directions of TTF will be discussed below.

TTF was first experimented in vitro on various cell lines to test the effects of magnetic fields on mitosis in 2004. These studies were based on the premise that the magnetic fields would interfere with mitosis by separating the charged tubulin dimers from the growing end of microtubules and mitotic spindle fibers [189]. The experimental analysis indicated common mechanisms across various GBM and breast cancer cells. The magnetic field affected the organelles by not allowing proper alignment for cytokinesis, thus rendering the dividing cells dead upon division [189]. Additional in vitro experiments to explain further mechanisms of this new therapy deduced that the magnetic field interfered with septin fibers and endoplasmic reticulum function. This led to the tumor cells undergoing stress and autophagy. The extent of protein manipulation and differences encountered by normal and malignant cells warrants further investigation [190]. The combination of TTF and TMZ led to increased ROS production and upregulation in heme oxygenase-1 (OH-1), resulting in cell apoptosis [191]. Although there are gaps in the mechanisms of TTF, the experimental studies provided insight into the mechanisms of TTF, leading to monotherapy as well as combination therapy clinical trials in GBM.

Clinical trials

The EF-11 phase III clinical trial tested the efficacy of TTF as monotherapy as compared to standard of care [188]. OS, the primary endpoint, was 6.6 vs 6.0 months. Other endpoints of TTF vs chemotherapy showed similar OS at 1-year but improved survival at 6 months, 21.4% vs 15.1%. Severe adverse events were 6% vs 16% for the TTF vs standard of care, respectively. Mild and moderate side effects consisted mostly of skin irritation and rash in areas under the transducers [188]. The outcome of the EF-11 trial in 2011 led to FDA approving the use of TTF therapy for refractory GBM [192]. The PRiDe trial was a follow-up study that analyzed the results of 457 patients from 91 US cancer centers [193, 194]. Patient demographics were similar to the EF-11 trial. However, TTF was initiated at an early time point. This trial showed OS of 9.6 months as compared to 6.6 months favoring the TTF arm. The following outcomes were noted for long-term follow-up: 1-year (44% vs 20%) and 2-year (30% vs 9%) survival rates were increased vs chemotherapy alone. Additionally, the second trial noted better compliance (>75% daily compliance vs <75% daily compliance) and this correlated with improved OS (13.5 months vs 4.0 months). A follow-up clinical trial is known as EF-14 compared TTF along with TMZ vs TMZ alone [188]. The primary endpoints, progression-free survival, were 6.7 vs 4.0 months, and OS, 20.9 vs 16.0 months. Skin toxicity continued to be an issue in the experimental group as compared to chemotherapy alone. In 2015, this trial led to the FDA approving TMZ and TTF combination therapy for newly diagnosed GBM.

Adverse effects and limitations

In the EF-11 trial, the only noted adverse effect of TTF treatment was skin irritation, presenting as a rash on the skin areas exposed to the cap transducers. This occurred in mild (6%) and moderate severity (16%) in the TTF-treated patients vs 0% of non-TTF-treated patients. The skin toxicity, which is reversible with no requirement for treatment, was preferred over the systemic adverse effects encountered with chemotherapy [188]. In the EF-14 trial, where patients would receive chemotherapy in both the experimental and control groups, it was found that the occurrence of systemic adverse effects was comparable, thus attributing the systemic toxicity to chemotherapy rather than TFF [187]. There was skin toxicity in 52% of TTF-TMZ treated patients as compared to no skin effects with TMZ alone. There was no change in the overall occurrence of seizures.

Adverse effects are not the only drawback to this new therapy, but the cost may play a major role. Prior to TMZ treatment as the first-line care, the treatment for brain tumors ranked 4th in average monthly cost at $8478. These expenses were mostly incurred by surgery and radiation therapy [6]. TMZ, an oral agent costs about $2677–$3544 USD per month in the early 2000s and currently is estimated to be about $1155 per month [195]. The average cost for TTF therapy is about $21,000 USD per month. Since a patient's average treatment lasts 4.1 months, the overall cost is around $86,000 USD [6]. Many healthcare insurance plans do not cover TTF treatment, based on the incremental cost-effective ratio of $800,000 USD to gain a quality life-year. Of note, TTF, as adjuvant therapy, still requires gross total resection, radiation, and TMZ, indicating the overall cost to be relatively high as compared to standard care.

Quality of life

TTF has no half-life since it requires an "ON" switch with the continuous application for an effect on GBM cells. Additionally, to undergo the therapy a shaven head is required in order to fit the cap for more than 18h/day and greater than 50% of the days per month [196]. Hair loss is a common systemic side effect of certain chemotherapeutic agents, however self-imposed head shaving may go against a patient's cultural, religious, or personal beliefs. With a low 2-year survival rate compared to other cancers, it is imperative that the therapy leads to an acceptable quality of life. Taking these concerns into account, including the knowledge that compliance increases treatment efficacy, TTF might appeal to certain determined patients and their families. Considering the benefit of progression-free and OS to be clinically significant, many patients appear to be committed to TTF treatment. As discussed above, there is no overall decrease in quality of life, except for skin discomfort.

As for any new therapy, skepticism to TTF is expected. Though most experts concluded that TTF clinical trials were statistically and clinically significant, study designs led some to question the results. One area of concern was the lack of a double-blinded trial with a "sham" device, with the transmitter emitting heat [197]. However, owing to the poor prognosis of GBM, there are ethical issues with sham control arms. The experts who are critical of the lack of a sham group deemed that the results of TTF could not be explained by a placebo effect. Other studies lacking placebo groups did not show improvement in OS as noted in the TTF trials. Even with skepticism in published TTF studies, some answers may be shed with ongoing long-term observational studies.

A major hindrance in the treatment of brain tumors is to have the drug penetrate the BBB. Added to this hindrance is the ease by which GBM cells gain resistance to the first-line drug, TMZ. TTF showed efficacy, regardless of MGMT status, which is a known determinant of GBM drug resistance [13]. TTF treatment bypasses the hindrance posed by the BBB, underscoring the focus on TTF treatment for drug-resistant GBM. TTF disrupts the DNA repair mechanisms in drug-resistant GBM cells. In this regard, it is possible that TTF treatment will target the G-CSCs. Several ongoing clinical trials incorporating TTF may reveal if this type of therapy could be the first true alternative for drug-resistant GBM [197].

Alternative delivery methods

As discussed above, a major impediment to GBM treatment is the presence of the BBB. The BBB has extensive tight junctions that prevent the transport of most chemotherapeutic drugs [198]. This hurdle will exist even if discoveries resulted in new and effective therapies. Hence, the science has a large focus on mechanisms to transport therapies across the BBB to directly deliver the drugs to the tumor.

Cell drug delivery methods

Various cellular therapies are currently being explored, particularly mesenchymal stem cells (MSCs) [199]. It has been demonstrated that MSCs are chemo attracted to tumorigenic and inflammatory regions and areas of tissue insult [200]. They can extravasate through the BBB to reach their target, where they release secretory substances [201]. They are also immune evasive since they lack MHC class II, thus "off-the-shelf" MSCs are available for allogeneic transplantation, making them good delivery vehicles for targeted drug delivery in GBM [202]. Various studies used MSCs to deliver microRNA (miRNA) mimics to target GBM cell lines. In one such study, MSCs were loaded with and delivered anti-miR-9 to the cancer cells via exosomes. Anti-miR-9 inhibited PTCH1 downregulation, thereby successfully chemosensitizing GBM cells to TMZ [158]. Other studies have engineered MSCs to deliver suicide genes in a rat glioma model [202]. While there are many advantages to using MSCs to deliver therapy, a major drawback is in the immunosuppressive action of MSCs via Tregs. MSCs can become tumor supportive rather than suppressive [199, 203]. To overcome this, a few GBM mouse studies have encapsulated MSCs in synthetic extracellular matrices [204]. One study used MSCs secreting a target drug such as IFN- β and a suicide gene, such as the Herpes simplex virus thymidine kinase-ganciclovir system (HSVtk-gcv) [205]. After delivery of IFN-β, MSCs engineered to express HSVtk can be activated by GCV, resulting in cell death [206, 207].

Nanotherapeutics

Currently, safe and effective nanotechnology-related therapies are in development for GBM. Such technology includes nanoparticles (NP) to target ABC transporter genes to reduce multidrug resistance. Investigators have proposed encapsulation of this ligand-modifying NP to prevent recognition by drug efflux transporters, consequently enhancing chemosensitivity [208, 209]. Another approach was to encapsulate miRNA or siRNA for the purpose of chemosensitizing GBM cells. Thus, solid lipid NPs (lipids pass through the BBB) were used to deliver miR-145 to inhibit GBM tumorigenicity [210]. SiRNA into NPs to target GLUT3 effectively decreased the target in the G-CSCs and decreased the tumor size [211]. While there is a need for additional validation and in vivo experiments, nanomedicine has the potential for future therapeutic delivery to GBM.

Other approaches

Immune-checkpoint (IC) inhibitors

Immune-checkpoint molecules lead to immunosuppression of CD8 T-cells resulting in immune-evasion of tumor cells. Other IC inhibitors such as those specific for PD-1/PD-L1, CTLA-4, and PDL-2 are possible treatment options [212]. The reports for such treatments mainly indicate negative prognosis by PD-L1 and TIM-3 expressions on GBM cells. Perhaps miRs or antibodies such as nivolumab (PD-1), atezolizumab (anti-PDL-1), or anti-TIM3, which are in ongoing trials, might show promise [213, 214]. Experimental mouse models with stereotactic radiosurgery tested PD-1 blockade alone, and in combination with TIM-3

[215, 216]. While preclinical studies indicate IC inhibitors as promising, to date, there is no record of success with IC.

Viral delivery

Similar to the use of herpes simplex to deliver the virus, there have been studies with other types of modified viruses to express tumor-attenuating genes. One study focused on the Zika virus (ZIKV) since it is an oncolytic flavivirus with a predilection for neuronal stem cells. ZIKV rapidly reduced the ability of G-CSCs to form spheres, indicating that the virus may target the most immature GBM cells. Although relative success was noted for ZIKV to reduce tumor burden in mouse models, ZIKV performance in humans remains a question with major concerns for safety [217].

It is abundantly clear from the above studies the struggles we face with finding an effective treatment for GBM. It is also clear that there will be no single treatment as an option but a combination of multiple strategies is the way forward. These new therapeutic approaches as well as possible new delivery systems have the potential for success and need further studies. There are ongoing studies to test novel combination therapies with effective delivery methods. More importantly, there is a need for further clinical trials to ensure the safety and efficacy of viral treatment in humans.

Conclusion

We concluded with a brief summary to show that despite overwhelming drug resistance for GBM, there are several ongoing clinical trials (Table 1). Patient variability, tumor heterogeneity, state of disease progression, and chemoresistance are all factors contributing to the complexity of GBM treatment. Despite advances in treatment, recurrence is almost inevitable and survival remains poor for most patients. TMZ resistance is a major clinical challenge in newly diagnosed and refractory GBM, which has prompted the investigation of numerous pathways and agents in order to develop effective therapeutic approaches for TMZ-resistant GBM. Although preclinical studies suggest promise by many compounds, these experimental outcomes thus far are not translated into successful clinical trial settings. However, there remains hope with extensive ongoing research. Improved strategies with lessons learned from failures could form the basis to overcome TMZ resistance, which may entail multimodal and patient-specific approaches.

References

[1] Arvold ND, Reardon DA. Treatment options and outcomes for glioblastoma in the elderly patient. Clin Interv Aging 2014;9:357–67.

[2] Abel TJ, Ryken T, Lesniak MS, Gabikian P. Gliadel for brain metastasis. Surg Neurol Int 2013;4(Suppl 4):S289–93.

[3] Omar AI, Mason WP. Temozolomide: the evidence for its therapeutic efficacy in malignant astrocytomas. Core Evid 2010;4:93–111.

[4] Castro BA, Aghi MK. Bevacizumab for glioblastoma: current indications, surgical implications, and future directions. Neurosurg Focus 2014;37(6), E9.

[5] Zhu P, Zhu JJ. Tumor treating fields: a novel and effective therapy for glioblastoma: mechanism, efficacy, safety and future perspectives. Chin Clin Oncol 2017;6(4):41.
[6] Bernard-Arnoux F, Lamure M, Ducray F, Aulagner G, Honnorat J, Armoiry X. The cost-effectiveness of tumor-treating fields therapy in patients with newly diagnosed glioblastoma. Neuro-Oncology 2016;18(8):1129–36.
[7] Carvalho BF, Fernandes AC, Almeida DS, Sampaio LV, Costa A, Caeiro C, et al. Second-line chemotherapy in recurrent glioblastoma: a 2-cohort study. Oncol Res Treat 2015;38(7–8):348–54.
[8] Lara-Velazquez M, Al-Kharboosh R, Jeanneret S, Vazquez-Ramos C, Mahato D, Tavanaiepour D, et al. Advances brain tumor surgery for glioblastoma in adults. Brain Sci 2017;7(12):166.
[9] Tseng WL, Hsu HH, Chen Y, Tseng SH. Tumor recurrence in a glioblastoma patient after discontinuation of prolonged temozolomide treatment. Asian J Neurosurg 2017;12(4):727–30.
[10] Lee SY. Temozolomide resistance in glioblastoma multiforme. Genes Dis 2016;3(3):198–210.
[11] Hsieh P, Yamane K. DNA mismatch repair: molecular mechanism, cancer, and ageing. Mech Ageing Dev 2008;129(7–8):391–407.
[12] Noonan EM, Shah D, Yaffe MB, Lauffenburger DA, Samson LD. O6-Methylguanine DNA lesions induce an intra-S-phase arrest from which cells exit into apoptosis governed by early and late multi-pathway signaling network activation. Integr Biol 2012;4(10):1237–55.
[13] Jiapaer S, Furuta T, Tanaka S, Kitabayashi T, Nakada M. Potential strategies overcoming the temozolomide resistance for glioblastoma. Neurol Med Chir 2018;58(10):405–21.
[14] Yu W, Zhang L, Wei Q, Shao A. O(6)-methylguanine-DNA methyltransferase (MGMT): challenges and new opportunities in glioma chemotherapy. Front Oncol 2019;9:1547.
[15] Weingart J, Grossman SA, Carson KA, Fisher JD, Delaney SM, Rosenblum ML, et al. Phase I trial of polifeprosan 20 with carmustine implant plus continuous infusion of intravenous O6-benzylguanine in adults with recurrent malignant glioma: new approaches to brain tumor therapy CNS consortium trial. J Clin Oncol 2007;25(4):399–404.
[16] Quinn JA, Jiang SX, Reardon DA, Desjardins A, Vredenburgh JJ, Rich JN, et al. Phase II trial of temozolomide plus o6-benzylguanine in adults with recurrent, temozolomide-resistant malignant glioma. J Clin Oncol 2009;27 (8):1262–7.
[17] Blumenthal DT, Rankin C, Stelzer KJ, Spence AM, Sloan AE, Moore Jr DF, et al. A phase III study of radiation therapy (RT) and O(6)-benzylguanine + BCNU versus RT and BCNU alone and methylation status in newly diagnosed glioblastoma and gliosarcoma: Southwest Oncology Group (SWOG) study S0001. Int J Clin Oncol 2015;20(4):650–8.
[18] Papachristodoulou A, Signorell RD, Werner B, Brambilla D, Luciani P, Cavusoglu M, et al. Chemotherapy sensitization of glioblastoma by focused ultrasound-mediated delivery of therapeutic liposomes. J Control Release 2019;295:130–9.
[19] Javanmard S, Loktionova NA, Fang Q, Pauly GT, Pegg AE, Moschel RC. Inactivation of O(6)-alkylguanine-DNA alkyltransferase by folate esters of O(6)-benzyl-2′-deoxyguanosine and of O(6)-[4-(hydroxymethyl)benzyl]guanine. J Med Chem 2007;50(21):5193–201.
[20] Mlala S, Oyedeji AO, Gondwe M, Oyedeji OO. Ursolic acid and its derivatives as bioactive agents. Molecules 2019;24(15):2751.
[21] Pathak AK, Bhutani M, Nair AS, Ahn KS, Chakraborty A, Kadara H, et al. Ursolic acid inhibits STAT3 activation pathway leading to suppression of proliferation and chemosensitization of human multiple myeloma cells. Mol Cancer Res 2007;5(9):943–55.
[22] Zhu Z, Du S, Ding F, Guo S, Ying G, Yan Z. Ursolic acid attenuates temozolomide resistance in glioblastoma cells by downregulating O(6)-methylguanine-DNA methyltransferase (MGMT) expression. Am J Transl Res 2016; 8(7):3299–308.
[23] Marziniak M, Meuth S. Current perspectives on interferon Beta-1b for the treatment of multiple sclerosis. Adv Ther 2014;31(9):915–31.
[24] Natsume A, Wakabayashi T, Ishii D, Maruta H, Fujii M, Shimato S, et al. A combination of IFN-beta and temozolomide in human glioma xenograft models: implication of p53-mediated MGMT downregulation. Cancer Chemother Pharmacol 2008;61(4):653–9.
[25] Wakabayashi T, Natsume A, Mizusawa J, Katayama H, Fukuda H, Sumi M, et al. JCOG0911 INTEGRA study: a randomized screening phase II trial of interferonbeta plus temozolomide in comparison with temozolomide alone for newly diagnosed glioblastoma. J Neuro-Oncol 2018;138(3):627–36.
[26] Minn AJ. Interferons and the immunogenic effects of cancer therapy. Trends Immunol 2015;36(11):725–37.

[27] Grogan PT, Sarkaria JN, Timmermann BN, Cohen MS. Oxidative cytotoxic agent withaferin A resensitizes temozolomide-resistant glioblastomas via MGMT depletion and induces apoptosis through Akt/mTOR pathway inhibitory modulation. Investig New Drugs 2014;32(4):604–17.
[28] Bi Y, Li H, Yi D, Bai Y, Zhong S, Liu Q, et al. Beta-catenin contributes to cordycepin-induced MGMT inhibition and reduction of temozolomide resistance in glioma cells by increasing intracellular reactive oxygen species. Cancer Lett 2018;435:66–79.
[29] Chen Z, Wei X, Shen L, Zhu H, Zheng X. 20(S)-ginsenoside-Rg3 reverses temozolomide resistance and restrains epithelial-mesenchymal transition progression in glioblastoma. Cancer Sci 2019;110(1):389–400.
[30] Yang HC, Wang JY, Bu XY, Yang B, Wang BQ, Hu S, et al. Resveratrol restores sensitivity of glioma cells to temozolamide through inhibiting the activation of Wnt signaling pathway. J Cell Physiol 2019;234(5):6783–800.
[31] Honari M, Shafabakhsh R, Reiter RJ, Mirzaei H, Asemi Z. Resveratrol is a promising agent for colorectal cancer prevention and treatment: focus on molecular mechanisms. Cancer Cell Int 2019;19:180.
[32] Okada M, Sato A, Shibuya K, Watanabe E, Seino S, Suzuki S, et al. JNK contributes to temozolomide resistance of stem-like glioblastoma cells via regulation of MGMT expression. Int J Oncol 2014;44(2):591–9.
[33] Strobel H, Baisch T, Fitzel R, Schilberg K, Siegelin MD, Karpel-Massler G, et al. Temozolomide and other alkylating agents in glioblastoma therapy. Biomedicine 2019;7(3):69.
[34] Torgovnick A, Schumacher B. DNA repair mechanisms in cancer development and therapy. Front Genet 2015;6:157.
[35] Morales J, Li L, Fattah FJ, Dong Y, Bey EA, Patel M, et al. Review of poly (ADP-ribose) polymerase (PARP) mechanisms of action and rationale for targeting in cancer and other diseases. Crit Rev Eukaryot Gene Expr 2014;24(1):15–28.
[36] Parrish KE, Cen L, Murray J, Calligaris D, Kizilbash S, Mittapalli RK, et al. Efficacy of PARP inhibitor rucaparib in orthotopic glioblastoma xenografts is limited by ineffective drug penetration into the central nervous system. Mol Cancer Ther 2015;14(12):2735–43.
[37] Kizilbash SH, Gupta SK, Chang K, Kawashima R, Parrish KE, Carlson BL, et al. Restricted delivery of talazoparib across the blood-brain barrier limits the sensitizing effects of PARP inhibition on temozolomide therapy in glioblastoma. Mol Cancer Ther 2017;16(12):2735–46.
[38] Gupta SK, Kizilbash SH, Carlson BL, Mladek AC, Boakye-Agyeman F, Bakken KK, et al. Delineation of MGMT hypermethylation as a biomarker for veliparib-mediated temozolomide-sensitizing therapy of glioblastoma. J Natl Cancer Inst 2016;108(5), djv369.
[39] Clarke MJ, Mulligan EA, Grogan PT, Mladek AC, Carlson BL, Schroeder MA, et al. Effective sensitization of temozolomide by ABT-888 is lost with development of temozolomide resistance in glioblastoma xenograft lines. Mol Cancer Ther 2009;8(2):407–14.
[40] Robins HI, Zhang P, Gilbert MR, Chakravarti A, de Groot JF, Grimm SA, et al. A randomized phase I/II study of ABT-888 in combination with temozolomide in recurrent temozolomide resistant glioblastoma: an NRG oncology RTOG group study. J Neuro-Oncol 2016;126(2):309–16.
[41] Su JM, Thompson P, Adesina A, Li XN, Kilburn L, Onar-Thomas A, et al. A phase I trial of veliparib (ABT-888) and temozolomide in children with recurrent CNS tumors: a pediatric brain tumor consortium report. Neuro-Oncology 2014;16(12):1661–8.
[42] Montaldi AP, Godoy PR, Sakamoto-Hojo ET. APE1/REF-1 down-regulation enhances the cytotoxic effects of temozolomide in a resistant glioblastoma cell line. Mutat Res Genet Toxicol Environ Mutagen 2015;793:19–29.
[43] Montaldi AP, Sakamoto-Hojo ET. Methoxyamine sensitizes the resistant glioblastoma T98G cell line to the alkylating agent temozolomide. Clin Exp Med 2013;13(4):279–88.
[44] Tang JB, Svilar D, Trivedi RN, Wang XH, Goellner EM, Moore B, et al. N-methylpurine DNA glycosylase and DNA polymerase beta modulate BER inhibitor potentiation of glioma cells to temozolomide. Neuro-Oncology 2011;13(5):471–86.
[45] Rinne M, Caldwell D, Kelley MR. Transient adenoviral N-methylpurine DNA glycosylase overexpression imparts chemotherapeutic sensitivity to human breast cancer cells. Mol Cancer Ther 2004;3(8):955–67.
[46] Fishel ML, He Y, Smith ML, Kelley MR. Manipulation of base excision repair to sensitize ovarian cancer cells to alkylating agent temozolomide. Clin Cancer Res 2007;13(1):260–7.
[47] Agnihotri S, Gajadhar AS, Ternamian C, Gorlia T, Diefes KL, Mischel PS, et al. Alkylpurine-DNA-N-glycosylase confers resistance to temozolomide in xenograft models of glioblastoma multiforme and is associated with poor survival in patients. J Clin Invest 2012;122(1):253–66.

[48] Everhard S, Tost J, El Abdalaoui H, Criniere E, Busato F, Marie Y, et al. Identification of regions correlating MGMT promoter methylation and gene expression in glioblastomas. Neuro-Oncology 2009;11(4):348–56.
[49] Bannister AJ, Kouzarides T. Regulation of chromatin by histone modifications. Cell Res 2011;21(3):381–95.
[50] Ecker J, Witt O, Milde T. Targeting of histone deacetylases in brain tumors. CNS Oncol 2013;2(4):359–76.
[51] Suraweera A, O'Byrne KJ, Richard DJ. Combination therapy with histone deacetylase inhibitors (HDACi) for the treatment of cancer: achieving the full therapeutic potential of HDACi. Front Oncol 2018;8:92.
[52] Hosein AN, Lim YC, Day B, Stringer B, Rose S, Head R, et al. The effect of valproic acid in combination with irradiation and temozolomide on primary human glioblastoma cells. J Neuro-Oncol 2015;122(2):263–71.
[53] Weller M, Gorlia T, Cairncross JG, van den Bent MJ, Mason W, Belanger K, et al. Prolonged survival with valproic acid use in the EORTC/NCIC temozolomide trial for glioblastoma. Neurology 2011;77(12):1156–64.
[54] Krauze AV, Myrehaug SD, Chang MG, Holdford DJ, Smith S, Shih J, et al. A phase 2 study of concurrent radiation therapy, temozolomide, and the histone deacetylase inhibitor valproic acid for patients with glioblastoma. Int J Radiat Oncol Biol Phys 2015;92(5):986–92.
[55] Galanis E, Anderson SK, Miller CR, Sarkaria JN, Jaeckle K, Buckner JC, et al. Phase I/II trial of vorinostat combined with temozolomide and radiation therapy for newly diagnosed glioblastoma: results of alliance N0874/ABTC 02. Neuro-Oncology 2018;20(4):546–56.
[56] Iwamoto FM, Lamborn KR, Kuhn JG, Wen PY, Yung WK, Gilbert MR, et al. A phase I/II trial of the histone deacetylase inhibitor romidepsin for adults with recurrent malignant glioma: North American brain tumor consortium study 03-03. Neuro-Oncology 2011;13(5):509–16.
[57] Lee EQ, Reardon DA, Schiff D, Drappatz J, Muzikansky A, Grimm SA, et al. Phase II study of panobinostat in combination with bevacizumab for recurrent glioblastoma and anaplastic glioma. Neuro-Oncology 2015;17(6):862–7.
[58] Gurbani SS, Yoon Y, Weinberg BD, Salgado E, Press RH, Cordova JS, et al. Assessing treatment response of glioblastoma to an HDAC inhibitor using whole-brain spectroscopic MRI. Tomography 2019;5(1):53–60.
[59] Regad T, Targeting RTK. Signaling pathways in cancer. Cancers (Basel) 2015;7(3):1758–84.
[60] Du Z, Lovly CM. Mechanisms of receptor tyrosine kinase activation in cancer. Mol Cancer 2018;17(1):58.
[61] Wieduwilt MJ, Moasser MM. The epidermal growth factor receptor family: biology driving targeted therapeutics. Cell Mol Life Sci 2008;65(10):1566–84.
[62] Pearson JRD, Regad T. Targeting cellular pathways in glioblastoma multiforme. Signal Transduct Target Ther 2017;2:17040.
[63] An Z, Aksoy O, Zheng T, Fan QW, Weiss WA. Epidermal growth factor receptor and EGFRvIII in glioblastoma: signaling pathways and targeted therapies. Oncogene 2018;37(12):1561–75.
[64] Reardon DA, Nabors LB, Mason WP, Perry JR, Shapiro W, Kavan P, et al. Phase I/randomized phase II study of afatinib, an irreversible ErbB family blocker, with or without protracted temozolomide in adults with recurrent glioblastoma. Neuro-Oncology 2015;17(3):430–9.
[65] Zahonero C, Aguilera P, Ramirez-Castillejo C, Pajares M, Bolos MV, Cantero D, et al. Preclinical test of dacomitinib, an irreversible EGFR inhibitor, confirms its effectiveness for glioblastoma. Mol Cancer Ther 2015;14(7):1548–58.
[66] Sepulveda-Sanchez JM, Vaz MA, Balana C, Gil-Gil M, Reynes G, Gallego O, et al. Phase II trial of dacomitinib, a pan-human EGFR tyrosine kinase inhibitor, in recurrent glioblastoma patients with EGFR amplification. Neuro-Oncology 2017;19(11):1522–31.
[67] Liu X, Chen X, Shi L, Shan Q, Cao Q, Yue C, et al. The third-generation EGFR inhibitor AZD9291 overcomes primary resistance by continuously blocking ERK signaling in glioblastoma. J Exp Clin Cancer Res 2019;38(1):219.
[68] Rajaratnam V, Islam MM, Yang M, Slaby R, Ramirez HM, Mirza SP. Glioblastoma: pathogenesis and current status of chemotherapy and other novel treatments. Cancer 2020;12(4):937.
[69] Dreier A, Barth S, Goswami A, Weis J. Cetuximab induces mitochondrial translocalization of EGFRvIII, but not EGFR: involvement of mitochondria in tumor drug resistance? Tumour Biol 2012;33(1):85–94.
[70] Chong DQ, Toh XY, Ho IA, Sia KC, Newman JP, Yulyana Y, et al. Combined treatment of nimotuzumab and rapamycin is effective against temozolomide-resistant human gliomas regardless of the EGFR mutation status. BMC Cancer 2015;15:255.
[71] Johns TG, McKay MJ, Cvrljevic AN, Gan HK, Taylor C, Xu H, et al. MAb 806 enhances the efficacy of ionizing radiation in glioma xenografts expressing the de2-7 epidermal growth factor receptor. Int J Radiat Oncol Biol Phys 2010;78(2):572–8.

[72] Reardon DA, Desjardins A, Vredenburgh JJ, O'Rourke DM, Tran DD, Fink KL, et al. Rindopepimut with bevacizumab for patients with relapsed EGFRvIII-expressing glioblastoma (ReACT): results of a double-blind randomized phase II trial. Clin Cancer Res 2020;26(7):1586–94.
[73] Schuster J, Lai RK, Recht LD, Reardon DA, Paleologos NA, Groves MD, et al. A phase II, multicenter trial of rindopepimut (CDX-110) in newly diagnosed glioblastoma: the ACT III study. Neuro-Oncology 2015; 17(6):854–61.
[74] Weller M, Butowski N, Tran DD, Recht LD, Lim M, Hirte H, et al. Rindopepimut with temozolomide for patients with newly diagnosed, EGFRvIII-expressing glioblastoma (ACT IV): a randomised, double-blind, international phase 3 trial. Lancet Oncol 2017;18(10):1373–85.
[75] Suryadevara CM, Desai R, Abel ML, Riccione KA, Batich KA, Shen SH, et al. Temozolomide lymphodepletion enhances CAR abundance and correlates with antitumor efficacy against established glioblastoma. Onco Targets Ther 2018;7(6), e1434464.
[76] Badhiwala J, Decker WK, Berens ME, Bhardwaj RD. Clinical trials in cellular immunotherapy for brain/CNS tumors. Expert Rev Neurother 2013;13(4):405–24.
[77] Norden AD, Schiff D, Ahluwalia MS, Lesser GJ, Nayak L, Lee EQ, et al. Phase II trial of triple tyrosine kinase receptor inhibitor nintedanib in recurrent high-grade gliomas. J Neuro-Oncol 2015;121(2):297–302.
[78] Gerstner ER, Eichler AF, Plotkin SR, Drappatz J, Doyle CL, Xu L, et al. Phase I trial with biomarker studies of vatalanib (PTK787) in patients with newly diagnosed glioblastoma treated with enzyme inducing anti-epileptic drugs and standard radiation and temozolomide. J Neuro-Oncol 2011;103(2):325–32.
[79] Kalpathy-Cramer J, Chandra V, Da X, Ou Y, Emblem KE, Muzikansky A, et al. Phase II study of tivozanib, an oral VEGFR inhibitor, in patients with recurrent glioblastoma. J Neuro-Oncol 2017;131(3):603–10.
[80] Batchelor TT, Mulholland P, Neyns B, Nabors LB, Campone M, Wick A, et al. Phase III randomized trial comparing the efficacy of cediranib as monotherapy, and in combination with lomustine, versus lomustine alone in patients with recurrent glioblastoma. J Clin Oncol 2013;31(26):3212–8.
[81] Brown N, McBain C, Nash S, Hopkins K, Sanghera P, Saran F, et al. Multi-center randomized phase II study comparing cediranib plus gefitinib with cediranib plus placebo in subjects with recurrent/progressive glioblastoma. PLoS One 2016;11(5), e0156369.
[82] Ellingson BM, Gerstner ER, Smits M, Huang RY, Colen R, Abrey LE, et al. Diffusion MRI phenotypes predict overall survival benefit from anti-VEGF monotherapy in recurrent glioblastoma: converging evidence from phase II trials. Clin Cancer Res 2017;23(19):5745–56.
[83] Jimenez-Pascual A, Siebzehnrubl FA. Fibroblast growth factor receptor functions in glioblastoma. Cell 2019; 8(7):715.
[84] Nelson KN, Meyer AN, Wang CG, Donoghue DJ. Oncogenic driver FGFR3-TACC3 is dependent on membrane trafficking and ERK signaling. Oncotarget 2018;9(76):34306–19.
[85] Zhang Y, Dube C, Gibert Jr M, Cruickshanks N, Wang B, Coughlan M, et al. The p53 pathway in glioblastoma. Cancer 2018;10(9):297.
[86] Lang FF, Bruner JM, Fuller GN, Aldape K, Prados MD, Chang S, et al. Phase I trial of adenovirus-mediated p53 gene therapy for recurrent glioma: biological and clinical results. J Clin Oncol 2003;21(13):2508–18.
[87] Kim SS, Rait A, Kim E, Pirollo KF, Chang EH. A tumor-targeting p53 nanodelivery system limits chemoresistance to temozolomide prolonging survival in a mouse model of glioblastoma multiforme. Nanomedicine 2015;11(2):301–11.
[88] Grillone A, Battaglini M, Moscato S, Mattii L, de Julian FC, Scarpellini A, et al. Nutlin-loaded magnetic solid lipid nanoparticles for targeted glioblastoma treatment. Nanomedicine (Lond) 2019;14(6):727–52.
[89] Villalonga-Planells R, Coll-Mulet L, Martinez-Soler F, Castano E, Acebes JJ, Gimenez-Bonafe P, et al. Activation of p53 by nutlin-3a induces apoptosis and cellular senescence in human glioblastoma multiforme. PLoS One 2011;6(4), e18588.
[90] Verreault M, Schmitt C, Goldwirt L, Pelton K, Haidar S, Levasseur C, et al. Preclinical efficacy of the MDM2 inhibitor RG7112 in MDM2-amplified and TP53 wild-type glioblastomas. Clin Cancer Res 2016;22(5):1185–96.
[91] Her NG, Oh JW, Oh YJ, Han S, Cho HJ, Lee Y, et al. Potent effect of the MDM2 inhibitor AMG232 on suppression of glioblastoma stem cells. Cell Death Dis 2018;9(8):792.
[92] Wu Q, Cao Z, Xiao W, Zhu L, Xie Q, Li L, et al. Study on therapeutic action and mechanism of TMZ combined with RITA against glioblastoma. Cell Physiol Biochem 2018;51(6):2536–46.
[93] Liu T, Zhang L, Joo D, Sun SC. NF-kappaB signaling in inflammation. Signal Transduct Target Ther 2017;2:17023.

[94] Yu Z, Chen Y, Wang S, Li P, Zhou G, Yuan Y. Inhibition of NF-kappaB results in anti-glioma activity and reduces temozolomide-induced chemoresistance by down-regulating MGMT gene expression. Cancer Lett 2018;428:77–89.

[95] Collins PE, Mitxitorena I, Carmody RJ. The ubiquitination of NF-kappaB subunits in the control of transcription. Cell 2016;5(2):23.

[96] Arkwright R, Pham TM, Zonder JA, Dou QP. The preclinical discovery and development of bortezomib for the treatment of mantle cell lymphoma. Expert Opin Drug Discovery 2017;12(2):225–35.

[97] Vlachostergios PJ, Hatzidaki E, Stathakis NE, Koukoulis GK, Papandreou CN. Bortezomib downregulates MGMT expression in T98G glioblastoma cells. Cell Mol Neurobiol 2013;33(3):313–8.

[98] Rahman MA, Gras Navarro A, Brekke J, Engelsen A, Bindesboll C, Sarowar S, et al. Bortezomib administered prior to temozolomide depletes MGMT, chemosensitizes glioblastoma with unmethylated MGMT promoter and prolongs animal survival. Br J Cancer 2019;121(7):545–55.

[99] Kong XT, Nguyen NT, Choi YJ, Zhang G, Nguyen HN, Filka E, et al. Phase 2 study of bortezomib combined with temozolomide and regional radiation therapy for upfront treatment of patients with newly diagnosed glioblastoma multiforme: safety and efficacy assessment. Int J Radiat Oncol Biol Phys 2018;100(5):1195–203.

[100] Soubannier V, Stifani S. NF-kappaB signalling in glioblastoma. Biomedicine 2017;5(2):29.

[101] Zanotto-Filho A, Braganhol E, Schroder R, de Souza LH, Dalmolin RJ, Pasquali MA, et al. NFkappaB inhibitors induce cell death in glioblastomas. Biochem Pharmacol 2011;81(3):412–24.

[102] Li M, Liang RF, Wang X, Mao Q, Liu YH. BKM120 sensitizes C6 glioma cells to temozolomide via suppression of the PI3K/Akt/NF-kappaB/MGMT signaling pathway. Oncol Lett 2017;14(6):6597–603.

[103] Cahill KE, Morshed RA, Yamini B. Nuclear factor-kappaB in glioblastoma: insights into regulators and targeted therapy. Neuro-Oncology 2016;18(3):329–39.

[104] Chung WJ, Sontheimer H. Sulfasalazine inhibits the growth of primary brain tumors independent of nuclear factor-kappaB. J Neurochem 2009;110(1):182–93.

[105] Takeuchi S, Wada K, Nagatani K, Otani N, Osada H, Nawashiro H. Sulfasalazine and temozolomide with radiation therapy for newly diagnosed glioblastoma. Neurol India 2014;62(1):42–7.

[106] Topacio BR, Zatulovskiy E, Cristea S, Xie S, Tambo CS, Rubin SM, et al. Cyclin D-Cdk4,6 drives cell-cycle progression via the retinoblastoma protein's C-terminal helix. Mol Cell 2019;74(4):758–70 [e4].

[107] Rayman JB, Takahashi Y, Indjeian VB, Dannenberg JH, Catchpole S, Watson RJ, et al. E2F mediates cell cycle-dependent transcriptional repression in vivo by recruitment of an HDAC1/mSin3B corepressor complex. Genes Dev 2002;16(8):933–47.

[108] Barroso-Sousa R, Shapiro GI, Tolaney SM. Clinical development of the CDK4/6 inhibitors ribociclib and abemaciclib in breast cancer. Breast Care (Basel) 2016;11(3):167–73.

[109] Yin L, Li H, Liu W, Yao Z, Cheng Z, Zhang H, et al. A highly potent CDK4/6 inhibitor was rationally designed to overcome blood brain barrier in gliobastoma therapy. Eur J Med Chem 2018;144:1–28.

[110] Kroemer G, Marino G, Levine B. Autophagy and the integrated stress response. Mol Cell 2010;40(2):280–93.

[111] Taylor MA, Das BC, Ray SK. Targeting autophagy for combating chemoresistance and radioresistance in glioblastoma. Apoptosis 2018;23(11 – 12):563–75.

[112] Golden EB, Cho HY, Jahanian A, Hofman FM, Louie SG, Schonthal AH, et al. Chloroquine enhances temozolomide cytotoxicity in malignant gliomas by blocking autophagy. Neurosurg Focus 2014;37(6), E12.

[113] Rosenfeld MR, Ye X, Supko JG, Desideri S, Grossman SA, Brem S, et al. A phase I/II trial of hydroxychloroquine in conjunction with radiation therapy and concurrent and adjuvant temozolomide in patients with newly diagnosed glioblastoma multiforme. Autophagy 2014;10(8):1359–68.

[114] Kim G, Ko YT. Small molecule tyrosine kinase inhibitors in glioblastoma. Arch Pharm Res 2020;43(4):385–94.

[115] Kerbel RS. Tumor angiogenesis. N Engl J Med 2008;358(19):2039–49.

[116] Diaz RJ, Ali S, Qadir MG, De La Fuente MI, Ivan ME, Komotar RJ. The role of bevacizumab in the treatment of glioblastoma. J Neuro-Oncol 2017;133(3):455–67.

[117] Ellis LM. Mechanisms of action of bevacizumab as a component of therapy for metastatic colorectal cancer. Semin Oncol 2006;33(5 Suppl 10):S1–7.

[118] Gilbert MR, Dignam JJ, Armstrong TS, Wefel JS, Blumenthal DT, Vogelbaum MA, et al. A randomized trial of bevacizumab for newly diagnosed glioblastoma. N Engl J Med 2014;370(8):699–708.

[119] Piccioni DE, Selfridge J, Mody RR, Chowdhury R, Li S, Lalezari S, et al. Deferred use of bevacizumab for recurrent glioblastoma is not associated with diminished efficacy. Neuro-Oncology 2014;16(6):815–22.
[120] Fujita K, Kubota Y, Ishida H, Sasaki Y. Irinotecan, a key chemotherapeutic drug for metastatic colorectal cancer. World J Gastroenterol 2015;21(43):12234–48.
[121] Wick W, Gorlia T, Bendszus M, Taphoorn M, Sahm F, Harting I, et al. Lomustine and bevacizumab in progressive glioblastoma. N Engl J Med 2017;377(20):1954–63.
[122] Tini P, Pirtoli L. Combining ipilimumab and bevacizumab in glioblastoma is really safe and effective? Clin Oncol (R Coll Radiol) 2016;28(10):663.
[123] Carter T, Shaw H, Mulholland P. Combining ipilimumab and bevacizumab in glioblastoma: is it really safe and effective? Author response. Clin Oncol (R Coll Radiol) 2016;28(10):664.
[124] Carter T, Shaw H, Cohn-Brown D, Chester K, Mulholland P. Ipilimumab and Bevacizumab in Glioblastoma. Clin Oncol (R Coll Radiol) 2016;28(10):622–6.
[125] Hasan S, Chen E, Lanciano R, Yang J, Hanlon A, Lamond J, et al. Salvage fractionated stereotactic radiotherapy with or without chemotherapy and immunotherapy for recurrent glioblastoma multiforme: a single institution experience. Front Oncol 2015;5:106.
[126] Wachsberger PR, Burd R, Cardi C, Thakur M, Daskalakis C, Holash J, et al. VEGF trap in combination with radiotherapy improves tumor control in u87 glioblastoma. Int J Radiat Oncol Biol Phys 2007;67(5):1526–37.
[127] de Groot JF, Piao Y, Tran H, Gilbert M, Wu HK, Liu J, et al. Myeloid biomarkers associated with glioblastoma response to anti-VEGF therapy with aflibercept. Clin Cancer Res 2011;17(14):4872–81.
[128] Holash J, Davis S, Papadopoulos N, Croll SD, Ho L, Russell M, et al. VEGF-trap: a VEGF blocker with potent antitumor effects. Proc Natl Acad Sci U S A 2002;99(17):11393–8.
[129] de Groot JF, Lamborn KR, Chang SM, Gilbert MR, Cloughesy TF, Aldape K, et al. Phase II study of aflibercept in recurrent malignant glioma: a North American brain tumor consortium study. J Clin Oncol 2011;29(19):2689–95.
[130] Loilome W, Joshi AD, Ap Rhys CM, Piccirillo S, Vescovi AL, Gallia GL, et al. Glioblastoma cell growth is suppressed by disruption of fibroblast growth factor pathway signaling. J Neuro-Oncol 2009;94(3):359–66.
[131] Marx GM, Pavlakis N, McCowatt S, Boyle FM, Levi JA, Bell DR, et al. Phase II study of thalidomide in the treatment of recurrent glioblastoma multiforme. J Neuro-Oncol 2001;54(1):31–8.
[132] Riva M, Imbesi F, Beghi E, Galli C, Citterio A, Trapani P, et al. Temozolomide and thalidomide in the treatment of glioblastoma multiforme. Anticancer Res 2007;27(2):1067–71.
[133] Baumann F, Bjeljac M, Kollias SS, Baumert BG, Brandner S, Rousson V, et al. Combined thalidomide and temozolomide treatment in patients with glioblastoma multiforme. J Neuro-Oncol 2004;67(1–2):191–200.
[134] Penas-Prado M, Hess KR, Fisch MJ, Lagrone LW, Groves MD, Levin VA, et al. Randomized phase II adjuvant factorial study of dose-dense temozolomide alone and in combination with isotretinoin, celecoxib, and/or thalidomide for glioblastoma. Neuro-Oncology 2015;17(2):266–73.
[135] Hassler MR, Sax C, Flechl B, Ackerl M, Preusser M, Hainfellner JA, et al. Thalidomide as palliative treatment in patients with advanced secondary glioblastoma. Oncology 2015;88(3):173–9.
[136] Stockhammer F, Misch M, Koch A, Czabanka M, Plotkin M, Blechschmidt C, et al. Continuous low-dose temozolomide and celecoxib in recurrent glioblastoma. J Neuro-Oncol 2010;100(3):407–15.
[137] Stupp R, Hegi ME, Neyns B, Goldbrunner R, Schlegel U, Clement PM, et al. Phase I/IIa study of cilengitide and temozolomide with concomitant radiotherapy followed by cilengitide and temozolomide maintenance therapy in patients with newly diagnosed glioblastoma. J Clin Oncol 2010;28(16):2712–8.
[138] Weller M, Nabors LB, Gorlia T, Leske H, Rushing E, Bady P, et al. Cilengitide in newly diagnosed glioblastoma: biomarker expression and outcome. Oncotarget 2016;7(12):15018–32.
[139] Eisele G, Wick A, Eisele AC, Clement PM, Tonn J, Tabatabai G, et al. Cilengitide treatment of newly diagnosed glioblastoma patients does not alter patterns of progression. J Neuro-Oncol 2014;117(1):141–5.
[140] Stupp R, Hegi ME, Gorlia T, Erridge SC, Perry J, Hong YK, et al. Cilengitide combined with standard treatment for patients with newly diagnosed glioblastoma with methylated MGMT promoter (CENTRIC EORTC 26071-22072 study): a multicentre, randomised, open-label, phase 3 trial. Lancet Oncol 2014;15(10):1100–8.
[141] Reardon DA, Cheresh D. Cilengitide: a prototypic integrin inhibitor for the treatment of glioblastoma and other malignancies. Genes Cancer 2011;2(12):1159–65.

[142] Roomi MW, Kalinovsky T, Rath M, Niedzwiecki A. Modulation of MMP-2 and MMP-9 secretion by cytokines, inducers and inhibitors in human glioblastoma T-98G cells. Oncol Rep 2017;37(3):1907–13.
[143] Agarwal S, Muniyandi P, Maekawa T, Kumar DS. Vesicular systems employing natural substances as promising drug candidates for MMP inhibition in glioblastoma: a nanotechnological approach. Int J Pharm 2018;551 (1–2):339–61.
[144] Perry MC, Demeule M, Regina A, Moumdjian R, Beliveau R. Curcumin inhibits tumor growth and angiogenesis in glioblastoma xenografts. Mol Nutr Food Res 2010;54(8):1192–201.
[145] Wang Y, Xing D, Zhao M, Wang J, Yang Y. The role of a single angiogenesis inhibitor in the treatment of recurrent glioblastoma multiforme: a meta-analysis and systematic review. PLoS One 2016;11(3), e0152170.
[146] Prieto-Vila M, Takahashi RU, Usuba W, Kohama I, Ochiya T. Drug resistance driven by cancer stem cells and their niche. Int J Mol Sci 2017;18(12):2574.
[147] Al-Hajj M, Wicha MS, Benito-Hernandez A, Morrison SJ, Clarke MF. Prospective identification of tumorigenic breast cancer cells. Proc Natl Acad Sci U S A 2003;100(7):3983–8.
[148] Yi Y, Hsieh IY, Huang X, Li J, Zhao W. Glioblastoma stem-like cells: characteristics, microenvironment, and therapy. Front Pharmacol 2016;7:477.
[149] Singh SK, Hawkins C, Clarke ID, Squire JA, Bayani J, Hide T, et al. Identification of human brain tumour initiating cells. Nature 2004;432(7015):396–401.
[150] Wang J, Sakariassen PO, Tsinkalovsky O, Immervoll H, Boe SO, Svendsen A, et al. CD133 negative glioma cells form tumors in nude rats and give rise to CD133 positive cells. Int J Cancer 2008;122(4):761–8.
[151] Cheray M, Begaud G, Deluche E, Nivet A, Battu S, Lalloue F, et al. Cancer stem-like cells in glioblastoma. In: De Vleeschouwer S, editor. Glioblastoma. Brisbane: AU; 2017.
[152] Fletcher JI, Haber M, Henderson MJ, Norris MD. ABC transporters in cancer: more than just drug efflux pumps. Nat Rev Cancer 2010;10(2):147–56.
[153] Zhou J, Wang CY, Liu T, Wu B, Zhou F, Xiong JX, et al. Persistence of side population cells with high drug efflux capacity in pancreatic cancer. World J Gastroenterol 2008;14(6):925–30.
[154] Gottesman MM, Fojo T, Bates SE. Multidrug resistance in cancer: role of ATP-dependent transporters. Nat Rev Cancer 2002;2(1):48–58.
[155] Singh A, Settleman J. EMT, cancer stem cells and drug resistance: an emerging axis of evil in the war on cancer. Oncogene 2010;29(34):4741–51.
[156] Fletcher JI, Williams RT, Henderson MJ, Norris MD, Haber M. ABC transporters as mediators of drug resistance and contributors to cancer cell biology. Drug Resist Updat 2016;26:1–9.
[157] Munoz JL, Rodriguez-Cruz V, Rameshwar P. High expression of miR-9 in CD133(+) glioblastoma cells in chemoresistance to temozolomide. J Cancer Stem Cell Res 2015;3:e1003.
[158] Munoz JL, Bliss SA, Greco SJ, Ramkissoon SH, Ligon KL, Rameshwar P. Delivery of functional anti-miR-9 by mesenchymal stem cell-derived exosomes to glioblastoma Multiforme cells conferred chemosensitivity. Mol Ther Nucleic Acids 2013;2, e126.
[159] Lee CAA, Banerjee P, Wilson BJ, Wu S, Guo Q, Berg G, et al. Targeting the ABC transporter ABCB5 sensitizes glioblastoma to temozolomide-induced apoptosis through a cell-cycle checkpoint regulation mechanism. J Biol Chem 2020;295(22):7774–88.
[160] Lai EC. Notch signaling: control of cell communication and cell fate. Development 2004;131(5):965–73.
[161] Kopan R. Notch signaling. Cold Spring Harb Perspect Biol 2012;4(10):a011213.
[162] Fan X, Khaki L, Zhu TS, Soules ME, Talsma CE, Gul N, et al. NOTCH pathway blockade depletes CD133-positive glioblastoma cells and inhibits growth of tumor neurospheres and xenografts. Stem Cells 2010;28 (1):5–16.
[163] Wang J, Wakeman TP, Lathia JD, Hjelmeland AB, Wang XF, White RR, et al. Notch promotes radioresistance of glioma stem cells. Stem Cells 2010;28(1):17–28.
[164] Takebe N, Harris PJ, Warren RQ, Ivy SP. Targeting cancer stem cells by inhibiting Wnt, Notch, and Hedgehog pathways. Nat Rev Clin Oncol 2011;8(2):97–106.
[165] Fan X, Matsui W, Khaki L, Stearns D, Chun J, Li YM, et al. Notch pathway inhibition depletes stem-like cells and blocks engraftment in embryonal brain tumors. Cancer Res 2006;66(15):7445–52.
[166] Farnie G, Willan PM, Clarke RB, Bundred NJ. Combined inhibition of ErbB1/2 and Notch receptors effectively targets breast ductal carcinoma in situ (DCIS) stem/progenitor cell activity regardless of ErbB2 status. PLoS One 2013;8(2), e56840.

[167] Choudhry Z, Rikani AA, Choudhry AM, Tariq S, Zakaria F, Asghar MW, et al. Sonic hedgehog signalling pathway: a complex network. Ann Neurosci 2014;21(1):28–31.
[168] Clement V, Sanchez P, de Tribolet N, Radovanovic I, Ruiz i Altaba A. HEDGEHOG-GLI1 signaling regulates human glioma growth, cancer stem cell self-renewal, and tumorigenicity. Curr Biol 2007;17(2):165–72.
[169] Bar EE, Chaudhry A, Lin A, Fan X, Schreck K, Matsui W, et al. Cyclopamine-mediated hedgehog pathway inhibition depletes stem-like cancer cells in glioblastoma. Stem Cells 2007;25(10):2524–33.
[170] Dijkgraaf GJ, Alicke B, Weinmann L, Januario T, West K, Modrusan Z, et al. Small molecule inhibition of GDC-0449 refractory smoothened mutants and downstream mechanisms of drug resistance. Cancer Res 2011;71 (2):435–44.
[171] Munoz JL, Rodriguez-Cruz V, Walker ND, Greco SJ, Rameshwar P. Temozolomide resistance and tumor recurrence: Halting the Hedgehog. Cancer Cell Microenviron 2015;2(2), e747.
[172] Kim WY, Wang X, Wu Y, Doble BW, Patel S, Woodgett JR, et al. GSK-3 is a master regulator of neural progenitor homeostasis. Nat Neurosci 2009;12(11):1390–7.
[173] Chenn A, Walsh CA. Regulation of cerebral cortical size by control of cell cycle exit in neural precursors. Science 2002;297(5580):365–9.
[174] Rossi M, Magnoni L, Miracco C, Mori E, Tosi P, Pirtoli L, et al. Beta-catenin and Gli1 are prognostic markers in glioblastoma. Cancer Biol Ther 2011;11(8):753–61.
[175] Heimberger AB, Suki D, Yang D, Shi W, Aldape K. The natural history of EGFR and EGFRvIII in glioblastoma patients. J Transl Med 2005;3:38.
[176] Emlet DR, Gupta P, Holgado-Madruga M, Del Vecchio CA, Mitra SS, Han SY, et al. Targeting a glioblastoma cancer stem-cell population defined by EGF receptor variant III. Cancer Res 2014;74(4):1238–49.
[177] Verhaak RG, Hoadley KA, Purdom E, Wang V, Qi Y, Wilkerson MD, et al. Integrated genomic analysis identifies clinically relevant subtypes of glioblastoma characterized by abnormalities in PDGFRA, IDH1, EGFR, and NF1. Cancer Cell 2010;17(1):98–110.
[178] Manoranjan B, Chokshi C, Venugopal C, Subapanditha M, Savage N, Tatari N, et al. A CD133-AKT-Wnt signaling axis drives glioblastoma brain tumor-initiating cells. Oncogene 2020;39(7):1590–9.
[179] Carra E, Barbieri F, Marubbi D, Pattarozzi A, Favoni RE, Florio T, et al. Sorafenib selectively depletes human glioblastoma tumor-initiating cells from primary cultures. Cell Cycle 2013;12(3):491–500.
[180] Sareddy GR, Kesanakurti D, Kirti PB, Babu PP. Nonsteroidal anti-inflammatory drugs diclofenac and celecoxib attenuates Wnt/beta-catenin/Tcf signaling pathway in human glioblastoma cells. Neurochem Res 2013;38 (11):2313–22.
[181] Beckmann A, Hainz N, Tschernig T, Meier C. Facets of communication: gap junction ultrastructure and function in cancer stem cells and tumor cells. Cancer 2019;11(3):288.
[182] Murphy SF, Varghese RT, Lamouille S, Guo S, Pridham KJ, Kanabur P, et al. Connexin 43 inhibition sensitizes chemoresistant glioblastoma cells to temozolomide. Cancer Res 2016;76(1):139–49.
[183] Munoz JL, Rodriguez-Cruz V, Greco SJ, Ramkissoon SH, Ligon KL, Rameshwar P. Temozolomide resistance in glioblastoma cells occurs partly through epidermal growth factor receptor-mediated induction of connexin 43. Cell Death Dis 2014;5, e1145.
[184] Mulkearns-Hubert EE, Torre-Healy LA, Silver DJ, Eurich JT, Bayik D, Serbinowski E, et al. Development of a Cx46 targeting strategy for cancer stem cells. Cell Rep 2019;27(4):1062–72 [e5].
[185] Hitomi M, Deleyrolle LP, Mulkearns-Hubert EE, Jarrar A, Li M, Sinyuk M, et al. Differential connexin function enhances self-renewal in glioblastoma. Cell Rep 2015;11(7):1031–42.
[186] Grek CL, Sheng Z, Naus CC, Sin WC, Gourdie RG, Ghatnekar GG. Novel approach to temozolomide resistance in malignant glioma: connexin43-directed therapeutics. Curr Opin Pharmacol 2018;41:79–88.
[187] Stupp R, Taillibert S, Kanner A, Read W, Steinberg D, Lhermitte B, et al. Effect of tumor-treating fields plus maintenance temozolomide vs maintenance temozolomide alone on survival in patients with glioblastoma: a randomized clinical trial. JAMA 2017;318(23):2306–16.
[188] Stupp R, Wong ET, Kanner AA, Steinberg D, Engelhard H, Heidecke V, et al. NovoTTF-100A versus physician's choice chemotherapy in recurrent glioblastoma: a randomised phase III trial of a novel treatment modality. Eur J Cancer 2012;48(14):2192–202.
[189] Kirson ED, Gurvich Z, Schneiderman R, Dekel E, Itzhaki A, Wasserman Y, et al. Disruption of cancer cell replication by alternating electric fields. Cancer Res 2004;64(9):3288–95.

[190] Gera N, Yang A, Holtzman TS, Lee SX, Wong ET, Swanson KD. Tumor treating fields perturb the localization of septins and cause aberrant mitotic exit. PLoS One 2015;10(5), e0125269.
[191] Akbarnejad Z, Eskandary H, Dini L, Vergallo C, Nematollahi-Mahani SN, Farsinejad A, et al. Cytotoxicity of temozolomide on human glioblastoma cells is enhanced by the concomitant exposure to an extremely low-frequency electromagnetic field (100Hz, 100G). Biomed Pharmacother 2017;92:254–64.
[192] Hottinger AF, Pacheco P, Stupp R. Tumor treating fields: a novel treatment modality and its use in brain tumors. Neuro-Oncology 2016;18(10):1338–49.
[193] Fabian D, Guillermo Prieto Eibl MDP, Alnahhas I, Sebastian N, Giglio P, Puduvalli V, et al. Treatment of glioblastoma (GBM) with the addition of tumor-treating fields (TTF): a review. Cancers 2019;11(2):174.
[194] Mrugala MM, Engelhard HH, Dinh Tran D, Kew Y, Cavaliere R, Villano JL, et al. Clinical practice experience with NovoTTF-100A system for glioblastoma: the patient registry dataset (PRiDe). Semin Oncol 2014;41(Suppl 6):S4–S13.
[195] Wasserfallen JB, Ostermann S, Leyvraz S, Stupp R. Cost of temozolomide therapy and global care for recurrent malignant gliomas followed until death. Neuro-Oncology 2005;7(2):189–95.
[196] Toms SA, Kim CY, Nicholas G, Ram Z. Increased compliance with tumor treating fields therapy is prognostic for improved survival in the treatment of glioblastoma: a subgroup analysis of the EF-14 phase III trial. J Neuro-Oncol 2019;141(2):467–73.
[197] Mehta M, Wen P, Nishikawa R, Reardon D, Peters K. Critical review of the addition of tumor treating fields (TTFields) to the existing standard of care for newly diagnosed glioblastoma patients. Crit Rev Oncol Hematol 2017;111:60–5.
[198] D'Alessio A, Proietti G, Sica G, Scicchitano BM. Pathological and molecular features of glioblastoma and its peritumoral tissue. Cancer 2019;11(4):469.
[199] Aleynik A, Gernavage KM, Mourad Y, Sherman LS, Liu K, Gubenko YA, et al. Stem cell delivery of therapies for brain disorders. Clin Transl Med 2014;3:24.
[200] Rivera-Cruz CM, Shearer JJ, Figueiredo Neto M, Figueiredo ML. The immunomodulatory effects of mesenchymal stem cell polarization within the tumor microenvironment niche. Stem Cells Int 2017;2017:4015039.
[201] Hersh DS, Wadajkar AS, Roberts N, Perez JG, Connolly NP, Frenkel V, et al. Evolving drug delivery strategies to overcome the blood brain barrier. Curr Pharm Des 2016;22(9):1177–93.
[202] Lee HK, Finniss S, Cazacu S, Bucris E, Ziv-Av A, Xiang C, et al. Mesenchymal stem cells deliver synthetic microRNA mimics to glioma cells and glioma stem cells and inhibit their cell migration and self-renewal. Oncotarget 2013;4(2):346–61.
[203] Ye Z, Wang Y, Xie HY, Zheng SS. Immunosuppressive effects of rat mesenchymal stem cells: involvement of CD4+CD25+ regulatory T cells. Hepatobiliary Pancreat Dis Int 2008;7(6):608–14.
[204] Mao AS, Ozkale B, Shah NJ, Vining KH, Descombes T, Zhang L, et al. Programmable microencapsulation for enhanced mesenchymal stem cell persistence and immunomodulation. Proc Natl Acad Sci U S A 2019;116 (31):15392–7.
[205] Wang J, Lu XX, Chen DZ, Li SF, Zhang LS. Herpes simplex virus thymidine kinase and ganciclovir suicide gene therapy for human pancreatic cancer. World J Gastroenterol 2004;10(3):400–3.
[206] Duebgen M, Martinez-Quintanilla J, Tamura K, Hingtgen S, Redjal N, Wakimoto H, et al. Stem cells loaded with multimechanistic oncolytic herpes simplex virus variants for brain tumor therapy. J Natl Cancer Inst 2014;106 (6), dju090.
[207] Choi SH, Stuckey DW, Pignatta S, Reinshagen C, Khalsa JK, Roozendaal N, et al. Tumor resection recruits effector T cells and boosts therapeutic efficacy of encapsulated stem cells expressing IFNbeta in glioblastomas. Clin Cancer Res 2017;23(22):7047–58.
[208] Markman JL, Rekechenetskiy A, Holler E, Ljubimova JY. Nanomedicine therapeutic approaches to overcome cancer drug resistance. Adv Drug Deliv Rev 2013;65(13–14):1866–79.
[209] Livney YD, Assaraf YG. Rationally designed nanovehicles to overcome cancer chemoresistance. Adv Drug Deliv Rev 2013;65(13–14):1716–30.
[210] Shi S, Han L, Gong T, Zhang Z, Sun X. Systemic delivery of microRNA-34a for cancer stem cell therapy. Angew Chem Int Ed Engl 2013;52(14):3901–5.

[211] Xu CF, Liu Y, Shen S, Zhu YH, Wang J. Targeting glucose uptake with siRNA-based nanomedicine for cancer therapy. Biomaterials 2015;51:1–11.
[212] Preusser M, Lim M, Hafler DA, Reardon DA, Sampson JH. Prospects of immune checkpoint modulators in the treatment of glioblastoma. Nat Rev Neurol 2015;11(9):504–14.
[213] Han S, Feng S, Xu L, Shi W, Wang X, Wang H, et al. Tim-3 on peripheral CD4(+) and CD8(+) T cells is involved in the development of glioma. DNA Cell Biol 2014;33(4):245–50.
[214] Bersanelli M, Buti S. From targeting the tumor to targeting the immune system: transversal challenges in oncology with the inhibition of the PD-1/PD-L1 axis. World J Clin Oncol 2017;8(1):37–53.
[215] Zeng J, See AP, Phallen J, Jackson CM, Belcaid Z, Ruzevick J, et al. Anti-PD-1 blockade and stereotactic radiation produce long-term survival in mice with intracranial gliomas. Int J Radiat Oncol Biol Phys 2013;86(2):343–9.
[216] Kim JE, Patel MA, Mangraviti A, Kim ES, Theodros D, Velarde E, et al. Combination therapy with anti-PD-1, anti-TIM-3, and focal radiation results in regression of murine gliomas. Clin Cancer Res 2017;23(1):124–36.
[217] Zhu Z, Gorman MJ, McKenzie LD, Chai JN, Hubert CG, Prager BC, et al. Zika virus has oncolytic activity against glioblastoma stem cells. J Exp Med 2017;214(10):2843–57.

CHAPTER

22

Understanding signal transduction pathways to overcome targeted therapy resistance in glioblastoma

Arata Tomiyama[a,b]

[a]Department of Neurosurgery, National Defense Medical College, Saitama, Japan [b]Department of Brain Disease Translational Research, Juntendo University Faculty of Medicine, Tokyo, Japan

Abstract

Glioblastoma (GBM) is the most common yet highly malignant brain tumor. Several surgical techniques and adjuvant therapies have made progressive advances in the treatment of GBM over the past few decades; however, the therapeutic outcomes remain poor because only limited tumor can be surgically resected to minimize damage to the surrounding normal brain tissue. Moreover, there is high resistance of the residual tumors against adjuvant chemotherapies. Therefore, the development of novel chemotherapeutic methods that are effective against residual refractory tumors is urgently needed for GBM treatment. To solve this problem, it is crucial to understand the molecular mechanisms underlying the chemoresistance in GBMs. Owing to the recent advances in high-throughput screening techniques for the genetic, epigenetic, and proteomic analyses of clinical GBM samples, many advances have been made to understand the intracellular molecular networks that regulate the chemoresistance of the GBMs. In fact, multiple candidates have been identified as molecular therapeutic targets for GBM chemotherapy. Thus, in this chapter, the recent knowledge about intracellular signaling networks that possibly contribute to the chemoresistance of GBMs is summarized.

Abbreviations

ABCB1	ATP-binding cassette subfamily B member 1
ABCG2	ATP-binding cassette subfamily G member 2
Akt	protein kinase B
APC	adenomatous polyposis coli
ATF6	activating transcription factor 6
ATM	ataxia-telangiectasia mutated
ATR	ataxia-telangiectasia and Rad3 related
Axin	Axin-1
AXL	Tyro3, Axl, and Mer (TAM) RTK family member AXL

Glioblastoma Resistance to Chemotherapy: Molecular Mechanisms and Innovative Reversal Strategies
https://doi.org/10.1016/B978-0-12-821567-8.00015-4

BBB	blood–brain barrier
BCRP	breast cancer resistance protein
BER	base-excision repair
BiP	binding immunoglobulin protein
BMI1	B lymphoma Mo-MLV insertion region 1 homolog
CBL	casitas B-lineage lymphoma
CCL2	chemokine (C—C motif) ligand 2
CCL20	chemokine (C—C motif) ligand 20
CCL5	chemokine (C—C motif) ligand 5
CCR6	chemokine receptor 6
CD133	cluster of differentiation 133
CD243	cluster of differentiation 243
CD44	cluster of differentiation 44
CDC25	cell division cycle 25C
CDK	cyclin-dependent kinase
CDKN2A	cyclin-dependent kinase inhibitor 2A
CHK1	checkpoint kinase 1
CHK2	checkpoint KINASE 2
CHOP	C/EBP HOMOLOGOUS PROTEIN
CK1	casein kinase 1
CNS	central nervous system
CSC	cancer stem-like cells
CXCL11	C-X-C motif chemokine 11
CXCL12	C-X-C motif chemokine 12
DDR	discoidin domain receptor
DMR	DNA damage response
DR5	death receptor 5
DSB	double-strand break
E2F	members of the transcription factor E2
ECM	extracellular matrix
EGFR	epidermal growth factor receptor
EMT	epithelial to mesenchymal transition
Eph receptor	erythropoietin-producing hepatocellular receptor
ERK1/2	mitogen-activated protein kinase 3/1
ERSR	endoplasmic reticulum stress responses
FGFR	fibroblast growth factor receptor
FZD	frizzled
GBM	glioblastoma
GDNF	glial cell line-derived neurotrophic factor
GDP	guanosine diphosphate
GP130	glycoprotein 130
GRB2	growth factor receptor-bound protein 2
GSK-3	glycogen synthase kinase-3
GSK-3B	glycogen synthase kinase 3 beta
GTP	guanosine triphosphate
HA	hyaluronic acid
HGF	hepatocyte growth factor
HGFR/c-MET	hepatocyte growth factor receptor
HH	hedgehog
HIFα	hypoxia-inducing factor α
IL6	interleukin-6
IRS-1	insulin receptor substrate 1
JAK	janus kinase

LATS1/2	large tumor suppressor 1/2
LTBP4	latent transforming growth factor beta-binding protein 4
MAPK	mitogen-activated protein kinase
Mcl-1	induced myeloid leukemia cell differentiation protein Mcl-1
MDM2	mouse double minute 2
MDM4	mouse double minute 4
MDR1	multidrug resistance protein 1
MEK1/2	mitogen-activated protein kinase 1/2
MGMT	O^6-methylguanine-DNA methyltransferase
MLH1	MutL homolog 1, colon cancer, nonpolyposis type 2 (*E. coli*)
MMR	mismatch repair
MSH2	MutS protein homolog 2
MSH6	MutS protein homolog 6
mTOR	mammalian target of rapamycin
mTORC1	mammalian target of rapamycin complex 1
NF1	neurofibromin 1
NF-κB	nuclear factor-kappa B
Notch1	notch homolog 1
Noxa	phorbol-12-myristate-13-acetate-induced protein 1
OXPHOS	oxidative phosphorylation
p21	cyclin-dependent kinase inhibitor 1
p53	tumor protein p53
PDGF	platelet-derived growth factor
PDGFRA	platelet-derived growth factor receptor A
PDK1	3-phosphoinositide-dependent protein kinase-1
PEA-15	phosphoprotein enriched in astrocytes 15
P-gp	permeability glycoprotein
PI3K	phosphoinositide-3 kinase
PIP_2	phosphatidylinositol-4,5-bisphosphate
PIP_3	phosphatidylinositol-3,4,5-triphosphate
PKCα	protein kinase Cα
PMS2	mismatch repair endonuclease
PP2A	Protein phosphatase 2A
Ptch1	patched-1
PTEN	phosphatase and tensin homolog
PUMA	p53 upregulated modulator of apoptosis
RAF	rapidly accelerated fibrosarcoma
RB1	retinoblastoma-associated protein
RHEB	Ras homolog enriched in brain
RICTOR	rapamycin-insensitive companion of mammalian target of rapamycin
RTK	receptor tyrosine kinase
SH2 domain	SRC homology 2 domain
SMAD2/3	cytosolic mothers against decapentaplegic homolog 2/3
Smo	smoothened
SNAIL	zinc finger protein SNAI1
SOS	son of sevenless
SOX2	SRY (sex-determining region Y)-box 2
SSB	single-strand break
STAT	signal transducer and activator of transcription
TAZ	WW domain-containing transcription regulator protein 1
TCGA	The Cancer Genome Atlas
TEAD	TEA domain family member
TGFβ	transforming growth factor beta

TGFβ1	transforming growth factor beta1
TGFβR	transforming growth factor beta receptor
TME	tumor microenvironment
TMZ	temozolomide
TSC1 (Hamartin)	tuberous sclerosis 1
TSC2 (Tuberin)	tuberous sclerosis 2 complex
TWIST1	twist-related protein 1
VEGF	vascular endothelial growth factor
WHO	World Health Organization
XIAP	X-linked inhibitor of apoptosis protein
YAP1	yes-associated-protein 1
ZEB1	zinc finger *E*-box-binding homeobox 1

Conflict of interest

No potential conflicts of interest were disclosed.

Introduction

Glioblastoma (GBM) is the most common intracranial malignancy arising from the brain parenchyma. In many cases, a surgical gross total resection is almost impossible to perform because of the highly infiltrative growth of the tumor and the necessity to minimize the damage to the surrounding normal brain tissues. Therefore, the response of chemoradiotherapy against the postoperative and inoperative cases is important to determine the prognosis in GBM cases; however, GBMs are highly refractory against these therapies, even with the current standard protocols [1, 2]. Because of the GBM's refractory nature, various molecular mechanisms that confer this resistance have been suggested, especially since the identification of the epidermal growth factor receptor (EGFR) amplification in the 1980s [3]. Further, the World Health Organization (WHO) has accumulated evidence regarding genetic alterations in GBM cases and has compiled them in the latest WHO 2016 tumor grading and molecular classification. The development of high-throughput genome-wide screening techniques has identified the involvement of gene mutations and epigenetic changes in the GBM cells and the molecular mechanisms involved in the refractory nature of the GBMs. Furthermore, recently developed analysis techniques, such as single-cell RNA sequencing, have revealed critical roles of extensive intratumoral heterogeneity and plasticity in the regulation of chemoresistance of the GBMs [4, 5]. Also, the molecular regulatory mechanisms of the tumor microenvironment (TME), GBM-associated cells, extracellular matrix (ECM), and tissue hypoxia and acidosis have also been implicated in conferring chemoresistance in GBMs [6–9]. Consequently, understanding the molecular signaling networks underlying GBM chemoresistance is of extreme importance in overcoming the refractory nature of these tumors.

Current status of chemotherapy in GBM treatment

Currently, the standard therapy against GBMs is the maximal surgical resection of the tumor, followed by irradiation and administration of the alkylating agent temozolomide (TMZ) [2]. Compared with other DNA damaging agents, TMZ possesses greater advantages, such as its oral availability, fewer side effects, good bioavailability because of the highest penetration across the blood–brain barrier (BBB), and best clinical outcome as a chemotherapeutic agent against the GBMs [10–12]. In addition to the standard therapy, other alkylating agents, such as nimustine hydrochloride, carmustine, and lomustine [13–15], and bevacizumab [an antagonistic antibody of the vascular endothelial growth factor (VEGF)] are also currently used in GBM chemotherapy [16, 17]. Furthermore, to improve the poor prognosis of patients with GBMs, numerous clinical trials have been conducted with newly developed chemotherapeutic agents; however, the results of most of them are still far from satisfactory [18–23]. In this chapter, the current knowledge about molecular signaling pathways that confer resistance against TMZ and other chemotherapeutic agents are described.

Molecular mechanisms of TMZ resistance in GBM cells

TMZ-induced DNA replication stress and double-strand breaks

The DNA damage response (DMR) is the most fundamental pathway in the chemotherapy-induced molecular machinery of the GBMs. TMZ treatment can trigger the DMR and cause cell cycle arrest or death of the GBMs. In normal cells, TMZ induces the methylation of adenine and guanine residues in the DNA (N^7-guanine, N^3-adenine, and O^6-guanine) followed by the base-excision repair (BER) and mismatch repair (MMR) of the DNA [10, 24]. Interestingly, the TMZ-mediated DNA O^6-methylation is the most important DNA modification in TMZ-induced cell death as it shows the highest toxicity, despite the lowest frequency of occurrence [10, 24]. The TMZ-mediated DNA modification of O^6-guanine causes abnormal pairing of this guanine with thymidine instead of cysteine; this is subsequently followed by the MMR system activation. However, the MMR system repeatedly removes only the newly synthesized strands, while the O^6-modified guanines are not affected. The MMR-mediated system continues its futile DNA repair cycle and finally induces a DNA replication folk stall, which subsequently triggers a single-strand break (SSB) followed by a double-strand break (DSB) of the DNA. This TMZ-induced DNA replication stress along with the DSBs finally activates the DMR in the cells to induce the cell cycle arrest or cell death of the GBMs [25, 26].

The O^6-methylguanine-DNA methyltransferase (MGMT), a DNA-repair enzyme encoded by the *MGMT* gene, is known to reverse this TMZ-induced DNA methylation of O^6 [27, 28]. In GBMs, the methylation of the *MGMT* promoter is frequently observed and the GBM cases with *MGMT* promoter methylation show good prognosis under treatment with TMZ when compared with *MGMT* unmethylated-promoter cases [29–31]. Therefore, methylation of the *MGMT* promoter possibly determines the prognosis in TMZ-treated GBMs. However, recent evidence has demonstrated that GBM cases with *MGMT* promoter methylation show a good prognosis even without TMZ treatment [32]. This suggests that the *MGMT* promoter methylation status may be a hallmark of a good prognosis of GBM cases, even without TMZ treatment.

Further, the downregulation of the MutS protein homolog 6 (MSH6) function is also considered critical in the molecular mechanism underlying TMZ resistance. A MutS protein homolog 2 (MSH2)-MSH6 heterodimer complex is first formed at a mismatched base that subsequently recruits the MLH1-PMS2 heterodimer to initiate the excision of the mismatched base and incorporate a new base [33]. In GBMs, the suppression of MSH6 expression is observed in the relapsed cases after primary TMZ-mediated therapy [34, 35]; moreover, a mutation of MSH6 contributes to the resistance against TMZ. In addition, the downregulation of *MSH2* expression in GBM cells results in the acquisition of resistance against TMZ treatment in vitro, and the lower *MSH2* expression in clinical GBM cases is reportedly correlated with a poor response in primary TMZ-mediated therapy [36]. These results reaffirm that the downregulation of MSH6 or MSH2 expression levels, and their functions, are critical for TMZ resistance in GBMs.

Recently, TMZ-induced N^7-methylation has also received attention as a regulatory factor of the therapeutic response of TMZ in GBM therapy. In MMR-downregulated TMZ-resistant GBMs, the inhibition of poly (ADP-ribose) polymerase 1 (a DNA-repair enzyme) restores the SSBs that were rapidly triggered by the TMZ-induced N^7-methylation-mediated BER pathway [37–40] and reportedly induces cell death of the GBMs effectively [41, 42]. In addition, the expression of alkylpurine-DNA-*N*-glycosylase, a TMZ-induced N^7- and N^3-modification repair enzyme, is reported as a potent prognostic marker of the GBMs as well [43]. These evidence suggest that the TMZ-induced N^7-, N^3-, and O^6-methylation modifications play significant roles in determining the response of TMZ in GBM therapy. The molecular networks underlying the TMZ-induced DSBs in GBMs are summarized in Fig. 1.

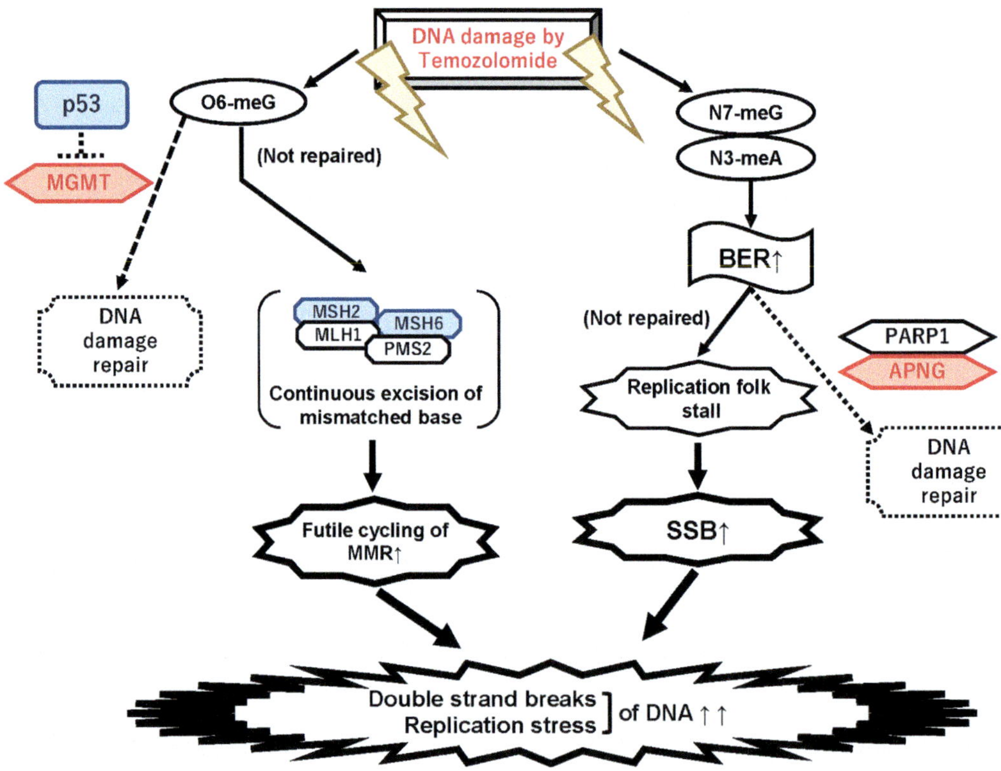

FIG. 1 Signal transduction by TMZ-induced DSBs and replication stress in the DNA. The solid lines and arrows indicate the signaling that promotes cell cycle arrest and/or cell death; whereas the dotted lines and arrows indicate the signaling that inhibits cell cycle arrest and/or cell death. The molecules whose expression and functions are upregulated and suppressed in the glioblastomas are marked with red and blue shading, respectively.

TMZ-induced DMR after replication stress and DSBs in the DNA

Ataxia-telangiectasia mutated (ATM) and ATM and RAD3-related-mediated (ATR) pathways

The TMZ-mediated replication stress and DSBs in the DNA stimulate the DMR activation. This process is orchestrated via a sequential activation of various molecules that trigger two pathways: the ataxia-telangiectasia mutated (ATM)-checkpoint kinase 2 (CHK2)-tumor protein p53 (p53)-mediated pathway (ATM-CHK2-p53) and ataxia-telangiectasia mutated RAD3-related (ATR)-checkpoint kinase 1(CHK1)-mediated pathway (ATR-CHK2 pathway) [44, 45]. After sensing the TMZ-triggered DNA damage and replication stress, both the cascades are activated and the transcription factor p53 (a master regulator of DMR) is stabilized and activated by ATM-, ATR-, CHK1-, and CHK2-mediated phosphorylation through various biological reactions [44–47]. It has been confirmed that point mutations of p53 frequently occur in GBM cases, which trigger the resistance against TMZ treatment (the detailed mechanism of p53 regulation in TMZ resistance in GBMs is given in "Regulation of p53 and its related molecules and pathways in TMZ-induced DMR in GBMs" Section) [48]. Further, CHK2 phosphorylates the protein tyrosine phosphatase, cell division cycle 25C (CDC25C); this induces the cell cycle arrest by a phosphorylation-mediated degradation of CDC25C [49, 50]. In addition, a dataset from The Cancer Genome Atlas (TCGA) demonstrated that heterozygous deletion of CHK2 is present in approximately 20% of all GBM cases [48], suggesting that CHK2 is a critical regulator of GBM cell proliferation during TMZ treatment.

In contrast, the TMZ-induced DNA replication stress triggers the activation of the ATR-CHK1 cascade [44, 51]. Importantly, a recent report demonstrated that TMZ strongly activates the ATR-CHK1-cyclin-dependent kinase 1 inhibitor (p21) cascade (ATR-CHK1-p21) in the GBM cells and this results in the arrest of cells in the G2/M phase. This further indicates that TMZ induces a persistent cell cycle arrest in the GBMs rather than causing their cell death via the hyperactivation of the cascade that results in growth suppression of the GBMs [52]. This may explain why TMZ possesses higher antitumor effects against the GBMs when compared with that of other anticancer drugs. These findings suggest that although the DMR-mediated signaling is complex and not as straightforward as mentioned above, the ATM- and/or ATR-dependent cascade activations may play pivotal roles in the TMZ resistance in GBMs. The molecular networks underlying ATR-, ATM-, and p53-mediated regulatory pathways of DMR by TMZ are summarized in Fig. 2.

Regulation of p53 and its related molecules and pathways in TMZ-induced DMR in GBMs

TMZ-induced ATM and ATR-dependent regulation of p53

Downstream of the ATM- and ATR-mediated DMR cascade induced by TMZ or other chemotherapeutic agents, the transcription factor p53 orchestrates various biological responses, such as autophagy, cell cycle arrest, cellular senescence, DNA repair, and cell death; these responses depend on its expression and activation levels. In its unstimulated state, the p53 protein expression is constantly suppressed by degradation caused by a continuous association with E3 ubiquitin ligase mouse double minute 2 (MDM2) at its N′-terminal transactivation domain [53]. Once DMR is activated by a drug treatment or other stimuli such

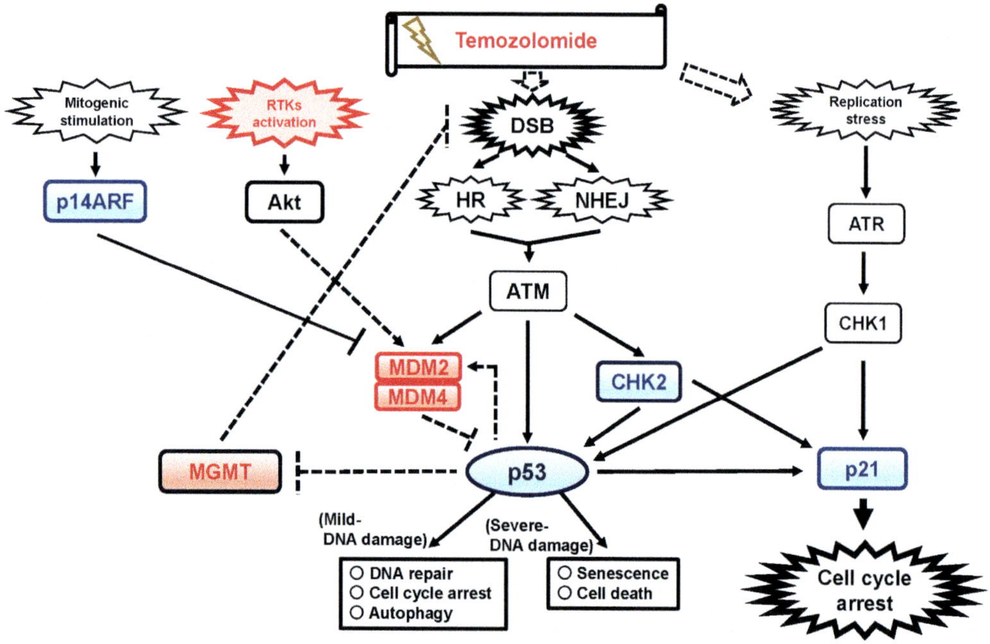

FIG. 2 Mechanism of temozolomide (TMZ)-induced double-strand breaks (DSBs) and replication stress in the DNA. The molecules whose expression and functions are upregulated and suppressed in the glioblastomas are marked with red and blue shading, respectively. The solid lines and arrows indicate the signaling that promotes DNA DSBs and/or replication stress; whereas the dotted lines and arrows indicate the signaling that inhibits DNA DSBs and/or replication stress.

as oxidative stress, irradiation, and ATM- and ATR-mediated phosphorylation, p53 is stabilized by dissociating from MDM2 and activated by forming tetramers in its active form; subsequently, it binds to the promoter regions of its target molecules [54]. A TCGA dataset revealed that one of the most frequently mutated genes in GBMs is *TP53* (p53-encoding gene) [48]. In particular, heterozygous point mutations of *TP53* are observed in about 30% of all GBM cases at multiple sites of its molecular domains according to the COSMIC database (https://cancer.sanger.ac.uk/cosmic). These are reported as a loss of function mutants of *TP53*; however, antagonism of wild-type p53, produced by a nonmutated allele during tetramer formation or gain of function by foaming beta sheet-like structure that evades proteolysis and forms intracellular aggregation, has also been reported as an additional oncogenic function of mutant p53 [55]. Therefore, *TP53* mutations in GBMs may serve as more powerful oncogenes when compared with the simple loss of function owing to mutation of a single allele.

TMZ-induced MDM2-dependent regulation of p53

MDM2 is an E3 ubiquitin ligase whose expression is positively regulated by phosphoinositide-3 kinase (PI3K)-protein kinase B (PKB-Akt) signaling [56]. It functions downstream of the plasma membrane-localized receptor tyrosine kinase (RTK)-mediated cell survival molecular cascade; this cascade is known to be aberrantly activated in the GBM cells (see "Chemoresistance of GBM induced by plasma membrane-localizing receptor-mediated survival signaling" Section). As described in "TMZ-induced ATM and ATR-dependent regulation of p53" Section, MDM2 associates with and antagonizes p53 by a ubiquitination-mediated p53 degradation [53]. Also, an amplification of MDM2 was discovered in GBM cases, especially in the wild-type p53 harboring cases, where it induced TMZ resistance by antagonizing p53 expression [57]. Importantly, MDM2 expression is upregulated by p53, which in turn blocks the expression and functions of p53 [58, 59]. Furthermore, MDM2 also possesses other functions that aid in cellular proliferation and survival [60, 61]. In addition, MDM4, another p53 regulator, which is structurally similar to MDM2 and lacks ubiquitin ligase activity, is also highly expressed in GBM tissues, interacts with the DNA binding domain of p53, and inhibits the association between p53 and its target promoter; this results in the suppression of p53 functions [48, 62, 63]. Lastly, MDM4 also forms a heterodimer with MDM2 that contributes toward the stable expression of MDM2 and degradation of MDM4 [53]. The molecular networks regarding TMZ-induced MDM2-MDM4-dependent regulation of p53 in GBMs are summarized in Fig. 2.

Regulation of TMZ-induced intrinsic and extrinsic cell death signaling downstream of p53

In general, the activation of p53 by DMR possibly induces tumor suppression by triggering cell cycle arrest and/or cell death [64, 65]. Therefore, inactivation of p53 by a *TP53* mutation is considered to protect the cancer cells from an anticancer agent-induced DMR-mediated cell cycle arrest or cell death. p53 activation causes cell death via a mitochondria-dependent intrinsic cascade and death receptor-mediated extrinsic pathway [64, 66]. A p53-dependent transcriptional upregulation of proapoptotic Bcl-2 family proteins, such as Bcl-2 homologous antagonist/killer, Bcl-2-associated X protein, phorbol-12-myristate-13-acetate-induced protein 1 (Noxa), and p53 upregulated modulator of apoptosis (PUMA) leads to the binding of p53 to their gene promoters or causes a p53-dependent activation of c-Jun N′-terminus kinase; consequently, this triggers the intrinsic mitochondria-dependent programmed cell death [64, 66]. Meanwhile, p53-dependent transcriptional upregulation of plasma membrane-localizing "death" receptors, such as Fas receptor, death receptor 5 (DR5), and p53 apoptosis effector related to PMP-22, contributes to a rapid extrinsic programmed cell death [64]. Both intrinsic and extrinsic programmed cell death cascades induce the caspase-dependent apoptosis and caspase-independent cell death, effectively. Consequently, a p53-dependent programmed cell death induced by TMZ is triggered by these multiple downstream signaling pathways. Further, cellular X-linked inhibitor of apoptosis protein (XIAP), a key regulator of both intrinsic and extrinsic programmed cell death signaling, suppresses the activation of caspase-3, -7, and -9 by triggering their ubiquitination-mediated degradation. In GBMs, an upregulated expression of XIAP is reportedly associated with cases that show poor prognosis [67, 68]. Further, the *XIAP* mRNA is stabilized by MDM2, its expression is induced by the RTK-PI3K-Akt pathway-mediated nuclear factor-kappa B (NF-κ B) [69], and MDM2 expression is induced by Akt [56]. Hence, these evidence suggests that TMZ-induced p53-dependent intrinsic/extrinsic cell death signaling in GBM cells is regulated mainly at the postmitochondrial level via a complicated Akt-, MDM2-, NF-κB-, and

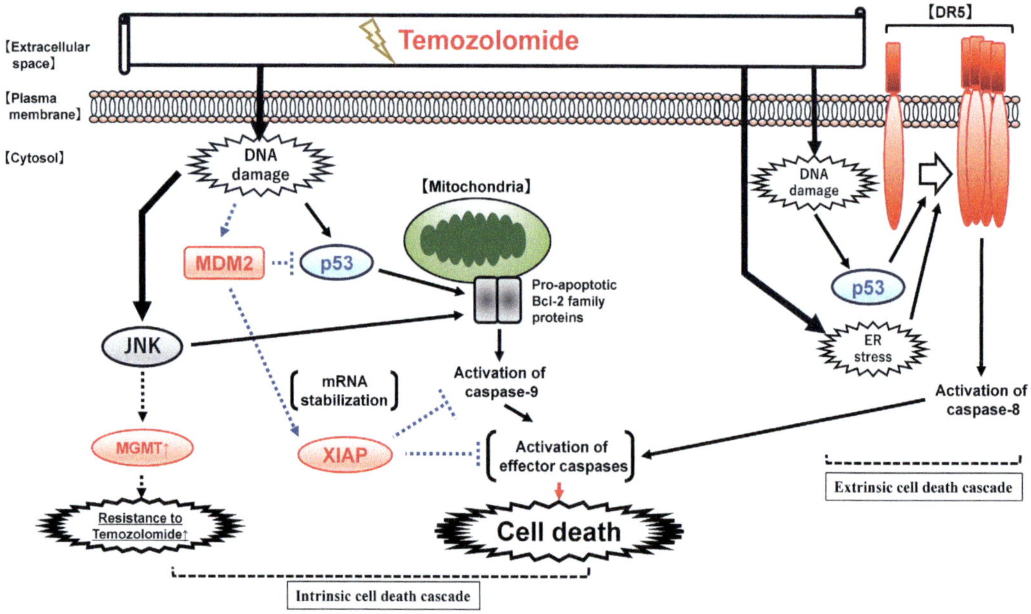

FIG. 3 The molecular machinery underlying the TMZ-induced cell death via intrinsic (mitochondria-mediated) and extrinsic (plasma membrane-localizing death receptor-triggered) cell death pathways. The molecules whose expression and/or functions are upregulated and suppressed in the glioblastomas are marked with red and blue shading, respectively. The solid lines and arrows indicate the signaling that promotes cell death; whereas the dotted lines and arrows indicate the signaling that inhibits cell death.

XIAP-mediated machinery. The molecular networks underlying p53-induced cell death are summarized in Fig. 3.

TMZ-induced regulation of p21 downstream of p53

The expression of p21, which induces cell cycle arrest and other biological responses, is upregulated by p53-dependent transcription [70]. p21 interacts with cyclin-dependent kinases (CDKs) to inhibit the cyclin/CDK interaction and induce G2 cell cycle arrest [70]. In fact, a lack of p21 upregulation was reportedly associated with a poor response to TMZ treatment in GBM cases [71]. On the other hand, an augmented expression of p21, which induces constitutive cell cycle arrest after TMZ treatment, could induce an effective growth suppression of the GBMs [52]. These evidence suggest that p53-dependent induction of p21 expression after TMZ treatment may be an essential therapeutic response and thus warrants further investigations.

TMZ-mediated regulation of MGMT expression downstream of p53

The key role of p53 in TMZ resistance in GBM cells at the epigenetic level is also known. MGMT expression is negatively regulated by p53 in GBM cells [72]. This implies that the inactivation of p53 induces MGMT expression and its downregulation contributes to TMZ resistance (Figs. 1 and 2).

TMZ-induced endoplasmic reticulum (ER) stress responses (ERSR) in GBMs

TMZ also reportedly induces ERSR in GBM cells [73, 74]. ERSR are generally triggered by the accumulation of misfolded proteins in the ER or by disruption of the intracellular calcium homeostasis. As a result, they activate various intracellular signaling pathways that consequently induce cell death and cell survival signaling [75, 76]. In the GBM cells, TMZ-induced upregulated expression levels of binding immunoglobulin protein (BiP, also known as GRP78), a representative pro-survival chaperone induced by ERSR, and C/EBP homologous protein (CHOP), a representative cell death mediator by ERSR, have been reported [74, 77]. Importantly, CHOP induces the upregulation of DR5 and triggers cell death via the extrinsic cell death cascade (Fig. 3). This suggests that TMZ can trigger the extrinsic as well as intrinsic cell death cascades [73]. Notably, the activation of activating transcription factor 6 (ATF6), an upstream regulator of BiP and CHOP, is specifically confirmed in GBM cells as a master regulator of ERSR [78]. In addition, the inhibition of BiP expression simultaneously with TMZ treatment results in the enhancement of TMZ toxicity in GBM cells [74, 79]. Furthermore, the TMZ-induced ERSR-dependent autophagy, which is considered to protect against the TMZ-induced GBM cell toxicity, has also been reported (see "TMZ-mediated regulation of autophagy in GBMs" Section) [73]. Thus, these insights indicate that TMZ-mediated ERSR may be responsible for TMZ resistance in GBM cells to a certain extent.

TMZ-mediated regulation of autophagy in GBMs

Autophagy is a stress response that is generally triggered by nutrient scarcity and various cytotoxic stresses [80, 81]. Consequently, the cells undergo autophagy and decompose their intracellular molecules via characteristic intracellular vesicles called autophagosomes and autolysosomes; this is followed by the recycling of the obtained amino acids for the synthesis of new intracellular molecules [80, 81]. While autophagy contributes toward cell survival during cellular stresses, in some situations it reportedly induces cell death as well [80, 81]. Chemotherapy-induced autophagy is classified into two groups: triggered by p53-dependent and p53-independent (e.g., ERSR) mechanisms. For instance, in GBMs treated with TMZ, the autophagy-inducing molecular signaling downstream of p53 is activated and plays critical roles in TMZ-induced GBM therapy [80]; whereas, TMZ-induced ERSR can also trigger autophagy via the activation of multiple cascades [73]. It is noteworthy that inhibition of autophagy in TMZ-treated GBM cells has resulted in the enhancement of TMZ-induced cell death [73, 82, 83]. Therefore, TMZ-mediated autophagy may also play protective roles in achieving TMZ-induced cell death.

Chemoresistance of GBM induced by plasma membrane-localizing receptor-mediated survival signaling

RTK and their related signaling-mediated chemoresistance

RTK

The RTK-mediated signaling is one of the fundamental intracellular signaling pathways responsible for the maintenance of cellular migration, invasion, proliferation, and survival.

In fact, the upregulated activation of RTK signaling is reportedly implicated in the oncogenesis of various cancers [84–86]. In general, the RTK ligands associate with the N′-terminal extracellular domain of their corresponding RTKs to form a self-dimer; this is followed by the binding of cytosolic adapter molecules to trigger a sequential activation of RTK downstream signaling [84–86]. On the other hand, the activation mutations of RTKs trigger self-dimerization or heterodimerization with other RTKs without association with their corresponding ligands; this results in a ligand-independent RTK activation and hypersensitivity to their ligands [84–86]. In addition, such hyperactivation of RTK-related signaling is known to trigger chemoresistance [84–86]. The GBMs are no exception to this as the first identified oncogenic mutation in them was the amplification of EGFR [3]. Subsequently, the mutational activation of other RTKs and downstream molecules has also been identified and highlighted as novel molecular targets in GBM therapy [84, 86]. Further, point mutations and oncogenic fusions of the RTKs have also been confirmed in GBMs by recent advances made in genome-wide screenings. Moreover, the downstream signaling of RTKs is associated with multiple cross talk signaling pathways and thus is not unidirectional because of negative feedback.

Chemoresistance induced by oncogenic activation of the RTKs

The amplification of N′-terminal (extracellular domain) truncation mutation (EGFRvIII) in EGFR is well known; that is, it is characterized by a lack of its extracellular domain [3, 87]. This EGFR mutant triggers the escape from activation-induced degradation via an impaired association with the E3-ubiquitin ligase, casitas B-lineage lymphoma, as well as through ligand-independent activation [88, 89]. In addition, TCGA has revealed multiple point mutations in the EGFR extracellular domain in GBM cases [48, 90]; however, the mechanisms underlying the pathogenesis of these EGFR point mutations are still not well elucidated. Another highly mutated RTK in GBM cases is the platelet-derived growth factor receptor A (PDGFRA). Approximately 10%–13% of GBM cases harbor this amplification (confirmed in TCGA), [91, 92] and 40% of the amplified PDGFRA are reported as PDGFRA$^{\Delta 8,\ 9}$, which is a deletion mutant of PDGFRA that lacks a part of its extracellular domain [93]. Similar to EGFRvIII, the PDGFRA$^{\Delta 8,\ 9}$ mutant triggers a ligand-independent activation and also possesses a higher transformation capacity [93]. Importantly, the oncogenic hyperactivation of the PDGFRA mutant facilitates the extracellular secretion of platelet-derived growth factor (PDGF), a ligand for PDGFRA; this forms an abnormal PDGFRA activation-PDGF secretion loop that further enhances the oncogenic capacity of the PDGFRA mutants, including their cellular survival and chemoresistance [91, 92]. Additionally, other hyperactivated RTKs associated with GBMs have been recently discovered. Their amplified forms occur with lower frequencies than that of EGFR and PDGFRA and include hepatocyte growth factor receptor (HGFR/c-MET); fibroblast growth factor receptor (FGFR) families; and Tyro-3, Axl, and Mer (TAM) RTK family member AXL (AXL) [48, 90, 94, 95]. Further, both GBM cells and intratumoral vascular endothelial cells express HGFR/c-MET in the GBM tissue; the autocrine loop of HGF-HGFR/c-MET is formed between these cells to induce further oncogenicity of the GBMs [96]. Recently, the upregulation of erythropoietin-producing hepatocellular receptor (Eph receptors) expression has also been reported in GBMs. The Eph receptors and Eph family receptor-interacting proteins (ephrins; ligands of Eph receptors) consist of multiple members: nine EphA receptors (EphA1–8 and Epha10), five EphB receptors (EphB1–4 and EphB6), five ephrin-A ligands, and three ephrin-B ligands. Some Eph receptors, such as EphA2, EphA3, and EphB2, are associated with poor prognosis in GBM cases and with undifferentiated phenotypes of the GBM stem cells. It is also suggested that these Eph receptors are involved in the regulation of chemoresistance of the GBMs. Another group of RTKs, the discoidin domain receptor (DDR) family, contributes to the oncogenic properties of the GBMs. In contrast to the other RTK families, the DDRs are generally activated by associating with ECM molecules, such as collagen, at the N′-terminal extracellular domains to conduct RTK signaling. Among the DDRs, DDR1 expression is elevated in GBM cases as per TCGA data; thus, DDR1 may participate in the modulation of GBM oncogenicity as well [97]. Moreover, less frequent recurrent oncogenic fusions of RTKs, such as EGFR, FGFRs, HGFR/c-MET, and neurotrophic tyrosine kinase receptor, have also been recognized as oncogenic in GBM cases. Although multiple RTKs with augmented oncogenic activation have been discovered in GBMs, most of the clinical trials targeting these RTKs have failed. This is probably because the downstream signaling of RTKs is common and not unidirectional because of negative feedback. Lastly, these downstream signaling pathways are classified into three pathways: the Ras-PI3K-AKT, Ras-MAPK, and JAK–STAT pathways. Various studies on these signaling pathways are ongoing and are further elaborated on in this section.

RTK-Ras-MAPK signaling

The RTK-Ras-MAPK cascade is one of the most essential molecular pathways involved in the maintenance of cellular functions such as differentiation, migration, invasion, proliferation, and survival. This cascade is initially triggered by the association of adapter proteins, such as the growth factor receptor-bound protein 2 (GRB2) or insulin receptor substrate 1 (IRS-1), to activate the RTKs; this is followed by the activation of a Ras-GTPase son of sevenless (SOS), which subsequently converts the GDP-bound inactive small GTPase Ras subfamilies (H-Ras, K-Ras, and N-Ras) to GTP-bound active forms [98]. The activated Ras subfamilies subsequently induce the dimerization and phosphorylation (activation) of the serine/threonine kinase known as rapidly accelerated fibrosarcoma (RAF; A-Raf, B-Raf, and C-Raf) by interacting with its N-terminal regulatory domain. Upon activation, RAF phosphorylates (hence activates) the downstream dual-specificity, serine/threonine kinases, mitogen-activated protein kinase 1/2 (MEK1/2); these subsequently phosphorylate and activate the mitogen-activated protein kinase 3/1 (ERK1/2) [99, 100]. Ultimately, the activated ERK1/2 activates the transcription factors by phosphorylation and shuttles into the nucleus from the cytosol to regulate important cellular functions by regulating the gene expression [101].

The chemoresistance induced by the oncogenic activation of the RTK-Ras-MAPK signaling

The oncogenic hyperactivation of RTK-Ras-MAPK cascade has been observed in many cancers [102]. In GBMs, the oncogenic *BRAF* gene (encoding B-Raf) point mutation, BRAF V600E, or the KIAA1549-BRAF fusion (the products are constitutively activated without association with Ras) are discovered at a relatively high rate (~20%) in pediatric GBMs and rarely in the adult ones (~1%–3%) [103–109]. On the other hand, neurofibromin 1 (NF1), a GTPase-activating protein that negatively regulates Ras, is highly mutated and thereby negatively regulated in the GBMs (approximately 20%) [48, 90]. Thus, the mutated NF1 in GBMs activates the Ras-MAPK pathway indirectly and this activated Ras-MAPK cascade accelerates GBM oncogenicity, survival, and chemoresistance [48, 90, 110, 111]. In malignant melanomas, a small molecule inhibitor of B-Raf-based chemotherapy, targeting BRAF V600E harboring tumors, has already been initiated; however, resistance against these B-Raf inhibitors has been pointed out. In these cases, the resistance is possibly provoked by an enhanced ERK1/2 activation via the disruption of multiple negative feedback loops, such as the ERK1/2-EGFR, ERK1/2-BRAF, ERK1/2-MEK1, and ERK1/2-dual-specificity phosphatases [112–114]. A cotreatment with ERK1/2 inhibitor is reported to be effective in overcoming this resistance [113, 114]. The molecular networks underlying the RTK-Ras-MAPK signaling in the GBMs are summarized in Fig. 4.

RTK-PI3K-Akt–mTORC1 (RTK-PKC-mTORC1) and RTK-PI3K-mTORC2 signaling pathways

The RTK-PI3K-Akt–mTOR and RTK-PKC-mTOR pathways also play crucial roles in the regulation of cellular biology. The activated RTKs phosphorylate (thus activate) PI3K by a sequential association with adapter or scaffold molecules (GRB2-SOS-Ras cascade); the activated-PI3K further phosphorylates and converts phosphatidylinositol-4,5-bisphosphate (PIP_2) to phosphatidylinositol-3,4,5-triphosphate (PIP_3) localized on the medial surface of the plasma membrane [115]. Then, PIP_3 recruits serine/threonine kinase 3-phosphoinositide-dependent protein kinase-1 (PDK1) and serine/threonine kinase Akt to the plasma membrane, resulting in a phosphorylation-mediated Akt activation by PDK1 [116]. Next, the activated Akt phosphorylates and inactivates tuberous sclerosis 1 (TSC1, hamartin)-tuberous sclerosis 2 (TSC2, tuberin) complex; this finally activates the mammalian target of rapamycin complex 1 (mTORC1) by inhibiting the GTP-binding protein Ras homolog enriched in brain (RHEB)-mediated suppression of mTORC1 [116, 117]. This complex is widely recognized as a master regulator of protein and lipid synthesis. Subsequently, these events result in the regulation of cell survival and intracellular energy metabolism/redox state. Another fundamental pathway regulated by the RTK-PI3K pathway is the RTK-PI3K-mTORC2 cascade [118, 119]. mTORC2 binds to ribosomes in a PIP_3 (PI3K)-dependent manner upon activation of the RTKs and activates AGC-subfamily kinases (Akt, serum and glucocorticoid-induced kinases, and PKCα), thereby resulting in cell survival and migration [118]. Furthermore, it is particularly noteworthy that mTOC2 facilitates glucose and lipid metabolisms by regulating the transcription factors c-Myc and sterol regulatory element-binding protein in glioblastomas [118, 119]. During cross talk with mTORC1, p70S6K (a downstream target of mTORC1) phosphorylates rapamycin-insensitive companion of mammalian target of rapamycin, i.e., RICTOR, an mTORC2 component, and suppresses its function [118–120]. In addition, a recent report revealed that RTKs can activate mTORC1 via the PKCα-dependent alternative pathway without activating the PI3K-Akt signaling [121]. In fact, p70S6K regulates IRS-1 by phosphorylation and thereby suppresses RTK signaling immediately after RTK activation [119, 122]. Collectively, these findings imply that RTK-mediated mTORC1 and mTORC2-related intracellular signaling pathways are regulated by a highly complex mechanism that includes a bypass or negative feedback loop.

Chemoresistance induced by oncogenic activation of the RTK-Akt–mTORC1 (RTK-PKC-mTORC1) and RTK-PI3K-mTORC2 signaling

In the RTK-PI3K-Akt-mTOR pathway, TCGA data of GBM cases have demonstrated that 18.3% of GBM exhibit somatic mutation of *PIK3CA* and *PIK3R1* genes, which encode the p85α

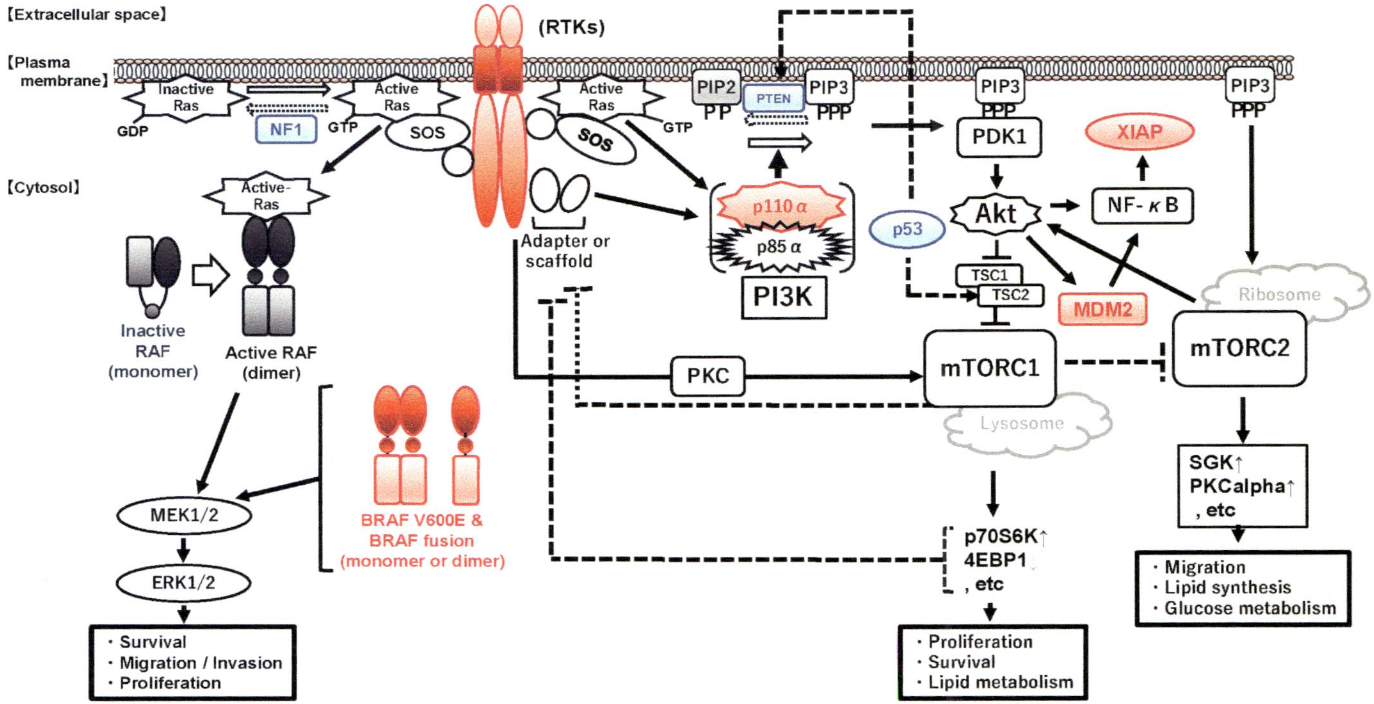

FIG. 4 The regulation of signal transduction associated with the receptor-type kinase (RTK)-Ras–Raf–MEK–ERK pathway and RTK-PI3K-Akt–mTOR (PI3K-PKC-mTOR) pathways. "P" labeling is attached to molecules that are phosphorylated as a result of signal transduction. The molecules whose expression and functions are upregulated and suppressed in the gliomas are marked with red and blue shading, respectively. The solid lines and arrows indicate signaling that promotes cell proliferation, migration, invasion, and survival; whereas the dotted lines and arrows indicate signaling that inhibits cell proliferation, migration, invasion, and survival.

subunit and the p110α subunit of PI3K, simultaneously. [48, 90] According to the COSMIC database, most *PIK3CA and PIK3R1* mutations in the GBMs are frequently discovered in the domain necessary for the interaction of the p85α subunit with the p110α subunit. It has been suggested that inhibition of p110α subunit activity by the p85α subunit is lost and the PI3K mutants become constitutively active, thereby resulting in sustained Akt activation. [48, 90, 123] In addition, phosphatase and tensin homolog (PTEN), a dual-specificity phosphatase that dephosphorylates PIP_3 and negatively regulates PI3K function, is frequently deleted in gliomas along with a loss of heterozygosity in chromosome 10, which is occurred mutually exclusive of *PIK3CA* and *PIK3R1* mutations. [48, 90, 124, 125] The loss of PTEN induces an enhanced production of PIP_3 and triggers enhancement of Akt activation; this also contributes to the elevation of GBM malignancy. Another phosphatase, protein serine/threonine phosphatase protein phosphatase 2 (PP2A), widely dephosphorylates serine/threonine kinases including ERK1/2 and Akt; importantly, Akt positively regulates PP2A activity [49, 126]. This implies that inhibition of Akt or its related molecules may result in rather an acceleration of ERK1/2 activity, and thereby induce the elevation of survival and chemoresistance. Thus, the protein phosphatases play key roles in the regulation of RTK-mediated chemoresistance. In the RTK-PKC-mTOR pathway, a recurrent oncogenic activation of PKCα has not been confirmed in GBMs [48, 90]. However, the RTK-PKC-mTOR signaling potentially bypasses the RTK-PI3K-mediated mTORC1 pathway and conducts RTK activation even when treated with PI3K or Akt inhibitors; thus, it contributes to chemoresistance even without its oncogenic activation. The molecular networks underlying the RTK-mTORC1 and RTK-mTORC2 signaling pathways in GBMs are summarized in Fig. 4.

RTK-JAK–STAT signaling

The cytosolic tyrosine kinase Janus kinase (JAK)-family proteins (JAK1/2) and transcription factor signal transducer and activator of transcription (STAT)-family proteins (STAT1–5/6A/6B) mediate the signaling axis that is activated downstream of the RTKs, especially downstream of cytokine receptors (a type of RTKs). [127, 128] The JAKs constitutively binds to the cytosolic domain of the RTKs and upon RTK activation, they become phosphorylated (activated) by their tyrosine residues and then phosphorylate specific cytosolic domains of the RTKs [127, 128]. Next, the STATs are tyrosine-phosphorylated by the activated JAKs and released into the cytosol. The released STATs form homodimers or heterodimers with other STATs via their SRC homology 2 domain (SH2 domain). Then, the dimerized STATs are further phosphorylated at their serine residue by ERK1/2 or PKCs and at tyrosine residues by cytosolic tyrosine kinase c-SRC and are shuttled into the nucleus [127–130]. The STATs eventually bind to the promoter region of the target genes and initiate transcription, resulting in cell migration, proliferation, survival, angiogenesis, and immunosuppression [127, 128, 131, 132]. The downstream signaling of STAT3 has been recently identified as a critical regulator of tumor biology; the master regulators of epithelial to mesenchymal transition [EMT; zinc finger *E*-box binding homeobox 1 (ZEB1), zinc finger protein SNAI1 (SNAIL), twist-related protein 1 (TWIST1)], antiapoptotic Bcl-2 family proteins, other apoptosis inhibitor proteins (B-cell lymphoma-extra large), induced myeloid leukemia cell differentiation protein Mcl-1 [(Mcl-1), survivin], and cancer stem-like cell markers [cluster of differentiation 133 (CD133), Kruppel-like factor 4], which potentially induce chemoresistance, are all transcriptionally upregulated by STAT3 [133, 134]. Hence, the RTK-JAK–STAT-mediated cascade constitutes complicated cross talk with other types of RTK-mediated signaling. Moreover, the activated-JAKs are also capable of activating the Ras-mediated pathway.

Chemoresistance induced by oncogenic activation of the RTK-JAK–STAT signaling

Although activation mutations of JAKs have been discovered in several cancers (especially in hematological malignancies), the genetic alteration-mediated enhanced activation of the JAK–STAT pathway is rare in GBMs [48, 135]. Nevertheless, the augmented activation of the JAK–STAT cascade by upstream hyperactivated RTKs reportedly contributes to the gain of chemoresistance in GBMs [136]. In GBMs with acquired TMZ resistance, an elevated activation of STAT3 has also been observed. Importantly, the inhibition of STAT3 activation resulted in the resensitization of the GBMs to chemotherapies [137]. Hence, the RTK-JAK–STAT pathway also contributes to the chemoresistance of the GBMs.

Integrin signaling-mediated chemoresistance of GBMs

Integrin signaling

Integrins are plasma membrane-localizing transmembrane receptors that regulate cell adhesion to the ECM and transduce intracellular signaling. Integrins consist of a heterodimer of α and β subunits (a total of 18 α and total 10 of β subunits); currently, 24 integrins, with different combinations of these subunits, exist [138, 139]. The integrins interact with ECM molecules, such as collagen, fibronectin, laminin, periostin, tenascin, thrombospondin, and vitronectin to form focal adhesions by clustering and to conduct intracellular signaling [138, 139]. Further, the integrins activate the PI3K-Akt–mTOR and Ras-MAPK signaling pathways either by themselves or by enhancing the RTK-mediated signaling. The integrins also specifically activate the intracellular signaling associated with cytoskeletal movements, for example, Rho-mediated signaling [138, 139]. Hence, the integrins are considered to induce tumorigenesis, invasion, and angiogenesis in multiple cancer systems because of the activation of the abovementioned molecular machinery.

Chemoresistance induced by oncogenic activation of integrin signaling

As mentioned above, integrins can activate the PI3K-Akt and Ras-MAPK pathways, both of which are downstream of RTK signaling and are essential for the survival of the GBM cells. Therefore, upregulation of integrin expression or activity may contribute to the chemoresistance of the GBMs. In fact, the integrins α2β1, α3β1, and α5β1 were reportedly overexpressed in multidrug resistant GBM cells [140]. In particular, integrin α5β1-Akt-PEA-15-mediated survivin-integrin α5β1 axis upregulation via p53 suppression has also been reported as essential for conferring chemoresistance [141]. In addition, a single-cell analysis of GBM samples clarified that integrin αv upregulation and integrin α7 downregulation were observed in TMZ-resistant clones and not in TMZ-sensitive clones [142]. Moreover, the tissue hypoxia-induced transforming growth factor-beta (TGFβ) signaling activation (see "TGFβ–transforming growth factor beta receptor (TGFβR) signaling-mediated chemoresistance of GBMs" Section) triggers the production of ECM periostin from the GBM and GBM-associated cells to activate the integrins; this induces the upregulation of GBM malignancy and tissue angiogenesis, which in turn contribute to the formation of GBM tissues and resistance against antiangiogenic therapy [143, 144]. These results suggest that integrins play essential roles in forming a TME, which are critical in determining therapeutic responses, and their upregulation may confer GBM chemoresistance as well.

TGFβ-transforming growth factor-beta receptor (TGFβR) signaling-mediated chemoresistance of GBMs

TGFβ-TGFβR signaling pathway

TGFβR is a receptor-type serine/threonine kinase that activates TGFβR-specific canonical and noncanonical signaling by associating their extracellular domains with the active forms of their ligand TGFβ (TGFβ1/2/3). In the canonical signaling pathway, by the association of TGFβ with TGFβRII, TGFβRII, and TGFβRI form a tetramer; this is followed by phosphorylation (activation) of both the receptors. The activated receptors phosphorylate cytosolic decapentaplegic homolog (SMAD) 2/3 proteins; subsequently, these SMAD2/3 proteins form a complex with SMAD4 [145–147]. This protein complex is translocated to the nucleus and acts as a transcriptional repressor with other cofactors to induce multiple cellular responses that regulate crucial cellular functions [145–147]. Importantly, SMAD2/3 proteins are also phosphorylated by MAPKs downstream of the RTKs (noncanonical TGF signaling) [148]. This means that the canonical and noncanonical TGFβ signaling pathways construct an intricate web with each other. The representative and most important cellular response caused by the activation of TGFβ-mediated canonical signaling is the EMT [149]. EMT causes changes in the cellular phenotype to achieve higher motility, lower intercellular adhesion, and lower cell division by transcriptional activation of distinct transcription factors, such as ZEB1, SNAIL, and TWIST1 [149].

Chemoresistance induced by oncogenic activation of TGFβ-TGFβR signaling

EMT also reportedly contributes to the acquisition of chemoresistance in GBMs. As noted above, TGF-mediated signaling is not only activated by TGFβR-mediated canonical signaling but also by the RTK noncanonical pathway. Therefore, in tumor cells with oncogenic activation of the RTKs, such as in GBMs, the RTKs can potentially enhance the expression of TGFβR-mediated canonical signaling-specific genes. In addition, frequent activation mutations of the latent transforming growth factor beta-binding protein 4 (*LTBP4*) gene (encodes the LTBP4 protein), which assists in the maturation of TGFβ and positively regulates TGFβ-mediated signaling, were observed in the relapsed GBM cases (after standard therapy) and not in the untreated cases [150]. The *LTBP4* mutations were correlated with higher LTBP4 expression levels; this indicates that LTBP4 induces the augmented activation of TGFβ signaling [150]. In addition, the involvement of TGFβ1-related signaling in upregulating the malignant behavior in relapsed GBM cases has also been reported [151–153]. Further, the interaction of the hypoxia-TGFβ1 axis with integrin signaling and periostin-dependent gain of GBM malignancy and chemoresistance has also been reported [143, 144]. Thus, TGFβ1-mediated signaling is considered to be hyperactivated in GBMs by multiple mechanisms and this could also trigger the chemoresistance of the GBMs.

Wnt signaling-mediated chemoresistance of GBMs

Wnt signaling pathway

The Wnt signaling pathway regulates complex signaling cascades that are associated with crucial cellular functions such as development, differentiation, migration, and proliferation [154, 155]. This signaling pathway is broadly classified into two: the canonical (β-catenin-dependent) and noncanonical (β-catenin-independent) Wnt pathways [154, 155]. In the canonical Wnt pathway, β-catenin, an essential protagonist of Wnt signaling, forms a protein complex composed of Axin-1 (Axin), casein kinase 1 (CK1), glycogen synthase kinase-3 (GSK-3), and adenomatous polyposis coli (APC) in its normal unstimulated state [154, 155]. In this state, β-catenin is continuously phosphorylated and undergoes proteasomal degradation, which results in its expressional suppression [154, 155]. However, when a Wnt ligand associates with Frizzled (FZD) receptor and low-density lipoprotein receptor-related protein (LRP) families, the CK1-GSK-3B-APC-β-catenin complex binds to the cytosolic domains of FZD and LRP; then, β-catenin dissociates from this complex to escape degradation [154, 155]. This stabilized β-catenin then shuttles into the nucleus and functions as a coactivator for the transcription factors of the T cell factor/lymphoid enhancer factor-1 family [154, 155].

Chemoresistance induced by oncogenic activation of Wnt signaling

Wnt signaling is also involved in the chemoresistance of cancer cells. An expressional upregulation of the drug efflux pump ABCB1 (induces the secretion of chemotherapeutic agents from cells) and stem cell marker cluster of differentiation 44 (CD44) due to hyperactivation of the canonical Wnt pathway was discovered in several cancers [156]. The CD44-mediated signaling transcriptionally upregulates Notch homolog 1 (Notch1) expression; thus a Wnt-CD44 mediated Notch1 upregulation may also participate in the Wnt-mediated chemoresistance. The overexpression of Frizzled-2 and downregulation of transcription factor LEF1 (Wnt pathway antagonist) were discovered in GBM cells with acquired TMZ resistance [157]. In addition, among the four molecular subtypes of glioblastoma (proneural, neural, classical, and mesenchymal; classified in TCGA), the expression of TCF4 and SRY (sex-determining region Y)-box 2 (SOX2), the activators of canonical Wnt pathway, were upregulated in the proneural type with the poorest prognosis [158]. Furthermore, as Akt phosphorylates and inactivates glycogen synthase kinase 3 beta (GSK-3B), it indicates that oncogenic RTK-mediated signaling possibly accelerates the canonical Wnt signaling as well [154, 159]. Therefore, it is suggested that the canonical Wnt pathway is involved in the chemoresistance of GBM cells.

Hedgehog signaling-mediated chemoresistance of GBMs

Hedgehog signaling pathway

The Hedgehog (HH) signaling pathway primarily regulates the development, homeostasis, and regeneration of organs during embryogenesis [160, 161]. The association of sonic hedgehog with its plasma membrane-localizing receptor Patched-1 (Ptch1) leads to a Ptch1-dependent inhibition of the transmembrane protein smoothened (Smo). As Smo is suppressed, it accumulates and initiates a downstream signaling cascade by the processing and activation of the following transcription factors: zinc finger proteins GLI1 (glioma-associated oncogene homolog; agonistic), GLI2 (agonistic/antagonistic), and GLI3 (antagonistic). [160, 161] The activated GLI factors form a molecular complex with a suppressor of fused homolog (SUFU) that is translocated into the nucleus to finally induce the expressions of Bcl-2, c-Myc, cyclin D1, GLI1, and Ptch1 by acting as their transcription factors [160, 161]. Additionally, the cross talk between the HH signaling and other signaling pathways, such as RTK-, Wnt-, and TGFβ-mediated signaling has also been reported [160, 161].

Chemoresistance induced by the oncogenic activation of HH signaling

An upregulated activation of the HH signaling pathway has been observed in GBMs; for instance, higher mRNA expression of *GLI1*, especially in PTEN-expressing cases, has been reported [162, 163]. In addition, the transcriptional products of GLI include antiapoptotic molecules and those required for the maintenance of cellular stemness; these molecules supposedly induce chemoresistance as the knockdown of GLI1 in GBM cells has indicated an enhancement in TMZ-induced cell death in vitro and suppression of tumorigenesis in vivo [163, 164]. Notably, GLI-dependent transcriptional products include MGMT, which is a DNA-repair enzyme that antagonizes TMZ-mediated DNA modifications (see "TMZ-induced DNA replication stress and double strand breaks" Section). [165] Therefore, these evidence indicate a possible contribution of HH signaling in the chemoresistance of GBMs.

Notch signaling-mediated chemoresistance of GBMs

Notch signaling pathway

The Notch signaling pathway normally regulates the cell fate during embryogenesis and maintains homeostasis in the adult organs. Notch signaling is activated by the association of plasma membrane-localizing transmembrane receptor Notch with its ligands (e.g., Delta-like or Jagged family members), which are expressed on the surface of other cells [166, 167]. Upon binding with its ligands, the extracellular domain of the Notch receptor is cleaved by the tumor necrosis factor-α-converting enzyme (S2 cleavage) followed by a gamma-secretase complex-mediated cleavage of the intracellular region of the Notch receptor (S3 and S4 cleavages) [166, 167]. Then, the Notch intracellular domain is released into the cytosol and translocated to the nucleus where it interacts with the transcription factor complex CBF1/Su(H)/Lag-1 [166, 167]. As a result, a Notch-mediated transcription is initiated and the expression of the target genes, such as B lymphoma Mo-MLV insertion region 1 homolog (BMI1), CD44, cyclin D3, hairy and enhancer of split 1 and 5, Myc, and p21 is triggered [166–168].

Chemoresistance induced by oncogenic activation of Notch signaling

The involvement of Notch signaling in the chemoresistance of cancer cells has also been reported. *Notch, DLL1,* and *JAG1* genes have been discovered in GBM cells and samples as overexpressed genes [169, 170]. In addition, Notch signaling induces the cancer stem cell markers, such as BMI1 and CD44, as described in the sections above. Importantly, a cotreatment of a gamma-secretase inhibitor, which blocks Notch pathway activation, with

TMZ resulted in the enhancement of TMZ-mediated cytotoxicity [171, 172]. These evidence suggest that enhanced Notch signaling may confer chemoresistance in GBM cells as well.

Chemoresistance of GBM induced by Hippo/Yes associated-protein 1 (YAP)- mediated survival signaling

Hippo/YAP1 signaling

The Hippo/YAP1 pathway mainly regulates the normal organ size by controlling the cell numbers and cell–cell and cell-ECM contact responses [173, 174]. The core machinery of this signaling pathway is the Mst1/2- protein salvador homolog 1 complex that is formed downstream of the extracellular stimulus. This complex phosphorylates and activates large tumor suppressor 1/2 (LATS1/2) kinases [173–175]. Then, the activated LATS1/2 kinases phosphorylate YAP1 and WW domain-containing transcription regulator protein 1 (TAZ), resulting in the association of YAP1/TAZ with 14-3-3 proteins followed by the proteasomal degradation of these molecules [173–175]. Once the Hippo pathway is inhibited, YAP1 and TAZ are activated and act as the transcriptional co-activators of transcription factor TEA domain family member (TEAD), thereby inducing the expression of genes that play essential roles in cell proliferation and survival [173–175]. In the Hippo/YAP1 pathway, other intracellular signaling pathways that play pivotal roles in cell proliferation and survival, such as PI3K-Akt, Ras-MAPK, TGFβ, and Wnt, are known to cross-react upstream and/or downstream of the Hippo/YAP pathway [173, 175]. Consequently, it has always been considered that the activation of the Mst-LATS-YAP/TAZ axis might be associated with the activation of other pathways [175].

The chemoresistance induced by oncogenic activation of Hippo/YAP signaling

In GBMs, the mRNA and protein levels of the Hippo effector YAP, and the mRNA and protein levels of the YAP target genes were reported as overexpressed [176, 177]. The inhibition of YAP1 expression resulted in the suppression of GBM cell proliferation in vitro [176]. In fact, the downregulation of the YAP1 inhibitor LATS1/2 was also discovered; this suggested that deregulated LATS1/2 activity partially participates in the upregulation of YAP1 expression in the GBMs [177]. In addition, a YAP1-mediated transcription has also been reported to contribute to the chemoresistance of cancer cells [174, 178]. Together, these previous findings suggest a pivotal role of the Hippo/YAP pathway in the chemoresistance of the GBM cells.

Chemoresistance of GBMs induced by p16INK4a/p14ARF-mediated survival signaling

p16INK4a/p14ARF signaling

p16INK4a (a product of the *CDKN2A* gene) and p14ARF (a splicing-isoform product of the *CDKN2A* gene) are acknowledged as cell cycle repressors. p16INK4a blocks the activity of the serine/threonine kinase cyclin-dependent kinase 4 (CDK4)-cyclin-dependent kinase 6 (CDK6)-cyclin D complex to suppress a CDK4/CDK6-mediated retinoblastoma-associated protein (RB1) phosphorylation [179, 180]. The unphosphorylated RB1 associates with members of the transcription factor E2 (E2F) family and inhibits an E2F-mediated translation, which results in G1/S cell cycle arrest [181, 182]. Following RB1 inactivation by phosphorylation, the E2F-mediated translation leads to an expression of cyclin E; then, cyclin E-serine/threonine kinase cyclin-dependent kinase 2 (CDK2) complex further phosphorylates RB1. By this process, RB1 is fully inactivated and the cell cycle progresses into the S phase [182]. On the other hand, p14ARF induces cell cycle arrest by activation of the p53-mediated pathway. Upon mitogenic stimulation, p14ARF forms a stable complex with the p53 inhibitor MDM2; this results in the inhibition of the p53-MDM2 association and is followed by p53 activation. As the cross talk between p16INK4a- and p14ARF-mediated signaling progresses, the activated p53 transcriptionally upregulates p21 expression; this blocks the binding of CDK4 and CKD6 with cyclins and leads to a G1/S cell cycle arrest. In contrast, the CDK4-cyclin D complex blocks the p21 functionality and thereby facilitates CDK2-cyclin E complex formation and progression of the cell cycle. In addition, the expression of p14ARF is induced by E2F; this suggests that E2F/p14ARF-mediated transcriptional regulation forms a self-inhibitory feedback loop. In this way, the signaling regulated by p16INK4a and p14ARF collaborate with each other and finally trigger the cell cycle arrest. Furthermore, there is cross talk between other signaling pathways as well; for instance, cyclin D1 expression is transcriptionally upregulated by RTK-MAPK signaling [183] and the GSK3β-mediated phosphorylation of cyclin D1 triggers the translocation of nuclear cyclin D1 into the cytoplasm, thereby resulting in the proteasomal degradation of cyclin D1 [184].

Chemoresistance induced by p16INK4a/p14ARF inactivation-mediated oncogenic activation of CDK/cyclin-RB1-E2F signaling

Mutations in *CDKN2A* and the encoding genes of the p16INK4a/p14ARF-mediated signaling-related molecules have been determined in glioma cases; TCGA data have demonstrated that *CDKN2A* is frequently homozygously deleted in GBMs (approximately 60% of all GBMs) and as a result, p16INK4a and p14ARF are also frequently inactivated by deletion [48, 185, 186]. In addition, CDK4 overexpression and RB1 deletion were also observed in TCGA data (approximately 15% of all GBMs) [48]. Since Akt phosphorylates and functionally blocks GSK3β [187], the hyperactivated RTK-mediated MAPK and Akt activation pathways also contribute to the cell cycle progression via the p16INK4a/p14ARF inactivation-E2F activation axis. In addition, an elevated expression of E2F in high-grade glioma cases, including GBMs, has been reported [188]. Thus, these results suggest that the oncogenic activation of the CDK/cyclins-RB1-E2F pathway via the suppression of p16INK4a/p14ARF and its related events could contribute to the enhanced proliferation and chemoresistance of the GBMs. The molecular networks regarding p16INK4a/p14ARF signaling are summarized in Fig. 5.

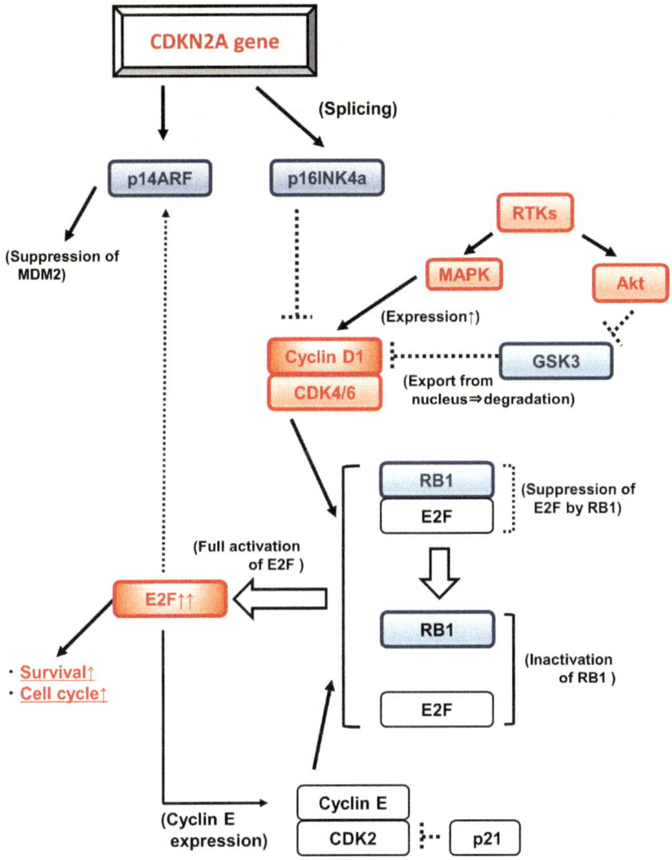

FIG. 5 The landscape of *CDKN2A* (p16INK4a/p14ARF)-induced molecular signals in gliomas. The molecules whose expression levels and functions are genetically upregulated and suppressed in the gliomas are marked with red and blue shading, respectively. The solid lines and arrows indicate the signaling that promotes cell cycle progression and/or survival; whereas the dotted lines and arrows indicate the signaling that inhibits cell cycle progression and/or survival.

Chemoresistance induced by cancer stem-like cells, TME, heterogeneity, and plasticity in GBMs

The GBM cancer stem-like cells (CSCs) are a minor population in the GBM tissues but play fundamental roles in the development of refractory GBMs [189]. The GBM CSCs possess cancer-initiating- and self-renewing-abilities and are associated with chemoresistance [189]. Further, the CSCs express specific molecules called stem cell markers, for instance, aldehyde dehydrogenase 1, CD133, and CD44 in the GBMs. Importantly, the CSCs also abundantly express molecules that excrete multiple substrates (including drugs) from the cells [189]. In GBMs, the CSCs mainly localize around the microenvironmental borders called niches in the GBM tissues; this uneven distribution supposedly contributes to the heterogeneity and plasticity of the GBMs [189]. In fact, recent investigations at the single-cell level have revealed that GBM is a highly heterogeneous tumor [4, 5]. In this case, "heterogeneity" can imply different meanings, such as histological, genetic/epigenetic, transcriptional, ECM, GBM-associated cell, and microenvironmental heterogeneity. These "heterogenetic" features of GBMs are different from that of other malignancies and thus are unique to the GBMs. In addition, single-cell analysis also revealed the plasticity of the GBM tissues. In GBMs, plasticity mainly indicates a phenotypic diversity (e.g., EMT), the transition of GBM cell differentiation from CSCs to differentiated cells because of TME heterogeneity (e.g., hypoxia), and interaction of the GBM-associated cells (e.g., astrocytes and microglia) in the GBM tissue. Thus, the heterogeneity and plasticity of the GBMs is considered to play fundamental roles in gaining chemoresistance and have been described in the following sections.

Mechanisms regulating the chemoresistance of GBM CSCs

CD133-mediated chemoresistance

As described above, the GBM CSCs express CSC-specific markers and the first candidate to be highlighted as a marker was CD133 or prominin-1 [190]. Currently, the significance of CD133 as a GBM CSC marker is not well acknowledged however it is still considered a prognostic marker of the GBMs [191, 192]. CD133 is a transmembrane glycoprotein that localizes at the plasma membrane, but its function is still not well elucidated. A recent study revealed that CD133 activates PI3K-Akt signaling via cytosolic SRC kinases and thus regulates the cell migration and invasion; further, it presents downstream of the RTKs that are aberrantly activated in GBMs (see "Chemoresistance of GBM induced by plasma membrane-localizing receptor-mediated survival signaling" Section) [193, 194]. On the other hand, CD133 also possibly regulates the SRC activity [195]. In addition, the genes related to Notch and HH pathways (see "Chemoresistance of GBM induced by plasma membrane-localizing receptor-mediated survival signaling" Section), both of which contribute to the chemoresistance in GBMs, were highly activated in the $CD133^+$ GBM population [194]; moreover, the expression of the angiogenic factor VEGF was also upregulated in the $CD133^+$ GBM population [196]. These instances indicate that CD133 also has the potential to contribute to chemoresistance in GBM cells.

CD44-mediated chemoresistance

In addition to CD133, CD44 is also a plasma membrane-localizing transmembrane glycoprotein that mainly regulates cellular adhesion along with ECM components, such as hyaluronic acid (HA) and osteopontin. Moreover, it is also considered a CSC-specific marker of the GBM cells [197, 198]. CD44 triggers multiple intracellular signaling pathways and thereby regulates various cellular functions; most functions lead to enhancement of stemness and the oncogenic properties of the cancer cells [197, 199]. For example, CD44 forms a heterodimer with plasma membrane-localizing receptors, such as RTKs, and activates their downstream signaling pathways [199]. CD44 also activates transcription factors, such as Sox2,

Nanog, Oct4, c-Jun, and HIF-2α [198, 200]. The activated RTKs and transcription factors induce chemoresistance via the upregulation of survival genes' expression levels.

Chemoresistance by ATP-binding cassette (ABC) subfamily-mediated extracellular transportation of drugs in GBMs

Another fundamental mechanism of drug resistance is the accelerated transportation of drugs from cells. The permeability glycoprotein [P-gp; also known as multidrug resistance protein 1 (MDR1), ABC subfamily B member 1 (ABCB1), or cluster of differentiation 243 (CD243)] is a transmembrane glycoprotein that belongs to the MDR/TAP subfamily of ABC transporters. The MDR/TAP subfamily proteins generally take charge of drug efflux and thereby may induce drug resistance [201, 202]. Like other MDR/TAP subfamily proteins, P-gp is an ATP-dependent drug pump that can transport multiple substrates from the cells [201, 202]. It is noteworthy that P-gp is highly expressed at the BBB, which is a molecular and physical barrier that protects the central nervous system (CNS) against circulating toxins and pathogens and contributes to neuroprotection [202–204]. In other words, P-gp also takes part in excretory roles by transporting anticancer agents from the CNS. P-gp expression is regulated by multiple intracellular signaling pathways, including the PI3K-Akt pathway, Wnt-β-catenin pathway, and MAPKs cascades, which are all aberrantly activated in the GBMs as described in "Chemoresistance of GBM induced by plasma membrane-localizing receptor-mediated survival signaling" Sectio [205–207]. Further, P-gp is highly expressed in the CSCs of the GBMs [208]. In addition, TMZ is one of the substrates of P-gp [209] and polymorphism of P-gp has been associated with a poor prognosis in GBM cases [210]. Furthermore, suppression of P-gp function has resulted in the enhancement of TMZ toxicity against the GBM cells in vitro [209, 211]. Thus, it is suggested that P-gp can confer resistance against chemotherapy of GBMs, particularly in the CSCs. Another ABC subfamily that is highly expressed in the GBM CSCs is the ABC subfamily G member 2 [ABCG2; alternatively referred to as the breast cancer resistance protein (BCRP)]. Similar to P-gp, ABCG2 is also an ATP-dependent pump that discharges intracellular xenobiotics into the extracellular space and is highly expressed at the BBB and in the GBM CSCs [212, 213]. In GBMs, the suppression of ABCG2 and P-gp functions reportedly enhance TMZ toxicity [214]; therefore, ABCG2 may also be involved in the positive regulation of GBM CSC chemoresistance.

Mechanisms of chemoresistance induced by the TME in GBMs

Mechanism of chemoresistance induced by ECM components

ECM is not only important for maintaining normal tissue homeostasis but is also crucial for regulating GBM chemoresistance by modulating the TME [215, 216]. In normal CNS tissues, the volume of ECM is 20% of the healthy brain. The ECM is mainly composed of fibronectin, laminin, and type IV collagen; however, their composition varies according to the CNS regions [8, 217, 218]. In GBMs, the ECM components are the same as those in a healthy brain, but also include increased levels of HA because of high levels of hyaluronan synthases

1, 2, and 3 [8, 219]. As described above, HA is a ligand of CD44 [197, 198]; this suggests that the upregulation of CD44-mediated signaling in GBM CSCs is enriched by extracellular HA. Further, the GBM cells produce and excrete osteopontin and milk fat globule-EGF factor 8 protein (Mfge8), both of which are known ligands of integrins and are capable of triggering angiogenesis and other oncogenic properties. Both are also known to transform the GBM-associated microglia into an amoeboid-like invasive phenotype via the activation of the integrin-PI3K-Akt cascade of the microglia [220–222]. In addition, glycoprotein fibulin-3, a component of the fibrillar ECM in connective tissues that is absent in the normal brain, has also been reported to contribute to the chemoresistance of the GBMs via metalloproteinase inhibitor 3 (TIMP3) suppression-mediated ADAM metallopeptidase domain 17 activation that ultimately activates the NF-κB [223] and paracrine-mediated Notch signaling pathways [224]. Furthermore, tenascin-C, another ECM component, is also associated with TMZ resistance in the GBM cells [225]. As described above, a hypoxia-induced periostin production from GBM cells is also involved in the resistance against antiangiogenic therapy and immunotherapy in GBM treatments [143, 144]. In fact, hypoxia-induced periostin is also capable of inducing infiltration of the GBM cell-associated macrophages and M2-polarization of the cells that trigger suppression of the antitumor immune responses [226]. Thus, these insights suggest that ECM may also participate in conferring the chemoresistance.

Mechanism of chemoresistance induced by hypoxia and acidosis

Hypoxia-inducing factor signaling

As described above, a hypoxic or acidic microenvironment is a key player in determining the heterogeneity and plasticity of the GBMs [6, 227–229]. Intracellular hypoxia causes deterioration of the mitochondrial oxidative phosphorylation (OXPHOS) process. Consequently, this induces the hyperactivation of glycolysis to compensate for the reduced energy production owing to OXPHOS loss [229–231]. Then, the augmented glycolysis triggers the overproduction of lactate, which induces an intracellular metabolic shift to acidosis [229–231]. Therefore, tissue hypoxia and acidosis are considered as a series of intracellular events [229–231]. Indeed, a fundamental role of tissue acidosis in the acquisition of enhanced malignancy and stemness in the GBM cells has been suggested [6, 229, 231]. The central player of this hypoxia-mediated biological response is the hypoxia-inducing factor α/β complex; the hypoxia-inducing factor α (HIFα; HIF-1α/2α/3α)-mediated pathway is generally activated by hypoxic conditions [232]. In normoxia, the HIFα is hydroxylated by prolyl-4-hydroxylase; this allows the binding of the E3 ubiquitin ligase von Hippel–Lindau tumor suppressor to HIFα and results in the continuous degradation of HIFα by the proteasome [232]. Once the cells are exposed to hypoxic conditions, HIFα is stabilized and accumulated by forming a heterodimer with Hif-1β and is translocated into the nucleus [232]. Then, this heterodimer binds to the consensus sequence in the DNA called the hypoxia-responsive element and functions as a transcription factor. As a result, the expression of hundreds of target molecules, including Bcl-2, EGF, VEGF, PDGF-B, and TWIST1 is triggered to finally induce angiogenesis, ECM remodeling, metabolic reprogramming, immunosuppression, cellular invasion, and survival [232]. The expression of HIFα is regulated at the mRNA level via transcription factors (NF-κB and STAT3) and at the protein level via the Akt–mTOR-dependent pathway [233, 234]. Moreover, HIF-1α is also known to activate the TGFβ signaling by upregulating the TGF β1 and TGFβ signaling-related molecules [235, 236].

Chemoresistance induced by HIF signaling in GBMs

Like other cancer systems, the essential role of a hypoxic microenvironment in the maintenance and expression of the CSCs in GBM tissues has been confirmed both in vitro and in vivo [237–239]. The HIFα positively regulates the GBM CSC's expressions; this suggests that HIFα participates in the regulation of chemoresistance via the maintenance of the GBM CSCs in intratumoral hypoxic regions such as hypoxic niches. Importantly, other oncogenic signaling pathways also contribute to the upregulation of HIFα expression and that of HIFα-mediated transcription products, such as ligands of RTKs, to further facilitate the activation of other oncogenic signaling pathways.

Mechanisms of chemoresistance induced by GBM-associated cells

GBM-associated astrocyte-mediated chemoresistance

The GBM-associated cells, such as astrocytes and microglia, are crucial components of the TME. The astrocytes around or adjacent to the GBM cells demonstrate a hypertrophic phenotype (reactive astrocytes) and are characterized by the upregulation of glial fibrillar acidic protein [240]. These reactive astrocytes interact with GBM cells by exchanging substances via extracellular excretion or transportation through gap junctions at the interface of both the cells [7, 240]. For example, the reactive astrocytes release TGFβ toward the GBM cells to induce their proliferative and invasive abilities [7, 241]. The reactive astrocytes also secrete glial cell line-derived neurotrophic factor (GDNF) to trigger the attraction of microglia toward the GBM cells [7, 242]. Further, the astrocytes secrete chemokine (C—C motif) ligand 20 (CCL20) and activate the chemokine receptor 6 (CCR6; plasma membrane-localizing receptor of CCL20)-mediated signaling in the GBM cells [7, 243]. This activated signaling enhances the HIF-1-mediated response via the subsequent activation of NF-κB and STAT3. The reactive astrocytes also release interleukin-6 (IL6) to activate the IL6-mediated signaling in the GBM cells [244]. Generally, IL6 associates with the plasma membrane-localizing receptors (IL6 receptor complex) and transduces the downstream signaling. The IL6 receptor complex is composed of an IL6 receptor subunit and IL6 signal transducer glycoprotein 130 (GP130). GP130 associates with IL6 and is phosphorylated and activates the JAK–STAT and Ras-MAPK signaling pathways [245, 246]. These IL6-mediated responses supposedly result in enhancing the gain of malignancy and conferring chemoresistance. In addition to these mechanisms, the gap junction-mediated signal transduction between the reactive astrocytes and GBM cells is also important. Through this signal transduction, calcium, mitochondria, and microRNAs (miRNAs), are possibly delivered to the GBM cells from astrocytes, which are then considered to induce protection from cell death [9, 247, 248].

GBM-associated microglia-mediated chemoresistance

Unlike the other glial cells, microglia arise from the yolk sac during embryogenesis and share a macrophage-like phenotype in response to stresses or pathogenetic stimuli that contribute to tissue plasticity [7, 249, 250]. Like normal macrophages, microglia can also polarize to the M1 (proinflammatory) and M2 (antiinflammatory) phenotypes in response to stimuli or the microenvironment to evoke their immunogenic functions [7, 250]. In GBMs, the microglia suppress their proinflammatory M1 polarization and polarize to their antiinflammatory M2 phenotype; this is because of the release of numerous soluble factors from the GBM cells, such as growth factors, cytokines, and chemokines, including CCL2, CCL5, C-X-C motif chemokine 11 (CXCL11), CXCL12, IL4, IL10, IL13, brain-derived neurotrophic factor, GDNF, Wnt5a, and miRNAs [7, 250]. These soluble factors activate the RTK downstream cascades and the G protein-coupled receptor signaling; this results in further M1 polarization and activation of the astrocytes [7, 250]. In turn, the activated/ M2-polarized astrocytes also secrete various soluble factors, including IL1, IL6, EGF, VEGF, TGFβ, and miRNAs to trigger the acquisition of further oncogenic properties of the GBM cells, including their chemoresistance [7, 250]. This positive feedback between the GBM cells and microglia possibly potentiates the chemoresistance of the GBM cells.

miRNA-mediated chemoresistance

miRNAs are short noncoding single-stranded RNA molecules that regulate gene expression in eukaryotic cells under physiological and pathogenetic conditions. The miRNAs regulate the expression of target genes by incompletely binding to the 3′-untranslated regions of the mRNAs to either suppress their translation or degrade the target mRNA itself [251]. The miRNA expression is endogenously regulated by primary intracellular signaling. The miRNAs may also be exogenously transferred into the GBM cells via exosomes or from other cells by gap junction-mediated cell–cell contact [247]; this results in further downstream signaling of the primary signaling pathways. In GBMs as well as in other cancers, multiple miRNAs regulate various oncogenic properties via the modulation of several molecular pathways. For example, miRNA-21, a representative oncogenic miRNA, is overexpressed in GBM tissues and cell lines and is considered to be involved in their oncogenesis and chemoresistance by targeting multiple molecules [252–254]. The target molecules of miRNAs, which are essential for chemoresistance of GBMs, are roughly classified as cell survival pathway-related molecules (e.g., downstream of RTK-related molecules) and cell death cascade-related molecules (e.g., proapoptotic Bcl-2 family proteins and DMR-related molecules, such as p53).

Conclusions and perspectives

As discussed, the molecular pathways governing the chemoresistance of the GBMs are various, complex, and intertwined. In addition, the deregulated molecular pathways induce chemoresistance by modulating target molecules, regulating related signaling pathways of the therapeutic drugs, and affecting the microenvironment of the tumor tissue; these also include the drug distribution in tumor tissues and the interaction of the GBM cells with the surrounding normal tissue-derived cells. Thus, these are the distinctive features that bestow the GBMs with their chemoresistance abilities. Therefore, it is necessary to consider the factors and associated molecular signaling pathways that affect the chemotherapeutic responses before selecting appropriate molecules as chemotherapeutic targets for the treatment of GBMs in the future.

References

[1] Gilbert MR, Wang M, Aldape KD, Stupp R, Hegi ME, Jaeckle KA, et al. Dose-dense temozolomide for newly diagnosed glioblastoma: a randomized phase III clinical trial. J Clin Oncol 2013;31:4085–91.

[2] Stupp R, Mason WP, van den Bent MJ, Weller M, Fisher B, Taphoorn MJ, et al. Radiotherapy plus concomitant and adjuvant temozolomide for glioblastoma. N Engl J Med 2005;352:987–96.

[3] Libermann TA, Nusbaum HR, Razon N, Kris R, Lax I, Soreq H, et al. Amplification, enhanced expression and possible rearrangement of EGF receptor gene in primary human brain tumours of glial origin. Nature 1985;313:144–7.

[4] Patel AP, Tirosh I, Trombetta JJ, Shalek AK, Gillespie SM, Wakimoto H, et al. Single-cell RNA-seq highlights intratumoral heterogeneity in primary glioblastoma. Science 2014;344:1396–401.

[5] Sottoriva A, Spiteri I, Piccirillo SG, Touloumis A, Collins VP, Marioni JC, et al. Intratumor heterogeneity in human glioblastoma reflects cancer evolutionary dynamics. Proc Natl Acad Sci U S A 2013;110:4009–14.

[6] Da Ros M, De Gregorio V, Iorio AL, Giunti L, Guidi M, de Martino M, et al. Glioblastoma chemoresistance: the double play by microenvironment and blood-brain barrier. Int J Mol Sci 2018;19:2879.

[7] Matias D, Balca-Silva J, da Graca GC, Wanjiru CM, Macharia LW, Nascimento CP, et al. Microglia/astrocytes-glioblastoma crosstalk: crucial molecular mechanisms and microenvironmental factors. Front Cell Neurosci 2018;12:235.

[8] Perrin SL, Samuel MS, Koszyca B, Brown MP, Ebert LM, Oksdath M, et al. Glioblastoma heterogeneity and the tumour microenvironment: implications for preclinical research and development of new treatments. Biochem Soc Trans 2019;47:625–38.

[9] Zhang X, Ding K, Wang J, Li X, Zhao P. Chemoresistance caused by the microenvironment of glioblastoma and the corresponding solutions. Biomed Pharmacother 2019;109:39–46.

[10] Ortiz R, Perazzoli G, Cabeza L, Jimenez-Luna C, Luque R, Prados J, et al. Temozolomide: an updated overview of resistance mechanisms, nanotechnology advances and clinical applications. Curr Neuropharmacol 2020;19:513–37.

[11] Villano JL, Seery TE, Bressler LR. Temozolomide in malignant gliomas: current use and future targets. Cancer Chemother Pharmacol 2009;64:647–55.

[12] Yung WK. Temozolomide in malignant gliomas. Semin Oncol 2000;27:27–34.

[13] Burri SH, Prabhu RS, Sumrall AL, Brick W, Blaker BD, Heideman Jr BE, et al. BCNU wafer placement with temozolomide (TMZ) in the immediate postoperative period after tumor resection followed by radiation therapy with TMZ in patients with newly diagnosed high grade glioma: final results of a prospective, multi-institutional, phase II trial. J Neuro-Oncol 2015;123:259–66.

[14] Happold C, Roth P, Wick W, Steinbach JP, Linnebank M, Weller M, et al. ACNU-based chemotherapy for recurrent glioma in the temozolomide era. J Neuro-Oncol 2009;92:45–8.

[15] McDermott M, Krouwer D, Prados M, Wilson C. Irradiation and CCNU in glioma therapy. J Neurosurg 1989;71:791–3.

[16] Khasraw M, Ameratunga M, Grommes C. Bevacizumab for the treatment of high-grade glioma: an update after phase III trials. Expert Opin Biol Ther 2014;14:729–40.

[17] Narayana A, Golfinos JG, Fischer I, Raza S, Kelly P, Parker E, et al. Feasibility of using bevacizumab with radiation therapy and temozolomide in newly diagnosed high-grade glioma. Int J Radiat Oncol Biol Phys 2008;72:383–9.

[18] de Groot J, Penas-Prado M, Alfaro-Munoz K, Hunter K, Pei BL, O'Brien B, et al. Window-of-opportunity clinical trial of pembrolizumab in patients with recurrent glioblastoma reveals predominance of immune-suppressive macrophages. Neuro-Oncology 2020;22:539–49.

[19] Le Rhun E, Taillibert S, Chamberlain MC. The future of high-grade glioma: where we are and where are we going. Surg Neurol Int 2015;6:S9–S44.

[20] Nabors LB, Fink KL, Mikkelsen T, Grujicic D, Tarnawski R, Nam DH, et al. Two cilengitide regimens in combination with standard treatment for patients with newly diagnosed glioblastoma and unmethylated MGMT gene promoter: results of the open-label, controlled, randomized phase II CORE study. Neuro-Oncology 2015;17:708–17.
[21] Reardon DA, Brandes AA, Omuro A, Mulholland P, Lim M, Wick A, et al. Effect of Nivolumab vs bevacizumab in patients with recurrent glioblastoma: the CheckMate 143 phase 3 randomized clinical trial. JAMA Oncol 2020;6:1003–10.
[22] Stupp R, Hegi ME, Gorlia T, Erridge SC, Perry J, Hong YK, et al. Cilengitide combined with standard treatment for patients with newly diagnosed glioblastoma with methylated MGMT promoter (CENTRIC EORTC 26071-22072 study): a multicentre, randomised, open-label, phase 3 trial. Lancet Oncol 2014;15:1100–8.
[23] Stupp R, Taillibert S, Kanner A, Read W, Steinberg D, Lhermitte B, et al. Effect of tumor-treating fields plus maintenance temozolomide vs maintenance temozolomide alone on survival in patients with glioblastoma: a randomized clinical trial. JAMA 2017;318:2306–16.
[24] Zhang J, Stevens MF, Bradshaw TD. Temozolomide: mechanisms of action, repair and resistance. Curr Mol Pharmacol 2012;5:102–14.
[25] Fu D, Calvo JA, Samson LD. Balancing repair and tolerance of DNA damage caused by alkylating agents. Nat Rev Cancer 2012;12:104–20.
[26] Wang JY, Edelmann W. Mismatch repair proteins as sensors of alkylation DNA damage. Cancer Cell 2006;9:417–8.
[27] Natarajan AT, Vermeulen S, Darroudi F, Valentine MB, Brent TP, Mitra S, et al. Chromosomal localization of human O6-methylguanine-DNA methyltransferase (MGMT) gene by in situ hybridization. Mutagenesis 1992;7:83–5.
[28] Tano K, Shiota S, Collier J, Foote RS, Mitra S. Isolation and structural characterization of a cDNA clone encoding the human DNA repair protein for O6-alkylguanine. Proc Natl Acad Sci U S A 1990;87:686–90.
[29] Hegi ME, Diserens AC, Gorlia T, Hamou MF, de Tribolet N, Weller M, et al. MGMT gene silencing and benefit from temozolomide in glioblastoma. N Engl J Med 2005;352:997–1003.
[30] Stupp R, Hegi ME, Mason WP, van den Bent MJ, Taphoorn MJ, Janzer RC, et al. Effects of radiotherapy with concomitant and adjuvant temozolomide versus radiotherapy alone on survival in glioblastoma in a randomised phase III study: 5-year analysis of the EORTC-NCIC trial. Lancet Oncol 2009;10:459–66.
[31] Uno M, Oba-Shinjo SM, Camargo AA, Moura RP, Aguiar PH, Cabrera HN, et al. Correlation of MGMT promoter methylation status with gene and protein expression levels in glioblastoma. Clinics (Sao Paulo) 2011;66:1747–55.
[32] Rivera AL, Pelloski CE, Gilbert MR, Colman H, De La Cruz C, Sulman EP, et al. MGMT promoter methylation is predictive of response to radiotherapy and prognostic in the absence of adjuvant alkylating chemotherapy for glioblastoma. Neuro-Oncology 2010;12:116–21.
[33] Stojic L, Brun R, Jiricny J. Mismatch repair and DNA damage signalling. DNA Repair (Amst) 2004;3:1091–101.
[34] Cahill DP, Levine KK, Betensky RA, Codd PJ, Romany CA, Reavie LB, et al. Loss of the mismatch repair protein MSH6 in human glioblastomas is associated with tumor progression during temozolomide treatment. Clin Cancer Res 2007;13:2038–45.
[35] Hunter C, Smith R, Cahill DP, Stephens P, Stevens C, Teague J, et al. A hypermutation phenotype and somatic MSH6 mutations in recurrent human malignant gliomas after alkylator chemotherapy. Cancer Res 2006;66:3987–91.
[36] McFaline-Figueroa JL, Braun CJ, Stanciu M, Nagel ZD, Mazzucato P, Sangaraju D, et al. Minor changes in expression of the mismatch repair protein MSH2 exert a major impact on glioblastoma response to temozolomide. Cancer Res 2015;75:3127–38.
[37] Almeida KH, Sobol RW. A unified view of base excision repair: lesion-dependent protein complexes regulated by post-translational modification. DNA Repair (Amst) 2007;6:695–711.
[38] Kim MY, Zhang T, Kraus WL. Poly(ADP-ribosyl)ation by PARP-1: 'PAR-laying' NAD+ into a nuclear signal. Genes Dev 2005;19:1951–67.
[39] Luo X, Kraus WL. On PAR with PARP: cellular stress signaling through poly(ADP-ribose) and PARP-1. Genes Dev 2012;26:417–32.
[40] Schreiber V, Dantzer F, Ame JC, de Murcia G. Poly(ADP-ribose): novel functions for an old molecule. Nat Rev Mol Cell Biol 2006;7:517–28.

[41] Cheng CL, Johnson SP, Keir ST, Quinn JA, Ali-Osman F, Szabo C, et al. Poly(ADP-ribose) polymerase-1 inhibition reverses temozolomide resistance in a DNA mismatch repair-deficient malignant glioma xenograft. Mol Cancer Ther 2005;4:1364–8.
[42] Curtin NJ, Wang LZ, Yiakouvaki A, Kyle S, Arris CA, Canan-Koch S, et al. Novel poly(ADP-ribose) polymerase-1 inhibitor, AG14361, restores sensitivity to temozolomide in mismatch repair-deficient cells. Clin Cancer Res 2004;10:881–9.
[43] Agnihotri S, Gajadhar AS, Ternamian C, Gorlia T, Diefes KL, Mischel PS, et al. Alkylpurine-DNA-N-glycosylase confers resistance to temozolomide in xenograft models of glioblastoma multiforme and is associated with poor survival in patients. J Clin Invest 2012;122:253–66.
[44] Caporali S, Falcinelli S, Starace G, Russo MT, Bonmassar E, Jiricny J, et al. DNA damage induced by temozolomide signals to both ATM and ATR: role of the mismatch repair system. Mol Pharmacol 2004;66:478–91.
[45] Zhang D, Tang B, Xie X, Xiao YF, Yang SM, Zhang JW. The interplay between DNA repair and autophagy in cancer therapy. Cancer Biol Ther 2015;16:1005–13.
[46] Levine AJ, Oren M. The first 30 years of p53: growing ever more complex. Nat Rev Cancer 2009;9:749–58.
[47] O'Connor MJ. Targeting the DNA damage response in cancer. Mol Cell 2015;60:547–60.
[48] Brennan CW, Verhaak RG, McKenna A, Campos B, Noushmehr H, Salama SR, et al. The somatic genomic landscape of glioblastoma. Cell 2013;155:462–77.
[49] Tomiyama A, Kobayashi T, Mori K, Ichimura K. Protein phosphatases-a touchy enemy in the battle against glioblastomas: a review. Cancers (Basel) 2019;11:241.
[50] Zannini L, Delia D, Buscemi G. CHK2 kinase in the DNA damage response and beyond. J Mol Cell Biol 2014;6:442–57.
[51] Smith J, Tho LM, Xu N, Gillespie DA. The ATM-Chk2 and ATR-Chk1 pathways in DNA damage signaling and cancer. Adv Cancer Res 2010;108:73–112.
[52] Aasland D, Gotzinger L, Hauck L, Berte N, Meyer J, Effenberger M, et al. Temozolomide induces senescence and repression of DNA repair pathways in glioblastoma cells via activation of ATR-CHK1, p21, and NF-kappaB. Cancer Res 2019;79:99–113.
[53] Wade M, Li YC, Wahl GM. MDM2, MDMX and p53 in oncogenesis and cancer therapy. Nat Rev Cancer 2013;13:83–96.
[54] Ashcroft M, Vousden KH. Regulation of p53 stability. Oncogene 1999;18:7637–43.
[55] Silva JL, Cino EA, Soares IN, Ferreira VF, de Oliveira GAP. Targeting the prion-like aggregation of mutant p53 to combat cancer. Acc Chem Res 2018;51:181–90.
[56] Zhou BP, Hung MC. Novel targets of Akt, p21(Cipl/WAF1), and MDM2. Semin Oncol 2002;29:62–70.
[57] Zhang Y, Dube C, Gibert Jr M, Cruickshanks N, Wang B, Coughlan M, et al. The p53 pathway in glioblastoma. Cancers (Basel) 2018;10:297.
[58] Barak Y, Gottlieb E, Juven-Gershon T, Oren M. Regulation of mdm2 expression by p53: alternative promoters produce transcripts with nonidentical translation potential. Genes Dev 1994;8:1739–49.
[59] Freedman DA, Wu L, Levine AJ. Functions of the MDM2 oncoprotein. Cell Mol Life Sci 1999;55:96–107.
[60] Feeley KP, Adams CM, Mitra R, Eischen CM. Mdm2 is required for survival and growth of p53-deficient cancer cells. Cancer Res 2017;77:3823–33.
[61] Teoh G, Urashima M, Ogata A, Chauhan D, DeCaprio JA, Treon SP, et al. MDM2 protein overexpression promotes proliferation and survival of multiple myeloma cells. Blood 1997;90:1982–92.
[62] Riemenschneider MJ, Buschges R, Wolter M, Reifenberger J, Bostrom J, Kraus JA, et al. Amplification and overexpression of the MDM4 (MDMX) gene from 1q32 in a subset of malignant gliomas without TP53 mutation or MDM2 amplification. Cancer Res 1999;59:6091–6.
[63] Venkatesan S, Lamfers ML, Dirven CM, Leenstra S. Genetic biomarkers of drug response for small-molecule therapeutics targeting the RTK/Ras/PI3K, p53 or Rb pathway in glioblastoma. CNS Oncol 2016;5:77–90.
[64] Haupt S, Berger M, Goldberg Z, Haupt Y. Apoptosis—the p53 network. J Cell Sci 2003;116:4077–85.
[65] Yoshida K, Miki Y. The cell death machinery governed by the p53 tumor suppressor in response to DNA damage. Cancer Sci 2010;101:831–5.
[66] Aubrey BJ, Kelly GL, Janic A, Herold MJ, Strasser A. How does p53 induce apoptosis and how does this relate to p53-mediated tumour suppression? Cell Death Differ 2018;25:104–13.
[67] Emery IF, Gopalan A, Wood S, Chow KH, Battelli C, George J, et al. Expression and function of ABCG2 and XIAP in glioblastomas. J Neuro-Oncol 2017;133:47–57.

[68] Tirapelli D, Lustosa IL, Menezes SB, Franco IM, Rodrigues AR, Peria FM, et al. High expression of XIAP and Bcl-2 may inhibit programmed cell death in glioblastomas. Arq Neuropsiquiatr 2017;75:875–80.
[69] Stehlik C, de Martin R, Kumabashiri I, Schmid JA, Binder BR, Lipp J. Nuclear factor (NF)-kappaB-regulated X-chromosome-linked iap gene expression protects endothelial cells from tumor necrosis factor alpha-induced apoptosis. J Exp Med 1998;188:211–6.
[70] Warfel NA, El-Deiry WS. p21WAF1 and tumourigenesis: 20 years after. Curr Opin Oncol 2013;25:52–8.
[71] Bocangel DB, Finkelstein S, Schold SC, Bhakat KK, Mitra S, Kokkinakis DM. Multifaceted resistance of gliomas to temozolomide. Clin Cancer Res 2002;8:2725–34.
[72] Sato A, Sunayama J, Matsuda K, Seino S, Suzuki K, Watanabe E, et al. MEK-ERK signaling dictates DNA-repair gene MGMT expression and temozolomide resistance of stem-like glioblastoma cells via the MDM2-p53 axis. Stem Cells 2011;29:1942–51.
[73] He Y, Su J, Lan B, Gao Y, Zhao J. Targeting off-target effects: endoplasmic reticulum stress and autophagy as effective strategies to enhance temozolomide treatment. Onco Targets Ther 2019;12:1857–65.
[74] Pyrko P, Schonthal AH, Hofman FM, Chen TC, Lee AS. The unfolded protein response regulator GRP78/BiP as a novel target for increasing chemosensitivity in malignant gliomas. Cancer Res 2007;67:9809–16.
[75] Chaudhari N, Talwar P, Parimisetty A, Lefebvre d'Hellencourt C, Ravanan P. A molecular web: endoplasmic reticulum stress, inflammation, and oxidative stress. Front Cell Neurosci 2014;8:213.
[76] Kim I, Xu W, Reed JC. Cell death and endoplasmic reticulum stress: disease relevance and therapeutic opportunities. Nat Rev Drug Discov 2008;7:1013–30.
[77] Weatherbee JL, Kraus JL, Ross AH. ER stress in temozolomide-treated glioblastomas interferes with DNA repair and induces apoptosis. Oncotarget 2016;7:43820–34.
[78] Dadey DY, Kapoor V, Khudanyan A, Urano F, Kim AH, Thotala D, et al. The ATF6 pathway of the ER stress response contributes to enhanced viability in glioblastoma. Oncotarget 2016;7:2080–92.
[79] Chen TC, Wang W, Golden EB, Thomas S, Sivakumar W, Hofman FM, et al. Green tea epigallocatechin gallate enhances therapeutic efficacy of temozolomide in orthotopic mouse glioblastoma models. Cancer Lett 2011;302:100–8.
[80] Escamilla-Ramirez A, Castillo-Rodriguez RA, Zavala-Vega S, Jimenez-Farfan D, Anaya-Rubio I, Briseno E, et al. Autophagy as a potential therapy for malignant glioma. Pharmaceuticals (Basel) 2020;13:156.
[81] Noguchi M, Hirata N, Tanaka T, Suizu F, Nakajima H, Chiorini JA. Autophagy as a modulator of cell death machinery. Cell Death Dis 2020;11:517.
[82] Jiang C, Shen F, Du J, Fang X, Li X, Su J, et al. Upregulation of CASC2 sensitized glioma to temozolomide cytotoxicity through autophagy inhibition by sponging miR-193a-5p and regulating mTOR expression. Biomed Pharmacother 2018;97:844–50.
[83] Kanzawa T, Germano IM, Komata T, Ito H, Kondo Y, Kondo S. Role of autophagy in temozolomide-induced cytotoxicity for malignant glioma cells. Cell Death Differ 2004;11:448–57.
[84] Nakada M, Kita D, Teng L, Pyko IV, Watanabe T, Hayashi Y, et al. Receptor tyrosine kinases: principles and functions in glioma invasion. Adv Exp Med Biol 2020;1202:151–78.
[85] Pottier C, Fresnais M, Gilon M, Jerusalem G, Longuespee R, Sounni NE. Tyrosine kinase inhibitors in cancer: breakthrough and challenges of targeted therapy. Cancers (Basel) 2020;12:731.
[86] Tomiyama A, Ichimura K. Signal transduction pathways and resistance to targeted therapies in glioma. Semin Cancer Biol 2019;58:118–29.
[87] Sugawa N, Ekstrand AJ, James CD, Collins VP. Identical splicing of aberrant epidermal growth factor receptor transcripts from amplified rearranged genes in human glioblastomas. Proc Natl Acad Sci U S A 1990;87:8602–6.
[88] Guo G, Gong K, Wohlfeld B, Hatanpaa KJ, Zhao D, Habib AA. Ligand-independent EGFR signaling. Cancer Res 2015;75:3436–41.
[89] Huang PH, Xu AM, White FM. Oncogenic EGFR signaling networks in glioma. Sci Signal 2009;2, re6.
[90] Cancer Genome Atlas Research N. Comprehensive genomic characterization defines human glioblastoma genes and core pathways. Nature 2008;455:1061–8.
[91] Hermanson M, Funa K, Hartman M, Claesson-Welsh L, Heldin CH, Westermark B, et al. Platelet-derived growth factor and its receptors in human glioma tissue: expression of messenger RNA and protein suggests the presence of autocrine and paracrine loops. Cancer Res 1992;52:3213–9.

[92] Lokker NA, Sullivan CM, Hollenbach SJ, Israel MA, Giese NA. Platelet-derived growth factor (PDGF) autocrine signaling regulates survival and mitogenic pathways in glioblastoma cells: evidence that the novel PDGF-C and PDGF-D ligands may play a role in the development of brain tumors. Cancer Res 2002;62:3729–35.
[93] Ozawa T, Brennan CW, Wang L, Squatrito M, Sasayama T, Nakada M, et al. PDGFRA gene rearrangements are frequent genetic events in PDGFRA-amplified glioblastomas. Genes Dev 2010;24:2205–18.
[94] Onken J, Vajkoczy P, Torka R, Hempt C, Patsouris V, Heppner FL, et al. Phospho-AXL is widely expressed in glioblastoma and associated with significant shorter overall survival. Oncotarget 2017;8:50403–14.
[95] Sadahiro H, Kang KD, Gibson JT, Minata M, Yu H, Shi J, et al. Activation of the receptor tyrosine kinase AXL regulates the immune microenvironment in glioblastoma. Cancer Res 2018;78:3002–13.
[96] Abounader R, Laterra J. Scatter factor/hepatocyte growth factor in brain tumor growth and angiogenesis. Neuro-Oncology 2005;7:436–51.
[97] Vehlow A, Klapproth E, Jin S, Hannen R, Hauswald M, Bartsch JW, et al. Interaction of discoidin domain receptor 1 with a 14-3-3-Beclin-1-Akt1 complex modulates glioblastoma therapy sensitivity. Cell Rep 2019;26:3672–83 [e3677].
[98] Pylayeva-Gupta Y, Grabocka E, Bar-Sagi D. RAS oncogenes: weaving a tumorigenic web. Nat Rev Cancer 2011;11:761–74.
[99] Lavoie H, Therrien M. Regulation of RAF protein kinases in ERK signalling. Nat Rev Mol Cell Biol 2015;16:281–98.
[100] Zhao Y, Adjei AA. The clinical development of MEK inhibitors. Nat Rev Clin Oncol 2014;11:385–400.
[101] Shaul YD, Seger R. The MEK/ERK cascade: from signaling specificity to diverse functions. Biochim Biophys Acta 1773;2007:1213–26.
[102] Downward J. Targeting RAS signalling pathways in cancer therapy. Nat Rev Cancer 2003;3:11–22.
[103] Horbinski C, Hamilton RL, Nikiforov Y, Pollack IF. Association of molecular alterations, including BRAF, with biology and outcome in pilocytic astrocytomas. Acta Neuropathol 2010;119:641–9.
[104] Jones DT, Kocialkowski S, Liu L, Pearson DM, Backlund LM, Ichimura K, et al. Tandem duplication producing a novel oncogenic BRAF fusion gene defines the majority of pilocytic astrocytomas. Cancer Res 2008;68:8673–7.
[105] Lawson AR, Tatevossian RG, Phipps KP, Picker SR, Michalski A, Sheer D, et al. RAF gene fusions are specific to pilocytic astrocytoma in a broad paediatric brain tumour cohort. Acta Neuropathol 2010;120:271–3.
[106] Nicolaides TP, Li H, Solomon DA, Hariono S, Hashizume R, Barkovich K, et al. Targeted therapy for BRAFV600E malignant astrocytoma. Clin Cancer Res 2011;17:7595–604.
[107] Schiffman JD, Hodgson JG, VandenBerg SR, Flaherty P, Polley MY, Yu M, et al. Oncogenic BRAF mutation with CDKN2A inactivation is characteristic of a subset of pediatric malignant astrocytomas. Cancer Res 2010;70:512–9.
[108] Schindler G, Capper D, Meyer J, Janzarik W, Omran H, Herold-Mende C, et al. Analysis of BRAF V600E mutation in 1,320 nervous system tumors reveals high mutation frequencies in pleomorphic xanthoastrocytoma, ganglioglioma and extra-cerebellar pilocytic astrocytoma. Acta Neuropathol 2011;121:397–405.
[109] Sievert AJ, Jackson EM, Gai X, Hakonarson H, Judkins AR, Resnick AC, et al. Duplication of 7q34 in pediatric low-grade astrocytomas detected by high-density single-nucleotide polymorphism-based genotype arrays results in a novel BRAF fusion gene. Brain Pathol 2009;19:449–58.
[110] Costa AA, Gutmann DH. Brain tumors in neurofibromatosis type 1. Neurooncol Adv 2019;1, vdz040.
[111] Lobbous M, Bernstock JD, Coffee E, Friedman GK, Metrock LK, Chagoya G, et al. An update on neurofibromatosis type 1-associated gliomas. Cancers (Basel) 2020;12:114.
[112] Luebker SA, Koepsell SA. Diverse mechanisms of BRAF inhibitor resistance in melanoma identified in clinical and preclinical studies. Front Oncol 2019;9:268.
[113] Saei A, Eichhorn PJA. Adaptive responses as mechanisms of resistance to BRAF inhibitors in melanoma. Cancers (Basel) 2019;11:1176.
[114] Yaeger R, Corcoran RB. Targeting alterations in the RAF-MEK pathway. Cancer Discov 2019;9:329–41.
[115] Fruman DA, Chiu H, Hopkins BD, Bagrodia S, Cantley LC, Abraham RT. The PI3K pathway in human disease. Cell 2017;170:605–35.
[116] Manning BD, Toker A. AKT/PKB Signaling: navigating the network. Cell 2017;169:381–405.
[117] Huang J, Manning BD. The TSC1-TSC2 complex: a molecular switchboard controlling cell growth. Biochem J 2008;412:179–90.

[118] Masui K, Cavenee WK, Mischel PS. mTORC2 in the center of cancer metabolic reprogramming. Trends Endocrinol Metab 2014;25:364–73.
[119] Saxton RA, Sabatini DM. mTOR signaling in growth, metabolism, and disease. Cell 2017;168:960–76.
[120] Treins C, Warne PH, Magnuson MA, Pende M, Downward J. Rictor is a novel target of p70 S6 kinase-1. Oncogene 2010;29:1003–16.
[121] Fan QW, Cheng C, Knight ZA, Haas-Kogan D, Stokoe D, James CD, et al. EGFR signals to mTOR through PKC and independently of Akt in glioma. Sci Signal 2009;2. ra4.
[122] Zhang J, Gao Z, Yin J, Quon MJ, Ye J. S6K directly phosphorylates IRS-1 on Ser-270 to promote insulin resistance in response to TNF-(alpha) signaling through IKK2. J Biol Chem 2008;283:35375–82.
[123] Liu S, Knapp S, Ahmed AA. The structural basis of PI3K cancer mutations: from mechanism to therapy. Cancer Res 2014;74:641–6.
[124] Li J, Yen C, Liaw D, Podsypanina K, Bose S, Wang SI, et al. PTEN, a putative protein tyrosine phosphatase gene mutated in human brain, breast, and prostate cancer. Science 1997;275:1943–7.
[125] Tohma Y, Gratas C, Biernat W, Peraud A, Fukuda M, Yonekawa Y, et al. PTEN (MMAC1) mutations are frequent in primary glioblastomas (de novo) but not in secondary glioblastomas. J Neuropathol Exp Neurol 1998;57:684–9.
[126] Ruvolo PP. Role of protein phosphatases in the cancer microenvironment. Biochim Biophys Acta Mol Cell Res 1866;2019:144–52.
[127] Rawlings JS, Rosler KM, Harrison DA. The JAK/STAT signaling pathway. J Cell Sci 2004;117:1281–3.
[128] Villarino AV, Kanno Y, O'Shea JJ. Mechanisms and consequences of Jak-STAT signaling in the immune system. Nat Immunol 2017;18:374–84.
[129] Cao X, Tay A, Guy GR, Tan YH. Activation and association of Stat3 with Src in v-Src-transformed cell lines. Mol Cell Biol 1996;16:1595–603.
[130] Ceresa BP, Horvath CM, Pessin JE. Signal transducer and activator of transcription-3 serine phosphorylation by insulin is mediated by a Ras/Raf/MEK-dependent pathway. Endocrinology 1997;138:4131–7.
[131] Chang N, Ahn SH, Kong DS, Lee HW, Nam DH. The role of STAT3 in glioblastoma progression through dual influences on tumor cells and the immune microenvironment. Mol Cell Endocrinol 2017;451:53–65.
[132] Piperi C, Papavassiliou KA, Papavassiliou AG. Pivotal role of STAT3 in shaping glioblastoma immune microenvironment. Cells 2019;8:1398.
[133] Jin W. Role of JAK/STAT3 signaling in the regulation of metastasis, the transition of cancer stem cells, and chemoresistance of cancer by epithelial-mesenchymal transition. Cells 2020;9:217.
[134] Li B, Huang C. Regulation of EMT by STAT3 in gastrointestinal cancer (review). Int J Oncol 2017;50:753–67.
[135] Hammaren HM, Virtanen AT, Raivola J, Silvennoinen O. The regulation of JAKs in cytokine signaling and its breakdown in disease. Cytokine 2018;118:48–63.
[136] Sang Y, Li Y, Song L, Alvarez AA, Zhang W, Lv D, et al. TRIM59 promotes gliomagenesis by inhibiting TC45 dephosphorylation of STAT3. Cancer Res 2018;78:1792–804.
[137] Kohsaka S, Wang L, Yachi K, Mahabir R, Narita T, Itoh T, et al. STAT3 inhibition overcomes temozolomide resistance in glioblastoma by downregulating MGMT expression. Mol Cancer Ther 2012;11:1289–99.
[138] Cooper J, Giancotti FG. Integrin signaling in cancer: mechanotransduction, stemness, epithelial plasticity, and therapeutic resistance. Cancer Cell 2019;35:347–67.
[139] Malric L, Monferran S, Gilhodes J, Boyrie S, Dahan P, Skuli N, et al. Interest of integrins targeting in glioblastoma according to tumor heterogeneity and cancer stem cell paradigm: an update. Oncotarget 2017;8:86947–68.
[140] Hikawa T, Mori T, Abe T, Hori S. The ability in adhesion and invasion of drug-resistant human glioma cells. J Exp Clin Cancer Res 2000;19:357–62.
[141] Renner G, Janouskova H, Noulet F, Koenig V, Guerin E, Bar S, et al. Integrin alpha5beta1 and p53 convergent pathways in the control of anti-apoptotic proteins PEA-15 and survivin in high-grade glioma. Cell Death Differ 2016;23:640–53.
[142] Meyer M, Reimand J, Lan X, Head R, Zhu X, Kushida M, et al. Single cell-derived clonal analysis of human glioblastoma links functional and genomic heterogeneity. Proc Natl Acad Sci U S A 2015;112:851–6.
[143] Huizer K, Zhu C, Chirifi I, Krist B, Zorgman D, van der Weiden M, et al. Periostin is expressed by pericytes and is crucial for angiogenesis in glioma. J Neuropathol Exp Neurol 2020;79:863–72.

[144] Park SY, Piao Y, Jeong KJ, Dong J, de Groot JF. Periostin (POSTN) regulates tumor resistance to antiangiogenic therapy in glioma models. Mol Cancer Ther 2016;15:2187–97.

[145] Colak S, Ten Dijke P. Targeting TGF-beta signaling in cancer. Trends Cancer 2017;3:56–71.

[146] Hill CS. Transcriptional control by the SMADs. Cold Spring Harb Perspect Biol 2016;8.

[147] Todaro GJ, De Larco JE, Fryling C, Johnson PA, Sporn MB. Transforming growth factors (TGFs): properties and possible mechanisms of action. J Supramol Struct Cell Biochem 1981;15:287–301.

[148] Massague J. Integration of Smad and MAPK pathways: a link and a linker revisited. Genes Dev 2003;17:2993–7.

[149] Lamouille S, Xu J, Derynck R. Molecular mechanisms of epithelial-mesenchymal transition. Nat Rev Mol Cell Biol 2014;15:178–96.

[150] Wang J, Cazzato E, Ladewig E, Frattini V, Rosenbloom DI, Zairis S, et al. Clonal evolution of glioblastoma under therapy. Nat Genet 2016;48:768–76.

[151] Elias MC, Tozer KR, Silber JR, Mikheeva S, Deng M, Morrison RS, et al. TWIST is expressed in human gliomas and promotes invasion. Neoplasia 2005;7:824–37.

[152] Mahabir R, Tanino M, Elmansuri A, Wang L, Kimura T, Itoh T, et al. Sustained elevation of snail promotes glial-mesenchymal transition after irradiation in malignant glioma. Neuro-Oncology 2014;16:671–85.

[153] Mehta S, Lo CC. Developmentally regulated signaling pathways in glioma invasion. Cell Mol Life Sci 2018;75:385–402.

[154] Patel S, Alam A, Pant R, Chattopadhyay S. Wnt signaling and its significance within the tumor microenvironment: novel therapeutic insights. Front Immunol 2019;10:2872.

[155] Zhan T, Rindtorff N, Boutros M. Wnt signaling in cancer. Oncogene 2017;36:1461–73.

[156] Kim JH, Park SY, Jun Y, Kim JY, Nam JS. Roles of Wnt target genes in the journey of cancer stem cells. Int J Mol Sci 2017;18:1604.

[157] Auger N, Thillet J, Wanherdrick K, Idbaih A, Legrier ME, Dutrillaux B, et al. Genetic alterations associated with acquired temozolomide resistance in SNB-19, a human glioma cell line. Mol Cancer Ther 2006;5:2182–92.

[158] Verhaak RG, Hoadley KA, Purdom E, Wang V, Qi Y, Wilkerson MD, et al. Integrated genomic analysis identifies clinically relevant subtypes of glioblastoma characterized by abnormalities in PDGFRA, IDH1, EGFR, and NF1. Cancer Cell 2010;17:98–110.

[159] Hermida MA, Dinesh Kumar J, Leslie NR. GSK3 and its interactions with the PI3K/AKT/mTOR signalling network. Adv Biol Regul 2017;65:5–15.

[160] Niyaz M, Khan MS, Mudassar S. Hedgehog signaling: an Achilles' heel in cancer. Transl Oncol 2019;12:1334–44.

[161] Sari IN, Phi LTH, Jun N, Wijaya YT, Lee S, Kwon HY. Hedgehog signaling in cancer: a prospective therapeutic target for eradicating cancer stem cells. Cells 2018;7:208.

[162] Chandra V, Das T, Gulati P, Biswas NK, Rote S, Chatterjee U, et al. Hedgehog signaling pathway is active in GBM with GLI1 mRNA expression showing a single continuous distribution rather than discrete high/low clusters. PLoS One 2015;10, e0116390.

[163] Xu Q, Yuan X, Liu G, Black KL, Yu JS. Hedgehog signaling regulates brain tumor-initiating cell proliferation and portends shorter survival for patients with PTEN-coexpressing glioblastomas. Stem Cells 2008;26:3018–26.

[164] Melamed JR, Morgan JT, Ioele SA, Gleghorn JP, Sims-Mourtada J, Day ES. Investigating the role of hedgehog/GLI1 signaling in glioblastoma cell response to temozolomide. Oncotarget 2018;9:27000–15.

[165] Wang K, Chen D, Qian Z, Cui D, Gao L, Lou M. Hedgehog/Gli1 signaling pathway regulates MGMT expression and chemoresistance to temozolomide in human glioblastoma. Cancer Cell Int 2017;17:117.

[166] Aster JC, Pear WS, Blacklow SC. The varied roles of notch in cancer. Annu Rev Pathol 2017;12:245–75.

[167] Venkatesh V, Nataraj R, Thangaraj GS, Karthikeyan M, Gnanasekaran A, Kaginelli SB, et al. Targeting notch signalling pathway of cancer stem cells. Stem Cell Investig 2018;5:5.

[168] Katoh M, Katoh M. Precision medicine for human cancers with notch signaling dysregulation (review). Int J Mol Med 2020;45:279–97.

[169] Bazzoni R, Bentivegna A. Role of notch signaling pathway in glioblastoma pathogenesis. Cancers (Basel) 2019;11:292.

[170] Xiu MX, Liu YM, Kuang BH. The oncogenic role of Jagged1/notch signaling in cancer. Biomed Pharmacother 2020;129:110416.

[171] Gilbert CA, Daou MC, Moser RP, Ross AH. Gamma-secretase inhibitors enhance temozolomide treatment of human gliomas by inhibiting neurosphere repopulation and xenograft recurrence. Cancer Res 2010;70:6870–9.

[172] Saito N, Fu J, Zheng S, Yao J, Wang S, Liu DD, et al. A high notch pathway activation predicts response to gamma secretase inhibitors in proneural subtype of glioma tumor-initiating cells. Stem Cells 2014;32:301–12.

[173] Zanconato F, Cordenonsi M, Piccolo S. YAP and TAZ: a signalling hub of the tumour microenvironment. Nat Rev Cancer 2019;19:454–64.

[174] Zhao Y, Yang X. The hippo pathway in chemotherapeutic drug resistance. Int J Cancer 2015;137:2767–73.

[175] Bao Y, Hata Y, Ikeda M, Withanage K. Mammalian hippo pathway: from development to cancer and beyond. J Biochem 2011;149:361–79.

[176] Wang Y, Pan P, Wang Z, Zhang Y, Xie P, Geng D, et al. Beta-catenin-mediated YAP signaling promotes human glioma growth. J Exp Clin Cancer Res 2017;36:136.

[177] Zhang H, Geng D, Gao J, Qi Y, Shi Y, Wang Y, et al. Expression and significance of Hippo/YAP signaling in glioma progression. Tumour Biol 2016;37:15665–76.

[178] Zanconato F, Cordenonsi M, Piccolo S. YAP/TAZ at the roots of cancer. Cancer Cell 2016;29:783–803.

[179] LaPak KM, Burd CE. The molecular balancing act of p16(INK4a) in cancer and aging. Mol Cancer Res 2014;12:167–83.

[180] Sherr CJ, Roberts JM. CDK inhibitors: positive and negative regulators of G1-phase progression. Genes Dev 1999;13:1501–12.

[181] Dyson NJ. RB1: a prototype tumor suppressor and an enigma. Genes Dev 2016;30:1492–502.

[182] Hinds PW, Mittnacht S, Dulic V, Arnold A, Reed SI, Weinberg RA. Regulation of retinoblastoma protein functions by ectopic expression of human cyclins. Cell 1992;70:993–1006.

[183] Lavoie JN, L'Allemain G, Brunet A, Muller R, Pouyssegur J. Cyclin D1 expression is regulated positively by the p42/p44MAPK and negatively by the p38/HOGMAPK pathway. J Biol Chem 1996;271:20608–16.

[184] VanArsdale T, Boshoff C, Arndt KT, Abraham RT. Molecular pathways: targeting the cyclin D-CDK4/6 axis for cancer treatment. Clin Cancer Res 2015;21:2905–10.

[185] Schmidt EE, Ichimura K, Reifenberger G, Collins VP. CDKN2 (p16/MTS1) gene deletion or CDK4 amplification occurs in the majority of glioblastomas. Cancer Res 1994;54:6321–4.

[186] Ueki K, Ono Y, Henson JW, Efird JT, von Deimling A, Louis DN. CDKN2/p16 or RB alterations occur in the majority of glioblastomas and are inversely correlated. Cancer Res 1996;56:150–3.

[187] Shaw M, Cohen P, Alessi DR. Further evidence that the inhibition of glycogen synthase kinase-3beta by IGF-1 is mediated by PDK1/PKB-induced phosphorylation of Ser-9 and not by dephosphorylation of Tyr-216. FEBS Lett 1997;416:307–11.

[188] Yu H, Li Z, Wang M. Expression and prognostic role of E2F transcription factors in high-grade glioma. CNS Neurosci Ther 2020;26:741–53.

[189] Gimple RC, Bhargava S, Dixit D, Rich JN. Glioblastoma stem cells: lessons from the tumor hierarchy in a lethal cancer. Genes Dev 2019;33:591–609.

[190] Singh SK, Clarke ID, Terasaki M, Bonn VE, Hawkins C, Squire J, et al. Identification of a cancer stem cell in human brain tumors. Cancer Res 2003;63:5821–8.

[191] Ahmed SI, Javed G, Laghari AA, Bareeqa SB, Farrukh S, Zahid S, et al. CD133 expression in glioblastoma multiforme: a literature review. Cureus 2018;10, e3439.

[192] Xu HS, Qin XL, Zong HL, He XG, Cao L. Cancer stem cell markers in glioblastoma—an update. Eur Rev Med Pharmacol Sci 2017;21:3207–11.

[193] Boivin D, Labbe D, Fontaine N, Lamy S, Beaulieu E, Gingras D, et al. The stem cell marker CD133 (prominin-1) is phosphorylated on cytoplasmic tyrosine-828 and tyrosine-852 by Src and Fyn tyrosine kinases. Biochemistry 2009;48:3998–4007.

[194] Wei Y, Jiang Y, Zou F, Liu Y, Wang S, Xu N, et al. Activation of PI3K/Akt pathway by CD133-p85 interaction promotes tumorigenic capacity of glioma stem cells. Proc Natl Acad Sci U S A 2013;110:6829–34.

[195] Chen YS, Wu MJ, Huang CY, Lin SC, Chuang TH, Yu CC, et al. CD133/Src axis mediates tumor initiating property and epithelial-mesenchymal transition of head and neck cancer. PLoS One 2011;6, e28053.

[196] Bao S, Wu Q, Sathornsumetee S, Hao Y, Li Z, Hjelmeland AB, et al. Stem cell-like glioma cells promote tumor angiogenesis through vascular endothelial growth factor. Cancer Res 2006;66:7843–8.

[197] Mooney KL, Choy W, Sidhu S, Pelargos P, Bui TT, Voth B, et al. The role of CD44 in glioblastoma multiforme. J Clin Neurosci 2016;34:1–5.
[198] Pietras A, Katz AM, Ekstrom EJ, Wee B, Halliday JJ, Pitter KL, et al. Osteopontin-CD44 signaling in the glioma perivascular niche enhances cancer stem cell phenotypes and promotes aggressive tumor growth. Cell Stem Cell 2014;14:357–69.
[199] Chen C, Zhao S, Karnad A, Freeman JW. The biology and role of CD44 in cancer progression: therapeutic implications. J Hematol Oncol 2018;11:64.
[200] Chanmee T, Ontong P, Kimata K, Itano N. Key roles of hyaluronan and its CD44 receptor in the stemness and survival of cancer stem cells. Front Oncol 2015;5:180.
[201] Robey RW, Pluchino KM, Hall MD, Fojo AT, Bates SE, Gottesman MM. Revisiting the role of ABC transporters in multidrug-resistant cancer. Nat Rev Cancer 2018;18:452–64.
[202] Wijaya J, Fukuda Y, Schuetz JD. Obstacles to brain tumor therapy: key ABC transporters. Int J Mol Sci 2017;18:2544.
[203] Decleves X, Fajac A, Lehmann-Che J, Tardy M, Mercier C, Hurbain I, et al. Molecular and functional MDR1-Pgp and MRPs expression in human glioblastoma multiforme cell lines. Int J Cancer 2002;98:173–80.
[204] Schinkel AH. P-glycoprotein, a gatekeeper in the blood-brain barrier. Adv Drug Deliv Rev 1999;36:179–94.
[205] Bargou RC, Jurchott K, Wagener C, Bergmann S, Metzner S, Bommert K, et al. Nuclear localization and increased levels of transcription factor YB-1 in primary human breast cancers are associated with intrinsic MDR1 gene expression. Nat Med 1997;3:447–50.
[206] Lim JC, Kania KD, Wijesuriya H, Chawla S, Sethi JK, Pulaski L, et al. Activation of beta-catenin signalling by GSK-3 inhibition increases p-glycoprotein expression in brain endothelial cells. J Neurochem 2008;106:1855–65.
[207] Yang JM, Vassil AD, Hait WN. Activation of phospholipase C induces the expression of the multidrug resistance (MDR1) gene through the Raf-MAPK pathway. Mol Pharmacol 2001;60:674–80.
[208] Nakai E, Park K, Yawata T, Chihara T, Kumazawa A, Nakabayashi H, et al. Enhanced MDR1 expression and chemoresistance of cancer stem cells derived from glioblastoma. Cancer Investig 2009;27:901–8.
[209] Munoz JL, Walker ND, Scotto KW, Rameshwar P. Temozolomide competes for P-glycoprotein and contributes to chemoresistance in glioblastoma cells. Cancer Lett 2015;367:69–75.
[210] Schaich M, Kestel L, Pfirrmann M, Robel K, Illmer T, Kramer M, et al. A MDR1 (ABCB1) gene single nucleotide polymorphism predicts outcome of temozolomide treatment in glioblastoma patients. Ann Oncol 2009;20:175–81.
[211] Tivnan A, Zakaria Z, O'Leary C, Kogel D, Pokorny JL, Sarkaria JN, et al. Inhibition of multidrug resistance protein 1 (MRP1) improves chemotherapy drug response in primary and recurrent glioblastoma multiforme. Front Neurosci 2015;9:218.
[212] Jin Y, Bin ZQ, Qiang H, Liang C, Hua C, Jun D, et al. ABCG2 is related with the grade of glioma and resistance to mitoxantone, a chemotherapeutic drug for glioma. J Cancer Res Clin Oncol 2009;135:1369–76.
[213] Mao Q, Unadkat JD. Role of the breast cancer resistance protein (BCRP/ABCG2) in drug transport—an update. AAPS J 2015;17:65–82.
[214] de Gooijer MC, de Vries NA, Buckle T, Buil LCM, Beijnen JH, Boogerd W, et al. Improved brain penetration and antitumor efficacy of temozolomide by inhibition of ABCB1 and ABCG2. Neoplasia 2018;20:710–20.
[215] Belousov A, Titov S, Shved N, Garbuz M, Malykin G, Gulaia V, et al. The extracellular matrix and biocompatible materials in glioblastoma treatment. Front Bioeng Biotechnol 2019;7:341.
[216] Henke E, Nandigama R, Ergun S. Extracellular matrix in the tumor microenvironment and its impact on cancer therapy. Front Mol Biosci 2019;6:160.
[217] de Gooijer MC, Guillen Navarro M, Bernards R, Wurdinger T, van Tellingen O. An experimenter's guide to glioblastoma invasion pathways. Trends Mol Med 2018;24:763–80.
[218] Lau LW, Cua R, Keough MB, Haylock-Jacobs S, Yong VW. Pathophysiology of the brain extracellular matrix: a new target for remyelination. Nat Rev Neurosci 2013;14:722–9.
[219] Koh I, Cha J, Park J, Choi J, Kang SG, Kim P. The mode and dynamics of glioblastoma cell invasion into a decellularized tissue-derived extracellular matrix-based three-dimensional tumor model. Sci Rep 2018;8:4608.
[220] Dai J, Peng L, Fan K, Wang H, Wei R, Ji G, et al. Osteopontin induces angiogenesis through activation of PI3K/AKT and ERK1/2 in endothelial cells. Oncogene 2009;28:3412–22.

[221] Ellert-Miklaszewska A, Wisniewski P, Kijewska M, Gajdanowicz P, Pszczolkowska D, Przanowski P, et al. Tumour-processed osteopontin and lactadherin drive the protumorigenic reprogramming of microglia and glioma progression. Oncogene 2016;35:6366–77.
[222] Raymond A, Ensslin MA, Shur BD. SED1/MFG-E8: a bi-motif protein that orchestrates diverse cellular interactions. J Cell Biochem 2009;106:957–66.
[223] Nandhu MS, Kwiatkowska A, Bhaskaran V, Hayes J, Hu B, Viapiano MS. Tumor-derived fibulin-3 activates pro-invasive NF-kappaB signaling in glioblastoma cells and their microenvironment. Oncogene 2017; 36:4875–86.
[224] Hu B, Nandhu MS, Sim H, Agudelo-Garcia PA, Saldivar JC, Dolan CE, et al. Fibulin-3 promotes glioma growth and resistance through a novel paracrine regulation of notch signaling. Cancer Res 2012;72:3873–85.
[225] Xia S, Lal B, Tung B, Wang S, Goodwin CR, Laterra J. Tumor microenvironment tenascin-C promotes glioblastoma invasion and negatively regulates tumor proliferation. Neuro-Oncology 2016;18:507–17.
[226] Guo X, Xue H, Shao Q, Wang J, Guo X, Chen X, et al. Hypoxia promotes glioma-associated macrophage infiltration via periostin and subsequent M2 polarization by upregulating TGF-beta and M-CSFR. Oncotarget 2016;7:80521–42.
[227] Dirkse A, Golebiewska A, Buder T, Nazarov PV, Muller A, Poovathingal S, et al. Stem cell-associated heterogeneity in glioblastoma results from intrinsic tumor plasticity shaped by the microenvironment. Nat Commun 2019;10:1787.
[228] Monteiro AR, Hill R, Pilkington GJ, Madureira PA. The role of hypoxia in glioblastoma invasion. Cells 2017;6:45.
[229] Poff A, Koutnik AP, Egan KM, Sahebjam S, D'Agostino D, Kumar NB. Targeting the Warburg effect for cancer treatment: ketogenic diets for management of glioma. Semin Cancer Biol 2019;56:135–48.
[230] Chiche J, Brahimi-Horn MC, Pouyssegur J. Tumour hypoxia induces a metabolic shift causing acidosis: a common feature in cancer. J Cell Mol Med 2010;14:771–94.
[231] Peppicelli S, Bianchini F, Calorini L. Dynamic scenario of metabolic pathway adaptation in tumors and therapeutic approach. Onco Targets Ther 2015;2:225–32.
[232] Lee P, Chandel NS, Simon MC. Cellular adaptation to hypoxia through hypoxia inducible factors and beyond. Nat Rev Mol Cell Biol 2020;21:268–83.
[233] D'Ignazio L, Bandarra D, Rocha S. NF-kappaB and HIF crosstalk in immune responses. FEBS J 2016;283:413–24.
[234] Hudson CC, Liu M, Chiang GG, Otterness DM, Loomis DC, Kaper F, et al. Regulation of hypoxia-inducible factor 1alpha expression and function by the mammalian target of rapamycin. Mol Cell Biol 2002;22:7004–14.
[235] Deng B, Zhu JM, Wang Y, Liu TT, Ding YB, Xiao WM, et al. Intratumor hypoxia promotes immune tolerance by inducing regulatory T cells via TGF-beta1 in gastric cancer. PLoS One 2013;8, e63777.
[236] Mingyuan X, Qianqian P, Shengquan X, Chenyi Y, Rui L, Yichen S, et al. Hypoxia-inducible factor-1alpha activates transforming growth factor-beta1/Smad signaling and increases collagen deposition in dermal fibroblasts. Oncotarget 2018;9:3188–97.
[237] Bar EE, Lin A, Mahairaki V, Matsui W, Eberhart CG. Hypoxia increases the expression of stem-cell markers and promotes clonogenicity in glioblastoma neurospheres. Am J Pathol 2010;177:1491–502.
[238] Bar EE. Glioblastoma, cancer stem cells and hypoxia. Brain Pathol 2011;21:119–29.
[239] Seidel S, Garvalov BK, Wirta V, von Stechow L, Schanzer A, Meletis K, et al. A hypoxic niche regulates glioblastoma stem cells through hypoxia inducible factor 2 alpha. Brain 2010;133:983–95.
[240] Charles NA, Holland EC, Gilbertson R, Glass R, Kettenmann H. The brain tumor microenvironment. Glia 2012;60:502–14.
[241] Zhang ZM, Yang Z, Zhang Z. Distribution and characterization of tumor-associated macrophages/microglia in rat C6 glioma. Oncol Lett 2015;10:2442–6.
[242] Ku MC, Wolf SA, Respondek D, Matyash V, Pohlmann A, Waiczies S, et al. GDNF mediates glioblastoma-induced microglia attraction but not astrogliosis. Acta Neuropathol 2013;125:609–20.
[243] Jin P, Shin SH, Chun YS, Shin HW, Shin YJ, Lee Y, et al. Astrocyte-derived CCL20 reinforces HIF-1-mediated hypoxic responses in glioblastoma by stimulating the CCR6-NF-kappaB signaling pathway. Oncogene 2018;37:3070–87.
[244] Chen W, Xia T, Wang D, Huang B, Zhao P, Wang J, et al. Human astrocytes secrete IL-6 to promote glioma migration and invasion through upregulation of cytomembrane MMP14. Oncotarget 2016;7:62425–38.

[245] West AJ, Tsui V, Stylli SS, Nguyen HPT, Morokoff AP, Kaye AH, et al. The role of interleukin-6-STAT3 signalling in glioblastoma. Oncol Lett 2018;16:4095–104.
[246] Yeung YT, McDonald KL, Grewal T, Munoz L. Interleukins in glioblastoma pathophysiology: implications for therapy. Br J Pharmacol 2013;168:591–606.
[247] Buruiana A, Florian SI, Florian AI, Timis TL, Mihu CM, Miclaus M, et al. The roles of miRNA in glioblastoma tumor cell communication: diplomatic and aggressive negotiations. Int J Mol Sci 2020;21:1950.
[248] Sin WC, Crespin S, Mesnil M. Opposing roles of connexin43 in glioma progression. Biochim Biophys Acta 1818;2012:2058–67.
[249] Ginhoux F, Lim S, Hoeffel G, Low D, Huber T. Origin and differentiation of microglia. Front Cell Neurosci 2013;7:45.
[250] Gutmann DH, Kettenmann H. Microglia/brain macrophages as central drivers of brain tumor pathobiology. Neuron 2019;104:442–9.
[251] Peng Y, Croce CM. The role of MicroRNAs in human cancer. Signal Transduct Target Ther 2016;1:15004.
[252] Kwak HJ, Kim YJ, Chun KR, Woo YM, Park SJ, Jeong JA, et al. Downregulation of Spry2 by miR-21 triggers malignancy in human gliomas. Oncogene 2011;30:2433–42.
[253] Li Y, Li W, Yang Y, Lu Y, He C, Hu G, et al. MicroRNA-21 targets LRRFIP1 and contributes to VM-26 resistance in glioblastoma multiforme. Brain Res 2009;1286:13–8.
[254] Papagiannakopoulos T, Shapiro A, Kosik KS. MicroRNA-21 targets a network of key tumor-suppressive pathways in glioblastoma cells. Cancer Res 2008;68:8164–72.

CHAPTER

23

Targeting the molecular mechanisms of glioma stem cell resistance to chemotherapy

Jawad Fares[a], *Deepak Kanojia*[a], *Alex Cordero*[a], *Ilya Ulasov*[b], *and Maciej S. Lesniak*[a]

[a]Department of Neurological Surgery, Feinberg School of Medicine, Northwestern University, Chicago, IL, United States [b]Group of Experimental Biotherapy and Diagnostics, Institute for Regenerative Medicine, Sechenov First Moscow State Medical University, Moscow, Russia

Abstract

Glioma stem cells (GSCs) are a subpopulation of glioblastoma cells that display high levels of cellular plasticity. They possess functional characteristics that permit sustained self-renewal, persistent proliferation, and tumor initiation. In addition, they have an intrinsic ability to invade through the brain parenchyma, evade the immune system, adapt to microenvironmental pressures, and promote angiogenesis. Adaptive interactions between GSCs and their biological niches in the brain microenvironment offer them a safe haven from drugs that cannot cross the blood-brain barrier. Furthermore, GSC inherent resistance to treatment renders standard of care therapy futile, leading to tumor progression and recurrence. As such, GSCs have been blamed for clinical deterioration and dismal survival outcomes in patients with glioma. Mechanisms of GSC resistance include O(6)-methylguanine-DNA methyltransferase (MGMT) promoter methylation status, enhanced repair of DNA damage, impaired induction of apoptosis, aberrant DNA checkpoint analysis, expression of ATP-binding cassette (ABC) transporters and multidrug resistance, and phenotypic plasticity. Understanding the molecular mechanisms of GSC chemoresistance allows the development of efficient therapeutic strategies to overcome it. Pro-differentiation therapies, including bone morphogenic proteins (BMPs) and posttranscriptional modification using microRNAs (miRNAs), drive GSCs into more differentiated states, sensitizing them to cytotoxic treatments. Targeting of the immunogenic, neuronal, hypoxic, and/or angiogenic niches of GSCs offer promising therapeutic potential. Intracellularly, inhibition of the cell signaling pathways that promote GSC survival and progression is essential for reversing GSC chemotherapeutic resistance. So far, the translation of monotherapeutic agents to the clinical scene failed to instigate the desired clinical benefit in patient outcomes. Therefore, the combination of therapeutic strategies holds promise as efforts to overcome GSC resistance continue to build. Technological advances in single-cell genomics and CRISPR-Cas9 genome-editing tools and derivatives offer better opportunities to achieve strides in precision cancer therapeutics.

Glioblastoma Resistance to Chemotherapy: Molecular Mechanisms and Innovative Reversal Strategies
https://doi.org/10.1016/B978-0-12-821567-8.00014-2

Abbreviations

ABC ATP-binding cassette
ADAM10 disintegrin and metalloproteinase domain-containing protein 10
AKT protein kinase B
ALDH1 aldehyde dehydrogenase 1
APE1 apurinic/apyrimidinic endonuclease
AurA aurora kinase A
Bak1 BCL2 antagonist/killer 1
Batf3 basic leucine zipper ATF-like transcription factor 3
BBB blood-brain barrier
Bcl2L12 Bcl2-like12
BMPR1a BMP type IA receptor
CA9 carbonic anhydrase 9
CAMTA1 calmodulin-binding transcription activator 1
CAR chimeric antigen receptor
CDK6 cyclin-dependent kinase 6
Chk checkpoint kinase
c-Met hepatocyte growth factor receptor
COX-2 cyclooxygenase
CSF1R colony-stimulating factor 1 receptor
CSMD1 CUB and sushi multiple domains 1
CTLA-4 cytotoxic T-lymphocyte-associated protein 4
ECM extracellular matrix
EGFR epidermal growth factor receptor
EMT epithelial-to-mesenchymal transition
EPHA6 erythropoietin-producing hepatocellular carcinoma receptor
ERK extracellular-signal-regulated kinase
EZH2 enhancer of zeste 2 polycomb repressive complex 2 subunit
FAK focal adhesion kinase
GFAP glial fibrillary acidic protein
GLUT3 glucose transporter 3
GSCs glioma stem cells
GSIs gamma secretase inhibitors
HDAC histone deacetylases
HES1 Hes family BHLH transcription factor 1
HIF hypoxia-inducible factor
HOXD10 homeobox D10
HSV herpes simplex virus
ID2 inhibitor of differentiation 2
IDH isocitrate dehydrogenase
IDO1 indoleamine 2,3-dioxygenase 1
IFNβ interferon-beta
IL13RA2 IL-13 receptor subunit alpha 2
IL-6 interleukin-6
INPP5F inositol polyphosphate-5-phosphatase F
JAK2 Janus kinase 2
MAPK mitogen-activated protein kinase

MCT	monocarboxylate transporter
MGMT	O(6)-methylguanine-DNA methyltransferase
MHC-I	major histocompatibility complex I
miRNA	microRNA
MMP-9	matrix metallopeptidase 9
MRP1	multidrug resistance protein 1
mTOR	mammalian target of rapamycin
MYC	MYC proto-oncogene
NCOR1	nuclear receptor corepressor 1
NF-κB	nuclear factor kappa B
NK	natural killer
NLGN3	neuroligin-3
OLIG2	oligodendrocyte transcription factor 2
PARP	poly ADP-ribose polymerase
PD-1	programmed death 1
PDGFRA	platelet derived growth factor receptor alpha
PDK1	phosphoinositide-dependent kinase-1
PD-L1	programmed death-ligand 1
PGK1	phosphoglycerate kinase 1
PI3K	phosphoinositide 3-kinase
PIAS3	protein inhibitor of activated STAT 3
PLAGL2	PLAG1 like zinc finger 2
PlGF	placental growth factor
PTEN	phosphatase and tensin homolog
PTPRZ1	protein tyrosine phosphatase receptor type Z1
RBPJ	recombining binding protein suppressor of hairless
ROS	reactive oxygen species
RTK	receptor tyrosine kinases
RTVP-1	related to testis-specific, vespid, and pathogenesis proteins 1
Shh	sonic hedgehog
SMO	smoothened
SOX2	SRY-Box transcription factor 2
STAT3	signal transducer and activator of transcription 3
Syk	spleen tyrosine kinase
TAMs	tumor-associated macrophages and myeloid cells
TGF-β	transforming growth factor beta
TNC	tenascin C
TNF-α	tumor necrosis factor alpha
TRIM8	tri-partite motif-containing protein 8
UPR	unfolded protein response
VEGF	vascular endothelial growth factor
WNT5A	Wnt family member 5A

Conflict of interest

No potential conflicts of interest were disclosed.

Introduction

Malignant glioma is the most common and lethal primary brain tumor. It has a dismal 5-year survival rate of only 5% and a median survival of less than 15 months [1, 2]. The current standard of care for malignant glioma involves maximal surgical resection followed by radiation and cytotoxic chemotherapy with temozolomide [3]. Tumor-treating fields represent a new therapeutic modality that extends overall patient survival to 20.9 months when used in combination with radiotherapy and chemotherapy [4]. Most recently, the extra removal of noncontrast-enhanced tumor tissue has shown to increase survival up to 37.3 months in certain populations [5]. Still, the presence of originally resistant malignant subpopulations of cells alongside acquired epigenetic and genetic alterations of naive tumor cells renders them able to survive therapeutic interventions [6]. This leads to full recurrence of the tumor with a more aggressive phenotype.

Genetic heterogeneity, which is at the root of tumor resistance and recurrence, is acquired through an evolutionary process. Description of glioma subclones carrying genetic amplifications of distinctive receptor tyrosine kinases (RTK), such as epidermal growth factor receptor (EGFR) and platelet-derived growth factor receptor alpha (PDGFRA) further endorsed this concept [7, 8]. As such, therapeutic agents that specifically target a single RTK would not be able to eradicate the whole tumor and, subsequently, nontargeted subclones would lead to tumor relapse [9]. Functional heterogeneity dictates the existence of a hierarchical model of tumorigenesis, in which specific subpopulations of glioma cells, named glioma stem cells (GSCs), are capable of self-renewal, multipotency, and tumorigenic induction [10–13] (Fig. 1). Single-cell RNA sequencing of primary tumor specimens captured the great extent of intratumoral heterogeneity and identified a slow-dividing quiescent population of stem-like cells in glioblastoma tumors [14–16]. The division of a GSC leads to the generation of two differentiated glioma cells or self-renewal to sustain the GSC pool. As such, GSCs have been directly correlated with the underlying recurrence in the face of aggressive multimodal therapies [17–19]. Studies on human tumor samples have shown that GSCs survive irradiation and chemotherapy better than non-GSC matched populations [20–24], leading to therapeutic resistance [25, 26]. Clinical observation of both recurrent tumors and tumors examined immediately after chemotherapy and radiotherapy have been shown to contain a higher proportion of GSCs [27]. Furthermore, the stemness of GSCs is a dynamic process influenced by factors in the microenvironment, such as hypoxia, intratumoral pH, and even therapeutic stress [18, 28–32]. Therefore, differentiated glioma cells can also acquire a GSC state in response to microenvironmental cues or therapeutic stress. Isolating GSCs can be done by neurosphere formation assays or through two-dimensional adherent culture of GSCs on poly-L-lysine/laminin-coated plates [33]. More recently, three-dimensional organoid culture systems of human glioblastomas recapitulated in vivo tumor growth, cancer stem cell heterogeneity, and hypoxic gradients [34]. In addition, CRISPR-Cas9 genomic editing helped generate glioblastoma organoid models of invasive phenotypes [35–37].

Isolation and detection of GSCs permitted the understanding of their biology. GSCs share a decent amount of markers with neural stem cells that include SRY-Box transcription factor 2 (SOX2), oligodendrocyte transcription factor 2 (OLIG2), MYC proto-oncogene (MYC), and Nestin, as well as cell surface markers, such as CD133, L1 Cell Adhesion Molecule

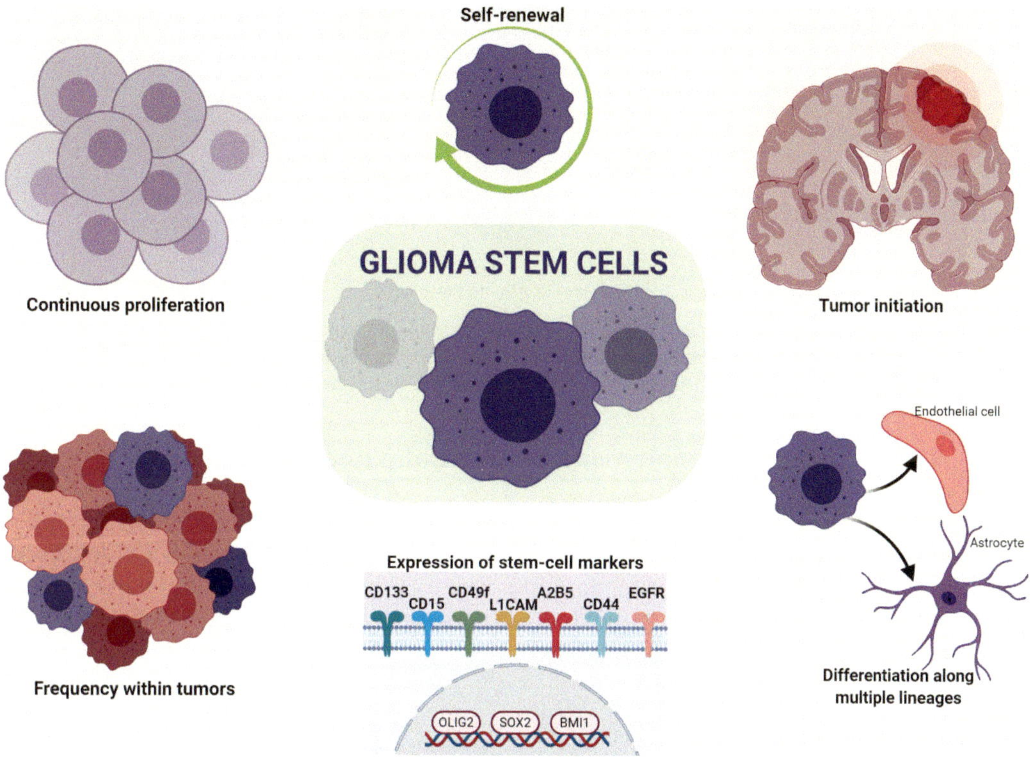

FIG. 1 Characteristics of glioma stem cells (GSCs). GSCs are defined by functional characteristics that govern their self-renewal, continuous proliferation, and tumor initiation capacity. In addition, the GSC frequency within tumors, the expression of a set of stem cell markers, and their ability to differentiate along multiple lineages are common features and characteristics of stem cells in general.

(L1CAM), CD44, and A2B5 [38]. Promising examples of other common markers include CD15/SSEA, Notch, enhancer of zeste 2 polycomb repressive complex 2 subunit (EZH2), signal transducer and activator of transcription 3 (STAT3), and a host of other surface moieties and transcription factors [9, 12, 39, 40].

GSCs have been blamed for clinical deterioration and dismal survival outcomes in patients with glioma (Fig. 2). Most therapeutic strategies are mainly focused on the direct ablation of GSCs by targeting cell surface markers and specific pathways that are required for maintaining GSC stemness. Other plausible approaches include specifically targeting the ability of GSCs to interact with their microenvironment, exploiting their dependency on angiogenesis, and halting their immune evasion mechanisms.

The elimination of GSCs is crucial in treating malignant glioma. This chapter explores the adaptation of GSCs to their microenvironment in the brain, the molecular mechanisms that govern GSC chemoresistance and malignant recurrence, and novel targeted therapies and effective molecular mechanisms that can be explored to overcome GSC chemoresistance.

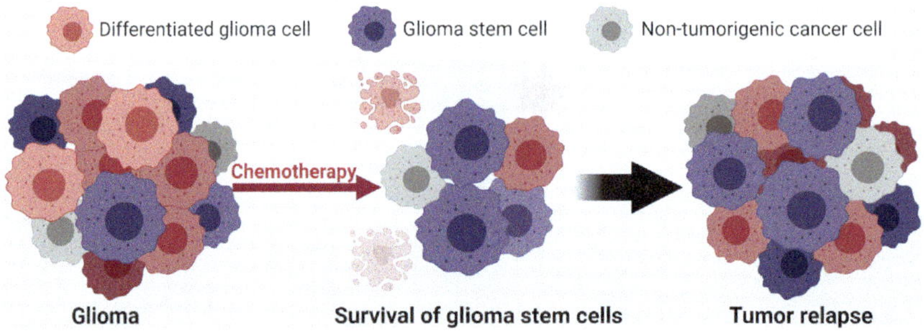

FIG. 2 Chemoresistance of glioma stem cells (GSCs). Chemotherapeutic treatment of malignant gliomas can effectively target differentiated and nontumorigenic cancer cells that are highly proliferative. Slowly dividing or quiescent GSCs remain unaffected, which later leads to tumor relapse and chemoresistance.

The adaptive microenvironment of glioma stem cells

GSCs have a synergistic, codependent relationship with their environment within the tumor, inhabiting specialized niches that have now been characterized. Stromal and cellular interactions are necessary to preserve stem-like features. In addition, interactions with the neuronal, immune, and perivascular niches are necessary to maintain survival and adaptation in the brain tumor microenvironment (Fig. 3).

Stromal interactions

Stromal interactions are required for GSC maintenance. Integrin α7, for instance, maintains GSC proliferation and invasiveness, which makes it a target for therapeutic intervention [41]. Furthermore, high expression of the extracellular matrix (ECM) glycoprotein tenascin C (TNC) in isocitrate dehydrogenase (IDH)-mutant glioma increases ECM stiffness, which leads to increased mechano-signaling through the oncogenic focal adhesion kinase (FAK) pathway [42]. In addition, TNC promotes the preservation of stemness through integrin-mediated Notch signaling [43, 44]. Moreover, mesenchymal GSCs, nontumorigenic stromal components of malignant gliomas, support GSCs through interleukin-6 (IL-6)/STAT3 signaling [45] and exosomal secretions of microRNA (miRNA) [46].

Cellular interactions

Transcriptomic and genomic mapping of malignant gliomas allowed functional characterization of their intratumoral heterogeneity [47]. Intercellular interactions play vital roles in maintaining the complex cellular systems of gliomas. Microtubes that connect glioma cells form a multicellular anatomic network that permits connexin-mediated communication between cells, provide tracts for invasion, and are required for maintenance of the tumor cell network after therapy [48]. Functional connexin-mediated gap junctions are essential for GSC self-renewal [49]. Communication via secreted factors is also important, as differentiated glioblastoma cells produce brain-derived neurotrophic factor to support the growth and survival of their stem cell counterparts [50].

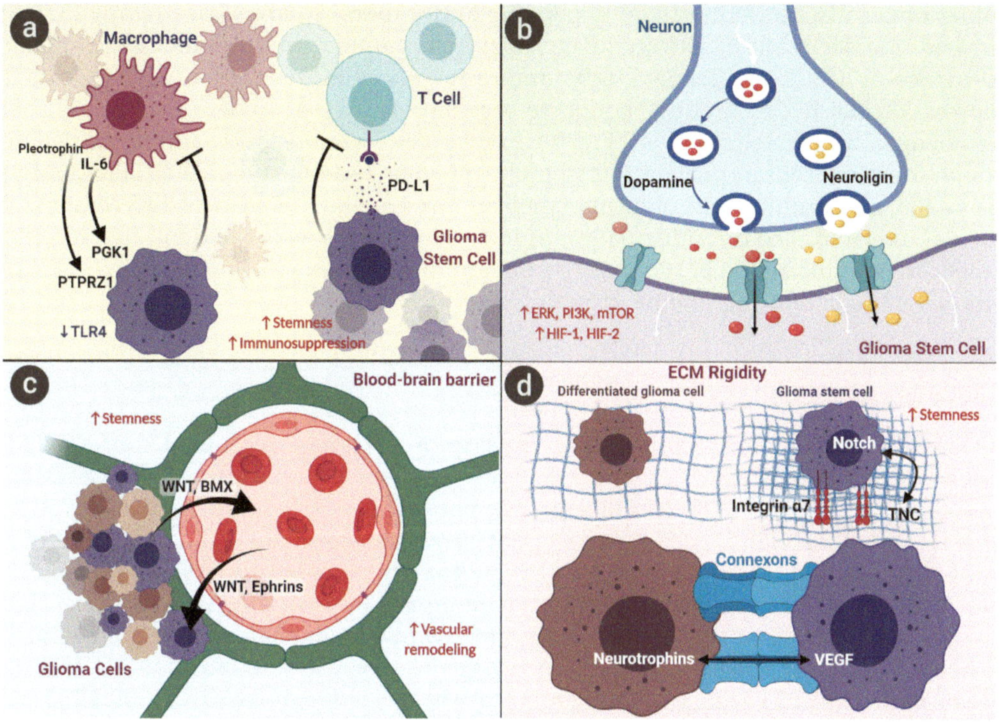

FIG. 3 The microenvironmental landscape of glioma stem cells (GSCs). Interactions of GSCs with their immunogenic niche leads to the immunosuppression of T cells through the PD-L1 pathway and the recruitment of tumor-associated macrophages that further promote immunosuppression and increase stemness of GSCs via the pleiotrophin-PTPRZ1 and the IL-6-PGK1 signaling pathways (A). The release of dopamine and neuroligin-3 in the synaptic cleft between neurons and GSCs further promotes the growth of GSCs and increases stemness through ERK-PI3K-mTOR signaling pathway. Neuronal interactions can also lead to the activation of HIF-1 and HIF-2 under normoxic conditions (B). Furthermore, GSCs form and remodel blood vessels using Wnt7, BMX, and ephrin signaling pathways (C). Stromal interactions via integrin α7 and TNC are required for GSC maintenance through the oncogenic FAK and Notch signaling pathways. Cellular interactions via connexons and tubules allow the exchange of factors, such as VEGF and neurotrophins, to support the growth and survival of GSCs (D).

The immunosuppressive niche

Immune editing of developing gliomas by both the innate and adaptive arms of the immune system drives the evolution of GSCs as they evade surveillance. Besides, intercellular communication and intracellular signaling of GSCs can shape their local immune environment.

GSCs mediate the formation of the immunosuppressive niche in glioma by secreting factors and altering cell surface markers [51]. Through aberrant presentation of antigens, GSCs cause anergy and induce the death of cytotoxic effector T cells and the proliferation of tumor-promoting regulatory T cells, further limiting a potential immune response [51]. A high expression of indoleamine 2,3-dioxygenase 1 (IDO1), an enzyme involved in tryptophan metabolism, is implicated in glioblastoma chemotherapeutic resistance via immunosuppression [52]. The expression of IDO1 in GSCs is higher than in normal glioblastoma cells, rendering IDO1 a valuable and important target for therapy [52]. Furthermore, the secretion of extracellular vesicles containing programmed death-ligand 1 (PD-L1) helps GSCs to escape T-cell immunity [53]. GSC-derived exosomes can also suppress T cells by promoting monocyte maturation [54]. Moreover, tumor-associated macrophages and myeloid cells (TAMs) induce GSC maintenance and tumorigenicity through the pleiotrophin-protein tyrosine phosphatase receptor type Z1 (PTPRZ1) signaling axis [55] and promote glycolytic metabolism through the IL-6-phosphoglycerate kinase 1 (PGK1) axis [56]. The downregulation of innate immune sensors like toll-like receptor 4 further prevents alteration of stemness by the immune system [57] and leads to the recruitment of TAMs, enhancing glioma growth [58, 59].

The neuronal niche

Dopaminergic signaling leading to the release of dopamine receptor D4 promotes the growth of GSCs and increases stemness through autophagic mechanisms and extracellular-signal-regulated kinase (ERK)-mTOR signaling pathways [60]. In addition, the activation of Notch receptors through Jagged promotes GSC invasion along unmyelinated white matter tracts by increasing the levels of SOX9 and SOX2 transcription factors in GSCs [61]. Furthermore, the dopamine D2 receptor has long been thought to be associated with neuronal signaling only; however, recent studies have supported its role in gliomagenesis [62, 63]. Notably, dopamine D2 receptor expression was increased in response to temozolomide treatment and elevated in CD133+ cells, promoting self-renewal and activation of HIF-1 and HIF-2 under normoxic conditions [63].

Advances in cancer neuroscience revealed that neuronal stimulation promotes glioma growth through the secretion of soluble mitogen neuroligin-3 (NLGN3), a synaptic adhesion molecule, to support FAK and phosphoinositide 3-kinase (PI3K)-mammalian target of rapamycin (mTOR) signaling [64, 65]. Microenvironmental NLGN3 is cleaved from both neurons and oligodendrocyte precursor cells via disintegrin and metalloproteinase domain-containing protein 10 (ADAM10) sheddase [65]. It induces transcriptional changes that include the upregulation of several synapse-related genes in glioma cells [65]. A better understanding of tumor-neuron communications allows the development of therapies that can target such interactions.

The hypoxic and perivascular niche

The hypoxic and perivascular niche is a common site for GSCs [51, 66]. This niche surrounds the poorly formed and friable blood vessels that infiltrate the glioma [66] and creates a microenvironment that has high interstitial pressure and poor delivery of oxygen and other nutrients. As such, drug penetration into the tumor is limited, promoting tumoral resistance to therapy [67].

GSCs are important for the recruitment of blood vessels to support continued tumor growth [68]. They invade along blood vessels using an Olig2-Wnt7 signaling axis to secure access to nutrients and a path for cell migration [69]. Evidence also increasingly suggests that GSCs can differentiate into vascular pericytes or endothelial-like cells to support vessel remodeling. GSC-derived pericytes depend primarily on transforming growth factor beta

(TGF-β) and BMX-tyrosine kinase activity to shape their microenvironment and maintain the integrity of the blood-tumor barrier interactions that can be targeted for therapeutic benefit [70–72]. GSC-derived endothelial-like cells rely on Wnt family member 5A (WNT5A) signaling to promote cellular lineage infidelity and acquisition of endothelial-like phenotypes, which promotes tumor neovascularization and invasion [73].

GSCs often prosper in zones of mild to moderate hypoxia [74]. They utilize hypoxia-inducible factor (HIF) signaling to preserve stem properties and drive survival and expansion [39, 74]. Hypoxia signaling pathways are abnormally activated due to the high expression of inhibitor of differentiation 2 (ID2) that prevents HIF-2α degradation [29]. Hypoxia further protects GSCs by limiting drug penetration and decreasing the production of reactive oxygen species (ROS), which are crucial to the success of currently available treatments. Moreover, hypoxia contributes to chemotherapeutic resistance by inducing O(6)-methylguanine-DNA methyltransferase (MGMT) expression in GSCs [75].

Mechanisms of glioma stem cell resistance to chemotherapy

Molecular mechanisms through which GSCs evade chemotherapeutic cytotoxicity comprise increased DNA and base excision repair mechanisms and increased DNA checkpoints that allow GSCs to circumvent the genetic damage instigated by therapeutic agents, such as temozolomide. In addition, GSC resistance to chemotherapy is also favored by the decreased bioavailability of chemotherapeutic factors in the brain, mainly due to the limited BBB permeability and the presence of ATP-binding cassette (ABC) transporter channels that prevent the accumulation of cytotoxic agents in the brain. Furthermore, the plasticity of GSCs, demonstrated by continued self-renewal and dedifferentiation, limits the efficacy of current therapeutic regimens. In combination, these mechanisms of resistance explain why current chemotherapeutic agents have been ineffective.

Advanced DNA repair system

The base excision repair system in cells corrects base pair aberrations identified by the mismatch repair system [76]. In normal cells, it is capable of recognizing and excising a variety of base alterations induced by chemotherapy [76]. Damaged base pairs are removed by their corresponding DNA glycosylase and/or apurinic/apyrimidinic endonuclease (APE1). Next, DNA polymerase and ligase proceed with strand repair and corrections [77]. APE1 has been shown to be overexpressed in GSCs [77]. Poly ADP-ribose polymerase (PARP-1), an effector of apoptosis, controls both the mismatch repair system and base excision repair and is upregulated in glioma [78]. Therefore, GSCs develop resistance to therapies that induce nucleic acid damage and can become more genetically heterogeneous (Fig. 4).

GSCs profit from robust DNA repair mechanisms that render alkylating agents, such as temozolomide, ineffective, and lead to poor patient survival [79]. Increased expression of the DNA repair protein MGMT in GSCs is an important factor in this process [20]. Temozolomide cytotoxicity depends on the mismatch repair system of glioma cells that recognize alkylated base pairs, including the O-6-methylguanine introduced by temozolomide, as well as other base pair mismatches, insertions, and deletions [80, 81]. If DNA damage is irreversible, the cell undergoes apoptosis via breakage of DNA double strands [81]. In GSCs,

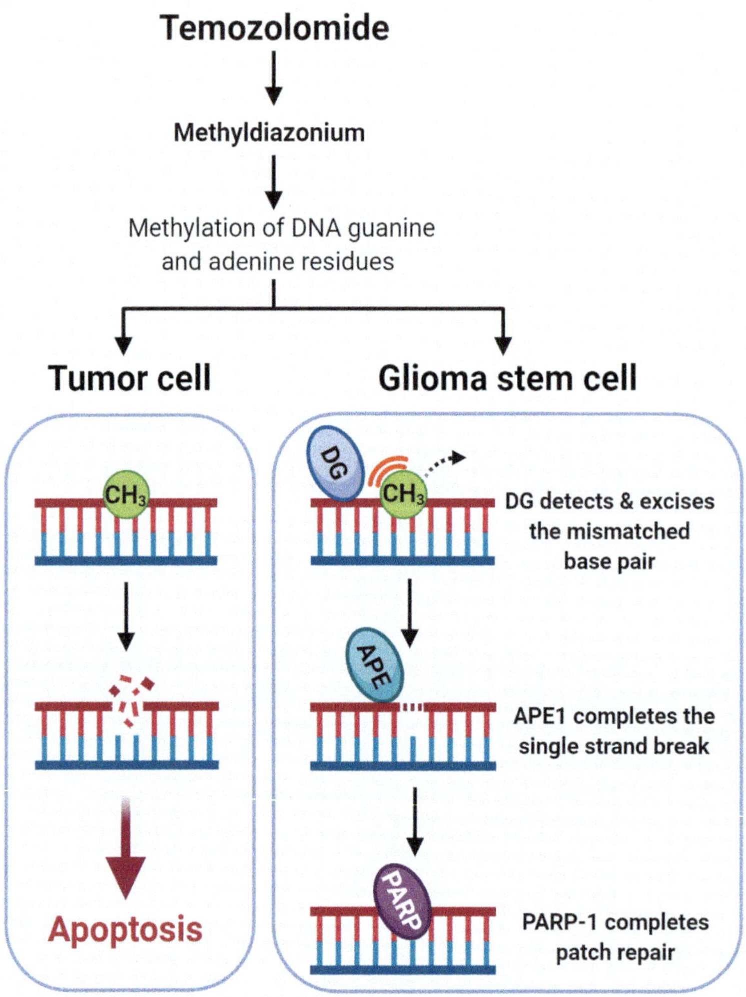

FIG. 4 The base excision repair system in glioma stem cells (GSCs). Upon the administration of temozolomide, it is converted into methyldiazonium, an active cation, which methylates the DNA guanine and adenine residues in tumor cells, leading to the accumulation of DNA strand breaks that eventually triggers apoptosis. In GSCs, the enhanced basic excision repair system detects and excises the offending methyl group by methylpurine DNA glycosylase (DG). Then, AP endonuclease (APE1) completes the single-strand break and signals for poly (ADP-ribose) polymerase (PARP) to complete patch repair. This leads to the chemoresistance of GSCs.

the mismatch repair system is not conserved, as it can reverse O-6-methylguanine alkylation and correct mismatches, additions, and deletions. This allows DNA synthesis and cell cycle progression and permits the continuous replication of cells (Fig. 5).

The cell cycle is controlled by different checkpoints that ensure its integrity. In GSCs, DNA checkpoints, particularly checkpoint kinases 1 (Chk1) and 2 (Chk2) prevent cell cycle

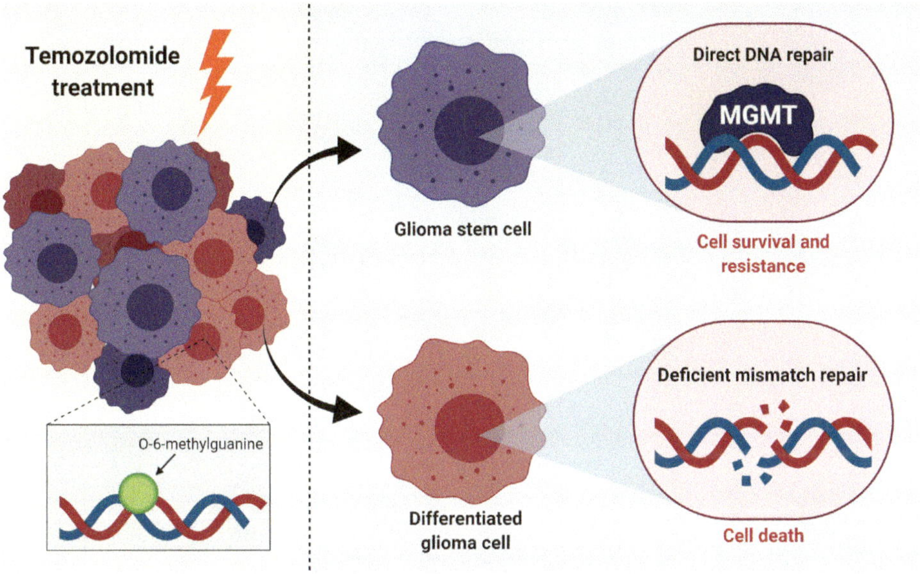

FIG. 5 The mismatch repair system in glioma stem cells (GSCs). Temozolomide treatment induces the methylation of DNA guanine residues at the O6 position (O-6-methylguanine). In GSCs, the expression of methylguanine methyltransferase (MGMT) neutralizes the O-6-methylguanine adduct and restores DNA integrity. This allows glioblastomas to evade chemotherapeutic toxicity. Deficient mismatch repair can lead to DNA strand breaks and cell death.

progression upon detection of DNA damage. Chk1 is active in the S and G2 phases, while Chk2 is activated in response to DNA damage detected at any point during the cell cycle [82]. Inherently, Chk1 and Chk2 activity is increased in GSCs, rendering them resistant to cell death enacted by DNA damage [80].

Limited bioavailability of chemotherapeutic agents

The BBB prevents more than 98% of all macromolecules from penetrating into the brain, including anticancer drugs [83, 84]. This makes it one of the most important factors responsible for chemotherapy failure in brain tumors. GSCs further overexpress ABC transporter channels at their surface [85, 86]. ABC channels actively transport molecules, including temozolomide, outside the cell, preventing the accumulation of cytotoxic agents inside GSCs [87, 88]. ABCG2, mainly, is further associated with subpopulations of cells that are stem-like and multidrug-resistant [85] (Fig. 6).

Increased phenotypic plasticity

Phenotypic plasticity in glioblastoma describes the ability of glioma cells to change from one differentiated state to another. The retroversion of differentiated glioblastoma cells to a more stem-like and progenitor state allows them to benefit from the GSC survival

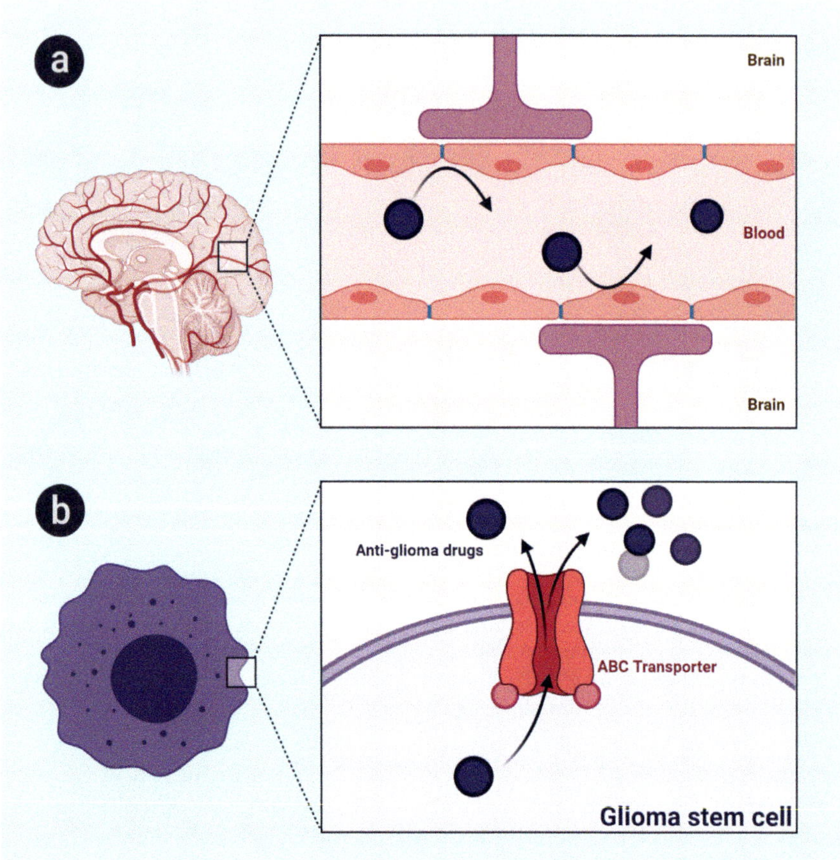

FIG. 6 Limited bioavailability of chemotherapeutic drugs leads to chemoresistance. The blood-brain barrier is a highly selective semipermeable border of endothelial cells that are supported by the end feet of astrocytes. It prevents drugs from nonselectively crossing into the brain where glioma cells reside (A). Glioma stem cells express an ABC transporter that prevents the accumulation of cytotoxic agents, like temozolomide, inside the cell and actively transports them out (B).

mechanisms and self-renewal capacities. This interchange between states can occur incidentally or can be induced by temozolomide therapy [18]. HIF2α, loss of phosphatase and tensin homolog (PTEN), and loss of von Hippel-Lindau have been proposed as markers of conversion into a more stem-like state [18, 89, 90].

Therapeutic strategies to overcome glioma stem cell resistance to chemotherapy

Glioblastoma is a highly heterogenic tumor, with GSCs being responsible for tumor progression, drug resistance, and relapse [91]. Overcoming the resistance mechanisms of GSCs can sensitize them to chemotherapeutic agents and improve patient outcomes. Several strategies have been deployed to counter GSC resistance. Therapies that promote GSC differentiation, target GSC hosting niches, and/or inhibit GSC intracellular signaling and survival pathways have all been explored.

Pro-differentiation therapies

Pro-differentiation therapies induce the progression of GSCs into more differentiated glioma cells, which sensitize them to cytotoxic treatments, decrease their ability to engraft, and/or trigger apoptotic processes [92–95]. Conversion to more differentiated glioma cells limits and, in some cases, reverses the negative immune effects mediated by GSCs [51]. To date inducers of cellular differentiation effective in GSCs include bone morphogenic proteins (BMPs) and posttranscriptional modification using microRNAs (miRNAs). Preclinical studies of these compounds have demonstrated their ability to decrease GSC neurosphere formation and tumor engraftment while also inducing apoptosis. With improved delivery and specificity, they may present promising translational therapeutic options.

Targeting bone morphogenic protein signaling

BMPs elicit a significant decrease in the stem-like and tumor-initiating glioma cell population [92]. They activate their cognate receptors and trigger the Smad signaling cascade in human glioblastoma cells [92] (Table 1).

BMP-2 signaling in GSCs regulates erythropoietin-producing hepatocellular carcinoma receptor (EPHA6) phosphorylation [96]. Overexpression of the EPHA6 receptor in GSCs results in the formation of a protein complex of EPHA6 receptor and ALK-2, a BMP type I receptor encoded by the *ACVR1* gene, and increases the susceptibility of GSCs to BMP-2-induced apoptosis [96]. ALK-2 has crucial roles in the induction of apoptosis in GSCs [97]. In addition, distal-less homeobox 2 (*DLX2*), a BMP target gene, promotes apoptosis and neural differentiation of GSCs [97]. Interestingly, valproic acid, an antiepileptic compound, induced *BMP-2*, *BMP-4*, *ACVR1*, and *DLX2* mRNA expression with a concomitant increase in phosphorylation of Smad1/5; silencing of *ALK-2* or *DLX2* expression partially suppressed it [97]. GBMP1a,

TABLE 1 Role of bone morphogenic proteins in glioma stem cells.

Bone morphogenic protein	Expression in glioma	Mechanism of action	References
BMP-2	Downregulated	Induction of apoptosis and differentiation via EPHA6 receptor and ALK-2	[96, 97]
BMP-4	Downregulated	Reduces GSC size and growth kinetics by activating Smad1/5/8 signaling	[92, 98]
BMP-7	Downregulated	Suppresses CD133, MGMT, ABCB1, and ABCG2 and induces Snail to promote the astrocytic differentiation of GSCs	[99, 100]
BMP-9	Downregulated	Activates Smads and suppresses the PI3K-AKT-MAPK and RhoA-Cofilin signaling pathways	[101]
BMPR1a	Upregulated	Decreases expression of CD133, Olig1, and Olig2 in tumor-associated astrocytes	[102]

a BMP-2-mimicking peptide, activated BMP signaling in glioblastoma cells through Smad1/5/8 phosphorylation [103]. GBMP1a treatment was found to be sufficient to enhance osteogenic differentiation of mesenchymal stem cells and to induce astroglial differentiation of GSCs, in vitro, sensitizing glioma cells to therapy [103]. Moreover, suppressing promoters of stemness through the BMP-2 signaling is important for the differentiation of GSCs into more chemotherapy-susceptible states. For instance, ID1 expression promotes self-renewal of GSCs by inhibiting BMP-receptor 2 expression through miR-17 and miR-20a, which are transcriptional targets of MYC [104]. ID1 simultaneously regulates stemness by increasing *Myc* expression and activating the Wnt and Shh signaling in GSCs [105]. Thus, inhibiting ID1 can be a novel therapeutic strategy of combinatorial targeting of both stemness and differentiation in glioblastoma.

BMP-4 induces the concomitant reduction in the size of the CD133+ GSC population and in the growth kinetics of glioblastoma cells [92]. Moreover, the overexpression of *BMP-4* in human neural stem cells drives them to trace GSCs both in vitro and in vivo, significantly inhibiting their invasive growth and promoting their differentiation and apoptosis by activating Smad1/5/8 signaling [98]. BMP-4 also drives the differentiation of GSCs and thus improves chemotherapeutic efficacy [106]. Nonviral, biodegradable polymeric nanoparticles that are BBB-permeable can be used to deliver *BMP-4* in a human primary malignant glioma model [107]. Targeted mesenchymal cells expressing the *BMP-4* plasmid DNA secrete BMP-4 growth factor while maintaining their multipotency and preserving their migration and invasion capacities [107]. Rats treated with these nanoparticles showed significantly improved survival compared to controls, demonstrating that synthetic polymeric nanoparticles are a safe and effective approach for stem cell-based cancer-targeting therapies [107].

Reduced endogenous BMP-7 signaling in GSCs may contribute to maintained stemness, epithelial-to-mesenchymal transition (EMT), and chemoresistant phenotype. *BMP-7* was identified as one of the top downregulated genes in MGMT-expressing GSCs [99]. Augmenting BMP-7 signaling in GSCs by exogenous BMP-7 treatment did not effectively limit GSC growth but markedly sensitized them to temozolomide treatment, leading to loss of self-renewal and migration capacities [99]. In addition, BMP-7 treatment induced senescence of GSCs in vitro and suppressed mRNA expression of *CD133*, *MGMT*, *ABCB1*, and *ABCG2*. In vivo, BMP-7 treatment significantly prolonged the survival time of animals after intracranial injection of GSCs, whereas a combination of BMP-7 and temozolomide further extended animal survival [99]. Moreover, BMP-7 induces the transcription factor Snail to promote the astrocytic differentiation of GSCs [100]. Snail, in turn, represses stemness in GSCs through Smad signaling and by decreasing the secretion of TGF-β1, a cytokine that promotes self-renewal and proliferation of GSCs [100].

BMP-9, known to be involved in GSC differentiation, invasion, and metastasis, is able to trigger the activation of Smads in patient-derived glioblastoma cells and to strongly inhibit proliferation and invasion by suppressing the PI3K-protein kinase B (AKT)-mitogen-activated protein kinase (MAPK) and RhoA-Cofilin pathways, respectively [101]. Intriguingly, BMP-9 treatment is sufficient to induce a strong differentiation of GSCs and to significantly impede glioblastoma cell transdifferentiation into tumor-derived endothelial cells [101]. Of note, BMP-9 mediates the inhibition of angiogenesis during glioblastoma tumor formation, rendering it a promising candidate to be further developed for the future therapeutic management of glioblastoma [101].

Despite the preclinical success, BMP induction therapy faces significant challenges. Chromatin accessibility mapping identified loci in GSCs that failed to alter in response to BMPs. These loci were enriched in SOX transcription factor-binding motifs [108]. SOX transcription factors, therefore, may limit differentiation commitment. Furthermore, BMP inhibition also represents a viable therapeutic approach in adult glioma. Tumor-associated astrocytes are affected by BMP signaling [102]. The deletion of the BMP type IA receptor (BMPR1a) in astrocytes increased median survival compared to mice with BMPR1a-intact astrocytes, in vivo [102]. In vitro analysis showed that deletion of BMPR1a in oncogenic astrocytes resulted in decreased cell proliferation, invasion, and migration, as well as increased expression of stemness markers [102]. In addition, inhibition of BMP signaling in murine cells and astrocytoma cells with a small molecule BMP receptor kinase inhibitor resulted in similar tumor suppressive effects in vitro [102].

Role of microRNAs in differentiation

miRNAs are small, noncoding RNAs that are 19–25 nucleotides in length and regulate gene expression by antisense complementarity to mRNA targets [109]. They can regulate cell differentiation, cell proliferation, apoptosis, stress resistance, stem cell maintenance, and metabolism [110–114]. A single miRNA may regulate the expression of many target genes, and one gene may be targeted by multiple miRNAs. miRNAs that downregulate the expression of oncogenes act as tumor suppressors; in contrast, miRNAs that downregulate the expression of tumor suppressor genes behave like oncogenes [115, 116]. miRNAs have been implicated in the pathogenesis of glioblastoma and many have been reported to promote GSC differentiation. Key miRNAs can have a broad impact on genes related to the neuronal differentiation of GSCs (Table 2).

Inhibiting oncogenic miRNAs

Using an in vitro system, a microarray-based high-throughput analysis determined the global expression changes of miRNAs during differentiation of GSCs [117]. The detected changes in differentiating glioma cells included the downregulation of hsa-miR-93 and hsa-miR-106a [117]. miR-9/9*, miR-17, miR-10b, and miR-21 were upregulated in human glioblastoma cells and tissues [109]. Inhibiting these miRNAs countered the resistance of GSCs in preclinical settings. For example, the suppression of miR-9/9* and miR-17 leads to the reduction of neurosphere formation and stimulates GSC differentiation. miR-9/9* signals through the stemness-related transcription factor SOX2 [118] and the calmodulin-binding transcription activator 1 (CAMTA1). CAMTA1 is also a target of miR-17 and has tumor suppressor properties [119]. miR-17 promotes the generation of GSCs through inhibiting PTEN [148] and activating the PI3K/Akt and STAT3 signaling pathways [120]. miR-10b is upregulated in GSCs and its inhibition strongly reduces the proliferation, invasion, and migration of GSCs by activating the expression of CUB and sushi multiple domains 1 (*CSMD1*) and homeobox D10 (*HOXD10*); *CSMD1* and *HOXD10* are tumor suppressor genes in GSCs [121, 122]. miR-1587 and miR-3620-5p were shown to increase the proliferation of GSCs. miR-1587, particularly, also inhibited the nuclear receptor corepressor 1 (*NCOR1*), a tumor suppressor of GSCs, after exosomal transfer [46]. miR-21 inhibition sensitizes GSCs to chemotherapy by directly inhibiting the expression of PTEN, which controls GSC senescence [120, 149, 150]. The mechanism of action involves targeting CDC25A, a direct target regulator of GSC proliferation

TABLE 2 Role of microRNAs in glioma stem cells.

miRNA	Expression in glioma	Mechanism of action	References
hsa-miR-93	Upregulated	Preserves stemness of GSCs	[117]
hsa-miR-106a	Upregulated	Preserves stemness of GSCs	[117]
miR-9/9*	Upregulated	Increases neurosphere formation through SOX2 signaling and CAMTA1	[118]
miR-17	Upregulated	Increases neurosphere formation through the inhibition of CAMTA1 and PTEN and activation of PI3K/Akt and STAT3 signaling pathways	[119, 120]
miR-10b	Upregulated	Inhibits the expression of tumor suppressor genes *CSMD1* and *HOXD10*	[121, 122]
miR-21	Upregulated	Increases the expression of CDC25A, a direct target regulator of GSC proliferation	[123, 124]
miR-1587	Upregulated	Inhibits *NCOR1*, a tumor suppressor of GSCs	[46]
miR-3620-5p	Upregulated	Promotes proliferation of GSCs	[46]
miR-125b	Upregulated	Enhances *Bak1* and *PIAS3* expression, promotes GSC fission, and activates the Wnt/β-catenin signaling pathway	[125–127]
CMV70-3P	Upregulated	Increases neurosphere formation through SOX2 expression	[128]
miR-29a	Downregulated	Promotes differentiation of GSCs	[117]
miR-29b	Downregulated	Promotes differentiation of GSCs	[117]
miR-221	Downregulated	Promotes differentiation of GSCs	[117]
miR-222	Downregulated	Promotes differentiation of GSCs	[117]
miR-34a	Downregulated	Inhibits the expression of *MET*, *CDK*, *Notch-1*, and *Notch-2*	[129–131]
miR-7-5p	Downregulated	Inhibits Yin Yang 1 and Notch signaling pathway	[132]
miR-203	Downregulated	Reduces self-renewal and increases expression of glial and neuronal differentiation markers	[133]
miR-137	Downregulated	Inhibits oncogenic proteins such as c-KIT, AKT2, CDC42, CDK6, RTVP-1, and TGF-β2	[95, 134–136]
miR-124	Downregulated	Inhibits CDK6 and reduces the expression of stem cell markers, such as BMI1, Nanog, and Nestin	[137, 138]
miR-128	Downregulated	Inhibits oncogenic proteins implicated in neuronal differentiation and gliomagenesis	[134]

TABLE 2 Role of microRNAs in glioma stem cells—cont'd

miRNA	Expression in glioma	Mechanism of action	References
miR-7	Downregulated	Inhibits oncogenic proteins implicated in neuronal differentiation and gliomagenesis	[134]
miR-181b	Downregulated	Decreases neurosphere formation	[139]
miR-182	Downregulated	Represses Bcl2L12, c-Met, and HIF2α	[140]
miR-218-5p	Downregulated	Reduces the expression of *A2B5, Nestin, PLAGL2, ALDH1, and SOX2*	[141]
miR-128	Downregulated	Inhibits BMI1 and SUZ12 to decrease neurospheres and reduce self-renewal	[142–145]
miR-218	Downregulated	Reduces stem cell markers, such as CD133, SOX2, Nestin, and BMI1 and inhibits the Wnt signaling pathway	[146, 147]

[123, 124]. Inhibition of miR-125b was reported to enhance the chemosensitivity of GSCs to temozolomide by targeting BCL2 antagonist/killer 1 (Bak1) and protein inhibitor of activated STAT 3 (PIAS3) [125, 126]. miR-125b is important for GSCs fission. As such, its suppression can sensitize GSCs to temozolomide in a limited fashion [125]. Combining a PI3K inhibitor with miR-125b inhibitor markedly decreased GSC proliferation through the inactivation of the Wnt/β-catenin pathway and further sensitized the GSCs to temozolomide [127]. CMV70-3P miRNA increases GSCs and promotes glioblastoma stemness [128]. Inhibition of CMV70-3P expression significantly attenuated the ability of primary glioma cells to proliferate and form neurospheres through SOX2 expression [128].

Promoting tumor-suppressing miRNAs

Several miRNAs have been found to be downregulated in glioblastoma compared to normal brain tissue. Therefore, their upregulation might dictate tumor suppressor effects on glioma progression.

The concomitant increase of key miRNAs in GSCs can be a useful therapeutic strategy in promoting differentiation. Using an in vitro system, a microarray-based high-throughput analysis determined the global expression changes of miRNAs during differentiation of GSCs [117]. This detected changes in differentiating GSCs, including the overexpression of miR-21, miR-29a, miR-29b, miR-221 and miR-222 [117]. The overexpression of miR-34a leads to an increase in GSC differentiation [151, 152]. miR-34a directly inhibits the expression of *MET*, cyclin-dependent kinases (*CDK*), *Notch-1*, and *Notch-2* in GSCs [129–131], which reduces the stemness and therapeutic resistance of GSCs [40]. miR-7-5p also suppresses stemness and enhances temozolomide sensitivity of GSCs by targeting Yin Yang 1 [132], an endogenous transcription factor that is associated with Notch signaling [153]. Moreover, miR-203 expression is reduced in human-derived CD133+ GSCs [133]. Transfecting it into GSCs reduces the capacity for self-renewal and increases the expression of glial and neuronal differentiation markers [133]. This, in turn, leads to reduced proliferation rates and an increase in the rate of apoptosis. A genome-wide target mapping in glioblastoma cells followed by an RNASeq analysis revealed that 1468 genes were inhibited by miR-137 [134]. Relevant targets include oncogenic proteins and key regulators of neurogenesis like c-KIT, AKT2, CDC42, cyclin-dependent kinase 6 (CDK6), and TGF-β2 [134, 135]. In addition to promoting differentiation of GSCs, miR-137 has been reported to inhibit the stemness of GSCs by targeting markers, such as related to testis-specific, vespid, and pathogenesis proteins 1 (RTVP-1) [95, 136]. Interestingly, several identified miR-137 targets are also regulated by miR-124, miR-128, and miR-7, which are equally implicated in neuronal differentiation and gliomagenesis [134]. Using mesenchymal stem cells that can migrate to the cancer site and exert antitumor effect has also been explored in glioblastoma. Wharton's jelly-stem cells can successfully deliver miR-124 to glioblastoma cells via dependent or exosome-independent processes [137]. Delivered miR-124 significantly decreases the activity of the target gene CDK6 and enhances the chemosensitivity of glioblastoma cells to temozolomide [137]. Moreover, the overexpression of miR-124 and the knockdown of Snail family transcriptional repressor 2 (*SNAI2*) inhibit the CD133+ cell subpopulation in neurospheres and reduces the expression of stem cell markers, such as BMI1, Nanog, and Nestin [138].

Targeting the stemness markers of GSCs through miRNAs is an important therapeutic strategy. miR-181b functions as a tumor suppressor in malignant glioma; its expression is significantly decreased in GSCs [139]. Overexpression of miR-181b decreases neurosphere formation and exerts a synergistic effect on temozolomide-induced inhibition of secondary neurospheres and on cell growth inhibition rates in glioblastoma [139]. miR-182 is a regulator of apoptosis, growth, and differentiation in glioblastoma. Repression of Bcl2-like12 (Bcl2L12), hepatocyte growth factor receptor (c-Met), and HIF2α are of central importance to miR-182 antitumor activity, as it results in enhanced GSC susceptibility to therapy and a decrease in neurosphere formation and stemness in vitro [140]. Synthesized gold nanoparticles covalently functionalized with mature miR-182 duplexes reduced glioblastoma burden and increased animal survival, indicating that miR-182 activities can be a novel strategy for therapeutic intervention in glioblastoma [140]. Furthermore, miR-218-5p inhibits the stem-like and invasive properties of GSCs. Its expression is lower in human GSCs than in noncancerous brain tissue and normal human astrocyte cells [141]. Overexpression of miR-218-5p inhibits glioma growth significantly by suppressing neurosphere formation and reducing the expression of stem cell markers A2B5, Nestin, PLAG1 like zinc finger 2 (PLAGL2), aldehyde dehydrogenase 1 (ALDH1), and SOX2 [141].

BMI1 is a polycomb group epigenetic gene silencer that is highly expressed in various types of human cancers [154] and required for the maintenance of adult stem cells in many organs [155–158]. The effects of BMI1 on glioblastoma stemness are reported to be associated with the upregulation of p21CIP1 and with decreased Akt activation [142]. miR-128 is generally downregulated in GSCs [142, 159–161]. It targets the polycomb complex protein BMI1 to inhibit glioblastoma proliferation and self-renewal [142]. It also targets SUZ12 that is essential for GSC maintenance [143–145]. The overexpression of miR-218 leads to a significant decrease in the volume of glioblastoma neurospheres and reduces the self-renewal capacity of GSCs. In addition, overexpression of miR-218 reduces stem cell markers, such as CD133, SOX2, Nestin, and BMI1, in glioblastoma neurospheres [146]. These findings suggest that miR-218 regulates the glioblastoma stemness through a BMI1-mediated epigenetic pathway. FZD4 and LEF1, which are involved in the Wnt pathway, are regulated by miR-218 and also affect glioblastoma stem cell stemness [147]. This indicates the potential and relevance of miR-218 as a novel class of therapeutic targets in glioblastoma.

Signaling regulators of glioma stem cell differentiation

The interaction of tumor cells with the ECM is an important aspect of therapeutic strategies. Fibronectin increases cellular differentiation of GSCs through the decrease in SOX-2 and Nestin expression levels, and the increased levels of glial fibrillary acidic protein (GFAP) and β-tubulin [162]. In addition, fibronectin suppresses p53-mediated apoptosis and upregulates P-glycoprotein expression, making GSCs more chemoresistant to alkylating agents such as carmustine [162].

Epigenetic activation of the PAX6-DLX5-WNT5A axis drives GSC differentiation and governs the diffuse spread of glioma cells throughout the brain parenchyma, contributing to the lethality of GBM [73].

Lysine demethylase KDM1A is overexpressed in glioblastoma as it modulates GSC stemness and differentiation. Pharmacological inhibition of KDM1A using NCL-1 and NCD-38 significantly reduced the cell viability and neurosphere formation and induced apoptosis of GSCs with little effect on differentiated cells [163]. Mechanistic studies showed that KDM1A inhibitors induce activation of the unfolded protein response (UPR) pathway [163]. KDM2B is also highly expressed in glioblastoma surgical specimens compared to the normal brain [164]. Targeting KDM2B decreased the GSC pool, which was potentiated in combination with etoposide or lomustine [164].

The blockade of the nuclear factor kappa B (NF-κB) pathway promotes terminal differentiation and senescence of GSCs. This effect was partly mediated by reduced levels of the NF-κB target cyclin D1 [165]. SN50, a cell-permeable peptide inhibitor of NF-κB, induces robust differentiation of human GSCs, causing loss of their oncogenic potential. Treatment with SN50 increases sensitivity to temozolomide and radiation [166]. This supports the rationale for therapeutic strategies aimed to promote premature senescence of differentiating GSCs by blocking key factors within the NF-κB pathway.

Senescent glioma cells also seem to play a role in promoting the malignant progression of glioblastoma through the release of proangiogenic factors, including vascular endothelial growth factor (VEGF), IL-6, and IL-8, that boost the tumor formation in vivo [189]. In response, GSCs downregulate telomerase, shorten telomeres, and eventually become senescent themselves [189]. As such, targeting senescent glioma cells might be an important strategy for therapeutic consideration.

Repositioning therapeutics for glioma stem cell differentiation

Drug repurposing for the treatment of malignant glioma is another way to find therapeutic targets for GSCs. Quetiapine, an atypical antipsychotic, can effectively suppress GSCs-initiated tumor growth, act synergistically with temozolomide on glioma suppression, and suppress effectively temozolomide-resistant GSCs [190]. The antiglioma effect of quetiapine was due to its actions of promoting the differentiation of GSCs into oligodendrocyte-like cells and its inhibitory effect on the Wnt/β-catenin signaling pathway [190]. Antidepressants, imipramine and amitriptyline, can also play a role as adjuvant glioblastoma therapy, as they possess the ability to switch phenotypes from GSC to non-GSC [191]. Retinoic acid is well known as a potent modulator of differentiation and proliferation in normal stem cells. GSC differentiation using all-trans retinoic acid in vitro induces therapy-sensitizing effects, impairs the secretion of angiogenic cytokines, and disrupts GSC motility [192]. Furthermore, ex vivo differentiation reduces tumorigenicity of GSCs, showing that all-trans retinoic acid treatment can induce antitumor effects in vivo [192].

Targeting glioma stem cell niches

Besides direct cellular manipulation, some have attempted to target GSCs by modifying the niches that they characteristically occupy. Targeting the immunogenic, neuronal, hypoxic, metabolic, and angiogenic niches destabilize GSCs within the brain microenvironment.

Targeting the immunogenic niche

Immune evasion mechanisms of GSCs can be mediated by specific protein expression patterns or through released immune-related factors. Targeting the promoters of immunogenic adaptation through immunotherapy, dendritic cell vaccines, or oncolytic viral therapy continues to be explored for better clinical outcomes (Fig. 7).

Programmed death-ligand 1 (PD-L1) is highly expressed on glioma-associated TAMs, which are known to promote glioma immune escape [193] and facilitate resistance by driving neovascularization and progression of glioblastoma [194–199]. A lipid nanoparticle formulation surface-functionalized with an anti-PD-L1 therapeutic antibody enabled effective and specific delivery of therapeutic payload to TAMs [193]. Specifically, encapsulation of dinaciclib, a cyclin-dependent kinase inhibitor, into PD-L1-targeted nanoparticles led to a robust depletion of TAMs and an attenuation of their immunosuppressive functions [193]. Furthermore, TAMs promote regulatory B-cell function by delivering microvesicles transporting membrane-bound PD-L1, able to be uptaken by tumoral B cells [200]. The transfer of functional PD-L1 via microvesicles conferred regulatory B cells the potential to suppress CD8+ T-cell activation and acquisition of an effector phenotype [200]. Local delivery of B cell-depleting anti-CD20 immunotherapy improved overall survival in vivo [200]. Moreover, GSCs secrete periostin for the recruitment of TAMs [58]. Silencing periostin in GSCs markedly reduces TAM quantities, inhibits tumor growth, and increases survival of mice bearing GSC-derived xenografts [58].

Therapeutic modulation of phagocytosis in glioblastoma can activate both innate and adaptive antitumor immunity. Phagocytosis of glioma cells is critical for the generation of antitumor immunity. However, glioblastoma cells can evade phagocytosis by upregulating antiphagocytosis molecule CD47. Combining CD47 blockade with temozolomide enhances T-cell priming and inhibits glioma growth but activates immune checkpoint [201]. Sequential administration of an anti-PD-1 antibody overcomes this potential adaptive resistance [201]. Furthermore, antioxidant agents that serve as nutritional supplements might affect tumor cell phagocytosis. Ocoxin oral solution, in particular, affects GSC properties through a direct effect on macrophage polarization in vitro and in vivo [202]. It inhibits the pro-tumoral features of M2 macrophages, thereby targeting the supportive immunogenic niche of GSCs [202]. In addition, targeting colony-stimulating factor 1 receptor (CSF1R), which controls the production, differentiation, and function of macrophages, has been shown to be effective in preclinical studies. Inhibition of CSF1R on TAMs has been shown to suppress the crosstalk between GSCs and TAMs [199, 203].

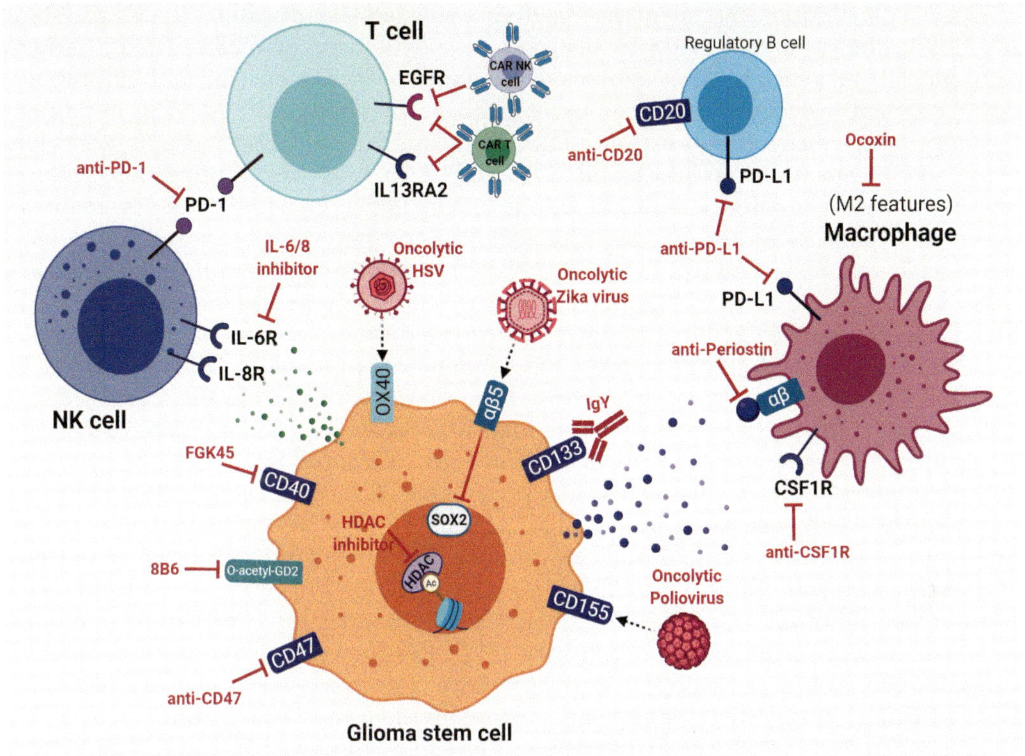

FIG. 7 Immunogenic therapies against glioma stem cells (GSCs). Targeting the immune evasion mechanisms of GSCs can be done by inhibiting protein surface markers, such as CD47, *O*-acetyl-GD2, CD40, OX40, CD133, and CD155. Inhibiting intracellular molecules such as HDAC and SOX2 can also potentiate immune activation. Immunotherapy can further target released immune-related factors, such as periostin, IL-6, and IL-8. Inhibiting CSF1R on TAMs can also restore the immunogenicity of macrophages, as well as targeting the M2 features via ocoxin. Anti-PD-L1 therapy can prevent the immunosuppression of macrophages. Anti-CD20 can remove the inhibitory effect of regulatory B cells on cytotoxic T cells, as wells as anti-PD-1 therapy can do on both T cells and NK cells. CAR-T cells and CAR NK cells targeting the EGFR and IL13RA2 on T cells can potentiate their immune cytotoxicity against GSCs.

Targeting tumor surface antigens of GSCs can be an effective strategy to trigger an adaptive immune response and activate cytotoxic T cells [204]. Downregulation of major histocompatibility complex I (MHC-I) expression was associated with Wnt pathway activation in GSCs [204]. Histone deacetylases (HDAC) inhibition improved MHC-I expression, which could be partly reverted by Wnt pathway activation [204]. The enhancement of T-cell immune response resulting from HDAC inhibition was dependent on the cytotoxic T-cell cognition of tumor antigens presented by MHC-I molecule in GSCs [204]. Furthermore, interferon-beta (IFNβ) can be a promising immunotherapeutic strategy for the treatment of glioma. Mesenchymal stem cells used as cellular vehicles to locally deliver IFNβ in glioma shows a remarkable inhibition on the glioma growth through increasing dendritic cells that express basic leucine zipper ATF-like transcription factor 3 (Batf3+) and promoting CD8+ T lymphocyte infiltration within the glioma [205]. Moreover, IFNβ can potentiate tumor debulking, a mainstay in glioblastoma treatment that results in a substantial reduction of TAMs and simultaneous recruitment of CD4/CD8 T cells [206]. Locally delivery of encapsulated mesenchymal stem cells that secrete IFNβ significantly enhances selective postsurgical infiltration of CD8 T cells and directly induces cell cycle arrest in glioblastoma cells [206]. In addition, targeting CD40, which is highly expressed in GSCs, using the anti-CD40 agonistic monoclonal antibody FGK45 increases apoptosis and CD4+ and CD8+ T-cell tumor infiltration [207]. This leads to significant and direct antitumor effects in vitro and in vivo [207]. Besides, analysis of patient-derived glioblastoma cells revealed that GSCs overexpressed *O*-acetyl-GD2 [208]. A monoclonal anti-*O*-acetyl-GD2 antibody 8B6 immunotherapy significantly increased temozolomide genotoxicity and tumor cell death by enhancing temozolomide tumor uptake [208]. Furthermore, the combination therapy decreased the expression of the GSC markers CD133 and Nestin and compromised GSC self-renewal capabilities [208]. An avian IgY immunotoxin against CD133+ GSCs reduced cell viability, in vitro [209]. Animals treated with IgY-abrin immunotoxin showed more than 50% decrease of tumor volume, suggesting IgY-abrin as a novel approach for the immunotherapy of glioblastoma [209].

Natural killer (NK) cells are functionally suppressed in the glioblastoma tumor microenvironment [210]. The survival and differentiation of GSCs are controlled through two distinct phenotypes of cytotoxic and noncytotoxic/split anergized NK cells, respectively [211]. Split anergized NK cells promote GSC differentiation but result in the resistance of differentiated glioma cells to NK cell-mediated cytotoxicity. Undifferentiated GSCs remained more susceptible to NK cell lysis [212]. Resistance to cytotoxicity is mediated by sustained release of IL-6 and IL-8 in the presence of decreased IFN-γ postdifferentiation [212]. As such, immunotherapies designed to retain the tumor killing function as well as IFN-γ secretion by the NK cells, while suppressing tumor-induced IL-6 and IL-8 secretion can be effective. Furthermore, blocking the programmed death 1 (PD-1)/B7-H1 pathway in NK cells promotes the NK cell-mediated immune response and the subsequent targeting of GSCs [213]. B7-H6, a newly identified member of the B7 costimulatory molecule family, is a crucial regulator of NK cell-mediated immunity. B7-H6 expression is upregulated in GSCs [214]. Knockdown of B7-H6 in GSCs led to the inhibition of cell proliferation, with a decrease in the expression of the oncogene *Myc* as well as inactivation of PI3K/Akt and ERK/MAPK signaling pathways [214]. Thus, B7-H6 can be a potential molecule for GSC targeted immunotherapy.

Immunotherapy effects, including induction of immunogenic cell death and depletion of regulatory T cells by temozolomide, remains limited [215, 216]. Although immune checkpoint inhibitors targeting cytotoxic T-lymphocyte-associated protein 4 (CTLA-4) and the PD-1/PD-L1 axis demonstrate remarkable efficacy in other cancers and in preclinical studies of glioblastoma, their role in patients remains to be determined pending completion of clinical trials. Thus, utilization of innate and adaptive immune responses to target GSCs and glioblastoma remains an active and fruitful area of the ongoing investigation.

The efficacy of some immunotherapeutic regimens against glioblastoma can possibly confer therapeutic effects against GSCs. Vaccination, chimeric antigen receptor (CAR) T-cell therapy, and oncolytic viral therapy have shown promise in preclinical settings of glioblastoma and early-phase clinical trials.

Vaccination has been considered one of the most promising approaches to improve the outcomes of patients with glioblastoma, although negative results from several phase II and phase III trials challenge the current concept of vaccination as a single-modality immunotherapy [215]. Treatment with dendritic cell vaccines containing glioblastoma-specific cytomegalovirus antigens in combination with immune adjuvants extended survival in a small cohort of patients [217]. Furthermore, treatment with autologous tumor lysate-containing dendritic cells may also be a feasible approach [218]. ICT-107, a patient-derived dendritic cell vaccine developed against six antigens highly expressed in GSCs [219], is currently under clinical evaluation for use in patients. In two early clinical trials, administration of personalized vaccines containing tumor neoepitopes elicited strong T-cell immunologic responses with evidence of immunologic memory and tumor-infiltrating capacity [220, 221]. Moreover, neural stem-like cells engineered from autologous patient-derived fibroblasts migrate toward glioblastoma tumors and deliver cytotoxic compounds to tumor cells in mouse models [222]. This modality has been investigated in early-phase clinical trials [223].

CAR T-cell therapy for glioblastoma depends on the identification of stably expressed and sufficiently tumor-specific antigens [215]. CAR-T cells against tumor-specific targets, including IL-13 receptor subunit alpha 2 (IL13RA2) [224] and EGFRvIII [225], display some efficacy in patients with recurrent glioblastoma without major limiting toxicities in phase 1 clinical trials. In preclinical studies, CAR NK cells targeting EGFR appear to prolong survival in xenograft-bearing mouse glioblastoma models [226]. In addition, a CAR-T cell tuned to recognize EGFR amplified expression levels in solid tumors showed efficient in vitro cytokine secretion and glioma cell lysis [227]. With additional IL-2 support, all in vivo tumors were eradicated without recurrence [227]. Furthermore, CAR-T cells directed against B7-H3, an immune checkpoint protein that is highly expressed in GSCs, effectively decreases neurospheres in vitro and in vivo [228]. Interestingly, CD28-co-stimulated CAR-T cells released more inflammatory cytokines [228]. Therefore, employing CAR-T cells against GSCs can be a plausible therapeutic strategy for future consideration.

Oncolytic viruses represent a new therapeutic modality with increasing promise. Improved oncolytic adenovirus treatment regimens have demonstrated efficacy in mouse models through stimulating an influx of T cells [229] and in human patients through altering the composition of TAMs [230]. Modified herpes simplex virus (HSV) combined with immune checkpoint inhibition increase CD8+ T-cell infiltration and transition to a more inflammatory antitumor macrophage phenotype [231]. Using syngeneic GSC orthotopic models, a curative immunotherapeutic strategy required an oncolytic HSV expressing IL-12 in combination with two checkpoint inhibitors, anti-PD-1 and anti-CTLA-4 [232]. Furthermore, combining oncolytic HSV expressing antiangiogenic cytokine IL12 with systemic VEGFR tyrosine kinase inhibitor, axitinib, synergistically enhanced antitumor efficacy against GSCs [233]. Moreover, oncolytic HSV expressing murine IL-12 alongside anti-CTLA-4 and anti-PD-1 antibodies was associated with macrophage influx and M1-like polarization, along with increased T effector to T regulatory cell ratios [231].

Other viruses have also been employed as oncolytic agents against glioma. Intratumoral injection of an oncolytic adenovirus expressing the immune costimulator OX40 ligand and an anti-PD-L1 antibody showed synergistic inhibition of GSCs and significantly increased survival in vivo [229]. Furthermore, a phase I trial of modified poliovirus therapy has demonstrated some efficacy in the treatment of recurrent glioblastoma, with evidence that a subset of patients may gain relatively long-term survival benefit [234]. Zika virus, the highly neurotropic flavivirus that causes microcephaly by killing neural precursor cells and other brain cells, also displays therapeutic oncolytic activity against GSCs. Zika virus preferentially infected and killed GSCs in a SOX2-dependent manner using integrin αvβ5 [235]. Inhibition of integrin αvβ5 reversed the Zika virus elimination of GSCs [235], suggesting that integrin αvβ5 is a functional cancer stem cell marker essential for glioblastoma maintenance and Zika virus infection, thereby providing a potential therapy for glioblastoma. In organoid and mouse models, Zika virus selectively ablates GSC populations, although concerns surrounding biosafety limit widespread adoption of this virus in clinical trials at the present time [236]. In the future, combining viruses with other immunotherapy modalities might exert proinflammatory responses that could potentially exploit GSCs.

Targeting the neuronal niche

The rise of the cancer neuroscience field facilitated the understanding of the bidirectional crosstalk between the brain tissue and glioblastoma cells [237, 238]. GSCs interact with their neurogenic niche via secreted signals and the expression of proteins that can be targeted therapeutically. ADAM10 inhibitors prevent the release of NLGN3, a synaptic adhesion molecule that promotes glioma proliferation, into the tumor microenvironment and robustly blocks glioblastoma growth, offering a promising target for therapeutic intervention [65]. Trifluoperazine, a dopamine receptor antagonist, can interfere with the phenotype conversion of glioma cells to GSCs by directing the loss of Nanog mRNA expression, which is usually induced by radiation [239]. Trifluoperazine also promotes the activation of glycogen synthase kinase 3 (GSK3), leading to a consecutive reduction in phosphorylated-Akt, SOX2, and β-catenin protein levels [239]. Continuous treatment with trifluoperazine and a single dose of radiation reduces the number of GSCs and prolonged survival in vivo [239].

Repositioning drugs used for neurological and psychiatric diseases can be an important tool against GSCs, since these agents will be able to cross the BBB. By screening chemical libraries, a nondepolarizing neuromuscular blocker, atracurium besylate, was identified as a small molecule, which effectively induces astroglial but not neuronal differentiation of GSCs [240]. Functionally, atracurium besylate irreversibly inhibits the clonogenicity of several patient-derived GSC neurosphere lines [240]. In addition, a second nondepolarizing neuromuscular blocker, vecuronium, also induced GSC astrocytic differentiation [240] (Fig. 8). Furthermore, brexpiprazole, a serotonin-dopamine activity modulator used in schizophrenia, can sensitize GSCs to Osimertinib, a third-generation EGFR inhibitor, via reduction of survivin [241]. Moreover, chlorpromazine, an antipsychotic, inhibits cytochrome *c* oxidase bearing the COX4-1 regulatory subunit from chemoresistant GSCs, without affecting other mitochondrial complexes [242]. This led to cell cycle arrest selectively in GSCs [242].

Targeting the glioma stem cell metabolism and the hypoxic niche

Oxidative stress is an important and clinically relevant component of temozolomide-induced therapeutic effects in GSCs, whereby cytotoxicity seems to be mediated by aldehydes resulting from lipid peroxidation. GSCs use ALDH1A3 to reduce the number of toxic aldehydes, becoming resistant to temozolomide therapy. As such, targeting ALDH1A3 in GSCs offers therapeutic benefit in patients with temozolomide-resistant glioblastoma [167]. Doxycycline, repurposed clinically as a safe mitochondrial inhibitor, can be used to target Forkhead Box M1 and mitochondrial biogenesis in GSCs [168, 169]. Moreover, GSCs reprogram their metabolic machinery and preferentially take up glucose to survive in environments with

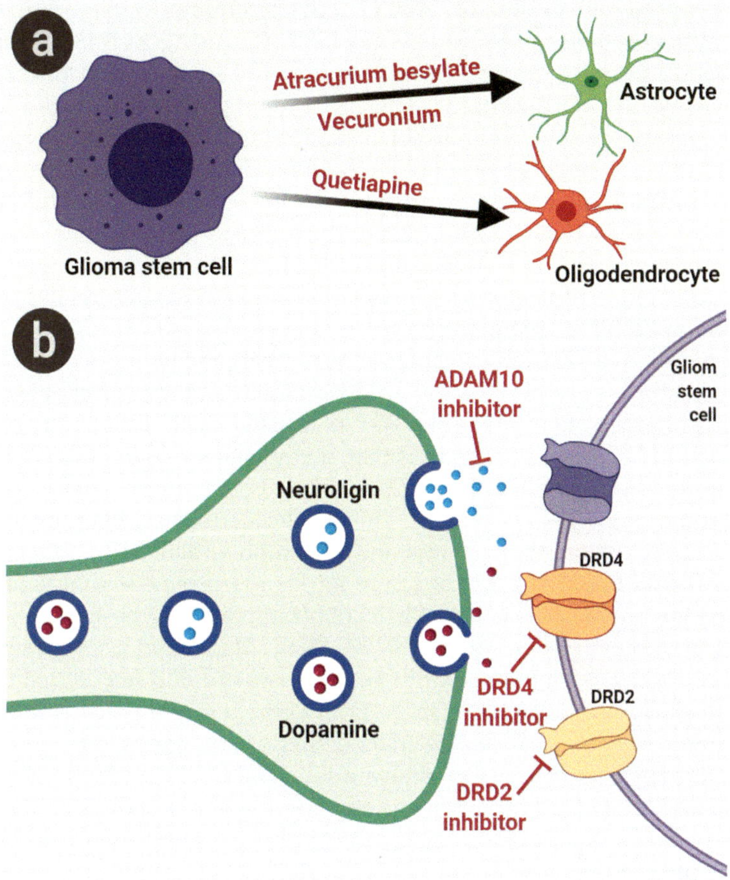

FIG. 8 Targeting the neurogenic niche of glioma stem cells (GSCs). Repurposing drugs, such as atracurium besylate and vecuronium, as adjuvant therapy for glioblastoma, can promote the differentiation of GSCs into astrocytes. Quetiapine, an atypical antipsychotic promotes the differentiation of GSCs into oligodendrocyte-like cells (A). Antidepressants, imipramine and amitriptyline, can also switch phenotypes from GSC to non-GSC. Targeting the neuroligin-3 signaling through ADAM10 inhibitors, as well as inhibiting dopaminergic signaling through DRD2 and DRD4 can be an effective therapeutic approach to destabilize the neuronal niche of GSCs (B).

limited nutrients by expressing the high-affinity glucose transporter 3 (GLUT3) [170]. GLUT3 thus represents a promising therapeutic target for potential selective inhibition of GSCs. Utilizing an anticancer compound library, R406 was identified as an active metabolite of spleen tyrosine kinase (Syk) inhibitors, with specific cytotoxicity against GSCs [171]. It induced a metabolic shift from glycolysis to oxidative phosphorylation and led to the subsequent production of excess ROS in GSCs [171]. Mechanistically, the anti-GSC effect of R406 was due to the disruption of Syk/PI3K/Akt signaling [171].

Photodynamic therapy is a form of phototherapy that involves light and a photosensitizing chemical substance that is used in conjunction with molecular oxygen to elicit cell death. It has been rendered as a promising adjunct for GSC treatment [172]. Some photosensitizers have been reported to be substrates of ABCG2, a membrane transporter that mediates chemoresistance. Nevertheless, the inhibition of ABCG2, using KO143, during photosensitizer accumulation and irradiation promises to restore full susceptibility of GSCs to photodynamic treatment [172]. Furthermore, ABCG2 downregulation in GSCs enhances the therapeutic efficacy of demethoxycurcumin, a curcuminoid found in turmeric, via apoptosis and G0/G1 cell cycle arrest in GSCs [173]. Notably, demethoxycurcumin increased ROS levels in GSCs, resulting in increased cytochrome C and caspase-3 activity [173]. Moreover, nanoparticles loaded with temozolomide can be effective as photothermal therapy agents in combination with hyperbaric oxygen therapy to sensitize GSCs to the combination treatment [174]. Moreover, the expression of stem-like cell markers and hypoxia-related molecules were decreased after combination treatment [174].

Metformin, an oral hypoglycemic agent, can be repositioned to induce cytotoxicity against GSCs. It has been shown to suppress spheroid formation and size, inhibit the expression of CD133 stemness marker, and attenuate TGF-β-induced EMT in glioma cells [176, 177]. Inhibiting the activity of yes-associated protein, a key effector of the Hippo pathway, regulates metformin-induced action against GSCs [176]. Furthermore, metformin inhibits the chloride current that depends on the function of chloride intracellular channel protein 1 in GSCs [178]. The combination of metformin and 9-cis retinoic acid further increases apoptosis in GSCs [179]. Moreover, adding metformin to the standard temozolomide regimen offers a synergistic response to overcome resistance in glioblastoma by increasing Bax/Bcl-2 ratio and decreasing the expression of GSC marker CD90 [180]. Metformin plus sorafenib is another combination that produces high levels of ROS, decreases efflux pump activity and generates synergistically high apoptotic rates in GSCs [177]. Reducing STAT3-phosphorylation, using the STAT3 inhibitor Stattic, enhances the effect of metformin against GSCs. The combination treatment of Stattic and metformin additively inhibited GSC proliferation [181]. Furthermore, in preclinical cancer models, dual inhibition of both glycolysis and oxidative phosphorylation, respectively, using the glucose analog 2-deoxyglucose and metformin, has been shown to effectively reduce tumor growth and dissemination [182]. Similarly, the biguanide drug phenformin leads to cancer cell death through metabolic targeting only when combined with genetic disruption of monocarboxylate transporters (MCT) involved in cancer metabolic symbiosis [183].

Overcoming hypoxia by increasing local tissue oxygen is a strategy that has been explored to overcome the chemoresistance of GSCs [243]. In addition, targeting HIFs and their downstream signaling factors continues to be explored as a therapeutic strategy to disrupt GSC chemoresistance and survival. Nimustine, a nitrosourea agent with antineoplastic activity, has antitumor activities in patients with malignant glioma [175]. In a hypoxic microenvironment, hyperbaric oxygen therapy could inhibit glioma cell proliferation and inflammatory cell infiltration, and exert a sensitizing effect on nimustine therapy partially through enhancing oxygen pressure in tumor tissues and lowering expression levels of HIF-1α, tumor necrosis factor alpha (TNF-α), IL-1β, VEGF, matrix metallopeptidase 9 (MMP-9), and NF-κB [175]. Other small molecule inhibitors have also shown promise in countering the hypoxic conditions that foster the proliferation of GSCs. Acriflavine is a small molecule that inhibits the binding between Basigin and MCT4, proteins that are important for mediating GSC signaling in hypoxia [184]. It inhibits the intratumoral expression of VEGF and tumor neovascularization [184]. Topotecan is a novel inhibitor of global SUMOylation. It also inhibits

CDK6 and HIF-1α, instigating pronounced changes in cell cycle progression and cellular metabolism in GSCs [185]. Polymeric nanoparticle-induced CXCR4-overexpressing human stem cells show enhanced migration in response to hypoxia and are efficient in penetrating the hypoxic core within GSC spheres [186]. Therefore, nanoparticles can be potent cellular vehicles for targeting tumor hypoxia. Temozolomide efficacy may be increased by targeting carbonic anhydrase 9 (CA9), a hypoxia-responsive gene important for maintaining the altered pH gradient of tumor cells [187]. CA12 also was shown to mediate p-glycoprotein resistance against temozolomide [244]. SLC-0111, an inhibitor of CA9 and CA12, used concurrently with temozolomide in patient-derived glioblastoma xenograft cells reduced cell growth and induced cell cycle arrest via DNA damage in vitro [187]. In addition, this treatment shifted tumor metabolism to a suppressed bioenergetic state in vivo [187]. SLC-0111 also inhibited the enrichment of GSCs after temozolomide treatment via suppressing CD133 expression and neurosphere formation capacity [187]. Combination of a cyclooxygenase (COX-2) inhibitor, NS-398, with carmustine showed a synergistic effect in both hypoxia and normoxia against chemoresistant GSCs [188]. Prostaglandin E2 levels, a manifestation of COX-2 activity, were increased during hypoxia but were reduced by the combination regimen during both hypoxia and normoxia [188]. The combination also resulted in a greater reduction of the cancer stem cell marker CD133 [188] (Table 3).

Targeting the angiogenic niche

The strategic positioning of GSCs in a perivascular niche dictates crosstalk between GSCs and endothelial cells of angiogenic vessels. Such interactions yield targets for therapeutic disruption. Inhibitors of mTOR, delta-like 4, laminins, nitric oxide, and sonic hedgehog (Shh) ligand have all been developed to reduce the survival of GSCs, with limited success in preclinical studies and early clinical trials [245–248]. Antiangiogenic therapies targeting VEGF, stromal-derived factor 1, and others combined with the current standard of care may be more relevant [249]. VEGF, specifically, has been recognized as a key regulator of angiogenesis [250]. This necessitated the development of several VEGF-targeting agents, such as bevacizumab [251]. Nevertheless, GSCs can develop resistance to bevacizumab, identifying the need for alternative angiogenic pathways as potential targets for therapy [252].

Tumor cells that survive after bevacizumab treatment tend to accumulate and surround remaining blood vessels [253]. As such, overcoming GSC resistance to chemotherapeutic agents and bevacizumab requires blocking the activation of alternative angiogenic pathways. Clinical trials utilizing combination therapies have been initiated using agents such as onartuzumab (an anti-c-Met antibody), an angiopoietin inhibitor, and an antiplacental growth factor (PlGF) antibody. Furthermore, combination therapies targeting the angiogenic niche along with other GSC survival mechanisms, such as tumor hypoxia and autophagy, were also employed in an attempt to achieve clinical benefit.

A phase I dose-escalation study of RO5323441, a novel anti-PlGF monoclonal antibody, with bevacizumab did not appear to improve outcomes in patients with recurrent glioblastoma [254]. A phase II, double-blinded, placebo-controlled trial of bevacizumab plus trebananib, a novel Fc fusion protein that sequesters angiotensin1/2, did not significantly improve progression-free survival or overall survival [255]. Bevacizumab and vorinostat, a histone deacetylase inhibitor, did not improve progression-free survival or overall survival [256]. Temsirolimus, an mTOR inhibitor, in combination with bevacizumab failed to detect

TABLE 3 Therapeutic strategies to target the glioma stem cell metabolism and hypoxia.

Therapy	Mechanism of action	References
ALDH1A3 inhibitor	Inhibits ALDH1A3 to prevent GSCs from reducing toxic aldehydes after temozolomide therapy	[167]
Doxycycline	Inhibits Forkhead Box M1 and mitochondrial biogenesis	[168, 169]
GLUT3 inhibitor	Inhibits preferential uptake of glucose by GSCs	[170]
R406	Inhibits the Syk/PI3K/Akt signaling pathway and increases the production of ROS	[171]
KO143	Inhibits ABCG2 during photosensitizer accumulation and irradiation	[172]
Demethoxycurcumin	Induces G0/G1 cell cycle arrest and apoptosis via increased ROS levels and increased cytochrome C and caspase-3 activity	[173]
Hyperbaric oxygen therapy + temozolomide	Decreases the expression of stem-like cell markers and hypoxia-related proteins	[174]
Hyperbaric oxygen therapy + nimustine	Enhances oxygen pressure and lowers expression of HIF-1α, TNF-α, IL-1β, VEGF, MMP-9, and NF-κB	[175]
Metformin	Suppresses spheroid formation and size, inhibits the expression of CD133, attenuates TGF-β-induced EMT, inhibits the Hippo oncogenic pathway, and decreases the chloride current in GSCs	[176–178]
Metformin + 9-cis retinoic acid	Synergistically increases apoptosis	[179]
Metformin + temozolomide	Increases Bax/Bcl-2 ratio and decreases the expression of GSC marker CD90	[180]
Metformin + sorafenib	Produces high levels of ROS, decreases efflux pump activity, and increases apoptosis in GSCs	[177]
Metformin + stattic	Reduces STAT3-phosphorylation and additively inhibits GSC proliferation	[181]
Metformin + 2-deoxyglucose	Inhibits both glycolysis and oxidative phosphorylation	[182]
Phenoformin + MCT inhibitor	Inhibits cancer metabolic symbiosis and increases apoptosis	[183]
Acriflavine	Inhibits the binding between Basigin and MCT4 and suppresses the intratumoral expression of VEGF and tumor neovascularization	[184]
Topotecan	Inhibits global SUMOylation, CDK6, and HIF-1α	[185]
Polymeric nanoparticle-induced CXCR4-overexpressing human stem cells	Penetrates the hypoxic core within GSC spheres	[186]
SLC-0111 + temozolomide	Inhibits CA9 and CA12, induces cell cycle arrest, decreases tumor metabolism, and suppresses CD133 expression and neurosphere formation capacity	[187]
NS-398 + carmustine	Reduces prostaglandin E2 levels and suppresses CD133	[188]

clinical benefit in patients with progressive glioblastoma [257]. Early trials on MLN0128, an oral inhibitor of TORC1/2 complexes, in combination with bevacizumab also did not exhibit survival benefit in patients with recurrent glioblastoma [258].

Nevertheless, some combination therapy trials involving antiangiogenic therapies have shown some promise. Antiangiogenic therapy is known to increase hypoxia in glioblastoma [252]. Evofosfamide is a hypoxia-activated prodrug that leads to the release of the alkylating agent bromo-isophosphoramide mustard. The released cytotoxin alkylates DNA, inducing intrastrand and interstrand crosslinks [259]. Early trials in combination with bevacizumab demonstrated some improvement in progression-free survival in patients with progressive glioblastoma, meriting further investigation [260]. Targeting endoglin, an endothelial cell membrane receptor, using TRC105 in combination with bevacizumab showed some clinical activity in bevacizumab-resistant cases [261].

Targeting intracellular signaling and survival pathways

Targeting intracellular signaling and cell survival pathways that drive therapeutic resistance of GSCs is vital for better patient survival and clinical outcomes (Fig. 9).

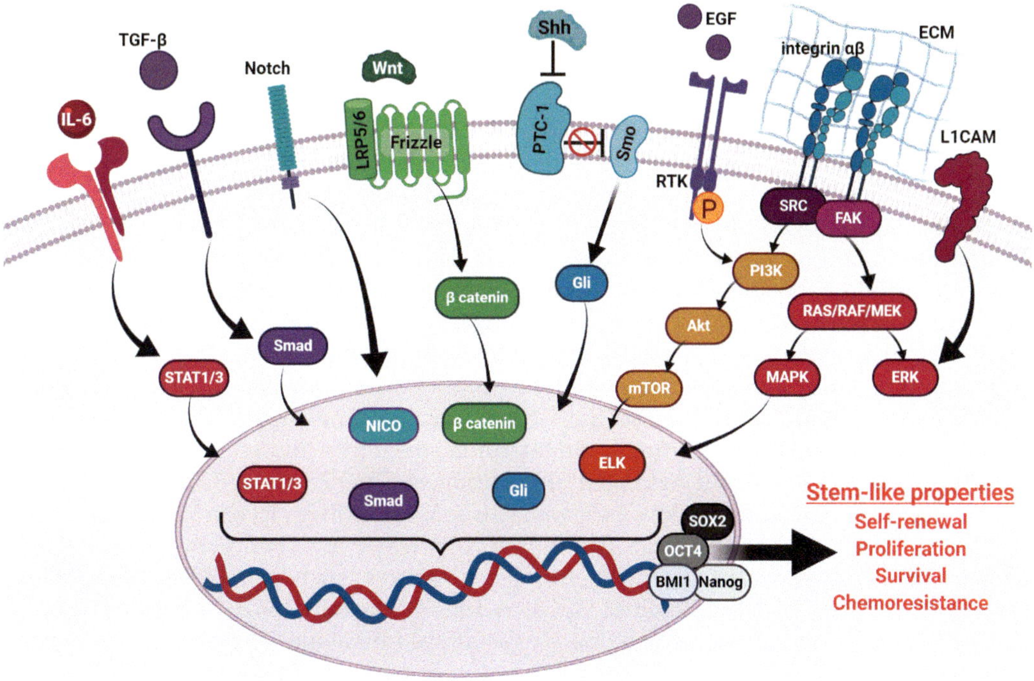

FIG. 9 The intracellular signaling pathways of glioma stem cells (GSCs). Deregulation of molecular signaling pathways in GSCs promotes their self-renewal, proliferation, survival, and chemoresistance.

Targeting the PI3K-Akt-mTOR pathway

Kinase inhibitors that affect the PI3K-Akt and MAPK pathways have shown important roles in inhibiting GSC proliferation. Sorafenib specifically targets GSCs, decreases stemness, halts cellular growth, and induces apoptotic cell death [262]. It targets several tyrosine kinase receptors involved in tumor growth progression and neoangiogenesis, such as VEGFR2 and VEGFR3, PDGFRβ, fibroblast growth factor receptor 1, CD135, RET, and c-Kit [263]. Furthermore, sorafenib directly inhibits the downstream serine/threonine kinase Raf, a key member of the MEK-ERK signaling pathway [263]. Specifically, sorafenib inhibits the antiapoptotic member of Bcl-2 family, Mcl-1, which induces apoptosis in GSCs [262]. Combining sorafenib with a histone deacetylase inhibitor and a bromodomain inhibitor further increases these apoptotic effects and result in a significant regression of glioblastoma in vivo [264]. Human mesenchymal stromal cells can be also be used to carry sorafenib to glioblastoma cells following intranasal administration [265]. However, the therapeutic effect is modest due to the protumorigenic properties of mesenchymal stem cells, which may limit the action of the released sorafenib. Suppression of RAF-MEK or PI3K synergizes with the cytotoxicity of receptor tyrosine kinase inhibitors in GSCs [266]. The combination of erlotinib and sorafenib against GSC11 cells synergistically induces apoptosis and autophagic cell death, as well as suppresses Akt and ERK signaling pathways. It also decreases nuclear pyruvate kinase isozymes M1/M2 and β-catenin expression in vitro [266]. Combination treatment further affects glucose metabolism, fatty acid metabolism, gene transcription, histone methylation, iron transport, stress response, cell cycle, and apoptosis [266]. Furthermore, constitutive ERK1/2 activation sustains a stem-like state in glioblastoma. Inhibition of ERK1/2 restores neurogenesis during murine astrocytoma formation, inducing neuronal differentiation in tumorspheres [267]. Differentiation of glioblastoma neurospheres following the inhibition of ERK1/2 necessitates the expression of miR-124 and the depletion of its target gene SOX9 [267]. Overexpression of miR124 depletes SOX9 and promotes a stem-like-to-neuronal transition, with reduced tumorigenicity and increased sensitivity to radiation [267]. A correlation analysis revealed that miR-124 expression promotes the differentiation of GSCs through the inhibition of the Nogo-NgR pathway [268]. Moreover, dual inhibition of the phosphoinositide-dependent kinase-1 (PDK1) and aurora kinase A (AurA) pathways, each one playing a pivotal role in cellular survival/migration/differentiation, represents an innovative strategy to overcome GSCs [269]. Ligand SA16 can inhibit both the PDK1 and the AurA pathways, thus inducing GSC differentiation and apoptosis [269]. In addition, GSCs can be targeted by mibefradil, which inhibits the T-type calcium channel Cav3.2 [270]. This calcium channel is highly expressed in human glioblastoma specimens and enriched in GSCs [270]. Mibefradil treatment or RNAi-mediated attenuation of Cav3.2 was sufficient to inhibit the growth, survival, and stemness of GSCs and also sensitized them to temozolomide chemotherapy by inhibiting prosurvival Akt-mTOR pathways and stimulating proapoptotic survivin and BAX pathways [270]. Other kinase inhibitors, such as OTS964, potentiate the cytotoxicity of temozolomide against GSCs, reducing the initial size and controlling the regrowth of GSC populations, which directly facilitates initial and long-term control of glioblastoma recurrence [271].

Targeting Notch signaling

Gamma secretase inhibitors (GSIs) are a class of drugs that gained attention during the past decade in cancer treatment. The use of GSIs for cancers is primarily based on the premise that GSIs act by inhibiting the cleavage of γ-secretase, which result in blocking Notch 1 signaling [272]. Modulation of the Notch pathway increases glioma cell sensitivity to temozolomide [273]. Notch blockade combined with radiation therapy and temozolomide further prolongs survival in glioblastoma by decreasing GSC growth and reducing GSC markers CD133, SOX2, and Nestin [274]. In a phase 0/I trial, Notch inhibition via the GSI RO4929097 was added to temozolomide and radiotherapy [275]. Immunohistochemistry of treated tumors showed a significant decrease in proliferation and in the expression of the Notch intracellular domain by tumor cells and blood vessels [275]. Nevertheless, recurrence occurred, associated with alterations in angiogenesis signaling pathways [275].

In GSCs, the Notch signaling pathway regulates self-renewal and drug resistance. However, blocking its activity with GSIs fails to suppress the growth of GSCs, as GSIs seem to be active in only a fraction of GSCs that exhibit constitutive Notch activity [276]. The loss of PTEN function leads to resistance to Notch inhibition, which causes the transfer of key oncogenic signaling from the Notch pathway to the PI3K pathway [276]. Notch regulates PTEN expression and the activity of the PI3K pathway in GSCs. GSIs attenuate the Notch pathway but increase PTEN expression via Hes family BHLH transcription factor 1 (HES1) [276]. Therefore, targeting both pathways is important to improve therapeutic efficacy in glioblastoma [276]. Isoliquiritigenin is a flavonoid that also downregulates Notch1 and HES1 in glioblastoma, leading to inhibition of GSC growth differentiation into astrocytes and neurons [277].

Downregulation of Akt is an important mechanism to sensitize GSCs to GSI treatment [87]. Farnesyltransferase inhibitors significantly improve sensitivity to GSIs. Different models show that the combination of farnesyltransferase inhibitors and GSIs, but not either agent alone, significantly reduces GSC growth [278]. Combinatorial effects are mediated, at least partially, through inhibition of Akt and cell cycle progression [278]. Recombining binding protein suppressor of hairless (RBPJ) is considered a central transcriptional mediator of Notch activity and is preferentially expressed in GSCs for self-renewal and tumor growth. Targeting RBPJ was more effective than Notch inhibitors in suppressing glioblastoma growth [279]. RBPJ controls MYC transcription through CDK9 [279]. Treatment with a bromodomain inhibitor decreases both MYC and RBPJ expression [279]. Furthermore, hypoxia can stimulate the activation of RBPJ in glioblastoma. RBPJ blockade can suppress glioblastoma invasion in GSCs present in the hypoxic niche [280]. Targeting RBPJ can be an effective anti-GSC therapy to impede invasive properties and induce chemosensitivity in GSCs [280].

Targeting Shh signaling

The Shh pathway is highly expressed in GSCs. Treating GSCs with the Shh inhibitor cyclopamine impaired new neurosphere formation, and glioblastoma cells injected intracranially following Shh blockade were no longer able to form tumors in athymic mice [281]. Combining micellarized cyclopamine with temozolomide exhibited synergistic cytotoxicity, correlating with their ability to induce apoptosis and eradicate GSCs [282]. In addition, the combination was accompanied by enhanced blockage of the Shh pathway [282].

Targeting the Shh pathway alongside other oncogenic pathways is an important strategy in glioblastoma treatment. The combined inhibition of PI3K-Akt-mTOR and Shh pathways was superior to single pathway inhibition in suppressing glioblastoma growth. NVP-LDE-225, inhibitor of smoothened (SMO) antagonist, synergizes with NVP-BEZ-235, dual inhibitor of PI3K and mTOR, in inhibiting the self-renewal capacity of GSCs and decreasing the expression of pluripotency maintaining factors (Nanog, c-Myc, Oct4, and SOX2), Musashi1, cyclin D1, and Bcl-2, and transcription and expression of zinc finger protein Gli. The combination further induces the expression of cleaved caspase-3, cleaved PARP and Bim, as well as inhibits EMT [283].

Therapeutic resistance of GSCs can be overcome by targeting influencers of the Shh pathway. Inhibition of HDAC6 in GSCs decreases cell proliferation and induces differentiation and apoptosis through the inactivation of the Shh-Gli1 signaling pathway [284]. Moreover, HDAC6 inhibition decreases DNA damage repair capacity of GSCs and leads to increased radiosensitivity. This provides a promising novel drug target to overcome GSC stemness and radioresistance [284]. The loss of cullin3 represents another common signaling node for controlling the activity of intracellular Wnt and Shh signaling pathways mediated by ID1. Cullin3 interacts with Gli2 and disheveled segment polarity protein 2 and induces their degradation via ubiquitination. ID1 activates GSC proliferation, self-renewal, and tumorigenicity by suppressing cullin3 ubiquitin ligase [105]. An ID1-high/Cullin-low expression signature correlates with a poor patient prognosis. Combined control of Gli2 and β-catenin effectively diminishes GSC properties [105]. The human sialidase, NEU4, is a regulator of neuronal differentiation and its overexpression promotes GSCs through the Shh and Wnt/β-catenin signaling pathways. NEU4 inhibition decreases the stem cell-like gene expression and markers of NANOG, OCT-4, SOX-2, CD133 expression and significantly decreases GSC survival [285]. Furthermore, Shh-Gli1 signaling regulates the expression of the stem cell marker BMI1 in glioma [286]. Knockdown of BMI1 or inhibition of Shh-Gli1 signaling using the inhibitor GANT61 decreases the expression of multidrug resistance protein 1 (MRP1), which implicates a role for Shh-Gli1-BMI1 cell signaling in the regulation of MRP1 and the promotion of chemoresistance in glioma [286]. The combination of GANT61 and temozolomide achieved a synergic effect as it drastically decreased the LD50 of temozolomide in vitro [286]. Moreover, polyethylenimine-spherical nucleic acid nanoparticles can be used against Gli1 to reduce the chemoresistance and stemness of GSCs [287]. Targeting the Shh pathway through miRNAs can also promote GSC differentiation. miR-326 is an inhibitor of SMO and their expression is inversely correlated in glioma [288]. miR-326 overexpression downregulated SMO and its downstream genes thus decreasing the activity of the hedgehog pathway [288]. This, in turn, decreased stemness and prompted differentiation in GSCs.

Targeting STAT3 signaling

STAT3 regulates diverse cellular processes, including cell growth, differentiation, and apoptosis, and is frequently activated during gliomagenesis. This makes it an ideal target for therapy. Napabucasin (BBI608), a novel small molecule inhibitor of STAT3, suppresses the expression of stemness-associated genes including STAT3 and inhibits the formation of GSCs in vitro [289]. Nonphosphorylated STAT3 inhibitor 31 is a BBB-permeable agent that strongly binds to the STAT3 protein, effectively suppresses STAT3 phosphorylation and its downstream transcriptional targets, and potently kills GSCs [290]. LLL12, a nonpeptide small molecule, is another potent inhibitor of STAT3 phosphorylation, which decreases cell viability, and induces apoptosis in GSCs with elevated levels of phosphorylated-STAT3 [291]. WP1193, a novel Janus kinase 2 (JAK2)/STAT3 pathway inhibitor, significantly decreased the proliferation of GSCs [292]. WP1193 suppresses phosphorylation of JAK2 and STAT3 with high potency and demonstrates a dose-dependent inhibition of proliferation and neurosphere formation of GSCs [292]. Resveratrol, a polyphenol in grapes, is reported to be a potent and nontoxic cancer-preventive compound. It suppresses tumorigenicity and enhances radiosensitivity in GSCs by inhibiting the STAT3 axis [293]. 4-Acetylantroquinonol B, a bioactive isolate of Antrodia cinnamomea, was found to inhibit GSC properties and tumorigenesis by disrupting the β-Catenin/TCF-1/STAT3 signaling axis [294]. Thus, projecting 4-acetylantroquinonol B as a potent therapeutic agent for anti-GBM target therapy [294].

In patients with recurrent and temozolomide-resistant glioma, overexpression of the transcription factor AP-2α was found to reduce stemness and sensitize GSCs to temozolomide by inhibiting the Nanog-SOX2-CD133 axis and the IL-6-STAT3 signaling pathway [295]. Consequently, this led to a decrease in MGMT and PD-L1 expression [295]. STAT3 inhibition leads to the upregulation of the histone H3K27me2/3 demethylase Jmjd3 (KDM6B), which slows GSC growth and neurosphere formation [296]. Targeting tripartite motif-containing protein 8 (TRIM8) is an alternative way to downregulate STAT3 signaling in GSCs. TRIM8 overexpression is highly correlated with stem cell markers p-STAT3, c-MYC, SOX2, Nestin, and CD133 [297]. TRIM8 activates STAT3 by suppressing the expression of PIAS3, an inhibitor of STAT3, through E3-mediated ubiquitination and proteasomal degradation [297]. IL6-activated STAT3 enhances B7-H4 expression, which correlates with poor survival outcomes in patients with malignant glioma [298]. CD133+ cells mediate immunosuppression by silencing B7-H4 expression on microglia, blocking effective T-cell immune responses [298]. Countering this phenomenon can be an effective strategy against GSCs. Tetraspanin CD9 is a GSC biomarker that is relevant to GSC maintenance. It stabilizes the IL-6 receptor glycoprotein 130 by preventing its ubiquitin-dependent lysosomal degradation to facilitate the STAT3 activation in GSCs [299]. Targeting CD9 potently inhibits GSC tumor growth [299]. Inositol Polyphosphate-5-Phosphatase F (INPP5F) is differentially expressed in GSCs isolated from glioma-bearing patients [300]. INPP5F inhibits STAT3 signaling through dephosphorylation, which suppresses self-renewal and proliferation potentials and reduces tumorigenicity of glioblastoma [300].

Although STAT3 inhibition can induce apoptosis in serum-derived glioblastoma cells, GSCs can develop resistance to induced apoptotic cell death [301]. Unfortunately, there are no successful STAT3 inhibitors currently in clinical trials. However, strong clinical evidence implicating STAT3 as a major factor in glioblastoma justifies the identification of safe and effective strategies for inhibiting STAT3 [302]. As such, combining STAT3 inhibitors with other chemotherapeutic agents can yield better results. The combination of minocycline, a tetracycline antibiotic, and STAT3 inhibitor synergistically reduces the cell viability of GSCs in vitro and shows clinical benefit in vivo [303]. Cardamonin is a novel inhibitor of STAT3 that locks the nuclear translocation and dimerization of STAT3 in CD133+ GSCs [304]. It inhibits proliferation and induces apoptosis in CD133+ GSCs. In combination with temozolomide, the apoptotic effect is further enhanced [304]. STX-

0119, another inhibitor of STAT3, is an effective therapeutic compound in combination with temozolomide plus radiation therapy against GSCs, even in temozolomide-resistant glioblastoma [305]. Notably, STX-0119 demonstrated strong inhibition of the expression of c-Myc, survivin, cyclin D1, HIF-1α, and VEGF [305]. In addition, it suppressed stem cell-associated genes CD44, Nanog, Nestin, and CD133 [305]. Ovatodiolide, a major bioactive diterpenoid isolate of *Anisomeles indica*, inhibits the oncogenicity and stemness of GSCs, as well as potentiate the anticancer effect of temozolomide, via dysregulation of the JAK2-STAT3 signaling pathway [306]. It induces apoptosis, reduces stemness markers, and inhibits tumor sphere formation [306].

Regulating autophagy

Autophagy is one of the many signaling mechanisms that regulate tumor growth. Accumulating evidence suggests that unique properties of stem cells depend on autophagy. As such, modulation of autophagy can increase the efficacy of anti-GSC therapeutic options [307].

Induction of cytotoxic autophagy can be an effective strategy against GSCs. Glioblastoma cells feature mTOR upregulation which relates to high stemness and chemoresistance. Administering mTOR inhibitor and autophagy inducer, rapamycin, stimulates gene transcription related to neuronal differentiation while inhibiting stemness related genes such as Nestin [308]. Rapamycin further increases βIII-tubulin, which leads GSCs to form pyramidal neuron-like shapes owing long branches, making the cells more susceptible to chemotherapy [308]. BIX01294, an inhibitor of histone methyltransferase, also induces autophagy-dependent differentiation of GSCs [309].

On the other hand, inhibition of cytoprotective autophagy is a potential method to overcome chemotherapeutic resistance in GSCs. Cytoprotective autophagy is a survival response that offers cancer cells the ability to buffer against starvation and evade apoptotic signals. Long noncoding RNA AC023115.3 is induced by cisplatin in human glioblastoma cells [310]. Elevated AC023115.3 promotes cisplatin-induced apoptosis by inhibiting autophagy, thereby improving chemosensitivity of GSCs to cisplatin by regulating the miR-26a-GSK3β-Mcl1 pathway [310]. In addition, inhibition of the Notch pathway in GSCs increases autophagy markers significantly [311]. By combining the late stage autophagy inhibitor chloroquine with the GSI MRK003, a significant induction in apoptosis and reduction in growth was noted as compared to Notch inhibition alone [311].

Conclusion

The progress made in understanding the epigenetic, metabolic, microenvironmental, and developmental foundations of GSCs allowed the understanding of how GSCs develop resistance to chemotherapy and the identification of potential targets for glioma therapy. New technologies, such as single-cell genomics and CRISPR-Cas9 genome-editing tools and derivatives, offer better capabilities to precisely target key regulators of stemness and resistance at the DNA levels; however, its application for mRNA inhibition remains unestablished. Organoid models for glioblastoma fail to recapitulate the serum and blood components that are necessary for GSC growth. Moreover, there has not been an effective translation of preclinical advancements to the clinical scene. Patient survival in glioblastoma remains dismal and the disease continues to be fatal despite the administration of various chemotherapeutic agents and the employment of different surgical and radiotherapy strategies.

In the era of immunotherapy, immune checkpoint inhibition and cell-based therapies, including CAR-T cells, must be investigated further in the coming years. Future immune-based strategies must focus on combinations of different immune-checkpoint inhibitors with different treatment modalities to reverse local immunosuppression in the microenvironment. Oncolytic viral therapy can show better clinical outcomes in combination with immune checkpoint inhibitors and targeting of other microenvironmental adaptive mechanisms, such as hypoxia and angiogenesis. Moreover, metabolic targeting of GSCs can further improve the efficacy of immunotherapy and radiotherapy. Cancer neuroscience is a rising field. It is now established that the nervous system, via electrochemical interactions, contributes to glioblastoma initiation, growth, recurrence, and chemoresistance. Proper understanding of the complex crosstalk between the nervous system and glioblastoma cells is necessary to elucidate therapeutic targets to improve patient outcomes. In addition, deciphering the interactions between the nervous system and the other microenvironmental niches is equally important to achieve strides in precision cancer therapeutics.

References

[1] Stupp R, Hegi ME, Mason WP, van den Bent MJ, Taphoorn MJB, Janzer RC, et al. Effects of radiotherapy with concomitant and adjuvant temozolomide versus radiotherapy alone on survival in glioblastoma in a randomised phase III study: 5-year analysis of the EORTC-NCIC trial. Lancet Oncol 2009;10:459–66.

[2] Ostrom QT, Gittleman H, Fulop J, Liu M, Blanda R, Kromer C, et al. CBTRUS statistical report: primary brain and central nervous system tumors diagnosed in the United States in 2008–2012. Neuro-Oncology 2015;17 (Suppl 4):iv1–iv62.

[3] Stupp R, Mason WP, van den Bent MJ, Weller M, Fisher B, Taphoorn MJB, et al. Radiotherapy plus concomitant and adjuvant temozolomide for glioblastoma. N Engl J Med 2005;352:987–96.

[4] Stupp R, Taillibert S, Kanner A, Read W, Steinberg D, Lhermitte B, et al. Effect of tumor-treating fields plus maintenance temozolomide vs maintenance temozolomide alone on survival in patients with glioblastoma: a randomized clinical trial. JAMA 2017;318:2306–16.

[5] Molinaro AM, Hervey-Jumper S, Morshed RA, Young J, Han SJ, Chunduru P, et al. Association of maximal extent of resection of contrast-enhanced and non-contrast-enhanced tumor with survival within molecular subgroups of patients with newly diagnosed glioblastoma. JAMA Oncol 2020;6:495–503.

[6] Gimple RC, Bhargava S, Dixit D, Rich JN. Glioblastoma stem cells: lessons from the tumor hierarchy in a lethal cancer. Genes Dev 2019;33:591–609.

[7] Little SE, Popov S, Jury A, Bax DA, Doey L, Al-Sarraj S, et al. Receptor tyrosine kinase genes amplified in glioblastoma exhibit a mutual exclusivity in variable proportions reflective of individual tumor heterogeneity. Cancer Res 2012;72:1614–20.

[8] Snuderl M, Fazlollahi L, Le LP, Nitta M, Zhelyazkova BH, Davidson CJ, et al. Mosaic amplification of multiple receptor tyrosine kinase genes in glioblastoma. Cancer Cell 2011;20:810–7.

[9] Auffinger B, Spencer D, Pytel P, Ahmed AU, Lesniak MS. The role of glioma stem cells in chemotherapy resistance and glioblastoma multiforme recurrence. Expert Rev Neurother 2015;15:741–52.

[10] Singh SK, Clarke ID, Terasaki M, Bonn VE, Hawkins C, Squire J, et al. Identification of a cancer stem cell in human brain tumors. Cancer Res 2003;63:5821–8.

[11] Galli R, Binda E, Orfanelli U, Cipelletti B, Gritti A, De Vitis S, et al. Isolation and characterization of tumorigenic, stem-like neural precursors from human glioblastoma. Cancer Res 2004;64:7011–21.

[12] Singh SK, Hawkins C, Clarke ID, Squire JA, Bayani J, Hide T, et al. Identification of human brain tumour initiating cells. Nature 2004;432:396–401.

[13] Yuan X, Curtin J, Xiong Y, Liu G, Waschsmann-Hogiu S, Farkas DL, et al. Isolation of cancer stem cells from adult glioblastoma multiforme. Oncogene 2004;23:9392–400.
[14] Patel AP, Tirosh I, Trombetta JJ, Shalek AK, Gillespie SM, Wakimoto H, et al. Single-cell RNA-seq highlights intratumoral heterogeneity in primary glioblastoma. Science 2014;344:1396–401.
[15] Tirosh I, Venteicher AS, Hebert C, Escalante LE, Patel AP, Yizhak K, et al. Single-cell RNA-seq supports a developmental hierarchy in human oligodendroglioma. Nature 2016;539:309–13.
[16] Filbin MG, Tirosh I, Hovestadt V, Shaw ML, Escalante LE, Mathewson ND, et al. Developmental and oncogenic programs in H3K27M gliomas dissected by single-cell RNA-seq. Science 2018;360:331–5.
[17] Huang Z, Cheng L, Guryanova OA, Wu Q, Bao S. Cancer stem cells in glioblastoma—molecular signaling and therapeutic targeting. Protein Cell 2010;1:638–55.
[18] Auffinger B, Tobias AL, Han Y, Lee G, Guo D, Dey M, et al. Conversion of differentiated cancer cells into cancer stem-like cells in a glioblastoma model after primary chemotherapy. Cell Death Differ 2014;21:1119–31.
[19] Ahmed AU, Auffinger B, Lesniak MS. Understanding glioma stem cells: rationale, clinical relevance and therapeutic strategies. Expert Rev Neurother 2013;13:545–55.
[20] Liu G, Yuan X, Zeng Z, Tunici P, Ng H, Abdulkadir IR, et al. Analysis of gene expression and chemoresistance of CD133+ cancer stem cells in glioblastoma. Mol Cancer 2006;5:67.
[21] Chen J, Li Y, Yu T-S, McKay RM, Burns DK, Kernie SG, et al. A restricted cell population propagates glioblastoma growth after chemotherapy. Nature 2012;488:522–6.
[22] Bao S, Wu Q, McLendon RE, Hao Y, Shi Q, Hjelmeland AB, et al. Glioma stem cells promote radioresistance by preferential activation of the DNA damage response. Nature 2006;444:756–60.
[23] De Bacco F, D'Ambrosio A, Casanova E, Orzan F, Neggia R, Albano R, et al. MET inhibition overcomes radiation resistance of glioblastoma stem-like cells. EMBO Mol Med 2016;8:550–68.
[24] Caragher SP, Sachdev S, Ahmed A. Radiotherapy and glioma stem cells: searching for chinks in cellular armor. Curr Stem Cell Rep 2017;3:348–57.
[25] Hardee ME, Marciscano AE, Medina-Ramirez CM, Zagzag D, Narayana A, Lonning SM, et al. Resistance of glioblastoma-initiating cells to radiation mediated by the tumor microenvironment can be abolished by inhibiting transforming growth factor-beta. Cancer Res 2012;72:4119–29.
[26] Cheng L, Wu Q, Huang Z, Guryanova OA, Huang Q, Shou W, et al. L1CAM regulates DNA damage checkpoint response of glioblastoma stem cells through NBS1. EMBO J 2011;30:800–13.
[27] Tamura K, Aoyagi M, Ando N, Ogishima T, Wakimoto H, Yamamoto M, et al. Expansion of CD133-positive glioma cells in recurrent de novo glioblastomas after radiotherapy and chemotherapy. J Neurosurg 2013;119:1145–55.
[28] Hjelmeland AB, Wu Q, Heddleston JM, Choudhary GS, MacSwords J, Lathia JD, et al. Acidic stress promotes a glioma stem cell phenotype. Cell Death Differ 2011;18:829–40.
[29] Lee G, Auffinger B, Guo D, Hasan T, Deheeger M, Tobias AL, et al. Dedifferentiation of glioma cells to glioma stem-like cells by therapeutic stress-induced HIF signaling in the recurrent GBM model. Mol Cancer Ther 2016;15:3064–76.
[30] Olmez I, Shen W, McDonald H, Ozpolat B. Dedifferentiation of patient-derived glioblastoma multiforme cell lines results in a cancer stem cell-like state with mitogen-independent growth. J Cell Mol Med 2015;19:1262–72.
[31] Safa AR, Saadatzadeh MR, Cohen-Gadol AA, Pollok KE, Bijangi-Vishehsaraei K. Glioblastoma stem cells (GSCs) epigenetic plasticity and interconversion between differentiated non-GSCs and GSCs. Genes Dis 2015;2:152–63.
[32] Kim R-K, Suh Y, Cui Y-H, Hwang E, Lim E-J, Yoo K-C, et al. Fractionated radiation-induced nitric oxide promotes expansion of glioma stem-like cells. Cancer Sci 2013;104:1172–7.
[33] Lee J, Kotliarova S, Kotliarov Y, Li A, Su Q, Donin NM, et al. Tumor stem cells derived from glioblastomas cultured in bFGF and EGF more closely mirror the phenotype and genotype of primary tumors than do serum-cultured cell lines. Cancer Cell 2006;9:391–403.
[34] Hubert CG, Rivera M, Spangler LC, Wu Q, Mack SC, Prager BC, et al. A three-dimensional organoid culture system derived from human glioblastomas recapitulates the hypoxic gradients and cancer stem cell heterogeneity of tumors found in vivo. Cancer Res 2016;76:2465–77.
[35] Ogawa J, Pao GM, Shokhirev MN, Verma IM. Glioblastoma model using human cerebral organoids. Cell Rep 2018;23:1220–9.
[36] Oldrini B, Curiel-Garcia A, Marques C, Matia V, Uluckan O, Grana-Castro O, et al. Somatic genome editing with the RCAS-TVA-CRISPR-Cas9 system for precision tumor modeling. Nat Commun 2018;9:1466.

[37] Chow RD, Guzman CD, Wang G, Schmidt F, Youngblood MW, Ye L, et al. AAV-mediated direct in vivo CRISPR screen identifies functional suppressors in glioblastoma. Nat Neurosci 2017;20:1329–41.

[38] Brescia P, Richichi C, Pelicci G. Current strategies for identification of glioma stem cells: adequate or unsatisfactory? J Oncol 2012;2012:376894.

[39] Mao X-G, Zhang X, Xue X-Y, Guo G, Wang P, Zhang W, et al. Brain tumor stem-like cells identified by neural stem cell marker CD15. Transl Oncol 2009;2:247–57.

[40] Wang J, Wakeman TP, Lathia JD, Hjelmeland AB, Wang X-F, White RR, et al. Notch promotes radioresistance of glioma stem cells. Stem Cells 2010;28:17–28.

[41] Haas TL, Sciuto MR, Brunetto L, Valvo C, Signore M, Fiori ME, et al. Integrin alpha7 is a functional marker and potential therapeutic target in glioblastoma. Cell Stem Cell 2017;21:35–50.e9.

[42] Miroshnikova YA, Mouw JK, Barnes JM, Pickup MW, Lakins Johnathan N, Kim Y, et al. Tissue mechanics promote IDH1-dependent HIF1α—tenascin C feedback to regulate glioblastoma aggression. Nat Cell Biol 2016;18:1336–45.

[43] Sarkar S, Mirzaei R, Zemp FJ, Wei W, Senger DL, Robbins SM, et al. Activation of NOTCH signaling by Tenascin-C promotes growth of human brain tumor-initiating cells. Cancer Res 2017;77:3231–43.

[44] Barnes JM, Kaushik S, Bainer RO, Sa JK, Woods EC, Kai F, et al. A tension-mediated glycocalyx-integrin feedback loop promotes mesenchymal-like glioblastoma. Nat Cell Biol 2018;20:1203–14.

[45] Hossain A, Gumin J, Gao F, Figueroa J, Shinojima N, Takezaki T, et al. Mesenchymal stem cells isolated from human gliomas increase proliferation and maintain stemness of glioma stem cells through the IL-6/gp130/STAT3 pathway. Stem Cells 2015;33:2400–15.

[46] Figueroa J, Phillips LM, Shahar T, Hossain A, Gumin J, Kim H, et al. Exosomes from glioma-associated mesenchymal stem cells increase the tumorigenicity of glioma stem-like cells via transfer of miR-1587. Cancer Res 2017;77:5808–19.

[47] Puchalski RB, Shah N, Miller J, Dalley R, Nomura SR, Yoon J-G, et al. An anatomic transcriptional atlas of human glioblastoma. Science 2018;360:660–3.

[48] Osswald M, Jung E, Sahm F, Solecki G, Venkataramani V, Blaes J, et al. Brain tumour cells interconnect to a functional and resistant network. Nature 2015;528:93–8.

[49] Hitomi M, Deleyrolle LP, Mulkearns-Hubert EE, Jarrar A, Li M, Sinyuk M, et al. Differential connexin function enhances self-renewal in glioblastoma. Cell Rep 2015;11:1031–42.

[50] Wang X, Prager BC, Wu Q, Kim LJY, Gimple RC, Shi Y, et al. Reciprocal signaling between glioblastoma stem cells and differentiated tumor cells promotes malignant progression. Cell Stem Cell 2018;22:514–528.e5.

[51] Wei J, Wu A, Kong L-Y, Wang Y, Fuller G, Fokt I, et al. Hypoxia potentiates glioma-mediated immunosuppression. PLoS One 2011;6, e16195.

[52] Ozawa Y, Yamamuro S, Sano E, Tatsuoka J, Hanashima Y, Yoshimura S, et al. Indoleamine 2,3-dioxygenase 1 is highly expressed in glioma stem cells. Biochem Biophys Res Commun 2020;524:723–9.

[53] Ricklefs FL, Alayo Q, Krenzlin H, Mahmoud AB, Speranza MC, Nakashima H, et al. Immune evasion mediated by PD-L1 on glioblastoma-derived extracellular vesicles. Sci Adv 2018;4, eaar2766.

[54] Domenis R, Cesselli D, Toffoletto B, Bourkoula E, Caponnetto F, Manini I, et al. Systemic T cells immunosuppression of glioma stem cell-derived exosomes is mediated by monocytic myeloid-derived suppressor cells. PLoS One 2017;12, e0169932.

[55] Shi Y, Ping Y-F, Zhou W, He Z-C, Chen C, Bian B-S-J, et al. Tumour-associated macrophages secrete pleiotrophin to promote PTPRZ1 signalling in glioblastoma stem cells for tumour growth. Nat Commun 2017;8:15080.

[56] Zhang Y, Yu G, Chu H, Wang X, Xiong L, Cai G, et al. Macrophage-associated PGK1 phosphorylation promotes aerobic glycolysis and tumorigenesis. Mol Cell 2018;71:201–215.e7.

[57] Alvarado AG, Thiagarajan PS, Mulkearns-Hubert EE, Silver DJ, Hale JS, Alban TJ, et al. Glioblastoma cancer stem cells evade innate immune suppression of self-renewal through reduced TLR4 expression. Cell Stem Cell 2017;20:450–461.e4.

[58] Zhou W, Ke SQ, Huang Z, Flavahan W, Fang X, Paul J, et al. Periostin secreted by glioblastoma stem cells recruits M2 tumour-associated macrophages and promotes malignant growth. Nat Cell Biol 2015;17:170–82.

[59] Otvos B, Silver DJ, Mulkearns-Hubert EE, Alvarado AG, Turaga SM, Sorensen MD, et al. Cancer stem cell-secreted macrophage migration inhibitory factor stimulates myeloid derived suppressor cell function and facilitates glioblastoma immune evasion. Stem Cells 2016;34:2026–39.

[60] Dolma S, Selvadurai HJ, Lan X, Lee L, Kushida M, Voisin V, et al. Inhibition of dopamine receptor D4 impedes autophagic flux, proliferation, and survival of glioblastoma stem cells. Cancer Cell 2016;29:859–73.
[61] Wang J, Xu S-L, Duan J-J, Yi L, Guo Y-F, Shi Y, et al. Invasion of white matter tracts by glioma stem cells is regulated by a NOTCH1-SOX2 positive-feedback loop. Nat Neurosci 2019;22:91–105.
[62] Marisetty AL, Lu L, Veo BL, Liu B, Coarfa C, Kamal MM, et al. REST-DRD2 mechanism impacts glioblastoma stem cell-mediated tumorigenesis. Neuro-Oncology 2019;21:775–85.
[63] Caragher SP, Shireman JM, Huang M, Miska J, Atashi F, Baisiwala S, et al. Activation of dopamine receptor 2 prompts transcriptomic and metabolic plasticity in glioblastoma. J Neurosci 2019;39:1982–93.
[64] Venkatesh HS, Johung TB, Caretti V, Noll A, Tang Y, Nagaraja S, et al. Neuronal activity promotes glioma growth through neuroligin-3 secretion. Cell 2015;161:803–16.
[65] Venkatesh HS, Tam LT, Woo PJ, Lennon J, Nagaraja S, Gillespie SM, et al. Targeting neuronal activity-regulated neuroligin-3 dependency in high-grade glioma. Nature 2017;549:533–7.
[66] Zhu TS, Costello MA, Talsma CE, Flack CG, Crowley JG, Hamm LL, et al. Endothelial cells create a stem cell niche in glioblastoma by providing NOTCH ligands that nurture self-renewal of cancer stem-like cells. Cancer Res 2011;71:6061–72.
[67] Rosso L, Brock CS, Gallo JM, Saleem A, Price PM, Turkheimer FE, et al. A new model for prediction of drug distribution in tumor and normal tissues: pharmacokinetics of temozolomide in glioma patients. Cancer Res 2009;69:120–7.
[68] Filatova A, Acker T, Garvalov BK. The cancer stem cell niche(s): the crosstalk between glioma stem cells and their microenvironment. Biochim Biophys Acta 1830;2013:2496–508.
[69] Griveau A, Seano G, Shelton SJ, Kupp R, Jahangiri A, Obernier K, et al. A glial signature and Wnt7 signaling regulate glioma-vascular interactions and tumor microenvironment. Cancer Cell 2018;33:874–889.e7.
[70] Cheng L, Huang Z, Zhou W, Wu Q, Donnola S, Liu JK, et al. Glioblastoma stem cells generate vascular pericytes to support vessel function and tumor growth. Cell 2013;153:139–52.
[71] Zhou W, Chen C, Shi Y, Wu Q, Gimple RC, Fang X, et al. Targeting glioma stem cell-derived pericytes disrupts the blood-tumor barrier and improves chemotherapeutic efficacy. Cell Stem Cell 2017;21:591–603.e4.
[72] Shi Y, Guryanova OA, Zhou W, Liu C, Huang Z, Fang X, et al. Ibrutinib inactivates BMX-STAT3 in glioma stem cells to impair malignant growth and radioresistance. Sci Transl Med 2018;10.
[73] Hu B, Wang Q, Wang YA, Hua S, Sauve C-EG, Ong D, et al. Epigenetic activation of WNT5A drives glioblastoma stem cell differentiation and invasive growth. Cell 2016;167:1281–1295.e18.
[74] Seidel S, Garvalov BK, Wirta V, von Stechow L, Schanzer A, Meletis K, et al. A hypoxic niche regulates glioblastoma stem cells through hypoxia inducible factor 2 alpha. Brain 2010;133:983–95.
[75] Pistollato F, Abbadi S, Rampazzo E, Persano L, Della Puppa A, Frasson C, et al. Intratumoral hypoxic gradient drives stem cells distribution and MGMT expression in glioblastoma. Stem Cells 2010;28:851–62.
[76] Johannessen T-CA, Bjerkvig R, Tysnes BB. DNA repair and cancer stem-like cells—potential partners in glioma drug resistance? Cancer Treat Rev 2008;34:558–67.
[77] Bobola MS, Blank A, Berger MS, Stevens BA, Silber JR. Apurinic/apyrimidinic endonuclease activity is elevated in human adult gliomas. Clin Cancer Res 2001;7:3510–8.
[78] Wharton SB, McNelis U, Bell HS, Whittle IR. Expression of poly(ADP-ribose) polymerase and distribution of poly(ADP-ribosyl)ation in glioblastoma and in a glioma multicellular tumour spheroid model. Neuropathol Appl Neurobiol 2000;26:528–35.
[79] Bartkova J, Hamerlik P, Stockhausen MT, Ehrmann J, Hlobilkova A, Laursen H, et al. Replication stress and oxidative damage contribute to aberrant constitutive activation of DNA damage signalling in human gliomas. Oncogene 2010;29:5095–102.
[80] Ropolo M, Daga A, Griffero F, Foresta M, Casartelli G, Zunino A, et al. Comparative analysis of DNA repair in stem and nonstem glioma cell cultures. Mol Cancer Res 2009;7:383–92.
[81] Yip S, Miao J, Cahill DP, Iafrate AJ, Aldape K, Nutt CL, et al. MSH6 mutations arise in glioblastomas during temozolomide therapy and mediate temozolomide resistance. Clin Cancer Res 2009;15:4622–9.
[82] Bartek J, Lukas J. Chk1 and Chk2 kinases in checkpoint control and cancer. Cancer Cell 2003;3:421–9.
[83] Fares J, Kanojia D, Rashidi A, Ulasov I, Lesniak MS. Genes that mediate metastasis across the blood-brain barrier. Trends Cancer 2020;6:660–76.
[84] Balca-Silva J, Matias D, Ad C, Sarmento-Ribeiro AB, Lopes MC, Moura-Neto V. Cellular and molecular mechanisms of glioblastoma malignancy: implications in resistance and therapeutic strategies. Semin Cancer Biol 2019;58:130–41.

[85] Li W-Q, Li Y-M, Tao B-B, Lu Y-C, Hu G-H, Liu H-M, et al. Downregulation of ABCG2 expression in glioblastoma cancer stem cells with miRNA-328 may decrease their chemoresistance. Med Sci Monit 2010;16:HY27–30.
[86] Xu Z-Y, Wang K, Li X-Q, Chen S, Deng J-M, Cheng Y, et al. The ABCG2 transporter is a key molecular determinant of the efficacy of sonodynamic therapy with photofrin in glioma stem-like cells. Ultrasonics 2013;53:232–8.
[87] Bleau A-M, Hambardzumyan D, Ozawa T, Fomchenko EI, Huse JT, Brennan CW, et al. PTEN/PI3K/Akt pathway regulates the side population phenotype and ABCG2 activity in glioma tumor stem-like cells. Cell Stem Cell 2009;4:226–35.
[88] Golebiewska A, Bougnaud S, Stieber D, Brons NHC, Vallar L, Hertel F, et al. Side population in human glioblastoma is non-tumorigenic and characterizes brain endothelial cells. Brain 2013;136:1462–75.
[89] Kanno H, Sato H, Yokoyama T-A, Yoshizumi T, Yamada S. The VHL tumor suppressor protein regulates tumorigenicity of U87-derived glioma stem-like cells by inhibiting the JAK/STAT signaling pathway. Int J Oncol 2013;42:881–6.
[90] Korkaya H, Paulson A, Charafe-Jauffret E, Ginestier C, Brown M, Dutcher J, et al. Regulation of mammary stem/progenitor cells by PTEN/Akt/beta-catenin signaling. PLoS Biol 2009;7, e1000121.
[91] Buruiana A, Florian ŞI, Florian AI, Timiş T-L, Mihu CM, Miclauş M, et al. The roles of miRNA in glioblastoma tumor cell communication: diplomatic and aggressive negotiations. Int J Mol Sci 2020;21.
[92] Piccirillo SGM, Reynolds BA, Zanetti N, Lamorte G, Binda E, Broggi G, et al. Bone morphogenetic proteins inhibit the tumorigenic potential of human brain tumour-initiating cells. Nature 2006;444:761–5.
[93] Gal H, Pandi G, Kanner AA, Ram Z, Lithwick-Yanai G, Amariglio N, et al. MIR-451 and Imatinib mesylate inhibit tumor growth of glioblastoma stem cells. Biochem Biophys Res Commun 2008;376:86–90.
[94] Ernst A, Campos B, Meier J, Devens F, Liesenberg F, Wolter M, et al. De-repression of CTGF via the miR-17-92 cluster upon differentiation of human glioblastoma spheroid cultures. Oncogene 2010;29:3411–22.
[95] Silber J, Lim DA, Petritsch C, Persson AI, Maunakea AK, Yu M, et al. miR-124 and miR-137 inhibit proliferation of glioblastoma multiforme cells and induce differentiation of brain tumor stem cells. BMC Med 2008;6:14.
[96] Raja E, Morikawa M, Nishida J, Tanabe R, Takahashi K, Seeherman HJ, et al. Tyrosine kinase Eph receptor A6 sensitizes glioma-initiating cells towards bone morphogenetic protein-induced apoptosis. Cancer Sci 2019;110:3486–96.
[97] Raja E, Komuro A, Tanabe R, Sakai S, Ino Y, Saito N, et al. Bone morphogenetic protein signaling mediated by ALK-2 and DLX2 regulates apoptosis in glioma-initiating cells. Oncogene 2017;36:4963–74.
[98] Liu S, Yin F, Zhao M, Zhou C, Ren J, Huang Q, et al. The homing and inhibiting effects of hNSCs-BMP4 on human glioma stem cells. Oncotarget 2016;7:17920–31.
[99] Tso JL, Yang S, Menjivar JC, Yamada K, Zhang Y, Hong I, et al. Bone morphogenetic protein 7 sensitizes O6-methylguanine methyltransferase expressing-glioblastoma stem cells to clinically relevant dose of temozolomide. Mol Cancer 2015;14:189.
[100] Caja L, Tzavlaki K, Dadras MS, Tan EJ, Hatem G, Maturi NP, et al. Snail regulates BMP and TGFbeta pathways to control the differentiation status of glioma-initiating cells. Oncogene 2018;37:2515–31.
[101] Porco E, Maule F, Boso D, Rampazzo E, Barbieri V, Zuccolotto G, et al. BMP9 counteracts the tumorigenic and pro-angiogenic potential of glioblastoma. Cell Death Differ 2018;25:1808–22.
[102] Hover LD, Owens P, Munden AL, Wang J, Chambless LB, Hopkins CR, et al. Bone morphogenetic protein signaling promotes tumorigenesis in a murine model of high-grade glioma. Neuro-Oncology 2016;18:928–38.
[103] Rampazzo E, Dettin M, Maule F, Scabello A, Calvanese L, D'Auria G, et al. A synthetic BMP-2 mimicking peptide induces glioblastoma stem cell differentiation. Biochim Biophys Acta Gen Subj 1861;2017:2282–92.
[104] Jin X, Jin X, Kim LJY, Dixit D, Jeon H-Y, Kim E-J, et al. Inhibition of ID1-BMPR2 intrinsic signaling sensitizes glioma stem cells to differentiation therapy. Clin Cancer Res 2018;24:383–94.
[105] Jin X, Jeon H-M, Jin X, Kim E-J, Yin J, Jeon H-Y, et al. The ID1-CULLIN3 axis regulates intracellular SHH and WNT signaling in glioblastoma stem cells. Cell Rep 2016;16:1629–41.
[106] Xi G, Best B, Mania-Farnell B, James CD, Tomita T. Therapeutic potential for bone morphogenetic protein 4 in human malignant glioma. Neoplasia 2017;19:261–70.
[107] Mangraviti A, Tzeng SY, Gullotti D, Kozielski KL, Kim JE, Seng M, et al. Non-virally engineered human adipose mesenchymal stem cells produce BMP4, target brain tumors, and extend survival. Biomaterials 2016;100:53–66.
[108] Caren H, Stricker SH, Bulstrode H, Gagrica S, Johnstone E, Bartlett TE, et al. Glioblastoma stem cells respond to differentiation cues but fail to undergo commitment and terminal cell-cycle arrest. Stem Cell Rep 2015;5:829–42.

[109] Gao X, Jin W. The emerging role of tumor-suppressive microRNA-218 in targeting glioblastoma stemness. Cancer Lett 2014;353:25–31.
[110] Ciafre SA, Galardi S, Mangiola A, Ferracin M, Liu CG, Sabatino G, et al. Extensive modulation of a set of microRNAs in primary glioblastoma. Biochem Biophys Res Commun 2005;334:1351–8.
[111] Chen X, Liang H, Zhang J, Zen K, Zhang C-Y. Horizontal transfer of microRNAs: molecular mechanisms and clinical applications. Protein Cell 2012;3:28–37.
[112] Floyd D, Purow B. Micro-masters of glioblastoma biology and therapy: increasingly recognized roles for microRNAs. Neuro-Oncology 2014;16:622–7.
[113] Ma R, Yan W, Zhang G, Lv H, Liu Z, Fang F, et al. Upregulation of miR-196b confers a poor prognosis in glioblastoma patients via inducing a proliferative phenotype. PLoS One 2012;7, e38096.
[114] Ma R, Jiang T, Kang X. Circulating microRNAs in cancer: origin, function and application. J Exp Clin Cancer Res 2012;31:38.
[115] Godlewski J, Newton HB, Chiocca EA, Lawler SE. MicroRNAs and glioblastoma; the stem cell connection. Cell Death Differ 2010;17:221–8.
[116] Esquela-Kerscher A, Slack FJ. Oncomirs—microRNAs with a role in cancer. Nat Rev Cancer 2006;6:259–69.
[117] Aldaz B, Sagardoy A, Nogueira L, Guruceaga E, Grande L, Huse JT, et al. Involvement of miRNAs in the differentiation of human glioblastoma multiforme stem-like cells. PLoS One 2013;8, e77098.
[118] Jeon H-M, Sohn Y-W, Oh S-Y, Oh S-Y, Kim S-H, Beck S, et al. ID4 imparts chemoresistance and cancer stemness to glioma cells by derepressing miR-9*-mediated suppression of SOX2. Cancer Res 2011;71:3410–21.
[119] Schraivogel D, Weinmann L, Beier D, Tabatabai G, Eichner A, Zhu JY, et al. CAMTA1 is a novel tumour suppressor regulated by miR-9/9* in glioblastoma stem cells. EMBO J 2011;30:4309–22.
[120] Moon S-H, Kim D-K, Cha Y, Jeon I, Song J, Park K-S. PI3K/Akt and Stat3 signaling regulated by PTEN control of the cancer stem cell population, proliferation and senescence in a glioblastoma cell line. Int J Oncol 2013; 42:921–8.
[121] Kamal M, Shaaban AM, Zhang L, Walker C, Gray S, Thakker N, et al. Loss of CSMD1 expression is associated with high tumour grade and poor survival in invasive ductal breast carcinoma. Breast Cancer Res Treat 2010;121:555–63.
[122] Sun L, Yan W, Wang Y, Sun G, Luo H, Zhang J, et al. MicroRNA-10b induces glioma cell invasion by modulating MMP-14 and uPAR expression via HOXD10. Brain Res 2011;1389:9–18.
[123] Shi L, Zhang J, Pan T, Zhou J, Gong W, Liu N, et al. MiR-125b is critical for the suppression of human U251 glioma stem cell proliferation. Brain Res 2010;1312:120–6.
[124] Li Y, Zhao S, Zhen Y, Li Q, Teng L, Asai A, et al. A miR-21 inhibitor enhances apoptosis and reduces G(2)-M accumulation induced by ionizing radiation in human glioblastoma U251 cells. Brain Tumor Pathol 2011;28:209–14.
[125] Chen J, Fu X, Wan Y, Wang Z, Jiang D, Shi L. miR-125b inhibitor enhance the chemosensitivity of glioblastoma stem cells to temozolomide by targeting Bak1. Tumour Biol 2014;35:6293–302.
[126] Shi L, Wan Y, Sun G, Zhang S, Wang Z, Zeng Y. miR-125b inhibitor may enhance the invasion-prevention activity of temozolomide in glioblastoma stem cells by targeting PIAS3. BioDrugs 2014;28:41–54.
[127] Shi L, Fei X, Wang Z, You Y. PI3K inhibitor combined with miR-125b inhibitor sensitize TMZ-induced anti-glioma stem cancer effects through inactivation of Wnt/beta-catenin signaling pathway. In Vitro Cell Dev Biol Anim 2015;51:1047–55.
[128] Ulasov IV, Kaverina NV, Ghosh D, Baryshnikova MA, Kadagidze ZG, Karseladze AI, et al. CMV70-3P miRNA contributes to the CMV mediated glioma stemness and represents a target for glioma experimental therapy. Oncotarget 2017;8:25989–99.
[129] Fan X, Khaki L, Zhu TS, Soules ME, Talsma CE, Gul N, et al. NOTCH pathway blockade depletes CD133-positive glioblastoma cells and inhibits growth of tumor neurospheres and xenografts. Stem Cells 2010;28:5–16.
[130] Zhen Y, Zhao S, Li Q, Li Y, Kawamoto K. Arsenic trioxide-mediated notch pathway inhibition depletes the cancer stem-like cell population in gliomas. Cancer Lett 2010;292:64–72.
[131] Fan X, Matsui W, Khaki L, Stearns D, Chun J, Li Y-M, et al. Notch pathway inhibition depletes stem-like cells and blocks engraftment in embryonal brain tumors. Cancer Res 2006;66:7445–52.
[132] Jia B, Liu W, Gu J, Wang J, Lv W, Zhang W, et al. MiR-7-5p suppresses stemness and enhances temozolomide sensitivity of drug-resistant glioblastoma cells by targeting Yin Yang 1. Exp Cell Res 2019;375:73–81.
[133] Deng Y, Zhu G, Luo H, Zhao S. MicroRNA-203 as a stemness inhibitor of glioblastoma stem cells. Mol Cell 2016;39:619–24.

[134] Tamim S, Vo DT, Uren PJ, Qiao M, Bindewald E, Kasprzak WK, et al. Genomic analyses reveal broad impact of miR-137 on genes associated with malignant transformation and neuronal differentiation in glioblastoma cells. PLoS One 2014;9, e85591.
[135] Fares J, Bou Diab Z, Nabha S, Fares Y. Neurogenesis in the adult hippocampus: history, regulation, and prospective roles. Int J Neurosci 2019;129:598–611.
[136] Bier A, Giladi N, Kronfeld N, Lee HK, Cazacu S, Finniss S, et al. MicroRNA-137 is downregulated in glioblastoma and inhibits the stemness of glioma stem cells by targeting RTVP-1. Oncotarget 2013;4:665–76.
[137] Sharif S, Ghahremani MH, Soleimani M. Delivery of exogenous miR-124 to glioblastoma multiform cells by Wharton's Jelly mesenchymal stem cells decreases cell proliferation and migration, and confers chemosensitivity. Stem Cell Rev Rep 2018;14:236–46.
[138] Xia H, Cheung WKC, Ng SS, Jiang X, Jiang S, Sze J, et al. Loss of brain-enriched miR-124 microRNA enhances stem-like traits and invasiveness of glioma cells. J Biol Chem 2012;287:9962–71.
[139] Li P, Lu X, Wang Y, Sun L, Qian C, Yan W, et al. MiR-181b suppresses proliferation of and reduces chemoresistance to temozolomide in U87 glioma stem cells. J Biomed Res 2010;24:436–43.
[140] Kouri FM, Hurley LA, Daniel WL, Day ES, Hua Y, Hao L, et al. miR-182 integrates apoptosis, growth, and differentiation programs in glioblastoma. Genes Dev 2015;29:732–45.
[141] Wu Z, Han Y, Li Y, Li X, Sun T, Chen G, et al. MiR-218-5p inhibits the stem cell properties and invasive ability of the $A2B5^{+}CD133^{-}$ subgroup of human glioma stem cells. Oncol Rep 2016;35:869–77.
[142] Godlewski J, Nowicki MO, Bronisz A, Williams S, Otsuki A, Nuovo G, et al. Targeting of the Bmi-1 oncogene/stem cell renewal factor by microRNA-128 inhibits glioma proliferation and self-renewal. Cancer Res 2008;68:9125–30.
[143] Peruzzi P, Bronisz A, Nowicki MO, Wang Y, Ogawa D, Price R, et al. MicroRNA-128 coordinately targets polycomb repressor complexes in glioma stem cells. Neuro-Oncology 2013;15:1212–24.
[144] Suva M-L, Riggi N, Janiszewska M, Radovanovic I, Provero P, Stehle J-C, et al. EZH2 is essential for glioblastoma cancer stem cell maintenance. Cancer Res 2009;69:9211–8.
[145] Natsume A, Ito M, Katsushima K, Ohka F, Hatanaka A, Shinjo K, et al. Chromatin regulator PRC2 is a key regulator of epigenetic plasticity in glioblastoma. Cancer Res 2013;73:4559–70.
[146] Tu Y, Gao X, Li G, Fu H, Cui D, Liu H, et al. MicroRNA-218 inhibits glioma invasion, migration, proliferation, and cancer stem-like cell self-renewal by targeting the polycomb group gene Bmi1. Cancer Res 2013;73:6046–55.
[147] Jin X, Jeon H-Y, Joo KM, Kim J-K, Jin J, Kim SH, et al. Frizzled 4 regulates stemness and invasiveness of migrating glioma cells established by serial intracranial transplantation. Cancer Res 2011;71:3066–75.
[148] Li H, Yang BB. Stress response of glioblastoma cells mediated by miR-17-5p targeting PTEN and the passenger strand miR-17-3p targeting MDM2. Oncotarget 2012;3:1653–68.
[149] Zhang S, Wan Y, Pan T, Gu X, Qian C, Sun G, et al. MicroRNA-21 inhibitor sensitizes human glioblastoma U251 stem cells to chemotherapeutic drug temozolomide. J Mol Neurosci 2012;47:346–56.
[150] Zhou X, Ren Y, Moore L, Mei M, You Y, Xu P, et al. Downregulation of miR-21 inhibits EGFR pathway and suppresses the growth of human glioblastoma cells independent of PTEN status. Lab Investig 2010;90:144–55.
[151] Li Y, Guessous F, Zhang Y, Dipierro C, Kefas B, Johnson E, et al. MicroRNA-34a inhibits glioblastoma growth by targeting multiple oncogenes. Cancer Res 2009;69:7569–76.
[152] Guessous F, Zhang Y, Kofman A, Catania A, Li Y, Schiff D, et al. microRNA-34a is tumor suppressive in brain tumors and glioma stem cells. Cell Cycle 2010;9:1031–6.
[153] Yeh T-S, Lin Y-M, Hsieh R-H, Tseng M-J. Association of transcription factor YY1 with the high molecular weight notch complex suppresses the transactivation activity of notch. J Biol Chem 2003;278:41963–9.
[154] Glinsky GV, Berezovska O, Glinskii AB. Microarray analysis identifies a death-from-cancer signature predicting therapy failure in patients with multiple types of cancer. J Clin Invest 2005;115:1503–21.
[155] Molofsky AV, Pardal R, Iwashita T, Park I-K, Clarke MF, Morrison SJ. Bmi-1 dependence distinguishes neural stem cell self-renewal from progenitor proliferation. Nature 2003;425:962–7.
[156] Raaphorst FM. Self-renewal of hematopoietic and leukemic stem cells: a central role for the polycomb-group gene Bmi-1. Trends Immunol 2003;24:522–4.
[157] Lessard J, Sauvageau G. Bmi-1 determines the proliferative capacity of normal and leukaemic stem cells. Nature 2003;423:255–60.
[158] Proctor E, Waghray M, Lee CJ, Heidt DG, Yalamanchili M, Li C, et al. Bmi1 enhances tumorigenicity and cancer stem cell function in pancreatic adenocarcinoma. PLoS One 2013;8, e55820.

[159] Fu J, Rodova M, Nanta R, Meeker D, Van Veldhuizen PJ, Srivastava RK, et al. NPV-LDE-225 (Erismodegib) inhibits epithelial mesenchymal transition and self-renewal of glioblastoma initiating cells by regulating miR-21, miR-128, and miR-200. Neuro-Oncology 2013;15:691–706.

[160] Shi Z-M, Wang J, Yan Z, You Y-P, Li C-Y, Qian X, et al. MiR-128 inhibits tumor growth and angiogenesis by targeting p70S6K1. PLoS One 2012;7, e32709.

[161] Papagiannakopoulos T, Friedmann-Morvinski D, Neveu P, Dugas JC, Gill RM, Huillard E, et al. Pro-neural miR-128 is a glioma tumor suppressor that targets mitogenic kinases. Oncogene 2012;31:1884–95.

[162] Yu Q, Xue Y, Liu J, Xi Z, Li Z, Liu Y. Fibronectin promotes the malignancy of glioma stem-like cells via modulation of cell adhesion, differentiation, proliferation and chemoresistance. Front Mol Neurosci 2018;11:130.

[163] Sareddy GR, Viswanadhapalli S, Surapaneni P, Suzuki T, Brenner A, Vadlamudi RK. Novel KDM1A inhibitors induce differentiation and apoptosis of glioma stem cells via unfolded protein response pathway. Oncogene 2017;36:2423–34.

[164] Staberg M, Rasmussen RD, Michaelsen SR, Pedersen H, Jensen KE, Villingshoj M, et al. Targeting glioma stem-like cell survival and chemoresistance through inhibition of lysine-specific histone demethylase KDM2B. Mol Oncol 2018;12:406–20.

[165] Nogueira L, Ruiz-Ontanon P, Vazquez-Barquero A, Lafarga M, Berciano MT, Aldaz B, et al. Blockade of the NFkappaB pathway drives differentiating glioblastoma-initiating cells into senescence both in vitro and in vivo. Oncogene 2011;30:3537–48.

[166] Zhang L, Ren X, Cheng Y, Liu X, Allen JE, Zhang Y, et al. The NFkappaB inhibitor, SN50, induces differentiation of glioma stem cells and suppresses their oncogenic phenotype. Cancer Biol Ther 2014;15:602–11.

[167] Wu W, Wu Y, Mayer K, von Rosenstiel C, Schecker J, Baur S, et al. Lipid peroxidation plays an important role in chemotherapeutic effects of temozolomide and the development of therapy resistance in human glioblastoma. Transl Oncol 2020;13:100748.

[168] De Luca A, Fiorillo M, Peiris-Pages M, Ozsvari B, Smith DL, Sanchez-Alvarez R, et al. Mitochondrial biogenesis is required for the anchorage-independent survival and propagation of stem-like cancer cells. Oncotarget 2015;6:14777–95.

[169] Gong A, Huang S. FoxM1 and Wnt/beta-catenin signaling in glioma stem cells. Cancer Res 2012;72:5658–62.

[170] Flavahan WA, Wu Q, Hitomi M, Rahim N, Kim Y, Sloan AE, et al. Brain tumor initiating cells adapt to restricted nutrition through preferential glucose uptake. Nat Neurosci 2013;16:1373–82.

[171] Sun S, Xue D, Chen Z, Ou-Yang Y, Zhang J, Mai J, et al. R406 elicits anti-Warburg effect via Syk-dependent and -independent mechanisms to trigger apoptosis in glioma stem cells. Cell Death Dis 2019;10:358.

[172] Abdel Gaber SA, Muller P, Zimmermann W, Huttenberger D, Wittig R, Abdel Kader MH, et al. ABCG2-mediated suppression of chlorin e6 accumulation and photodynamic therapy efficiency in glioblastoma cell lines can be reversed by KO143. J Photochem Photobiol B 2018;178:182–91.

[173] Chen L, Shi L, Wang W, Zhou Y. ABCG2 downregulation in glioma stem cells enhances the therapeutic efficacy of demethoxycurcumin. Oncotarget 2017;8:43237–47.

[174] Zeng X, Wang Q, Tan X, Jia L, Li Y, Hu M, et al. Mild thermotherapy and hyperbaric oxygen enhance sensitivity of TMZ/PSi nanoparticles via decreasing the stemness in glioma. J Nanobiotechnol 2019;17:47.

[175] Lu Z, Ma J, Liu B, Dai C, Xie T, Ma X, et al. Hyperbaric oxygen therapy sensitizes nimustine treatment for glioma in mice. Cancer Med 2016;5:3147–55.

[176] Yuan X, Wei W, Bao Q, Chen H, Jin P, Jiang W. Metformin inhibits glioma cells stemness and epithelial-mesenchymal transition via regulating YAP activity. Biomed Pharmacother 2018;102:263–70.

[177] Aldea MD, Petrushev B, Soritau O, Tomuleasa CI, Berindan-Neagoe I, Filip AG, et al. Metformin plus sorafenib highly impacts temozolomide resistant glioblastoma stem-like cells. J BUON 2014;19:502–11.

[178] Gritti M, Wurth R, Angelini M, Barbieri F, Peretti M, Pizzi E, et al. Metformin repositioning as antitumoral agent: selective antiproliferative effects in human glioblastoma stem cells, via inhibition of CLIC1-mediated ion current. Oncotarget 2014;5:11252–68.

[179] Songthaveesin C, Sa-Nongdej W, Limboonreung T, Chongthammakun S. Combination of metformin and 9-cis retinoic acid increases apoptosis in C6 glioma stem-like cells. Heliyon 2018;4, e00638.

[180] Valtorta S, Lo Dico A, Raccagni I, Gaglio D, Belloli S, Politi LS, et al. Metformin and temozolomide, a synergic option to overcome resistance in glioblastoma multiforme models. Oncotarget 2017;8:113090–104.

[181] Leidgens V, Proske J, Rauer L, Moeckel S, Renner K, Bogdahn U, et al. Stattic and metformin inhibit brain tumor initiating cells by reducing STAT3-phosphorylation. Oncotarget 2017;8:8250–63.

[182] Farge T, Saland E, de Toni F, Aroua N, Hosseini M, Perry R, et al. Chemotherapy-resistant human acute myeloid leukemia cells are not enriched for leukemic stem cells but require oxidative metabolism. Cancer Discov 2017;7:716–35.

[183] Marchiq I, Albrengues J, Granja S, Gaggioli C, Pouysségur J, Simon M-P. Knock out of the BASIGIN/CD147 chaperone of lactate/H+ symporters disproves its pro-tumour action via extracellular matrix metalloproteases (MMPs) induction. Oncotarget 2015;6.

[184] Voss DM, Spina R, Carter DL, Lim KS, Jeffery CJ, Bar EE. Disruption of the monocarboxylate transporter-4-basigin interaction inhibits the hypoxic response, proliferation, and tumor progression. Sci Rep 2017;7:4292.

[185] Bernstock JD, Ye D, Gessler FA, Lee Y-J, Peruzzotti-Jametti L, Baumgarten P, et al. Topotecan is a potent inhibitor of SUMOylation in glioblastoma multiforme and alters both cellular replication and metabolic programming. Sci Rep 2017;7:7425.

[186] Jiang X, Wang C, Fitch S, Yang F. Targeting tumor hypoxia using nanoparticle-engineered CXCR4-overexpressing adipose-derived stem cells. Theranostics 2018;8:1350–60.

[187] Boyd NH, Walker K, Fried J, Hackney JR, McDonald PC, Benavides GA, et al. Addition of carbonic anhydrase 9 inhibitor SLC-0111 to temozolomide treatment delays glioblastoma growth in vivo. JCI Insight 2017;2.

[188] Jalota A, Kumar M, Das BC, Yadav AK, Chosdol K, Sinha S. A drug combination targeting hypoxia induced chemoresistance and stemness in glioma cells. Oncotarget 2018;9:18351–66.

[189] Ouchi R, Okabe S, Migita T, Nakano I, Seimiya H. Senescence from glioma stem cell differentiation promotes tumor growth. Biochem Biophys Res Commun 2016;470:275–81.

[190] Wang Y, Huang N, Li H, Liu S, Chen X, Yu S, et al. Promoting oligodendroglial-oriented differentiation of glioma stem cell: a repurposing of quetiapine for the treatment of malignant glioma. Oncotarget 2017;8:37511–24.

[191] Bielecka-Wajdman AM, Lesiak M, Ludyga T, Sieron A, Obuchowicz E. Reversing glioma malignancy: a new look at the role of antidepressant drugs as adjuvant therapy for glioblastoma multiforme. Cancer Chemother Pharmacol 2017;79:1249–56.

[192] Campos B, Wan F, Farhadi M, Ernst A, Zeppernick F, Tagscherer KE, et al. Differentiation therapy exerts antitumor effects on stem-like glioma cells. Clin Cancer Res 2010;16:2715–28.

[193] Zhang P, Miska J, Lee-Chang C, Rashidi A, Panek WK, An S, et al. Therapeutic targeting of tumor-associated myeloid cells synergizes with radiation therapy for glioblastoma. Proc Natl Acad Sci U S A 2019;116:23714–23.

[194] Badie B, Schartner J. Role of microglia in glioma biology. Microsc Res Tech 2001;54:106–13.

[195] Hussain SF, Yang D, Suki D, Grimm E, Heimberger AB. Innate immune functions of microglia isolated from human glioma patients. J Transl Med 2006;4:15.

[196] Kennedy BC, Maier LM, D'Amico R, Mandigo CE, Fontana EJ, Waziri A, et al. Dynamics of central and peripheral immunomodulation in a murine glioma model. BMC Immunol 2009;10:11.

[197] Li W, Graeber MB. The molecular profile of microglia under the influence of glioma. Neuro-Oncology 2012;14:958–78.

[198] Wagner S, Czub S, Greif M, Vince GH, Suss N, Kerkau S, et al. Microglial/macrophage expression of interleukin 10 in human glioblastomas. Int J Cancer 1999;82:12–6.

[199] Pyonteck SM, Akkari L, Schuhmacher AJ, Bowman RL, Sevenich L, Quail DF, et al. CSF-1R inhibition alters macrophage polarization and blocks glioma progression. Nat Med 2013;19:1264–72.

[200] Lee-Chang C, Rashidi A, Miska J, Zhang P, Pituch KC, Hou D, et al. Myeloid-derived suppressive cells promote B cell-mediated immunosuppression via transfer of PD-L1 in glioblastoma. Cancer Immunol Res 2019;7:1928–43.

[201] von Roemeling CA, Wang Y, Qie Y, Yuan H, Zhao H, Liu X, et al. Therapeutic modulation of phagocytosis in glioblastoma can activate both innate and adaptive antitumour immunity. Nat Commun 2020;11:1508.

[202] Hernandez-SanMiguel E, Gargini R, Cejalvo T, Segura-Collar B, Nunez-Hervada P, Hortiguela R, et al. Ocoxin modulates cancer stem cells and M2 macrophage polarization in glioblastoma. Oxidative Med Cell Longev 2019;2019:9719730.

[203] Quail DF, Bowman RL, Akkari L, Quick ML, Schuhmacher AJ, Huse JT, et al. The tumor microenvironment underlies acquired resistance to CSF-1R inhibition in gliomas. Science 2016;352, aad3018.

[204] Yang W, Li Y, Gao R, Xiu Z, Sun T. MHC class I dysfunction of glioma stem cells escapes from CTL-mediated immune response via activation of Wnt/beta-catenin signaling pathway. Oncogene 2020;39:1098–111.

[205] Mao J, Cao M, Zhang F, Zhang J, Duan X, Lu L, et al. Peritumoral administration of IFNbeta upregulated mesenchymal stem cells inhibits tumor growth in an orthotopic, immunocompetent rat glioma model. J Immunother Cancer 2020;8.

[206] Choi SH, Stuckey DW, Pignatta S, Reinshagen C, Khalsa JK, Roozendaal N, et al. Tumor resection recruits effector T cells and boosts therapeutic efficacy of encapsulated stem cells expressing IFNbeta in glioblastomas. Clin Cancer Res 2017;23:7047–58.
[207] Shoji T, Saito R, Chonan M, Shibahara I, Sato A, Kanamori M, et al. Local convection-enhanced delivery of an anti-CD40 agonistic monoclonal antibody induces antitumor effects in mouse glioma models. Neuro-Oncology 2016;18:1120–8.
[208] Fleurence J, Bahri M, Fougeray S, Faraj S, Vermeulen S, Pinault E, et al. Impairing temozolomide resistance driven by glioma stem-like cells with adjuvant immunotherapy targeting O-acetyl GD2 ganglioside. Int J Cancer 2020;146:424–38.
[209] Chavez-Cortez E-G, Vargas Felix G, Rangel Lopez E, Sotelo J, Martinez-Canseco C, Perez-de la Cruz V, et al. Production and evaluation of an avian IgY immunotoxin against CD133+ for treatment of carcinogenic stem cells in malignant glioma: IgY immunotoxin for the treatment of glioblastoma. J Oncol 2019;2019:2563092.
[210] Fares J, Fares MY, Fares Y. Natural killer cells in the brain tumor microenvironment: defining a new era in neuro-oncology. Surg Neurol Int 2019;10:43.
[211] Tseng H-C, Bui V, Man Y-G, Cacalano N, Jewett A. Induction of split anergy conditions natural killer cells to promote differentiation of stem cells through cell-cell contact and secreted factors. Front Immunol 2014;5:269.
[212] Kozlowska AK, Tseng H-C, Kaur K, Topchyan P, Inagaki A, Bui VT, et al. Resistance to cytotoxicity and sustained release of interleukin-6 and interleukin-8 in the presence of decreased interferon-gamma after differentiation of glioblastoma by human natural killer cells. Cancer Immunol Immunother 2016;65:1085–97.
[213] Huang BY, Zhan YP, Zong WJ, Yu CJ, Li JF, Qu YM, et al. The PD-1/B7-H1 pathway modulates the natural killer cells versus mouse glioma stem cells. PLoS One 2015;10, e0134715.
[214] Chen H, Guo Y, Sun J, Dong J, Bao Q, Zhang X, et al. Preferential expression of B7-H6 in glioma stem-like cells enhances tumor cell proliferation via the c-Myc/RNMT axis. J Immunol Res 2020;2020:2328675.
[215] Lim M, Xia Y, Bettegowda C, Weller M. Current state of immunotherapy for glioblastoma. Nat Rev Clin Oncol 2018;15:422–42.
[216] Fares J, Fares MY, Fares Y. Immune checkpoint inhibitors: advances and impact in neuro-oncology. Surg Neurol Int 2019;10:9.
[217] Mitchell DA, Batich KA, Gunn MD, Huang M-N, Sanchez-Perez L, Nair SK, et al. Tetanus toxoid and CCL3 improve dendritic cell vaccines in mice and glioblastoma patients. Nature 2015;519:366–9.
[218] Prins RM, Soto H, Konkankit V, Odesa SK, Eskin A, Yong WH, et al. Gene expression profile correlates with T-cell infiltration and relative survival in glioblastoma patients vaccinated with dendritic cell immunotherapy. Clin Cancer Res 2011;17:1603–15.
[219] Phuphanich S, Wheeler CJ, Rudnick JD, Mazer M, Wang H, Nuno MA, et al. Phase I trial of a multi-epitope-pulsed dendritic cell vaccine for patients with newly diagnosed glioblastoma. Cancer Immunol Immunother 2013;62:125–35.
[220] Hilf N, Kuttruff-Coqui S, Frenzel K, Bukur V, Stevanovic S, Gouttefangeas C, et al. Actively personalized vaccination trial for newly diagnosed glioblastoma. Nature 2019;565:240–5.
[221] Keskin DB, Anandappa AJ, Sun J, Tirosh I, Mathewson ND, Li S, et al. Neoantigen vaccine generates intratumoral T cell responses in phase Ib glioblastoma trial. Nature 2019;565:234–9.
[222] Bago JR, Okolie O, Dumitru R, Ewend MG, Parker JS, Werff RV, et al. Tumor-homing cytotoxic human induced neural stem cells for cancer therapy. Sci Transl Med 2017;9.
[223] Portnow J, Synold TW, Badie B, Tirughana R, Lacey SF, D'Apuzzo M, et al. Neural stem cell-based anticancer gene therapy: a first-in-human study in recurrent high-grade glioma patients. Clin Cancer Res 2017;23:2951–60.
[224] Brown CE, Alizadeh D, Starr R, Weng L, Wagner JR, Naranjo A, et al. Regression of glioblastoma after chimeric antigen receptor T-cell therapy. N Engl J Med 2016;375:2561–9.
[225] O'Rourke DM, Nasrallah MP, Desai A, Melenhorst JJ, Mansfield K, Morrissette JJD, et al. A single dose of peripherally infused EGFRvIII-directed CAR T cells mediates antigen loss and induces adaptive resistance in patients with recurrent glioblastoma. Sci Transl Med 2017;9.
[226] Han J, Chu J, Keung Chan W, Zhang J, Wang Y, Cohen JB, et al. CAR-engineered NK cells targeting wild-type EGFR and EGFRvIII enhance killing of glioblastoma and patient-derived glioblastoma stem cells. Sci Rep 2015;5:11483.
[227] Ravanpay AC, Gust J, Johnson AJ, Rolczynski LS, Cecchini M, Chang CA, et al. EGFR806-CAR T cells selectively target a tumor-restricted EGFR epitope in glioblastoma. Oncotarget 2019;10:7080–95.
[228] Nehama D, Di Ianni N, Musio S, Du H, Patane M, Pollo B, et al. B7-H3-redirected chimeric antigen receptor T cells target glioblastoma and neurospheres. EBioMedicine 2019;47:33–43.

[229] Jiang H, Rivera-Molina Y, Gomez-Manzano C, Clise-Dwyer K, Bover L, Vence LM, et al. Oncolytic adenovirus and tumor-targeting immune modulatory therapy improve autologous cancer vaccination. Cancer Res 2017;77:3894–907.
[230] van den Bossche WBL, Kleijn A, Teunissen CE, Voerman JSA, Teodosio C, Noske DP, et al. Oncolytic virotherapy in glioblastoma patients induces a tumor macrophage phenotypic shift leading to an altered glioblastoma microenvironment. Neuro-Oncology 2018;20:1494–504.
[231] Saha D, Martuza RL, Rabkin SD. Macrophage polarization contributes to glioblastoma eradication by combination immunovirotherapy and immune checkpoint blockade. Cancer Cell 2017;32:253–267.e5.
[232] Saha D, Martuza RL, Rabkin SD. Oncolytic herpes simplex virus immunovirotherapy in combination with immune checkpoint blockade to treat glioblastoma. Immunotherapy 2018;10:779–86.
[233] Saha D, Wakimoto H, Peters CW, Antoszczyk SJ, Rabkin SD, Martuza RL. Combinatorial effects of VEGFR kinase inhibitor axitinib and oncolytic virotherapy in mouse and human glioblastoma stem-like cell models. Clin Cancer Res 2018;24:3409–22.
[234] Desjardins A, Gromeier M, Herndon 2nd JE, Beaubier N, Bolognesi DP, Friedman AH, et al. Recurrent glioblastoma treated with recombinant poliovirus. N Engl J Med 2018;379:150–61.
[235] Zhu Z, Mesci P, Bernatchez JA, Gimple RC, Wang X, Schafer ST, et al. Zika virus targets glioblastoma stem cells through a SOX2-integrin alphavbeta5 axis. Cell Stem Cell 2020;26:187–204.e10.
[236] Zhu Z, Gorman MJ, McKenzie LD, Chai JN, Hubert CG, Prager BC, et al. Zika virus has oncolytic activity against glioblastoma stem cells. J Exp Med 2017;214:2843–57.
[237] Yu K, Lin C-CJ, Hatcher A, Lozzi B, Kong K, Huang-Hobbs E, et al. PIK3CA variants selectively initiate brain hyperactivity during gliomagenesis. Nature 2020;578:166–71.
[238] Venkatesh HS, Morishita W, Geraghty AC, Silverbush D, Gillespie SM, Arzt M, et al. Electrical and synaptic integration of glioma into neural circuits. Nature 2019;573:539–45.
[239] Bhat K, Saki M, Vlashi E, Cheng F, Duhachek-Muggy S, Alli C, et al. The dopamine receptor antagonist trifluoperazine prevents phenotype conversion and improves survival in mouse models of glioblastoma. Proc Natl Acad Sci U S A 2020;117:11085–96.
[240] Spina R, Voss DM, Asnaghi L, Sloan A, Bar EE. Atracurium Besylate and other neuromuscular blocking agents promote astroglial differentiation and deplete glioblastoma stem cells. Oncotarget 2016;7:459–72.
[241] Suzuki S, Yamamoto M, Sanomachi T, Togashi K, Sugai A, Seino S, et al. Brexpiprazole, a serotonin-dopamine activity modulator, can sensitize glioma stem cells to osimertinib, a third-generation EGFR-TKI, via survivin reduction. Cancers (Basel) 2019;11.
[242] Oliva CR, Zhang W, Langford C, Suto MJ, Griguer CE. Repositioning chlorpromazine for treating chemoresistant glioma through the inhibition of cytochrome c oxidase bearing the COX4-1 regulatory subunit. Oncotarget 2017;8:37568–83.
[243] Liong C, Ortiz D, Ao-ieong E, Navati MS, Friedman JM, Cabrales P. Localized increase of tissue oxygen tension by magnetic targeted drug delivery. Nanotechnology 2014;25:265102.
[244] Salaroglio IC, Mujumdar P, Annovazzi L, Kopecka J, Mellai M, Schiffer D, et al. Carbonic anhydrase XII inhibitors overcome P-glycoprotein-mediated resistance to temozolomide in glioblastoma. Mol Cancer Ther 2018;17:2598–609.
[245] Teodorczyk M, Schmidt MHH. Notching on cancer's door: notch signaling in brain tumors. Front Oncol 2014;4:341.
[246] Galan-Moya EM, Le Guelte A, Lima Fernandes E, Thirant C, Dwyer J, Bidere N, et al. Secreted factors from brain endothelial cells maintain glioblastoma stem-like cell expansion through the mTOR pathway. EMBO Rep 2011;12:470–6.
[247] Lathia JD, Li M, Hall PE, Gallagher J, Hale JS, Wu Q, et al. Laminin alpha 2 enables glioblastoma stem cell growth. Ann Neurol 2012;72:766–78.
[248] Ulasov IV, Nandi S, Dey M, Sonabend AM, Lesniak MS. Inhibition of sonic hedgehog and notch pathways enhances sensitivity of CD133(+) glioma stem cells to temozolomide therapy. Mol Med 2011;17:103–12.
[249] Ricci-Vitiani L, Pallini R, Biffoni M, Todaro M, Invernici G, Cenci T, et al. Tumour vascularization via endothelial differentiation of glioblastoma stem-like cells. Nature 2010;468:824–8.
[250] Chung AS, Ferrara N. Developmental and pathological angiogenesis. Annu Rev Cell Dev Biol 2011;27:563–84.
[251] Calabrese C, Poppleton H, Kocak M, Hogg TL, Fuller C, Hamner B, et al. A perivascular niche for brain tumor stem cells. Cancer Cell 2007;11:69–82.
[252] Tamura R, Tanaka T, Miyake K, Yoshida K, Sasaki H. Bevacizumab for malignant gliomas: current indications, mechanisms of action and resistance, and markers of response. Brain Tumor Pathol 2017;34:62–77.

[253] Tamura R, Tanaka T, Miyake K, Tabei Y, Ohara K, Sampetrean O, et al. Histopathological investigation of glioblastomas resected under bevacizumab treatment. Oncotarget 2016;7:52423–35.
[254] Lassen U, Chinot OL, McBain C, Mau-Sorensen M, Larsen VA, Barrie M, et al. Phase 1 dose-escalation study of the antiplacental growth factor monoclonal antibody RO5323441 combined with bevacizumab in patients with recurrent glioblastoma. Neuro-Oncology 2015;17:1007–15.
[255] Lee EQ, Zhang P, Wen PY, Gerstner ER, Reardon DA, Aldape KD, et al. NRG/RTOG 1122: a phase 2, double-blinded, placebo-controlled study of bevacizumab with and without trebananib in patients with recurrent glioblastoma or gliosarcoma. Cancer 2020;126:2821–8.
[256] Ghiaseddin A, Reardon D, Massey W, Mannerino A, Lipp ES, Herndon 2nd JE, et al. Phase II study of bevacizumab and vorinostat for patients with recurrent World Health Organization Grade 4 malignant glioma. Oncologist 2018;23:157-e21.
[257] Lassen U, Sorensen M, Gaziel TB, Hasselbalch B, Poulsen HS. Phase II study of bevacizumab and temsirolimus combination therapy for recurrent glioblastoma multiforme. Anticancer Res 2013;33:1657–60.
[258] Nayak L, Hays J, Muzikansky A, Gaffey S, Do K, Hilton J, et al. ATNT-18A phase I study of MLN0128 and bevacizumab in patients with recurrent glioblastoma and other solid tumors. Neuro-Oncology 2015;17, v14.
[259] Weiss GJ, Infante JR, Chiorean EG, Borad MJ, Bendell JC, Molina JR, et al. Phase 1 study of the safety, tolerability, and pharmacokinetics of TH-302, a hypoxia-activated prodrug, in patients with advanced solid malignancies. Clin Cancer Res 2011;17:2997–3004.
[260] Brenner A, Zuniga R, Sun JD, Floyd J, Hart CP, Kroll S, et al. Hypoxia-activated evofosfamide for treatment of recurrent bevacizumab-refractory glioblastoma: a phase I surgical study. Neuro-Oncology 2018;20:1231–9.
[261] Ahluwalia MS, Rogers LR, Chaudhary RT, Newton HB, Seon BK, Jivani MA, et al. A phase 2 trial of TRC105 with bevacizumab for bevacizumab refractory glioblastoma. J Clin Oncol 2016;34:2035.
[262] Carra E, Barbieri F, Marubbi D, Pattarozzi A, Favoni RE, Florio T, et al. Sorafenib selectively depletes human glioblastoma tumor-initiating cells from primary cultures. Cell Cycle 2013;12:491–500.
[263] Wurth R, Barbieri F, Florio T. New molecules and old drugs as emerging approaches to selectively target human glioblastoma cancer stem cells. Biomed Res Int 2014;2014:126586.
[264] Zhang Y, Ishida CT, Ishida W, Lo S-FL, Zhao J, Shu C, et al. Combined HDAC and bromodomain protein inhibition reprograms tumor cell metabolism and elicits synthetic lethality in glioblastoma. Clin Cancer Res 2018;24:3941–54.
[265] Clavreul A, Pourbaghi-Masouleh M, Roger E, Lautram N, Montero-Menei CN, Menei P. Human mesenchymal stromal cells as cellular drug-delivery vectors for glioblastoma therapy: a good deal? J Exp Clin Cancer Res 2017;36:135.
[266] Shingu T, Holmes L, Henry V, Wang Q, Latha K, Gururaj AE, et al. Suppression of RAF/MEK or PI3K synergizes cytotoxicity of receptor tyrosine kinase inhibitors in glioma tumor-initiating cells. J Transl Med 2016;14:46.
[267] Sabelstrom H, Petri R, Shchors K, Jandial R, Schmidt C, Sacheva R, et al. Driving neuronal differentiation through reversal of an ERK1/2-miR-124-SOX9 axis abrogates glioblastoma aggressiveness. Cell Rep 2019; 28:2064–2079.e11.
[268] Meng Y, Shang F, Zhu Y. miR-124 participates in the proliferation and differentiation of brain glioma stem cells through regulating Nogo/NgR expression. Exp Ther Med 2019;18:2783–8.
[269] Daniele S, Sestito S, Pietrobono D, Giacomelli C, Chiellini G, Di Maio D, et al. Dual inhibition of PDK1 and aurora kinase a: an effective strategy to induce differentiation and apoptosis of human glioblastoma multiforme stem cells. ACS Chem Neurosci 2017;8:100–14.
[270] Zhang Y, Cruickshanks N, Yuan F, Wang B, Pahuski M, Wulfkuhle J, et al. Targetable T-type calcium channels drive glioblastoma. Cancer Res 2017;77:3479–90.
[271] Sugimori M, Hayakawa Y, Tamura R, Kuroda S. The combined efficacy of OTS964 and temozolomide for reducing the size of power-law coded heterogeneous glioma stem cell populations. Oncotarget 2019;10:2397–415.
[272] Golde TE, Koo EH, Felsenstein KM, Osborne BA, Miele L. Gamma-secretase inhibitors and modulators. Biochim Biophys Acta 2013;1828:2898–907.
[273] Clement V, Sanchez P, de Tribolet N, Radovanovic I, Ruiz i Altaba A. HEDGEHOG-GLI1 signaling regulates human glioma growth, cancer stem cell self-renewal, and tumorigenicity. Curr Biol 2007;17:165–72.
[274] Yahyanejad S, King H, Iglesias VS, Granton PV, Barbeau LMO, van Hoof SJ, et al. NOTCH blockade combined with radiation therapy and temozolomide prolongs survival of orthotopic glioblastoma. Oncotarget 2016;7:41251–64.

[275] Xu R, Shimizu F, Hovinga K, Beal K, Karimi S, Droms L, et al. Molecular and clinical effects of notch inhibition in glioma patients: a phase 0/I trial. Clin Cancer Res 2016;22:4786–96.
[276] Saito N, Hirai N, Aoki K, Suzuki R, Fujita S, Nakayama H, et al. The oncogene addiction switch from NOTCH to PI3K requires simultaneous targeting of NOTCH and PI3K pathway inhibition in glioblastoma. Cancers (Basel) 2019;11.
[277] Lin Y, Sun H, Dang Y, Li Z. Isoliquiritigenin inhibits the proliferation and induces the differentiation of human glioma stem cells. Oncol Rep 2018;39:687–94.
[278] Ma Y, Cheng Z, Liu J, Torre-Healy L, Lathia JD, Nakano I, et al. Inhibition of farnesyltransferase potentiates NOTCH-targeted therapy against glioblastoma stem cells. Stem Cell Rep 2017;9:1948–60.
[279] Xie Q, Wu Q, Kim L, Miller TE, Liau BB, Mack SC, et al. RBPJ maintains brain tumor-initiating cells through CDK9-mediated transcriptional elongation. J Clin Invest 2016;126:2757–72.
[280] Maciaczyk D, Picard D, Zhao L, Koch K, Herrera-Rios D, Li G, et al. CBF1 is clinically prognostic and serves as a target to block cellular invasion and chemoresistance of EMT-like glioblastoma cells. Br J Cancer 2017;117:102–12.
[281] Mao P, Joshi K, Li J, Kim S-H, Li P, Santana-Santos L, et al. Mesenchymal glioma stem cells are maintained by activated glycolytic metabolism involving aldehyde dehydrogenase 1A3. Proc Natl Acad Sci U S A 2013;110:8644–9.
[282] Liu Y-J, Ma Y-C, Zhang W-J, Yang Z-Z, Liang D-S, Wu Z-F, et al. Combination therapy with micellarized cyclopamine and temozolomide attenuate glioblastoma growth through Gli1 down-regulation. Oncotarget 2017;8:42495–509.
[283] Nanta R, Shrivastava A, Sharma J, Shankar S, Srivastava RK. Inhibition of sonic hedgehog and PI3K/Akt/mTOR pathways cooperate in suppressing survival, self-renewal and tumorigenic potential of glioblastoma-initiating cells. Mol Cell Biochem 2019;454:11–23.
[284] Yang W, Liu Y, Gao R, Yu H, Sun T. HDAC6 inhibition induces glioma stem cells differentiation and enhances cellular radiation sensitivity through the SHH/Gli1 signaling pathway. Cancer Lett 2018;415:164–76.
[285] Silvestri I, Testa F, Zappasodi R, Cairo CW, Zhang Y, Lupo B, et al. Sialidase NEU4 is involved in glioblastoma stem cell survival. Cell Death Dis 2014;5, e1381.
[286] Shahi MH, Farheen S, Mariyath MPM, Castresana JS. Potential role of Shh-Gli1-BMI1 signaling pathway nexus in glioma chemoresistance. Tumour Biol 2016;37:15107–14.
[287] Melamed JR, Ioele SA, Hannum AJ, Ullman VM, Day ES. Polyethylenimine-spherical nucleic acid nanoparticles against Gli1 reduce the chemoresistance and stemness of glioblastoma cells. Mol Pharm 2018;15:5135–45.
[288] Du W, Liu X, Chen L, Dou Z, Lei X, Chang L, et al. Targeting the SMO oncogene by miR-326 inhibits glioma biological behaviors and stemness. Neuro-Oncology 2015;17:243–53.
[289] Han D, Yu T, Dong N, Wang B, Sun F, Jiang D. Napabucasin, a novel STAT3 inhibitor suppresses proliferation, invasion and stemness of glioblastoma cells. J Exp Clin Cancer Res 2019;38:289.
[290] Haftchenary S, Luchman HA, Jouk AO, Veloso AJ, Page BDG, Cheng XR, et al. Potent targeting of the STAT3 protein in brain cancer stem cells: a promising route for treating glioblastoma. ACS Med Chem Lett 2013;4:1102–7.
[291] Ball S, Li C, Li P-K, Lin J. The small molecule, LLL12, inhibits STAT3 phosphorylation and induces apoptosis in medulloblastoma and glioblastoma cells. PLoS One 2011;6, e18820.
[292] Sai K, Wang S, Balasubramaniyan V, Conrad C, Lang FF, Aldape K, et al. Induction of cell-cycle arrest and apoptosis in glioblastoma stem-like cells by WP1193, a novel small molecule inhibitor of the JAK2/STAT3 pathway. J Neuro-Oncol 2012;107:487–501.
[293] Yang Y-P, Chang Y-L, Huang P-I, Chiou G-Y, Tseng L-M, Chiou S-H, et al. Resveratrol suppresses tumorigenicity and enhances radiosensitivity in primary glioblastoma tumor initiating cells by inhibiting the STAT3 axis. J Cell Physiol 2012;227:976–93.
[294] Liu H-W, Su Y-K, Bamodu OA, Hueng D-Y, Lee W-H, Huang C-C, et al. The disruption of the beta-catenin/TCF-1/STAT3 signaling axis by 4-acetylantroquinonol B inhibits the tumorigenesis and cancer stem-cell-like properties of glioblastoma cells, in vitro and in vivo. Cancers (Basel) 2018;10.
[295] Huang W, Zhong Z, Luo C, Xiao Y, Li L, Zhang X, et al. The miR-26a/AP-2alpha/Nanog signaling axis mediates stem cell self-renewal and temozolomide resistance in glioma. Theranostics 2019;9:5497–516.
[296] Sherry-Lynes MM, Sengupta S, Kulkarni S, Cochran BH. Regulation of the JMJD3 (KDM6B) histone demethylase in glioblastoma stem cells by STAT3. PLoS One 2017;12, e0174775.
[297] Zhang C, Mukherjee S, Tucker-Burden C, Ross JL, Chau MJ, Kong J, et al. TRIM8 regulates stemness in glioblastoma through PIAS3-STAT3. Mol Oncol 2017;11:280–94.

[298] Yao Y, Ye H, Qi Z, Mo L, Yue Q, Baral A, et al. B7-H4(B7x)-mediated cross-talk between glioma-initiating cells and macrophages via the IL6/JAK/STAT3 pathway lead to poor prognosis in glioma patients. Clin Cancer Res 2016;22:2778–90.
[299] Shi Y, Zhou W, Cheng L, Chen C, Huang Z, Fang X, et al. Tetraspanin CD9 stabilizes gp130 by preventing its ubiquitin-dependent lysosomal degradation to promote STAT3 activation in glioma stem cells. Cell Death Differ 2017;24:167–80.
[300] Kim HS, Li A, Ahn S, Song H, Zhang W. Inositol polyphosphate-5-phosphatase F (INPP5F) inhibits STAT3 activity and suppresses gliomas tumorigenicity. Sci Rep 2014;4:7330.
[301] Sherry MM, Reeves A, Wu JK, Cochran BH. STAT3 is required for proliferation and maintenance of multipotency in glioblastoma stem cells. Stem Cells 2009;27:2383–92.
[302] Chang N, Ahn SH, Kong D-S, Lee HW, Nam D-H. The role of STAT3 in glioblastoma progression through dual influences on tumor cells and the immune microenvironment. Mol Cell Endocrinol 2017;451:53–65.
[303] Chang C-H, Liu W-T, Hung H-C, Gean C-Y, Tsai H-M, Su C-L, et al. Synergistic inhibition of tumor growth by combination treatment with drugs against different subpopulations of glioblastoma cells. BMC Cancer 2017;17:905.
[304] Wu N, Liu J, Zhao X, Yan Z, Jiang B, Wang L, et al. Cardamonin induces apoptosis by suppressing STAT3 signaling pathway in glioblastoma stem cells. Tumour Biol 2015;36:9667–76.
[305] Ashizawa T, Miyata H, Iizuka A, Komiyama M, Oshita C, Kume A, et al. Effect of the STAT3 inhibitor STX-0119 on the proliferation of cancer stem-like cells derived from recurrent glioblastoma. Int J Oncol 2013;43:219–27.
[306] Su Y-K, Bamodu OA, Tzeng Y-M, Hsiao M, Yeh C-T, Lin C-M. Ovatodiolide inhibits the oncogenicity and cancer stem cell-like phenotype of glioblastoma cells, as well as potentiate the anticancer effect of temozolomide. Phytomedicine 2019;61:152840.
[307] Ulasov I, Fares J, Timashev P, Lesniak MS. Editing cytoprotective autophagy in glioma: an unfulfilled potential for therapy. Trends Mol Med 2020;26:252–62.
[308] Ferrucci M, Biagioni F, Lenzi P, Gambardella S, Ferese R, Calierno MT, et al. Rapamycin promotes differentiation increasing betaIII-tubulin, NeuN, and NeuroD while suppressing nestin expression in glioblastoma cells. Oncotarget 2017;8:29574–99.
[309] Ciechomska IA, Przanowski P, Jackl J, Wojtas B, Kaminska B. BIX01294, an inhibitor of histone methyltransferase, induces autophagy-dependent differentiation of glioma stem-like cells. Sci Rep 2016; 6:38723.
[310] Ma B, Yuan Z, Zhang L, Lv P, Yang T, Gao J, et al. Long non-coding RNA AC023115.3 suppresses chemoresistance of glioblastoma by reducing autophagy. Biochim Biophys Acta, Mol Cell Res 2017;1864:1393–404.
[311] Natsumeda M, Maitani K, Liu Y, Miyahara H, Kaur H, Chu Q, et al. Targeting notch signaling and autophagy increases cytotoxicity in glioblastoma neurospheres. Brain Pathol 2016;26:713–23.

CHAPTER

24

Immunotherapy for glioblastoma as a means to overcome resistance to standard therapy

Kristen A. Batich[a,b], *John H. Sampson*[a,b,c], *Mustafa Khasraw*[a,b,c], *Henry S. Friedman*[a,b,c], *and David M. Ashley*[a,b,c]

[a]Preston Robert Tisch Brain Tumor Center at Duke, Duke University Medical Center, Durham, NC, United States [b]Department of Neurosurgery, Duke University Medical Center, Durham, NC, United States [c]Duke Cancer Institute, Duke University Medical Center, Durham, NC, United States

Abstract

High-grade gliomas arising in the central nervous system portend an aggressive course of neurologic devastation with a dismal prognosis for nearly all patients. Treatment strategies for these cancers have yielded minimal headway in the ability to slow the progression of the disease as well as prevent tumor recurrence, which invariably occurs. In this chapter, we discuss the evolution of standard therapies for high-grade gliomas anaplastic astrocytoma (WHO grade III) and glioblastoma (WHO grade IV). This chapter delineates why the successes of immunotherapy for other cancers has not translated in the same magnitude of responses for gliomas. The recently uncovered mechanisms of posttreatment resistance and alterations of the immune landscape are also discussed herein.

Abbreviations

APC	antigen-presenting cell
BBB	blood-brain barrier
BCNU	carmustine
CAM	cell adhesion molecule
CAR	chimeric antigen receptor
CCNU	lomustine
CDLN	cervical draining lymph node

Glioblastoma Resistance to Chemotherapy: Molecular Mechanisms and Innovative Reversal Strategies
https://doi.org/10.1016/B978-0-12-821567-8.00018-X

c-Kit	stem cell factor receptor
CNS	central nervous system
COL-IV	collagen-type IV
CRC	colorectal cancer
CSF	cerebrospinal fluid
CTLA-4	cytotoxic T lymphocyte-associated antigen 4
CX3CL1	fractalkine
DC	dendritic cell
DLN	draining lymph node
ECM	extracellular matrix
EGFRvIII	epidermal growth factor receptor variant III
EphA2	erythropoietin-producing hepatocellular carcinoma A2
FDA	Food and Drug Administration
FGFR	FGF receptor
Flt-1	vascular endothelial growth factor receptor-1
GBM	glioblastoma
GLASS	Glioma Longitudinal AnalySiS
HER2	human epidermal growth factor receptor 2
HIF-1α	hypoxia-inducible factor 1-alpha
HLA	human leukocyte antigen
IDO	indoleamine 2,3-dioxygenase
IGFR	IGF receptor
IL-13Rα2	interleukin receptor 13Rα2
KDR	kinase insert domain receptor
LEC	lymphatic endothelial cell
MET	scatter factor/hepatocyte growth factor receptor
MGMT	O-6-methylguanine-DNA methyltransferase
MHC	major histocompatibility complex
MLV	meningeal lymphatic vessel
MMP	matrix metalloprotease
MMR	mismatch repair
NO	nitric oxide
OS	overall survival
PCV	procarbazine, lomustine, vincristine
PD-1	programmed death-1
PDGFR	platelet-derived growth factor receptor
PD-L1	PD1 ligand 1
PFS	progression-free survival
PI3K	phosphatidylinositol 3-kinase
PLGF	placental growth factor
ROS	reactive oxygen species
RT	radiation therapy
RTOG	Radiation Therapy Oncology Group
SASP	senescence associated secretory phenotype
TAM	tumor-associated macrophage
TGF-β	transforming growth factor-beta
TH	T-helper
TIL	tumor infiltrating leukocyte
TIMP	tissue inhibitor of metalloprotease
TMB	tumor mutational burden
TME	tumor microenvironment
TMZ	temozolomide
TREG	T regulatory cell
VEGF	vascular endothelial growth factor
VEGFR	vascular endothelial growth factor receptor

Conflict of interest

No potential conflicts of interest were disclosed.

Introduction: Evolution of therapy for glioblastoma

Glioblastoma (GBM), the most common primary malignant brain tumor in adults and most aggressive, remain uniformly lethal with a median overall survival (OS) of <21 months, despite maximal surgical resection, high-dose radiation, and dose-limited chemotherapy [1–3]. Even with optimal treatment, nearly all GBMs will recur with a median time to progression from first-line treatment of around 6.9 months [1, 4]. Thus, the field is in desperate need of more durable treatments. As with other rare and aggressive cancers, however, reaching such treatments would require further depth in the understanding of this disease.

All the therapies that have demonstrated a survival benefit for high-grade gliomas in Phase III trials, including radiation, chemotherapy [temozolomide (TMZ), procarbazine, lomustine, vincristine (PCV)], and tumor-treating fields (TTFs), are based on relatively nonspecific targeting of proliferating cells. Recent advances in the molecular pathogenesis of gliomas as well as the distinction between primary and secondary GBM [5] suggest divergent pathways of their natural history and have unveiled some potential reasons for the failure of certain therapies.

The high-grade gliomas (anaplastic astrocytoma and GBM) carry a nearly uniformly dismal outcome. For GBM, even the most optimistic predictions of long-term survival of greater than 5 years after diagnosis only approach 1% or 2% [6], although higher 5-year survival rates have been recently published in Phase 3 studies [7]. The median survival of patients with newly diagnosed GBM has improved slightly in the last decade and now ranges between 15 and 18 months [8, 9]. Because of the dismal prognosis, patients with these tumors are treated early in the course of the disease with aggressive multimodal regimens that comprise surgical resection, radiotherapy, and chemotherapy. A timeline of approved therapies for primary and recurrent GBM is shown in Fig. 1. Today, unfortunately, the choice of drug therapy for primary and recurrent GBM is limited to a handful of agents.

Conventional chemotherapies

Nitrosourea compounds are well established in glioma therapy, as they are capable of penetrating the brain through an intact blood-brain barrier (BBB). Chemotherapy for GBM has evolved since the 1970s when researchers first reported data on the efficacy of carmustine (BCNU) [10], a nitrosourea compound that crosses the BBB and targets glioma cells directly [10–12]. Shortly after, another nitrosourea compound known as lomustine (CCNU) had demonstrated benefit in glioma responses. Oral lomustine received approval in 1976 [13], and intravenous carmustine received approval in 1977 [14] for use as single agents or in combination with other approved chemotherapeutic agents in patients with primary or metastatic brain tumors who had already undergone surgery or radiation therapy [15]. Carmustine wafers are synthetic biodegradable polymers loaded with carmustine. This product was first approved in 1996 for the treatment of recurrent GBM as an adjunct to surgery and was subsequently approved in 2003 for the first-line treatment of high-grade gliomas as an adjunct to

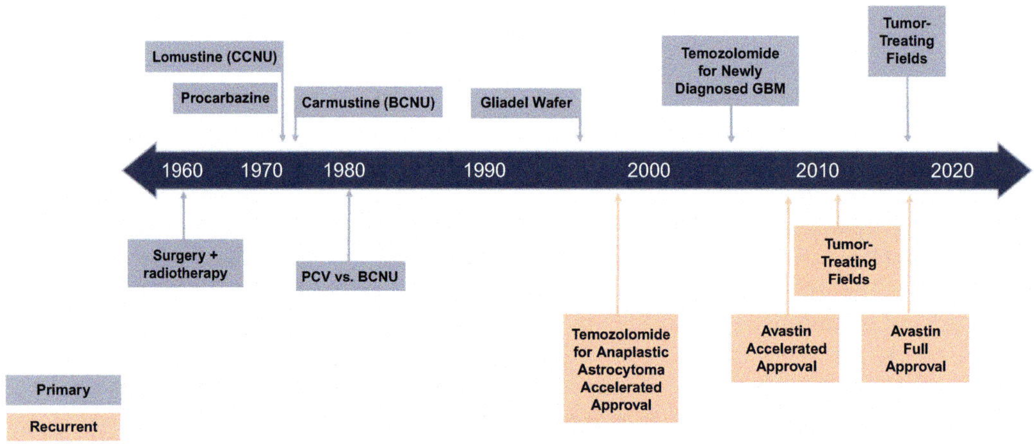

FIG. 1 Milestones in approved therapies for primary and recurrent glioblastoma. This timeline shows some of the milestones in the treatment of glioblastoma. In the 1970s radiation therapy was shown to improve outcome. This was soon followed by the observation that nitrosoureas can add additional benefit, although the impact on overall survival is minimal. The early 1980s CT imaging was used to describe response and progression. MR imaging became available in the mid-1980s and in the early 1990s some standard criteria were adopted to help clinical trialists assess response in investigative studies. Gliadel Wafers were approved as a delivery vehicle of BCNU for patients undergoing a second surgery, and then temozolomide was approved for relapsed gliomas. After the seminal randomized clinical trial showing benefit of concurrent temozolomide with radiotherapy for primary GBM, concurrent and adjuvant temozolomide was approved for primary disease. The use of tumor-treating fields with standard of care RT+TMZ was studied in the past 10 years and in 2017 was approved for primary disease. *PCV*, procarbazine, lomustine, and vincristine.

surgery and radiation [16, 17]. While treated patients demonstrated a longer OS compared to placebo controls, approval was based on a reduction in systemic toxicity using this locally applied therapy. In a Phase III trial, 240 newly diagnosed adults undergoing resection of any type of malignant glioma were randomly assigned to placement of up to eight carmustine wafers or a placebo, followed by standard RT. Patients receiving the carmustine polymer had only a modest 2-month increase in median survival, although statistically significant (13.9 vs 11.6 months). When the analysis was restricted to patients specifically with GBM tumors, the difference in survival was not statistically significant. Moreover, toxicities with carmustine polymers were similar to the placebo with an additional increase in the incidence of cerebrospinal fluid leakage and intracranial hypertension compared to placebo [18].

In the late 1970s, Walker and others demonstrated the efficacy of radiation therapy for brain cancers [14, 15, 19]. Shortly after radiation therapy was implemented for GBM treatment, standard practice shifted toward combining chemotherapy with radiation therapy, given the observation of this combination resulting in a longer time to progression [20]. It was not until two decades later, however, that the use of the oral DNA alkylating chemotherapy drug TMZ was granted accelerated approval by the US Food and Drug Administration (FDA) [21, 22]. Currently, TMZ is the preferred FDA-approved chemotherapeutic agent for GBM [23, 24]. One feature enabling its success is that TMZ dosing and frequency can be tailored based on patient characteristics and tolerability (standard dose 150–200 mg/m^2 for 5 days of a 28-day cycle, metronomic (or daily) TMZ at 50 mg/m^2 daily for 1 month cycles).

The combination of nitrosoureas with TMZ had been studied as well, based on the premise of synergistic effects of DNA damage [25, 26]. Fortunately, these studies did not reveal significant toxicity, but because of single-arm designs with notably small sample sizes in each one, interpretations of any clinically meaningful benefit for a combinatorial chemotherapy have yet to be unveiled.

Targeted and antiangiogenic agents

Because of the extensive endothelial proliferation that characterizes GBM and the discovery of vascular endothelial growth factor (VEGF) in solid tumors and its significance in the angiogenesis of tumor growth, targeting VEGF in GBM began in the early 2000s and was first employed in the recurrent setting [27]. Angiogenesis is fueled by several proangiogenic factors in high-grade gliomas, among which VEGF is dominant. Six VEGF isoforms (VEGF-A through -E and placental growth factor) and other biologically active VEGF variants generated by alternative gene splicing or protease cleavage, are secreted by tumor cells, infiltrating inflammatory cells and platelets and are harbored in the extracellular matrix [28–30]. Anti-VEGF-targeted therapies for high-grade gliomas are shown in Fig. 2. Glial tumors exhibit a prototypic "angiogenic switch" whereby a transformation from a low-grade to high-grade tumor necessitates induction of proangiogenic mediators and new blood vessel formation [31]. Furthermore, increased VEGF expression predicts glioma aggressiveness and poorer outcome [32]. VEGF expression in GBM is most concentrated adjacent to areas of necrosis and hypoxia, including cellular pseudopalisades at the tumor leading edge [31, 33–36].

The first documented use of bevacizumab in patients with GBM was a small series of patients with recurrent GBM treated by Stark-Vance [37]. Bevacizumab is a recombinant humanized monoclonal IgG1 antibody that has been tested in high-grade gliomas. It selectively binds to and neutralizes the biologic activity of human VEGF by interfering with the binding of its receptors vascular endothelial growth factor receptor-1 (Flt-1) and kinase insert domain receptor (KDR) on the surface of endothelial cells. Currently, bevacizumab is the only antiangiogenic therapy approved for recurrent GBM or grade III anaplastic astrocytoma. In 2009, the FDA granted accelerated approval for bevacizumab as a single agent for patients with GBM with progressive disease following prior RT and TMZ therapy, and full approval was granted in 2017 [38, 39].

Following observations that bevacizumab improved outcomes when administered with chemotherapy for colorectal, breast, lung, and pancreatic cancers [40], an initial evaluation of bevacizumab with the topoisomerase-1 inhibitor irinotecan (Camptosar; Pfizer) reported a dramatic rate of radiographic response among patients with recurrent GBM [37]. These results led to a Phase II study of this regimen for recurrent malignant glioma patients [41, 42]. Irinotecan was included in this regimen because it has modest activity as a salvage agent among recurrent malignant glioma patients, including a 5%–15% radiographic response rate and a median progression-free survival (PFS) of nearly 12 weeks [43–45]. Next, the addition of bevacizumab to lomustine chemotherapy was tested for recurrent GBM [46–48], and although treatment with lomustine plus bevacizumab prolonged PFS, it did not confer an OS advantage over monotherapies in patients with progressive GBM [49].

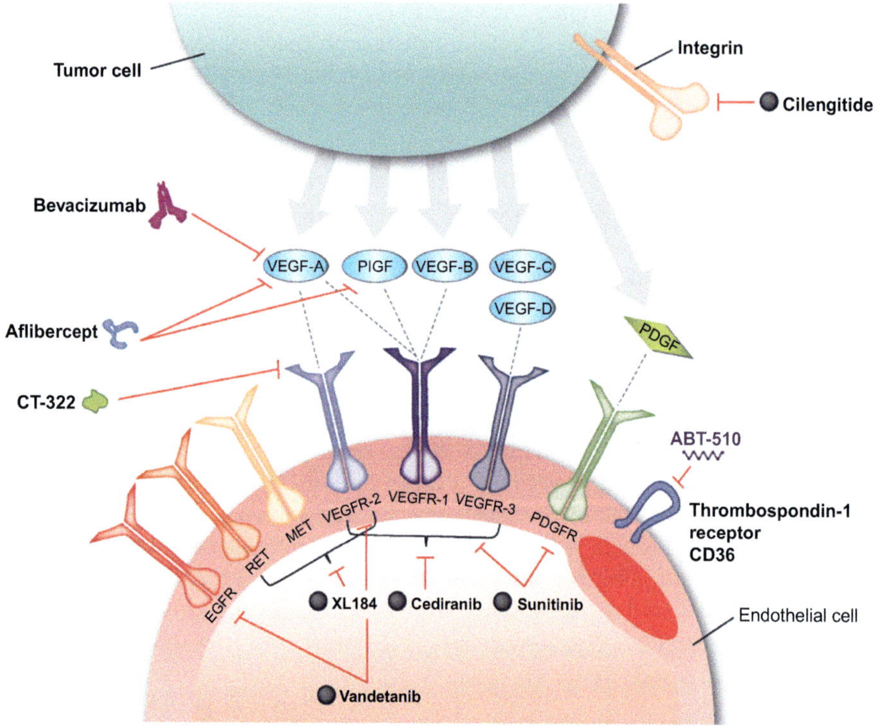

FIG. 2 Molecular targets of antiangiogenic agents in GBM. Cilengitide is a cyclic peptide that binds to and inhibits the activities of the a(*v*)b(3) and a(*v*)b(5) integrins. Bevacizumab is a humanized monoclonal IgG1 antibody that binds to and inhibits VEGF-A. Aflibercept is a fusion protein that binds all isoforms of VEGF-A, as well as PlGF. Cediranib, sunitinib, vandetanib, XL184, and CT-322 are multireceptor tyrosine kinase inhibitors. ABT-510 is a nonapeptide that targets the thrombospondin-1 receptor CD36. *EGFR*, epidermal growth factor receptor; *PDGFR*, platelet-derived growth factor receptor; *PlGF*, placental growth factor; *VEGF-A*, vascular endothelial growth factor A; *VEGFR*, vascular endothelial growth factor receptor. *Reproduced with permission from Beal K, Abrey LE, Gutin PH. Antiangiogenic agents in the treatment of recurrent or newly diagnosed glioblastoma: analysis of single-agent and combined modality approaches. Radiat Oncol 2011;6:2.*

Bevacizumab treatment for patients with newly diagnosed GBM was studied in two Phase III clinical trials. In the Radiation Therapy Oncology Group (RTOG) 0825 study, 637 patients were randomized to TMZ-based chemoradiation with or without bevacizumab. Patients in the bevacizumab arm started the agent 4 weeks into the radiation therapy protocol and continued for 6–12 cycles of maintenance therapy. The primary endpoints of these trials were to evaluate OS and PFS. Results showed a median OS of 15.7 months with the addition of bevacizumab compared to 16.1 months without the treatment. PFS improved slightly in the bevacizumab arm (10.7 vs 7.3 months, $p=0.004$), but the difference did not meet the predetermined level of statistical significance ($p=0.002$) [50]. The European trial (AVAglio trial) involved 921 patients and revealed a modest improvement in PFS (10.6 months in the bevacizumab arm vs 6.2 months in the placebo arm) but virtually identical OS (16.8 months vs 16.7 months, respectively) [51]. With respect to quality metrics, the AVAglio trial reported an improvement in quality of life. This was not evident in the RTOG 0825 study, which even suggested a negative impact on neurocognitive functions. In both studies, patients who had progressed were eligible to continue treatment with bevacizumab, based on the discretion of the treating investigators. What is quite compelling, from an exploratory analysis of the AVAglio study population, is that the subset of patients who did not receive a second line of therapy beyond progression had a significant extension in PFS and OS compared to those randomized to placebo [52]. Without the effect of previously confounded results from crossover in this study (48.2%) and in RTOG 0825 (48.3%), this analysis suggested that the addition of bevacizumab to the standard of care for newly diagnosed GBM appeared to prolong PFS and OS in those who never received second-line therapy.

In summary, there is evidence that bevacizumab prolongs PFS in newly diagnosed and recurrent GBM, with a potential impact on quality of life. It is unclear if subsets of people with GBM may benefit from antiangiogenic therapies, nor their utility in other glioma histologies [53]. Bevacizumab has seen increasing use for the management of radiation necrosis and as a steroid-sparing agent [54].

Therapies for GBM discussed thus far take a relatively nonselective approach to glioma cell death. The rationale for employing the immune system to target GBM is based on the premise that the inherent biologic specificity of the immune system can meet the need for a more precise therapy. Furthermore, the surveillance and multiple arms of the immune system offer an additional weapon of cell memory that has demonstrated activity even within the once-believed immunoprivileged central nervous system (CNS) [55, 56]. The next section encompasses the rationale underlying immune system targeting of GBM and how the integration of such strategies with prior approved therapies has uncovered mechanisms of resistance.

Immune surveillance and immunotherapy strategies in translation

Immunotherapeutic strategies are being evaluated in isolation and integrated with currently approved therapies for GBM at an accelerated rate. Immunotherapy for regarded "cold" brain tumors has largely been propelled by treatment failures of standard therapy regimens. Such prior shortcomings can be attributed to the high-grade invasiveness and microscopic dissemination at the time of diagnosis increased resistance to radiotherapy and TMZ therapy [57], the impracticality of total surgical resection, and the comparative intolerance of the normal brain and bone marrow for cytotoxic therapies.

The term immunotherapy refers to any approach that seeks to mobilize or manipulate the immune system for a therapeutic benefit. In this regard, there are numerous strategies for overcoming resistance the immune system may hold in targeting cancer. These include nonspecific activation of the immune system with microbial components or cytokines, antigen-specific adoptive immunotherapy with antibodies or T cells and active immunotherapy, which is commonly a direct "vaccination" or administration of vehicles specific to the tumor. The main limitation of using antibodies is that target proteins must be present in the extracellular milieu.

By contrast, T cells may target intracellular and extracellular antigens, which are presented at the cell surface in complexes with major histocompatibility complex (MHC) class I and class II, respectively [58]. The identification of defined tumor antigens, which are unique or shared proteins expressed on the surface of tumor cells, prompted the development of adoptive T-cell therapy. This approach has transcended into the use of genetically engineered T cells known as T cells with chimeric antigen receptors (CARs) [59]. To date, four GBM antigens have been targeted by CAR T cells in clinical studies, three of which have recently released results from clinical trials: epidermal growth factor receptor variant III (EGFRvIII) [60], human epidermal growth factor receptor 2 (HER2) [61], and interleukin receptor 13Rα2 (IL-13Rα2) [62]. Targeting erythropoietin-producing hepatocellular carcinoma A2 (EphA2) [63] is also in clinical trials, but results have not yet been released. Other CAR strategies for GBM, including chlorotoxin-directed lysis (NCT04214392) [64, 65] and tumor-associated antigen targeting of glioma cell proteoglycans [66], are also underway.

Predominantly, CAR-T therapies have been employed in the recurrent setting for GBM, which carries its own hurdles to effective immune responses. Yet, one of the most attractive immunotherapy strategies is vaccination. Active immunotherapy via vaccination is aimed at eliciting a specific de novo immune response against select tumor antigens. This form of therapy is expected to induce both therapeutic T-cell immunity in the form of tumor-specific effector T cells [67] and protective T-cell immunity in the form of tumor-specific memory T cells that can control tumor recurrence [68].

Early studies in mice revealed that the immune system can recognize and reject tumors [69]. Once thought to be siloed, the central nervous system and brain parenchyma itself has received considerable attention in detailing the pathways for immune cell entry and egress. Studies using models of multiple sclerosis, stroke, and pathogen-mediated encephalitis elucidated key afferent pathways into the brain whereby immune cells enter via (1) the choroid plexus into the cerebrospinal fluid (CSF) [70, 71], (2) meningeal afferent blood vessels into the subarachnoid space and Virchow Robin space [72, 73], and (3) across the BBB via selectin and integrin ligand-receptor interactions [70]. Antigen-presenting cells have previously been thought to exit the brain via perivascular spaces across BBB endothelial cells into the draining lymph nodes (DLNs) via afferent blood vessels. Not too long ago, however, the idea that the brain is an immune-privileged site dominated the field mainly due to the apparent lack of a classical lymphatic drainage system. Indeed, the network of lymphatic drainage from the brain by no means classical. Experimental evidence has disputed the immune privilege hypothesis, showing that tracers and proteins injected into the brain parenchyma and the CSF made their way to the brain's cervical draining lymph nodes (CDLNs) [74–76]. Intraparenchymal macromolecules were shown to exit the parenchyma mainly via perivascular efflux into the CSF [77], while the CSF itself was reported to exit the CNS mainly through arachnoid granulations into the venous system [78] and along the neural sheaths of cranial nerves (particularly the olfactory nerves), using the perineural lymphatics and return to the ethmoid/nasal mucosa and superficial CDLNs [75, 79]. A composite illustration of currently known mechanisms of antigen drainage from the brain to CDLNs as well as tolerance mechanisms against CNS tumor antigens is shown in (Fig. 3).

Lymphatic drainage in the CNS to secondary lymphoid organs

The meningeal lymphatic vessels (MLVs) have recently been characterized as an important drainage route for CNS- and CSF-derived soluble molecules and for meningeal immune cells

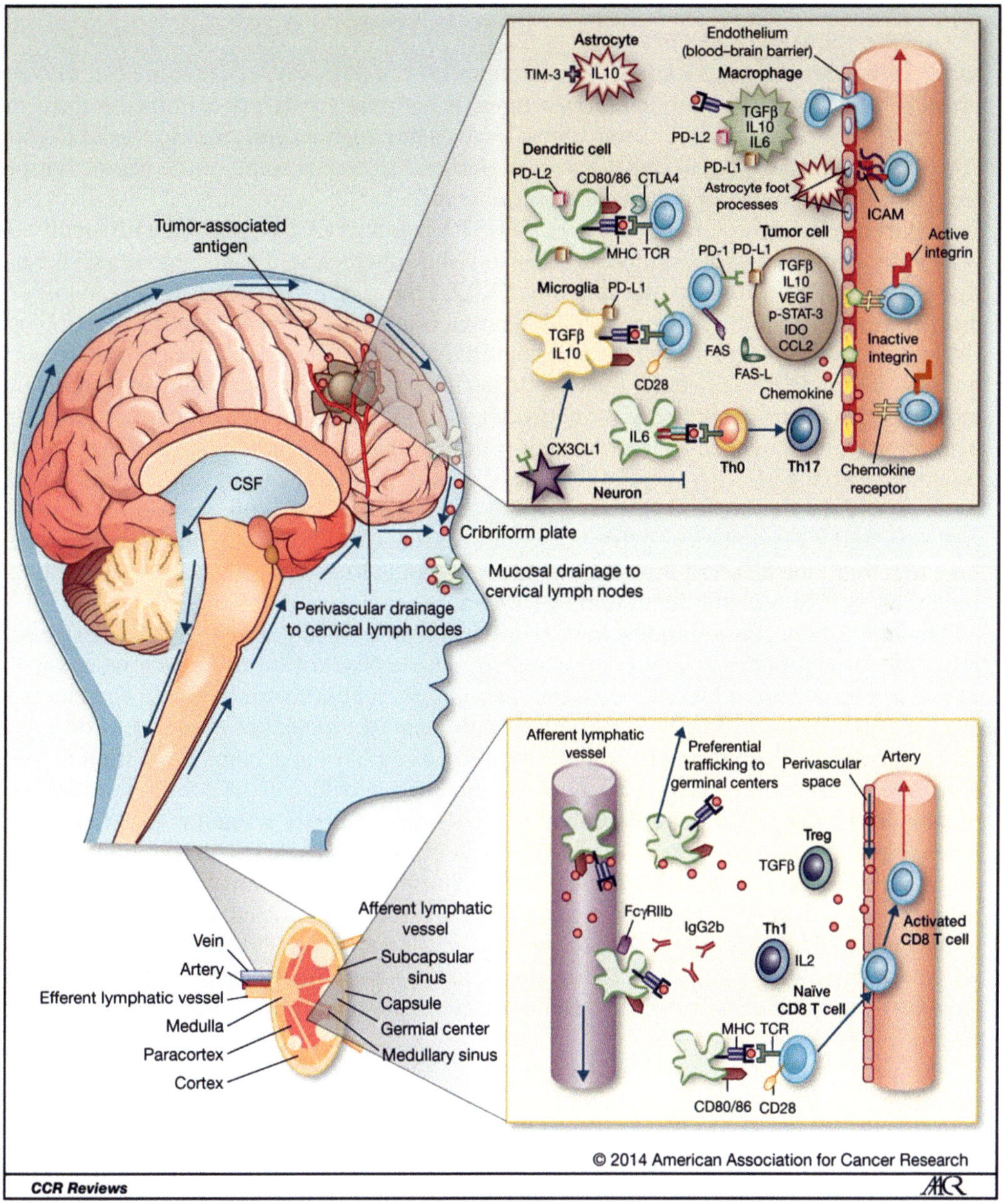

FIG. 3 Summary of adaptive immune response activation and tolerance mechanisms against CNS tumor antigens. Antigens drain to cervical lymph nodes along two distinct pathways: (i) soluble antigens and activated APCs pass through the cribriform plate and are conveyed to cervical lymph nodes via lymphatic pathways draining the nasal mucosa, (ii) soluble antigens in interstitial fluid drain along capillary wall basement membranes and between smooth muscle cells in the tunica media of muscular arteries to reach the cervical lymph nodes. In the cervical lymph nodes, CNS APCs preferentially migrate to germinal centers, stimulating B-cell predominant responses. The presence of CD32 (FcγRIIB) and Tregs producing TGF-β, further blunt the immune response. Activated T cells acquire CNS tropism via expression of α4 and β1 integrins. In the CNS vasculature, these integrins bind cell adhesion molecules (CAMs) expressed on the surface of vascular endothelial cells, mediating T-cell egress. Once in the tumor microenvironment, expression of immune checkpoint ligands on a variety of cell types indicates that immune checkpoint signaling may play a central role in the suppression of cytotoxic effector function. Inhibitory cytokines, including TGF-β, IL10 are also produced by a variety of cell types, including microglia, astrocytes, tumor cells, and immigrating macrophages. CX3CL1 (fractalkine) expressed by neurons inhibits activation of microglia. Furthermore, skewing toward Th17 differentiation limits the maturation of CD8 T cells from Th0 cells entering the inflamed CNS. This combination of immunosuppressive mechanisms both intrinsic and extrinsic to the tumor cells themselves is likely to subserve blunted immune responses against glioma antigens. *Reproduced with permission from Jackson CM, Lim M, Drake CG. Immunotherapy for brain cancer: recent progress and future promise. Clin Cancer Res 2014;20:3651–9.*

to directly enter the deep CDLNs. In searching for the pathways of immune cell movement and CSF drainage throughout the meninges, a lymphatic network within the dura mater along the superior sagittal and transverse sinuses has been recently rediscovered in rodents [74, 80] and humans [81, 82], and has been shown to possess some of the classic lymphatic functions. Like conventional lymphatic endothelial cells (LECs), meningeal LECs express surface markers CD31, VEGFR3, Prox1, PDPN, LYVE-1, and CCL21, and they efficiently drain both soluble molecules and immune cells from the subarachnoid space into the CDLNs. Under pathologic conditions, while ablation of MLVs reduced the inflammatory responses in an animal model of multiple sclerosis [73], dysfunction of MLVs propagated the pathology of Alzheimer's disease and age-associated cognitive decline [83, 84]. This means that MLVs play an important role in connecting the CNS to peripheral immunity and thus represent a promising therapeutic target not only for neurodegenerative and neuroinflammatory diseases but also for immune awakening and targeting brain tumors. Interestingly, findings that were obtained with the MLVs located at the dorsal part of the skull [74, 80] were recently opposed by observations that instead the basal MLVs at the base of the skull are a superior drainage pathway and morphologically distinct from dorsal MLVs [79, 83, 85, 86]. Taking these findings together, our directed immunotherapies continue to evolve in order to reflect the most potent aspects of an antiglioma immune response.

The majority of tumor models in mice are intracranially implanted tumor lines, ones that share the same genetic backbone as the host for the purposes of studying interactions with the native immune system. We are now seeing that implanted intracranial tumors can induce extensive remodeling of MLVs, depending on the site of inoculation. When tumor is placed more dorsally beneath the dura or in the striatum and pharmacochemical ablation of the dorsal MLVs is tested, there is a significant reduction in dendritic cell (DC) trafficking from brain tumors to the deep CDLNs [87]. Although DCs represent only a small population of leukocytes, they are the most powerful antigen-presenting cells (APC) with the unique ability to activate naive T cells [88]. As sentinel members of the innate immune system, DCs respond to antigens and molecules containing pathogen-associated molecular patterns or damage-associated molecular patterns by the generation of protective cytokines [89]. As members of the acquired immune system, DCs respond to these harmful molecules by efficient antigen uptake, processing and presentation, and hence DCs are crucial in the initiation of adaptive immune responses.

We previously introduced that high-grade gliomas undergo an "angiogenic switch" whereby a hypoxia-rich environment is established with the induction of VEGF analogs [90]. Hypoxia, a prominent feature of GBM [36, 91], enhances the expression and stabilization of hypoxia-inducible factor 1-alpha (HIF-1α), which acts as a transcription factor to activate a myriad of target genes that regulate tumor angiogenesis, migration and survival, including VEGF and VEGFRs [92]. Aberrant activation of multiple growth factor receptors in malignant glioma, including EGFR [93, 94], platelet-derived growth factor receptor (PDGFR) [95], scatter factor/hepatocyte growth factor receptor (MET) [96], IGF receptor (IGFR) [97, 98], stem cell factor receptor (c-Kit) [99], and FGF receptor (FGFR) [100, 101], also increases VEGF activity. Dysregulated signaling of the phosphatidylinositol 3-kinase (PI3K/Akt) and the Ras/MAPK pathways, which occur frequently among GBM, also augment VEGF expression [94, 102]. Expression of VEGFRs (VEGFR-1, VEGFR-2, and VEGFR-3) and coreceptors, including the neuropilins, although very low in the normal brain, are also markedly increased in high-grade gliomas [33–35, 103, 104]. Ligand binding activates VEGFRs, triggering a downstream intracellular signaling cascade that promotes endothelial cell proliferation, survival, activation, invasion, migration, and permeability [105]. VEGF signaling further fortifies the angiogenic response by triggering endothelial cell nitric oxide synthesis [106], and by mobilizing bone marrow-derived endothelial cell progenitors [107, 108].

Because GBM is both heavily vascularized and necrotic/hypoxic, there is a constant push-and-pull dynamic within a confined intracranial space. The newly formed blood vessels are leaky with a disrupted BBB, which leads to the formation of peritumoral edema and increased intracranial pressure [109]. The outflow of CSF is accelerated under conditions of acute rises in intracranial pressure such as the introduction of large CSF infusion volumes [110, 111]. In the context of established high-grade gliomas, however, preclinical evidence suggests a disruption in CSF drainage whereby the outflow of CSF is decreased in glioma-bearing mice compared to nontumor bearing mice. Surprisingly, in some glioma-bearing mice, there was high tracer signal in the sacral spinal region with tracer transport to the iliac lymph nodes and more tracer transport to systemic blood, indicating that CSF may be redirected in glioma-bearing mice to spinal lymphatic outflow pathways after the blockage of cranial CSF outflow routes [112].

With reduced CSF outflow, transport of GBM-derived antigens to DLNs is also likely be minimal, compromising the activation and proliferation of antitumor T cells, known as T-cell priming, eventually setting the stage for GBM to root and to progress. The expression of the lymphangiogenic factor VEGF-C has been reported to be very low in GBM [113] compared to healthy controls, which may also hinder the expansion of lymphatic vessels that are draining the CNS. The VEGF-C pathway has major implications for effective antigen drainage and exit through the CNS. In a recent study, ectopic expression of VEGF-C using adenoviral delivery into the CSF of glioma-bearing mice promoted enhancing priming of CD8+ T cells in the deep CDLNs, migration of CD8+ T cells into the tumor and unprecedented clearance of GBM with evidence of immunologic memory [114]. VEGF-C overexpression in brain tumor lines resulted in activation of the CCR7-CCL21 pathway, which is a vital pathway for DC homing to lymph nodes [87]. Consistently, mice bearing tumors with VEGF-C overexpression had decreased tumor volumes, and this benefit was abrogated in the absence of MLVs. By reopening these channels with VEGF-C, immune surveillance appeared to be reinstated.

Overcoming limited migration of DC therapies

There are a variety of factors presumed to be important for realizing DC vaccine efficacy, including the state of maturation of injected DCs [115–117], the route of injection [118, 119], and the efficiency of DC migration to vaccine site-draining lymph nodes. A crucial feature of DCs is their ability to migrate at various stages of maturation to allow appropriate antigen capture and presentation. Initially, DC precursors migrate from the bone marrow to peripheral tissues, where they reside in an immature state and act as sentinels for incoming pathogens and antigens. Once a DC encounters antigen at sites of peripheral inflammation, it undergoes a series of chemokine receptor changes to relocate to secondary lymphoid organs such as local DLNs. There in the lymph nodes, incoming DCs will encounter naive and central memory T cells and present antigen with concomitant costimulatory signals, initiating a cascade of T-cell priming for this specific antigen of interest [88]. The ability of DCs to follow this highly orchestrated migration pattern is reliant on chemokines being expressed at the right time. Chemokines bind to G-protein-coupled receptors that trigger intracellular signaling pathways involved in cell motility and activation [120].

Chemokines and their DC receptors play a key role in governing DC trafficking to lymph nodes via afferent lymphatic vessels [121, 122] and from blood [123]. As such, these axes have become relevant targets for immunotherapy. Based on the natural ability of DCs to mature and migrate to lymph nodes upon stimulation, mature DCs generated ex vivo have been pulsed with tumor antigens and injected back into patients as a form of antitumor immunotherapy. The number of DCs injected and the proportion that migrates to draining lymph nodes are likely important factors, as these parameters directly affect the magnitude of T-cell priming. Monitoring the in vivo migration of labeled DCs in patients with advanced melanoma has revealed that a maximum of 4% of intradermally injected DCs were able to reach regional lymphatics and home to local lymph nodes [124]. Similarly, de Vries et al. have found that DC migration efficiency in patients was severely limited and that the highest levels of migration were dependent on DC maturation state [117]. Preclinical investigations have revealed that preconditioning the vaccine site with inflammatory cytokines or mature DCs significantly increases the migration of a subsequent DC vaccine to draining lymph nodes, which then proportionally increased the magnitude of the CD4+ T-cell response [121]. Furthermore, distinct maturation and activation signals (PGE-2) [125] might induce the preferential expression of CCR7 by DCs, thereby increasing the capacity of these DCs to respond to the appropriate lymph node homing ligands CCL19 and CCL21.

A key parameter in driving DC migration to lymph nodes is the route of administration of DC vaccines. Antigen-loaded DCs might prime T-cell responses regardless of the route of injection, but the quality of responses is drastically affected. This has been shown by the induction of mainly T-helper (Th)-1-cell responses after intradermal and intralymphatic administration compared to nonpolarized T cell and antibody responses after intravenous administration [126]. Furthermore, recent studies in mice demonstrated that the induction of tissue-specific immunity was associated with the tissue origin of DCs, as distinct lineages will have differing migratory capacities with specificity for certain tissues [127]. In a mouse model of melanoma, both intravenous and subcutaneous immunization with peptide-pulsed DCs induced peptide-specific memory T cells in the spleen and enabled the control of metastasis in the lungs. However, whereas subcutaneous immunization also induced memory T cells in the lymph nodes, allowing subsequent protection against subcutaneously growing tumors, intravenous immunization had failed to do so [127]. Therefore, DCs can prime T cells with different homing capacities. For translational application in DC cancer vaccine trials, the consequence of such an instructive role of DCs is that, for example, intravenous administration of a DC vaccine against a melanoma would be unlikely to induce skin-homing effector cells. Certain trials are evaluating the ability of a high number of autologous DCs directly injected into the lymph nodes, bypassing the migratory steps from peripheral intradermal sites, to stimulate highly potent T-cell responses in the setting of newly diagnosed (NCT00323115) [128] and recurrent (NCT00766753) [129] GBM. Unfortunately, the laborious process of intranodal vaccinations has likely prevented the scalability of this translational application.

Clinical trials with corroborating preclinical studies have demonstrated that even in the steady state, there is a significant inefficiency in LN homing, with generally less than 5% of injected DCs actually reaching the vaccine site-draining lymph nodes [117, 118, 130]. Moreover, T-cell priming against glioma antigens is more successful in distant, nondraining lymph nodes compared to tumor-draining nodes [131, 132]. Preclinical investigations have revealed that preconditioning the vaccine site with inflammatory cytokines or mature DCs significantly increases the migration of a subsequent DC vaccine to DLNs [121]. By preconditioning the vaccine site with resultant increases in DC migration, the number of dividing antigen-specific T cells in DLNs was increased, and this was proportional to the absolute numbers of DCs reaching DLNs.

Our own studies have uncovered the relative inefficiency of intradermally administered DCs in reaching vaccine site-draining lymph nodes. This limited migration is apparent with vaccination of DCs into the intradermal compartment, whereby administered cells are required to home toward afferent lymphatic vessels in the skin and passage toward skin site-draining lymph nodes. Although increased DC migration to lymph nodes induces greater effector T-cell responses, whether the enhanced migration resulted in improved immune responses and reduced tumor burden had yet to be answered. For this reason, we assessed the impact of DC migration on clinical outcomes in patients with newly diagnosed GBM. In the preclinical setting, a variety of adjuvants have been employed to induce local inflammation with the goal of enhancing the immunogenicity of administered tumor-specific DCs or tumor-derived peptides [116, 133, 134]. In our studies, the use of Td unilaterally at the vaccine site produced a recall response that increased the bilateral migration of our tumor-specific DC vaccine in patients with GBM and in mice. Moreover, patients randomized to Td showed a statistically significant increase in PFS and OS even compared to those receiving unpulsed DCs [135, 136]. As a result, our current DC vaccine trials for GBM employ a mode of vaccine site preconditioning to ensure superior LN homing.

The field has uncovered a wealth of information on the biology of gliomas and its utilization of the CNS to remain relatively undetected by the immune system. Aside from mechanisms of altering the ability of immune cells to enter and exit efficiently, there exist mechanisms of direct immune suppression including MHC I downregulation [137], recruitment and proportional increases in T regulatory cells (Tregs) [138], secretion of IL-10 [139], indoleamine 2,3-dioxygenase (IDO) [140], and transforming growth factor-beta (TGF-β) [140, 141]. With all of this knowledge about barriers to realizing the full potential of immunotherapy, DCs and T cells are now being equipped with the right machinery to realize their functional goal as a defense against diseases of the brain and spinal cord. Therapies such as receptor stabilization for integrin or chemokine receptor expression are one line of investigation. The remaining strategies entail maturation cocktails with toll-like receptor activating adjuvants [116] and immunocytokines [142] for more durable, memory-inducing antitumor responses.

Immune checkpoint inhibitors in glioma

Immune checkpoint blockade targeting the programmed death-1 (PD-1)/PD1 ligand 1 (PD-L1) axis or cytotoxic T lymphocyte-associated antigen 4 (CTLA-4) has shown remarkable successes in lung, melanoma, head and neck, and renal cancers [143–146], with the increasing number of approvals by the US FDA. However, GBM has not seen such success with checkpoint blockade when used as monotherapies in preclinical models and in clinical trials [147, 148]. The Phase III trial CheckMate-143 compared the anti-PD-1 therapy nivolumab with bevacizumab in the treatment of recurrent GBM found no benefit in OS [148]. Next, the Phase III CheckMate-498 study (NCT02617589) showed similarly disappointing results in newly diagnosed patients GBM (of note, this study population was selected for only O-6-methylguanine-DNA methyltransferase (*MGMT*) promoter unmethylated patients) [149]. The CheckMate-548 (NCT02667587) [150], which is evaluating nivolumab plus TMZ in patients with *MGMT*-methylated GBM, is still pending. The field has been forced to dig deeper as to why these strategies failed with primary brain tumors.

Strategies that promote T-cell responses such as activated DCs [151] or radiation [152, 153] can potentiate anti-PD-1 therapy against GBM, suggesting that the barriers in the treatment of primary CNS tumors lie in T-cell priming. In patients with GBM receiving neoadjuvant anti-PD-1 therapies, expression of VEGF-C was highly correlated with T-cell infiltration [154]. Antiglioma T-cell responses in this preclinical study were further enhanced with checkpoint blockade, where anti-PD-1 alone had marginal effects on survival from established tumors, a combination treatment with VEGF-C mRNA and anti-PD-1 antibody resulted in regression of established gliomas. There was evidence of durable memory using this combination strategy in rechallenge of tumor to the contralateral hemisphere and transfer of host T cells to naïve hosts [114]. Given the importance of MLVs for successful T-cell priming, another set of investigators utilized VEGF-C overexpression with anti-PD-1/CTLA-4 blockade. This translated into significantly longer survival times against established tumors, and once CDLNs were removed, the survival benefit had been abolished [87]. Now with the updated understanding of these entry and exit pathways, investigators utilizing immunotherapies are undertaking methodologies to alter the machinery of adoptively transferred immune cell subsets or the expression of cognate receptors for purposes of counteracting tumor-derived immune suppression and evasion.

Mechanisms of chemotherapy resistance and implications for immune activation

GBM classically presents as a genetically heterogeneous disease with a varying degree of gene amplification, promoter methylation, deletion, and point mutations. Moreover, these tumors exhibit a relatively low tumor mutational burden (TMB) [155, 156]. For reference, the median TMB in newly diagnosed glioma is around 1.5 mutations/megabase compared to >10 mutations/megabase in melanoma or colorectal cancers [157]. Clonal selection of driver mutations occurs early on in oncogenesis, and it is the later acquisition of numerous passenger mutations that results in such intratumoral heterogeneity. Functional heterogeneity of glioma cells is not only evident by genetic makeup but also by nongenetic alterations such as stemness features and those interactions with the microenvironment [158]. A number of recent studies demonstrated intratumoral heterogeneity in GBM by showing regional variation and cell-to-cell differences in the expression of receptor tyrosine kinases [159–161]. Also, recent studies of GBM geography demonstrate maps of regional copy number and gene expression differences that importantly shed light on mechanisms of intratumoral clonal evolution in space and time and show that individual tumors can contain cells corresponding to different GBM subgroups [162].

Before treatment, GBM has a neutral evolutionary pattern, whereby mutational variants expand at similar rates and accumulate [163]. However, once chemotherapy or radiotherapy is introduced, there is a considerable expansion of subclones, and this is driven by selective pressures of the therapy itself. For low-grade gliomas, a hypermutation phenotype induced by treatment with TMZ is observed in up to 60% of tumors at recurrence [164]. There exist two main pathways to hypermutation in gliomas: a de novo pathway associated with constitutional defects in DNA polymerase and mismatch repair (MMR) genes, and a more common posttreatment pathway, associated with acquired resistance driven by MMR defects in chemotherapy-sensitive gliomas that recur after treatment with the chemotherapy drug TMZ [165]. Specifically, the mutational signature of posttreatment hypermutated gliomas can be recapitulated by TMZ-induced damage in cells with MMR deficiency.

Hypermutations in GBM still occur at a relatively low frequency of around 10% of patients after exposure to TMZ [166]. One study prospectively sequenced single GBM clones and demonstrated that single tumors contain clones that respond differently to TMZ. Interestingly, in this study, clonal analysis of the *MGMT* enzyme that removes the placed alkyl groups did not correlate with TMZ responsiveness, suggesting that new biomarkers of drug responsiveness are still needed [167].

These changes following standard of care therapy have a number of major implications for the fate of immunotherapy. There have been previous observations of subsets of patients with recurrent tumors who respond to immune checkpoint inhibitors [148, 168]. This has prompted an investigation into predictive biomarkers of immune checkpoint response [169]. High TMB is associated with response to checkpoint inhibitors in several cancer types but not in gliomas [170, 171]. Conversely, hypermutated gliomas with MMR deficiency have been less responsive to PD-1 blockade compared to gliomas with a lower TMB [165]. MMR deficiency recently emerged as an indicator of response to PD-1 blockade in patients with other solid cancers [172, 173] leading to the first tissue-agnostic cancer-drug approval by the US FDA for use of the PD-1 blocker pembrolizumab in patients with MMR-deficient cancers. In GBM, recurrent defects in the MMR pathway drive hypermutation and acquired TMZ resistance in chemotherapy-sensitive gliomas. However, the hypermutation phenotype and linked MMR deficiencies in gliomas are unique. The first distinction is that in colorectal cancers (CRCs) and others, MMR inactivation occurs early in tumor progression, whereas in posttreatment gliomas it arises late. The second distinction is that the response of nonglioma MMR-deficient tumors (such as CRC) to PD-1 blockade is based on their increased neoantigen burden and immune infiltration. This failed to occur in analyses of glioma specimens. In this study, MMR-deficient CRCs exhibited significantly more infiltrating T cells than their MMR-proficient counterparts, yet both MMR-deficient and MMR-proficient glioma samples lacked significant T-cell infiltrates. Similarly, there was no relative increase in neoantigen expression between de novo and posttreatment MMR-deficient gliomas [165]. Gliomas might therefore differ from other cancers on which the approval was based, and these differences might influence immune recognition of tumors and the response to immunotherapy.

TMB has also correlated with the level of preexisting antitumor adaptive immunity [174–176]. In a recurrent GBM population, a nonsignificant inverse relationship between TMB and radiographic/histological responses to PD-1 blockade was evident [177]. In fact, a recent analysis of three immunotherapy clinical trial cohorts for GBM has unveiled that patients with recurrent GBM tumors with a lower TMB share an enriched inflammatory gene signature relative to recurrent GBM tumors with higher TMB levels [170, 177, 178]. Notably, within the recurrent GBM cohort treated with poliovirus therapy and checkpoint blockade, patients with the low TMB phenotype associated with long-term survival [179]. These observations may in part explain the poor performance of checkpoint blockade in primary or recurrent GBM. A great emphasis must be placed that gliomas do not share the same relationships with the immune system as other solid tumors. Since we now know that chemotherapy can drive the acquisition of hypermutated populations without promoting a response to PD-1 blockade, caution will need to be exercised in the clinical application of TMB and MSI when predicting responses to checkpoint blockade.

Immunoediting is a process whereby the immune system eliminates tumor cells carrying recognizable epitopes on the surface of immune and tumor cells expressing human leukocyte antigen (HLA) molecules that are required for T-cell recognition. Immunoediting over time can select for antigen-negative, resistant subclones [180]. Evidence of this process has been shown in several cancer types, including glioma [181–183]. In comparing newly diagnosed GBM specimens with those at recurrence from the same patients, the Glioma Longitudinal AnalySiS (GLASS) Consortium found evidence in support of immunoediting in gliomas, suggesting that active immunosurveillance is ongoing in some adult patients [166]. Immunoediting can be evident in tumors with high cytolytic activity and low levels of immunosuppressive activity [184]. Hypermutated tumors, which have high loads of neoantigens, have also been associated with highly cytolytic microenvironments [183]. However, the Consortium did not observe any differences in the observed-to-expected neoantigen ratio between hypermutated recurrent tumors and their primary original tissue, nor were differences in neoantigen burden observed between hypermutated and nonhypermutated recurrent tumors. Yet, there was a strong association with the total number of unique HLA loci in a patient, similar to findings in lung cancer [185]. In addition, supporting the possibility of immunoediting in some patients is a recent study that examined the whole exome and transcriptome following αPD-1 therapy in recurrent glioma samples. Those tumors defined as responsive were associated with a more diverse pattern of evolution, possibly due to the elimination of neoantigens [177]. Several strategies seek to overcome these inherent mechanisms of resistance. For example, attempts to identify patients with hypermutated tumors at diagnosis are underway, in an effort to enrich for patients who may respond to immune checkpoint inhibition (NCT03718767) [186].

Conclusion

Efforts devoted to understanding the biology of gliomas and specifically high-grade gliomas have truly changed the landscape of our priorities in choosing the most optimal treatment. Not too long ago, cytotoxic chemotherapy agents were the mainstay for GBM treatment with the major advantage being their penetration of the BBB and activity on glial cells. Radiation therapy soon became a mainstay and was utilized in conjunction with chemotherapy given superior results of DNA damage with a combined modality. Now that radiation therapy and chemotherapy is the standard of care backbone following resection, the time has come to delve into how our chemoradiation strategies alter the landscape of these tumors and alter the full potential of our vaccines and adoptive T-cell strategies. The standard therapies for GBM, principally radiation and chemotherapy, are both profoundly immunosuppressive and thwart many immunotherapeutic attempts. Several studies have utilized the lymphopenic effect of chemotherapies to foster homeostatic proliferation of cancer vaccines [187, 188], but we are now witnessing how serial cycles of chemotherapy amidst vaccination, in turn, will dampen such immune responses [189]. The strong evidence of unmethylated *MGMT* status and the unfavorable response to TMZ lend an opportunity to understand how this population might respond differently to immune-activating strategies, ones without the abrogating effect of serial chemotherapy cycles.

As we are integrating a standard of care therapies with investigative immunotherapies, we are only now uncovering detrimental, additive, and synergistic outcomes. The integration of serial DC vaccines with serial chemotherapy cycles and the observed detriment to de novo immunity with this combination is dramatically changing the landscape of clinical trial design. Even the eligibility criteria based on *MGMT* methylation status, hypermethylated tumors, and TMB is directing enrollment toward more immune-based upfront therapies. VEGF inhibitors for GBM has been characterized to have both immune-activating and immune-dampening effects [190, 191], and some anti-VEGF strategies are being employed within clinical protocols to curtail the immune propagation of CNS inflammation and unforgiving neurologic deficits.

Preclinical studies and human sequencing data are indicating that the baseline inflammatory status of primary and recurrent GBM differentially can predict their susceptibility to immunotherapy. As more and more immunotherapies are being tested for primary and recurrent tumors, we need to pay more attention to the posttreatment alterations described: TMB, MMR defects, neoantigen expression of remaining subclones, and adaptive changes of the glial lymphatic system. By improving our ability to predict glial-derived immunosuppression at the time of and following the first-line immunotherapies, our targeting of resistant subclones would be more accurate and direct. The past 5 years of rigorous sequencing data have unveiled tumor-specific changes from chemotherapy. Now knowing that TMZ can evoke a hypermutated phenotype and resistant clones, should we even be considering additional chemotherapies at that time? Is there a molecular basis to say that these subclones would respond accordingly? With these questions in mind, the field has devoted multidisciplinary efforts to develop predictive biomarkers of immune-based therapeutic efficacy in GBM.

Radiation therapy is known to have lasting effects on the structure and milieu of the GBM microenvironment and in some studies facilitates tumor aggressiveness upon recurrence (Fig. 4). Such changes include hypoxia, innate immune activation, and extracellular matrix changes. We are starting to understand the effects that alkylating and nitrosourea chemotherapies have on the features of recurrent tumors, but the field desperately requires improved modeling of radiation-induced changes and what characterizes a "resistant" tumor. Hypoxic conditions facilitate tumor radioresistance and recurrence [192]. Radiation-induced stabilization and activation of HIF-1 elicits protective processes through regulating downstream target genes, such as nitric oxide (NO), which can stimulate immunosuppressive and antiapoptotic responses [193]. Vascular remodeling is a hallmark of ischemia-reperfusion injury, and radiation is known to affect vascular integrity causing vasculopathy, vascular

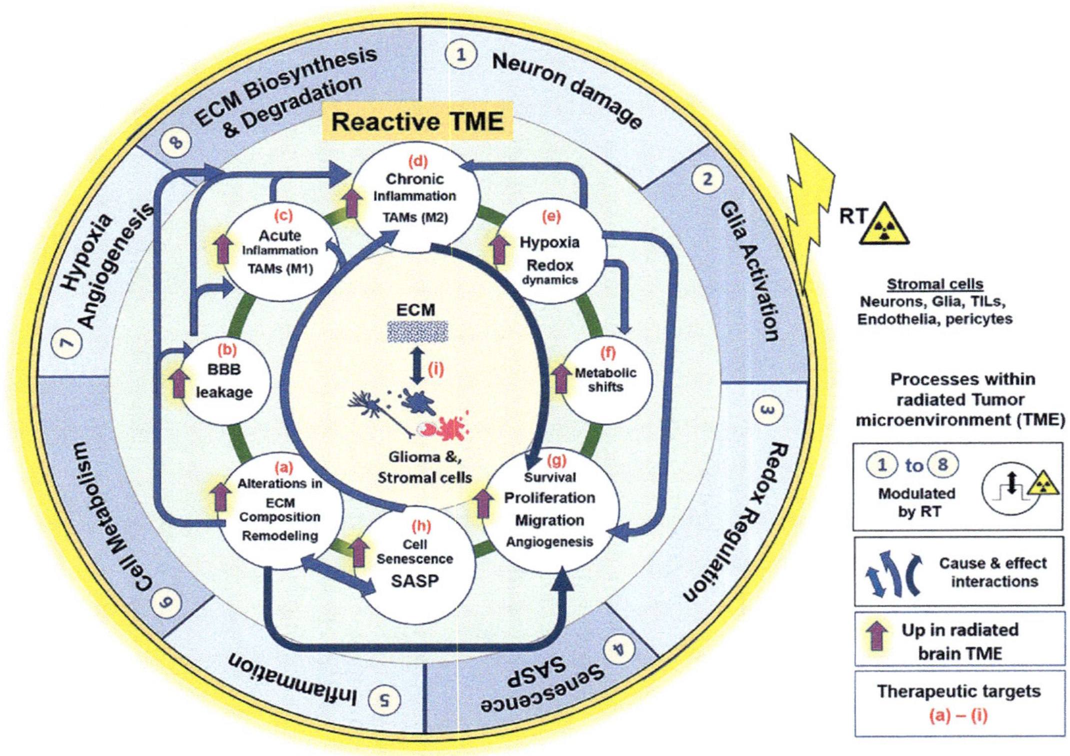

FIG. 4 Effects of RT on the glioblastoma (GBM) tumor microenvironment (TME). The ECM and its interaction with cellular components such as glia (microglia and astrocytes), glioma cells, endothelia, pericytes, and peripherally derived tumor-infiltrating leukocytes (TILs), play a central role in the GBM TME, which contributes to tumor cell survival, proliferation, migration, and invasion. The illustration represents the key pathophysiological processes and their interactions within the radiated TME. The outer *blue circle* represents biological phenomenon (1–8) that are directly impacted by RT. The inner *green circle* represents the reactive GBM TME, with its various processes, alterations or adaptive mechanisms that are upregulated in effect of RT, described from (a–i). *Blue arrows* indicate the "cause and effect" interactions between these processes facilitating GBM pathology and, *dark blue arrows* signify the primary role of the respective alterations in facilitating cell motility and invasiveness and thus aggressive tumor recurrence. Radiation injury leads to neuronal damage, overactivation of M1 microglia, and elicits an acute inflammatory response, with high ROS production in neurons and glial cells. There is alteration in ECM composition, and ECM-cell interactions. MMP/TIMP imbalance degrades Col-IV in basement membranes, which leads to blood-brain barrier leakage. Pronounced inflammation causes infiltration of leukocytes (monocyte-derived cell populations, macrophages), which along with activated microglia form the TAMs. TME of progressive tumors favor M2 phenotype and establishment of chronic inflammation. Redox dysregulation in effect of RT causes exacerbation of SASP phenotype and tumor cell adaptive processes, like hypoxia, metabolic shifts, and redox regulation (ROS/NO production), leading to neoangiogenesis and ECM remodeling. These alterations collectively make the TME permissive to glioma cell migration and invasion, thereby contributing to resistance and an aggressive tumor recurrence. All these biological processes in the reactive TME are potential therapeutic targets for improved glioblastoma care, with having ECM-cell interactions central to the manifestation of each of these phenomena. *BBB*, blood-brain barrier; *Col-IV*, collagen-type IV; *ECM*, extracellular matrix; *M1/M2*, proinflammatory or immune suppressive phenotypes of TAMs; *MMP*, matrix metalloproteases; *NO*, nitric oxide; *ROS*, reactive oxygen species; *RT*, radiation therapy; *SASP*, senescence-associated secretory phenotype; *TAM*, tumor-associated macrophages; *TILs*, tumor-infiltrating leukocytes; *TIMP*, tissue inhibitor of metalloproteases, respectively. *Reproduced with permission from Gupta K, Burns TC. Radiation-induced alterations in the recurrent glioblastoma microenvironment: therapeutic implications. Front Oncol 2018;8.*

depletion, hypoxia, and neoangiogenesis [194]. Levels of VEGF and angiotensin are increased in GBM postradiation, contributing to angiogenesis and tumor growth [195], while SDF promotes vasculogenesis [196]. In contrast to intracranial metastatic tumors, which can be eliminated via localized or whole-brain radiation, glioma stem cells survive radiation therapy, and eventual recurrence is fostered by radiation-induced changes in the tumor microenvironment [195, 197]. Radiation induces changes in the extracellular matrix composition that may facilitate tumor cell infiltration and pro-migratory tumor cell phenotypes [198, 199].

The invasive nature of GBM is induced by radiation and facilitates distant failures in the event of prolonged local control. Given this debilitating but consistent observation, the use of systemic radiosensitizers in combination with radiotherapy is being explored to enhance the effects of radiation with the aims of lowering radiation and chemotherapy doses with improved efficacy for GBM [200, 201]. Radiation-induced extracellular matrix alterations for GBM infiltration are potentially important therapeutic targets, and targeting of cytoskeletal dynamics, as well as microglial-matrix interactions that lead to pro-tumorigenic phenotypes in GBM, are being investigated [202].

Despite gross total resection, high dose radiation, and dose-limited chemotherapy for adults with GBM, median survival in unselected populations has been essentially frozen at 15 months for decades. Even the aggressive implementation of novel therapies such as antiangiogenic agents, viral vectors [203, 204], and tumor-treating fields [3, 205] have provided small benefits when compared to the advances in immunotherapy being seen in other cancers. A major driver for barely moving the needle lies in the relatively small number of patients with this diagnosis and its dire consequences. Thus, clinical testing of new agents is handled with small, nonrandomized studies that generally thwart the reliable and reproducible establishment of efficacy. Moreover, rare cancers like GBM are often aggressive which produces a narrow therapeutic window, therapeutic nihilism, and reluctance to enroll patients in placebo-controlled trials. As a result, among ~200 registered GBM trials identified between 2005 and 2015, nearly half were single-arm studies with sample sizes less than 50 total patients, almost 90% were open-label, and only 40% had a reportable randomization scheme [206]. These constraints have made demonstrating efficacy for novel agents, namely immunotherapies, difficult in GBM. In the end, the safest and most potent strategies will be those that take a multimodal, Phase-dependent approach to eradicate GBM.

References

[1] Stupp R, Mason WP, van den Bent MJ, Weller M, Fisher B, Taphoorn MJB, Belanger K, Brandes AA, Marosi C, Bogdahn U, Curschmann J, Janzer RC, Ludwin SK, Gorlia T, Allgeier A, Lacombe D, Cairncross JG, Eisenhauer E, Mirimanoff RO, European Organisation for Research and Treatment of Cancer Brain Tumor and Radiotherapy Groups; National Cancer Institute of Canada Clinical Trials Group. Radiotherapy plus concomitant and adjuvant temozolomide for glioblastoma. N Engl J Med 2005;352:987–96.

[2] Stupp R, Hegi ME, Mason WP, van den Bent MJ, Taphoorn MJ, Janzer RC, Ludwin SK, Allgeier A, Fisher B, Belanger K, Hau P, Brandes AA, Gijtenbeek J, Marosi C, Vecht CJ, Mokhtari K, Wesseling P, Villa S, Eisenhauer E, Gorlia T, Weller M, Lacombe D, Cairncross JG, Mirimanoff RO, European Organisation for Research and Treatment of Cancer Brain Tumour and Radiation Oncology Groups; National Cancer Institute of Canada Clinical Trials Group. Effects of radiotherapy with concomitant and adjuvant temozolomide versus radiotherapy alone on survival in glioblastoma in a randomised phase III study: 5-year analysis of the EORTC-NCIC trial. Lancet Oncol 2009;10:459–66.

[3] Stupp R, Taillibert S, Kanner A, Read W, Steinberg D, Lhermitte B, Toms S, Idbaih A, Ahluwalia MS, Fink K, Di Meco F, Lieberman F, Zhu JJ, Stragliotto G, Tran D, Brem S, Hottinger A, Kirson ED, Lavy-Shahaf G, Weinberg U, Kim CY, Paek SH, Nicholas G, Bruna J, Hirte H, Weller M, Palti Y, Hegi ME, Ram Z. Effect of tumor-treating fields plus maintenance temozolomide vs maintenance temozolomide alone on survival in patients with glioblastoma: a randomized clinical trial. JAMA 2017;318:2306–16.
[4] Majewska P, Ioannidis S, Raza MH, Tanna N, Bulbeck H, Williams M. Postprogression survival in patients with glioblastoma treated with concurrent chemoradiotherapy: a routine care cohort study. CNS Oncol 2017;6:307–13.
[5] Louis DN, Perry A, Reifenberger G, von Deimling A, Figarella-Branger D, Cavenee WK, Ohgaki H, Wiestler OD, Kleihues P, Ellison DW. The 2016 World Health Organization classification of tumors of the central nervous system: a summary. Acta Neuropathol 2016;131:803–20.
[6] Scott JN, Rewcastle NB, Brasher PM, Fulton D, MacKinnon JA, Hamilton M, Cairncross JG, Forsyth P. Which glioblastoma multiforme patient will become a long-term survivor? A population-based study. Ann Neurol 1999;46:183–8.
[7] Poon MTC, Sudlow CLM, Figueroa JD, Brennan PM. Longer-term (≥2 years) survival in patients with glioblastoma in population-based studies pre- and post-2005: a systematic review and meta-analysis. Sci Rep 2020;10:11622.
[8] Darefsky AS, King Jr JT, Dubrow R. Adult glioblastoma multiforme survival in the temozolomide era: a population-based analysis of Surveillance, Epidemiology, and End Results registries. Cancer 2012;118:2163–72.
[9] Koshy M, Villano JL, Dolecek TA, Howard A, Mahmood U, Chmura SJ, Weichselbaum RR, McCarthy BJ. Improved survival time trends for glioblastoma using the SEER 17 population-based registries. J Neurooncol 2012;107:207–12.
[10] Wilson CB, Boldrey EB, Enot KJ. 1,3-bis (2-chloroethyl)-1-nitrosourea (NSC-409962) in the treatment of brain tumors. Cancer Chemother Rep 1970;54:273–81.
[11] Walker MD. Chemotherapy: adjuvant to surgery and radiation therapy. Semin Oncol 1975;2:69–72.
[12] Walker MD, Hurwitz BS. BCNU (1,3-bis(2-chloroethyl)-1-nitrosourea; NSC-409962) in the treatment of malignant brain tumor—a preliminary report. Cancer Chemother Rep 1970;54:263–71.
[13] Parney IF, Chang SM. Current chemotherapy for glioblastoma. Cancer J 2003;9:149–56.
[14] Walker MD, Alexander Jr E, Hunt WE, MacCarty CS, Mahaley Jr MS, Mealey Jr J, Norrell HA, Owens G, Ransohoff J, Wilson CB, Gehan EA, Strike TA. Evaluation of BCNU and/or radiotherapy in the treatment of anaplastic gliomas. A cooperative clinical trial. J Neurosurg 1978;49:333–43.
[15] Walker MD, Green SB, Byar DP, Alexander Jr E, Batzdorf U, Brooks WH, Hunt WE, MacCarty CS, Mahaley Jr MS, Mealey Jr J, Owens G, Ransohoff 2nd J, Robertson JT, Shapiro WR, Smith Jr KR, Wilson CB, Strike TA. Randomized comparisons of radiotherapy and nitrosoureas for the treatment of malignant glioma after surgery. N Engl J Med 1980;303:1323–9.
[16] Brem H, Piantadosi S, Burger PC, Walker M, Selker R, Vick NA, Black K, Sisti M, Brem S, Mohr G, et al. Placebo-controlled trial of safety and efficacy of intraoperative controlled delivery by biodegradable polymers of chemotherapy for recurrent gliomas. The Polymer-brain Tumor Treatment Group. Lancet 1995; 345:1008–12.
[17] Valtonen S, Timonen U, Toivanen P, Kalimo H, Kivipelto L, Heiskanen O, Unsgaard G, Kuurne T. Interstitial chemotherapy with carmustine-loaded polymers for high-grade gliomas: a randomized double-blind study. Neurosurgery 1997;41:44–8. discussion 8–9.
[18] Westphal M, Hilt DC, Bortey E, Delavault P, Olivares R, Warnke PC, Whittle IR, Jääskeläinen J, Ram Z. A phase 3 trial of local chemotherapy with biodegradable carmustine (BCNU) wafers (Gliadel wafers) in patients with primary malignant glioma. Neuro Oncol 2003;5:79–88.
[19] Salazar OM, Rubin P, Feldstein ML, Pizzutiello R. High dose radiation therapy in the treatment of malignant gliomas: final report. Int J Radiat Oncol Biol Phys 1979;5:1733–40.
[20] Fine HA, Dear KB, Loeffler JS, Black PM, Canellos GP. Meta-analysis of radiation therapy with and without adjuvant chemotherapy for malignant gliomas in adults. Cancer 1993;71:2585–97.
[21] Brada M, Hoang-Xuan K, Rampling R, Dietrich PY, Dirix LY, Macdonald D, Heimans JJ, Zonnenberg BA, Bravo-Marques JM, Henriksson R, Stupp R, Yue N, Bruner J, Dugan M, Rao S, Zaknoen S. Multicenter phase II trial of temozolomide in patients with glioblastoma multiforme at first relapse. Ann Oncol 2001;12:259–66.
[22] Yung WK, Albright RE, Olson J, Fredericks R, Fink K, Prados MD, Brada M, Spence A, Hohl RJ, Shapiro W, Glantz M, Greenberg H, Selker RG, Vick NA, Rampling R, Friedman H, Phillips P, Bruner J, Yue N, Osoba D, Zaknoen S, Levin VA. A phase II study of temozolomide vs. procarbazine in patients with glioblastoma multiforme at first relapse. Br J Cancer 2000;83:588–93.

[23] Alifieris C, Trafalis DT. Glioblastoma multiforme: pathogenesis and treatment. Pharmacol Ther 2015;152:63–82.
[24] Frosina G. The glioblastoma problem: targeting by combined medicinal chemistry approaches. Curr Med Chem 2015;22:2506–24.
[25] Herrlinger U, Rieger J, Koch D, Loeser S, Blaschke B, Kortmann RD, Steinbach JP, Hundsberger T, Wick W, Meyermann R, Tan TC, Sommer C, Bamberg M, Reifenberger G, Weller M. Phase II trial of lomustine plus temozolomide chemotherapy in addition to radiotherapy in newly diagnosed glioblastoma: UKT-03. J Clin Oncol 2006;24:4412–7.
[26] Herrlinger U, Tzaridis T, Mack F, Steinbach JP, Schlegel U, Sabel M, Hau P, Kortmann RD, Krex D, Grauer O, Goldbrunner R, Schnell O, Bähr O, Uhl M, Seidel C, Tabatabai G, Kowalski T, Ringel F, Schmidt-Graf F, Suchorska B, Brehmer S, Weyerbrock A, Renovanz M, Bullinger L, Galldiks N, Vajkoczy P, Misch M, Vatter H, Stuplich M, Schäfer N, Kebir S, Weller J, Schaub C, Stummer W, Tonn JC, Simon M, Keil VC, Nelles M, Urbach H, Coenen M, Wick W, Weller M, Fimmers R, Schmid M, Hattingen E, Pietsch T, Coch C, Glas M. Lomustine-temozolomide combination therapy versus standard temozolomide therapy in patients with newly diagnosed glioblastoma with methylated MGMT promoter (CeTeG/NOA-09): a randomised, open-label, phase 3 trial. Lancet 2019;393:678–88.
[27] Erdem-Eraslan L, van den Bent MJ, Hoogstrate Y, Naz-Khan H, Stubbs A, van der Spek P, Böttcher R, Gao Y, de Wit M, Taal W, Oosterkamp HM, Walenkamp A, Beerepoot LV, Hanse MC, Buter J, Honkoop AH, van der Holt B, Vernhout RM, Sillevis Smitt PA, Kros JM, French PJ. Identification of patients with recurrent glioblastoma who may benefit from combined bevacizumab and CCNU therapy: a report from the BELOB trial. Cancer Res 2016;76:525–34.
[28] Kowanetz M, Ferrara N. Vascular endothelial growth factor signaling pathways: therapeutic perspective. Clin Cancer Res 2006;12:5018–22.
[29] Park JE, Keller GA, Ferrara N. The vascular endothelial growth factor (VEGF) isoforms: differential deposition into the subepithelial extracellular matrix and bioactivity of extracellular matrix-bound VEGF. Mol Biol Cell 1993;4:1317–26.
[30] Möhle R, Green D, Moore MA, Nachman RL, Rafii S. Constitutive production and thrombin-induced release of vascular endothelial growth factor by human megakaryocytes and platelets. Proc Natl Acad Sci U S A 1997;94:663–8.
[31] Kaur B, Tan C, Brat DJ, Post DE, Van Meir EG. Genetic and hypoxic regulation of angiogenesis in gliomas. J Neurooncol 2004;70:229–43.
[32] Leon SP, Folkerth RD, Black PM. Microvessel density is a prognostic indicator for patients with astroglial brain tumors. Cancer 1996;77:362–72.
[33] Plate KH, Breier G, Weich HA, Mennel HD, Risau W. Vascular endothelial growth factor and glioma angiogenesis: coordinate induction of VEGF receptors, distribution of VEGF protein and possible in vivo regulatory mechanisms. Int J Cancer 1994;59:520–9.
[34] Samoto K, Ikezaki K, Ono M, Shono T, Kohno K, Kuwano M, Fukui M. Expression of vascular endothelial growth factor and its possible relation with neovascularization in human brain tumors. Cancer Res 1995;55:1189–93.
[35] Schmidt NO, Westphal M, Hagel C, Ergün S, Stavrou D, Rosen EM, Lamszus K. Levels of vascular endothelial growth factor, hepatocyte growth factor/scatter factor and basic fibroblast growth factor in human gliomas and their relation to angiogenesis. Int J Cancer 1999;84:10–8.
[36] Rong Y, Durden DL, Van Meir EG, Brat DJ. 'Pseudopalisading' necrosis in glioblastoma: a familiar morphologic feature that links vascular pathology, hypoxia, and angiogenesis. J Neuropathol Exp Neurol 2006;65:529–39.
[37] Stark-Vance V. Bevacizumab and CPT-11 in the treatment of relapsed malignant glioma. Neuro Oncol 2005;7:369.
[38] Cohen MH, Shen YL, Keegan P, Pazdur R. FDA drug approval summary: bevacizumab (Avastin) as treatment of recurrent glioblastoma multiforme. Oncologist 2009;14:1131–8.
[39] Friedman HS, Prados MD, Wen PY, Mikkelsen T, Schiff D, Abrey LE, Yung WK, Paleologos N, Nicholas MK, Jensen R, Vredenburgh J, Huang J, Zheng M, Cloughesy T. Bevacizumab alone and in combination with irinotecan in recurrent glioblastoma. J Clin Oncol 2009;27:4733–40.
[40] Jain RK, Duda DG, Clark JW, Loeffler JS. Lessons from phase III clinical trials on anti-VEGF therapy for cancer. Nat Clin Pract Oncol 2006;3:24–40.
[41] Vredenburgh JJ, Desjardins A, Herndon 2nd JE, Marcello J, Reardon DA, Quinn JA, Rich JN, Sathornsumetee S, Gururangan S, Sampson J, Wagner M, Bailey L, Bigner DD, Friedman AH, Friedman HS. Bevacizumab plus irinotecan in recurrent glioblastoma multiforme. J Clin Oncol 2007;25:4722–9.

[42] Desjardins A, Reardon DA, Herndon 2nd JE, Marcello J, Quinn JA, Rich JN, Sathornsumetee S, Gururangan S, Sampson J, Bailey L, Bigner DD, Friedman AH, Friedman HS, Vredenburgh JJ. Bevacizumab plus irinotecan in recurrent WHO grade 3 malignant gliomas. Clin Cancer Res 2008;14:7068–73.
[43] Friedman HS, Petros WP, Friedman AH, Schaaf LJ, Kerby T, Lawyer J, Parry M, Houghton PJ, Lovell S, Rasheed K, Cloughsey T, Stewart ES, Colvin OM, Provenzale JM, McLendon RE, Bigner DD, Cokgor I, Haglund M, Rich J, Ashley D, Malczyn J, Elfring GL, Miller LL. Irinotecan therapy in adults with recurrent or progressive malignant glioma. J Clin Oncol 1999;17:1516–25.
[44] Batchelor TT, Gilbert MR, Supko JG, Carson KA, Nabors LB, Grossman SA, Lesser GJ, Mikkelsen T, Phuphanich S. Phase 2 study of weekly irinotecan in adults with recurrent malignant glioma: final report of NABTT 97-11. Neuro Oncol 2004;6:21–7.
[45] Prados MD, Lamborn K, Yung WK, Jaeckle K, Robins HI, Mehta M, Fine HA, Wen PY, Cloughesy T, Chang S, Nicholas MK, Schiff D, Greenberg H, Junck L, Fink K, Hess K, Kuhn J. A phase 2 trial of irinotecan (CPT-11) in patients with recurrent malignant glioma: a North American Brain Tumor Consortium study. Neuro Oncol 2006;8:189–93.
[46] Wick W, Gorlia T, Bendszus M, Taphoorn M, Sahm F, Harting I, Brandes AA, Taal W, Domont J, Idbaih A, Campone M, Clement PM, Stupp R, Fabbro M, Le Rhun E, Dubois F, Weller M, von Deimling A, Golfinopoulos V, Bromberg JC, Platten M, Klein M, van den Bent MJ. Lomustine and bevacizumab in progressive glioblastoma. N Engl J Med 2017;377:1954–63.
[47] Heiland DH, Masalha W, Franco P, Machein MR, Weyerbrock A. Progression-free and overall survival in patients with recurrent Glioblastoma multiforme treated with last-line bevacizumab versus bevacizumab/lomustine. J Neurooncol 2016;126:567–75.
[48] Taal W, Oosterkamp HM, Walenkamp AM, Dubbink HJ, Beerepoot LV, Hanse MC, Buter J, Honkoop AH, Boerman D, de Vos FY, Dinjens WN, Enting RH, Taphoorn MJ, van den Berkmortel FW, Jansen RL, Brandsma D, Bromberg JE, van Heuvel I, Vernhout RM, van der Holt B, van den Bent MJ. Single-agent bevacizumab or lomustine versus a combination of bevacizumab plus lomustine in patients with recurrent glioblastoma (BELOB trial): a randomised controlled phase 2 trial. Lancet Oncol 2014;15:943–53.
[49] Song J, Xue YQ, Zhao MM, Xu P. Effectiveness of lomustine and bevacizumab in progressive glioblastoma: a meta-analysis. Onco Targets Ther 2018;11:3435–9.
[50] Gilbert MR, Dignam JJ, Armstrong TS, Wefel JS, Blumenthal DT, Vogelbaum MA, Colman H, Chakravarti A, Pugh S, Won M, Jeraj R, Brown PD, Jaeckle KA, Schiff D, Stieber VW, Brachman DG, Werner-Wasik M, Tremont-Lukats IW, Sulman EP, Aldape KD, Curran Jr WJ, Mehta MP. A randomized trial of bevacizumab for newly diagnosed glioblastoma. N Engl J Med 2014;370:699–708.
[51] Chinot OL, Wick W, Mason W, Henriksson R, Saran F, Nishikawa R, Carpentier AF, Hoang-Xuan K, Kavan P, Cernea D, Brandes AA, Hilton M, Abrey L, Cloughesy T. Bevacizumab plus radiotherapy-temozolomide for newly diagnosed glioblastoma. N Engl J Med 2014;370:709–22.
[52] Chinot OL, Nishikawa R, Mason W, Henriksson R, Saran F, Cloughesy T, Garcia J, Revil C, Abrey L, Wick W. Upfront bevacizumab may extend survival for glioblastoma patients who do not receive second-line therapy: an exploratory analysis of AVAglio. Neuro Oncol 2016;18:1313–8.
[53] Ameratunga M, Pavlakis N, Wheeler H, Grant R, Simes J, Khasraw M. Anti-angiogenic therapy for high-grade glioma. Cochrane Database Syst Rev 2014;(9):CD008218. https://doi.org/10.1002/14651858.CD008218.pub3.
[54] Zhuang H, Shi S, Yuan Z, Chang JY. Bevacizumab treatment for radiation brain necrosis: mechanism, efficacy and issues. Mol Cancer 2019;18:21.
[55] Owens T, Renno T, Taupin V, Krakowski M. Inflammatory cytokines in the brain: does the CNS shape immune responses? Immunol Today 1994;15:566–71.
[56] Stevens A, Klöter I, Roggendorf W. Inflammatory infiltrates and natural killer cell presence in human brain tumors. Cancer 1988;61:738–43.
[57] Hegi ME, Diserens AC, Gorlia T, Hamou MF, de Tribolet N, Weller M, Kros JM, Hainfellner JA, Mason W, Mariani L, Bromberg JE, Hau P, Mirimanoff RO, Cairncross JG, Janzer RC, Stupp R. MGMT gene silencing and benefit from temozolomide in glioblastoma. N Engl J Med 2005;352:997–1003.
[58] Townsend AR, Gotch FM, Davey J. Cytotoxic T cells recognize fragments of the influenza nucleoprotein. Cell 1985;42:457–67.
[59] Choi BD, Curry WT, Carter BS, Maus MV. Chimeric antigen receptor T-cell immunotherapy for glioblastoma: practical insights for neurosurgeons. Neurosurg Focus 2018;44, E13.

[60] O'Rourke DM, Nasrallah MP, Desai A, Melenhorst JJ, Mansfield K, Morrissette JJD, Martinez-Lage M, Brem S, Maloney E, Shen A, Isaacs R, Mohan S, Plesa G, Lacey SF, Navenot JM, Zheng Z, Levine BL, Okada H, June CH, Brogdon JL, Maus MV. A single dose of peripherally infused EGFRvIII-directed CAR T cells mediates antigen loss and induces adaptive resistance in patients with recurrent glioblastoma. Sci Transl Med 2017;9(399): eaaa0984. https://doi.org/10.1126/scitranslmed.aaa0984.

[61] Ahmed N, Brawley V, Hegde M, Bielamowicz K, Kalra M, Landi D. HER2-specific chimeric antigen receptor-modified virus-specific T cells for progressive glioblastoma: a phase 1 dose-escalation trial. JAMA Oncol 2017;3:1094–101.

[62] Brown C, Alizadeh D, Starr R, Weng L, Wagner J, Naranjo A. Regression of glioblastoma after chimeric antigen receptor T-cell therapy. N Engl J Med 2016;375:2561–9.

[63] Yi Z, Prinzing BL, Cao F, Gottschalk S, Krenciute G. Optimizing EphA2-CAR T Cells for the adoptive immunotherapy of glioma. Mol Ther Methods Clin Dev 2018;9:70–80.

[64] US National Library of Medicine, ClinicalTrials.gov; 2020.

[65] Wang D, Starr R, Chang WC, Aguilar B, Alizadeh D, Wright SL, Yang X, Brito A, Sarkissian A, Ostberg JR, Li L, Shi Y, Gutova M, Aboody K, Badie B, Forman SJ, Barish ME, Brown CE. Chlorotoxin-directed CAR T cells for specific and effective targeting of glioblastoma. Sci Transl Med 2020;12(533):eaaw2672. https://doi.org/10.1126/scitranslmed.aaw2672.

[66] Pellegatta S, Savoldo B, Di Ianni N, Corbetta C, Chen Y, Patané M, Sun C, Pollo B, Ferrone S, DiMeco F, Finocchiaro G, Dotti G. Constitutive and TNFα-inducible expression of chondroitin sulfate proteoglycan 4 in glioblastoma and neurospheres: Implications for CAR-T cell therapy. Sci Transl Med 2018;10(430):eaao2731. https://doi.org/10.1126/scitranslmed.aao2731.

[67] Yu JS, Liu G, Ying H, Yong WH, Black KL, Wheeler CJ. Vaccination with tumor lysate-pulsed dendritic cells elicits antigen-specific, cytotoxic T-cells in patients with malignant glioma. Cancer Res 2004;64:4973–9.

[68] Klebanoff CA, Gattinoni L, Torabi-Parizi P, Kerstann K, Cardones AR, Finkelstein SE, Palmer DC, Antony PA, Hwang ST, Rosenberg SA, Waldmann TA, Restifo NP. Central memory self/tumor-reactive CD8+ T cells confer superior antitumor immunity compared with effector memory T cells. Proc Natl Acad Sci U S A 2005;102:9571–6.

[69] Foley EJ. Antigenic properties of methylcholanthrene-induced tumors in mice of the strain of origin. Cancer Res 1953;13:835–7.

[70] Sagar D, Foss C, El Baz R, Pomper MG, Khan ZK, Jain P. Mechanisms of dendritic cell trafficking across the blood-brain barrier. J Neuroimmune Pharmacol 2012;7:74–94.

[71] De Laere M, Berneman ZN, Cools N. To the brain and back: migratory paths of dendritic cells in multiple sclerosis. J Neuropathol Exp Neurol 2018;77:178–92.

[72] Arjmandi A, Liu K, Dorovini-Zis K. Dendritic cell adhesion to cerebral endothelium: role of endothelial cell adhesion molecules and their ligands. J Neuropathol Exp Neurol 2009;68:300–13.

[73] Louveau A, Herz J, Alme MN, Salvador AF, Dong MQ, Viar KE, Herod SG, Knopp J, Setliff JC, Lupi AL, Da Mesquita S, Frost EL, Gaultier A, Harris TH, Cao R, Hu S, Lukens JR, Smirnov I, Overall CC, Oliver G, Kipnis J. CNS lymphatic drainage and neuroinflammation are regulated by meningeal lymphatic vasculature. Nat Neurosci 2018;21:1380–91.

[74] Louveau A, Smirnov I, Keyes TJ, Eccles JD, Rouhani SJ, Peske JD, Derecki NC, Castle D, Mandell JW, Lee KS, Harris TH, Kipnis J. Structural and functional features of central nervous system lymphatic vessels. Nature 2015;523:337–41.

[75] Kida S, Pantazis A, Weller RO. CSF drains directly from the subarachnoid space into nasal lymphatics in the rat. Anatomy, histology and immunological significance. Neuropathol Appl Neurobiol 1993;19:480–8.

[76] Cserr HF, Harling-Berg CJ, Knopf PM. Drainage of brain extracellular fluid into blood and deep cervical lymph and its immunological significance. Brain Pathol 1992;2:269–76.

[77] Iliff JJ, Wang M, Liao Y, Plogg BA, Peng W, Gundersen GA, Benveniste H, Vates GE, Deane R, Goldman SA, Nagelhus EA, Nedergaard M. A paravascular pathway facilitates CSF flow through the brain parenchyma and the clearance of interstitial solutes, including amyloid β. Sci Transl Med 2012;4:147ra11.

[78] Go KG, Houthoff HJ, Hartsuiker J, Blaauw EH, Havinga P. Fluid secretion in arachnoid cysts as a clue to cerebrospinal fluid absorption at the arachnoid granulation. J Neurosurg 1986;65:642–8.

[79] Ma Q, Ineichen BV, Detmar M, Proulx ST. Outflow of cerebrospinal fluid is predominantly through lymphatic vessels and is reduced in aged mice. Nat Commun 2017;8:1434.

[80] Aspelund A, Antila S, Proulx ST, Karlsen TV, Karaman S, Detmar M, Wiig H, Alitalo K. A dural lymphatic vascular system that drains brain interstitial fluid and macromolecules. J Exp Med 2015;212:991–9.
[81] Absinta M, Ha SK, Nair G, Sati P, Luciano NJ, Palisoc M, Louveau A, Zaghloul KA, Pittaluga S, Kipnis J, Reich DS. Human and nonhuman primate meninges harbor lymphatic vessels that can be visualized noninvasively by MRI. Elife 2017;6:e29738. https://doi.org/10.7554/eLife.29738.
[82] Johnston M, Zakharov A, Papaiconomou C, Salmasi G, Armstrong D. Evidence of connections between cerebrospinal fluid and nasal lymphatic vessels in humans, non-human primates and other mammalian species. Cerebrospinal Fluid Res 2004;1:2.
[83] Ahn JH, Cho H, Kim J-H, Kim SH, Ham J-S, Park I, Suh SH, Hong SP, Song J-H, Hong Y-K, Jeong Y, Park S-H, Koh GY. Meningeal lymphatic vessels at the skull base drain cerebrospinal fluid. Nature 2019;572:62–6.
[84] Da Mesquita S, Louveau A, Vaccari A, Smirnov I, Cornelison RC, Kingsmore KM, Contarino C, Onengut-Gumuscu S, Farber E, Raper D, Viar KE, Powell RD, Baker W, Dabhi N, Bai R, Cao R, Hu S, Rich SS, Munson JM, Lopes MB, Overall CC, Acton ST, Kipnis J. Functional aspects of meningeal lymphatics in ageing and Alzheimer's disease. Nature 2018;560:185–91.
[85] Ma Q, Decker Y, Müller A, Ineichen BV, Proulx ST. Clearance of cerebrospinal fluid from the sacral spine through lymphatic vessels. J Exp Med 2019;216:2492–502.
[86] Ma Q, Ries M, Decker Y, Müller A, Riner C, Bücker A, Fassbender K, Detmar M, Proulx S. Rapid lymphatic efflux limits cerebrospinal fluid flow to the brain. Acta Neuropathol 2019;137(1):151–65. https://doi.org/10.1007/s00401-018-1916-x.
[87] Hu X, Deng Q, Ma L, Li Q, Chen Y, Liao Y, Zhou F, Zhang C, Shao L, Feng J, He T, Ning W, Kong Y, Huo Y, He A, Liu B, Zhang J, Adams R, He Y, Tang F, Bian X, Luo J. Meningeal lymphatic vessels regulate brain tumor drainage and immunity. Cell Res 2020;30:229–43.
[88] Palucka K, Banchereau J. Cancer immunotherapy via dendritic cells. Nat Rev Cancer 2012;12:265–77.
[89] Granucci F, Zanoni I, Ricciardi-Castagnoli P. Central role of dendritic cells in the regulation and deregulation of immune responses. Cell Mol Life Sci 2008;65:1683–97.
[90] Reardon DA, Wen PY, Desjardins A, Batchelor TT, Vredenburgh JJ. Glioblastoma multiforme: an emerging paradigm of anti-VEGF therapy. Expert Opin Biol Ther 2008;8:541–53.
[91] Kaur B, Khwaja FW, Severson EA, Matheny SL, Brat DJ, Van Meir EG. Hypoxia and the hypoxia-inducible-factor pathway in glioma growth and angiogenesis. Neuro Oncol 2005;7:134–53.
[92] Zagzag D, Zhong H, Scalzitti JM, Laughner E, Simons JW, Semenza GL. Expression of hypoxia-inducible factor 1α in brain tumors: association with angiogenesis, invasion, and progression. Cancer 2000;88:2606–18.
[93] Maity A, Pore N, Lee J, Solomon D, O'Rourke DM. Epidermal growth factor receptor transcriptionally up-regulates vascular endothelial growth factor expression in human glioblastoma cells via a pathway involving phosphatidylinositol 3'-kinase and distinct from that induced by hypoxia. Cancer Res 2000;60:5879–86.
[94] Pore N, Liu S, Haas-Kogan DA, O'Rourke DM, Maity A. PTEN mutation and epidermal growth factor receptor activation regulate vascular endothelial growth factor (VEGF) mRNA expression in human glioblastoma cells by transactivating the proximal VEGF promoter. Cancer Res 2003;63:236–41.
[95] Ma D, Nutt CL, Shanehsaz P, Peng X, Louis DN, Kaetzel DM. Autocrine platelet-derived growth factor-dependent gene expression in glioblastoma cells is mediated largely by activation of the transcription factor sterol regulatory element binding protein and is associated with altered genotype and patient survival in human brain tumors. Cancer Res 2005;65:5523–34.
[96] Abounader R, Laterra J. Scatter factor/hepatocyte growth factor in brain tumor growth and angiogenesis. Neuro Oncol 2005;7:436–51.
[97] Clarke K, Smith K, Gullick WJ, Harris AL. Mutant epidermal growth factor receptor enhances induction of vascular endothelial growth factor by hypoxia and insulin-like growth factor-1 via a PI3 kinase dependent pathway. Br J Cancer 2001;84:1322–9.
[98] Trojan J, Cloix JF, Ardourel MY, Chatel M, Anthony DD. Insulin-like growth factor type I biology and targeting in malignant gliomas. Neuroscience 2007;145:795–811.
[99] Sun L, Hui AM, Su Q, Vortmeyer A, Kotliarov Y, Pastorino S, Passaniti A, Menon J, Walling J, Bailey R, Rosenblum M, Mikkelsen T, Fine HA. Neuronal and glioma-derived stem cell factor induces angiogenesis within the brain. Cancer Cell 2006;9:287–300.
[100] Li VW, Folkerth RD, Watanabe H, Yu C, Rupnick M, Barnes P, Scott RM, Black PM, Sallan SE, Folkman J. Microvessel count and cerebrospinal fluid basic fibroblast growth factor in children with brain tumours. Lancet 1994;344:82–6.

[101] Stefnik DF, Rizkalla LR, Soi A, Goldblatt SA, Rizkalla WM. Acidic and basic fibrolast growth factors are present in glioblastoma multiforme. Cancer Res 1991;51:5760–5.

[102] Pelloski CE, Lin E, Zhang L, Yung WKA, Colman H, Liu J-L, Woo SY, Heimberger AB, Suki D, Prados M, Chang S, Barker FG, Fuller GN, Aldape KD. Prognostic associations of activated mitogen-activated protein kinase and Akt pathways in glioblastoma. Clin Cancer Res 2006;12:3935–41.

[103] Vajkoczy P, Farhadi M, Gaumann A, Heidenreich R, Erber R, Wunder A, Tonn JC, Menger MD, Breier G. Microtumor growth initiates angiogenic sprouting with simultaneous expression of VEGF, VEGF receptor-2, and angiopoietin-2. J Clin Invest 2002;109:777–85.

[104] Grau SJ, Trillsch F, Herms J, Thon N, Nelson PJ, Tonn JC, Goldbrunner R. Expression of VEGFR3 in glioma endothelium correlates with tumor grade. J Neurooncol 2007;82:141–50.

[105] Dvorak HF. Vascular permeability factor/vascular endothelial growth factor: a critical cytokine in tumor angiogenesis and a potential target for diagnosis and therapy. J Clin Oncol 2002;20:4368–80.

[106] Papapetropoulos A, García-Cardeña G, Madri JA, Sessa WC. Nitric oxide production contributes to the angiogenic properties of vascular endothelial growth factor in human endothelial cells. J Clin Invest 1997;100:3131–9.

[107] Rajantie I, Ilmonen M, Alminaite A, Ozerdem U, Alitalo K, Salven P. Adult bone marrow-derived cells recruited during angiogenesis comprise precursors for periendothelial vascular mural cells. Blood 2004;104:2084–6.

[108] Asahara T, Murohara T, Sullivan A, Silver M, van der Zee R, Li T, Witzenbichler B, Schatteman G, Isner JM. Isolation of putative progenitor endothelial cells for angiogenesis. Science 1997;275:964–7.

[109] Gleiser C, Wagner A, Fallier-Becker P, Wolburg H, Hirt B, Mack AF. Aquaporin-4 in astroglial cells in the CNS and supporting cells of sensory organs—a comparative perspective. Int J Mol Sci 2016;17:1411.

[110] McComb JG. Recent research into the nature of cerebrospinal fluid formation and absorption. J Neurosurg 1983;59:369–83.

[111] Boulton M, Armstrong D, Flessner M, Hay J, Szalai JP, Johnston M. Raised intracranial pressure increases CSF drainage through arachnoid villi and extracranial lymphatics. Am J Physiol 1998;275:R889–96.

[112] Ma Q, Schlegel F, Bachmann SB, Schneider H, Decker Y, Rudin M, Weller M, Proulx ST, Detmar M. Lymphatic outflow of cerebrospinal fluid is reduced in glioma. Sci Rep 2019;9:14815.

[113] Jenny B, Harrison JA, Baetens D, Tille JC, Burkhardt K, Mottaz H, Kiss JZ, Dietrich PY, De Tribolet N, Pizzolato GP, Pepper MS. Expression and localization of VEGF-C and VEGFR-3 in glioblastomas and haemangioblastomas. J Pathol 2006;209:34–43.

[114] Song E, Mao T, Dong H, Boisserand LSB, Antila S, Bosenberg M, Alitalo K, Thomas J-L, Iwasaki A. VEGF-C-driven lymphatic drainage enables immunosurveillance of brain tumours. Nature 2020;577:689–94.

[115] Nair S, McLaughlin C, Weizer A, Su Z, Boczkowski D, Dannull J, Vieweg J, Gilboa E, Nair S, McLaughlin C, Weizer A, Su Z, Boczkowski D, Dannull J, Vieweg J, Gilboa E. Injection of immature dendritic cells into adjuvant-treated skin obviates the need for ex vivo maturation. J Immunol 2003;171:6275–82.

[116] Prins RM, Craft N, Bruhn KW, Khan-Farooqi H, Koya RC, Stripecke R, Miller JF, Liau LM. The TLR-7 agonist, imiquimod, enhances dendritic cell survival and promotes tumor antigen-specific T cell priming: relation to central nervous system antitumor immunity. J Immunol 2006;176:157–64.

[117] De Vries I, Krooshoop D, Scharenborg N, Lesterhuis W, Diepstra J, Van Muijen G, Strijk S, Ruers T, Boerman O, Oyen W, Adema G, Punt C, Figdor CG. Effective migration of antigen-pulsed dendritic cells to lymph nodes in melanoma patients is determined by their maturation state. Cancer Res 2003;63:7–12.

[118] Eggert A, Schreurs M, Boerman O, Oyen W, de Boer A, Punt C, Figdor C, Adema G. Biodistribution and vaccine efficiency of murine dendritic cells are dependent on the route of administration. Cancer Res 1999;59:3340–5.

[119] Lesterhuis WJ, de Vries IJ, Schreibelt G, Lambeck AJ, Aarntzen EH, Jacobs JF, Scharenborg NM, van de Rakt MW, de Boer AJ, Croockewit S, van Rossum MM, Mus R, Oyen WJ, Boerman OC, Lucas S, Adema GJ, Punt CJ, Figdor CG. Route of administration modulates the induction of dendritic cell vaccine-induced antigen-specific-T cells in advanced melanoma patients. Clin Cancer Res 2011;17:5725–35.

[120] Moser B, Wolf M, Walz A, Loetscher P. Chemokines: multiple levels of leukocyte migration control☆. Trends Immunol 2004;25:75–84.

[121] Martin-Fontecha A, Sebastiani S, Hopken UE, Uguccioni M, Lipp M, Lanzavecchia A, Sallusto F. Regulation of dendritic cell migration to the draining lymph node: impact on T lymphocyte traffic and priming. J Exp Med 2003;198:615–21.

[122] Gunn MD, Kyuwa S, Tam C, Kakiuchi T, Matsuzawa A, Williams LT, Nakano H. Mice lacking expression of secondary lymphoid organ chemokine have defects in lymphocyte homing and dendritic cell localization. J Exp Med 1999;189:451–60.

[123] Diacovo TG, Blasius AL, Mak TW, Cella M, Colonna M. Adhesive mechanisms governing interferon-producing cell recruitment into lymph nodes. J Exp Med 2005;202:687–96.
[124] Verdijk P, Aarntzen EH, Lesterhuis WJ, Boullart AC, Kok E, van Rossum MM, Strijk S, Eijckeler F, Bonenkamp JJ, Jacobs JF, Blokx W, Vankrieken JH, Joosten I, Boerman OC, Oyen WJ, Adema G, Punt CJ, Figdor CG, de Vries IJ. Limited amounts of dendritic cells migrate into the T-cell area of lymph nodes but have high immune activating potential in melanoma patients. Clin Cancer Res 2009;15:2531–40.
[125] Scandella E, Men Y, Gillessen S, Förster R, Groettrup M. Prostaglandin E2 is a key factor for CCR7 surface expression and migration of monocyte-derived dendritic cells. Blood 2002;100:1354–61.
[126] Fong L, Brockstedt D, Benike C, Wu L, Engleman EG. Dendritic cells injected via different routes induce immunity in cancer patients. J Immunol 2001;166:4254–9.
[127] Mullins DW, Sheasley SL, Ream RM, Bullock TN, Fu YX, Engelhard VH. Route of immunization with peptide-pulsed dendritic cells controls the distribution of memory and effector T cells in lymphoid tissues and determines the pattern of regional tumor control. J Exp Med 2003;198:1023–34.
[128] Fadul CE, Fisher JL, Hampton TH, Lallana EC, Li Z, Gui J, Szczepiorkowski ZM, Tosteson TD, Rhodes CH, Wishart HA, Lewis LD, Ernstoff MS. Immune response in patients with newly diagnosed glioblastoma multiforme treated with intranodal autologous tumor lysate-dendritic cell vaccination after radiation chemotherapy. J Immunother 2011;34:382–9.
[129] Okada H, Kalinski P, Ueda R, Hoji A, Kohanbash G, Donegan TE, Mintz AH, Engh JA, Bartlett DL, Brown CK, Zeh H, Holtzman MP, Reinhart TA, Whiteside TL, Butterfield LH, Hamilton RL, Potter DM, Pollack IF, Salazar AM, Lieberman FS. Induction of CD8+ T-cell responses against novel glioma-associated antigen peptides and clinical activity by vaccinations with {alpha}-type 1 polarized dendritic cells and polyinosinic-polycytidylic acid stabilized by lysine and carboxymethylcellulose in patients with recurrent malignant glioma. J Clin Oncol 2011;29:330–6.
[130] Eggert AAO, van der Voort R, Torensma R, Moulin V, Boerman OC, Oyen WJG, Punt CJA, Diepstra H, de Boer AJ, Figdor CG, Adema GJ. Analysis of dendritic cell trafficking using EGFP-transgenic mice. Immunol Lett 2003;89:17–24.
[131] Ohlfest JR, Andersen BM, Litterman AJ, Xia J, Pennell CA, Swier LE, Salazar AM, Olin MR. Vaccine injection site matters: qualitative and quantitative defects in CD8 T cells primed as a function of proximity to the tumor in a murine glioma model. J Immunol 2013;190:613–20.
[132] Munn DH, Mellor AL. The tumor-draining lymph node as an immune-privileged site. Immunol Rev 2006;213:146–58.
[133] Bliss J, Maylor R, Stokes K, Murray KS, Ketchum MA, Wolf SF. Interleukin-12 as vaccine adjuvant. Characteristics of primary, recall, and long-term responses. Ann N Y Acad Sci 1996;795:26–35.
[134] Chagnon F, Tanguay S, Ozdal OL, Guan M, Ozen ZZ, Ripeau JS, Chevrette M, Elhilali MM, Thompson-Snipes LA. Potentiation of a dendritic cell vaccine for murine renal cell carcinoma by CpG oligonucleotides. Clin Cancer Res 2005;11:1302–11.
[135] Mitchell DA, Batich KA, Gunn MD, Huang MN, Sanchez-Perez L, Nair SK, Congdon KL, Reap EA, Archer GE, Desjardins A, Friedman AH, Friedman HS, Herndon 2nd JE, Coan A, McLendon RE, Reardon DA, Vredenburgh JJ, Bigner DD, Sampson JH. Tetanus toxoid and CCL3 improve dendritic cell vaccines in mice and glioblastoma patients. Nature 2015;519:366–9.
[136] Batich KA, Mitchell DA, Healy P, Herndon JE, Sampson JH. Once, twice, three times a finding: reproducibility of dendritic cell vaccine trials targeting cytomegalovirus in glioblastoma. Clin Cancer Res 2020;26:5297–303.
[137] Zagzag D, Salnikow K, Chiriboga L, Yee H, Lan L, Ali MA, Garcia R, Demaria S, Newcomb EW. Downregulation of major histocompatibility complex antigens in invading glioma cells: stealth invasion of the brain. Lab Invest 2005;85:328–41.
[138] Fecci PE, Mitchell DA, Whitesides JF, Xie W, Friedman AH, Archer GE, Herndon 2nd JE, Bigner DD, Dranoff G, Sampson JH. Increased regulatory T-cell fraction amidst a diminished CD4 compartment explains cellular immune defects in patients with malignant glioma. Cancer Res 2006;66:3294–302.
[139] Qi L, Yu H, Zhang Y, Zhao D, Lv P, Zhong Y, Xu Y. IL-10 secreted by M2 macrophage promoted tumorigenesis through interaction with JAK2 in glioma. Oncotarget 2016;7:71673–85.
[140] Zhai L, Lauing KL, Chang AL, Dey M, Qian J, Cheng Y, Lesniak MS, Wainwright DA. The role of IDO in brain tumor immunotherapy. J Neurooncol 2015;123:395–403.

[141] Roy LO, Poirier MB, Fortin D. Differential expression and clinical significance of transforming growth factor-beta isoforms in GBM tumors. Int J Mol Sci 2018;19:1113.

[142] Weiss T, Puca E, Silginer M, Hemmerle T, Pazahr S, Bink A, Weller M, Neri D, Roth P. Immunocytokines are a promising immunotherapeutic approach against glioblastoma. Sci Transl Med 2020;12, eabb2311.

[143] Larkin J, Chiarion-Sileni V, Gonzalez R, Grob JJ, Cowey CL, Lao CD, Schadendorf D, Dummer R, Smylie M, Rutkowski P, Ferrucci PF, Hill A, Wagstaff J, Carlino MS, Haanen JB, Maio M, Marquez-Rodas I, McArthur GA, Ascierto PA, Long GV, Callahan MK, Postow MA, Grossmann K, Sznol M, Dreno B, Bastholt L, Yang A, Rollin LM, Horak C, Hodi FS, Wolchok JD. Combined nivolumab and ipilimumab or monotherapy in untreated melanoma. N Engl J Med 2015;373:23–34.

[144] Motzer RJ, Escudier B, McDermott DF, George S, Hammers HJ, Srinivas S, Tykodi SS, Sosman JA, Procopio G, Plimack ER, Castellano D, Choueiri TK, Gurney H, Donskov F, Bono P, Wagstaff J, Gauler TC, Ueda T, Tomita Y, Schutz FA, Kollmannsberger C, Larkin J, Ravaud A, Simon JS, Xu L-A, Waxman IM, Sharma P. Nivolumab versus everolimus in advanced renal-cell carcinoma. N Engl J Med 2015;373:1803–13.

[145] Garon EB, Rizvi NA, Hui R, Leighl N, Balmanoukian AS, Eder JP, Patnaik A, Aggarwal C, Gubens M, Horn L, Carcereny E, Ahn M-J, Felip E, Lee J-S, Hellmann MD, Hamid O, Goldman JW, Soria J-C, Dolled-Filhart M, Rutledge RZ, Zhang J, Lunceford JK, Rangwala R, Lubiniecki GM, Roach C, Emancipator K, Gandhi L. Pembrolizumab for the treatment of non–small-cell lung cancer. N Engl J Med 2015;372:2018–28.

[146] Ferris RL, Blumenschein Jr G, Fayette J, Guigay J, Colevas AD, Licitra L, Harrington K, Kasper S, Vokes EE, Even C, Worden F, Saba NF, Iglesias Docampo LC, Haddad R, Rordorf T, Kiyota N, Tahara M, Monga M, Lynch M, Geese WJ, Kopit J, Shaw JW, Gillison ML. Nivolumab for recurrent squamous-cell carcinoma of the head and neck. N Engl J Med 2016;375:1856–67.

[147] Khasraw M, Reardon DA, Weller M, Sampson JH. PD-1 inhibitors: do they have a future in the treatment of glioblastoma? Clin Cancer Res 2020;26:5287–96.

[148] Filley AC, Henriquez M, Dey M. Recurrent glioma clinical trial, CheckMate-143: the game is not over yet. Oncotarget 2017;8:91779–94.

[149] US National Library of Medicine, ClinicalTrials.gov; 2015.

[150] US National Library of Medicine, ClinicalTrials.gov; 2016.

[151] Garzon-Muvdi T, Theodros D, Luksik AS, Maxwell R, Kim E, Jackson CM, Belcaid Z, Ganguly S, Tyler B, Brem H, Pardoll DM, Lim M. Dendritic cell activation enhances anti-PD-1 mediated immunotherapy against glioblastoma. Oncotarget 2018;9:20681–97.

[152] Zeng J, See AP, Phallen J, Jackson CM, Belcaid Z, Ruzevick J, Durham N, Meyer C, Harris TJ, Albesiano E, Pradilla G, Ford E, Wong J, Hammers HJ, Mathios D, Tyler B, Brem H, Tran PT, Pardoll D, Drake CG, Lim M. -Anti-PD-1 blockade and stereotactic radiation produce long-term survival in mice with intracranial gliomas. Int J Radiat Oncol Biol Phys 2013;86:343–9.

[153] Kim JE, Patel MA, Mangraviti A, Kim ES, Theodros D, Velarde E, Liu A, Sankey EW, Tam A, Xu H, Mathios D, Jackson CM, Harris-Bookman S, Garzon-Muvdi T, Sheu M, Martin AM, Tyler BM, Tran PT, Ye X, Olivi A, Taube JM, Burger PC, Drake CG, Brem H, Pardoll DM, Lim M. Combination therapy with anti-PD-1, anti-TIM-3, and focal radiation results in regression of murine gliomas. Clin Cancer Res 2017;23:124–36.

[154] Cloughesy TF, Mochizuki AY, Orpilla JR, Hugo W, Lee AH, Davidson TB, Wang AC, Ellingson BM, Rytlewski JA, Sanders CM, Kawaguchi ES, Du L, Li G, Yong WH, Gaffey SC, Cohen AL, Mellinghoff IK, Lee EQ, Reardon DA, O'Brien BJ, Butowski NA, Nghiemphu PL, Clarke JL, Arrillaga-Romany IC, Colman H, Kaley TJ, de Groot JF, Liau LM, Wen PY, Prins RM. Neoadjuvant anti-PD-1 immunotherapy promotes a survival benefit with intratumoral and systemic immune responses in recurrent glioblastoma. Nat Med 2019;25:477–86.

[155] Hodges TR, Ott M, Xiu J, Gatalica Z, Swensen J, Zhou S, Huse JT, de Groot J, Li S, Overwijk WW, Spetzler D, Heimberger AB. Mutational burden, immune checkpoint expression, and mismatch repair in glioma: implications for immune checkpoint immunotherapy. Neuro Oncol 2017;19:1047–57.

[156] Bouffet E, Larouche V, Campbell BB, Merico D, de Borja R, Aronson M, Durno C, Krueger J, Cabric V, Ramaswamy V, Zhukova N, Mason G, Farah R, Afzal S, Yalon M, Rechavi G, Magimairajan V, Walsh MF, Constantini S, Dvir R, Elhasid R, Reddy A, Osborn M, Sullivan M, Hansford J, Dodgshun A, Klauber-Demore N, Peterson L, Patel S, Lindhorst S, Atkinson J, Cohen Z, Laframboise R, Dirks P, Taylor M, Malkin D, Albrecht S, Dudley RW, Jabado N, Hawkins CE, Shlien A, Tabori U. Immune checkpoint inhibition for hypermutant glioblastoma multiforme resulting from germline biallelic mismatch repair deficiency. J Clin Oncol 2016;34:2206–11.

[157] Chalmers ZR, Connelly CF, Fabrizio D, Gay L, Ali SM, Ennis R, Schrock A, Campbell B, Shlien A, Chmielecki J, Huang F, He Y, Sun J, Tabori U, Kennedy M, Lieber DS, Roels S, White J, Otto GA, Ross JS, Garraway L, Miller VA, Stephens PJ, Frampton GM. Analysis of 100,000 human cancer genomes reveals the landscape of tumor mutational burden. Genome Med 2017;9:34.
[158] Dirkse A, Golebiewska A, Buder T, Nazarov PV, Muller A, Poovathingal S, Brons NHC, Leite S, Sauvageot N, Sarkisjan D, Seyfrid M, Fritah S, Stieber D, Michelucci A, Hertel F, Herold-Mende C, Azuaje F, Skupin A, Bjerkvig R, Deutsch A, Voss-Böhme A, Niclou SP. Stem cell-associated heterogeneity in Glioblastoma results from intrinsic tumor plasticity shaped by the microenvironment. Nat Commun 2019;10:1787.
[159] Little SE, Popov S, Jury A, Bax DA, Doey L, Al-Sarraj S, Jurgensmeier JM, Jones C. Receptor tyrosine kinase genes amplified in glioblastoma exhibit a mutual exclusivity in variable proportions reflective of individual tumor heterogeneity. Cancer Res 2012;72:1614–20.
[160] Snuderl M, Fazlollahi L, Le LP, Nitta M, Zhelyazkova BH, Davidson CJ, Akhavanfard S, Cahill DP, Aldape KD, Betensky RA, Louis DN, Iafrate AJ. Mosaic amplification of multiple receptor tyrosine kinase genes in glioblastoma. Cancer Cell 2011;20:810–7.
[161] Szerlip NJ, Pedraza A, Chakravarty D, Azim M, McGuire J, Fang Y, Ozawa T, Holland EC, Huse JT, Jhanwar S, Leversha MA, Mikkelsen T, Brennan CW. Intratumoral heterogeneity of receptor tyrosine kinases EGFR and PDGFRA amplification in glioblastoma defines subpopulations with distinct growth factor response. Proc Natl Acad Sci U S A 2012;109:3041–6.
[162] Sottoriva A, Spiteri I, Piccirillo SGM, Touloumis A, Collins VP, Marioni JC, Curtis C, Watts C, Tavaré S. Intratumor heterogeneity in human glioblastoma reflects cancer evolutionary dynamics. Proc Natl Acad Sci U S A 2013;110:4009–14.
[163] Muscat AM, Wong NC, Drummond KJ, Algar EM, Khasraw M, Verhaak R, Field K, Rosenthal MA, Ashley DM. The evolutionary pattern of mutations in glioblastoma reveals therapy-mediated selection. Oncotarget 2018;9:7844–58.
[164] Johnson BE, Mazor T, Hong C, Barnes M, Aihara K, McLean CY, Fouse SD, Yamamoto S, Ueda H, Tatsuno K, Asthana S, Jalbert LE, Nelson SJ, Bollen AW, Gustafson WC, Charron E, Weiss WA, Smirnov IV, Song JS, Olshen AB, Cha S, Zhao Y, Moore RA, Mungall AJ, Jones SJM, Hirst M, Marra MA, Saito N, Aburatani H, Mukasa A, Berger MS, Chang SM, Taylor BS, Costello JF. Mutational analysis reveals the origin and therapy-driven evolution of recurrent glioma. Science 2014;343:189–93.
[165] Touat M, Li YY, Boynton AN, Spurr LF, Iorgulescu JB, Bohrson CL, Cortes-Ciriano I, Birzu C, Geduldig JE, Pelton K, Lim-Fat MJ, Pal S, Ferrer-Luna R, Ramkissoon SH, Dubois F, Bellamy C, Currimjee N, Bonardi J, Qian K, Ho P, Malinowski S, Taquet L, Jones RE, Shetty A, Chow K-H, Sharaf R, Pavlick D, Albacker LA, Younan N, Baldini C, Verreault M, Giry M, Guillerm E, Ammari S, Beuvon F, Mokhtari K, Alentorn A, Dehais C, Houillier C, Laigle-Donadey F, Psimaras D, Lee EQ, Nayak L, McFaline-Figueroa JR, Carpentier A, Cornu P, Capelle L, Mathon B, Barnholtz-Sloan JS, Chakravarti A, Bi WL, Chiocca EA, Fehnel KP, Alexandrescu S, Chi SN, Haas-Kogan D, Batchelor TT, Frampton GM, Alexander BM, Huang RY, Ligon AH, Coulet F, Delattre J-Y, Hoang-Xuan K, Meredith DM, Santagata S, Duval A, Sanson M, Cherniack AD, Wen PY, Reardon DA, Marabelle A, Park PJ, Idbaih A, Beroukhim R, Bandopadhayay P, Bielle F, Ligon KL. Mechanisms and therapeutic implications of hypermutation in gliomas. Nature 2020;580:517–23.
[166] Barthel FP, Johnson KC, Varn FS, Moskalik AD, Tanner G, Kocakavuk E, Anderson KJ, Abiola O, Aldape K, Alfaro KD, Alpar D, Amin SB, Ashley DM, Bandopadhayay P, Barnholtz-Sloan JS, Beroukhim R, Bock C, Brastianos PK, Brat DJ, Brodbelt AR, Bruns AF, Bulsara KR, Chakrabarty A, Chakravarti A, Chuang JH, Claus EB, Cochran EJ, Connelly J, Costello JF, Finocchiaro G, Fletcher MN, French PJ, Gan HK, Gilbert MR, Gould PV, Grimmer MR, Iavarone A, Ismail A, Jenkinson MD, Khasraw M, Kim H, Kouwenhoven MCM, LaViolette PS, Li M, Lichter P, Ligon KL, Lowman AK, Malta TM, Mazor T, McDonald KL, Molinaro AM, Nam D-H, Nayyar N, Ng HK, Ngan CY, Niclou SP, Niers JM, Noushmehr H, Noorbakhsh J, Ormond DR, Park C-K, Poisson LM, Rabadan R, Radlwimmer B, Rao G, Reifenberger G, Sa JK, Schuster M, Shaw BL, Short SC, Smitt PAS, Sloan AE, Smits M, Suzuki H, Tabatabai G, Van Meir EG, Watts C, Weller M, Wesseling P, Westerman BA, Widhalm G, Woehrer A, Yung WKA, Zadeh G, Huse JT, De Groot JF, Stead LF, Verhaak RGW, GLASS Consortium. Longitudinal molecular trajectories of diffuse glioma in adults. Nature 2019;576:112–20.
[167] Meyer M, Reimand J, Lan X, Head R, Zhu X, Kushida M, Bayani J, Pressey JC, Lionel AC, Clarke ID, Cusimano M, Squire JA, Scherer SW, Bernstein M, Woodin MA, Bader GD, Dirks PB. Single cell-derived clonal analysis of human glioblastoma links functional and genomic heterogeneity. Proc Natl Acad Sci U S A 2015;112:851–6.

[168] Sharma P, Allison JP. The future of immune checkpoint therapy. Science 2015;348:56–61.

[169] Havel JJ, Chowell D, Chan TA. The evolving landscape of biomarkers for checkpoint inhibitor immunotherapy. Nat Rev Cancer 2019;19:133–50.

[170] Samstein RM, Lee CH, Shoushtari AN, Hellmann MD, Shen R, Janjigian YY, Barron DA, Zehir A, Jordan EJ, Omuro A, Kaley TJ, Kendall SM, Motzer RJ, Hakimi AA, Voss MH, Russo P, Rosenberg J, Iyer G, Bochner BH, Bajorin DF, Al-Ahmadie HA, Chaft JE, Rudin CM, Riely GJ, Baxi S, Ho AL, Wong RJ, Pfister DG, Wolchok JD, Barker CA, Gutin PH, Brennan CW, Tabar V, Mellinghoff IK, DeAngelis LM, Ariyan CE, Lee N, Tap WD, Gounder MM, D'Angelo SP, Saltz L, Stadler ZK, Scher HI, Baselga J, Razavi P, Klebanoff CA, Yaeger R, Segal NH, Ku GY, DeMatteo RP, Ladanyi M, Rizvi NA, Berger MF, Riaz N, Solit DB, Chan TA, Morris LGT. Tumor mutational load predicts survival after immunotherapy across multiple cancer types. Nat Genet 2019;51:202–6.

[171] Marabelle A, Fakih M, Lopez J, Shah M, Shapira-Frommer R, Nakagawa K, Chung HC, Kindler HL, Lopez-Martin JA, Miller Jr WH, Italiano A, Kao S, Piha-Paul SA, Delord JP, McWilliams RR, Fabrizio DA, Aurora-Garg D, Xu L, Jin F, Norwood K, Bang YJ. Association of tumour mutational burden with outcomes in patients with advanced solid tumours treated with pembrolizumab: prospective biomarker analysis of the multicohort, open-label, phase 2 KEYNOTE-158 study. Lancet Oncol 2020;21:1353–65.

[172] Le DT, Uram JN, Wang H, Bartlett BR, Kemberling H, Eyring AD, Skora AD, Luber BS, Azad NS, Laheru D, Biedrzycki B, Donehower RC, Zaheer A, Fisher GA, Crocenzi TS, Lee JJ, Duffy SM, Goldberg RM, de la Chapelle A, Koshiji M, Bhaijee F, Huebner T, Hruban RH, Wood LD, Cuka N, Pardoll DM, Papadopoulos N, Kinzler KW, Zhou S, Cornish TC, Taube JM, Anders RA, Eshleman JR, Vogelstein B, Diaz LA. PD-1 blockade in tumors with mismatch-repair deficiency. N Engl J Med 2015;372:2509–20.

[173] Marabelle A, Le DT, Ascierto PA, Di Giacomo AM, De Jesus-Acosta A, Delord JP, Geva R, Gottfried M, Penel N, Hansen AR, Piha-Paul SA, Doi T, Gao B, Chung HC, Lopez-Martin J, Bang YJ, Frommer RS, Shah M, Ghori R, Joe AK, Pruitt SK, Diaz Jr LA. Efficacy of pembrolizumab in patients with noncolorectal high microsatellite instability/mismatch repair-deficient cancer: results from the phase II KEYNOTE-158 study. J Clin Oncol 2020;38:1–10.

[174] Schumacher TN, Schreiber RD. Neoantigens in cancer immunotherapy. Science 2015;348:69–74.

[175] Goodman AM, Kato S, Bazhenova L, Patel SP, Frampton GM, Miller V, Stephens PJ, Daniels GA, Kurzrock R. Tumor mutational burden as an independent predictor of response to immunotherapy in diverse cancers. Mol Cancer Ther 2017;16:2598–608.

[176] Hellmann MD, Nathanson T, Rizvi H, Creelan BC, Sanchez-Vega F, Ahuja A, Ni A, Novik JB, Mangarin LMB, Abu-Akeel M, Liu C, Sauter JL, Rekhtman N, Chang E, Callahan MK, Chaft JE, Voss MH, Tenet M, Li XM, Covello K, Renninger A, Vitazka P, Geese WJ, Borghaei H, Rudin CM, Antonia SJ, Swanton C, Hammerbacher J, Merghoub T, McGranahan N, Snyder A, Wolchok JD. Genomic features of response to combination immunotherapy in patients with advanced non-small-cell lung cancer. Cancer Cell 2018;33:843–852.e4.

[177] Zhao J, Chen AX, Gartrell RD, Silverman AM, Aparicio L, Chu T, Bordbar D, Shan D, Samanamud J, Mahajan A, Filip I, Orenbuch R, Goetz M, Yamaguchi JT, Cloney M, Horbinski C, Lukas RV, Raizer J, Rae AI, Yuan J, Canoll P, Bruce JN, Saenger YM, Sims P, Iwamoto FM, Sonabend AM, Rabadan R. Immune and genomic correlates of response to anti-PD-1 immunotherapy in glioblastoma. Nat Med 2019;25:462–9.

[178] Desjardins A, Gromeier M, Herndon 2nd JE, Beaubier N, Bolognesi DP, Friedman AH, Friedman HS, McSherry F, Muscat AM, Nair S, Peters KB, Randazzo D, Sampson JH, Vlahovic G, Harrison WT, McLendon RE, Ashley D, Bigner DD. Recurrent glioblastoma treated with recombinant poliovirus. N Engl J Med 2018;379:150–61.

[179] Gromeier M, Brown MC, Zhang G, Lin X, Chen Y, Wei Z, Beaubier N, Yan H, He Y, Desjardins A, Herndon II JE, Varn FS, Verhaak RG, Zhao J, Bolognesi DP, Friedman AH, Friedman HS, McSherry F, Muscat AM, Lipp ES, Nair SK, Khasraw M, Peters KB, Randazzo D, Sampson JH, McLendon RE, Bigner DD, Ashley DM. Very low tumor mutation burden identifies inflamed recurrent glioblastomas responsive to cancer immunotherapy. Nat Commun 2020;12:352.

[180] Dunn GP, Old LJ, Schreiber RD. The three Es of cancer immunoediting. Annu Rev Immunol 2004;22:329–60.

[181] Marty Pyke R, Thompson WK, Salem RM, Font-Burgada J, Zanetti M, Carter H. Evolutionary pressure against MHC class II binding cancer mutations. Cell 2018;175:416–428.e13.

[182] McGranahan N, Rosenthal R, Hiley CT, Rowan AJ, Watkins TBK, Wilson GA, Birkbak NJ, Veeriah S, Van Loo P, Herrero J, Swanton C. Allele-specific HLA loss and immune escape in lung cancer evolution. Cell 2017;171:1259–1271.e11.

[183] Wang Q, Hu B, Hu X, Kim H, Squatrito M, Scarpace L, deCarvalho AC, Lyu S, Li P, Li Y, Barthel F, Cho HJ, Lin YH, Satani N, Martinez-Ledesma E, Zheng S, Chang E, Sauvé CG, Olar A, Lan ZD, Finocchiaro G, Phillips JJ, Berger MS, Gabrusiewicz KR, Wang G, Eskilsson E, Hu J, Mikkelsen T, DePinho RA, Muller F, Heimberger AB, Sulman EP, Nam DH, Verhaak RGW. Tumor evolution of glioma-intrinsic gene expression subtypes associates with immunological changes in the microenvironment. Cancer Cell 2017;32:42–56.e6.

[184] Rooney MS, Shukla SA, Wu CJ, Getz G, Hacohen N. Molecular and genetic properties of tumors associated with local immune cytolytic activity. Cell 2015;160:48–61.

[185] Rosenthal R, Cadieux EL, Salgado R, Bakir MA, Moore DA, Hiley CT, Lund T, Tanić M, Reading JL, Joshi K, Henry JY, Ghorani E, Wilson GA, Birkbak NJ, Jamal-Hanjani M, Veeriah S, Szallasi Z, Loi S, Hellmann MD, Feber A, Chain B, Herrero J, Quezada SA, Demeulemeester J, Van Loo P, Beck S, McGranahan N, Swanton C, Swanton C, Jamal-Hanjani M, Veeriah S, Czyzewska-Khan J, Johnson D, Laycock J, Rosenthal R, Gorman P, Hynds RE, Wilson G, Birkbak NJ, Watkins TBK, McGranahan N, Escudero M, Stewart A, Van Loo P, Rowan A, Hiley C, Abbosh C, Goldman J, Stone RK, Denner T, Ward S, Nye E, Joshi K, Ben Aissa A, Wong YNS, Georgiou A, Quezada S, Hartley JA, Lowe HL, Herrero J, Lawrence D, Hayward M, Panagiotopoulos N, Falzon M, Borg E, Marafioti T, Janes SM, Forster M, Ahmad T, Lee SM, Papadatos-Pastos D, Carnell D, Mendes R, George J, Ahmed A, Taylor M, Choudhary J, Summers Y, Califano R, Taylor P, Shah R, Krysiak P, Rammohan K, Fontaine E, Booton R, Evison M, Crosbie P, Moss S, Joseph L, Bishop P, Quinn AM, Doran H, Leek A, Harrison P, Moore K, Waddington R, Novasio J, Blackhall F, Rogan J, Smith E, Dive C, Tugwood J, Brady G, Rothwell DG, Pierce J, Gulati S, Naidu B, Langman G, Trotter S, Bancroft H, Kerr A, Kadiri S, Middleton G, Djearaman M, Fennell D, Shaw JA, Le Quesne J, Moore DA, Nakas A, Rathinam S, Monteiro W, Marshall H, Nelson L, Riley J, Primrose L, Martinson L, Anand G, Khan S, Nicolson M, Kerr K, Palmer S, Remmen H, Miller J, Buchan K, Chetty M, Gomersall L, Lester J, Morgan F, Adams H, Davies H, Kornaszewska M, Attanoos R, Lock S, MacKenzie M, Wilcox M, Bell H, Hackshaw A, Ngai Y, Smith S, Gower N, Ottensmeier C, Chee S, Johnson B, Alzetani A, Shaw E, Lim E, De Sousa P, Barbosa MT, Bowman A, Jordan S, Rice A, Raubenheimer H, Bhayani H, Hamilton M, Mensah N, Ambrose L, Devaraj A, Chavan H, Nicholson AG, Lau K, Sheaff M, Schmid P, Conibear J, Ezhil V, Prakash V, Russell P, Light T, Horey T, Danson S, Bury J, Edwards J, Hill J, Matthews S, Kitsanta Y, Suvarna K, Fisher P, Shackcloth M, Gosney J, Feeney S, Asante-Siaw J, Ryanna K, Dawson A, Tuffail M, Bajaj A, Brozik J, Walter H, Carey N, Price G, Gilbert K, Webb J, Patel A, Chaturvedi A, Granato F, Baker K, Carter M, Priest L, Krebs MG, Lindsay C, Gomes F, Chemie F, George R, Patrini D, Khiroya R, Shaw P, Skrzypski M, Sunderland MW, Reading James L, Beastall C, Mangal N, Peggs K, Lim E, Al-Bakir M, Navani N, Scarci M, Ensell L, Biswas D, Salgado R, Razaq M, Nicod J, Beck S, Lopez S, Huebner A, Dietzen M, Mourikis T, Adefila-Ideozu T, Begum S, Klein H, Mani A, Carvalho S, Kaniu D, Realingo C, Malima M, Booth S, Lim L, Rao J, Tenconi S, Socci L, Kibutu F, Agyemang M, Young R, Blyth KG, Dick C, Kirk A, Kidd A, The TRACERx Consortium. Neoantigen-directed immune escape in lung cancer evolution. Nature 2019;567:479–85.

[186] US National Library of Medicine, ClinicalTrials.gov; 2018.

[187] Dudley ME, Wunderlich JR, Robbins PF, Yang JC, Hwu P, Schwartzentruber DJ, Topalian SL, Sherry R, Restifo NP, Hubicki AM, Robinson MR, Raffeld M, Duray P, Seipp CA, Rogers-Freezer L, Morton KE, Mavroukakis SA, White DE, Rosenberg SA. Cancer regression and autoimmunity in patients after clonal repopulation with antitumor lymphocytes. Science 2002;298:850–4.

[188] Asavaroengchai W, Kotera Y, Mule JJ. Tumor lysate-pulsed dendritic cells can elicit an effective antitumor immune response during early lymphoid recovery. Proc Natl Acad Sci U S A 2002;99:931–6.

[189] Batich KA, Reap EA, Archer GE, Sanchez-Perez L, Nair SK, Schmittling RJ, Norberg P, Xie W, Herndon II JE, Healy P, McLendon RE, Friedman AH, Friedman HS, Bigner D, Vlahovic G, Mitchell DA, Sampson JH. Long-term survival in glioblastoma with cytomegalovirus pp65-targeted vaccination. Clin Cancer Res 2017;23:1898–909.

[190] Soubéran A, Brustlein S, Gouarné C, Chasson L, Tchoghandjian A, Malissen M, Rougon G. Effects of VEGF blockade on the dynamics of the inflammatory landscape in glioblastoma-bearing mice. J Neuroinflammation 2019;16:191.

[191] Turkowski K, Brandenburg S, Mueller A, Kremenetskaia I, Bungert AD, Blank A, Felsenstein M, Vajkoczy P. VEGF as a modulator of the innate immune response in glioblastoma. Glia 2018;66:161–74.

[192] Rey S, Schito L, Koritzinsky M, Wouters BG. Molecular targeting of hypoxia in radiotherapy. Adv Drug Deliv Rev 2017;109:45–62.

[193] Kim YH, Yoo KC, Cui YH, Uddin N, Lim EJ, Kim MJ, Nam SY, Kim IG, Suh Y, Lee SJ. Radiation promotes malignant progression of glioma cells through HIF-1alpha stabilization. Cancer Lett 2014;354:132–41.
[194] Short SC, Buffa FM, Bourne S, Koritzinsky M, Wouters BG, Bentzen SM. Dose- and time-dependent changes in gene expression in human glioma cells after low radiation doses. Radiat Res 2007;168:199–208.
[195] Kioi M, Vogel H, Schultz G, Hoffman RM, Harsh GR, Brown JM. Inhibition of vasculogenesis, but not angiogenesis, prevents the recurrence of glioblastoma after irradiation in mice. J Clin Invest 2010;120:694–705.
[196] Januel E, Ursu R, Alkhafaji A, Marantidou A, Doridam J, Belin C, Levy-Piedbois C, Carpentier AF. Impact of renin-angiotensin system blockade on clinical outcome in glioblastoma. Eur J Neurol 2015;22:1304–9.
[197] Rizzo AE, Yu JS. Radiation therapy for glioma stem cells. Adv Exp Med Biol 2015;853:85–110.
[198] Cordes N, Meineke V. Cell adhesion-mediated radioresistance (CAM-RR). Extracellular matrix-dependent improvement of cell survival in human tumor and normal cells in vitro. Strahlenther Onkol 2003;179:337–44.
[199] Wild-Bode C, Weller M, Rimner A, Dichgans J, Wick W. Sublethal irradiation promotes migration and invasiveness of glioma cells: implications for radiotherapy of human glioblastoma. Cancer Res 2001;61:2744–50.
[200] Chang JE, Khuntia D, Robins HI, Mehta MP. Radiotherapy and radiosensitizers in the treatment of glioblastoma multiforme. Clin Adv Hematol Oncol 2007;5:894–902. 7-15.
[201] Peters KB, Kirkpatrick JP, Batinic-Haberle I, Affronti ML, Woodring S, Iden D, Lipp ES, Boyd K, Healy P, Herndon J, Spasojevic I, Penchev S, Gad S, Silberstein D, Johnson MO, Randazzo D, Desjardins A, Friedman HS, Ashley DM, Crapo J. First in human clinical trial of a metalloporphyrin dual radioprotectant and radiosensitizer, BMX-001, in newly diagnosed high-grade glioma undergoing concurrent chemoradiation. Int J Radiat Oncol Biol Phys 2019;105:E106.
[202] Gandalovičová A, Rosel D, Fernandes M, Veselý P, Heneberg P, Čermák V, Petruželka L, Kumar S, Sanz-Moreno V, Brábek J. Migrastatics-anti-metastatic and anti-invasion drugs: promises and challenges. Trends Cancer 2017;3:391–406.
[203] Chiocca EA, Nassiri F, Wang J, Peruzzi P, Zadeh G. Viral and other therapies for recurrent glioblastoma: is a 24-month durable response unusual? Neuro Oncol 2019;21:14–25.
[204] Kaufmann JK, Chiocca EA. Glioma virus therapies between bench and bedside. Neuro Oncol 2014;16:334–51.
[205] Green LM. Tumor treating fields substantially improve survival in GBM. OncLive; 2017.
[206] Cihoric N, Tsikkinis A, Minniti G, Lagerwaard FJ, Herrlinger U, Mathier E, Soldatovic I, Jeremic B, Ghadjar P, Elicin O, Lössl K, Aebersold DM, Belka C, Herrmann E, Niyazi M. Current status and perspectives of interventional clinical trials for glioblastoma—analysis of ClinicalTrials.gov. Radiat Oncol 2017;12:1.

CHAPTER

25

Role of microRNA therapy in presensitizing glioblastoma cells to temozolomide treatment

Ramasamy Paulmurugan[a] *and Tarik F. Massoud*[b]

[a]Cellular Pathway Imaging Laboratory (CPIL), Molecular Imaging Program at Stanford (MIPS), Department of Radiology, Stanford University School of Medicine, Palo Alto, CA, United States
[b]Laboratory of Experimental and Molecular Neuroimaging (LEMNI), Molecular Imaging Program at Stanford (MIPS), Department of Radiology, Stanford University School of Medicine, Stanford, CA, United States

Abstract

MicroRNAs are small noncoding RNAs of 18–22 bases that regulate gene expression in cells. MicroRNA-mediated reprogramming approaches have shown effective therapeutic enhancement of cancers to chemo- and radiotherapies. In this chapter, we focus specifically on the use of microRNAs for improving TMZ therapy in glioblastoma.

Abbreviations

ANP32A	acidic nuclear phosphoprotein 32 family member A
BER	base excision repair
BBB	blood-brain barrier
Bcl-2	B-cell lymphoma 2
CA9	carbonic anhydrase 9
CCND1	cyclin D1
CDC25C	cell division cycle 25C
CDK	cyclin dependent kinase
EGFR	epidermal growth factor receptor
FAT4	FAT atypical cadherin 4
GBM	glioblastoma
GSCs	glioma stem cells
HIF3A	hypoxia inducible factor 3 alpha subunit

Glioblastoma Resistance to Chemotherapy: Molecular Mechanisms and Innovative Reversal Strategies
https://doi.org/10.1016/B978-0-12-821567-8.00024-5

HOXD10	homeobox D10
IGFBP3	insulin like growth factor binding protein 3
LRRFIP1	LRR binding FLII interacting protein 1
MGMT	O^6-methylguanine-DNA methyltransferase
MiRNAs	MicroRNAs
MSI1	musashi RNA binding protein 1
PAMAM	poly-amido amine
PDCD4	programmed cell death 4
PLGA	poly(lactic-*co*-glycolic acid)
PLK1	polo like kinase 1
PTEN	phosphatase and tensin homolog
qRT-PCR	real-time quantitative reverse transcription PCR
RECK	reversion inducing cysteine rich protein with Kazal motifs
RhoC	Ras homolog gene family, member C
RISC	RNA induced silencing complex
RNA	ribonucleic acid
RPS6KB1	ribosomal protein S6 kinase B1
SiRNA	small interfering RNA
SMARCA4	SWI/SNF related, matrix associated, actin dependent regulator of chromatin, subfamily A, member 4
SOX9	SRY-box transcription factor 9
SPRY2	sprouty RTK signaling antagonist 2
TMZ	temozolomide
TGF-β	transforming growth factor beta
VEGF	vascular endothelial growth factor
WEE1	wee1 G2 checkpoint kinase

Conflict of interest

No potential conflicts of interest were disclosed.

Introduction

Glioblastoma (GBM) is an aggressive form of astrocytoma with high morbidity and mortality [1]. Despite aggressive treatment strategies including surgical resection, radiotherapy, and adjuvant chemotherapy using temozolomide (TMZ), GBM remains an aggressive malignancy with poor prognosis [2]. Hence, there is a need to develop more effective molecular therapeutic strategies targeting tumor proliferation, invasion, and chemoresistance. Dysregulated expression of several microRNAs (miRNAs) has been associated with oncogenesis of many cancers including GBM, affecting proliferation, apoptosis, invasion, migration, and resistance to therapy. Oncogenic miRNAs (oncomiRs) promote carcinogenesis by targeting cell cycle regulators and proapoptotic genes. Inhibition of such endogenous oncomiRs using sequence-specific antisense oligonucleotides has been shown as a promising molecularly targeted anticancer therapy for GBM. Similarly, tumor suppressor miRNAs are downregulated in cancer. Supplementation of tumor suppressor miRNAs by delivering synthetic miRNA mimics that reprogram cancer cells by blocking endogenous oncogenes and antiapoptotic genes has shown therapeutic improvement in preclinical animal models of human GBM.

TMZ is the main drug currently used for the treatment of GBM in the clinic. TMZ extends survival of patients with GBM, but in most cases, it eventually demonstrates resistance. TMZ is an alkylating prodrug that induces DNA damage by producing O^6-methylguanine. But O^6-methylguanine-DNA methyltransferase (MGMT) is expressed at high level in GBM, which repairs the main cytotoxic damage generated in DNA by TMZ. Hence, TMZ therapy becomes inefficient while creating drug resistance in GBM after initial treatment using TMZ. In addition, mismatch repair and base excision repair (BER) by the endogenous DNA repair mechanism also contribute to TMZ resistance. TMZ treatment also induces self-protective autophagy, a mechanism by which tumor cells develop resistance to TMZ treatment. Emerging evidence demonstrates that a small population of cells expressing markers such as CD133, CD15/SSEA, and CD44 [3–6] has been identified as stem cells, also called GBM stem cells (GSCs), contributing to drug resistance and tumor recurrence owing to their ability for self-renewal and invasion into neighboring tissues. Some molecules present within tumors maintain stem cell properties of GSCs. Other molecules or signaling pathways regulate stemness and influence MGMT activity, making these GSCs attractive therapeutic targets. Treatments targeting these molecules and pathways resulting in suppression of stemness in GSCs can decrease MGMT activity and improve therapeutic outcomes in GBM. Recently, some novel therapeutic strategies, targeted molecules, immunotherapies, and miRNAs have provided new potential treatments for highly resistant GBM cases. In this chapter, we highlight the current status of different resistance mechanisms, novel strategies of miRNA therapeutics for enhancing the effect of TMZ, and emerging therapeutic approaches to eliminate GSCs, all with the aim to produce a successful GBM treatment. We discuss the future directions for basic and clinical research to achieve therapeutic miRNAs as novel molecularly targeted adjuvant therapies against GBM [7].

Role of microRNAs in cells

MiRNAs are endogenously expressed small noncoding RNAs of ~18–22nt in sizes that play an important role in gene regulatory functions in animals and plants by targeting mRNAs through degradation or translational repression. MiRNAs with their unique property of targeting many cellular mRNAs of genes that are involved in various cellular pathways serve as an important therapeutic target for treating different cellular diseases including cancer. Compared to siRNA- and shRNA-mediated gene silencing, miRNAs are biologically compatible since the delivered miRNA mimics can enter the endogenous biogenesis pathways of miRNAs for further processing in delivering functional effect, which is considered more specific. Further, each miRNA can target more than one cellular mRNAs (~80–100 targets) to control multiple steps involved in a cellular pathway or multiple cellular pathways involved in regulating cellular property. Since this field is relatively young, the list of human miRNAs added to the library is constantly increasing, and up to now around 2000 miRNAs have been considered as well-validated candidates with detailed functions reported for their cellular regulatory mechanisms [8].

MiRNAs in eukaryotic cells are mainly encoded by nuclear DNA. They are expressed in cells similar to mRNA transcription. RNA polymerase II enzyme transcribes miRNAs to generate primary miRNA transcripts, these pri-miRNAs fold into stem-loop structures that are cleaved by the microprocessor complex (Drosha and DGCR8) to generate long hairpin miRNA precursors called pre-miRNAs, which are of 60–120 nucleotides in length. These pre-miRNAs are then transported from the nucleus to the cytoplasm by exportin-5 complex where they are further processed by the RNase III enzyme along with dicer, to cleave and produce mature miRNA duplexes. These mature miRNA duplexes undergo unwinding and the guide (antisense) strands are selected by argonaute for integration into the RNA-induced silencing complex (RISC). Within RISC, the antisense strand of the miRNA binds with the complementary mRNA molecule, resulting in degradation and/or translational repression of target genes (Fig. 1) [9, 10].

Regulatory role of miRNAs in cancer

The dysregulated expression of miRNAs has been observed in every cancer. Maintaining cellular homeostasis is crucial to keep cells at their constant metabolic status while maintaining genomic integrity. Any change in regulatory functions that are indirectly controlled by miRNAs can determine the fate of cells. Recent analysis of miRNA profiles in 15 epithelial cancers for their expression status has identified tumor suppressive and oncogenic miRNAs in human cancers. In that study [11], the authors analyzed the cancers comprising 7316 clinical samples from the Cancer Genome Atlas. The association of miRNA expression and target regulation with the phenotypic hallmarks of cancer was assessed. This study utilized penalized regression techniques to integrate transcriptomic, methylation, and mutation data, to find evidence for a complex map of interactions underlying the relationship of miRNA regulation and the hallmarks of cancer. The results highlight a high redundancy for the oncomiR-1 cluster of oncogenic miRNAs, in particular hsa-miR-17-5p. In addition, the study also revealed the presence of extensive miRNA regulation on tumor suppressor genes such as PTEN, FAT4, and CDK12, uncovering an alternative mechanism of repression in the absence of mutation, methylation, or copy number changes (Fig. 2) [11].

While the network of oncomiRs and tumor suppressor miRNAs regulates gene expression, the presence of mutation-assisted hyper-activation of oncogenes along with nonfunctional mutation in tumor suppressor genes further increase in the loss of cellular senescence and result in oncogenesis. It has been observed that more than 50% of DNA coding for miRNAs are located in cancer-associated genomic regions or in fragile sites [12, 13]. This observation suggests that miRNAs may play a more important role in the pathogenesis of wide range of human cancers than originally projected. Overexpression of miRNAs in cancers, such as miR-17–92, may function as oncomiRs and promote cancer development by negatively regulating tumor suppressor genes and/or genes that control cell differentiation and apoptosis. Similarly, underexpressed miRNAs in cancers, such as let-7, function as tumor suppressor genes and may inhibit cancers by regulating oncogenes and/or genes that control cell differentiation and apoptosis. These observations clearly predict that miRNA therapy could be a powerful tool for cancer prevention and therapeutic applications [14]. Moreover, the synthesis and delivery of synthetic miRNA mimics and antisense miRNAs are much simpler compared to the delivery of full-length genes to achieve therapeutic functions.

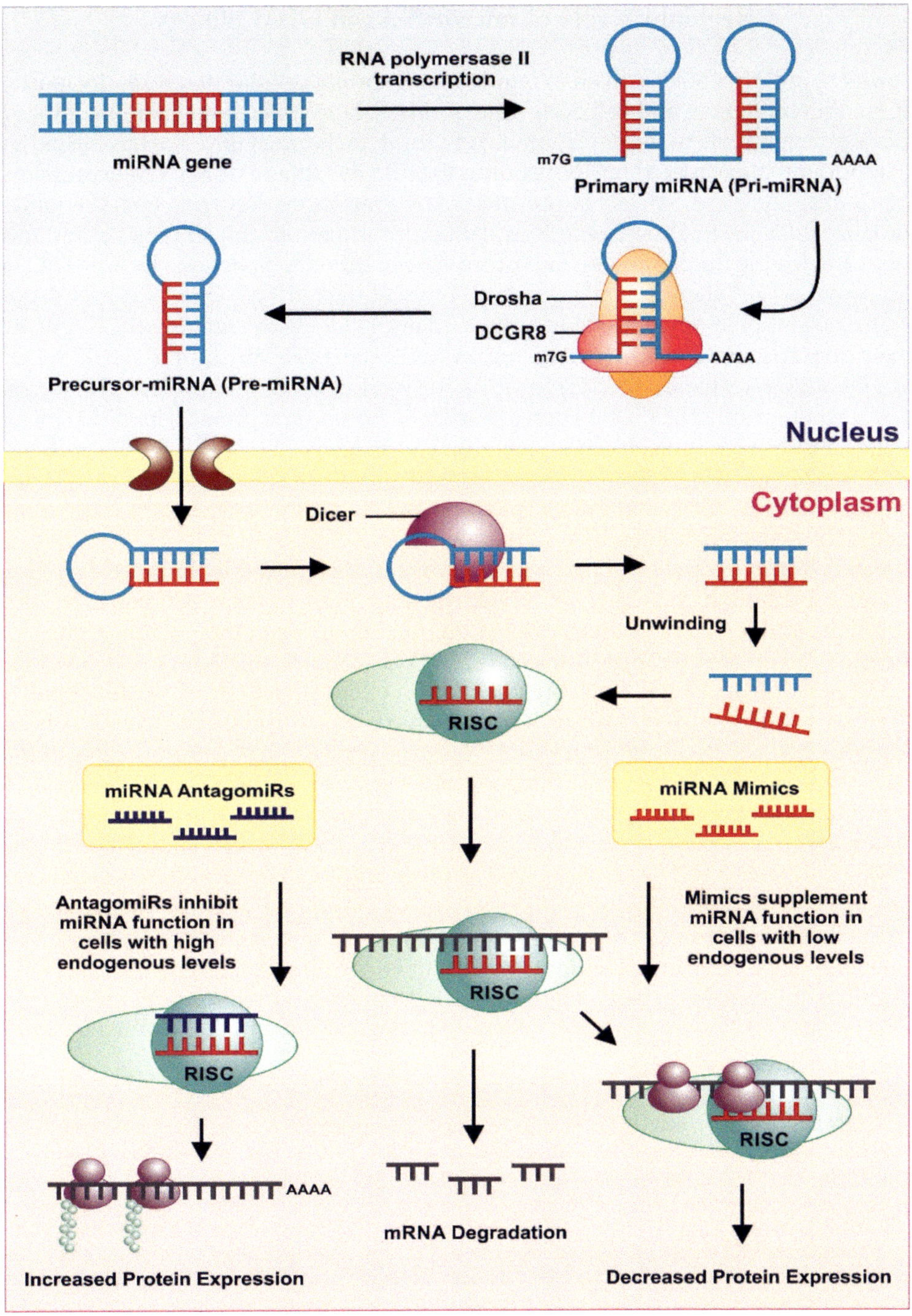

FIG. 1 Biogenesis and function of miRNAs. MiRNA genes are transcribed by RNA polymerase II (Pol II) to form a miRNA precursor called primary miRNA (pri-miRNA). The pri-mRNA is then cleaved by the RNase III Drosha and DiGeorge syndrome critical region 8 (DGCR8), to form a hairpin-shaped precursor miRNA (pre-miRNA) of

(Continued)

Regulatory role of microRNAs in GBM biology

In general, miRNAs are involved in regulation of normal cellular processes, including proliferation, differentiation, and cell cycle regulation [9]. However, dysregulation of these vital processes can contribute to the development of cancer, including GBM. We have listed a number of important dysregulated miRNAs linked to GBM in Table 1 [10, 35]. The expression level of various miRNAs in cancer cells can be affected by aberrations occurring in the genome, such as chromosomal abnormalities, genetic mutations, or mutations in components and machineries that are driving the expression and processing of miRNA biogenesis include DICER and RISC complexes [9]. When comparing normal brain tissue to GBM, about 256 miRNAs of relevance to the pathobiology of GBM have been identified to be overexpressed. A few of these routinely pop-up in the literature, including: miRNA-10b, miR-15b, the miRNA-17–92 cluster, miRNA-21, miRNA-93, miRNA-125b, miRNA-100, miRNA-381, and miRNA-542 [23, 36, 37]. Similarly, 95 other miRNAs have been reported to be underexpressed in GBM, including: miRNA-7, miRNA-29, miRNA-34a, miRNA-101, miRNA-128, miRNA-137, miRNA-138, and miRNA-218 [38–40]. Some important miRNAs that are dysregulated in GBM are also shown in Fig. 3. This list is continually growing as new miRNAs with their roles in various pathways have constantly been discovered.

As stated above, a number of molecular pathways are affected by miRNA dysregulation in cancer, contributing to malignant phenotypes such as evasion of apoptosis, aberrant proliferation, and drug resistance, all owing to the upregulation of oncogenes or the downregulation of tumor suppressor genes [41]. For instance, miRNA-21 is known to have a critical role in GBM, as it has widespread effects on cell death and drug resistance pathways. It is the most common upregulated miRNA in various cancers (in more than 50% of cancers) including GBM [42]. In addition, miRNA-21 overexpression regulates antiapoptotic pathways by targeting p53, TGF-β, and the apoptotic signaling pathways of mitochondria that lead to tumor development, evasion, drug resistance, and metastasis [43]. MiRNA-21 also downregulates the tumor suppressor genes PTEN (phosphatase and tensin homolog), PDCD4 (programmed cell death 4), as well as other targets such as ANP32A (acidic nuclear phosphoprotein 32 family member A), SMARCA4 (SWI/SNF related, matrix associated, actin-dependent regulator of chromatin, subfamily A, member 4), SPRY2 (sprouty RTK signaling antagonist 2), IGFBP3 (insulin-like growth factor-binding protein 3), LRRFIP1 (LRR binding FLII interacting protein 1), and RECK (reversion-inducing cysteine-rich protein with Kazal motifs). Taken together, miRNA-21 has the potential to influence cell cycle

FIG. 1, **CONT'D** approximately 70 nucleotides in length. The pre-miRNA is exported from the nucleus into the cytoplasm. The RNase III enzyme Dicer processes the pre-miRNA in the cytoplasm, resulting in miRNA duplex of 22 nucleotides in length. The miRNA duplex unwinds and then one of the strands is incorporated into the RNA-induced silencing complex (RISC). MiRNAs-RISC complex recognize and bind with the mRNA based on seed complementation (using 2–7 nucleotides of miRNA) and result in either mRNA degradation or translational inhibition. AntagomiRs are antisense oligonucleotides, with nucleotides complementary to the endogenous miRNA. Adding exogenous antagomiRs inhibits the complementary endogenous miRNA from degrading mRNA, resulting in increased protein expression. Adding exogenous miRNA mimics results in supplementing the endogenous miRNA expression, resulting in increased mRNA degradation or translational repression. Abbreviation: m7G: 7-methylguanosine (a modified form of guanosine attached to the 5′ ends of mRNAs) [10].

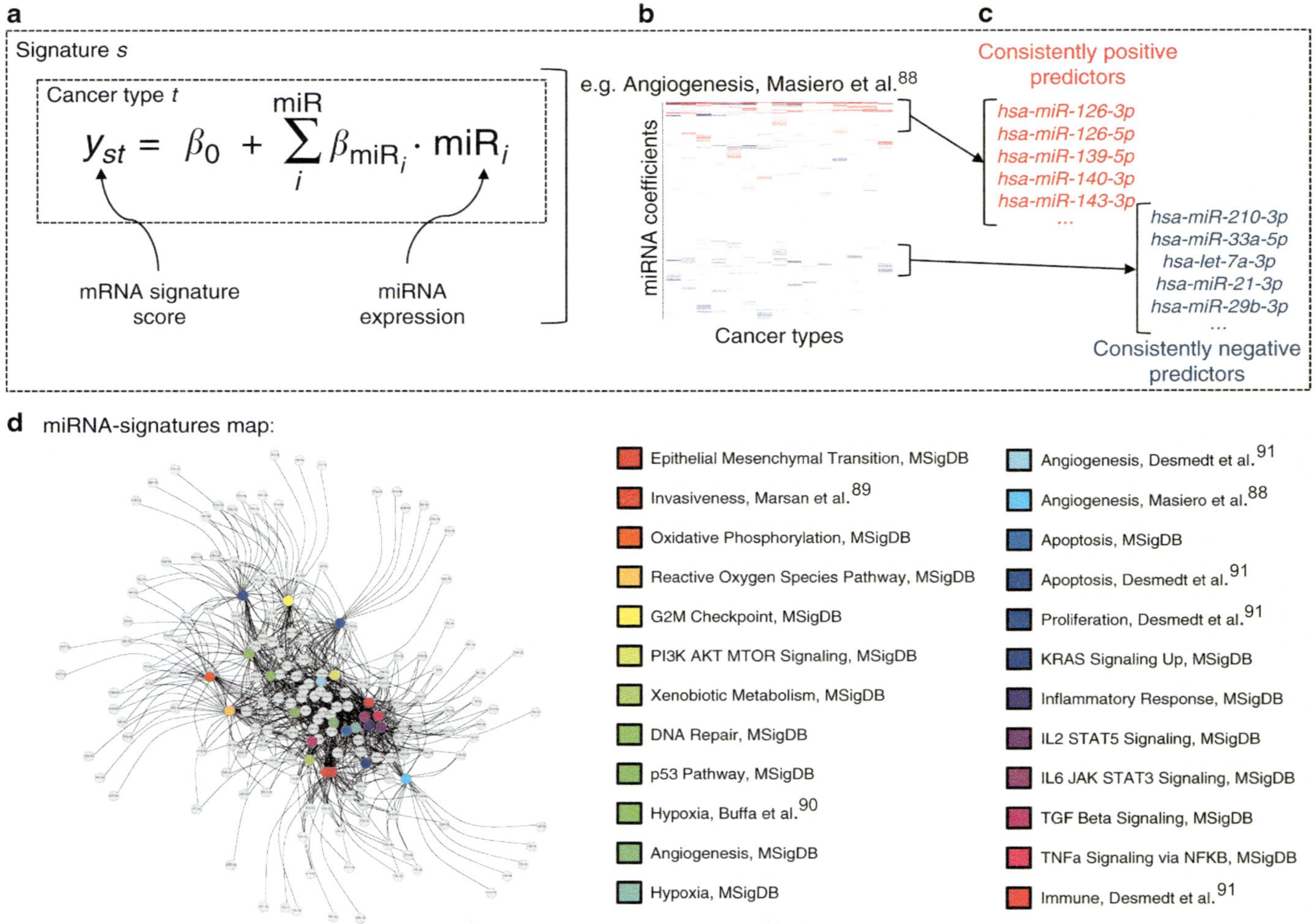

FIG. 2 The penalized regression analysis used to integrate transcriptomic, methylation, and mutation data, to find evidence for a complex map of interactions underlying the relationship of miRNA regulation and the hallmarks of cancer [11]. (A) Overview of the linear model used in the fitting, for each gene signature and cancer type under consideration. (B) Example of a heatmap depicting the values of the coefficients identified for the miRNA predictors *(rows)*, across cancer types *(columns)* for our previously developed angiogenesis signature. (C) Consistently positive and negatively ranking miRNA coefficients, identified as statistically significant by the rank product statistic, are taken as the positive and negative hallmark-associated miRNA for each hallmark signature. (D) Network map of signatures *(colored circles)* and their positively associated miRNA *(gray circles)*, connected by edges when an association was found, highlighting strong interconnectivity between distinct molecular signatures [11].

TABLE 1 Dysregulated expression of small noncoding RNAs (miRNAs) in GBM.

MicroRNAs	Expression status	Targets	Pathway involved
miR-9/miR-9*	Upregulated	CAMTA1, SOX2 [15, 16]	Proliferation; self-renewal
miR-10b	Upregulated	HOXD10 [1, 17, 18]	Invasion
miR-15b	Upregulated	CCNE1 [19]	Proliferation, cell cycle
miR-17	Upregulated	POLD2, CTGF [20, 21]	Angiogenesis, growth, viability, apoptosis, proliferation
miR-18a	Upregulated	Smad4, CTGF [20, 21]	Angiogenesis, growth, apoptosis, viability, proliferation
miR-20a	Upregulated	TGFβ-RII, CTGF [20, 22]	Angiogenesis, tumor growth
miR-21	Upregulated	PTEN, RECK, PDCD4, Caspase, SPRY2, TIMP3 [22, 23]	Invasion, apoptosis, chemoresistance, proliferation, tumor growth
miR-25	Upregulated	Mdm21, TSC1 [24]	Tumor growth
miR-26a/miR-26b	Upregulated	PTEN, IFNB [24]	Apoptosis, proliferation, tumor growth
miR-92	Upregulated	CTGF [24]	Viability, proliferation
miR-93	Upregulated	Integrin-β8 [24]	Angiogenesis, tumor growth, proliferation
miR-125b	Upregulated	Bmf [25]	Invasion, apoptosis, proliferation, migration, tumor volume
miR-221/miR-222	Upregulated	PUMA, P57, P27, Akt, PTPμ [26]	Proliferation, invasion, apoptosis, migration, cell cycle
miR-381	Upregulated	LRRC4 [25]	Proliferation, tumor volume
miR-7	Downregulated	EGFR, IRS1, IRS2 [2, 27]	Proliferation, invasion, migration, viability, radiosensitivity
miR-29b	Downregulated	PDPN [22, 28, 29]	Apoptosis, invasion, proliferation
miR-101	Downregulated	EZH2 [22, 28, 29]	Angiogenesis, invasion, migration, viability
miR-128	Downregulated	EGFR, Bmi-1 [24]	Proliferation, self-renewal, tumor volume, angiogenesis
miR-137	Downregulated	CDK6, Msi1, EZH2 [22, 29, 30]	Proliferation, cell cycle, invasion, migration
miR-138	Downregulated	Msi1 [22]	Proliferation
miR-146b-5p	Downregulated	EGFR [17]	Invasion, migration, proliferation
miR-181a/miR-181b/miR-181d	Downregulated	Bcl-2, K-Ras [24], Cdc42, RhoA and N-cadherin [31]	Apoptosis, radiosensitivity, invasion, proliferation, tumor volume
miR-218	Downregulated	IKK-β [30]	Invasion
miR-326	Downregulated	Notch-1, Notch-2, PKM2 [32]	Apoptosis, invasion, viability, proliferation, tumor growth
miR-195/miR-455/miR-10a	Down/upregulated	Drug resistance [33]	Apoptosis, invasion, viability, proliferation, tumor growth
miR-17-92	Down/upregulated	ant-EGFRvIII [34]	T-cell-mediated cytotoxicity

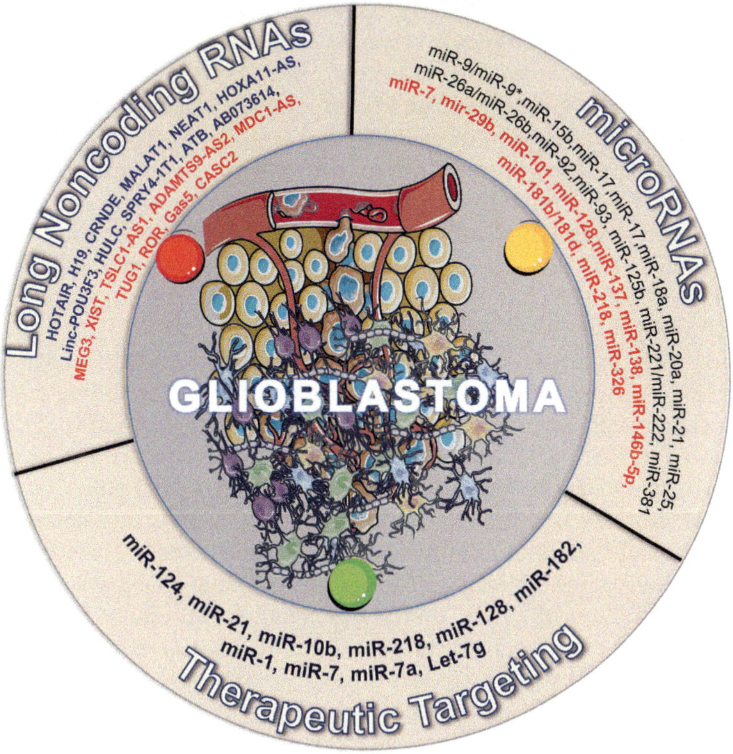

FIG. 3 Regulatory ncRNAs in GBM pathogenesis, and therapeutic targets for GBM. Schematic demonstration of important ncRNAs dysregulated in GBM, and targeting them as therapeutic agents for GBM treatment. The RNAs highlighted in black (upregulated) and red (downregulated) differentiates the expression levels of these RNAs in GBM tissues compared to control tissues. Small ncRNAs (miRNAs) and lncRNAs tested for GBM therapy are listed [10].

progression, invasion, and metastasis [27, 44–50]. In addition, overexpression of miRNA-21 has been found to significantly inhibit the effect of the chemotherapeutic drug TMZ by downregulating the proapoptotic proteins Bax and caspase-3, and upregulating the antiapoptotic Bcl-2 protein [51].

Another miRNA that has been associated with higher-grade gliomas is miRNA-10b. MiR-10b directly suppresses HOXD10, an upstream target of urokinase receptor and RhoC [52, 53]. Overexpression of miRNA-10b causes enhanced growth, invasion, and angiogenesis [53]. Angiogenesis is one of the important hallmarks of tumor growth, which also contributes to hypoxia, poor delivery of therapeutic drugs to the tumors, and motility of tumor cells for metastasis. Highly vascularized tumors have become more proliferative and invasive owing to the enhanced availability of oxygen and nutrients. Further, neovascularization aids in cell migration and metastasis [54]. MiRNAs which are involved in promoting angiogenesis are called angiomiRs [17]. An important miRNA that promotes angiogenesis is miRNA-296, which is upregulated in endothelial cells that express VEGF. Recently, it has been shown that hypoxia was associated with expression of certain miRNAs that led to enhanced angiogenesis. For example, miRNA-210-3p targets HIF3A (hypoxia inducible factor 3 alpha subunit), a negative regulator of hypoxic response, which downregulates VEGF (vascular endothelial growth factor). Thus, overexpression of miRNA-210-3p enhances the expression of HIF, VEGF, and CA9 (carbonic anhydrase 9) and leads to neovascularization [55].

In contrast to the overexpression of miR-210 that contributes to change in VEGFR signaling, downregulation of miRNA-7, an important tumor suppressor miRNA, contributes to oncogenesis in GBM. MiRNA-7 directly targets EGFR and represses the Akt signaling pathway in GBM. Thus, underexpression of miRNA-7 and miR-146 enhances proliferation and invasiveness of tumor cells [28, 56, 57]. Another miRNA that represses the expression of EGFR in tumor cells is miRNA-128 [24]. Hence, oncogenesis and drug resistance enhances when multiple miRNAs target the same cell-signaling pathways.

In addition to EGFR target genes associated with proliferation [WEE1 (wee1 G2 checkpoint kinase) [19] and MSI1 (Musashi RNA-binding protein 1)], RPS6KB1 (ribosomal protein S6 kinase B1) is associated with angiogenesis and also dysregulated in GBM [15]. MiRNA-34a is also downregulated in GBM and targets CDK6 (cyclin-dependent kinase 6) and CCND1 (cyclin D1) genes involved in the proliferative signaling (cyclin/CDK) pathway [34, 58]. Another example is that of miRNA-100, the expression of which is downregulated in GBM. It suppresses proliferation and induces apoptosis and cell cycle arrest in tumor cells by targeting multiple tumor-related genes including PLK1 (polo-like kinase 1). Expression of miRNA-100 downregulates the function of PLK1, thereby increasing expression of p53 and enhancing the radiosensitization of tumor cells. It targets CDC25C (cell division cycle 25C) expression and induces G2/M phase arrest [59]. Even though miR-100 expression is slightly upregulated in low-grade glioma compared to normal brain tissues, its expression is downregulated in GBM.

Microarray-based profiling of miRNA expression in GBM identified miRNA-21 and miRNA-10b as significantly overexpressed oncomiRs. Functional studies have identified miRNA-21 as an oncogene controlling tumor invasion, apoptosis, cell proliferation, and chemoresistance [26, 60–62]. MiRNA-10b is another miRNA highly expressed in metastatic breast cancer, pancreatic adenocarcinomas, and GBM. It is an oncogenic miRNA expressed specifically in glioma tumors but not in normal brain cells. MiRNA-10b inhibition has been shown to compromise cell proliferation, survival, migration, and invasion. Recently, coinhibition of miRNA-21 and miRNA-10b has been shown to induce cell cycle arrest, reduce migration, and enhance apoptosis in glioma cells. The treatment of GBM using a combination of antimiRNA-10b and antimiRNA-21 blocked the function of endogenous miRNA-21 and miRNA-10b in glioma cells and enhanced therapeutic response of glioma cells to TMZ [20].

A significant limitation of GBM therapy is that many chemotherapeutic drugs fail to cross the blood-brain barrier (BBB). Systemic drug toxicity also remains a serious concern and the cause of many constraints when attempting to increase the dosage for most drugs, including the standard anti-GBM drug, TMZ. To minimize unwanted off-target toxicity and to maximize the effect on GBM, a specific strategy that drives GBM biology may provide potentially a more meaningful and molecularly targeted avenue to treat GBM. MiRNAs are regulatory molecules that can serve as a novel molecularly targeted therapeutic agent that can independently or in combination with TMZ or other chemotherapeutic agents can improve clinical GBM therapy in future.

MiRNAs role in therapeutic sensitization of GBM to chemotherapy

TMZ is the main chemotherapy currently used for clinical treatment. As stated above, many other drugs that are effective in inducing cancer cell apoptosis have been avoided owing to dose-limited toxicity when treating GBM patients. The BBB limits the amount of drugs that can be delivered to GBM. Hence, to achieve therapeutic doses of drugs at the target tumor site, it is necessary to deliver a high amount of systemic dose. However, the use of higher doses generates higher levels of toxicity to other organs, and this eliminates many cytotoxic agents from use for GBM therapy. Modulation of miRNAs in GBM has the potential to presensitize cancer cells to chemotherapy. Hence, several other drugs that originally showed high potential in treating cancer but have been stopped owing to dose-limiting toxicity can reemerge as repurposed drugs for GBM therapy by combination with miRNA therapy. Along these lines, several studies are currently underway to reevaluate the potential of many drugs as combination therapies with miRNAs. We have evaluated miRNAs (miRNA-100 and antimiRNA-21) in combination with doxorubicin in GBM cells of different p53 status (p53-null, wild type, and mutant) and found significant levels of enhancement in apoptosis with use of low doses of drugs. Slowly this field is progressing as it explores other drugs that can potentially change the availability of chemotherapies for GBM beyond TMZ.

The role of miRNAs in therapeutic sensitization of GBM to TMZ

The limitations of TMZ include: (1) ideally, it is necessary to use very high doses of the drug for therapy; (2) the mechanism (methylation of nucleoside to produce O^6-methylguanine) by which this drug induces cell death; and (3) the pH-sensitive nature of this prodrug into an active form that results in premature inactivation of the potency of the drug in inducing the metabolic process. All these major downsides of TMZ can be improved by delivering the drug locally to GBMs by taking advantage of recent advancements in nanotechnology, as well as by presensitizing the tumor cells using p53-refolding drugs or miRNA-targeted reprogramming approaches. Here, we specifically focus on explaining miRNA-targeted therapies that are currently under investigation for sensitizing GBM cells to TMZ therapy.

Dysregulated miRNAs in GBM and response to TMZ

The first study on the role of miRNAs in TMZ resistance in GBM was established by artificially creating a GBM model where TMZ-resistant GBM cells (U251) were evaluated for dysregulated miRNAs by comparing with the parental cell line. The study identified that three miRNAs, miRNA-195, miRNA-455-3p, and miRNA-10a(*), were dysregulated in TMZ-resistant U251 cells compared to the parental cells. The study further evaluated suppression of miRNA-455-3p and miRNA-10a functions by using antimiRNAs, and supplementing miRNA-195 function by delivering miRNA-195 mimic, and found moderate improvement in TMZ therapy in resistant cells [63]. In a similar fashion, miRNA-21 overexpression was explored in a series of artificially generated clones of TMZ-resistant U54MG GBM cells and it was found that inhibition of overexpressed miRNA-21 improved cellular apoptosis in TMZ-resistant clones of U54MG cells [22]. Later, miRNA-17-92 was identified as important T cell markers downregulated in patient with GBM. Hence, expression of these miRNAs along with chimeric antigen-expressing T cells improved cytotoxicity in U87 GBM cells stably expressing mutant EGFRvIII (U87-EGFRvIII), and improved survival of mice bearing U87-EGFRvIII tumor xenografts by enhancing tumor sensitivity to TMZ [16].

It is well known that MGMT impacts response of GBM cells to TMZ in patients. MiRNAs targeting MGMT expression and their response to TMZ have been explored in GBM. The two miRNAs (miRNA-221 and miRNA-222) that are directly involved in MGMT regulation are found to be overexpressed in GBM. The suppression of miRNA-221 and miRNA-222 has resulted in improving TMZ therapy, but also showed enhanced DNA damage from some unknown mechanism. Hence, this may result in poor prognosis and induce oncogenesis rather than improving therapy outcome [31]. Similarly, miRNA-181a/b/c/d subunits have been downregulated in GBM, and contribute resistance to TMZ therapy. Supplementation of miRNA-181 subunits has improved TMZ therapy by modulating Rap1B-mediated cytoskeleton remodeling and related molecular (Cdc42, RhoA, and N-cadherin) changes in GBM [64]. In addition to the abovementioned miRNAs, differential expression of miRNA-200a-3p and miRNA-21 has been observed in grade II, III, and IV gliomas. It is well known that miRNA-21 contributes to drug resistance, invasion, and metastasis. The results of this study show that miRNA-200a-3p regulates MGMT expression and promotes tumor responsiveness to TMZ [21].

Synthetic miRNA mimics and antisense miRNAs therapeutically modulate cellular genes and facilitate anticancer effects

Since the expression levels of several miRNAs are dysregulated in cancer and contribute to tumor cell resistance to anticancer therapy, metastasis, and invasion, the use of synthetic miRNA mimics to supplement tumor suppressor miRNAs, and antisense miRNAs to block the function of oncomiRs, have been extensively explored in the literature. The synthetic miRNAs mimics and antisense miRNAs can be prepared using various base-modified oligonucleotides to improve the stability, but delivering these synthetic oligonucleotides to target tumors in vivo is challenging. Hence, various nanotechnology-based delivery approaches have been explored, including the use of liposomes, cationic lipids, chitosan, and polymer nanoparticles. However, owing to toxicity, many of the nanoformulations are not suitable for clinical applications except the polymer-based nanoparticles that we briefly discuss in the following section.

PLGA nanoparticles for biocompatible and clinically feasible delivery of therapeutic miRNAs for GBM therapy

For therapeutic strategies targeting the brain, the presence of the BBB presents a significant challenge. Systemically delivered antisense miRNAs distribute broadly into most tissues except the CNS owing to the presence of BBB [65]. The BBB consists of specialized endothelial cells and astrocyte end feet that line capillaries and prevent the nonregulated delivery of blood-borne chemicals and therapeutics to the brain. Hence, it is important to identify a nontoxic, biocompatible, and clinically translatable delivery approaches to deliver therapeutic drugs and small RNAs to the brain to treat GBM and other diseases. The unique size, shape, and surface properties of nanoparticles make them attractive drug and gene delivery carriers to penetrate the BBB into the brain. Indeed, cationic lipid-based and poly(amido amine) dendrimers (PAMAM)-based nanocarriers have been shown as efficient gene and miRNA delivery vehicles to the brain [66]. However, a major drawback with cationic delivery vehicles is their toxicity, poor miRNA loading efficiency, and insufficient nonsustained delivery of miRNAs at the target site [29, 67, 68]. Compared to most other types of delivery vehicles, PLGA nanocarriers have significantly greater advantages, such as biocompatibility, ability to encapsulate hydrophobic and hydrophilic drugs, sustained release of encapsulated molecules over time, long-term stability of encapsulated molecules, functionalizable external surface for targeting, and FDA approval for their use. Hence, we here focus on studies that used PLGA nanoparticles for efficient intracellular delivery of miRNAs while allowing sustained release of the encapsulated miRNAs. This results in prolonged suppression of endogenous oncomiRs while supplementing apoptotic and cell death pathways in GBM cells. We had previously demonstrated similar physicochemical advantages of PLGA nanoparticles in delivery of miRNAs to GBM [69].

Antisense-microRNA-21 sensitizes GBM to TMZ

Since TMZ remains the standard clinically available chemotherapy for GBM the current clinical success is hindered by its ineffectiveness with high IC_{50} values for TMZ (in the millimolar range), and moreover, the recent studies have shown that GBM is resistant to the cytotoxic effects of TMZ [70, 71]. Hence, we tested antisense-miRNA-21 targeting oncomiR-21 in combination with TMZ for GBM therapy. MicroRNA-21 is an oncomiR overexpressed in GBM, thus controlling important aspects of GBM biology. Recent studies have identified miRNA-21 overexpression in low-grade astrocytomas, anaplastic astrocytomas, and GBMs, offering potential opportunities for improving targeted GBM treatment [43, 72, 73]. Antisense miRNA-mediated knockdown of endogenous miRNA-21 has been reported to impair tumorigenic properties of cancer cells expressing miRNA-21 at high levels [33, 73]. TMZ blocks the function of endogenously overexpressed miRNA-21. The delivery of antimiRNA-21 can reprogram GBM cells and sensitize them to TMZ therapy. Hence, we evaluated this approach for the first time by delivering antimiRNA-21 in GBM cells. We used PLGA nanoparticles carrying antisense miRNA-21 in GBM cells and tested their response to TMZ treatment. PLGA nanoparticles encapsulating antisense miRNA-21 showed effective intracellular delivery and sustained silencing of miRNA-21 function in GBM cells with different phenotypes, such as U87MG (p53-wt), LN229 (p53-mt), and T98G (polyploidy) cells. The delivery of antisense miRNA-21 significantly reduced the number of viable cells, and increased (1.6-fold) cell cycle arrest at G2/M phase upon TMZ treatment in U87MG cells. This delivery of antimiRNA-21 restored the expression of the miRNA-21 target genes PTEN (by 67%) and caspase-3 (by 15%) upon cotreatment with TMZ. This promising PLGA nanoparticle-based platform for antisense miRNA-21 delivery to GBM is an effective cotherapeutic strategy in cell culture [69].

MiRNA-21 is one of the first miRNAs to be described as an oncomiR. It is overexpressed in a wide range of cancers [74], and has been identified as an oncogene controlling apoptosis, cell proliferation, and migration [26, 60–62]. Recently, using high-throughput miRNA profiling, miRNA-21 has been shown to be strongly overexpressed in nearly all analyzed human GBM specimens [33]. Knockdown of miRNA-21 also results in significant reduction in cell viability in GBM cell lines [33]. The exact mechanisms that control these changes were not completely understood previously. However, considering the general chemoresistance of GBM to its standard treatment using TMZ, and the significant role of miRNA-21 in cancer cell biology, the question arises as to whether a potential beneficial therapeutic effect may be possible by adding miRNA-21 knockdown to the therapeutic response of GBM cells to TMZ. Recently, antisense miRNA-assisted miRNA-21 inhibition has been shown to enhance the chemosensitivity of TMZ-resistant GBM cells [22, 75, 76]. Our study of delivering antisense-miRNA-21 also showed that the recovery of apoptotic PTEN and caspase is the major mechanism responsible for the improved therapeutic efficiency of GBM cells to TMZ [69].

Commercially available transfection agents, such as Oligofectamine, and polycationic dendrimer-based delivery vehicles form stable polyamino lipoplexes with sense and antisense miRNAs, to then deliver them across cell membranes. However, such agents are not suitable for in vivo applications owing to degradation by circulating serum nucleases, systemic toxicity, off-target effects, and rapid renal clearance. Despite efforts to improve the plasma stability of antisense miRNAs, therapeutic applications require an effective delivery system to protect the miRNAs from systemic degradation and achieve localized concentration at the target site. PLGA nanoparticles are attractive platforms for encapsulation of antisense miRNAs and their sustained release at the target site. Hence, we used PLGA-based nanoparticle vehicle to encapsulate antisense miR-21 and deliver it in a "Trojan horse" manner to GBM cells in culture. In addition, miRNA-21 is a commonly overexpressed miRNA in many cancers, and is associated with genes that regulate cell proliferation, apoptosis, invasiveness, and drug resistance. This strategy can also be extended to other cancers.

Targeting miRNA-21 in combination with miRNA-10b improves GBM response to TMZ therapy

Even though, silencing the function of oncomiR miRNA-21 improves GBM response to TMZ therapy, targeting the aggressiveness, heterogeneity, metastasis, and the invasive nature of GBM requires multiple cellular targets to be modulated to potentially improve therapeutic outcome. Hence, following the miRNA-21-targeted therapeutic study in GBM cells and its evaluation against TMZ, we further expanded our investigations to a combination therapy targeting miRNA-21 and miRNA-10b, the two important oncomiRs dysregulated in various cancers, including GBM. There are more than 300 miRNAs that are reported as either upregulated or downregulated in GBM, and studies have shown that many of the established GBM cell lines exhibit elevated miRNA-21 and miRNA-10b levels [23, 30, 33, 72].

Microarray-based profiling of miRNA expression in GBMs has identified miRNA-21 and miRNA-10b (miR-10b) as oncomiRs that are significantly overexpressed in GBM [30, 33, 72]. Indeed, miRNA-21 was one of the first oncomiRs studied in various cancers including GBM, controlling tumor invasion, apoptosis, cell proliferation, and chemoresistance [26, 33, 51, 61, 62, 74]. MiRNA-10b is another miRNA highly expressed in metastatic breast cancer, pancreatic adenocarcinomas, and GBM [30, 72, 77, 78]. It is also an oncogenic miRNA that is specifically expressed in glial tumors but not in normal brain cells [30]. MiRNA-10b inhibition has been shown to compromise cell proliferation, survival, migration, and invasion in GBM [30, 32]. Recently, coinhibition of miRNA-21 and miRNA-10b has been shown to induce cell cycle arrest, and reduce migration and apoptosis in GBM cells [25]. The exact mechanisms controlling these changes are not understood fully. Considering the continued abysmal clinical prognosis of GBM, its general chemoresistance to standard treatment using TMZ, and the significant role of miRNAs in cancer cell biology, we considered antisense oligonucleotide-assisted downregulation of miRNA-21 and miRNA-10b function in GBM cells might enhance their therapeutic response to TMZ.

Antisense oligonucleotide-mediated knockdown of either endogenous miRNA-21 or miRNA-10b has been shown to impair oncogenic properties of cancer cells [25, 26, 30, 32, 33, 43, 51, 62, 73]. In agreement with prior studies, we observed a significant expression of miRNA-21 and miRNA-10b in U87 MG, LN229, and T98G GBM cells [20]. Compared to neuronal cells (HCN-2), GBM cells exhibited up to 10-fold increase in endogenous miRNA-21 levels and a nearly 10^{10}-fold increase in miRNA-10b expression. However, we contend that measuring the fold change in the expression of endogenous oncomiRs is only a relative measurement of their quantity because the target miRNAs could be either overexpressed in GBM cells compared to control cells, or not expressed in control cells (in the case of miR-10b, this manifests as a huge-fold value).

As for antisense-miRNA-21 delivery in our previous study, we delivered PLGA nanoparticles carrying antisense miRNA-21 and antisense miRNA-10b to GBM cells. Coinhibition of miRNA-21 and miRNA-10b significantly reduced the number of viable cells by 25%, and increased (2.9-fold) cell cycle arrest at G2/M phase upon TMZ treatment in U87-MG cells. The results of that study support that controlled delivery of nanoparticles and the associated concurrent silencing of miRNA-21 and miRNA-10b prior to TMZ treatment can be an effective molecularly targeted therapeutic strategy for GBM in cell culture. Its application in vivo can be challenging but potentially achievable using custom mixing of different proportions of miRNA-loaded nanoparticles for delivery.

Molecular treatment tailored to intracellular levels of endogenous miRNA targets for enhancing TMZ therapy

The sustained release of miRNAs from PLGA nanoparticles is advantageous for delivering continuous effects. In a recent study, it has been shown that continuous, local, and targeted intratumoral infusion of antimiRNA-10b is an effective therapeutic strategy for malignant brain tumors [79]. For simultaneous silencing of endogenous miRNA-21 and miRNA-10b, which are expressed at significantly different levels in GBM cells, it would be imperative to study the efficacy of nanoparticles to intracellularly deliver and sustainably release sufficient copies of each antisense oligonucleotide to silence the endogenous targets without leaving any free targets that can show some level of oncomiR function. In one of our recent studies, we investigated the percentage uptake of miRNA-542-3p loaded nanoparticles in

U87-MG cells as a function of their incubation concentration using qRT-PCR. Though the percentage of uptake remained low at all concentrations of nanoparticles studied (from estimating the extracellular levels of nanoparticles), the absolute number of our nontargeted nanoparticles internalized by cells increased with increasing concentrations of nanoparticles in the culture medium. Our miRNA loading, release, and cell internalization results showed that the nontargeted PLGA nanoparticles used in our study can still accumulate in sufficient quantities inside GBM cells for subsequent efficient downregulation, but the information provided significant merits because it highlighted the desirability for additional chemical modifications to actively target nanoparticles to the cells of interest to increase even more the ratio of nanoparticles internalized by the cells without the need to use high concentrations of nanoparticles in the incubation medium. Functionalizing the nanoparticles would also be more relevant in the clinical setting, where systemically injected nanoparticles can be targeted to accumulate in the tumor.

Owing to the significantly higher copy number of miRNA-21 molecules (>25,000) compared to miRNA-10b (<1500) in GBM cells, we investigated the concentration of antimiR-21 required to significantly downregulate miRNA-21. We used U87-MG cells, since they express the highest number of miRNA-21 copies per cell (60,000/cell). We thus incubated U87-MG cells with increasing concentrations of antimiR-21 (0–25 pmol) encapsulated in PLGA nanoparticles, having shown previously that PLGA nanoparticles alone do not induce toxicity at the concentrations used for the study (0–20 μg/mL) [80]. We observed a significant downregulation (71%) with 6.25 pmol of antimiR-21, and the level of downregulation was sustained at higher concentrations. Once we identified this effective antimiR-21 concentration, we coincubated U87-MG cells with antimiR-21 plus antimiR-10b loaded PLGA nanoparticles to achieve simultaneous suppression of endogenous miRNA-21 and miRNA-10b. We observed significant downregulation of miRNA-21 (50%; $P < 0.01$) and miRNA-10b (55%; $P < 0.01$) when cells were cotransfected with both antisense miRNAs. It bears repeating that in the absence of targeting, the percentage cellular uptake of our nonfunctionalized PLGA nanoparticles was low, which resulted in the need to incubate cells with higher concentrations of nanoparticles to bring about effective miRNA-21 and miRNA-10b downregulation. With further improvements in the design of our nanoparticle delivery system, especially with surface functional groups targeting cell surface receptors, the cellular uptake of PLGA nanoparticles was significantly improved to achieve even more precise delivery of antisense miRNA molecules tailored to copy numbers of endogenous targets.

MiRNAs targeting GSCs to achieve GBM therapy

Cancer cell resistance to chemotherapy and recurrence have been highly related to stemness of cancer cells, and a small population cancer stem cells present within the tumor. While rapidly proliferating cancer cells respond to chemotherapy, the cancer stem cells and cancer cells with the activated stemness and their quiescent nature have shown resistance to treatment. Hence, therapies targeting stemness is an important focus in successful cancer therapy. MiRNAs targeting GSCs have been considered attractive targets for GBM therapy. Recently, it has been shown that miRNA-451 is regulated in vivo by the AMPK pathway, where AMPK/miRNA-451 loop has the ability to switch between proliferative and migratory pattern of glioma cells behavior. This study postulated that AMPK/miRNA-451 negative reciprocal feedback loop allows GBM cells/GSCs to adapt to the tumor "ecosystem" by metabolic and behavioral flexibility, and that disruption of such a loop reduces invasiveness and diminishes therapy resistance [81]. Similarly, a few other studies have revealed that miRNA-101 was significantly downregulated in human cancers including GBM, and was associated with aggressive tumor cell proliferation and stem cell self-renewal. But the underlying molecular mechanism of the deregulation of miRNA-101 in GBM remained largely unknown. Hence, a study by Liu et al. found that miR-101 could inhibit the proliferation and invasion of glioma cells both in vitro and in vivo by directly targeting SOX9, a transcription factor responsible for self-renewal in cells [25]. The study found that silencing of SOX9 or the overexpression miRNA-101 had similar effects on glioma cell proliferation and invasion. Further evaluations indicated that miRNA-101 altered SOX9 expression by directly targeting of 3′UTR. The findings of this study suggested that miRNA-101 regulates glioma proliferation, migration, and invasion via directly downregulating SOX9 both in vitro and in vivo, and miRNA-101 could be a potential therapeutic target for future glioma treatment [82].

Similarly, it has been previously shown by us and others that miRNA-100 decreases GBM growth by PLK1/p53 signaling to enhance GBM response to TMZ therapy. Recently, it has been found that miRNA-100 supplementation in GBM targets multiple cellular pathways such as SMARCA5 and its downstream target STAT3 (known as mTOR-STAT3-Notch pathway), ErbB3 (Her3), and its downstream AKT and ERK signaling pathways, while nestin and L1CAM tumor-initiating cells (CSCs) in GBM [18]. While several miRNA markers have constantly been reported for targeting CSCs, to our knowledge there are no reports of CSC-targeting miRNAs in combination with TMZ were used for GBM therapy, except our report of miRNA-100-loaded PLGA nanoparticles delivered via the intranasal route and evaluated with TMZ in vivo [83].

MicroRNAs targeting multiple cellular mechanisms as potential future therapies for GBM

The promising regulatory role of miRNAs in cellular pathways and the adoption of synthetic miRNA mimics and antisense miRNAs have opened a new area of molecularly targeted therapy for cancer. Even though several studies have reported the use of independent miRNAs for their therapeutic role in cancer, including GBM, either alone or in combination with chemo- and radiotherapies, only a few studies including ours have explored more than one miRNA in anticancer therapies. Currently, we are working on multiple miRNAs systematically identified based on their roles in multiple cellular pathways, and the cellular pathways important for cancer therapy in combination with chemotherapy (Fig. 4). We also selected miRNAs based on cancer types and the oncogenic process that are important for each cancer type. This selection process also includes CSCs and immune cytokine signaling as our primary focus, since these pathways are important for activating endogenous immune signaling to take care of anticancer immune process by the host immune system. That is a natural process in cancer therapy, where most of the anticancer therapies only help in initiating the therapy, while the host immune system takes over to deliver the outcome as either beneficial or detrimental.

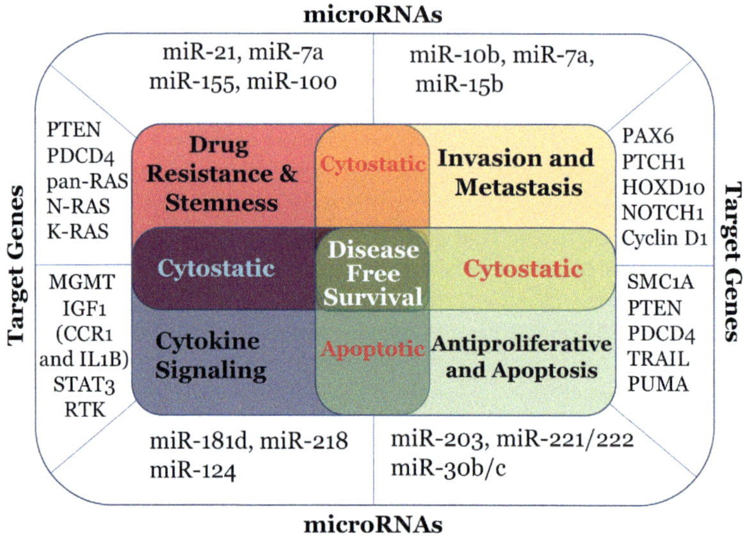

FIG. 4 Hypothetical model showing the selection process for the identification of multiple microRNAs targeting four important cellular mechanisms (cytokine signaling, invasion and metastasis, apoptosis, and stemness) important for enhancing anticancer chemotherapy in cancer.

Conclusion and future perspectives

The last two decades have been considered "the era of noncoding RNA biology" [84]. Recent technological advances in the next generation sequencing have discovered several new noncoding RNAs and their functional roles in various cellular regulatory mechanisms [10]. Since the expression of these RNAs is dysregulated in different cellular diseases, including cancer, they are currently recognized as potential therapeutic targets for cancer. Currently, several additional investigations are underway to find more miRNAs and their roles in various cellular pathways. This process allows identification of miRNAs involved in multiple signaling pathways that can be combined to deliver multiple miRNAs simultaneously targeting many cellular genes to provide cumulative therapeutic effects. This process has the potential to change the future of cancer therapy in the clinic. In addition, the novel development in nanoparticle-based drug and small RNA delivery systems further facilitates miRNA therapeutics feasibility in the clinic. Based on the phrase by Walter Gilbert in 1986, "The RNA World," which was further stressed by the Nobel laureate Sidney Altman in 1989 [85–87], the recent developments in noncoding RNA biology are changing our understanding of cellular regulatory mechanisms originally thought to be controlled by proteins, and now expanded to RNAs. Excitingly, and of great relevance to addressing the multiple daunting challenges posed by GBM, perhaps we are about to witness a resurgence of the phrase "The RNA World" in a completely different but potentially clinically useful guise that highlights the preeminent role of all types of RNAs, biological and therapeutic, in GBM.

References

[1] Wen PY, Kesari S. Malignant gliomas in adults. N Engl J Med 2008;359:492–507.
[2] Stewart LA. Chemotherapy in adult high-grade glioma: a systematic review and meta-analysis of individual patient data from 12 randomised trials. Lancet 2002;359:1011–8.
[3] Singh SK, Hawkins C, Clarke ID, Squire JA, Bayani J, Hide T, et al. Identification of human brain tumour initiating cells. Nature 2004;432:396–401.
[4] Ogden AT, Waziri AE, Lochhead RA, Fusco D, Lopez K, Ellis JA, et al. Identification of A2B5+CD133- tumor-initiating cells in adult human gliomas. Neurosurgery 2008;62:505–14 [discussion 514-505].
[5] Son MJ, Woolard K, Nam DH, Lee J, Fine HA. SSEA-1 is an enrichment marker for tumor-initiating cells in human glioblastoma. Cell Stem Cell 2009;4:440–52.
[6] Anido J, Saez-Borderias A, Gonzalez-Junca A, Rodon L, Folch G, Carmona MA, et al. TGF-beta receptor inhibitors target the CD44(high)/Id1(high) Glioma-initiating cell population in human Glioblastoma. Cancer Cell 2010;18:655–68.
[7] Jiapaer S, Furuta T, Tanaka S, Kitabayashi T, Nakada M. Potential strategies overcoming the temozolomide resistance for glioblastoma. Neurol Med Chir (Tokyo) 2018;58:405–21.
[8] Alles J, Fehlmann T, Fischer U, Backes C, Galata V, Minet M, et al. An estimate of the total number of true human miRNAs. Nucleic Acids Res 2019;47:3353–64.
[9] Shea A, Harish V, Afzal Z, Chijioke J, Kedir H, Dusmatova S, et al. MicroRNAs in glioblastoma multiforme pathogenesis and therapeutics. Cancer Med 2016;5:1917–46.
[10] Paulmurugan R, Malhotra M, Massoud TF. The protean world of non-coding RNAs in glioblastoma. J Mol Med (Berl) 2019;97:909–25.
[11] Dhawan A, Scott JG, Harris AL, Buffa FM. Pan-cancer characterisation of microRNA across cancer hallmarks reveals microRNA-mediated downregulation of tumour suppressors. Nat Commun 2018;9:5228.
[12] Calin GA, Sevignani C, Dumitru CD, Hyslop T, Noch E, Yendamuri S, et al. Human microRNA genes are frequently located at fragile sites and genomic regions involved in cancers. Proc Natl Acad Sci U S A 2004;101:2999–3004.
[13] Drakaki A, Iliopoulos D. MicroRNA gene networks in oncogenesis. Curr Genomics 2009;10:35–41.
[14] Zhang B, Pan X, Cobb GP, Anderson TA. microRNAs as oncogenes and tumor suppressors. Dev Biol 2007;302:1–12.
[15] Shi ZM, Wang J, Yan Z, You YP, Li CY, Qian X, et al. MiR-128 inhibits tumor growth and angiogenesis by targeting p70S6K1. PLoS One 2012;7, e32709.
[16] Ohno M, Ohkuri T, Kosaka A, Tanahashi K, June CH, Natsume A, et al. Expression of miR-17-92 enhances antitumor activity of T-cells transduced with the anti-EGFRvIII chimeric antigen receptor in mice bearing human GBM xenografts. J Immunother Cancer 2013;1:21.
[17] Wang S, Olson EN. AngiomiRs—key regulators of angiogenesis. Curr Opin Genet Dev 2009;19:205–11.
[18] Alrfaei BM, Clark P, Vemuganti R, Kuo JS. MicroRNA miR-100 decreases glioblastoma growth by targeting SMARCA5 and ErbB3 in tumor-initiating cells. Technol Cancer Res Treat 2020;19, 1533033820960748.
[19] Wuchty S, Arjona D, Li A, Kotliarov Y, Walling J, Ahn S, et al. Prediction of associations between microRNAs and gene expression in Glioma biology. PLoS One 2011;6, e14681.
[20] Ananta JS, Paulmurugan R, Massoud TF. Tailored nanoparticle codelivery of antimiR-21 and antimiR-10b augments glioblastoma cell kill by temozolomide: toward a "personalized" anti-microRNA therapy. Mol Pharm 2016;13:3164–75.
[21] Berthois Y, Delfino C, Metellus P, Fina F, Nanni-Metellus I, Al Aswy H, et al. Differential expression of miR200a-3p and miR21 in grade II-III and grade IV gliomas: evidence that miR200a-3p is regulated by O (6)-methylguanine methyltransferase and promotes temozolomide responsiveness. Cancer Biol Ther 2014;15:938–50.
[22] Wong ST, Zhang XQ, Zhuang JT, Chan HL, Li CH, Leung GK. MicroRNA-21 inhibition enhances in vitro chemosensitivity of temozolomide-resistant glioblastoma cells. Anticancer Res 2012;32:2835–41.
[23] Moller HG, Rasmussen AP, Andersen HH, Johnsen KB, Henriksen M, Duroux M. A systematic review of microRNA in glioblastoma multiforme: micro-modulators in the mesenchymal mode of migration and invasion. Mol Neurobiol 2013;47:131–44.
[24] Papagiannakopoulos T, Friedmann-Morvinski D, Neveu P, Dugas JC, Gill RM, Huillard E, et al. Pro-neural miR-128 is a glioma tumor suppressor that targets mitogenic kinases. Oncogene 2012;31:1884–95.

[25] Dong CG, Wu WK, Feng SY, Wang XJ, Shao JF, Qiao J. Co-inhibition of microRNA-10b and microRNA-21 exerts synergistic inhibition on the proliferation and invasion of human glioma cells. Int J Oncol 2012;41:1005–12.
[26] Asangani IA, Rasheed SA, Nikolova DA, Leupold JH, Colburn NH, Post S, et al. MicroRNA-21 (miR-21) post-transcriptionally downregulates tumor suppressor Pdcd4 and stimulates invasion, intravasation and metastasis in colorectal cancer. Oncogene 2008;27:2128–36.
[27] Zhou X, Ren Y, Moore L, Mei M, You Y, Xu P, et al. Downregulation of miR-21 inhibits EGFR pathway and suppresses the growth of human glioblastoma cells independent of PTEN status. Lab Invest 2010;90:144–55.
[28] Katakowski M, Zheng X, Jiang F, Rogers T, Szalad A, Chopp M. MiR-146b-5p suppresses EGFR expression and reduces in vitro migration and invasion of glioma. Cancer Invest 2010;28:1024–30.
[29] Ewert KK, Zidovska A, Ahmad A, Bouxsein NF, Evans HM, McAllister CS, et al. Cationic liposome-nucleic acid complexes for gene delivery and silencing: pathways and mechanisms for plasmid DNA and siRNA. Top Curr Chem 2010;296:191–226.
[30] Gabriely G, Yi M, Narayan RS, Niers JM, Wurdinger T, Imitola J, et al. Human glioma growth is controlled by microRNA-10b. Cancer Res 2011;71:3563–72.
[31] Quintavalle C, Mangani D, Roscigno G, Romano G, Diaz-Lagares A, Iaboni M, et al. MiR-221/222 target the DNA methyltransferase MGMT in glioma cells. PLoS One 2013;8, e74466.
[32] Guessous F, Alvarado-Velez M, Marcinkiewicz L, Zhang Y, Kim J, Heister S, et al. Oncogenic effects of miR-10b in glioblastoma stem cells. J Neurooncol 2013;112:153–63.
[33] Chan JA, Krichevsky AM, Kosik KS. MicroRNA-21 is an antiapoptotic factor in human glioblastoma cells. Cancer Res 2005;65:6029–33.
[34] Sun F, Fu H, Liu Q, Tie Y, Zhu J, Xing R, et al. Downregulation of CCND1 and CDK6 by miR-34a induces cell cycle arrest. FEBS Lett 2008;582:1564–8.
[35] Schraivogel D, Weinmann L, Beier D, Tabatabai G, Eichner A, Zhu JY, et al. CAMTA1 is a novel tumour suppressor regulated by miR-9/9* in glioblastoma stem cells. EMBO J 2011;30:4309–22.
[36] Ivo D'Urso P, Fernando D'Urso O, Damiano Gianfreda C, Mezzolla V, Storelli C, Marsigliante S. miR-15b and miR-21 as circulating biomarkers for diagnosis of glioma. Curr Genomics 2015;16:304–11.
[37] Xia HF, He TZ, Liu CM, Cui Y, Song PP, Jin XH, et al. MiR-125b expression affects the proliferation and apoptosis of human glioma cells by targeting Bmf. Cell Physiol Biochem 2009;23:347–58.
[38] Winter J, Diederichs S. MicroRNA biogenesis and cancer. Methods Mol Biol 2011;676:3–22.
[39] Song L, Huang Q, Chen K, Liu L, Lin C, Dai T, et al. miR-218 inhibits the invasive ability of glioma cells by direct downregulation of IKK-beta. Biochem Biophys Res Commun 2010;402:135–40.
[40] Sha HH, Wang DD, Chen D, Liu SW, Wang Z, Yan DL, et al. MiR-138: a promising therapeutic target for cancer. Tumour Biol 2017;39, 1010428317697575.
[41] Messaoudi K, Clavreul A, Lagarce F. Toward an effective strategy in glioblastoma treatment. Part II: RNA interference as a promising way to sensitize glioblastomas to temozolomide. Drug Discov Today 2015;20:772–9.
[42] Krichevsky AM, Gabriely G. miR-21: a small multi-faceted RNA. J Cell Mol Med 2009;13:39–53.
[43] Papagiannakopoulos T, Shapiro A, Kosik KS. MicroRNA-21 targets a network of key tumor-suppressive pathways in glioblastoma cells. Cancer Res 2008;68:8164–72.
[44] Gaur AB, Holbeck SL, Colburn NH, Israel MA. Downregulation of Pdcd4 by mir-21 facilitates glioblastoma proliferation in vivo. Neuro Oncol 2011;13:580–90.
[45] Yang CH, Yue J, Pfeffer SR, Fan M, Paulus E, Hosni-Ahmed A, et al. MicroRNA-21 promotes glioblastoma tumorigenesis by down-regulating insulin-like growth factor-binding protein-3 (IGFBP3). J Biol Chem 2014;289:25079–87.
[46] Li Y, Li W, Yang Y, Lu Y, He C, Hu G, et al. MicroRNA-21 targets LRRFIP1 and contributes to VM-26 resistance in glioblastoma multiforme. Brain Res 2009;1286:13–8.
[47] Kwak HJ, Kim YJ, Chun KR, Woo YM, Park SJ, Jeong JA, et al. Downregulation of Spry2 by miR-21 triggers malignancy in human gliomas. Oncogene 2011;30:2433–42.
[48] Chen Y, Liu W, Chao T, Zhang Y, Yan X, Gong Y, et al. MicroRNA-21 down-regulates the expression of tumor suppressor PDCD4 in human glioblastoma cell T98G. Cancer Lett 2008;272:197–205.
[49] Abba M, Patil N, Allgayer H. MicroRNAs in the regulation of MMPs and metastasis. Cancers (Basel) 2014;6:625–45.
[50] Zhou X, Zhang J, Jia Q, Ren Y, Wang Y, Shi L, et al. Reduction of miR-21 induces glioma cell apoptosis via activating caspase 9 and 3. Oncol Rep 2010;24:195–201.

[51] Shi L, Chen J, Yang J, Pan T, Zhang S, Wang Z. MiR-21 protected human glioblastoma U87MG cells from chemotherapeutic drug temozolomide induced apoptosis by decreasing Bax/Bcl-2 ratio and caspase-3 activity. Brain Res 2010;1352:255–64.
[52] Sasayama T, Nishihara M, Kondoh T, Hosoda K, Kohmura E. MicroRNA-10b is overexpressed in malignant glioma and associated with tumor invasive factors, uPAR and RhoC. Int J Cancer 2009;125:1407–13.
[53] Sun L, Yan W, Wang Y, Sun G, Luo H, Zhang J, et al. MicroRNA-10b induces glioma cell invasion by modulating MMP-14 and uPAR expression via HOXD10. Brain Res 2011;1389:9–18.
[54] Wurdinger T, Tannous BA. Glioma angiogenesis: towards novel RNA therapeutics. Cell Adh Migr 2009;3:230–5.
[55] Agrawal R, Pandey P, Jha P, Dwivedi V, Sarkar C, Kulshreshtha R. Hypoxic signature of microRNAs in glioblastoma: insights from small RNA deep sequencing. BMC Genomics 2014;15:686.
[56] Kefas B, Godlewski J, Comeau L, Li Y, Abounader R, Hawkinson M, et al. microRNA-7 inhibits the epidermal growth factor receptor and the Akt pathway and is down-regulated in glioblastoma. Cancer Res 2008;68:3566–72.
[57] Liu Z, Liu Y, Li L, Xu Z, Bi B, Wang Y, et al. MiR-7-5p is frequently downregulated in glioblastoma microvasculature and inhibits vascular endothelial cell proliferation by targeting RAF1. Tumour Biol 2014;35:10177–84.
[58] Lukiw WJ, Cui JG, Li YY, Culicchia F. Up-regulation of micro-RNA-221 (miRNA-221; chr Xp11.3) and caspase-3 accompanies down-regulation of the survivin-1 homolog BIRC1 (NAIP) in glioblastoma multiforme (GBM). J Neurooncol 2009;91:27–32.
[59] Li C, Gao Y, Zhang K, Chen J, Han S, Feng B, et al. Multiple roles of MicroRNA-100 in human cancer and its therapeutic potential. Cell Physiol Biochem 2015;37:2143–59.
[60] Meng F, Henson R, Wehbe-Janek H, Ghoshal K, Jacob ST, Patel T. MicroRNA-21 regulates expression of the PTEN tumor suppressor gene in human hepatocellular cancer. Gastroenterology 2007;133:647–58.
[61] Si ML, Zhu S, Wu H, Lu Z, Wu F, Mo YY. miR-21-mediated tumor growth. Oncogene 2007;26:2799–803.
[62] Zhu S, Wu H, Wu F, Nie D, Sheng S, Mo YY. MicroRNA-21 targets tumor suppressor genes in invasion and metastasis. Cell Res 2008;18:350–9.
[63] Ujifuku K, Mitsutake N, Takakura S, Matsuse M, Saenko V, Suzuki K, et al. miR-195, miR-455-3p and miR-10a (*) are implicated in acquired temozolomide resistance in glioblastoma multiforme cells. Cancer Lett 2010;296:241–8.
[64] She X, Yu Z, Cui Y, Lei Q, Wang Z, Xu G, et al. miR-181 subunits enhance the chemosensitivity of temozolomide by Rap1B-mediated cytoskeleton remodeling in glioblastoma cells. Med Oncol 2014;31:892.
[65] Geary RS, Norris D, Yu R, Bennett CF. Pharmacokinetics, biodistribution and cell uptake of antisense oligonucleotides. Adv Drug Deliv Rev 2015;87:46–51.
[66] Zhu Y, Liu C, Pang Z. Dendrimer-based drug delivery systems for brain targeting. Biomolecules 2019;9:790.
[67] Pecot CV, Calin GA, Coleman RL, Lopez-Berestein G, Sood AK. RNA interference in the clinic: challenges and future directions. Nat Rev Cancer 2011;11:59–67.
[68] Safinya CR, Ewert KK, Majzoub RN, Leal C. Cationic liposome-nucleic acid complexes for gene delivery and gene silencing. New J Chem 2014;38:5164–72.
[69] Ananta JS, Paulmurugan R, Massoud TF. Nanoparticle-delivered antisense MicroRNA-21 enhances the effects of temozolomide on glioblastoma cells. Mol Pharm 2015;12:4509–17.
[70] Bocangel DB, Finkelstein S, Schold SC, Bhakat KK, Mitra S, Kokkinakis DM. Multifaceted resistance of gliomas to temozolomide. Clin Cancer Res 2002;8:2725–34.
[71] Gaspar N, Marshall L, Perryman L, Bax DA, Little SE, Viana-Pereira M, et al. MGMT-independent temozolomide resistance in pediatric glioblastoma cells associated with a PI3-kinase–mediated HOX/stem cell gene signature. Cancer Res 2010;70:9243–52.
[72] Ciafre SA, Galardi S, Mangiola A, Ferracin M, Liu CG, Sabatino G, et al. Extensive modulation of a set of microRNAs in primary glioblastoma. Biochem Biophys Res Commun 2005;334:1351–8.
[73] Corsten MF, Miranda R, Kasmieh R, Krichevsky AM, Weissleder R, Shah K. MicroRNA-21 knockdown disrupts glioma growth in vivo and displays synergistic cytotoxicity with neural precursor cell delivered S-TRAIL in human gliomas. Cancer Res 2007;67:8994–9000.
[74] Volinia S, Calin GA, Liu CG, Ambs S, Cimmino A, Petrocca F, et al. A microRNA expression signature of human solid tumors defines cancer gene targets. Proc Natl Acad Sci U S A 2006;103:2257–61.
[75] Qian X, Ren Y, Shi Z, Long L, Pu P, Sheng J, et al. Sequence-dependent synergistic inhibition of human glioma cell lines by combined temozolomide and miR-21 inhibitor gene therapy. Mol Pharm 2012;9:2636–45.

[76] Zhang S, Wan Y, Pan T, Gu X, Qian C, Sun G, et al. MicroRNA-21 inhibitor sensitizes human glioblastoma U251 stem cells to chemotherapeutic drug temozolomide. J Mol Neurosci 2012;47:346–56.
[77] Bloomston M, Frankel WL, Petrocca F, Volinia S, Alder H, Hagan JP, et al. MicroRNA expression patterns to differentiate pancreatic adenocarcinoma from normal pancreas and chronic pancreatitis. JAMA 2007;297:1901–8.
[78] Ma L, Teruya-Feldstein J, Weinberg RA. Tumour invasion and metastasis initiated by microRNA-10b in breast cancer. Nature 2007;449:682–8.
[79] Teplyuk NM, Uhlmann EJ, Gabriely G, Volfovsky N, Wang Y, Teng J, et al. Therapeutic potential of targeting microRNA-10b in established intracranial glioblastoma: first steps toward the clinic. EMBO Mol Med 2016;8:268–87.
[80] Ananta JS, Paulmurugan R, Massoud TF. Temozolomide-loaded PLGA nanoparticles to treat glioblastoma cells: a biophysical and cell culture evaluation. Neurol Res 2016;38:51–9.
[81] Ogawa D, Ansari K, Nowicki MO, Salinska E, Bronisz A, Godlewski J. MicroRNA-451 inhibits migration of Glioblastoma while making it more susceptible to conventional therapy. Noncoding RNA 2019;5:25.
[82] Liu N, Zhang L, Wang Z, Cheng Y, Zhang P, Wang X, et al. MicroRNA-101 inhibits proliferation, migration and invasion of human glioblastoma by targeting SOX9. Oncotarget 2017;8:19244–54.
[83] Sukumar UK, Bose RJC, Malhotra M, Babikir HA, Afjei R, Robinson E, et al. Intranasal delivery of targeted polyfunctional gold-iron oxide nanoparticles loaded with therapeutic microRNAs for combined theranostic multimodality imaging and presensitization of glioblastoma to temozolomide. Biomaterials 2019;218:119342.
[84] Bartoszewski R, Sikorski AF. Editorial focus: entering into the non-coding RNA era. Cell Mol Biol Lett 2018;23:45.
[85] Neveu M, Kim HJ, Benner SA. The "strong" RNA world hypothesis: fifty years old. Astrobiology 2013;13:391–403.
[86] Cech TR. The RNA worlds in context. Cold Spring Harb Perspect Biol 2012;4, a006742.
[87] Shampo MA, Kyle RA, Steensma DP. Sidney Altman—Nobel laureate for work with RNA. Mayo Clin Proc 2012;87, e73.
[88] Masiero M, Simões FC, Han HD, Snell C, Peterkin T, Bridges E, et al. A core human primary tumor angiogenesis signature identifies the endothelial orphan receptor ELTD1 as a key regulator of angiogenesis. Cancer Cell 2013;24:229–41.
[89] Marsan M, Van den Eynden G, Limame R, Neven P, Hauspy J, Van Dam PA, et al. A core invasiveness gene signature reflects epithelial-to-mesenchymal transition but not metastatic potential in breast cancer cell lines and tissue samples. PLoS One 2014;9:e89262.
[90] Buffa FM, Harris AL, West CM, Miller CJ. Large meta-analysis of multiple cancers reveals a common, compact and highly prognostic hypoxia metagene. Br J Cancer 2010;102:428–35.
[91] Desmedt C, Haibe-Kains B, Wirapati P, Buyse M, Larsimont D, Bontempi G, et al. Biological processes associated with breast cancer clinical outcome depend on the molecular subtypes. Clin Cancer Res 2008;14:5158–65.

CHAPTER

26

Drug penetration through the blood–brain barrier after radiotherapy: New approaches to bypass glioblastoma chemoresistance

Iacopo Sardi[a], *Milena Guidi*[a], *Laura Giunti*[b], *Anna Lisa Iorio*[a], *Giacomo Casati*[a], *Maria Luigia Censullo*[a], *and Barbara Rombi*[c]

[a]Neuro-Oncology Unit, Department of Pediatric Oncology, Meyer Children's Hospital, Florence, Italy [b]Medical Genetics Unit, Meyer Children's Hospital, Florence, Italy [c]Trento Proton Therapy Center, Azienda Provinciale per i Servizi Sanitari (APSS), Trento, Italy

Abstract

Glioblastoma (GBM) is one of the most aggressive tumors known. For many years, the most varied therapeutic strategies have been adopted without obtaining significant prognostic results. Intrinsic resistance of GBM is due to the interaction between tumor and microglia cells, especially in a hypoxic microenvironment. Instead, the blood–brain barrier (BBB) plays an essential role of extrinsic resistance to certain drugs preventing their release into the brain. Although in vitro and in vivo studies show that doxorubicin is a highly toxic chemotherapy agent for glioma cells, its effectiveness in the treatment of GBM is limited by efflux pumps that actively expel this agent at the BBB level. Realistically, mechanical damage induced by radiation therapy temporarily modifies the BBB making it more permeable to anthracyclines. Whole brain radiotherapy (WBRT) could be a novel approach to alter the whole BBB allowing effective drugs, such as doxorubicin, to reach diffuse GBM stem cells at therapeutic doses.

Abbreviations

BBB	blood–brain barrier
CNS	central nervous system
DOX	doxorubicin
GBM	glioblastoma multiforme
OS	Overall survival

Glioblastoma Resistance to Chemotherapy: Molecular Mechanisms and Innovative Reversal Strategies
https://doi.org/10.1016/B978-0-12-821567-8.00019-1

PFS Progression-free survival
RTP radiotherapy
TMZ temozolomide
WBRT whole brain radiotherapy

Conflict of interest

No potential conflicts of interest were disclosed.

Introduction

Glioblastoma (GBM) is the commonest and the most lethal primary brain tumor in adults but conversely in children, it accounts for no more than 3%–15% of primary central nervous system (CNS) tumors [1, 2]. Despite the current therapies that associate radiotherapy (RTP) with chemotherapy after surgery prolonging survival, the prognosis is still dismal [2].

In the last decade, there has been enormous progress in developing new targeted therapies for non-CNS solid tumors [3]. Conversely, for brain tumors, strong chemoresistance remains the main factor responsible for the failure to respond to several antineoplastic agents. Ongoing preclinical and clinical studies have, therefore, focused on identifying the diverse mechanisms of chemoresistance in GBM in order to develop new effective treatment strategies [4].

The resistance mechanisms are to be found in molecular and genetic causes, including tumor microenvironment, dysregulation of apoptosis and angiogenesis, overexpression of drug efflux transporter pumps such as P-glycoprotein, unusual behavior of glioblastoma stem cells (GSCs), and high activity of DNA repair system. GBM has the characteristic of creating an immunosuppressive microenvironment by the secretion of cytokines and chemokines laying the foundations for an escape from host immune surveillance and invasion [2, 5]. Intriguingly, knowledge of microenvironment genesis could be very useful for the development of immune-targeted therapies.

Moreover, the endothelial cells of blood–brain barrier (BBB) are crucial in the creation of and excluding the vast majority of cancer therapeutics from normal brain [3]. The BBB is composed of drug efflux ABC transporters that regulate the entry of every molecule and chemical. Therefore, several studies have been aimed at modulating the ABC transporters to bypass drug efflux mechanisms and to test drugs associations that could increase BBB permeability [6]. Many anticancer agents have significantly impaired distribution into normal brain parenchyma. In fact, over the past 30 years many strategies have been planned such as modification of drugs, high dose of chemotherapy, and use of drug carriers [2, 7].

The anthracyclines are currently indicated for several tumors, but what is their role in GBM therapy? Doxorubicin (DOX) is not common used for CNS tumors due to its low BBB penetration; however, new strategies have been investigated, especially pharmacological and mechanical modulation of the BBB [8, 9]. One hypothesis is that temporary destruction of the BBB after irradiation facilitates the entry into nervous tissue of DOX. The demonstration of this BBB modification is represented by the damage, which appears months or years after RTP, sometimes visible as radiation-associated syndrome. Recently, the whole brain radiotherapy (WBRT) is back in vogue again for therapy of diffuse or multifocal GBMs [10].

This chapter provides an overview on how the microenvironment and molecular characteristics operate in GBM chemoresistance, offering the opportunities to implement treatments for GBM patients using novel approaches.

The molecular mechanisms and genetics of chemoresistance in glioblastoma

GBM is the most aggressive brain tumors and, during its tumorigenesis and progression, it develops multiple-drug resistance (MDR) to conventional treatments. It has an innate genetic and cellular heterogeneity and contains cancer stem cell (CSCs), responsible of radiation and chemotherapies resistance and tumor relapse [11–14]. In fact, CSCs show an increased expression of drug efflux ABC transporters (P-gp, breast cancer resistance protein, BCRP1/ABCG2, and MRP1) that extrude xenobiotics and endogenous compounds out of the cell [15], a robust DNA damage response [16–21], a deregulated modulation of transcriptional machineries, and reduced apoptosis.

ABC transporters are membrane transporter glycoproteins and they have a physiological role in the transport of many types of products, included the principal chemotherapeutic agents, and in the maintaining of the isolation of the BBB [22]. High expression of ABC transporters is present in resistant tumors, determining decreased antineoplastic drugs accumulated in cancer cells. Alteration of DNA repair mechanisms [DNA damage response (DDR)] can represent another genetic mechanism involved in MDR. In nontumoral cell, it is implicated in the maintenance of genetic stability but, in cancer cell, it is culpable in the removal of DNA lesions caused by anticancer agents and responsible for the development of MDR and tumor relapse.

Currently, ionizing radiation (IR) and temozolomide (TMZ) are the first-line chemotherapy for GBM's treatment. IR produces a base damage or single-strand breaks (SSBs) repaired by base excision repair (BER) and SSBs repair mechanisms, respectively and double-strand breaks (DSBs) repaired by homologous recombination (HR) and nonhomologous end-joining (NHEJ) mechanisms.

TMZ acts by methylation DNA at N3 position of adenine and at the O6 and N7 positions of guanine [23]. BER repairs N7 and N3 methylguanine instead, O6-methylguanine DNA methyltransferase (MGMT) repairs O6-methylguanine [24, 25]. MGMT promoter methylation is present in GBM [26, 27], and is associated with an increase in TMZ sensitivity, favorable clinical response to TMZ, and consequent prolonged survival [27–29]. The absence of MGMT brings the unrepaired 06-meG to incorrectly pair with thymidine resulting in inappropriate actives the mismatch repair system (MMR). This phenomenon is called the "futile cycle" of MMR cycle that brings to DSBs formation, arrest of the cycle arrest and apoptosis [30]. Deficiency of MMR [31, 32] can produce TMZ resistance of glioma cells. In fact, in cases of recurrent exposure to TMZ, it is possible to stimulate the development of somatic MSH6 mutations with the generation of TMZ-resistant clones [32] that could induce tumor recurrence [33].

Recently, to better understand the acquired mechanisms of drug resistance in GBM, a new research approach has been possible to simulate the cell microenvironment and its in vivo cell behaviors through the development of three-dimensional (3D) culture (3D GBM organoids). These may be used to obtain reliable transcriptomic profiles for the identification of novel molecular targets in clinical treatment of GBM with the aim to development an anti-GBM drug screening [34].

MicroRNAs (miRNAs), short endogenous single-stranded noncoding RNAs molecules, play an important role in drug sensitivity/resistance phenomenon in different cancers included GBM [35–43]. It is known that miRNAs are involved in acquired TMZ resistance phenomenon and that GBM patients and GBM cell lines with acquired TMZ resistance have a typical miRNAs expression profile [44, 45]. MiRNAs are also involved in MDR through increase of the expression level of drug transporter genes in tumors included GBM [46–48], and modulating specific signal pathway(s) that control ABCB1/P-gp-mediated MDR-related genes [49] or pathway(s) implicated in cellular damage repair provoked using cytotoxic therapy or cell proliferation signaling pathways altered in GBM [50–53]. MiRNAs can modulate chemoresistance through their release into exosomes, small extracellular vesicles, and contain DNA, RNAs, miRNAs, growth factors, and enzymes released from tumors. These exosomes play a role in cell–cell communication and are involved in different physiological and pathological processes [54, 55].

Emerging studies show strong evidences on the oncogenic role a new class of RNAs: long noncoding RNA (LncRNAs) (RNAs with length ranges between 200 nt and 1 kb), and circular RNA (circRNAs), an endogenous circular noncoding RNAs. In particular, a LncRNA MALAT1 shows expression level increased in TMZ-resistant GBM cells and it binds miR-101. The ratio between MALAT1 and miR-101 regulates chemoresistance of GBM cells [56].

Alterations in the BBB permeability to increase the action of doxorubicin

A prerequisite for the efficacy of antineoplastic agents is that they reach an efficacious concentration inside the cancerous cells. Unfortunately, in CNS tumors, this phenomenon is hampered by the presence of the BBB, a dynamic interface that separates the brain from the circulatory system.

Different invasive methods have been tested to increase drugs delivery into the brain; these methods include brain microdialysis, intracerebral implantation, and intraventricular delivery [57]. However, these approaches could cause damage to the surrounding healthy tissues and severe side effects in patients. On the other hand, various noninvasive approaches have been explored to improve drug delivery through BBB and to limit side effects [5]. The BBB is composed by a set of endothelial cells with tight junctions and other cells of neurovascular unit, which controls and limits the passage of numerous molecules within the brain, thus representing a major physical and physiological hurdle for CNS drug delivery.

Drug delivery across the BBB is meanly mediated by the overexpression of ATP-binding cassette (ABC) transporters [58]. In all, 48 types of ABC transporter proteins have been currently identified in humans; among them P-glycoprotein (P-gp), breast cancer resistance protein (BCRP), and multidrug resistance-associated proteins (MRPs) are involved in the absorption, distribution, and toxicity profile of many drugs.

Veringa et al. have screened a group of primary GBM cell cultures, for their sensitivity to different drugs. Results have shown that DOX, a classic anticancer agent, displays a higher cytotoxic effect compared to the commonly used TMZ and to other consolidated drugs and new molecules present in the panel [59]. DOX is a historical antineoplastic agent belonging to the chemical class of anthracycline. DOX binds to nucleic acids, presumably by specific intercalation of the planar anthracycline nucleus with the DNA double helix inhibiting topoisomerases II. DOX is currently indicated by the US Food and Drug Administration for numerous oncological diseases as leukemia, lymphoma, neuroblastoma, sarcomas, breast, and ovarian carcinoma and gastric carcinoma. However, DOX is a P-gp substrate, which actively expels anthracyclines at the BBB level, thus limiting its effectiveness in the treatment of CNS tumors.

Pharmacological approaches have been developed to overcome GBM DOX resistance. Literature data concerning the effect of opioids on the BBB permeability have reported a reversible modulation of the BBB using morphine. Combinational administration involving this opioid agent might act as permeability enhancer for chemotherapy agents [60–62]. Morphine is the most used drug for pain management in oncological patients, but data regarding the effect of this drug on tumor cells growth are numerous and controversial [63–67]. On the other hand, in vivo studies have suggested that stress associated with high doses of morphine administration or withdrawal can influence BBB function [60, 68, 69] and allow the accumulation and efficacy of anticancer drugs [58]. These controversial data made it necessary to clarify the effect of morphine treatments on GBM tumor cells growth. In vitro results showed that exposure to micromolar concentrations of morphine did not influence cells growth, indicating that this agent does not represent a proliferative stimulus for the tested GBM cells [6].

Our group has further investigated the possible interaction between the opioid agent and the efflux protein P-gp for better characterizing the molecular mechanism underlying the effect of morphine on BBB permeability. For the first time, our in vitro results demonstrated that morphine is an effective P-gp inhibitor, suggesting that the opioid agent could enhance the effects of chemotherapeutic drugs in brain tissue through reduction of the ATP-binding cassette transporter function [6].

To explore the role of morphine on DOX brain uptake in vivo, adult rats were pretreated with morphine (10 mg/kg, three times over 24 h) before injection of DOX (12 mg/kg) and quantitative analysis of DOX levels in brain tissues was performed using mass spectrometry. The obtained data indicated that morphine pretreatment was able to allow DOX penetration inside all brain areas, including cerebral hemispheres, cerebellum, and brainstem. However, DOX toxicity has to be considered as a potential limitation to its unrestricted use. So, the anthracycline levels were quantified in plasma, heart, and kidney of treated rats. Results indicated that coadministration of morphine did not increase DOX levels in either tissue 1 h after administration [8].

Afterward, antitumor efficacy of DOX plus morphine treatment has been evaluated in an orthotopic GBM xenograft model. Luminescent GBM cells (U87MG-luc) were injected in the left lobe of the brain of immunodeficient mice. After tumor implantation, animals were treated with DOX (5 and 2.5 mg/kg, weekly) with or without morphine pretreatment (10 mg/kg, weekly). Bioluminescence imaging was used to monitor tumor growth and response to therapy. Data demonstrated that 5 mg/kg DOX resulted in a significant tumor regression, while the lower dose (2.5 mg/kg) was not effective. However, if combined with morphine, the group treated with 2.5 mg/kg DOX showed a tumor growth decrease. The average bioluminescence observed in the 2.5 mg/kg DOX plus morphine group was fivefold lower than the group treated using 2.5 mg/kg DOX alone ($P = 0.0053$) and eightfold lower than the vehicle group ($P = 0.0004$).

In addition, an in vitro BBB model (composed by MDCKII cells transfected with human MDR1 gene encoding for P-gp) has been used to explore the role of morphine on DOX uptake. This experiment showed that DOX concentration increased in MDCKII-P-gp transfected cells in the presence of morphine, with a significantly higher intracellular level of the anthracycline comparing control group vs DOX plus morphine group ($P < 0.05$) [70]. It is therefore conceivable that morphine, trough P-gp inhibition, may induce a transient alteration of the permeability of BBB, enhancing the spread of drugs normally unable to cross the BBB, even if administered to a low-dose regimen. These results suggest that DOX, either alone or in combination with morphine, appears to be effective in controlling GBM growth in an orthotopic xenograft mouse model.

It is becoming clear that combination or sequential treatment modalities, which target multiple pathways, may lead to better control of aberrant cell proliferation. Recently, a new pharmacological approach combining DOX and rapamycin, a potent inhibitor of mTOR that is largely involved in GBM proliferation and progression has been explored [71]. In vitro results demonstrated the cytotoxic effect of rapamycin on GBM cell lines (A172, U87MG, and T98G) with different degrees of sensibility. In these cell lines, excluding T98G (a DOX-resistant cell line [72]), DOX effectiveness was appreciable only after 48h of treatment indicating that rapamycin blocks cells proliferation in a short period, while the DOX cytotoxic mechanism requires more time to induce GBM cells death. A stimulating result was obtained in T98G cells, prolonging treatments at 72h. T98G cells remained resistant to DOX but showed a high sensitivity when cotreated with DOX and rapamycin, suggesting that rapamycin probably sensitizes T98G cells to DOX.

Literature data about the interaction between rapamycin and P-gp, the major responsible of drug efflux at BBB level, are currently conflicting. Different studies reported that rapamycin and its analog, tacrolimus, tested at 2.5 μM concentration, were able to modulate Pgp and BCRP activity [73, 74]. On the other hand, in vitro data on fibrosarcoma, leukemia, and immune cells reported that rapamycin did not modulate Pgp, MRP-1, or BCRP at its clinically achievable concentration of 0.25 μM. The same result was obtained also using a rapamycin analog, tacrolimus, at its clinically achievable concentration of 0.08 μM.

Accordingly, Iorio et al. also reported comparable data evaluating the effect of rapamycin on DOX uptake by using an in vitro model of BBB. DOX is a P-gp substrate and, making use of its autofluorescence, they have cytometrically quantified DOX intracellular accumulation in the presence of various doses of rapamycin, in MDCKII P-gp transfected cells. No differences in DOX signal were observed comparing all treated samples vs controls, or vs DOX alone. This result indirectly demonstrated that rapamycin (at tested concentrations) has no effect on DOX-P-gp-mediated transport across the BBB.

These results indicated that the affinity and modulatory effect of rapamycin on the ABC transporter was strongly related to the used dosage of the mTOR inhibitor. The efficacy of the cotreatment rapamycin plus DOX was also tested in a xenograft mouse model of human GBM. A strong efficacy of 5 mg/kg rapamycin treatment was reported on cell proliferation, tumor volume inhibition, and survival [75]. In vivo results have indicated that all treatments had noteworthy effects on tumor volume inhibitions, with different degrees of significance. The most effective approaches resulted rapamycin plus 5 mg/kg DOX and rapamycin plus 2.5 mg/kg DOX. These results were also supported by immunohistological analysis. Nuclear Ki-67 immunopositivity was higher in rapidly growing malignant tumors than in their slowly growing benign counterparts [76–78]. Brain samples were stained with hematoxylin–eosin or Ki67 immunostaining, in order to calculate the proliferation index within the tumor mass. Rapamycin plus 2.5 mg/kg DOX samples presented a proliferation index comparable to rapamycin treatment, while rapamycin plus 5 mg/kg DOX resulted the best treatment with the lowest value of proliferation and a statistically significant difference compared to controls [79].

Intriguingly, these in vitro and in vivo data not only strongly support the effectiveness of DOX in GBM therapy but also underline the need to develop a combinational approach that improves the anthracycline delivery across the BBB, limiting the efflux activity of the ABC transporters, reducing drug dosage and related side effects.

Consolidated and new treatments for glioblastoma

Over the years, GBM has been studied extensively and hundreds of clinical trials have been undertaken. Current strategies of treatment are surgical removal of tumor, chemotherapy, and radiation therapy. It has been shown that a gross total removal seems to increase survival and determine a better response to radio- and chemotherapy. However, given the infiltrating nature of GBM, a complete resection of the tumor is not always possible [80, 81].

Starting in 1960, several phase II randomized controlled trials had been evaluated. Radiation therapy has been shown to be the first postsurgical adjuvant therapy, and in the following years some radiosensitizers such as misonidazole and metronidazole have been investigated without significant benefits. Nitrosureas have played an important role in GBM's history CCNU and BCNU in addition to radiation therapy have been shown to increase survival [82]. A double-blind, randomized phase III trial was conducted to test the efficacy of carmustine (BCNU) wafers (Gliadel) after intraoperative diagnosis of malignant glioma. However, Gliadel plus TMZ and RTP were associated with significant toxicities in approximately 40% of patients including myelosuppression, infections, and neurological deficits. Therefore, this therapy never acquired universal acceptance [83, 84].

In 2005, Stupp and coworkers made a very important contribution to the treatment of newly diagnosed GBM. They described the use of TMZ, given concomitantly to postoperative RTP and in adjuvant phase. TMZ had to be administrated concomitantly with RTP at 75 mg/m^2 daily, followed by 150–200 mg/m^2 for 5 days every 4 weeks for a total of 6 months. This strategy had efficacy similar to the BCNU and less toxicity; the median survival was 14.6 months with RTP + TMZ and 12.1 months with RTP alone [85].

TMZ is an imidazotetrazine derivative of the alkylating agent dacarbazine. Circulating, it gets converted to monomethyl triazone imidazole carboxamide (MTIC) and its activity interferes with DNA replication; with this conformation it can cross BBB.

At the same time, the Swiss group studied the role of promoter methylation of the enzyme O-6-metillguanine-DNA methyltransferase (MGMT) in GBM patients treated with TMZ [27]. This enzyme removes the genetic alteration produced by alkylants, thus protecting the neoplastic cell from the lethal damage induced by these drugs. Based on the induced MGMT protection mechanisms, Wen et al. analyzed its role in TMZ chemoresistance of GBM [86]. A phase III study evaluated the role of bevacizumab in combination with TMZ. The dose used to the treatment group of bevacizumab was 10 mg/kg i.v. for 2 weeks, plus RTP (2 Gy 5 days a week) and oral TMZ (75 mg/m^2/day) for 6 weeks. After 28 days, treatment cycle changed and maintenance dose was administered as follows: bevacizumab (10 mg/kg i.v. every 2 weeks) plus TMZ (150–200 mg/m^2/day for 5 days) was continued for 6-week cycle followed by bevacizumab monotherapy (15 mg/kg i.v. every 3 week). This adjuvant therapy did not improve the survival rate in patients with GBM. The median of progression free survival (PFS) was 12.6 months [87].

The experience that the addition of bevacizumab to RTP plus TMZ in pediatric patients with newly diagnosed nonbrainstem high-grade glioma (HERBY trial) reported similar results not improving event-free survival in group treated with bevacizumab [88].

Antiepileptic drugs have also been studied. Hosein et al. conducted an in vitro study, and through retrospective analyses reported that the valproic acid (VPA) in combination with TMZ and RTP improved the survival rate from 13.96 to 17.35 months. Also, levetiracetam has been evaluated. The experiments were carried out in GBM cells, and there was a greater inhibitory effect than TMZ alone [89].

The role of histone deacetylases (HDACs) was investigated. It could be implicated in the regulation of cell survival, proliferation, tumor cell differentiation, cell cycle arrest, and apoptosis. Galanis E. described the drug vorinostat, a histone deacetylase (HDAC) inhibitor that has radiosensitizing properties. The maximum tolerated dose (MTD; phase I) and efficacy (phase II) of vorinostat combined with standard chemoradiation in newly diagnosed GBM were evaluated. However, the results obtained were not optimal. The MTD was defined at 300 mg/day, vorinostat had acceptable tolerability but the overall survival (OS) at 15 months was 55.1% and median OS was 16.1 months. These data opened the possibility for the future trials [90].

Over the years, many molecules have been used as curcumin, an ingredient of the Indian spice turmeric. In a xenograft mouse models, it has been demonstrated that TMZ and curcumin have synergetic action by enhancing apoptosis. The combination of these two molecules resulted in a more pronounced inhibitory effect blocking of AKT/mTOR signaling [91].

One of the first molecules designed to reduce resistance to TMZ was O6-benzylguanine (O6-BG); it is a potent inactivator of O6-alkylguanine-DNA alkyltransferase (AGT), a single-turnover DNA repair protein that is expressed in all tissues. In the past years, studies on O6-BG maximal tolerated dose and toxicity were considered [92].

Interestingly, resveratrol, a type of natural compound produced by many plants in response to injury, turned out to enhance TMZ cytotoxic effect by ROS-dependent AMPK-TSC mTOR signaling pathway [93]. Shikonin is a component of a Chinese herbal medicine with various biological activities such as apoptosis and necroptosis in human glioma cells. This natural compound and its derivatives revealed strong cytotoxicity EGFR-expressing cell lines [94]. In vivo and vitro studies investigated melatonin because it inhibits cell proliferation [95].

Amplification and/or overexpression of EGFR is a common molecular alteration of classical and neural GBM subtypes (approximately 40% of cases). A phase II study combined erlotinib with RTP and TMZ showing a better survival than historical controls without significant correlation with overexpressed, amplified, or mutated EGFR [96]. Later, a phase III study (OSAG 101-BSA-05) was conducted to evaluate efficacy of nimotuzumab, a monoclonal antibody against EGFR added to the Stupp regimen. This trial failed to show statistically significant advantage in terms of extension of PFS and OS. In addition, EGFR amplification did not correlate with clinical efficacy of nimotuzumab [97].

Due to the presence of IL-13 receptor subunit-α2 (IL-13Rα2) and EGFR variant III (EGFRvIII) on GBM's cells, the study of engineered chimeric antigen receptor (CAR)-modified T cells has started. These cells can be infused into the patient to target whichever surface-exposed tumor antigen is specified by the CAR's extracellular targeting moiety of antigens. The first studies were carried out on IL-13 receptor because it is extremely specific on tumor cells and with limited expression on normal brain cells. With trials termed "IL-13 zetakine" that expected postresection intracavitary infusions of CAR cell, Brown et al. demonstrated antiglioma activity and a rapid increase in necrotic tumor with good tolerance [98]. Good results have been achieved by treating a 50-year-old patient with recurrent multifocal GBM, with a total of 16 intracranial infusions. This patient experienced regression of all intracranial and spinal tumors lasting for 7.5 months [99].

With reference to the EGFR gene, around 40% of all newly diagnosed GBMs carry amplification of the gene, and about 50% of EGFR-amplified GBMs contain constitutively active and oncogenic EGFRvIII; therefore, both vaccine and CAR T-cell therapies against EGFRvIII have been developed. Johnson et al. described the first phase I trial in which 10 patients were treated using infused EGFRvIII-directed CAR T cells, demonstrating a good efficacy and no tumor regression [100].

Currently, the biggest problem with the use of these molecules is the tumor microenvironment and tumor's metabolic landscape. It seems that these can change after using the CAR cell and they can develop mechanisms of adaptive immune resistance in GBM including activation of mitogen-activated protein kinase signaling and glioma stem cell stimulation [101].

The use of vaccines for now has also shown limitations. A trial of phase III that uses rindopepimut (peptide cancer vaccine which targets a mutant protein called EGFRvIII) with TMZ for patients with newly diagnosed, demonstrated not only the same outcome of standard treatment but also a high rate (57%) of EGFRvIII antigen loss in posttreatment tissue [102].

To date, there are several current ongoing trials of CAR T-cell therapies for GBM, however, the challenges are still numerous [101]. Kast and coworkers evaluated targeting multiple active growth pathways with a multidrug oral treatment to achieve cure or long-term growth suppression of GBM. The CAALA regimen was proposed to improve outcomes by enabling a more effective surgery and RTP approach. By combining 5-ALA (5-aminolevulinic acid) with ciprofloxacin, deferiprone, an iron-chelating drug, 5-fluorouracil (5-FU) and febuxostat, a xanthine oxidase inhibitor, the researchers were able to indicate findings of both increasing intraoperative fluorescence identification of malignant tissue, and increasing intraoperative photodynamic treatment effectiveness [103].

Ionizing radiation interferes with BBB permeability by facilitating the passage of drugs

Radiation therapy has the advantage of temporarily destroying the BBB and, therefore, making the action of chemotherapy more effective. On the other hand, this phenomenon may exacerbate radiation-induced brain toxicity, acute edema, and a later cognitive impairment. An acute edema due to a BBB permeability alteration is a constant toxic effect after irradiation. Dexamethasone is a common antiedema drug able to reduce inflammation acting on capillary permeability [104, 105].

Ionizing radiation induces an inflammatory response in the nervous system through the stimulation of astrocytes and microglial cells through inflammatory cytokines, including tumor necrosis factor-alpha (TNFa) [106]. The radiation-induced BBB damage seems to be exacerbated by TNFa action. Consequently, the inhibition of TNFa may be key to abrogate BBB damage and to inhibit astrogliosis. In vivo studies showed that, following irradiation, a significant microglia response may be observed at 24 and 48h. Interestingly, it was significantly inhibited when animals receiving an antibody treatment with antimouse TNF-a mAb [107].

A pilot study on various CNS tumors (6/14 were gliomas) showed that the BBB permeability around neoplastic lesions was approximately 20%. After a total dose of 30Gy irradiation, the permeability by ^{99M}Tc-GH imaging increased to 75%. Interestingly, 8months after irradiation, BBB recovered entirely and permeability levels in normal tissue and in the tumor had returned to pretreatment rate [108].

The radiation-associated cognitive syndrome results from microvascular and parenchymal damage. In vitro and in vivo experiments revealed the endothelial damage might occur months to years after RTP leading to poor perfusion of normal CNS tissue. This response phenomenon could cause the radiation-associated cognitive syndrome through local demyelination [10].

Although the main studies on radiation effects on CNS have been principally limited to higher single-dose irradiation, an important study on 56 GBM patients demonstrated that fractionated radiotherapy (FRT) can open the BBB and improve the effects of CCNU chemotherapy. This group of patients irradiated with 40Gy to the whole brain and boosted at 50–70Gy to the tumor administered at 2Gy per fraction had a significantly a better prognosis compared to patients treated with RTP alone ($P=0.0021$) [109]. Interestingly, animal model studies on the BBB alterations induced by FRT established that BBB permeability did not increase until 90days after RTP, while microglia response and brain microvasculature activity increased significantly, it was already starting at 60days. Moreover, irradiation caused a change in astrocyte morphology and an increase in cell numbers in the cerebral cortex [110, 111].

Multifocal GBM (M-GBM) represents the most deadly subgroup of GBMs. Almost half the M-GBMs belonged to the mesenchymal subtype [112]. Whole brain radiotherapy WBRT should be considered as the clinical target volume in M-GBM such as gliomatosis cerebri for reaching the distant infiltrating disease, invisible to MRI scans. Recently, a monocentric retrospective study evaluated the efficacy and safety of upfront WBRT and TMZ according to the Stupp program in 11M-GBM patients [113]. Despite their response, they had a median OS and PFS of 10 and 5months, respectively, consistent with the findings of the previous reports, a multivariate analysis of 50M-GBMs showed that no significant differences were found in the time-to-progression between 3D conformal RTP and WBRT. The authors showed that focal RTP plus concomitant and adjuvant TMZ seems more effective and less toxic compared to WBRT that always results in a cognitive and quality of life worsening [114]. The experience of the Mayo Clinic College of Medicine on gliomatosis cerebri patients showed that median PFS was longer by 12months (16.5 vs 4.5months, $P<0.01$) and median OS was longer by 21months (27.5 vs 6.5months, $P<0.01$) in patients treated with WBRT. In all, 19 patients (51%) received some form of chemotherapy [115].

Despite these poorly interpretable results due to the low number of cases, there are currently at least three ongoing studies for newly diagnosed GBM using WBRT (NCT01822275, NCT02590263, NCT03746080). Particularly of interest is the use of low-dose WBRT to treat microscopic CNS diffuse GBM cells outside of the focal radiation dose to the tumor mass consisting of 46 Gy and a 2.5-cm margin on T1 enhancement for a 14-Gy boost.

Based on the experimental results on temporary BBB damage after radiotherapy, WBRT seems a valid approach to treat a highly aggressive disease resistant to consolidated therapies, facilitating antineoplastic action of chemotherapy.

Preclinical trials have demonstrated that DOX likely has a good antineoplastic activity against human gliomas. Veringa et al. analyzed various established anticancer agents as well as new agents such as small molecules and monoclonal antibodies against primary GBM cell cultures. They found that DOX and other few drugs resulted in the highest levels of apoptosis [59]. DOX, when delivered locally, had a dose-dependent antitumor activity in a gliosarcoma orthotopic xenograft model. Rats treated with DOX polymers had a better outcome. The median survival for DOX group was 45 days compared with 21 days of the control group ($P < 0.0001$) [116]. After a prolonged exposure to anthracyclines, GBM stem cells displayed a higher apoptotic rate compared to the commonly used TMZ and etoposide [117].

Unfortunately, the therapeutic efficacy of DOX is lower in brain tumors because of its poor penetration through the BBB, mediated by multidrug resistance efflux transporters. In vivo studies in rats treated with DOX after a pharmacological modulation of BBB with a morphine pretreatment had significantly prolonged survival [8]. Our experience with the T98G DOX-resistant GBM cell line revealed an increased apoptosis rate after a prolonged (96 h) exposure to DOX, while shorter times (24–72 h) did not yield any statistical differences compared to controls [40]. A phase II trial investigated the efficacy and toxicity of liposomal DOX combined with TMZ and RTP on 22 recurrent GBMs. The results showed that the treatment was well tolerated and the toxicity was acceptable [118]. Phase II clinical studies have investigated the use of pegylated liposomal DOX combined with TMZ and RTP in patients with newly diagnosed GBM. The combination seems well tolerated without adding significant clinical benefit to TMZ alone [119, 120].

Our experience with using a prolonged exposure to DOX after RTP for a heterogeneous group of high-risk gliomas of pediatric age (ClinicalTrials.gov Identifier: NCT02758366) concluded that a 96-h DOX infusion was well tolerated without any particular mild toxicity. The pilot study provided three courses of prolonged DOX infusion after RTP on three DIPG patients highlighted unacceptable hematological toxicity in these children. Therefore, the study was amended to reduce toxicity using a single cycle of DOX. In agreement with other recent studies, our preliminary results showed that DOX was still well tolerated.

Conclusion

GBM chemoresistance is related to the interaction between tumor and microglia cells, especially in a hypoxic microenvironment. The BBB is critical to protect brain parenchyma, but it also represents the major hurdle for the delivery of drugs agents into the brain. Although TMZ is an established drug in the treatment of GBM, in vitro and in vivo studies show that DOX is a highly toxic chemotherapy agent against glioma cells. Unfortunately, MDR mechanisms, including P-gp, actively expel DOX at the BBB level, thus limiting its effectiveness in the treatment of CNS tumors.

Mechanical damage induced by radiation therapy temporarily modifies the BBB, making it more permeable to anthracyclines. This could allow for better control of the disease locally, but the microscopic CNS diffuse GBM cells outside of the focal radiation dose could be the cause of relapse. An approach to be taken into consideration to alter the whole BBB is WBRT, not only for the direct effect of radiation on all diffuse GBM stem cells but also because drugs such as DOX can effectively target widespread minimal disease.

Acknowledgment

This work was supported by Fondazione Anna Meyer—Firenze.

References

[1] Thomas AA, Brennan CW, DeAngelis LM, Omuro AM. Emerging therapies for glioblastoma. JAMA Neurol 2014;71:1437–44.
[2] Zhang X, Ding K, Wang J, Li X, Zhao P. Chemoresistance caused by the microenvironment of glioblastoma and the corresponding solutions. Biomed Pharmacother 2019;109:39–46.
[3] Sarkaria JN, Hu LS, Parney IF, Pafundi DH, Brinkmann DH, Laack NN, Giannini C, Burns TC, Kizilbash SH, Laramy JK, Swanson KR, Kaufmann TJ, Brown PD, Agar NYR, Galanis E, Buckner JC, Elmquist WF. Is the blood–brain barrier really disrupted in all glioblastomas? A critical assessment of existing clinical data. Neuro-Oncology 2018;20:184–91.
[4] Sun S, Wong TS, Zhang XQ, Pu JK, Lee NP, Day PJ, Ng GK, Lui WM, Leung GK. Protein alterations associated with temozolomide resistance in subclones of human glioblastoma cell lines. J Neuro-Oncol 2012;107:89–100.
[5] da Ros M, De Gregorio V, Iorio AL, Giunti L, Guidi M, de Martino M, Genitori L, Sardi I. Glioblastoma chemoresistance: the double play by microenvironment and blood-brain barrier. Int J Mol Sci 2018;19:2879.
[6] Iorio AL, da Ros M, Genitori L, Lucchesi M, Colelli F, Signorino G, Cardile F, Laffi G, de Martino M, Pisano C, Sardi I. Tumor response of temozolomide in combination with morphine in a xenograft model of human glioblastoma. Oncotarget 2017;8:89595–606.
[7] Taal W, Oosterkamp HM, Walenkamp AM, Dubbink HJ, Beerepoot LV, Hanse MCJ, Buter J, Honkoop AH, Boerman D, de Vos FYF, Dinjens WNM, Enting RH, Taphoorn MJB, van den Berkmortel FWPJ, Jansen RLH, Brandsma D, Bromberg JEC, van Heuvel I, Vernhout RM, van der Holt B, van den Bent MJ. Single-agent bevacizumab or lomustine versus a combination of bevacizumab plus lomustine in patients with recurrent glioblastoma (BELOB trial): a randomised controlled phase 2 trial. Lancet Oncol 2014;15:943–53.
[8] Sardi I, la Marca G, Giovannini MG, Malvagia S, Guerrini R, Genitori L, Massimino M, Aricò M. Detection of doxorubicin hydrochloride accumulation in the rat brain after morphine treatment by mass spectrometry. Cancer Chemother Pharmacol 2011;67:1333–40.
[9] Villodre ES, Kipper FC, Silva AO, Lenz G, da Costa Lopez PL. Low dose of doxorubicin potentiates the effect of temozolomide in glioblastoma cells. Mol Neurobiol 2018;55:4185–94.
[10] Doolittle ND, Anderson CP, Bleyer WA, Cairncross JG, Cloughesy T, Eck SL, Guastadisegni P, Hall WA, Muldoon LL, Patel SJ, Peereboom D, Siegal T, Neuwelt EA. Importance of dose intensity in neuro-oncology clinical trials: summary report of the sixth annual meeting of the blood-brain barrier disruption consortium. Neuro-Oncology 2001;3:46–54.
[11] Reya T, Morrison SJ, Clarke MF, Weissman IL. Stem cells, cancer, and cancer stem cells. Nature 2001;414:105–11.
[12] Facchino S, Abdouh M, Bernier G. Brain cancer stem cells: current status on glioblastoma multiforme. Cancer (Basel) 2011;3:1777–97.
[13] Chen J, Li Y, Yu TS, McKay RM, Burns DK, Kernie SG, Parada LF. A restricted cell population propagates glioblastoma growth after chemotherapy. Nature 2012;488:522–6.
[14] Ehtesham M, editor. Stem cell biology in neoplasms of the central nervous system, vol. 853. Switzerland: Springer International Publishing; 2015.

[15] Salmaggi A, Boiardi A, Gelati M, Russo A, Calatozzolo C, Ciusani E, Sciacca FL, Ottolina A, Parati EA, La Porta C, Alessandri G, Marras C, Croci D, De Rossi M. Glioblastoma-derived tumorospheres identify a population of tumor stem-like cells with angiogenic potential and enhanced multidrug resistance phenotype. Glia 2006; 54:850–60.
[16] Caldera V, Mellai M, Annovazzi L, Monzeglio O, Piazzi A, Schiffer D. MGMT hypermethylation and MDR system in glioblastoma cancer stem cells. Cancer Genomics Proteomics 2012;9:171–8.
[17] Squatrito M, Holland EC. DNA damage response and growth factor signaling pathways in gliomagenesis and therapeutic resistance. Cancer Res 2011;71:5945–9.
[18] Atkins RJ, Ng W, Stylli SS, Hovens CM, Kaye AH. Repair mechanisms help glioblastoma resist treatment. J Clin Neurosci 2015;22:14–20.
[19] Erasimus H, Gobin M, Niclou S, Van Dyck E. DNA repair mechanisms and their clinical impact in glioblastoma. Mutat Res Rev Mutat Res 2016;769:19–35.
[20] Schmalz PGR, Shen MJ, Park JK. Treatment resistance mechanisms of malignant glioma tumor stem cells. Cancer (Basel) 2011;3:621–35.
[21] Alexander BM, Pinnell N, Wen PY, D'Andrea A. Targeting DNA repair and the cell cycle in glioblastoma. J Neuro-Oncol 2012;107:463–77.
[22] Dean M, Fojo T, Bates S. Tumour stem cells and drug resistance. Nat Rev Cancer 2005;5:275–84.
[23] Denny BJ, Wheelhouse RT, Stevens MF, Tsang LL, Slack JA. NMR and molecular modeling investigation of the mechanism of activation of the antitumor drug temozolomide and its interaction with DNA. Biochemistry 1994;33:9045–51.
[24] Zhukovskaya N, Rydberg B, Karran P. Inactive O6-methylguanine-DNA methyltransferase in human cells. Nucleic Acids Res 1992;20:6081–90.
[25] Tano K, Shiota S, Collier J, Foote RS, Mitra S. Isolation and structural characterization of a cDNA clone encoding the human DNA repair protein for O6-alkylguanine. Proc Natl Acad Sci USA 1990;87:686–90.
[26] Weller M, Stupp R, Reifenberger G, Brandes AA, Van Den Bent MJ, Wick W, Hegi ME. MGMT promoter methylation in malignant gliomas: ready for personalized medicine? Nat Rev Neurol 2010;6:39–51.
[27] Hegi ME, Diserens AC, Gorlia T, Hamou MF, de Tribolet N, Weller M, Kros JM, Hainfellner JA, Mason W, Mariani L, Bromberg JEC, Hau P, Mirimanoff RO, Cairncross JG, Janzer RC, Stupp R. MGMT gene silencing and benefit from temozolomide in glioblastoma. N Engl J Med 2005;352:997–1003.
[28] Esteller M, Hamilton SR, Burger PC, Baylin SB, Herman JG. Inactivation of the DNA repair gene O6-methylguanine-DNA methyltransferase by promoter hypermethylation is a common event in primary human neoplasia. Cancer Res 1999;59:793–7.
[29] Martinez R, Schackert G, Yaya-Tur R, Rojas-Marcos I, Herman JG, Esteller M. Frequent hypermethylation of the DNA repair gene MGMT in long-term survivors of glioblastoma multiforme. J Neuro-Oncol 2007;83:91–3.
[30] Hirose Y, Berger MS, Pieper RO. p53 effects both the duration of G2/M arrest and the fate of temozolomide-treated human glioblastoma cells. Cancer Res 2001;61:1957–63.
[31] Friedman HS, Kerby T, Calvert H. Temozolomide and treatment of malignant glioma. Clin Cancer Res 2000; 6:2585–97.
[32] Cahill DP, Levine KK, Betensky RA, Codd PJ, Romany CA, Reavie LB, Batchelor TT, Futreal PA, Stratton MR, Curry WT, Iafrate AJ, Louis DN. Loss of the mismatch repair protein MSH6 in human glioblastomas is associated with tumor progression during temozolomide treatment. Clin Cancer Res 2007;13:2038–45.
[33] Yip S, Miao J, Cahill DP, Iafrate AJ, Aldape K, Nutt CL, et al. MSH6 mutations arise in glioblastomas during temozolomide therapy and mediate temozolomide resistance. Clin Cancer Res Off J Am Assoc Cancer Res 2009;15:4622–9.
[34] Chaicharoenaudomrung N, Kunhorm P, Promjantuek W, Rujanapun N, Heebkaew N, Soraska N, Noisa P. Transcriptomic profiling of 3D glioblastoma tumoroids for the identification of mechanisms involved in anticancer drug resistance. In Vivo 2020;34:199–211.
[35] Xu Y, Xia F, Ma L, Shan J, Shen J, Yang Z, Liu J, Cui Y, Bian X, Bie P, Qian C. MicroRNA-122 sensitizes HCC cancer cells to adriamycin and vincristine through modulating expression of MDR and inducing cell cycle arrest. Cancer Lett 2011;310:160–9.
[36] Sun C, Li N, Yang Z, Zhou B, He Y, Weng D, Fang Y, Wu P, Chen P, Yang X, Ma D, Zhou J, Chen G. miR-9 regulation of BRCA1 and ovarian cancer sensitivity to cisplatin and PARP inhibition. J Natl Cancer Inst 2013;105:1750–8.
[37] Shen R, Wang Y, Wang CX, Yin M, Liu HL, Chen JP, Han JQ, Wang WB. MiRNA-155 mediates TAM resistance by modulating SOCS6-STAT3 signalling pathway in breast cancer. Am J Transl Res 2015;7:2115–26.

[38] Dong Z, Ren L, Lin L, Li J, Huang Y, Li J. Effect of microRNA-21 on multidrug resistance reversal in A549/DDP human lung cancer cells. Mol Med Rep 2015;11:682–90.

[39] Wong STS, Zhang XQ, Zhuang JT, Chan HL, Li CH, Leung GKK. MicroRNA-21 inhibition enhances in vitro chemosensitivity of temozolomide-resistant glioblastoma cells. Anticancer Res 2012;32:2835–41.

[40] Giunti L, da Ros M, Vinci S, Gelmini S, Iorio AL, Buccoliero AM, Cardellicchio S, Castiglione F, Genitori L, de Martino M, Giglio S, Genuardi M, Sardi I. Anti-miR21 oligonucleotide enhances chemosensitivity of T98G cell line to doxorubicin by inducing apoptosis. Am J Cancer Res 2014;5:231–42.

[41] Bai Y, Liao H, Liu T, Zeng X, Xiao F, Luo L, Guo H, Guo L. MiR-296-3p regulates cell growth and multi-drug resistance of human glioblastoma by targeting ether-à-go-go (EAG1). Eur J Cancer 2013;49:710–24.

[42] Qian Z, Zhou S, Zhou Z, Yang X, Que S, Lan J, Qiu Y, Lin Y. miR-146b-5p suppresses glioblastoma cell resistance to temozolomide through targeting TRAF6. Oncol Rep 2017;38:2941–50.

[43] Tian T, Mingyi M, Qiu X, Qiu Y. MicroRNA-101 reverses temozolomide resistance by inhibition of GSK3β in glioblastoma. Oncotarget 2016;7:79584–95.

[44] Ujifuku K, Mitsutake N, Takakura S, Matsuse M, Saenko V, Suzuki K, Hayashi K, Matsuo T, Kamada K, Nagata I, Yamashita S. MiR-195, miR-455-3p and miR-10a(*) are implicated in acquired temozolomide resistance in glioblastoma multiforme cells. Cancer Lett 2010;296:241–8.

[45] Slaby O, Lakomy R, Fadrus P, Hrstka R, Kren L, Lzicarova E, Smrcka M, Svoboda M, Dolezalova H, Novakova J, Valik D, Vyzula R, Michalek J. MicroRNA-181 family predicts response to concomitant chemoradiotherapy with temozolomide in glioblastoma patients. Neoplasma 2010;57:264–9.

[46] Munoz JL, Rodriguez-Cruz V, Ramkissoon SH, Ligon KL, Greco SJ, Rameshwar P. Temozolomide resistance in glioblastoma occurs by miRNA-9-targeted PTCH1, independent of sonic hedgehog level. Oncotarget 2015;6:1190–201.

[47] Feng R, Dong L. Knockdown of microRNA-127 reverses adriamycin resistance via cell cycle arrest and apoptosis sensitization in adriamycin-resistant human glioma cells. Int J Clin Exp Pathol 2015;8:6107–16.

[48] Blower PE, Chung JH, Verducci JS, Lin S, Park JK, Dai Z, Liu CG, Schmittgen TD, Reinhold WC, Croce CM, Weinstein JN, Sadee W. MicroRNAs modulate the chemosensitivity of tumor cells. Mol Cancer Ther 2008;7:1–9.

[49] Sui H, Fan ZZ, Li Q. Signal transduction pathways and transcriptional mechanisms of ABCB1/Pgp-mediated multiple drug resistance in human cancer cells. J Int Med Res 2012;40:426–35.

[50] Ohgaki H, Kleihues P. Genetic pathways to primary and secondary glioblastoma. Am J Pathol 2007; 170:1445–53.

[51] Mayo LD, Dixon JE, Durden DL, Tonks NK, Donner DB. PTEN protects p53 from Mdm2 and sensitizes cancer cells to chemotherapy. J Biol Chem 2002;277:5484–9.

[52] Tanaka K, Babic I, Nathanson D, Akhavan D, Guo D, Gini B, Dang J, Zhu S, Yang H, De Jesus J, Amzajerdi AN, Zhang Y, Dibble CC, Dan H, Rinkenbaugh A, Yong WH, Vinters HV, Gera JF, Cavenee WK, Cloughesy TF, Manning BD, Baldwin AS, Mischel PS. Oncogenic EGFR signaling activates an mTORC2-NF-κB pathway that promotes chemotherapy resistance. Cancer Discov 2011;1:524–38.

[53] Kefas B, Godlewski J, Comeau L, Li Y, Abounader R, Hawkinson M, Lee J, Fine H, Chiocca EA, Lawler S, Purow B. MicroRNA-7 inhibits the epidermal growth factor receptor and the Akt pathway and is down-regulated in glioblastoma. Cancer Res 2008;68:3566–72.

[54] Zocco D, Ferruzzi P, Cappello F, Kuo WP, Fais S. Extracellular vesicles as shuttles of tumor biomarkers and anti-tumor drugs. Front Oncol 2014;4:267.

[55] Mc Namee N, O'Driscoll L. Extracellular vesicles and anti-cancer drug resistance. Biochim Biophys Acta Rev Cancer 2018;1870:123–36.

[56] Janaki Ramaiah M, Divyapriya K, Kartik Kumar S, Rajesh YBRD. Drug-induced modifications and modulations of microRNAs and long non-coding RNAs for future therapy against glioblastoma multiforme. Gene 2020;723:144126.

[57] Arvanitis CD, Ferraro GB, Jain RK. The blood-brain barrier and blood-tumour barrier in brain tumours and metastases. Nat Rev Cancer 2020;20:26–41.

[58] Gottesman MM, Fojo T, Bates SE. Multidrug resistance in cancer: role of ATP-dependent transporters. Nat Rev Cancer 2002;2:48–58.

[59] Veringa SJ, Biesmans D, van Vuurden DG, Jansen MH, Wedekind LE, Horsman I, Wesseling P, Vandertop WP, Noske DP, Kaspers GJ, Hulleman E. In vitro drug response and efflux transporters associated with drug resistance in pediatric high grade glioma and diffuse intrinsic pontine glioma. PLoS ONE 2013;8, e61512.

[60] Sharma HS, Ali SF. Alterations in blood-brain barrier function by morphine and methamphetamine. Ann N Y Acad Sci 2006;1074:198–224.

[61] Yousif S, Saubaméa B, Cisternino S, Marie-Claire C, Dauchy S, Scherrmann JM, Declèves X. Effect of chronic exposure to morphine on the rat blood-brain barrier: focus on the P-glycoprotein. J Neurochem 2008; 107:647–57.

[62] Dagenais C, Graff CL, Pollack GM. Variable modulation of opioid brain uptake by P-glycoprotein in mice. Biochem Pharmacol 2004;67:269–76.

[63] Amini-Khoei H, Hosseini MJ, Momeny M, Rahimi-Balaei M, Amiri S, Haj-Mirzaian A, Khedri M, Jahanabadi S, Mohammadi-Asl A, Mehr SE, Dehpour AR. Morphine attenuated the cytotoxicity induced by arsenic trioxide in H9c2 cardiomyocytes. Biol Trace Elem Res 2016;173:132–9.

[64] Cao LH, Li HT, Lin WQ, Tan HY, Xie L, Zhong ZJ, Zhou JH. Morphine, a potential antagonist of cisplatin cytotoxicity, inhibits cisplatin-induced apoptosis and suppression of tumor growth in nasopharyngeal carcinoma xenografts. Sci Rep 2016;6:1–10.

[65] Lazarczyk M, Matyja E, Lipkowski AW. A comparative study of morphine stimulation and biphalin inhibition of human glioblastoma T98G cell proliferation in vitro. Peptides 2010;31:1606–12.

[66] Bimonte S, Barbieri A, Rea D, Palma G, Luciano A, Cuomo A, Arra C, Izzo F. Morphine promotes tumor angiogenesis and increases breast cancer progression. Biomed Res Int 2015;2015:161508.

[67] Ge ZH, Wang ZX, Yu TL, Yang N, Sun Y, Hao CL, Sun LX. Morphine improved the antitumor effects on MCF-7 cells in combination with 5-fluorouracil. Biomed Pharmacother 2014;68:299–305.

[68] Barenholz YC. Doxil®–the first FDA-approved nano-drug: lessons learned. J Control Release 2012;160:117–34.

[69] Gentile E, Cilurzo F, Di Marzio L, Carafa M, Ventura CA, Wolfram J, Paolino D, Celia C. Liposomal chemotherapeutics. Future Oncol 2013;9:1849–59.

[70] da Ros M, Iorio AL, Consolante D, Cardile F, Muratori M, Fantappiè O, Lucchesi M, Guidi M, Pisano C, Sardi I. Morphine modulates doxorubicin uptake and improves efficacy of chemotherapy in an intracranial xenograft model of human glioblastoma. Am J Cancer Res 2016;6:639–48.

[71] Duzgun Z, Eroglu C. Biray Avci. Role of mTOR in glioblastoma. Gene 2016;575:187–90.

[72] Sardi I, Fantappiè O, la Marca G, Giovannini MG, Iorio AL, da Ros M, Malvagia S, Cardellicchio S, Giunti L, de Martino M, Mazzanti R. Delivery of doxorubicin across the blood-brain barrier by ondansetron pretreatment: a study in vitro and in vivo. Cancer Lett 2014;353:242–7.

[73] Arceci RJ, Stieglitz K, Bierer BE. Immunosuppressants FK506 and rapamycin function as reversal agents of the multidrug resistance phenotype. Blood 1992;80:1528–36.

[74] Gupta A, Dai Y, Vethanayagam RR, Hebert MF, Thummel KE, Unadkat JD, Ross DD, Mao Q. Cyclosporin A, tacrolimus and sirolimus are potent inhibitors of the human breast cancer resistance protein (ABCG2) and reverse resistance to mitoxantrone and topotecan. Cancer Chemother Pharmacol 2006;58:374–83.

[75] Arcella A, Biagioni F, Oliva MA, Bucci D, Frati A, Esposito V, Cantore G, Giangaspero F, Fornai F. Rapamycin inhibits the growth of glioblastoma. Brain Res 2013;1495:37–51.

[76] Tüttenberg J, Fink W, Back W, Wenz F, Schadendorf D, Thomé C. A rare case of primary sellar melanoma: case report. J Neurosurg 2004;100:931–4.

[77] Rousseau A, Kujas M, van Effenterre R, Boch AL, Carpentier A, Leroy JP, Poirier J. Primary intracranial myopericytoma: report of three cases and review of the literature. Neuropathol Appl Neurobiol 2005;31:641–8.

[78] Ohsie SJ, Sarantopoulos GP, Cochran AJ, Binder SW. Immunohistochemical characteristics of melanoma. J Cutan Pathol 2008;35:433–44.

[79] Iorio AL, Da Ros M, Pisano C, de Martino M, Genitori L, Sardi I. Combined treatment with doxorubicin and rapamycin is effective against in vitro and in vivo models of human glioblastoma. J Clin Med 2019;8:331.

[80] Lacroix M, Abi-Said D, Fourney DR, Gokaslan ZL, Shi W, DeMonte F, Lang FF, McCutcheon IE, Hassenbusch SJ, Holland E, Hess K, Michael C, Miller D, Sawaya R. A multivariate analysis of 416 patients with glioblastoma multiforme: prognosis, extent of resection, and survival. J Neurosurg 2001;95:190–8.

[81] ALA-Glioma Study Group, Stummer W, Reulen HJ, Meinel T, et al. Extent of resection and survival in glioblastoma multiforme: identification of and adjustment for bias. Neurosurgery 2008;62:564–76.

[82] Adamson C, Kanu OO, Mehta AI, Di C, Lin N, Mattox AK, Bigner DD. Glioblastoma multiforme: a review of where we have been and where we are going. Expert Opin Investig Drugs 2009;18:1061–83.

[83] Westphal M, Ram Z, Riddle V, Hilt D, Bortey E, Executive Committee of the Gliadel ® Study Group. Gliadel ® wafer in initial surgery for malignant glioma: long-term follow-up of a multicenter controlled trial. Acta Neurochir 2006;148:269–75.

[84] Westphal M, Hilt DC, Bortey E, Delavault P, Olivares R, Warnke PC, Whittle IR, Jääskeläinen J, Ram Z. A phase 3 trial of local chemotherapy with biodegradable carmustine (BCNU) wafers (Gliadel wafers) in patients with primary malignant glioma. Neuro-Oncology 2003;5:79–88.
[85] Stupp R, Mason WP, van den Bent MJ, Weller M, Fisher B, Taphoorn MJB, Belanger K, Brandes AA, Marosi C, Bogdahn U, Curschmann J, Janzer RC, Ludwin SK, Gorlia T, Allgeier A, Lacombe D, Cairncross JG, Eisenhauer E, Mirimanoff RO. Radiotherapy plus concomitant and adjuvant temozolomide for glioblastoma. N Engl J Med 2005;352:987–96.
[86] Wen PY, Kesari S. Malignant gliomas in adults. N Engl J Med 2008;359:492–507.
[87] Chinot OL, Wick W, Mason W, Henriksson R, Saran F, Nishikawa R, Carpentier AF, Hoang-Xuan K, Kavan P, Cernea D, Brandes AA, Hilton M, Abrey L, Cloughesy T. Bevacizumab plus radiotherapy-temozolomide for newly diagnosed glioblastoma. N Engl J Med 2014;370:709–22.
[88] Grill J, Massimino M, Bouffet E, Azizi AA, McCowage G, Cañete A, Saran F, Le Deley MC, Varlet P, Morgan PS, Jaspan T, Jones C, Giangaspero F, Smith H, Garcia J, Elze MC, Rousseau RF, Abrey L, Hargrave D, Vassal G. Phase II, open-label, randomized, multicenter trial (HERBY) of bevacizumab in pediatric patients with newly diagnosed high-grade glioma. J Clin Oncol 2018;36:951–8.
[89] Hosein AN, Lim YC, Day B, Stringer B, Rose S, Head R, Cosgrove L, Sminia P, Fay M, Martin JH. The effect of valproic acid in combination with irradiation and temozolomide on primary human glioblastoma cells. J Neuro-Oncol 2015;122:263–71.
[90] Galanis E, Anderson SK, Miller CR, Sarkaria JN, Jaeckle K, Buckner JC, Ligon KL, Ballman KV, Moore Jr DF, Nebozhyn M, Loboda A, Schiff D, Ahluwalia MS, Lee EQ, Gerstner ER, Lesser GJ, Prados M, Grossman SA, Cerhan J, Giannini C, Wen PY. Alliance for Clinical Trials in Oncology and ABTC. Phase I/II trial of vorinostat combined with temozolomide and radiation therapy for newly diagnosed glioblastoma: results of Alliance N0874/ABTC 02. Neuro Oncol 2018;20:546–56.
[91] Zanotto-Filho A, Braganhol E, Klafke K, Figueiró F, Terra SR, Paludo FJ, Morrone M, Bristot IJ, Battastini AM, Forcelini CM, Bishop AJR, Gelain DP, Moreira JCF. Autophagy inhibition improves the efficacy of curcumin/temozolomide combination therapy in glioblastomas. Cancer Lett 2015;358:220–31.
[92] Quinn JA, Jiang SX, Reardon DA, Desjardins A, Vredenburgh JJ, Rich JN, Gururangan S, Friedman AH, Bigner DD, Sampson JH, McLendon RE, Herndon Jr JE, Walker A, Friedman HS. Phase I trial of temozolomide plus O^6-benzylgua- nine 5-day regimen with recurrent malignant glioma. Neuro-Oncology 2009;11:556–61.
[93] Yuan Y, Xue X, Guo RB, Sun XL, Hu G. Resveratrol enhances the antitumor effects of temozolomide in glioblastoma via ROS-dependent AMPK-TSC-mtor signaling pathway. CNS Neurosci Ther 2012;18:536–46.
[94] Zhao Q, Kretschmer N, Bauer R, Efferth T. Shikonin and its derivatives inhibit the epidermal growth factor receptor signaling and synergistically kill glioblastoma cells in combination with erlotinib. Int J Cancer 2015;137:1446–56.
[95] Martin V, Sanchez-Sachez AM, Herrera F, Gomez-Manzano C, Fueyo J, Alvarez-Vega MA, Antolín I, Rodriguez C. Melatonin-induced methylation of the ABCG2/BCRP promoter as a novel mechanism to overcome multidrug resistance in brain tumors stem cell. Br J Cancer 2013;108:2005–12.
[96] Prados MD, Chang SM, Butowski N, DeBoer R, Parvataneni R, Carliner H, Kabuubi P, Ayers-Ringler J, Rabbitt J, Page M, Fedoroff A, Sneed PK, Berger MS, McDermott MW, Parsa AT, Vandenberg S, James CD, Lamborn KR, Stokoe D, Haas-Kogan DA. Phase II study of erlotinib plus temozolomide during and after radiation therapy in patients with newly diagnosed glioblastoma multiforme or gliosarcoma. J Clin Oncol 2009;27:579–84.
[97] Westphal M, Heese O, Steinbach JP, Schnell O, Schackert G, Mehdorn M, Schulz D, Simon M, Schlegel U, Senft C, Geletneky K, Braun C, Hartung JG, Reuter D, Metz MW, Bach F, Pietsch T. A randomised, open label phase III trial with nimotuzumab, an anti-epidermal growth factor receptor monoclonal antibody in the treatment of newly diagnosed adult glioblastoma. Eur J Cancer 2015;51:522–32.
[98] Brown CE, Badie B, Barish ME, Weng L, Ostberg JR, Chang WC, Naranjo A, Starr R, Wagner J, Wright C, Zhai Y, Bading JR, Ressler JA, Portnow J, D'apuzzo M, Forman SJ, Jensens MC. Bioactivity and safety of IL13Rα2-redirected chimeric antigen receptor CD8+ T cells in patients with recurrent glioblastoma. Clin Cancer Res 2015;21:4062–72.
[99] Brown CE, Alizadeh D, Starr R, Weng L, Wagner JR, Naranjo A, Ostberg JR, Blanchard MS, Kilpatrick J, Simpson J, Kurier A, Priceman SJ, Wang X, Harshbarger TL, D'Apuzzo M, Ressler JA, Jensen MC, Barish ME, Chen M, Portnow J, Forman SJ, Badie B. Regression of glioblastoma after chimeric antigen receptor t-cell therapy. N Engl J Med 2016;375:2561–9.
[100] Johnson DB, Frampton GM, Rioth MJ, Yusko E, Xu Y, Guo X, Ennis RC, Fabrizio D, Chalmers ZR, Greenbowe J, Ali SM, Balasubramanian S, Sun JX, He Y, Frederick DT, Puzanov I, Balko JM, Cates JM, Ross JS, Sanders C,

Robins H, Shyr Y, Miller VA, Stephens PJ, Sullivan RJ, Sosman JA, Lovly CM. Targeted next generation sequencing identifies markers of response to PD-1 blockade. Cancer Immunol Res 2016;4:959–67.

[101] Bagley SJ, Desai AS, Linette GP, June CH, O'Rourke DM. CAR T-cell therapy for glioblastoma: recent clinical advances and future challenges. Neuro-Oncology 2018;20:1429–38.

[102] Weller M, Butowski N, Tran DD, Recht LD, Lim M, Hirte H, Ashby L, Mechtler L, Goldlust SA, Iwamoto F, Drappatz J, O'Rourke DM, Wong M, Hamilton MG, Finocchiaro G, Perry J, Wick W, Green J, He Y, Turner CD, Yellin MJ, Keler T, Davis TA, Stupp R, Sampson JH. Rindopepimut with temozolomide for patients with newly diagnosed, EGFRvIII-expressing glioblastoma (ACT IV): a randomised, double-blind, international phase 3 trial. Lancet Oncol 2017;18:1373–85.

[103] Kast RE, Skuli N, Sardi I, Capanni F, Hessling M, Frosina G, Kast AP, Karpel-Massler G, Halatsch ME. Augmentation of 5-Aminolevulinic acid treatment of glioblastoma by adding ciprofloxacin, deferiprone, 5-fluorouracil and febuxostat: the CAALA regimen. Brain Sci 2018;8:203.

[104] Jarden JO, Dhawan V, Poltorak A, Posner JB, Rottenberg DA. Positron emission tomographic measurement of blood-to-brain and blood-to-tumor transport of ^{82}Rb: the effect of dexamethasone and whole brain radiation therapy. Ann Neurol 1985;18:636–46.

[105] D'Avella D, Cicciarello R, Albiero F, Mesiti M, Gagliardi ME, Russi E, d'Aquino A, Tomasello F, d'Aquino S. Quantitative study of blood-brain barrier permeability changes after experimental whole-brain radiation. Neurosurgery 1992;30:30–4.

[106] Hwang SY, Jung JS, Kim TH, Lim SJ, Oh ES, Kim JY, Ji KA, Joe EH, Cho KH, Han IO. Ionizing radiation induces astrocyte gliosis through microglia activation. Neurobiol Dis 2006;21:457–67.

[107] Wilson CM, Gaber MW, Sabek OM, Zawaski JA, Merchant TE. Radiation-induced astrogliosis and blood-brain barrier damage can be abrogated using anti-TNF treatment. Int J Radiat Oncol Biol Phys 2009;74:934–41.

[108] Qin S, Wise M, Cobbold SP, Leong L, Kong YC, Parnes JR, Waldmann H. Induction of tolerance in peripheral T cells with monoclonal antibodies. Eur J Immunol 1990;20:2737–45.

[109] Qin D, Ou G, Mo H, Song Y, Kang G, Hu Y, Gu X. Improved efficacy of chemotherapy for glioblastoma by radiation-induced opening of blood-brain barrier: clinical results. Int J Radiat Oncol Biol Phys 2001;5:959–62.

[110] Brown WR, Thore CR, Moody DM, Robbins ME, Wheeler KT. Vascular damage after fractionated whole-brain irradiation in rats. Radiat Res 2005;164:662–8.

[111] Yuan H, Gaber MW, Boyd K, Wilson CM, Kiani MF, Merchant TE. Effects of fractionated radiation on the brain vasculature in a murine model: blood-brain barrier permeability, astrocyte proliferation, and ultrastructural changes. Int J Radiat Oncol Biol Phys 2006;66:860–6.

[112] Liu Q, Liu Y, Li W, Wang X, Sawaya R, Lang FF, Yung WK, Chen K, Fuller GN, Zhang W. Genetic, epigenetic, and molecular landscapes of multifocal and multicentric glioblastoma. Acta Neuropathol 2015;130:587–97.

[113] Lahmi L, Idbaih A, Rivin Del Campo E, Hoang-Xuan K, Mokhtari K, Sanson M, Canova CH, Carpentier A, Jacob J, Maingon P, Feuvret L. Whole brain radiotherapy with concurrent temozolomide in multifocal and/or multicentric newly diagnosed glioblastoma. J Clin Neurosci 2019;68:39–44.

[114] Showalter TN, Andrel J, Andrews DW, Curran Jr WJ, Daskalakis C, Werner-Wasik M. Multifocal glioblastoma multiforme: prognostic factors and patterns of progression. Int J Radiat Oncol Biol Phys 2007;69:820–4.

[115] Chen S, Tanaka S, Giannini C, Morris J, Yan ES, Buckner J, Lachance DH, Parney IF. Gliomatosis cerebri: clinical characteristics, management, and outcomes. J Neuro-Oncol 2013;112:267–75.

[116] Lesniak MS, Upadhyay U, Goodwin R, Tyler B, Brem H. Local delivery of doxorubicin for the treatment of malignant brain tumors in rats. Anticancer Res 2005;25:3825–31.

[117] Eramo A, Ricci-Vitiani L, Zeuner A, Pallini R, Lotti F, Sette G, Pilozzi E, Larocca LM, Peschle C, De Maria R. Chemotherapy resistance of glioblastoma stem cells. Cell Death Differ 2006;13:1238–41.

[118] Chua SL, Rosenthal MA, Wong SS, Ashley DM, Woods AM, Dowling A, Cher LM. Phase 2 study of temozolomide and Caelyx in patients with recurrent glioblastoma multiforme. Neuro-Oncology 2004;6:38–43.

[119] Beier CP, Schmid C, Gorlia T, Kleinletzenberger C, Beier D, Grauer O, Steinbrecher A, Hirschmann B, Brawanski A, Dietmaier C, Jauch-Worley T, Kölbl O, Pietsch T, Proescholdt M, Rümmele P, Muigg A, Stockhammer G, Hegi M, Bogdahn U, Hau P. RNOP-09: pegylated liposomal doxorubicine and prolonged temozolomide in addition to radiotherapy in newly diagnosed glioblastoma-a phase II study. BMC Cancer 2009;9:308.

[120] Ananda S, Nowak AK, Cher L, Dowling A, Brown C, Simes J, Rosenthal MA, Cooperative Trials Group for Neuro-Oncology (COGNO). Phase 2 trial of temozolomide and pegylated liposomal doxorubicin in the treatment of patients with glioblastoma multiforme following concurrent radiotherapy and chemotherapy. J Clin Neurosci 2011;18:1444–8.

CHAPTER

27

Nanoparticles beyond the blood-brain barrier for glioblastoma

Lynn Jena[a], *Nicholas J. Dunne*[b,c,d,e,f,g,h], *and Helen O. McCarthy*[a,i]

[a]School of Pharmacy, Queen's University Belfast, Belfast, United Kingdom [b]School of Mechanical and Manufacturing Engineering, Dublin City University, Dublin, Ireland [c]Centre for Medical Engineering Research, School of Mechanical and Manufacturing Engineering, Dublin City University, Dublin, Ireland [d]Department of Mechanical and Manufacturing Engineering, School of Engineering, Trinity College Dublin, Dublin, Ireland [e]Advanced Manufacturing Research Centre (I-Form), School of Mechanical and Manufacturing Engineering, Dublin City University, Dublin, Ireland [f]Advanced Materials and Bioengineering Research Centre (AMBER), Royal College of Surgeons in Ireland and Trinity College Dublin, Dublin, Ireland [g]Advanced Processing Technology Research Centre, Dublin City University, Dublin, Ireland [h]Trinity Centre for Biomedical Engineering, Trinity Biomedical Sciences Institute, Trinity College Dublin, Dublin, Ireland [i]School of Chemical Sciences, Dublin City University, Dublin, Ireland

Abstract

Glioblastoma (GBM) is an incurable highly malignant cancer of the central nervous system (CNS). GBM is a complex cancer, which involves multiple molecular pathways and genetic mutations that propagate tumor growth. To prolong tumor control and patient survival, novel therapeutic strategies are required. The blood-brain barrier (BBB) is the protective barrier of the CNS and acts to maintain the brain microenvironment. It is a highly organized structure that impedes the delivery of many therapeutic agents to the brain. The emerging strategies which have been developed to circumvent the BBB for effective GBM therapy are detailed in this chapter. Invasive and noninvasive methods which can temporarily disrupt the overall organization and normal structure of the BBB in order to promote the penetration of nanoparticles to the CNS are described along with an evaluation of the key delivery technologies.

Glioblastoma Resistance to Chemotherapy: Molecular Mechanisms and Innovative Reversal Strategies
https://doi.org/10.1016/B978-0-12-821567-8.00027-0

Abbreviations

ABC	adenosine triphosphate-binding cassette
Ag	silver
AJ	adheren junction
ALS	amyotrophic lateral sclerosis
AMT	adsorptive-mediated transcytosis
Au	gold
BBB	blood-brain barrier
BCB	blood-cerebrospinal fluid barrier
BCEC	brain capillary endothelial cell
BCRP	breast cancer resistance protein
BEC	brain endothelial cell
CBB	CSF-brain barrier
CBSA	cationic bovine serum albumin
CNS	central nervous system
CPP	cell-penetrating peptide
CSF	cerebrospinal fluid
CYP	cytochrome P450
DC-Chol	3-(*N*-(*N*′, *N*′-dimethylaminoethane) carbamoyl) cholesterol
DOTAP	1,2-dioleoyl-3-trimethylammonium-propane
DOX	doxorubicin
DOX-CL	DOX-loaded cationic liposome
DSS-DAK-PEG-OPSS	disuccinimidyl suberate-diamino ketal-PEG-orthopyridyl disulfide
EPC	egg phosphatidylcholine
EPR	enhanced permeability and retention
EVG	elvitegravir
FUS	focused ultrasound
GBM	glioblastoma multiforme
HAS	human serum albumin
HFE	high-frequency electroporation
IgM	immunoglobulin (Ig) M antibody
ISF	interstitial fluid
LAT1	L-type amino-acid transporter
LDL	low-density lipoprotein
L-DOPA	levodopa
LfR	lactoferrin
LNP	lipid NP
LRP-1	lipoprotein receptor-related protein 1
MAN	p-aminophenyl-α-D-manno-pyranoside
MGMT	O6-methylguanine-DNA methyltransferase
MMP	matrix metalloproteinases
MPS	mononuclear phagocyte system
MRI	magnetic resonance imaging
MRP	multidrug resistance protein
NP	nanoparticle
NVU	neurovascular unit
OX-26	anti-Tf receptor antibody IgG2a
PC	phosphocholine
PEF	pulsed electric field
PEG	polyethylene glycol
PEG2000-DSPE	PEG-distearoylphosphosphatidylethanolamine
PEG-PHDCA	methoxypolyethyleneglycol cyanoacrylate-co-hexadecyl cyanoacrylate

P-gp	P-glycoprotein
PLA	poly-lactic acid
PLGA-PLL	poly(lactic-co-glycolic acid)
PLL	poly-L-lysine
PMNP	paramagnetic NP
POH	perillyl alcohol
PVA	poly(vinyl alcohol)
RAP	receptor associated protein
RGD	arginine-glycine-aspartic acid
RMT	receptor-mediated transport
RVG	rabies virus glycoprotein
SA	sialic acid
siRNA	small interfering RNA
SPION	super paramagnetic iron oxide NP
SREBP-2	sterol regulatory element-binding protein-2
TAT	HIV virus trans-activator of transcription
TEB	poly [triphenylamine-4-vinyl-(P-methoxy-benzene)]
TEER	transendothelial cell electrical resistance
Tf	transferrin
TJ	tight junction
TME	tumor microenvironment
TMZ	temozolomide
VE	vascular endothelial
WGA	wheat germ agglutinin
WHO	World Health Organization

Conflict of interest

No potential conflicts of interest were disclosed.

Introduction

Glioblastoma (GBM) is the most common primary tumor of the central nervous system (CNS), comprising more than 60% of all gliomas [1]. The World Health Organization (WHO) classify GBM as a grade IV highly malignant tumor, derived from glial cells [2]. The current standard of care in the treatment of GBM includes maximal safe resection, followed by concomitant radiotherapy and chemotherapy in which temozolomide (TMZ) is the main chemotherapeutic drug [3]. Other adjuvant therapies include carmustine, procarbazine, lomustine, and vincristine [4]. Despite these aggressive therapies, the median survival time is currently 14.6 months, with an overall 5-year survival rate of less than 5% [5]. The main challenges faced in the treatment of GBM are determined by the location of the tumor and the failure of current therapies to overcome the cellular and genetic heterogeneity of tumor cell populations. For example, the high activity levels of the DNA repair protein O6-methylguanine-DNA methyltransferase (MGMT) in tumor cells create a resistant phenotype by inhibiting the therapeutic effect of alkylating agents such as TMZ, thereby preventing DNA damage and cytotoxicity [6]. Although surgical approaches, radiotherapy, and adjuvant chemotherapy have shown gradual improvements in survival and quality of life for those with GBM, the prognosis is still poor. Drug delivery is impeded by a number of physiological barriers which include the blood-brain barrier (BBB), presence of efflux pumps and the consequential low drug concentration at the tumor site [7].

Until recently, it was reported that the majority of small-molecule drugs and nearly all macromolecular drugs such as therapeutic antibodies, nucleic acids, and recombinant proteins cannot cross the BBB [8]. However, the development of novel drug delivery systems and alternative strategies that can effectively transport therapeutic agents into the CNS have opened up new avenues in the treatment of GBM [9].

Nanomaterials can be employed as versatile drug transportation systems across the BBB, which can deliver the loaded therapeutic agents into the CNS. Many nanomaterials, including inorganic nanoparticles (NPs), polymeric NPs, and liposomes, have demonstrated unparalleled advantages, such as high drug loading content, controllable drug release, excellent stability, capabilities for passive and active targeting, biocompatibility, and biodegradability [10]. In addition, strategies have bypassed uptake by the mononuclear phagocyte system (MPS) in order to prolong blood circulation [11]. This significantly enhances the BBB transport of NPs, resulting in improved drug accumulation within the CNS. This chapter provides an overview of the delivery obstacles and discusses the key strategies designed to transport therapeutic cargoes across the BBB.

Barriers to delivery

The BBB

Unlike other solid cancers, GBM rarely colonize regions external to the CNS [12]. The unique properties of the cranial microenvironment support growth and invasion within that niche as evidenced by the low incidence of extracranial metastasis. The BBB is a highly selective barrier composed of a tightly packed layer of endothelial cells that surround cerebral vasculature and the spinal cord [13]. The BBB acts as a physical barrier that separates the CNS from peripheral tissues and maintain homeostasis [14]. While the BBB is critical for regulating transport across the CNS, it also represents a significant hindrance to drug transport into the brain. As it stands, very few CNS conditions including epilepsy, chronic pain, depression, and schizophrenia are treatable using small molecule drugs [15]. Therefore, understanding the BBB is imperative to developing novel strategies to improve drug delivery into the CNS. Generally, while drug delivery tends to focus on the vascular BBB due to the large surface area, it must be noted other barriers such as the blood-cerebrospinal fluid (CSF) barrier (BCB) and CSF-brain barrier (CBB) exist [16]. In contrast to the tightly knit BBB, the BCB and CBB are "leaky" in nature, indicating a possible route for drug delivery into the brain [17]. However, despite the transport of the drug into the CSF, the penetration of substances from CSF to brain parenchyma is facilitated through diffusion. The large distances create a diffusion barrier that remains an obstacle for effective drug delivery. This can lead to rapid clearance of the therapeutic agent by glial cells before it can effectively permeate into the brain parenchyma.

In the BBB, brain endothelial cells (BECs), which include pericytes, vascular smooth muscle cells, astrocytes, microglia, and neurons, are located between the brain parenchyma and the vascular system [18]. Interactions between these components have led to the concept of

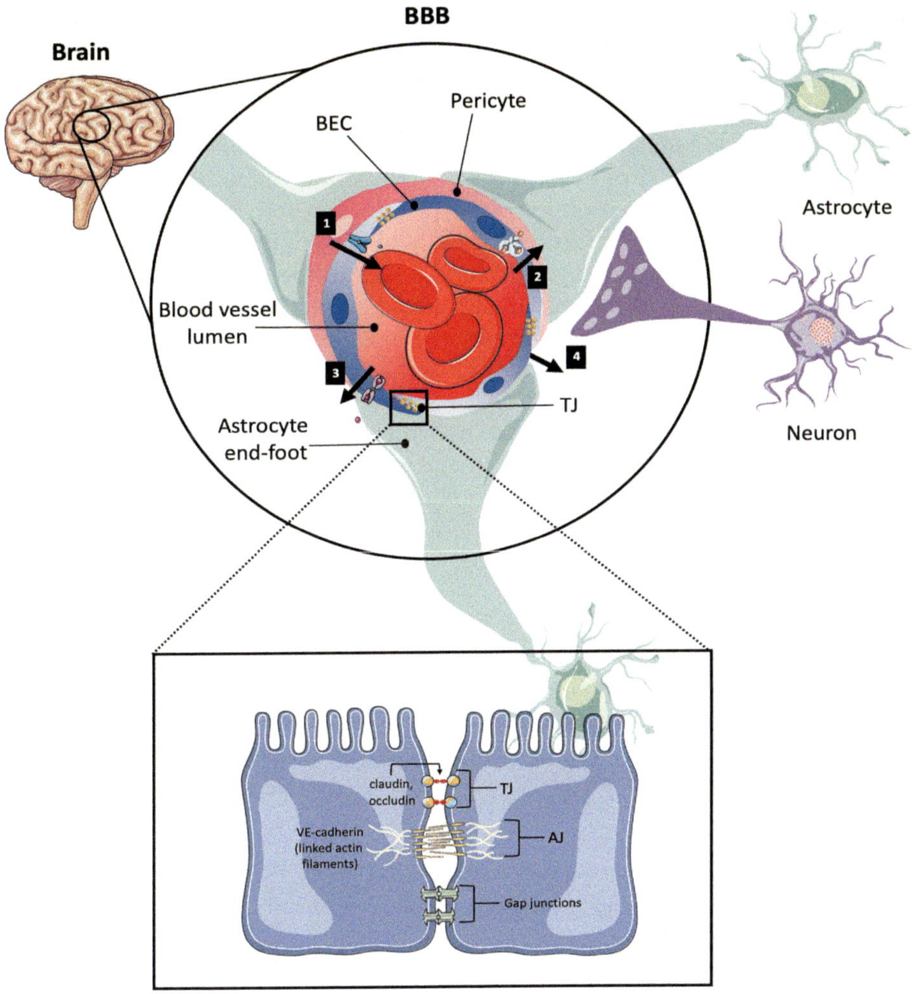

FIG. 1 Schematic representation of the cell types that form the NVU of the BBB and some of the mechanisms which govern transport across the BBB, (1) Efflux pump transporters, (2) receptor transport, (3) transporter proteins, (4) passive diffusion. Tight junction (TJ); adherens junction (AJ); and brain endothelial cells (BECs).

the neurovascular unit (NVU) (Fig. 1), where each cell type contributes to BBB function and regulates the expression of the BBB phenotype across BECs [13]. These specialized cells include astrocytes which mediate the transport of nutrients and waste products between the brain capillaries and neurons while modulating neuronal activity and synaptic transmission [19]. The membrane potentials and ionic concentrations of a group of 50–100 coupled astrocytes are regulated in a coordinated manner where ions and small molecules travel between astrocytes via junctions [20]. Furthermore, astrocytes demonstrate antioxidant effects where during oxidative stress, ischemia, and inflammation, astrocytes upregulate anaerobic glycolysis to provide adequate ATP for cell survival or glutamate uptake under hypoxia [21, 22].

Pericytes reside on the abluminal surface of the endothelium of the basement membrane and regulate the blood flow in the brain capillary through contraction and relaxation [23]. These cells play an important role in the establishment of the BBB and provide structural support for the mature BBB while imparting phagocytic activity [24]. The presence of paracellular junctions known as tight junctions (TJs), adheren junctions (AJs), or gap junctions stringently regulate the extracellular environment of the neural tissue thereby protecting the CNS from harmful molecules and pathogens in the blood [25]. In addition, the lack of fenestration and pinocytotic vesicles effectively precludes paracellular transport across the endothelium [26].

TJs confer paracellular diffusion and induce a high transendothelial cell electrical resistance (TEER) thus impeding ions and small charged molecules from crossing the BBB to a greater extent than AJs [27]. AJs are ubiquitous in the vasculature, sit abluminal to TJs and consist primarily of vascular endothelial (VE)-cadherin. AJs mediate cell-cell adhesion and bind to the actin cytoskeleton via catenins thus contributing to barrier integrity [28]. Furthermore, the TJ complex includes two classes of transmembrane molecules namely claudins and occludins. These transmembrane proteins interact and form seals in the gap between adjacent BECs via accessory proteins called zonulae occludens [26]. Interestingly, the "tight" properties such as TJs, efflux, and nutrient transporters are enhanced in BECs compared to peripheral endothelial cells where "leaky" properties such as transcytosis are predominant and in which BECs lack [13].

Therefore, BBB drug transport is predominantly via transcellular diffusion through the membranes [29]. Circulating molecules are transported into the brain via lipid-mediated transport of small nonpolar molecules by passive diffusion or energy-dependent transport mechanisms [30]. Transporters are expressed on the luminal membrane and use ATP to transport highly selective substrates down a concentration gradient into the brain, thus delivering essential nutrients to the target site [31].

Exploiting endogenous transport routes across the BBB

Despite the restricted transport of molecules across the BBB, the brain has a high demand for energy and nutrients [32]. Therefore, multiple mechanisms exist that mediate the uptake of specific endogenous substances. Some of these provide a useful basis for the development of novel strategies for drug delivery. Under physiological conditions, substances may cross the BBB by passive diffusion in nonenergy dependent manner or energy-dependent processes which include receptor-mediated transport (RMT), and adsorptive-mediated transcytosis (AMT) (Fig. 2) [33].

The majority of large molecules and 98% of all small molecules do not cross the BBB [8]. Of the small percentage which does, transport is predominantly through spontaneous passive diffusion, down a concentration gradient. For example, despite TMZ being able to cross the BBB relatively well via diffusion, it is estimated that <1% of TMZ administered reaches the brain [34]. It is a nonsaturable pathway where transport is influenced by the drug chemistry [35].

Transcellular transport is limited to small molecules such as O_2, CO_2, caffeine, barbiturates and ethanol which can diffuse through the cell membrane [33]. It is generally reported that drugs with a positive charge, a low molecular weight (<400–600 Da), and a high lipophilic partition rate favor crossing luminal and abluminal cell membranes through this mechanism;

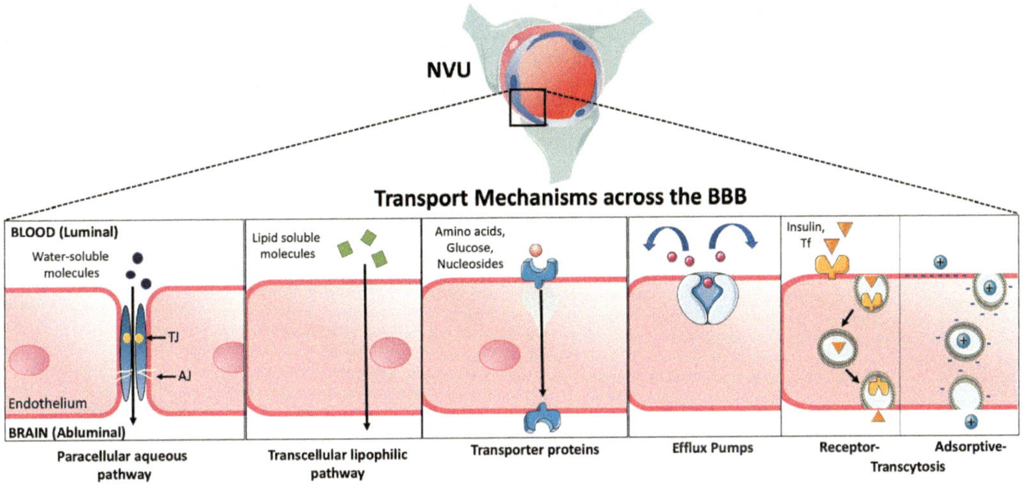

FIG. 2 Schematic representation of the blood-brain barrier and the transport mechanisms.

the drug must be able to partition from the BBB into the aqueous environment of the interstitial fluid (ISF) in order to reach the target site and exert a therapeutic effect [8, 35]. There is a clear correlation between increased lipophilicity (LogP) and the rate and extent of penetration into the brain as observed by the penetration of blood gases and several drugs such as anesthetics [36]. Opioids are the most widely reported example of how increased lipophilicity can facilitate transport across the BBB; heroin is a lipophilic diacyl derivative of morphine and is able to cross the BBB with a potency twofold greater than morphine [37].

Yet, while lipid solubility can increase the transport rate across the BBB, lipophilic drugs are more likely to pass through lipid bilayers and therefore more likely to distribute to tissues with a high adipose density which can lead to an increased incidence of off target effects [38]. A high degree of lipophilicity can result in a high rapid metabolic turnover associated with low solubility and poor absorption [26].

The transcytosis of small polar molecules such as glucose, amino acids, organic anions and cations, nucleosides and other biologically important molecules, occurs through carrier-mediated transport systems which are expressed at both the apical/luminal (blood side) and abluminal (brain side) membranes of BBB endothelial cells [39]. Typically, the apical/luminal side contains glycoproteins and proteoglycans, such as syndecan-1, -2, -4, and glypican which facilitates the transport of hydrophilic compounds across the BBB [40, 41]. In contrast, the abluminal membrane has been reported to contain more caveolae, a membrane microdomain that is rich in sphingomyelin and glycosphingolipids which are necessary for intracellular vesicle trafficking [40].

Macromolecules, such as proteins, peptides, and many NP therapeutics are transported across the BBB through both specific and nonspecific interactions with proteins and receptors expressed on the surface of the BECs where endocytosis and transcytosis facilitate transport (Table 1) [53]. Subsequently, efficient GBM delivery methods have been developed which utilize receptor- or adsorption-mediated endocytic transport where it has been estimated that 10%–15% of all proteins in the NVU are transporters [54].

TABLE 1 Specific transport proteins on the BBB.

Transport-mediated	Active efflux-mediated	Receptor-mediated
Glucose transporter [42]	Organic anionic transporter [43]	Tf receptor [44]
Cationic amino acid transporter, member 1 [45]	Organic anionic transporter polypeptide [43]	Insulin receptor [42]
Large neutral amino acid transporter, member 1 [42]	Adenosine triphosphate-binding cassette (ABC) transporter, subfamily B (P-gp, MRPs) [46, 47]	Insulin-like growth receptor [48]
Monocarboxylic acid transporter [42, 43]	ABC transporter, subfamily G [49]	Low-density lipoprotein receptor [44]
Nucleoside transporter [50]	Glutamic acid, amino acid transporter [42]	Diphtheria toxin receptor [51]
Choline transporter [50]	Taurine transporter [52]	Leptin receptor [42]
		Neonatal FC receptor [48]
		Scavenger receptor call B type [48]
		Nicotinic acetylcholine receptor [48]

Although lipophilic molecules permeate across the BBB via the transcellular route, hydrophilic molecules, i.e., cationic NPs, exploit the adsorption-mediated transcytosis pathway [55]. AMT requires a cell surface ligand interaction which is triggered through electrostatic interaction between the positively charged molecules or proteins and negatively charged lipid membrane [56]. It is a nonspecific process where molecules interact with the cell surface membrane and concentrate at the site prior to internalization through endocytosis [57]. Subsequently, vesicular transport occurs within the cell, resulting in the eventual release of the molecule or protein into the brain parenchyma [58]. Cationic proteins and peptides such as cationic bovine serum albumin (CBSA) and the HIV virus trans-activator of transcription (TAT) have been used for AMT brain drug delivery.

Lu et al. synthesized cationic albumin-conjugated pegylated (maleimide-PEG, MW 3400 Da) NPs where CBSA was functionalized to the NP surface. The presence of CBSA significantly increased the cell uptake of encapsulated 6-coumarin in rat brain capillary endothelial cells, where the fluorescent intensity was up to twofold higher compared to unmodified NPs after 4 h incubation in vitro [59]. Furthermore, the apparent BBB permeability of CBSA-NPs was up to 9.0-fold higher than unmodified NPs. In vivo, brain tissue distribution was enhanced as evidenced by a 2.3-fold in the CBSA-NPs treatment group compared to unmodified NPs [60].

Interestingly, the surface characteristics of CBSA-NP were significantly different from those of the conventional cationic liposomes or NPs. CBSA-NPs had a negative zeta potential, however, authors proposed that the net electrical charge in the local CBSA microenvironments remained positively charged. It was suggested with CBSA protruding on the external surface of the NP, this conformation enabled electrostatic binding to the negatively charged residues on BBB.

In another study, Li et al. used a combination of two biotin labeled peptides, Angiopep-2 (ANG: TFFYGGSRGKRNNFKTEEY) and Angiopep-2—TAT (biotin ANG-TAT: TFFYGGSRGKRNNFK[Biotin]TEEYGRKKRRQRRRPPQQ), to deliver paclitaxel across the BBB [61]. In vitro, TAT conjugation to ANG-TAT paclitaxel (40 μM) demonstrated increased cellular uptake in U87MG GBM cells with a peak uptake of ~18 μM compared to ANG-TAT (20 μM) or ANG at ~8 and 6 μM, respectively at the 8 h time point. In vivo, ANG-TAT was injected into the tail vein of CD-1 nude mice carrying intracerebral implants of U87MG GBM cells and demonstrated the greatest total brain uptake at a total concentration of 17.5 μg after 5 min with corresponding increased mouse survival at 35 days compared to ANG and free paclitaxel at 12.5 μg (27 days) and ~4 μg (20 days) respectively. However, all biological membranes are negatively charged, therefore one of the inherent limitations of adsorptive-mediated targeting is poor selectivity which is associated with toxicity issues [48, 56].

Receptor-mediated transcytosis offers specific targeting where vesicular trafficking is induced upon a ligand binding to a site-specific domain. Typically, a circulating ligand binds to the receptor expressed on the apical surface of the plasma membrane, forming receptor-ligand complexes [62, 63]. The cargo is processed into the intracellular compartment via membrane invagination leading to the eventual formation of an endosome. While some endosomes fuse with lysosomes, a fraction of the endocytic vesicles are shuttled to the basolateral side and exocytosis releases the cargo into the brain parenchyma [53, 63].

For example, iron-loaded transferrin (Tf) binds to the corresponding Tf receptor on the BBB. Tf (78 kDa) is the main protein responsible for the regulation and distribution of serum iron, utilized for various biological processes such as DNA synthesis, cellular metabolism and proliferation [64, 65]. Tf reversibly binds two atoms of ferric iron (Fe^{3+}) at a high affinity which triggers a conformational change that is required for selective recognition by the Tf receptor. Upon receptor binding, internalization is mediated through clathrin-mediated endocytosis [64]. The vesicle-based mechanism allows for transport of a wide range of heterogeneous proteins and molecules into the brain, including Tf, insulin, leptin, growth factors, and plasma proteins [57, 62]. Tf receptors are ubiquitously expressed on the surface of all cells, which can lead to indiscriminate binding of Tf-ligand bound particles across all tissues with preferential localization determined by the NP pharmacokinetics [66]. However, many cancer therapies which utilize Tf as a targeting ligand, rely upon the overexpression of Tf receptors as observed in cancer. For example, experimental studies have analyzed the expression of Tf receptors in GBM cancer and found increased expression on both proliferating and malignant cells compared to normal brain tissue [67, 68]. Furthermore, clinical studies have revealed that Tf receptor-1 expression was associated with poorer outcomes in GBM [69].

It is thought that iron transport across the BBB occurs through two possible mechanisms; iron is dissociated from Tf within the low pH endosomal compartments of the brain endothelial cells before release on the abluminal face or the Tf complex dissociates from the Tf receptor within endosomes [70]. Yet, it remains clear that the Tf receptor must transcytose from blood-to-brain across the BBB to release its cargo.

Clark et al. prepared Tf-containing 80-nm gold (Au) NPs with an acid-cleavable disuccinimidyl suberate-diamino ketal-PEG-orthopyridyl disulfide (DSS-DAK-PEG-OPSS) linkage between the Tf ligand and the NP core to facilitate RMT across the BBB [71]. Authors exploited the rapid acidification of the endosome, which occurs during endocytosis, to cleave the Tf ligand from the NP in order to facilitate free movement of the NP into the brain parenchyma. The acid-sensitive DAK was selected as it has a hydrolysis half-life of 60 min at pH 5.5 but more than 24 h at pH 7.4 and 37°C, maintaining the stability of the functionalized Tf during systemic circulation, but able to efficiently cleave in the acidic vesicles during the transcytosis. NPs were systemically administered to BALB/c mice via tail vein injection and allowed to circulate for 12 h. In vitro and in vivo results showed the formulation with the most Tf ligands, cleavable 200Tf-C, had the highest uptake into brain cells. For example, 93% and 94% of cleavable 120Tf-C and 200Tf-C formulations, respectively, crossed the in vitro bEnd.3 cell BBB model within 2 h compared with noncleavable 200Tf-N at 47%. Furthermore, cleavable 200Tf-C NPs had the highest brain penetration with 2.7-fold increase observed compared to the noncleavable formulation in vivo. Such results demonstrate how BBB transport mechanisms can be utilized to enhance drug delivery.

The low-density lipoprotein receptor-related protein 1 (LRP-1) is another highly expressed receptor present in brain cells which has been utilized for drug delivery [72]. The 19-mer Angiopep-2 peptide was developed as a specific ligand for LRP-1 [73]. LRP-1 facilitates the transcytosis of ligands such as lactoferrin, melanotransferrin, and receptor-associated proteins, across the BBB. This highlights LRP-1 as a potential target moiety for drug delivery systems, where it can be exploited for BBB targeting, especially in the treatment of GBM [61]. Angiopep-2 has exhibited significant potential for the mediation of brain-targeted drug delivery transport. Competitive binding by α2-macroglobulin, a specific ligand for LRP-1, can block the accumulation of Angiopep-2 thereby indicating Angiopep-2 facilitates brain-targeted drug delivery through LRP-1-mediated transcytosis [61, 74]. Notably, brain entry via a low-density lipoprotein (LDL) receptor-mediated route has shown to harbor a potentially higher RMT capacity since the rate of brain uptake of intravenously administered melanotransferrin (P97) and receptor-associated protein (RAP), two ligands for LRP-1, greatly exceeds that of Tf [62]. However, due to the limited number of the receptors on the cell surface, RMT remains a rate-limited process [57].

To enhance RMT across the BBB, Guo et al. mediated the upregulation of LRP-1 on BMECs through the delivery of statins-loaded, Angiopep-2-bound, poly(lactic-*co*-glycolic acid)-poly-L-lysine (PLGA-PLL) polymeric NPs (S@A-NPs) [75]. Statins have been shown to increase the expression of LDL receptors and LRP-1 on endothelial cells. Statins act to inhibit cholesterol synthesis which corresponds to the increased expression of sterol regulatory element-binding protein-2 (SREBP-2), a transcription factor that activates the *Ldlr* genes which subsequently mediates the expression on LRP-1 [76, 77].

Thereafter, Angiopep-2 effectively facilitated RMT across the BBB. Systemic administered S@A-NPs loaded with doxorubicin (DOX) lengthened median survival of mice bearing brain metastases to 59.5 days compared to mice treated with free DOX, S@A-NPs and saline at 52, 40, and 39.5 days, respectively. In another study, Lu et al. synthesized fluorescent polymeric, poly [Triphenylamine-4-vinyl-(P-methoxy-benzene)] (TEB) NPs, surface modified with three types of ligands—Tf, LRP and lactoferrin (LfR), through covalent linking [78]. Ligand-decorated TEB-NPs exhibited selective and enhanced internalization into bEnd.3 cells where the fluorescence intensities of ligand modified TEB-NPs were significantly higher. The transportation efficiencies of TEB-NP-Tf, TEB-NP-LRP and

TEB-NP-LfR were 28.0%, 25.7%, and 29.0%, respectively, in contrast to 7.5% observed by unmodified NPs. Therefore, transcytosis presents an attractive transport route for targeted nanomedicine strategies across the BBB.

Efflux pumps

In addition to the occlusive nature of the BBB provided by TJs, efflux transporter proteins expressed at the BBB present significant problems for the treatment of GBM with current chemotherapeutics [79].

ABC transporters are a superfamily of ATP-dependent proteins which are widely recognized as important determinants for the transport of endogenous and exogenous molecules in the CNS to regulate homeostasis [80]. Changes in the level of expression and functionality of these transporters influence the efficacy and safety of the drugs transported [81].

Among ABC transporters, P-glycoprotein (P-gp) is the most common of the BBB efflux pumps and is responsible for the active transport of low molecular weight hydrophobic chemotherapeutics out of the brain [82]. P-gp resides only on the apical membrane of endothelial cells, which modulates drug transport in a unidirectional manner [83]. Typically, endogenous substrates of P-gp are nonpolar and less amphiphilic molecules such as lipids, steroid hormones and cytokines [84]. Moreover, small molecule chemotherapeutic drugs such as TMZ, methotrexate, paclitaxel and anthracyclines, have been proven to be the substrates of P-gp [85]. The brain-to-blood efflux caused by P-gp, significantly impedes the rate of drug uptake by the BBB. For example, TMZ treatment is able to extend the median survival of GBM patients by 2.5 months which indicates the therapeutic agent is able to effectively cross the BBB [86]. However, Gooijer et al. used drug transporter knockout mouse models to demonstrate how P-gp and Breast Cancer Resistance Protein (BCRP) collectively limit the brain penetration of this drug [87]. All mice were given 50mg/kg TMZ via tail vein injection; the concentration of TMZ increased 1.5-fold with the combined deletion of both transporters and a similar increase was observed when TMZ was administered to wild-type mice in combination with the dual P-gp/BCRP inhibitor, elacridar. This study highlights how efflux transporters impede the delivery of conventional chemotherapeutics and the presence of P-gp efflux pumps present in the TME also contribute to a poor response to GBM therapy [88].

In addition, the expression of ABC transporters holds relevance in the variability of the interindividual response to treatment in GBM [89]. It has been reported 30% of the population overexpress P-gp while 25% of the population under express P-gp, resulting in treatment response variation between individuals [35]. Other ABC efflux transporters include the ABCC multidrug resistance protein (MRP) transporter subfamily, which has five members (ABCC1-5) and BCRP [90]. These transporters play a critical role in BBB function and are known to contribute significantly to the inherent or acquired multidrug resistance phenotype of cancer cells and the BBB. In addition to chemotherapeutic agents, BCRP substrates include glutathione and steroid hormones while substrates of MRPs are more polar molecules such as sulfate-conjugated molecules or endogenous metabolites [50]. Overlapping substrate profiles have been recognized in MRP, P-gp and BCRP transporters and are attributed to the effective clearing of small molecule chemotherapeutics drugs away from the cerebrum which has major clinical significance for resistance to chemotherapy in the treatment of GBM [85].

Therefore, modulation of ABC efflux transporters at the BBB through the use of efflux inhibitors to compete with endogenous substrates could provide a novel strategy to enhance drug delivery across the BBB and generate new therapeutic options for drug-resistant CNS diseases such as GBM.

Emerging delivery strategies to mitigate the BBB

Recent advances in molecular biology and drug development have provided a greater understanding of GBM progression, supporting the development of new classes of therapeutic agents that can circumvent the BBB. Generally, two main drug delivery strategies have been identified; BBB disruption through physical opening of the BBB (paracellular) and transport across the BBB (transcellular) [91].

Novel pharmacological methods can be identified as BBB disruption techniques or non BBB disruptions techniques (Table 2).

BBB disruption techniques

BBB disruption refers to an increase in leakiness due to a reduction in barrier tightness [92]. The loss of BBB integrity enables the entry of cytokines and immune cells into the CNS, which causes alterations in the cerebral microenvironment [13]. Moreover, the temporal reversible disruption of the BBB through chemical or physical strategies has been exploited to allow therapeutic agents to penetrate the brain and reach the GBM tumor site [93].

Direct injection or convection-enhanced delivery are some of the strategies which have been employed to improve drug concentrations at the tumor site, however, associated safety concerns limit clinical application due to the need for open surgery [94]. Surgical risks include infection and hemorrhage which can lead to secondary inflammation and further damage to the BBB, with the leakage of plasma proteins and neurotoxic substances [95].

Considering the importance of the BBB and how it maintains homeostasis in the brain, barrier disruption aims to affect TJ integrity and endothelial structure but equally needs to be reversible and fine-tuned. The increased permeability of the BBB can be advantageous for drug delivery, however, the fundamental properties of the BBB must be maintained, as there

TABLE 2 Methods employed to circumvent the BBB.

BBB disruption techniques	Non BBB disruptive techniques
• Osmotic opening of BBB	• Intranasal delivery
• Disruption of the BBB by focused ultrasound (FUS) and microbubbles	• Prodrug analogs
• Nonthermal reversible electroporation	• Efflux pump inhibitors
	• NP-based delivery systems

is the potential extravasation of circulating factors that can disturb homeostasis and cause toxicity of neurons [96].

Osmotic opening

Traditional approaches to transiently alter BBB integrity involve the injection of a hyperosmotic solution such as mannitol prior to the administration of a therapeutic drug [97]. Hyperosmotic solutions act to induce BEC shrinkage which consequently leads to an increase in vascular leakage within the brain parenchyma [98]. The use of an osmotic agent is estimated to increase drug delivery by 10–100 times in comparison to delivering the drug alone, where the barrier can remain open for up to 3h [99]. TJs at the cerebrovascular endothelium are reversibly opened under osmotic stress, and this opening can facilitate entry of water-soluble drugs, proteins, and other solutes of interest into the brain for experimental or therapeutic purposes [100]. Rapoport et al. discovered that when hypertonic solutions of electrolytes and nonelectrolytes of minimal osmolarities were applied topically for 10min to the pial surface of the rabbit cerebral cortex, the BBB was reversibly opened as evidenced by extravasation of Evans blue-albumin while more lipid-soluble nonelectrolytes damaged the barrier irreversibly [101, 102].

Recently, it was shown osmotic disruption of the BBB, achieved via intracarotid injection of mannitol, could be exploited to facilitate drug delivery into the brain. Godinho et al. administered 25% mannitol through the right carotid artery of Sprague-Dawley rats, causing osmotic disruption prior to treatment with hydrophobic small interfering RNA (siRNA), previously modified with phosphocholine (PC)-docosahexanoic acid [103].

Mannitol caused osmotic shrinkage of BECs and transient widening of the TJs to facilitate the delivery of PC-docosahexanoic acid-hydrophobic siRNAs to the brain which do not readily cross the BBB when administered alone by intravenous injection. Delivery was most concentrated in proximal regions of the brain vascular tree and authors found that by increasing the hydrophobicity of this biological therapeutic, enhanced retention of siRNA was observed in the brain without compromising the therapeutic action.

Likewise, Choi et al. found combined osmotic treatment with mannitol and TMZ increased the BBB permeability. The treatment was found to increase the concentration of Evans blue dye within the stroke lesion site from ~1 to 6 µg of dye per mg of tissue, and also reduced the expression of the TJ protein, occludin, by ~30% with regard to relative protein expression in rat brain microvessels [104].

Many clinical trials have been conducted where the use of a hyperosmotic solution based on mannitol or NaCl enhances the delivery of a therapeutic agent to the brain tumor to significantly enhance overall patient survival [3, 99]. For example, Guillaume et al. reported significant tumor reduction and prolonged progression-free survival in a phase I trial; BBB/BBTB disruption was induced through a mannitol infusion in 13 patients with malignant gliomas followed by intra-arterial administration of melphalan/carboplatin [105]. However, widespread clinical application of osmotic disruption of the BBB is limited; it is generally considered an invasive procedure as it requires repeated hospitalization, sedation, and adverse effects include severe systemic toxicity, strokes, seizures, and potential new tumor formation [97, 106].

Although physical opening of the BBB has yielded significant enhancement of drug delivery, as of yet, there are no strategies to monitor BBB disruption in real time. Kiviniemi et al.

used direct-current electroencephalography to monitor BBB opening, induced by carotid or vertebral artery mannitol infusion [107]. Such considerations are required to ensure optimal patient safety.

Disruption of BBB by FUS and microbubbles

Recently, promising studies have indicated that ultrasound can be employed to locally deliver a gene or drug to a specific area in the brain [108]. Ultrasound waves can be focused using either single element concave transducers or electronically controlled phased arrays, in which the energy becomes concentrated over a focal spot, resulting in a high-intensity field [109]. Long periods of exposure of biological tissues to FUS increases the absorption of the acoustic energy by the surrounding tissues, which is converted to thermal energy, leading to a temperature rise in that area [110]. This temperature rise is also known as "hyperthermia" which can persist for up to several hours [110, 111]. Consideration must be given to the temperatures reached and the duration of exposure as previous studies have indicated that temperatures in the range of 43°C–46°C have detrimental effects on the brain tissue [111]. Early application of high-intensity ultrasound on the BBB was accompanied by neuronal and glial damage [112, 113]. However, after reducing the acoustic intensity, BBB opening was still observed but without the previously associated damage [114, 115].

It was found that the dual activity of FUS and microbubbles, which are 1–10 μm diameter lipid or protein-encapsulated gas spheres, enabled the reduction of the acoustic intensity needed to produce BBB opening when combined with FUS [116]. Typically during treatment, microscopic gas bubbles are injected into the blood at the same time ultrasound begins. Microbubbles transform kinetic energy from the traveling ultrasound wave to the local microenvironment, due to the high compressible properties and tendency to oscillate under ultrasound [117]. Notably, this leads to mechanical disruption and temporary permeabilization of neighboring cells and tissue, such as the BBB by widening interendothelial gaps and TJs [118]. The BBB opening is immediate and reversible, lasting for several hours after the sonication, to facilitate the penetration of the drug molecules [119].

Zhao et al. synthesized microbubbles with a mean diameter of 1546.8 ± 114.2 nm which were able to disrupt the BBB with ultrasound and authors found a higher microbubble dose was associated with more profound BBB disruption [120]. Moreover, when the dose increased from 2.0×10^7 to 3.0×10^7 microbubbles, Evans blue extravasation in the striatum increased by over 100 μg/g tissue, which was more prominent than in the cortex with neuronal damage and loss only seen in the higher microbubble dose treatment group. One major concern of BBB disruption is the risk of nonrecovery. Recovery occurred within 6 h and 4 h in the cortex and striatum, respectively, indicating that the effects of BBB disruption were reversible.

Microbubbles offer the potential for ligand-receptor targeting, payload conjugation and act as delivery vehicles for therapeutic agents, similar to liposomal, NP- or polymer-based vehicles [117]. In addition, in situ targeting with the use of FUS enables NP transport across the paracellular route due to the transient disruption of intercellular junctions. For example, Lin et al. utilized FUS and microbubbles to enhance the delivery of DOX-loaded cationic liposomes (DOX-CLs) [121]. FUS + DOX-CLs exhibited the greatest growth inhibition in vitro and in vivo as shown through magnetic resonance imaging (MRI) and histological analysis. Moreover, a survival time of 81.2 days was observed in a rat C6-glioma model, compared with the DOX-CLs, DOX, and FUS group alone at 35, 30.4, and 22.3 days, respectively.

Nonhuman primate studies with rhesus macaques have shown that microbubble-enhanced FUS can successfully induce local BBB opening with minimal side effects, even with repeated opening [122, 123]. Therefore, the safety of FUS application in glioma therapy appears to be promising. Neuronavigation systems minimize off target effects by guiding the FUS for precise BBB opening and facilitate monitoring of BBB opening in real time [124–126]. Abrahao et al. conducted the first-in-human trial of BBB opening using magnetic resonance-guided FUS in patients with the CNS disease, amyotrophic lateral sclerosis (ALS) [127]. Successful BBB opening was achieved in all subjects, with BBB recovery within 24h and no serious adverse effects.

Currently, FDA-approved commercial microbubbles include Definity, Optison, Sonazoid, and SonoVue [128]. FUS has also shown great promise through the development of Optune. This device obtained FDA approval in 2015 where it creates low-intensity and alternating tumor-treating fields, as a combined treatment together with TMZ for newly diagnosed GBM [129]. Furthermore, a Phase 1/2a clinical trial due for completion in 2020 is underway where patients with recurrent GBM are treated with a new-generation SonoCloud prior to carboplatin chemotherapy (NCT03744026) [130]. The commercialization of these newly established therapies highlight the clinical relevance of FUS+microbubbles in GBM shows the promise of such treatment strategies for GBM, with the consistent development and approval of new products.

Nonthermal reversible electroporation

Despite displaying the capability to safely and reversibly disrupt the BBB, FUS is limited by the small area the focal beam can target, leading to prolonged treatment times for larger treatment areas while osmotic disruption using agents such as mannitol may lead to adverse effects such as abnormal neuron function [131]. Alternatively, pulsed electric fields (PEFs), including electroporation, provide another strategy for drug delivery across the BBB.

PEFs are short, intense electrical pulses that have been used for tumor ablation. During electroporation, PEFs are applied across cells or tissue, resulting in destabilization of the electrical potential across the cell membrane [132]. The membrane potential exceeds a critical threshold value, resulting in the formation of aqueous pores in the lipid bilayer, which enhance the permeability of the cell membrane [133].

Electroporation can be either reversible, where membrane pores reseal over time, or irreversible where the induced permanent defects in the cell membrane result in cell death [134]. Notably, even if with reversible electroporation, cell death remains a possibility as the temporary disruption during the electroporation process can lead to apoptosis [131]. Reversible electroporation is an attractive strategy for drug delivery as it can be used in a controlled manner to temporarily disrupt the cell membrane as well as TJs for gene or drug delivery [135]. More importantly, this indicates that PEFs can potentially enhance the permeability of both the transcellular or paracellular routes across the BBB.

Hjouj et al. demonstrated that it is possible to permeabilize BECs in the BBB in a controlled manner, therefore highlighting the feasibility of using PEFs for clinical application [136]. Authors studied the reversible and irreversible electroporation thresholds of the BBB in male Sprague-Dawley rats via MRI. Authors found reversible disruption occurred between electric field intensities of 330 and 500 V/cm using a total number of 50–70 μs pulses at a frequency of 4 Hz, resulting in greater BBB disruption volumes larger than the volume of tissue damage.

Other studies have investigated the effects of PEFs on the integrity of the BBB. In a similar manner, Lorenzo et al. delivered high-frequency electroporation (HFE) at a voltage-to-distance ratio of 600 V/cm to cause BBB disruption in a rat brain model [137]. MRI was used to verify disruption and authors were able to demonstrate BBB disruption for 1–72 h after HFE with BBB recovery observed at the 96 h time point, Furthermore, 1200 V/cm resulted in widespread necrosis, while 600 V/cm induced BBB disruption with minimal associated tissue damage. Electroporation represents a promising field for BBB disruption and in combination with a chemotherapeutic shows potential clinical translation.

Non BBB disruptive techniques

Intranasal delivery

Intranasal inhalation is a promising drug delivery route that can bypass the BBB. Intranasally administered drugs can penetrate the CNS through the olfactory or trigeminal routes through intracellular and extracellular pathways [138]. During intracellular uptake, the drug is taken up by olfactory or trigeminal sensory neurons which transmit to the olfactory bulb or to the pons while in extracellular uptake, the drug passes through TJs, paracellular cleft, the lamina propria and perineural space to reach the subarachnoid space [118, 138]. Furthermore, when administered through the respiratory route, the drug molecules can reach the brainstem [118].

Wang et al. demonstrated that when administered intranasally, a combined therapy of bortezomib with Perillyl alcohol (POH) (also known as NEO100) could exert antitumor effects in vivo against orthotopic GBM [139]. POH is a natural compound found in many essential oils such as peppermint and spearmint where the amphipathic character makes it readily soluble in the lipid bilayers of gliomas cells, leading to effective POH delivery into these cells [140]. It affects multiple cell growth regulation processes which are proposed to lead to anticancer activity [141]. Authors found brain tumor-bearing mice in the bortezomib +NEO100 treatment group had significantly longer survival times, as compared to mice that received bortezomib or NEO100 alone. Moreover, bortezomib concentrations in the brain were also substantially higher in the combined treatment group with intranasal administration in contrast to intravenous delivery [139]. For example, the concentration of bortezomib postintravenous administration only reached ~0.03 μg/mL compared to ~3 μg/mL postintranasal administration after 1 h. This study demonstrated that the noninvasive intranasal administration of a NEO100-based formulation can facilitate effective brain delivery of a BBB-impermeable drug.

Clinical studies have been performed in Brazil where positive results led to the first Phase 1/2A clinical trials in the United States for a synthetic GMP grade NEO100 (Neonc Technologies, Inc.) (ClinicalTrials.gov identifier: NCT02704858) for the treatment of patients with recurrent GBM [142]. Initiated in 2016, the treatment protocol consists of POH being inhaled four times daily over a period of 6 months. The main aim of Phase 1 is to determine the safety and tolerability of intranasal NEO100 and identify the Maximum Tolerable Dose. Meanwhile, the main aim of Phase 2a is to determine the progression-free survival rate at 6 months. Although the study is expected to conclude in October 2020, this study represents the development of NEO100 as a novel, intranasally administered therapy for malignant glioma patients, who no longer respond to other types of therapy [142].

Furthermore, intranasal administration can be utilized for NP CNS delivery. Owing to the small size, when administered into the nasal cavity NPs are further transported by olfactory and trigeminal nerves of nasal mucosa into the CNS [138]. Chu et al. developed ephrin type-A receptor 3 tyrosine kinase antibody-modified PLGA NPs to target GBM through intranasal delivery [143]. In vivo imaging and distribution studies in glioma-bearing rats showed high accumulation in glioma tissue 4h postadministration with significantly higher tumor cell apoptosis ($P < 0.01$) than other formulations. Furthermore, treatment prolonged the median survival time of glioma-bearing rats, 1.37-fold, up to 26 days compared to PLGA NPs. Intranasal administration reduces peripheral distribution and showed the greatest accumulation in the brain, especially at the tumor site, with a small amount of fluorescence detected in the lungs. Meanwhile, after intravenous administration, the fluorescence was distributed throughout the body, mostly in the liver, indicating that nasal administration is a better delivery strategy for GBM treatment. Intranasal delivery exploits the unique anatomical structure between the olfactory and trigeminal nerves of nasal mucosa and the CNS to provide successful drug targeting which then eliminates the risk of surgery and minimizes off site effects to normal tissue [138].

Prodrug analogs

A pragmatic approach to improve drug delivery across the BBB is by increasing the lipophilicity of polar compounds. The lipid composition of the membrane of human BECs is approximately 33% phosphatidylcholine, 25% phosphatidylethanolamine, 17% sphingomyelin, 11% phosphatidylserine and 4.8% phosphatidylinositol, which therefore favors the transport of lipid-soluble drugs across the BBB [33].

Van de Waterbeemd et al. compared the CNS penetration of the hydrophilic drug, sulpiride, with the more lipophilic D2 antagonists, remoxipride, haloperidol, and thioridazine [144]. A higher free (unbound) plasma concentration of sulpiride was required to diffuse into the CSF compared to the other lipophilic compounds which were in close equilibria between the plasma and the CSF.

Although increased hydrophobicity enhances BBB diffusion, in practice, a high degree of lipophilicity leads to low plasma solubility, lung and liver uptake, and nonspecific binding to plasma proteins which reduces bioavailability and subsequently leads to lower brain uptake [145]. In particular, lipophilic compounds are often substrates for cytochrome P450 and other enzymes, thus increasing metabolic rates and further reducing bioavailability [146].

One possible strategy to circumvent these issues is to improve brain uptake via prodrug formation. Prodrugs are pharmacologically inactive compounds that result from chemical modifications of biologically active species, designed to improve the physicochemical properties [147]. Postadministration, the prodrug is brought closer to the receptor site where it is converted to the active form, usually via a single activating step [148]. For example, esterification or amidation of hydroxy-, amino-, or carboxylic acid-containing drugs, may greatly enhance lipid solubility thus enhance transport across the BBB [149]. Generally, esters are preferred to amides due to the superior hydrolysis rate that provides maximal CNS drug distribution from a systemically administered dose [149, 150]. Once in the CNS, hydrolysis of the modifying group will release the active compound so it can exert an effect. Placzek et al. used a prodrug strategy to increase the permeability of the active drug, sobetirome, across the BBB [150]. The carboxylic acid moiety was esterified so once across the BBB nonspecific esterases could cleave the prodrug ester, hence releasing the active drug sobetirome in the CNS. A significant twofold increase of brain sobetirome level was observed with ethanolamine ester prodrug. In addition to improving BBB penetration, the ester prodrug effectively masked sobetirome in circulation and peripheral organs thus reducing the systemic drug clearance and potentially increasing the therapeutic index.

Another prodrug approach to enhance the CNS transport is the development of bioprecursor prodrugs. Bioprecursors involve bioreversible chemical manipulation of the parent drug molecule without the linkage of an auxiliary moiety [149]. Levodopa (L-DOPA), a prodrug of dopamine, is one of the most well-known bioprecursors which cross the BBB [151]. L-DOPA acts as substrate for the neutral L-Type amino-acid transporter (LAT1) expressed at the BBB [149]. Once it is transported into the brain, L-DOPA is decarboxylated to dopamine by aromatic amino-acid decarboxylase, enabling the pharmacological effects to be realized [152].

Unfortunately, prodrugs are not without limitations. Although increased lipophilicity may improve movement across the BBB, such compounds also have an increased propensity to nonselectively bind to other tissues [26]. This often leads to off target toxicity as observed with potent drugs. Additionally, enhanced efflux activity may occur due to lipophilic binding to P-gp transporters, resulting in poor drug retention in the brain and reduced biological action [153].

Notably, genetic polymorphisms often lead to interpatient variability in drug response, with alterations in the pharmacokinetics and pharmacodynamics profile due to differences in enzyme production and receptor expression [154]. Cytochrome P450 (CYP) represent an enzyme superfamily involved in the oxidative metabolism of structurally diverse molecules such as drugs and fatty acids [155]. For example, polymorphism in the *CYP2D6* gene is commonly observed which leads to significantly different phenotypic drug responses with up to a 10-fold difference in the required dose to achieve the same plasma concentration in different individuals [156, 157].

This demonstrates that the prodrug strategy alone is not sufficient to achieve CNS delivery but rather are targeted approach is required.

Efflux pump inhibitors

Due to the location of ABC transporters at the luminal site of BECs on the BBB, the multispecificity and potency of these transporter proteins offers a desirable inhibition target to enhance BBB entry of therapeutics into the brain. The overexpression of ABC transporters at the BBB indicates pharmacological modulation of these transporters could form a novel strategy in GBM treatment [80].

It has been shown that inhibition of efflux transporters at the BBB can enhance the brain concentration of the substrate drugs which can provide a therapeutic advantage. As part of a larger study, Dagenais et al. used P-gp competent wild-type mice and Mdr1a/1b(−) knockout mice naturally lacking Mdr1to determine the effect of efflux pump inhibition [158]. A significant increase in the brain uptake of loperamide and verapamil in the absence and presence of efflux pump inhibitor, quinidine, was observed when both compounds were administered through in situ mouse brain perfusion. Furthermore, when the concentration of quinidine was increased, the brain uptake of loperamide and verapamil was increased by factors of 9 and 5.1, respectively, indicative of full P-gp inhibition. However, quinidine is known as a first-generation P-gp inhibitor [153]. First-generation inhibitors are characterized by poor specificity, therefore, require higher doses to reverse multidrug resistance and associated toxicity [80, 159]. Second-generation P-gp inhibitors which include valspodar and biricodar are generally more selective and are better tolerated although use is limited due to unpredictable pharmacokinetics and interactions with other transporter proteins. Third-generation P-gp inhibitors, such as zosuquidar, tariquidar, laniquidar, and elacridar have a high potency and specificity for P-gp with demonstrated efficacy [80, 160, 161]. Tang et al. coadministered cabazitaxel with elacridar [162]. It was observed that elacridar increased the brain accumulation of cabazitaxel 9.6-fold in wild-type mice where similar concentrations observed for ABCB1a/1b;ABCG2 −/− mice treated with or without elacridar. The data from this study indicate that intravenous elacridar coadministration could potentially inhibit the activity of mouse ABCB1 in the BBB, and lead to increased concentrations of cabazitaxel concentrations in the brain for better therapeutic efficacy.

Nanoparticle-based delivery systems

The unique properties of nanomaterials such as biocompatibility, reduced size, prolonged blood circulation, and minimal toxicity have been exploited in the development of emerging drug delivery platforms [10]. Generally, systemic administration is the attractive route for NP administration as it is minimally invasive [163]. NPs are transported via blood circulation until the BBB is reached. NPs <200 nm are able to effectively cross the BBB and extravasate from the leaky vasculature to reach the tumor site where retention occurs in the tumor due to the "enhanced permeability and retention" (EPR) effect [118, 164, 165]. Once in the tumor microenvironment, NP delivery system must be able to enter cells and overcome endosomal entrapment of the cargo to exert a therapeutic effect [58].

The transport of NPs across biological barriers is typically governed by size-dependent uptake where AMT and RMT facilitate uptake through endocytosis [166]. Caveolin-mediated endocytosis occurs when plasma membrane invaginations of about 80 nm in size are produced to form lipid rafts [167, 168]. After this initial step, caveolin vesicles fuse together to generate caveosomes, which are trafficked into the intracellular compartment but are uniquely are able to avoid fusion with lysosomes [169].

Clathrin-mediated endocytosis, occurs in clathrin-enriched areas of the cell membrane, forming endocytic vesicles up to 200 nm in diameter [167]. Once inside the cell, the clathrin-coated vesicles fuse together to form an early endosome that progresses to a late endosome, with a corresponding decrease in the vesicular pH [170]. Finally, the vesicles fuse with lysosomes, causing degradation of the attached cargo [171]. Therefore, in order to be a successful delivery vehicle, NPs should be able to circumvent the endosome to avoid cargo degradation.

In addition, studies have shown that cationic delivery vehicles are effective at cell membrane permeation due to the electrostatic interactions that occur between the cationic drug delivery system and the negatively charged cell membranes to trigger AMT [56]. However, there is a major limitation to the use of cationic NPs for systemic administration and brain delivery; nonspecific uptake by peripheral tissues and the binding of serum proteins due to the charged surface can lead to an increased risk of adverse events and reduce the therapeutic capacity of the drug [172]. Furthermore, systemic administration can lead to opsonization which occurs when the cationic particle becomes coated with nonspecific complement factors, fibrinogen, or IgG known as opsonin proteins [173]. This is a biological process that occurs in the blood to make foreign molecules visible to phagocytic cells such as monocytes, macrophages, neutrophils, and dendritic cells, of the MPS [174]. After opsonization, phagocytic cells engulf and ultimately remove the foreign materials such as NPs from the bloodstream [175]. The nonspecific binding of serum proteins can also lead to the formation of a "protein corona" which can lead to agglomeration and reduce the drug bioavailability.

Serum protein adsorption is commonly regulated by modifying the NP surface with polyethylene glycol (PEG) to enhance the colloidal stability and prolong the circulation time and prevent MPS clearance [176]. PEG can be synthesized in a linear or branched polymer configuration to generate a hydrated protective layer that imparts a steric barrier to the surface of NPs [177]. The presence of a large excluded volume sterically impedes NPs from interacting with proximate NPs and circulating opsonin proteins, acting to mask the surface charge. These properties helped minimize the nonspecific serum protein adsorption and off-target binding, thereby improving distribution to the brain tissue. Calvo et al. prepared poly (methoxypolyethyleneglycol cyanoacrylate-co-hexadecyl cyanoacrylate) (PEG-PHDCA) and non PEGylated PHDCA NPs and found the blood concentration of PEG-PHDCA NPs was 26.92% and 25.11% higher in the PEGylated treatment group at 1h and 4h postintravenous administration, respectively, compared to the unmodified group in OF1 mice [178]. This facilitates higher chances of NP entry and retention in tumor tissue through the EPR effect.

However, while PEGylation prolongs circulation time, when the PEGylated molecule reaches the cell surface, the bulky nature of PEG can sterically hinder uptake. Cruje et al. reported a higher cell uptake was observed when Au NPs were grafted with lower PEG surface densities and shorter chain lengths all cancer cell lines [179]. For example, high-density 2kDa PEG Au NPs (HP2K-GNP) demonstrated NP uptake just below 60% in MDA-MB-231 cells while low-density 2kDa-PEG Au NPs (LP2K-GNP) had an uptake of ~75%. Furthermore, when PEG was increased to 5kDa and grafted at a low surface density ~15% NP uptake was observed in the same cell line. However, non PEGylated Au NPs demonstrated the highest cell uptake in Hela, MDA-MB-231 and MCF-7 cells compared to PEGylated counterparts, demonstrating how PEG can impede drug delivery.

Novel strategies have been developed which facilitate targeted delivery while evading the uptake problems often exhibited by PEGylation in which the addition of a targeting ligand can improve uptake. Gosk et al. prepared anti-Tf receptor antibody IgG2a (OX-26) OX-26-PEG-liposomes which were administered to P18 rats by an in situ perfusion method [180]. It was observed that the uptake of OX-26-PEG-liposomes by BECs was two times higher than observed in the PEG-liposome. After infusion for 15min, the uptake rate of the targeted liposomes increased to 16 times higher than that of pure PEG-liposome with no associated toxicity. Additionally, Bruun et al. developed targeted cationic lipid NPs (LNP) through the surface conjugation of angiopep to target the LRP-1 expressed on the BBB [181]. In addition, a matrix metalloproteinases (MMP)-cleavable PEG moiety was used to conceal the overall positive surface charge and to exploit the high level of MMPs often expressed in the tumor microenvironment (TME) and inflammatory BBB conditions. Authors found cleavage of the PEG moiety increased uptake 10-fold in the two MMP-2/9 expressing brain cell lines, bEnd.3

and U87MG. Functionalization of the NP surface with different targeted ligands as well as the use of cleavable moieties helps to improve transport across the BBB [93]. Active targeting aims to increase selectivity and cell internalization while reducing the risk of adverse events. In addition, modifications can also impart controlled drug release which offers a new perspective on drug delivery through certain triggers, i.e., pH, light, and temperature control [182].

Many studies have focused on the development of nanocarrier drug delivery systems including liposomes, polymers, and cell-penetrating peptides (CPPs) for the delivery of therapeutic agents in the treatment of GBM. However, not all of these therapies will translate for clinical application as the presence of the BBB and tumor heterogeneity impede the delivery of many formulations in vivo [165].

The parameters of the main nano-drug delivery vehicles required for successful drug delivery across the BBB are discussed below.

Liposomes

Liposomes are desirable delivery systems due to the flexible structure, capabilities for transporting and protecting a wide range of therapeutic hydrophilic/hydrophobic biomolecules, biocompatibility, and biodegradability, and easy manipulation of composition [183]. Liposomes are colloidal carriers made of concentric membrane bilayers containing biocompatible phospholipids that enclose an aqueous core [184]. The amphiphilic phospholipid bilayer of liposomes closely resembles the mammalian cell membrane, facilitating effective cell-liposome interactions for cellular uptake [185]. Furthermore, the versatile structure enables the loading of both hydrophobic molecules such as DNAs or RNAs (within the phospholipid bilayer) and hydrophilic drugs (in the center core) to alter the pharmacokinetics and biodistribution [183].

Surface-modified brain-targeted liposomes can be transported across the intact BBB by RMT or AMT for the treatment of different grades of GBM tumors [186, 187].

Ying et al. demonstrated the BBB crossing of synthesized egg phosphatidylcholine (EPC), cholesterol, and PEG-distearoylphosphosphatidylethanolamine (PEG_{2000}-DSPE) liposomes loaded with the model drug, daunorubicin, for the treatment of gliomas [188]. The formulated liposomes were surface modified with p-aminophenyl-α-D-manno-pyranoside (MAN) and Tf which have a specific affinity for BBB receptors, GLUT1 and Tf receptor, respectively. Daunorubicin liposomes were delivered into the cranium of C6-glioma bearing Sprague-Dawley rats via tail vein injection. 1 week posttreatment, the median survival time of rats treated with daunorubicin liposome modified with MAN and Tf was significantly longer at 22 days than that of rats treated with physiological saline (13 days), free daunorubicin (17 days), daunorubicin liposomes (18 days), daunorubicin liposomes modified with MAN (19 days) and daunorubicin liposomes modified with TF (18 days), respectively. Furthermore, tumor growth inhibitory effect of daunorubicin liposome modified with MAN and Tf after crossing the BBB has significantly enhanced up to 64% compared to other treatment groups. The dual-targeting effect of the liposomes enabled effective BBB transport, in order to reach the tumor site.

Moreover, liposomes help to increase drug bioavailability by protecting the therapeutic cargo from degradation by the immune system and biological enzymes. Xiao et al. prepared wheat germ agglutinin (WGA)-modified vinorelbine cationic liposomes, 100–120 nm in size, for the treatment of glioma. WGA was modified on the liposomal surface as a targeting moiety to facilitate RMT across the BBB with cationic 3-(*N*-(*N*′, *N*′-dimethylaminoethane) carbamoyl) cholesterol (DC-Chol) [189]. The chemotherapeutic drug, vinorelbine, was encapsulated within the liposomes and in vitro and in vivo studies exhibited BBB and tumor targeting capabilities. A C6-glioma-bearing mice model was established to demonstrate the targeting and therapeutic properties of the liposomes in the brain where mice survival ranged from 15 to 22 days in the saline control group, 18–26 days for free vinorelbine, 21–29 days for vinorelbine liposomes, 22–34 days for vinorelbine cationic liposomes, 23–33 days for WGA modified vinorelbine liposomes and 26–42 days for WGA modified vinorelbine cationic liposomes. WGA modified vinorelbine cationic liposomes provided the optimal therapeutic efficacy with negligible side effects.

However, the relationship between the physicochemical properties and immunogenicity of liposomal delivery systems has been an area of concern. Kedmi et al. analyzed the toxicity of prepared cationic liposomes composed of the positive lipid, 1,2-dioleoyl-3-trimethylammonium-propane (DOTAP), in animal models [190]. The intravenous injection of cationic liposomes (+54.3 mV) in C57BL/6 mice led to significantly increased liver enzymes in serum by three- to sixfold compared to neutral and negatively charged (−59.2 mV) liposomes, indicating induced hepatotoxicity. Treatment with cationic liposomes provoked a strong pro-inflammatory response by inducing the expression of Th1 cytokines (IL-2, IFNγ and TNFα) which was 10- to 75-fold higher than observed in control groups. Furthermore, the body weight of C57BL/6 mice decreased by up to 5.5%. While cationic liposomes are promising vehicles for gene delivery, consideration of the potential inflammatory response is critical for improving therapeutic benefits and avoiding immune toxicity. Some studies have reported a strong immune response upon repeated injection, where PEGylated particles often lose the long-circulating characteristic in a phenomenon referred to as accelerated blood clearance (ABC) [191, 192]. Wang et al. reported the ABC phenomenon in PEGylated liposomes was due to the selective binding of anti-PEG immunoglobulin (Ig) M antibody (anti-PEG IgM) to the PEG surface, where binding was more pronounced when PEG was in the "brush" formation, subsequently leading to complement activation [192]. Thereafter, opsonization by C3 fragments occurs following the second dose of PEGylated liposomes which in turn leads to uptake by the MPS and reducing the circulation time of PEGylated liposomes with concomitant increased liver uptake [191, 192].

Circulating anti-PEG antibodies have been found in healthy human subjects and are thought to be induced due to exposure to PEG-containing food and cosmetics [193]. Chanan-Khan et al. reported PEGylated liposomal formulation of DOX, Doxil/Caelyx induced moderate to severe hypersensitivity reactions in 13 out of 29 patients and activated complement in 21 out of 29 patients [194]. It can be assumed that naturally occurring anti-PEG antibodies in circulation, prime an immune response against treatment with PEGylated materials, leading to induced hypersensitivity reactions and reduced therapeutic efficacy.

Nevertheless, the commercialization of liposomal formulations has facilitated the development of glutathione PEGylated liposomal DOX (2B3-101) as a new treatment option for patients with brain cancer, based on already marketed Doxil/Caelyx [195]. The drug has an additional glutathione coating to enhance drug delivery across the BBB. It was found the brain-to-blood ratio of DOX was 4.8-fold higher after administration of 2B3-101 as compared with generic Caelyx highlighting the effectiveness of a targeted approach [196]. Moreover,

TABLE 3 Examples of clinically approved NP therapies.

Drug name	NP type/drug	Approved indication	Size	Surface charge	Crosses BBB?	References
Doxil (Caelyx) (Janssen) FDA approval: 1995	PEGylated liposomal DOX	Ovarian cancer HIV-associated Kaposi's sarcoma Multiple myeloma	~100 nm	−24.89 mV	No	[198–201]
Abraxane (Celgene) FDA approval: 2005 EMA approval: 2008	Albumin-particle bound paclitaxel	Advanced nonsmall cell lung cancer Primary metastatic pancreatic cancer Secondary metastatic breast cancer	~130 nm	~ −8.1 mv	Unknown	[200, 202–204]
Myocet (Teva) EMA approval: 2000	Non PEGylated liposomal DOX	Primary treatment of metastatic breast cancer	~190 nm	N/A	Not enough data to support	[199, 205, 206]
DaunoXome (Galen) FDA approval: 1996	Non PEGylated liposomal daunorubicin	HIV-associated Kaposi's sarcoma		Near neutral	No	[199, 205, 207]

when taken to Phase 1/2a clinical trials 2B3-101 was deemed safe and well tolerated with no neuro- or cardiotoxicity observed in patients [197]. Indeed, liposomal formulations have paved the way for the wider application of nanomedicine in the clinical setting for a range of cancers (Table 3).

Polymeric nanoparticles

Polymeric NPs can be made up of various biodegradable polymers which can help to effectively deliver the drug to a target site and increase the therapeutic efficacy. Polymeric NPs are simple to synthesize, are stable, and can extend blood circulation times as well as the drug release profile [208]. The polymers are typically biocompatible, biodegradable, and nontoxic to biological tissues [209]. For example, chitosan is a natural polysaccharide, derived from the exoskeleton of arthropods by the deacetylation of chitin and is capable of crossing the BBB [210]. Chitosan has minimal toxic properties and biocompatibility which makes it an excellent polymer for drug delivery use. Other biocompatible and biodegradable polymers such as poly-lactic acid (PLA) and PLGA are also commonly used where tunable properties are observed that lead to more controllable drug release kinetics and enhance the encapsulation efficiency of the drug cargo [211]. For example, Gelperina et al. prepared PLGA/poly(vinyl alcohol)(PVA) and PLGA/human serum albumin (HSA) NPs with surfactants poloxamer 188 or polysorbate 80, to enable efficient brain delivery [212]. High loading efficiencies of up to 75% and 97% DOX were achieved in the DOX-PLGA/PVA and DOX-PLGA/HSA NP formulations, respectively. Furthermore, drug release was measured in vitro and 1h postinjection, the percentage of free drug released reached 60% (DOX-PLGA/PVA) or 40% (DOX-PLGA/HSA), highlighting that the different formulation composition is a significant parameter that can enhance the pharmacokinetic profile of the drug.

Gong et al. utilized polymers to improve the drug delivery of the antiretroviral, elvitegravir (EVG) into the brain [213]. EVG was successfully encapsulated into positively charged NPs PLGA NPs, ~135nm in size. Using an in vitro BBB model (bEnd.3/C8-D1A astrocytes), the formulated NPs exhibited greater penetration compared to free EVG drug in a dose-dependent manner. In addition, the presence of the P-gp inhibitor elacridar further increased the in vitro BBB penetration for free EVG while the penetration of PLGA-EVG NPs showed no change in the presence and absence of the P-gp inhibitor. This indicated that while EVG transport across the CNS is partly modulated by P-gp efflux transportation, the efflux transporter P-gp may be bypassed by effective reformulation of EVG into polymeric NPs. Furthermore, brain delivery of specific ligand-conjugated NPs was greatly enhanced compared to unmodified NPs. Tosi et al. surface-modified PLGA NPs with a BBB-penetrating, similopioid peptide (g7) for BBB crossing and with a sialic acid (SA) residue to bind to brain receptors [214]. In vivo biodistribution was determined in Wistar-Hannover rats, where approximately 4%–6% of the injected NP dose was able to reach the brain in the SA-g7-PLGA-NP after 1h and the concentration was maintained over a 5h period. In contrast, 13% of the injected dose was observed in the g7-NP treatment group at 1.5h, however, the concentration decreased over where less than 0.7% of the injected dose was detected 24h after administration.

Other targeting moieties have been used to aid transport across the BBB for the treatment of GBM. For example, Kuo and Chen cross-linked lactoferrin and folic ligands on PLGA NPs and reported up to a twofold increase in the permeability of etoposide-loaded NPs across the BBB which in turn resulted in a twofold tumor suppression compared to the free etoposide treated group at 48h [215].

Some drawbacks associated with the formulation of polymeric NPs include the presence of residual organic solvents after preparation and the production of toxic monomers which can elicit an immunogenic response when administered [216]. Nondegradable polymers tend to accumulate in the liver and spleen which can lead to toxicity. Serious allergic reactions have been reported when NPs are delivered systemically including anaphylaxis, pulmonary and cerebral edema, volume overload, and platelet dysfunction [211]. Yang and Lai reported the existence of anti-PEG antibodies in about 42% of patients with no history of treatment with PEGylated products in healthy blood donors [217]. The higher prevalence of preexisting anti-PEG antibodies in healthy individuals was attributed to the frequent use of PEG-containing ingredients commonly used in processed food, personal care, cosmetics, and household cleaning products. The presence of preexisting anti-PEG antibodies might trigger further immunogenic responses when human subjects receive PEGylated therapeutics, thereby hindering clinical progress [211, 217]. Although, preparation methods for polymeric NPs are well-established, toxicity concerns must be addressed to permit wider clinical application.

Metallic gold nanoparticles

Metal oxides are commonly used in the diagnosis of diseases (fluorescent dyes, MRI contrast agents) and as drug delivery systems (photosensitizers) [218, 219]. Metallic NPs can resonate in the magnetic field, thus delivering energy directly to the target cancer cells [220]. Metal-based NPs include Au, silver (Ag), and zinc oxide NPs [221]. Au NPs are most commonly used as radiosensitizers for the treatment of solid tumors where radiation is applied and cancer cell death is induced by double-stranded DNA breaks [222]. Au has a high atomic number, showing high contrast and k-edge (~81 keV) and has preferential accumulation in tumors which makes it an ideal drug delivery system [223]. Sonavane et al. showed Au NPs with a hydrodynamic size of 15–50 nm could permeate effectively across the BBB [224].

Similar to many other NP-delivery systems, AuNPs have low specificity due to the lack of a targeted moiety, however, surface functionalization has enabled targeted delivery into the CNS [225]. Gonzalez-Carter and Porter functionalized the surface of multibranched nanoflower-like AuNPs with L-DOPA (L-DOPA-AuNFs) for the treatment of Parkinson's disease [226]. In an in vitro human BBB model (hCMEC/D3 monolayers), L-DOPA-AuNFs efficiently penetrated the BBB (37.9% of the injected dose) compared to similarly sized and shaped AuNFs functionalized with a nontargeting ligand. Furthermore, efficient BBB transcytosis occurred with minimal cell accumulation, which was indicative of efficient endothelial escape. Khongkow et al. encapsulated Au NPs with rabies virus glycoprotein (RVG)-targeted exosomes [225]. Exosomes are natural nanosized vesicles produced and secreted by cells, which have been utilized as a drug delivery system for low molecular weight drugs. Au NPs coated with RVG-exosome demonstrated effective BBB transport; 20% of particles had crossed the BBB at 24 h in an in vitro BBB coculture model (bEnd.3/ALT cells). This was also demonstrated in vivo where Au NPs coated with RVG-exosome had transport across BBB as determined by the fluorescence intensity of a fluorescent marker.

Magnetic particles

Although many studies have demonstrated the transport of metal NPs across the BBB after systemic administration, the percentage of the injected dose that reaches the target site is low compared to organs such as the liver and spleen [227]. The development of magnetic NPs has enabled enhanced and targeted drug delivery across the BBB in which magnetic NPs can be used to deliver therapeutic molecules, i.e., siRNA, cDNA, and polypeptides [228]. Magnetic NPs can also improve pharmacokinetics and lower toxicity [229]. Other applications of magnetic NPs include diagnostics and contrast agents for MRI, and therapy via magnetic fluid hyperthermia [230].

Magnetic NPs typically consist of an iron oxide metal core which imparts magnetic properties due to the presence of an unpaired electron [231]. Iron is a popular core material due to the associated low toxicity and metabolic elimination through the endogenous iron metabolic pathway where it is converted into a more stable form for safe elimination [232]. Multiple coating is used to enhance delivery and provide protection of the drug using nontoxic biocompatible materials such as polysaccharides (dextrin), PEG, synthetic polymer, phospholipids and proteins [233]. In particular, the transport of magnetic NPs across the BBB can be enhanced by the application of an extracranially applied magnetic force for targeted therapy of macromolecules to the brain [234]. Magnetic NPs can be divided into two subdivisions; paramagnetic NPs (PMNPs) which are greater than 100 nm and superparamagnetic iron oxide NPs (SPIONs) which are less than 100 nm. SPIONs are highly attractive for drug delivery [235].

The physical properties impart a high magnetic susceptibility in the presence of an external magnet in contrast to the weaker magnetic susceptibility of PMNPs due to the larger diameter, over large distances [231]. SPIONs are readily retained within tissues and undergo biocompatible degradation resulting in a good candidate for both targeting purposes and conjugation with biotherapeutics [235]. Fu et al. prepared targeted fluorescent SPION NPs which were delivered via tail vein injection into mice bearing a U87MG GBM xenograft model [236]. Particles were able to permeate the BBB and together with magnetic targeting, tumors regressed at a faster rate compared to when the magnetic force was withheld as observed by a tumor signal decay half-life of 0.853 and 6.197 days, respectively.

The surface coat enables the attachment of ligands or monoclonal antibodies such as anti-Tf receptor antibodies with a specific affinity for BECs which minimizes off target side effects [231].

Thomsen et al. encapsulated fluorescent red magnetic NPs in liposomes which were subsequently conjugated to antibodies, targeting the rat Tf receptor (OX26), to form magnetic immunoliposomes [237]. Enhanced uptake of magnetic liposomes by rat brain capillary endothelial cells (BCECs) was observed in vitro. The uptake of pure magnetic NPs increased 4.5-fold when encapsulated into a liposome and when OX26 monoclonal antibodies were conjugated to magnetic liposomes, uptake increased 7.3-fold and 1.6-fold compared to pure magnetic NPs and magnetic liposomes, respectively. In situ brain perfusion on P16 Wistar rats revealed enhanced uptake of magnetic OX26 immunoliposomes when compared to naked magnetic NPs or unmodified magnetic liposomes. Fluorescence detection with laminin highlighted the basal membrane and revealed that magnetic OX26 immunoliposomes were confined to the interior of BCECs. When an external magnetic force was applied, OX26 immunoliposomes permeated the intraparenchymal capillaries in a distinct perivascular labeling fluorescence distribution across the vascular bed, indicative of transport across the BBB through both BCECs and the basal membrane.

Similarly, magnetic NPs have been used in vivo to target gliomas. Marie et al. utilized magnetic liposome NPs to selectively target GBM in mice orthotopically implanted with U87MG glioma cells [238]. An eternal magnetic field was applied on the head of the mice to concentrate the magnetic NPs at the tumor site and in vivo tracking was achieved by quantitative MRI. Histological analysis using confocal fluorescence microscopy confirmed the significant accumulation of magnetic NPs in the cancerous tissue with effective internalization observed in endothelial and GBM cells.

Low toxicity is commonly observed with SPIONs as it has been reported that the iron oxide core is degraded to low molecular weight iron which is effectively cleared by brain macrophages, neurons, and microglia [235]. However, more work is needed to elucidate the effects on neuron activity. The high oxidative activity of the ferrous metabolite formed could ultimately cause neurodegeneration or a physiological mechanism could be triggered to remove excess iron and undergo incorporation in the body's iron stores [235, 239].

Peptide-based particles

CPPs are short cationic and/or amphipathic peptides that have been used to facilitate ubiquitous drug delivery across lipid bilayers [240]. CPPs undergo direct translocation across membranes to deliver the transported cargo into the intracellular compartment with limited associated toxicity in the absence of receptor recognition [241]. Peptides have garnered interest as delivery systems due to the safety, low immunogenicity, and high encapsulation efficiency profile commonly seen [242]. The TAT peptide is the most commonly used brain delivery peptide vehicle which is characterized by high permeability across nearly all cell membranes by AMT [118]. For example, the surface of liposomal encapsulated DOX was functionalized with the cell-penetrating TAT peptide (AYGRKKRRQRRR), to exploit AMT drug delivery for GBM treatment [243]. The TAT peptide effectively penetrated the BBB, to deliver DOX to C6 glioma-bearing rats. Qin et al. reported that the median survival of rats treated with DOX-TAT-LIP was significantly longer at 89 days than those of rats treated with physiological saline or DOX at 54 days and 32 days, respectively [243]. Another common targeting peptide for drug delivery across the BBB is the tripeptide, arginine-glycine-aspartic acid (RGD). It specifically binds αvβ3 which is overexpressed on tumor cells to facilitate BBB crossing via integrin-mediated transcytosis [244]. Typically, cyclic RGD peptides are preferred as this class of peptides is less prone to enzyme peptidolytic cleavage [245]. Belhadj et al. developed a multifunctional liposomal glioma-targeted drug delivery system (c(RGDyK)/pHA-LS) modified with cyclic RGD (c(RGDyK)) and p-hydroxybenzoic acid (pHA) in which c(RGDyK) could target integrin $\alpha_v\beta_3$ overexpressed on the BBB and tumor barrier and pHA could target dopamine receptors on the BBB [246]. Findings showed the fluorescence signal c(RGDyK)/pHA-LS group was much stronger at 4h postinjection in U87MG tumor-bearing BALB/c mice compared with controls, indicating BBB transport. Likewise, Liu et al. conjugated c(RGDfK) to the arginine-rich CPP, R8, to develop the multifunctional peptide, R8-RGD [247]. Authors reported a twofold increase in the cellular uptake of R8-RGD liposomes compared to R8 alone in vitro. Furthermore, in vivo studies demonstrated that R8-RGD-paclitaxel liposomes could be efficiently delivered into the brain after intravenous systemic administration in C6 glioma-bearing mice. Particles demonstrated BBB crossing and therapeutic efficacy where the paxcitaxel-R8-RGD-lipo group prolonged the medium survival to 48 days, significantly longer than control groups with no reported toxicity.

Moreover, cationic peptides have the ability to package anionic cargo such as nucleic acids and small molecules drugs in NPs and aid delivery into cells. McCarthy et al. developed the cationic amphipathic RALA peptide (WEARLARALARALARHLARALARALRACEA) for the delivery of anionic cargo and improved intracellular delivery [248–256]. It was recognized that amphipathicity promoted interaction with negatively charged cell membranes, mediated through heparan sulfate proteoglycans, to facilitate translocation of therapeutic cargo as supported by Chen et al. [254, 257] Furthermore, the cationic charge facilitates electrostatic interaction with anionic cargo where cargo is complexed into peptide NPs <200nm [254]. Unfortunately, for many nonviral delivery systems, the endosomal escape process is very inefficient, resulting in the entrapment of most of the cargo inside the endosome [258]. At the low pH of the mature endosome (pH < 5), many drug molecules become degraded, hindering the accumulation of therapeutic concentrations at the active site [170]. However, the pH-responsive fusogenic activity of RALA enables the release of complexed cargo before it can be degraded. As the endosome becomes more acidic, RALA basic amino acid residues in the peptide sequence become protonated which is associated with an increase in alpha-helicity [254]. It is proposed the protonation of the histidine amino residue contributes to the "proton sponge effect" and amphipathic characteristics of the peptide lead to endosomal fusion, enabling the release of the cargo. The unique characteristics of RALA make it an ideal drug delivery system for transport across the BBB; the cationic charge facilitates AMT where the charged particle interacts with the membrane of BECs and enables the delivery of anionic molecules. Additionally, the pH-responsive nature of RALA initiates endosomal escape during endocytosis, enhancing the intracellular availability of the drug.

Furthermore, while CPPs selectively pass the BBB, not all CPPs cross the BBB to the same extent where different uptake patterns have been observed based on CPP characteristics. Stalmans et al. investigated the in vivo BBB transport of five different CPPs with different cell-penetrating ability; pVEC, SynB3, Tat 47–57, transportan 10 (TP10) and TP10-2 [259]. TP10 and pVEC had the highest cellular uptake with a cell-penetrating response of 1.641 and 1.318. From this data, authors distinguished three groups of CPPs; the cationic-amphipathic peptide pVEC group showed a rapid brain influx, short cationic CPPs SynB3 and Tat 47–57 had relatively high initial brain influx rates, while the third group consisting of the transportan analogs, TP10 and TP10-2, showed either no or slow brain influx, respectively. The initial two cationic groups are composed of arginine-rich peptides, while the transportans only contain lysine-residues in their sequences with a reduced charge density of 19% in contrast to 44%, 50%, and 73% of pVEC, SynB3 and Tat 47–57, respectively. The results demonstrated arginine-rich CPPs had more effective and rapid BBB transport. Furthermore, authors found while all peptides have a random coil structure in solution, pVEC adopts a β-sheet structure and TP10 becomes α-helical, while SynB3 and Tat 47–57 remain randomly coiled at the membrane interface. This suggests the arginine-rich nature and alpha-helical structure of RALA are superior characteristics for BBB transport.

Therefore, while CPP properties of peptides do not directly infer BBB-penetrating ability, consideration must be given to other factors which can influence uptake. Similar to many cationic compounds, delivery can be nonselective and bioavailability is typically reduced due to the adsorption of serum proteins and MPS clearance in the systemic circulation [206]. However, the potential to conjugate a targeting moiety to the RALA peptide could be utilized to promote CNS targeting.

Conclusion

Despite advances in treatments, an unmet need persists for novel treatments for invasive GBM which ultimately improves patient outcomes. Here, we have described the current biological barriers which impede drug delivery and evolving strategies to overcome these barriers for GBM treatment (Table 4). While the intracranial delivery of therapeutics is promising, significant optimization is required before translation into the clinic as it has been reported only 0.7% of the administered dose of NPs reach solid tumors [273]. So far, liposomes have been the most successful nanomaterial-based formulation for clinical application, leading the way for FDA-approved nanomedicines. The success is predominantly governed

TABLE 4 Summary table showing the advantages and limitations of each delivery strategy in GBM.

Strategies	Advantages	Limitations	References
Osmotic opening	• Cargo size is not limited so facilitates the delivery of a wide range of agents • Reversible	• Risk of long-term damage—uncontrolled entry of non CNS molecules into the brain parenchyma during temporary disruption • Requires minimum osmolarity and duration of infusion • Quick reversal of permeability	[101]
FUS and microbubbles	• Noninvasive • Reversible • Can be used to enhance brain uptake of a wide variety of drugs	• Expensive, • Can overheat the brain • Risk of long-term damage—uncontrolled entry of non CNS molecules into the brain parenchyma during temporary disruption	[97, 260]
Nonthermal reversible electroporation	• Reversible • Specific targeting • Rapid healing	• Risk of thermal damage to extracellular structures • Risk of pores within the membrane not closing	[261]
Intranasal	• Rapid • Direct so avoid first pass metabolism, large surface area available for drug transport	• Clearance by mucus layer and cilia within nose • Poor drug delivery	[262, 263]
Prodrug analogs	• Tissue targeted delivery of lipophilic molecules	• Poor biological activity	[149, 264]
Efflux pump inhibitors	• Can be used to enhance brain uptake of a wide variety of drugs	• Reduce the BBB natural defense from toxins • Normally well tolerated may become neurotoxic in the absence of ABC transporters at the BBB	[80]
Liposomes	• Enhance drug efficacy	• Potentially toxic to membranes and immunogenic	[184]
Polymeric nanoparticles	• Controlled drug release is possible • Numerous biocompatible polymers have been approved for biomedical applications	• Clearance by MPS • Inflammatory response • Degradation pathway	[163, 265, 266]
Metallic Au nanoparticles	• Biocompatibility • Modifiable • Can be used for both diagnosis and treatment	• Low brain targeting • Stability issues during storage	[267, 268]

Continued

TABLE 4 Summary table showing the advantages and limitations of each delivery strategy in GBM—cont'd

Strategies	Advantages	Limitations	References
Magnetic particles	• Large surface area • Easy synthesis • Low cytotoxicity	• Expensive cost of synthesis material • Guided mobility dependent om external source • Limited scale up production	[221, 269, 270]
Peptide-based particles	• Drug is able to cross the BBB • Low toxicity and immunogenic profile • Effective and safe	• Nontargeted cationic NPs lack specificity • Potential for opsonization and removal from systematic circulation	[271, 272]

by the versatility where both hydrophobic and hydrophilic drugs can be encapsulated and the structural similarity to mammalian cell membranes [185].

However, no nanomedicine is yet approved for the primary treatment of GBM which is made more difficult due to the presence of the BBB and the aggressive phenotype of GBM. Novel approaches involve targeting overexpressed proteins and receptors on BECs to promote drug delivery across the BBB associated with minimal off target effects. Furthermore, combination therapies with conventional chemotherapies have already proved successful in GBM treatment in vitro and in vivo using various NP formulations. To this end, clinical translation is very much a possibility for the strategies described.

References

[1] Taylor OG, Brzozowski JS, Skelding KA. Glioblastoma Multiforme: an overview of emerging therapeutic targets. Front Oncol 2019;9(963):1–11.
[2] Louis DN, Perry A, Reifenberger G, von Deimling A, Figarella-Branger D, Cavenee WK, et al. The 2016 World Health Organization classification of tumors of the central nervous system: a summary. Acta Neuropathol 2016;131(6):803–20.
[3] Fernandes C, Costa A. Current standards of care in glioblastoma therapy. In: Glioblastoma. Codon Publications; 2017. p. 197–241.
[4] Parasramka S, Talari G, Rosenfeld M, Guo J, Villano JL. Procarbazine, lomustine and vincristine for recurrent high-grade glioma. Cochrane Database Syst Rev 2017;7(7):CD011773.
[5] Davis M. Glioblastoma: overview of disease and treatment. Clin J Oncol Nurs 2016;20(5):S2–8.
[6] Yu W, Zhang L, Wei Q, Shao A. O6-Methylguanine-DNA methyltransferase (MGMT): challenges and new opportunities in glioma chemotherapy. Front Oncol 2020;9(1547):1–11.
[7] Hersh DS, Wadajkar AS, Roberts N, Perez JG, Connolly NP, Frenkel V, et al. Evolving drug delivery strategies to overcome the blood brain barrier. Curr Pharm Des 2016;22(9):1177–93.
[8] Pardridge WM. Drug transport across the blood–brain barrier. J Cereb Blood Flow Metab 2012;32(11):1959–72.
[9] Carlsson SK, Brothers SP, Wahlestedt C. Emerging treatment strategies for glioblastoma multiforme. EMBO Mol Med 2014;6(11):1359–70.
[10] Patra JK, Das G, Fraceto LF, Campos EVR, Rodriguez-Torres MDP, Acosta-Torres LS, et al. Nano based drug delivery systems: recent developments and future prospects. J Nanobiotechnol 2018;16(1):71.
[11] Gustafson HH, Holt-Casper D, Grainger DW, Ghandehari H. Nanoparticle uptake: the phagocyte problem. Nano Today 2015;10(4):487–510.
[12] D'Alessio A, Proietti G, Sica G, Scicchitano BM. Pathological and molecular features of glioblastoma and its peritumoral tissue. Cancers (Basel) 2019;11(4):469.

[13] Daneman R, Prat A. The blood-brain barrier. Cold Spring Harb Perspect Biol 2015;7(1):1–23.
[14] Serlin Y, Shelef I, Knyazer B, Friedman A. Anatomy and physiology of the blood–brain barrier. Semin Cell Dev Biol 2015;38:2–6.
[15] Pardridge WM. The blood-brain barrier: bottleneck in brain drug development. NeuroRX 2005;2(1):3–14.
[16] Sharma G, Lakkadwala S, Modgil A, Singh J. The role of cell-penetrating peptide and transferrin on enhanced delivery of drug to brain. Int J Mol Sci 2016;17(6):806.
[17] Strazielle N, Ghersi-Egea J-F. Potential pathways for CNS drug delivery across the blood-cerebrospinal fluid barrier. Curr Pharm Des 2016;22(35):5463–76.
[18] Haddad-Tóvolli R, Dragano NRV, Ramalho AFS, Velloso LA. Development and function of the blood-brain barrier in the context of metabolic control. Front Neurosci 2017;11(224):1–12.
[19] García-Cáceres C, Balland E, Prevot V, Luquet S, Woods SC, Koch M, et al. Role of astrocytes, microglia, and tanycytes in brain control of systemic metabolism. Nat Neurosci 2019;22(1):7–14.
[20] Verkhratsky A, Nedergaard M. Physiology of astroglia. Physiol Rev 2018;98(1):239–389.
[21] Stobart JL, Anderson CM. Multifunctional role of astrocytes as gatekeepers of neuronal energy supply. Front Cell Neurosci 2013;7(38):1–21.
[22] Xu S, Lu J, Shao A, Zhang JH, Zhang J. Glial cells: role of the immune response in ischemic stroke. Front Immunol 2020;11(294):1–16.
[23] Zhou Y, Peng Z, Seven ES, Leblanc RM. Crossing the blood-brain barrier with nanoparticles. J Control Release 2018;270:290–303.
[24] Haseloff RF, Blasig IE, Bauer H-C, Bauer H. In search of the astrocytic factor(s) modulating blood–brain barrier functions in brain capillary endothelial cells in vitro. Cell Mol Neurobiol 2005;25(1):25–39.
[25] Fu BM. Experimental methods and transport models for drug delivery across the blood-brain barrier. Curr Pharm Biotechnol 2012;13(7):1346–59.
[26] Pajouhesh H, Lenz GR. Medicinal chemical properties of successful central nervous system drugs. NeuroRX 2005;2(4):541–53.
[27] Romero IA, Radewicz K, Jubin E, Michel CC, Greenwood J, Couraud P-O, et al. Changes in cytoskeletal and tight junctional proteins correlate with decreased permeability induced by dexamethasone in cultured rat brain endothelial cells. Neurosci Lett 2003;344(2):112–6.
[28] Wettschureck N, Strilic B, Offermanns S. Passing the vascular barrier: endothelial signaling processes controlling extravasation. Physiol Rev 2019;99(3):1467–525.
[29] Mikitsh JL, Chacko AM. Pathways for small molecule delivery to the central nervous system across the blood-brain barrier. Perspect Med Chem 2014;6:11–24.
[30] Saraiva C, Praça C, Ferreira R, Santos T, Ferreira L, Bernardino L. Nanoparticle-mediated brain drug delivery: overcoming blood-brain barrier to treat neurodegenerative diseases. J Control Release 2016;235:34–47.
[31] Abbott NJ, Patabendige AAK, Dolman DEM, Yusof SR, Begley DJ. Structure and function of the blood–brain barrier. Neurobiol Dis 2010;37(1):13–25.
[32] Grabrucker AM, Ruozi B, Belletti D, Pederzoli F, Forni F, Vandelli MA, et al. Nanoparticle transport across the blood brain barrier. Tissue Barriers 2016;4(1):e1153568.
[33] Wong AD, Ye M, Levy AF, Rothstein JD, Bergles DE, Searson PC. The blood-brain barrier: an engineering perspective. Front Neuroeng 2013;6(7):1–22.
[34] Bouzinab K, Summers H, Zhang J, Stevens MFG, Moody CJ, Turyanska L, et al. In search of effective therapies to overcome resistance to Temozolomide in brain tumours. Cancer Drug Resist 2019;2:1018–31.
[35] Banks WA. Characteristics of compounds that cross the blood-brain barrier. BMC Neurol 2009;9(Suppl. 1):5–9.
[36] Tosi G, Bortot B, Ruozi B, Dolcetta D, Vandelli MA, Forni F, et al. Potential use of polymeric nanoparticles for drug delivery across the blood-brain barrier. Curr Med Chem 2013;20(17):2212–25.
[37] Schaefer CP, Tome ME, Davis TP. The opioid epidemic: a central role for the blood brain barrier in opioid analgesia and abuse. Fluids Barriers CNS 2017;14(1):32.
[38] Kok-Yong S, Lawrence L. Drug distribution and drug elimination. In: Basic pharmacokinetic concepts and some clinical applications. Intechopen; 2015.
[39] Redzic Z. Molecular biology of the blood-brain and the blood-cerebrospinal fluid barriers: similarities and differences. Fluids Barriers CNS 2011;8(1):3.
[40] Worzfeld T, Schwaninger M. Apicobasal polarity of brain endothelial cells. J Cereb Blood Flow Metab 2016; 36(2):340–62.

[41] Reitsma S, Slaaf DW, Vink H, van Zandvoort MAMJ, MGA oE. The endothelial glycocalyx: composition, functions, and visualization. Pflugers Arch 2007;454(3):345–59.
[42] Barar J, Rafi MA, Pourseif MM, Omidi Y. Blood-brain barrier transport machineries and targeted therapy of brain diseases. Bioimpacts 2016;6(4):225–48.
[43] Tsuji A. Small molecular drug transfer across the blood-brain barrier via carrier-mediated transport systems. NeuroRX 2005;2(1):54–62.
[44] John R, Vaswani H, Dandekar P, Devarajan PV. Brain cancer receptors and targeting strategies. In: Targeted intracellular drug delivery by receptor mediated endocytosis. AAPS Advances in the Pharmaceutical Sciences Series. Springer; 2019.
[45] Smith QR. Transport of glutamate and other amino acids at the blood-brain barrier. J Nutr 2000;130(4):1016S–22S.
[46] Miller DS. Regulation of ABC transporters at the blood-brain barrier. Clin Pharmacol Ther 2015;97(4):395–403.
[47] Agarwal S, Uchida Y, Mittapalli RK, Sane R, Terasaki T, Elmquist WF. Quantitative proteomics of transporter expression in brain capillary endothelial cells isolated from P-glycoprotein (P-gp), breast cancer resistance protein (Bcrp), and P-gp/Bcrp knockout mice. Drug Metab Dispos 2012;40(6):1164–9.
[48] Gao H. Progress and perspectives on targeting nanoparticles for brain drug delivery. Acta Pharm Sin B 2016;6(4):268–86.
[49] Dauchy S, Dutheil F, Weaver RJ, Chassoux F, Daumas-Duport C, Couraud P-O, et al. ABC transporters, cytochromes P450 and their main transcription factors: expression at the human blood-brain barrier. J Neurochem 2008;107(6):1518–28.
[50] Sanchez-Covarrubias L, Slosky LM, Thompson BJ, Davis TP, Ronaldson PT. Transporters at CNS barrier sites: obstacles or opportunities for drug delivery? Curr Pharm Des 2014;20(10):1422–49.
[51] Gaillard PJ, Brink A, de Boer AG. Diphtheria toxin receptor-targeted brain drug delivery. Int Congr Ser 2005;1277:185–98.
[52] Kang Y-S, Ohtsuki S, Takanaga H, Tomi M, Hosoya K, Terasaki T. Regulation of taurine transport at the blood-brain barrier by tumor necrosis factor-α, taurine and hypertonicity. J Neurochem 2002;83(5):1188–95.
[53] Villaseñor R, Lampe J, Schwaninger M, Collin L. Intracellular transport and regulation of transcytosis across the blood–brain barrier. Cell Mol Life Sci 2019;76(6):1081–92.
[54] Cardoso I, Santos LM, Alemi M. Modulating role of TTR in Aβ toxicity, from health to disease. In: Exploring new findings on amyloidosis. InTech; 2016. p. 18.
[55] Beloqui A, des Rieux A, Préat V. Mechanisms of transport of polymeric and lipidic nanoparticles across the intestinal barrier. Adv Drug Deliv Rev 2016;106:242–55.
[56] Hervé F, Ghinea N, Scherrmann J-M. CNS delivery via adsorptive transcytosis. AAPS J 2008;10(3):455–72.
[57] Xiao G, Gan L-S. Receptor-mediated endocytosis and brain delivery of therapeutic biologics. Int J Cell Biol 2013;2013:1–14.
[58] Jena L, McErlean E, McCarthy H. Delivery across the blood-brain barrier: nanomedicine for glioblastoma multiforme. Drug Deliv Transl Res 2020;10(2):304–18.
[59] Lu W, Zhang Y, Tan Y-Z, Hu K-L, Jiang X-G, Fu S-K. Cationic albumin-conjugated pegylated nanoparticles as novel drug carrier for brain delivery. J Control Release 2005;107(3):428–48.
[60] Lu W, Wan J, She Z, Jiang X. Brain delivery property and accelerated blood clearance of cationic albumin conjugated pegylated nanoparticle. J Control Release 2007;118(1):38–53.
[61] Li Y, Zheng X, Gong M, Zhang J. Delivery of a peptide-drug conjugate targeting the blood brain barrier improved the efficacy of paclitaxel against glioma. Oncotarget 2016;7(48):79401–7.
[62] Lajoie JM, Shusta EV. Targeting receptor-mediated transport for delivery of biologics across the blood-brain barrier. Annu Rev Pharmacol Toxicol 2015;55(1):613–31.
[63] Fung KYY, Fairn GD, Lee WL. Transcellular vesicular transport in epithelial and endothelial cells: challenges and opportunities. Traffic 2018;19(1):5–18.
[64] Tortorella S, Karagiannis TC. Transferrin receptor-mediated endocytosis: a useful target for cancer therapy. J Membr Biol 2014;247(4):291–307.
[65] Yiannikourides A, Latunde-Dada G. A Short review of Iron metabolism and pathophysiology of iron disorders. Medicines 2019;6(3):85.
[66] Shen Y, Li X, Dong D, Zhang B, Xue Y, Shang P. Transferrin receptor 1 in cancer: a new sight for cancer therapy. Am J Cancer Res 2018;8(6):916–31.

[67] Calzolari A, Larocca LM, Deaglio S, Finisguerra V, Boe A, Raggi C, et al. Transferrin receptor 2 is frequently and highly expressed in glioblastomas. Transl Oncol 2010;3(2):123–34.
[68] Recht L, Torres CO, Smith TW, Raso V, Griffin TW. Transferrin receptor in normal and neoplastic brain tissue: implications for brain-tumor immunotherapy. J Neurosurg 1990;72(6):941–5.
[69] Rosager AM, Sørensen MD, Dahlrot RH, Hansen S, Schonberg DL, Rich JN, et al. Transferrin receptor-1 and ferritin heavy and light chains in astrocytic brain tumors: expression and prognostic value. PLoS One 2017;12(8):e0182954.
[70] Paterson J, Webster CI. Exploiting transferrin receptor for delivering drugs across the blood-brain barrier. Drug Discov Today Technol 2016;20:49–52.
[71] Clark AJ, Davis ME. Increased brain uptake of targeted nanoparticles by adding an acid-cleavable linkage between transferrin and the nanoparticle core. Proc Natl Acad Sci U S A 2015;112(40):12486–91.
[72] Kanekiyo T, Bu G. The low-density lipoprotein receptor-related protein 1 and amyloid-B clearance in Alzheimer's disease. Front Aging Neurosci 2014;6(93):1–12.
[73] Bertrand Y, Currie J-C, Poirier J, Demeule M, Abulrob A, Fatehi D, et al. Influence of glioma tumour microenvironment on the transport of ANG1005 via low-density lipoprotein receptor-related protein 1. Br J Cancer 2011;105(11):1697–707.
[74] Sagare AP, Deane R, Zlokovic BV. Low-density lipoprotein receptor-related protein 1: a physiological Aβ homeostatic mechanism with multiple therapeutic opportunities. Pharmacol Ther 2012;136(1):94–105.
[75] Guo Q, Zhu Q, Miao T, Tao J, Ju X, Sun Z, Li H, Xu G, Chen H, Han L. LRP1-upregulated nanoparticles for efficiently conquering the blood-brain barrier and targetedly suppressing multifocal and infiltrative brain metastases. J Control Release 2019;303:117–29
[76] Shinohara M, Tachibana M, Kanekiyo T, Bu G. Role of LRP1 in the pathogenesis of Alzheimer's disease: evidence from clinical and preclinical studies. J Lipid Res 2017;58(7):1267–81.
[77] Rashid S, Curtis DE, Garuti R, Anderson NN, Bashmakov Y, Ho YK, et al. Decreased plasma cholesterol and hypersensitivity to statins in mice lacking Pcsk9. Proc Natl Acad Sci U S A 2005;102(15):5374–9.
[78] Lu Q, Cai X, Zhang X, Li S, Song Y, Du D, et al. Synthetic polymer nanoparticles functionalized with different ligands for receptor-mediated transcytosis across the blood–brain barrier. ACS Appl Bio Mater 2018;1 (5):1687–94.
[79] Karkan D, Pfeifer C, Vitalis TZ, Arthur G, Ujiie M, Chen Q, et al. A unique carrier for delivery of therapeutic compounds beyond the blood-brain barrier. PLoS One 2008;3(6):e2469.
[80] Löscher W, Potschka H. Blood-brain barrier active efflux transporters: ATP-binding cassette gene family. NeuroRX 2005;2(1):86–98.
[81] Shukla S, Ohnuma S, Ambudkar SV. Improving cancer chemotherapy with modulators of ABC drug transporters. Curr Drug Targets 2011;12(5):621–30.
[82] Godbey WT. Cellular transport. In: An introduction to biotechnology. Elsevier Ltd. 2014. p. 35–64.
[83] Choi YH, Yu A-M. ABC transporters in multidrug resistance and pharmacokinetics, and strategies for drug development. Curr Pharm Des 2014;20(5):793–807.
[84] Davis TP, Sanchez-Covarubias L, Tome ME. P-glycoprotein trafficking as a therapeutic target to optimize CNS drug delivery. Adv Pharmacol 2014;71:25–44.
[85] Wang D, Wang C, Wang L, Chen Y. A comprehensive review in improving delivery of small-molecule chemotherapeutic agents overcoming the blood-brain/brain tumor barriers for glioblastoma treatment. Drug Deliv 2019;26(1):551–65.
[86] Stupp R, Hegi ME, Mason WP, van den Bent MJ, Taphoorn MJ, Janzer RC, et al. Effects of radiotherapy with concomitant and adjuvant temozolomide versus radiotherapy alone on survival in glioblastoma in a randomised phase III study: 5-year analysis of the EORTC-NCIC trial. Lancet Oncol 2009;10(5):459–66.
[87] de Gooijer MC, de Vries NA, Buckle T, Buil LCM, Beijnen JH, Boogerd W, et al. Improved brain penetration and antitumor efficacy of temozolomide by inhibition of ABCB1 and ABCG2. Neoplasia 2018;20(7):710–20.
[88] Yergeri M, Kapse-Mistry S, Srivastava R, Govender T. Nanodrug delivery in reversing multidrug resistance in cancer cells. Front Pharmacol 2014;5(159):1–22.
[89] López-Fernández LA. ATP-binding cassette transporters in the clinical implementation of pharmacogenetics. J Perinat Med 2018;8(4):40.
[90] Whitlock BD, Leslie EM. Efflux transporters in anti-cancer drug resistance: Molecular and functional identification and characterization of multidrug resistance proteins (MRPs/ABCCs). In: Drug efflux pumps in cancer

resistance pathways: from molecular recognition and characterization to possible inhibition strategies in chemotherapy. Elsevier; 2020. p. 31–65.
[91] Madsen SJ, Hirschberg H. Site-specific opening of the blood-brain barrier. J Biophotonics 2010;3(5–6):356–67.
[92] Małkiewicz MA, Szarmach A, Sabisz A, Cubała WJ, Szurowska E, Winklewski PJ. Blood-brain barrier permeability and physical exercise. J Neuroinflammation 2019;16(1):15.
[93] Deng CX. Targeted drug delivery across the blood–brain barrier using ultrasound technique. Ther Deliv 2010; 1(6):819–48.
[94] Mainprize T, Lipsman N, Huang Y, Meng Y, Bethune A, Ironside S, et al. Blood-brain barrier opening in primary brain tumors with non-invasive MR-guided focused ultrasound: a clinical safety and feasibility study. Sci Rep 2019;9(1):321.
[95] Keep RF, Zhou N, Xiang J, Andjelkovic AV, Hua Y, Xi G. Vascular disruption and blood–brain barrier dysfunction in intracerebral hemorrhage. Fluids Barriers CNS 2014;11(1):18.
[96] Obermeier B, Daneman R, Ransohoff RM. Development, maintenance and disruption of the blood-brain barrier. Nat Med 2013;19(12):1584–96.
[97] Parodi A, Rudzińska M, Deviatkin A, Soond S, Baldin A, Zamyatnin A. Established and emerging strategies for drug delivery across the blood-brain barrier in brain cancer. Pharmaceutics 2019;11(5):245.
[98] Al-Ahmady ZS. Selective drug delivery approaches to lesioned brain through blood brain barrier disruption. Expert Opin Drug Deliv 2018;15(4):335–49.
[99] Rodriguez A, Tatter S, Debinski W. Neurosurgical techniques for disruption of the blood–brain barrier for glioblastoma treatment. Pharmaceutics 2015;7(3):175–87.
[100] Dong X. Current strategies for brain drug delivery. Theranostics 2018;8(6):1481–93.
[101] Rapoport S. Osmotic opening of the blood–brain barrier: principles, mechanism, and therapeutic applications. Cell Mol Neurobiol 2000;20:217–30.
[102] Rapoport S, Hori M, Klatzo I. Testing of a hypothesis for osmotic opening of the blood-brain barrier. Am J Physiol 1972;223(2):323–31.
[103] Godinho BMDC, Henninger N, Bouley J, Alterman JF, Haraszti RA, Gilbert JW, et al. Transvascular delivery of hydrophobically modified siRNAs: gene silencing in the rat brain upon disruption of the blood-brain barrier. Mol Ther 2018;26(11):2580–91.
[104] Choi C, Kim HM, Shon J, Park J, Kim H-T, Oh S-H, et al. Additional increased effects of mannitol-temozolomide combined treatment on blood-brain barrier permeability. Biochem Biophys Res Commun 2018;497(2):769–75.
[105] Guillaume DJ, Doolittle ND, Gahramanov S, Hedrick NA, Delashaw JB, Neuwelt EA. Intra-arterial chemotherapy with osmotic blood-brain barrier disruption for aggressive oligodendroglial tumors: results of a phase I study. Neurosurgery 2010;66(1):48–58.
[106] Marchi N, Angelov L, Masaryk T, Fazio V, Granata T, Hernandez N, et al. Seizure-promoting effect of blood? Brain barrier disruption. Epilepsia 2007;48(4):732–42.
[107] Kiviniemi V, Korhonen V, Kortelainen J, Rytky S, Keinänen T, Tuovinen T, et al. Real-time monitoring of human blood-brain barrier disruption. PLoS One 2017;12(3):e0174072.
[108] Fisher DG, Price RJ. Recent advances in the use of focused ultrasound for magnetic resonance image-guided therapeutic nanoparticle delivery to the central nervous system. Front Pharmacol 2019;10(1348):1–14.
[109] Hynynen K, Jones RM. Image-guided ultrasound phased arrays are a disruptive technology for non-invasive therapy. Phys Med Biol 2016;61(17):R206–48.
[110] Prada F, Kalani MYS, Yagmurlu K, Norat P, Del Bene M, DiMeco F, et al. Applications of focused ultrasound in cerebrovascular diseases and brain tumors. Neurotherapeutics 2019;16(1):67–87.
[111] Fan C-H, Yeh C-K. Microbubble-enhanced focused ultrasound-induced blood–brain barrier opening for local and transient drug delivery in central nervous system disease. J Med Ultrasound 2014;22(4):183–93.
[112] Patrick JT, Nolting MN, Goss SA, Dines KA, Clendenon JL, Rea MA, Heimburger RF. Ultrasound and the blood-brain barrier. Adv Exp Med Biol. 1990;267:369–81.
[113] Vykhodtseva NI, Hynynen K, Damianou C. Pulse duration and peak intensity during focused ultrasound surgery: theoretical and experimental effects in rabbit brain in vivo. Ultrasound Med Biol 1994;20(9):987–1000.
[114] Su W-S, Tsai M-L, Huang S-L, Liu S-H, Yang F-Y. Controllable permeability of blood-brain barrier and reduced brain injury through low-intensity pulsed ultrasound stimulation. Oncotarget 2015;6(39):42290–9.
[115] Mesiwala AH, Farrell L, Wenzel HJ, Silbergeld DL, Crum LA, Winn HR, et al. High-intensity focused ultrasound selectively disrupts the blood-brain barrier in vivo. Ultrasound Med Biol 2002;28(3):389–400.

[116] Azad TD, Pan J, Connolly ID, Remington A, Wilson CM, Grant GA. Therapeutic strategies to improve drug delivery across the blood-brain barrier. Neurosurg Focus 2015;38(3):E9.
[117] Song K-H, Harvey BK, Borden MA. State-of-the-art of microbubble-assisted blood-brain barrier disruption. Theranostics 2018;8(16):4393–408.
[118] Bors L, Erdő F. Overcoming the blood–brain barrier. Challenges and tricks for CNS drug delivery. Sci Pharm 2019;87(1):6.
[119] Chen K-T, Wei K-C, Liu H-L. Theranostic strategy of focused ultrasound induced blood-brain barrier opening for CNS disease treatment. Front Pharmacol 2019;10(86):1–14.
[120] Zhao B, Chen Y, Liu J, Zhang L, Wang J, Yang Y, et al. Blood-brain barrier disruption induced by diagnostic ultrasound combined with microbubbles in mice. Oncotarget 2018;9(4):4897–914.
[121] Lin Q, Mao K-L, Tian F-R, Yang J-J, Chen P-P, Xu J, et al. Brain tumor-targeted delivery and therapy by focused ultrasound introduced doxorubicin-loaded cationic liposomes. Cancer Chemother Pharmacol 2016;77(2):269–80.
[122] Downs ME, Buch A, Sierra C, Karakatsani ME, Chen S, Konofagou EE, et al. Long-term safety of repeated blood-brain barrier opening via focused ultrasound with microbubbles in non-human primates performing a cognitive task. PLoS One 2015;10(5):e0125911.
[123] McDannold N, Arvanitis CD, Vykhodtseva N, Livingstone MS. Temporary disruption of the blood-brain barrier by use of ultrasound and microbubbles: safety and efficacy evaluation in rhesus macaques. Cancer Res 2012;72(14):3652–63.
[124] Jones RM, O'Reilly MA, Hynynen K. Transcranial passive acoustic mapping with hemispherical sparse arrays using CT-based skull-specific aberration corrections: a simulation study. Phys Med Biol 2013;58(14):4981–5005.
[125] Wei K-C, Tsai H-C, Lu Y-J, Yang H-W, Hua M-Y, Wu M-F, et al. Neuronavigation-guided focused ultrasound-induced blood-brain barrier opening: a preliminary study in swine. Am J Neuroradiol 2013;34(1):115–20.
[126] Wu S-Y, Aurup C, Sanchez CS, Grondin J, Zheng W, Kamimura H, et al. Efficient blood-brain barrier opening in primates with neuronavigation-guided ultrasound and real-time acoustic mapping. Sci Rep 2018;8(1):7978.
[127] Abrahao A, Meng Y, Llinas M, Huang Y, Hamani C, Mainprize T, et al. First-in-human trial of blood–brain barrier opening in amyotrophic lateral sclerosis using MR-guided focused ultrasound. Nat Commun 2019;10(1):4373.
[128] Frinking P, Segers T, Luan Y, Tranquart F. Three decades of ultrasound contrast agents: a review of the past, present and future improvements. Ultrasound Med Biol 2020;46(4):892–908.
[129] Fabian D, Guillermo Prieto Eibl M, Alnahhas I, Sebastian N, Giglio P, Puduvalli V, et al. Treatment of glioblastoma (GBM) with the addition of tumor-treating fields (TTF): a review. Cancers (Basel) 2019;11(2):174.
[130] ClinicalTrials.gov [Internet]. Available from https://clinicaltrials.gov/ct2/show/NCT03744026.
[131] Bonakdar M, Wasson EM, Lee YW, Davalos RV. Electroporation of brain endothelial cells on chip toward permeabilizing the blood-brain barrier. Biophys J 2016;110(2):503–13.
[132] Al-Sakere B, André F, Bernat C, Connault E, Opolon P, Davalos RV, et al. Tumor ablation with irreversible electroporation. PLoS One 2007;2(11):e1135.
[133] Kotnik T, Rems L, Tarek M, Miklavčič D. Membrane electroporation and electropermeabilization: mechanisms and models. Annu Rev Biophys 2019;48(1):63–91.
[134] Sharabi S, Bresler Y, Ravid O, Shemesh C, Atrakchi D, Schnaider-Beeri M, et al. Transient blood–brain barrier disruption is induced by low pulsed electrical fields in vitro: an analysis of permeability and trans-endothelial electric resistivity. Drug Deliv 2019;26(1):459–69.
[135] Shi J, Ma Y, Zhu J, Chen Y, Sun Y, Yao Y, et al. A review on electroporation-based intracellular delivery. Molecules 2018;23(11):3044.
[136] Hjouj M, Last D, Guez D, Daniels D, Sharabi S, Lavee J, et al. MRI study on reversible and irreversible electroporation induced blood brain barrier disruption. PLoS One 2012;7(8):e42817.
[137] Lorenzo MF, Thomas SC, Kani Y, Hinckley J, Lee M, Adler J, et al. Temporal characterization of blood–brain barrier disruption with high-frequency electroporation. Cancers (Basel) 2019;11(12):1850.
[138] Gänger S, Schindowski K. Tailoring formulations for intranasal nose-to-brain delivery: a review on architecture, physico-chemical characteristics and mucociliary clearance of the nasal olfactory mucosa. Pharmaceutics 2018;10(3):116.
[139] Wang W, Swenson S, Cho H-Y, Hofman FM, Schönthal AH, Chen TC. Efficient brain targeting and therapeutic intracranial activity of bortezomib through intranasal co-delivery with NEO100 in rodent glioblastoma models. J Neurosurg 2020;132(3):959–67.

[140] Bruinsmann FA, Richter Vaz G, de Cristo Soares Alves A, Aguirre T, Raffin Pohlmann A, Stanisçuaski Guterres S, et al. Nasal drug delivery of anticancer drugs for the treatment of glioblastoma: preclinical and clinical trials. Molecules 2019;24(23):4312.
[141] da Fonseca C, Khandelia H, Salazar M, Schönthal A, Meireles O, Quirico-Santos T. Perillyl alcohol: dynamic interactions with the lipid bilayer and implications for long-term inhalational chemotherapy for gliomas. Surg Neurol Int 2016;7(1):1.
[142] ClinicalTrials.gov [Internet]. Available from https://clinicaltrials.gov/ct2/show/NCT02704858.
[143] Chu L, Wang A, Ni L, Yan X, Song Y, Zhao M, et al. Nose-to-brain delivery of temozolomide-loaded PLGA nanoparticles functionalized with anti-EPHA3 for glioblastoma targeting. Drug Deliv 2018;25(1):1634–41.
[144] van de Waterbeemd H, Smith DA, Jones BC. Lipophilicity in PK design: methyl, ethyl, futile. J Comput Aided Mol Des 2001;15:273–86.
[145] Waterhouse R. Determination of lipophilicity and its use as a predictor of blood–brain barrier penetration of molecular imaging agents. Mol Imaging Biol 2003;5(6):376–89.
[146] Kiani YS, Jabeen I. Lipophilic metabolic efficiency (LipMetE) and drug efficiency indices to explore the metabolic properties of the substrates of selected cytochrome P450 isoforms. ACS Omega 2020;5(1):179–88.
[147] Rautio J, Meanwell NA, Di L, Hageman MJ. The expanding role of prodrugs in contemporary drug design and development. Nat Rev Drug Discov 2018;17(8):559–87.
[148] Caruso G, Merlo L, Caffo M. Brain drug-delivery attempts. In: Innovative brain tumor therapy. Woodhead Publishing; 2014. p. 35–52.
[149] Rautio J, Kumpulainen H, Heimbach T, Oliyai R, Oh D, Järvinen T, et al. Prodrugs: design and clinical applications. Nat Rev Drug Discov 2008;7(3):255–70.
[150] Placzek AT, Ferrara SJ, Hartley MD, Sanford-Crane HS, Meinig JM, Scanlan TS. Sobetirome prodrug esters with enhanced blood–brain barrier permeability. Bioorg Med Chem 2016;24(22):5842–54.
[151] Haddad F, Sawalha M, Khawaja Y, Naijar A, Karaman R. Dopamine and levodopa prodrugs for the treatment of Parkinson's disease. Molecules 2017;23(1):40.
[152] Hauser RA. Levodopa: past, present, and future. Eur Neurol 2009;62(1):1–8.
[153] Amin ML. P-glycoprotein inhibition for optimal drug delivery. Drug Target Insights 2013;7:27–34.
[154] Zanger UM, Schwab M. Cytochrome P450 enzymes in drug metabolism: regulation of gene expression, enzyme activities, and impact of genetic variation. Pharmacol Ther 2013;138(1):103–41.
[155] Kumar S. Engineering cytochrome P450 biocatalysts for biotechnology, medicine and bioremediation. Expert Opin Drug Metab Toxicol 2010;6(2):115–31.
[156] Ahmed S, Zhou Z, Zhou J, Chen S-Q. Pharmacogenomics of drug metabolizing enzymes and transporters: relevance to precision medicine. Genomics Proteomics Bioinformatics 2016;14(5):298–313.
[157] Miksys SL, Tyndale RF. Drug-metabolizing cytochrome P450s in the brain. J Psychiatry Neurosci 2002;27 (6):406–15.
[158] Dagenais C, Graff CL, Pollack GM. Variable modulation of opioid brain uptake by P-glycoprotein in mice. Biochem Pharmacol 2004;67(2):269–76.
[159] Dewanjee S, Dua T, Bhattacharjee N, Das A, Gangopadhyay M, Khanra R, et al. Natural products as alternative choices for P-glycoprotein (P-gp) inhibition. Molecules 2017;22(6):871.
[160] Patel KJ, Tannock IF. The influence of P-glycoprotein expression and its inhibitors on the distribution of doxorubicin in breast tumors. BMC Cancer 2009;9(1):356.
[161] Dash RP, Jayachandra Babu R, Srinivas NR. Therapeutic potential and utility of elacridar with respect to P-glycoprotein inhibition: an insight from the published in vitro, preclinical and clinical studies. Eur J Drug Metab Pharmacokinet 2017;42(6):915–33.
[162] Tang SC, Kort A, Cheung KL, Rosing H, Fukami T, Durmus S, et al. P-glycoprotein, CYP3A, and plasma carboxylesterase determine brain disposition and oral availability of the novel taxane cabazitaxel (jevtana) in mice. Mol Pharm 2015;12(10):3714–23.
[163] Chenthamara D, Subramaniam S, Ramakrishnan SG, Krishnaswamy S, Essa MM, Lin F-H, et al. Therapeutic efficacy of nanoparticles and routes of administration. Biomater Res 2019;23(1):20.
[164] Masserini M. Nanoparticles for brain drug delivery. ISRN Biochem 2013;2013:1–18.
[165] Golombek SK, May J-N, Theek B, Appold L, Drude N, Kiessling F, et al. Tumor targeting via EPR: strategies to enhance patient responses. Adv Drug Deliv Rev 2018;130:17–38.

[166] He Q, Liu J, Liang J, Liu X, Li W, Liu Z, et al. Towards improvements for penetrating the blood–brain barrier—recent progress from a material and pharmaceutical perspective. Cell 2018;7(4):24.
[167] El-Sayed A, Harashima H. Endocytosis of gene delivery vectors: from clathrin-dependent to lipid raft-mediated endocytosis. Mol Ther 2013;21(6):1118–30.
[168] Lajoie P, Nabi IR. Regulation of raft-dependent endocytosis. J Cell Mol Med 2007;11(4):644–53.
[169] Kiss AL, Botos E. Endocytosis via caveolae: alternative pathway with distinct cellular compartments to avoid lysosomal degradation? J Cell Mol Med 2009;13(7):1228–37.
[170] Hu Y-B, Dammer EB, Ren R-J, Wang G. The endosomal-lysosomal system: from acidification and cargo sorting to neurodegeneration. Transl Neurodegener 2015;4(1):18.
[171] Huotari J, Helenius A. Endosome maturation. EMBO J 2011;30(17):3481–500.
[172] Blanco E, Shen H, Ferrari M. Principles of nanoparticle design for overcoming biological barriers to drug delivery. Nat Biotechnol 2015;33(9):941–51.
[173] Wani TU, Raza SN, Khan NA. Nanoparticle opsonization: forces involved and protection by long chain polymers. Polym Bull 2020;77(7):3865–89.
[174] Salmaso S, Caliceti P. Stealth properties to improve therapeutic efficacy of drug nanocarriers. J Drug Deliv 2013;2013:1–19.
[175] Nie S. Understanding and overcoming major barriers in cancer nanomedicine. Nanomedicine 2010;5(4):523–8.
[176] Oh JY, Kim HS, Palanikumar L, Go EM, Jana B, Park SA, et al. Cloaking nanoparticles with protein corona shield for targeted drug delivery. Nat Commun 2018;9(1):4548.
[177] Suk JS, Xu Q, Kim N, Hanes J, Ensign LM, Sciences H, et al. PEGylation as a strategy for improving nanoparticle-based drug and gene delivery. Adv Drug Deliv Rev 2016;99:28–51.
[178] Calvo P, Gouritin B, Chacun H, Desmaële D, D'Angelo J, Noel J-P, Georgin D, Fattal E, Andreux JP, Couvreur P. Long-circulating pegylated polycyanoacrylate nanoparticles as new drug carrier for brain delivery. Pharm Res 2001;18:1157–66.
[179] Cruje C, Chithrani DB. Polyethylene glycol density and length affects nanoparticle uptake by cancer cells. J Nanomed Res 2014;1(1):1–6.
[180] Gosk S, Vermehren C, Storm G, Moos T. Targeting anti—transferrin receptor antibody (OX26) and OX26-conjugated liposomes to brain capillary endothelial cells using in situ perfusion. J Cereb Blood Flow Metab 2004;24(11):1193–204.
[181] Bruun J, Larsen TB, Jolck RI, Eliasen R, Holm R, Gjetting T, et al. Investigation of enzyme-sensitive lipid nanoparticles for delivery of siRNA to blood—brain barrier and glioma cells. Int J Nanomedicine 2015;10:5995–6008.
[182] Bardania H, Tarvirdipour S, Dorkoosh F. Liposome-targeted delivery for highly potent drugs. Artif Cells Nanomed Biotechnol 2017;45(8):1478–89.
[183] Sercombe L, Veerati T, Moheimani F, Wu SY, Sood AK, Hua S. Advances and challenges of liposome assisted drug delivery. Front Pharmacol 2015;6(286):1–13.
[184] Bozzuto G, Molinari A. Liposomes as nanomedical devices. Int J Nanomedicine 2015;10:975–99.
[185] Beltrán-Gracia E, López-Camacho A, Higuera-Ciapara I, Velázquez-Fernández JB, Vallejo-Cardona AA. Nanomedicine review: clinical developments in liposomal applications. Cancer Nanotechnol 2019;10(1):11.
[186] Karim R, Palazzo C, Evrard B, Piel G. Nanocarriers for the treatment of glioblastoma multiforme: current state-of-the-art. J Control Release 2016;227:23–37.
[187] Koukourakis MI, Koukouraki S, Fezoulidis I, Kelekis N, Kyrias G, Archimandritis S, et al. High intratumoural accumulation of stealth® liposomal doxorubicin (Caelyx®) in glioblastomas and in metastatic brain tumours. Br J Cancer 2000;83(10):1281–6.
[188] Ying X, Wen H, Lu W-L, Du J, Guo J, Tian W, et al. Dual-targeting daunorubicin liposomes improve the therapeutic efficacy of brain glioma in animals. J Control Release 2010;141(2):183–92.
[189] Xiao Y, Cheng L, Xie H, Ju R, Wang X, Fu M, et al. Vinorelbine cationic liposomes modified with wheat germ agglutinin for inhibiting tumor metastasis in treatment of brain glioma. Artif Cells Nanomed Biotechnol 2018;46 (s.3):S524–37.
[190] Kedmi R, Ben-Arie N, Peer D. The systemic toxicity of positively charged lipid nanoparticles and the role of toll-like receptor 4 in immune activation. Biomaterials 2010;31(26):6867–75.
[191] Santos S. The use of coating agents to enhance liposomes blood circulation time. J Mol Pharm Org Process Res 2015;3(1):1000e120.

[192] Deng Y, Wang C, Cheng X, Su Y, Pei Y, Song Y, et al. Accelerated blood clearance phenomenon upon cross-administration of PEGylated nanocarriers in beagle dogs. Int J Nanomedicine 2015;10:3533–45.
[193] Shiraishi K, Yokoyama M. Toxicity and immunogenicity concerns related to PEGylated-micelle carrier systems: a review. Sci Technol Adv Mater 2019;20(1):324–36.
[194] Chanan-Khan A, Szebeni J, Savay S, Liebes L, Rafique NM, Alving CR, et al. Complement activation following first exposure to pegylated liposomal doxorubicin (Doxil®): possible role in hypersensitivity reactions. Ann Oncol 2003;14(9):1430–7.
[195] Birngruber T, Raml R, Gladdines W, Gatschelhofer C, Gander E, Ghosh A, et al. Enhanced doxorubicin delivery to the brain administered through glutathione PEGylated liposomal doxorubicin (2B3-101) as compared with generic Caelyx,®/Doxil®—a cerebral open flow microperfusion pilot study. J Pharm Sci 2014;103(7):1945–8.
[196] Gaillard PJ, Appeldoorn CCM, Dorland R, van Kregten J, Manca F, Vugts DJ, et al. Pharmacokinetics, brain delivery, and efficacy in brain tumor-bearing mice of glutathione Pegylated liposomal doxorubicin (2B3-101). PLoS One 2014;9(1):e82331.
[197] Kerklaan BM, Jager A, Aftimos P, Dieras V, Altintas S, Anders C, et al. NT-23 * phase 1/2A study of glutathione pegylated liposomal doxorubicin (2B3-101) in breast cancer patients with brain metastases (BCBM) or recurrent high grade gliomas (HGG). Neuro Oncol 2014;16(Suppl 5):v163.
[198] Ali SM, Sheikh S, Ahmad A, Ahmaf MU, Chen P, Paithankar M, et al. Bioequivalence study of pegylated doxorubicin hydrochloride liposome (PEGADRIA) and DOXIL® in ovarian cancer patients: physicochemical characterization and pre-clinical studies. J Nanomed Nanotechnol 2016;7(2):1000361.
[199] Aronson JK. Anthracyclines—liposomal formulations. In: Meyler's side effects of drugs. Elsevier Science & Technology; 2016. p. 522–6.
[200] Anselmo AC, Mitragotri S. Nanoparticles in the clinic: an update. Bioeng Transl Med 2019;4(3).
[201] BC Cancer. Drug database: doxorubicin pegylated liposomal. Provincial Health Services Authority; 2020.
[202] Zhao M, Lei C, Yang Y, Bu X, Ma H, Gong H, et al. Abraxane, the nanoparticle formulation of paclitaxel can induce drug resistance by up-regulation of P-gp. PLoS One 2015;10(7):e0131429.
[203] Yuan H, Guo H, Luan X, He M, Li F, Burnett J, et al. Albumin nanoparticle of paclitaxel (Abraxane) decreases while taxol increases breast cancer stem cells in treatment of triple negative breast Cancer. Mol Pharm 2020;17(7):2275–86.
[204] BC Cancer. Drug database: paclitaxel, nanoparticle, albumin-bound (nab). Provincial Health Services Authority; 2020.
[205] Yeh M-K, Chang H-I, Cheng M-Y. Clinical development of liposome based drugs: formulation, characterization, and therapeutic efficacy. Int J Nanomedicine 2011;7:49–60.
[206] Vieira D, Gamarra L. Getting into the brain: liposome-based strategies for effective drug delivery across the blood-brain barrier. Int J Nanomedicine 2016;11:5381–414.
[207] Bulbake U, Doppalapudi S, Kommineni N, Khan W. Liposomal formulations in clinical use: an updated review. Pharmaceutics 2017;9(4):12.
[208] Sardoiwala MN, Kaundal B, Choudhury SR. Development of engineered nanoparticles expediting diagnostic and therapeutic applications across blood–brain barrier. In: Handbook of nanomaterials for industrial applications. Elsevier; 2018. p. 696–709.
[209] Calzoni E, Cesaretti A, Polchi A, Di Michele A, Tancini B, Emiliani C. Biocompatible polymer nanoparticles for drug delivery applications in cancer and neurodegenerative disorder therapies. J Funct Biomater 2019;10(1):4.
[210] Mahmoud BS, AlAmri AH, McConville C. Polymeric nanoparticles for the treatment of malignant gliomas. Cancers (Basel) 2020;12(1):175.
[211] Rai R, Alwani S, Badea I. Polymeric nanoparticles in gene therapy: new avenues of design and optimization for delivery applications. Polymers (Basel) 2019;11(4):745.
[212] Gelperina S, Maksimenko O, Khalansky A, Vanchugova L, Shipulo E, Abbasova K, et al. Drug delivery to the brain using surfactant-coated poly(lactide-co-glycolide) nanoparticles: influence of the formulation parameters. Eur J Pharm Biopharm 2010;74(2):157–63.
[213] Gong Y, Chowdhury P, Nagesh PKB, Rahman MA, Zhi K, Yallapu MM, et al. Novel elvitegravir nanoformulation for drug delivery across the blood-brain barrier to achieve HIV-1 suppression in the CNS macrophages. Sci Rep 2020 Dec 2;10(1):3835.
[214] Tosi G, Vergoni AV, Ruozi B, Bondioli L, Badiali L, Rivasi F, et al. Sialic acid and glycopeptides conjugated PLGA nanoparticles for central nervous system targeting: in vivo pharmacological evidence and biodistribution. J Control Release 2010;145(1):49–57.

[215] Kuo Y-C, Chen Y-C. Targeting delivery of etoposide to inhibit the growth of human glioblastoma multiforme using lactoferrin- and folic acid-grafted poly(lactide-co-glycolide) nanoparticles. Int J Pharm 2015;479 (1):138–49.
[216] Ribeiro AM, Veiga F, Figueiras A. Biodegradable polymeric nanostructures: design and advances in oral drug delivery for neurodegenerative disorders. In: Nanostructures for oral medicine. Elsevier; 2017. p. 61–86.
[217] Yang Q, Lai SK. Anti-PEG immunity: emergence, characteristics, and unaddressed questions. Wiley Interdiscip Rev Nanomed Nanobiotechnol 2015;7(5):655–77.
[218] Estelrich J, Sánchez-Martín MJ, Busquets MA. Nanoparticles in magnetic resonance imaging: from simple to dual contrast agents. Int J Nanomedicine 2015;10(1):1727–41.
[219] Sun J, Kormakov S, Liu Y, Huang Y, Wu D, Yang Z. Recent progress in metal-based nanoparticles mediated photodynamic therapy. Molecules 2018;23(7):1704.
[220] Sawicki K, Czajka M, Matysiak-Kucharek M, Fal B, Drop B, Męczyńska-Wielgosz S, et al. Toxicity of metallic nanoparticles in the central nervous system. Nanotechnol Rev 2019;8(1):175–200.
[221] Khan I, Saeed K, Khan I. Nanoparticles: properties, applications and toxicities. Arab J Chem 2019;12(7):908–31.
[222] Joh DY, Sun L, Stangl M, Al Zaki A, Murty S, Santoiemma PP, et al. Selective targeting of brain tumors with gold nanoparticle-induced radiosensitization. PLoS One 2013;8(4):e62425.
[223] Borran AA, Aghanejad A, Farajollahi A, Barar J, Omidi Y. Gold nanoparticles for radiosensitizing and imaging of cancer cells. Radiat Phys Chem 2018;152:137–44.
[224] Sonavane G, Tomoda K, Makino K. Biodistribution of colloidal gold nanoparticles after intravenous administration: effect of particle size. Colloids Surf B Biointerfaces 2008;66(2):274–80.
[225] Khongkow M, Yata T, Boonrungsiman S, Ruktanonchai UR, Graham D, Namdee K. Surface modification of gold nanoparticles with neuron-targeted exosome for enhanced blood–brain barrier penetration. Sci Rep 2019;9(1):8278.
[226] Gonzalez-Carter DA, Ong ZY, McGilvery CM, Dunlop IE, Dexter DT, Porter AE. L-DOPA functionalized, multi-branched gold nanoparticles as brain-targeted nano-vehicles. Nanomed Nanotechnol Biol Med 2019;15(1):1–11.
[227] Sintov AC, Velasco-Aguirre C, Gallardo-Toledo E, Araya E, Kogan MJ. Metal nanoparticles as targeted carriers circumventing the blood–brain barrier. Int Rev Neurobiol 2016;130:199–227.
[228] Jiang Y, Huo S, Hardie J, Liang X-J, Rotello VM. Progress and perspective of inorganic nanoparticle-based siRNA delivery systems. Expert Opin Drug Deliv 2016;13(4):547–59.
[229] Saucier-Sawyer JK, Deng Y, Seo Y-E, Cheng CJ, Zhang J, Quijano E, et al. Systemic delivery of blood–brain barrier-targeted polymeric nanoparticles enhances delivery to brain tissue. J Drug Target 2015;23(7–8):736–49.
[230] Ferreira M, Sousa J, Pais A, Vitorino C. The role of magnetic nanoparticles in cancer nanotheranostics. Materials (Basel) 2020;13(2):266.
[231] Wahajuddin A. Superparamagnetic iron oxide nanoparticles: magnetic nanoplatforms as drug carriers. Int J Nanomedicine 2012;7:3445–71.
[232] Arami H, Khandhar A, Liggitt D, Krishnan KM. In vivo delivery, pharmacokinetics, biodistribution and toxicity of iron oxide nanoparticles. Chem Soc Rev 2015;44(23):8576–607.
[233] Wankhede M, Bouras A, Kaluzova M, Hadjipanayis CG. Magnetic nanoparticles: an emerging technology for malignant brain tumor imaging and therapy. Expert Rev Clin Pharmacol 2012;5(2):173–86.
[234] Fang K, Liu P, Dong S, Guo Y, Cui X, Zhu X, et al. Magnetofection based on superparamagnetic iron oxide nanoparticle-mediated low lncRNA HOTAIR expression decreases the proliferation and invasion of glioma stem cells. Int J Oncol 2016;49(2):509–18.
[235] Thomsen LB, Thomsen MS, Moos T. Targeted drug delivery to the brain using magnetic nanoparticles. Ther Deliv 2015;6(10):1145–55.
[236] Fu A, Wilson RJ, Smith BR, Mullenix J, Earhart C, Akin D, et al. Fluorescent magnetic nanoparticles for magnetically enhanced cancer imaging and targeting in living subjects. ACS Nano 2012;6(8):6862–9.
[237] Thomsen LB, Linemann T, Birkelund S, Tarp GA, Moos T. Evaluation of targeted delivery to the brain using magnetic immunoliposomes and magnetic force. Materials (Basel) 2019;12(21):3576.
[238] Marie H, Lemaire L, Franconi F, Lajnef S, Frapart Y, Nicolas V, et al. Superparamagnetic liposomes for MRI monitoring and external magnetic field-induced selective targeting of malignant brain tumors. Adv Funct Mater 2015;25(8):1258–69.
[239] Mackenzie EL, Iwasaki K, Tsuji Y. Intracellular Iron transport and storage: from molecular mechanisms to health implications. Antioxid Redox Signal 2008;10(6):997–1030.

[240] Borrelli A, Tornesello AL, Tornesello ML, Buonaguro FM. Cell penetrating peptides as molecular carriers for anti-cancer agents. Molecules 2018;23(2):295.
[241] Ruseska I, Zimmer A. Internalization mechanisms of cell-penetrating peptides. Beilstein J Nanotechnol 2020;11:101–23.
[242] Bruno BJ, Miller GD, Lim CS. Basics and recent advances in peptide and protein drug delivery. Ther Deliv 2013;4(11):1443–67.
[243] Qin Y, Chen H, Zhang Q, Wang X, Yuan W, Kuai R, et al. Liposome formulated with TAT-modified cholesterol for improving brain delivery and therapeutic efficacy on brain glioma in animals. Int J Pharm 2011;420 (2):304–12.
[244] Israel LL, Galstyan A, Holler E, Ljubimova JY. Magnetic iron oxide nanoparticles for imaging, targeting and treatment of primary and metastatic tumors of the brain. J Control Release 2020;320:45–62.
[245] Bellis SL. Advantages of RGD peptides for directing cell association with biomaterials. Biomaterials 2011;32 (18):4205–10.
[246] Belhadj Z, Zhan C, Ying M, Wei X, Xie C, Yan Z. Multifunctional targeted liposomal drug delivery for efficient glioblastoma treatment. Oncotarget 2017;8(40):66889–900.
[247] Liu Y, Ran R, Chen J, Kuang Q, Tang J, Mei L, et al. Paclitaxel loaded liposomes decorated with a multifunctional tandem peptide for glioma targeting. Biomaterials 2014;35(17):4835–47.
[248] Bennett R, Yakkundi A, McKeen HD, McClements L, McKeogh TJ, McCrudden CM, et al. RALA-mediated delivery of FKBPL nucleic acid therapeutics. Nanomedicine (Lond) 2015;10:1–30.
[249] Sathy BN, Olvera D, Gonzalez-Fernandez T, Cunniffe GM, Pentlavalli S, Chambers P, et al. RALA complexed α-TCP nanoparticle delivery to mesenchymal stem cells induces bone formation in tissue engineered constructs in vitro and in vivo. J Mater Chem B 2017;5:1753–64.
[250] McCrudden CM, McBride JW, McCaffrey J, Ali AA, Dunne NJ, Kett VL, et al. Systemic RALA/iNOS nanoparticles: a potent gene therapy for metastatic breast Cancer coupled as a biomarker of treatment. Mol Ther - Nucleic Acids 2016;2017(6):249–58.
[251] Massey AS, Pentlavalli S, Cunningham R, McCrudden CM, McErlean EM, Redpath P, et al. Potentiating the anticancer properties of bisphosphonates by Nanocomplexation with the cationic amphipathic peptide. RALA Mol Pharm 2016;13(4):1217–28.
[252] Cole G, McCaffrey J, Ali AA, McBride JW, McCrudden CM, Vincente-Perez EM, et al. Dissolving microneedles for DNA vaccination: improving functionality via polymer characterization and RALA complexation. Hum Vaccin Immunother 2017;13(1):50–62.
[253] Mulholland EJ, Ali A, Robson T, Dunne NJ, McCarthy HO. Delivery of RALA/siFKBPL nanoparticles via electrospun bilayer nanofibres: an innovative angiogenic therapy for wound repair. J Control Release 2019;316:53–65.
[254] McCarthy HO, McCaffrey J, McCrudden CM, Zholobenko A, Ali AA, McBride JW, et al. Development and characterization of self-assembling nanoparticles using a bio-inspired amphipathic peptide for gene delivery. J Control Release 2014;189:141–9.
[255] Cole G, Ali AA, McErlean E, Mulholland EJ, Short A, McCrudden CM, et al. DNA vaccination via RALA nanoparticles in a microneedle delivery system induces a potent immune response against the endogenous prostate cancer stem cell antigen. Acta Biomater 2019;96:480–90.
[256] McCulloch A, Bennie L, Coulter JA, McCarthy HO, Dromey B, Grimes DR, et al. Nuclear uptake of gold nanoparticles deduced using dual-angle X-ray fluorescence mapping. Part Part Syst Charact 2019;36 (9):1900140.
[257] Chen CJ, Tsai KC, Kuo PH, Chang PL, Wang WC, Chuang YJ, et al. A heparan sulfate-binding cell penetrating peptide for tumor targeting and migration inhibition. Biomed Res Int 2015;2015:237969.
[258] Pei D, Buyanova M. Overcoming endosomal entrapment in drug delivery. Bioconjug Chem 2018;30(2):273–83.
[259] Stalmans S, Bracke N, Wynendaele E, Gevaert B, Peremans K, Burvenich C, et al. Cell-penetrating peptides selectively cross the blood-brain barrier in vivo. PLoS One 2015;10(10):e0139652.
[260] Abdul Razzak R, Florence GJ, Gunn-Moore FJ. Approaches to CNS drug delivery with a focus on transporter-mediated transcytosis. Int J Mol Sci 2019;20(12):3108.
[261] Maor E, Ivorra A, Rubinsky B. Non thermal irreversible electroporation: novel Technology for Vascular Smooth Muscle Cells Ablation. PLoS One 2009;4(3):e4757.
[262] Kanazawa T, Kaneko M, Niide T, Akiyama F, Kakizaki S, Ibaraki H, et al. Enhancement of nose-to-brain delivery of hydrophilic macromolecules with stearate- or polyethylene glycol-modified arginine-rich peptide. Int J Pharm 2017;530(1–2):195–200.

[263] Erdő F, Bors LA, Farkas D, Bajza Á, Gizurarson S. Evaluation of intranasal delivery route of drug administration for brain targeting. Brain Res Bull 2018;143:155–70.
[264] Zawilska JB, Wojcieszak J, Olejniczak AB. Prodrugs: a challenge for the drug development. Pharmacol Rep 2013;65(1):1–14.
[265] Wait SD, Prabhu RS, Burri SH, Atkins TG, Asher AL. Polymeric drug delivery for the treatment of glioblastoma. Neuro Oncol 2015;17(Suppl 2):ii9–ii23.
[266] Chan JM, Valencia PM, Zhang L, Langer R, Farokhzad OC. Polymeric nanoparticles for drug delivery. Methods Mol Biol 2010;624:163–75.
[267] Lesniak MS. Blood-brain barrier permeable gold nanoparticles: an efficient delivery platform for enhanced malignant glioma therapy and imaging. Small 2014;10(24):5137–50.
[268] Epanchintseva AV, Poletaeva JE, Pyshnyi DV, Ryabchikova EI, Pyshnaya IA. Long-term stability and scale-up of noncovalently bound gold nanoparticle-siRNA suspensions. Beilstein J Nanotechnol 2019;10:2568–78.
[269] Giustini AJ, Petryk AA, Cassim SA, Tate JA, Baker I, Hoopes PJ. Magnetic nanoparticle hyperthermia in cancer treatment. Nano Life 2013;1:1–23.
[270] Guo T, Lin M, Huang J, Zhou C, Tian W, Yu H, et al. The recent advances of magnetic nanoparticles in medicine. J Nanomater 2018;2018:1–8.
[271] Heitz F, Morris MC, Divita G. Twenty years of cell—penetrating peptides: from molecular mechanisms to therapeutics. Br J Pharmacol 2009;157(July 2008):195–206.
[272] Kristensen M, Birch D, Nielsen HM. Applications and challenges for use of cell-penetrating peptides as delivery vectors for peptide and protein cargos. Int J Mol Sci 2016;17(2):185.
[273] Formicola B, Dal Magro R, Montefusco-Pereira CV, Lehr C-M, Koch M, Russo L, et al. The synergistic effect of chlorotoxin-mApoE in boosting drug-loaded liposomes across the BBB. J Nanobiotechnol 2019;17(1):115.

CHAPTER

28

Small molecules targeting misfolded mutants of p53 as a rescue strategy to improve glioblastoma chemotherapy

Arutselvan Natarajan[a,*], *Husam A. Babikir*[b,*], *Ramasamy Paulmurugan*[a], *and Tarik F. Massoud*[b]

[a]Cellular Pathway Imaging Laboratory (CPIL), Molecular Imaging Program at Stanford (MIPS), Department of Radiology, Stanford University School of Medicine, Palo Alto, CA, United States
[b]Laboratory of Experimental and Molecular Neuroimaging (LEMNI), Molecular Imaging Program at Stanford (MIPS), Department of Radiology, Stanford University School of Medicine, Stanford, CA, United States

Abstract

p53 is a transcription factor that activates many genes involved in essential maintenance of genetic stability. *P53* is the most frequently mutated gene in human cancers. One-third of these p53 mutations are structural, resulting in mutants with disrupted protein conformations. Here, we review current progress in p53-targeted drug development. These are strategies to reactivate wild-type function of a misfolded mutant p53. Most p53-targeted drugs are still at early stages of development and many of them are progressing slowly toward clinical implementation. There are significant challenges to meet before clinical translation of new antimisfolding p53-targeted drugs.

Abbreviations

ARF	p53-alternative reading frame
DBD	DNA-binding domain
GBM	glioblastoma multiforme
GSCs	GBM stem cells
INK4	family of cyclin-dependent kinase inhibitors
MAPK	mitogen-activated protein kinase

*These authors shared first authorship.

Glioblastoma Resistance to Chemotherapy: Molecular Mechanisms and Innovative Reversal Strategies
https://doi.org/10.1016/B978-0-12-821567-8.00017-8

MDM2	mouse double minute 2 homolog
MGMT	methyl guanine-DNA-methyl transferase
NF1	neurofibromatosis type 1
Nutlin-3	4,5-dihydroimidazoline
PI3K	phosphoinositide 3-kinase
PRIMA-1	2,2-bis (hydroxymethyl)-1-azabicyclo (2-2-2) octan-3-one or APR-246
RITA	2,5-*bis*-(5-hydroxymethyl-2-thienyl)-furan
TMZ	temozolomide

Conflict of interest

The authors declare no competing interests.

Introduction

p53 protein is the guardian of the genome. It is a hub for numerous signaling pathways. It functions primarily as a transcription factor to activate different genes involved in crucial biological events such as cell cycle arrest, DNA repair, apoptosis, and senescence [1]. Therefore, the deleterious effects of mutant p53 lead to perturbations in fundamental cellular processes such as transcriptional regulation, metabolic pathways, and microRNA maturation. Importantly, therefore, many alterations in the p53 network can lead to cancer. Indeed, *P53* has been recognized as the most frequently mutated gene in human cancer; it is inactive to varying degrees owing to different mutations in over 50% of cancer patients [2]. These *P53* mutations are reported in almost every type of cancer, with different frequencies according to the type of cancer. For instance, high-grade serous ovarian carcinoma has a 96% frequency of *P53* mutations, while hematopoietic malignancies have only 10% [3].

A significant number of such *P53* mutations are structural, resulting in a mutant protein molecule that harbors a residue altering the overall protein conformation, in contradistinction to contact point mutations where the mutant protein maintains wild-type folding but the mutation affects DNA binding. Over the past decade and a half, p53-directed drug development has seen significant progress, with many discoveries of lead compounds and some drugs reaching clinical trials. However, there is still no effective p53-targeted therapy available for clinical use to date [1].

Here, we review available research in one aspect of p53-targeted drug development, that is, therapeutic strategies to reactivate the wild-type function of structurally mutant p53. We critically examine the currently known misfolding mutations in p53, and the lead drug compounds that can help counter these oncogenic structural changes by restoring protein folding and function to p53. Lastly, we review the current hurdles and limitations that slow down the clinical translation of antimisfolding p53-targeted drugs.

History of the tumor suppressor p53

Studies in the 1950s showed the ability of some viruses, e.g., SV40 and adenoviruses, to induce cancer when injected into animals. In the late 1970s, the interaction between SV40 T-antigen and an unknown new protein of molecular weight 53 kDa was reported; this protein was then named p53 [4]. In the following decade, p53 was recognized as a product of an oncogene. However this definition was imprecise and the result of a fallacy; the p53 cDNA used in all early cloning experiments was mutated, and therefore, it encoded a mutant protein. Subsequent studies revealed the role of the wild-type version of p53 and showed it as a transcription factor regulating different genes involved in important biological events such as cell cycle arrest, DNA repair, apoptosis, and senescence [1].

p53 regulation

In normal physiological processes, wild-type p53 expression is tightly regulated and maintained at low levels by its negative regulator, the E3 ubiquitin protein ligase, mouse double minute 2 homolog (MDM2) [5]. Conversely, under genotoxic stresses, activation of p53 occurs through posttranslational modifications such as phosphorylation, methylation, and acetylation. Once activated, p53 induces upregulation of its target genes, including that for MDM2, which then binds to the N-terminus of p53 to result in its ubiquitination and proteasomal degradation [6].

p53 structure

The human *P53* gene encodes a protein of 393 amino acids that are highly conserved through evolution. It is functionally active as a homotetramer made of four identical monomers. Structurally, the p53 full-length protein comprises five different functional domains: (1) the transactivation domain at the N-terminus, (2) the proline-rich domain, (3) the DNA-binding domain (core domain), (4) the oligomerization domain (tetramerization domain), and (5) the regulatory domain at the extreme carboxyl terminus (Fig. 1). The transactivation domain is crucial for p53 stability; it is the p53 region where the negative regulator MDM2 binds and ultimately represses the p53 transcriptional activity [7]. The proline-rich domain is indispensable, involved in p53-dependent apoptosis [8]. The DNA-binding domain stretches between residues 94 and 292, and is made of a central immunoglobulin-like β-sandwich scaffold and other structural units that together shape the DNA-binding surface. Two other structural elements that contribute to the DNA-binding surface are a loop-sheet-helix motif and two large loops (L2 and L3). This p53 architecture is well maintained by a zinc ion that binds to Cys176 and Cys238 [9]. The p53 consensus DNA sequences, where the core domain binds, incorporate two unconnected PuPuPuC(A/T)(A/T)GPyPyPy palindromes (Pu denotes purine, while Py is pyrimidine) separated by a 0–13 base pair spacer [10]. The tetramerization domain of p53 is found at its C-terminus and contributes to formation of the p53 homotetramer, leading to elevation of p53 transcriptional activity.

The p53 pathway

p53 is present at low levels during normal cellular conditions, but once activated, it acts as a transcription factor that binds to promoters of downstream target genes implicated in various cellular pathways and networks, including cell cycle arrest, DNA repair, and apoptosis. Upon exposure of a cell to stress, especially stress affecting DNA, p53 levels increase and there are

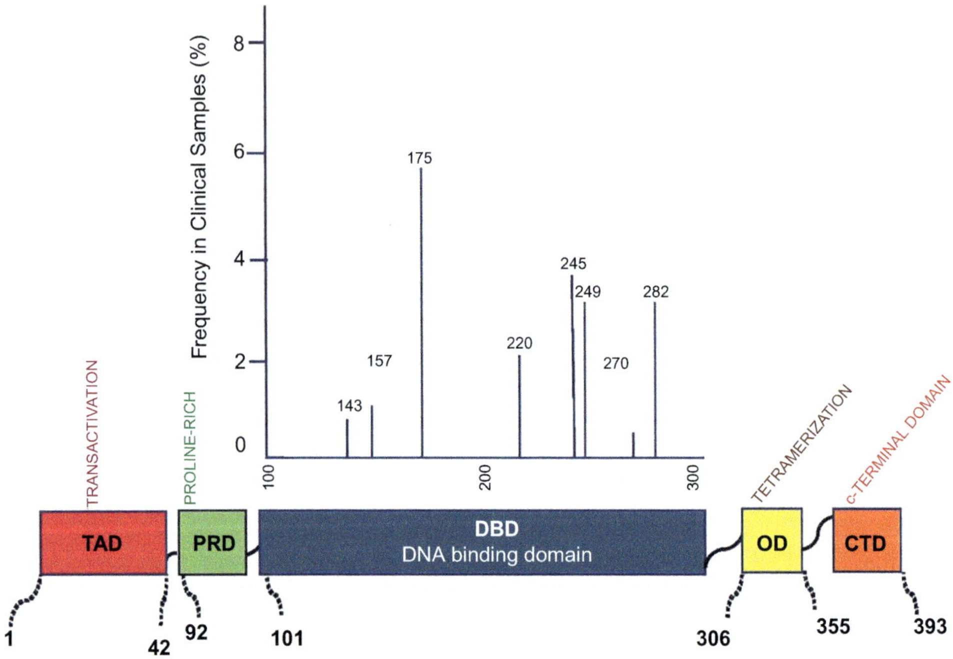

FIG. 1 Schematic representations depicting the domain structure of p53. The positions of the hot structural p53 mutations and their frequencies of clinical presentation are shown on the DNA binding domain of p53 protein.

associated alterations in p53 posttranslational modifications. Activated p53 binds to a conserved sequence in the open reading frame of *GADD45*, *P21*, and *MDM2* genes to induce their expression. However, both *P21* and *MDM2* can be activated in the absence of DNA damage, while induced p53 activates *GADD45* only in the presence of DNA damage [11]. Expressed p21 and GADD45 bind to different cyclin genes leading to the inhibition of the tumor suppressor Rb [12].

p53-Dependent apoptosis is mediated primarily through proapoptotic proteins such as Bax, Puma, and Noxa. However, antiapoptotic proteins, including Bcl2 and Mcl1, also greatly affect the p53 apoptotic pathway. Elevated levels of p53 spontaneously enhance the expression of proapoptotic proteins Noxa, Puma, and Bax. This proapoptotic activity subsequently results in the inhibition of antiapoptotic proteins [13].

A further mechanism consequent to p53 activation is senescence, which can be regarded as a different form of cell cycle arrest. It is initiated by various stresses, including DNA damage and oxidative stress. This process is mediated by the cell cycle inhibitor p21, which leads to the induction of an immune response that causes tumor clearance through activated macrophages. Triggering of macrophage induction is associated with "eat me" signaling, causing macrophage engulfment and elimination of senescent cells [14].

Overall, p53 activity leads to different biological responses that ultimately favor various cellular fates over others by regulating cycle arrest, apoptosis, or senescence. In that regard, crucial questions arise: First, what drives cells to choose a specific pathway? Second, what factors impact the choice of cells to die or to undergo cell cycle arrest? The first aspect determining cellular fate upon p53 activation is promoter selectivity. For example, binding of p53 to its recognition sites at promoters of apoptosis-inducing genes such as *BAX* requires high-affinity binding, when compared to p53 recognition sites at promoters of cell cycle-arrest genes such as *P21* [15]. Posttranslational modifications through phosphorylation, methylation or acetylation also influence p53 pathways and triggering of p53 transcriptional activity [16]. Moreover, interaction of p53 with other binding partners greatly affects its choice in prompting apoptosis or cell cycle arrest; this is supported by the vital role of p53 family members, p63 and p73, in activation of the proapoptotic genes *NOXA* and *BAX* [17]. Interestingly, the level of p53 protein itself can also determine cell fate, with low p53 amounts leading to cell cycle arrest while high levels promote apoptosis [18]. Additionally, the biological response to genetic restoration of wild-type p53 relies on tumor type. As such, reactivation of p53 in lymphomas results in apoptosis while restoration of p53 in sarcomas and hepatocellular carcinomas triggers cell senescence [19].

p53 mutants

The *P53* is the most frequently mutated gene in human cancer, with its function abolished or altered in about half of cancers. Unlike other tumor suppressors, p53 is principally affected by missense mutations that are mostly clustered in the DNA-binding domain. Different phenotypes related to p53 missense mutations have been studied. Gain-of-function (GOF) is observed in some cancers, but complete loss-of-function (LOF) is one of the hallmarks of p53 mutations that affect cancer response to chemoradiation. There is in fact a strong correlation between p53 mutation type and cancer prognosis, as supported by a large clinical study in 1794 patients with breast cancer [20]. Notably, mutant p53 protein has a half-life in cells that is 4 to 20 times longer than in the wild-type version [21].

Based on protein crystal structure, p53 mutations have been divided into two major categories: one is p53 contact mutations, where the mutant protein maintains the wild-type conformation but carries a mutation that affects DNA binding property (that is, these mutations inactivate p53 by replacing residues for essential contacts with its DNA response elements), such as in mutants carrying single amino acid changes at R248Q, R248W, R273H, and R273C residues. The other category is p53 structural mutations, where the p53 molecule contains a residue that changes the overall architecture and conformation, as reported in mutants carrying the R175H, G245S, R249S, and R282W residues [22] (Fig. 1). Further, the presence of structural mutations in the p53 core domain can change its function owing to improper protein folding, eliminating the ability of p53 to bind and activate its target genes. Protein misfolding is not noticeable in mutant p53 molecules harboring contact mutations, since the DNA binding domain is always properly folded [23].

p53 mutations are the main culprits in abolishing the oncosuppressive feature of p53, resulting in its LOF. p53 controls transcription activity as a homotetramer. Therefore, expression consequent to hetero-oligomerization of both wild-type and mutant p53 within p53 heterozygous cancers inactivates the wild-type version expressed from the normal allele and effectively transcribes the mutant p53. This is a phenomenon termed the "dominant-negative mechanism" [24]. Furthermore, mutant p53 not only inhibits the expression of the wild-type form of p53, but it also eliminates the biological role of other p53 family members through oligomerization [25]. The accumulation of p53 in cancer cells is good evidence of p53 failure as a transcription factor activating *MDM2* expression. High levels of mutant p53 described in malignant cells resist degradation by different oncogenic pathways [26]. In vivo and in vitro studies have also shown that mutations in the *P53* gene not only induce LOF, but they can also enhance GOF [27]. In addition, in vivo studies have highlighted the role of mutant p53 in promoting tumor growth and high metastatic rates [28]. Interestingly, p53 mutants (p53-R175H, p53-R273H, and p53-D281G) can function as unique transcription factors that target their own set of genes. This is supported by microarray studies showing the upregulation of some genes involved in cellular growth including *cyclin B2*, *angiopoietin 1*, and *MCL1*. None of these genes is activated by wild-type p53, which is explained by the lack of wild-type p53 binding sites on the sequences of these genes [29]. Other genes activated by mutant p53 include *MDR1* and *BAG1*, both involved in development of drug resistance in cancer cells expressing mutant p53 [29]. Importantly, mutant p53 affects activation of the oncogene *c-MYC*. The N-terminus region of mutant p53 is principally responsible for upregulation of *c-MYC* and other mutant p53 target genes [30]. Taken together, GOF consequent to mutant p53 significantly contributes to cancer cell pathology by promoting cell growth, hampering cell death, and increasing drug resistance.

Distribution and frequency of p53 mutations and their clinical relevance

Mutations in the *P53* gene are reported in almost all cancer types [3]. Kandoth et al. used exome DNA sequencing to compare the incidence of p53 mutations in 12 different cancer types [31]. *P53* was the most mutated gene in 10 types, and another two cancers (renal and colorectal) had lesser mutation frequencies [31]. p53 mutations are more common in solid cancers than in hematological malignancies. Mutations in p53 are also regarded as a typical hallmark of the most difficult-to-treat malignancies including lung, breast, and ovarian cancers [32, 33]. In triple-negative breast cancer, p53 is mutated in 65%–80% of biopsy samples. Recent studies have found that p53 is mutated in 12% of luminal A type tumors, 29% of luminal B type tumors, 72% of HER2-enriched tumors, and 80% of basal-type tumors [33, 34]. Some p53 mutations are specifically associated with exposure to carcinogens—hepatocellular carcinoma often contains a somatic mutation at codon 249 of *P53* where Arg is mutated to Ser (R249S) [35]. The expression of R249S in this cancer is attributed to aflatoxin B1 exposure, a mycotoxin produced by *Aspergillus* [36]. Similarly, p53 mutations are frequent in cancers related to tobacco. There is also strong evidence that p53 mutations in lung cancers (G to T transversions) are directly linked to DNA damage by cigarette smoke [37] (Fig. 1).

p53 with single allele, both allele and null mutants

The status of the p53 genotype is a key factor in sensitivity to different anticancer drugs. Multiple strategies to activate p53 function have been investigated, depending on the homotetramerization or heterotetramerization of p53 in different types of cancer. In cancers containing wild-type p53 (without p53 mutations in both alleles), the function of p53 is frequently impaired owing to enhanced activity of the E3 ubiquitin protein ligase MDM2 and MDMX proteins, or

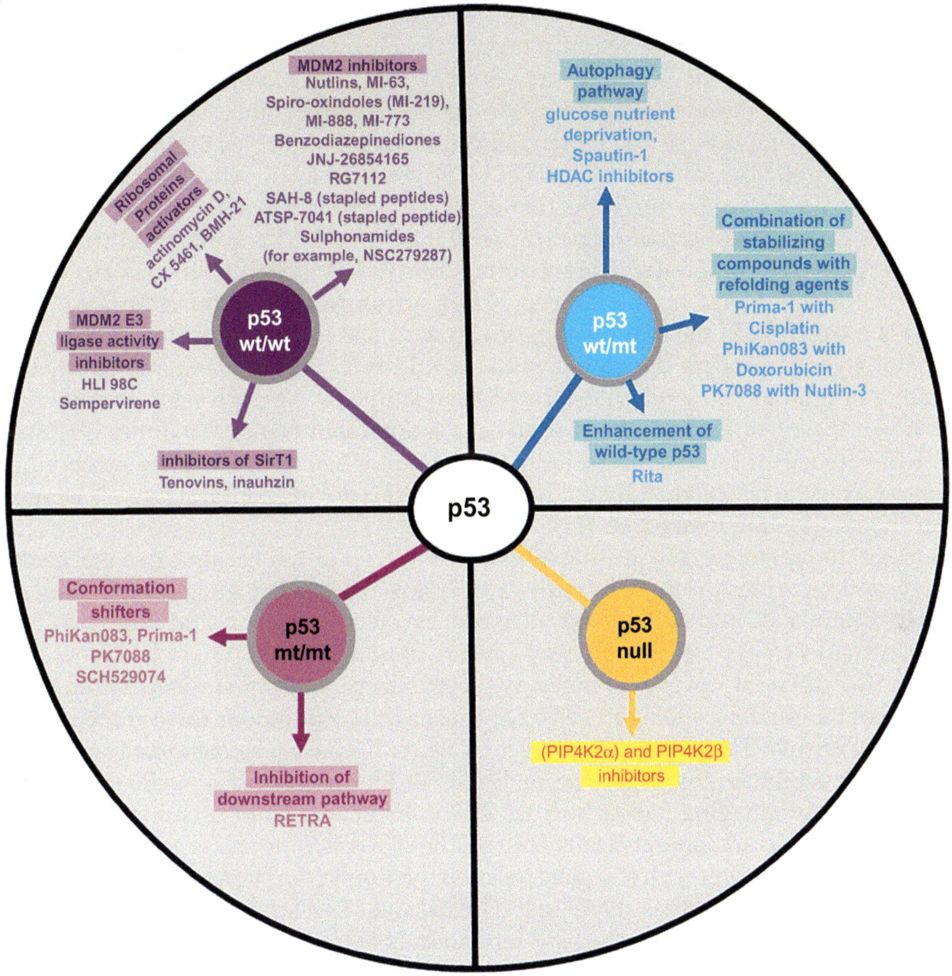

FIG. 2 Schematic diagram illustrating different phenotypes of p53 in cancer cells, and drugs that target p53 protein and recover its function through various mechanisms. The upper left quarter *(purple circle)* illustrates the strategies employed to enhance the function of p53 in cancer cells expressing wild-type p53. The mechanisms targeting the heterozygosity of p53 (single allele mutant) are shown in the upper right quarter and are depicted in *blue*. The *pink circle* represents tumor cells harboring double p53 mutant alleles and different strategies used to restore the wild-type function of p53. The lower right quarter demonstrates cancer cells with deletion of *P53* gene *(yellow circle)*.

loss of expression of the natural MDM2 inhibitor (Fig. 2). Normally, both MDM2 and MDMX bind to the N-terminal transactivation domain of p53 and block its transcriptional function [38]. Overexpression of MDM2 and MDMX occurs after posttranslational modifications or enhanced mRNA stability, leading to inactivation of p53 function in different cancer types. Two main approaches to enhance the activity of wild-type p53 have been considered. The first one increases the level of wild-type p53 by disrupting the MDM2-p53 interaction, which eventually reduces the proteasomal degradation of p53 by MDM2. Nutlins, benzodiazepinediones, and spirooxindoles have been lead compounds in promotion of wild-type p53 by diminishing proteasomal degradation [19, 39]. The other mechanism to increase wild-type p53 activity is mediated by deactivation of enzymes that negatively impact p53 regulation through posttranslational modifications. Tenovin 1 and tenovin 6 are two such molecules. They inhibit sirtuin enzymes involved in destabilization of p53 by its deacetylation [40] (Fig. 2).

In cancers carrying double p53 mutant alleles, different strategies to stabilize mutant p53 have been explored, including restoration of wild-type p53 function by refolding agents such as Prima-1 (p53 reactivation and induction of massive apoptosis) and PhiKan083 [41, 42], or inhibition of downstream signaling pathways of mutant p53. This strategy addresses inhibition of oncogenic effects of mutant p53 resulting from its interaction with downstream targets such as p63 or p73. Based on this concept, RETRA (reactivation of transcriptional reporter activity) is a small molecule that disrupts the mutant p53 and p73 interaction [43].

In heterozygous cancers (bearing a mutant/wild-type allele combination), heterotetramerization of mutant p53 with wild-type p53 can suppress the expression of the wild-type p53 that is transcribed from the normal allele in a dominant-negative manner [24]. Accordingly, RITA (reactivation of p53 and induction of tumor cell apoptosis) is a compound that stabilizes wild-type p53 by directly binding to p53 and inducing conformational changes that abolish the p53-MDM2 interaction [44].

A further therapeutic approach that targets heterozygous cancers aims to enhance mutant p53 degradation through autophagy pathways. This is mediated by glucose restriction, which especially diminishes mutant p53 and increases the stabilization of wild-type p53 [45]. Another important strategy for heterozygous cancers combines two different therapeutic agents targeting two different oncogenic pathways. For example, reconstitution of wild-type p53 function from a mutated allele can sensitize cancer cells and render them more responsive to alkylating agents. This approach succeeded when Prima-1 was combined with cisplatin in lung cancer cell lines [46]. Furthermore, PhiKan083 has also worked well when combined with other DNA-damaging agents such as doxorubicin and temozolomide [47]. Similarly, PK7088 was found to act synergistically with Nutlin-3 in different cancer cell lines [48] (Fig. 2).

In homozygous cancers with *P53* gene deletion (p53-null), two types of kinases, phosphatidylinositol 5-phosphate 4-kinase type-2α (PIP4K2α) and PIP4K2β, were found to be crucial for proliferation of p53-null tumor cells [49]. Inhibition of these two enzymes blocks the growth of cancer cells lacking p53.

Therapeutic strategies targeting p53

Studies in mouse models have shown that restoration of p53 function is essential for suppression of many cancers. This notion has led to concerted efforts to develop efficient strategies for reactivation of p53 function. These strategies are of two types: mechanisms that mainly target mutant p53, e.g., enhancement of p53 structural stability, promotion of mutant p53 proteasomal degradation, and inhibition of protein-protein interactions [50]. The second aims to restore p53 function by activating wild-type p53, primarily through inhibition of MDM2 or MDMX (p53 negative regulators). Nutlins and SAH-8 are good examples of MDM2 inhibitors that target p53 upstream regulators by ribosomal protein activators such as actinomycin D [51]. In this chapter, we mainly focus on structural mutations of p53 protein, and examine strategies that aim to stabilize mutant p53 structure to restore its wild-type function.

Therapeutic approaches to restore wild-type p53 function of mutant p53 protein via restoration of structural stability

Loss of wild-type p53 function initiated by destabilizing structural mutations can be reversed by introducing second-site point mutations. For example, NMR structural analysis has shown that introduction of an H168R mutation into mutant p53 protein harboring the R249S mutation can rescue the structural changes induced by this mutant [52].

A further mechanism to stabilize mutant p53 structure is by binding of a short nine-residue peptide, CDB3, to the p53 core domain. CDB3 is synthesized based on the co-crystal structure of p53 with its binding partner ASPP (apoptosis stimulating proteins of p53). CDB3 shifts the folding equilibrium toward the native state by increasing the melting temperature of wild-type p53 and R249S mutant [53, 54]. It has been shown that the DNA binding domain of wild-type p53 unfolds rapidly at body temperature. CDB3 stabilizes the structural conformation of p53 by enhancing the thermodynamic and kinetic stability of both wild-type and mutant p53.

Aside from these designated peptides, different chemical compounds have been used to stabilize p53 structure. Some dock into the cavity created by a specific mutation. For example, PhiKan083 and PK7088 are two molecules that efficiently bind to the mutation-induced cleft of Y220C, thus restoring the wild-type p53 structure [41, 55]. Other compounds such as Prima-1 and Mira-1 operate in different ways; they modify cysteine residues in wild-type and mutant p53, which promotes the structural stability of p53 protein [56]. Another therapeutic approach to reactivating mutant p53 entails the use of zinc, since efficient folding of p53 protein relies on the presence of zinc ion in the middle of the p53 beta sandwich [22]. Some mutations in p53, such as R175H and G245S, affect zinc binding, which leads to improper p53 folding. Moreover, the absence of metallothioneins, zinc-storing proteins, can result in conformational changes and aggregation of p53 [57]. To avoid this, supplementation with zinc is sufficient to restore proper folding of p53 [58]. NSC319726 can also enhance p53 folding by chelating zinc and promoting its binding to mutant p53 [59].

Considerations and principles of drugs that rescue mutant p53 structure

Many small molecules have been dedicated and used to restore the wild-type conformation of mutant p53. Initial biophysical characterization of any new compound is the first and main principle that should be considered when designing these molecules. Indeed, the structure of any new drug interaction with p53 should be validated through biophysical methods such as X-ray crystallography or NMR. For instance, structural analysis has confirmed the specific binding of PhiKan083 to p53-Y220C and provided valuable information about the binding affinity of this drug [41]. Biological response is also important in validating p53 reactivation by new compounds. The specificity of a new drug can be verified for its activity in p53-mutant cells rather than p53-null or p53-wild-type cells; for example, Prima-1 exerts its antitumor activity primarily through p53-dependent pathways in pancreatic cancer cells [60]. Also, the biological response should be validated in animal models. The efficacy of some new drugs in reactivating mutant p53 may be associated with elevation of specific biomarkers, e.g., serum levels of MIC1 protein in mice [61]. Toxicity in animal models should also be evaluated. Mutant p53 reactivators should not be toxic to animals expressing p53 mutations in their germline, and neither to wild-type animals. Moreover, the therapeutic index of a mutant p53 compound is crucial to its success as a potential therapeutic agent. For example, Mira-1 is very effective as a p53 conformation shifter in inducing antimyeloma activity in tumor-bearing mice without resulting in toxicity to normal tissues [62].

Assay systems to identify new drugs stabilizing p53 protein structure

Based on structural studies, almost one-third of p53 mutations lead to significant conformation changes [55]. To restore the wild-type structure of p53, effective assay systems should be used to identify unique molecules that stabilize it. One of the commonly used strategies is use of X-ray crystallography and NMR to localize a specific pocket within a mutant p53 protein created by a structural mutation, and then to design a new molecule that firmly binds to the mutation site but also weakly to wild-type p53 [63]. As such, Boeckler et al. adopted a structure-based in silico screening analysis of the crystal structure of mutant p53-Y220C. They screened more than two million compounds to identify an suitable molecule that can fill the surface crevice created by the Y220C mutation [55, 64], and found a carbazole derivative (PhiKan083) as a unique compound that strongly binds to the mutation-induced cavity, thus raising the melting temperature of mutant p53-Y220C protein and eventually leading to its structural stability [55]. Based on the crystal structure of the p53 core domain, computational methods have found another cavity between loop L1 and sheet S3 [65]. Accordingly, an ensemble-based virtual screening was implemented for 1324 compounds, identifying stictic acid as a potential molecule to stabilize mutant p53. Apart from the use of an in silico approach, a recent study has adopted phage display technology to generate reactivating peptides [66]. These potential peptides are randomly present on phages [67], and have been selected among others according to their ability to induce proper folding of mutant p53 [66]. Cell-based screening assays have also been used to select small molecules that rescue mutant p53 in tumor cells. For example, several compounds from the National Cancer Institute/Developmental Therapeutics Program (NCI/DTP) have been screened and Prima-1 was identified as a molecule that improves the folding of mutant p53 protein and inhibits the growth of cancer cells in a mutant p53-dependent manner [68]. Another approach to identify small molecules that restore p53 function uses a DNA binding assay [69]. Hence, Demma et al. screened a directed chemical library of quinazolines and identified SCH529074 as a small molecule that promotes DNA binding activity and stabilizes the wild-type conformation of mutant p53 [70].

We recently developed and characterized a new generalizable biosensor assay to identify compounds that rescue misfolded proteins. We used an in vivo protein-fragment-assisted complementation assay (PCA) based on a split bioluminescence reporter complementation technology, to image protein folding and misfolding in cells and in mice [47]. Using the same strategy, Paulmurugan et al. have optimized a unique p53 molecular imaging complementation biosensor to accelerate preclinical testing of drugs that target and correct aberrant p53 misfolding in glioblastoma cells [47] (Fig. 3). We successfully monitored the impact of PhiKan083 or SCH529074 as compounds that restore mutant p53 structure and downstream

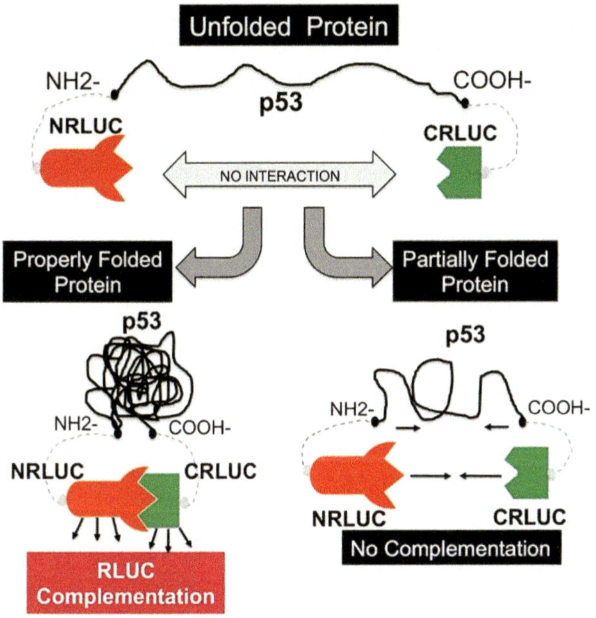

FIG. 3 Principles of p53 molecular imaging complementation biosensor assay to monitor p53 folding status using bioluminescence imaging.

function. The implementation of this p53 molecular imaging biosensor in the drug discovery process will likely pave the way for numerous translational applications, such as in the future construction of cell- and molecular imaging-based high-throughput screening platforms to discover new or repurposed anticancer drugs targeting misfolded p53.

Drugs that rescue the deleterious effect of p53 structural mutations (Tables 1 and 2)

Prima-1/Prima-1^{MET}

Prima-1 was identified in a cell-based screen of a selection of compounds from the NCI library [46]. It was observed to inhibit the growth of cancer cells harboring mutant p53 but had minor effects on cancer cell lines expressing wild-type or deleted p53 (null). This anticancer effect of Prima-1 has been validated in different cancer cell lines including myeloma, lung, pancreas, and breast cancers [60, 78]. To improve the anticancer activity and membrane permeability of Prima-1, Bykov et al. added a methyl group at the end of the original structure and named the new compound Prima-1^{MET} [82]. The mechanism of action of Prima-1 or Prima-1^{MET} in reactivating mutant p53 depends on one of their decomposition products called methylene quinuclidinone (MQ), which functions as the active form of Prima-1. MQ has a chemically active double bond that covalently reacts with thiol groups in mutant p53, to alkylate its cysteine residues [56]. This alkylation leads to the formation of disulfide bonds that can further influence the stability, structure, and function of p53 protein [56, 83, 84].

TABLE 1 Compounds that restore the wild-type conformation of mutant p53.

Compound	Discovery assay system	p53 targeting/ binding region	Structure/peptide sequence	References
Prima-1	Cell-based screening assay	Thiol groups at DBD	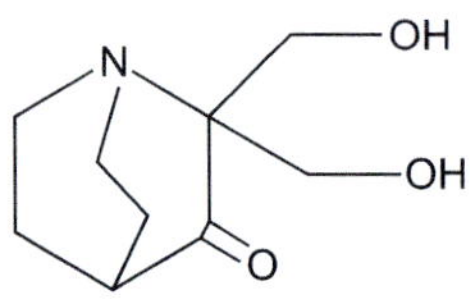	[68]
Stima-1	Cell-based screening assay	Cysteine residues in DBD	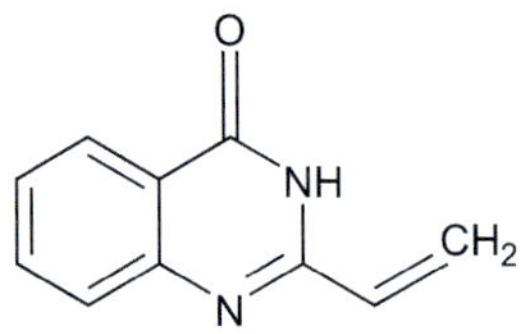	[71]
CP-31398	In silico screening	R249S, R273H mutations at DBD	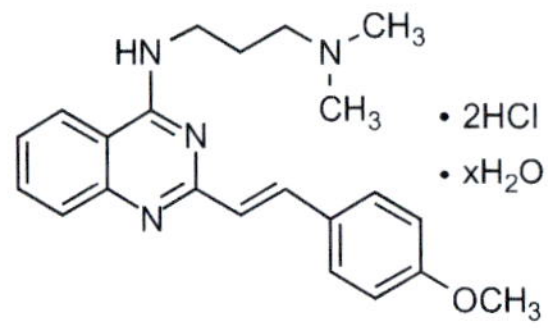	[72]
NSC319726/ZMC 1 (zinc metallochaperone-1)	Cell-based screening assay	R175H, R273H mutations at DBD	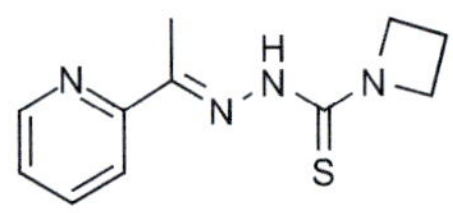	[59]
CDB3	In silico	DBD	REDEDEIEW	[54]
PhiKan083	Structure-based in silico screening	Y220C mutation at DBD	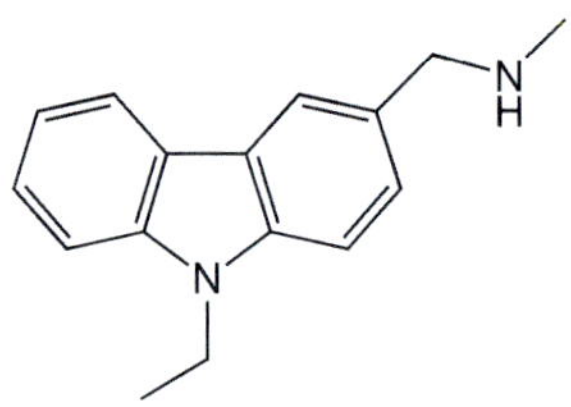	[55]
SCH529074	P53 DNA binding assay	R175H, R273H, R249S at DBD	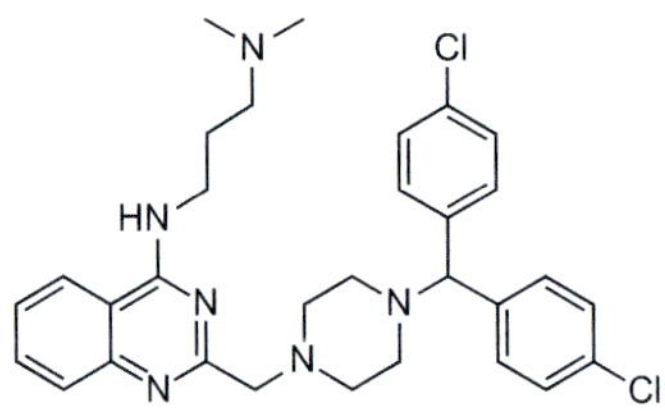	[70]
Mira-1	Cell-based screening assay	Cysteine residues in DBD	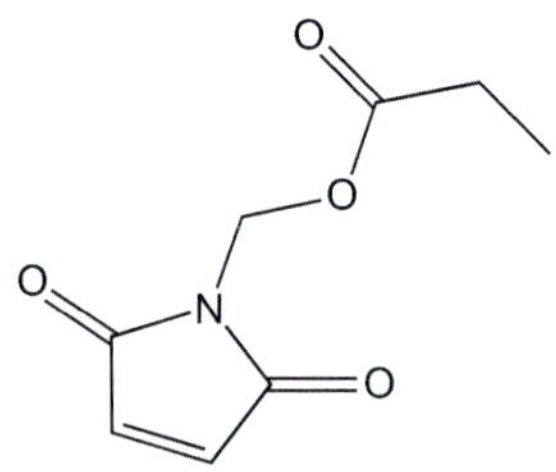	[62]

TABLE 1 Compounds that restore the wild-type conformation of mutant p53—cont'd

Compound	Discovery assay system	p53 targeting/ binding region	Structure/peptide sequence	References
Small peptides pCAP	Phage display	DBD	Different peptides sequences	[66]
Peptide 46	In silico	DBD and C-terminal	GSRAHSSHLKSKKGQSTSRHKK	[73]
Stictic acid	Ensemble-based virtual screening	Cys124 at DBD	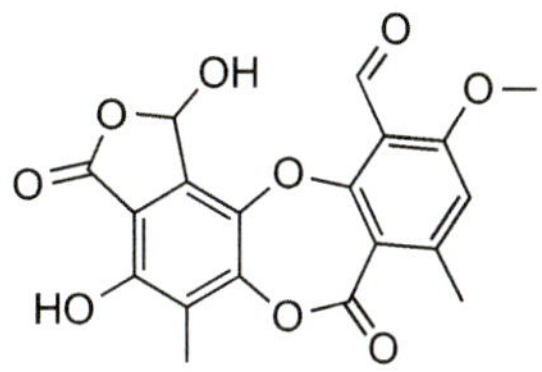	[65]

DBD, *DNA-binding domain.*

TABLE 2 Anticancer activity of compounds targeting structural mutations of p53 protein.

Compound	Mechanism of action	Validation in cancer cell lines	Validation in animal models	Stage of development
Prima-1	Bind to thiol group and alkylate the cysteine residues	Myeloma [74], lung [75], pancreas [60], breast cancer [76], Saos-2 cells [77]	1. Xenografts of BT-474, HCC-1428, and T47-D cells in mice [78] 2. Xenograft of DLD-1 cells in CRC mouse model [79]	Completed Phase-1 clinical trial [80] Phase-II proof of concept initiated by Aprea (https://ClinicalTrials.gov identifier: NCT03268382)
Mira-1	Covalently modify cysteine residues in DBD	Multiple myeloma (MM) [62], the human osteosarcoma and lung adenocarcinoma cell lines [46] colon carcinoma cell line [81]	Mouse xenograft model of multiple myeloma [62]	Preclinical
SCH529074	Restore sequence-specific DNA binding	Colorectal tumor cell line [70]	Human DLD-1 colorectal cancer xenograft model [70]	Preclinical
Stictic acid	Modify the cysteine 124 residue at DBD and thus stabilizes L1/S3 pocket	Human osteosarcoma cells [65]	NA	Preclinical

Continued

TABLE 2 Anticancer activity of compounds targeting structural mutations of p53 protein—cont'd

Compound	Mechanism of action	Validation in cancer cell lines	Validation in animal models	Stage of development
PhiKan083	Bind to p53-Y220C specific cavity and raise the p53 melting temperature	NA	NA	NA
NSC319726/ZMC1	Facilitate zinc binding and redox changes	Ovarian cancer cell lines and mouse embryonic fibroblasts (MEFs) [59]	Xenografts tumor assays from human tumor cell lines, TOV112D, H460 and MDAMB468 [59]	Preclinical
Stima-1	Modify cysteines in the p53 core domain	The human osteosarcoma and lung adenocarcinoma cell lines [71]	NA	Preclinical

To validate these conformational changes induced by Prima-1, conformation-specific antibodies have been used in immunofluorescence experiments. The PAb1620 antibody has been used to recognize the wild-type conformation of p53, while PAb240 antibody is a robust tool to identify mutant p53 protein structure [66]. Both of them are highly effective in verifying the ability of Prima-1 to restore wild-type structure [42, 78]. Rather than only inhibiting cell proliferation and inducing apoptosis in different cancer cell lines, Prima-1^{MET} also demonstrates an efficient anticancer activity against various xenograft tumors [78, 79, 85]. Thus, Prima-1^{MET} was the first compound targeting structural mutations of p53 to proceed to in-human clinical trials. It was successful in a Phase I clinical trial, exhibiting some efficacy in patients with hematological malignancies and prostate cancer (https://ClinicalTrials.gov identifier: NCT00900614) [80]. Further, a Phase II proof-of-concept study has been initiated by Aprea AB (Stockholm, Sweden) where platinum was used in combination with Prima-1 in ovarian cancer patients to achieve a better therapeutic effect (https://ClinicalTrials.gov identifier: NCT03268382).

PhiKan083, PK7088, and PK11007

PhiKan083 and PK7088 differ in their chemical structures—PK7088 is a pyrazole and PhiKan083 is a carbazole derivative. Both compounds have been designed from an in silico screen analysis of the p53-Y220C crystal structure using computational and fragment screening [48, 55]. The hotspot p53 mutation Y220C is reported in approximately 75,000 new cancer cases every year [48]. It disturbs the wild-type conformation of p53 by creating a cavity on the protein surface [55]. Either PhiKan083 or PK7088 can fit well into this cavity, and consequently, they raise the melting temperature of mutant p53 leading to restoration of the wild-type conformation [55]. Consistent with this reactivation of mutant p53, a recent study found that PK7088 induced growth inhibition and apoptosis in cancer cell lines carrying the Y220C mutation [48]. However, these observations have not yet been validated in animal models. Regarding PhiKan083, Paulmurugan et al. have recently investigated its potential anticancer activity in cancer cell lines harboring the p53-Y220C mutation [47]. They observed a synergistic treatment effect when combining a low concentration of PhiKan083 with either temozolamide or doxorubicin in Ln229 glioblastoma cells expressing different p53 mutations. A significantly increased number of apoptotic cells was observed in Ln229 cells stably expressing the p53-Y220C mutation compared to parental Ln229 cells. Recently, a 2-sulfonylpyrimidines named PK11007 was identified as a mild thiol alkylator with anticancer activity in several cell lines, especially those harboring mutant p53. PK11007 stabilizes p53 in vitro via selective alkylation of two surface-exposed cysteines without compromising its DNA binding activity. Another recent study showed that targeting mutant p53 with PK11007 is a potential approach for treating p53-mutated breast cancer, including the subgroup with TN disease.

SCH529074

In an effort to identify small molecules that reactivate mutant p53 and restore its DNA binding activity, Demma et al. used a sensitive p53 DNA binding assay to screen a directed chemical library of quinazolines [70]. SCH529074, which belongs to a class of piperazinylquinazolines, was identified as an effective compound that restores the wild-type function of two p53 mutations: the contact point mutation R273H and the structural mutation R249S [70]. Consistent with the conformational changes induced by SCH529074, immunoprecipitation experiments using monoclonal antibody 1620 have confirmed the ability of SCH529074 to restore the epitope for the binding of PAb1620 conformation-specific antibody in cells harboring the p53 mutation. Thus, SCH529074 was found to be very effective in inhibiting proliferation of mutant p53-expressing cell lines either by inducing cell cycle arrest or apoptosis [70]. Furthermore, in vivo efficacy of SCH529074 was determined in a tumor xenograft model, where it was shown that it is a potent compound in reducing tumor growth [70].

NSC319726/ZMC1, NSC319725, and NSC328784

Zinc is crucial to stabilize p53 by enhancing proper folding of the p53 DNA binding domain. Lack of zinc causes deleterious changes in the structure of p53, leading to its aggregation in cells [57, 58]. Of note, some specific mutations in p53 protein, such as C242S, H179R, C176F, and R175H, can impair zinc binding; these mutations are commonly present in different cancers [63]. Hence, restoration of proper folding of mutant p53 by enhancing zinc binding is an attempted therapeutic approach in cancer treatment. Zinc metallochaperones are small molecules that function as transportation vehicles to carry zinc into cells. NSC319726/ZMC1 is a zinc metallochaperone identified through screening; it selectively kills p53-R175H cancer cells but with less impact on cells expressing wild-type p53 [86]. The anticancer activity of ZMC1 has been attributed to two factors: zinc binding and redox changes [59, 86, 87]. NSC319726/ZMC1 was found to suppress cell growth and induce apoptosis in cancer cells expressing the p53-R175H mutation. Moreover, the in vivo effects of

ZMC1 have been verified in xenograft models where it was shown that ZMC1 could significantly suppress the growth of tumor xenografts using TOV112D cell lines possessing the p53-R175H mutation [59]. Interestingly, ZMC1 was found to be toxic to mice carrying the R175H mutation in their germline but not to wild-type mice. This confirms ZMC1 selectivity in a 175-allele-specific mutant p53-dependent manner [59]. To confirm the restoration of the wild-type structure of mutant p53 when using ZMC1, immunofluorescence analysis using p53 wild-type-specific antibody (PAb1620) and p53 mutant-specific antibody (PAb240) has revealed the induction of conformation changes in cells harboring the p53-R175H mutation following treatment with this drug [59].

In addition to NSC319726 (ZMC1), another two compounds belonging to the same class of thiosemicarbazones (TSC), NSC319725 and NSC328784, were examined in a number of cell-based assays for their ability to reactivate mutant p53, and were consequently found to be powerful in restoring the wild-type function of mutant p53 and behaving similarly to NSC319726 (ZMC1) in both biophysical and cell-based assays.

Stictic acid

A structural analysis of p53 protein shows that Cys124 is a crucial amino acid for p53 stabilization. It is found at the center of a transiently open binding pocket between loop L1 and sheet S3 of the p53 DNA-binding domain [65]. Mutation of Cys124 to alanine eliminates the reactivation effects promoted by Prima-1 in Saos-2 cells harboring an R175H mutation. Ensemble-based virtual screening was performed to identify compounds that bind to the L1/S3 pocket, identifying 45 compounds that bind efficiently [65]. Among these, stictic acid emerged as a potential compound that restores the wild-type function of mutant p53. It has been found to effectively induce expression of *P21* and *PUMA* in cell lines carrying p53 mutations [65].

Small peptides

Rather than using low molecular-weight chemical compounds, small peptides have emerged as good candidates to restore the wild-type function of mutant p53. They can be identified using different strategies, e.g., peptide 46 was identified in an in silico screen. It binds to the carboxy-terminal residues 361–382 of p53 and functions by restoring wild-type folding [73]. Other small peptides were identified by phage display technology, with several successfully converting mutant p53 to a form exhibiting wild-type folding [66]. The anticancer activity of these peptides was validated in vivo, where they were found to enhance apoptosis selectively in mutant p53-expressing cells [66].

Limitations of compounds targeting structural mutations of p53

To date, only limited success has been achieved despite the huge efforts exerted to develop unique compounds targeting structural mutations of p53. Most mutant p53-reactivating compounds available so far are progressing slowly toward clinical trials; presently only Prima-1 has successfully completed Phase I trials [80]. One compound targeting Chk1 in p53-deficient triple-negative breast cancer was found to be effective in vitro and in animal models, but it failed to achieve any success in human trials [88, 89].

Different factors negatively contribute to halting the clinical translation of mutant p53-based compounds, including observed strong impact on safety, high levels of toxicity as a result of weak specificity toward p53, nonsignificant effects on restoring p53 function, and the inability to penetrate cells owing to multiple drug resistance mechanisms [90]. Regarding small peptides as potential candidates to restore wild-type function of mutant p53, overcoming susceptibility to proteolytic degradation is considered a major challenge before possible clinical applications. Moreover, delivery is recognized as one of the main limitations of peptide-based therapies, and this drawback is also related to fast renal clearance from the circulation [66].

A significant obstacle associated with identifying new molecules targeting structural mutations of p53 is the choice of screening methods available currently. For example, screening approaches based on p53 growth inhibition or other phenotypic assays may identify compounds that function upstream of p53 and that may help in its reactivation. This is supported by observations that some of the identified compounds inhibit tumor growth through p53-independent mechanisms [91]. Moreover, compounds arising from screening protocols based on structural analysis may reactivate mutant p53 in a pathway(s) that does not involve post-translational modification. Therefore, more investigations will be required to confirm the role of these compounds in triggering p53 protein posttranslational modifications.

Structural mutations of p53 destabilize the protein structure kinetically and thermodynamically. According to the Law of Mass Action, ligand binding to an unfolded form of p53 will convert it to the native state and thus restore its wild-type conformation. Based on this concept, various mutant p53-reactivating compounds have been designed and synthesized. Despite their good biological activities, some of these compounds lack full biophysical characterization. In fact, it was found that many of these compounds do not even physically interact with p53, but act instead via different pathways [92]. One of these compounds is CP-31398, which has been described as a mutant p53 conformation shifter [72, 93]. However, studies that have used different biophysical approaches to validate this stabilization of p53 by CP-31398 were not successful [94]. Additionally, CP-31398 was found to act as a DNA intercalator that destabilizes the p53-cDNA complex in vitro [94]. All in all, the anticancer activity of CP-31398 appears attributed to a pathway that does not involve physical binding to p53.

Synergizing the effect of drugs targeting mutant p53

To overcome some of the obstacles mentioned above, combining different therapeutic agents could be usefully adopted to enhance the clinical efficacy intended from promoting the pharmacological reactivation of mutant p53 pathways. One example is the combination of Prima-1, targeting mutant p53, with cisplatin, which is a DNA alkylator that inhibits the mutant p53-related pathway. It was found that mutant p53 increases chemoresistance to cisplatin [90], and therefore, targeting the cause of this chemoresistance (mutant p53) using Prima-1 results in an effective combination therapy. Therefore, a combination treatment using cisplatin and Prima-1 was indeed shown to result in synergistic anticancer activity in SCID mice carrying human tumor xenografts [82]. Moreover, outstanding synergistic effects were observed with Prima-1 in combination with doxorubicin or other DNA-damaging drugs in ovarian cancer cells [95]. Prima-1 also exhibited profound synergistic effects against pancreatic cancer cells when it was combined with the MDM2 inhibitor Nutlin-3 [60]. In triple-negative breast cancer cells, mutant p53 activates proteasome gene transcription, leading to resistance to the proteasome inhibitor carfilzomib. Combining Prima-1 and the proteasome inhibitor carfilzomib is a good strategy to overcome this chemoresistance and to enhance the therapeutic efficacy against solid tumors and metastases [76]. Much more work is necessary to further understand how drug combinations using Prima-1 or other mutant p53 activators can cause synergistic therapeutic effects. Future research will be important to address questions regarding combination strategies and how they can induce the transcription of proapoptotic downstream genes, or to enhance cell death through inhibition of anti-apoptotic proteins.

Conclusion

Over the past four decades, *P53* has emerged as one of the most important genes in human cancer. Enormous efforts have been made to study mutant p53. There is general agreement that "mutant p53" is not just one protein, but is in fact numerous protein variants that lead to a wide range of oncogenic responses. Consequently, mutant p53 presents us with an attractive target for therapy, certainly requiring more focused efforts in future drug discovery projects to achieve suitable anticancer compound "hits." However, several important challenges and hurdles should be overcome before effective translation of research observations to clinical practice. It is notable that most p53-related drugs are still at early stages of development and many of them are progressing slowly toward clinical use. Currently, Prima-1 is the only drug targeting mutant p53 that has completed phase I/II clinical trials. On the other hand, many compounds have failed to reach clinical stages. Chk1 inhibitor is a good example of this, as it has exhibited efficacy in in vitro and mouse model experiments but failed to achieve a significant response in humans [89, 96]. Further investigations are required on different drug combinations that enhance the effectiveness of mutant p53 therapeutics and prevent issues related to toxicity of p53 modulator monotherapy. Moreover, new avenues in p53 research, such as p53-mediated microRNA regulation and links with stem cell biology, should be explored to strengthen the field of p53-targeted therapy .

References

[1] Mantovani F, Walerych D, Sal GD. Targeting mutant p53 in cancer: a long road to precision therapy. FEBS J 2017;284:837–50. https://doi.org/10.1111/febs.13948.
[2] Vogelstein B, Lane D, Levine AJ. Surfing the p53 network. Nature 2000;408:307–10. https://doi.org/10.1038/35042675.
[3] Rivlin N, Brosh R, Oren M, Rotter V. Mutations in the p53 tumor suppressor gene: important milestones at the various steps of tumorigenesis. Genes Cancer 2011;2:466–74. https://doi.org/10.1177/1947601911408889.
[4] Eliyahu D, Raz A, Gruss P, Givol D, Oren M. Participation of p53 cellular tumour antigen in transformation of normal embryonic cells. Nature 1984;312:646–9.

[5] Pant V, Lozano G. Limiting the power of p53 through the ubiquitin proteasome pathway. Genes Dev 2014;28:1739–51. https://doi.org/10.1101/gad.247452.114.
[6] Honda R, Tanaka H, Yasuda H. Oncoprotein MDM2 is a ubiquitin ligase E3 for tumor suppressor p53. FEBS Lett 1997;420:25–7.
[7] Chen J, Marechal V, Levine AJ. Mapping of the p53 and mdm-2 interaction domains. Mol Cell Biol 1993;13:4107–14.
[8] Sakamuro D, Sabbatini P, White E, Prendergast GC. The polyproline region of p53 is required to activate apoptosis but not growth arrest. Oncogene 1997;15:887–98. https://doi.org/10.1038/sj.onc.1201263.
[9] Hainaut P, Mann K. Zinc binding and redox control of p53 structure and function. Antioxid Redox Signal 2001;3:611–23. https://doi.org/10.1089/15230860152542961.
[10] el-Deiry WS, Kern SE, Pietenpol JA, Kinzler KW, Vogelstein B. Definition of a consensus binding site for p53. Nat Genet 1992;1:45–9. https://doi.org/10.1038/ng0492-45.
[11] Xiao G, Chicas A, Olivier M, Taya Y, Tyagi S, Kramer FR, et al. A DNA damage signal is required for p53 to activate gadd45. Cancer Res 2000;60:1711–9.
[12] Harper JW, Adami GR, Wei N, Keyomarsi K, Elledge SJ. The p21 Cdk-interacting protein Cip1 is a potent inhibitor of G1 cyclin-dependent kinases. Cell 1993;75:805–16.
[13] Chen L, Willis SN, Wei A, Smith BJ, Fletcher JI, Hinds MG, et al. Differential targeting of prosurvival Bcl-2 proteins by their BH3-only ligands allows complementary apoptotic function. Mol Cell 2005;17:393–403. https://doi.org/10.1016/j.molcel.2004.12.030.
[14] Lujambio A, Akkari L, Simon J, Grace D, Tschaharganeh DF, Bolden JE, et al. Non-cell-autonomous tumor suppression by p53. Cell 2013;153:449–60. https://doi.org/10.1016/j.cell.2013.03.020.
[15] Schlereth K, Charles JP, Bretz AC, Stiewe T. Life or death: p53-induced apoptosis requires DNA binding cooperativity. Cell Cycle 2010;9:4068–76. https://doi.org/10.4161/cc.9.20.13595.
[16] Lane DP, Brown CJ, Verma C, Cheok CF. New insights into p53 based therapy. Discov Med 2011;12:107–17.
[17] Flores ER, Tsai KY, Crowley D, Sengupta S, Yang A, McKeon F, et al. p63 and p73 are required for p53-dependent apoptosis in response to DNA damage. Nature 2002;416:560–4. https://doi.org/10.1038/416560a.
[18] Kracikova M, Akiri G, George A, Sachidanandam R, Aaronson SA. A threshold mechanism mediates p53 cell fate decision between growth arrest and apoptosis. Cell Death Differ 2013;20:576–88. https://doi.org/10.1038/cdd.2012.155.
[19] Xue W, Zender L, Miething C, Dickins RA, Hernando E, Krizhanovsky V, et al. Senescence and tumour clearance is triggered by p53 restoration in murine liver carcinomas. Nature 2007;445:656–60. https://doi.org/10.1038/nature05529.
[20] Olivier M, Langerød A, Carrieri P, Bergh J, Klaar S, Eyfjord J, et al. The clinical value of somatic TP53 gene mutations in 1,794 patients with breast cancer. Clin Cancer Res 2006;12:1157–67. https://doi.org/10.1158/1078-0432.CCR-05-1029.
[21] Finlay CA, Hinds PW, Tan TH, Eliyahu D, Oren M, Levine AJ. Activating mutations for transformation by p53 produce a gene product that forms an hsc70-p53 complex with an altered half-life. Mol Cell Biol 1988;8:531–9.
[22] Cho Y, Gorina S, Jeffrey PD, Pavletich NP. Crystal structure of a p53 tumor suppressor-DNA complex: understanding tumorigenic mutations. Science 1994;265:346–55.
[23] Bullock AN, Henckel J, Fersht AR. Quantitative analysis of residual folding and DNA binding in mutant p53 core domain: definition of mutant states for rescue in cancer therapy. Oncogene 2000;19:1245–56. https://doi.org/10.1038/sj.onc.1203434.
[24] Strano S, Dell'Orso S, Mongiovi AM, Monti O, Lapi E, Di Agostino S, et al. Mutant p53 proteins: between loss and gain of function. Head Neck 2007;29:488–96. https://doi.org/10.1002/hed.20531.
[25] Cadwell C, Zambetti GP. The effects of wild-type p53 tumor suppressor activity and mutant p53 gain-of-function on cell growth. Gene 2001;277:15–30.
[26] Suh YA, Post SM, Elizondo-Fraire AC, Maccio DR, Jackson JG, El-Naggar AK, et al. Multiple stress signals activate mutant p53 in vivo. Cancer Res 2011;71:7168–75. https://doi.org/10.1158/0008-5472.CAN-11-0459.
[27] Muller PA, Vousden KH. Mutant p53 in cancer: new functions and therapeutic opportunities. Cancer Cell 2014;25:304–17. https://doi.org/10.1016/j.ccr.2014.01.021.
[28] Lang GA, Iwakuma T, Suh YA, Liu G, Rao VA, Parant JM, et al. Gain of function of a p53 hot spot mutation in a mouse model of Li-Fraumeni syndrome. Cell 2004;119:861–72. https://doi.org/10.1016/j.cell.2004.11.006.

[29] Scian MJ, Stagliano KE, Ellis MA, Hassan S, Bowman M, Miles MF, et al. Modulation of gene expression by tumor-derived p53 mutants. Cancer Res 2004;64:7447–54. https://doi.org/10.1158/0008-5472.CAN-04-1568.
[30] Frazier MW, He X, Wang J, Gu Z, Cleveland JL, Zambetti GP. Activation of c-myc gene expression by tumor-derived p53 mutants requires a discrete C-terminal domain. Mol Cell Biol 1998;18:3735–43.
[31] Kandoth C, McLellan MD, Vandin F, Ye K, Niu B, Lu C, et al. Mutational landscape and significance across 12 major cancer types. Nature 2013;502:333–9. https://doi.org/10.1038/nature12634.
[32] Peifer M, Fernandez-Cuesta L, Sos ML, George J, Seidel D, Kasper LH, et al. Integrative genome analyses identify key somatic driver mutations of small-cell lung cancer. Nat Genet 2012;44:1104–10. https://doi.org/10.1038/ng.2396.
[33] Network CGA. Comprehensive molecular portraits of human breast tumours. Nature 2012;490:61–70. https://doi.org/10.1038/nature11412.
[34] Shah SP, Roth A, Goya R, Oloumi A, Ha G, Zhao Y, et al. The clonal and mutational evolution spectrum of primary triple-negative breast cancers. Nature 2012;486:395–9. https://doi.org/10.1038/nature10933.
[35] Gouas D, Shi H, Hainaut P. The aflatoxin-induced TP53 mutation at codon 249 (R249S): biomarker of exposure, early detection and target for therapy. Cancer Lett 2009;286:29–37. https://doi.org/10.1016/j.canlet.2009.02.057.
[36] Bressac B, Kew M, Wands J, Ozturk M. Selective G to T mutations of p53 gene in hepatocellular carcinoma from southern Africa. Nature 1991;350:429–31. https://doi.org/10.1038/350429a0.
[37] Pfeifer GP, Denissenko MF, Olivier M, Tretyakova N, Hecht SS, Hainaut P. Tobacco smoke carcinogens, DNA damage and p53 mutations in smoking-associated cancers. Oncogene 2002;21:7435–51. https://doi.org/10.1038/sj.onc.1205803.
[38] Li Q, Lozano G. Molecular pathways: targeting Mdm2 and Mdm4 in cancer therapy. Clin Cancer Res 2013;19:34–41. https://doi.org/10.1158/1078-0432.CCR-12-0053.
[39] Saha MN, Qiu L, Chang H. Targeting p53 by small molecules in hematological malignancies. J Hematol Oncol 2013;6:23. https://doi.org/10.1186/1756-8722-6-23.
[40] Lain S, Hollick JJ, Campbell J, Staples OD, Higgins M, Aoubala M, et al. Discovery, in vivo activity, and mechanism of action of a small-molecule p53 activator. Cancer Cell 2008;13:454–63. https://doi.org/10.1016/j.ccr.2008.03.004.
[41] Bauer MR, Jones RN, Baud MG, Wilcken R, Boeckler FM, Fersht AR, et al. Harnessing fluorine-sulfur contacts and multipolar interactions for the design of p53 mutant Y220C rescue drugs. ACS Chem Biol 2016;11:2265–74. https://doi.org/10.1021/acschembio.6b00315.
[42] Bykov VJ, Issaeva N, Shilov A, Hultcrantz M, Pugacheva E, Chumakov P, et al. Restoration of the tumor suppressor function to mutant p53 by a low-molecular-weight compound. Nat Med 2002;8:282–8. https://doi.org/10.1038/nm0302-282.
[43] Kravchenko JE, Ilyinskaya GV, Komarov PG, Agapova LS, Kochetkov DV, Strom E, et al. Small-molecule RETRA suppresses mutant p53-bearing cancer cells through a p73-dependent salvage pathway. Proc Natl Acad Sci U S A 2008;105:6302–7. https://doi.org/10.1073/pnas.0802091105.
[44] Issaeva N, Bozko P, Enge M, Protopopova M, Verhoef LG, Masucci M, et al. Small molecule RITA binds to p53, blocks p53-HDM-2 interaction and activates p53 function in tumors. Nat Med 2004;10:1321–8. https://doi.org/10.1038/nm1146.
[45] Rodriguez OC, Choudhury S, Kolukula V, Vietsch EE, Catania J, Preet A, et al. Dietary downregulation of mutant p53 levels via glucose restriction: mechanisms and implications for tumor therapy. Cell Cycle 2012;11:4436–46. https://doi.org/10.4161/cc.22778.
[46] Bykov VJ, Issaeva N, Zache N, Shilov A, Hultcrantz M, Bergman J, et al. Reactivation of mutant p53 and induction of apoptosis in human tumor cells by maleimide analogs. J Biol Chem 2005;280:30384–91. https://doi.org/10.1074/jbc.M501664200.
[47] Paulmurugan R, Afjei R, Sekar TV, Babikir HA, Massoud TF. A protein folding molecular imaging biosensor monitors the effects of drugs that restore mutant p53 structure and its downstream function in glioblastoma cells. Oncotarget 2018;9:21495–511. https://doi.org/10.1007/s00109-016-1437-9.
[48] Liu X, Wilcken R, Joerger AC, Chuckowree IS, Amin J, Spencer J, et al. Small molecule induced reactivation of mutant p53 in cancer cells. Nucleic Acids Res 2013;41:6034–44. https://doi.org/10.1093/nar/gkt305.
[49] Emerling BM, Hurov JB, Poulogiannis G, Tsukazawa KS, Choo-Wing R, Wulf GM, et al. Depletion of a putatively druggable class of phosphatidylinositol kinases inhibits growth of p53-null tumors. Cell 2013;155:844–57. https://doi.org/10.1016/j.cell.2013.09.057.

[50] Maslon MM, Hupp TR. Drug discovery and mutant p53. Trends Cell Biol 2010;20:542–55. https://doi.org/10.1016/j.tcb.2010.06.005.
[51] Lohrum MA, Ludwig RL, Kubbutat MH, Hanlon M, Vousden KH. Regulation of HDM2 activity by the ribosomal protein L11. Cancer Cell 2003;3:577–87.
[52] Nikolova PV, Wong KB, DeDecker B, Henckel J, Fersht AR. Mechanism of rescue of common p53 cancer mutations by second-site suppressor mutations. EMBO J 2000;19:370–8. https://doi.org/10.1093/emboj/19.3.370.
[53] Issaeva N, Friedler A, Bozko P, Wiman KG, Fersht AR, Selivanova G. Rescue of mutants of the tumor suppressor p53 in cancer cells by a designed peptide. Proc Natl Acad Sci U S A 2003;100:13303–7. https://doi.org/10.1073/pnas.1835733100.
[54] Friedler A, Hansson LO, Veprintsev DB, Freund SM, Rippin TM, Nikolova PV, et al. A peptide that binds and stabilizes p53 core domain: chaperone strategy for rescue of oncogenic mutants. Proc Natl Acad Sci U S A 2002;99:937–42. https://doi.org/10.1073/pnas.241629998.
[55] Boeckler FM, Joerger AC, Jaggi G, Rutherford TJ, Veprintsev DB, Fersht AR. Targeted rescue of a destabilized mutant of p53 by an in silico screened drug. Proc Natl Acad Sci U S A 2008;105:10360–5. https://doi.org/10.1073/pnas.0805326105.
[56] Lambert JM, Gorzov P, Veprintsev DB, Soderqvist M, Segerback D, Bergman J, et al. PRIMA-1 reactivates mutant p53 by covalent binding to the core domain. Cancer Cell 2009;15:376–88. https://doi.org/10.1016/j.ccr.2009.03.003.
[57] Loh SN. The missing zinc: p53 misfolding and cancer. Metallomics 2010;2:442–9. https://doi.org/10.1039/c003915b.
[58] Pintus SS, Ivanisenko NV, Demenkov PS, Ivanisenko TV, Ramachandran S, Kolchanov NA, et al. The substitutions G245C and G245D in the Zn(2+)-binding pocket of the p53 protein result in differences of conformational flexibility of the DNA-binding domain. J Biomol Struct Dyn 2013;31:78–86. https://doi.org/10.1080/07391102.2012.691364.
[59] Yu X, Vazquez A, Levine AJ, Carpizo DR. Allele-specific p53 mutant reactivation. Cancer Cell 2012;21:614–25. https://doi.org/10.1016/j.ccr.2012.03.042.
[60] Izetti P, Hautefeuille A, Abujamra AL, de Farias CB, Giacomazzi J, Alemar B, et al. PRIMA-1, a mutant p53 reactivator, induces apoptosis and enhances chemotherapeutic cytotoxicity in pancreatic cancer cell lines. Invest New Drugs 2014;32:783–94. https://doi.org/10.1007/s10637-014-0090-9.
[61] Ray-Coquard I, Blay JY, Italiano A, Le Cesne A, Penel N, Zhi J, et al. Effect of the MDM2 antagonist RG7112 on the P53 pathway in patients with MDM2-amplified, well-differentiated or dedifferentiated liposarcoma: an exploratory proof-of-mechanism study. Lancet Oncol 2012;13:1133–40. https://doi.org/10.1016/S1470-2045(12)70474-6.
[62] Saha MN, Chen Y, Chen MH, Chen G, Chang H. Small molecule MIRA-1 induces in vitro and in vivo anti-myeloma activity and synergizes with current anti-myeloma agents. Br J Cancer 2014;110:2224–31. https://doi.org/10.1038/bjc.2014.164.
[63] Joerger AC, Fersht AR. Structure-function-rescue: the diverse nature of common p53 cancer mutants. Oncogene 2007;26:2226–42. https://doi.org/10.1038/sj.onc.1210291.
[64] Joerger AC, Ang HC, Fersht AR. Structural basis for understanding oncogenic p53 mutations and designing rescue drugs. Proc Natl Acad Sci U S A 2006;103:15056–61. https://doi.org/10.1073/pnas.0607286103.
[65] Wassman CD, Baronio R, Demir Ö, Wallentine BD, Chen CK, Hall LV, et al. Computational identification of a transiently open L1/S3 pocket for reactivation of mutant p53. Nat Commun 2013;4:1407. https://doi.org/10.1038/ncomms2361.
[66] Tal P, Eizenberger S, Cohen E, Goldfinger N, Pietrokovski S, Oren M, et al. Cancer therapeutic approach based on conformational stabilization of mutant p53 protein by small peptides. Oncotarget 2016;7:11817–37. https://doi.org/10.18632/oncotarget.7857.
[67] FitzGerald K. In vitro display technologies—new tools for drug discovery. Drug Discov Today 2000;5:253–8.
[68] Bykov VJ, Issaeva N, Selivanova G, Wiman KG. Mutant p53-dependent growth suppression distinguishes PRIMA-1 from known anticancer drugs: a statistical analysis of information in the National Cancer Institute database. Carcinogenesis 2002;23:2011–8.
[69] Demma MJ, Wong S, Maxwell E, Dasmahapatra B. CP-31398 restores DNA-binding activity to mutant p53 in vitro but does not affect p53 homologs p63 and p73. J Biol Chem 2004;279:45887–96. https://doi.org/10.1074/jbc.M401854200.

[70] Demma M, Maxwell E, Ramos R, Liang L, Li C, Hesk D, et al. SCH529074, a small molecule activator of mutant p53, which binds p53 DNA binding domain (DBD), restores growth-suppressive function to mutant p53 and interrupts HDM2-mediated ubiquitination of wild type p53. J Biol Chem 2010;285:10198–212. https://doi.org/10.1074/jbc.M109.083469.

[71] Zache N, Lambert JM, Rokaeus N, Shen J, Hainaut P, Bergman J, et al. Mutant p53 targeting by the low molecular weight compound STIMA-1. Mol Oncol 2008;2:70–80. https://doi.org/10.1016/j.molonc.2008.02.004.

[72] Foster BA, Coffey HA, Morin MJ, Rastinejad F. Pharmacological rescue of mutant p53 conformation and function. Science 1999;286:2507–10.

[73] Selivanova G, Iotsova V, Okan I, Fritsche M, Strom M, Groner B, et al. Restoration of the growth suppression function of mutant p53 by a synthetic peptide derived from the p53 C-terminal domain. Nat Med 1997;3:632–8.

[74] Tessoulin B, Descamps G, Moreau P, Maiga S, Lode L, Godon C, et al. PRIMA-1Met induces myeloma cell death independent of p53 by impairing the GSH/ROS balance. Blood 2014;124:1626–36. https://doi.org/10.1182/blood-2014-01-548800.

[75] Zandi R, Selivanova G, Christensen CL, Gerds TA, Willumsen BM, Poulsen HS. PRIMA-1Met/APR-246 induces apoptosis and tumor growth delay in small cell lung cancer expressing mutant p53. Clin Cancer Res 2011;17:2830–41. https://doi.org/10.1158/1078-0432.CCR-10-3168.

[76] Walerych D, Lisek K, Sommaggio R, Piazza S, Ciani Y, Dalla E, et al. Proteasome machinery is instrumental in a common gain-of-function program of the p53 missense mutants in cancer. Nat Cell Biol 2016;18:897–909. https://doi.org/10.1038/ncb3380.

[77] Lambert JM, Moshfegh A, Hainaut P, Wiman KG, Bykov VJ. Mutant p53 reactivation by PRIMA-1MET induces multiple signaling pathways converging on apoptosis. Oncogene 2010;29:1329–38. https://doi.org/10.1038/onc.2009.425.

[78] Liang Y, Besch-Williford C, Hyder SM. PRIMA-1 inhibits growth of breast cancer cells by re-activating mutant p53 protein. Int J Oncol 2009;35:1015–23.

[79] Li XL, Zhou J, Chan ZL, Chooi JY, Chen ZR, Chng WJ. PRIMA-1met (APR-246) inhibits growth of colorectal cancer cells with different p53 status through distinct mechanisms. Oncotarget 2015;6:36689–99. https://doi.org/10.18632/oncotarget.5385.

[80] Lehmann S, Bykov VJ, Ali D, Andren O, Cherif H, Tidefelt U, et al. Targeting p53 in vivo: a first-in-human study with p53-targeting compound APR-246 in refractory hematologic malignancies and prostate cancer. J Clin Oncol 2012;30:3633–9. https://doi.org/10.1200/JCO.2011.40.7783.

[81] Bou-Hanna C, Jarry A, Lode L, Schmitz I, Schulze-Osthoff K, Kury S, et al. Acute cytotoxicity of MIRA-1/NSC19630, a mutant p53-reactivating small molecule, against human normal and cancer cells via a caspase-9-dependent apoptosis. Cancer Lett 2015;359:211–7. https://doi.org/10.1016/j.canlet.2015.01.014.

[82] Bykov VJ, Zache N, Stridh H, Westman J, Bergman J, Selivanova G, et al. PRIMA-1(MET) synergizes with cisplatin to induce tumor cell apoptosis. Oncogene 2005;24:3484–91. https://doi.org/10.1038/sj.onc.1208419.

[83] Camacho CJ, Thirumalai D. Modeling the role of disulfide bonds in protein folding: entropic barriers and pathways. Proteins 1995;22:27–40. https://doi.org/10.1002/prot.340220105.

[84] Kaar JL, Basse N, Joerger AC, Stephens E, Rutherford TJ, Fersht AR. Stabilization of mutant p53 via alkylation of cysteines and effects on DNA binding. Protein Sci 2010;19:2267–78. https://doi.org/10.1002/pro.507.

[85] Krayem M, Journe F, Wiedig M, Morandini R, Najem A, Sales F, et al. p53 Reactivation by PRIMA-1(Met) (APR-246) sensitises (V600E/K)BRAF melanoma to vemurafenib. Eur J Cancer 2016;55:98–110. https://doi.org/10.1016/j.ejca.2015.12.002.

[86] Yu X, Blanden AR, Narayanan S, Jayakumar L, Lubin D, Augeri D, et al. Small molecule restoration of wildtype structure and function of mutant p53 using a novel zinc-metallochaperone based mechanism. Oncotarget 2014;5:8879–92. https://doi.org/10.18632/oncotarget.2432.

[87] Yu Y, Kalinowski DS, Kovacevic Z, Siafakas AR, Jansson PJ, Stefani C, et al. Thiosemicarbazones from the old to new: iron chelators that are more than just ribonucleotide reductase inhibitors. J Med Chem 2009;52:5271–94. https://doi.org/10.1021/jm900552r.

[88] Ma CX, Cai S, Li S, Ryan CE, Guo Z, Schaiff WT, et al. Targeting Chk1 in p53-deficient triple-negative breast cancer is therapeutically beneficial in human-in-mouse tumor models. J Clin Invest 2012;122:1541–52. https://doi.org/10.1172/JCI58765.

[89] Ma CX, Ellis MJ, Petroni GR, Guo Z, Cai SR, Ryan CE, et al. A phase II study of UCN-01 in combination with irinotecan in patients with metastatic triple negative breast cancer. Breast Cancer Res Treat 2013;137:483–92. https://doi.org/10.1007/s10549-012-2378-9.

[90] Bossi G, Lapi E, Strano S, Rinaldo C, Blandino G, Sacchi A. Mutant p53 gain of function: reduction of tumor malignancy of human cancer cell lines through abrogation of mutant p53 expression. Oncogene 2006;25: 304–9. https://doi.org/10.1038/sj.onc.1209026.

[91] Peltonen K, Colis L, Liu H, Jaamaa S, Moore HM, Enback J, et al. Identification of novel p53 pathway activating small-molecule compounds reveals unexpected similarities with known therapeutic agents. PLoS One 2010;5: e12996. https://doi.org/10.1371/journal.pone.0012996.

[92] Wiman KG. Strategies for therapeutic targeting of the p53 pathway in cancer. Cell Death Differ 2006;13: 921–6. https://doi.org/10.1038/sj.cdd.4401921.

[93] Wischhusen J, Naumann U, Ohgaki H, Rastinejad F, Weller M. CP-31398, a novel p53-stabilizing agent, induces p53-dependent and p53-independent glioma cell death. Oncogene 2003;22:8233–45. https://doi.org/10.1038/sj.onc.1207198.

[94] Rippin TM, Bykov VJ, Freund SM, Selivanova G, Wiman KG, Fersht AR. Characterization of the p53-rescue drug CP-31398 in vitro and in living cells. Oncogene 2002;21:2119–29. https://doi.org/10.1038/sj.onc.1205362.

[95] Mohell N, Alfredsson J, Fransson Å, Uustalu M, Byström S, Gullbo J, et al. APR-246 overcomes resistance to cisplatin and doxorubicin in ovarian cancer cells. Cell Death Dis 2015;6:e1794. https://doi.org/10.1038/cddis.2015.143.

[96] Origanti S, Cai SR, Munir AZ, White LS, Piwnica-Worms H. Synthetic lethality of Chk1 inhibition combined with p53 and/or p21 loss during a DNA damage response in normal and tumor cells. Oncogene 2013;32:577–88. https://doi.org/10.1038/onc.2012.84.

CHAPTER

29

p53 supplementation as a targeted cancer gene therapy for glioblastoma

Uday Kumar Sukumar[a], Tarik F. Massoud[b], and Ramasamy Paulmurugan[a]

[a]Cellular Pathway Imaging Laboratory (CPIL), Molecular Imaging Program at Stanford (MIPS), Department of Radiology, Stanford University School of Medicine, Palo Alto, CA, United States
[b]Laboratory of Experimental and Molecular Neuroimaging (LEMNI), Molecular Imaging Program at Stanford (MIPS), Department of Radiology, Stanford University School of Medicine, Stanford, CA, United States

Abstract

TP53, also referred to as the "elephant gene," encodes for the proapoptotic protein p53 that guards the genome from mutations and aberrations. It is also one of the most commonly deregulated tumor suppressor genes in cancer. Nearly 84% of glioblastoma (GBM) patients have a dysregulated p53-ARF-MDM2 pathway, which is also evident in 94% of GBM cell lines adopted for in vitro assays. The majority of p53 dysregulation stems from mutations in TP53 and other regulatory proteins that are involved in its biogenesis feedback loop. GBM carries different grades of such mutations and exhibits higher cell invasion, migration, proliferation, evasion from apoptosis, and cancer cell stemness, which ultimately gives raise to resistance to standard therapeutic approaches. Expression of these mutations correlates with worse prognosis, highlighting their potential importance as markers and targets for GBM therapy. Thus, restoration of wild-type p53 function or supplementation of p53 selectively to GBM cells can sensitize the cells to subsequent chemotherapy/radiation therapy by promoting apoptosis, cell cycle arrest, and senescence. In light of the significance of this strategy for the current clinical practices for GBM therapy, in this chapter we will elaborate upon different strategies adopted for restoration or supplementation of wild-type p53 function in GBM, and their clinical potential for efficient therapeutic management. We focus on cancer gene therapy for supplementation of p53 as a strategy to improve chemo- and radiotherapies in GBM.

Abbreviations

BBB blood-brain barrier
CFDA China Food and Drug Administration
CSCs cancer stem cells

Glioblastoma Resistance to Chemotherapy: Molecular Mechanisms and Innovative Reversal Strategies
https://doi.org/10.1016/B978-0-12-821567-8.00013-0

DBD	DNA-binding domain
EPR	enhanced permeability and retention effect
GBM	glioblastoma
GCV	ganciclovir
GOF	adenovirus
HSV1-TK	Herpes Simplex Virus Thymidine Kinase
MDM2	murine double minute 2
MGMT	O6-methylguanine DNA-methyltransferase
mut-p53	mutant p53
ROS	reactive oxygen species
TfR	transferrin receptor
TMZ	temozolomide
wt-p53	wild type p53

Conflict of interest

No potential conflicts of interest were disclosed.

Role of p53 in cell cycle regulation

p53 is a protein of 53 kDa encoded by 393 amino acids that act as the "guardian of genome" [1]. It plays a vital role as a transcriptional regulator of cellular homeostasis in the event of stress or cellular damage [2]. It acts as a centerpiece of the cellular signaling network to control cell proliferation, cell cycle arrest, senescence, apoptosis, and autophagy [3]. It controls the genomic integrity of cells (Fig. 1). p53 is a transcription factor that forms a homotetramer composed of specific domains for interaction with specific nucleotide sequences in the open reading frame of regulatory proteins to drive the expression [6]. The p53 expression level in a cell is constantly kept in check by ubiquitination and degradation by MDM2 and MDM4 (MDMX) regulatory proteins. On the other hand, phosphorylation of amino-terminus serine promotes its stabilization by disrupting its interaction with the negative regulators MDM2 and MDM4 [7]. Such a feedback-regulated expression level of p53 in normal cells enables them to carefully regulate cellular homeostasis based on stress factors and prevents from the development of "abnormal cells" and their propagation. The term abnormal cells here refers to cells experiencing nutrient deprivation, hypoxia, oxidative stress, DNA damage-assisted hyperproliferation signals, ribonucleotide depletion, replicative stress, and telomere attrition. In the event of such abnormalities, p53 instigates signaling cascades that orchestrate apoptosis or autophagy and eliminate defective cells to avoid cellular oncogenesis [8]. Thus, by mediating selective elimination of abnormal cells, p53 inhibits progression of mutant or dysregulated cells and maintains an active checkpoint, and is thus called the "guardian of genome."

TP53 mutation and its relevance to glioblastoma development

As evident from the state of uncontrolled proliferation, invasion, and metastasis of cancer cells, it is certain that such cancer cells have to overcome the p53-mediated checkpoint to thrive with uncontrolled proliferation and maintain tumor growth in patients. If p53 was functional in cancer cells, genes that are involved in cell cycle arrest, apoptosis,

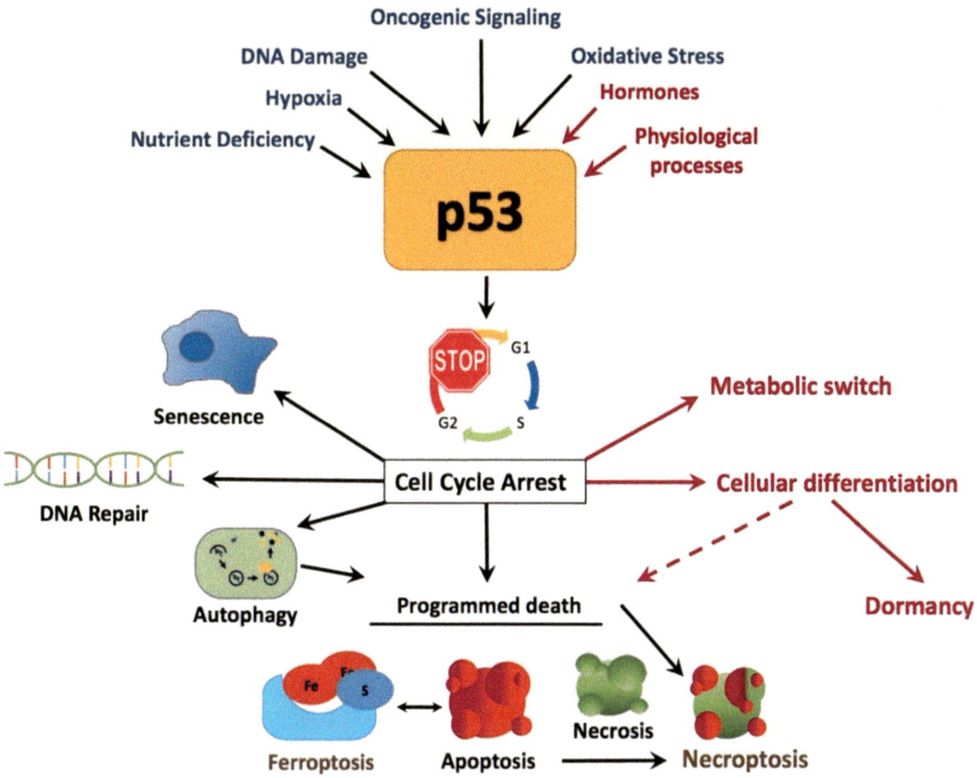

FIG. 1 The canonical and noncanonical tumor suppressor roles of p53. In cells, p53 is activated by a range of stress signals. These activators include nutrient stress, hypoxic conditions, activation of oncogenes, DNA damage, and oxidative stress from reactive oxygen species (ROS). As a result of all these activities, the function of p53 increases tremendously in stressed cells. Classical or canonical responses of p53 include transcriptionally and translationally regulate downstream target gene expression to control cell cycle arrest, repair the DNA damage, and place the cells in a state of senescence or induce apoptosis [4]. The noncanonical pathways include cell metabolism, autophagy, and necrosis [5].

differentiation, and senescence would be actively expressed, thereby eliminating these cancer cells. However, it is evident that is not the case from uncontrolled proliferation of cancer cells. Thus, the p53 pathway is altered or dysregulated in most human cancers, and, as a result of which, they grow and divide, with consequent accumulation of mutations that characterize their abnormal metabolism, membranes, cytoskeletal proteins, and morphology [9]. As an outcome of the suppressive role of p53 in cancer cells survival, proliferation, and progression, it has earned the label of a "tumor suppressor gene." Owing to its crucial role in apoptosis and cell growth control, and considering that its pathway is altered in the majority of human cancers, p53 is one of the most appealing targets for developing cancer therapies. To date, several strategies have been explored to target p53 and its pathways [10]. The majority of sporadic cancers carry TP53 mutations. For instance, germline TP53 mutations are causative of the Li Fraumeni syndrome, a rare familial cancer predisposition, further supporting a crucial role of p53 mutations in driving gliomagenesis [11]. As a result of TP53 mutations, there is loss of wild-type p53 (wt-53) function, which presents a favorable status for cancer development. Such mutations can be broadly categorized into three different types, i.e., loss of function, gain of function, and dominant-negative function [12]. Most of p53 mutations are associated with the DNA-binding domain of the protein, and arise owing to missense codon reading resulting in single amino acid substitution within the DNA-binding domain [13]. As a result of a mutational amino acid in the DNA-binding region, p53 loses its ability to regulate expression of canonical p53 target genes that are involved in cellular homeostasis. Some of the mutant variants exert trans-dominant repression over the wild-type counterpart, especially in the case of cells with single-allele mutations [14].

Despite rapid strides being made in cancer therapeutics, glioblastoma (GBM) still remains one of the deadliest cancers in humans with a median survival of 12–15 months [15, 16]. The conventional therapeutic strategies in clinical practice include radiation, surgery, and chemotherapy, which only pushes the average survival of patients with GBM to nearly 15 months and offers no cure [17]. The major challenges to these therapeutic approaches arise from hallmark features of GBM, which include uncontrolled cellular proliferation, diffuse infiltration, a propensity for necrosis, robust angiogenesis, strong resistance to apoptosis, and rampant genomic instability [18]. The diffuse and infiltrative nature of GBM leaves no defined tumor margin and, thus, leaves the patients with minimal scope for complete surgical resection of tumor.

The resistance of GBM to therapies has been ascribed, among many other factors discussed in this book, to hallmark characteristics of intratumoral heterogeneity. This implies the presence of different cell populations with distinct genetic mutations or expression profiles within a tumor [17]. Such heterogeneity plays a crucial role in tumor growth progression and resistance to conventional therapeutic approaches. The source of heterogeneity lies in genetic alterations in tumor suppressor genes and amplification of oncogenes. Among these genetic alterations, tumor suppressor gene p53 is the most aberrant gene in astrocytic gliomas [19]. The wt-p53 is lost in 50% of grade II and grade III astrocytomas, whereas an additional 10%–15% carry a mutation in one of the alleles for p53 [20]. The prevalence of p53 mutations also differs based on molecular subtypes of GBM, which includes proneural, mesenchymal, neural, and classical subtypes, with 54%, 32%, 21%, and 0% probabilities for p53 mutations, respectively [21]. As outlined earlier, p53 mutations/deletions can be categorized into different types that lead to loss of function and dominant-negative function. The obvious outcome of p53 loss of function is elimination of its crucial role as a tumor suppressor gene, thus favoring progression of abnormal GBM cells. On the other hand, reports indicate that some mutant forms of p53 are able to inactivate the endogenous wt-p53 protein in a dominant-negative fashion by forming a heterotetramer complex. These findings suggest that dominant-negative mutations in one p53 allele should be sufficient to inactivate p53 function in a cell by activation of two main negative regulators, murine double minute 2 and 4 (MDM2 and MDM4). Such dominant-negative p53 mutations are also reported to be associated with early onset of GBM [22].

Correlation of TP53 status to grades of gliomas

The predominant region of mutations in TP53 in GBM is missense mutations ascribed to "six hotspots" that lie in either DNA contact regions or conformational regions determining

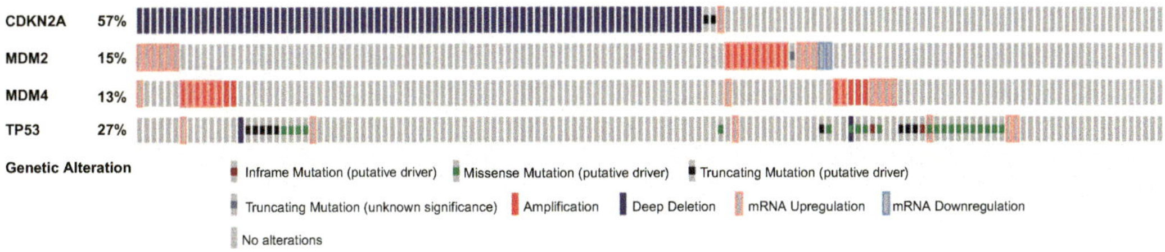

FIG. 2 The p53 pathway is highly deregulated in GBM. The most common mutations of the p53 pathway are missense mutations in TP53, deletions of CDKN2A/ARF, and/or amplifications of MDM2 and MDM4. These mutations often lead to diminished tumor suppressor activity [25]. *Adapted from cBioportal.*

native folding for wt-p53 function [23]. There is growing evidence suggesting that mutant p53 (mut-p53) protein is not only highly expressed in GBM but also possesses oncogenic functions beyond those acquired through loss of wt-p53 [24]. Mutation in TP53 has been found to have strong correlation with GBM progression, invasion, antiapoptotic, and proliferative phenotype. It also predisposes the patients to progression from low-grade to high-grade lesions (Fig. 2) [26]. It has been found that even within a single GBM tumor, heterogeneous regions of low-grade (well-differentiated) and high-grade (poorly differentiated) lesions with distinct p53 mutation patterns have been found to coexist [27]. In light of p53 role as a tumor suppressor gene and prevalence of its mutation in high-grade GBM, determination of p53 status can be of vital clinical importance to decide specific therapeutic approaches for GBM. A study by Pollack et al. suggested that patients with p53-positive malignant gliomas of childhood might have both shorter time to progression and survival [28]. However, no appropriate multivariate analyses (including tumor grade, age, performance status, and extent of surgical resection) were performed. Although preexisting literature strongly implies the role of p53 status in GBM progression, there is no clear correlation between the p53 status of GBM with respect to different grades of GBM. In a recent study, among 35 clinically annotated human primary GBM samples, 10 (29%) tumors registered prototypical TP53 mutations [29]. Likewise, a study by Shiraishi et al. indicated that the p53 mutation not only contributes to tumorigenesis in gliomas (as an early event) but also to progression to malignancy (as a late event) [30]. A systematic review with multivariate comparisons (considering factors such as grade, age, and performance status) should enable us to judge the value of p53 as a predictor of response to treatment and prognostic factor for survival in gliomas.

TP53-based therapeutic strategies for glioblastoma

It is clear from extensive studies that GBM cells are associated with high prevalence of p53 mutations. This has prompted the therapeutic concept of restoring or supplementing wt-p53 to GBM cells to suppress their proliferation and progression, and to sensitize them to concomitant chemotherapy (Fig. 3). Such approaches to restore the tumor suppressive function of p53 have already been explored earlier and proven to be promising for GBM treatment. In the following section, we will revisit different strategies used to restore wt-p53 function in

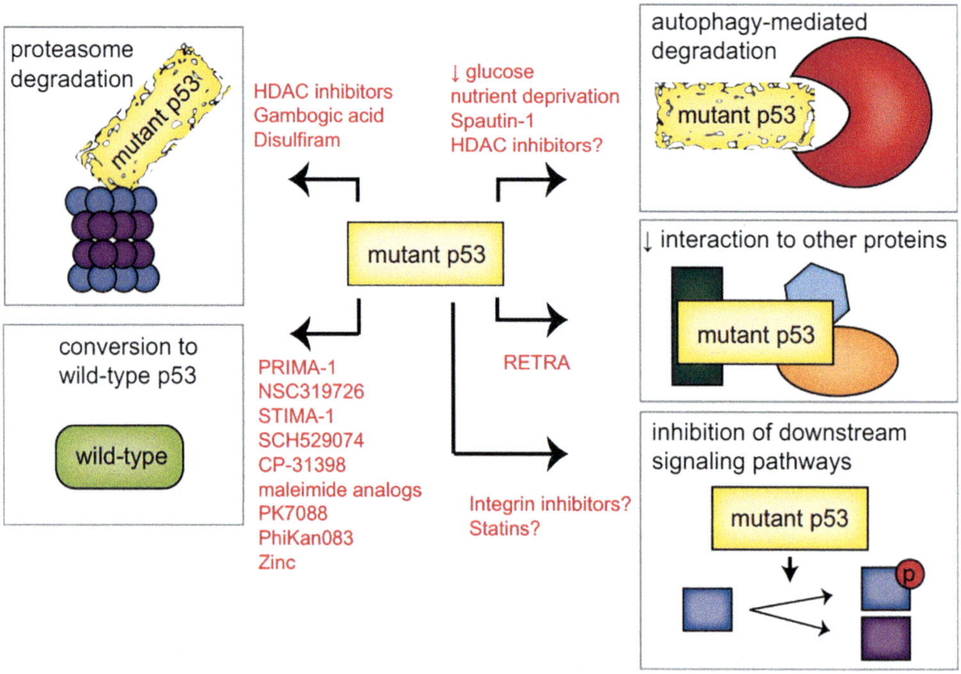

FIG. 3 Strategies that are currently being explored to target mutant p53 to achieve wt-p53 function. Depicted in red are schematics of the strategies that are currently being explored to target cancers expressing mutant-p53 protein. These strategies include promotion of mutant-p53 degradation through the proteasome and autophagy pathways, restoration of wild-type p53 activity, interference with the interaction of mutant-p53 with other proteins, and interference in signaling pathways downstream of mutant-p53 protein. *Reproduced with permission from Muller PA, Vousden KH. Mutant p53 in cancer: new functions and therapeutic opportunities. Cancer Cell 2014;25:304–317.*

GBM, to impose cellular proapoptotic and antiproliferative effects [31]. Aside from direct supplementation of the p53 gene by different gene delivery approaches, other strategies aimed at regulating p53 expression levels include: inhibition of the MDM2/p53 complex to prevent p53 from its degradation (possible only for GBM with single-allele mutation), small molecule-based approaches to reactivate mut-p53, and inhibition of gain of function [32]. While many of these strategies are still in early phases of clinical trials or in preclinical development for GBM therapy, these treatments may allow for the effective targeting of p53-mediated therapeutic enhancement. Although currently no therapeutic approaches for GBM based on supplementing the p53 gene have progressed beyond phase II trials, it was already been translated to the clinic for the treatment of other cancers, such as head and neck cancer (Gendicine). This strategy is thus believed to hold some potential and promise for GBM therapy [33]. However, the ability to selectively deliver p53 to GBM cells to induce apoptosis while avoiding any damage to healthy brain cells remains elusive. Apart from designing rational carriers for targeted delivery of p53 to GBM, rendering them with specific attributes to enable crossing of the blood-brain barrier (BBB) has imposed additional significant challenges.

Nanotechnology provides promising tools to tackle biological barriers effectively because they offer possibilities for surface functionalization of nanoparticles with multiple ligands that enable targeting and crossing of the BBB. Nanoparticles also have the ability to diffuse through the leaky vasculature of tumor tissues by the enhanced permeability and retention effect (EPR) [34]. This cancer-specific attribute is characterized by poor lymphatic drainage allowing the accumulation of nanoparticles to reach much higher concentrations than those found in plasma. Thus, effective p53 therapy for GBM can be accomplished by using such nanomaterials as they can improve the delivery of p53 across the BBB, penetrate into brain tissue and reach the tumor tissue at therapeutic concentrations, and at the same time avoid side effects to normal brain tissue. Along those lines, one such promising study with clinical relevance was reported by Kim et al., where an antitransferrin receptor-based tumor-targeting immunoliposome nanocomplex could efficiently cross the BBB and deliver p53 selectively to target GBMs and their cancer stem cells (CSCs) implicated in recurrence and treatment resistance. Transferrin receptor-mediated transcytosis of nanocarriers downregulated MGMT and induced apoptosis in intracranial GBM xenografts [35].

Restoration of p53 function by small molecules

After success against a wide range of tumors, small molecule and peptide inhibitors of p53 regulatory proteins have gained attention for GBM therapy. As in the case of other cancers, most TP53 alterations in GBM are missense mutations in the DNA-binding domain (DBD), leading to inhibition of transcription factor activity. Importantly, mut-p53 protein is highly expressed in GBM and numerous studies have demonstrated that mut-p53 possesses oncogenic functions beyond those acquired through loss of wt-p53. These p53 mutants have been designated gain-of-function (GOF) mut-p53. GOF mut-p53 is associated with disruption of the MDM2/p53 negative feedback loop and often leads to enhanced proliferation, migration, invasion, and resistance to chemotherapy. The mutations responsible for GOF mut-p53 are not localized within the DNA bonding domain, instead they arise from dysregulation of p53 posttranslation modification at different residues [36]. Consequently, efforts have been made to develop drugs that could reactivate wt-p53 by changing the conformation of the mut-p53 protein. A wide range of compounds have been developed for restoring wt-p53 function from mut-p53 in different cancers, including GBM. These are discussed in Chapter 28 of this book.

p53-targeted cancer gene therapy

As p53 function is lost in many cancers, it is logical to attempt restoring p53 function by reintroducing wt-p53 protein. It was demonstrated for the first time in the late 1980s that transfection of the wt-p53 gene into a variety of human cancer cells could induce apoptosis and growth inhibition [37]. It has been shown that reactivation of p53 function in murine cancer model systems triggers a curative response. This has spurred numerous efforts to develop p53-based cancer therapies, including for GBM. In early attempts, p53 gene therapy was explored extensively using adenovirus vectors, which were later improved by developing oncolytic virus engineered to replicate and kill only p53 defective cells. It has been shown that adenovirus-mediated p53 may enhance the radioresponse of GBM cells using wt-p53, and this radiosensitization may at least partially involve a restoration of the propensity for apoptosis [38].

Adenovirus-mediated delivery of wild-type p53

Virus-based gene therapy has emerged as an efficient carrier for horizontal gene transfer. The tropism of some herpesviruses, adenoviruses, and paramyxoviruses has been exploited to deliver genes to neural and GBM cells [39]. The first clinical study on engineered viral gene delivery to GBM was carried out in the early 1990s, where transduced cells were used to deliver engineered retroviruses into the stroma of GBM [40]. The initial attempts at virus-mediated gene delivery were confined to suicide genes such as AdV-TK and cytosine deaminase, which severely limited the efficacy and raised safety concerns in the subsequent clinical trials [41]. As an alternative, researchers have turned their attention toward virus-mediated delivery of tumor suppressor genes such as p53/MDM2, p16/Rb, BCL2, and PTEN to GBM [42]. The first study on the delivery of tumor suppressor gene p53 under the control of a CMV promoter used a nonreplicating adenovirus serotype 5 as a carrier and established promising outcomes in in vitro and in vivo experiments. Upon restoration of wt-p53 function in cells infected with virus, there was increased apoptosis of cells in vitro, which manifested as reduced tumorigeneses in subsequent in vivo study. Following these promising outcomes, adenoviral-based p53 therapies have been tested in clinical trials [43, 44]. One phase I study involved 15 patients with recurrent GBM, but a maximum tolerated dose was not reached. The authors found that intratumoral Ad-p53 led to expression of functional p53 protein but only within close proximity to the injection site. This phase I study was not powered to examine overall survival; however, survival in these patients was similar to historical norms. One patient remained alive at the time of publication more than 3 years following treatment. Thus, Ad-p53 administration was shown to be safe with demonstrated expression near the injection site; the authors felt that future research should focus on generating more widespread expression [33]. Although effective in pretransduced cells, from the clinical translational standpoint, inoculation of the same vector in GBM-bearing mice did not achieve sufficient gene transfer to manifest inhibition of cell proliferation and induction of apoptosis in intracranial tumor growth. The delivered p53-mediated cell cycle arrest and DNA repair, or induced apoptosis, senescence, and/or autophagy, depending upon cellular stress conditions. These observations led to a conclusion that these effects are not solely dependent on the restoration of a functional copy of p53.

Engineered viruses with conditional replication and retargeting have revamped the specificity and targetability of viruses for gene delivery [45]. Furthermore, the robust bystander effects and cellular immune responses to infected cells extend an advantage over other particle carriers for gene delivery. However, viral carriers are still associated with challenges such as limited infection, rapid clearance by immune cells, and limited spread across physical barriers. As a result of these drawbacks, nonviral carriers, specifically targeted nanoparticles, have become a subject of interest in GBM gene therapy.

Nanoparticles as carriers for p53 gene therapy

The use of nanotechnology, that is, manipulation of submicron-sized materials, to deliver genetic material into tumor cells is a relatively novel strategy that remains largely experimental. The additional advantage of these carriers is permeation across the BBB using versatile surface modifications, better diffusion and cellular uptake in the tumor mass, and passive tumor-targeting mechanisms relying on the EPR effect [46]. The only nanocarriers that have reached the clinical stage in GBM are the artificial lipid-based microvesicles called liposomes. A liposomal vector was devised in the early 2000s to carry a plasmid coding for HSV1-TK, which was given to patients with recurrent GBM in a phase I/II trial via intratumoral infusion, followed by the administration of the prodrug ganciclovir for 14 days [47]. Radiographic response was observed in most patients in this clinical trial, ranging from focal effects to 50% reduction in tumor volume, without major adverse events. A p53 gene delivery by such nanocarriers is still limited to preclinical studies, some of which are discussed in the following section.

In 2018 and 2019, Kim et al. improved the efficacy of an IL2 immune checkpoint inhibitor in GBM by restoring p53 functions through the delivery of TP53 gene via a novel cationic liposome (SGT-53) [48, 49]. The SGT nano-delivery liposome-based carrier was composed of a wt-p53 plasmid DNA payload encapsulated within a cationic liposome, the surface of which was decorated with an antitransferrin receptor (TfR) single-chain antibody fragment (scFv) as the tumor-targeting moiety. The liposomal nanocarrier SGT-53 was designed to target tumor cells via the TfR, which is highly expressed on the surface of many tumor cells, thus facilitating receptor-mediated endocytosis and enhancing the targeting performance of the nanoparticles. The same p53-loaded lipid carrier has entered phase II clinical trial recently as a combination therapy with the conventional clinically used drug for GBM, i.e., temozolomide (TMZ) [50]. Similarly, Tofilon et al. demonstrated that adenovirus-mediated delivery of p53 may enhance the radioresponse of brain tumor cells containing wt-p53, and that this radiosensitization may involve converting from a clonogenic to the more sensitive apoptotic form of cell death [38]. Although the mechanism underlying this enhanced apoptotic susceptibility is unknown, the AdSCMV-p53-infected cells have a higher level of p53 protein, which increases apoptosis further after irradiation, and this exogenous p53 is transcriptionally active.

The chemoresistance to standard treatment for GBM has been linked to a p53-specific mutation in GBM cells through increased expression of O6-methylguanine DNA-methyltransferase (MGMT), an enzyme that repairs DNA damage caused by TMZ [51]. The application of magnetic nanoparticles to deliver TP53 as a gene of interest to GBM (U87) cells across a simulated BBB model has been attempted in vitro. The selective activation of magnetic nanocarriers in GBM cells triggered an apoptotic cascade following targeted supplementation of p53 [52]. A novel RDP-p53 fusion protein was also investigated by a research group for antiproliferation of cancerous neural cell as an effective approach of targeted p53 protein delivery to treat GBM [53]. The results showed selective uptake of RDP-p53 by these cells, followed by initiation of cell apoptotic pathways and cell cycle arrest at the G2/M phase in in vitro assays.

Alternatively, another research group designed peptides against the p53 inhibitors MDMX and MDM2, delivered to GBM using RGD-targeted micelles. The micelles could reach an intracranial tumor and exert potent p53-dependent antiproliferative activity, and also sensitized the tumors to subsequent TMZ treatment [54]. With the advent of molecules capable of interacting with wt-p53 and other p53 regulatory proteins, a whole new class of small molecules has emerged to regulate levels of negative p53 feedback factors or those which could refold mut-p53 into functional p53 confirmation. Limitations of tumor suppressor-based gene therapy have included poor gene transfer, as mentioned, lack of a bystander effect, and potential resistance arising from the inherent genetic heterogeneity within GBM [55]. However, transduction of tumor suppressor genes, such as p53, may present an excellent opportunity for combinatorial therapy since they could resensitize the cells to radiation and chemotherapy.

p53 with other suicide genes as combination therapy

Despite the high efficiency of cancer cell transduction achieved by viral vectors and the spread of injected vectors beyond the site of intratumoral injection, it is unlikely that every tumor cell will be transduced by this method. In vivo studies of intratumoral injections of Ad-p53 into subcutaneous tumors in mice showed a certain degree of regression exceeding that originally predicted by the transduction efficiency of the vector, suggesting the involvement of some bystander effects. This phenomenon, which is critical in achieving a widespread antitumor effect, was first noted in vivo in brain tumor cells transduced with the HSV1-TK gene and then exposed to ganciclovir (Cytovene). The ganciclovir is nontoxic to normal cells, but the transduced viral thymidine kinase converts the drug into cytotoxic ganciclovir triphosphate that interferes with DNA synthesis and leads to death of the transduced tumor cells [56]. Trask et al. conducted a phase I study in which a replication deficient Ad bearing the HSV1-TK gene driven by the Rous sarcoma virus (RSV) promoter (Adv.RSVtk) was injected intratumorally followed by ganciclovir (GCV) treatment [57]. The virus was safely tolerated even at the highest dose, but failed to achieve significant tumor cell death. A similar study using HSV-TK gene, this time with the CMV promoter, was conducted by Sandmair et al. [58]. There was an increase in frequency of epileptic attacks and a "tendency" for improved survival was reported. Another trial was conducted by controlling the HSV-TK gene using the Ad major late promoter (MLP) [59]. The study concluded that patients tolerated the virus and the subsequent GCV treatment, but did not proceed further as it involved an extensive procedure of 50 intraoperative wound-bed injections (0.2 mL each), and at the same time this could not achieve significant improvement in patient survival. Similar studies by Germano et al. and Immonen et al. using a similar Ad vector showed little to moderate improvement in survival rates and moderate tolerability [60, 61]. A replication competent version of this Ad in GBMs showed increased cytotoxicity, reduced tumor volume, and increased survival in a mouse xenograft model.

Clinical status of p53 cancer gene therapy

A phase I study using intratumoral adenovirus-mediated p53 transfection in recurrent GBM patients demonstrated apoptosis, p21 upregulation, and minimal toxicity. Intratumoral injection of Ad-p53 allowed for exogenous transfer of the p53 gene and expression of functional p53 protein. However, at the dose and schedule evaluated, transduced cells were only found within a short distance of the injection site. Progression-free survival was 13 weeks, overall survival was 43 weeks, and three patients were alive at time of publication [62]. However, p53 distribution did not disperse in the intratumoral mass resulting in some tumor cells expressing wt-p53, which could have altered survival. Although toxicity was minimal, widespread distribution still remained a significant goal. While tumor cells containing a mut-p53 allele are typically more sensitive to the apoptotic effects of Ad-p53, the majority of patients in their study had tumors containing cells with wt-p53, which are often less sensitive to Ad-p53-mediated apoptosis. Likewise, in phase I clinical study of Advexin as a monotherapy, 15 patients with recurrent GBMs received Advexin intratumorally or into the postresection cavity [63]. Although clinical toxicity was minimal, again the distribution of Advexin in tumor cells was limited to the areas near the injection site. These findings suggested that intratumoral injection of Ad-p53 was well tolerated in GBM patients, but the widespread distribution of Ad-p53 vector was necessary for inducing profound antitumor effects in brain tumors. Another study utilizing conditionally replicating adenoviruses, which only replicate in p53 dysfunctional cells, showed efficacy in GBM as well as other tumors as a proof-of-principle. In that study, median time to progression after adenovirus treatment was 46 days while median survival was 6.2 months, with three patients alive at 19 months [64]. Along the same lines, Gendicine (recombinant human p53 adenovirus) was developed by Shenzhen SiBiono GeneTech Co. Ltd., which was approved in 2003 by the China Food and Drug Administration (CFDA) as a first-in-class gene therapy product to treat head and neck cancer, and entered the commercial market in 2004 [33]. As an alternative approach, use of mesenchymal stem cells and neural stem cell vehicles for viruses are also being developed to target GBM but have not reached clinical development. Here, cellular agents would be used to target tumor cells and transmit oncolytic viral particles [65]. A variety of p53 transfection trials are also currently in phase 1 clinical trial in GBM but have not been published (https://clinicaltrials.gov).

Future perspectives and therapeutic potentials

Despite such progress of molecules in other tumor models, preclinical and clinical trial development of such compounds in GBM has been limited (https://clinicaltrials.gov). It would not be unreasonable to see future experimental protocols and agents that show progress in other cancer models to be rapidly investigated in GBM, especially owing to the significant morbidity and mortality of this brain malignancy. Although TP53 is an established tumor suppressor gene which averts propagation of abnormal or damaged cells by inducing cell cycle arrest and apoptosis, its role in GBM is still a subject of controversy. Some research groups have claimed that p53 is not associated with GBM patient survival and no direct correlation could be deduced [66–68]. On the other hand, others propose that p53 alteration is a negative prognostic factor and, in some instances, acts as positive prognostic factor [69–71]. Such observations are further compounded by the intricacies of p53 signaling pathways, isoforms, and mutant heterogeneity that limit the understanding of how this pathway regulates gliomagenesis. In GBM, the exact timing of TP53 mutation and coexistence of subpopulations with heterogeneous TP53 status has raised concerns for targeted therapeutic approaches. The other subject of contention arises from growing evidence in support of the claim that not only loss of wt-p53 function but also gain-of-function oncogenic mut-p53 can also enhance GBM proliferation and progression. Above all, further research along these lines is expected to address crucial gaps in establishing a strong correlation of GBM stages with respect to TP53 status, which can ultimately enable us to arrive at efficient cancer management strategies. Apart from adopting conventional knockout mouse models, the use of knock in mouse models to understand the role of GOF mut-p53 can also provide some new insights. Successful development of new-generation drugs aimed at p53 reactivation will supplement the range of chemotherapeutic agents and their combined use will improve the efficacy of therapy.

References

[1] Saha T, Guha D, Manna A, Panda AK, Bhat J, Chatterjee S, et al. G-actin guides p53 nuclear transport: potential contribution of monomeric actin in altered localization of mutant p53. Sci Rep 2016;6:32626.
[2] Hu W, Feng Z, Levine AJ. The regulation of multiple p53 stress responses is mediated through MDM2. Genes Cancer 2012;3:199–208.
[3] Sui X, Han W, Pan H. p53-induced autophagy and senescence. Oncotarget 2015;6:11723–4.
[4] Moulder DE, Hatoum D, Tay E, Lin Y, McGowan EM. The roles of p53 in mitochondrial dynamics and cancer metabolism: the pendulum between survival and death in breast cancer? Cancers (Basel) 2018;10:189.
[5] Hager KM, Gu W. Understanding the non-canonical pathways involved in p53-mediated tumor suppression. Carcinogenesis 2014;35:740–6.
[6] Sullivan KD, Galbraith MD, Andrysik Z, Espinosa JM. Mechanisms of transcriptional regulation by p53. Cell Death Differ 2018;25:133–43.
[7] Shadfan M, Lopez-Pajares V, Yuan ZM. MDM2 and MDMX: alone and together in regulation of p53. Transl Cancer Res 2012;1:88–9.
[8] Bieging KT, Mello SS, Attardi LD. Unravelling mechanisms of p53-mediated tumour suppression. Nat Rev Cancer 2014;14:359–70.
[9] Sever R, Brugge JS. Signal transduction in cancer. Cold Spring Harb Perspect Med 2015;5:a006098.
[10] Almazov VP, Kochetkov DV, Chumakov PM. The use of p53 as a tool for human cancer therapy. Mol Biol (Mosk) 2007;41:947–63.
[11] Malkin D. Li-fraumeni syndrome. Genes Cancer 2011;2:475–84.
[12] Muller PA, Vousden KH. Mutant p53 in cancer: new functions and therapeutic opportunities. Cancer Cell 2014;25:304–17.
[13] Baugh EH, Ke H, Levine AJ, Bonneau RA, Chan CS. Why are there hotspot mutations in the TP53 gene in human cancers? Cell Death Differ 2018;25:154–60.
[14] Vieler M, Sanyal S. p53 isoforms and their implications in cancer. Cancers (Basel) 2018;10:288.
[15] Omuro A, DeAngelis LM. Glioblastoma and other malignant gliomas: a clinical review. JAMA 2013;310:1842–50.
[16] Tamimi AF, Juweid M. Epidemiology and outcome of glioblastoma. In: De Vleeschouwer S, editor. Glioblastoma. Brisbane (AU): Codon Publications; 2017:143.
[17] Davis ME. Glioblastoma: overview of disease and treatment. Clin J Oncol Nurs 2016;20:S2–8.
[18] Pedrote MM, Motta MF, Ferretti GDS, Norberto DR, Spohr T, Lima FRS, et al. Oncogenic gain of function in glioblastoma is linked to mutant p53 amyloid oligomers. iScience 2020;23:100820.
[19] Rivlin N, Brosh R, Oren M, Rotter V. Mutations in the p53 tumor suppressor gene: important milestones at the various steps of tumorigenesis. Genes Cancer 2011;2:466–74.
[20] von Deimling A, Eibl RH, Ohgaki H, Louis DN, von Ammon K, Petersen I, et al. p53 mutations are associated with 17p allelic loss in grade II and grade III astrocytoma. Cancer Res 1992;52:2987–90.
[21] Backes C, Harz C, Fischer U, Schmitt J, Ludwig N, Petersen BS, et al. New insights into the genetics of glioblastoma multiforme by familial exome sequencing. Oncotarget 2015;6:5918–31.
[22] Shi D, Gu W. Dual roles of MDM2 in the regulation of p53: ubiquitination dependent and ubiquitination independent mechanisms of MDM2 repression of p53 activity. Genes Cancer 2012;3:240–8.
[23] Leroy B, Anderson M, Soussi T. TP53 mutations in human cancer: database reassessment and prospects for the next decade. Hum Mutat 2014;35:672–88.
[24] Shetzer Y, Molchadsky A, Rotter V. Oncogenic mutant p53 gain of function nourishes the vicious cycle of tumor development and cancer stem-cell formation. Cold Spring Harb Perspect Med 2016;6:a026203.
[25] Gao J, Aksoy BA, Dogrusoz U, Dresdner G, Gross B, Sumer SO, et al. Integrative analysis of complex cancer genomics and clinical profiles using the cBioPortal. Sci Signal 2013;6:pl1.
[26] Zhang Y, Dube C, Gibert Jr M, Cruickshanks N, Wang B, Coughlan M, et al. The p53 pathway in glioblastoma. Cancers (Basel) 2018;10:297.

[27] Testa U, Castelli G, Pelosi E. Genetic abnormalities, clonal evolution, and cancer stem cells of brain tumors. Med Sci (Basel) 2018;6:85.

[28] Pollack IF, Finkelstein SD, Woods J, Burnham J, Holmes EJ, Hamilton RL, et al. Expression of p53 and prognosis in children with malignant gliomas. N Engl J Med 2002;346:420–7.

[29] Zheng H, Ying H, Yan H, Kimmelman AC, Hiller DJ, Chen AJ, et al. p53 and Pten control neural and glioma stem/progenitor cell renewal and differentiation. Nature 2008;455:1129–33.

[30] Shiraishi S, Tada K, Nakamura H, Makino K, Kochi M, Saya H, et al. Influence of p53 mutations on prognosis of patients with glioblastoma. Cancer 2002;95:249–57.

[31] Forte IM, Indovina P, Iannuzzi CA, Cirillo D, Di Marzo D, Barone D, et al. Targeted therapy based on p53 reactivation reduces both glioblastoma cell growth and resistance to temozolomide. Int J Oncol 2019;54:2189–99.

[32] Sabapathy K, Lane DP. Therapeutic targeting of p53: all mutants are equal, but some mutants are more equal than others. Nat Rev Clin Oncol 2018;15:13–30.

[33] Zhang WW, Li L, Li D, Liu J, Li X, Li W, et al. The first approved gene therapy product for cancer Ad-p53 (Gendicine): 12 years in the clinic. Hum Gene Ther 2018;29:160–79.

[34] Mahmoud BS, AlAmri AH, McConville C. Polymeric nanoparticles for the treatment of malignant gliomas. Cancers (Basel) 2020;12:175.

[35] Kim SS, Rait A, Kim E, Pirollo KF, Nishida M, Farkas N, et al. A nanoparticle carrying the p53 gene targets tumors including cancer stem cells, sensitizes glioblastoma to chemotherapy and improves survival. ACS Nano 2014;8:5494–514.

[36] Nguyen TA, Menendez D, Resnick MA, Anderson CW. Mutant TP53 posttranslational modifications: challenges and opportunities. Hum Mutat 2014;35:738–55.

[37] Lane DP, Cheok CF, Lain S. p53-based cancer therapy. Cold Spring Harb Perspect Biol 2010;2:a001222.

[38] Lang FF, Yung WK, Raju U, Libunao F, Terry NH, Tofilon PJ. Enhancement of radiosensitivity of wild-type p53 human glioma cells by adenovirus-mediated delivery of the p53 gene. J Neurosurg 1998;89:125–32.

[39] Nandi S, Lesniak MS. Adenoviral virotherapy for malignant brain tumors. Expert Opin Biol Ther 2009;9:737–47.

[40] Kwiatkowska A, Nandhu MS, Behera P, Chiocca EA, Viapiano MS. Strategies in gene therapy for glioblastoma. Cancers (Basel) 2013;5:1271–305.

[41] Hossain JA, Marchini A, Fehse B, Bjerkvig R, Miletic H. Suicide gene therapy for the treatment of high-grade glioma: past lessons, present trends, and future prospects. Neurooncol Adv 2020;2:vdaa013.

[42] Guo XE, Ngo B, Modrek AS, Lee WH. Targeting tumor suppressor networks for cancer therapeutics. Curr Drug Targets 2014;15:2–16.

[43] Senzer N, Nemunaitis J, Nemunaitis M, Lamont J, Gore M, Gabra H, et al. p53 therapy in a patient with Li-Fraumeni syndrome. Mol Cancer Ther 2007;6:1478–82.

[44] Lang FF, Bruner JM, Fuller GN, Aldape K, Prados MD, Chang S, et al. Phase I trial of adenovirus-mediated p53 gene therapy for recurrent glioma: biological and clinical results. J Clin Oncol 2003;21:2508–18.

[45] Hajeri PB, Sharma NS, Yamamoto M. Oncolytic adenoviruses: strategies for improved targeting and specificity. Cancers (Basel) 2020;12:1504.

[46] Golombek SK, May JN, Theek B, Appold L, Drude N, Kiessling F, et al. Tumor targeting via EPR: strategies to enhance patient responses. Adv Drug Deliv Rev 2018;130:17–38.

[47] Reszka RC, Jacobs A, Voges J. Liposome-mediated suicide gene therapy in humans. Methods Enzymol 2005;391:200–8.

[48] Kim SS, Harford JB, Moghe M, Rait A, Chang EH. Combination with SGT-53 overcomes tumor resistance to a checkpoint inhibitor. Onco Targets Ther 2018;7:e1484982.

[49] Kim SS, Rait A, Kim E, Pirollo KF, Chang EH. A tumor-targeting p53 nanodelivery system limits chemoresistance to temozolomide prolonging survival in a mouse model of glioblastoma multiforme. Nanomedicine 2015;11:301–11.

[50] Yingchoncharoen P, Kalinowski DS, Richardson DR. Lipid-based drug delivery systems in cancer therapy: what is available and what is yet to come. Pharmacol Rev 2016;68:701–87.

[51] Zhang X, Zhang W, Cao WD, Cheng G, Zhang YQ. Glioblastoma multiforme: molecular characterization and current treatment strategy (Review). Exp Ther Med 2012;3:9–14.

[52] Eslaminejad T, Nematollahi-Mahani SN, Ansari M. Glioblastoma targeted gene therapy based on pEGFP/p53-loaded superparamagnetic iron oxide nanoparticles. Curr Gene Ther 2017;17:59–69.

[53] Wu J, Zhang E, Fu A. A novel cell-permeable RDP-p53 fusion protein for specific inhibition on the growth of cancerous neural cells. Drug Deliv 2016;23:2464–70.

[54] Chen X, Tai L, Gao J, Qian J, Zhang M, Li B, et al. A stapled peptide antagonist of MDM2 carried by polymeric micelles sensitizes glioblastoma to temozolomide treatment through p53 activation. J Control Release 2015;218:29–35.
[55] Hong YK, Joe YA, Yang YJ, Lee KS, Son BC, Jeun SS, et al. Potentials and limitations of adenovirus-p53 gene therapy for brain tumors. J Korean Med Sci 2000;15:315–22.
[56] Culver KW, Ram Z, Wallbridge S, Ishii H, Oldfield EH, Blaese RM. In vivo gene transfer with retroviral vector-producer cells for treatment of experimental brain tumors. Science 1992;256:1550–2.
[57] Trask TW, Trask RP, Aguilar-Cordova E, Shine HD, Wyde PR, Goodman JC, et al. Phase I study of adenoviral delivery of the HSV-tk gene and ganciclovir administration in patients with current malignant brain tumors. Mol Ther 2000;1:195–203.
[58] Sandmair AM, Loimas S, Puranen P, Immonen A, Kossila M, Puranen M, et al. Thymidine kinase gene therapy for human malignant glioma, using replication-deficient retroviruses or adenoviruses. Hum Gene Ther 2000;11:2197–205.
[59] Smitt PS, Driesse M, Wolbers J, Kros M, Avezaat C. Treatment of relapsed malignant glioma with an adenoviral vector containing the herpes simplex thymidine kinase gene followed by ganciclovir. Mol Ther 2003;7:851–8.
[60] Immonen A, Vapalahti M, Tyynela K, Hurskainen H, Sandmair A, Vanninen R, et al. AdvHSV-tk gene therapy with intravenous ganciclovir improves survival in human malignant glioma: a randomised, controlled study. Mol Ther 2004;10:967–72.
[61] Germano IM, Fable J, Gultekin SH, Silvers A. Adenovirus/herpes simplex-thymidine kinase/ganciclovir complex: preliminary results of a phase I trial in patients with recurrent malignant gliomas. J Neurooncol 2003;65:279–89.
[62] Chen GX, Zhang S, He XH, Liu SY, Ma C, Zou XP. Clinical utility of recombinant adenoviral human p53 gene therapy: current perspectives. Onco Targets Ther 2014;7:1901–9.
[63] Tazawa H, Kagawa S, Fujiwara T. Advances in adenovirus-mediated p53 cancer gene therapy. Expert Opin Biol Ther 2013;13:1569–83.
[64] Foreman PM, Friedman GK, Cassady KA, Markert JM. Oncolytic virotherapy for the treatment of malignant glioma. Neurotherapeutics 2017;14:333–44.
[65] Parker Kerrigan BC, Shimizu Y, Andreeff M, Lang FF. Mesenchymal stromal cells for the delivery of oncolytic viruses in gliomas. Cytotherapy 2017;19:445–57.
[66] Saito T, Hama S, Kajiwara Y, Sugiyama K, Yamasaki F, Arifin MT, et al. Prognosis of cerebellar glioblastomas: correlation between prognosis and immunoreactivity for epidermal growth factor receptor compared with supratentorial glioblastomas. Anticancer Res 2006;26:1351–7.
[67] Stark AM, Hugo HH, Witzel P, Mihajlovic Z, Mehdorn HM. Age-related expression of p53, Mdm2, EGFR and Msh2 in glioblastoma multiforme. Zentralbl Neurochir 2003;64:30–6.
[68] Newcomb EW, Cohen H, Lee SR, Bhalla SK, Bloom J, Hayes RL, et al. Survival of patients with glioblastoma multiforme is not influenced by altered expression of p16, p53, EGFR, MDM2 or Bcl-2 genes. Brain Pathol 1998;8:655–67.
[69] Birner P, Piribauer M, Fischer I, Gatterbauer B, Marosi C, Ungersbock K, et al. Prognostic relevance of p53 protein expression in glioblastoma. Oncol Rep 2002;9:703–7.
[70] Soini Y, Niemela A, Kamel D, Herva R, Bloigu R, Paakko P, et al. p53 immunohistochemical positivity as a prognostic marker in intracranial tumours. APMIS 1994;102:786–92.
[71] Jaros E, Perry RH, Adam L, Kelly PJ, Crawford PJ, Kalbag RM, et al. Prognostic implications of p53 protein, epidermal growth factor receptor, and Ki-67 labelling in brain tumours. Br J Cancer 1992;66:373–85.

Index

Note: Page numbers followed by *f* indicate figures and *t* indicate tables.

A

ABCB1 protein, 249–250, 251–253*t*
ABCC1 protein, 250–253
ABCG2 protein, 254
Abemaciclib, 45, 514–521*t*, 525
ABT 414, 25
Accelerated blood clearance (ABC), 728
  transporters, 690–692, 717
Active transport, 244–245
Acute lymphoblastic leukemia (ALL), 455
Adenovirus-mediated delivery, 779–782
Adenovirus p53, 514–521*t*, 523–524
Adjuvant chemotherapy, 668
Adoptive T-cell immunotherapy, 641–642, 650–651
Adsorptive-mediated transcytosis (AMT), 712
Aerobic glycolysis, 273–274
Afatinib, 513–522, 514–521*t*
AKT inhibitors, 55
Aldehyde dehydrogenase 1 family member A3 (ALDH1A3), 217, 385–386
Alkylating agent chemoresistance
  defective DNA repair systems, 210
  miRNAs, 210
  p53, 209
Alkylpurine-DNA-*N*-glycosylase (APNG), 266, 474, 482, 485–486
Allogeneic CAR T cells, 462–463
α-disintegrin and metalloproteinase (ADAM), 295
Anaplastic astrocytomas (grade III), 269–270
Angiogenesis, 3–4, 431–433, 525–526, 639–641, 643*f*, 652*f*, 675–676
  glioma stem-like cells (GSCs), role of, 139–140
Angiogenic switch, 639, 644–645
AngiomiRs, 675–676
Angiopoietin-1 and 2, 431–433
Anion superoxide, 275
Anthracyclines, 690, 692–695, 699
Antiangiogenic therapies (AAT), 432*f*, 434–443
Antiapoptotic proteins, 751
Antibody-drug conjugates (ADCs), 254, 256
Antigen-presenting cells (APCs), 463
Anti PD1 antibodies, 26
Anti-PD L1 antibodies, 26
Antisense lncRNAs, 418
Antisense MicroRNA-21, 679–680
Antitransferrin receptor (TfR), 780–781
Antitumor vaccines, 522
Anti-VEGF antibodies, 435–436
Apoptosis, 191–192, 206–208, 207*f*, 382–383, 774, 779, 782–783
  Bcl-2 family proteins, 208
  and chemoresistance, 208
  cross talk, 211–213, 213*f*
  extrinsic pathway of, 206–207, 207*f*
  inhibitors of apoptosis proteins (IAPs), 208
  intrinsic pathway of, 206, 207*f*
  loss, 493
  mediators, 207
  poly (ADP-ribose) polymerases (PARPs) family protein, 207–208
  regulation, 499
Apurinic/apyrimidinic endonuclease-1 (APE1), 512
Arsenic trioxide, 514–521*t*, 524–525
Artesunate effect, 500
Aspirin, 514–521*t*
Astrocytes, 710–712
Astrocytic neural cell, 320
Ataxia telangiectasia and Rad3-related (ATR) kinases, 190–191, 375, 553
Ataxia-telangiectasia mutated (ATM) pathway, 190–191, 553
ATF6 pathway, 224
α-thalassemia/mental retardation syndrome X-linked (ATRX) protein, 11
Atherosclerosis, 219, 273
ATP-binding cassette (ABC)
  ABCB1 protein, 249–250, 251–253*t*
  ABCC1 protein, 250–253
  ABCG2 protein, 254
  breast cancer resistance protein (BCRP), 254
  clinical value, 254–256, 255*f*
  MDR1, 249–250, 251–253*t*
  multidrug resistance-associated protein 1 (MRP1), 250–253
  P-glycoprotein (P-gp), 249–250, 251–253*t*
  proteins

ATP-binding cassette (ABC) *(Continued)*
functions of, 247, 247*t*
structure of, 246, 246*f*
transporters, 142
in glioblastoma, 248–254, 249*f*
mechanisms of action, 247–248, 248*f*
ATP-binding cassette superfamily G member 2 (ABCG2), 190
Aurora A, 46
Autophagic cell death (ACD), 211–213
Autophagy, 525, 620
anticancer activity mechanisms of alkylating agents, 215–217
chemoresistance to alkylating drugs in GBM, 217–218, 218*f*
cross talk, 211–213, 213*f*
in glioblastoma promotion and suppression, 213–218, 215*t*
initiation and overcome chemoresistance in GBM, 214–215, 216*t*
pathway, 210–213, 212*f*
TMZ-mediated regulation of, 557
Autophagy-related gene (ATG) proteins, 211–213
Axicabtagene ciloleucel, 455
AZD9291, 513–522, 514–521*t*

B

BafilomycinA1 (BafA1), 218
BAL101553, 45–46
Base-excision repair (BER) mechanism, 143, 266, 509–511, 691
B-cell lymphoma 2 (BCL-2), 191, 206
family proteins, 208
in chemoresistance of glioma cells, 209
Belinostat, 41
Benzodiazepinediones, 753–754
Bevacizumab, 25, 435–439, 526–527, 639–641, 640*f*
Bidirectional lncRNAs, 418
Bioengineered glioblastoma, 227–229, 228*f*
Biological hallmarks
altered metabolism, 5–6
angiogenesis, 3–4
cancer stem-like cells, 7
heterogeneity, 7
immunosuppressive microenvironment, 6–7
local invasion, 4–5
microvascular hyperplasia, 3–4
pseudopalisading necrosis, 3
Biomarkers, 284–285
Biomimetic systems, 487
Bioprecursor prodrugs, 724
Bispecific T-cell engager (BiTE), 461–462
BKM120, 514–521*t*, 524–525, 528
Blood–brain barrier (BBB), 138, 189–190, 205, 244–245, 245*f*, 637–638, 690, 710–718, 711*f*, 776–777
brain endothelial cells (BECs), 710–712
disruption techniques, 718–734
FUS and microbubbles, 719–720
nonthermal reversible electroporation, 720–721
osmotic opening, 718–722
drug delivery strategies, 717–718, 718*t*
efflux pumps, 712–717
endogenous transport routes, 710–717, 713*f*, 714*t*
endothelial cells of, 690
non-BBB disruptive techniques, 721–722
efflux pump inhibitors, 723–724
intranasal delivery, 722–734
nanoparticle-based delivery systems, 724–725
prodrug analogues, 722–723
permeability
alterations, 692–695
ionizing radiation, 697–699
recurrent glioblastoma, 390
Bone marrow-derived cells (BMDCs), 431–433
Bone marrow-derived progenitor cells, 437–438
Bone morphogenic protein (BMP), 168–169, 599–601
Bortezomib (BTZ), 514–521*t*, 524
Brain capillary endothelial cells (BCECs), 732
Brain endothelial cells (BECs), 710–712
Brain extracellular matrix, 338–339
Brain matrix regulation, 340–341
Brain tumor-initiating cells (BTICs), 188
Brain tumors, 136. *See also* Glioblastoma (GBM)
Brain tumor tissue-engineered models, 229–232, 230*f*
Breast cancer, 323–324
Breast cancer resistance protein (BCRP), 254, 717
Breast tumor cells, 323–324
Buparlisib, 54
Buthionine sulfoximine (BSO), 279
Bystander effects, 349–351

C

Cancer-associated fibroblasts (CAFs), 389
Cancer cells, 210
Cancer gene therapy, p53 function, 779
Cancer Genome Atlas, 670
Cancers, 219, 752–754, 761
Cancer stem cells (CSCs), 7, 188, 192, 229–232, 320, 324, 528, 569–574, 691
CD44-mediated chemoresistance, 570–571
CD133-mediated chemoresistance, 570
cellular intratumoral heterogeneity, 188
hypoxia and, 493–494
marker, 493–494
Cancer susceptibility candidate 2 (CASC2), 413–415
Cancer therapy, 279, 683

Cannabis sativa, 384
Carbonic anhydrase IX (CAIX), 461
Carboplatin, 24
Carboxyfluorescein succinimidyl ester (CFSE), 326–327
Cardiomyopathy, 219
Cardiovascular disorders, 273
Carmustine (BCNU), 24
Carmustine (BCNU) wafer, 24, 637–638
Carnosic acid (CA), 217
CARs. *See* Chimeric antigen receptors (CARs)
Caspases, 382–383
Cataracts, 219
Cationic bovine serum albumin (CBSA), 714
Cationic lipids, 678
CC-115, 39
CCL5 protein, 302
CCNU. *See* Lomustine (CCNU)
CDK/cyclin-RB1-E2F signaling, 570
CDK4/6 inhibitors, 43–45
CD44-mediated chemoresistance, 570
CD133 protein, 138–139, 493–494
Cediranib, 514–521*t*, 523
Celecoxib, 527
Cell cycle checkpoints, 143–144
Cell cycle inhibitor p21, 751–752
Cell cycle regulation, 43
  p53 role in, 774
Cell drug delivery methods, 533–535
Cell metabolism, 385–386
Cell-penetrating peptides (CPPs), 733–734
Cell quiescence, 323–324
Cell survival signals, 529
  gap junctional intercellular communication (GJIC), 530–531
  notch signaling, 529–531
  sonic hedgehog signaling (SHH), 529–530
  WNT-PI3K-AKT signaling, 530
Cellular senescence, 349–350
Cemiplimab, 26
Central nervous system (CNS), 204, 642–645
Cerebrospinal fluid (CSF), 138, 642, 645
Cervical draining lymph nodes (CDLNs), 642–644
Cetuximab, 514–521*t*, 522
C6-glioma-bearing mice model, 727–728
Chaperon-mediated autophagy (CMA), 210–211, 281–283, 283*t*, 283*f*
Checkpoint inhibitors, 647–648
Checkpoint kinases 1 (Chk1), 375
Checkpoint kinases 2 (Chk2), 375
Chemokines, 645–646
Chemoradiotherapy, 324
Chemoresistance, 208
  apoptosis and, 208
  by ATP-binding cassette (ABC) subfamily-mediated extracellular transportation, 571
  GBM-associated cells, 573–574
  in glioblastoma, 209–210
  glioblastoma, factors involved in
    apoptosis regulation, 499
    mismatch repair (MMR) pathway, 498–499
    O^6-methylguanine- DNA methyltransferase (MGMT), 498
    other factors, 499
  in glioma cells, 209
  HIF signaling, 573
  mechanisms
    hypoxia and acidosis, 572–573
    tumor microenvironment (TME), 571–574
  receptor tyrosine kinase (RTK), 558–563
Chemotherapeutic agents, 597
Chemotherapy
  carmustine (BCNU) wafer, 24
  conventional, 637–639
  in GBM treatment, 550–551
  nitrosourea-based therapy, 24
  nonalkylating chemotherapy, 24
  resistance, 322
    mechanisms, 648–650
  temozolomide (TMZ), 23
  with temozolomide (TMZ), 10
  tumor microenvironment (TME), 171
Chimeric antigen receptors (CARs), 641–642
Chimeric antigen receptor (CAR) T-cell immunotherapies
  hematologic malignancies, 455
  immune checkpoint inhibitors, 463–464
  intrinsic dysfunction, 459–460
  intrinsic T-cell fitness, 462–463
  therapeutic vaccination, 463
  tumor heterogeneity, 458
  tumor-specific targets, 458
  tumor trafficking, 458–459, 462
  viral therapies, 463
Chitosan, 678
Chloroethylating nitrosoureas (CNUs), 372
Chloroquine (CQ), 384, 514–521*t*, 525–526
Chromatin, 40
cis-lncRNAs, 420
Cisplatin, 214–215
c-Jun NH2-terminal kinases (JNK), 511
Clathrin-mediated endocytosis, 725
Claudins, 712
CNS. *See* Central nervous system (CNS)
Colorectal cancers (CRCs), 649

Complete resection of enhancing tumor (CRET), 138
Contrast-enhancing tumor (CET), 138
Convection-enhanced delivery, 57
Conventional chemotherapies, 637–639
CRISPER/Cas9, 501, 503
CRISPR-Cas9 system, 463–464
Cross talk between UPR, autophagy and apoptosis, 221, 222*f*
CSF. *See* Cerebrospinal fluid (CSF)
Curcumin, 514–521*t*, 524–525, 528
CXCR4, 382
CXCR2 receptor, 440–443
Cyclin-dependent kinase inhibitor, p21 (CDKN1A), 325
Cyclooxygenases (COXs), 273
Cytochrome P450 (CYP), 273, 724
Cytokine-induced immunosuppression, 295–302
  interleukin-6 (IL-6), 297–298
  interleukin-8 (IL-8), 301–302
  interleukin-1β (IL-1β), 299–301
  other cytokines, 301–302
  signal transducer and activator of transcription 3 (STAT3) signaling, 297–298
  sphingolipid rheostat, 299–301
  T cell paralysis, 301
  transforming growth factor-β (TGF-β), 301
Cytokine-release syndrome (CRS), 455–456
Cytokines, 460
Cytoprotective autophagy, 620
Cytoskeletal remodeling, 5
Cytotoxic resistance, 320
Cytotoxic sensitivity, in glioblastoma cells, 327–330, 329–330*f*

D

Dacomitinib, 513–522, 514–521*t*
Dactolisib, 56
Definity, 721
Dendritic cell (DC) therapies, 645–647
Detox machinery, 277
Dexamethasone, 697–698
Diabetes, 219, 273
Dichloroacetate (DCA), 385–386
Diffuse large B-cell lymphoma (DLBCL), 455
Dimethylaminomicheliolide (DMAMCL), 385–386
Direct proapoptotic agents, 48–49
Direct reactive oxygen species effects, 275–277, 276*f*
DLNs. *See* Draining lymph nodes (DLNs)
DNA checkpoints, 190–191
DNA damage response (DMR), 551, 553–554
DNA protein kinase (DNA-PK) inhibitors, 39
DNA repair mechanisms, 508–509
DNA synthesis inhibition, 103
Dominant-negative mechanism, 753
Double strand breaks, 551–552
Dovitinib, 435–436
Doxorubicin (DOX), 690, 692–695, 699
  antineoplastic activity, 699
  antitumor efficacy of, 693
  morphine role on, 693–694
  therapeutic efficacy, 699
Draining lymph nodes (DLNs), 642, 644, 647
Drug efflux pumps, 190
Drug efflux transporters, 529–531
Drug resistance, 268–269

E

Efflux pumps, 380–381, 712–717
  inhibitors, 723–724
EGFR mutation variant III (EGFRvIII), 387–388
EGFRvIII chimeric antigen receptor T cells, 305, 457
Elacridar, 724–725
Electroporation, 721
Elongation factor (EF-1α), 282
Elvitegravir (EVG), 730
Endoplasmic reticulum (ER), 206
  stress, 219–220, 220*f*, 268
Endoplasmic reticulum stress responses (ERSR), TMZ-induced ERSR, 557
Endothelial cells (ECs), 194, 244–245, 437–438
Endothelial progenitor cells (EPCs), 4
Enhanced permeability and retention (EPR) effect, 725, 777–778, 780
Enzyme induced anticonvulsant drugs (EIACDs), 526
Ependymal neural cell, 320
EPIC1, 412
Epidermal growth factor receptor (EGFR), 8–9, 214, 271, 387–388, 499–500, 500*t*, 513–526
  amplification of, 696
  mutations, 499–501
  therapeutic resistance, 165
  tyrosine kinase pathway, 24–25
Epigenetic targeting, 511–512
Epithelial to mesenchymal transition (EMT), 280–281, 323–324, 347–348
Erlotinib, 513–522, 514–521*t*
Exosomal mRNA profiles
  biomarker, 482
  parent cellular status, 480–482
  TMZ treatment efficacy in clinical samples, 482–486
Exosomes, 475, 486
  secretion, 386–387
Extracellular matrix (ECM), 4–5, 194, 492–493
Extracellular microenvironment, 345
Extrachromosomal DNA, 10

F

Fas-associated death domain (FADD), 206–207
Fas ligand (Fas-L), 206–207
Fatty acid synthase (FAS), 274
Fibroblast growth factor (FGF)-inhibitors, 526–527
Fibroblast growth factor receptor (FGFR), 523
First-generation P-gp inhibitor, 724–725
Fluorescein isothiocyanate (FITC), 463
Fluorescence-activated cell sorting (FACS), 323–324
Fractionated radiotherapy (FRT), 698
Functional heterogeneity, 590
Futile cycle, 266–267, 691

G

Gain-of-function (GOF), 752
Galectin-1, 299, 389
Gamma secretase inhibitors (GSIs), 617
Ganciclovir (GCV) treatment, 781–782
Gap junctional intercellular communication (GJIC), 530–531
GBM stem cells (GSCs), 254, 668–669
Gefitinib, 513–522, 514–521*t*
GEMMs. *See* Genetically engineered mouse models (GEMMs)
Gendicine, 782
Gene expression analysis, 325
Gene expression profiles, 500–501, 501*t*
Gene therapy, 103
  p53 protein, 780
Genetically engineered mouse models (GEMMs), 501–503
Genetic heterogeneity, 590
Genetic polymorphisms, 724
Genotoxic stresses, 750
GJIC. *See* Gap junctional intercellular communication (GJIC)
Gliadel wafers, 508
Glial fibrillary acidic protein (GFAP), 282, 502
Glioblastoma (GBM), 136–137, 204, 244, 265–266, 320, 403–405, 492
  antiangiogenic therapies (AAT), 435
  antisense-microRNA-21, 679–680
  ATP-binding cassette (ABC) transporters in, 248–254, 249*f*
  blood–brain barrier (BBB) permeability
    alterations, 692–695
    ionizing radiation, 697–699
  cell quiescence, 323–324
  chemoresistance in, 209–210
  chemotherapy, 82–83
  cytokine-induced immunosuppression (*see* Cytokine-induced immunosuppression)
  cytotoxic sensitivity in, 327–330, 329–330*f*
  evolution of therapy for, 637–641
    conventional chemotherapies, 637–639
    targeted and antiangiogenic agents, 639–641
  genetics of chemoresistance, 691–692
  genomic classification of, 214
  history of, 18–23
    pre-Stupp era, 19
  and hypoxia-inducible factor-1α (HIF-1α) activity, 280–281, 281*f*
  molecular drivers of quiescence, 324–325
  molecular events, 83–84
  molecular mechanisms, 691–692
  proliferative potential in, 327–330, 329–330*f*
  quiescence and treatment resistance, 324
  quiescent cells, 326–327
  and reactive oxygen species (ROS), 271–272
    therapeutic activity in, 278–279
    therapeutic target for, 284–285
  resistance mechanisms in, 321–322
    chemotherapy resistance, 322
    resistance to radiation, 321–322
  secretome, 295
  standard of care, 20–23, 337
  standard treatment, 430–431
  subtypes, 205
  therapeutic approaches, 96
  therapy, 682–683
  treatment
    chemotherapy in, 550–551
  treatments, 82, 695–697
  vital dye retention, 327–330, 329–330*f*
Glioblastoma multiforme (GBM)
  advanced imaging, 487
  chemotherapy monitoring, 474
  delivery methods, 713
  treatment of, 709–710
Glioblastoma stem cells (GSCs), 369–381
  chemoresistance mechanisms, 371
    aberrant DNA repair, 372–375
    cell cycle checkpoints deregulation, 375
    deregulation, 375–379
    efflux pumps, 380–381
  functional and molecular properties, 371
  vs. neural stem cells (NSCs), 369–370, 370*t*
Glioma Longitudinal AnalySiS (GLASS) Consortium, 650
Gliomas, 205, 320
  cells, 274
    invasion, 4–5
  WHO classification, 403, 430–431
Glioma stem cells (GSCs), 7, 137, 192–193
  adaptive microenvironment, 592–595, 593*f*
  angiogenic niche, 613–615

Glioma stem cells (GSCs) *(Continued)*
cellular interactions, 592
characteristics, 590, 591*f*
chemoresistance, 591, 592*f*
heterogeneity and plasticity, impact of, 141–142
hypoxic and perivascular niche, 594–595
targeting, 610–613
identification, 138–139
immunosuppressive niche, 593–594
targeting, 606–610
intracellular signaling pathways, 615–620, 615*f*
autophagy regulation, 620
Notch signaling, 617
PI3K-Akt–mTOR, 616
Shh signaling pathway, 617–618
STAT3 signaling, 618–620
isolation and detection, 590–591
neuronal niche, 594
targeting, 610
role
in angiogenesis, 139–140
in immune evasion, 140–141
in resistance to hypoxia, 139
specific molecular pathways, 142, 142*f*
ATP-binding cassette (ABC) transporter channel, 142
base excision repair (BER) mechanism, 143
cell cycle checkpoints, 143–144
methylguanine-DNA-methyltransferase (MGMT), 143
mismatch repair(MMR) mechanism, 143
stromal interactions, 592
surgical resection, 137–138
temozolomide (TMZ) bioavailability, 138
Glucose oxidase (GOX), 273
Glucose-regulated protein (GRP), 219
Glucose transporter type (GLUT)-1, 280–281
Glutamine, 6
Glutathione (GSH), 278
Glyceraldehyde-3-phosphate dehydrogenase (GAPDH), 280–281
Glycerol-3-phosphate dehydrogenase 1 (GPD1), 192–193
GRP78, 219, 223–224
GSCs. *See* Glioblastoma stem cells (GSCs)

H

HCT116-p21-null colon cancer cells, 325
Hedgehog (HH) signaling-mediated chemoresistance
oncogenic activation, 566
signaling pathway, 102, 565–566
Hematologic malignancies, 455
Heme oxygenase (HO), 273
Heme oxygenase-1 (HO-1), 278
Heterogeneity, 7, 776
Heterozygous cancers, 754–756
Hexokinase 2 (HK2) expression, 280–281, 385–386
HIF signaling, 573
High frequency electroporation (HFE), 721–722
Hippo/yes associated-protein 1 (YAP)-mediated survival signaling chemoresistance
oncogenic activation, 567
pathway, 567
Histone acetyltransferases (HATs), 512
Histone deacetylase inhibitors (HDACs), 40–41, 696
Histone deacetyltransferases (HDACs), 512
Histone methylase inhibitors, 42
Homeostasis, 339
Homozygous cancers, 756
HOTAIR, 414
HrasV12 model, 502
Human epidermal growth factor receptor 2 (HER2) CAR T cells, 457–458
Human P53 gene, 750–751
Hydrogen peroxide, 277
Hydroxychloroquine (HCQ), 514–521*t*, 525–526
Hypermutated tumors, 650
Hypermutations, 649
Hyperosmotic solutions, 719
Hyperthermia, 720
Hypoxia, 139, 157–158, 229–232, 280, 389, 644–645
in glioma cells, 209
Hypoxia-inducible factor (HIF-2α), 229–232
Hypoxia inducible factor-1 (HIF-1), 348–349
Hypoxia-inducible factor-1α (HIF-1α), 280–281, 281*f*, 431–433
Hypoxia inducible factor 3 alpha subunit (HIF3A), 675–676
Hypoxia inducible factors (HIFs), 493
signaling, 572–573
Hypoxia response elements (HRE), 280
Hypoxia, tumor microenvironment component, 493–494
cancer stem cell, 493–494
Hypoxic-inducible factor-1 (HIF1), 322

I

IL-CXCR2 pathway, 440–443
IL13Rα2 CAR T cells, 456
Imetelstat, 39
Immune activation, implications for, 648–650
Immune cells, 193–194
Immune-checkpoint (IC) inhibitors, 306, 306*t*, 460, 463–464, 534–535, 647–648
Immune diseases, 219
Immune evasion, glioma stem-like cells (GSCs), 140–141
Immunoediting, 650
Immunologic cell death, 214

Immunomagnetic exosome RNA (iMER) technology
advantages, 486
components, 475, 486
exosome enrichment and RNA analysis, 478–480, 479–480*f*
exosome mRNA profiles
parent cellular status, 480–482, 483*f*
TMZ treatment efficacy, in clinical samples, 482–486, 485*f*
treatment efficacy, biomarker for, 482, 484*f*
microfluidic chip operation, 476–478, 477*f*
Immunostimulatory cytokines, 460
Immunosuppressive microenvironment, 6–7
Immunotherapy, 26, 641–642
Indirect reactive oxygen species effects, 275–277, 276*f*
Indoleamine 2,3-dioxygenase (IDO), 302–305
Indoleamine 2,3-dioxygenase 1 (IDO1), 593–594
Inhibitors of apoptosis proteins (IAPs), 208
Integrase inhibitors, 527–528
Integrin signaling-mediated chemoresistance
integrin signaling, 563
oncogenic activation, 563–564
Intensity-modulated radiation therapy (IMRT), 321
Interferon-β (IFN-β), 510
Intergenic lncRNAs, 418
Interleukin-6 (IL-6), 297–298
Interleukin-8 (IL-8), 301–302
Interleukin-1β (IL-1β), 299–301
Interstitial fluid (ISF), 712–713
Intranasal delivery, 722–734
Intratumoral heterogeneity (ITH), 186, 368–369
cellular and functional, 188
genetic and transcriptomic, 187
treatment resistance, molecular mechanisms of, 188–192
apoptosis, 191–192
DNA checkpoints, 190–191
drug efflux pumps, 190
O6-methylguanine-DNA methyltransferase (MGMT), 189
Intratumoral hypoxia, 341–342
Intronic lncRNAs, 418
In vitro 3D model, 494
Ionizing radiation (IR), 345, 691
IRE1, 224
Irinotecan, 24
Ischemia, 219
Isocitrate dehydrogenase (IDH), 8, 205, 270–271
ITH. *See* Intratumoral heterogeneity (ITH)

J

JNK. *See* c-Jun NH2-terminal kinases (JNK)

K

K-M enhancer, 267
Kynurenine pathway, 302–305

L

Laniquidar, 724–725
Lapatinib, 513–522, 514–521*t*
Lenvatinib, 514–521*t*, 523
Leukapheresis, 455
Levodopa (L-DOPA), 724
LINC00021, 409
Lipid peroxidation, 276
Lipoprotein receptor-related protein 1 (LRP-1), 716
Liposomes, 678, 725–734, 729*t*, 735–736*t*
Lipoxygenases (LOXs), 273
LncRNA 003092.1, 415
Lnc-TALC, 413–414
Local invasion, 4–5
Lomustine (CCNU), 24, 637–638
Lon noncoding RNAs (lncRNAs), 405–413
classification, 418, 420
functions
gene expression regulation, 419
nucleus, 419
mechanisms of actions, 419–420
Loss-of-function (LOF), 752
Lymphatic drainage, 642–645
Lymphatic endothelial cells (LECs), 642–644
Lymphodepleting chemotherapy, 459–460

M

M2698, 55
mAb806, 460–461
Macroautophagy, 210–211
Macromolecules, 713
Magnetic particles, 731
MALAT1, 412
Malignant glioma, 590, 695
Malignant recurrence, 137
Matrix metalloproteinases (MMP) inhibitors, 295, 528
MDR1, 249–250, 251–253*t*
Mebendazole, 45
Melanoma, 323–324
Meningeal lymphatic vessels (MLVs), 642–644
Mesenchymal stem cells (MSCs), 533, 782
Metabolism, 5–6
Metabolomics profiling, 119
Metallic gold nanoparticles, 729–730
Methylene quinuclidinone (MQ), 759
O^6-Methylguanine-DNA methyltransferase (MGMT), 10, 23, 143, 189, 267, 453–454, 498, 551, 691, 781
expression downstream of p53, 556

O^6-Methylguanine-DNA methyltransferase (MGMT)-targeted therapies, 509
c-Jun NH2-terminal kinases (JNK), 511
interferon-β (IFN-β), 510
mismatch repair (MMR), 509–510
O6-benzylguanine (O6-BG), 510
O6-(4-bromothenyl) guanine (O6-BTG), 510
ursolic acid (UA), 510
withaferin A (WA), 510–511
Microarray technology, 117–118
Microautophagy, 210–211
Microbubbles, 720
Microglial activation, 340–341
MicroRNA-21, 680–681
MicroRNAs (miRNAs), 601–605, 692
in alkylating agent chemoresistance, 210
chemoresistance, 574
modulation, 104–105
MicroRNAs (MiRNAs) therapy
biogenesis and function of, 671–672*f*
in cancer, regulatory role of, 670, 673*f*
in cells, 669–670
in GBM biology, regulatory role of, 672–676, 674*t*, 675*f*
multiple cellular mechanisms for GBM, 683, 684*f*
therapeutic sensitization of GBM
chemotherapy, 677
temozolomide (TMZ), 677–682
Microvascular hyperplasia, 3–4
Mira-1, 757
MiR-155 HG, 409
miRNA-10b, 680–681
Mismatch repair (MMR) mechanism, 143, 509–510, 649
futile cycle, 691
pathway, 498–499
proteins, 499
Mitochondria, 385
Mitochondrial membrane depolarization, 268
Mitochondrial outer membrane permeabilization (MOMP), 206
Mitochondrial permeability transition pore (MPTP), 268
Mitochondrial porins, 274
MK2206, 55
MMAC1 gene, 205
Molecular detection technologies, 487
Molecular pathology, 7
epigenetic markers
a-thalassemia/mental retardation syndrome X-linked (ATRX) protein, 11
MGMT methylation, 10
telomerase reverse transcriptase (TERT), 11
molecular markers
epidermal growth factor receptor (EGFR), 8–9
extrachromosomal DNA, 10
isocitrate dehydrogenase, 8
phosphate and tensin homolog (PTEN), 9
phosphoinositide 3-kinase (PI3K)/AKT pathway, 9
p53 pathway, 9
1p/19q codeletion, 8
Ras/Raf/MAPK pathway, 9–10
retinoblastoma (Rb) pathway, 9
Monoamine oxidase B-activated pro-drug effect, 500
Monocyte chemoattractant protein 1 (CCL2), 299
Mononuclear phagocyte system (MPS), 710
Morphine, 693
Mouse double minute 2 homolog (MDM2), 270, 554–555
Mouse models
genetically engineered mouse models (GEMMs), 501–503
transplantation models, 500–501
mTORC1/2 inhibitors, 55–56
mTOR-STAT3-Notch pathway, 683
Multidrug resistance (MDR), 267
Multidrug resistance-associated protein 1 (MRP1), 190, 250–253
Multidrug resistance-associated proteins (MRPs), 692
Multidrug resistance protein 1 (MDR1), 190
Multifocal GBM (M-GBM), 698
Murine double minute (MDM2), 524
Myeloidderived suppressor cells (MDSCs), 388–389

N

Nanocarrier technologies, 254
Nanoparticle-based delivery systems, 724–725
liposomes, 725–734, 729*t*, 735–736*t*
magnetic particles, 731
metallic gold nanoparticles, 729–730
peptide-based particles, 731–732
polymeric nanoparticles, 727–729
Nanoparticles, 777–778, 780
Nanotherapeutics, 534
National Cancer Institute/Developmental Therapeutics Program (NCI/DTP), 757
NEAT1, 412–413
Neddylation pathway, 50–51
NEO100, 722
Neoplastic transformation, 273
Nephrotic syndrome, 219
Neural stem cells (NSCs), 369–370, 370*t*, 375–377, 782
Neuroblastoma in vitro models, 227–229, 228*f*
Neurodegenerative diseases, 273
Neurodegenerative disorders, 219
Neurofibromin 1 (NF1), 214
Neurosphere assay, 494
Neurosurgery, 137–138
Neurovascular unit (NVU), 710–712
Neutrophils, 389

NFκB tumor suppressor, 169–170, 523–524
NF1 tumor suppressor, 501
Niche interactions, 193–194
Nicotinamide adenine dinucleotide phosphate (NADPH), 269–270, 277
Nimotuzumab, 514–521*t*, 522
Nimustine (ACNU), 24
Nintedanib, 435–436, 514–521*t*, 523
Nitrosourea-based therapy, 24
Nitrosureas, 637–638, 695
Nivolumab, 26
Nonalcoholic steatosis, 219
Nonalkylating chemotherapy, 24
Non-BBB disruptive techniques, 721–722
Noncoding RNA epigenetics, 416–417
Nonhomologous end joining (NHEJ), 372
Nonrandom chromosome segregation, 379
Nonselective autophagy, 210–211
Nonthermal reversible electroporation, 720–721
Nontoxic fluorescent dyes, 326–327
Notch intracellular domain (NICD), 529
Notch pathway, 325, 529–531
Notch signaling-mediated chemoresistance
  oncogenic activation, 566–567
  pathway, 566
NSC319725, 763–764
NSC328784, 763–764
NSC319726/ZMC1, 763–764
Nuclear dot protein 52 (NDP52), 210–211
Nucleotide-binding domains (NBD), 246
NUMB, 377
Nutlins, 753–754

O

O6-benzylguanine (O6-BG), 510
O6-(4-bromothenyl) guanine (O6-BTG), 510
Occludins, 712
OLIG2, 382
Oligodendroglial neural cell, 320
On-chip nanoscale biosensing technologies, 475, 486–487. *See also* Immunomagnetic exosome RNA (iMER) technology
Oncogenesis, 320
Oncogenic activation, 558–563
  of CDK/cyclin-RB1-E2F signaling, 568
  of HH signaling, 566
  of hippo/YAP signaling, 567
  of integrin signaling, 563–564
  of notch signaling, 566–567
  of RTK-Akt–mTORC1 signaling, 560–562
  of RTK-JAK–STAT signaling, 563
  of RTKPI3K-mTORC2 signaling, 560–562
  of RTK-Ras-MAPK signaling, 559–560
  of TGFβ-TGFβR signaling, 564
  of Wnt signaling, 565
Oncogenic miRNAs (oncomiRs), 668
1p/19q codeletion, 8
Opioids, 712–713
Opsonin proteins, 725–726
Optison, 721
Oregon Green, 326–327
  cell entry and activation, 328*f*
  dye fluorescence, 329*f*
Osimertinib, 52–53, 514–521*t*
Osmotic opening techniques, 718–722
Ovarian cancer, 323–324
Overall survival (OS), 267
Oxidative stress, 273, 385–386
OX-26-PEG-liposomes, 726–727

P

P53, 208
  in alkylating agent chemoresistance, 209
Palbociclib, 43–45, 514–521*t*, 525
Pan-Bcl2 inhibitor gossypol, 217
Pancreatic cancer, 323–324
Pancreatic ductal adenocarcinoma (PDAC), 461
Panitumumab, 514–521*t*, 522
Panobinostat, 41
Pan-PI3K inhibitors, 54
Paramagnetic nanoparticles (PMNPs), 731–732
PARP inhibitor, 500–501
Patient-derived cell lines (PDCL), 256
Patient-derived xenograft (PDX) model, 500–501
PATZ1, 382
PAXIP1-AS1, 408
Pazopanib, 514–521*t*, 523
PDGFR, 165–166
Pembrolizumab, 26
Pentose phosphate pathway (PPP), 269–270
Peptide-based particles, 731–732
Perillyl alcohol (POH), 722
Perineuronal nets (PNNs), 339
Periodic acid-Schiff (PAS) staining, 433–434
Perivascular niche, 389
Peroxiredoxins (PRDXs), 277
Pevonedistat, 50–51
p21 factor, 325, 556
P-glycoprotein (P-gp), 249–250, 251–253*t*, 717
Phenotypic plasticity, 381–382, 597–598
PhiKan083, 762–763
Phosphatase and tensin homolog (PTEN), 9, 191, 214, 270
Phosphoinositide 3-kinase (PI3K)/AKT pathway, 9
Phosphoionositide 3-kinase (PI3K/0) inhibitor, 462–463
Phospholipase C, 164–170
Phosphorylate retinoblastoma protein (pRb), 524–525

Photodynamic therapy, 321, 611–612
PI3K/AKT/mTOR pathway, 53
PI3K/mTOR pathway, 26
PI3K pathway, 56
p16INK4a/p14ARF-mediated survival signaling
CDK/cyclin-RB1-E2F signaling, 568
hemoresistance, 567–568
PK7088, 762–763
PK11007, 762–763
Platelet-derived growth factor receptor (PDGFR) alpha, 214
Poly ADP-ribose polymerase (PARP-1), 372, 374
Polycomb group of proteins, 415–416
Polyethylene glycol (PEG), 726
Polymeric nanoparticles, 678, 727–729
Poly(lactic-*co*-glycolic acid) (PLGA) nanoparticles, 678–679
Poly (ADP-ribose) polymerases (PARPs) family protein, 207–208
Polyunsaturated fatty acids (PUFAs), 276
Positron emission tomography (PET), 249–250
p53 protein, 750, 760–762*t*
anticancer activity, 761–762*t*
ATR-dependent regulation, 553–554
cancer gene therapy, 779
adenovirus-mediated delivery, 779–782
clinical status, 782
nanoparticles, 780
p53 with other suicide genes, 780–782
canonical and noncanonical tumor suppressor roles, 774, 775*f*
correlation, 774–776
distribution and frequency of, 754
drugs targeting effect, 762–763
gene therapy, 780
limitations of, 759–762
mutants, 753–754
mutation, 774
pathway, 9, 751–753
principles of, 757–758
regulation, 751
restoration, 777–779
role, 774
with single allele, both allele and null mutants, 754–756, 755*f*
structural mutations, 759–764
structure, 751, 752*f*, 758–759, 759*f*
therapeutic potentials, 782–783
therapeutic strategies, 756, 776–777
TMZ-induced regulation
ATM and ATR-dependent regulation, 553–554
intrinsic and extrinsic cell death signaling downstream, 555–556
MGMT expression downstream, 556
mouse double minute 2 (MDM2), 554–555
p21 downstream, 556
tumor suppressor, 499, 501, 523
tumor suppressor, history of, 750–751
wild-type p53 function, 757
pre-miRNAs, 669–670
Prima-1/Prima-1^{MET}, 759–762
Primary glioblastoma (pGBM), 187
Primary miRNA (pri-miRNA), 671–672*f*
Proapoptotic proteins, 751
Procaspase 3 (PC3), 49
Pro-differentiation therapies, 599–606
Prodrug analogues, 722–723
Programmed cell death (PCD), 206, 208
Programmed death-ligand 1 (PD-L1), 306–307
Progression-free survival (PFS), 267
Proinflammatory responses, 229–232
Proliferative potential, in glioblastoma cells, 327–330, 329–330*f*
Proneural-to-mesenchymal transition (PMT), 192, 382
Protein corona, 725–726
Protein kinase RNA-like ER kinase (PERK), 224
Protein misfolding, 752–753
Protein oxidation, 275
Protein phosphatase 2A (PP2A), 48
Proteomics, 117
Proton beam therapy, 321
Pseudopalisading necrosis, 3
Psoriasis, 219
Pulmonary fibrosis, 219
Pulse-chase assays, 323
Pulsed electric fields (PEFs), 721
Pyruvate dehydrogenase kinase (PDK), 274

Q

Quiescence, 323
molecular drivers of, 324–325
and treatment resistance, 324
Quiescent cells, 323–324, 326–327
Quiescent GBM tumor cells, 325–326

R

Radiation-induced bystander effects (RIBE), 350–351
Radiation therapy (RT), 278–279, 321–322, 337, 345–351, 508, 638–639, 651–653, 652*f*, 668, 690, 695–699
tumor microenvironment (TME), 170
Random chromosome segregation, 379, 380*f*
Rapamycin, 694–695
RAS/MEK/ERK pathway, 56–57
Ras/Raf/MAPK pathway, 9–10
Reactivation of p53 and induction of tumor cell apoptosis (RITA), 754–756

Reactive oxygen species (ROS), 213–214, 271–272
and chaperone mediated autophagy (CMA), 281–283, 283*t*, 283*f*
direct, 275–277, 276*f*
indirect, 275–277, 276*f*
inflammation, 266
metabolism and mitochondrial dysfunction, 273–275
overview, 285–286
scavenging, modulation of, 277–278
therapeutic activity in GBM, 278–279
Receptor-mediated transcytosis, 715
Receptor-mediated transport (RMT), 712
Receptor tyrosine kinases (RTKs), 206, 512–513, 557–558
chemoresistance, 558–563
RTK-Akt–mTORC1, 560–562
RTK-JAK–STAT signaling, 562–563
RTK-PI3K-Akt–mTORC1 pathway, 560
RTKPI3K-mTORC2 signaling, 560–562
RTK-PI3K-mTORC2 signaling pathway, 560
RTK-Ras-MAPK signaling, 559–560
epidermal growth factor receptor (EGFR), 163
inhibitors, 435–436
JAK/STAT, 164
PI3K/AKT/MTOR, 164
RAS/MAPK/ERK, 163–164
recurrent glioblastoma, 387–388
Recurrent glioblastoma (rGBM), 187, 368–369
extracellular signaling pathways, 386
blood-brain barrier (BBB), 390
exosome secretion, 386–387
perivascular niche, 389
receptor tyrosine kinases (RTKs), 387–388
tumor microenvironment (TME), 388–389
intrinsic and extrinsic mechanisms, 381
apoptosis escape, 382–383
autophagy, 383–385
cell metabolism, 385–386
phenotypic plasticity, 381–382
transdifferentiation, 381–382
Redox cell homeostasis, 273, 278
Redox signaling, 349
Resistance, 270–271
cell-intrinsic and -extrinsic mechanisms
cell state and proneural-to-mesenchymal transition, 192–193
immune cell, 193–194
niche interactions, 193–194
Retinoblastoma (Rb) pathway, 9
Ribociclib, 514–521*t*, 525
Rindopepimut, 436, 514–521*t*, 522
RNA G-quadruplexes (RG4), 415
RNA-induced silencing complex (RISC), 669–670
The RNA World, 684
Romidepsin, 41
Rous sarcoma virus (RSV) promoter, 781–782
RTKs. *See* Receptor tyrosine kinases (RTKs)
Rucaparib effect, 500–501

S

SCH529074, 763
Sebaceous tumors, 323–324
Second-generation P-gp inhibitors, 724–725
Secretome, 295
Self-eating processes, 211–213, 213*f*
Self-killing processes, 211–213, 213*f*
Semaphorin7A (SEMA7A), 386–387
Senescence-associated secretory phenotype (SASP), 350
Sense-overlapping lncRNAs, 418
SGT-53, 514–521*t*, 523–524
SHH. *See* Sonic hedgehog signaling (SHH)
shRNA library screening, 119
Signaling effectors, 377
Signaling pathway inhibitors, 101–103
Signaling pathways, 278
Signal transducer and activator of transcription 3 (STAT3) signaling pathways, 191–192, 297–298, 299*t*
Simvastatin, 225–226
Single-cell RNA sequencing, 590
Single nucleotide polymorphisms (SNPs), 249
Small peptides, 764
SNHG12, 410
SNHG15, 411
Solid tumors, 6
Solute carrier (SLC), 244–245
Sonazoid, 721
Sonic hedgehog signaling (SHH), 162, 529–530
SonoVue, 721
Sorafenib, 514–521*t*, 523, 530
SOX2OT, 408–409
Sphere-forming efficiency (SFE), 377, 378*f*
Sphingolipid rheostat, 299–301, 300*f*
Sphingosine kinase (SK), 300–301, 300*f*
Sphingosine 1-phosphate (S1P), 300–301
Spirooxindoles, 753–754
Spliced XBP-1, 221
Standard of care (SoC) treatment, 186
STAT3. *See* Signal transducer and activator of transcription 3 (STAT3) signaling pathways
Stereotactic radiosurgery (SRS), 321
Stictic acid, 764
Stupp protocol, 265–266
Sulfasalazine, 514–521*t*, 525
Sunitinib, 514–521*t*, 523
Superoxide anion, 274, 277

Superparamagnetic iron oxide nanoparticles (SPIONs), 731–732
Surgical resection, 337, 668
"Switchable" CAR T cells, 461
Synthetic lethality, 57, 321
Synthetic MicroRNAs mimics, 678
Synthetic Notch receptor (SynNotch) receptor, 461

T

Tariquidar, 724–725
Tazemetostat, 42
T-cell intrinsic dysfunction, 459–460
T cell paralysis, 301
Telomerase inhibition, 39
Telomerase reverse transcription (TERT), 11
  mutations, 7
  promoter, 205
Temozolomide (TMZ), 23, 35–36, 143, 186, 189, 215–216, 244, 265–266, 320, 337, 459–460, 474, 482–487, 485*f*, 498, 508, 651, 668–669, 677–678, 691–693, 695–696, 698–700, 709–710, 780–781
  alkylating agents, 372
  bioavailability, 138
  chemotherapy, 23
  DNA replication stress, 551–552
  FDA-approved chemotherapeutic agent, 638–639
  GBM chemoresistance, affect, 225–226
  mechanism of action, 266–271, 269*f*
  resistance, 266–267
  therapy, 681–682
Temozolomide (TMZ) resistance, 403–404
  acquired cells, 88–95
  antiangiogenic therapies, 526
    FGF-inhibitors and combinations, 526–527
    integrase inhibitors, 527–528
    matrix metalloproteinases (MMP) inhibitors, 528
    VEGF-inhibitors and VEGF-trap, 526–528
  basic excision repair (BER)-targeted therapies, 509–511
  cancer stem cells, 404, 405*f*
  cell drug delivery methods, 533–535
  DNA repair mechanisms, 508–509
  epigenetic targeting, 511–512
  genome-wide screening modulators, 117–120
  immune-checkpoint (IC) inhibitors, 534–535
  MGMT-targeted therapies (*see* O^6-Methylguanine-DNA methyltransferase (MGMT)-targeted therapies)
  molecular mechanisms
    DNA damage response (DMR), 551
    MutS protein homolog 6 (MSH6), 551–552
    O^6-methylguanine-DNA methyltransferase (MGMT), 551
    TMZ-induced N^7-methylation, 552
  nanotherapeutics, 534
  receptor tyrosine kinases (RTKs), 512–513
    autophagy, 525
    epidermal growth factor receptor (EGFR), 513–526
    fibroblast growth factor receptor (FGFR), 523
    NFκB tumor suppressor, 523–524
    phosphorylate retinoblastoma protein (pRb), 524–525
    p53 tumor suppressor, 523
    vascular endothelial growth factor receptor (VEGFR), 513–522
  research models, 84–85, 88–91
  single modulator, 116–117
  targeting cancer stem cells, 528
    cell survival signals (*see* Cell survival signals)
    drug efflux transporters, 529–531
  tumor treating fields (TTF), 531
    adverse effects, 532
    clinical trials, 531–533
    limitations, 532
    quality of life, 532–533
  viral delivery, 534–535
  in vitro models, 114–115
  in vivo models, 115–116
Tenascin-C, 299
TEP1 gene, 205
TGFβ-transforming growth factor-beta receptor (TGFβR) signaling-mediated chemoresistance
  oncogenic activation, 564
  signaling pathway, 564
The Cancer Genome Atlas (TCGA), 42–43, 141, 187
Third-generation P-gp inhibitors, 724–725
Three-dimensional (3D) conformal radiation therapy, 321
3-Methyladenine (3-MA), 218
Thrombospondin-1 (TSP1), 348–349
Tight junctions (TJs), 710–712
Tisagenlecleucel, 455
Tissue-engineered tumor models, 226–227
Tivozanib, 435–436, 514–521*t*, 523
TMB. *See* Tumor mutational burden (TMB)
TNF receptor-associated death domain (TRADD), 206–207
TNF-related apoptosisinducing ligands (TRAIL), 206–207
TP73-AS1, 406–407
Tp53-induced glycolysis and apoptosis regulator (TIGAR), 269–270
TP53 mutations, 190
Transdifferentiation, 381–382
Transendothelial cell electrical resistance (TEER), 712
Transferrin (Tf), 715

Transforming growth factor-β (TGF-β), 166–168, 301
trans-lncRNAs, 420
Transmembrane domain (TMD), 246
Transmembrane helices, 246
Transplantation models, 500–501
Trebananib, 436–437
Triple-negative breast cancer, 753
Trojan horse, 680
Tryptophan metabolism, 302–305, 304*f*
TTF. *See* Tumor treating fields (TTF)
Tumor-associated macrophages (TAMs), 6–7, 193, 350, 386–387
Tumor heterogeneity, 569–574
Tumor hypoxia, 341–342, 348–349
Tumor immune microenvironment, 194
Tumor microenvironment (TME), 155–156, 192–194, 569–574
  chemotherapy, 171
  radiotherapy, 170
  recurrent glioblastoma, 388–389
  targeting, 171–172
Tumor mutational burden (TMB), 6, 648–651
Tumor necrosis factor (TNF) receptor, 206–207, 698
Tumor, node, metastasis (TNM), 2
Tumor plasticity, 569–574
Tumor progression, 320
Tumor protein 53 (TP53), 214
Tumor spheroid/organoids, 226
Tumor suppressor protein (Tp53), 269–270, 750–751
Tumor treating fields (TTF), 20–23, 508, 531
Two dimensional (2D) assay, 492
2D monolayer culture system, 494
Tyrosine kinase pathway
  epidermal growth factor receptor (EGFR) pathway, 24–25
  met, 26
  PI3K/mTOR pathway, 26
  vascular endothelial growth factors (VEGF), 24–25

U

Ubiquitin-proteasome pathway, 49–50
Ultrasound-mediated BBB opening (UMBO), 254, 256
Unfolded protein response (UPR), 219–220, 220*f*, 268
  chemotherapy induced cell death in GBM, 224–225
  GBM-induced chemoresistant to TMZ, 223–224
  in glioblastoma chemoresistance, 221–223, 223*f*
Ursolic acid (UA), 510
US Food and Drug Administration (FDA), 265–266

V

Valproate, 40
Valproic acid (VA), 512
Vascular adenosine triphosphatase (V-ATPase) inhibitor, 218
Vascular endothelial growth factor (VEGF), 166, 639, 644–645, 675–676
  inhibitors, 526–528
  tyrosine kinase pathway, 24–25
Vascular endothelial growth factor A (VEGF-A), 139–140
Vascular endothelial growth factor receptor (VEGFR), 214, 513–522
Vascular mimicry (VM), 432*f*, 433–434, 438–439
Vasculogenesis, 4, 431–433
Vatalanib, 436–437, 514–521*t*, 523
VEGF-VEGFR pathway, 435
Veliparib, 38, 500
Vessel cooption, 4
Viral delivery, 534–535
Viral therapies, 463
Virus-based gene therapy, 779
Vistusertib, 55–56
Vital dye retention, in glioblastoma cells, 327–330, 329–330*f*
Voltage-dependent anion channels (VDACs), 274
Vorinostat, 40–41
VXM01 vaccine, 514–521*t*, 523

W

Warburg effect, 273–274
Whole brain radiotherapy (WBRT), 690, 698–700
Wild-type p53 (wt-p53), 774, 776–779, 778*f*
Withaferin A (WA), 510–511
WNT-PI3K-AKT signaling, 530
Wnt signaling-mediated chemoresistance
  oncogenic activation, 565
  signaling pathway, 565
World Health Organization (WHO), 204, 709–710
wt-p53. *See* Wild-type p53 (wt-p53)

X

Xenotransplantation, 500–501
XIST, 414
X-ray crystallography, 757

Z

Zika virus (ZIKV), 534–535
Zinc metallochaperones, 764
Zonulae occludens, 712
Zosuquidar, 724–725
ZSTK474, 54